International Scientific and Practical Conference: Water Power Energy Forum 2018

IOP Conference Series: Earth and Environmental Science
Volume 288

Kazan, Russia
29 October – 2 November 2018

ISBN: 978-1-5108-9330-6
ISSN: 1755-1307

Printed from e-media with permission by:

Curran Associates, Inc.
57 Morehouse Lane
Red Hook, NY 12571

Some format issues inherent in the e-media version may also appear in this print version.

Institute of Physics
Dirac House, Temple Back
Bristol BS1 6BE UK

Phone: 44 1 17 929 7481
Fax: 44 1 17 920 0979

techtracking@iop.org

Additional copies of this publication are available from:

Curran Associates, Inc.
57 Morehouse Lane
Red Hook, NY 12571 USA
Phone: 845-758-0400
Fax: 845-758-2633
Email: curran@proceedings.com
Web: www.proceedings.com

TABLE OF CONTENTS

WATER TREATMENT: PROCESSES AND EQUIPMENT

INNOVATIVE AQUA BIO TECHNOLOGIES AT ENERGY FACILITIES

ENERGY AND RESOURCE SAVING

HEAT AND MASS TRANSFER THROUGH VARIOUS SURFACES, INTENSIFICATION OF HEAT AND MASS TRANSFER PROCESSES. HEAT SUPPLY

IOP Conf. Series: Earth and Environmental Science **288** (2019) 011001 doi:10.1088/1755-1315/288/1/011001

International Water Power Engineering Forum - 2018 (WPEF-2018)

Abstract

The International Water Power Engineering Forum - 2018 (WPEF-2018) was held in Kazan State Energy University (Kazan), Russia, from October 29 to November 2, 2018.

10th anniversary Water Power Engineering Forum was organized by Kazan State Power Engineering University, National Research University "MPEI", JSC "NPK - Mediana - Filter" with sponsorship from JSC NPK - Mediana - Filter.

The International Water Power Engineering Forum - 2018 was of considerable scientific and practical interest.

WPEF-2018 is of considerable scientific and practical interest. It contributes to the development of advanced experience in the implementation and operation of world water treatment technologies, wastewater treatment and disposal, ensuring optimal water-chemical regimes, improvement of chemical and technological monitoring systems, development of aqua biotechnology and aquaculture. Issues related to heat supply, energy and resource saving were covered as well. The conference participants were the representatives of academic science, university science and members of the largest energy enterprises.

The proceedings are a compilation of the accepted papers and represent an achievement of the conference. This book covers 2 chapters: Chapter 1 Water treatment, aqua biotechnologies and Chapter 2 Thermotechnical Engineering.

Peer review statement

All papers published in this volume of *IOP Conference Series: Earth and Environmental Science* have been peer reviewed through processes administered by the proceedings Editors. Reviews were conducted by expert referees to the professional and scientific standards expected of a proceedings journal published by IOP Publishing.

IOP Conf. Series: Earth and Environmental Science **288** (2019) 012001 doi:10.1088/1755-1315/288/1/012001

Innovative technologies for industrial wastewater treatment

O.V. Atamanova[1], **E.I. Tichomirova** [1], **N.A. Politayeva** [2], **A.L. Podolsky** [1], **M.V. Istrashkina** [1]

[1]Yuri Gagarin State Technical University of Saratov, 77 Polytekhnicheskaya St., Saratov, Russia

[2]Saint Petersburg Polytechnic University of Peter the Great, 29 Polytekhnicheskaya St., St. Petersburg, Russia

O_V_Atamanova@mail.ru

Abstract. We justified the need to purify wastewater, coming from industrial enterprises, from aromatic compounds using sorption technologies. The objects and methods of our research are described in this paper. In the process of studying adsorption mechanisms of aromatic compounds on bentonites of various modifications in static and dynamic conditions, the basic characteristics of adsorption processes were established. Bentonite, modified by carbon nanotubes after thermal treatment, proved to be the most effective with respect to organic compounds. Three most effective compositions of multicomponent adsorption filter for wastewater treatment from nitro and amino compounds are described. We developed the technological design of a local water treatment plant designed for purifying the wastewater, coming from industrial plants, from aromatic compounds. Multicomponent adsorption filters for wastewater treatment have been developed and patented.
Key words: adsorption, bentonite, aromatic amine compounds, hydrodynamic conditions

Introduction

Modern production of paint, varnish, textile, rubber, etc., is a potential cause for pollution of surface water bodies and groundwater, since wastewaters of those industries are characterized by increased content of organic impurities, including toxic content [1]. At present, studies of wastewater treatment from pollutants represented by aromatic compounds acquire particular urgency.

Complete removal of organic matter from wastewater is a difficult task [1], despite the fact that, at present, there is a wide range of biological, physicochemical and other methods for water purification. Each of these methods has its drawbacks, which substantially narrows the limits of their applicability.

Adsorption methods have a number of advantages over other existing methods of water treatment. The main advantages of these methods include simplicity of instrumentation, the ability to extract many organic substances from water, including biologically stable ones, which cannot be removed by other methods [2]. Another significant advantage is an absence of secondary impurities [2-3].

Modern problems of industrial ecology in the field of cleaning contaminated waters in both water supply and wastewater systems include the need to develop approaches to improving adsorption technologies and the composition of sorbents by means of their chemical modification [4]. Currently, the Scientific and Educational Center "Industrial Ecology" at Yuri Gagarin State Technical University of Saratov (SSTU) works on developing innovative technologies for wastewater treatment from nitro and amino compounds within the framework of the State Task of the Russian Federation Ministry of

Education and Science, grant No. 5.3922.2017/PCh: "Development of environmentally friendly energy-saving technologies for the integrated treatment of water contaminated as a result of natural and man-made emergencies for local water treatment plants in problem regions of the Russian Federation".

Objects and methods of research

The objects of investigation were bentonites modified by various methods, including high-temperature heat treatment under various modes (with preheating or burning directly at a given temperature); modification of the bentonite structure by organic components - carbon nanotubes (CNT) and glycerol, either individually or in combination with each other. Bentonite modification was carried out in collaboration with our partners from scientific-industrial enterprise "LISSKON".

Studies on adsorption efficiency were conducted both for individual modifications of bentonites and for multicomponent adsorption filtering loads, including various modifications of bentonite and other sorbents (AN-17-8 anion exchange resin, ASKG silica gel, zeolite, etc.). In our experiments, the control was represented by BAU-A activated carbon. The major methods of our study included spot test reactions, photometry, titration, electron microscopy, and statistical data treatment.

We selected nitro and amino compounds as model pollutants in our study. The presence of those was detected in sewage waters of paint and textile production enterprises of Saratov Region [5]. These included o-toluidine, o-phenylenediamine and p-dinitrobenzene. In the course of laboratory studies of multicomponent adsorption filters' effectiveness, complex model solutions were used, comprising of various nitro and amino aromatic compounds, such as p-dinitrobenzene, m-aminophenol, p-nitrophenol, p-nitroaniline, o-toluidine, and others.

Results and Discussion

The results of our research revealed the features of the adsorption processes of the studied pollutants by various bentonite modifications.

Studying the adsorption mechanism of nitro and amino compounds

Justification of the adsorption capacity of modified bentonites presumed studying their adsorption characteristics in laboratory conditions. Estimation of sorption efficiency parameters was carried out by establishing equilibrium concentrations of organic substances. In each experiment, an equilibrium concentration of organic substances was determined from the steady-state values of each studied substance concentration over time.

The results of conducted studies of organic substances' adsorption on bentonites of various modifications under static conditions made it possible to establish the characteristics of the adsorption process, such as static exchange capacity (SEC), K_d, S, which allowed the selection of the adsorbent under specific conditions of predominance of polluting organic matter in wastewater [6].

Analysis of the adsorption isotherms of o-toluidine and o-phenylenediamine on modified and unmodified bentonites under static conditions showed that this process was most accurately described by the Freundlich isotherm. Thus, we concluded that the adsorption of o-toluidine and o-phenylenediamine on the studied sorbents proceeded according to the ion-exchange mechanism. Addition of glycerol to bentonite modified by CNT somewhat reduced the effectiveness of the adsorbate interaction with aluminosilicate and nanotubes, which could be explained by the intrinsic adsorption of glycerol on the adsorbent and reduction of the adsorption centers' numbers [7].

Analysis of the adsorption isotherms of p-dinitrobenzene on the studied bentonites under static conditions showed that this process was most accurately described by Langmuir isotherms. In this case, monomolecular layers of the adsorbate were formed on the adsorbent surface. The highest adsorption efficiency of p-dinitrobenzene was manifested by the bentonite modified by CNT [8].

The established efficiency parameters of the interaction of o-toluidine, o-phenylenediamine and p-dinitrobenzene ions with the studied modifications of bentonites under dynamic conditions were dynamic exchange capacity (DEC), K_d, and S. It is worth noting that DEC of the studied bentonites

IOP Conf. Series: Earth and Environmental Science **288** (2019) 012001 doi:10.1088/1755-1315/288/1/012001

with respect to ions of organic substances was decreasing in the following series: p-dinitrobenzene > o-toluidine> o-phenylenediamine.

The adsorption effectiveness of organic substances' ions under dynamic conditions was the highest for the bentonite modified with CNT. It was slightly less for the bentonite modified with glycerol, and even less for the bentonite modified with glycerol and CNT. The lowest values were found for unmodified granular bentonite.

The results of investigating the adsorption mechanism of composite adsorbents with respect to the studied organic substances confirmed the expediency of using the proposed adsorbents as the constituents of sorption filters for treating the wastewaters of chemical industry enterprises.

Developing the composition of multicomponent adsorption filters

Comparison of the efficiencies of water purification from various nitro and amino compounds made it possible to single out the following compositions of the filtering loads for multicomponent adsorption filters (the list of filter layers is presented in a top-down direction):

– Zeolite; bentonite modified with CNT and glycerol after thermal treatment at 550 °C with a gradual increase in temperature; organobentonite TU 952752-2000; bentonite modified with CNT and glycerol after heat treatment at a temperature of 550 °C; silica gel brand ASKG; Bentonite after heat treatment at a temperature of 800 °C. The efficiency of water treatment was 98.9%;

– Bentonite modified with CNT thermally treated at a temperature of 550 °C; AB-17-8 anion-exchange resin; ASKG silica gel. The efficiency of this composition was 98.8%;

– Peat; zeolite; organobentonite TU 952752-2000; bentonite after heat treatment at 650 °C; bentonite modified with glycerol after heat treatment at 650 °C. The efficiency of water treatment was 95.4%.

Therefore, we recommend these filtering load compositions for using in multicomponent adsorption filters designed for industrial sewage treatment from nitro and amino compounds.

Developing innovative technological designs for wastewater treatment from aromatic compounds

Multicomponent adsorption filters (MAF) with described filtering load compositions are advisable to use at local wastewater treatment plants. The proposed technological design of the local wastewater treatment station is shown in Fig. 1.

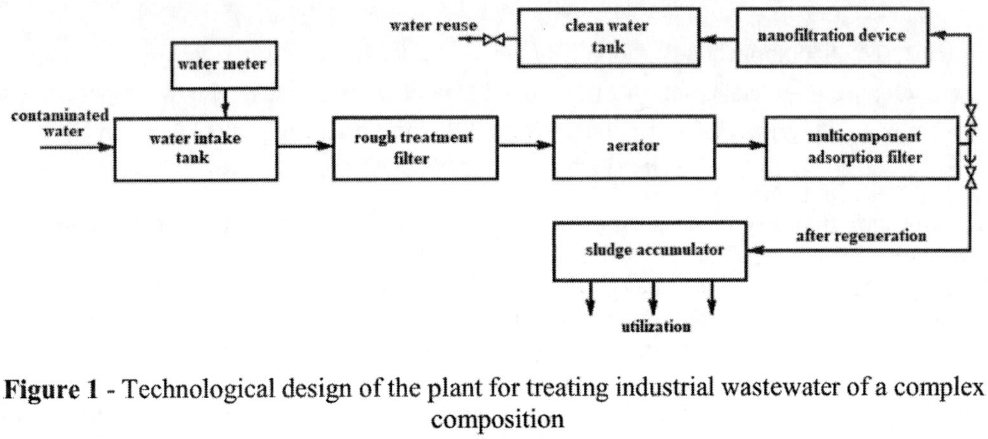

Figure 1 - Technological design of the plant for treating industrial wastewater of a complex composition

The technology of the proposed design presumes the following stages of water purification:

- Wastewater enters the receiving tank, neutralized by chemical interaction with substances (mixing acidic wastewater with alkaline waters, addition of the reagent in required concentrations). Volume of incoming sewage is measured by water meters. Further, settling of water in the water intake tank takes place;

- Wastewater neutralized and suspended from mechanical impurities is subjected to rough cleaning on filters;
- At the next stage, preliminarily treated wastewater is subjected to aeration using an oxygen compressor;
- Then, water is deferred, softened and purified from organic substances while passing through a multicomponent adsorption filter with a complex filtering load consisting of various modified bentonites;
- Further on, purified water passes the post-treatment stage in the nanofiltration block, dripping through the membrane filters;
- Treated water enters a storage tank for clean water, where its volume is measured, and then discharges into an open reservoir;
- Filtering load, including several types of sorbents, which are placed in a multicomponent adsorption filter, periodically undergoes regeneration;
- Sorbent, completely saturated with contaminants, is utilized with lowering its hazard class and its further use in construction.

A sorbent of a multilayer filtering load is placed in each of the MAF cassettes. The number of withdrawable cassettes can be from 3 to 7 pieces (Fig. 2).

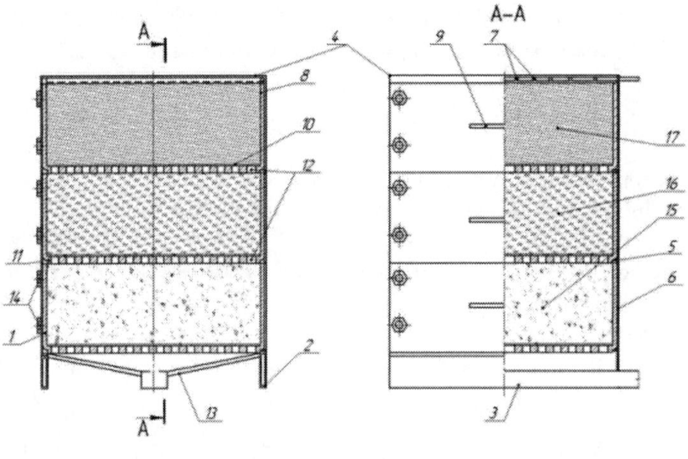

Figure 2 - Multicomponent adsorption filter with retractable cassettes: 1 - cassette; 2 - support; 3 - branch pipe; 4 – metal case; 5 - guides; 6 - side wall; 7 - perforated tube; 8 - upper cassette; 9 - bracket; 10 - grid; 11 - distributing device; 12 - short branch pipes; 13 - catchment tray; 14 - bolt; 15 ... 17 - sorbents [9]

The adsorption filter [9] includes cassettes located one above the other, filled with various kinds of sorbents, and a branch pipe for the drainage of purified water. Cassettes are made of high-strength plastic and placed in a metal case; they extend along the guides, fixed to the side walls of the housing. In the ceiling part of the metal case, there are perforated tubes, through the holes of which water to be purified is fed into the upper cassette. The lower part of each cassette is equipped with a grid and a distributing device, guiding the water to be purified from the upper cassette into the lower cassette through short branches. In the bottom of the metal case, there is a catchment tray for purified water, to which a branch pipe for draining purified water is attached.

The team of scientists at the Scientific and Educational Center "Industrial Ecology", SSTU, proposed and patented several innovative designs of adsorption filters for local wastewater treatment plants [10]. The results of industrial tests conducted at the scientific-industrial enterprise "LISSKON" confirmed the feasibility of the proposed technological solutions for industrial wastewater treatment.

Conclusions

We proposed and investigated the principle of combining various sorbents in a single multicomponent filter, based on their selection for various groups of organic contaminants and on the ability of these contaminating substances to ionize in an aqueous environment.

The parameters, characterizing the adsorption efficiency of organic compounds by unmodified bentonite and bentonite modifications under different static and dynamic conditions, were identified. Bentonite modified with CNT proved to be the most effective with respect to organic compounds. Its parameters with respect to o-toluidine were: SEC = 58.42 meq / g, K_d = 0.07 mg / l, S = 73.03%. Its parameters with respect to o-phenylenediamine amounted to: SEC= 51.31 meq / g, K_d = 0.05 mg / l, S = 64.14%. For p-dinitrabenzene, SEC = 1.56 meq / g, K_d = 0.14 mg / l, S = 83.57%.

Adsorption of o-toluidine and o-phenylenediamine on the studied bentonites proceeded according to the ion-exchange mechanism. The adsorption of p-dinitrobenzene under static conditions was described by the Langmuir isotherm.

We developed and tested innovative designs of multicomponent adsorption filters for water treatment from a complex of organic pollutants [9, 10], designed to provide comprehensive purification of wastewaters, contaminated with a complex of organic compounds. The expediency of using the proposed multicomponent adsorption filters was confirmed by the results of production tests at a scientific-industrial enterprise "LISSKON".

Acknowledgements and Conflict of Interests

The authors are grateful to the scientific-industrial enterprise LISSKON located in the city of Saratov, Russia, for using their laboratory space and equipment for conducting the research. The study was supported by the grant No. 5.3922.2017/PCh from the Russian Federation Ministry of Education. The authors state no conflict of interests involved.

References

[1] Grushko Ya M 1982 Vrednyye organicheskiye soyedineniya v promyshlennykh stochnykh vodakh: Spravochnik. 2-e izd.. pererab. i dop. (L.: Khimiya) 216 s (Rus.)

[2] Ochistka i ispolzovaniye stochnykh vod v promyshlennom vodosnabzhenii / A.M. Koganovskiy [i dr.] 1983 (M.: Khimiya) 288 s (Rus.)

[3] Veselovskaya EV 2000 Issledovaniye protsessa adsorbtsii korrozionnoaktivnykh soyedineniy kondensata TES na modifitsirovannoy uglerodsoderzhashchey poverkhnosti // Prioritetnyye napravleniya razvitiya energetiki na poroge XXI veka i puti ikh resheniya: Materialy Vseros. konf. (Novocherkassk: YuRGTU) s 79-80 (Rus.)

[4] Dytnersky Yu I 1995 Processes and equipment of chemical technology: textbook for universities. Part 2. Mass exchange processes and equipment (Moscow: Chemistry) 368 p

[5] Sovershenstvovaniye sorbtsionnykh metodov ochistki zagryaznennykh prirodnykh i stochnykh vod: kollektivnaya monografiya / Pod red. dokt. biol. nauk. professora Tikhomirovoy E I 2017 (Saratov: SGTU imeni Gagarina Yu.A.) 154 s (Rus.)

[6] Alekseeva A A, Fazullin D D, Kharlyamov D A, Mavrin G V, Stepanova S V, Shaikiev I G and Shaimardanova A S 2016 The use of leaves of different tree species as a sorption material for extraction of heavy metal ions from aqueous media (International Journal Of Pharmacy & Technology) vol 8 No 2 p 14375-14391

[7] Tikhomirova E I, Atamanova O V, Kosarev A V [i dr.] 2018 Issledovaniye mekhanizma adsorbtsii orto-fenilendiamina na bentonitakh v staticheskikh usloviyakh (Fundamentalnyye issledovaniya) No 1 s 18-23 (Rus.)

[8] Kosarev A V, Istrashkina M V, Tikhomirova E I [i dr.] 2018 Kharakteristiki adsorbtsii o-toluidina na modifitsirovannykh formakh bentonita v zadache ochistki stochnykh vod (Sovremennyye naukoyemkiye tekhnologii) No 2 s 53-58 (Rus.)

[9] Atamanova O V, Kosarev A V, Tikhomirova E I i Istrashkina M V 2017 Patent na poleznuyu model No 174230 (19) RU (11) 174230 (13) U1. Sorbtsionnyy filtr (RF) zayavka No

2017115281 zayavleno 28.04.2017 opubl. 09.10.2017 Byul. «Izobreteniya. Poleznyye modeli» No 28 6 s (Rus.)

[10] Atamanova O V, Kosarev A V, Tikhomirova E I i Istrashkina M V 2017 Patent na poleznuyu model No 169004 (19) RU (11) 169004 (13) U1. Sorbtsionnyy filtr (RF) zayavka No 2016141893 zayavleno 25.10.2016 opubl. 01.03.2017 Byul. «Izobreteniya. Poleznyye modeli» No 7 7 s (Rus.)

IOP Conf. Series: Earth and Environmental Science **288** (2019) 012002 doi:10.1088/1755-1315/288/1/012002

Research and development of acid corrosion inhibitor for single stage process of isoprene synthesis

[1]Burganov R.T., [1]Shepelin V.A., [1]Gilmullina A.R., [1]Kovrizhnyh E.A., [2]Khasanova D.I., [3]Dikaryova U.M.,

[1]PJSC Nizhnekamskneftekhim, 423570, Nizhnekamsk, Tatarstan Republic

[2]National Research University, Moscow Power Engineering University

[3]JSC "Fine Chemicals R&D Center"

BurganovRT@nknh.ru

Abstract. New acid corrosion inhibitor OPC-2000 has been developed for a single stage process of isoprene synthesis. Gravimetric method has found high effectiveness of the inhibitor in condi- tions of dilute ortho-phosphoric and sulfuric acids towards alloys Incoloy 825 and 12X18H10T being applied within the range of 10-16 g/dm^3. According to electrochemical investigation, OPC- 2000 is an anodic class inhibitor. It has high dispersing properties (no sediment formation in reac- tors), the fact being confirmed in the course of PJSC Nizhnekamsneftekhim's industrial trials while a single stage of isoprene synthesis.

Acid corrosion of chemical and petrochemical facilities is one of the major challenge of industries and one of the most important factors, determining the run life of process facility.

One of the most effective ways of protection from the acid corrosion is using inhibitors, generally im- proving mechanical properties of steels and alloys [1].

Ortho-phosphoric acid is used as a catalyst in the single stage process of isoprene synthesis which, due to entry of the process materials (*i.e.* hydrocarbons) becomes very corrosive, making the run life of Incoloy 825 reactors very short and raising the risk of emergency shut-down of production facilities. In spite of high corrosion resistance of this alloy, the run life of the reactors is not more than two years, as a result. Incoloy 825 is an alloy of nickel (38-46%), chromium (19-23%), iron (28-31%), and molybdenum, titani- um and copper. Considering this fact, metal salts soluble in these conditions should be used as inhibitors of the acid corrosion due to employment of these salts (salts of chromium, copper, nickel, molybdenum) as alloying components for the major reactor materials [1-3].

A great number of gravimetric corrosion tests in different acid conditions were carried out to determine the optimal composition of inhibitor and its dosing. As a result of the research, the inhibiting composition of OPC-2000 grade [4] with the most effective and optimal composition and its formulation was devel- oped at R&D of PJSC Nizhnekamskneftekhim.

The work depicts an influence of the new acid corrosion inhibitor of OPC-2000 grade in conditions of pure 6% sulfuric and 6% ortho-phosphoric acids containing added quantities of 2% TMC (trimethyl car- binol) and 0.2% formaldehyde. The research was carried out with specimens made of alloy Incoloy 825 and stainless steel 12X18H10T, as this steel is one of the most commonly used, strong and less expensive constructive materials.

Metal specimens of 20x10x3 mm size were kept in solutions of corrosion media containing the inhibitors under study as well as without using the latter. Protecting effectiveness of the inhibitor was determined as loss of mass of the metal specimen being kept in corrosion medium for 6 hours, at temperature 150°C. High temperature corrosion tests were carried out in a reaction vessel namely, in bomb calorimeter of a self-sealing type. Electrochemical studies were carried out using potentiostat PI-50-1.1 in the standard three-electrode electrochemical cell (V = 0.08 dm3) having cathode and anode compartments separated and naturally aerated at 50 °C [5-6]. The metal specimens were analyzed by scanning electron microsco- py with the help of microscope (SEM, HITACHITM-3030) and accelerating electric potential 15 kW in the StandartMode, using special current-conducting adhesive tape as a support for the tested specimen (ConductingDouble-sidedTape). In parallel, we detected elements composing the surface of the speci- mens using the method of energy-dispersive microanalysis and spectrometer Quantax 70 (Bruker).

To evaluate the influence of admixtures on corrosive activity of phosphoric acid, some preliminary tests were performed employing pure ortho-phosphoric acid.

Table 1 – Inhibitor Effectiveness towards Stainless Steel 12X18H10T and Alloy Incoloy 825 @ 150°C. Exposure Period – 6 Hours.

Tested Medium	Concentration of Inhibitor, g/dm3	Corrosion Rate, mm/year	Inhibitor Effectiveness, Z, %
	12X18H10T		
6% solution of H_3PO_4 (control)	-	0.0915	-
+ OPC-2000	5	0.0365	60
	10	0.0000	100
	16	0.0000	100
	Incoloy 825		
6% solution of H_3PO_4 (control)	-	0.0398	-
+ OPC-2000	5	0.0000	100
	10	0.0000	100
	16	0.0000	100

According to Table 1, OPC-2000 provides 100% protection both for stainless steel & for Incoloy 825 within the medium of diluted ortho-phosphoric acid.

As a result of the laboratory studies, OPC-2000 was found to have high dispersing properties, as no sedi- ment formation was faced upon heating, all the solutions remaining clear (Figure 1).

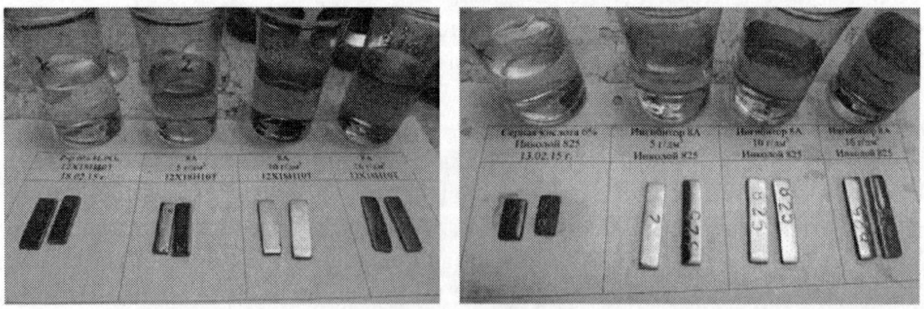

Figure 1 – Appearance of Stainless Steel 12X18H10T and Incoloy 825 Specimens after 6-hours Test in 6% Ortho-Phosphoric Acid

Besides, the inhibitor was studied in more aggressive conditions of sulfuric acid to evaluate its protective properties. The results are summarized in Table 2 and in the Figure 2.

Table 2 – Inhibitor Effectiveness towards Stainless Steel 12X18H10T and Incoloy 825 @ 150°C. Exposure Period – 6 hours.

Tested Medium	Concentration of Inhibitor, g/dm^3	Corrosion Rate, mm/year	Effectiveness of Inhibitor, Z, %	Inhibition Factor, γ
		12X18H10T		
6% solution of H2SO4 (control)	-	57.3289	-	-
	5	47.4247	17	1.2
+OPC-2000	10	27.0550	53	2.1
	16	0.1727	99	332.0
		Incoloy 825		
6% solution of H2SO4 (control)	-	0.4252	-	-
	5	0.0000	100	-*
+OPC-2000	10	0.0000	100	-*
	16	0.0000	100	-*

Note * - It is not possible to calculate γ (inhibition factor) at 100% protection (Formula 3).

The most effective dosage of OPC-2000 inhibitor for Incoloy 825 in 6 % sulfuric acid is 5-16 g/dm^3 (Ta- ble 2). The optimal dosage of OPC-2000 for the stainless steel in 6% sulfuric acid is 16 g/dm^3, and the rate of corrosion shows 332 time decrease.

High effectiveness of the studied inhibitor for the Incoloy 825 protection makes possible to mitigate nega- tive environmental impact being used in smaller dosages (10 g/dm^3).

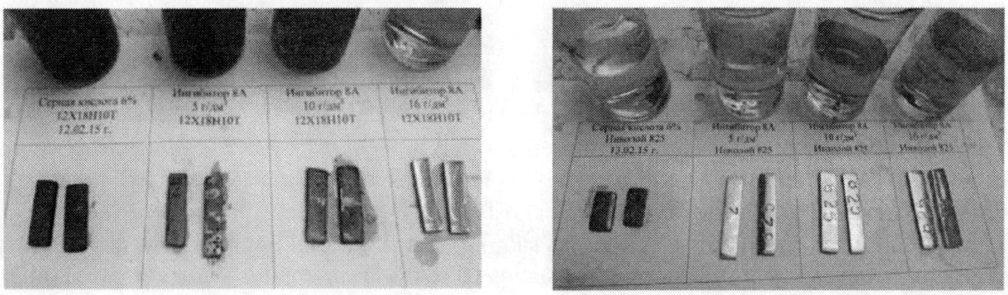

Figure 2 - Stainless Steel 12X18H10T and Incoloy 825 Samples after 6-hours Test in 6% Sulfuric Acid

The gravimetrical and electrochemical polarization measurements of inhibitor OPC-2000 show good con- sistency, *i.e.* high effectiveness of inhibitor in acid conditions is confirmed by polarization curves. Anode polarization curve, plotted without inhibitor (control), has a classic appearance with all standard regions of this curve, namely: intensive anodic dissolution (I), active-passive state (II), passive region (III) and transpassive region (IV – the region where well-soluble metal complex with higher oxidation number is formed).

The anode polarization curve (Figure3), plotted against the values for the stainless steel 12X18H10T in 6% solution of ortho-phosphoric acid without the inhibitor shows very low current density in potentials area from 650 to 750 mW, meaning passive state of the steel surface due to protective film formed to- gether with oxygen (*i.e.* oxide formation). The anodic passivation region of the stainless steel when the latter is in passive state is expanded from 250 to 620 mW at optimal dosage of inhibitor. Such expansion of the region in the presence of OPC-2000 is associated with protecting film formed against higher amount of the passivating substance on the surface and probably, improvement of protecting properties of the passivation layer formed by chemical absorption of the salts of alloying metal components. Shifting of 2,2*curves towards positive values witnesses the deceleration of anode metal dissolution, meaning that corrosion rate is slowdown by anode pattern.

As for the new inhibitor OPC-2000 for Incoloy 825 protection, it shows similar behavior. Most likely, that OPC-2000 inhibits hydrogen charging by several ways, namely: barrier film formation and changing in rate-controlling step of the hydrogen release or by film formation with simultaneous hydrogenation of some inhibitor's molecule part.

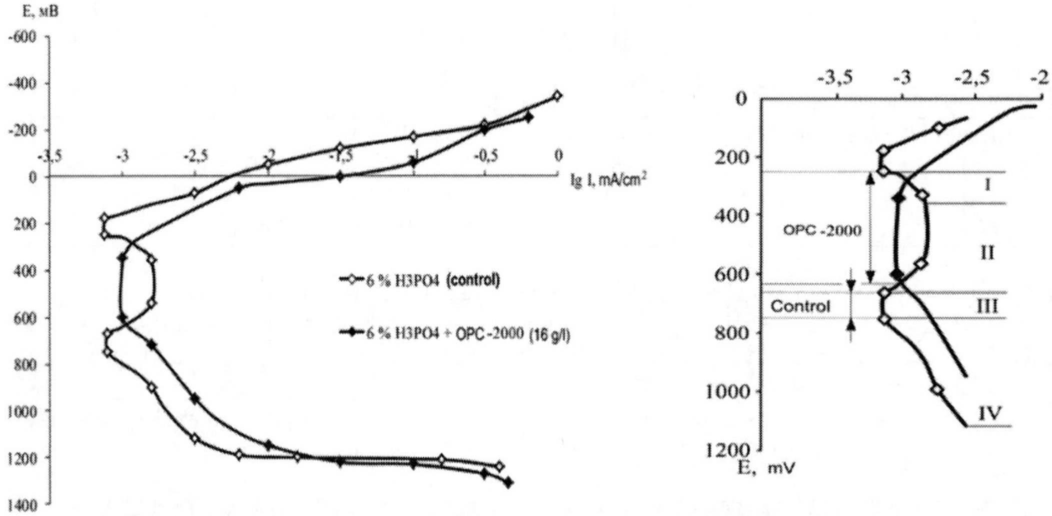

Figure 3 – Cathode and Anode Curves of Stainless Steel 12X18H10T in 6% Ortho-Phosphoric Acid at 50 °C

Comparative examination of samples photomicrograph, received by the scan electron microscope (SEM, HITACHITM-3030), shows the most vivid picture taken for alloy 12X18H10T in conditions of 6% sulfu- ric acid, as the latter promotes the most intensive corrosion processes. According to our observations, the surface of alloy 12X18H10T sample, tested in conditions without the inhibitor, was covered with crumbly crystal deposits, consisting of 70% iron oxide, with corrosion pits being detected underneath. The surface of the specimen, tested in the presence of inhibitor OPC-2000, had also spongy structure, but free from any visual signs of corrosion (Figure 4).

As for Incoloy 825 specimens, tested in 6% sulfuric acid with the inhibitor and without it, no deposits were found. The difference was in number of surface pittings formed with and without the inhibitor. Lines and groves that can be observed on the surface are caused by abrasion treatment of metal specimens before the corrosion studies. Similar situation is observed for the investigations of Incoloy 825 and 12X18H10T in 6% ortho-phosphoric acid.

Thus, this recently developed corrosion inhibitor OPC-2000, representing a multicomponent mixture of inorganic salts (chromium, copper, nickel, molybdenum), showed high effectiveness in Incoloy 825 and 12X18H10T alloys protection tested in 6% ortho-phosphoric acid and in more corrosion active medium of 6% sulfuric acid. According to the results electrochemical studies, OPC-2000 inhibitor belongs to an- odic type. Tests with added process materials (*i.e.* trimethyl carbinol and formaldehyde) have confirmed high effectiveness of inhibitor OPC-2000 being used at optimal levels of 10-16 g/ dm^3. OPC-2000 pos- sesses high dispersing properties, proved by the absence of the sediments formation upon heating as the solutions remained clear.

The industrial trials of a single stage isoprene synthesis at PJSC Nizhnekamsneftekhim have confirmed the inhibitor effectiveness.

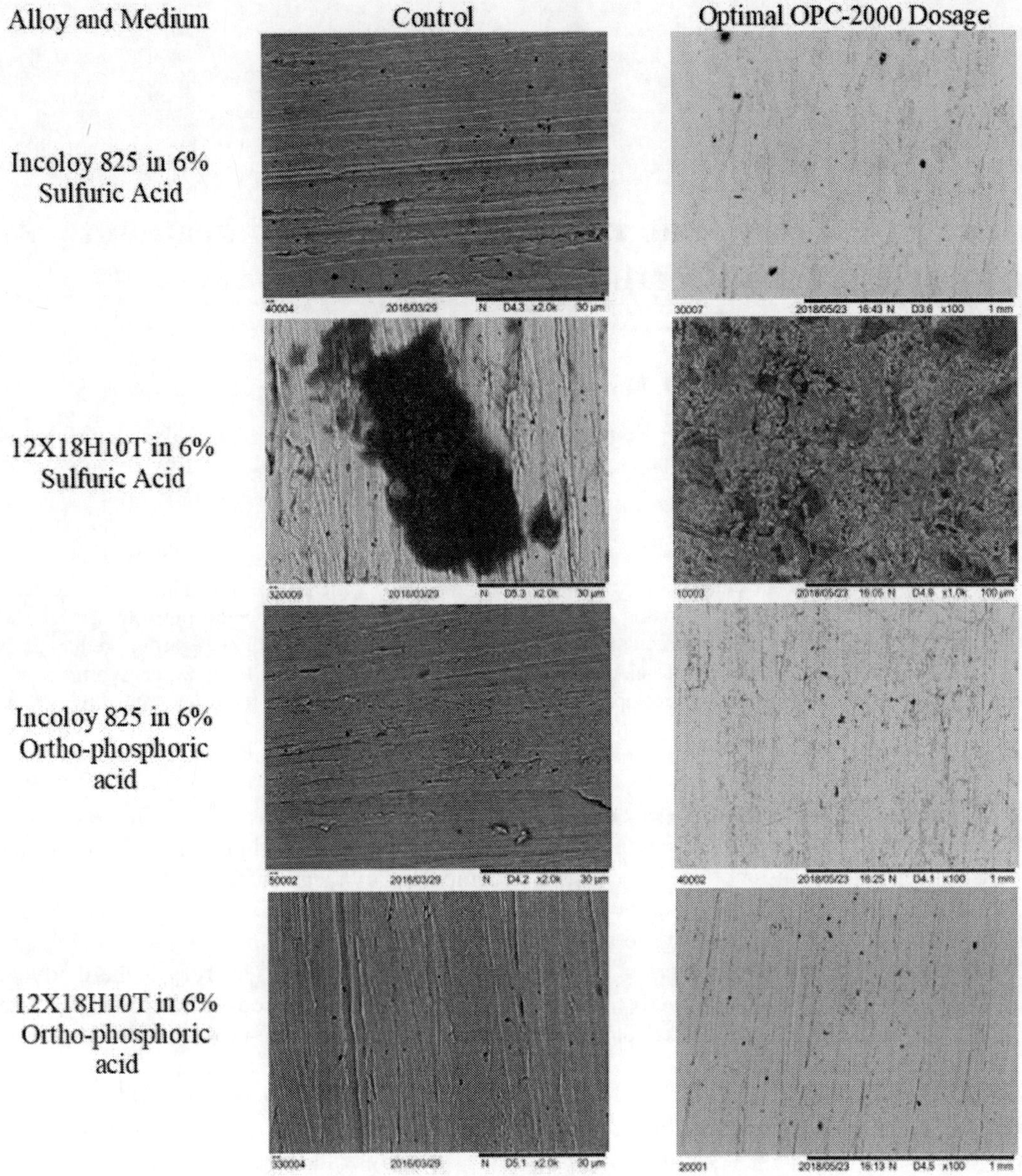

Figure 4 –Specimens' Picture after Testing (Scanning Microscopy)

Reference

[1] Ivanov E.S. *Metal Corrosion Inhibitors in Acid Media.* – M.: Metallurgiya, 1986.

[2] Ugrumov O.V., Ivshin Ya.V. *Nitrogen-Phosphorous-Containing Corrosion Inhibitors for Oil-Field Equipment.* – Kazan: Kazan State University, 2009.

[3] Bregman J.I. *Corrosion Inhibitors; eds. Professor Antropov L.I.* – L.: Khimia, 1966.

[4] *Patent RF No.* 22593569, 2015 *Inhibiting Composition for Metals against Acid Corrosion Reshetnikov S.M. Acid corrosion inhibitors for metals.* – L.: Khimia, 1986.

[5] Measuring Method No. 226. *Practice for High Temperature Corrosion Studies.*

[6] Freyman L.I., Makarov V.A., Bryksin I.E. *Potensiostatic Methods in Corrosion Studies and Electrochemical Protection.* – L.: Khimia, 1972.

IOP Conf. Series: Earth and Environmental Science **288** (2019) 012003 doi:10.1088/1755-1315/288/1/012003

Problems of energy and resource efficiency in heat and power engineering: water treatment

E.N. Bushuev[1], B.M. Larin[1], E.A. Karpychev[1]

[1]Ivanovo State Power Engineering University, 153003, Russian Federation, Ivanovo, Rabfakovskaya St., 34

admin@xxte.ispu.ru

Abstract. An analysis is submitted of the main developments in energy and resource efficiency both in currently operating water treatment systems and those being put into service. A big potential for energy efficiency exists in reverse osmosis installations regarding reduced source water consumption and also disposal of concentrate forming in the water treatment system cycle. A promising development in energy saving is reducing the temperature of raw water warming.

1. Energy and resource efficiency at thermal power plants (TPP)

The principal aim of energy and resource efficiency at currently operating thermal power plants is to reduce specific fuel consumption. This is often accompanied by measures to reduce operating costs, including those relating to repairs and operating personnel. The purpose of such measures is usually to increase the profitability of a power generating facility.

Modern heat and power engineering development is moving towards increased thermodynamic cycle efficiency both by changing over to binary (double) cycles in combined cycle gas turbine (CCGT), and by increasing the initial parameters of main steam in once-through ultra supercritical (USC) boilers using solid fuel.

Increased power generation unit efficiency is achieved by improving the thermodynamic cycle, optimizing plant cycle, improving main and auxiliary equipment and raising turbine inlet steam parameters. Table 1 shows data for possible increased efficiency with increased thermodynamic cycle parameters.

The problems of heat engineering development in this direction are related to the need for close coordination and control in the operation of different kinds of equipment (combined cycle units), construction materials, developed heat exchange surfaces, water quality (USC and CCGT units).

The potential for improved energy efficiency at coal-burning condensation power plants may be related to changing over to new steam turbine plants with ultra supercritical steam parameters. Under the leading role of the All-Russian Heat Engineering Institute (AHEI), a 600MW power generation unit was developed in 2007-2008 with a once-through boiler operating on Kuzbas grades G and D bituminous coal at 28.4 MPa (Pp-1860-28,4-600 CS). Boiler outlet steam parameters meet the following conditions: pressure – 28.4 MPa, temperature – 600 °C [2]. Calculated analysis was performed for once-through overheating boiler operation (Pp-1000-25-585) with circulating boiling layer and oxyfuel firing, and the possibility of using this boiler in big coal-fired power generation units with USC once-through boilers was demonstrated [3]. Data from the Russian Central Engineering Research Institute (CERI) are given for new heat resistant chrome steel grades 12KH10M1V1MFBR and 10KH9V2MFBR for turbine rotors, boiler piping, and steam piping for USC power generation units implemented in Russian industry [4]. With parameters increased from standard p_0 = 25 MPa, t_0 =545°C to p_0 = 30 МПа, t_0 = 700°C, power generation efficiency is increased by approximately 45%. Plant cycle optimization and improvements in main and auxiliary equipment allow a further 2.4% increase in efficiency.

Table 1. Potential for improving thermodynamic cycle parameters [1]

Measure	Increase in efficiency, %
Increasing main steam temperature by 1 °C	0.02
Increasing main steam pressure by 1 MPa	0.1
Increasing re-superheating temperature by 1 °C	0.015
Using second resuperheating	1.2
Lowering condensate pressure by 1 kPa	1
Increasing feed water temperature by 1°C	0.02

At present, highly maneuverable combined cycle power generation units burning natural gas with efficiency of more than 55-57% are widely used.

Apart from improved economic indicators in power generation, increased power plant efficiency is one of the principal ways to reduce harmful atmospheric emissions.

2. Energy and resource efficiency in water treatment systems

The tasks of achieving energy and resource efficiency in water treatment, cooling and water-chemical regime systems in heat power equipment main and auxiliary circuits at thermal power stations are al- ways relevant, and increasingly so with growing electric and heat power consumption. The problems of fulfilling these tasks are related to the need to ensure standard water quality with limited discharge of waste water, and in the final analysis are determined by the cost of technological measures.

Water treatment plant (WTP) at thermal power plants in an overwhelming majority of cases are represented by traditional installations for pretreatment (clarifiers and mechanical filters) and ion exchange filters (softening and demineralization). The last 10 years have seen the active introduction of reverse osmosis systems (ROS) at thermal power plants, usually, however, with the retention of ion exchange filters.

Energy efficiency for water treatment facilities at thermal power plants is determined by reducing own electric power consumption and heat consumption for source water heating. Electric power consumption may be reduced by adjusting design, for example, by reducing the number of high output pumps and providing their electric motors with frequency drives. The number of operating pumps may be reduced, for example, by bypassing decarbonizers in water demineralization circuits with filter chains during lime softening of water in clarifiers. Heat energy consumption may also be reduced by adjusting design. Thus, in some cases source water for WTP is drawn from the return line of circulating water piping, that is, steam turbine condensate cooling system. At some thermal power plants, the condensers are provided with «built in» loops, ensuring independent source water heating with heat from steam condensation. Moreover, in all these cases reduced water consumption for own needs and wastewater flow in WTP leads to reduced wastewater losses.

The problems of implementing the above and other energy efficiency measures are usually related to the need to change (complicate) the technological designs of currently operating equipment, sometimes with the necessity of installing automatic process monitoring (including chemical) and control.

The main criteria regarding the choice of water demineralization processes are economic. A comparison of demineralization operating costs demonstrates the advantage of chemical methods – ion exchange at strong acid anion concentrations ($[Cl^-]+[SO_4^{2-}]$) of less than 3 mmol/l [5]. Such low mineralized water predominates in Russia's central and northern regions.

In addition to developing promising water treatment technologies for thermal power plants, it is also necessary to achieve the highest possible indicators for currently operating natural water demineralizers. Energy and resource efficiency with traditional chemical demineralizers may be achieved by implementing the following measures [6]:

- stepwise regeneration of first stage H-cation exchange filters with increasing sulfuric acid concentration;
- regeneration of second stage OH-anion exchange filters with 6-12 hour maturation of NaOH solution in the filter after passing 65–70 % of working fluid volume;
- periodic chemical treatment of anion exchange in first stage filters by saline-alkaline or acidic-saline regeneration.

Changing over to promising ion exchange technologies in express (high velocity) filters ensures energy and resource efficiency, but requires thorough automated monitoring and control of equipment operation.

Membrane water treatment technologies have significant advantages over ion exchange according to environmental indicators (Fig. 1), but require more thorough water pretreatment.

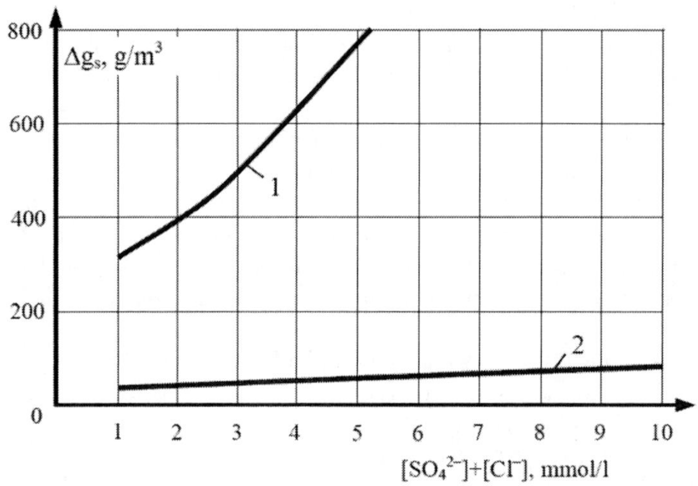

Fig. 1. Relative salt discharge without taking into account salt discharge from water supplied with source water (Δg_s) depending on method of desalting and mineralizing source water: 1 – chemical desalting based on countercurrent ion exchange filters; 2 – desalting based on reverse osmosis with further desalting in ion exchange filters

Water treatment systems based on membrane technologies are characterized by high consumption of waste water (up to 50% of system capacity).

Reverse osmosis systems have big resource efficiency potential and possibilities for lower source water requirements, also disposal of concentrates forming in the water treatment cycle. Using a number of design adjustments for reverse osmosis systems, such as multicascade and mixing of water flows, permits a reduction in own requirements (up to 30%) with negligible deterioration in the quality of the resulting permeate [7, 8]. Reduced consumption of concentrates leads to reduced concomitant heat loss.

A number of design adjustments are proposed, envisaging partial softening of reverse osmosis system concentrate in an H-carboxyl filter [9]. The performed calculations show that using the proposed technologies enables up to 80% of concentrate to be returned to the system inlet. Adding processed concentrate to ROS source water reduces the likelihood of carbonate deposits forming on the membrane, enabling us to simplify, and often dispense with source water treatment.

Processed concentrate may also be used for other purposes, for example, after softening in a carboxyl H-cation exchanger and decarbonization to replenish the heating system.

Regeneration of a H-carboxyl cation exchanger is performed by stoichiometric consumption of sulfuric acid. The resulting discharge is diverted into a gypsum crystallizer; the resulting precipitate may be dehydrated at a slime compacting station. The product obtained in this way meets the material re- quirements for gypsum binding substances.

One of the important issues to be resolved in the design and operation of reverse osmosis systems is choosing a rational treatment of source water to prevent the formation of sediment on membrane surfaces. For this purpose, the following methods of treating source water are commonly used [7]:

- acidification;
- softening with ion exchange (Na-cation exchange or H-carboxyl) filters;
- addition of antiscalant;
- nano-filtration.

Fig. 2 shows the estimated relation between consumption of reagents to achieve a Langelier index (LSI) of not more than 0.7 and the degree of source water concentration at reverse osmosis system inlet, with various kinds of source water pretreatment. The resulting increased quantity of discharged salts is shown in Fig. 3.

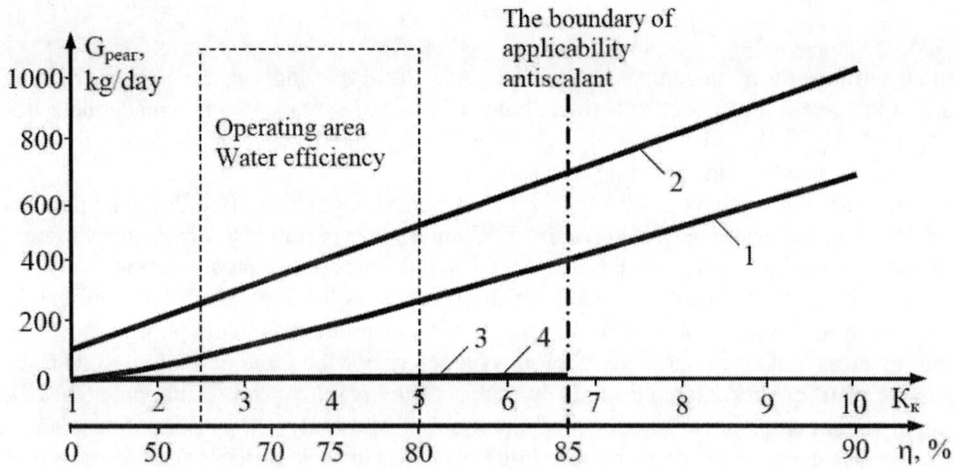

Fig. 2. Dependence of reagent consumption for source water treatment at reverse osmosis system inlet on permeate concentration ($K_к$) and discharge (water efficiency η):
1 – sulfuric acid during acidification; 2 – table salts with countercurrent Na-cation exchange;
3 – addition of antiscalant; 4 – reagents for periodic washing (rinsing) of nano-filtration modules

Analysis of the results of numeric study (examination) shows maximum consumption and thus maximum quantity of discharged salts occur with Na-cation exchange, although this is the most reliable method of preventing the formation of harness cation deposits on membrane surfaces. The least consumption of reagents occurs when water is treated with antiscalant. The toxicity of these substances during discharge of reverse osmosis system concentrate should be borne in mind, however.

An important energy efficiency issue in water treatment systems is the possibility of lowering the temperature at which source water is heated.

This temperature may be lowered by using flocculants at the pretreatment stage. These are already being used at most of water treatment systems, with, however, “out of date” polyacrylamide (PAA) or other flocculants that do not always achieve the required results.

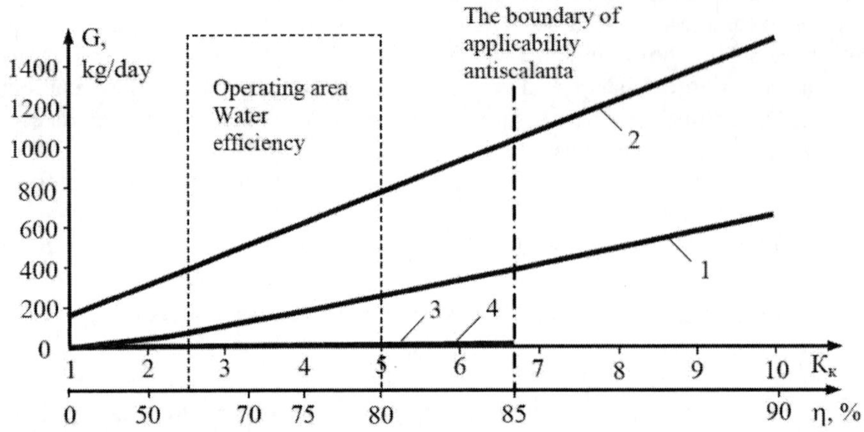

Fig. 3. Dependence of daily discharged ions on permeate concentration ($K_к$) and discharge (water efficiency η) and type of source water treatment at reverse osmosis system inlet:
1 – acidification with sulfuric acid; 2 – Na-cation exchange; 3 – addition of antiscalant;
4 – nano-filtering

Inadequate efficiency of flocculants may be due to the following causes: improper preparation, storage and transportation of flocculants; bad choice of points for reagent entry; inadequate flocculant mixing; unsuitability of flocculant grade for clarifier hydraulic and slime modes; excessive heating of source water (not related to lime softening modes).

Heating source water to above 25 °C does not always have a positive effect for coagulation modes. On the one hand, source water temperatures of 25 °C and higher permit faster hydrolysis of coagulants and lower water viscosity, but on the other hand intensify the process of gas formation.

There is the notion of «cold coagulation», which presupposes heating water to a significantly lower temperature, even to the point of treating the water without preheating. This permits not only lower thermal energy costs, but also improved hydraulic characteristics for slime in relation to its flocculence actually in the clarifier. For example, rapid hydrolysis may result in the formation of heavy slime, containing in its structure huge quantities of microbubbles, mainly CO_2. Using flocculants in this mode permits concentration of the slime into large objects with a significant surface area, and hence high lifting capacity. This leads to foaming in a significant part of the slime on the surface of the water catchment spouts, poor water settling in the slime compactor, and consequently, increased extent of continuous blowing and lower «cut-off» water quality. « Cold» coagulation has a positive effect on the hydraulic mode of clarifiers, and specifically helps to suppress floating slime by the action of released gases.

Water treatment at lower temperatures is also possible in ultrafiltration systems. Research shows [10] that although the coagulation process in cold water (at temperatures of 8–11 °C) is worse than in heated water, treatment by the ultrafiltration method remains stable with only a small increase in coagulant dosage. Water treatment at lower temperatures increases water viscosity, resulting in better membrane retaining ability, which gives the ultrafiltration method an advantage over traditional pretreatment methods, where lower water temperature leads to larger doses of reagents and aluminium slip into clarified water. Lower clarified water temperatures have a favourable impact on the work of reverse osmosis membranes.

3. Conclusion

High energy and resource efficiency indicators at thermal power plants may thus be achieved by various technological measures, including equipment modernization, improved efficiency of power generation units, and others. In addition to this, there is a need for modernization of auxiliary equipment, particularly water treatment systems. Furthermore, high energy and resource efficiency indicators may be achieved both at currently operating and new water treatment systems at thermal power plants.

References

[1] Tumanovskii A.G., Shvarts A.L., Somova E.V., Verbovetskii E.K., Avrutskii G.D., Ermakova S.V., Kalugin R.N., Lazarev M.V. Review of the coal-fired, over-supercritical and ultrasupercritical steam power plants // Thermal Engineering. 2017. Vol. 64. № 2. pp. 83–96.

[2] Tumanovskii A.G., Shvarts A.L., Verbovetskii E.Kh., Tugolukov E.A., Smyshlyaev A.A., Nesiolovskii O.V., Petrova N.V. A coal-fired boiler for a new-generation power unit for ultrasupercritical steam conditions // Thermal Engineering. 2009. Vol. 56. № 6. pp. 447-455.

[3] Supranov V.M., Ryabov G.A., Mel'Nikov D.A. Studying the possibility and advisability of using a Pp-1000-25-585 circulating fluidized-bed boiler in the oxyfuel combustion mode // Thermal Engineering. 2011. Vol. 58. № 7. pp. 593-601.

[4] Dub A.V., Skorobogatykh V.N., Shchenkova I.A. New heat-resistant chromium steels for a promising objects of power engineering // Thermal Engineering. 2008. Vol. 55. № 7. pp. 594-601.

[5] Yurchevskii E.B., Larin B.M. Development, study, and introduction of water-treatment equipment with improved environmental characteristics // Thermal Engineering. 2005. Vol. 52. № 7. pp. 532-538.

[6] Larin B.M., Bushuev E.N., Larin A.B., Karpychev E.A., Zhadan A.V. Improvement of water treatment at thermal power plants // Thermal Engineering. 2015. Vol. 62. № 4. pp.286-292.

[7] Panteleev A.A., Riabchikov B.E., Khoruzhii O.V., Gromov S.L., Sidorov A.R. Membrane technologies in the industrial water treatment –Moscow: DeLi plus, 2012. 429 p.

[8] Bushuev E.N., Eremina N.A., Zhadan A.V. Analysis of Water Treatment Modern Technology at Heat Power Plants // Vestnik of Ivanovo State Power Engineering University. 2013. №1. pp. 8-14.

[9] Yurchevskii E.B., Solodiannikova Yu. V., Pichugina M.A., Kavun O. Yu. Vortex reactors in the schemes of utilization of concentrate of reverse osmosis units at Thermal power plant and Nuclear power plant // Tyazheloe mashinostroenie. 2016. № 6. pp. 9-14.

[10] Zhadan A.V., Bushuev E.N. Rationale for selecting water treatment schemes at thermal power plants based on ultrafiltration technique // Vestnik of Ivanovo State Power Engineering Uni versity. 2014. №2. pp. 5-9.

IOP Conf. Series: Earth and Environmental Science **288** (2019) 012004 doi:10.1088/1755-1315/288/1/012004

Potentiometric analysis of metastable carbonate water coolants of thermal power plants and heat supply systems

A A Chichirov, A A Filimionova, B A Gilfanov, A A Gafiatullina and N D Chichirova

Kazan state power engineering university, Russia, 420066 Kazan, Krasnoselskaya, 51

b_gilfanov@bk.ru

Abstract. The paper describes a new method for studying the structure of metastable carbonate solutions, based on obtaining pK- and pH-spectra of solutions from potentiometric titration data. The decoding method is used to obtain the equations for calculating the coordinates of the spectra. In the production of thermal and electrical energy in modern thermal power plants (TPP) operate with water systems that have a constantly producing phase instability. Such systems include (1) water systems in intermediate stages of water treatment, in particular lime-coagulated or clarified water after pre-treatment, (2) cooling water in the circulating cooling system (COO) of the station, (3) water heat transfer medium in conjugated heat networks , (4) boiler water. All these systems are characterized by the presence of a stabilized (metastable) solid phase, mainly calcium carbonate, capable of forming a crystalline phase of calcite under certain conditions. The crystallization process is undesirable, since it can lead to dense deposits on the heat exchange surfaces. Approximately the instability of such systems is characterized by the value of ΔQw - by changing the total alkalinity of water in time (hours, days). The method of quantitative determination of stabilized calcium carbonate in water systems of TPPs is currently lacking. Development of a technique for potentiometric analysis of carbonate-alkaline solutions.

The experimental part

The initial solutions Na2CO3, NaHCO3, CaCl2, NaOH were prepared on the basis of bidistillate and dry chemically pure substances. Model solutions were obtained by mixing the starting solutions in a certain proportion. About 10 minutes after the preparation, potentiometric titration of model solutions with 0.1 N HCl solution was carried out. Titration was carried out in a potentiometric cell, closed from the atmosphere, with a working glass electrode and a silver chloride reference electrode. During the titration, the solution was stirred with a magnetic stirrer. Before and after the experiment, the electrode was calibrated using standard buffer solutions.

To find the equivalence point, a differential curve is frequently constructed in the coordinates δE / δV - V, (δpH / δV - V). The maximum of the obtained curve indicates the point of equivalence, and the reading along the abscissa axis corresponding to this maximum gives the volume of the titrant consumed for titration up to the point of equivalence. The definition of the equivalence point along the differential curve is much more accurate than the simple E-V (pH-V) dependence.

The obtained results were presented in the form of dependences in the coordinates pH-V (HCl), pH-f, δpH / δf-f, and δpH / f-pH, q-pKa. Here, f = nt / n0 is the degree of titration, which is equal to the ratio of the quantity of the currently tipped substance (nt) to the original quantity (n0), pKa is the acid dissociation constants. The obtained experimental curves were compared with the results of

IOP Conf. Series: Earth and Environmental Science **288** (2019) 012004 doi:10.1088/1755-1315/288/1/012004

calculation of chemical equilibria in model solutions using the previously written applied computer program (PP) "Structure Analysis and Classification of Water" [2 ± 5].

Results and discussion

The formation of calcium carbonate can occur according to different mechanisms, depending on the total concentration of calcium ions with (Ca2 +) in solution [3]:

with (Ca2 +) ≥ 10 mM - spontaneous precipitation;

with (Ca2 +) ≥ 3 mM - delayed homogeneous formation of nucleation centers (embryos);

with (Ca2 +) ≥ 2.5 mM - deposition on solid microparticles;

with (Ca2 +) ≤ 2,5 mM - formation of supersaturated solution and subsequent precipitation of precipitate.

In this case supersaturated (metastable) solutions retain phase stability up to several days. A study of the kinetics of the process showed [4, 5] that the precipitation of CaCO3 proceeds in two stages: the formation of calcium carbonate and its subsequent crystallization. Moreover, the first stage - the formation of salt as a result of ionic transformations - proceeds rapidly during a time of 1 to 10 minutes.

The structure of metastable solutions was studied by the potentiometric method. For guaranteed entry into the region of supersaturated solutions, the measurements were started 10 minutes after the initial solutions were drained. In alkaline carbonate solutions that do not contain hardness ions, there are three potential jumps - at pH 3.5 ÷ (I), pH 8.2 (II) and pH 10 ÷ 11. When calcium ions with a concentration of more than 1 mM are introduced into the solution, which corresponds to an excess of PR (CaCO3), there are four jumps.

On the concentration-logarithmic diagram calculated by means of PP (Fig. 1), it is possible to visually determine the coordinates of the titration curves and the reactions corresponding to them.

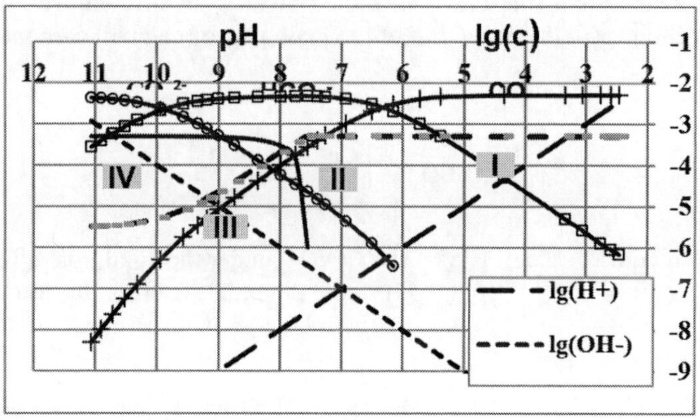

Figure1. Concentration-logarithmic diagram of CaCl2 solution (1 mM), Na2CO3 (10 mM), NaOH (1mM).

For the I point of equivalence (pH 3 ÷ 5), the potential-determining reaction-neutralization of bicarbonate ions

$$H^{+} + HCO_{3}^{-} \Leftrightarrow H_{2}CO_{3}, \qquad (1)$$

which is characterized by the dissociation constant of weak carbonic acid in the first stage (K1)

$$K_1 = \gamma_1^2 \frac{[H^+]\cdot[HCO_3^-]}{[H_2CO_3]}. \tag{2}$$

At the equivalence point: [HCO3-] = [H +]

For the II equivalence point (pH 7 ÷ 8,2), the potential-determining reaction is the carbonate-bicarbonate-ion transition, including ions, formed during the dissolution and dissociation of calcium carbonate. The diagram (Figure 1) shows that in the region between the II and III equivalence points, complete dissolution of CaCO3 occurs.

$$H^+ + CO_3^{2-} \Leftrightarrow HCO_3^{2-} \tag{3}$$

$$CaCO_3 \Leftrightarrow Ca^{2+} + CO_3^{2-}. \tag{4}$$

The equilibrium is characterized by a dissociation constant of carbonic acid in the second stage (K2) and PR (CaCO3)

$$K_2 = \gamma_2 \frac{[H^+]\cdot[CO_3^{2-}]}{[HCO_3^-]}. \tag{5}$$

At the equivalence point [CO32-] = [CO2] or max [HCO3-].

For the III equivalence point (pH 6.5 ÷ 8.2), the potential-determining reaction is the carbonate-bicarbonate-ion transition, with the exception of the main part of calcium carbonate, which remains in the insoluble state

$$\Pi P_{CaCO_3} = \left[Ca^{2+}\right]\cdot\left[CO_3^{2-}\right]. \tag{6}$$

In the third point of equivalence of [Ca2 +] = [CO32-] under the condition with (CO2) ≥ s (Ca2 +).

For the IV equivalence point (pH 10.5 ÷ 11.5), the potential-determining reaction-neutralization of hydroxyl ions:

$$OH^- + HCO_3^- \Leftrightarrow H_2O + CO_3^{2-}. \tag{7}$$

In the fourth point of equivalence, [OH-] = [HCO3-].

In the equations (2) and (5), the ion activities are equated to their concentrations or to the coefficients $\gamma = 1$.

The results show that in the pK spectrum only the maximum of the peak is observed in the region of the characteristic value of pK0 (coincides with pK2). The peaks of other peaks lie in the region pK = pH ($\alpha = 1/2$). Therefore, for practical purposes, the use of pH values ($\alpha = 1/2$) seems preferable to us.

In Fig. 4 represents the experimental and calculated theoretical pH spectra of the solution under study. Good agreement with the concentration-logarithmic diagram indicates the correctness of the correlation of peaks and reactions. The number of peaks and their coordinates (pHeq) unambiguously characterize the composition and structure of the solution.

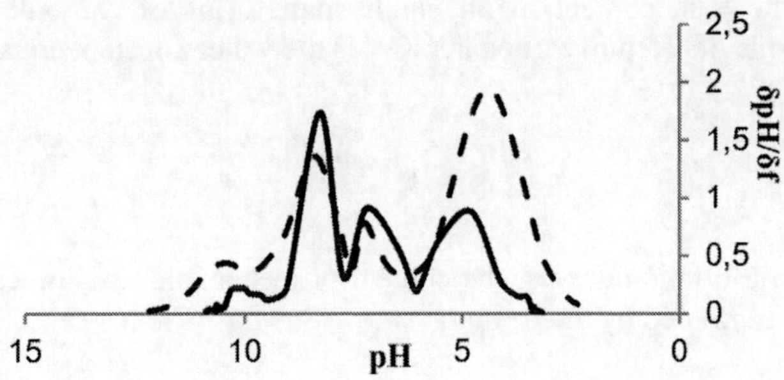

Figure 2. The pH spectrum of a model solution of CaCl2 (2 mM), Na2CO3 (10 mM), NaOH (2 mM).

For an approximate calculation of the characteristics of potentiometric curves and pH spectra, we use the basic equation obtained by summing the expressions (2), (5), (6).

$$a_{\mathrm{H}^+}^2 \cdot \frac{\sqrt{\Pi\mathrm{P}}}{K_1 \cdot K_2} + a_{\mathrm{H}^+} \cdot \frac{\sqrt{\Pi\mathrm{P}}}{\gamma_1 \cdot K_2} + \frac{\sqrt{\Pi\mathrm{P}}}{\gamma_2} = c_{\mathrm{CO}_2}; \tag{8}$$

Taking the activity coefficients γ = 1, we obtain an expression for the concentration of hydrogen ions:

$$\left[\mathrm{H}^+\right] = \frac{-\dfrac{\sqrt{\Pi\mathrm{P}_{KK}}}{K_2} \pm \sqrt{\dfrac{\Pi\mathrm{P}_{KK}}{K_2^2} - 4 \cdot \dfrac{\sqrt{\Pi\mathrm{P}}}{K_1 K_2} \cdot \left(\sqrt{\Pi\mathrm{P}_{KK}} - c_{\mathrm{CO}_2}\right)}}{2 \cdot \dfrac{\sqrt{\Pi\mathrm{P}}}{K_1 \cdot K_2}}. \tag{9}$$

For regions near the peaks on the curve of the pH spectrum, after appropriate simplifications, we obtain expressions for finding the coordinates of the spectrum (Table 1).

Table 1. Characteristics of pH-spectra in systems CO32- - OH - - Ca2 +

Peak	Value pH	Formula for calculating the pH of ECB	
		$Na_2CO_3 + NaOH$	$Na_2CO_3 + NaOH + CaCl_2$
1	3-5	$\frac{1}{2}\left(\mathrm{p}K_1 - \lg c_{\mathrm{CO}_2}\right)$	$\frac{1}{2}\left(\mathrm{p}K_1 - \lg c_{\mathrm{CO}_2}\right)$
2	7-8,2	$\frac{1}{2}\left(\mathrm{p}K_1 + \mathrm{p}K_2\right)$	$\frac{1}{2}\left(\mathrm{p}K_1 + \mathrm{p}K_2 - \lg C_{\mathrm{Ca}^{2+}} - \frac{1}{2}\mathrm{p}\Pi\mathrm{P}\right)$
3	6,5-9,5	-	$\mathrm{p}K_2 - \lg \cdot \left(c_{\mathrm{CO}_2} - c_{\mathrm{CaCO}_3}\right) - \frac{1}{2}\mathrm{p}\Pi\mathrm{P}$
4	10,5-11,5	$\frac{1}{2}\left(\mathrm{p}K_w + \mathrm{p}K_2 + \lg c_{\mathrm{CO}_2}\right)$	$\frac{1}{2} \cdot \left(\mathrm{p}K_w + \mathrm{p}K_2 + \lg \cdot \left(c_{\mathrm{CO}_2} - c_{\mathrm{CO}_3}\right)\right)$

Here c (CO2) is the total content of all equilibrium forms of carbonic acid; with (CaCO3) - concentration of all forms of calcium carbonate; KW is the water constant equal to [H +] · [OH-].

$$\mathbf{ПР}_{\mathbf{CaCO}_3} = \mathbf{5 \cdot 10^{-9}}.$$

By the potentiometric titration curves, the content of metastable calcium carbonate in the solution can be quantitatively determined by the formula:

$$c(CaCO_3) = c(HCl) \cdot (V_{III}(HCl) - V_{II}(HCl))/V\text{р-ра}. \quad (10)$$

In a series of 20 experiments with metastable solutions, a good agreement was obtained between the experimental and calculated values of the CaCO3 content. The standard deviation was ± 15 rel. %, which can be considered acceptable for metastable equilibrium and small concentrations.

It should be noted that in the chemical laboratories of TPP, the hydrate alkalinity, acidity and carbon dioxide content are determined by titration in the presence of indicators of phenolphthalein and methylorange, which change their color depending on the pH of the solution. Moreover, the phenolphthalein color transition occurs in the pH range of 8.5 ÷ 10.0. Meanwhile, as can be seen from the pH spectrum of a metastable solution, a part of the carbonate bound to calcium is not titrated. But this part of carbonates is titrated in the area of methylorange (pH 3,0 ÷ 4,0). Thus, a systematic error occurs.

The results in the article allow us to draw some conclusions:

- Potentiometric and pH-spectroscopic analysis of metastable alkaline solutions containing carbonates and stiffness ions allows qualitative and quantitative determination of their structure and ion-molecular composition, including stabilized calcium carbonate;

- the method of potentiometric titration can be used for the analysis of metastable solutions, including turbid or colored, which is difficult or impossible, since the use of color indicators gives a systematic error.

References

[1] Garmash A V, Vorobyova O N, Kudryavtsev A V and Danchenko N N 1998 *Journal of Analytical Chemistry* **4** 411-417

[2] Chichirov A A, Chichirova N D, Khusainov R R, Filimonov A G and Filippov I E 2006 *Energysaving and watertreatment* **2** 31-34

[3] Filimonov A G, Chichirov A A, Gainutdinova L I and Chichirova N D 2012 *Energy problems* **5-6** 11-16

[4] Boskey A L, Posner A S J/ Phys. Chem. 1973. v.77, p. 2313.

[5] Khoroshilov A V, Semenova I V and Simonova S V 2004 *Energysaving and watertreatment* **3** 20-22

IOP Conf. Series: Earth and Environmental Science **288** (2019) 012005 doi:10.1088/1755-1315/288/1/012005

Purification of acidic highly mineralized spent regeneration solution of Novo-Salavatskaya TTP

A.A.Chichirov, A.Yu Vlasova., S.M. Vlasov, A.V. Pechenkin

Kazan state power engineering university, Russia, 420066 Kazan, Krasnoselskaya, 51

vlasovsm@list.ru, vlasovaay@mail.ru

Annotation. The results of laboratory studies and industrial experiments on the neutralization and purification of highly mineralized acid waste regeneration solutions of the ion-exchange water-preparation plant H-cationite filters are given. As the reagent the carbonate sludge from the pre-cleaning of the water treatment plant was used. A sludge was used in a dry dehydrated form. During a laboratory study, it was shown that there are a significant sulfate content reduction and a pH increase. The sulfate determination was carried out by turbidimetric method using a Shimadzu UV-1800 spectrophotometer. The water treatment sludge using at Novo-Salavatskaya TPP is able to reduce the sulfate content in acid waste water from 16.8 g/l to 1 g/l. This way of neutralizing can reduce these penalty sanctions for thermal power plants for wastewater discharge.

Introduction

The creation of thermal power plants (TPPs) with a minimum harmful impact on the environment (environmentally friendly TPPs) is without a doubt one of the most important scientific and technological task. With limited natural fresh water and constant deterioration of the water sources condition while the requirements stricter of regulatory authorities to the discharged sewage quality, the assessment of the TPPs environmental impact becomes one of the main criteria for the operation effectiveness.

Based on laboratory studies and industrial experiments a new method for purifying highly mineralized wastewater from sulfate-containing components is proposed, on the example of Novo-Salavatskaya CHP.

Theoretical part

The main problems of water use in thermal power plants are:

- large volume of highly mineralized wastewater (WW) from water treatment plants (WTP). On averafe, their volume is up to 25% of the chemically desalted water (CDW) production;
- the sulphate content in the WW WTP is highly overestimated, which caused the huge fines presence;
- high chemical reagents specific consumption including the sulfuric acid for the regeneration (110 g/geq);
- problems with wastewater treatment plants and waste storage, while increasing production of chemically desalted water (CDW). Lack of space and equipment.

A radical way to reduce water consumption and protect water bodies from pollution is the water losses reduction and the reuse of sewage in the TPP technological cycles. The experience shows that with repeated use wastewater, in most cases, a smaller depth of purification and therefore less cost than with WW discharge into water bodies are required.

IOP Conf. Series: Earth and Environmental Science **288** (2019) 012005 doi:10.1088/1755-1315/288/1/012005

To reduce water consumption and WW discharge, the most promising directions are:
- the maximum application of recycling water management systems;
- the re-consecutive water use in several technological cycles;
- the water treatment method application, as a result of which WW are not formed at all or can be used in other cycles directly or after appropriate treatment;
- the allocation and use of valuable substances contained in WW production .

"Traditional" scheme, pre-cleaning-ionite desalination, includes bleach with liming and coagulation or only with coagulation, mechanical filters (MF) with appropriate loading, one or two ion-exchange (IE) stages, figure 1. This scheme is considered "classical" for the water preparation for steam and condensate losses replenishment on power units with type boilers. Therefore, utilization of highly mineralized acid spent regeneration solutions is of great importance in the water treatment system.

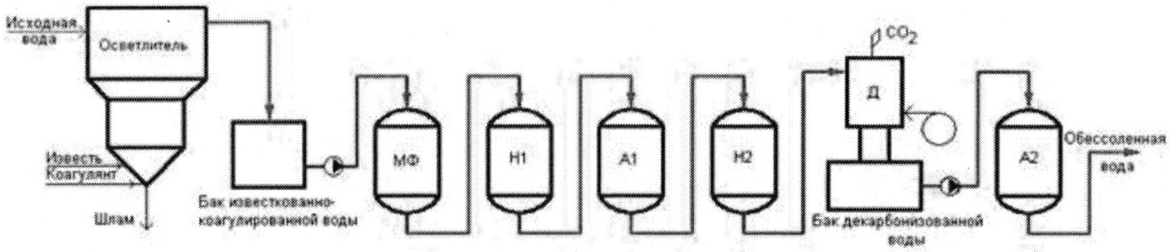

Figure 1. Water desalination classical scheme

When pre-cleaning, slurry water with a solids concentration of 5 to 50 kg/m^3 is formed. Further clarified water after pre-cleaning passes the ion-exchange treatment. After the ion-exchange part chemically demineralized water is obtained. As ion-exchange filters, strongly and weakly acidic cation exchangers H+ and Na +, strongly and weakly basic anion exchangers in OH- or salt form are used.

The H-cationite filters regeneration is carried out by inorganic sulfuric acid of various concentrations (from 0.8 to 4%). To ensure the necessary depth of the cation exchange regeneration, the regeneration solution consumption in most cases is 2-3 times higher than the stoichiometric amount. As a result of chemical purification, sewage is formed, which contains all the reagents excess, and also Ca 2+ and Mg 2+, which enter it during the regeneration process. In the process, a large number of acidic, highly mineralized wastewaters are formed. There are many difficulties in processing and disposal of there wastes, since the sulfate content is very high.

There are various methods for wastewater treatment, but each method has its drawbacks: the use of expensive reagents, the water purification insufficient, the used reagents toxicity and the inability to use on an industrial scale. Therefore, a new technology is needed that would satisfy all the requirements for wastewater treatment.

Description of the principle technological scheme of the WTP at the "Novo-Salavatskaya" TPP

The wastewater treatment plant is designed to wastewater treat from mechanical impurities and petroleum products to a content of: mechanical impurities – no more than 3 mg/dm^3, petroleum products – no more than 0.3 mg/dm^3 with their supply to the cooling tower or to the sewage treatment plant of "GazpromneftekhimSalavat".

On the pre-cleaning, which is carried out by the liming coagulation method with iron salts, carbonate sludge is formed as a solid waste. Its composition is determined by the component composition of the natural water and is approximately the same for TPPs in the central Russian Federation part terms of dry matter, the sludge consists of approximately 80% calcium carbonate (CaCo3), 15% magnesium hydroxide (Mg(OH)2), 2% iron hydroxide, the remainder is other impurities (SiO2, organic substances). WTP slurry is an alkaline calcium-containing reagent. Consequently, its expected action is similar to the lime.

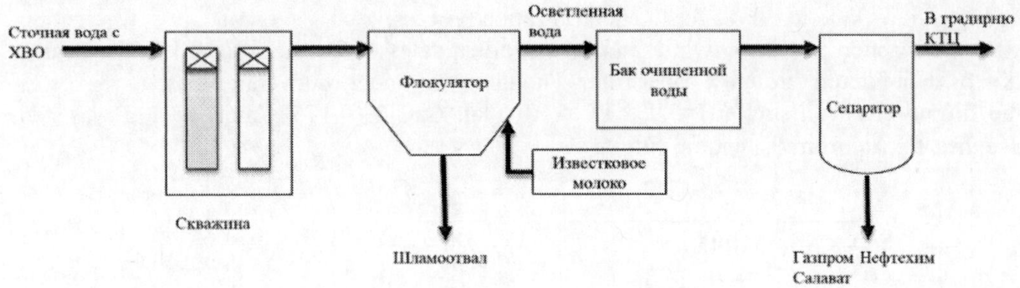

Figure 2. Schematic diagram of the wastewater treatment plant at the Novo-Salavatskaya TPP.

Laboratory research

For laboratory experiments was provided from Novo-Salavatskaya TPP. In the form of a calcium-containing reagent, the slurry waters if thermal power plants were used. The sludge content in the slurry water is about 2%

The Novo-Salavatskaya TPP provided different water kinds: slime water; water from the waching of the H-cation filter first stage; water from the washing of the Na-cation filter first stage; water from the neutralizer tank.

In the laboratory, the analysis of the provided water was carried out, during which the following water characteristics were measured: pH, conductivity, the sulphate content. Table 1 shows the results. The following equipment was used: UV spectrophotometer Shimadzu-1800, conductivity meter HannaHI 9033. The method sulfates determination was the turbodimetric method.

Table 1. Waters characteristics of the Novo-Salavatskaya TPP.

Name	Appearance	pH	Conductivity	Sulphate condent
Sludge water	Color: yellow-brown. There is a dispersed phase, at the bottom of the bottle a brown precipitate is observed. The slurry content in the water is approximately 2%.	10,29	0,23	0,04
The first stage of the H-cation filter	Color: colorless, transparent. At the bottom of the bottle a white precipitate is observed	1,13	More than 20	16,8
The first stage of the Na-cation filter	Color: colorless, transparent.	8,65	More than 20	0,16
Neitralizer tank	Color: yellowish color. The suspension is observed.	6,54	4,21	4,2

Before the research beginning, spectra of the investigated waters were taken with a UV spectrophotometer.

We took 100 ml of water from the first stage of the H-cation filter, in which the sulfate content was 0.175 mol/l. To neutralize sulphates, it is necessary to take sludge in a minimum molar ratio of 1:1. If you take the sludge as CaCO3 (100 g/mole), then for 100 ml of the prepared solution you need to take a slurry of 1.75 g / 100 ml.

Water from the first stage of the H-cation filter has a pH=1.13 units. The neutralization reaction was carried out with mixing in a glass beaker in which electrodes for measuring total salinity (TS) were placed. Sludge was added in two divided doses of 1.75 g of slurry per 100 ml of water.

IOP Conf. Series: Earth and Environmental Science **288** (2019) 012005 doi:10.1088/1755-1315/288/1/012005

Figure 3 shows kinetic curves for the neutralization process.

After the reaction passing a white and brown precipitates could be observed in the beaker. /the solution os then filtered through a paper filter. The filter was under room conditions.

The filtrate characteristics pH=5.93, TS=5.87 g/kg. CaCl2 (0.175 g/mol) or 1 g was added to remove the sulphate from the neutral filtrate 1.

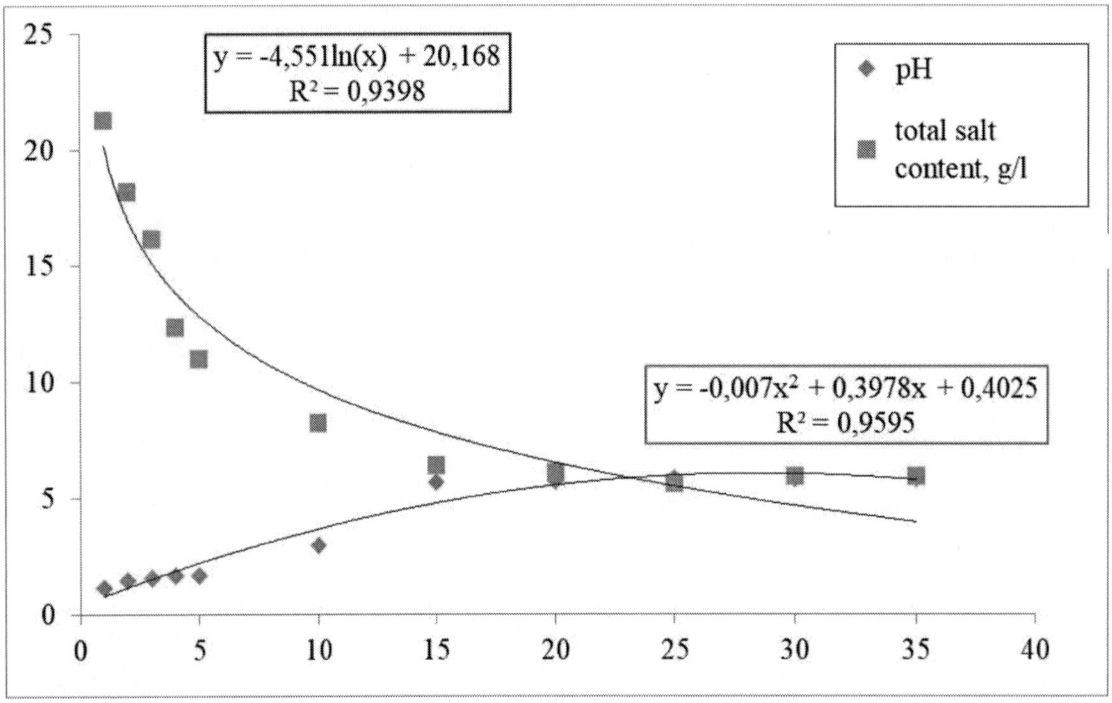

Figure 3. The change in the hydrogen index and total salt content during neutralization of the water from the first stage of the H-cation filter by pre-cleaning sludge .

The reaction kinetics are given in Table 2.

Table 2. The reaction kinetics for treating water from the first stage of the H-cation filter Further, the calcium chloride was added to the filtrage-1.

τ, min	1	2	3	4	5	10	15	20	25	30	35
pH	1,15	1,46	1,58	1,65	1,67	3,0	5,67	5,75	5,85	5,85	5,87
total salt content, g/l	21,3	18,2	16,18	12,34	11,00	8,23	6,43	6,12	5,62	5,96	5,93

The reaction kinetics are given in Table 3 and Figure 4.

Table 3. The filtrare-1 treatment reaction kinetics.

τ, min	1	2	3	4	5	6
pH	5,97	5,99	6,0	6,01	6,02	6,02
total salt content, g/l	More than 20					

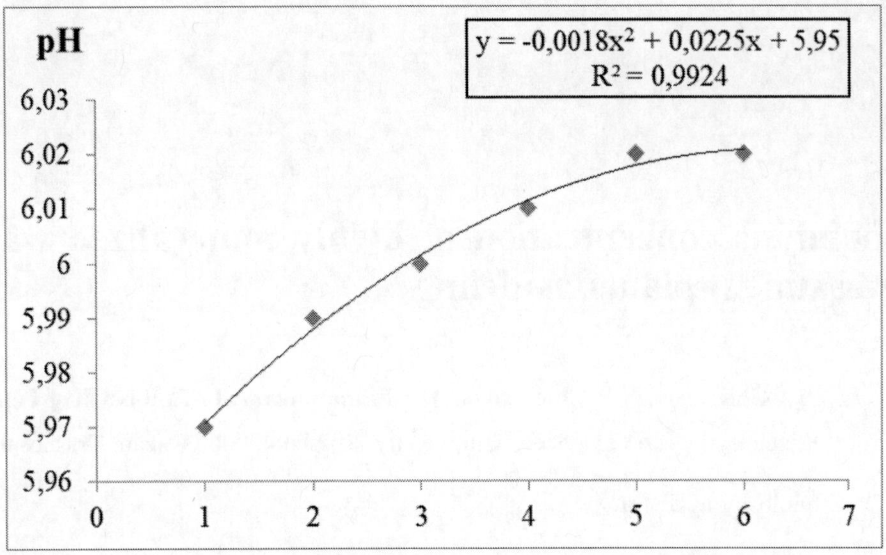

Figure 4. Change in the hydrogen index of the filtrate-1

The filtrate 1 was left for a day in a glass. During this time, a while precipitate was formed. Then the solution was filtered (filtrate-2) and the residual amount of sulfates. Has determined thus, the final results of the experiment are summarized in Table 4.

Table 4. The experiment results

Water	Sulfate content, g/l
first stage of the H-cation filter	16,8
filtrate -1	6,14
filtrate -2	0,98

References

[1] Vlasova, A.Y., Chichirova, N.D., Chichirov, A.A., Filimonova, A.A., Vlasov, S.M. Resource-saving technology for neutralization and purification of acidic and hard-concentrated, liquid waste of the ion-exchange water treatment plant of tppscomplex of water for drinkable small settlements. Water and Ecology, 2017, № 2 (70), p. 3-17

IOP Conf. Series: Earth and Environmental Science **288** (2019) 012006 doi:10.1088/1755-1315/288/1/012006

Electrodialysis concentration of highly mineralized wastes of water treatment plants modeling

A A Chichirov, N D Chichirova, A A Filimonova, A I Minibaev, L I Tolmachev

Kazan state power engineering university, Russia, 420066 Kazan, Krasnoselskaya, 51

minibaev-a@list.ru

Abstract. Studies on electrodialysis concentration/desalination of concentrated salt solutions (NaCl) as a model of soft highly mineralized liquid waste have been carried out. The apparatus of the EMA line manufactured by JSC «Membranines Technologijos LT» were used as electro-membrane devices. Apparatus are distinguished by reliable operation in concentrated mineral solutions under high currents. IONSEP-HC/MC ion-selective cation- and anion-exchange membranes intended for work with highly mineralized sewage was used. Studies have shown the possibility of utilization of concentrated electrolyte solutions, as models of highly mineralized VPU waste with high efficiency. To a residual electrolyte concentration in dialysate at a level of 0.2-0.25 g/l, the unit cost of electricity for salt transfer is approximately 1 kWh/kg, the desalination ratio is 100-150. When the concentration of electrolyte in dialysate is less than 0.1 g/l, the process practically stops due to the increasing influence of the reverse diffusional salt transfer. The calculated maximum salt (NaCl) concentration in the concentrate is 15-16 mass. %

In the production of pure and deionized water from a natural water by ion exchange, electrodialysis or reverse osmosis methods a bulk amount of highly mineralized wastewaters (regenerates and brine) forms, that generates secondary pollution problems.

Thermal power plants (TPP) are one of the main sources of highly mineralized effluent [1]. The main source of highly drains on TPP are water treatment plants (WTP), mainly ion-exchange baromembrane desalting plants. At present, at most TPP acidic highly mineralized effluents from cationite filters and alkaline highly mineralized effluents from anionite filters are mutually neutralized, and the final highly mineralized discharge is dumped. In some countries, the method of utilization of highly mineralized effluent by evaporation with the disposal of solid salts is used, which cannot be called successful. In this case all valuable chemical components of the effluent are irretrievably lost.

In solving the problem, it is of interest to use electrodialysis concentration of liquid wastes prior to neutralization to a level suitable for reuse in the cycle of TPP. In this soft liquid wastes containing mainly sodium salts and alkaline liquidwastes of ion exchange WTP are of great interest, since they do not contain hardness ions, and sodium salts and alkalis are valuable reagents themselves. There is experience, mainly foreign, of using electrodialysis for concentrating reverse osmosis concentrates with obtaining a salt masterbatch suitable for evaporation with obtaining dry salts [2, 3].

The tasks of the work included testing the operability of the electrodialysis method using promising industrial electro -membrane devices and membranes.

Methods and objects

Investigations were carried out on EMA-400/2 electromembrane apparatus in the version of two – channel assembly with flushing (near-electrode) chambers. The main characteristics of the apparatus are shown in Table 1.

Table 1. Characteristics of the EMA-400/2 apparatus (JSC «Membranines Technologijos LT»)

The number of cameras on 1 path	200
Bandwidth per path	
- working range	3.5-5.0 m^3/h
- maximum flow	6.0 m^3/h
Maximum voltage on the device (U_{max})	494 V
Maximum current per device (I_{max})	28.7 A
Chamber thickness (distance between membranes)	1 mm
Working area of membranes	24 dm^2
Path length in the active zone	7.9 dm

The apparatus were installed heterogeneous membrane produced by IONTECH (table 2).

Table 2. Membrane characteristics.

Model	IONSEP-HC-C	IONSEP-HC-A
Thickness, mm	0.42	0.42
Water content, %	40	40
Ion exchange capacity, mol/kg	2.2	2
Burst strength, MPa	0.6	0.6
Water permeation time, min	45	90
Solution diffusion index x10^{-3} mmol, NaCl/(cm^2·h·mol/L)	3.2	3.7
Selective ratio, %	92	92
Surface electric resistance, Ω·cm^2	10-12	10-13

The experimental plant included two EMA-400/2 units connected sequentially with a parallel flow of dialysate and concentrate, intermediate recirculation tanks with a volume of 500 l and pumps with variable frequency drive ohm. Power supply apparatus – independent from two DC power sources.

Concentrated NaCl solutions as models of soft highly mineralized liquid waste were used. Studies were performed in the recirculation mode until the complete development of the initial solutions. In the experiments, the feed rates of the concentrate and dialysate were varied. Conducted constant monitoring of current (I), voltage (U), temperature and salt content of the solutions. Then, using known methods, the process efficiency was calculated – current efficiency (η), specific consumption of electrical energy for salt transfer (w_s). In addition, parasitic side processes were calculated: reverse diffusion, salt transfer and osmotic water transfer. The salt flow through a membrane pair (j_s, *mol-eq/m^2·s*) was determinedby the change in salt concentration in circulation storage tanks and by the change in salt concentration in the concentrate and dialysate at the inlet/outlet of the apparatus.

Results and discussion

The experimental results were processed in the framework of the limit electrodialysis concentration model [4]. According to the model, the flows salt and water through the membranes include diffusion and osmotic ones:

$$j_s = -P_s(c_К - c_Д) + \eta \frac{i}{F}, \quad (1)$$

$$j_w = P_w(c_К - c_Д) + t_w \frac{i}{F}, \quad (2)$$

where j_s and j_w – respectively, flows of salt and water into the concentration chambers; $P_s = P_s^a + P_s^k$, $P_w = P_w^a + P_w^k$ – diffusion and osmotic permeability of the membrane pair, respectively; $t_w = t_w^a + t_w^k$ – the transfer number of water through a membrane pair; $\eta = 1 - t_+^a + t_-^k$ – current output for the transfer of cations (c) and anions (a); $c_к$ and $c_д$ – salt concentration in the concentration and desalting chambers of, respectively i – current density, F – Faraday number.

In the model, it was assumed that in concentrated solutions the electroosmotic transfer of free water is negligible (water is transferred predominantly as part of hydration shells of ions). Therefore, the number of water transfer was represented as $t_w = t_+^a + t_-^k = h\eta$, where h – the number of salt hydration.

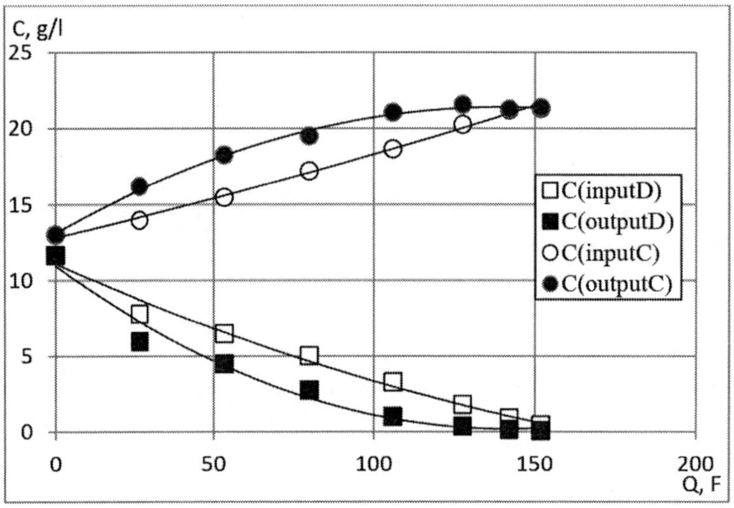

Figure. 1. Dynamics of salt concentration in concentrate and dialysate lines at a circulation rate om EMP 5000 l/h

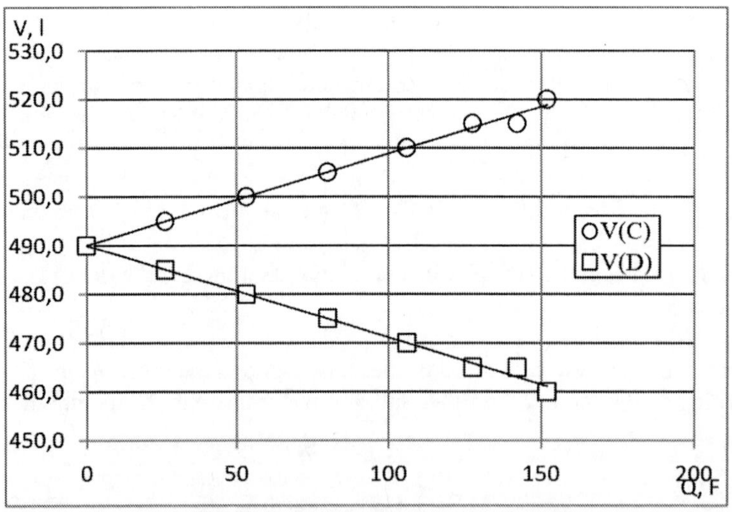

Figure. 2. The change in the volume of concentrate and dialysate at the rate of circulation of each solution 5000 l/h

All four phenomenological parameters of the model: P_s, P_w, η and h can be determined by linearizing equations 1 and 2, dividing their right and left parts by Δc:

$$\frac{j_s}{\Delta c} = -P_s + \frac{\eta i}{F\Delta c}, \tag{3}$$

$$\frac{j_w}{\Delta c} = P_w + \frac{h\eta i}{F\Delta c}. \tag{4}$$

The experimental data were processed in coordinates $\frac{j_s}{\Delta c} - \frac{i}{\Delta c}$ and $\frac{j_w}{\Delta c} - \frac{i}{\Delta c}$ using the method of least squares to determine the transport characteristics of the membrane pair.

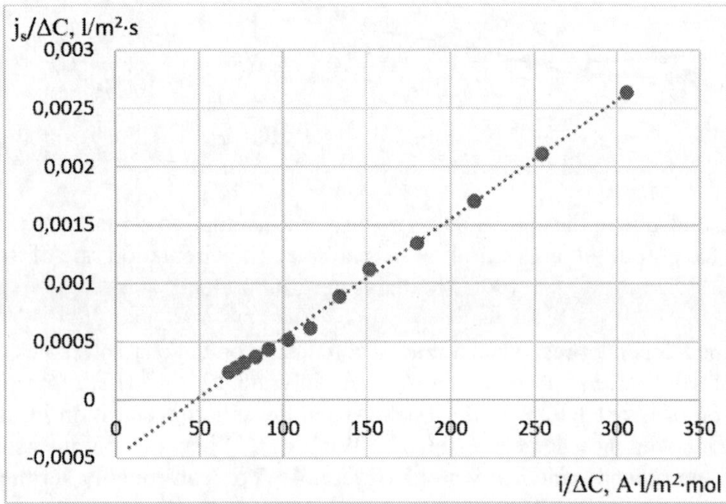

Figure. 3. The salt flow of through the membrane pair, depending on the current density at a circulation rate of 1000 l/h.

Characteristics of mass transfer processes in the concentration of model solutions on a cascade of two EMAs with a sweep in the circulation rate of dialysate and concentrate are presented in table 3.

Table 3. The transport characteristics of the IONSEP-HC-C/IONSEP-HC-A membrane pair in EMA-400/2 apparatus, depending on the circulation rate.

№	$\vartheta_д$, L/h	$\vartheta_к$, L/h	$P_s \cdot 10^7$, m/s	η	$P_w \cdot 10^5$, m/s	t_w, mol/F	h, $\frac{\text{mol } H_2O}{\text{mol salt}}$
1	2000	2000	1.96	0.682	-	-	-
2	6000	6000	-	0.75	-	9.65	12.9
3	4,000	4,000	1.86	0.715	0.5	10.62	14.85
4	3000	3000	2.24	0.71	0.56	9.17	12.9
5	5000	5000	4.33	0.9245	0.41	11.1	12.0
6	1000	1000	4.034	1.023	1.46	20.7	20.2
7	2000	100 0	4.85	0.982	1.54	14.57	14.8
8	1000	60 0	3.9	0.907	0.7	8.44	9.31

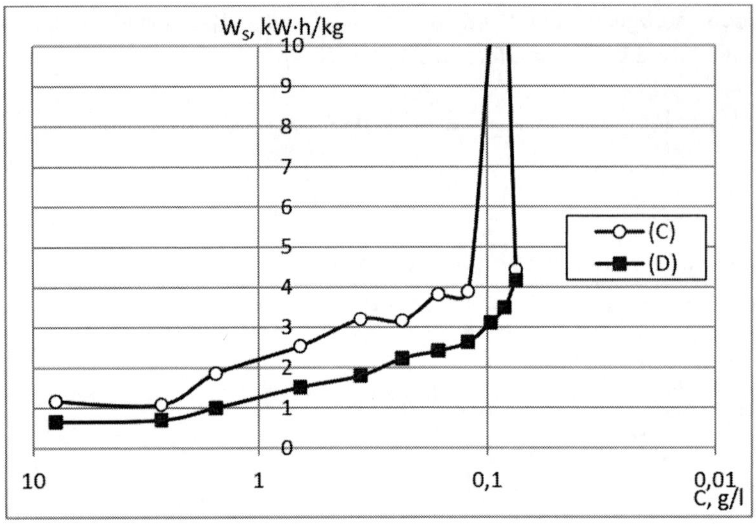

Figure. 4. Dependence of the specific electric power consumption on salt transfer through the membranes on the output salt concentration in dialysate. The circulation rate of the concentrate – 1000 l/h, dialysate – 2000 l/h

From the obtained dependences, we can draw a conclusion on the characteristics of the process.

In the range of salt concentrations in transferred dialysate 12-1 g/l (i.e. more than 1 g/l) specific energy consumption of about 1 to Wh·h/kg t salt. When the salt concentration in the dialysate is less than 1 g/l efficiency does not decrease (up to 2 Wh·h/kg). When the output salt concentration in dialysate ≈ 0.1 g/l the process practically stops (figure. 4). You can roughly set the minimum output concentration of salt in dialysate at 0.2-0.25 g/l in which the high efficiency is maintained. By ratnost desalting ($\frac{C_c}{C_d}$) it will be 100-150.

From the transport characteristics, the calculated maximum concentration of salt in the concentrate will be 15.4±1 mass. % or 2.9 M.

The linear flow rate of solutions in the chambers of the apparatus varies from the minimum in the middle part of the chamber, where the gap width is 40 cm to the maximum at the entrance and exit of the chamber, where the gap width is 2. Calculated average speed data are presented in Table 4.

Table 4. The average linear flow rate in the apparatus of the EMA-400/2.

Volumetric circulation rate (υ), l/h	Minimum linear velocity in chambers, cm/s	Maximum linear velocity in chambers, cm/s
1000	0.85	14.0
6000	5.0	82.5

The obtained data indicate that water speeds on average exceed critical values [5] even at low space velocities. circulation (costs) up to 1000 l/h in the streams saved turbulent mode.

From the obtained conclusion it follows that it is possible to obtain finished products – partially desalinated water and concentrate in one pass at a low flow rate of solutions (2000-1000 l/h), which is confirmed experimentally. At the same time, significant heating of the solutions (more than 5 °C) is not checked.

Conclusion

Studies conducted on promising industrial electro-membrane devices equipped with cation-and anion-exchange membranes resistant to high current loads have shown the possibility of utilization of concentrated electrolyte solutions, as models of highly mineralized VPU waste with high efficiency.

To a residual electrolyte concentration in dialysate at a level of 0.2-0.25 g/l, the unit cost of electricity for salt transfer is approximately 1 kWh/kg, the desalination ratio is 100-150. When the concentration of electrolyte in dialysate is less than 0.1 g/l, the process practically stops due to the increasing influence of the reverse diffusional salt transfer. The calculated maximum salt (NaCl) concentration in the concentrate is 15-16 mass. %

The work was performed under the financial support of applied scientific research and experimental developments by the Ministry of Education and Science of the Russian Federation in the scope of the implementation of the Federal Target Program “Research and Development in the Priority Fields of the Development of the Russian Science and Technology Complex for 2014–2020” (agreement no. 14.577.21.0238 of October 3, 2016, stage 3, unique identifier of the project RFMEFI57716X0238).

References

[1] Kopylov A.S., Lavygin V.M., Ochkov V.F. Water treatment in the energy of Textbook. allowance 2nd ed., stereotypical. 2006 MEI M. 309 p.

[2] McGovern R.K., S.M. Zubair, J.H. Lienhard V. Hybrid electrodialysis reverse osmosis system for saline brines. IDA Journal of Desalination and Water Reuse. Volume 6, 2014. Pages 15-23.

[3] Victor F. Medina, Jared L. Johnson, Scott A. Waisner Journal of Environmental Engineering. Vol. 141, Issue 7

[4] Demin A.V., Zabolotsky V.I. Verification of the model of the maximum electrodialysis concentration of the electrolyte solution. Electrochemistry, 2008, vol. 44, No. 9, p. 1140-1146

[5] Pilat B.V. Basics of electrodialysis. - M .: Avvalon, 2004. - 456 p.

IOP Conf. Series: Earth and Environmental Science **288** (2019) 012007 doi:10.1088/1755-1315/288/1/012007

Atomic force microscopy and IR spectrometry application in detecting the type and nature of contaminants on reverse osmosis membrane elements

N D Chichirova, A A Chichirov and S R Saitov

Kazan state power engineering university, Russia, 420066 Kazan, Krasnoselskaya, 51

caapel@mail.ru

Abstract. The main, the most acute problem of barometric technology today is reagent provision. For its stable and trouble-free operation, a complex of expensive and unique chemical reagents (biocides, antiscales, washing compositions) is required. Each manufacturer develops his own chemical composition for a certain membrane type. This leads to a significant cost rise of reagents, as well as creates a dependence of the consumer technology from a particular supplier. However, these reagents do not always meet the quality requirements - despite the guarantees provided by manufacturers, the equipment fails before the due service life as a result of deposits in the pores and on the membrane surfaces. To develop a more effective washing composition, it is required to determine the type and consist of these deposits. The report illustrates the methods for determining the consist and type of deposits on reverse osmosis membranes. The solution of the problem of reagent provision for the example of the wastewater treatment plant at the Ashalchinsky field of the extra-viscous oil of PJSC "Tatneft" is presented.

Introduction

The experience of the vapor gravitational technology application obtained by PJSC "Tatneft" at the Ashalchinskoye field of super viscous oil showed that the use of steam for its extraction poses a problem both in preparation and disposal of a water large amount [1].

By the end of 2014, the volume of wastewater chemical water treatment (CWT) of the Ashalchinsky field was about 8,000 m^3 per day.

The problem solving of the region's environmental tension was achieved through the a sewage treatment plant (STP) commissioning based on baromembrane technologies. However, during the STP operation, a number of problems arose: firstly, the washing composition declared by the manufacturer did not cope with its main task - washing the membranes away from salt deposits - as a result premature failure of the reverse osmosis unit STP membrane elements. Occurs secondly, the composition claimed by the manufacturer contains components other than the components of the CWT boiler plant compositions at the Ashalchinsky and North Ashalchinsky fields, which causes the customer to purchase components from various sources.

In the third, the washing-out composition of the CWT has unique reagents in its composition, because of which the consumer is forced to apply for reagents strictly to the manufacturer.

The development of a unified composition of a more efficient washing composition, consisting of available chemical reagents, would solve the above problems [2].

Wastewater treatment unit boiler "Ashalchi" Ashalchinsky extra-viscous oil fields of PJSC

"Tatneft" STP boiler "Ashalchi" is designed to process the wastewater treatment the CWT boiler at Ashalchinsky field and re-use them for the CWT boiler North Ashalchinsky field.

The STP scheme includes a submerged ultrafiltration unit, a reverse osmosis unit, and a sodium cationing unit (figure 1).

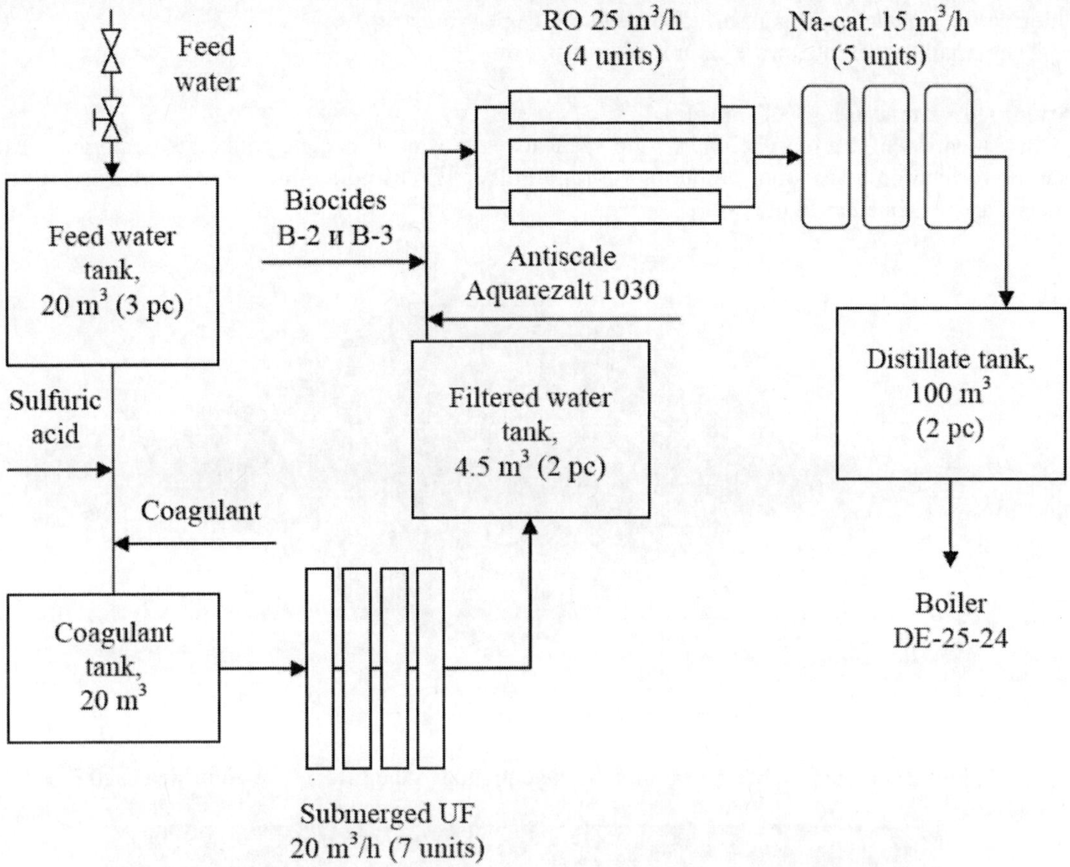

Figure 1. Scheme CWT STP.

The reverse osmosis consists of four reverse osmosis (RO) units "Sharya MP-25-RO", three of which are workers, one is a backup one. The capacity of each installation is permeate 25 m^3/h. The estimated flow rate for the initial water is 38.5 m^3/h. Work on manufacturing and installation of "Sharya MP-25-RO" was carried out by the enterprise "BIOTECH-PROGRESS" in 2015. The installation itself was commissioned in May 2016. In the summer of 2017, the roll elements were replaced because of their premature failure. One of these elements samples was used for research.

Selection and preparation of reverse osmosis membrane material samples

The initial material for the investigation of the deposit composition was the spent cellulose acetate membranes of the reverse osmotic roll filtration element NanoRO K 8040-C of the Sharya MP-25-RO unit for processing of the Ashalchi boiler plant effluents.

RO NanoRO K 8040-C consists of 21 cellulose acetate membranes and a perforated tube. Samples were selected from 5 points along the length and 4 points along the widths 1, 8, 15 and 21 of the membrane layer. A total of 80 samples. The dimensions of the samples are 30 × 40 mm.

Further sample preparation was carried out:

- 2 samples (at the membrane unit inlet and outlet, 21 layers) were soaked in a distillate - for subsequent examination with a scanning electron microscope;
- 5 samples (along the length of the filter, 15 layer) were soaked in 4% citric acid solution and 1% acetic acid solution for 15 hours to obtain acidic extracts (for UF spectrometry);
- 5 samples (along the length of the filter, 15 layer) were soaked in a 1% solution of sodium chloride for 15 hours to obtain salt extracts (for UF spectrometry).

The remaining samples were examined in dry form.

Atomic force microscopy of samples

Investigation of the morphology of the spent reverse osmosis acetate cellulose membrane surface was performed on a scanning probe microscope Solver HV using atomic force microscopy. Surface images are presented in figure 2-4.

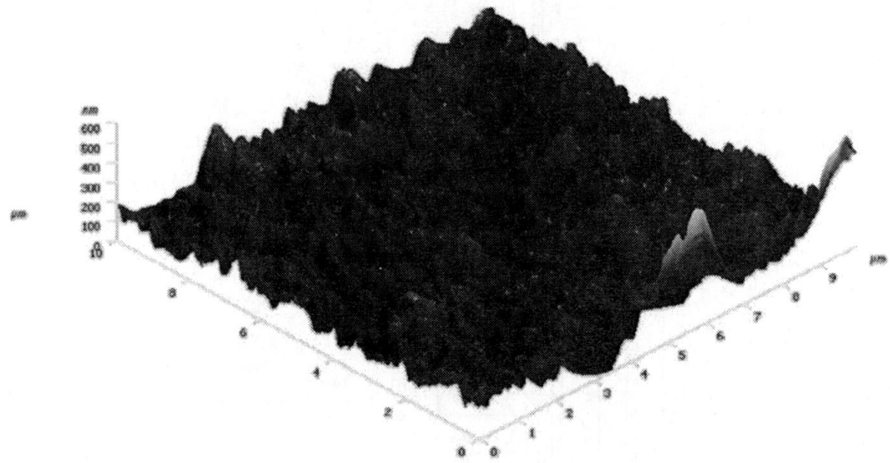

Figure 2. 3D surface of first sample (at the entrance to the filter). The zone size is 10 μm.

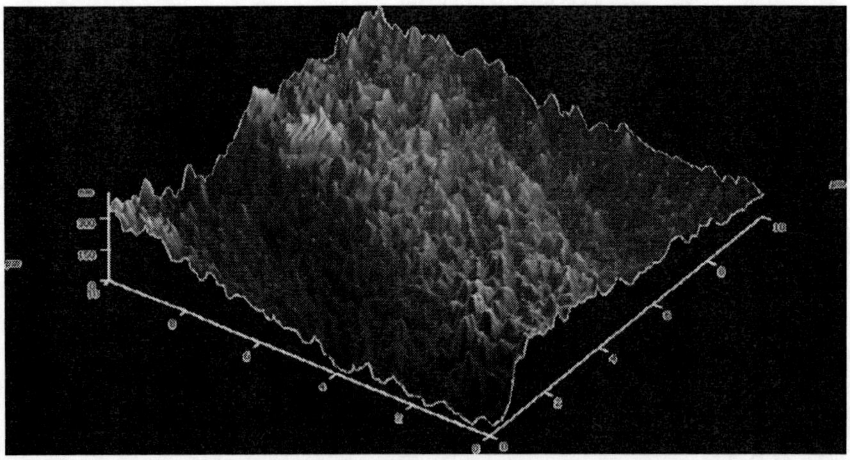

Figure 3. 3D surface of third sample. The zone size is 10 μm.

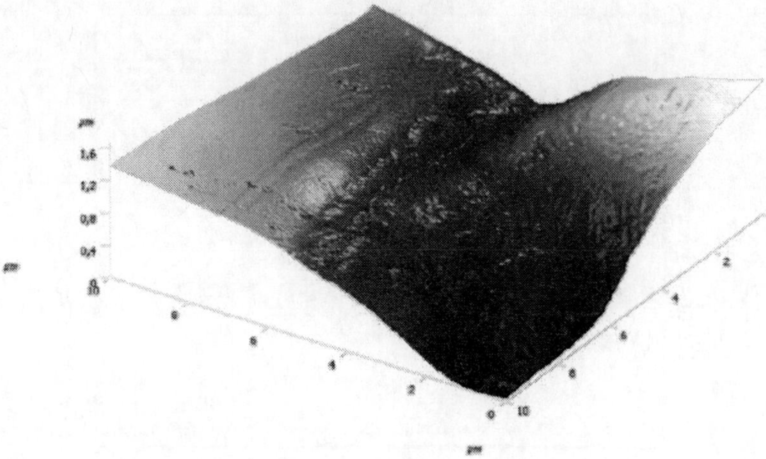

Figure 4. 3D surface of fifth sample (at the out to the filter). The zone size is 10 μm.

A total of 5 dry samples were scanned (along the length of the filter, 21 layers).

Analyzing the pictures, you can see that the closer the sample is to the exit from the device, the smoother the profile of its surface becomes.

Investigation of reverse-osmotic acetate cellulose membrane surfaces by IR spectrometry

Two dry samples were used for the studies: at the inlet and outlet of the filter, 8 membrane layer.

Investigation of infrared spectra was carried out using the IR spectrometer ALPHA (ALPHA-E module) by Bruker Optics (Germany) using the ATR (attenuated total reflection) method. Infrared spectra (figure 5, 6) are taken in the middle infrared range using the OPUS software. The natural surface of samples on a ZnSe crystal is studied. The area of the sample contact with the crystal is 19.6 mm^2.

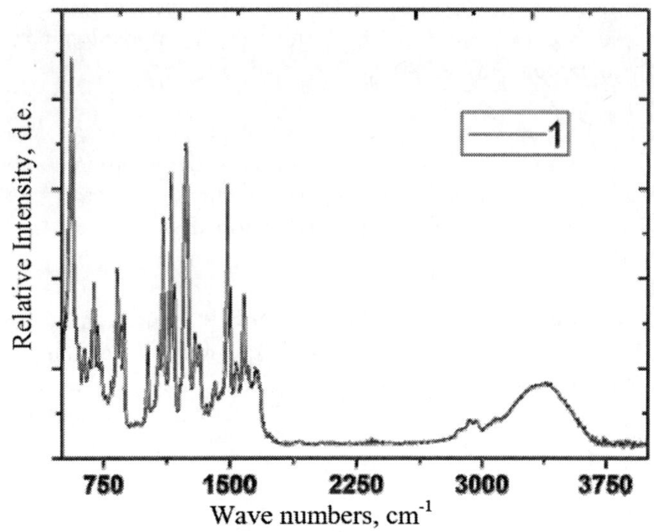

Figure 5. IR spectrum of the membrane sample at the inlet to the filter.

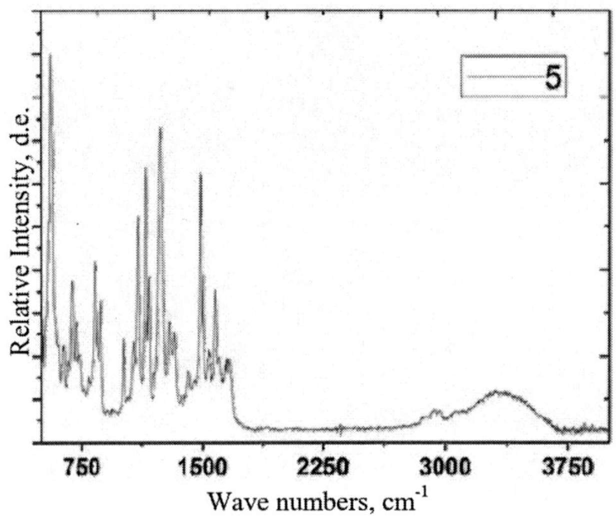

Figure 6. IR spectrum of the membrane sample at the outlet of the filter.

Measurement mode:

- Resolution - 2 cm^{-1}.
- Sample scan time - 50 scans.
- Background scan time - 50 scans.
- Measuring range - 500 ... 4000 cm^{-1}.

The intensity peaks presented on figures 5, 6 correspond to a certain set of chemical components, of which the membranes are deposited. Interpretation of atomic force microscopy results and infrared spectrometry of membrane samples allowed to determine the consist and character of the deposits, which, in turn, made it possible to select the optimal consist of the more efficient washing composition for the previously described effluent treatment unit. The composition components were unified, generally available chemical reagents.

Conclusion

On the basis of information of the feed water parameters, the washing composition consist, and the operating equipment a regime map for the management of water-chemical regime of the baromembrane unit for medium-pressure boiler was developed.

Acknowledgments

The work was financially supported by the Ministry of Education and Science of the Russian Federation of applied scientific research and experimental development (Agreement No. 14.577.21.0238 of 03.10.2016) within the framework of the Federal Target Program "Research and development in priority areas for the development of Russia's scientific and technological complex for 2014-2020 ". Id: RFMEFI57716X0238.

References

[1] Sakhabutdinov R Z, Gubaidulin F R, Loyko A V, Itskov S V and Buslaev E S 2014 Development of technology that envisages processing of waste waters after chemical-water treatment of boilers to produce steam at Ashalchinsky field *Equipment and technologies for oil and gas industry* 5 35-39

[2] Chichirova N D *et al* 2017 *J. Phys.: Conf. Series* 891 012276

IOP Conf. Series: Earth and Environmental Science **288** (2019) 012008 doi:10.1088/1755-1315/288/1/012008

Method Of Aluminum Salts Extraction From Wastewater Using Desalination Technology: Syr Darya River Case Study[1]

Lyazzat Ryskhanovna Junussova[1], **Stanislav Viktorovich Chicherin**[2]

[1]Non-profit JSC Almaty University of Power Engineering and Telecommunications, Almaty, Kazakhstan
[2]Omsk State Transport University (OSTU), Omsk, Russia

man_csv@hotmail.com

Abstract. The Aral Sea Region is characterized by being a very demanding consumer of industrial wastewater treatment. We studied the effects of various pretreatment methods on aluminum production. This method can be used as an alternative one in industrial wastewater treatment. The aim of multi-year study is to explore a nanofiltration as a one of the most attractive technologies for this application since nanofiltration membranes can retain ions and small organic molecules from an aqueous solution. But it is also very challenging due to the presence of salts and operating problems such as fouling, salt deposition, etc. Result becomes available thank to experimental set-up based on an ion exchange and a reverse osmosis. Results are also verified by evaluation tests and field applications showing their usage and wastewater conversion efficiency. The technological scheme includes combined water treatment plant using desalination technology and additional block functioning for an aluminum production and a temporary storage. Aluminum is to be produced of the Syr Darya river rich of it with the help of cutting-edge technology of oxide film removal and cartridge packing. This work evaluates the use of aluminum accumulated in the reaction unit by means of aluminum powder plus water module to reduce the pollutant contents of industrial wastewater quickly and effectively. The invention makes it possible to reduce the cost of aluminum by 3-4 times. Energy consumption of the desalination process was also discussed. Extracted aluminum is a superior material for industry. The high activity of aluminum to water is prohibited by a thin layer of aluminum oxide on its surface. The water treatment system with a combined phase purification and desalination may contribute up to 50 per cent to meeting the reduction of energy consumption for end product (aluminum).
Keywords: Water Treatment, Water-Chemical Regimes, Ion Exchange, Desalination, Ultrafiltration, Membrane, Multicam, Electrodialysis, Sewage, Wastewater, Saline, Reverse Osmosis, Salt Content, Initial, Final.

Water treatment system has been installed and tested on a combined phase purification and desalination using membrane technology capacity of $1m^3$.

Description of the development of the experimental setup

Experimental combined water treatment plant -1 m^3 / h, mounted on the base of the "AUPET" the department "Industrial power" laboratory " Water treatment installation and water chemistry."

The technological scheme includes:

1. Blok- Pretreatment consists of an ultrafiltration module;

[1] The Authors thank the Government of the Russian Federation for its financial support of the Decree No. 860 August, 29 2017

2. Blok- The reaction unit for extracting aluminum from water to produce hydrogen: aluminum powder + water module;
3. Blok - hydride accumulator for hydrogen storage module.

Technological scheme of the proposed installation of the experimental setup is shown in fig.

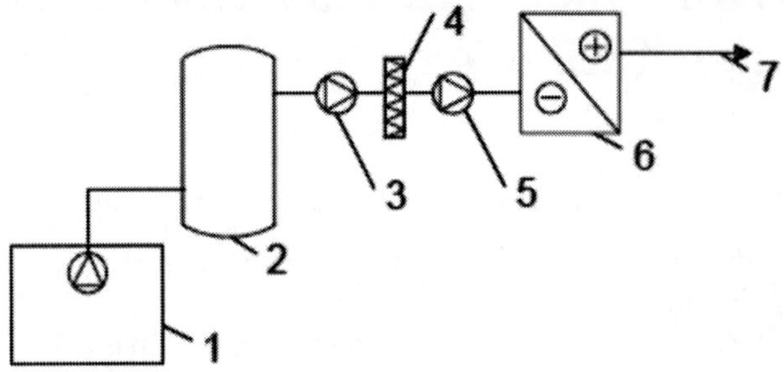

Fig. 1. The technological scheme of the designed compound water treatment plant:
1 – Well, 2 – Storage tank, 3, 5 – Centrifugal pump, 4 – Pretreatment consists of an ultrafiltration module, 6 – Reaction unit for extracting aluminum from water to produce hydrogen, 7 – Supply line of treated water to consumers

The results of the studies are shown in Fig. 2-4.

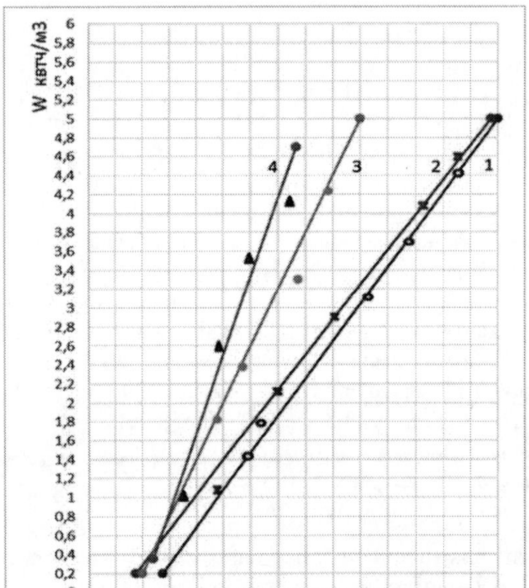

Fig. 2. Specific electricity consumption for various options of electrodialysis water desalination (Damkler, $\frac{d}{1-\alpha}$, dependence)
Lines 1, 2, 3, 4 represent the content of recoverable salts in desalinated water:
1 - Ca2+; 2 - Al2+; 3 - Cl-; 4 - SO42-

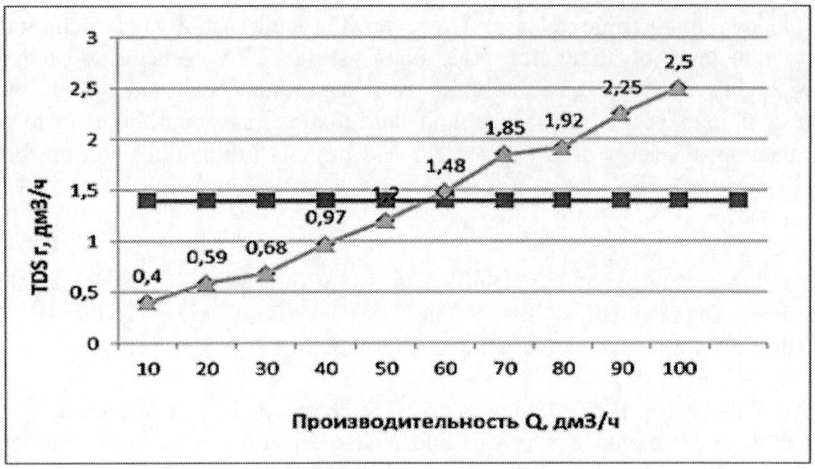

Fig. 3. The total amount of dissolved substances as opposite of the installation flow rate

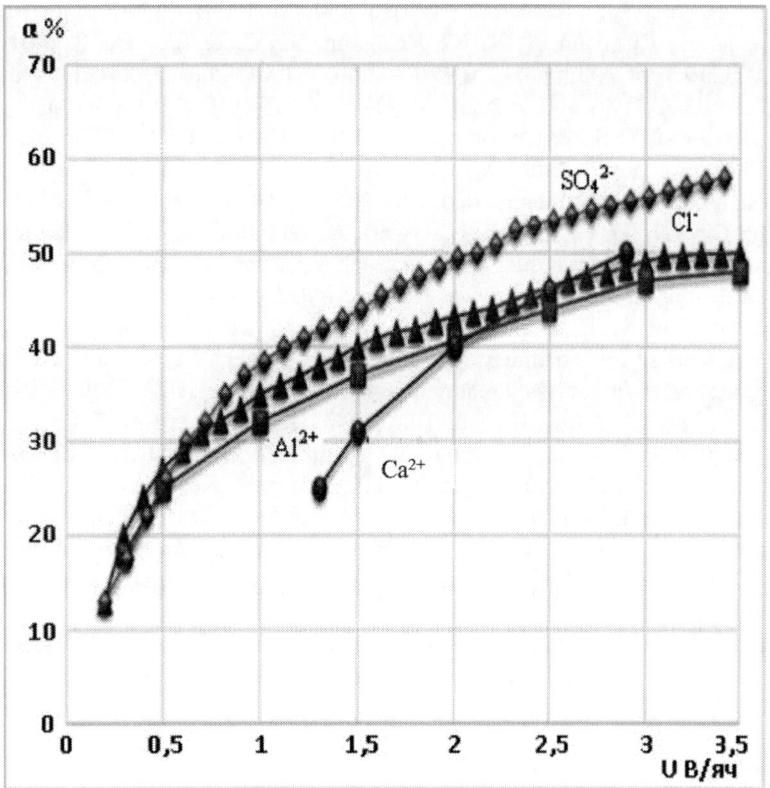

Fig. 4. The variation of water desalting by separate components on the value of input voltage U (V/cell)

Meantime, a new technology based on using electro-dialysis unit responsible for electrolytic extraction of hydrogen is developed. Hydrogen itself is an ideal energy source due to being environmentally friendly. Although the storage and transportation of hydrogen gas currently is still challenging. In addition, the product of aluminum reacted with water is Al(OH)3, which is easily recycled to produce aluminum by known exothermic release. However, the high activity of aluminum to water is prohibited by a thin layer of aluminum oxide on its surface. The generation of hydrogen can be achieved using highly corrosive basic solution such as sodium hydroxide. Thus, water treatment system with a combined phase purification and desalination may contribute up to 50 per cent to meeting the reduction of energy consumption for end product (aluminum) and up to 50 per cent to hydrogen as a byproduct.

References

[1] Ananieva O.A., Kunitsyna T.E., Shilina A.S., Milinchuk V.K. Hydrogen Production by Chemical Decomposition of Saline Water (Poluchenie vodoroda khimicheskim razlozheniem mineralizovannoi vody). Alternative Energy and Ecology (ISJAEE). 2012;(5-6):140-144. (In Russ.)

[2] Viana, M. M., Rangel, B. F., Júnior, A. N., Melchert, M. B. M., & Dweck, J. (2015). Semi-pilot scale sewage sludge pyrolysis and characterization of obtained fractions by thermal analysis. Journal of Thermal Analysis and Calorimetry, 123(2), 981–991. doi:10.1007/s10973- 015-4821-8

[3] Cao, B., Zhang, W., Wang, Q., Huang, Y., Meng, C., & Wang, D. (2016). Wastewater sludge dewaterability enhancement using hydroxyl aluminum conditioning: Role of aluminum speciation. Water Research, 105, 615–624. doi:10.1016/j.watres.2016.09.016

[4] Il'in, A. A., Rumyantsev, R. N., Veisgaim, V. V., & Il'in, A. P. (2016). Mechanochemical Oxidation of Aluminum for Production of Its Oxides, Hydroxides and Hydrogen. Russian Journal of Physical Chemistry A, 90(4), 764–770. doi:10.1134/s0036024416040105

[5] Razavi-Tousi, S. S., & Szpunar, J. A. (2014). Role of Ball Milling of Aluminum Powders in Promotion of Aluminum-Water Reaction to Generate Hydrogen. Metallurgical and Materials Transactions E, 1(3), 247–256. doi:10.1007/s40553-014-0024-7

[6] Sahu, O., Rao, D. G., Gopal, R., Tiwari, A., & Pal, D. (2017). Treatment of wastewater from sugarcane process industry by electrochemical and chemical process: Aluminum (metal and salt). Journal of Water Process Engineering, 17, 50–62. doi:10.1016/j.jwpe.2017.03.005

[7] Van der Bruggen, B., Canbolat, Ç. B., Lin, J., & Luis, P. (2017). The Potential of Membrane Technology for Treatment of Textile Wastewater. Sustainable Membrane Technology for Water and Wastewater Treatment, 349–380. doi:10.1007/978-981-10-5623-9_13

[8] Sun, Y., Zhu, C., Zheng, H., Sun, W., Xu, Y., Xiao, X., … Liu, C. (2017). Characterization and coagulation behavior of polymeric aluminum ferric silicate for high- concentration oily wastewater treatment. Chemical Engineering Research and Design, 119, 23– 32. doi:10.1016/j.cherd.2017.01.009

[9] Senatov S.N. About the Concept of Water Treatment for Thermal Networks (O kontseptsii vodopodgotovki dlya teplovykh setei). Energysaving and Watertreatment. 2016;3(101):3-7. (In Russ.)

[10] Chichirova N.D., Akhmetova I.G. Traditional and Modern Solutions of Methods of Water Treatment of Industrial Power Systems (Traditsionnye i sovremennye resheniya metodov vodopodgotovki promyshlennoi teploenergetiki). Energysaving and Watertreatment. 2016;3(101):8-14. (In Russ.)

[11] Chicherin S. V. The Rapid Assessment the Status of Heat Distribution System Technique (Metodika dlya obobshchennoi otsenki sostoyaniya truboprovodov teplovykh setei). NDT World. 2017;4(20);66-68. DOI: 10.12737/article_5a2fc155a8c669.22998743

[12] Razikov Z. A., Yunusov M. M. Assessment of Heavy Metal Pollution in Waters of the Syr Darya (in the Republic of Tajikistan) (Otsenka urovnya zagryazneniya vody r. Syrdar'ya tyazhelymi metallami (na territorii respubliki Tadzhikistan)). XXI century. Technosphere safety. 2016;4(1):48-53. (In Russ.)

[13] Barrera-Díaz, C., Varela-Guerrero, V., Gonzalez-Rivas, N., & Ureña-Núñez, F. (2014). Double aluminum recovery and its reuse in wastewater treatment. International Journal of

Environmental Science and Technology, 12(9), 2979–2986. doi:10.1007/s13762-014-0716-7

[14] Perelygin, Y. P., Kolchugina, I. G., Rashevskaya, I. V., Flyagin, A. A. (2017). Electrical Conductivity of Aqueous Solutions Containing Sulfuric Acid and Sulfate Aluminum, Copper or Cadmium. University Proceedings. Volga Region. Natural Sciences, 2(18), 37-43. doi:10.21685/2307-9150-2017-2-519.

[15] Kolesnikov A. V., Krutchkova L., Sinitsina A. E., Kisilenko P. N. Features of Electroflotation Extraction of Difficultly Connected Aluminum Compounds from Wastewater Waters. Journal Adnvances in Chemistry and Chemical Technology. 2017; 6-31(187);43-45

[16] Chorghe, D., Sari, M. A., & Chellam, S. (2017). Boron removal from hydraulic fracturing wastewater by aluminum and iron coagulation: Mechanisms and limitations. Water Research, 126, 481–487. doi:10.1016/j.watres.2017.09.057

[17] Elabbas, S., Ouazzani, N., Mandi, L., Berrekhis, F., Perdicakis, M., Pontvianne, S., … Leclerc, J.-P. (2016). Treatment of highly concentrated tannery wastewater using electrocoagulation: Influence of the quality of aluminium used for the electrode. Journal of Hazardous Materials, 319, 69–77. doi:10.1016/j.jhazmat.2015.12.067

[18] Junussova L.R. Comparative Performance Analysis of Baromembranes Desalination. Issues of Energy and Resource Saving. 2014;(1-2), 254-257. (In Russ.)

[19] Vurdova N.G., Fomicheva V.T. Electro-dialysis of natural and waste water. Moscow, ASV Publ. 2001. 119 p.

[20] Choi E. – Y., Choi J. – H., Moon S. – H. An elektrodialiysis Model for Determination of the Optimal Current Density // Conference Proceedings. – May 4 – 6, 2002, Sharm El – Sheikh, Egypt, v. 1, p. 305 -310.

[21] Mukhiddinov D.N., Junussova L.R. Improving the Quality of Desalinated Water with Combined Water Treatment Plant Boiler (Uluchshenie kachestva obessolennoi vody na kombinirovannoi VPU kotel'nykh). Alternative Energy and Ecology (ISJAEE). 2015;(23):167- 176. (In Russ.) DOI:10.15518/isjaee.2015.23.021

[22] Chicherin S.V., Glukhov S.V. Centralized Heat Supply System and Techniques of Planning Its Loads // Urban Construction and Architecture. 2017. V.7, 3. Pp. 129-136. DOI: 10.17673/Vestnik.2017.03.22.

[23] Latosov, E., Volkova, A., Siirde, A., Kurnitski, J., & Thalfeldt, M. (2017). Methodological Approach to Determining the Effect of Parallel Energy Consumption on District Heating System. Environmental and Climate Technologies, 19(1). doi:10.1515/rtuect-2017-0001

[24] Bogdanović, B., Brand, R. A., Marjanović, A., Schwickardi, M., & Tölle, J. (2000). Metal-doped sodium aluminium hydrides as potential new hydrogen storage materials. Journal of Alloys and Compounds, 302(1-2), 36–58. doi:10.1016/s0925-8388(99)00663-5

[25] Kalinkina N. M., Belkina N. A., Sidorova A. I., Galibina N. A., Nikerova K. M. Bioassay of Lake Onego bottom sediments toxicity based on their chemical composition and deepwater macrozoobenthos state. Ecology Principles. 2017;1(22);81-103.

PROCESSING OF ALKALINE WASTEWATER OF TPP EVAPORATIVE WATER TREATMENT PLANT WITH ELECTROMEMBRANE METHODS

A A Filimonova, N D Chichirova, A A Chichirov, A I Minibaev, R V Buskin

Kazan state power engineering university
Russia, 420066 Kazan, Krasnoselskaya , 51

e-mail: aachichirova@mail.ru

Abstract. The thermal desalination technology is considered one of the most environmentally friendly and perfect for the preparation of steam turbine additional water. The purging water of the evaporative unit is an alkaline, highly mineralized, intensely colored solution, in which sodium salts (sulfates, chlorides) and alkali are mainly contained. The total salt content exceeds 30 g/kg. There is significant content of organic substances, silicates. A technology has been developed for processing of a thermo-desalting complex excess purge water by electromembrane methods to produce alkali and softened water. A schematic solution of the electromembrane utilization of purge water excess from an evaporator water treatment plant is presented.

Water treatment schemes using thermal desalination technology are currently used at a number of thermal power plants (TPPs), both in the Russian Federation and abroad, namely, Saransk TPP, Kazan TPP-3, Rostov TPP-2, Omsk TPP-5, Surgut state district power station, etc. The main elements of such schemes are surface type evaporators, during which operation purging water of high mineralization is formed. Analysis of thermal desalination plant operation shows that their use allows to significantly reduce the source water consumption, reduce the waste water discharge and the chemical reagent consumption which largely fits into concept of drainage and wastelessness of industrial and energy enterprises companies adopted in recent years [1, 2]. Within this concept, a set of technological solutions for rational water use was developed Kazan TPP-3 with the introduction of a low-waste water treatment plant on wastewater heat-treated additional water for boilers, as well as sludge and chemical reagents suitable for further use in the technological process [3].

Table 1. Average data on the composition of evaporative water treatment plant purging water. [4].

Index	Concentration
pH	13,2-13,7
Alkalinity	95-100 mg-eq/dm^3
Total hardness	0,45-0,85 mg-eq/dm^3
Carbonate hardness	0,25-0,54 mg-eq/dm^3
Total salt content	10-45 g/l
Na^+	19-41 g/dm^3
SO_4^{2-}	9900-20009 mg/dm^3
Cl^-	2150-4500 mg/dm^3
Cu^{2+}	0,25-1,15 mg/dm^3
SiO_2	280-440 mg/dm^3
Fe^{n+}	9,8-1,7 mg/dm^3
Suspended substances	96-147 mg/dm^3

In the process of the described technology implementing, an excessive amount of purging water of the evaporative plant is formed, an average 0,8-1,0 m^3/h (Table 1).

Purge water is a highly mineralized solution that contains mainly sodium salts (chlorides, sulfates) and a large number of insoluble impurities – iron, copper, silicon compounds. In addition, natural high molecular weight organic compounds are present. These impurities are mostly in the colloidal state, or go into a colloidal state during the water treatment. Hence the need is to separate small and insoluble impurities before and during processing. In the purge water there are also organic substances of natural origin. This is evidenced by the color of the water – dark brown. There is no chemical analysis for organic substances content [5].

It should be noted a large amount of alkali (4-8 g/l, depending on the month) in the evaporators purge water. This fact is of fundamental importance for a technical solution (scheme) of purge water recycling.

Water treatment schemes with significant or complete wastewater elimination of can be realized with the electromembrane device help [6]. Taking into account the given data on the purge water composition and the requirements for the resulting solutions, electromembrane treatment can be realized successively in steps on three types electric-membrane apparatus:

1) alkali separation from salts;
2) salts decomposition into alkali and acid;
3) water desalination and alkaline solution concentration.

The investigations were carried out using EMA-100 experimental-laboratory electromembrane apparatus, developed by us [7], with a maximum chamber number of 100. Using membranes were of the «Shchekinoazot» (Russia).

The purge water Scheme of the three-stage treatment of is shown in Fig. 1.

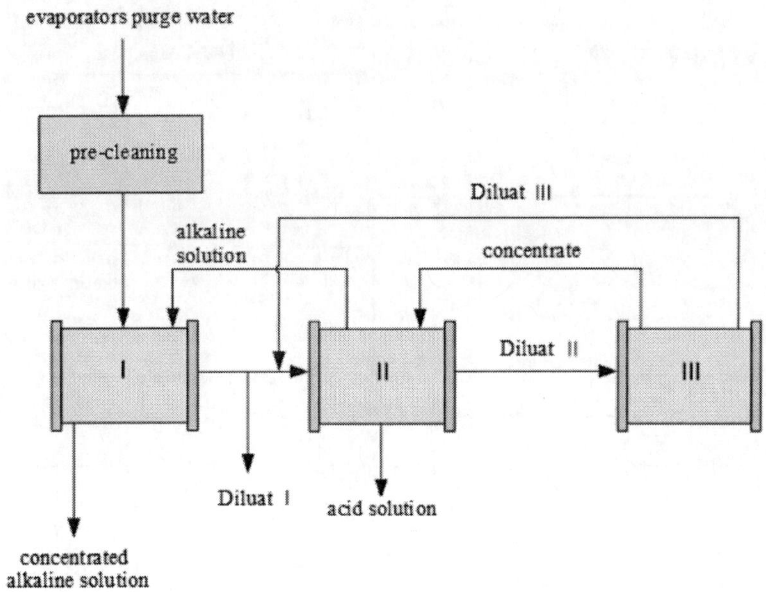

Fig. 1. Utilization of evaporator purging using electromembrane modules for alkaline and acidic solution obtaining (shortened): I – electromembrane alkali separation; II – bipolar electromembrane apparatus; III – electromembrane concentration, desalination.

At the first-stage, a partial alkali separation from the initial solution in the EMA with cation and anion exchange membranes occurs. The driving force of the process is the higher hydroxyl ion mobility in comparison with sulfate and chloride ions. Since the process selectivity is not high, it is possible to as produce an alkaline solution containing the initial solution salts as the product.

The EMA of the first stage produces a concentrated alkaline solution and diluat I. The latter is a more diluat solution of the starting salts and the remaining alkali. Diluat I is the starting solution for the second-stage EMA.

The EMA of the second-stage is assembled with bipolar membranes and serves to separate the solution of salt into alkaline and acidic solutions. As products in the second-stage, a diluat II is formed, which is a more diluat solution of the starting salts, unconcentrated alkali solutions and acids mixtures. Diluat II is sent to the EMA of the third-stage, the alkaline solution is concentrated on the first-stage or to the EMA alkali concentrator. An acidic solution containing a sulfuric, hydrochloric and nitric acid mixture, is sent to the consumer.

On third-stage EMA, the process of the diluat II concentration-desalination is carried out to produce partially desalted water with a salt concentration of about 0,3 g/l (diluat III) and concentrate. Diluat III is used to prepare alkaline and acidic solutions in the second stage of EMA.

The presented scheme is flexible enough. It is possible to gradually reduce the steps, starting with the last. If you remove the EMA third-stage (concentration-desalination), partially desalinated water for second-stage can be taken from the water treatment plant.

Equivalent water amount in the diluat II form (sodium salt solution) is directed to make up the heating network. That is, water is exchanged between water treatment and electromembrane plants. With the reduction of the third- and second- stages simultaneously an alkaline solution and diluat I can obtained on the electromembrane apparatus of the first-stage. Alkaline solution is sent to concentration or directly to the consumer. Diluat I (saline solution) can be used for the regeneration of Na-cation exchanger filters, for the heating system replenishing or for the evaporator replenishing.

To obtain a cleaner alkali and a more concentrated acid solution, it is necessary to install one more type of EMA concentrator (Fig. 2).

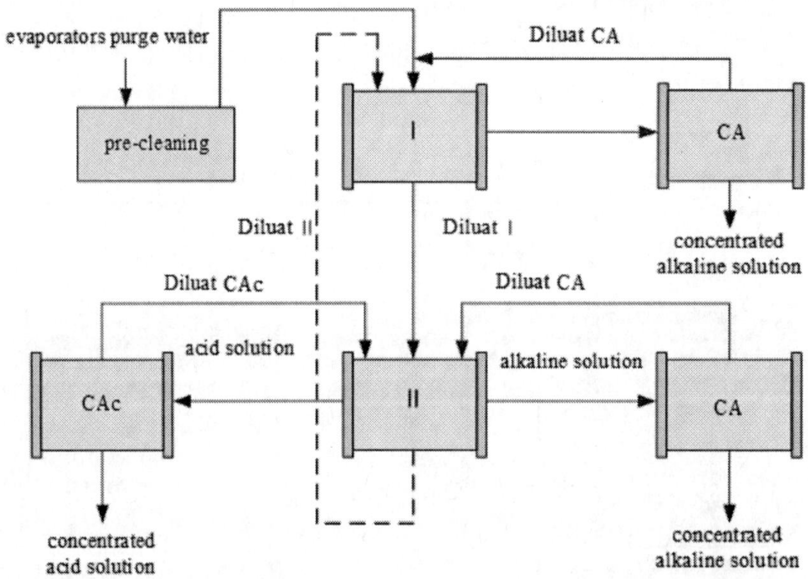

Fig. 2. A complete scheme for the utilization of evaporator purge using electromembrane devices to produce pure alkali and a acid mixture: I – electromembrane alkali separation; II – bipolar electromembrane apparatus for the salt decomposition; CA – electromembrane alkali concentration; CAc – electromembrane concentration of acids.

As EMA products, alkaline and acidic solutions are obtained. Depending on the presence step of alkali step electromembrane concentration, step as well as the degree of salts decomposition in bipolar

EMA, the alkali concentration in the alkaline solution will be from 4 to 8 mass. %. As impurities, the initial sodium sulfate and sodium chloride salts are present up to 10 % of the base material.

The second product – an acidic solution contains the initial salt and a mixture of sulfuric with inclusions of hydrochloric and nitric acids. Depending on the degree of salts decomposition at the second-stage of electromembrane treatment, the acids concentration can vary from 0 to 3 %. In the extreme case, when the salts are completely decomposed, approximately equal amounts of alkaline and acidic solutions are obtained.

If it is decided to limit the alkaline and saline solutions production, a second-stage electromembrane treatment of the purge water is possible (Fig. 3). The scheme includes the alkaline solution the separating and concentrating steps.

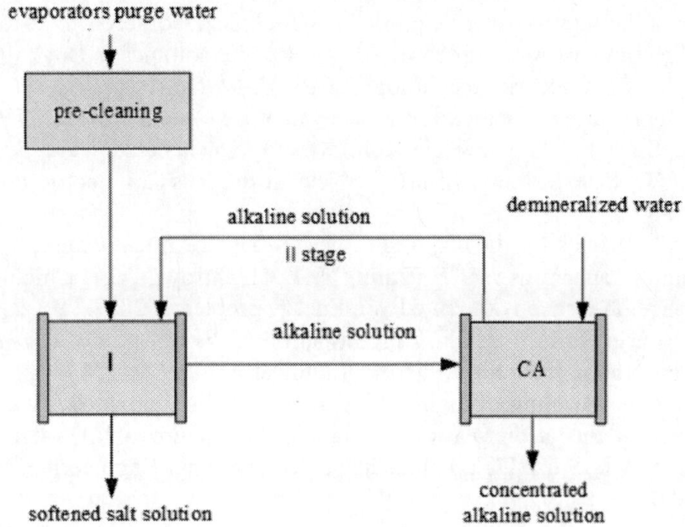

Fig. 3. Scheme of purge utilization with pure concentrated alkali and saline solution obtaining: IV – electromembrane preparation of alkali of the second-stage.

The resulting concentrated alkaline solution (concentrate) is sufficiently high purity and can be used in the plant cycle to regenerate the anion exchanger filters. The softened salt solution can be fed to the heating system.

Thus, it is economically and technically simple to obtain concentrated alkaline and soft salt solutions by electromembrane methods, with their further utilization in the station cycle.

At the same time, the acid production in its pure form is difficult. It is possible to obtain only the acids mixture (sulfuric, hydrochloric and nitric). Further separation seems economically inexpedient. This limits the scope data the acid mixture application in the plants process cycle.

Conclusions

1. The composition data of the evaporative water treatment plant purge water are analyzed. A large alkali amount in the evaporative plant purging water is of fundamental importance for the choice of the technical solution for purging. The electromembrane technology using represents an opportunity for purging water separation into separate components and their subsequent utilization in the technological cycle.

2. In accordance with the electromembrane processing features and limitations, the requirements for the conditions and quality of the treated water and the obtained solutions are formulated.

3. The schemes of evaporators purging water electromembrane processing have been developed, allowing to allocate valuable raw materials reusable in the production cycle:

- scheme for alkaline and acid solutions obtaining from the evaporators purging water;
- shortened scheme for alkaline and acid solutions obtaining;
- scheme for alkaline and softened solution obtaining;

4. Relatively low operating costs make the most expedient use of the scheme of purging water utilization with concentrated alkaline and softened salt solutions obtaining, which are used in the station cycle. For industrial development, the described scheme is recommended.

References

[1] Technologies of water water washing at TES // Sedlov A.S., Potapkina E.N., Grazhdanov A.A., Gordeev A.A., Tomashevskaya M.S. // In the collection: The State and Prospects for the Development of Electrical and Heat Technologies of the XVIIIth Benardos's Readings: Proceedings of the International Scientific and Technical Conference. 2015. pp. 75-78.
[2] The main directions of waste-free and low-waste technologies // Boyko N.I., Odaryuk V.A., Safonov A.V. // Civil security technology. 2015. V. 12. No. 1 . pp. 68-72.
[3] Experience of creating a complex low-waste water use system at Kazan TPP-3 // Fardiev I.Sh., Zakirov I.A., Silov I.Yu., Galiev I.I., Korolev A.G., Shishchenko V.V., Sedlov A.S., Ilina I.P., Sidorova S. .V., Khaziakhmetova F.R. // New in the Russian electric power industry. 2009. No. 3. pp. 30-37.
[4] Technological schemes for the disposal of wastewater treatment evaporator installations using electromembrane apparatus // T.F. Vafin, N.D. Chichirova, A.A. Chichirov, I.A. Zakirov // Izvestiya Vysshikh Uchebnykh Zavedenii. Energy problems. 2012. № 1-2. pp. 182-186.
[5] Sedlov A.S., Larin B.M., Ilina I.P. // Investigation of the removal of organic substances in the distillate of an evaporative unit // Thermal Engineering. - 1999. - №7. - pp. 16-19
[6] Electromembrane recycling of highly mineralized alkaline blowdown water from evaporative water treatment plants at thermal power stations // Chichirova N.D., Chichirov A.A., Lyapin A.I., Minibaev A.I., Silov I.Y., Tolmachev L.I. // Thermal Engineering. 2016. T. 63. No. 14. pp. 1046-1050.
[7] No. 121500 patent for a utility model "Plant for processing industrial wastewater and producing a concentrated alkaline solution and softened salt solution" // T.F. Vafin, A.A. Chichirov // FGBOU VPO "KGEU". 2012. №30.

Acknowledgments

The work was financially supported by the Ministry of Education and Science Russian Federation, applied scientific research and experimental development (Agreement No. 14.577.21.0238 03.10.2016) within the framework of the Federal Program "Research and development in priority areas for the development of Russia's scientific and technological complex for 2014-2020 Years ". The unique identifier PNIER RFMEFI57716X0238.

Investigation of calcium carbonate precipitation in the presence of scale inhibitors

N P Kakurkin, A A Malashenko, N E Pudova, E Yu Perunova
D.Mendeleev University of Chemical Technology of Russia

npk@muctr.ru

Abstract. The work presents a study on the precipitation of calcium carbonate in the presence of four inhibitors: aminotris(methylenephosphonic acid) (ATMP), 1-hydroxyethane-1,1 diphosphonic acid (HEDP), polyacrylate (PA), copolymer of acrylic and allylsulfonic acids (PAC). Using the laboratory methods for evaluating the effect of antiscalants for inhibition, the following order in terms of decreasing effectiveness HEDP > ATMP > PAS > PA at a concentration of 4 ppm has been observed. It was revealed that polymeric antiscalants can effectively cope with the suppression of the precipitation of calcium carbonate at higher concentrations than phosphorus-containing preparations. It has been shown that in the presence of antiscalants, a change in the structure and modification of $CaCO_3$ occurs.

1. Introduction

Calcium carbonate is one of the most commonly encountered scale deposits in reverse osmosis systems. The deposition of $CaCO_3$ and other poorly soluble components leads to significant negative consequences, among them a decrease in system performance and inter-regeneration period of the reverse osmosis unit and, in the most difficult cases, irreversible membrane failure [1-2].

One of the ways to prevent the formation of deposits on the surface of membranes is the using of the scale inhibitors or antiscalants. As a rule, modern antiscalants consist of various classes of organic compounds, among them the most common inhibitors based on phosphonic acids (for example, HEDP, ATMP) [3-4]. However, the growing environmental problems associated with the pollution of surface water sources with phosphorus and nitrogen-containing compounds contributed to the development of new ecofriendly inhibitors of scale [5-6].

In this work, a comparison was made of classical phosphorus-containing antiscalants and polymers based on polyacrylic acid (PAA) and polyacrylate with sulfo group (PAS).

2. Experimental

2.1. Materials

$CaCl_2$, Na_2CO_3. All chemicals are of analytical grade, also was used deionized water. Phosphonate solutions of antiscalants were obtained from solid aminotris(methylenephosphonic acid) ATMP (98%) and 1-hydroxyethane 1,1- diphosphonic acid HEDP (98%). New polymer antiscalants polyacrylic acid PA and copolymer of acrylic and allylsulfonic acids were supplied in the form of solutions.

2.2. Test methods

The system of calcium carbonate was obtained by draining in a stoichiometric ratio of solutions of calcium chloride and sodium carbonate, previously thermostatted at 25 ° C. The system was mixed for 30 seconds, which ensured the uniform distribution of the reagents in the solution and increased the rate of formation of the precipitate.

$$CaCl_2+Na_2CO_3 \rightarrow CaCO_3\downarrow+2NaCl$$

The concentration of $CaCO_3$ was 250 ppm, the pH of the solution was 10.2. The concentration of the inhibitor was 4 ppm of the original product, then the concentration was increased to 10 ppm.

The effectiveness of inhibitors to prevent the precipitation of calcium carbonate was determined using two express laboratory techniques. The first method was based on determining the content of dissolved Ca^{2+} in the system after interacting with the antiscalant [7-9], based on which the degree of inhibition was calculated by the formula:

$$I,\% = \frac{[\mathrm{Ca}]\mathrm{final} - [\mathrm{Ca}]\mathrm{blank}}{[\mathrm{Ca}]\mathrm{init} - [\mathrm{Ca}]\mathrm{blank}} * 100$$

where [Ca] final and [Ca]blank are final calcium concentrations with and without the presence of an inhibitors and [Ca]init is the initial calcium concentration.

The second laboratory technique consisted in visual observation of the solutions using an optical microscope with the possibility of photo fixation. The photographs were transferred to a black and white colour and using a special program we determined the percentage of a photo filled with color - m,% (Fig.1).

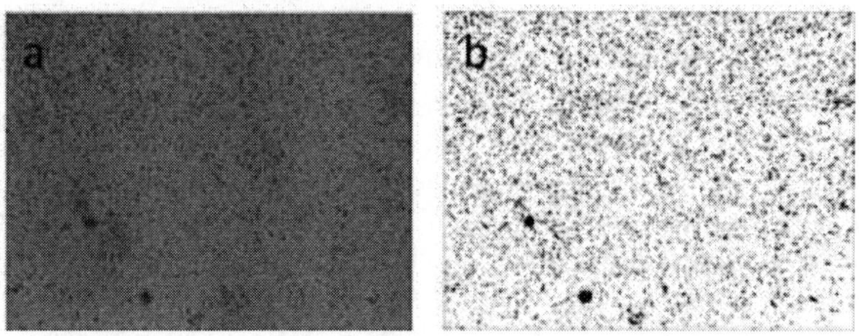

Fig.1. Images of $CaCO_3$ obtained using an optical microscope (a), (b) – images transferred to a black and white colour

This indicator is selected as a parameter characterizing the amount of precipitate formed. For this method, the degree of inhibition was defined as:

$$I,\% = \frac{[\mathrm{m}]\mathrm{sample} - [\mathrm{m}]\mathrm{control}}{[\mathrm{m}]\mathrm{sample}} * 100$$

where [m]sample is percentage of the photo filled with color without the presence of an inhibitors, [m]control is percentage of the photo filled with color with the presence of an inhibitors.

The time of the study in both cases was 1 hour.

The solid phase of the deposited component was investigated using a scanning electron microscopy (SEM) Microscope JEOL JSM-6510 LV.

3. Results and discussion

3.1. Determination of the effectiveness of scale inhibitors

The study of the effectiveness of antiscalants using two laboratory methods showed that phosphorus-containing inhibitors HEDP and ATMP completely suppress the precipitation of calcium carbonate in the chosen conditions.

The polymeric antiscalants PA and PAS were less effective, and the degree of inhibition for these antiscalants varied depending on the method of investigation used (table 1). Differences in I,%, depending on the research method, in our opinion, are caused by the limit ability of an optical micriscope to capture solid particles less than a micron in size. To study the effect of the dose of polymeric inhibitors on the precipitation of calcium carbonate, the antiscalant concentration was increased to 10 ppm.

Table1.
Results for the determination of the degree of inhibition of calcium carbonate precipitation

Inhibitor	Inhibition (%)			
	Liquid phase analysis		Microscopic observation	
	concentration, ppm			
	4	10	4	10
HEDP	100 ± 8	—	100 ± 13	—
ATMP	95 ± 7	—	100 ± 13	—
PAS	66 ± 5	90 ± 7	98 ± 13	100 ± 13
PA	60 ± 5	77± 6	78 ± 10	90 ± 11

3.2 Solid phase investigation.

The SEM images of calcium carbonate are presented in Figures 2, 3. $CaCO_3$ particles, obtained in experiments without an inhibitor (Fig.2), were similar in size and shape, and the most stable phase of calcium carbonate, calcite, prevailed in the scale.

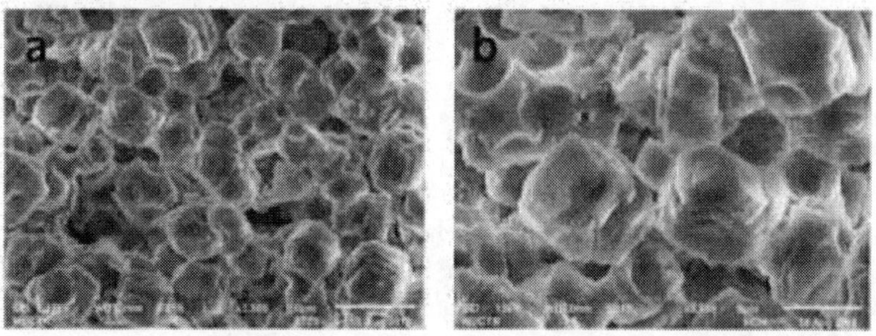

Fig.2. SEM images of $CaCO_3$ crystals: (a) - (x2500), (b) - (x5000)

When inhibitors were used a significant decrease of calcium carbonate amount and particle size were observed. Moreover, the more effective the inhibitor was, the more the structure of the $CaCO_3$ precipitate was changed. The smallest changes were recorded when using a polyacrylate-based inhibitor (Fig. 3a). In the presence of PAS, an aggregate of calcium carbonate particles was formed, consisting of both rhombohedron calcite and round-shaped crystals characteristic of a metastable vaterite (Fig. 3b). The greatest changes were recorded when using HEDP and ATMP (Fig. 3c-d), while the sediment lost its crystalline structure and became friable, which leads to simpler removal of it from the surface of the reverse osmosis membrane.

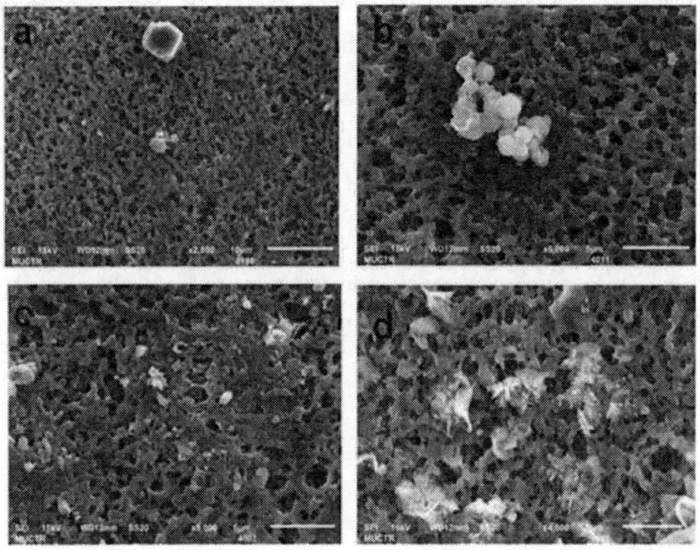

Fig.3. SEM images of CaCO3 in the presents of inhibitors. (a) – PA, (b) – PAS, (c) – ATMP, (d) – HEDP

4. Conclusion

1. Using two express laboratory techniques, different types of scale inhibitors to prevent the precipitation of calcium carbonate were compared. A number of the effectiveness of antiscalants at a concentration of 4 ppm is evaluated as HEDP> ATMP> PAS> PA.
2. The relative effectiveness of polymeric antiscalants at a concentration of 10 ppm was shown.
3. It is defined that the addition of inhibitors leads to a change in the structure of the precipitate of calcium carbonate.
4. It was revealed that the more effective the antiscalant is, the more the sediment structure changes towards the formation of a less stable and easier to remove modification of calcium carbonate.

References

[1] Amjad Z and P Koutsoukos 2014 *Desalination* **335** pp 55–63
[2] Tang Y *et al* 2008 *Desalination* **228** pp 55–60
[3] Belarbi Z *et al* 2014 *Journal of Crystal Growth* **386** pp 208–214
[4] Lui G *et al* 2014 *Designed Monomers and Polymers* **20** pp 397–405
[5] Kumar M *et al* 2015 *Environmental Technology* **36** pp 1281-1290
[6] Lui Y *et al* 2014 *Appl. Polym. Sci.* **131** pp 39792/1-39792/12
[7] Wang *et al* 2015 *Desalination and Water Treatment* **53** pp 3491–3498
[8] Al-Roomi Y M *et al* 2015 *Desalination* **375** pp 138–148
[9] Amjad Z *et al* 2014 *Int. J. Corros. Scale Inhib.* **3** pp 35–47

IOP Conf. Series: Earth and Environmental Science **288** (2019) 012011 doi:10.1088/1755-1315/288/1/012011

Biofouling on water intakes and fight against them

Kalinichenko M. Ya.

North-Caucasus Federal University, Stavropol, st. Pushkin, 1

mix2710@mail.ru

Abstract. Relevance of the research is caused by high costs of maintenance for Dreissena removal from water inlets of a water intake system.
In this regard, the article is directed to the development of the technology and the technological scheme of biofouling prevention on water intakes.
The main approach to this problem research is the development of the technology and the technological scheme of biofouling prevention on water intakes.

1. Introduction

Pipelines on water intakes (especially on the regulated sources) are subject to internal fouling by hydrobionts most of which are often Dreissena mollusks. Fouling often happens to be considerable. It leads to critical losses of pressure in the water intake soaking-up system and threat of pump stations laying-off. Driyssena larvae seldom move independently in the water supply system, generally under the water flow influence.

The Dreissena layer on internal walls of pipelines reaches 7...10 cm, and the mass of fouling up to 7 kg/m^2. At such fouling the hydraulic resistance of pipelines significantly increases. It attracts additional expenses of the electric power on water supply. In this regard fight with Dreissena on the operating water intakes is to be considered not only as an instrument for ensuring of uninterrupted water supply, but also as a measure of economy of the electric power. So it is seen to be important to prevent entry of hydrobionts in water intakes.

The scale of annual expenses for fight against Dreissena is impressive. According to the experts of the U. S. Congress additional costs of maintenance for Dreissena removal were more than 3 billion dollars in 1993-99. On average, a total of 376,000 $ is spent on protection against biofouling of one power station in the United States, and 822,000 $ for a nuclear power plant.

2. Formulation of the problem.

Small microorganisms (fungi and bacteria) begin the process of new surfaces colonization. They cover very quickly all available surfaces of pipelines, forming a thin film similar to a transparent gel.

Biofilm reduces the efficiency of heat transfer process and promotes large microorganisms (mollusks and algae) attachment.

Weak tendrils of larvae cannot be fixed on the walls of the tubes without biofilm and will be carried away outside by water flow.

As it is known, the bacterial membrane is semipermeable.

If you electrically "charge" the bacteria flowing through the pipe-line, then the water molecules align themselves around it, forming several ordered layers of pure (on the molecular level) water (Figure 1) around the bacterium.

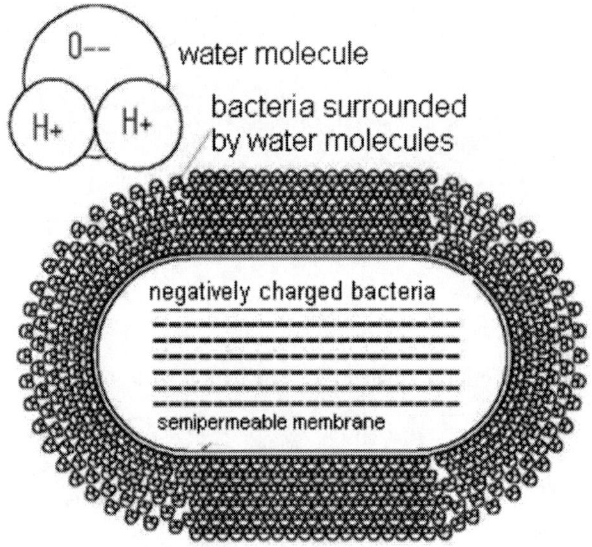

Figure 1 – Scheme of a bacteria "charging", where water molecules are polarly aligned

There is osmotic pressure that makes water molecules penetrate the membrane into the bacterium from all sides. It instantly leads to rupture of its membrane.

3. Discussion of the results

Internal surfaces of the pipelines are proposed to impart a weak positive charge. Due to the attraction of water molecules to charged surfaces, several stable layers are formed from identically oriented water molecules (Fig. 2).

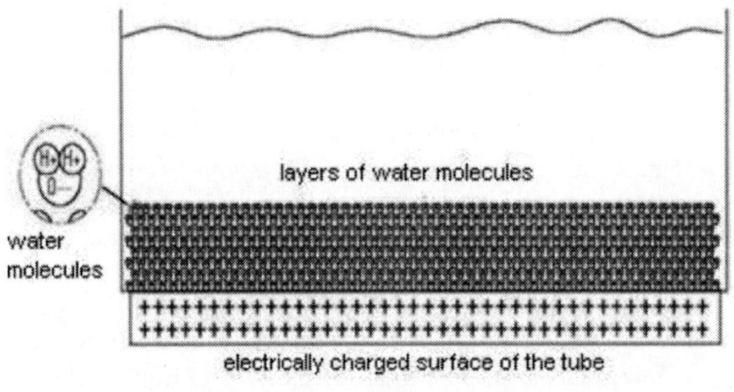

Figure 2 – Scheme of microbial fixation preventing on the walls of pipes

The proposed scheme of preventing water fouling is shown in Figure 3.

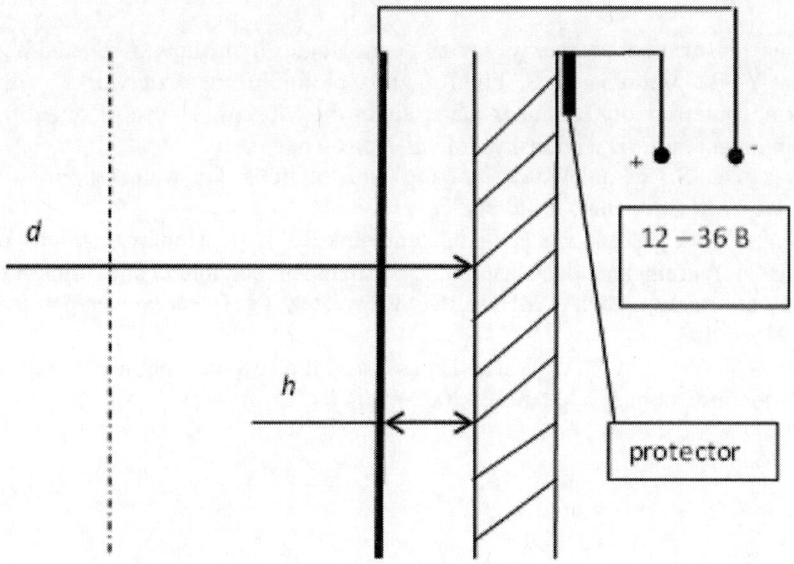

Figure 3 – Diagram of the device for preventing the biofouling of water intake

Thus, several layers of pure (on the molecular level) water are created on the surfaces of the pipelines. They represent a stable barrier to fungi and bacteria, which prevents the microorganisms from fixing on the surface.

Consequently, suppressing the bacteria themselves and not allowing them to fixate on the walls of the pipelines, biofilm formation is prevented. Protector provides protection of anode fracture of the pipeline.

Similar to bacteria and particles of pollution, the Dreissena larvae and other microorganisms are also electrically charged. As a result, the larva cells begin to experience a high external osmotic pressure and, as already shown above, they will be surrounded by pure water layers (Figure 4).

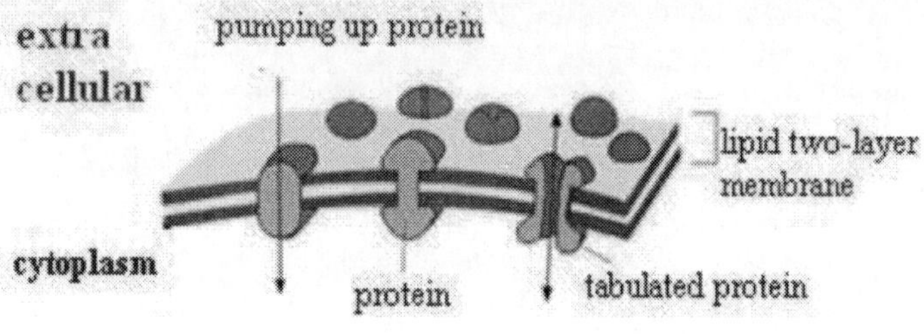

Figure 4 – Scheme of vital activity suppression of Dreissena larvae

As a result, the larvae metabolic processes will be disturbed. The cell, trying to stabilize the internal pressure, will continuously eject its cytoplasm into external environment, receiving only pure water instead of nutrients.

At the same time, some of the cells immediately die of osmotic pressure drop, while others cannot normally function.

4. Conclusion.

Physical ways to prevent fouling by Dreissena appear to be promising because of their non-reagency and safety for the environment.

The required DC power for processing will be about 1 kW at a current density of 0.5 A/dm^2.

References

[1] http://www.plasma.com.ua/energy_saving_technologies/hydroflow/product85.html.

[2] Bagotsky Y. B., Velmina E. S. Fight against biofouling at waterworks. – Improvement of drinking water quality (Seminar materials in the Moscow House of scientific and technical promotion n.a. F.E. Dzerzhinsky). M., 1977, 160 c.

[3] Obrazovsky A. S., et al. Water intaking constructions for water supply from superficial sources. – M.: Strojizdat, 1976, 368 c.

[4] Ryabchenko V. A., Rusanova N. A., Korobejnikova L. I. Modern methods of fight against biological fouling and deposits in the systems of economic and drinking water supply. Survey information COSTI MHUI RSFSR, "Water Supply and Sewerage" series, it. 1 (32). – M. 1976, 64c.

[5] Yakovlev S.V, Voronov Y.V. Water disposal and the sewage treatment / Textbook for higher education institutions: – M.: ACB, 2002 – 704 c.

IOP Conf. Series: Earth and Environmental Science **288** (2019) 012012 doi:10.1088/1755-1315/288/1/012012

Reagent approach for silicate deposits prevention in waste heat boilers

Khasanova D.I., Panteleev A.A.

Federal State Budgetary Educational Institution of Higher Education, Moscow Power Engineering University

Abstract. New reagent composition with optimal phosphate:thermal polymer ratio has been developed and studied for steam-and-condensate channel protection and its corrosion prevention as well as for on-line washing to remove old silicate deposits formations. Optimal dosage and water-chemistry conditions for a medium pressure steam boiler with specified quality of feed water have been selected. The causes of silicate deposits in medium pressure waste-heat boiler have been analyzed. The reagent efficiency has been evaluated while industrial applications.

Radiant surface of steam boilers are heated actively with flare system causing high local heating. So, even in case of a short decrease of heat-transfer between the boiler's wall and boiling water, the temperature of a steam-generating tube will exceed visibly the temperature of boiling water. The deposits formed on internal surfaces though being of low thickness but due to their low heat-conductivity can cause blisters, air leakage and explosions of water tubes enduring high heat-stress as a result of higher thermal resistance and significant local metal overheat. Besides, scale precipitation from the boiler unit can get into turbine together with steam, causing clogging of nozzle and paddle channels, making narrower their cross-section and resulting in pressure drop of the disks and diaphragm cascades. All these things cause low efficiency of the turbine, even the scale clogging being not severe.

Scale precipitation generally occurs due to deterioration of water environment of the unit, *i.e.* inefficient demineralization, low-quality condensate purification, condenser leakage and presence of non- boiled water as a result, omissions in boiler feed water inhibition *etc.* [1-2]

The most spread scales found in boilers and turbine flow channels are sulphates, calcium and magnesium carbonates, ferric & copper oxides, alumina and silicon oxide, the silicate deposits being the most challenging due to their low solubility and thermal conductivity. The scales of this kind are difficult to remove by chemical washing, the mechanical approach being usually in use. The most effective environment for silicate deposits cleaning is hydrofluoric acid, but its application is usually avoided by the reason of toxic properties. So, it was a challenge to find a solution for the "on-line" scale washing.

Silicate sediments found in a boiler have heat conductivity 50-500 times lower comparing to steel conductivity. Consequently, this scale sometimes promotes a 100 ^{0}C and even higher overheat of boiler' structure, so ductile properties of metal are affected adversely. Operating the boilers with scale being present leads to inevitable economic losses driven by fuel rate increase. Depending on deposits type, this unreasonable losses cause higher fuel consumption, making practically a several percentage increase per 1 mm of deposits (acc. to some literature data, this fuel increase amounts to 8-10% per 1 mm of deposits) due to higher heat losses along with exit gases. Table 1 gives data on thermal conductivity of some alloys and the most spread deposits [3-4].

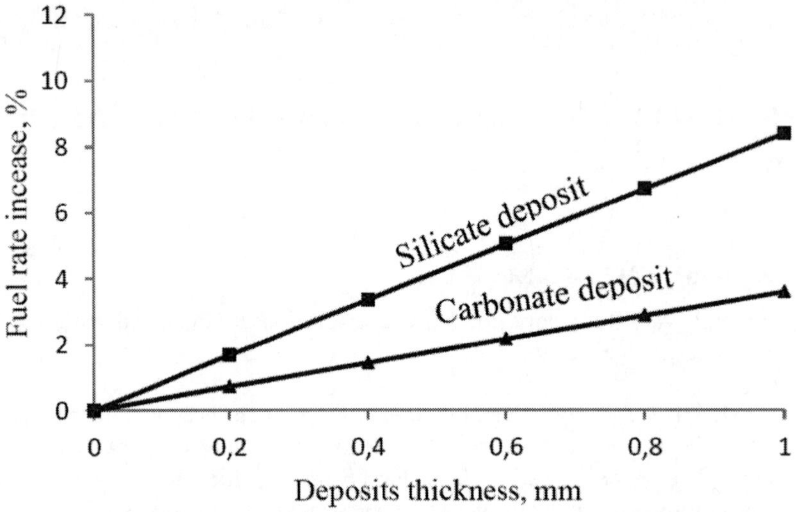

Fig. 1 – Comparative Analysis of Silicate & Carbonate Deposits.

Table 1. Thermal Conductivity Factor

Name	Thermal Conductivity Factor, Wt/(m·degree)	Note
Carbon Steel	48.0	-
12X18H9T	14.5	-
Brass	92-134	-
Hematite Fe_2O_3, porosity 0%	6.3	Scale is formed in high temperate and low pH conditions
Magnetite Fe_3O_4, porosity 0%	3.5	Main part of protection layer
Calcium Sulphate (gypsum) $CaSO_4 \cdot 2H_2O$	1.6	Firm and easy sticking scale
Calcium Carbonate $CaCO_3$	1.2	Scale (the most spread composition)
Silicate scale (xonotlite): $5CaO \cdot 5SiO_2 \cdot H_2O$	0.8	Very firm scale

Deposits being formed in boilers and heated with waste nitrous gases employed by nitrogen fertilizer industries, can cause lower capacity of the unit or unscheduled shut-down meaning reduce of the principal product and a failure of a production plan.

Thus, one nitric acid factory faced some problems in the drums of nitrous-gases heat-waste boilers (KN 80/40) caused by lower heat removal related to scale precipitations on a heat-transfer surface.

Inspection of drums' internals in waste-heat boiler (boiler's separating drums of nitric acid Workshop KN-80/40) has detected both drums' and cyclones' inner surfaces of salt compartment to be covered with grey and beige deposits of 1 cm thickness; white deposits have been found in clean compartments of the boiler, predominantly on the surfaces below the line of liquid and vapour phase separation; as for the grids in clean compartments, 3 mm- thick deposits of grey and beige colour have been detected (Figure 2).

According to performed examinations, the scale has the following composition: silicates, being the main deposit and amounting to 33.4-50.0%; magnesium oxide - 2.9-36.9%; calcium oxide – 2.2-43.8%; iron oxide – 1.2 – 4.4% (Table 2).

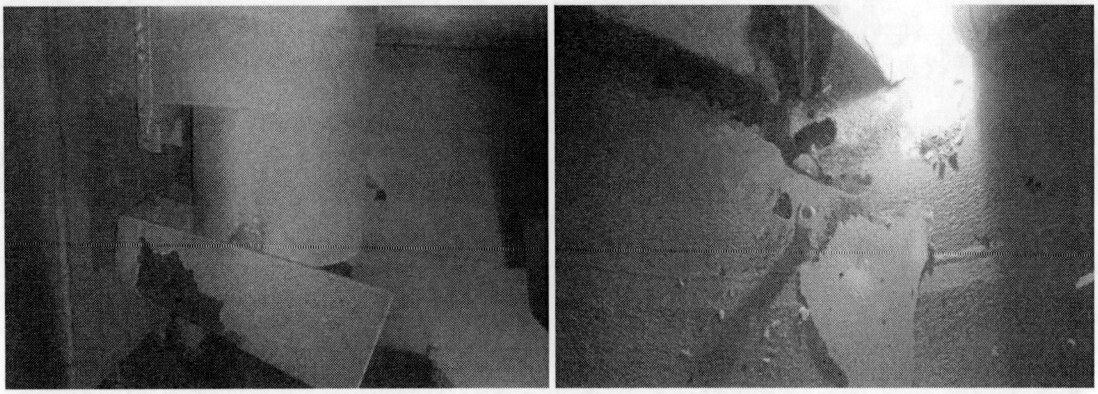

Figure 2 –Picture of Waste-Heat Boiler KN-80/40 (Salt Compartment, 2nd Unit) at Nitric Acid Workshop

Table 2 – Analysis of Scale Samples

Sampling Point	Phase Composition (Matrix)	Main Chemical Components
Unit 1 Salt Compartment	$Ca_6Si_6O17(OH)_2$	SiO_2 – 50% CaO – 43.8% MgO- 2.9%
Unit 1 Clean Compartment	$Mg_3(Si_2O_5)(OH)_4$ $Ca_6Si_6O17(OH)_2$ - admixture Fe_2O_3 – admixture	SiO_2 – 30.9% CaO – 4.9% MgO- 29.3% Fe_2O_3 – 8.6%
Unit 2 Salt Compartment	$Mg_3(Si_2O_5)(OH)_4$ $Ca_6Si_6O17(OH)_2$ - примесь Fe_2O_3 – admixture	SiO_2 – 33.4% CaO – 6.8% MgO- 30.4% Fe_2O_3 – 3.6%
Unit 2 Clean Compartment	$Ca_6Si_6O17(OH)_2$ $Mg_3(Si_2O_5)(OH)_4$	SiO_2 – 42% CaO – 34.7% MgO- 20.7% Fe_2O_3 – 1.4%

The case being studied, it was found out that the main cause for the deposits of such kind was employment of partially demineralized (instead of completely demineralized) water as a feed water for the boiler. This water was determined to contain significant amount of silicon compounds entering the process along with natural water as well as non-reactive silicon. Sediments of such kind can be prevented by using demineralized water for feeding the boiler.

Due to inefficiency of the demineralizing unit failing to feed waste-heat boilers with absolutely minerals-free water, it was offered to use reagent complex, *i.e.* sediments dispersants, to minimize the deposits in the drums of nitrous gases waste-heat boilers.

A great number of laboratory experiments was carried out, the broad range of concentration and ratios of alkaline phosphates (ortho-, polyphosphates) and water soluble polymers were studied. The optimal phosphate:thermal polymer ratio and a new complex reagent MF-RSB-35 were developed as a result.

The aim of this work is to study AquaComplex MF-RSB-35 complex reagent for feed water treatment to improve boilers efficiency and reliability, and to minimize corrosion and/or scale and silicon sediments troubles. Optimal reagent dosage and chemistry conditions for medium pressure steam boiler fed with water of specified quality were determined by this study.

The reagent AquaComplex MF-RSB-35 is a balanced mixture of alkaline phosphates and polymeric dispersants inhibiting scale precipitation and corrosion inside a boiler. MF-RSB-35 contains special dispersing polymer to reduce accumulation of silica and sodium silicate. Phosphates being a part of the composition are included to inhibit boiler corrosion and scales precipitation. As for the water- soluble polymer incorporated into the composition, it represents a carboxylic copolymer namely, a non- ionic thermal polymer with molecular weight of 5000 and the following structural formula (Figure 3).

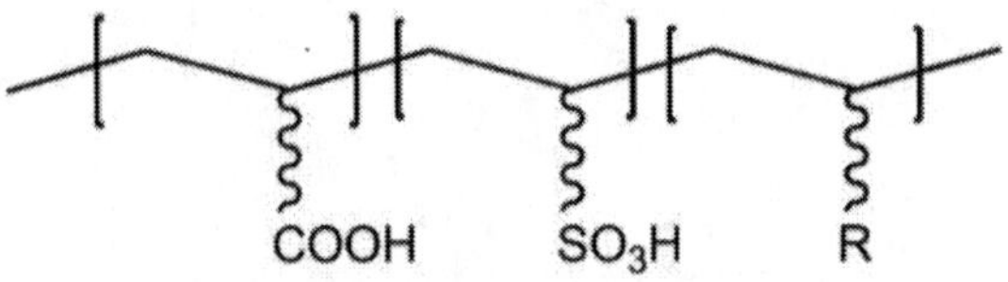

Figure 3 – Thermal Polymer Structural Formula

The performance range of the reagent under study is assumed to be within 10-100 mg/dm^3 depending on feed water quality and the boiler operating conditions. To monitor the dependency of major feed water specifications on reagent dosing rate correctly, it was decided to keep the purging volume the same. To achieve maximum "on-line" purging of the boiler, it was settled to perform pilot testing at maximum possible purging, *i.e.* 6 % of the units capacity. Besides, a maximum concentration was added while initial feeding, namely 100 mg/dm^3, each further dosage making up 10 mg/dm^3 proportionally to the feed water volume.
As it is shown in Fig. 4, pH values of feed and boiler water are relatively stable while treatment procedure, although being along the lower limit and slightly below target values.

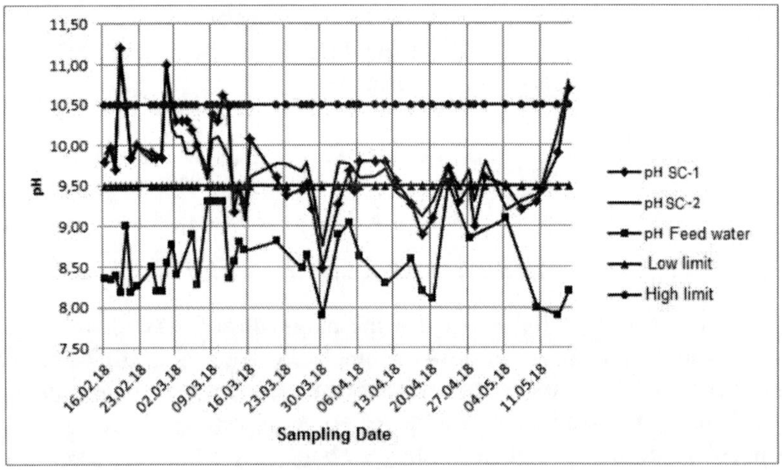

Figure 4 – Feed and Boiler Water pH Values of while Testing

The level of orthophosphate within the boiler water is a main criterion characterizing the concentration of the inhibitor. Water soluble polymer is contained in a certain ratio towards orthophosphate, so knowing phosphates concentration, the content of polymer in a boiler water can be calculated accordingly. As it is shown by the diagram of Fig. 5, this value was above the normal getting stable within specification limits afterwards. As for the reagent dosage, it was adjusted depending on boiler water exponents during the test.

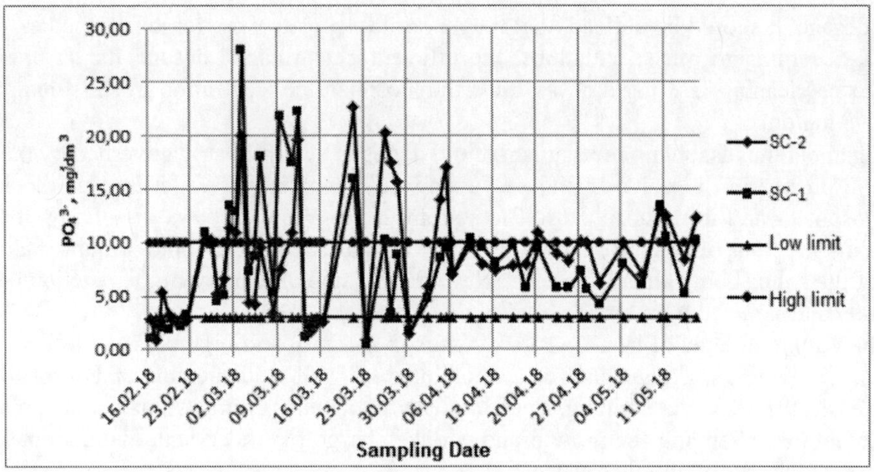

Figure 5 – Orthophosphate Fluctuations in Feed and Boiler Water while Testing.

Figure 6 contains diagram of silicates fluctuations in feed and boiler water (Salt Compartments 1 and 2). The silicate content in feed water varies within 3.5-5.5 mg/dm^3, being practically 180-800 mg/dm^3 in boiler water which is two orders of magnitude higher. The second salt compartment (SC-2) has the concentration several times higher, comparing to the first one (SC-1). As for the boiler's vaporization coefficient, it is about 10-20 (judging by the iron salts contained in feed and boiler water). Consequently, the silicate content in purging boiler water is to be within 35-110 mg/dm^3 at relevant vaporization coefficient. Taking into consideration much higher silicate concentrations present in boiler water (not proportional to the mentioned vaporization coefficient), we supposed these silicates come from the older boiler deposits while washing. Higher silicates content in boiler purging water at enhanced purging is an evident of scale dissolving and deposits removal from the boiler. This idea is proved by the fact of better heat removal, the temperature difference showed a 6-12°C increase thus, reaching the normal values. The test being completed, the equipment was opened and inspected, and we didn't find any sediments that had been previously present.

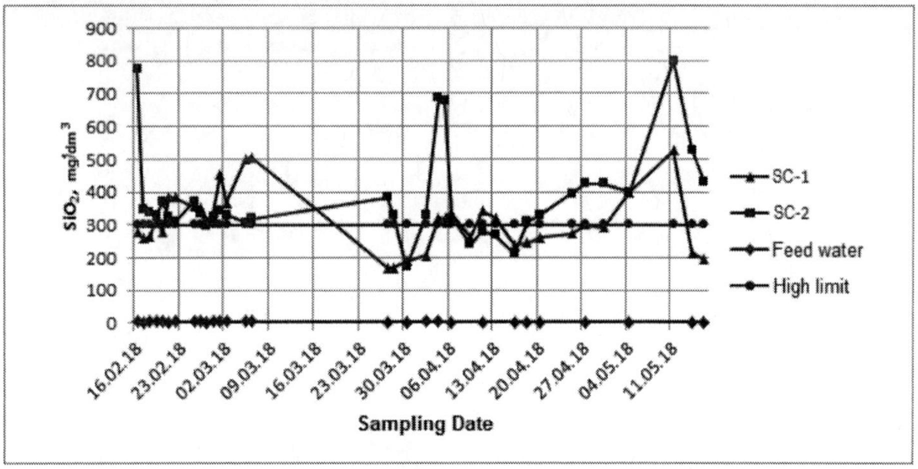

Figure 6 – Silicates Fluctuations in Feed and Boiler Water while Testing.

Basing on the quality indices of water in salt compartment, we decreased the reagent supply while washing. Actually, the optimal reagent content was 2.5-3 times lower comparing to theoretically calculated dosage. It should be noticed, that the more old deposits were removed, the less reagent was needed. So, according to our speculations, the efficient performance dosage for an operated boiler following to its cleaning is 3 mg/dm^3, as for orthophosphate concentration in salts compartments, it makes up 5-6 mg/dm^3.

Speaking about the dissolving mechanism of old deposits, it is probably governed by the carboxylic groups being present in the polymer and preventing the formation of water-insoluble compounds of calcium phosphates and iron phosphates. The sulfonic acid group supports dissolving of the existing silicate sediments and inhibits formation of new crystals, and non-ionic groups facilitate steric repulsion of the solids being already formed. Nevertheless, the chemistry of the reaction needs further detailed study and some additional research.

Thus, the minimum level of iron-based sediments in medium and high pressure boilers is achieved by the special polymers making a part of the inhibitor, as well as dispersing and suspending oxides, that can be easily removed via draining. The polymers that control deposits formation, are adsorbed on metal oxide entities, changing the growth and surface charge of the crystal, and alkaline phosphates diminish corrosion activity of boiler water.

Employment of reagent complex MF-RSB-35 for internal boiler water treatment provides sediments control and their minimization reduces the areas of hydroxide concentrations, thus ensuring the heat-exchange at maximal efficiency and eliminating the problems of tubes overheating enduring high temperature impact.

Reference

[1] M. Yovchev. Corrosion in Heat and Nuclear Power Equipment. – Moscow: Energoatomizdat, 1988.

[2] A.A. Kot, Z.V. Deeva. Water Chemistry Specifications for CHP High-Power Generating Sets. – Moscow:-"Energia", 1978.

[3] M.S. Shkrob, V.I. Vulfson. Water Preparation, Specification and Control at Steam Power Plants. – Moscow: "Energia", 1978.

[4] V.E. Doroschuk, V.B. Rubin. Water Chemistry Specifications and Reliability of 500 and 800 MWt Power Generating Sets. – Moscow: "Energoizdat", 1981

THE TECHNOLOGY OF OBTAINING MODIFIED SORBENTS BASED ON SILICATE PRODUCTION WASTE

V.A. Kulagin, O.G. Dubrovskaya, D.N. Gudkov, A.I. Matyushenko
Siberian Federal University, 79 Pr. Svobodny, Krasnoyarsk, Russia

v.a.kulagin@mail.ru

Abstract. *The article deals with the problem of rational use of waste of enterprises for the production of silicate cement. Currently the production gliezh waste (burnt clay) does not find its application and is disposed of. A method of obtaining a sorbent based on the gliezh waste is suggested, the aluminium oxide content in which ranges from 15 to 40 % and silicon oxide – from 55-75 %. Pre-burnt clay is treated with sulphuric acid, and then calcined in a muffle furnace at t = 300 °C. The resultant sorbent has high rates of wastewater treatment from heavy metals.*
Key words: sorbent, sorption material, secondary material resource, modified sorbent, wastewater treatment.

1. Introduction.

Waste water of metallurgical and thermal power plants often exceeds the maximum permissible concentrations of harmful substances. The search for effective methods to control harmful emissions into the environment is an urgent task of the modern world. One of these methods is the use of sorbents. They are often used in heat power plants for purification of petroleum products [1–6]. The heavy metal ions, ammonium salts, phosphates, chlorides and other compounds contained in them also cause environmental damage to the environment. Therefore, the use of sorbents helps to significantly reduce the level of hazardous substances to the maximum permissible concentrations. At the same time, methods and technologies of modification of the sorption material are constantly being searched for in order to improve the efficiency of wastewater treatment and obtain economic benefits [7–13]. Thermal and chemical types of sorption material treatments, as well as their combination, are widely used.

The search for ways to modify sorbents is often associated with the use of production waste [1, 6, 7, 11–15]. Environmental friendliness of the solution is achieved not only by cleaning water environments, but also by reducing the impact of waste materials on nature [13]. In this paper the technology of the gliezh modification, a residue from the production of silica brick is considered.

2. Materials of research

A sample of raw material based on the gliezh for the production of sorbents with the selective ability to extract heavy metals has been studied. The general characteristics are given in Table 1, where the main parameters are the sorption capacity and the degree of extraction of the pollutant.

Table 1. General characteristics of the studied sample

Indicators	The studied sample
Base	Gliezh
Type	1.15-1.3 mm free-form granules, white - cream color
Sorption capacity (absorption capacity) for copper, mg / g	15.79–26.7
Application temperature, °C	+4...+25
Degree of heavy metal extraction	80.5 %
Cu	62 %
Fe	86,6 %
Pb	92.9 %
Maximum dose of sorbent, g/ l	5.0–12.8
Dose of discharged sorbent, mg/l	0.83

The gliezh-based clays used for the aluminum sulfate production are to contain a significant amount of kaolinite and are to be easily decomposed by acids. Kaolinite clays containing 15–40 % Al_2O_3, 55-75% SiO_2, 1-2% Fe(III) 0,5–1 % Fe(II) meet these requirements. As a sample, grey clay was used, the main fraction of which is SiO_2 quartz (d = 3.33, 2.45, 1.81 Å), and diffraction peaks relating to kaolinite $Al_2O_3{\cdot}2SiO_2{\cdot}2H_2O$ (d = 7.10, 3.56 Å) and to feldspar with d = 3.23, 3.18 Å are observed. In addition, impurities of muscovite are available in clays (d = 9.9, 4.95 Å). A diffraction pattern, made with the help of the D8 ADVANCE diffractometer (Bruker, Germany) is shown in Fig. 1. The results of the thermal analysis carried out with the help of the synchronous thermal analysis STA 449 F1 Jupiter (NETZSCH, Germany) device are presented in Fig. 2.

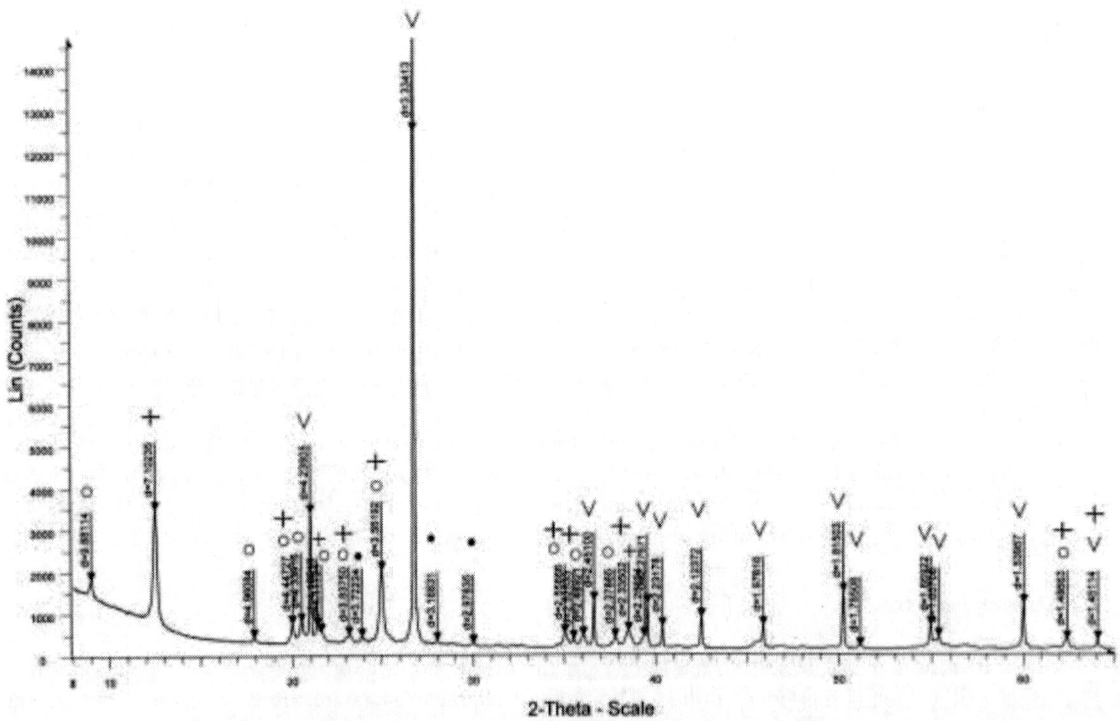

Fig. 1. The diffraction pattern of the initial grey clay sample:
v – Quartz (SiO_2), + – Kaolinit, • – Feldspar, o – Muscovit

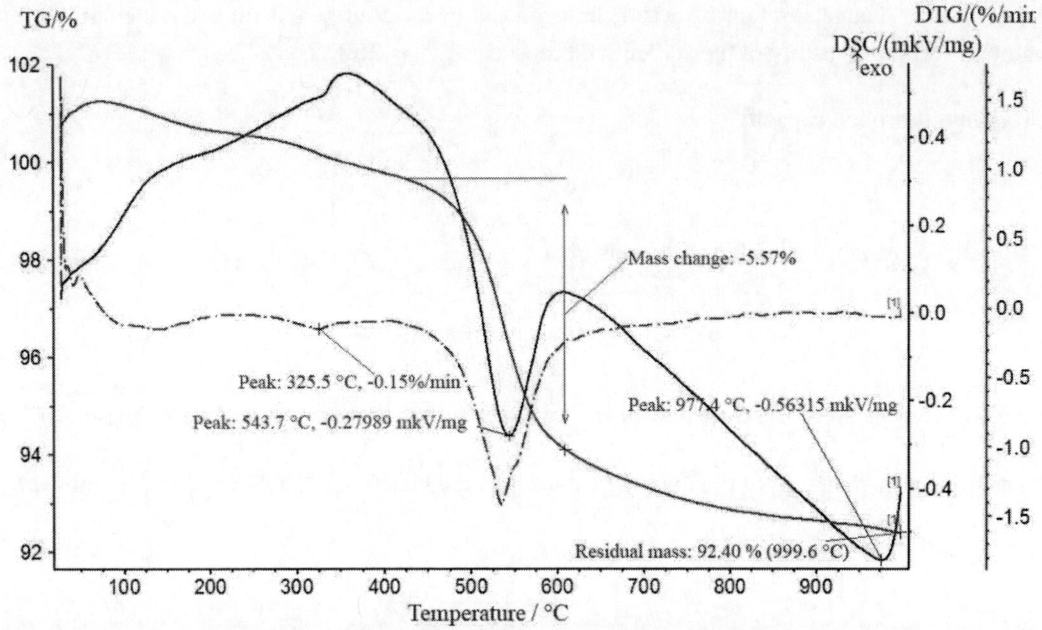

Fig. 2. Thermogram of the initial sample of grey clay

In this case deep endoeffect at 500–600 °C is observed in the DSC curve related to the Al_2O_3 $2SiO_2{\cdot}2H_2O$ kaolinite – there is an allocation of water from the hydroxyl groups, the process of the kaolinite amorphization is taking place and the mass of the sample (TG curves) is reduced by 5.57 %. The content of the kaolinite calculated according to the graphs, is about 39.9 %. The clay chemical composition is given in Table 2.

Table 2. Chemical composition of the initial sample, mass %

	Sample	SiO_2	Al_2O_3	Fe_2O_3	CaO	MgO	SO_3	Na_2O	K_2O	TiO_2
Grey clay	7.65	68.02	19.90	0.90	0.19	1.09	0.12	0.13	0.98	0.1

3.The experiment methodology

A portion of the clay is ground and calcined in a muffle furnace at a temperature of 750-800 C for one hour. The resultant cake is cooled, crushed and then treated in sulphuric acid for leaching of Al2O3. Concentrated sulfuric acid is added in excess (105% of the stoichiometric ratio), then diluted with water and stirred at a temperature of ≈ 100 C. The resulting solution is decanted; the sludge is washed with hot water to extract aluminum sulfate. Wash water is added to the solution, filtered, evaporated in a water bath until a thick mass is formed. Then the thickened mass is calcined in a muffle furnace at a temperature of 300 C, obtaining a white crumbly mass consisting almost of pure aluminum sulfate.
The calculation of the characteristics of sorption materials was carried out by the following method:
1. The maximum dose of sorbent is determined by the formula:

$$D_{sb}^{\max} = \frac{C_{en} - C_{ex}}{a_{sb}^{\min}},$$

where C_{en}, C_{ex} – concentration of petroleum products in the source and filtered water, , $a_{sb}^{\min}$ – the minimum sorption capacity of the sorbent, defined as $a_{sb}^{\min} = 253\sqrt{C_{ex}}$;

2. Maximum sorption capacity

$$a_{sb}^{\max} = 253\sqrt{C_{en}}.$$

3. Dose of sorbent discharged from the adsorber

$$D_{sb} = \frac{C_{en} - C_{ex}}{K_{sb} \cdot a_{sb}^{\max}}$$

where K_{sb}– the degree of exhaustion of the capacity of the sorbent. According to the experimental data K_{sb}= 0,5.

4. The approximate height of the load, which ensures the cleaning of the drains, is determined by the formula:

$$\mathrm{H} = \frac{D_{sb}^{\max} \cdot t_{ds}^{op}}{k_{ds} \cdot a_{sb}^{\max}}$$

where t_{ds}^{op} – the approximate duration of operation of the plant to the skip; k_{ds} –the specified degree of exhaustion of the sorbent capacity.

5. The degree of extraction of metals is defined as

$$R = (C_1 - C_2)/C_1 \cdot 100\ \%,$$

where C_1 is the initial concentration of metal in the solution; C_2 is the concentration of metals in the filtrate.

4. Results of research and discussion

The diffractogram and the thermogram $Al_2(SO_4)_3$ are shown in Fig. 3 and 4, respectively. The sludge remaining after decanting and washing with water is quartz containing minor impurities of calcium, aluminum and iron compounds (0.1 %). The diffractogram of the sludge is shown in Fig. 5.

Thus, the presented clays are suitable for the production of coagulant – aluminum sulfate and sludge used to produce microspheres. The results of the use of the modified sorbent are presented in Table 3. The degree of extraction of metals is given in Table 4. The number of experimental lines-6. The error of experimental studies is 5 %, which characterizes a sufficiently high degree of reliability of the results.

Table 3. The results of experimental research

Pollutants	Substance concentration, mg/l		
	Before treatment	After treatment with small particle gliezh in suspended layer	After treatment with large particle gliezh filtering
Ph	7–7.5	8–10	9–10
Copper (Cu)	5	1.8	1.9
Iron (Fe)	13.5	0.15	1.8
Plumbum (Pb)	0.7	0.002	0.05

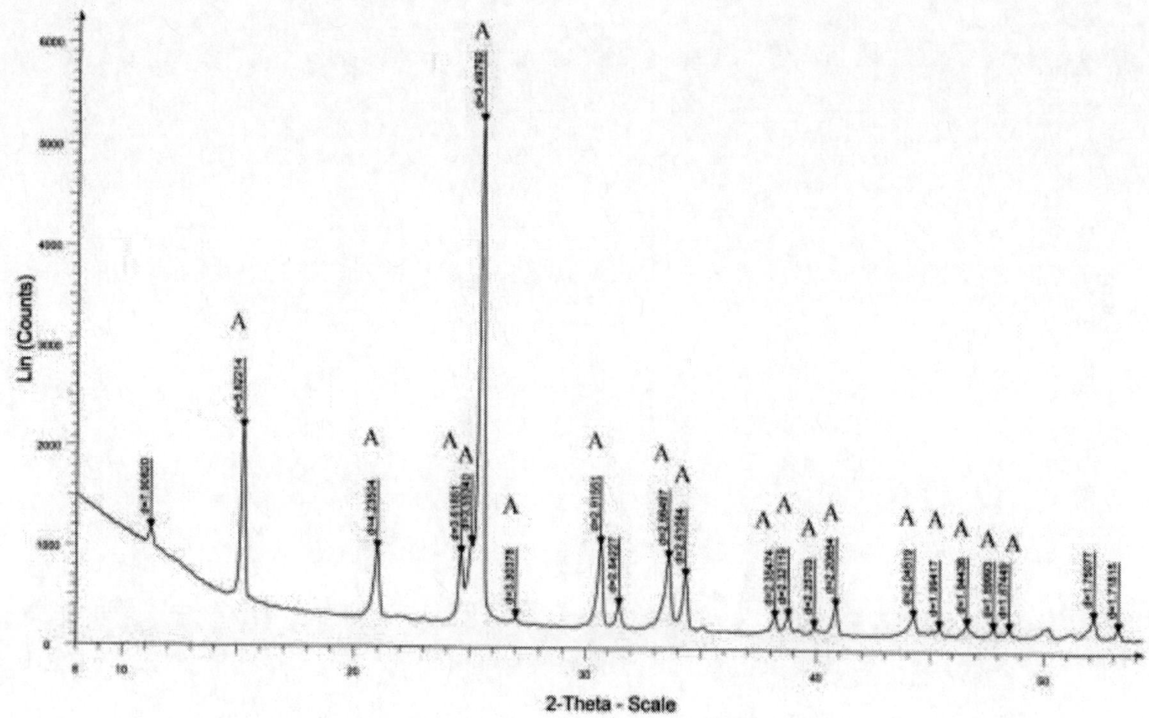

Fig. 3. The diffraction pattern of aluminium sulphate: A – $Al_2(SO_4)_3$

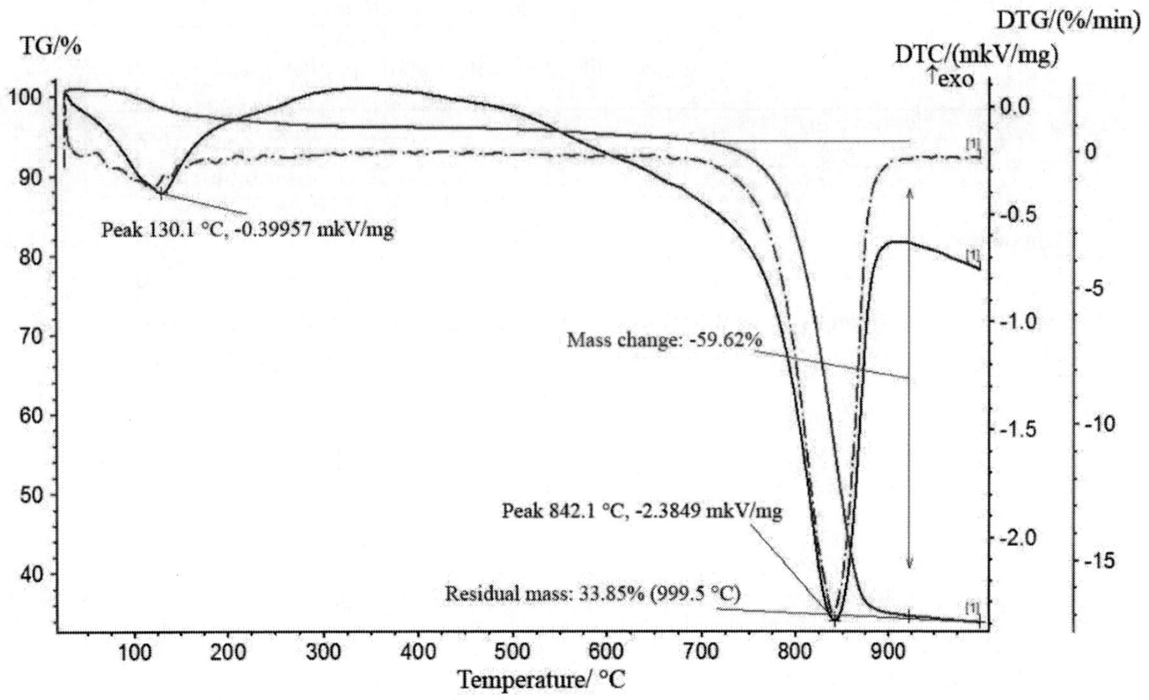

Figure 4. Thermogram of $Al_2(SO_4)_3$ aluminum sulfate

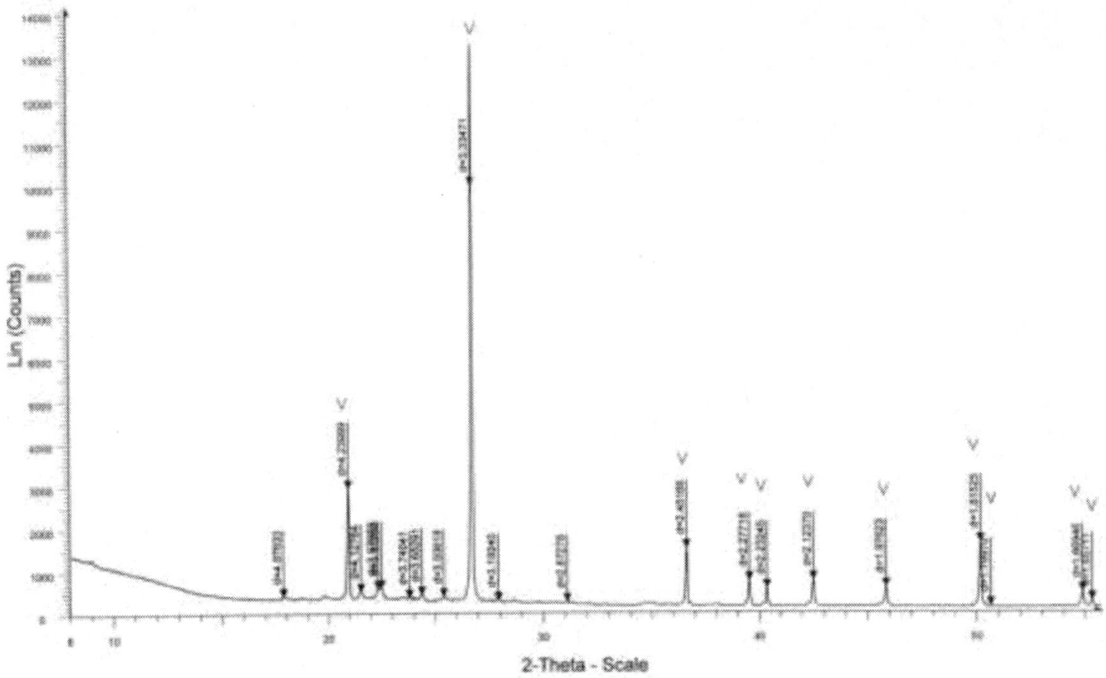

Fig. 5. The sludge after the decomposition of the clay with sulfuric acid: v – Quartz (SiO_2)

Table 4. The degree of extraction of metals

Pollutants	The degree of extraction of metals, %	
	After treatment with small particle gliezh in suspended layer	After treatment with large particle gliezh filtering
Copper (Cu)	64	62
Iron (Fe)	98.9	86.7
Plumbum (Pb)	99.7	92.9

5. Conclusion

It is obvious that sorbents on the basis of gliezh are possible to use in the treatment of wastewater from heavy metals (waste water of metallurgical complex), in the treatment of waste water of the energy complex, as well as for the oil refining industry at the stage of reservoir water preparation.

References

[1] Nikolaeva L A and Iskhakova R Ya 2017 *Thermal Engineering* No 6 pp 72–78
[2] Tihomirova E I, Atamanova O V, Zamatyrina V A and Koshelev A V 2016 *ELPIT Academic herald* No 1 pp. 80–85
[3] Privalova N M, Dvadnenko M V, Nekrasova A A and Privalov D M 2017 *KubGAU research journal* No 127(03) pp 1–10
[4] Popova T V and Privalova N M 2016 *Research collection* No 2 pp 1–16
[5] Zamatyrina V A, Tihomirova E I 2016 *Proceedings of Samara research center* No 18(2) pp 369–372
[6] Hala Ahmed Hegazi 2013 *HBRC Journal*. № 9 276–282
[7] Fuks S L, Hitrin S V, Vasilevich A D, Tetesova A A, Pshenichnikova L N and Nagovicyna O A 2017 *Advanced science* No 2 pp 1–12

[8] Lihareva O B, Ivanov M G and Matern A I 2008 *Chemistry in the interests of sustainable development* No 16 pp 415–420
[9] Gurova TV and Shubin I N 2015 *Research problems: modern technologies and technical progress* No 4 pp 44–47
[10] Polunina I A, Vysockij V V, Senchihin I N., Polunin K E, Goncharova I S, Petuhova G A and Buryak A K 2017 *Colloid journal* No 79(2) pp 192–197
[11] Dubrovskaya O G, Kulagin V A, Kurilina T A and Li Feng-Chen 2017 *J. Sib. Fed. Univ. Eng. technol*. No 10(5) pp 621–630
[12] Aseel M Aljeboree, Abbas N Alshirifi and Ayad F Alkaim Arabian 2017 *Journal of Chemistry*. No 10 pp 3381–3393
[13] Barakat M A 2011 *Arabian Journal of Chemistry*. No 4 pp 361–377
[14] Kyrii S O, Kosogina I V, Astrelin I M and Kyrienko O Yu 2015 *Наукові вісті НТУУ "КПІ"*. No 6 pp 95–101
[15] Sabino De Gisi, Giusy Lofrano, Mariangela Grassi and Michele Notarnicola 2016 *Sustainable Materials and Technologies* No 9 pp 10–40
[16] Yarusova S B, Gordienko P S, Buravlev I Yu, Zhevtun I G and Narbutovich A A 2018 *Developments of modern natural science* No 3 pp 39–43

Acknowledgment

The reported study was funded by Russian Foundation for Basic Research, Government of Krasnoyarsk Territory, Krasnoyarsk Region Science and Technology Support Fund to the research projects № 17-48-240386 p_a and 16-41-242156 p_офи_м.

Highly efficient catalysts from waste water sludge of thermal power plants and waste of machine-building industries for gas purification from nitrogen oxides

A. F. Mahotkin [1], I. A. Mahotkin [2], A. E. Suharnikov [3]

[1,2]Kazan Scientific Research Technological University, Russia,K. Marx Str., 68 420015 Kazan, Tatarstan, Russia.

[3]Kazan State University of Architecture and Engineering, Russia, Zelenaya Str., 1 420043 Kazan, Tatarstan, Russia.

oxzkstu@kstu.ru, Sukcharnikov@yandex.ru

Annotation. Relatively cheap and high-efficient catalysts for purification of waste gases of thermal power plants from nitrogen oxides have been developed. Pilot batches of catalysts were manufactured. The analysis of the results of research and testing of new catalysts in industrial conditions of thermal power plants of chemical enterprises.

When operating thermal power plants on fuel oil on the surface of boilers, a precipitate of incomplete combustion products is usually formed. Washing coppers from the sediment was previously carried out with water. At the same time, waste water was sent to the sludge collector. The study of the sediment composition in the sludge collector of Zainsk State district power plant showed that the main components of the sludge are: incomplete combustion products, iron oxides and vanadium pentoxide. The latter substances are known as effective components for the preparation of catalysts [1]. Drying, sifting, grinding, mixing with clay of the Kuganak Deposit, annealing and pressing allowed to turn the toxic sludge of the Zainsk State district power plant into a non-toxic high-efficient catalyst for selective catalytic purification of gases from nitrogen oxides. Fabrication and study of catalysts was carried out at the Kazan scientific research technological University, Bugulminsky porcelain factory. New catalysts have been tested in industrial conditions for purification of waste gases from nitrogen oxides at three thermal power plants in the cities: Zainsk, Naberezhnye Chelny, Moscow [2]. Modern sanitary standards for purification of waste gases from nitrogen oxides by ammonia at a temperature of 350 °C have been achieved. Ammonia is dosed with an excess of 30-40 %. The maximum degree of purification of exhaust gases is achieved on the granular catalyst. Minimum hydraulic resistance is achieved on a honeycomb catalyst. The required phase contact time in the catalyst bed is about one second.

The total amount of sludge in the sludge collector of Zainsk State district power plant alone reaches 15 000 tons, including vanadium pentoxide in the sludge about 1000 tons. It should be noted that this amount of valuable raw materials is sufficient for the manufacture of catalysts to solve the environmental problem of purification of waste gases from nitrogen oxides of all thermal power plants and all chemical enterprises of the Republic of Tatarstan. The technology of manufacturing new catalysts is relatively simple and can be implemented in red brick factories. The term of reliable operation of the catalyst is three years.

The creation of a modern production of high-performance catalyst in Zainsk would represent a solution to a complex scientific, technical and environmental problem of great importance for the Republic of Tatarstan. From the slurry Zainsk SDPP can be made highly efficient catalysts for the sum more than two billions roubles.

However, at present, the sludge collector has been eliminated. The sludge of Zainsk State district power plant is almost completely covered by the land and solid waste of the city. The Republic was left without cheap valuable raw materials for the manufacture of high-performance catalyst.

Further development of the effective scientific direction on creation of catalysts on the basis of processing of waste of various productions allowed to receive one more type of the simplest cheap and effective catalyst in Kazan scientific research technological University. The study showed that the effective catalyst for the reduction of nitrogen oxides with ammonia is carbon steel shavings taken from scrap metal. Simple rusty twisted shavings at 350 °C and phase contact time for one second provides effective cleaning of waste gases from nitrogen oxides with ammonia to modern sanitary standards. The period of reliable operation of the chip as a catalyst is one year. In the early stages of the study, we took care of ways to prepare the active surface of the chip. Tests in industrial conditions of power plant - 11 of Moscow and chemical enterprises in the cities of Solikamsk, Sterlitamak showed that preliminary preparation of the chip surface is not required. Conventional rusty and even oily twisted carbon steel chips taken from scrap metal in an industrial catalytic gas treatment reactor are rapidly converted into a highly efficient catalyst. Tested in Moscow at the power plant – 11 was conducted with the flow rate of the gas flow of 60 000 m^3/h, in Solikamsk and Sterlitamak tests conducted with the flow rate of the gas flow 30 000 m^3/h. the Hydraulic resistance of the layer of chips was in the range of 50-100 mm water column.

Among the developed catalysts, granulated iron-oxide catalysts have maximum efficiency in the conditions of thermal power plants and chemical enterprises [2]. Pilot tests of iron-oxide catalyst for purification of waste gases from nitrogen oxides were carried out in the conditions of the Moscow power plant – 11 at a gas flow rate of 60 000 m^3/h and a temperature of 350 °C. Comparison of the tested catalysts at a cost, starting from the cheapest, looks the following way: the turnings of carbon steel, zhelezovanadievoj granular catalyst from slurry waste Zainsk SDPP, cell catalyst from slurry waste Zainsk SDPP, plate, the catalyst sheet of carbon steel, the catalyst in the form of pipes made of carbon steel, zhelezobaktery granular catalyst. Therefore, the rational use of waste from various industries for the manufacture of catalysts can give a significant economic effect.

References

[1] Khodakov Yu.S. Nitrogen oxides and heat power engineering: problems and solutions. Monograph. Moscow:AST-M, 2001, 432 p

[2] Makhotkin I. A. Purification of gas emissions from vapors, aerosols and dust of toxic substances: abstract of dissertation ... kand. techn. sciences: 03.02.08. Kazan: KAZAN state technical University, 2011, p. 20

IOP Conf. Series: Earth and Environmental Science **288** (2019) 012015 doi:10.1088/1755-1315/288/1/012015

Waste gas treatment technology against dust in vortex apparatuses with closed water circulation

A. F. Mahotkin [1], I. A. Mahotkin [2], A. V. Starkova [3]

[1,2,3] Kazan Scientific Research Technological University, Russia,K. Marx Str., 68, 420015 Kazan, Tatarstan, Russia.

oxzkstu@kstu.ru

Abstract. This article covers the development results and manufacturing application of high technology of water scrubbing for waste gases against dust for environmental management of finely broken stone of limestone wastes burning in rotary kilns. The followings are obtained under production conditions firstly: current sanitary standards for gas cleaning, comprehensive utilization of trapped slurry and closed water circulation.

It has been established that limestone blocks are crushed into pieces of 80 mm diameter to make quicklime by limestone burning in shaft kilns [1]. However waste substances in the form of finely broken stone are appeared during a crushing process. It is not allowed to burn finely broken stone in shaft kilns. Quantity of resulting finely broken stone is large. For limestone burning rotary kilns find an application even though technical and economic features of a rotary kiln is a little bit worse than a shaft kiln but it will be balanced out by a disposal of a large amount of limestone wastes. Temperature for limestone burning is 1100-1200°C. When a rotary kiln is operated a huge amount of dust is appeared. Major constituents of dust are lime and limestone particles. There are usually two stages for waste gas treatment. Dry cyclone separator is used in the first stage and wet scrubbing is used in the second stage. Irrigated spraying Venturi tubes with cyclones are most popular in the second stage. However spraying Venturi tubes do not meet current sanitary standards for gas treatment against fine dust of lime and limestone. That’s why it is required to apply additional apparatuses for purification after Venturi tubes. In fact, additional apparatuses with gas filtration through porous membrane have a high performance [3] but they are not reliable due to dust clogs the filter way too fast in short time of operation. Electrostatic precipitators are more reliable but more costly. - For instance electrostatic precipitators are used in limestone burning stage during caustic ash production [2]. An essential fault of a theory of gas scrubbing against dust is high water flow rate [2]. The last mentioned is required an additional stage for waste water treatment.

It's becoming apparent that priority is a development and a manufacturing application of fundamentally new enhanced apparatuses for waste gas scrubbing against dust with water closed circulation. This problem is particularly sensitive not only for construction material production but and for modern large-scale production of caustic ash, solid-fuel combined heat and power stations as well [4].

Key reason for poor performance of dry cyclone separators for fine dust trapping is that when a particle size is reduced a force of friction relation of gas flow to centrifugal force acting on a particle shall be increased. For this reason small particles are absorbed by gas flow and do not sediment in cyclone separator. Small particles are rotated along with gas flow and sub-micron particles even move toward to central line of apparatus. Particle-size study for lime and limestone dust made clear that maximum of particle weight concentration is related to diameter of particles 3-5 microns. However number of sub-micron particles are much more than number of relatively large particles. As a result several motion paths of particles are appeared when gas flow rotates. Relatively large particles are pressed by centrifugal force to sides. Relatively small particles come to equilibrium state between friction force of gas flow to a particle and centrifugal force. These particles gyrate. Simultaneously the third motion path of sub-micron particles is formed. Sub-micron particles are spirals from sides toward a central line of apparatus. Inertial settling to sides is impossible for sub-micron particles. All these particles go out of cyclone separator and will go out of any other dry apparatus with gas flow rotation. Gas rotation amplification or reduction in diameter of apparatus do not make possible to separate dust sub-micron particles from gas. A particulate emission from a dry cyclone separator or from a dry vortex apparatus will always be above allowable sanitary standards. Accordingly an effective way to improve an efficiency of a gas scrubbing operation against a fine dust is a dampening of dry particles by liquid to form more solid wetted particles followed by their capture by rotating gas-liquid flow. This method can be organized as in both a cyclone separator and in a dedicated vortex apparatuses. Research of wetting actions for lime and limestone dust particles has testified that each particle gets wet immediately. Wetting time is about 0.01second. For this reason it is required to increase a possibility of particle collision between liquid and a dust particle inside a vortex gas scrubbing.

In a positive liquid separation from wetted collected particles full liquid flow can be returned to circulation in closed water circulation. Accordingly in development of vortex apparatuses for high-efficient waste gas scrubbing against dust the following tasks become important: development of an interreacting method for gas and liquid inside apparatus, development of additional apparatuses for water closed circulation, and development of method for comprehensive utilization of collected wetted particles.

In the furtherance of these goals several plants in Russian Federation manufactured pilot-plant equipment as per our projects. These equipment are dedicated to investigate complex dry and of wetting actions for lime and limestone dust particles later on rotary kilns [5]. Key pilot plants are the following: cyclone separator for dry dust separation and installed one by one vortex apparatuses for wet gas scrubbing. Wet gas scrubbing plant was sprinkled by water with closed water circulation. Water is fed to a plant in amount required only for a dampening of particles and evaporation. All collected dust is returned to construction material production. The high degree of a waste gas treatment against lime and limestone dust has been firstly achieved in production. This degree was 99.9 %. Designed industrial plants are simple and operationally reliable. New technology of gas treatment with water closed circulation against dust has found an application in different chemical industry sectors. Flow resistance of plant is range within 500-600 mmAq. New apparatuses are not clogged with sludge and reliable operate without repair for some years.

References

[1] Monastyrev A. V. Manufacture of lime. M.: Higher school, 1986. 192 p.
[2] Uzhov V. N. Cleaning of industrial gases from dust. M.: Chemistry, 1981. 392 p.
[3] Zaitsev D. I., Weaver G. A., Stoev, N. D. The production of soda. M.: Chemistry, 1986. 312 p.
[4] Power System: Mezhvuzovski. thematic. collection is scientific. works / Moscow.energetic.int (technic.un-t) Kazansk.phil. Kazan, 1997. 128 p.
[5] Makhotkin I. A. Purification of gas emissions from vapors, aerosols and dust of toxic substances: abstract of dissertation ... kand. techn. sciences: 03.02.08. Kazan: KAZAN state technical University, 2011. 20 p.

Technology for cleaning off-gases from dust in vortex devices with closed water circulation

A. F. Mahotkin [1], I. A. Mahotkin [2], A. V. Starkova [3]

[1,2,3] Kazan Scientific Research Technological University, Russia,K. Marx Str., 68, 420015 Kazan, Tatarstan, Russia.

oxzkstu@kstu.ru

Annotation. The results of the development and implementation in the production of highly efficient technology of wet cleaning of waste gases from dust to solve environmental problems of the processes of burning waste of small limestone in rotary kilns. For the first time in industrial conditions is reached: modern sanitary standards of cleaning gases, full utilization of collected slurry and closed-loop water circulation.

It is known that to obtain lime by firing limestone in mine furnaces, limestone blocks are first crushed to pieces with a diameter of about 80 mm [1]. However, if the shredding waste is generated in the form of small stone. Small stone annealing in shaft furnaces impossible. The amount of fine stone formed is large. For the firing of small limestone stone rotary kilns are used, although the technical and economic indicators of the rotary kiln of small stone are slightly worse than the mine furnace, but this is compensated by the elimination of large amounts of waste of small limestone. The process of firing of small limestone stone is carried out at a temperature of 1100-1200 °C. A large amount of dust is generated during the operation of the rotary kiln. The main components in the dust are lime and limestone particles. Waste gas cleaning is usually carried out in two stages. At the first stage, a dry cyclone is used, and at the second stage, a wet method is used. The greatest distribution in the second stage received irrigated spray Venturi pipe with the cyclones. However, Venturi spray pipes do not provide modern sanitary standards for cleaning gases from fine dust of lime and limestone. Therefore, after irrigated Venturi pipes with cyclones it is necessary to use additional devices. It is known that additional devices with gas filtration through porous partitions are highly effective [2], but they are quickly clogged with dust and therefore are not reliable. More reliable are electric filters. However, electrostatic precipitators are very expensive. For example, electrostatic precipitators are used at the stage of calcination of limestone in the soda ash production [3]. A significant disadvantage of the method of wet cleaning of gases from dust is a large water consumption [3]. The latter requires the use of an additional stage of wastewater treatment.

It becomes obvious that an urgent task is to develop and introduce into production fundamentally new more effective devices for wet cleaning of waste gases from dust with a closed water circulation. This problem is acute not only for the production of building materials, but also for the modern large-scale production of soda ash, as well as for thermal power plants operating on solid fuel [4].

The main reason for the low efficiency of dry cyclones for the capture of fine dust is that with a decrease in the particle size, the ratio of the friction force of the gas flow to the centrifugal force acting on the particle increases. Therefore, small particles are carried away by the gas flow and do not settle in the cyclone. Small particles rotate together with the gas flow, and especially small even tend to the axis of the device. The study of the dispersed composition of lime and limestone dust showed that the maximum weight concentration of particles refers to the particle diameter of 3-5 microns. However, the number of particularly small particles is much larger than the number of relatively large ones. This leads to the fact that the rotation of the gas flow produces several trajectories of the particles. Relatively large particles are pressed by centrifugal forces to the periphery. Relatively small particles enter into a state of equilibrium between the friction forces of the gas flow on the particle and the centrifugal forces. These particles rotate in a circle. At the same time, a third form of the trajectory of very small particles is formed. Particularly small particles of the device move in a spiral from the periphery to the axis of the device. For very small particles, centrifugal deposition to the periphery is not possible. All of these particles will fly out of the standard cyclone and will fly out of any other dry devices where the gas flow is rotating. Increasing the intensity of the gas twist, or reducing the diameter of the device does not allow to separate from the gas very small dust particles. The emission of particles from a dry cyclone, or from a dry vortex apparatus will always be higher than the permissible sanitary norms. Therefore, an effective way to improve the efficiency of the gas cleaning process from particularly fine dust is wetting the dry particle with a liquid to form more massive wetted particles, followed by the capture of wetted particles by a rotating gas-liquid flow. This method can be organized both in a cyclone and in special vortex devices. The study of the process of wetting the dust particles of lime and limestone showed that each particle is wetted with water almost instantly. The wetting time is about 0.01 seconds. Therefore, inside the vortex apparatus of wet cleaning of gases from dust, it is necessary to increase the probability of collision of dust particles with the liquid.

With a reliable separation of the liquid from the wetted trapped particles, the entire liquid flow can be returned to the circulation in a closed water cycle. Therefore, when creating vortex devices for highly efficient wet cleaning of waste gases from dust, the following tasks become relevant: development of a method of interaction between gas and liquid inside the apparatus, development of additional devices for closed water circulation and development of a method for the complete utilization of the trapped wetted particles.

To solve these problems at several domestic production facilities on our projects, pilot plants have been created to study the processes of complex dry and wet cleaning of exhaust gases from lime and limestone dust after rotating furnaces [5]. The main apparatus pilot plant were: a cyclone for trapping the dry dirt and sequentially installed vortex apparatus for the wet purification of gases. The installation of wet gas cleaning was irrigated with water with a closed water rotation. Water is fed into the plant in the amount necessary only for wetting the particles and evaporation. The caught dust all comes back to production of construction materials. In industrial conditions, for the first time, a high degree of purification of exhaust gases from lime and limestone dust was achieved, which reached more than 99.9 %. The developed industrial installations are simple and reliable in operation. The new technology of gas cleaning from dust with a closed water circulation has found application in different branches of chemical industry. The hydraulic resistance is in the range of 500-600 mm of water. art. New devices are not clogged with sediment and reliably operate without repair for several years.

Reference

[1] Monastyrev A. V. Manufacture of lime. M.: Higher school, 1986. 192 p.
[2] Uzhov V. N. Cleaning of industrial gases from dust. M.: Chemistry, 1981. 392 p.
[3] Zaitsev D. I., Weaver G. A., Stoev, N. D. The production of soda. M.: Chemistry, 1986. 312 p.
[4] Power System: Mezhvuzovski. thematic. collection is scientific. works /Moscow.energetic.int (technic.un-t) Kazansk.phil. Kazan, 1997. 128 p.
[5] Makhotkin I. A. Purification of gas emissions from vapors, aerosols and dust of toxic substances: abstract of dissertation ... kand. techn. sciences: 03.02.08. Kazan: KAZAN state technical University, 2011, p. 20

Adsorption of industrial wastewater from oil products with application of mathematical modeling

L.A. Nikolaeva, R. Ya. Iskhakova

51, Krasnoselskaya str., Kazan, 420066

larisanik16@mail.ru

Abstract. The granulated hydrophobic sorption material created on the basis of the chemical water treatment sludge of thermal power plants "SM-5" is studied. The information basis obtained in the experiments on the kinetics of adsorption makes it possible to model the dynamics of adsorption in a fixed bed of "SM-5" granules.

Nowadays one of the leading branches of the industrial sector of the Republic of Tatarstan is the chemical and petrochemical industry, where a significant amount of wastewater is formed. The more efficient and environmentally friendly methods of wastewater treatment are required because of the increasing the requirements for the permissible discharge of wastewater.

Different methods of wastewater treatment are used to reduce the anthropogenic impact on water resources. One of the methods of the dissolved oil wastewater treatment is adsorption on activated carbons of different types. Industrial-produced sorbents are expensive (several hundred thousand rubles per ton). Therefore, developments of finding the new cheap sorption materials based on waste production and modeling these processes are of great scientific and practical importance.

In the early works [1, 2], granulated hydrophobic sorption material "SM-5" was obtained. This material was create on the basis of the chemical water treatment sludge of thermal power plants and was used for the oily wastewater treatment at the enterprises of the chemical and petrochemical complex. The creation of granules was consisted of the following stages: heat treatment at 700 ° C for 60 minutes, granulation of material to optimal particle size of 0.5-2.5 mm, impregnation with 5% aqueous emulsion "GKZh-94M" at a ratio of 1: 2 with a binder, drying at 150 ° C for 60 minutes. The bulk density of resulting material was a 536 kg / m^3, a total pore volume of 0.57 cm^3 / g, an average granule size of 0.5 to 2.5 mm, an abrasion resistance of 68% (mass), a moisture capacity of 0,9% (mass), buoyancy - 99% (mass) / 96 h.

The mechanism and kinetic of dissolved oil adsorption onto the "SM-5" granules under static and dynamic conditions are studied. The adsorption isotherm under static conditions is obtained. The efficiency of oily wastewater treatment under dynamic conditions simulating an industrial adsorption filter is determined.

The model case of isothermal adsorption without temperature effects is considered. A model solution with an oil product concentration (OP) of 1.35 mg / dm^3 is passed through the adsorbent layer of length x. This concentration is the average at the inlet to the adsorption filters in wastewater treatment systems from the OP. The height of the "SM-5"layer is 20 cm, the weight is 54.38 g, the filtering speed is 3.5 m / h. Inadequate cleaning efficiency of OP is fixed at a concentration of 0.1 mg / dm^3. The initial concentration is denoted by C_0, the specific water velocity per unit section of the adsorbent is denoted by w. These parameters are kept constant. In addition, we assume that the concentration of OP in the adsorbent before the start of the experiment is zero.

The basic system of assumptions is formulated. The change of the flow density of wastewater due to the loss of OP can be neglected. Flow in the filter has the same direction at a constant speed. The flow of wastewater through the filter will be considered as a pseudobinary mixture consisting of water and OP. The sorption layer is considered to be a continuous porous body. The presence of only OP in wastewater leads to a decrease of sorption of other components. Components with a less than 0.1% are not considered.

In this case, the dynamics of adsorption can be described by the equation of the material balance between the solid and liquid phases, the kinetics of the impurity transfer process from the liquid to the adsorbent grains and the adsorption isotherm.

The total material balance describing the adsorption process in the form of a nonlinear partial differential equation is considered (1) [3]:

$$\rho_b \frac{\partial a}{\partial \tau} + w\frac{\partial c}{\partial x} + \varepsilon\frac{\partial c}{\partial \tau} - \varepsilon D^* \frac{\partial^2 c}{\partial x^2} = 0 \quad (1)$$

a –adsorption, kg / kg; c – current concentration of OP in the flow, kg / m^3; τ – time, s; w – velocity of the liquid flow, m/s; x – coordinate depending on the height of the adsorbent bed, m; ε - porosity of the sorbent; D*– the longitudinal diffusion coefficient with molecular diffusion and convective mixing along the layer, m^2 / s; ρ_b – bulk density, kg / m^3.

The equation of the kinetic of the adsorption process is written in the form (2):

$$\rho_b \frac{\partial a}{\partial \tau} = \beta_i (c - c^*), \quad (2)$$

β_i – mass-transfer coefficient, s^{-1}; c* – equilibrium OP concentration, kg / m^3. The equation of the adsorption isotherm is in the form (7):

$$a = f(c^*) \quad (3)$$

Differential equations of mass transfer in a granular layer are solved with initial and boundary conditions: $\tau = 0$; $0 \le x \le L$; c = 0; a = 0; T = To = const. The boundary conditions for x = 0 are written in the form: $\tau > 0$; x = L; c = c0 = const; a = a (τ); T = const.

The obtained data on the adsorption capacity were used to construct the kinetic curve of adsorption (Fig. 1).

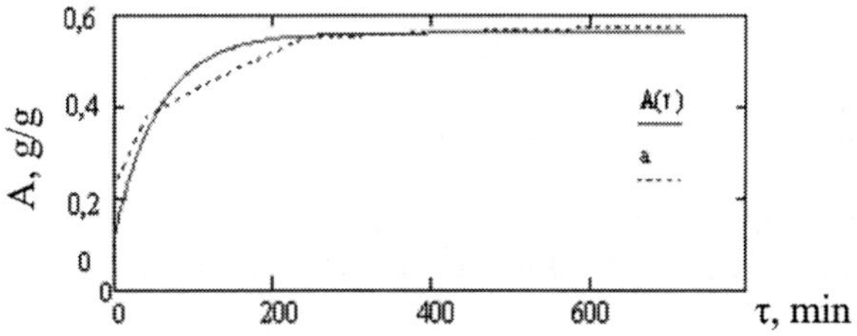

Figure 1 – Kinetic curve of OP adsorption on SM-5 material: a - experimental data, A (τ) - approximation by exponential dependence

The mathematical model equations were solved by the iterative method.

The solution of the system of differential equations (1) - (3) is the determination of the OP concentration in the flow.

In each differential element of the SM-5 layer a surface that reflect the change in concentration both in the layer and in time is formed during the calculation of material balance of the adsorption process. In calculations absorption kinetic of the OP from the wastewater is considered (Fig. 2). The concentration dependences of the adsorbed OP in the sorbent layer is identified from the particular solution of the equations system.

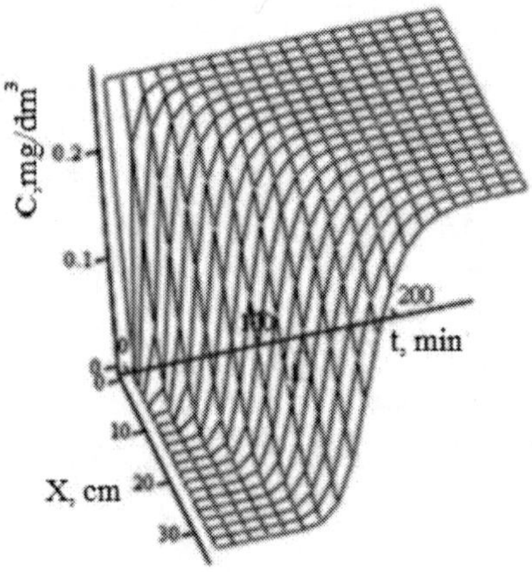

Figure 2 – Concentration dependence on layer height and contact time

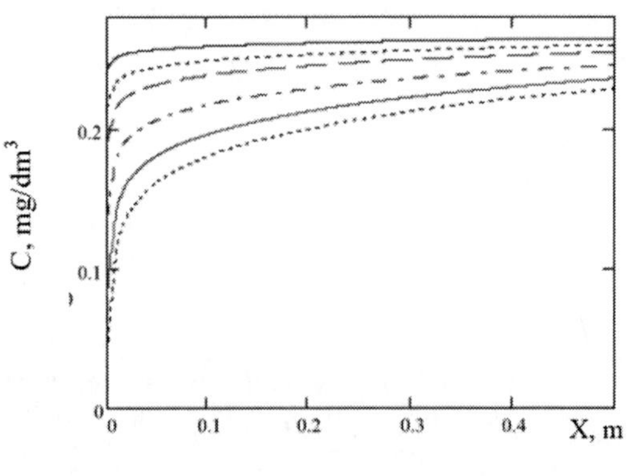

Figure 3 – Concentration distribution of the OP in the sorbent layer

The adsorption isotherm has a convex form (the L-type isotherm). The convex isothermal line is of type I according to the BET classification and indicates the presence of micropores in the adsorbent.

In the case of longitudinal diffusion two factors have a place simultaneously. The first factor causes dilution of the adsorption front. The second factor is related to the type of isotherm. The coefficient D* is found experimentally.

The convex adsorption isotherm shows the adsorption of OP on the surface of the sludge granules. Let's consider the motion of two points of the adsorption front C_1 and C_2, with $C_2 > C_1$. The derivative of the isotherm is $f'(C_1) > f'(C_2)$, so the point with the concentration of C_2 should move faster than the point with the concentration C_1. But the saturation of the layer at the C_2 concentration should be preceded by the saturation of the layer $C_1 > C_2$. As a result, the front along the layer should become stronger.

But these simulations refer to the model situation in the reactor of ideal displacement without the front blurring. In practice, the final adsorption rate and the inhomogeneity of the layer, inevitably leads to a smearing of the adsorption front. In this case the adsorption kinetics is defined by the equations:

$$\rho_b \frac{da}{d\tau} = \beta_{liq}(c - c^*) , \quad (4)$$

$$\rho_b \frac{da}{d\tau} = \beta_{sol}(a^* - a) . \quad (5)$$

β_{liq} , β_{sol} – coefficients of external and internal mass transfer; c –is the concentration in the flow outside the grain; c * – the concentration at the external surface, which is in equilibrium with the amount of adsorption in the grain volume; a – adsorption; a^* – equilibrium adsorption in the flow.

When the process is limited by external diffusion formula (4) and when the process is limited by internal diffusion formula (5) are used.

The information basis obtained in the experiments on the kinetics of adsorption makes it possible to model the dynamics of adsorption in a fixed bed of "SM-5" granules. It is implemented on the basis of the material balance equations in partial derivatives and the subsequent approximation using the built-in functions of various computational programs.

Reference

[1] Nikolaeva L.A., Golubchikov M.A. The study of the oil products sorption for power plants wastewaters y a modified sludge of clarifiers // Thermal Engineering. 2012. № 5. P. 59.

[2] Nikolaeva L.A., Golubchikov M.A. Purification of industrial wastewater from oil products by modified sorption materials based on carbonate sludge // Water supply and sanitary technique. 2016. №7. P.51-58.

[3] Romankov P.G., Frolov V.F. Mass exchange processes of chemical technology. L .: Chemistry, 1990. - 384 p.

IOP Conf. Series: Earth and Environmental Science **288** (2019) 012018 doi:10.1088/1755-1315/288/1/012018

Purification of water sources from oil contamination by hydrophobic carbonate sludge

L.A. Nikolaeva, D.A. Khamzina

420066, Republic of Tatarstan, Kazan, Krasnoselskaya St. 51

larisanik16@mail.ru

Abstract. The paper presents laboratory and experimental studies on the purification of surface water sources from oil contamination by sorption material on the basis of production waste. Energy waste is a slurry of chemical water treatment, which is formed during liming and coagulation at the stage of preliminary purification of natural water. Based on the sludge, the hydrophobic sorption material "MGSM" was developed. The following requirements are imposed on the sorption materials that allow, at low costs, the most effective cleaning of the water surface from petroleum products: hydrophobicity, oil capacity, buoyancy, ease of utilization from the water surface, efficient work in a different temperature range, resistance to water disruption, non-toxicity, optimal price. The material obtained has these requirements. The technical characteristics of the developed sorption material were determined: oil capacity, wetting contact angle, an express analysis of the quality of the aqueous extract of the hydrophobic sorbent developed for the acute lethal toxicity of fish of the *Poecilliareticulata Pet.* and crustaceans *Daphniamagna Str.* A technological scheme for utilization of oil-contaminated hydrophobic sorption material as a secondary energy resource in industrial enterprises is proposed, a calculation of the volume of air and combustion products during its combustion, calculation of harmful emissions, thermal calculation, and prevented environmental damage.

Ecological situation in the country is determined by the enterprises of the heat and energy complex and the oil industry, chemical, petrochemical and machine-building industries, the construction complex and agriculture. The existing structure of industry, obsolete technologies form a wide range of environmental problems and sharply exacerbate the ecological situation.

Oil pollution is a particular danger to the environment. Increasingly, accidents occur in the extraction, transportation and refining of oil.

In this case, the method of localization of oil contamination by booms and liquidation of oil and oil products by sorption materials, developed on the basis of production wastes, is less resource and energy-consuming.

For the purification of oil and petroleum products use mechanical, physicochemical, chemical and biological methods. One effective method is adsorption - this method allows the purification of natural and waste water from oil and petroleum products to the required value without secondary water pollution. The efficiency of purification depends on the correspondence of the parameters of the porous structure of the sorbent to the size of the impurity molecules.

In the literature [1], waste from the wood processing and cellulose industry - from 88 to 94%, from sunflower to 79%, wheat husk - up to 82%, etc., has been proposed as a sorbent.

Previously, [2] a hydrophobic sorption material for aqueous media from oil contamination, a modified hydrophobic sorption material ("MGSM"), was developed on the basis of chemical water treatment sludge. The sorbent is obtained by modifying the powdered chemical water treatment sludge (wetness of the sludge $W = 3\%$, size of the particle fraction $r = 0.09\text{-}0.5$ mm) with a 100% silicon organic liquid "Silor" [3] at a volume ratio of liquid and solid phases of 1: 5 "Silor" is a waste product of chemical destruction: silicone rubber compounds, sealants, compounds formed in the process of manufacturing

IOP Conf. Series: Earth and Environmental Science **288** (2019) 012018 doi:10.1088/1755-1315/288/1/012018

rubber products based on silicone rubbers. Heat treatment is carried out at 150 ° C for one hour to establish a constant mass. The developed material is called the modified hydrophobic sorption material "MGSM" [4]. The technological characteristics of "MGSM", which are presented in Table. 1.

Table 1 - Technological characteristics of "MГСМ"

Characteristic	Value
Particle size, mm	0,09-0,5
Bulk density, kg / m3	790
Oil intensity (relative to the Shiyskoye field oil), y / y	0,95
Humidity,%	3
Specific surface, m2 / g	65,8
Total pore volume, cm3 / g	0,8
Water absorption,%	1,3
Buoyancy, mass% (96 hours)	98-99
Adsorption capacity by iodine, mg / g	14
Adsorption capacity by methylene blue, mg / g	40

To determine the degree of hydrophobicity of "MGSM", the contact angle of wetting the surface of "MGSM" was determined, which was 115 °. A laboratory experiment on the use of "MGSM" for cleaning the aquatic environment from oil contamination was carried out. 0,06 g "MГСМ" is placed in a container with water of 1 liter, which is artificially contaminated with oil. The time of contact of "MGSM" with oil is 24 hours, while the value of the total initial concentration of oil products is 100 mg / dm3, the area of the oil slick is 39.5 cm2. After

24 hours "MGSM", which is saturated with NP, is collected from the water surface by means of a special device, which is a grid with a cell size of 0.5 x 0.5 mm fixed on a metal frame. The water after the treatment is analyzed for the residual content of the NP. The residual concentration of petroleum products was 0.008 (± 0.03) mg / dm3.

Express - control of water quality for acute lethal toxicity of the guppy culture of *Poecilliareticulate Pet.* and crustaceans *Daphniamagna Str.* for carrying out ecological examination of the technology of surface water purification from oil contamination with a new hydrophobic sorption material [5].

In Table. 2 shows water quality indicators, the analysis of which revealed that "MGCM" does not create secondary pollution of the aquatic environment. At the same time, the efficiency of cleaning oil contamination "MГСМ" was more than 90%.

Table 2 - Water quality indicators after oil products purification with the material "MGSM"

Indicator name	Indicators after purification	The value of MPC*
SiO_3^{2-}, mg / dm^3	0,03	10,00
J_o, mg-equ / dm^3	0,55	7,00
Zh^{2+}Ca, mg-eqv / dm^3	0,25	-
Ok, $mgKMnO_4$ / dm^3	0,75	5,00
$Щ_o$, meq / dm^3	0,35	-
Fe_o, mg / dm^3	0,04	0,30

* MPC of substances in water of water objects of domestic and drinking and cultural-household water use (SanPin 2.1.4.10749 - 01)

In table 2 shows water quality indicators, the analysis of which revealed that "MGCM" does not create secondary pollution of the aquatic environment. At the same time, the efficiency of treatment for oil contamination of "MGSM" was more than 90%.

To determine the effectiveness of cleaning

water environment from NP in full-scale conditions, an experiment was carried out to remove oil contamination 5 mm thick on the pond-settler of JSC "Mari Pulp and Paper Mill" [6]. The studies were carried out using a frame made of expanded polystyrene with a size of 40x20 cm, which was covered with a polymer cloth and impregnated with silicone liquid "Silor". The quantity of "MGSM" necessary for purification is calculated - 343.7 g. "MGSM " completely remains on the surface and is collected by special devices. The efficiency of purification from NP "MGSM" was more than 99%.

After carrying out the laboratory and full-scale experiment, a basic scheme for cleaning the pond-settler from oil and oil products "MGSM", which is shown in Fig. 1.

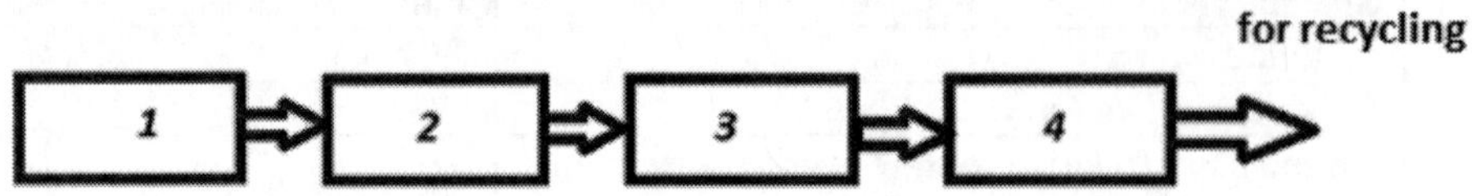

Figure 1. Scheme of cleaning from oil pollution of the pond-settler "MGSM":
1 - concentration of oil pollution, 2 - spraying of "MGSM", 3 - collection of saturated "МГСМ", 4 - utilization of spent "МГСМ".

For the localization of oil pollution, boom bins of constant buoyancy are used, the brand of booms is selected depending on the size of the oil pollution. For the application of "MGSM" to the region of localized NP, an autonomous sprayer

PC-1. In real conditions, it is difficult to determine the amount of "MGSM", which is necessary for the complete absorption of NP, so "MGSM" is sprayed with excess. According to the laboratory experiment, the time of contact of "MGSM" with the oil product to its full saturation is 30 minutes. Silicone-organic liquid "Silor" gives the waterproofing effect of "MGSM" and reduces the risks of drowning saturated "MGSM" to a minimum. The saturated "MGSM" remains completely on the surface and is easily assembled by the NP-2 oil skimmer.

After oil contamination is cleaned, the sorbed material produced by "sorbent material" is formed "MGSM", which is proposed to be used as a fuel resource in the same enterprise.

Based on the biotesting conducted, it was established that the water extract of the lubricated material "MGSM" does not have an acute toxic effect on fish of the *Poecilliareticulate Pet.* and crustaceans *Daphniamagna Str.* The obtained results confirm that the lubricated "MGSM" is practically non-hazardous and refers to V hazard class for the environment.

Damage caused to the settling pond due to the spill of the NP, which amounted to 2674.82 thousand rubles, was calculated, while the enlarged ecological and economic assessment of the damage prevented as a result of a spill of 1 tonne of oil will amount to 723.4 thousand rubles. The principal scheme of the solid waste incineration line for the utilization of the lubricated "MGSM" was modernized. The thermal calculation of the boiler was carried out during the combustion of the lubricated "MGSM": the efficiency of the boiler was 85.2%; the total fuel consumption is 0.336 kg/s.

References

[1] L.A. Nikolaeva Energy-Resource-saving Technology for Utilizing Oiled Sludge for Chemical Purification of Thermal Power Plants for Use as an Alternative Source of Solid Fuel. L.A. Nikolaeva, D.A. Khamzina // Problems of Energy. 2016. №1. Pp. 51-53.

[2] N.A. Sobgayda Use of production waste as adsorbents of petroleum products / N.A. Sobgaida, L.N. Olshanskaya., K.N. Makarova, Yu.A. Makarova // Ecology and industry of Russia. - 2009. №1. Pp. 36-38.

[3] F.M. Palyutin The use of silicone fluid "Silor" in the oil industry / F.M. Palyutin, VA Baburina, AS Romachin // Bulletin of the Kazan Technological University. №2. P.114-116.

[4] L.A. Nikolaeva, D.A. Khamzina. Investigation of chemical water treatment sludge as an oil sorbent in the purification of water objects // Bulletin of the Technological University. 2018. №1. Pp. 200-204.

[5] RD 52.24.635-2002 Observation of toxic pollution of bottom sediments in freshwater ecosystems based on biotesting.

[6] Mari Pulp and Paper Combine / http://www.marbum.ru/.

Pollution monitoring and optimization of technology of membrane water desalting of production installation

Odoyevtseva M.V., Makarenko E.A.

Volzhsky Branch of the National Research University «Moscow Power Engineering Institute» 69 Lenin St., Volzhsky, Volgograd region, 404110, Russia

vfmei@vfmei.ru

Summary. The majority of installations of the reverse osmosis take root into production without preliminary approbation. In this regard optimization of the equipment operation is performed in the conditions of commercial operation. The possible solution of this problem is use of test membrane installations of JSC NPK Mediana-Filter for evaluating intensity of pollution of membrane elements and the directed choice of the washing solutions of production installations.

Introduction

One of the main factors influencing stability, efficiency and operation period of membranes is their pollution by inorganic, organic and biological deposits [1]. Chemical cleaning of reverse osmosis installations is an integral part of technology of the reverse osmosis (ROI). Especially significant it is in the conditions of the changing quality of initial water, unstable work of system of retraining and also in the multistage systems of the reverse osmosis working with the high-concentrated solutions of salts [2].

As the majority of installations of the reverse osmosis are introduced into production without preliminary approbation, optimization of operation of the equipment are performed in the conditions of industrial operation. Therefore application of test installations for the purpose of optimization of technological parameters of the operating of reverse osmosis installations is the relevant direction of researches [3].

Results of researches

Test installation of JSC NPK Mediana-Filter allow:

- to define primary deposits on membrane elements;
- to pick up the optimum cleaning solutions;
- to define an order and working parameters of cycles of cleaning and number of their repetitions.

The hydraulic scheme of the test two-level installation of the reverse osmosis is provided on figure 1.

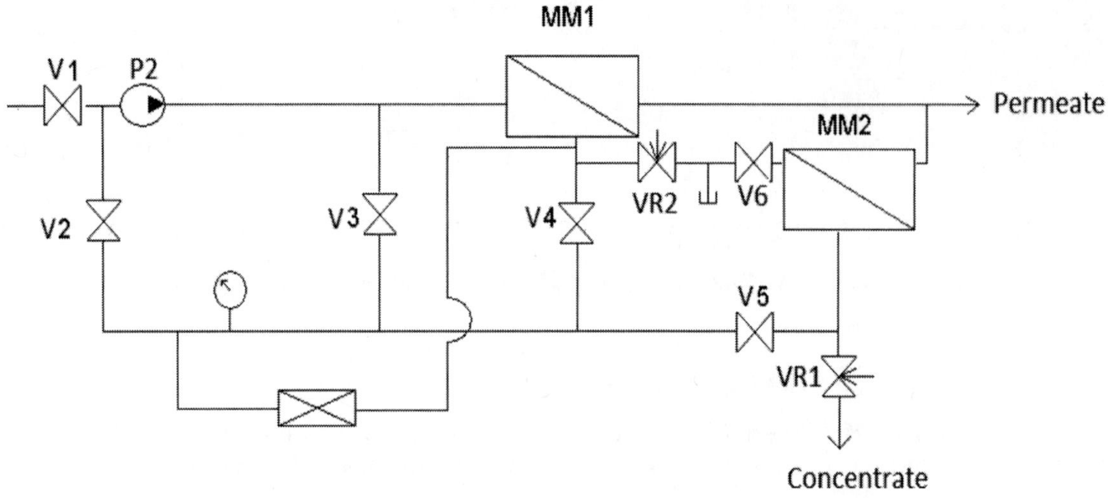

Figure 1. The hydraulic scheme of two-level installation of the reverse osmosis.
Control panel pump H2; BP1 adjusting valve for installation of pressure of the feeding solution on the membrane devices MM1 and MM2

The laboratory and test installation "worked" with production installation of the reverse osmosis.

The hydraulic scheme of the production installation of the reverse osmosis (IRO) is provided on figure 2.

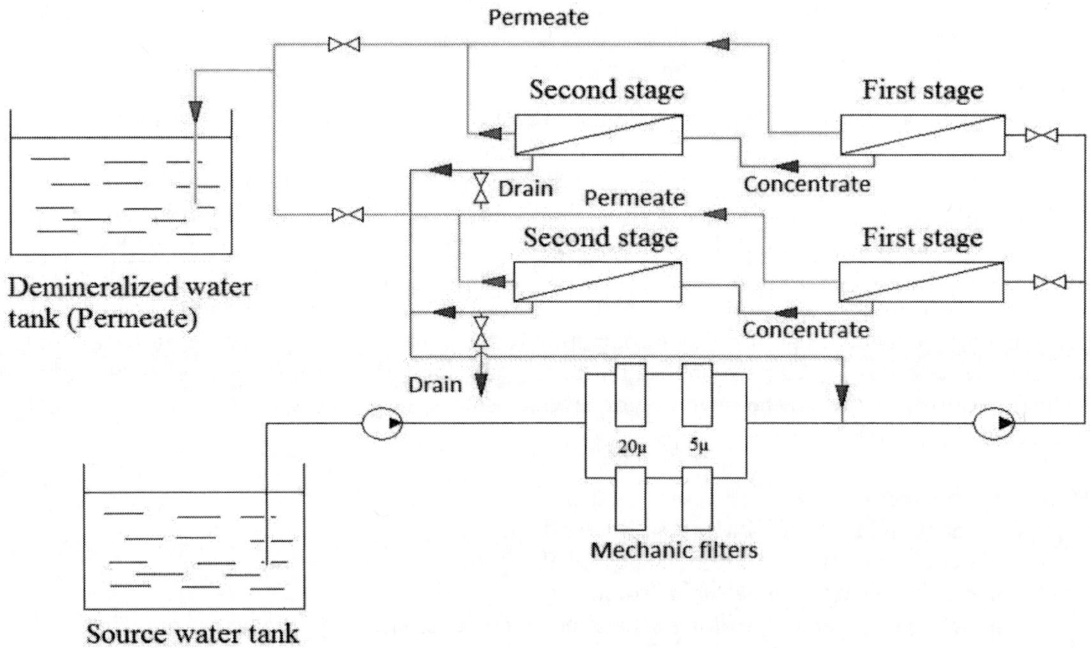

Figure 2. The hydraulic scheme of a production installation of the reverse osmosis

The test installation was connected in two points of production installation: to initial solution after five micron filters and on a concentrate after the first step.

The results of tests at connection of the test installation after five micron filters are given in table 1.

Table 1. The results of tests at connection of test installation after five micron filters

Indicators	Initial solution	concentrate	Permeate	Initial solution	concentrate	Permeate
	Beginning of work			Completion of work		
SDI	6	9	–	6	10	–
Жо, mg-ekv/dm3	3,40	4,60	0,02	4,50	6,40	0,02
ЖСа, mg-ckv/dm3	2,40	3,40	0,01	3,00	4,80	0,01
ЖMg, mg-ekv/dm3	1,00	1,20	0,01	1,50	1,60	0,01
Що, mg-ekv/dm3	2,6	3,4	0,1	3,0	4,8	0,2
Chlorides, mg/dm3	13,0	22,0	2,0	1,0	32,5	0,4
Salinity, mg/dm3	733	1000	9	523	770	6
pH	7,4	7,6	5,1	7,7	7,7	5,8
Expense, l/h	62,8	43,4	19,4	37,7	23,3	14,4
Conversion, %		31			38	
Selectivity of membranes %		98,7			98,9	

In the course of operation of the test installation conversion, selectivity of membranes and technological indicators of the permeate have slightly changed (table 1). Productivity of the test installation after its operation within 45 days (for 12 hours a day) on a concentrate decreased twice. At deterioration in technological indicators of a concentrate by 1,4-1,6 times the chemical cleaning of membranes is carried out. The results of the analysis of tests of the washing solutions are given in table 2.

Table 2. The structure of pollution of the washing solutions of membrane elements of the test installation

Indicator/solution	The acid washing solution	Alkaline washing solution
pH	2,5	12,0
Fe is the general, mg/dm3	0,2	0,1
SiO2, mg/dm3	0	0
The weighed substances, mg/dm3	2	7

For realization of a chemical cleaning the membrane element have been connected to the station of a chemical cleaning (figure 3).

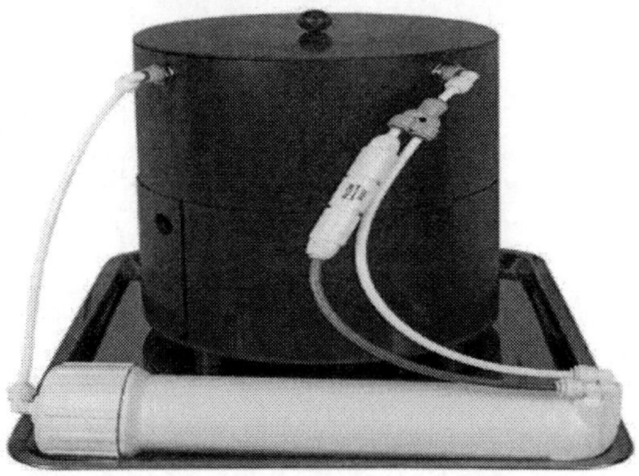

Figure 3. The station of a chemical cleaning of JSC NPK Mediana-Filter

The chemical cleaning was carried out in several steps:

– circulation of alkaline MF-A-T10 solution of JSC NPK Mediana-Filter through a membrane element;

– soaking of a membrane element with the subsequent circulation of alkaline solution through a membrane element;

– circulation of acid MF-B-L20 solution of JSC NPK Mediana-Filter through a membrane element;

– soaking of a membrane element with the subsequent circulation of acid solution through a membrane element;

– washing of membranes the distilled water to values pH =6-7 after circulation of alkaline and acid solutions.

The tests' results at connection of the test installation on a concentrate after the first step are given in table 3.

Table 3. The results of tests at connection of the test installation on a concentrate of the first step

Indicators	Initial solution	concentrate	permeate	Initial solution	concentrate	permeate
	Beginning of work			Completion of work		
Жо, mg-ekv/dm3	17,00	19,00	0,07	17,00	24,00	0,30
ЖCa, mg-ekv/dm3	11,00	14,50	0,03	11,00	18,50	0,30
ЖMg, mg-ekv/dm3	5,50	4,50	0,04	5,50	5,50	0,03
Що, mg-ekv/dm3	11,0	13,5	0,3	11,0	15,0	0,3
Chlorides, mg/dm3	102,0	124,0	1,4	102.0	150,0	3,5
Salinity, mg/dm3	1437,5	1725,0	11,5	1437,5	1950,0	11,5
pH	7,7	7,7	5,3	7,7	8,0	6,4
Expense, l/h	41,5	31,0	10,5	39,8	33,4	8,9
Conversion, %		25			21	
Selectivity of membranes %		98,6			96,6	

During the operation the following parameters of work of the test installation have been traced: expenses of streams, pressure on mechanical filters, the size of conversion and selectivity of membrane elements.

The criteria to carrying out a chemical cleaning have been: decline in production of test installation on a permeate for 9%; decrease in selectivity of a membrane element by 2% and deterioration of the permeate.

The results of researches of tests of solutions after a chemical cleaning are given in table 4.

Table 4. The structure of pollution of the washing solutions of membrane elements of test installation

Structure of pollution	Unit of measurements	Acid solution	Alkaline solution
1.Calcium	mg/dm3	–	–
2. Sulfates	mg/dm3	26	–
3. Iron	mg/dm3	0,24	–
4. sio_2	mg/dm3	–	3,2

The researches of tests of solutions after a chemical cleaning have revealed that on membrane elements of ttheest installation and, therefore, on membranes of the second stage of production UOO silicon oxide deposits prevail. The received results demonstrate expediency of reduction of chemical

IOP Conf. Series: Earth and Environmental Science **288** (2019) 012019 doi:10.1088/1755-1315/288/1/012019

cleanings of membranes with acid solutions. The reduction of duration of contact of membranes with the acid washing solution will provide increase in duration of their work on the water treatment equipment of the metallurgical enterprise at stable quality of the desalinated water.

The results of tests at connection of the test installation on a concentrate after the first step after a chemical cleaning are given in table 5.

Table 5. The parameters of tests after chemical cleaning at connection of the test installation on a concentrate of the first step

Indicators	Concentrate after the first steps of the production installation	Permeate from the test installation	Concentrate from the test installation
$Ж_o$, mg-ekv/dm3	18,00	0,14	23,00
$Ж_{Ca}$, mg-ekv/dm3	14,00	0,10	17,00
$Ж_{Mg}$, mg-ekv/dm3	4,00	0,04	6,00
$Щ_o$, mg-ekv/dm3	13,0	0,5	18,0
Chlorides, mg/dm3	128,5	2,2	155,0
Salinity, mg/dm3	1260,0	12,5	1680,0
pH	7,55	5,85	7,80
Expense, l/h	36,4	10,2	26,2
Conversion, %	–	28,7	
Selectivity ofmembranes, %		98,6	

After a chemical cleaning the selectivity of membranes and quality of the permeate have been restored (table 5). By results of researches it is established that in the course of work as production UOO the second stage is most polluted. It is promoted high extent of saturation of a concentrate of the first stage of UOO compounds of iron and oxide of silicon and the overestimated conversion of production installation (80%).

Conclusions

As a result of researches the efficiency of application the test installation of JSC NPK Mediana-Filter for definition of primary deposits on membrane elements of installation of the reverse osmosis of metallurgical production is shown.

Need of decrease in size of conversion of production installation to 65-75% and change of the modes of carrying out chemical cleaning due to reduction of number of cleanings with acid solutions is revealed. These actions will provide reduction of deposits of oxide of silicon and iron on membrane elements, will increase the term of operation of membrane elements of the water treatment equipment of the metallurgical enterprise at stable quality of the desalinated water.

References

[1] Panteleev A.A., Ryabchikov B E, Horuzhy O.V., Gromov S.L. and Sidorov A P 2012 Technology of membrane division in industrial water treatment (Moscow: DeLi plus) p 429
[2] Kocharov R.G. 2007 Theoretical bases of the reverse osmosis(Moscow: RHTU of D.I. Mendeleyev) p 143
[3] Svittsov A.A. 2007 Introduction to membrane technology (Moscow: DeLi Print) p 208

IOP Conf. Series: Earth and Environmental Science **288** (2019) 012020 doi:10.1088/1755-1315/288/1/012020

About the results of experimental study performed on a "decarbonizer – vacuum deaerator" system

M E Orlov, V I Sharapov

Ulyanovsk State Technical University, Severniy Venets Street, 32, Ulyanovsk, Russia

mi5h@mail.ru, vlad-sharapov2008@yandex.ru

Abstract. The paper discloses the results obtained in experimental study, implemented at an operating water treatment unit of the thermal power plant. The study was carried out via using the method of a complete factorial experiment. As a result of the experiments, the dependencies between objective functions and a number of controlled parameters have been determined. Regression equations have been derived and cross-sections of the response surfaces have been constructed in a natural scale, that adequately describe the processes of corrosive gases desorption within the "decarbonizer - vacuum deaerator" system. The data obtained allows one to assess the minimal technologically acceptable level, when it is possible to lower the temperature of the treated water and, consequently, to reduce the energy costs for its heating, and to specify the conditions when the reduction of the energy costs for supplying air to decarbonizer is possible.

Introduction

The reliability and economic viability of the heat source itself and the heat supply system as a whole depend on the selection of a rational operating mode of water treatment equipment. Decarbonizers and vacuum deaerators serve as the main water treatment equipment for the removal of corrosive gases at most thermal power plants and large boiler houses [1].

To obtain a proper description of the physico-chemical processes, taking place in water treatment units, and to create energy-efficient technologies for water degassing processes control, the experimental studies have been conducted by the authors, jointly with colleagues, at water treatment units of various heat sources [2, 3, 4]. The need to conduct the experimental study is caused by the absence of complex characteristics of the "decarbonizer - vacuum deaerator" system performance, depending on the controlled parameters of the hydraulic, thermal and chemical modes.

Experimental installation

The experimental study was carried out several years ago at an operating water treatment unit of one of the local thermal power plants in Ulyanovsk city [5]. The unit in question includes three parallel- connected packed decarbonizers with a capacity of 550 t/h and two horizontal jet-type vacuum de- aerators of the DV-800 type with a capacity of 800 t/h (Figure 1). Heat and mass transfer devices of this type are most often used for the anticorrosion treatment of makeup water of large heat supply systems. The air is supplied to decarbonizers via centrifugal fans of the C4-76 No.8 type with a nominal capacity of 14000 m^3/h and a head pressure of 1.4 kPa.

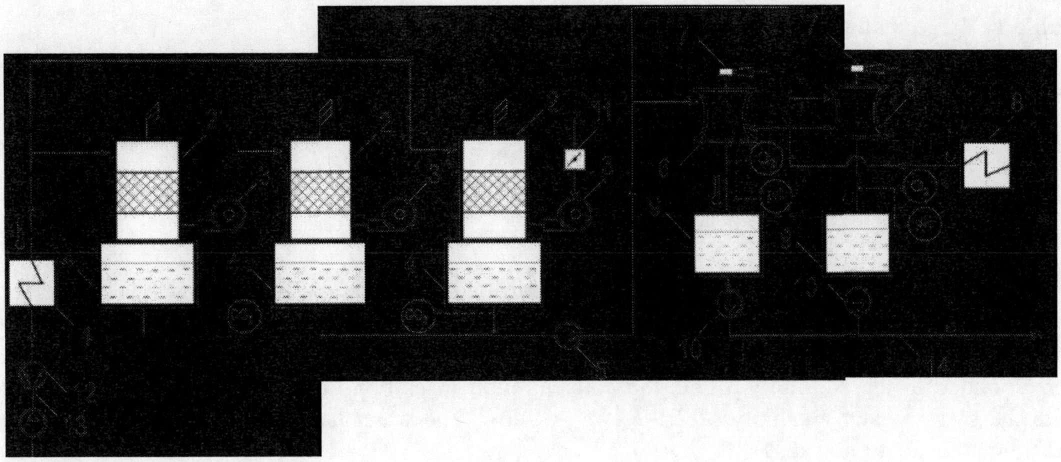

Figure 1. Schematic diagram of the experimental water treatment unit: 1 – source water heater; 2 – the decarbonizer; 3 – the fan; 4 – decarbonized water tank; 5 – decarbonized water pump; 6 – the vacuum de-aerator; 7 – the ejector; 8 – the heating agent heater; 9 – makeup water storage tank; 10 – the heat supply network feeding pump; 11 – the control damper; 12 – the flow-meter; 13 – source water pipeline; 14 – makeup water pipeline

The residual concentration of oxygen in de-aerated water was measured using an automatic portable analyzer, the oxygen analyzer "MARK-301T" manufactured by LLV "Vzor" (Nizhny Novgorod). The measurements were made at a sample temperature of 30 ± 5°C. In order to control the oxygen analyzer readings, chemical analyzes of oxygen content in de-aerated water were regularly performed using the "methylene blue" indicator. The pH value of water was determined electrometrically using a standard laboratory pH-meter of the "pH-340" type (Gomel). The content of carbon dioxide was determined by titration with a caustic soda in the presence of phenolphthalein in accordance with operating instructions. Measurements of flow-rates and temperatures were carried out with standard measuring equipment.

Methodology for performing the experiment

The "decarbonizer-vacuum deaerator" system testing was carried out via using the method of a complete factorial experiment of the 2^3 type. During this method implementation, the experiment follows a certain strategy, the controlled parameters vary simultaneously, and calculation of statistical estimates is performed in accordance with a strictly formalized procedure, which significantly reduces the number of necessary experiments to be carried out and improves the accuracy of the study [2, 6]. Estimates of the regression coefficients are independent of one another, which allow one to determine the strength and the nature of influence of the parameters and their interaction and to use the obtained model for analysis and optimization of the studied object.

The residual content of carbon dioxide in decarbonized water (objective function Y_1, mg/dm^3) and the pH value of decarbonized water (Y_2) were taken as the parameters of decarbonizers efficiency (objective functions) to be determined. When evaluating the performance of vacuum de-aerators, the residual content of oxygen dissolved in de-aerated water (Y_3, μg/dm^3) and the pH value of deaerated water (Y_4) were used as objective functions.

During study of the efficiency of desorption of the corrosive gases O_2 and CO_2 dissolved in makeup water, the influence of the three controlled regime factors, namely, the flow rate of source water G_w, source water temperature t_w, and flow-rate of air supplied to decarbonizer D_a, was estimated quantitatively. The values of the controlled parameters are given in Table 1 below.

Table 1. Values of the controlled parameters.

Parameter for constructing a regression equation	Controlled parameters		
	G_w, t/h	t_w, °C	D_a, %
Basic zero level X_{io}	380	13	75
Variability interval λ_i	140	5	25
Notation in a normalized form X_i	X_1	X_2	X_3

The intervals of the maximum change in the controlled parameters were as follows: the flow rate of source water was 240-520 t/h; water temperature was 8-18°C; the flow-rate of air was 50-100% of the nominal capacity of the decarbonizer by air. The initial content of oxygen in source tap water corresponded to the equilibrium one, the concentration of carbon dioxide was 12.76-13.20 mg/dm^3 with the source water alkalinity of 2.1-2.2 mg-eq/dm^3. The water heating magnitude in the vacuum deaerator was kept fixed at 15 ± 2°C.

The experimental study results

The key feature of the studies was the fact that for the first time it became possible to assess the impact of the flow-rate of air supplied to decarbonizer on carbon dioxide desorption. The change in flow-rate of air was implemented by means of a control damper 11, installed in the air duct before the fan 3 (Figure 1). It was assumed that when the damper is fully opened, the nominal fan capacity is $D_a = 100\%$ ($X_3 = +1$). The flow-rate of air supplied to decarbonizer was measured using a preliminary calibrated hand-held anemometer. Several de-carbonization modes were examined at different degrees of opening of the damper, installed on the suction pipe of the fan. As a result, it was managed to specify not only the impact of the flow-rate of air on the efficiency of CO_2 desorption, but also to obtain analytical dependencies that allow determining the actual flow-rate of air used for de-carbonization depending on the flow rate of water under treatment [4].

Based on the results of testing the "decarbonizer - vacuum deaerator" system, the following regression equations, that adequately describe the system operation, were obtained:

$$Y_1 = 3{,}6 + 0{,}17X_1 - 0{,}5X_2 - 0{,}33X_3 - 0{,}14X_2X_3, \quad (1)$$

$$Y_2 = 7{,}98 - 0{,}02X_1 + 0{,}06X_2 + 0{,}04X_3, \quad (2)$$

$$Y_3 = 164 + 51X_1 - 61X_2, \quad (3)$$

$$Y_4 = 8{,}25 - 0{,}06X_1 + 0{,}10X_2 + 0{,}03X_3 - 0{,}04X_1X_2. \quad (4)$$

The regime factors X_i are included into the regression equations in the normalized form and, with their substitution in (1) – (4), can take values from –1 to +1. For the practical use of equations, including construction of graphs - diagrams of the modes of operation for decarbonizers and de-aerators, it is necessary to convert the values of factors from the normalized form to a natural scale, which, to a certain degree, complicates the calculations and is not always convenient in engineering practice. It is much more convenient to use the graphical interpretation of regression equations and mathematical models, constructed in a natural scale

$$Y_1 = 4{,}33657 + 0{,}00121x_1 - 0{,}016x_2 + 0{,}00136x_3 - 0{,}00112x_2x_3; \quad (5)$$

$$Y_2 = 7{,}75829 - 0{,}00014x_1 + 0{,}012x_2 + 0{,}0016x_3; \quad (6)$$

$$Y_3 = 184{,}17121 + 0{,}36414x_1 - 12{,}2x_2; \quad (7)$$

$$Y_4 = 7{,}78057 - 0{,}00031x_1 + 0{,}04171x_2 + 0{,}0012x_3 - 0{,}00006x_1x_2. \quad (8)$$

In equations (5) – (8), x_1, x_2, x_3 are the values of the corresponding controlled parameters in their natural form.

In order to visualize the results of the experiment using the regression equations (1) – (4), calculations for the array of the values of objective functions for various combinations of regime factors were made and the cross-sections of the response surfaces in two-dimensional space were obtained (Figures 2 – 5).

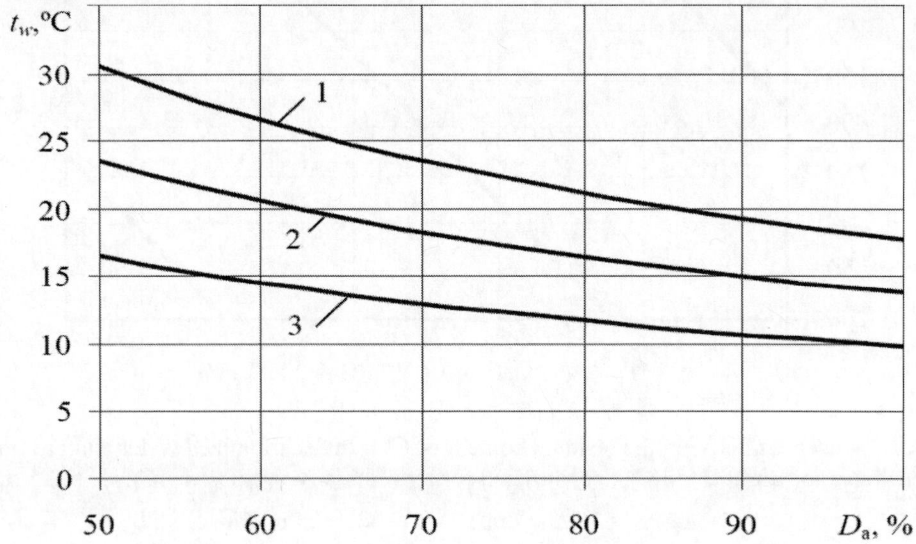

Figure 2. Dependences between the residual content of CO_2 in decarbonized water and the temperature of the treated water and the flow rate of air supplied to decarbonizer at the flow rate of source water G_w = 240 t/h, constructed for representative values: 1 - C_{CO2} = 2.5 mg/dm3; 2 - C_{CO2} = 3.0 mg/dm3; 3 - C_{CO2} = 3.5 mg/dm3.

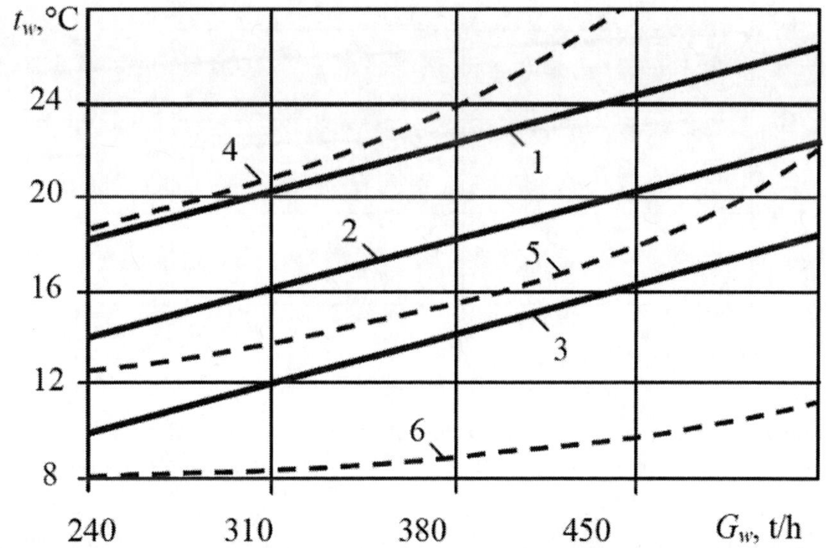

Figure 3. Dependences between the residual content of O_2 and the pH value of deaerated water and the flow rate and temperature of the treated water at D_a = 100%D_a^{nom}, constructed for representative values: 1 - $CO_2^{ост}$ = 50 µg/dm3; 2 - $CO_2^{ост}$ = 100 µg/dm3; 3 - $CO_2^{ост}$ = 150 µg/dm3; 4 - pH = 8.5; 5 - pH = 8.33; 6 - pH = 8.2.

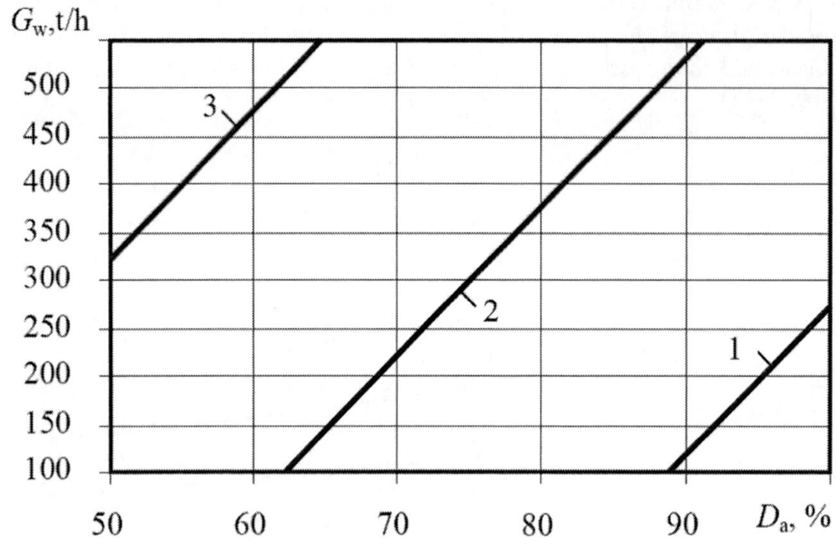

Figure 4. Dependences between the residual content of CO_2 in decarbonized water and the flow rates of treated water and air at t_w = 18°C, constructed for representative values: 1 - C_{CO_2} = 2.5 mg/dm^3; 2 - C_{CO_2} = 3.0 mg/dm^3; 3 - C_{CO_2} = 3.5 mg/dm^3.

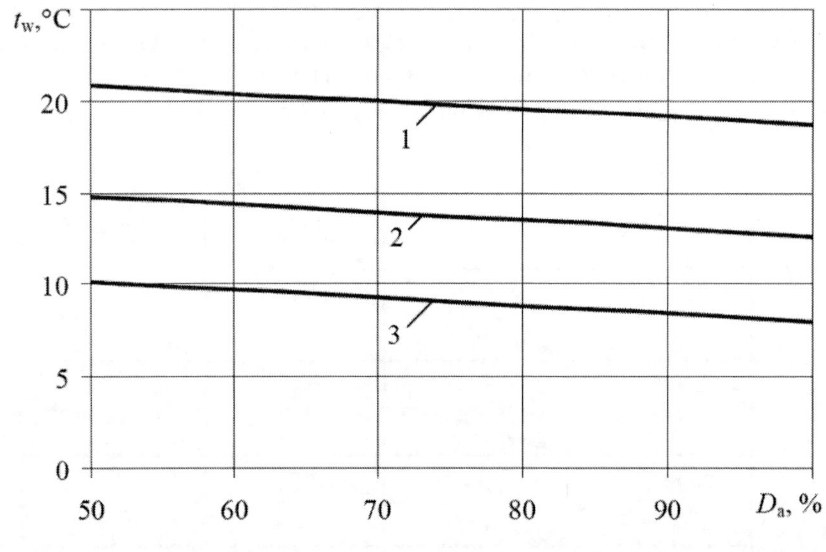

Figure 5. Dependences between the pH value of deaerated water and the flow rate and temperature of treated water at the flow rate of source water G_w = 240 t/h, constructed for representative values: 1 - pH = 8.5; 2 - pH = 8.33; 3 - pH = 8.2

Graphical dependencies (Figures 2 – 5) are based on calculation, but they are constructed on the basis of the experimentally obtained models of operation for decarbonizers and vacuum deaerators (1) – (4). The accuracy of predicting the values of the objective function using the diagrams is the same as when using the experimentally obtained regression equations. This accuracy is determined by the values of the variances of reproducibility and adequacy.

It can be seen from the diagrams that despite the low temperature level of makeup water treatment processes, the required efficiency of carbon dioxide desorption in decarbonizers (Y_1 = 3 mg/dm^3) and vacuum de-aerators (Y_4 = 8.33) is achieved within a wide range of studied modes. This can be explained by a sufficiently high bicarbonate alkalinity of the treated water.

The required efficiency of dissolved oxygen desorption (Y_3 = 50 μg/dm^3) is achieved only at the lower level of the unit's capacity and at the upper level of the temperature of the treated water, within the range of the studied intervals of the parameters change [4].

When setting up, maintaining and analyzing the operation of decarbonizers and vacuum deaerators, one can use both the regression equations and the polynomial dependencies and their graphical interpretation in the form of two-dimensional cross-sections of the response surfaces.

Conclusion

- The study of the "decarbonizer - vacuum deaerator" system, which is a part of the water treatment unit of one of the urban thermal power plants, was carried out via using the method of a complete factorial experiment.
- Based on the experimental data, multifactor regression equations were derived and cross sections of the response surfaces of objective functions in two-dimensional space were constructed, that adequately describe the processes of water degassing in a water treatment unit at low coolant temperatures.
- The mathematical models and diagrams obtained, characterizing the operation of the treatment unit for preparing makeup water of the heat supply network, allow one to assess the minimal technologically acceptable level, when it is possible to lower the temperature of the treated water and, consequently, to reduce the energy costs for its heating.
- The equations and graphical dependencies, describing the efficiency of carbon dioxide desorption from makeup water allow one, as well, to specify the minimal technologically acceptable level and the conditions when the reduction of the energy costs for supplying air to decarbonizer is possible.

References

[1] Sharapov V I 1994 Current problems in the use of vacuum deaerators for open heat-supply systems *Thermal Engineering* vol 41 **8** pp 635-639.
[2] Sharapov V I 1980 Analisis of operation of a vacuum deaerator of makeup water with the aid of multifactorial experiment *Thermal Engineering* vol 27 **3**.
[3] Sharapov V I, Deryabin A N, Orlov M E, Sivukhina M A and Tsyura D V 2000 Eksperimental'noe issledovanie ustanovki dlya podpitki sistemy teplosnabzheniya *Energosberezhenie* **1** pp 90-91.
[4] Sharapov V I, Fetkullov M R and Tsyura D V 2005 Ob ehnergeticheskoj ehffektivnosti upravleniya deaeratorami TEC po neskol'kim parametram *Problemy energetiki. Izvestiya vuzov* **3-4** pp 46-54.
[5] Orlov M E and Sharapov V I 2016 Eksperimental'noe issledovanie i matematicheskoe modelirovanie processov desorbcii korrozionno-aktivnyh gazov v vodopodgotovitel'nyh ustanovkah teplofikacionnyh sistem *Energosberezhenie i vodopodgotovka* **1** pp 3-8.
[6] Girshfel'd V Y, Akimenkova V M and Kulikov V E 1976 Ispytanie teploenergeticheskogo oborudovaniya i obrabotka rezul'tatov s ispol'zovaniem metodov planirovaniya eksperimenta *Teploenergetika* **2** pp 38-41.

IOP Conf. Series: Earth and Environmental Science **288** (2019) 012021 doi:10.1088/1755-1315/288/1/012021

Technology of utilization of liquid wastes generated during the preparation of main oil pipelines to transport diesel fuel with produce cheap steam or warm heat carrier

G.I. Pavlov, O.R. Sitnikov, Y.I. Khakimzyanova, V.A. Otradnova

Kazan National Research Technical University named after A.N. Tupolev – KAI, Russian Federation, Kazan, Karl Marx st., 10

pavlov16@mail.ru

Abstract. The existing technology of cleaning internal surfaces of oil pipelines from oil residues and solid sediments leads to the formation of a large volume of liquid hydrocarbons contaminated with mechanical particles and various resins. Currently, in the oil industry, mobile steam generator plants of PPU 1600/100 are used to steal sections of oil pipelines. The standard nozzle burner used in the PPU 1600/100 does not allow to burn effectively liquid waste similar to contaminated liquid hydrocarbons formed during the cleaning of the inner walls of the main oil pipelines. The article presents the results of operation of steam generator sets of PPU 1600/100 with a nozzle- free burner device VGU-1.

The increase in demand for petroleum products on the domestic market and competitiveness on foreign trading floors contributed to the construction and commissioning in some regions of the Russian Federation of a number of large oil refineries. The increase in the output of automotive fuels implies an increase in the volume of transportation of petroleum products. The most effective means of delivering oil and oil products from the producer to the consumer is the main pipeline transport. Despite the existing network of oil product pipelines, the difficulties in ensuring the throughput of existing oil product pipelines arise more often [1].

The optimal solution to the problem of ensuring the volume of transportation of petroleum products is the transfer of the main oil pipelines to the main oil product pipelines.

The need to transfer part of the oil pipelines for transportation of commercial petroleum products is due to a number of reasons [2]. The technology of transfer provides for the cleaning of the internal surface of oil pipelines from oil residues and asphalt and resin forming substances by a chemical method. For these purposes, special solvents are widely used, effectively removing oil residues and solid deposits from internal surfaces. In research [1] results of practical works on realization of this purification technology are given. The results of the control of the quality of the passed inspection lot of diesel fuel on the cleaned section of the pipeline show the effectiveness of the proposed purification technology.

The technology of cleaning the inner surface of the oil pipeline from oil residues and solid sediments by dissolving them in a special solvent, as well as the need to use a control lot of passed fuel, leads to the formation of a large volume of liquid hydrocarbons contaminated with mechanical particles and various resins. Of course, these fluids require disposal. Reuse is possible after regeneration. The regeneration of used solvent and diesel fuel is financially expensive [3]. A large part of the costs are for transportation of liquid waste to oil refineries by rail or road. These costs can be reduced if a part of the waste is used as a source of thermal energy to produce steam or a hot heat carrier. In addition, according to the proposed technology, cleaning of the inner part of the body of linear valves is carried out by steaming with disassembly [2]. Currently, in the oil industry, for the production of steam, mobile steam generator plants of PPU 1600/100 are widely used, in which the working fuel is diesel commercial fuel. The nozzle burner used in PPU 1600/100 does not allow to burn effectively liquid waste similar to contaminated liquid hydrocarbons formed during the cleaning of the inner walls of main oil pipelines.

There are several reasons for this:

1. Mechanical particles contained in liquid waste often cause contamination of the calibrated holes of the nozzle block, thereby violating the process of spraying the hydrocarbon liquid. Removal of solid particles from the liquid fraction requires the use of special technologies.
2. The high content of resinous compounds leads to the coking of the injector, leading its operation disrupted.
3. The fuel supply system of the PPU 1600/100 does not provide for controlling the pressure of the liquid fuel in front of the injector. The power of the burner is regulated by the alternate connection of three injectors located on the nozzle block. Regular nozzles are designed for diesel fuel and the use of other flammable liquids with a different viscosity is not provided.
4. The standard ignition system does not ensure the safe ignition of flammable liquids with a low flash point.

Because of the above-mentioned disadvantages, it is not possible to use energy-intensive liquid combustible wastes as a fuel in steam generators with standard burners. Another combustion principle is needed, according to which the fuel must be converted to the steam-gas state without spraying through the nozzle. The burner device, in which this principle is implemented, was developed by the specialists of the Department of Power Plants and Jet Engines of the Kazan National Research Technical University named after A.N. Tupolev. The essence of the work of the burner device is that the transformation of the liquid fraction into a steam-gas phase (sputtering, evaporation, thermal decomposition) takes place in a vortex flow. Moreover, the flow is interrupted in an orderly manner, creating a pulsation of the gas-air flow with a certain frequency. The supply of combustible liquid to the vortex zone is not carried out through a calibrated wound (nozzle), but through a tube having a diameter dozen times larger than the diameter of the nozzle. According to this principle, it is equally effective to burn both commercial fuels and tarred liquid combustible waste without fine cleaning [4].

At present, 72 modernized mobile steam generators with a nozzle-free burner unit VGU-1, developed under patent No. 2508501, are successfully operating in the Tatspetstransport MC.

Figure 1 shows the structural diagrams of a steam boiler with a nozzle burner block and a nozzle-free burner unit.

From Fig. 1 it can be seen that during spraying, the torch of the flame is elongated and the flame reaches the cold walls of the heat exchanger. From the theory of combustion, it is known that during the "freezing" of products of thermal decomposition of hydrocarbon combustibles, combustion processes are violated.

The upgraded mobile steam generator set was operated on a wide range of liquid combustible materials, such as waste from the petrochemical industry, petroleum products, low-grade furnace fuel.

a) with a nozzle burner block b) with a nozzle-free burner unit

Fig. 1. Diagram of a steam boiler

It was found that the concentration of harmful substances in gas emissions, such as oxides of nitrogen, carbon monoxide, unburned hydrocarbons are within acceptable limits. A comparative analysis of the content of harmful emissions in combustion products of diesel fuel shows that their concentrations during operation of the PPU in the nominal (nozzle) burner block are significantly worse than when the PPU is operating at the nozzle-free burner unit.

PPU 1600/100 with a certain periodicity is subject to routine maintenance. One of the types of work is the inspection and verification of the condition of the coil. Fig. 2 shows photographs of some sections of the coil heat exchanger during operation of the PPU on the nozzle burner block. Traces of carbon formation are clearly visible on them.

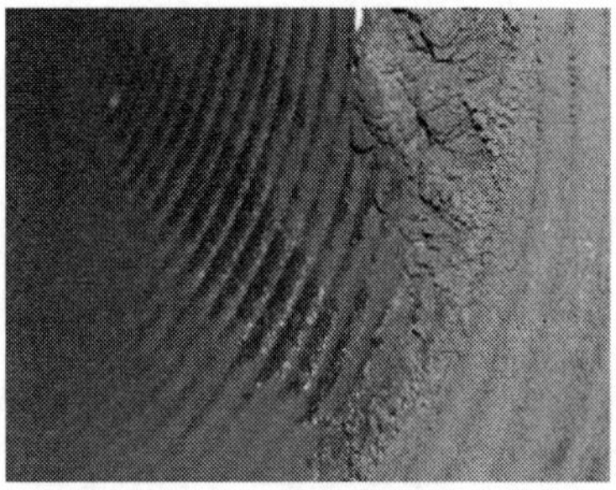

a) the inner surface of the coil

b) the outer surface of the coil

Fig. 2. Sections of coil heat exchanger

To the causes of poor fuel combustion and carbon formation when using a standard burner should be included: a long spray torch; heterogeneity of the spray of the centrifugal nozzle; coking nozzles; poor mixing of evaporation products with air in the combustion zone; narrow range of power regulation.

It is noted that the use of the nozzle-free burner assembly avoids carbon formation on the walls of the heat exchanger. This is facilitated by the technological features of fuel combustion in VGU-1: the fuel enters the cavity of the coil heat exchanger in a gasified form, which rapidly burns in an oxidizing environment (created by supplying secondary air).

In work [4] economic indicators obtained during the operation of the PPU 1600/100 with the modernized burner device VGU-1 are given. Waste fuel was used as fuel.

Polluted solvent, diesel fuel and their mixtures formed during the purification of the main oil pipelines are physicochemically close to the fuel waste. Based on this, it can be concluded that these flammable liquids are a promising fuel for PJSC Transneft. Burning these wastes can reduce significantly the energy resources expended in the maintenance of main pipelines.

References

[1] Revel-Muroz P.A. 2015 Transfer for transportation of diesel fuel of the oil pipeline and equipment used at the facilities of OAO AK Transneft *Science and technology of pipeline transport of oil and non-oil products* Vol.2(18) (Moscow: NIITNiNP) p.16-20.

[2] Chentsov A.N. 2014 Experience of experimental and practical measures for the preparation of the linear part of the pipeline to transport diesel fuel of the ecological class 5 according to TR TS 013/2011 *Science and technology of pipeline transport of oil and non-oil products* Vol. 3(15) (Moscow: NIITNiNP) p.32-38.

[3] Gafurov A.M., Б. Osipov B.M., Gafurov N.M., Gatina R.Z. 2016 A method of utilization of thermal waste energy industry to generate electricity. *Proceedings of the higher educational institutions. Energy sector problems* Vol.11-12 (Kazan: KSPEU) p.36-42.

[4] Pavlov G.I. 2017 About the technical and economic indicators of the upgraded mobile steam generator plant PPU 1600/100 *Oil and gas business* Vol.3 (Ufa: UGNTU).

IOP Conf. Series: Earth and Environmental Science **288** (2019) 012022 doi:10.1088/1755-1315/288/1/012022

Luminescent control of polycyclic aromatic hydrocarbons content in water systems

O A Plotnikova, E I Tihomirova

Politekhnicheskaya st., 77, SSTU, Saratov, Russia

plotnikova_o_a@mail.ru, tichomirova_ei@mail.ru

Abstract. To control the content of polycyclic aromatic hydrocarbons in aqueous media and to ensure effective sewage treatment of natural contaminated waters, a luminescent solid-substrate method was applied. This method permits combining the preceding sorption concentration of substances on a cellulose matrix, modified with a surfactant, and analyzing directly in the solid phase of the sorbent. Optimal characteristics of luminescent control of ecotoxicant content before and after water purification were determined

Introduction

Providing the required quality of drinking water as well as water for residential use, and compliance with sanitary and environmental quality standards for wastewaters requires monitoring and quality control of water both in water supply sources and at wastewater treatment facilities. Increasing the monitoring quality of water bodies for maintaining proper levels of public health is facilitated by creating new effective technologies for water quality control.

The article is devoted to the problems of water purification and to development of novel technologies for water quality control. Pollution adversely affects the natural sources of fresh water, which in turn has a negative impact on human health. It is crucial to identify the composition of all pollutants in the water body, which is not possible without using innovative methods for water quality monitoring.

The choice of cleaning method and water quality control requires conducting water analyses. The required stage of water analysis includes determination concentrations of particularly dangerous highly toxic substances, which can potentially cause considerable damage to human health. PAHs occupy one of the leading positions in the list of priority pollutants. Control of PAHs content in various environmental objects is mandatory. Most PAHs are carcinogenic and mutagenic substances that can accumulate in the organisms and pose a hazard even at trace concentrations in the environment [1-2]. Standard procedures for determination of PAHs in environmental objects are based mainly on chromatographic methods with various detectors [3-4]. Although the analytical effectiveness of these methods is widely recognized, the PAHs analysis in this case is characterized as laborious, relatively expensive and time-consuming process. For these reasons, it is of great interest to develop simpler and more sensitive methods for the detection and determination of PAHs. Luminescent methods are especially prospective for these purposes [5].

The goal of this study involved development of new technologies for adsorption water treatment from complex composition based on the use of modified bentonites.

Methods for Water Quality Control

We selected the luminescent assay as priority pollution control technique for ecotoxicants in the water. Luminescent methods have a number of advantages for determining the PAHs in aqueous media. First, many of the PAHs representatives are capable of fluorescing, and therefore the use of luminescent methods allows obtaining the best results of the analysis. Secondly, it is well known that these methods are among the most sensitive methods of analysis, which is especially important for identifying trace concentrations of substances in environmental objects.

However, it should be taken into account that the luminescence specters of the solutions of most organic substances are represented by broad, blurry bands. It is possible to simplify the identification of pollutants via significant narrowing of the spectrum bands, which can be achieved either by significant decrease in temperatures [6] or by using water-micellar solutions [7].

Another effective method for improving the characteristics of radiative processes in molecules is immobilization of a luminophore on a solid substrate [8-11]. This phenomenon is based on the method of solid-substrate luminescence (SSL), which enables combining preceding sorption concentrating and analysis of substances directly in the sorbent phase. Sorption of luminophores is carried out on various luminescent matrices, the most common of which is cellulose matrix [12-13]. It is well known that signal intensity of SSL luminophores on the cellulose matrix is dependent on the presence of various substances: alkalis, salts, and surfactants. They can significantly reduce or increase the intensity of luminescence. On the basis of luminescent methods, the development of sensors and screening systems for water quality control is conducted [10, 15-16].

Hence, we used the SSL method with analytical signal registration on modified cellulose matrix as the basis for original effective and highly sensitive technique of determining the PAHs.

We took Fluka-manufactured Purum brand pyrene, the least toxic PAHs, as a model compound for our experimental study. The fluorescence spectrum of pyrene was observed in the wavelength range 360-400 nm. As the matrix modifying agent, a cationic surfactant, cetyltrimethylammonium bromide (CTAB) was used. We used cellulose matrix of Red Tape brand for sorption concentrating (solid phase extraction) and applied luminescence research. Sorption of PAHs was carried out dynamically. With this goal in mind, the solution was filtered for 30 min through a sorbent layer in a plastic cartridge. Then the sorbent was dried for 15 minutes at a temperature of 80° C. PAHs luminescence spectra were recorded on a Perkin-Elmer fluorescence spectrometer LS 55.

Results of Studies on Luminescent Method of Water Quality Control

In the course of our experimental studies on the development of an effective method of water quality control for the content of ecotoxicants, we selected optimal characteristics of the luminescent determination of PAHs on a cellulose matrix.

We established that fluorescence intensity of pyrene, the PAHs representative, on the matrix is higher than in the aqueous solution, from which its sorption was conducted. We explain this finding by mobility loss of pyrene molecules during the sorption and by increase in the probability of radiative luminescent transitions (fluorescence) from the first singlet excited state to the ground state.

However, we discovered that an efficiency of sorption by a hydrophilic matrix of hydrophobic PAHs is not high enough. It can be increased by modifying the surface of the matrix with surfactants. It is known that water-micellar solutions in luminescent assay are widely used [7]. However, there are only a few studies, in which surfactants were used to observe luminescence on solid matrices [14]. We have experimentally confirmed that pyrene sorption from water-micellar solutions of surfactants, contributed to concentrating of PAHs in the surface layer of the sorbent. Such modification of the matrix surface allows improving analytical characteristics of the method.

It is known, that the sorption of cationic surfactants on the negatively charged matrix is more effective compared with anionic surfactants. Perhaps, that is due to the fact that cationic surfactants interact better with a hydrophilic matrix, the structure of which includes a hydroxyl group, -OH. We confirmed this hypothesis experimentally. Significant increase in pyrene fluorescence intensity with increasing concentration of the cationic surfactant in the solution is observed during the pyrene sorption from aqueous cationic surfactant on cellulose matrix. It was established experimentally, that the sorption of luminophore on a modified cellulose matrix could significantly increase fluorescence signal intensity.

The fluorescence spectra of pyrene on a surfactant-modified cellulose matrix were analyzed. Linear dependences of pyrene fluorescence intensity on surfactant concentrations in the solution were constructed. It was found that maximum fluorescence signal of pyrene on a solid matrix was observed at concentrations of the cationic surfactant CTAB near the critical concentration of micelle formation (9.5 · 10 - 4 M).

Limit of pyrene fluorescent detection on the modified cellulose matrix is 4.9 ng/ml, range of detectable contents is 10 - 200 ng/ml, relative standard deviation Sr = 0.04.

However, pyrene phosphorescence was observed neither by using CTAB, nor in solution, nor on the cellulose matrix. Phosphorescence allows analysing mixtures of PAHs with higher selectivity. The maximums of phosphorescence spectra of individual PAHs are more spaced in wavelength than in the fluorescence spectra.

Probably, the electrostatic repulsion of like-charged cations of heavy atoms, required to observe PAHs phosphorescence and of cationic surfactants take place.

Therefore, in the case of observing phosphorescence, it is expedient to use sodium dodecyl sulfate as concentrating and modifying agent.

Effectiveness of using luminescent assay for PAH content control in aqueous media

Sorption concentrating in combination with the measurement of an analytical signal on the surface of a sorbent allows reducing the detection limit of polluting components by several orders of magnitude. Using the sorbent modified with the surfactant allows increasing the effectiveness of concentrating the pollutants on the sorbent and sensitivity of the method. It has been experimentally established that the sorption of luminophores on a cellulose matrix modified by micellar nanosystems causes a significant increase in fluorescence signal intensity. Optimal characteristics of a luminescent solid-substrate method for controlling PAHs content in aqueous media are optimally chosen to ensure purification and quality control of sewage or natural waters

However, it should be noted that it is necessary to take into account possible composition of analyzed waters while using our original highly sensitive luminescent assay, because presence of interfering components (for example, heavy metals) may affect the results of luminescence analysis. In some cases of analyzing highly contaminated water, it is recommended to conduct a routine stage of sample preparation, including preliminary separation of the components and cleaning of the sample. Our future study will be devoted to the development of a solid-substrate luminescent method for determination of heavy metals and their concentrations in polluted aqueous media.

The results were obtained in the framework of implementing Federal grant No. 5.3922.2017/PCh of the Russian Ministry of Education

Reference

[1] Skupińska K, Misiewicz I and Kasprzycka-Guttman T Polycyclic aromatic hydrocarbons: physicochemical properties, environmental appearance and impact on living organisms 2004 *Acta Pol Pharm* **61** 233–40.

[2] Abdel-Shafy H I and Mansour M. S A review on polycyclic aromatic hydrocarbons: source, environmental impact, effect on human health and remediation 2016 *Egyptian Journal of Petroleum* **25** 107–23

[3] Wise S A, Sander L C and Schantz M M Analytical methods for determination of polycyclic aromatic hydrocarbons (PAHs) — A historical perspective on the 16 U.S. EPA priority pollutant PAHs 2015 *Polycyclic Aromatic Compounds* **35** 187–47

[4] Poster D L, Schantz M M, Sander L C and Wise S. A. Analysis of polycyclic aromatic hydrocarbons (PAHs) in environmental samples: a critical review of gas chromatographic (GC) methods 2006 *Anal Bioanal Chem* **386** 859–81

[5] Morales T V, Esponda S M, Rodríguez J J S, Aaron S E and Aaron J J Luminescence methods for study and determination of pollutants in the environment 2010 *Macedonian Journal of Chemistry and Chemical Engineering* **29** 1–42

[6] Wang H, Campiglia A D Direct determination of benzo[a]pyrene in water samples by a gold nanoparticle-based solid phase extraction method and laser-excited time-resolved Shpol'skii spectrometry 2010 *Talanta* **83** 233–40

[7] Santana Rodríguez J J, Halko R, Betancort Rodríguez J R and Aaron J J Environmental analysis based on luminescence in organized supramolecular systems 2006 *Anal Bioanal Chem* **385** 525–45

[8] Bortolato S A, Arancibia J A and Escandar G M A novel application of nylon membranes to the luminescent determination of benzo[a]pyrene at ultra trace levels in water samples 2008 *Anal Chim Acta* **613** 218–27

[9] Whitcomb J L and Campiglia A D Screening potential of solid-phase extraction and solid surface room temperature fluorimetry for polycyclic aromatic hydrocarbons in water samples 2001 *Talanta* **55** 509–18

[10] Whitcomb J L, Bystol A J and Campiglia A D Time-resolved laser-induced fluorimetry for screening polycylic aromatic hydrocarbons on solid-phase extraction membranes 2002 *Anal Chim Acta* **464** 261–72

[11] Dmitrienko S G, Gurariy E Y, Nosov R E and Zolotov Y A Solid-phase extraction of polycyclic aromatic hydrocarbons from aqueous samples using polyurethane foams in connection with solid-matrix spectrofluorimetry 2001 *Analytical Letters* **34** 425–38

[12] Ackerman A H and Hurtubise R J Methods for coating filter paper for solid-phase microextraction with luminescence detection and characterization of the coated filter paper by infrared spectrometry 2002 *Anal Chim Acta* **474** 77–89

[13] Hurtubise R J, Thompson A L and Hubbard S E Solid-phase room-temperature phosphorescence 2005 *Analytical Letters* **38** 1823–45

[14] Fernández-Sánchez J F, Segura Carretero A, Cruces-Blanco C and Fernández-Gutiérrez A Highly sensitive and selective fluorescence optosensor to detect and quantify benzo[a]pyrene in water samples 2004 *Anal Chim Acta* **506** 1–7

[15] Delgado B, Pino V, Ayala J H, Gonzalez V and Afonso A M Coupling micelle-mediated extraction using mixtures of surfactants and fluorescence measurements with a fiber-optic for the screening of PAHs in seawater 2005 *Analyst* **130** 571–77

[16] Fernández-Sánchez J F, Segura Carretero A, Cruces-Blanco C and Fernández-Gutiérrez A The development of solid-surface fluorescence characterization of polycyclic aromatic hydrocarbons for potential screening tests in environmental samples 2003 *Talanta* **60** 287–93

IOP Conf. Series: Earth and Environmental Science **288** (2019) 012023 doi:10.1088/1755-1315/288/1/012023

Influence of wheat husk modification on sorption properties of heavy metal ions

N. Politaeva[1], Y. Smyatskaya[1], E. Safonova[1], N. Barsukova[1]

[1] Higher School of Biotechnology and Food Technology, Peter the Great St. Petersburg Polytechnic University, 194021, Russia, St.Petersburg, Novorossiyskaya, 48-50

politaevana1971@gmail.com

Abstract. This article focuses on sorption properties of sorbents based on modified wheat husk. It presents kinetic characteristics and isotherms of sorbents sorption. We studied the influence of pH environment on the efficiency of wastewater treatment from heavy metal ions.

Introduction

Nowadays the problem of waste management is of urgent interest, and one of the main directions is wastewater purification [1–12]. Choosing a method of wastewater treatment is crucial not only for cleaning efficiency, but also for the cost of sorption material [13–24]. Thus production of sorbents based on waste has several advantages. Two problems are solved simultaneously: water purification and waste disposal. Availability and low cost of sorbents on the basis of agro-industrial complex waste attract attention of many Russian and foreign scientists. However, such sorbents often do not have high sorption characteristics, so the work associated with their modification and increase in physical, chemical, mechanical and adsorption properties is highly demanded.

Authors [25] studied the possibility of using waste of linen processing industry (short flax fiber) as a sorbent of zinc ions. It was shown that plasma modification of short linen fiber leads to an increase in its limiting sorption capacity in relation to zinc ions by 3 - 3.5 times. Modified husk of ripened sunflower seeds was used as sorbent for ions of iron and cadmium in [26].

To increase sorption properties of rice and buckwheat husks, scientists of Kazan State Technological University [27] suggest using special plasma plant for waste treatment. Sorbent is heated in the high-frequency plasma torch in an inert gas environment. The degree of purification of sewage from petroleum products with the help of such modified sorbents reaches 99% compared to no more than 70% without such treatment.

The waste of agricultural production - buckwheat husk is used in combination with polyurethane foam [28]. The composite material is used to collect oil, oil products and oils from the water surface.

Other authors propose the use of pine nuts shells as a raw material for the production of sorbent [29]. Researchers of Malaysian Institute of Chemical and Bioengineering Technology suggest using cheap sorbents for wastewater treatment such as tea leaves and coffee leaves that can simultaneously extract heavy metals (nickel, cadmium, chromium, zinc and lead) from polluted waters, and are also suitable for the elimination of turbidity [30].

Pakistani scientists suggest using leaves of the ficus as a sorbent for treating sewage from a number of organic and inorganic compounds [31]. Due to the composition of wood (cellulose, hemicellulose, pectin, lignin) and leaves (chlorophyll, carotene, anthocyanin, tannin), that contain functional carboxyl, carbonyl, amino and nitro groups, high sorption characteristics for heavy metals are achieved.

The purpose of this work was to study the effect of modifying the waste of agricultural processing - wheat husks on their sorption properties with respect to heavy metal ions (HMI).

Experimental part

In this work we used wheat husk and its modified analogues as sorbents, namely:

- Sorbent P-1 - wheat husk without modification. The sorbent consisted of oily yellowish husk particles with a size of about 4-5 mm in length and 1.5-2.5 mm in width.
- Sorbent P-2 - wheat husk after heat treatment at a temperature of 300 °C for 20 minutes in a special form with limited air access. These heat treatment conditions make it possible to obtain a sorbent with higher sorption properties [32]. The sorbent was a coarse black powder.
- Sorbent P-3 – wheat husk activated in a 0.2 N solution of hydrochloric acid. To do this, 25 g of wheat was boiled for 10 minutes in 500 ml of acid, then washed with distilled water until negative reaction to chloride ions (qualitative reaction with a 10% solution of silver nitrate). To wash 2.5 g of sorbent, 200 ml of water was required. The sorbent was then dried at 110° C for 1 hour. The sorbent became white without visible traces of vegetable fats.
- Sorbent P-4 - wheat husks heat-treated in the same way as P-2 sorbent, and then activated by hydrochloric acid as P-3 sorbent. The sorbent was a fine powder of black color.

To study the efficiency of wastewater treatment from HMI, each sorbent in an amount of 10 g was placed in 100 ml of a model solution and held for 24 hours. A solution containing a mixture of cadmium, zinc and lead ions with concentrations of 10·10-3 gl-1 was used as a model solution. The residual concentration was determined by the voltammetric method using the AKV-07 MK device manufactured by Aquilon. For final (Cfin) and initial (Cini) concentrations, the efficiency (E) of model wastewater treatment was calculated according to the formula:

$$E = \frac{Cini - Cfin}{Cini} * 100\%$$

The results of the calculations are given in Table 1

Table 1 Efficiency of wastewater treatment from HMI with sorbents based on wheat husks modified by different methods. Cinitial = 10 mg / l

HMI	P-1		P-2		P-3		P-4	
	Cfin mg/l	E,%	Cfinmg/l	E,%	Cfin mg/l	E,%	Cfin mg/l	E,%
Pb^{2+}	0.68	93.2	0.538	94.6	0.415	95.8	0.101	98.9
Cd^{2+}	0.57	94.3	0.324	96.8	0.268	97.3	0.130	98.6
Zn^{2+}	0.19	98.1	0.051	99.5	0.068	99.3	0.020	99.8

The analysis of the obtained data indicates that the highest purification efficiency with respect to the HMI was shown by P-4 sorbent. Sorbents P-2 and P-3 had similar values. The effectiveness of sorbents depends on their sorption properties, which, as our studies have shown, increase with modification.

The sorption capacity of the sorbent depends largely on the amount and size of the pores. The number of micropores is characterized by iodine absorption [33]. From the degree of filling of the pores of the sorbent with iodine molecules, it is possible to determine the number of pores with a size of about 1.0 nm. Sorbents from P-1 to P-4 were investigated according to the method described in [33]. The sorption capacity for iodine (Table 2) was determined by the formula:

$$E \equiv \frac{(V1 - V2) * 12.7}{M}$$, (mg/ dm3)

Here V1 is volume of sodium thiosulfate, used for fitration of the "zero" sample; V2 is volume of sodium thiosulfate, used for fitration of the sample after treatment with a sorbent; 12.7 is volume of iodine solution taken to clarify the sample; M is the mass of the sorbent (0.25 g).

Table 2 Sorption capacity (E) of sorbents for iodine

sorbent	P-1	P-2	P-3	P-4
$E{\cdot}10^{-3}$, $gr{\cdot}l^{-1}$	63.5	15.2	58.4	45.6

Sorption agents P-1 and P-3 (correspondingly 65.3·10-3 g l-1 and 58.4·10-3 g l-1) possess the largest value of sorption capacity for iodine. Apparently, these sorbents have a larger number of pores with dimensions of about 1 nm. During the heat treatment of wheat husks the process of coking of particles is likely to occur. This process produces pores with large dimensions, which, accordingly, leads to a decrease in the sorption capacity of the material by iodine (E = 15.2·10-3 g l-1). According to the procedure [33], it is also possible to determine the sorption capacity of methylene blue, which characterizes the presence of mesopores (~ 1.5 - 1.7 nm). None of the sorbents investigated by us discolored the solution of methylene blue. Consequently, the number of pores with such a size in the structure of materials is minimal and our sorbents are basically microporous (pore size ≤ 1 nm).

Basic information about sorption properties of the material and the nature of sorption makes it possible to obtain the shape of the curve of the sorption isotherm. The sorption isotherm for liquid media characterizes the dependence of the sorption capacity on the concentration at a constant temperature. According to the shape of the isotherm, one can also judge the presence and number of macro- and micropores. For the construction of sorption isotherms, it is necessary to establish the time to achieve sorption equilibrium.

To determine the time of sorption equilibrium and the value of sorption capacity, the process of sorption of HMI under static conditions from aqueous solutions of lead sulphate was studied with stirring and thermostating of the solution (T = 293 K). To obtain the kinetic curves of sorption, a sample of m = 0.1 g of P-1 - P-4 sorbents and 10 ml (V) of an aqueous solution of lead sulphate with an initial concentration of C0 = 10·10-3 g l-1 was placed in a series of tubes and aged from 5 minutes and up to 48 hours. After certain time intervals, the solution was separated from the sorbent by filtration and the current concentration of lead ions Ct was determined by voltammetric method using the AKV-07 MK instrument. The sorption capacity Ai of the sorbent at each given period of time was calculated by the formula [33]:

$$Ai = \frac{Co - Ct}{m} * V$$

Based on the results of calculations, kinetic curves of sorption were constructed (Fig. 1)

Under steady-state equilibrium conditions, the equilibrium concentration of metal ions in the solution Cp was determined in the system, and the equilibrium sorption capacity Ap was calculated by the formula:

$$Ap = \frac{Co - Cp}{m} * V$$

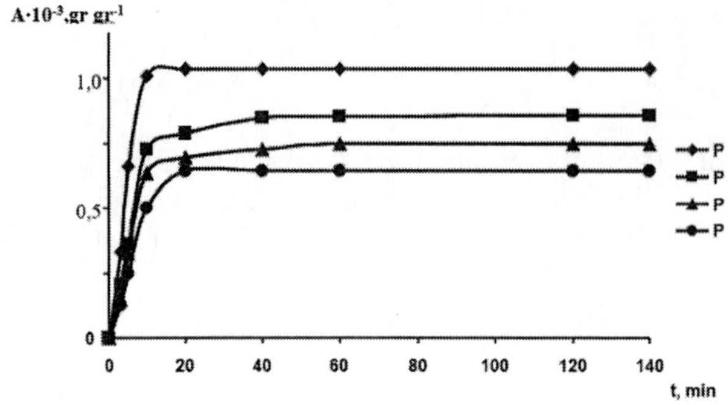

Fig.1. The kinetic curves of the sorption of Pb (II) Sorbents based on modified wheat husk

Important characteristics of sorbents are the amount of sorption capacity and the time to achieve sorption equilibrium. Kinetic studies have shown that in the sorption of Pb2+ sorbents on the basis of wheat husks, the equilibrium in the system is established first for P-1 and P-4 sorbents (in 20 min), and the equilibrium sorption capacity increases for sorbents in the series:

$$P\text{-}1(A_p=12.9) < P\text{-}2\ (A_p=15.0) < P\text{-}3 - 1(A_p=17.1) < P\ \text{-}4\ (A_p=20.7)$$

For the construction of sorption isotherms, weighed samples of P-1, P-3 sorbents were placed in a series of tubes and 10 ml of a lead sulfate solution with initial concentrations of 5, 10, 20, 30, 50, 100 mg / dm^3 was poured into them until equilibrium was reached. The solution was then separated from the sorbent by filtration and the final concentration of the HMI was determined therein (Fig. 2).

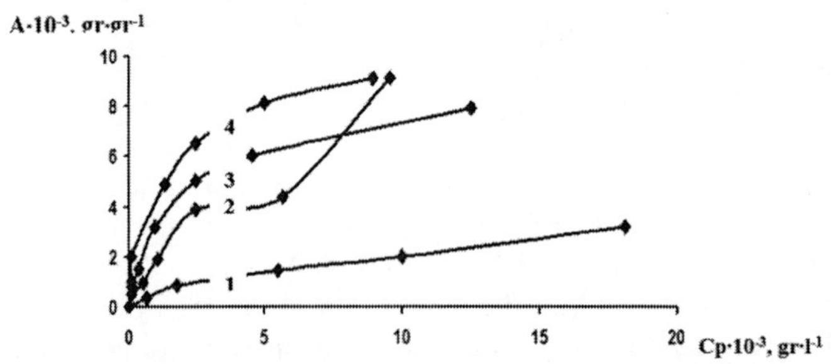

Fig. 2. Sorption isotherms of Pb (II) ions of sorbents on the basis of modified wheat husk: 1) P-1; 2) P-2; 3) P-3; 4) P-4

According to the BET theory [33] (Braunauer, Emmett and Teller), the shape of isotherms 1, 3, 4 in Fig. 2 indicates the presence of micropores in the sorbents. Steepness of the isotherm 4 and 3 characterizes the presence of ultramicropores. The curved shape of isotherm 2 is typical for sorbents with mixed micro- and macroporosity, which confirms our explanation of the minimum value for iodine absorption by P-2 sorbent (EP-2 = 15.2 mg·l^{-1}). The presence of porosity in sorbents indicates the possibility of physical sorption of HMI from solutions.

From the literature data it is known that the sorption value determined by the HMI depends on the pH of the environment, which significantly affects the sorption processes of the HMI present in solutions in the form of various complex particles.

The stability of metal complexes decreases in the case of deviation of the pH of the environment from neutral reaction: at low pH, because proton competes with the metal ion for the binding sites in the molecules, and at high values, because of competition between the hydroxyl group and the ligand.

IOP Conf. Series: Earth and Environmental Science **288** (2019) 012023 doi:10.1088/1755-1315/288/1/012023

To determine optimal conditions for sorption by pH, model solutions containing copper, lead, zinc and cadmium ions with an initial concentration of $10 \cdot 10^{-3}$ g l^{-1} were prepared, the pH of which was adjusted to pH = 2.0; 6.0; 8.0; 10.0. The necessary pH was adjusted with solutions (0.1 N) of sodium hydroxide or sulfuric acid. P-2 sorbent was used at the rate of 10 g of sorbent per 100 ml of the model solution. Solutions with the sorbent were held for 12 hours with stirring and thermostating (T = 293 K).

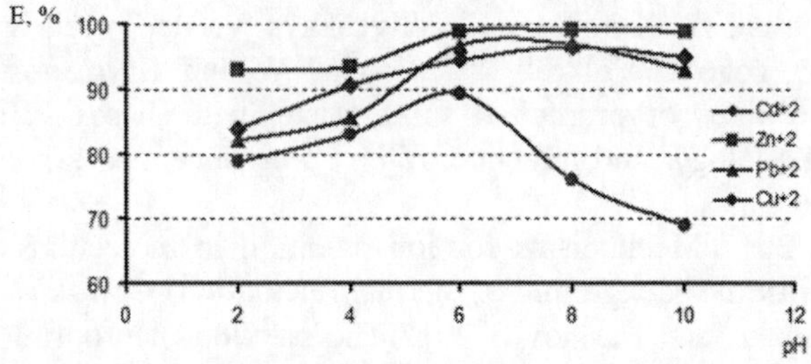

Fig.3. Efficiency of purification from HMI by sorbent P-2, depending on the pH of the solution

It can be seen from the Fig.3 that the most effective is the purification from zinc and cadmium ions in the range 6.0-10.0, from lead ions at pH=6-8, from copper ions at pH=6.0. In solutions with higher pH values (8, 10) copper precipitates a s copper hydroxide.

The mechanism of sorption process by proposed sorbents on the basis of modified wheat husk can be explained as follows. The main composition of wheat husks is fiber and cellulose. Cellulose is a natural polymer, the elementary links of which $-C_6H_{10}O_5$ - are combined into long linear macromolecules with the help of a glucose bond or an oxygen bridge -O-. A characteristic feature of cellulose is the presence in each elementary unit of three hydroxyl groups -OH-. The functional hydroxyl group is able to interact with the HMI, keeping them in the pores and on the surface of the sorbent. Consequently, the mechanism of cleaning HMI with these sorbents occurs not only due to physical sorption but also chemisorption. Polymer macromolecules interact with each other using van der Waals forces and hydrogen bonds, due to the presence of hydroxyl groups [34]. After thermal treatment, the polymer is oxidized and carbonized, which increases the physical adsorption of HMI with P-2 sorbent in comparison to P-1 sorbent.

Wheat husk also contains hydrophobic components: lipids (ether extract), lignin and waxy substances that determine the hydrophobic properties of P-1 sorbent (unprocessed husk of wheat), which makes it difficult to adsorb HMI from aqueous solutions, because of poor wettability, slowing down penetration of the solution into the interior of the sorbent. After acid treatment (sorbent P-3), in addition to polymer oxidation, hydrophobic substances are removed, which leads to an increase in the porosity and hydrophilicity of the sorbent. This increases the availability of functional groups, which increases chemisorption and the ability to bind HMI. Therefore after complex thermal and acid modification (sorbent P-4) its adsorption and chemisorption properties improve.

Thus, obtained data show that modifying the husk of wheat leads not only to physical properties change (porosity, dispersity, bulk density), but also causes changes in the chemical composition of the material. We carried out physico-chemical studies using infrared spectroscopy (IRS) (Fourier spectrometer Infralum FT-801 in tablets with KBr) of P-1 and P-2 sorbents. Samples analysis showed that P-1 sorbent contained cellulose, ester, protein and silicon compounds. After heat treatment at 300 ° C, the absorption ester and protein groups disappeared from the P-2 composition and the absorption bands of the silicon compounds increased. At the same time, bands were found that indicate the appearance of carbon in the P-2 sorbent formed during coking due to the conversion of organic constituents.

Carried out research allowed us to establish kinetic dependences of sorption processes by HMI sorbents on the basis of the initial and modified husk of wheat. Modification by heat treatment and hydrochloric acid promotes an increase in the porosity and hydrophilicity of the material and, as a result, increases the sorption capacity.

References

[1] Bukova M, Bondal A, Skvortsova O, Nikonova O, Kholodiakov A, Guseva I, Makarova T and Mirschel W 2016 Reclamation of the illegal dump for sustainable development the environment in Sverdlovo of Leningrad Oblast', Russia *MATEC Web Conf.* **73**

[2] Orlova T, Melnichuk A, Klimenko K, Vitvitskaya V, Popovych V, Dunaieva I, Terleev V, Nikonorov A, Togo I, Volkova Y, Mirschel W and Garmanov V 2017 Reclamation of landfills and dumps of municipal solid waste in a energy efficient waste management system: Methodology and practice *IOP Conference Series: Earth and Environmental Science* vol 90

[3] Shevkunov S V 2017 Mechanisms for ion retention in molecular water clusters in a planar nanopore against the background of thermal fluctuations *Colloid J.* **79** 399–413

[4] Braila N, Kirilkina A and Barinov S 2017 The solution of proofing pie for the waste sorting plant *IOP Conference Series: Earth and Environmental Science* vol 90

[5] Fedyukhin A, Sultanguzin I, Gyul'Maliev A and Sergeev V 2017 Biomass pyrolysis and gasification comprehensive modeling for effective power generation at combined cycle power plant *Eurasian Chem. J.* **19** 245–53

[6] Andrianova M and Bondarenko E 2016 Changing of Cations Concentrations in Waters of Polluted Urban River *MATEC Web of Conferences* vol 53

[7] Novikov D, Molodkina L, Chusov A and Vedmetskii Y 2015 Electrokinetic and electroconductivity properties of filtering material aqualat *Procedia Engineering* vol 117pp 264–72

[8] Chusov A N, Maslikov V I, Molodtsov D V, Zhazhkov V V and Riabuokhin O A 2015 Assessment of zonal distribution of methane on MSW landfills in northern regions for its usage in local power engineering *Mag. Civ. Eng.* **58** 44–55

[9] Bazarnova J G, Politayeva N A and Bozhuk S G 2017 Innovative technologies secondary use of processed active source Proceedings of the 2017 International Conference "Quality Management, Transport and Information Security, Information Technologies", IT and QM and IS 2017 pp 471–6

[10] Matveyeva A N, Pakhomov N A and Murzin D Y 2016 Recycling of Wastes from the Production of Alumina-Based Catalyst Carriers *Ind. Eng. Chem. Res.* **55** 9101–8

[11] Ryzhakova M G, Maslikov V I, Chusov A N and Korablev V V 2014 The environmental problem of household hazardous waste generation and treatment *Appl. Mech. Mater.* **675–677** 761–9

[12] Rudskoi A I and Kokorin V N 2013 Compaction of heterophase mechanical mixtures in production processes of utilizing industrial wastes (final tailings) at enterprises of ferrous metallurgy *Russ. J. Non-Ferrous Met.* **54** 518–21

[13] Arakcheev E N, Brunman V E, Brunman M V, Volkov A N, D'yachenko V A, Kochetkov A V and Petkova A P 2016 Automated production system for reagents used in water and wastewater processing *Russ. Eng. Res.* **36** 745–50

[14] Alshabab M S, Andrianova M and Alsalloum D 2016 Modification of Wastewater Treatment Technology at Cottonseed Oil Plant *MATEC Web of Conferences* vol 53

[15] Umanets V, Kalitova L, Kalitov D, Chusov A and Umanets E 2015 Conditions for safe disposal of industrial wastewater into the deep aquifer *International Multidisciplinary Scientific GeoConference Surveying Geology and Mining Ecology Management, SGEM* vol 1pp 135–44

[16] Simonenko E, Gomonov A, Rolle N and Molodkina L 2015 Modeling of H2O2 and UV oxidation of organic pollutants at wastewater post-treatment vol 117(Elsevier Ltd)pp 342–9

[17] Petkova A P, Brunman V E, Brunman M V, Konyashin A V and Arakcheev E N 2016 Reagent

production technology for disinfection and treatment of drinking water and wastewater *Ecol. Ind. Russ.* **20** 10–5

[18] Polyakova I V, Ezhova N M, Osipenko A A and Pisarev O A 2015 Polymer sorbent with the properties of an artificial cholesterol receptor *Russ. J. Phys. Chem. A* **89** 312–5

[19] Polyakova I V, Osipenko A A, Borovikova L N, Ezhova N M, Pisarev O A, Vlasova E N and Volchek B Z 2015 Synthesis and properties of polymeric and organo-inorganic amphiphilic sorbents molecularly imprinted with cholesterol *Russ. J. Appl. Chem.* **88** 1617–26

[20] Sobgaida N A and Kolesnikova M A 2006 Carbon sorbents for water purification from oil products

[21] Morozova V Y, Podvyaznikov M L and Samonin V V 2015 Effect of microscopic additions of fullerenes on the absorption capacity of inorganic sorbents for d elements *Russ. J. Appl. Chem.* **88** 1612–6

[22] Polyakova I V, Borovikova L N, Kipper A I and Pisarev O A 2017 Complex-Emulsion synthesis of organo-inorganic amphiphilic sorbents with specific affinity for glucose *Russ. J. Appl. Chem.* **90** 901–9

[23] Kolosentsev S D, Solovei V N, Kiseleva V L, Spiridonova E A and Samonin V V 2015 Effect of nitrogen- and sulfur-containing modifying additives on porous structure and sorption properties of carbon adsorbents *Russ. J. Appl. Chem.* **88** 430–5

[24] Leshchinskaya A P, Ezhova N M and Pisarev O A 2015 Synthesis and sorption properties of polymeric sorbents molecularly imprinted with uric acid *Russ. J. Appl. Chem.* **88** 820–5

[25] Nikiforova T E, Bagrovskaya N, Kozlov V and Natareev S V 2008 Sorption properties of plasma-modified flax fibers *Russ. J. Appl. Chem.* **81** 1175–9

[26] Zemnukhova L A, Tomshich S V, Shkorina E D and Klykov A G 2004 Polysaccharides from buckwheat production wastes *J. Appl. Chem.* **7** 1192–6

[27] Shchepakin M B, Gafarov I G, Mishynin G M and Israfilov I H 2000 Ecological-technological complex for purification of hydrosphere from oil and oil-products *Ecol. Ind. Russ.* **11** 41–4

[28] Zavyalov V S 2006 Sorbtion capacity of materials in relation to oil products *Ecol. Ind. Russ.* **8** 7–9

[29] Sabrina K and I. S H 2008 Tea waste as low cost adsorbent for removal of heavy metals and turbidity from synthetic wastewater *Proceedings International Conference on Environmental Research and Technology* pp 38–41

[30] Mozumder S I, Khan M R and Islam A 2008 Kinetics and mechanism of Cr(VI) adsorption onto tea-leaves waste *Asia-Pacific J. Chem. Eng.* **3** 452–8

[31] Qaiser S, Saleemi A R and Muhammad M A 2007 Heavy metal uptake by agro based waste materials *Electron. J. Biotechnol.* **10** 23–8

[32] Sobgaida N A, Olshanskaya L N and Makarova Y A 2008 j *Proceedings of the 6th Scientific and Practical Conference "Tatishchev Readings: Actual Problems of Science and Practice"* (Togliatti) pp 338–42

[33] Smirnov A D 1982 *Sorption purification of water* (Leningrad: Chemistry)

[34] Bryk M T 1989 *Destruction of the filled polymers* (Moscow: Chemistry)

IOP Conf. Series: Earth and Environmental Science **288** (2019) 012024 doi:10.1088/1755-1315/288/1/012024

Modern water treatment system for high-quality water used in the energy industry

E Rakhmetova[1], A Kozodaev[2], A Zavyalova[3], O Kondrateva[3], N Zvonkova[3]

[1]University of Genoa, via all'Opera Pia, 15 – DIME (prof. Roberto Revetria), Genova (GE), 16145, Italy

[2]Bauman Moscow State Technical University, 2-nd Baumanskaya Str., 5/1, Moscow, 105005, Russia

[3]National Research University "Moscow Power Engineering Institute", Krasnokazarmennaya Str., 14, Moscow, 111250, Russia

KondratyevaOYe@mpei.ru

Abstract. This article provides the detailed optimization processes and enhancement, in the field of water treatment systems, for the energy sector needs. The developing of the unique water purification system design based on modern techniques for equipment is presented according to ASTM standards. As well as the project proposes a new methodology for complex analysis of quality data after each purification section in order to manage all possible need in the water for the power plant. The analysis allows producing the more accurate selection of the filter systems, thereby prevent failures of such sophisticated and expensive equipment as reverse osmosis and continuous electrodeionization.

Introduction

The main purposes of introducing optimization processes and enhancement in the field of water treatment systems (WTS) for the energy sector needs [1] is a prevention of corrosion and boilers deterioration [2, 3] thus ensuring an increase of power units efficiency and protection of components, minimization of blowdown and reduction in the use of chemical reagents. The building scheme of the WTS is the general plan, which contains the full project definitions (as requirements program, treatment scheme, phasing of construction, estimation of an investment cost) and also is important for international decisions, including such considerations as the project mandate for organizational, technical, contractual and financial aspect. [1, 4] Improving energy efficiency in the WTS allows reducing total demand, avoiding the risk of brownouts or blackouts during high energy demand periods. Water efficiency strategies reduce the risk of water shortages, helping to ensure a reliable and continuous water supply. [4, 5]

At the current moment, widescale projects for the implementation of new high-tech combined cycle gas turbine (CCGT) plants implement in the power systems of the Russian Federation. In this regard, the requirements for the boiler feed water of the steam-turbine unit also increase: instead of the previously demanded electrical conductivity of the feed water for supercritical power units less than 0.3 μS/cm, combined-cycle power units put forward a requirement with an electrical conductivity of less than 0.2 μS/cm. [6]

Traditionally, for ensuring a quality of the boiler feed water simplified water treatment is used, for instance, Raw water reservoir - Dry filtration – Rapid bulk filtration - Adsorption - Softening - UV-disinfection – Microfiltration - Clear water reservoir. [1, 7, 8] However, increased requirements for desalinated filtrate will cause a decrease in a filter run that finally will lead to increase in operational expenses and decrease in reliability of plant operation.

Firstly, the ultra-pure water related to the electronics and microelectronics industries. The system developed within the framework of this project is aimed at providing all kinds of water consumption on the power plant, not only needs in the feed water, but maintenance of component parts of power units and electronic components by this reduce the total investment cost. The limits on the impurities are related to current contamination specifications were obtained by analytical methods (these processes are described in part 3), laboratory or on-line instrumentation control are also envisaged in building design (see part 2) and are used in accordance with current industry practice. The stage for water selection is defined in accordance with the specific manufacturing process and the water-quality recommendations apply at the point of distribution.

There are several international professional organizations that set technical standards for the quality of ultrapure water: International Organization for Standardization, American Society of Testing Materials, Semiconductor Equipment and Materials Institute, etc. In general, these organizations specify four types of water with three additional degrees of purity, described in the normative document requirements for deionized water. For a long time in Russia, the requirements for ultrapure water are regulated by OST 11.029.003-80. However, due to large-scale changes in the energy and electronics industries over the past decades, these requirements for the quality of de-ionized water are considered obsolete. Increasingly, engineers of modern Russian water treatment companies in the design of purification systems are guided by the standard ASTM D5127–13 «Standard Guide for Ultra-Pure Water Used in the Electronics and Semiconductor Industries». [9].

Development of the treatment scheme of WTS

The treatment scheme is embedded in the general planning process, first of all, is based on the data on the quality of raw water, requirements for purification water and requested process flow diagrams (PDF) of water consumption. Comparison of the laboratory analysis of initial water to requirements is a prime preparatory design stage of the station of water treatment. This project proposes a unique system of complex calculation and analysis of quality data for all purification sections in order to manage different needs described in part 1. Application of this way of technological development allows specifying most necessary and enough an inventory thanks to what it is possible to reduce overall dimensions of the station of water treatment and to considerably reduce prime cost of the project [1, 8].

In the presented article, we consider the full technological process of the ultra-pure WTS for the nuclear power station needs. The raw water source is the system of the central water supply. The treatment scheme comprising PDF is based on laboratory control data according to requirements of the ASTM D-5127-13 to type E-1 and is represented in Figure 1.

The primary section of the WTS utilizes the pretreatment and desalination processes. The purified water at the first stage passes a mesh water filter (MF) with the spectrum of filtration ranges of 100 µm. The water produced in the primary section has a purity equivalent to Type E-4 and is stored prior to use or next additional purification.

The secondary purification section is dedicated to bringing the effluent of the primary section up to the purity level of Type E-3. This section includes removal of organic and biological material and as well as includes additional desalination processes. This section includes removal of organic and biological material, with sand as a foundation in all cases, and as well as includes additional desalination processes:

- inorganic particle removing filters (FF), principally Fe (regenerative);
- adsorption filters (AF) with the activated carbon for chlorine, its derivatives and organic impurities removing(regenerative);

- water softening by two-bed ion-exchange filters (SF) with synthetic materials (NaCl for a regeneration).

Water treatment by the adsorption method was widely adopted in the production of water treatment and also processes of receiving valuable components of waste products. [10] Additionally, ultrafine particulate removal is implied in this treatment section. A significant part of the clear water flow from this section goes to energy production department (where be used as feed water) directly thus a degradation of the resistivity is avoided.

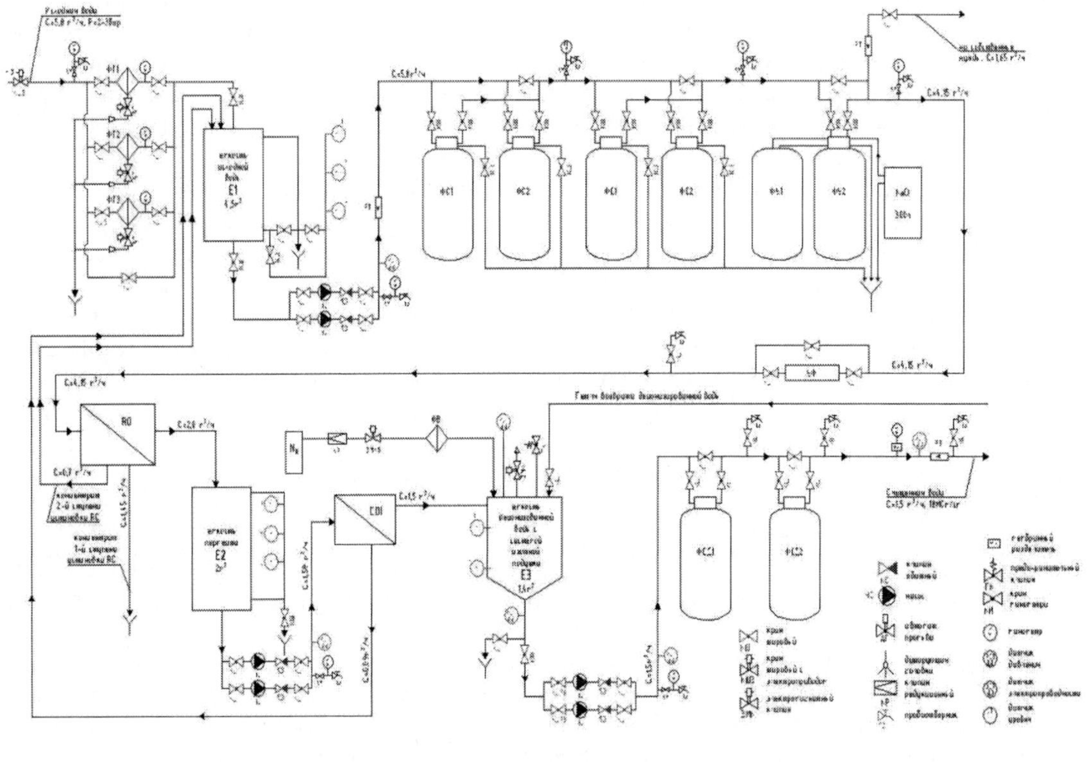

Figure 1. The scheme of the ultra-pure water treatment

Tertiary filtration or final removal process increases the purity to the requirements of type E-2 or E-1. This section includes ultraviolet light (UV) (254 nm), reverse osmosis, continuous electrodeionization and mixed-bed ion-exchange filters (MIF). The water from the tertiary section of the system is used with increased requirements for storage and is recirculated continuously (will be described in part 4).

Experience with ultraviolet radiation shows that if the dose of actual irradiation is not lower than a certain value for a particular water quality, then a sustainable decontamination effect is guaranteed. In the world practice, the minimum radiation dose requirements range from 16 to 40 mJ/cm^2. Removal of biological and organic contaminants is an important adjunct to any system used to prepare ultra-pure water. Dissolved organic compounds can accumulate in the system during the process as well as being present in the original water. Ultraviolet irradiation at 254 nm significantly reduces the growth of organisms by dislocating the DNA base pairs. This process prevents the bacteria replication, furthermore, total organic carbon (TOC) is reduced.

RO and EDI belong to the most advanced technologies in the production of ultrapure water, due to the ability to recycle some portion of the wastewater also. In presented RO system the water purification follows the next scheme: the first stage permeate goes to the second stage of RO, while the first stage concentrate fed back into the feed water supply to be recycled through the RO system to save water; the second stage concentrate goes to drain.

In accordance with the PDF shown in Figure 1, after the RO phase, the final water purification is carried out using electrodeionization (EDI) - the process of continuous demineralization of water using ion-exchange resins, ion-selective membranes and a constant electric field. Electrodeionization modules combine the advantages of speed and efficiency of ion exchange with the absence of a laborious and dangerous for the health and environment stage of regeneration of ion-exchange resins with acid and alkali. Modern installations of electrodeionization of water make it possible to obtain water with a resistivity of 18 MOhm*cm. In the process of EDI, about 5-7% of the drainage water (concentrate) is formed, which is sent to the clean water reservoir (type E-4)

Quality analysis of the feed water according to requirements

The application of reverse osmosis and electrodeionization, in industrial water treatment for needs of the energy sector, is not so widespread therefore there are simplified schemes with bunk filters are used. The choice of the membrane depends upon thereof the pore size, characteristics of RO system and the size of particles to be removed. This is due to the limited life of RO membranes and the high cost of components in that the membrane surface is exposed to organic, colloidal and mechanical deposits. However, often the real reason for reducing the life of the plant and premature deterioration of membranes are design errors. [12] Accordingly conduction of a qualitative analysis of all stages of water treatment that precede reverse osmosis is a necessity, and in case of unsatisfactory water quality, use additional methods of cleaning, especially in the presence of surfactants and petroleum products. [13]

Table 1 presents the water quality indicators obtained by calculation, including using software Integrated Membranes Solutions Design Software.

Table 1. The change in water quality after each type of treatment

Parameter	Unit	**Quality data of the raw water, treated water and requirements**						
		Raw water	**Clear water**					**Requirements**
			MF	**FF**	**AF**	**SF**	**RO***	
TOC	µg/L	3,0	0,6	0,2	0,1		0,000	< 2
Turbidity	µg/L	0,96	0,38	0,08	0,05		0,000	< 0,1
Alkalinity	mmol/L	1,42		1,2		0,25	0,298***	< 0,1
Hardness	mmol/L	1,92		1,8	1,3	0,1	0,002	< 0,02
Iron	µg/L	0,05		0,01			0,000	< 0,02
Manganese	µg/L	0,08 ± 0,02		0,03			0,001	< 0,05
Dissolved oxygen	µg/L	3,52			1,6		0,1	10
Ammonium	µg/L	0,143			0,1		0,01	< 0,1
Chloride	µg/L	4,66			1,2	1,5	0,000	< 0,02
Nitrate	µg/L	0,55				0,1	0,002	< 0,05
Calcium	µg/L	19,38				2,0	0,001	< 0,02
Sulfate	µg/L	26,06 ± 5,21					0,001	< 0,05
Magnesium	µg/L	9,22				1,2	0,000	< 0,05
Residue after evaporation	µg/L	133				116,5	0,42	< 0,5
Sodium	µg/L	6,291				28,82	0,112***	< 0,05
Resistivity, 20°C	On-line					165**	1,2	18,1

* – The parameters were calculated automatically in the program IMS Design Software for a two-stage RO with 12 membranes such as Hydranautics CPA2-4040 for the flow of 4.15 m^3/h.
** – The value was obtained automatically in the software after input of the original data.
*** – The required values are achieved by means of final filtration at the electrodeionization unit and mixed-bed demineralization filters.

Storage and distribution systems

The storage of high-quality water during production is very important. In proportion to the area and time of contact between the water, the air and the containment materials, the impurities penetrate into the clear water. It is necessary to provide a WTS with a loopdistribution design. Particular emphasis must be placed upon the atmosphere above the water, since those gases may contaminate the water with biological, organic, inorganic, and particulate impurities. [9]

In the considered project, high-purity nitrogen is used to blanket the atmosphere above the ultra-pure water stored in the container after deionization stage. As the distribution system presents a large area of contact between the water and pipes and biological impurities tend to accumulate in stagnant water, for those reasons, the flow of clear water is maintained on a continuous 24-h basis. To maintain the required quality E-1 additional mixed-bed ion-exchange (nonregenerative) systems is required. MIFs are shown in the treatment scheme and are a part of the loopdistribution.

Conclusion

Summarizing the issues and results discussed in the provided article:

– The unique water treatment scheme provides all kinds of water consumption on the power plant, as the feed water, high-quality water for the maintenance of component parts and electronic components of power units were developed and presented in Figure 1.

– By this design, there is the total investment cost of WTS for the plant, in general, is reduced.

– The changes in the main water quality indicators were determined after each stage, that allows determining a water take-off location for specific needs and simplifies the on-line instrumentation control, the results are presented in Table 1.

– The loopdistribution design and recycle system were envisaged as well.

References

[1] Moel P J, Verberk J Q and Van Dijk J C 2007 *Drinking water: principles and practices* (Singapore: World Scientific)
[2] Larionov K, Gvozdjakov D, Zenkov A and Zaytsev V 2016. Assessment of efficiency of regenerative heating system of feed water for high temperature turbine plant on decentralized power engineering facilities *EPJ Web of Conferences* **110** 01034
[3] U.S. Environmental protection agency 2013 *Energy efficiency in water and wastewater facilities* (Washington: Local Government Climate and Energy Strategy Series)
[4] Rashidi M M, Aghagoli A and Ali M 2014 Thermodynamic analysis of a steam power plant with double reheat and feed water heaters *Advances in Mechanical Engineering* **6** 940818
[5] Mrzljak V, Poljak I and Mrakovčić T 2017 Energy and exergy analysis of the turbo-generators and steam turbine for the main feed water pump drive on LNG carrier *Energy Conversion and Management* **140** pp 307-23
[6] Jurby watertech international 2011 Jurby news *Jurby watertech international* **31** pp 2-7
[7] Belikov S E 2007 *Water treatment* (Moscow: Akva-Term)
[8] Sutherland K 2008 Water filtration: bulk water filtration techniques *Filtration & separation* **10** pp 17-9
[9] Standard AS. D5127–13 2013 *Standard Guide for Ultra Pure Water Used in the Electronics and Semiconductor Industries. ASTM International, West Conshohocken, PA*
[10] Ksenofontov B S, Kozodayev A S, Taranov R A, Vinogradov M S, Balina A A and Petrova E V 2013 Elaboration of the fundamentals of the technology of extracting valuable components

from ash-slag heaps of heat and power generation *Ecology and industry of Russia* **8** pp 9-12

[11] Rakhmetova E R, Tkachenko Yu L 2017 Application of software for the analysis of the danger of the system for the preparation of ultrapure water for the production of microelectronics at the design stage *The Future of Mechanical Engineering in Russia Conf* pp 405-8.

[12] Werber J R, Deshmukh A and Elimelech M 2016 The critical need for increased selectivity, not increased water permeability, for desalination membranes *Environmental Science & Technology Letters* **4** pp 112-20.

[13] Ksenofontov B S, Kozodayev A S and Taranov R A 2013 Wastewater treatment of food industry enterprises from surfactants and fats by flotation *Ecology and industry of Russia* **11** pp 4-7

[14] Mallevialle J, Odendaal P E and Wiesner M R 1996 *Water treatment membrane processes* (USA: American Water Works Association)

[15] Asonov A M 2007 *Calculation of cation exchangers for water softening in industrial water treatment* (Ekaterinburg: USURTU)

IOP Conf. Series: Earth and Environmental Science **288** (2019) 012025 doi:10.1088/1755-1315/288/1/012025

Definition of an optimum method of water preparation for needs of the enterprises

R.R.Rotach[1], E.A.Akhmetov[2], S.O. Gaponenko[2]

[1] Russia, Kazan, LLC "KER-Engineering"

[2] Russia, Kazan, Kazan state power engineering university

rita-nurkaeva@mail.ru

Summary. Results of technical and economic calculations for various methods of water desalting are presented in article. The prime cost of the desalted water received by each of the considered methods is calculated (in costs of thermal and electric energy). On the basis of the obtained data conclusions are drawn on applicability of each of the considered water desalting methods.

Research objective

The research objective consisted in definition of the most optimum and economic method of preparation of water for needs of the enterprise. The following methods of water desalting have been considered: electrodialysis, method of the return osmosis, evaporator of instant boiling up.

Description of a research objects

Technical and economic calculations of electrodialysis installation (EDI), installation of the return osmosis (RO) and the evaporator of instant boiling up (EIB) with a consumption of water from 10 to 100 t/h have been made.

For each of devices the following indicators have been calculated:

– for electrodialysis installation – consumption of electric energy;

– for reverse-osmotic installation and the evaporator of instant boiling up – consumption of thermal and electric energy.

– for all installations calculation of cost of receiving the desalted water has been made (in costs only of thermal and electric energy).

Table 1. Salinity of initial water (Kazan)

Cation	Maintenance of ions		Anion	Maintenance of ions	
	m/kg	mg-eq/kg		m/kg	mg-eq/kg
Ca^{2+}	72	3,6	SO_4^{2-}	140	2,92
Mg^{2+}	14,5	1,21	Cl^-	15	0,42
Na^+	10,15	0,44	HCO_3^-	140,3	2,30
K^+	10,15	0,26			
Total	106,8	5,51	Total	295,3	5,64

Basic data for calculation of electrodialysis installation were the following indicators [4]:

– ionic composition of initial water, mg/kg;
– a consumption of initial water before giving on installation: G: 10, 25, 50, 75, 100 t/h;
– salinity of initial water (Table 1);
– characteristics of the used ionite membranes. In EDI ionite membranes of the Russian production of the MK and MA types 0,7 mm thick, are used by the size 500×1500 mm. A pro-masonry framework are made of a paronite 0,8 mm thick. In membrane space turbulize grids – separators from the viniplast are laid.

Basic data for calculation of the evaporator of instant boiling up [4]:

– consumption of unboiled water G: 10, 25, 50, 75, 100 t/h;
– reference temperature of unboiled water t_s=10°C;
– water temperature before an entrance to the installation t_m=30°C;
– the specific heat of initial water, c=4,19 kJ/kg ·°C;
– temperature of the heating steam t_{s1}=130°C.
– temperature of the initial clarified water (the summer mode): t_{s1}=30 °C;
– temperature of boiling of solution in the last step: t_b=47 °C.

Basic data for calculation of reverse-osmotic installation [4]:

– consumption of unboiled water G: 10, 25, 50, 75, 100 t/h;
– reference temperature of unboiled water t_s=10°C;
– water temperature before an entrance to the installation t_m=30°C;
– the specific heat of initial water, c=4,19 kJ/kg ·°C;
– temperature of the heating steam t_{s1}=130°C.

Calculation of consumption of electric energy for reverse-osmotic installation has been executed in the ROSA program.

The ROSA program is developed by the DOW CHEMICAL company for calculation of reverse-osmotic and nanofiltrational installations with application of the membrane FILMTEC elements.

For calculation of prime cost the following data have been used:

– cost of electric energy: C_{el}= 4,3 rub/kW · h;
– for heating of initial water before an entrance on installations it is used pairs of low parameters from a technological production line. Steam parameters are: p_{s1}=4 atm.; t_m=130 °C. The cost of steam is: 67,98 rub/t.

Results of numerical researches and their discussion

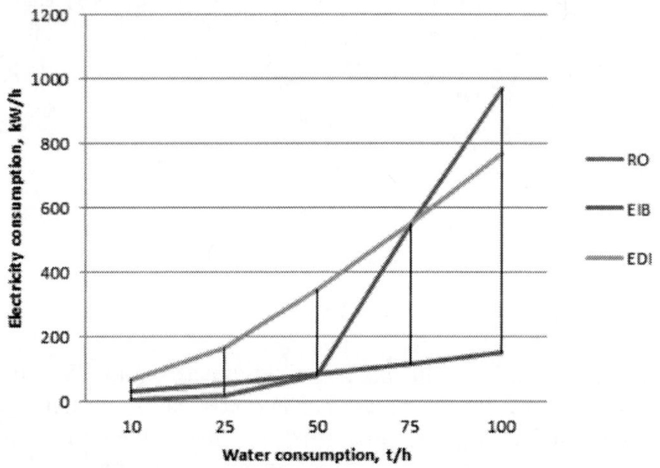

Fig. 1. Dependence of consumption of electric energy on a water consumption

According to the diagram submitted in fig. 1 it is visible that with increase in a consumption of initial water, consumption of electric energy increases during using of all considered methods. However the smallest consumption is the share of the evaporator of instant boiling up. The greatest consumption of electric energy is the share of electrodialysis installation at water expenses, it is less than 75 t/h. At expenses, it is more than 75 t/h, the greatest electricity consumption is the share of installation of the return osmosis. Also it is worth noticing that when using reverse-osmotic installations at water expenses less than 50 t/h electricity consumption is lower, than during using evaporators. However at expenses more than 50 t/h at the return osmosis electricity consumption sharply increases as a result of use of high-pressure pumps for water transmission through membranes [5].

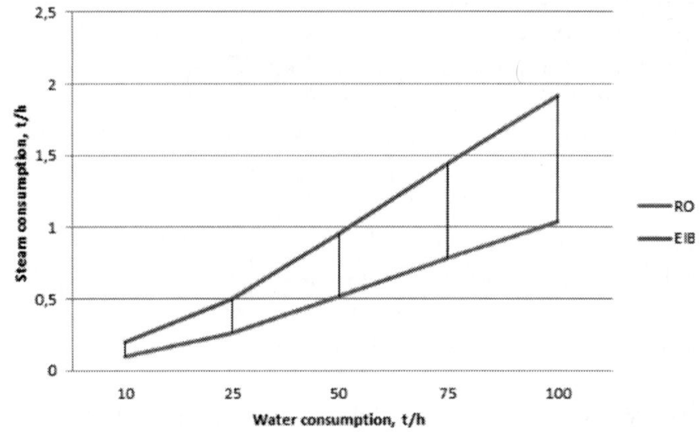

Fig. 2. Dependence of thermal energy consumption on a water consumption

On the diagram submitted in fig. 2 it is visible that thermal energy consumption increases during using both methods of a desalting. However during using evaporators of instant boiling up of heat it is consumed more, than at the return osmosis owing to more high reference temperature [5].

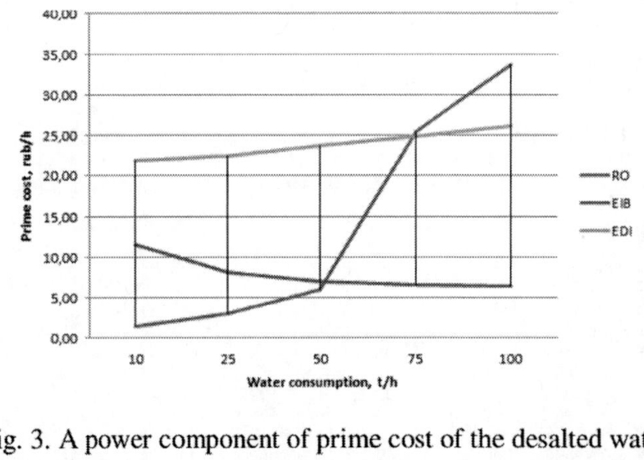

Fig. 3. A power component of prime cost of the desalted water

According to the schedule submitted in fig. 3 it is possible to draw the following conclusions: at expenses of water from 10 to 75 t/h the greatest prime cost of the desalted water is the share of electrodialysis installations, and at expenses, more than 75 t/h – of the return osmosis installations. At the return osmosis sharp increase in prime cost at water expenses is observed, it is more than 50 t/h. The smallest values of prime cost fall on a method of the return osmosis at expenses of water from 10 to 50 t/h. It is necessary to notice that for the evaporator of instant boiling up at increase in a water consumption, the prime cost decreases [6].

Conclusions

Thus, it is possible to consider reliable that at expenses, from 10 to 50 t/h are more economic to apply a method of the return osmosis to a water desalting, and at expenses from 50 to 100 t/h – the evaporator of instant boiling up.

However in view of the high cost of chemical reagents for washing of reverse-osmotic membranes, the prime cost of the desalted water with method of the return osmosis can grow considerably. Thereof the most optimum method of an water desalting is the evaporator of instant boiling up in all considered range of expenses of water.

References

[1] Kostrikin Yu.A., Meshchersky N.A., Korovin O.V. Water treatment and water mode of power facilities of low and average pressure. M.: Energy. 2010. – 252 pages.

[2] Hodzhis T. Technology of pure receiving// Pure locations and technological environments. – 2003. - No. 3. – p. 13-19.

[3] Bagrov O.N. The systems of complete reverse water-supply in nonferrous metallurgy. - M.: Metallurgy, 2008. - 143 pages.

[4] Akhmetov E.A., Rotach R.R. Comparing of energetic expenses in case of different methods of an water desalting for technological needs of the enterprises// International scientific and technical conference "Innovative machine-building technologies, the equipment and materials – 2016" - Kazan, 2016 p. 265-269.

[5] Rotach R.R. Water desalting methods for needs of the enterprises// The Nineteenth Postgraduate and master seminar devoted to Day of power engineering / KGEU. – Kazan 2015 – p. 87-88.

[6] Rotach R.R. Akhmetov E.A. Economic effect during heating feedwater in installation of the reverse osmosis// The Eleventh international youth scientific conference " To scientific progress creativity of young people" / VSTU. – Yoshkar-Ola, 2016. – P. 269-270.

[7] Gorodovih V.Ye., Obraztsov S.V., Kaplin A.A. Methods of deep electrochemical water purification on alternating asymmetrical current// Materials of a scientific and practical conference. Research methods in chemistry and chemical technology. - Tomsk, 2006. – P. 88-96.

IOP Conf. Series: Earth and Environmental Science **288** (2019) 012026 doi:10.1088/1755-1315/288/1/012026

Energy, mass-exchange and hydrodynamic efficiency of degassers at low-temperature deaeration of water for thermal power plants

V.I. Sharapov, E.V. Mingaraeva

Ulyanovsk, st. Severnyy Venets, 32

vlad-sharapov2008@yandex.ru

Abstract. A new technology for low-temperature degassing of the make-up water was developed using natural gas as a desorbing agent. The calculation of the energy efficiency of the developed technology for a typical power unit of the heat power plants (HPPs) is performed. Theoretically necessary and actual specific costs of natural gas are estimated to ensure the normative quality of the deaeration. The calculation of the hydrodynamic characteristics of the atmospheric deaerator at work on natural gas is presented.

A new technology for low-temperature deaeration of make-up water in a heating network

The main means of removing corrosive gases (O_2иCO_2) from water at thermal power plants is thermal deaeration. One of the most important ways to increase the energy efficiency of deaeration at CHP is the use of heating mediums with low potential [1].

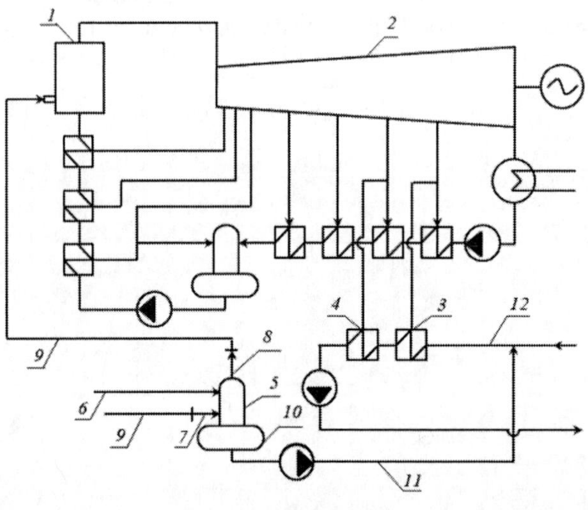

Fig. 1.Flow diagram of degasation of make up water of the heating supply system by natural gas. (1) steam boiler, (2) turbine, (3, 4) lower and upper network heaters, (5) deaerator, (6) pipeline of influent water, (7, 8) inlet and outlet branch for desorbent, (9) gas pipeline, (10) accumulator tank, (11) pipeline for make up water of the heating supply system, and (12) return network pipeline

Consider a new technology of low-temperature deaeration of make-up water of the heating network using natural gas as a desorbing agent, supplied to the boilers of the power plant. Natural gas does not contain oxygen and carbon dioxide in its composition.(Fig. 1) [2].

Natural gas after the reducing facilities has a very low, often negative temperature. Due to this, deaeration is carried out at relatively low temperatures (10-30°C). Mixing of cold deaerated make-up water with reverse network water leads to a significant decrease in the temperature of the return network water before the lower network heater, an increase in the generation of electricity by thermal consumption and, consequently, to an increase in the economics of operation of the thermal power plant.

Energy efficiency of the HPP during low-temperature deaeration of the makeup water of the heat network with natural gas

The indicators of the energy efficiency of the operation of thermal power plants are calculated by the method of the specific generation of electricity for heat consumption due to the selection of steam for heating the heat carriers in the thermal scheme of water treatment[3].

Power N_{ch}^{lnh},kW, developed by the turbine on thermal consumption due to the selection of steam from the turbine to the heating of water in the lower network heater:

$$N_{ch}^{lnh}=\frac{G_{n.w}(h_o-h_{lnh})(\tau_1-\tau_2)c}{h_{lnh}-h_{c_{lnh}}}\eta_e\eta_m, \quad (1)$$

where$G_{n.w}$ – consumption of network water through network heaters, t/h; τ_1– temperature of network water supplied to the consumer after heating in network heaters, °C; τ_2– return water temperature after mixing with cold deaerated make-up water, °C; c – specific heat capacity of water, kJ/(kg·°C); h_o– enthalpy of fresh steam, kJ/kg; h_{lnh}– enthalpy of the steam of the lower selection of the turbine, kJ/kg; $h_{c_{lnh}}$– enthalpy of the condensate of the lower network heater, kJ/kg; $\eta_e\eta_m$– electric and mechanical efficiency of the turbogenerator.

The power produced by the steam of the turbine selections, which is spent for the regenerative heating of steam condensate, used for heating the streams of network water in the lower network heater:

$$N_{reg}^{lnh}=\frac{G_{n.w}(\tau_1-\tau_2)\left(h_{f.w}-h_{c_{lnh}}\right)\left(h_o-h_{reg}^{lnh}\right)c}{\left(h_{lnh}-h_{c_{lnh}}\right)\left(h_{reg}^{lnh}-h_{f.w}\right)}\eta_e\eta_m, \quad (2)$$

whereh_{reg}^{lnh}=0,5(h_o+h_{lnh}) – enthalpy of conditional equivalent regenerative selection, kJ/kg.

The calculation for the upper network heater is similar.

The increase in fuel consumption for additional steam generation in the boiler with increasing steam flow to the lower and upper network heaters and lowering the enthalpy of this steam is calculated by the form:

$$\Delta B_{add}=\frac{\Delta D(h_o-h_{f.w})}{Q_{c.f}\eta_b}, \quad (3)$$

whereΔD– increase in steam consumption when the temperature of the network water changes, t/h;$Q_{c.f}$– combustion heat of conventional fuel, kJ/kg; $\eta_b$is the boiler efficiency.

The annual savings of the conventional fuel at the HPP using the new technology is:

$$\Delta B=\left(\left(\Delta N_{ch}+\Delta N_{reg}\right)\left(b_e^c-b_e^h\right)10^{-3}-\Delta B_{add}\right)n_h, \quad (4)$$

where b_e^c, b_e^h are specific consumptions of the equivalent fuel for power output in the condensation and heat extraction modes, respectively, kg/(kW·h); n_h – number of hours of use of the turbine.

The calculation showed that on an installation with a turbine T-100-130, a boiler with a steam capacity of 500 t/h and a feed water consumption of a heat network of 800 t/h, an annual savings of about 4587 t of fuel is achieved.

Mass exchange efficiency at low-temperature deaeration of the make-up water of the heating network

To investigate the possibility of using the technology of low-temperature deaeration of water, it is necessary to estimate the mass-exchange efficiency of the deaerator operation when using natural gas as the desorbing medium.

The basis of the method for determining the theoretically necessary specific consumption of natural gas to remove dissolved oxygen from the water d_{gas}^{min}, kg/t, is the solution of the balance equations for the processes of mass transfer and heat transfer during deaeration, provided that an equilibrium between the phases is reached at the exit from the deaerator [4]. With certain assumptions, it can be assumed that the largest mass-transfer efficiency of the deaerator is achieved by ensuring the standard quality of the deaeration with the minimum possible costs of the desorbing agent and the evaporation leaving the deaerator.

The equation for the material balance of the deaeration can be written in the form:

$$G_{i.w}X_{i.w}+D_{gas}Y_{gas}=G_{d.w}X_{d.w}+D_{vent}Y_{vent}, \tag{5}$$

where $G_{i.w}$ and $G_{d.w}$ is the consumption of the initial and deaerated waters, kg/h; D_{gas} is the consumption of natural gas supplied in the deaerator, kg/h; D_{vent} is the discharge of the deaerator vent steam (mixture of corrosive gases liberated from water and natural gas), kg/h; $X_{i.w}$ and $X_{d.w}$ are oxygen concentrations in the water at the inlet and outlet of the deaerator; and Y_{gas} and Y_{vent} are the oxygen contents in the natural gas at the dearator inlet and in the vented steam at the deaerator outlet.

According to the Dalton law, the total pressure of the gas and gas–steam mixtures is equal to the sum of partial pressures of gases and steams composing the mixture. From the Henry law, it follows that the concentration of the gas dissolved in water is proportional to the partial pressure of this gas over the water surface.

Oxygen concentration Y_{gas} in gas at the deaerator input is practically zero. The oxygen concentration in the vented steam leaving the deaerator depends on the flow chart of the water and steam motion in the apparatus. At the countercurrent, mole fraction Y_{vent} of O_2 in the gas-steam mixture is

$$Y_{vent}=K_H^{O_2}X_{i.w}/p, \tag{6}$$

where $K_H^{O_2}$ is the Henry coefficient (phase equilibrium constant for oxygen), Pa; and p is the pressure in the deaerator, Pa.

At the countercurrent of water and natural gas in the deaerator, the minimum consumption of natural gas is

$$D_{gas}^{min}=G_{i.w}\frac{p}{K_H^{O_2}}\frac{X_{i.w}-X_{d.w}}{X_{i.w}}, \tag{7}$$

and its specific value is

$$d_{gas}^{min}=\frac{D_{gas}^{min}}{G_{i.w}}. \tag{8}$$

The calculation of the theoretically necessary specific consumption of natural gas in the low-temperature degassing of water at the HPP showed that about 1 m^3 of natural gas is required per 1 ton of water.In the actual operating conditions of the deaerator, the required specific gas consumption for deaerating should be taken as 3-5 times as much as theoretically necessary (by analogy with thermal deaerators operating on a steam). Ensuring the necessary consumption of natural gas for deaeration of water in thermal power plants does not present any difficulties. With deaeration of 800 t/h of make up water, the required gas flow will be 2400-4000 m^3/h, while a steam boiler with a steam capacity of 500 t/h requires about 40000 m3/h of gas, i.e. deaeration will require no more than one tenth of the gas flow to the boiler [5].

Hydrodynamic efficiency of degassers working on natural gas

Calculation of the hydrodynamic characteristics of low-temperature deaeration of water using natural gas as a stripping medium is made for a jet-bubbling atmospheric deaerator DA-25 with an un-submerged perforated bubble-plate installed in the lower part of the deaeration column.

Hydrodynamic operating conditions of the deaerator for effective deaeration [6]:

1. Maintaining the desired gas velocities in the holes of the bubble-plate.
2. The presence of a gas cushion under the bubbling sheet, due to which the sheet is not immersed in water.
3. Absence of entrainment of droplets from the deaerator column.

The use of an un-submerged principle of bubbling, according to which water on a perforated sheet is continuously and repeatedly treated with gas supplied under the sheet and passing through the holes in it, is most effective when operating bubbling deaerators.

Under the sheet a gas layer ("cushion") is formed, which prevents dropping of liquid through the holes of the sheet. The hydrodynamic stability of un-submerged bubble sheet is determined by the rate of gas flow through the holes.

The minimum required speed may be determined by the empirical form [7-9].

$$w_{\min}=\frac{20{,}6}{\sqrt{\rho_{gas}}}, \tag{9}$$

whereρ_{gas}– the gas density under the sheet, kg/m^3.

The height of the gas cushion under the sheet is

$$h=2\cdot\sqrt[3]{\frac{\sigma^2}{\left(g\rho_w-g\rho_{gas}\right)^2 d}+\frac{\zeta w^2\rho_{gas}}{2g\left(\rho_w-\rho_{gas}\right)}}. \tag{10}$$

whereσ– coefficient of surface tension of water-gas system, kg/m; ρ_w – the density of water, kg/m^3; *d*– the diameter of the holes in the perforated sheet, m; *w* – the rate of passage through the holes of the sheet, m/s; ζ=1,9…2,0 – the coefficient of hydraulic resistance of the perforated sheet.

To assess the presence or absence of entrainment of drops from the deaerator is necessary to determine the gas velocity in the column deaerator [9, 10]. Sustainable mode downward flow exists at gas velocities of about 15-30 m/s, above which the entrainment of droplets [11].

In traditional thermal deaerators, used as the working medium is water steam, which condenses in the process of deaeration, flow rate of steam in the jet of the column is small and the danger of entrainment of droplets is practically absent. In deaerator using natural gas as a stripping agent the gas velocity and the risk of entrainment of drops from the column of deaeration to the gas pipeline before boiler is slightly higher, since natural gas by deaeration is not condenses.

The gas velocity in the column deaerator is

$$w_{col}=\frac{G_{gas}}{S}, \tag{11}$$

whereG_{gas}– gas consumption, m^3/h;S– section area, m^2.

As a result of calculation according to formulas (9) - (11) with a specific gas consumption of 3-5 m^3 / t, ie 3-5 times more than theoretically necessary, the following hydrodynamic characteristics of the deaerator DA-25 were determined:

1. The estimated gas velocity in the holes of the bubble sheet w_{min}=57,58 m/s.
2. The height of the gas cushion under the bubble sheet h=0,025 m.
3. The gas velocity in the column of deaerator w_{col}=0,13 m/s. Consequently, the entrainment of droplets is impossible, as in deaerator using steam as the stripping agent.

Thus, the operation of the deaerator jet-bubbling atmospheric deaerator on natural gas provided the required hydrodynamic conditions of operation of jet and bubbling steps of deaerating column.

Conclusions

1. A new technology for low-temperature deaeration of the make-up water of the heat network of thermal power plants using natural gas as a desorbing medium was developed.
2. The calculation of the energy efficiency of the new technology has been performed, showing that only one installation with a T-100-130 turbine and a 500 t/h steam generating boiler achieves an annual savings of about 4587 tons of conditional fuel.
3. Normative mass-exchange efficiency of deaerators using natural gas as stripping medium supplied to the burners of boilers of thermal power plants is ensured with a relatively low gas consumption for deaeration. Theoretically necessary and actual specific consumption of natural gas for low-temperature deaeration of water are estimated.
4. When operating a jet-bubbling deaerator with the use of natural gas as a working medium, all the necessary hydrodynamic conditions for the operation of the jet and bubbling deaeration stages are performed.

References

[1] Sharapov VI 1994 Current Problems in the Use of vacuum deaerators for Open Heat-Supply Systems*Thermal Engineering***41**pp 635-639.
[2] SharapovV I, PazushkinaO V and KudryavtsevaE VPatent RF № 2537656 Sposob raboty teplovoy elektricheskoy stantsii.
[3] Sharapov V I 2015Metodika otsenki energeticheskoy effektivnosti strukturnykh izmeneniy v teplovykh skhemakh TES *TrudyAkedemenergo*No 2 pp 27-37.
[4] Sharapov V I and Tsyura D V2003 *Termicheskie Deaeratory* (Ul'yanovsk: Ulyanovsk state technical university) p 560.
[5] Sharapov V I, Pazushkina O V and Kudryavtseva E V2016 Energy-Effective Method for Low-Temperature Deaeration of Make-up Water on the Heating Supply System of Heat Power Plants *Thermal Engineering***63** No. 1 pp 56-60.
[6] Ramm V M 1976*Absorbtsiya Gazov* (Moscow: Khimiya) p 656.
[7] Kutateladze S S and Styrikovich M A 1976*Gidrodinamika Gazozhidkostnykh Sistem* (Moscow: Energiya) p 296.
[8] Oliker I I and Permyakov V A1971 Termicheskaya Deaeratsiya Vody na Teplovykh Elektrostantsiyakh (Leningrad: Energiya)p 184.
[9] Brodov YuM, Aronson KE, Ryabchikov AYu and Nirenshteyn MA 2008*Spravochnik Po Teploobmennym Apparatam Paroturbinnykh Ustanovok* (Moscow: Izdatel'skiy dom MEI) p 480.
[10] Vodopodgotovitel'noe oborudovanie. Informatsionno-spravochnyy katalog. Saratov:OAO «Sarenergomash» 2007 p. 166.
[11] Kasatkin A G 1973 *Osnovnye Protsessy I Apparaty Khimicheskoy Tekhnologii* (Moscow: Khimiya) p 752.

IOP Conf. Series: Earth and Environmental Science **288** (2019) 012027 doi:10.1088/1755-1315/288/1/012027

Studying the sorption of heavy metal ions by materials based on food industry waste

Y. Smyatskaya[1], N. Politaeva [1], L. Olshanskaya [2], E. Tatarinceva [2], S. Efremova[3]

[1]Higher School of Biotechnology and Food Technology, Peter the Great St. Petersburg Polytechnic University, 194021, Russia, St.Petersburg, Novorossiyskaya, 48-50

[2]Engels Technological Institute (branch) of Saratov State Technical University named after Yuri Gagarin, Engels, Saratov region, Russia, 413100, Pl. Liberty, 17

[3]Penza State Technological University (440039, Russia, Penza city, 1a/11 Baidukova psg./Gagarina st.)

Makarovayulia169@mail.ru

Abstract. The present study treats on cleaning wastewater originated from industrial buildings using sorption materials. As sorption materials for the recovery of heavy metal ions, industry wastes were used: threshing millet, cotton fiber, chitosan. The basic sorption characteristics such as purification efficiency, sorption capacity, specific surface and sorption capacity on water were considered. Conditions and modes of heat treatment of sorption materials (temperature and time) are described. A combined use of materials in the form of a layer-by-layer filter and granules is proposed. The efficiency of cleaning materials is 93 to 99.9%, the sorption capacity is from 6.9 to 50 mg/g.

Introduction

Competent design of the sewerage system from industrial enterprises allows one to extend the service life of sewer facilities. Sewerage network, pumping stations and facilities for cleaning shop floor drains that are placed at the industrial site, refer to the system on-site sewage enterprises. Wastewater discharge from the territory of industrial enterprises is often carried out on a separate sewerage system, i.e. rainwater and industrial fused separately. For some enterprises with appropriate technical and economic justification the combined sewage system can be accepted [1].

In those enterprises where the production waste water in composition and nature of contamination close to domestic waters, satisfied with the overall sewer network to share their leads. Rainwater is discharged by drains. If you have conditionally-net production of water, they come in a water recycling or diverted watercourses. In this case, the separate sewerage system has two networks: a network of polluted industrial and domestic waters and a network of conditionally clean industrial and rainwater. If industrial wastewater is not take together with household due to the fact that they contain specific pollution (fats, oil products, heavy metal ions, etc.), which can cause irregularities in the sewer, arrange for local treatment facilities (e.g., girolomoni, septic tanks, filter for wool, neutralization installations, etc.), having them in the shops or near them.

In modern conditions, when water shortage is acute in many regions of the country, the urgent task is to create closed water supply systems with utilization of all water treatment products.

It is possible to prevent pollutants from entering the General sewage system by means of local wastewater treatment systems based on different methods: chemical, physico-chemical and biological [2-4]. One of the most effective methods is the sorption method of wastewater treatment, which allows to remove ions of heavy metals. Various materials (activated carbons, zeolite, silica gel, peat) are used as sorbents and find application in various technological solutions [5-7]. Ash is used for effective removal of lead, cadmium, copper and zinc from polluted water [8].

In recent years, much attention has been paid to the use of low-cost adsorbents from plant waste to remove Cu (II) and Zn (II) from polluted waters. In article [9] sorption characteristics of two types of the wastes received as a result of agriculture are studied: soy bran and mustard husk. The influence of contact time, initial concentration of metal ions, pH, sorbent mass and temperature on the adsorption capacity of materials as sorbents is investigated. Article [10] describes the use of olive seeds for extraction of iron, lead, copper.

This papers [11-12] presents the first results concerning the possible use of cotton stalks and apricot seeds as biosorbents of heavy metals. These materials were chosen due to its high availability and basic cellulose structure. The heavy metals selected as sorbates were Cu and Pb. The purpose of this paper was to study adsorption. The use of various agricultural wastes to remove heavy metal ions of these from aqueous solution under various experimental conditions. The influence of various operating parameters on biosorption, such as initial pH and metal concentration, adsorbent amount and adsorbent particle size, was studied and optimal experimental conditions were determined. Another purpose of this study was to test the toxic effect of heavy metals on the soil bacterium Pseudomonas aeruginosa ATCC 10145 before and after biosorption with agricultural waste.

For the treatment of wastewater from organic and inorganic dyes are used disaccharides [13].

The authors [14-15] show the possibility of using barley grain shells for sorption from water surfaces of oil and oil products. It was found that the modification of the shells with weak sulfuric acid solutions leads to the removal of the hydrophilic component from the material, which helps to reduce water absorption by 16.5% and thus increase the sorption capacity in relation to oil and petroleum products by 30 %.

The aim of this work was to develop sorption materials that allow highly efficient extraction of heavy metal ions from sewage effluents of industrial enterprises.

Methods

The objects of the study were agricultural waste i.e. carbonized waste of millet (CWM); cotton fiber (CF) - waste of weaving cotton fabrics; chitosan (waste from processed crustaceans).

As method for obtaining sorption materials, heat treatment is used in a muffle furnace (SNOL brand) in a special unit with limited oxygen access.

The thermal treatment modes for the threshing millet waste was a temperature of 300°C for 20 minutes, and for the weaving cotton fabrics wastes a 450°C temperature during 8 minutes (TCF).

The layer filter was made as follows: the materials were placed in a special polymeric housing in such a way that the outer layers contained TCF, and the inner layer - (CWM) sorbent.

The granular materials made of chitosan and CWM were prepared according to the following procedure:

Sample 1 - A mixture is prepared for the preparation of the granules. 40 g of chitosan was added to 960 g of acetic acid 3%. The mixture is stirred for 4-5 hours until the chitosan is completely dissolved. The obtained gel-like mixture is the poured through a syringe into a 5% solution of sodium hydroxide (NaOH). The formed granules are held for 24 hours in an alkali solution (NaOH), followed by washing with water of a 7.0-7.5 pH.

Sample 2- A CWM additive in an amount of 20% by weight is added to the resulting mixture of chitosan in acetic acid (described above for Sample 1). The mixture is thoroughly mixed and poured through a syringe into a 5% solution of sodium hydroxide (NaOH).

The content of heavy metal ions in the sample wastewater was monitored voltammetrically using the analyzer of the voltammetric model AKV-07 MK.

Results and Discussion

In [16-19], the authors selected the optimal conditions for the thermal treatment of agricultural waste. As a result of heat treatment, a carbonized waste of grind mill (CWM). The initial threshing of millet contains a large amount of organic substances (cellulose, lignin, cellulose, etc.), with heat treatment, a porous material containing carbon, which determines sorption properties of sorbents. For a given sorption material, the specific surface area is: S= 188 m^2/g, the total pore volume in water is V = 0.3 cm^3 /g [20].

To determine the efficiency of wastewater purification from IHM (Ions of heavy metals) (Pb(II), Cd(II), Zn(II), Cu(II) C initial = 10 mg/dm^3) using (CWM), 20 grams per liter of model wastewater solution were added to wastewater and static sorption was carried out for 20 minutes at constant stirring. (CWM) was separated from the water by filtration. Based on the final concentrations found in the solution, the wastewater purification efficiency and the CWM sorption capacity were determined (Table 1).

Table 1. Final concentration (C final), efficiency of clearing (E, %) IHM from sewage (C initial = 10 mg/l)

IHM	C final, mg/l	E, %	A, mg/g
		Sorbent CWM	
Pb(II)	0.68	93.0	15
Cd(II)	0.57	94.0	16
Zn(II)	0.19	98.0	18
Cu(II)	0.12	98.0	14
		Sorbent TCF.	
Pb(II)	0.54	94.5	6.9
Cd(II)	0.17	98.3	14.4
Zn(II)	0.39	96.0	13.2
Cu(II)	0.01	99.9	15.1

The next object of research as a sorption material (SM) for wastewater treatment from heavy metal ions was cotton fiber waste of weaving cotton fabrics. It is formed during the mill process while weaving the thread of the weft with the warp thread. In the article, the optimal conditions for the heat treatment for cotton fibre were established.

Treatment was carried out in a special cell without air access. Under these conditions, heat-treated cotton fiber (TCF) is formed which forms a high sorption characteristics, due to mechanical sorption. The main characteristics for cotton fibre: the specific surface area is - Ssp = 50.8 m2/g, the total pore volume in water is V pore = 0.8 cm3 /g, for TCF: the specific surface is - Ssp = 1700 m2/g, the total pore volume by water Vpor = 1.4 cm3/g,

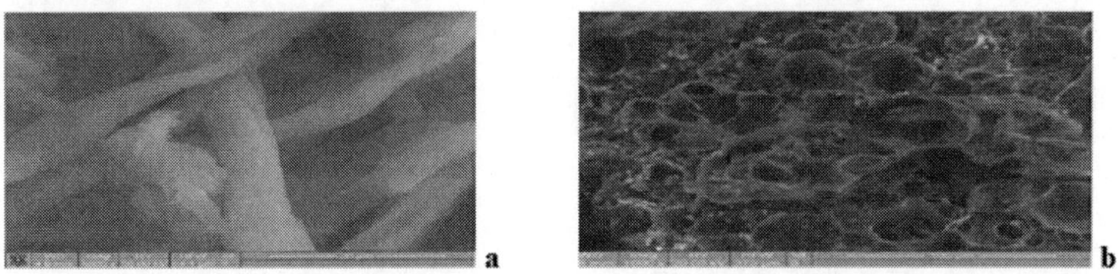

Figure 1. Morphology of the surface of TCF (a), CWM (b)

The obtained TCF was investigated for the purification of wastewater from the IHM in the static regime (Table 1). The time for achieving the sorption equilibrium was 30 min, during which the sorbent was kept at the optimum ratio of 1g/dm3. Based on the final concentrations found in the solution, the purification efficiency of wastewater and the sorption capacity of TCF were determined (Table 1). Morphology of the surface is show the figure 1.

When using powdered CWM as a sorbent, fine particles of CWM are carried away with the water flow and further pollutes the water. For convenient use of powdered CWM, sorption materials of CWM and TCF are proposed to be laid in layers [19].

The article [19] studied the ratio of fibrous and cellulose-containing waste (20: 80% by weight). External fibrous layers play the role of a framework for lose, powdery sorbents, and when the wastewater is filtered. The filter was analyzed for its ability to absorb the IHM dynamically. It turned out that the efficiency of purification from IHM is up to 93%. It has been established that the calculated values of the total sorption capacity (A, mg/g) for IHM increase in the series: Zn(II) (A = 13.6) < Cd(II) (A = 12.8) < Pb(II) (A = 12.4) (Fig. 2).

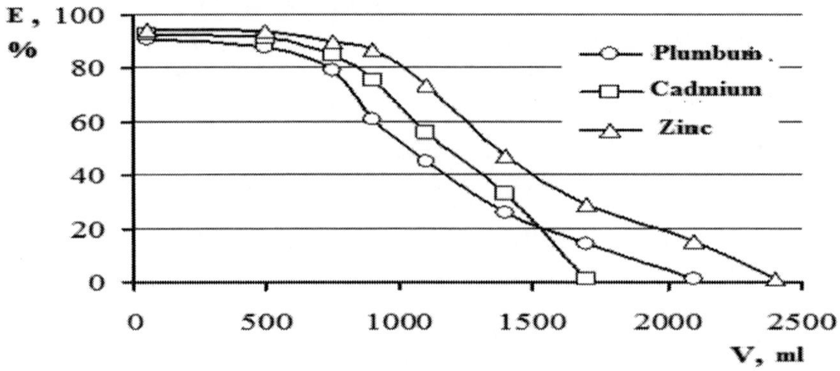

Figure 2. Dependence of the efficiency of IHM purification from the volume of the filtered wastewater through the filter (m = 5 g)

The next technical solution of the powdered material is its granulation. For the production of granules as a binder, proposed to use chitosan, which also has sorption properties. The paper [20] shows that powdered chitosan is used to purify wastewater from IHM. The obtained granular sorption composite materials (granules) of the following composition: 1 - chitosan; 2 - chitosan + CWM;

To study the sorption properties of graphs in an amount of 20 g/l, the addition of lead ions with initial concentration of 30 mg/l to model solutions containing lead ions and the sorption process carried out for 60 min under static conditions. Then, the model solutions were filtered, and the residual content of lead ions was analyzed. At the final and initial concentrations, the efficiency and sorption capacity of cleaning model solutions are calculated. The results are shown in Table 2.

Table 2. Influence of granules composition on sorption properties

Granules composition	C initial, mg/l	C final, mg/l	E,%	A, mg/g
Sample 1 (chitosan)	30	6.34	78.0	43
Sample 2 (chitosan + CWM)	30	0.41	98.6	50

From Table 2, it can be seen that the addition of CWM to chitosan granules makes it possible to increase the purification efficiency from 78 to 98% and the sorption capacity from 43 to 50 mg /g.

When studying the morphology of the surface of sample 2 with a CWM content of 20% (Figure 3), it can be seen that sample have an inhomogeneous structure, with a portion of the porous surface. With CWM content, the granule surface has pores with dimensions of 4-12 nm and 20-24 nm (Figure 3). The porous surface structure explains the high efficiency of the sample 2, in which the extraction of the IHM occurs due to physical adsorption into the pores.

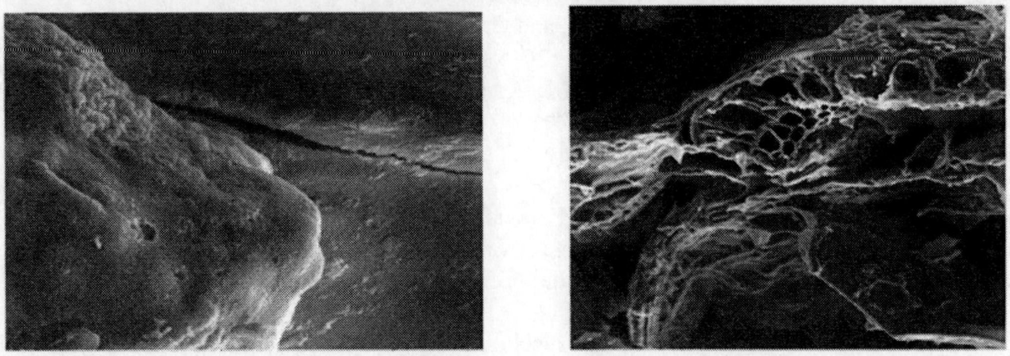

Figure 3. Morphology of the surface of granulated samples x 100; A-sample 1 (chitosan), sample 2 (chitosan -CWM)

Conclusion

All the above SM have their advantages and disadvantages: CWM and TCF are cheap materials with high sorption characteristics, but additional modifications are required for practical application. The design of the layer-by-layer filter makes it possible to clean the wastewater in a dynamic mode.

Granulated SM also have a high purification efficiency and CWM additive allows to reduce the cost of a material based on chitosan and to provide convenience of use.

As a result of the study, the following conclusions were drawn

1. Methods for obtaining SM from wastes and the possibility of cleaning industrial sewage with the use of obtained SM has been shown.

2. The main parameters of the SM were determined:

- CWM in relation to the IHM: sorption equilibrium time (20 min), optimum mass ratio of SM to the volume of model wastewater (20 g/dm^3), purification efficiency of wastewater (from 93 to 98%) and maximum sorption capacity (14-18 mg/g).
- TCF in relation to IHM: sorption equilibrium time (20 min), optimum mass ratio of SM to the volume of model wastewater (20 g/dm3), purification efficiency of wastewater (from 94.5 to 99.9%) and maximum sorption capacity (6, 9-15.1 mg/g).
- Filter based on CWM and TCF in relation to IHM: sorption equilibrium time (20 min), optimal mass ratio of SM to volume of model wastewater (20 g/dm3), purification efficiency of wastewater (up to 93%) and maximum sorption capacity (12.4 -13.6 mg/g).
- Granulated SM based on chitosan and chitosan with additive CWM in relation to IHM: sorption equilibrium time (20 min), optimal mass ratio of CM to volume of model wastewater (20 g/dm3), purification efficiency of wastewater (from 78 to 98.6%) and maximum sorption capacity (43-50 mg/g).

References

[1] Galkin Yu.A. 2012 Vestnik SGASU Town planning and architecture. No 4 (8) Pp. 64-70
[2] Chechevichkin V. and Vatin N 2014 Applied Mechanics and Materials P. 409-415.
[3] Molodkina L.M. 2010 Methods of purification of drinking, natural and waste water (SPb.: Politekhn. un-ta) p 275
[4] Filatov Ye.G. 2015 Izvestiya VUZ. Applied chemistry and biotechnology No 2 (13) Pp. 97-109.

[5] Vukojević Medvidović, A. Daković, M. Ugrina, M. Trgo, I. Nuić and M. Marković, 2017 Technologica Acta Vol 10 No 2 P 9–13.

[6] Chechevichkin, A.V., Vatin, N.I., Samonin, V.V. and Grekov, M.A. 2017Magazine of Civil Engineering.76 (8) P. 201-213

[7] Vatin, N.I., Chechevichkin, V.N., Chechevichkin, A.V., Shilova, Y. and Yakunin, L.A. 2014 Applied Mechanics and Materials 587-589 P. 565-572

[8] Harja M, Buema G, Mircea Sutiman D. and Cretescu I. 2013 Chemical Papers 67 (5) 497–508

[9] Humelnicu D, Ignat M. and Doroftei F. 2015 Environ Monit Assess. P. 187- 222

[10] Alslaili T. M., Abustan, I., Ahmad M. A. and Foul A. A. 2014 Desalination and Water Treatment. No. 52 P. 7887–7897.

[11] Kahraman S, Dogan N. and Erdemoglu S. 2008 J. Environment and Pollution Vol. 34 Nos 1/2/3/4 P275-283

[12] Bhatnagar A., Sillanpää M. and Witek-Krowiak A. 2015 Chem. Eng. J. P. 244–271.

[13] Yelnikov D. A. Wastewater treatment from synthetic and organic dyes by disaccharide production waste. avtoref. dis. ... kand. Tech.sci. : 03.02.08 D. A. Yelnikov. Penza 2013 20 p.

[14] Shaykhiyev I. Kondalenko O. A. and Trushkov S. M 2010 Exposition.Oil.Gas No 5 Pp. 46-50

[15] Shaykhiyev I. G., Stepanova S. V. and Kondalenko (Galblaub) O. A. 2011 Vestnik Kazanskogo tekhnologicheskogo universiteta. No 15 Pp. 244-251

[16] Kondalenko O. A., Stepanova S. V., Shaykhiyev I. G. 2012. Safety in the technosphere. No 6 Pp. 57-62.

[17] Sobgayda N.A., Olshanskaya L.N. and Makarova Yu.A. 2009 Chemical and oil and gas machine building No. 9 P.43-45

[18] Sobgayda N.A., Olshanskaya L.N., Makarova Yu.A, and Popova N.Ye 2009 Ecological problems of large cities : sb. nauch. tr. Ch. 1. (Saratov: Sarat. gos. tech. un-ta) Pp. 321–324

[19] Sobgayda N.A., Olshanskaya L.N. and Makarova Yu.A. 2012 Ecology of production No 3 Pp.68 - 71

[20] Taranovskaya Ye. 2015 Energy and resource saving of the environment: sb. dokl. Mezhdunar. nauch.-tekhn. konf.6 v 2 ch. (Belgorod: BGTU) Ch. II Pp. 118-122

IOP Conf. Series: Earth and Environmental Science **288** (2019) 012028 doi:10.1088/1755-1315/288/1/012028

Water purification from phenol on materials from subtropical crops' wastes

Vladimir Somin, Larisa Komarova

Barnaul, prospekt Lenina, 46

vladimir_somin@mail.ru

Abstract. The article considers the possibility of using materials on the bases of wastes of subtropical crops for water purification from phenol. The modification of banana peel, orange peel and pineapple peel by chemical and thermal methods was carried out. The static and dynamic capacity of obtained sorbents was determined; their ability to regeneration was estimated.

Introduction

Phenol is one of the most widespread pollutants that transfer into the natural water bodies with discharge of sewage waters from various enterprises. It leads to the toxic influence on organisms and to the significant change in the regime of biogenous elements and dissolved gases (oxygen, carbon dioxide). The self-purification of water bodies from phenol is rather slow, therefore it is necessary to prevent their transfer into the environment.

Sorption materials, which are in a dissolved condition, are used for the phenol removal. These sorption materials can be obtained from various wastes: wood shavings [1], wastes from processing of the buckwheat husks and sunflower husks [2-4], wastes from processing of subtropical crops in their growing regions. This approach will allow recovering wastes and receiving sorbents for water purification.

The aim of this article is to obtain sorbents for water purification from phenols from garden wastes: orange peel, coconut shell, banana and pineapple peel.

Experimental procedure

The research was carried out on phenols' simulated solutions with concentration from 10 to 2000 mg/l. The definition of phenols was conducted by means of photocolorimetric method using 4-aminoantipyrine. The method is based on the formation of phenol coloured compounds with 4-aminoantipyrine by potassium ferricyanide [K3Fe(CN)6]. This method allows determining the concentration of phenols in water to 5 mg/l; a sample must be diluted before carrying-out of an analysis of solution with stronger concentration.

Research of sorption kinetics

Initially, the kinetic characteristics of sorption on the native orange peel were researched. For this purpose, the solutions with a phenol concentration of 100 mg/l were used, in which 1 gram of peel was placed. Upon the expiry of certain time, the concentration of phenols in water was determined. The results are shown in Figure 1.

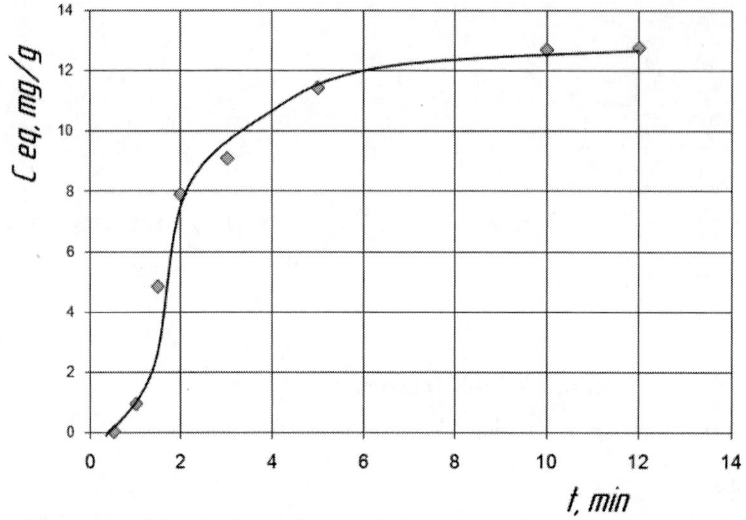

Figure 1 – Kinetic dependence of phenol sorption on orange peel

Figure 1 shows that the dynamic equilibrium in the system is established fairly quickly, at average in 10 minutes. At the same time, the maximal sorption capacity was 12,8 mg/g.

Research of sorption statics

The researched raw materials for the increase in sorption properties were modified by a sodium hydroxide solution with concentration of 500 mg/l and by pyrolysis method at a temperature of 600 °C. Sorption isotherms, shown in Figures 2-5, were obtained for all materials.

Figure 2 shows phenol sorption isotherms on the native and modified banana peel. A banana peel, modified by pyrolysis method has the maximal static capacity (17 mg/g), the maximal static capacity for the native material is 9,5 mg/g.

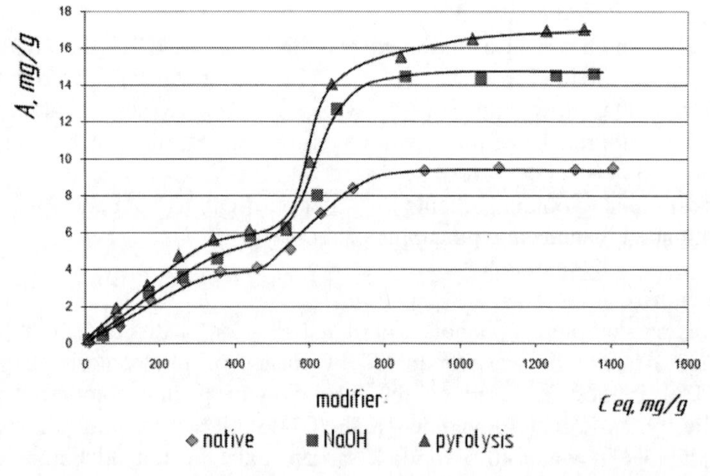

Figure 2 – Isotherms of phenol sorption by native and modified peel the banana

Sorption isotherms for orange peel are presented in Figure 3. It is stated that modification by pyrolysis increases the capacity to 16 mg/g in comparison with native peel (10 mg/g).

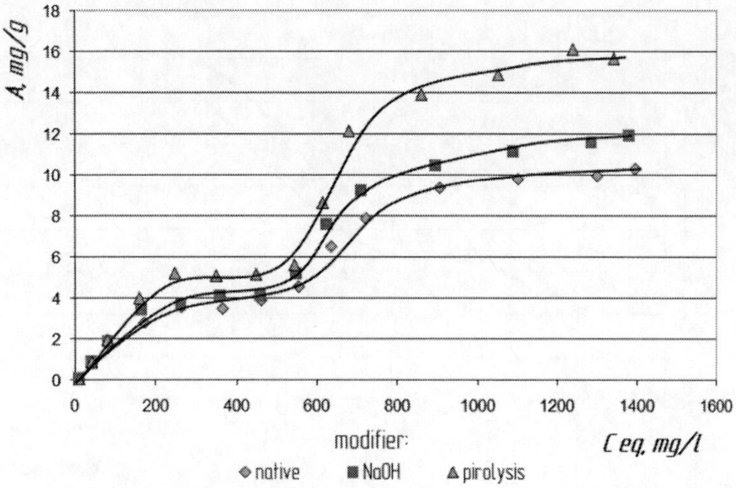

Figure 3 – Isotherms of phenol sorption by native and modified peel the orange

For pineapple peel, the similar research has shown that the peel modified by pyrolysis has the maximal statistic sorption capacity (11.1 mg/g) (Figure 4). The sorption characteristics of the material in the native form are 8.7 mg/g.

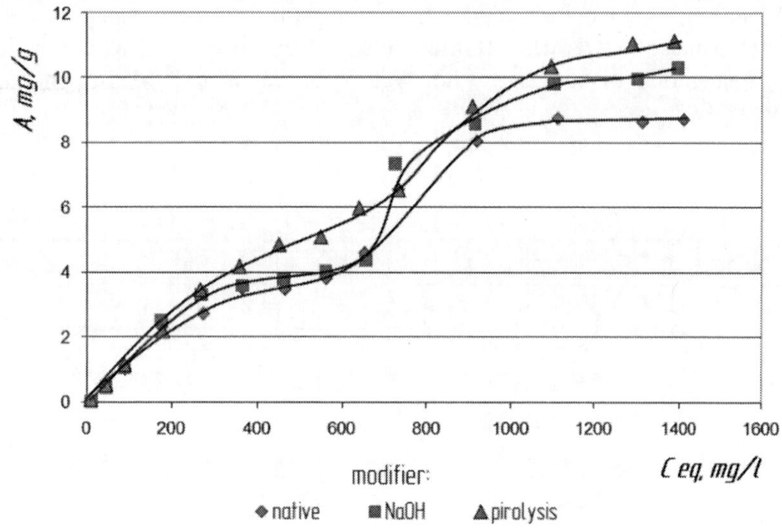

Figure 4 – Isotherms of phenol sorption by native and modified peel the pineapple

For coconut shell, the research was conducted only with native raw materials and raw materials modified by pyrolysis, since the original fraction was very finely dispersed (fine) and it was impossible to conduct pyrolysis processing. The results are shown in Figure 5. As it can be seen, modified and native coconut shell has almost the same sorption capacity – 7.6 and 7.9 mg/g respectively. This indicates that it is inexpedient to conduct modification by this method.

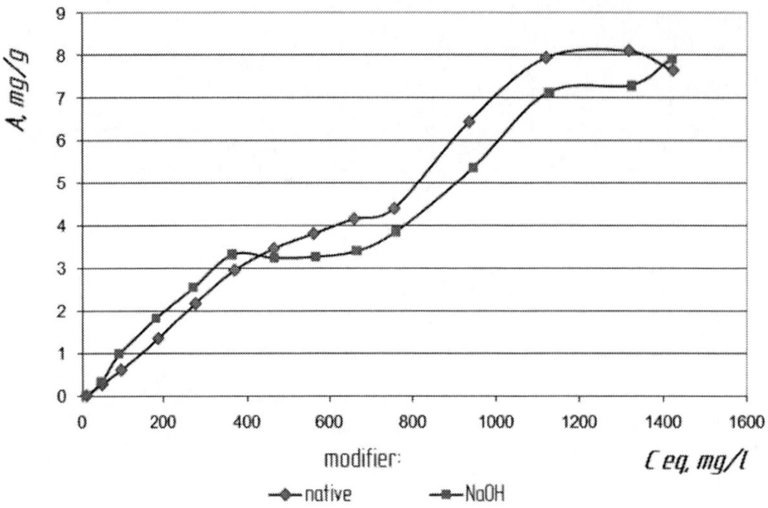

Figure 5 – Isotherms of phenol sorption by native and modified coconut shell

Research of sorption dynamics

Sorbent from pyrolysed banana showed maximal sorption capacity (17 mg/g) in static conditions. For this reason it was chosen for practical use as a sorption material for purifying water from phenol.

During the experiment, the dependence of purification efficiency on the used solution volume was obtained (Figure 6). As it can be seen, the maximum purification efficiency on a fresh material is 76%. The efficiency reduced to 22% after 10 liters of solution were passed. After purification of 18 liters of water, the regeneration of sorbent by hot water was conducted. Maximum purification efficiency decreased to 31%.

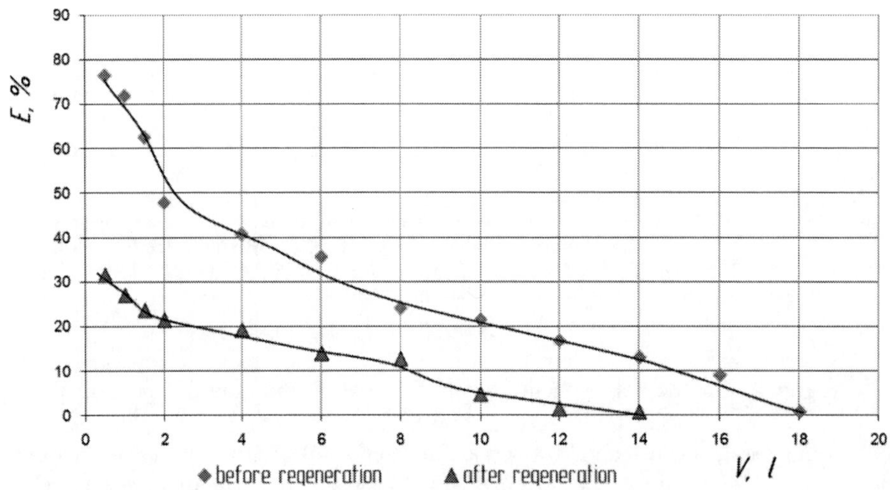

Figure 6 – Dependence of the removal efficiency of phenol from water (E,%) on the volume of the solution (V, l)

The dynamic capacity of the sorbet was calculated. It is 8 and 3 mg/g respectively for sorbent before and after regeneration. As a result, conduction of such type of regeneration is inexpedient. To sum it up, the conducted research has shown that sorbents based on proceedings of subtropical crops' wastes can be used for water purifications from petroleum products. Banana and orange peel modified by pyrolysis has presented the best sorption characteristics. They can be recommended for using in the regions of subtropical crops' growth.

References

[1] Somin V.A., Osokin V.M. (2013) *Issledovaniya po polucheniyu novyh sorbentov iz rastitel'nogo syr'ya dlya ochistki vody* [Studies on obtaining new sorbents from plant raw materials for water purification]. *Polzunovsky Vestnik*, no. 1, pp. 280-282.

[2] Somin V.A., Betc S.A., Komarova L.F. (2015) *Sorbciya fenola iz vody materialami na osnove rastitel'nogo syr'ya* [Sorption of phenol from water by materials based on plant raw materials]. *Vodoochistka*, no. 5-6, pp. 76-80.

[3] Somin V.A., Betc S.A., Komarova L.F. (2016) *Ispol'zovanie sorbentov na osnove prirodnogo syr'ya dlya ochistki fenolsoderzhashchih vod* [Use of sorbents on the basis of natural raw material for phenol-containing water purification] *Ecology and Industry of Russia*, vol. 20, no. 12, pp. 14-17.

[4] Galimova R.Z., SHajhiev I.G., Almazova G.A. (2017) *Osobennosti ochistki poverhnostnyh i stochnyh vod ot fenola s ispol'zovaniem othodov sel'skohozyajstvennogo proizvodstva* [Peculiarities of surface and sewage water purification from phenol using agricultural wastes]. Proceedings of the *Mezhdunarodnaya molodezhnaya nauchnaya konferenciya «Aktual'nye voprosy ehkologii volzhskogo bassejna»* (*Russia, Tolyatti, March 15□17, 2017*) (eds. Senatora S.A., Muhortovoj O.V., Saksonova S.V.), Tolyatti: Kassandra, pp. 87-90.

Methods of assessing the health impacts of chemicals ingested with drinking water

N V Stepanova and S F Fomina

Kazan Federal University, Institute of Fundamental Medicine and Biology, K. Marx street, 76, 420008, Kazan, Russia

stepmed@mail.ru

Abstract. Assessment of non-carcinogenic and carcinogenic risks of per oral intake of chemicals with drinking water for children aged 3-6 year sold living in 4 districts (zones) of the city of Kazan was carried out. A quantitative characteristic of risk was performed with application of regional exposure factors (REF) at the median level (Me) and the 95-thpercentile (95P). Analysis of the risk data showed that only 5 substances (magnesium, nitrates (in NO_3), fluorides, oil products, and chloroform) made from 62.5% to 89.8% of contribution to total risk value in all zones. General toxic effects for critical organs and systems were due to oil products (from 29 to 54.7%) in the 2nd and the 4th zones, chloroform (from 10 to 30.6%) and nitrates (from 12.8 to 35.9%) in all zones, magnesium (up to 11.2%) in the 3rd zone, and fluorides (from 13.7 to 14.3%) in the 1st and the 3rd zones. The unacceptable risk level (HI_{Me} = 11.8 and 9.9; $HI_{95\text{-th Perc}}$ = 14.8 and 12.5) was identified in the 2nd and the 4th zones with a mixed type of water in the utility and drinking water supply. The ingestion of chemicals with drinking water in different zones of the city of Kazan indicates an alarming and high level of non-carcinogenic and carcinogenic risks for the child population health.
Keywords: health risk assessment, exposure factors, drinking water, children.

Introduction

Monitoring of water ecosystems and water quality comprises an overall ecological assessment of the state and changes of water ecosystems for a variety of major aspects including ecological and sanitary characteristics of the water quality, bioproductivity, the flora and fauna protection, and valuable hydrobiocoenoses, ecological-toxicological and radio-ecological situation [1].The major trends to increase the effectiveness of using the resources of the multi-purpose water supply reservoirs include optimization of hydrological, hydrochemical and hydrobiological, the provision of water quality in accordance with regulatory requirements of the water users and consumers [2, 3, 4].

The analysis of modern stage of the drinking water quality should take into account and reflect national peculiarities of the drinking water supply within the state sand the region. According to the 4-th Guidelines on the drinking water quality of the World Health Organization (WHO) an approach based on risk assessment should be used for justification of management decision son provision of the drinking water safety [5]. The process must include the assessment of both the health risk, and risk management, and encompass all stages of the water supply system: from the water source, the distribution network, to the place of its consumption by humans. The quality of drinking waters depends largely on the quality of natural surface water source (the Volga River), which deteriorated in recent years in the region of the city of Kazan: the water is characterized as "dirty". Water supply to the citizens of Kazan is carried out from the "Volzhsky" surface intake, the underground water intakes and the artesian boreholes. The "Volzhsky" water intake provides more than 90% of the city population including the Kirovsky (the 1stzone) and Vakhitovsky districts (the 3rdzone) with drinking water. The population of the Sovetsky district (the 2ndzone) uses drinking water of a mixed character ("Volzhsky" water intake and underground water sources Aki, Azinoand Solidarnost). The Privolzhsky(the Volga) district of the city (the 4thzone) is provided with mixed water from the water in takes "Mirny", "Tankodrom", and "Volzhsky".

Materials and methods

The assessment of non-carcinogenic risk on ingestion of chemicals with drinking water was carried out for children aged 3-6 year sold living in 4 districts (zones) of the city of Kazan. The selection of research zones was made according to the type of the water supply source in these districts. The assessment of non-carcinogenic risk was per formed according to the values of the upper limit of the 95% CI of the results of the research carried out on the basis of an accredited laboratory of the Federal State-Funded Healthcare Institution "The Center of Hygiene and Epidemiology in the Republic of Tatarstan". The main document in the RF, which regulates the assessment of the population health risk, is "The Guidelines on Human Health Risk Assessment from Environmental Chemicals" P 2.1.10.1920–04 [6]. The study of the contaminant toxicity was carried on the basis of chronic daily substance intake (per oral route). The characteristic of general toxic effects was performed based on the hazard quotients (HQ) of certain substance sand total hazard indices (HI) for substances with unidirectional mechanism of action. The calculation of average daily dose (ADD) of per oral intake of chemicals with drinking water (mg·kg^{-1}·day^{-1}) was carried out according to formula 1 [6].

$$\text{ADD} = \frac{(Cw \times V \times EF \times ED)}{BW \times AT \times 365}, \qquad (1)$$

- ADD – average daily dose of ingestion with drinking water (mg·kg^{-1}·day^{-1}).
- CW – substance concentration in water, (mg·L^{-1}).
- V – amount of water intake (L·day^{-1}).
- EF – exposure frequency, (days·year^{-1}).
- ED – exposure duration (years).
- BW – body weight (kg).
- AT – averaging exposure time, years (for non-carcinogens, AT = ED × 365 days).

Non-carcinogenic risk was assessed according to hazard quotient (HQ) for each substance with application of the regional exposure factor sat the median level (Me) –usual exposure range, and the 95-thpercentile (P_{95}) –maximum reasonable exposure (formula 2).

$$HQ = \frac{ADD}{RfD}, \qquad (2)$$

- RfD–reference dose (safe exposure level) for each of the substances (μg·kg^{-1}·day^{-1}) [6]. Total hazard quotients (HI) were calculated according to formula 3.

$$THI = \Sigma HQ, \qquad (3)$$

To assess carcinogenic risk, the lifetime average daily doses (LADD), the carcinogenic potential factors (SFo), and age-dependent adjustment factors (ADAF) were used [7]. Regional exposure factors (REF) at the median level (Me) and the level of the 95-th percentile (95P) were determined based on the results of the questionnaire survey of parents, grandmothers and babysitters [8]. Non-carcinogenic and carcinogenic risks were assessed for the regional exposure factor sat the median level (Me), (usual

IOP Conf. Series: Earth and Environmental Science **288** (2019) 012029 doi:10.1088/1755-1315/288/1/012029

exposure range), and the level of the 95-th percentile (P95, maximum reasonable exposure). The THI values from 1.1 to 3.0 were taken for a tolerable level of non-carcinogenic effects; the range of THI values from 3 to 6 was regarded as an alarming risk level and THI values higher than 6 – as a high one. The TCR value equal to $1.0x10^{-5}$wasa tolerable risk level for carcinogenic effects [6]. The reference values for calculating the impact (exposure) assessment and non-carcinogenic risk assessment are given in Table 1.

Table 1 Concentration of chemical compounds in drinking water in the city of Kazan in the zones under study(mg/l)

Chemicals	CAS	Limit of detection	MPC, mg/l	RFD, mg/kg	Upper bound of 95% CI			
					1st zone	2nd zone	3rd zone	4th zone
Aluminum	7429-90-5	0.05	0.2	1	0.373	0.42	0.4	0.58
Barium	7440-39-3	0.01	4	0. 2	0.024	0.034	0.045	0.024
Iron	7439-89-6	0.1	0,3	0.3	0.8	1.71	1.9	0.7
Magnesium	7439-95-4	1	50	11	24.3	63.2	85.05	47.4
Nitrates (in NO_3)	14797-55-8	0.2	45	1.6	9.8	24.93	26	58.79
Nitrites (in NO_2)	14797-65-0	0.003	3.3	0.1	0.05	0.2	0.4	0.2
Sulphates	7440-43-9	0.0003	0.001	0.0005	0.0007	0.0006	0.0006	-
Cadmium	7439-96-5	0.01	0.1	0.14	-	0.131	-	002
Manganese	7439-92-1	0.05	0.01	0.024	0.007	0.0036	0.0076	0.004
Lead		0.01	7	0.6	1.01	0.64	0.92	0.68
Strontium	7440-50-8	0.02	1	0.019	0.021	0.017	0.015	0.028
Copper	7440-66-6	0.2	1	0.04	0.031	0.062	0.09	0.143
Zinc	16984-48-8	200	1.5	0.06	0.296	0.471	0.57	0.384
Oil products (total)		0.005	0.1	0.03	0.0172	1.993	0.1	1.01
Chloroform	67-66-3	0.001	0.1	0.01	0.106	0.119	0.147	0.115

Results and discussion

The results of non-carcinogenic risk assessment on peroral intake of chemicals with drinking water showed that the value of total risk indicated the alarming and unacceptable risk levels in all zones (Table 2).

Table 2. Hazard quotients (HQ) for substances with unidirectional mechanism of action

Substances	Hazard quotient (HQ)							
	1st zone		2nd zone		3rd zone		4th zone	
	Me	95-th Perc	Me	95-th Perc	Me	95-th Perc	Me	95-th Perc
Magnesium	0.21	027	0.56	0.70	0.75	0.94	0.42	0.53
Nitrates (in NO3)	0.59	0.75	1.51	1.90	1.58	1.98	3.57	4.48
Fluorides	0.48	0.60	0.76	0.96	0.92	1.16	0.62	0.78
Oil products (total)	0.06	0.07	6.45	8.10	0.32	0.41	3.27	4.11
Chloroform	1.03	1.29	1.16	1.45	1.43	1.79	1.12	1.40
HI	4.24	5.33	11.79	14.8	6.69	8.4	10.93	13.7

Analysis of the risk data showed that only 5 substances (magnesium, nitrates (in NO3), fluorides, oil products and chloroform made from 62.5% to 89.8% of contribution to total risk value in all zones. The hazard quotient for the rest of substances made less than 0.5.

The unacceptable level of total risk (HI_{Me}= 11.8 and 10.93; $HI_{95\text{-thPerc}}$ = 14.8 and 13.7) was identified in the 2nd and the 4th zones at the median level and the level of the 95-th percentile (Perc) having a mixed source of the drinking water supply. The assessment of values of the total hazard indices calculated on the basis of the regional exposure factors (the median concentration sand the 95-

thpercentile (Perc) for the child population of the city of Kazan indicated (excluding the 1st zone) the exceedance of the upper limit of the reference level (3.0) for three systems: the blood, the kidneys, and the cardiovascular system (Table 3).

Table 3. Total hazard indices (HI) for substances with unidirectional mechanism of action

HI for certain systems	Calculation of HI with application of regional exposure factors (Me)				Calculation of HI with application of regional exposure factors (95% perc)			
	1st zone	2nd zone	3rd zone	4th zone	1st zone	2nd zone	3rd zone	4th zone
HI the blood	2.14	3.63	4.25	5.28	2.68	4.56	5.34	6.63
HI the kidneys	1.25	7.77	1.93	4.42	1.58	9.76	2.42	5.55
HI the CVS	0.63	1.56	1.64	3.60	0.79	1.96	2.06	4.52
THI total	4.24	11.79	6.69	10.93	5.33	14.81	8.40	13.73

The major contribution to development of general toxic effects on the critical organs and systems given in Table 3, is made by oil products (from 29 to 54.7%) in the 2nd and the 4th zones, chloroform (from 10 to 30.6%) and nitrates (from 12.8 to 35.9%) in all zones, magnesium (up to 11.2%) in the 3rd zone, and fluorides (from 13.7 to 14.3%). In the 1st andthe 3rd zones. An unacceptable risk level (HI_{Me}= 11.8 and 9.9; $HI_{95\text{-thPerc}}$ = 14.8 and 12.5) was identified in the 2nd andthe 4th zones with mixed type of water in the utility and drinking water supply. The assessment of carcinogenic risk was carried out for 3 substances belonging, in accordance with IARC classification, to group1– the carcinogens known to man (cadmium), togroup2A- possible carcinogens (lead) and to group 2B – possible carcinogens (chloroform), which are found in drinking water (Table 4) [7, 9].

Table 4. Total carcinogenic risk (TCR) for the child population health on exposure to carcinogens in drinking water

Substance	SFo	CR, (Me)				CR, (95% perc)			
		1st zone	2nd zone	3rd zone	4th zone	1st zone	2nd zone	3rd zone	4th zone
Cd	0.38	6.7E-06	5.69E-06	5.69E-06	0	8.34E-06	7.15E-06	7.15E-06	0
Pb	0.047	8.2E-06	4.2E-06	8.9E-06	4.7E-06	1.03E-05	5.3E-06	1.2E-05	5.9E-06
($CHCl_3$)	0.0061	5.4E-06	6.04E-06	7.46E-06	5.8E-06	6.76E-06	7.59E-06	9.37E-06	7.33E-06
TCR		2.0E-05	1.6E-05	2.21E-05	1.1E-05	2.54E-05	2.0E-05	2.78E-05	1.3E-05

The highest carcinogenic risk level was registered in the 1st and the 3rd zones at the level of the 95 Perc. The value of TCR on exposure to carcinogens in water was tolerable only in the 4th zone. Analysis of risk levels with application of local factor sand age differences on exposure to chemicals ingested per orally with drinking water showed that the values of total carcinogenic risk (TCR) on exposure to three drinking water carcinogens – cadmium, lead and chloroform calculated with the account of ADAF exceeded the recommended risk levels in the 1-3 zones.

Conclusions

In our opinion, the ingestion of oil products is associated with pollution of the surface water supply source (the Volga River) and requires the detailed study. The researchers of the Kazan Federal University showed that the pollution of waters with suspended solid sand oil products was due to effluents from industrial and public utility enterprises, and the surface runoff from the urbanized territory [3, 4, 10]. A high proportion of unsatisfactory samples of iron, nitrates, fluorides, oil products and chloroform is responsible for development of non-carcinogenic effects on the kidneys, the cardiovascular system and the blood system and unacceptable carcinogenic risk for the child population health.

Both in Russia, and in the majority of cases abroad, the projects of risk assessment are, as a rule, of local character, and set a task of predicting the effects of the habitat pollution, which is unfavorable for the health, within the boundaries of certain cities/towns or districts [7, 10-19]. Such projects allow solving the important problems of risk management at the local level and under the conditions of the city of Kazan.

Acknowledgments

This work was funded by the subsidy allocated to Kazan Federal University for the state assignment in the sphere of scientific activities 19.9777.2017/8.9

Reference

[1] Kalaida M L 2003 Ecological assessment of the Kuibyshev reservoir under the conditions of anthropogenic impac (Kazan: Kazan State Power Engineering University) p 135

[2] Shagidullin R R, Latypova V Z, Nikitin O V and Yakovleva O G 2011 The evaluation of technogenic load of enterprises wastewaters on the Kuibyshev reservoir. *Georesources* **2(38)** 24-27

[3] Minakova E A, Mukhametshin F F and Shlychkov A P 2015 Water Consumption and Water Discharge in the Republic of Tatarstan in the modern period. *Journal of Ecology andIndustrial Safety* **1-2 (63-64)** 19-21

[4] Musin R Kh, Kurlyanov N A, Faizrakhmanova Z G and Musina R Z 2014 On the quality of waters used in the drinking water supply systems of the city of Kazan. *Georesources* **3 (58)** 9- 16.

[5] 2011 Guidelines for Drinking-Water Quality 4th edition *Guidelines World Health Organization (WHO)* (Gutenberg: WHO) p 564

[6] Rakhmanin J A et al 2004 *Guidelines for health risk assessment for the population on exposure to chemical substances polluting the environment (P 2.1.10.1920-04).* (Moscow: Federal Center of the State Committee for Sanitary and Epidemiological Control) p 143

[7] Stepanova N V, Valeeva E R and Fomina S F 2015 *Hygiene and sanitation* **5** 56 – 61

[8] Stepanova N V, Valeeva E R, Fomina S F and Ziyatdinova A I 2016 Assessment of non-carcinogenic risk for the child population health on consumption of drinking water. *Hygiene and Sanitation.* **95 (11)** 1079-83

[9] U S EPA 2011 *Exposure Factors Handbook (Final Report)* (Washington: U.S. Environmental Protection Agency, DC) EPA/600/R-09/052F

[10] 2008 WHO/SDE/WSH/05.08/123 Petroleum Products in Drinking-water. Background document for development of WHO *Guidelines for Drinking-water Quality* (Switzerland: WHO) p 20

[11] AbRazak N H, Praveena S M, Aris A Z and Hashim Z 2015 *J. Epidemiol. Glob. Health* **5**.10.1016/j.jegh.2015.04.003

[12] Andra S S and Makris K C 2014 *Water Health* **12** 81 – 93

[13] Chowdhury S and Hall K 2010 *Environmental International* **36** 453 – 60

[14] Gopal K, Tripathy S S, Bersillon J L and Dubey S P 2007 *J. Hazard Mater* **140** 1 – 6

[15] Whitaker H, Nieuwenhuusen M J, Best N, Fawell J, Gowers A and Elliot P 2003 *Journal of Exposure Analysis and Environmental Epidemiology***13** 17 – 23

[16] Villanueva C M, Kogevinas M, Cordier S, Templeton M R, Vermeulen R Nuckols J R, Nieumenhuijsen M J and Levallois P 2014 *Environmental Health Perspectives* **122** 10.1289/ehp.1206229

[17] Margot T B and Foos B P 2009 *Hum. Ecol. Risk Assess* **15** 923 – 47

[18] Karim Z, Mumtaz M and Kamal T 2011 *J of Chemical Society of Pakistan* **33** 215 – 9

[19] Basu M, Gupta S K, Singh G and Mukhopadhyay U 2011 *Environ Monit Assess* **178** 121– 34

IOP Conf. Series: Earth and Environmental Science **288** (2019) 012030 doi:10.1088/1755-1315/288/1/012030

Purification of natural water and wastewater from petroleum and petroleum products by sorption materials on a basis of industrial waste

E A Tatarintseva[1], I V Dolbnya[1], E A Bukharova[1], L N Olshanskaya[1], N A Politaeva[2]

[1]Yuri Gagarin State Technical University of Saratov, the Russian Federation

[2]Peter the Great St. Petersburg Polytechnic University

tatarinceva-elen@mail.ru

Abstract. Results of experimental research of sorption purification of wastewater and surface water from petroleum and petroleum products by sorption materials on a basis of industrial waste have been given. Polyethylene terephthalate, ferritized galvanic sludge (FGS), oxidized and thermally expandable graphite have been chosen as the components of the sorption materials (SM). The sorption material on the basis of FGS has magnetic properties and it allows to extract the sorbent from water with the help of a magnetic separation. Physical and chemical properties of SM (buoyancy, water adsorption, petroleum capacity), kinetic regularities of the sorption of petroleum products by the sorption materials have been studied. Specific surface area, porosity, and particles dispersion have been determined. A possibility of using sorbents to collect petroleum products from the water surface has been shown.
Keywords: polyethylene terephthalate, oxidized and thermally expanded graphite, ferritized galvanic sludge, waste utilization, water purification from petroleum products, sorption, sorption materials.

Introduction

Pollution by petroleum and petroleum products is one of the most widespread, dangerous ecological contaminants in the world, which today threatens to water resources. Petroleum leakage has a hugely detrimental impact on marine and freshwater ecology. Petroleum spots slow the flow of oxygen into the water, reduce the passage of sunlight, weakening the process of photosynthesis in plants and plankton. The vital activity is changing for many species of animals and plants with petroleum pollution. The main reasons for petroleum pollution of water are deterioration of petroleum pipelines, safety violation, illegal tie-ins and accidents. In the world, the biggest disasters happen as a result of accidents in the water areas. About 1,5 millions of cubic meters of petroleum and petroleum products get into the water every year. The amount of environmental and economic damage from spills of petroleum and petroleum products depends on many factors, such as a type of spilled petroleum products, a condition of the affected ecosystem, meteorological conditions, marine and river flows, seasons, the condition of local fisheries and tourism etc.

Also, the most acute problem facing humanity is the problem of the accumulation of industrial and domestic waste, which are stored in landfills. Modern scientific and technological progress is associated with a constant acceleration in the rates of consumption of water, land and material

IOP Conf. Series: Earth and Environmental Science **288** (2019) 012030 doi:10.1088/1755-1315/288/1/012030

resources. Annually the hundreds of millions of tons of various products of anthropogenic human activity are emitted into the environment. The most dangerous among them are galvanic sludge (GS), polymer waste, petroleum products (PP). The environment pollution by petroleum and petroleum products has the negative impact on the quantitative and qualitative indicators of water and soils. Therefore, the use of industrial waste in the creation of new sorption materials, which have low cost and high efficiency, in cleaning wastewater and surface waters from PP is an urgent task and has scientific and practical value.

The scientific significance of the solution of this problem is substantiated and experimentally confirmed by the possibility of obtaining sorption materials for purification of sewage and surface waters from petroleum products with the use of polyethylene terephthalate, thermally expanded and oxidized graphite, ferritized galvanic sludge, not previously used as composite materials. For the first time, the study of the regularities of adsorption of petroleum and PP has been carried out, possible ways of increasing the sorption capacity and efficiency of water purification by modifying the structure of sorbents, their chemical composition, etc have been proposed. Methods for obtaining sorbents based on industrial wastes for the recovery of pollutants have been developed. Purification of water can include mechanical and physicochemical methods. The choice of the method and technology of spillage of petroleum and petroleum products determines the efficiency of purification. Among a wide variety of methods of water purification from PP, the sorption method is the most appropriate. With the help of SM it is possible to achieve cleaning efficiency up to 80-99 %. An important factor is the possibility of recovery of petroleum products and multiple use of sorbents.

In connection with this, the research tasks were: the development of compositions and technologies for the production of sorption materials based on industrial wastes; the study and assessment of mechanical, physical and chemical and sorption properties of the sorbents obtained, the sorption purification of model solutions and wastewater from industrial products from petroleum products using developed sorption materials in a static mode, the study of the possibility of removing petroleum products from the water surface.

The results of experimental research

It is known [1] that finely dispersed polyolefin powders are used in the liquidation of petroleum spills and petroleum products. They are effective sorbents of these pollutants.

Therefore, the task of creating a fine-dispersed porous polymer sorption material from polyethylene terephthalate (PET) waste (packaging, bottles, tare, etc.) was solved. The method of preparation is based on the phase-inverting transition of the polymer from the liquid state to the solid state [2, 3]. The solvent mixture of benzyl alcohol and dibutyl phthalate plasticizer was heated, then ground PET was added and mixed until complete dissolution of PET to form a homogeneous mixture. Cooling the mixture to room temperature results in the precipitation of a solid product (call it PSM-1 - polymeric sorption material), which was combined with acetone in a ratio 1:2. The pores are formed as a result of leaching molecules of benzyl alcohol and dibutyl phthalate from PSM-1 with acetone. This method makes it possible to obtain a finely dispersed material with mesoporous structure.

Figure 1. Tablets sorption materials TS-1 and TS-2. **Figure 2.** Composite magnetic sorption material (CSM).

The obtained polymeric sorption material PSM-1 can be used for sorption purification of waters from petroleum products, as well as a binder in the production of tablet sorption materials based on thermally expanded graphite (TS-1) and oxidized graphite (TS-2), figure 1 [4, 5], as well as a composite sorption material with magnetic properties based on ferritized galvanic sludge (CSM), figure 2. CSM has magnetic properties, the residual magnetic induction (B_r) is equal to 0,05 T. Ferritized galvanic sludge (FGS) presents ferrites of different heavy metals, which included in the composition of the sludge. Ferrites have magnetic properties, and they determine their further useful applying in various spheres [6, 7, 8, 9, 10, 11]. The use of ferritized galvanic sludge as a magnetic constituent makes it possible to utilize industrial waste and to derive on their basis sorbents that can be extracted from the aqueous medium upon the completion of the sorption process by means of magnetic separation without additional capital and energy costs.

The main indicators of sorption materials used to collect petroleum products from the surface of reservoirs are petroleum capacity, buoyancy and water adsorption [12]. The properties of sorption materials are presented in table 1.

Table 1. Properties of sorbents.

	PSM-1	TS-1	TS-2	CSM
Water adsorption, g/g	0,5	0,2	0,2	0,2
Buoyancy (96 hours), %	80	100	100	100
Oil capacity, g/g	3,0	3,5	7,5	4,0
Petroleum capacity, g/g	4,5	9,0	15,0	3,5

The buoyancy of PSM-1 is determined by the ratio of open and closed pores, which are formed during the course of the formation of the structure. The high buoyancy of TS-1, TS-2 and CSM is related not only to the high hydrophobicity of the surface of these sorption materials, but also to their structure. Depending on the weather conditions, when localizing the petroleum slick, the time reserve should not exceed 24-72 hours from the time of the spill without the danger of causing significant harm to the environment. Thus, the use of developed sorption materials, which retain buoyancy for a long time, allows carrying out measures to collect petroleum from the water surface.

The economic efficiency of sorption technology, including the specific consumption of sorbent, is largely determined by the morphology of the surface and the porous structure of the sorption material [13]. Microstructural studies of sorption materials have shown the presence of an inhomogeneous surface of the sorbent with the presence of a large number of pores and depressions of various shapes and sizes, which are the most important factors ensuring a strong retention of the sorbate on the surface and in the volume of the sorbent.

Such characteristics of sorbents as specific surface area and porosity of the material play an important role in the sorption purification of water. Porosity is described by such characteristics as the determination of the total pores volume and pore size distribution. It was found that the structure of TS-2 has the largest total pore volume compared to PSM-1 and TS-1 (table 2).

Analysis of the distribution of pores by size showed that mesopores prevail in all sorption materials, which are convenient for layerwise penetration of adsorbed molecules of petroleum products.

CSM does not have porosity and the specific surface area is provided by the dispersity of the particles. The distribution of the particles of the sorbent CSM by their sizes has the following values: 10 μm – 2 %, from 10 to 100 μm – 15,22 %, from 100 to 1000 μm – 81,02 %.

Sorption properties of the obtained materials in relation to petroleum and petroleum products were studied. It has been experimentally established that the highest rate of sorption of petroleum products in both model systems and in sewage waters of the industrial enterprise «Petroleum refining plant of Saratov» (the Russian Federation) is observed in the first 10 minutes, then the sorption rate decreases. An increase in the sorption time of more than 30 minutes is impractical.

Table 2. Structural characteristics of sorption materials.

Sorption material	Pores radius, nm	Percentage ratio of total pore volume, %	Specific surface area, m^2/g	Total pore volume, cm^3/g
	till 2	6,38		
PSM-1	2 till 50	89,01	20,4	0,046
	over 50	4,61		
	till 2	8,41		
TS-1	2 till 50	67,62	33,9	0,096
	over 50	23,97		
	till 2	8,83		
TS-2	2 till 50	69,61	42,3	0,126
	over 50	21,56		
CSM	-	-	0,12	-

Figure 3 shows the kinetic curves of sorption of petroleum products for TS-1 and TS-2. The weight of the sorbent introduced was 0,5 g per 100 ml of the model system for TS-1 and 0,3 g per 100 ml of the system for TS-2 at t = 25 °C. It has been experimentally established that the greatest recovery of PP for TS-1 is observed in the first 30 minutes and for TS-2 in 20 minutes. A shorter time to achieve sorption equilibrium is characteristic of TS-2, which is explained by a more porous surface, which facilitates the penetration of petroleum products into the depth of the sorption material [14].

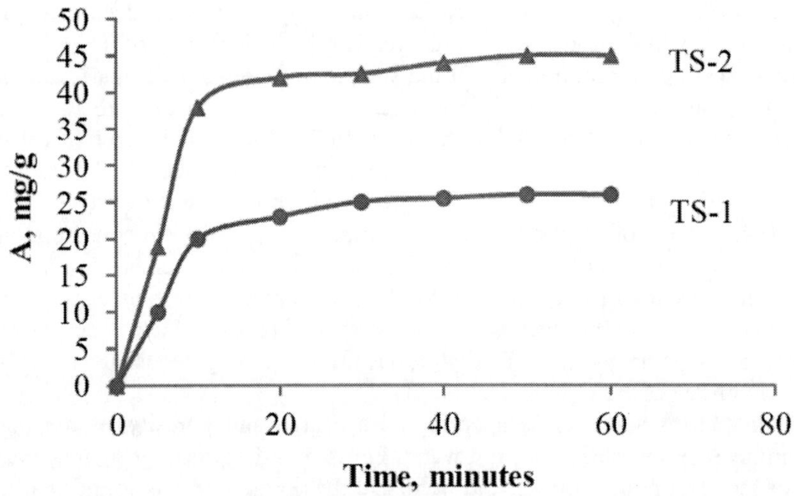

Figure 3. Kinetic curves of sorption of industrial oil I-20A by the sorption materials TS-1 and TS-2 ($C_{initial}$ = 121,7 mg/l), A – sorption capacity, mg/g, $C_{initial}$ – initial concentration, mg/l.

The analysis of the sorption curves showed that the maximum sorption of industrial oil by the sorption materials PSM-1, TS-1 and TS-2 happens in the first minutes, after which the sorbents are able to retain the sorbate, and it can be due not only to good hydrophobicity and oleophilicity of the obtained sorbents but to their structure [12].

It is known that the size of molecules of petroleum products fluctuates within 5-10 nm, and PSM-1 is mesoporous (the radius of prevailing pores is 2-50 nm) [15]. In addition, when solid oleophilic particles of polymer sorption material PSM-1 come into contact with a thick petroleum film around the sorbent particles, micelles form interacting with each other to form a specific network structure

[12]. This leads to a marked increase in the viscosity of the petroleum-water suspension, and at high concentrations of PSM-1 in water, dense conglomerates of petroleum contamination are observed.

When using TS-1 and TS-2, the placement of petroleum products is also possible in the voids of the sorption material due to capillary forces and oleophilicity. With a large thickness of the petroleum film, an effective introduction of the petroleum pollutant into the porous zone of tableted sorbents occurs. The use of PSM-1 is recommended for the liquidation of petroleum spill in the water bodies as filler for mats or sleeves.

The most promising direction of using magnetic SM is the collection of spilled petroleum products from the surface of water with subsequent extraction by magnetic separation. A series of experiments with the use of CSM for the sorption of petroleum and petroleum products with different film thickness (1-5 mm) was carried out. The contact of the petroleum film with the surface of the sorbents is the largest and prevents contact of SM with water. It is established that the maximum dynamic sorption capacity of the CSM is achieved at a film thickness of ≈ 3 mm and for CSM it is equals to g/g: 0,5 (kerosene); 1,1 (petroleum) and 1,2 (industrial oil). The efficiency of the purification of the water surface when using the sorption materials is determined by the amount of the sorbent used and it can reach values of 99,9 %.

The effective intercalation of petroleum and petroleum products into the porous granular sorbent zone occurs with a greater thickness. Moreover, the formation of micelles around the sorbent particles occurs upon contact of the solid oleophilic particles of CSM with thick petroleum film. These micelles interact with each other with the formation the specific net structure, which leads to a noticeable increase in the viscosity of the petroleum-water suspension, which can be easily removed by a magnet. Table 3 presents the comparative characteristics of the developed SM with those available on the market.

Table 3. Comparative characteristics of sorbents.

Brand of the sorbent	Material	Appearance of the sorbent	Cost per 1 kg, RUB	Petroleum capacity, kg/kg	Water adsorption, kg/kg
PSM-1	PET	powder	238	1,7-4,5	0,5
TS-2	Graphite	tablet	296	11-15	0,2
CSM	FGS and PSM-1	granules	153	3,0-4,0	0,2
Turbo-Jet	Peat	coarse-grained placer	370	2-3,6	2,0
OPUB	Polypropylene	granules	460	1,6-5,0	0,8
Unipolymer	Carbamide-formaldehyde resin	crumb	420	20-40	4,6-10,0
STRG	Graphite	powder	560-700	20-40	0,2

The adsorbents have the low cost, high sorption capacity, high mechanical strength and they are easily regenerated. The absorbed substances are extracted by the method of centrifugation or vacuum filters after the completion of the sorption process. It allows to reuse petroleum products in the industry as well as to regenerate sorbents with the possibility of their reuse. After the sorption/desorption ability (at least 5 cycles of regeneration), the materials are disposed of utilization by pyrolysis with obtaining the thermal energy or as resinous additives in asphalt mixtures at the production of paving.

Conclusion

The conducted research has shown the perspective of the application of the developed sorption materials on the basis of industrial waste for the liquidation of the petroleum products spills from the water surface. It is established that the SM data possess mechanical strength, good sorption capacity for petroleum products, hydrophobicity, buoyancy, and also the possibility of their regeneration and reuse.

References

[1] Samoylov N A, Khlestky R N, Shemetov A V et al 2001 *Sorption method of liquidation of emergency petroleum spills and petroleum products* (Moscow: Chemistry) p 150

[2] Mizerovsky L N and Pochivalov K V 2007 On the applicability of the phase rule to amorphous-crystalline polymer-liquid systems *Proceedings of universities Chemistry and chemical technology* 50 issue 3 pp 72–78

[3] Konyushenko V P et al Polyester composition for protective coating *Patent of the Russian Federation* no. 2381243 2010

[4] Bukharova E A, Tatarintseva E A and Zabedkov S L 2014 *Perspectives of the use of thermally expanded graphite compounds obtained by electrochemical synthesis in the production of sorbents for wastewater treatment* Scientific perspectives of the XXI century Achievements and prospects of the new century: materials of III International Scientific-practical Conference Part 1 Engineering (Novosibirsk: International Scientific Institute «Educatio») pp 17–21

[5] Bukharova E A and Tatarintseva E A 2014 *Application of composite sorbent on the basis of thermally expanded graphite in environmental technologies* Ecology and rational nature management of agro-industrial regions II International Youth Scientific Conference Part 2 (Belgorod: Publishing House of BSTU) pp 28–32

[6] Pritosiwi Gumelar 2012 *Removal of metal ions from synthetic and galvanic wastewater by their incorporation into ferrites:* Doktor-Ingenieurin genehmigte Dissertation (Jakarta) p 194

[7] Heuss-Aßbichler S, Huber A L and John M 2016 *Recovery of heavy metals from industrial wastewater – is it worth it?* Fifth International Conference on Industrial & Hazardous Waste Management: Proceedings CRETE 2016 (Chania – Crete – Greece, 27–30 September 2016) p 10

[8] Heuss-Aßbichler S, John M and Huber A L 2016 A new procedure for recovering heavy metals in industrial wastewater *WIT Transactions on Ecology and The Environment* Vol 202 pp 85–96

[9] Song X et al 2010 Removal of heavy metals by the ferrite process *Conference on Environmental Pollution and Public Health* pp 1030–32

[10] Harikishore Kumar Redd D and Yeoung-Sang Yun 2016 Spinel ferrite magnetic adsorbents: Alternative future materials for water purification? *Coordination Chemistry Reviews* Vol 315 pp 90–111

[11] Dolbnya I V, Tatarintseva E A, Kozmich C V and Ol'shanskaya L N 2016 Ferritisation as a method of electroplating sludge deactivation *East European Scientific Journal* 8 pp 109-113

[12] Veprikova E V, Tereshchenko E A, Chesnokov N V et al 2010 Features of water purification from petroleum products using petroleum sorbents, filtering materials and activated carbons *Journal of Siberian Federal University Chemistry* 3 pp 283–304

[13] Koganovskiy A M, Klimenko N A, Levchenko T M et al 1990 *Adsorption of organic substances from water* (Leningrad: Chemistry) p 256

[14] Bukharova E A, Tatarintseva E A, Serebryakov A V, Nagar Y N 2015 Obtaining polymeric compositional sorbents on a basis of thermally expanded and oxidized graphite for the purification of water objects from petroleum products *Water Purification Water treatment Water supply* 4 (88) pp 20–25

[15] Krasnov K S 2001 *Physical chemistry* (Moscow: High School) p 512

IOP Conf. Series: Earth and Environmental Science **288** (2019) 012031 doi:10.1088/1755-1315/288/1/012031

System of special water purification of spent fuel pool of Beloyarskaya Nuclear Power Plant unit BN-800

A I Valtseva and N V Valtsev

Federal State Autonomous Educational Institution of Higher Education «Ural Federal University named after the first President of Russia B.N. Yeltsin», 620002, Yekaterinburg, 19 Mira Street, Russia

Alex-Liga@yandex.ru

Abstract. The problem of handling of nuclear power plants radioactive waste has a significant impact on the development of the nuclear industry. An effective strategy for radioactive nuclear waste should include all of the components and processing steps since waste formation to final disposal, taking into account emergence of new technologies, changes in the regulatory authority requirements, the duration of the time intervals separating the initial and the final stage of the treatment process [1]. After unloading from the reactor core spent fuel assemblies are placed in spent fuel pools for removing of residual heat and radioactivity decrease. After one year of conditioning radioactivity is reduced by about 10 times, after 5 years radioactivity is reduced by 35 times. After 3-5 years the spent fuel assemblies can be transported to a centralized storage facility or sent for processing. However, the maintaining of necessary conditions for temporary storage is an important task that is carried out by special water treatment systems of the spent fuel pool [2]. This article considers the water purification system of the spent fuel pool of the Beloyarskaya nuclear power plant, which is the flagship of Russian nuclear power industry.

Introduction

The current stage of the development of nuclear power engineering is characterized by an improvement of economical and operational characteristics of nuclear power plants and by an increasing of the safety of all power plant's systems [3]. Nuclear safety of NPP is associated usually with a reactor unit, but the safe operation of the plant involves the appropriate level of nuclear and radiation safety at all stages of the internal fuel cycle from the loading of fresh nuclear fuel into the core and until the irradiated fuel is removed from the plant's territory (when it is provided by the fuel cycle strategy) [4].

When the spent fuel assemblies are removed from the reactor, they are placed into the water that serves as a shield for radiation and as a cooler. Storage in pools is used only as a temporary solution until the place for final disposal is found. As an alternative to storage in pools some of the materials can be stored on the surface in concrete or steel containers, called "dry containers" [9]. This storage is also considered as temporary. Spent fuel assemblies recovered from the reactor core are stored in the spent fuel pool at the power plant for 5-10 years to reduce the heat release and for short-lived fission products decay [5]. This operation, which is mandatory for all nuclear power plants, completes the fuel cycle of the reactor [2].

Sources of radioactive waste generation

During the operation of nuclear power plant various radioactive waste are formed, among the others a large amount of liquid radioactive wastes, which include: low-salt radioactive liquid medium (e.g. water of spent fuel storage pool); liquid radioactive medium from the reactor compartment; liquid radioactive medium from active laundry.

The purposes of cleaning systems of radioactive medium are to protect the environment and to maintain safe and reliable operation of the spent fuel storage pool that is ensured by the correct water chemistry of spent fuel pool [6].

Special water purification system FAL-4 is designed to purify pool water in order to achieve and maintain required water chemistry for unit 4 of Beloyarskaya nuclear power plant. In this system the ion exchange method, which is the easiest in terms of circuit and equipment, is utilized, it has a purification factor of 10^2.

Technical description of the FAL-4 system

As noted above, the FAL-4 system is designed for cleaning spent fuel pool to ensure water transparency, reduce its activity and maintain required water chemistry while storing spent fuel assemblies. The system is designed for cleaning of 60 m^3/h of pool water in order to maintain the quality standards given in Table 1 and also to ensure the reception of 300 m^3 of water from one of the pool compartments when it is taken out for repair.

Table 1. Standards of water quality of the spent fuel pool at unit's work in the nominal mode

Characteristic	Value
Transparency with respect to demineralized water, %	95
pH value	5,6÷7,5
Electrical conductivity, not more than, μs/cm	5,0
Chloride content, not more than, μg/kg	100
Activity, not more than, BK/kg	$3{,}7\times10^3$

The system consists of mechanical filters, H-cation filters, OH-anion filters, discrete material catchers, pre-repair drain tank of spent fuel pool and its pumps, pipelines, shutoff, nonreturn and safety valves, instrumentation (see Fig .1).

To ensure the maintenance of required water chemistry for storage of spent fuel assemblies the pool water is regularly fed for purification by pool cooling system pumps. Passing in parallel through the mechanical filters the water is purified from mechanical impurities and enters the ion exchange filters: H-cation and OH-anion, where it is cleaned of dissolved impurities and radionuclides, and returns back to the spent fuel pool. The technical characteristics of the filters are shown in Table 2.

Table 2. Technical characteristics of filters

Characteristic	Mechanical filter	H-cationic filter	OH-anionite filter
Operating pressure, MPa	0,6	0,6	0,6
Operating temperature, °C	45	45	45
Productivity, m^3/h	30	60	60
Volume of filter material, m^3	1,8	2,7	2,7
Durability of ion exchangers, years	2	2	2

In the case of carry-over of ion-exchange material due to integrity violation of the drainage system of filters descrete material catchers are provided. The loading of filtering materials is carried out using a mobile hydraulic monitor.

IOP Conf. Series: Earth and Environmental Science **288** (2019) 012031 doi:10.1088/1755-1315/288/1/012031

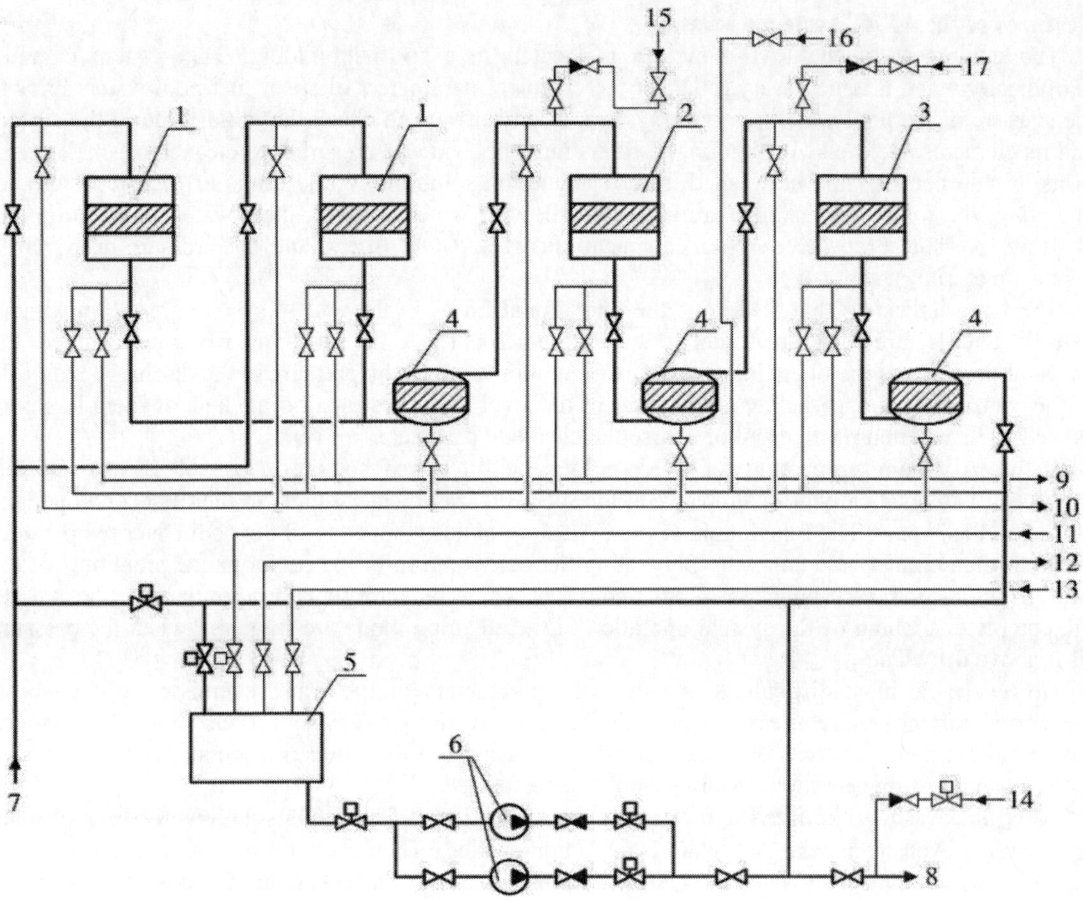

Figure 1. Technological diagram of special water purification of spent fuel pool: 1 – mechanical filter; 2 – H-cation filter; 3 – OH-cation filter; 4 – discrete material catcher; 5 – pre-repair drain tank of spent fuel pool; 6 – pre-repair drain tank of spent fuel pool pump; 7 – from spent fuel pool; 8 – to spent fuel pool; 9 – filter`s emptying to special drain; 10 – to LWR storage; 11 – from trap water filters; 12 – blowing off the tank to special gas treatment system; 13 – decontamination fluids; 14 – from water purification system; 15 – acid solution; 16 – alkali solution; 17 – water for auxiliary needs to filters.

In the case of taking one of the compartments of spent fuel pool out of service for repair, the water of the compartment to be disposed into the pre-repair drain tank of spent fuel pool. The water can be discharged into the tank through the cleaning system or bypassing it. After checking and confirming the required quality of water in the pre-repair drain tank the filling of the pool compartment after repair is carried out by the pre-repair drain tank pumps.

Apart it should be noted that in order to provide a long service life and high reliability of the equipment, fittings and pipelines when selecting their structural materials the influence of the following factors was taken into account: the influence of the chemical composition of the contacting media on the corrosive destruction of structural materials, working medium parameters (pressure, temperature), environmental parameters, as well as the practicability of the technology of manufacturing of equipment and pipelines without defects and in accordance with the requirements of technical conditions. Based on these factors, the main material used in the manufacture of equipment is austenitic stainless steel [7].

Features of the 4 FAL system operation

The equipment of the FAL-4 system is located in a special building. This system operates periodically when it is necessary to adjust the chemical parameters of spent fuel pool water. Prior to the start-up of the pool water treatment system, it is necessary to check the initial state of the system and its elements. Before switching on the filters must be loaded with sorbents, cleaned of small items; in its turn sorbents should be washed from regeneration solutions. While filters of FAL-4 system are operating, the following readings are inspected: flow of water to be cleaned, water temperature and pressure, pressure drop across the mechanical and H and OH filters, and the pressure drop across discrete material catchers.

The main indicators that determine the operational limits in the FAL-4 system are the maximum value of temperature of the spent fuel pool water, which is fed to the filters, the maximum value of the pressure drop across the operating filters, the minimum value of the pressure in the discharge manifold of the centrifugal pumps; the minimum value of the level in the pre-repair drain tank of spent fuel pool as well as its maximum value and of course the chemical parameters of water.

If the maximum temperature of water before the filters reaches more than 60 °C an alarm is triggered to automatically close the valve at the spent fuel pool water supply to the filters and open the valves on the bypass. The functioning of the FAL-4 system is stopped until the pool water temperature drops to the nominal value, thereby preventing the deterioration of the performance properties of the filter materials and the production of purified water of unsatisfactory quality. Similar interlocks up to the complete stopping of the system operation in the cleaning mode are triggered when the pressure rises above 0.1 MPa.

However, the most difficult is to adjust the parameters of the water chemistry. If the above mentioned parameters are rejected from the standardized values, the system filters should be removed from working and operations on backwashing and washing of the filters, regeneration of the filtering material and post-regeneration washing should be performed.

After backwashing the filtering material is regenerated with 5% HNO3 solutions for filters loaded with cation exchangers and 5% NaOH for anionic once. Then it is necessary to perform post-regeneration washing with water for auxiliary needs. Discharge of regeneration and washing water is carried out through the drain collector, which is disconnected from the system of special canalization and switched to the collector of hydraulic discharge into the tank of medium-radioactive sorbents. As a result of the regeneration, the dynamic exchange capacity of the filter media is restored, the washed filters can be put into operation and the functioning of the system can continue.

It should be noted that when the resource of the sorbent is exhausted and if it is not possible to restore the dynamic exchange capacity of the filter during regeneration, the filter load is hydraulically discharged to the medium-radioactive sorbent tank and the filters are loaded with fresh sorbent using hydrotransport [8].

The factors that may result in malfunction of FAL-4 system are failures of pre-repair drain tank of spent fuel pool pump, depressurization of the tank, violation of the integrity of the drainage system of the filters and de-energization of the system.

The system provides redundancy of the pumping equipment and manual duplication of the electric drives of the shutoff valves, therefore, the failure of the active elements does not lead to the failure of the system's functions.

Conclusion

The control system of FAL-4 special water purification system provides high operation stability and serves to protect the system during operation irregularities and in emergency situations. All components, as well as the system itself, belong to the class of repairable, recoverable items. The service life of the system is 40 years.

The technical and organizational decisions taken to ensure the safety of the operation of the FAL-4 system have been approved by previous design experience, testing, research, and confirmed by the operating experience of such systems at Russian NPPs [9].

The water purification system of the spent fuel pool meets the principles and criteria of radiation safety, prescriptions and rules in the field of nuclear energy.

References

[1] Asmolov VG 2007 Thermal Engineering vol 5 p 2
[2] Beckman I N Nuclear Industry (Moscow: Moscow State University Press) p 19
[3] Frolov I E 2011 Problems of forecasting vol 3 p 12–20
[4] Frolov K V 2005 Mechanical engineering (Moscow: MPEI Press) vol 4-25 p 944
[5] Kütt M, Freiβ F, Englert M 2014 Science and Global Security vol 22 p 108
[6] Kütt M, Freiβ F, Englert M 2014 Science and Global Security vol 22 p 115
[7] Patel S. 2016 Power / http://www.powermag.com/beloyarsk-nuclear-power-plant-unit-4-sverdlovsk-oblast-russia/?printmode=1
[8] Poltarakov G I, Vodyankin R E and Kuzmin A V 2011 Bulletin of the Tomsk Polytechnic University vol 319 p 13-16
[9] Rybin A A Momot O A 2017 Journal of Physics: Conference Series vol 781 article number 012013

IOP Conf. Series: Earth and Environmental Science **288** (2019) 012032 doi:10.1088/1755-1315/288/1/012032

Mathematical calculation of additional cooling methods to achieve the established temperature regime at the reservoir-cooler of Zainskaya state district power plant

A S Vinogradov, A M Gribkov, A A Filimonova, N D Chichirova, A A Chichirov

Kazan state power engineering university Russia, 420066 Kazan, Krasnoselskaya, 51

aachichirova@mail.ru

Abstract. A natural and technical complex, including the industrial part - the turbine condenser and cooling system, and the natural part - the cooling reservoir with the surrounding atmosphere, is one of the options for the operation of the circulating system technical water supply of thermal power station and nuclear power station. For cooling equipment Zainskaya state district power plant as a pond-cooler is used Zainskoe reservoir. According to the available data on the reservoir-cooler of Zainskaya state district power plant, the average water temperature in comparison with the natural temperature of the water body in summer months exceeds the current requirements, which negatively affects the functioning of the reservoir and the operation of the power plant itself. The mathematical analysis of the reservoir cooling capacity and the technical justification for the introduction of additional methods for achieving the optimum temperature regime at the point of heat exchange water discharge of the Zainskaya state district power plant are carried out in the article. According to the obtained results, the expected water temperature in the hottest month of the year as a result of wastewater discharge will not exceed a critical temperature maximum of 28 ° C.

The Zainskoe reservoir was created on the Stepnoi Zai River in 1965 as a cooling reservoir at Zainskaya state district power plant. Zainskaya state district power plant with a capacity of 2,200 MW is located on the left bank of the reservoir. Technical water supply of the state district power station is carried out according to the circulating scheme with the circulating water cooling in the reservoir. Heated water comes through two outflow channels. There is also a winter outlet channel from which the warmed waters are discharged into the middle part of the reservoir. The water intake for supply to the station is carried out from its middle part by three shore pumping stations, which allows the most intensive using the reservoir water area for heated water circulation.

Zainskoye reservoir performs the following functions of a diversified body of water [1]: reservoir-cooler state district power plant, recreational zone of Zainsk and Zainsky municipal district, fishery. The reservoir is not used for drinking water supply in adjacent settlements.

During the existence of the reservoir, a considerable amount of secondary deposits accumulated, which led to a reduction in its volume, the formation of extensive shallow-water areas, and active overgrowth by higher aquatic vegetation, which in turn leads to a decrease in heat exchange and intense thermal contamination of the reservoir [2].

The temperature regime formation of the reservoir is conditioned by weather conditions and the introduction of the heat in the reservoir with the circulating cooling water heated at the state district power plant. In the summer, the reservoir experiences the greatest heat load due to high outdoor temperatures and the maximum number of plant equipment operating, determined by the complex summer operating conditions of the Russian energy system. Recent years studies found that during the summer period it is not possible to achieve the necessary cooling effect of the circulating water, on which the actual power of the power plant depends [3,4]. There several reasons can be, for example, the reduction of evaporative cooling from the water surface [5]. The degree of influence of the state district power plant activity on the reservoir, including water temperature changes in it, has not previously been evaluated.

The hydrological characteristics of the reservoir, when sewage enters in it after being used for cooling needs of the Zainskaya state district power plant equipment, must satisfy the following requirements:

- the water temperature should not increase in comparison with the natural temperature of the water body by more than 5 °C, with a total temperature increase of no more than 28 °C in summer and 8 °C in winter;
- the dissolved oxygen concentration in the water should be at least 4-6 mg/dm^3 [6-8].

According to Zainskaya state district power plant temperature measurements at the reservoir-cooler of Zainskaya state district power plant, the average water temperature in comparison with the natural temperature of the water body exceeds the current requirements in the summer months, which negatively affects the reservoir functioning and the operation of the power plant itself. According to the actual measurements, the maximum water temperature in the reservoir during the operation of state district power plant in the summer months can reach 35 °C.

Mathematical analysis of the reservoir-cooler efficiency at Zainskaya state district power plant

The reservoir-cooler calculation was carried out on the basis of [9]. Methodical instructions apply to fully loaded cooling ponds (specific heat load above 150 W/m2), specially designed for cooling circulating water in the circulating water supply systems of TPS.

To evaluate the efficiency of the circulating water cooling, it is recommended to use the temperature distribution parameter P_T and the utilization factor K_{ev}:

$$K_{ev} = \frac{Fact}{F} \tag{1}$$

The total area of the reservoir-cooler F for today is 15.36 km^2. According to the initial project, the reservoir area was 20.45 km^2. The active cooling area in summer is F_{act} 8.7 km^2 [10]. Then Ken = 0.566.

The parameter P_T reflects the temperature decrease of the cooled circulating water relative to the average temperature level in the cooling pond and is determined by the formula:

$$P_T = \frac{T_s - T_{int}}{\Delta T} \tag{2}$$

where T_S is the mean free surface (mid-surface) temperature of the reservoir-cooler, °C; T_{int} - water temperature at the water intake of TPP, °C; $\Delta T = \Delta t_c$ - water temperature at the input and output from the capacitors (for July 2016 Δt_c = 7.8 °C). The average surface temperature T_S shall be taken as the average over the reservoir at the sampling points on 07/07/2016 from the temperature measurement schedule in the Zain reservoir. The average temperature at the water intake T_{int} will be taken as average at the shore pumping stations for the same day. If T_S = 25.4 °C, T_{int} = 23.9 °C, then P_T = 0.19.

According to the obtained values of the efficiency indicators for cooling reservoirs, we determine [9] that the Zainskoye reservoir refers to an elongated narrow type with a longitudinal direction of flow in one part and a stagnant zone in another. The water outlet and water intake are located so that the circulation zone occupies only a part of the reservoir area.

Next, we find the thermal load on the cooling pond. The technical water supply flow for the turbine K-200-130, operating at Zainskaya state district power plant, is 25,000 m^3/h [11]. There are 11 turbines in operation at the moment. Hence the estimated water flow maximum is:

$$V_м = 25000 \text{ м}^3/\text{h} \cdot 11 = 275\ 000 \text{ м}^3/\text{h} = 6{,}6 \cdot 10^6 \text{ м}^3/\text{day} \quad (3)$$

The maximum monthly average water temperature to the capacitors inlet for August is t_{c1} = 29.6 °C, at the outlet t_{c2} = 37.6 °C (average for August 2016). Heating water in turbines condensers is taken according to actual data:

$$\Delta t_c = 37{,}6 - 29{,}6 = 8{,}0 \text{ °C} \quad (4)$$

The mass water flow taking into account the water density at these temperatures is:

$$G = \Delta V_м * \rho = 275\ 000 \text{ м}^3/\text{h} * 994{,}26 \text{ кg/м}^3 = 273420 \text{ ton/h} \quad (5)$$

The heat amount supplied by the station's technical water in the water tank-cooler ΔQ is:

$$\Delta Q = G * \Delta t_c = 273\ 420 \text{ ton/h} * 8{,}0 \text{ °C} = 607\ 600 \text{ kcal/s} = 2\ 545\ 844 \text{ kJ/s} = 2546 \text{ MW} \quad (6)$$

The difference in the outlet and inlet water temperature in the reservoir changes on average by 2.34 °C (according to temperature measurements). If we take into account that the water flow in the river (flow effect) in July is equal to 6.38 m^3/s (according to state district power plant), due to the influence from the Stepnaya Zai river flow, 62,6 MW/s of heat is taken from the reservoir. In this period, the forecast load (according to the retrospective analysis of actual regimes) was 15% of the estimated maximum station load:

$$15\% * \Delta Q = 0{,}15 * 2546 \text{ MW} = 381{,}9 \text{ MW} \quad (7)$$

The natural heat loss from the Stepnaya Zai River running water of (62.6 MW/s) is 16.4% of the predicted thermal discharge of the station, i.e. the efficiency of the reservoir-cooler at this time was:

$$\eta_{ef} = 100\ \% - 16{,}4\ \% = 83{,}6\ \% \quad (8)$$

Since the total area of the reservoir is F = 15.4 km^2, then the specific heat load without insolation is:

$$Q_{rel} = \Delta Q / F = 165{,}3 \text{ W/м}^2 \quad (9)$$

Q_{rel} is more than 150 W/m^2 and the application of [9] is legitimate.

Let's perform an approximate evaluation of the cooled circulating water temperature. Using the nomogram of the reservoir-cooler calculating [9], we will find the temperature increasing in the cooling pond when the state district power plant works. We get that at a certain equilibrium temperature, T_{eq} (18.8 ° C for August), wind speed W (2.5 m/s) and temperature difference in turbine condensers ΔT (8.0 °C), the maximum overheating at the water intake δT_{max} in relation to the temperature T_{eq} is 7.4 °C.

So, under maximum load, state district power plant can ensure the water temperature in the reservoir is not higher than 28 °C in the summer months only if the water temperature at the entrance to the reservoir is not more than 20.6 °C. If the water temperature is 28 °C, the station will have to reduce the load to 0 with the termination of the thermal discharge completely.

In order to increase the load on state district power plant under these conditions, it is necessary to use complex technical measures for additional cooling, for example, air cooling or others.

Technical measures to achieve the established temperature regime at the reservoir-cooler of Zainskaya state district power plant

For air cooling, it is proposed to use the areas of supply and discharge channels, and on the ground along their banks to place spray devices.

It is necessary to calculate under what weather conditions the water temperature in the discharge channels can be reduced to 28 °C spray devices using. The average monthly relative humidity of air at 3 p.m. of the warmest month is 55%. We take the same values for the Zainsky reservoir. In ideal conditions, the cooling zone width should be taken as equal to the heating of the water in the condenser, i.e. $\Delta t = 8$ °C.

However, we can choose the cooling zone width of spray units only 4 °C, as the minimum possible, based on the following limitations:

1. it is economically unprofitable to cool down by less than 4 °C;
2. It is possible to cool more with the help of spray devices, but at a low ambient temperature, since the cooling capacity of the spray devices depends on the climatic conditions (air temperature and humidity).

Thus, for a given cooling zone width at 4 °C, to cool the water to the required water temperature at a discharge of 28 °C at a fixed humidity of 55%, the outside air temperature should not be above 27.5 °C (according to the nomogram for the temperature calculating of the chilled water in the spray unit and the psychrometric table) [12]. Taking into account the wind corrections, the shape of the spray pool, the water pressure before the nozzles, we get that the outside air temperature at 55% humidity should not be above 30 °C.

Average for 2015 - 2017 years in Zainsk, the duration of periods with temperatures above 30 °C is 72 hours per year. In this period, the spray devices operation will be inefficient, because they will not be able to cool the water by 4 °C and, consequently, the required water temperature at the discharge of 28 °C will not be reached.

Also, the spatter devices operation is impractical at water temperatures at the inlet of steam turbine condensers below 20 °C, when even with a maximum station load due to water heating at 8 °C, the temperature at the discharge will always be below 28 °C. The frequency of such modes is 6544 hours per year. The rest of the time - 2216 hours a year, spray devices are in demand to reduce the water temperature at the discharge in the reservoir below 28 °C.

Thus, the estimated time of spray devices using will be the difference in the duration of their demand period for the initial water temperature (2216 h/year) and the period of their inefficiency due to too high outside air temperatures (72 h/year) - 2,144 hours per year.

To cover the remaining small period of excessively high outside air temperatures (72 h/year), it is proposed to supplement the spray devices operation with the installation of a permanently operating ejection device in the discharge channel. According to our calculations, such an ejector will reduce the water temperature in the channel by approximately 1 °C. Since this is not enough, it is proposed to install 10 submerged axial pumps parallel to the ejector with a capacity of 20,000 m^3/h.

The estimated operating time of the mixing pumps work will be 3 days per year (72 hours), and the ejector should function continuously. The pumps installation will allow, together with the ejector, to reduce the water temperature in the discharge channel by 3 °C.

Conclusions and recommendations

1. At present, the Zainskoye cooling reservoir does not meet the requirements for the temperature regime, which provides a general increase in temperature to no more than 28 °C in summer and 8 °C in winter in comparison with the natural temperature of the water body;
2. To achieve a temperature regime in the summer months in accordance in the reservoir-cooler and ensure the full station operation with the required capacity, it is proposed to implement the following technical measures:

- To install spray units on the banks of the supply and discharge channels. The estimated time of sprays using will be 2144 hours per year. In the period of the highest outdoor temperatures (72 h/year), spray units can not independently provide sufficient the discharged water cooling, but they can be used in combination with additional cooling methods (dilution, ejection, etc.). Spray devices also saturate the water with oxygen, which allows a radical solution to the problem with oxygen deficiency in waste water.

- To install a simplified ejection device of the proposed design and 10 mixing pumps in the discharge channel. The estimated operating time of the mixing pumps will be 3 days per year (72 hours), and the ejector should function continuously.

References

[1] Amendments to the territorial planning scheme of the Zainsky municipal district of the Tatarstan Republic. Materials of the project substantiation, executed by "Tatinvest civic project" commissioned by the Executive Committee of Zainsky municipal district 5506-PZ 2012
[2] Ivanov D V, Shagidullin R R, Ziganshin I I, Osmelkin E V 2011 *Scientific notes of Kazan University. Natural Sciences series* 153 190-202
[3] Pryanishnikov F D, Beiner N V, Beiner P S 2014 *Power engineering. News of higher educational institutions and power associations* 1 82-88
[4] Saha P, Aksan N , Andersen J, Yan J, Simoneau J P, Leung L, Bertrand F, Aoto K, Kamide H 2013 *Nuclear Engineering and Design* 264 3-23
[5] Antonova L N, Kanuk G I, Michaiski D V, Omelchenko L N, Fokina A R 2012 *Eastern European Journal of Advanced Technology* 8 42-46
[6] Protection Rules of waters surface approved by the USSR State Committee for Nature Protection 22.03.1991 № 03-13/57-442.
[7] The Water Code of the Russian Federation No. 74-ФЗ of 3.06.2006
[8] Hygienic requirements for the protection of water surface SanPiN 2.1.5.980-00 2000 (Moscow)
[9] Methodical instructions on technological calculations of reservoirs-coolers. RD 153-34.2-21.144-2003.
[10] Development of the draft rules for the use of the Zainskoe reservoir P-13-79 2014 (St. Petersburg: BE Vedeneev VNIIG)
[11] Rizhkin V Ya 1967 *Thermal power plants* (Moscow: Energia) p 400
[12] Berman L D 1949 *Evaporative cooling of circulating water* (Moscow: Gosenergoizdat) p 440

Perspective direction of small rivers water purification from ions of heavy metals

G.G. Yagafarova S.V. Leontieva I.U. Gabitova D.R. Zinnatullina

Russian Federation, Republic of Bashkortostan, Ufa, Cosmonauts, 1

dianka_zinnatullina@mail.ru

Abstract. The article is devoted to the actual problem of purification of small rivers from ecotoxicants. In this paper we consider the biological method of cleaning small rivers with biofilters. The degree of bioaccumulation of heavy metals by water plants by Elodea canadensis and Cladophora aegagropila on the example of the river Shugurovka of the Republic of Bashkortostan is estimated.

Streams and small rivers predominate in the hydrographic network of any water basin. The main feature of the small rivers formation is their close relationship with the landscape of the basin. At present their condition as a result of a sharply increased anthropogenic load is estimated as catastrophic. Despite the prevalence of the water bodies pollution with heavy metals problem there are not many research related to the cleaning of small rivers in particular with bioligical methods of purification from heavy metals [1]. In addition because of the low ability to self purification of small rivers pollutants from them subseqently fall into larger rivers. The studied water body is the Shugurovka river is of interest because melting and storm water from petrochemical and oil refineries located near its bed along the terrain are sources of various ecotoxicants including heavy metal ions [2]. As shown by the results of the environmental monitoring carried out by the authors in the water of the river exceedances of MPC for copper strontium iron and other ions of heavy metals. The result of the biotest testify to the high toxicity if water Shugurovka practically along the whole length of the river although the sources of pollution do not exceed the permissible standards [3].

The purpose of this work was to study the biological method of cleaning small rivers from heavy metal ions using the example of the river Shugurovka.

Purification of small rivers and water bodies is proposed to be carried out by applying special designs of biofilters where as bioaccumulators of toxic ecotoxicants including heavy metals, water vegetation Elodea canadensis and Cladophora aegagropila.

For this purpose were investigated two types of water plants Elodea canadensis and Cladophora aegagropila. Previously conducted studies indicate their ability to accumulate various organic ecotoxicants [2-4].

Elodea is a kind of hydrocharis water plant lives in fresh water reservoirs with a water temperature of 16-24 °C suffers a prolonged temperature drop to 12 °C. Rigidity and hydrogen indicator for Elodea does not matter it grows both in very soft and hard water. Catching sunlight Elodea carries out photosynthesis however can withstand and moderate shading.

Cladophora aegagropila is a colony if green filamentous algae. Their filaments are arranged radially forming a ball. Lives in brackish water bodies with a water temperature 14-18 °C neutral pH 5,5-6,0. By its functionality Cladophora is an excellent mechanical and biological filter for water. In natural water bodies dwells on the bottom in shaded places.

Studies were conducted with contaminated water taken from the Shugurovka river in a similar pattern.

On the ability of algae to accumulate ions of heavy metals were judged by the reducation in their content in purified water and indirectly by the change in its toxicity.

Analyzes were conducted in an accredited laboratory for certified methods.

The content of strontium manganese iron at concentrations less than 0,05 mg/dm² was determined by atomic absorption spectrometry based on measuring the resonant absorption of light by free atoms of the metal being when light passes through an atomic vapor of the test sample formed in the flame on an atomic absorption spectrometer SHIMADZU [8].

The content of copper manganese ions at concentrations above 0,05 mg/dm² was determined by atomic absorption spectrometry with electrothermal atomization based on measuring the resonance absorption of light by free atoms of the element being determined when light passes through the atomic vapor of the sample being studied which is formed in a graphite atomizer on atomic absorption spectrophotometers Varian Spectr AA - 220 and SHIMADZU according ERD FL 14.1:2:4.140-98, M-03-505-119-08 [6].

Biotesting of the Shugurovka river water samples was carried out at the test object of the paramecium candatum cultured in a laboratory using standard techniques [9]. The essence of the method lies in the response of the body's test to the action of the toxicant with a change in behavioral responses. The perception if chemicals in paramecium occurs at the receptor level which explains the high sensitivity and speed of response to the effects of ecotoxicant. The criterion of toxicity functioning is the toxicity index the sighificative difference in the number of paramecium observed in the control and study sample [6].

Results of studying the process of heavy metal ions accumulation in seaweed in the Shugurovka river water are shown in figures 1-4.

The greater accumulating ability from the studied algae is possessed by Cladophora. While the content of copper ions decreased from 0,0014 to 0,0010 mg/l , manganese from 0,061 to 0,010 mg/l, iron from 0,11 to 0,06 mg/l, strontium from 1,00 to 0,61 mg/l, nickel from 0,0059 to 0,0051 mg/l.

In the joint cultivation of Elodea and Cladophora the accumulation of heavy metal ions us higher than when they are separately cultivated. At the same time it was possible to achieve the purification if river water to an environmentally safe level. The concentrations of all the metals studied decreased to below the MPC.

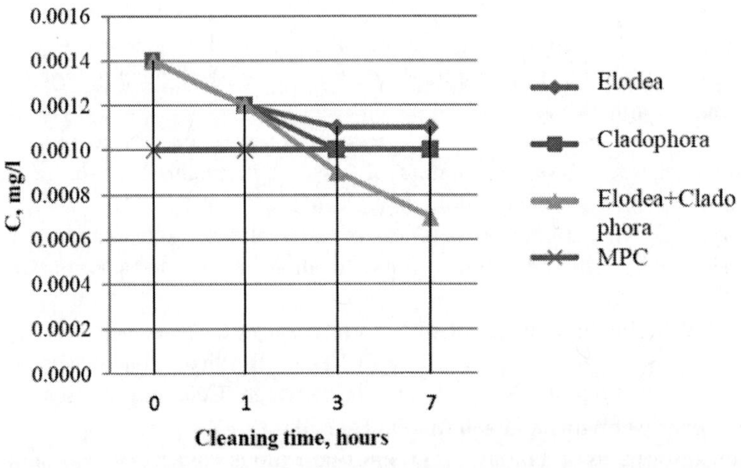

Figure 1 – Dynamics of accumulation of copper ions in the river water by algae

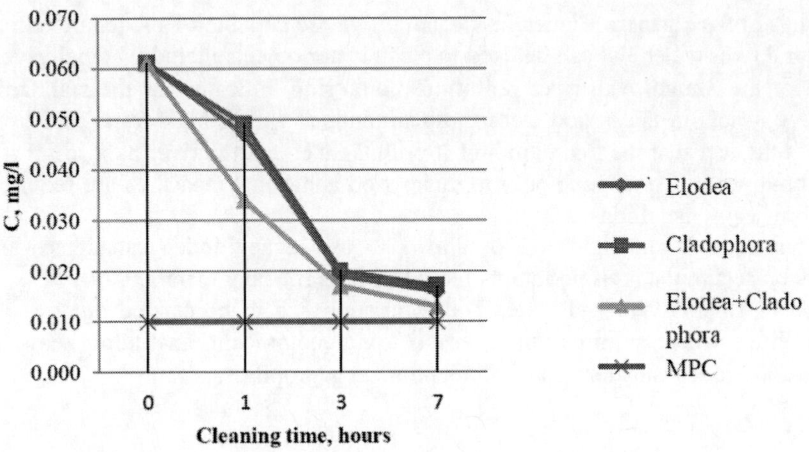

Figure 2 – Dynamics of accumulation of manganese ions in the river water by algae

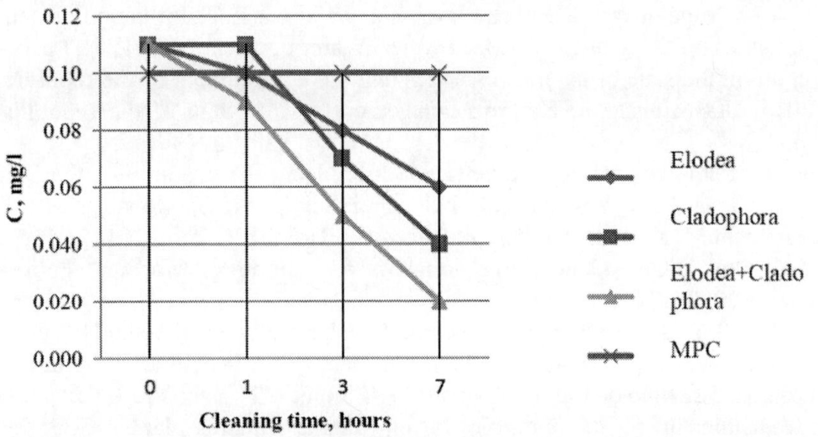

Figure 3 – Dynamics of accumulation of iron ions in the river water by algae

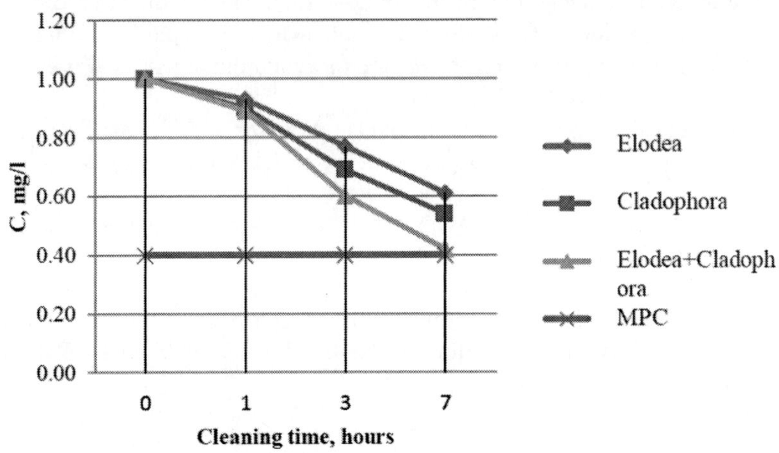

Figure 4 – Dynamics of accumulation of strontium ions in the river water by algae

Thus the algae of the genera Elodea and Cladophora are capable of bioaccumulating heavy metal ions from natural river water and can be used to clean water objects including small rivers.

The result of the Shugurovka river pollution monitoring indicate that the standard values of the MAC for heavy metals in the period from the beginning of spring high water and to winter. This is connected with the fact that the main amount if pollutants enters the river as a result if surface runoff with melt and rainwater. Correspondingly in this period additional measures are required to purify the river water from heavy metal ions.

Thus the temperature and light regime allows to use algae Elodea canadensis and Cladophora aegagropila as bioaccumulators if pollutants including ions if heavy metals.

Purification of small rivers and water bodies is supposed to be carried out by applying special designs of biofilters where as bioaccumulators if toxic ecotoxicants including heavy metals is used aquatic vegetation Elodea canadensis and Cladophora aegagropila.

Reference

[1] Assessment if the environmental status of the Shugurovka river. Magasumova A.T. Safarova V.I. Smirnova T.P. Khatmullina R.M. Stupun A P. Fatianova E.V. Water, chemistry and ecology 2011 №11 p.97-101.

[2] Safarov A.M. Yagafarova G.G. Leontieva S.V. Safarova R.V. Influence of petrochemical and all refineries on the quality of surface water. Water purification 2015 №7 p.19-22.

[3] State report on the state of natural resources and the environment of the republic Bashkortostan in 2015. Ufa Ministry of Nature Management and ecology of the republic Bashkortostan 2016.

[4] Monitoring of small rivers ecotoxicants pollution using the example of the Shugurovka river. S.V.Leontieva, G.G.Yagafarova, N.I. Fatikhova, I.U.Gabitova, N.S. Zinoveva, I.A. Fedorova, G.M. Kuznetsova, D.I. Yagafarova. PH PNRPU Perm 2017, p.116-125.

[5] SS 51592-2000 Water. General requirements for sampling. Moscow. Publishing house of standards 2000 p.31.

[6] ERD FL 14.1:2:4.128-98 Quantative chemical analysis of water. Method of measuring the mass concentration of petroleum products in samples of natural drinking wastewater by fluorometric method on the liquid analyzer "Fluorat -92".

[7] Bioaccumulation of organic pollution using higher aquatic plants. G.G. Yagafarova N.I. Fatikhova, I.A. Sukhareva, S.V. Leontieva, L.F. Korzhova, V.V. Mikulik. Bulletin of the technological university. Kazan 2017 V.20 № 1 p.169-172.

[8] Sukhareva I.A. Yagafarova G.G. Wastewater treatment from oil and petroleum products using algae Elodea. The perspective if the nation 2016. The collection of reports of IV the international competition of scientific research works 10.10.2016. Volume III (Natural and technical sciences). Scientific editor doctor of economic sciences professor A.V. Gumerov Kazan Ltd. "Raketa Soyuz" 2016, 150 p.

[9] Fatikhova N.I. Yagafarova G.G. Korzhova L.F. Leontieva S.V. Iagafarova D.I. Wastewater treatment from phenolic compounds using algae Cladophora aegagropila. Bulletin of the Kazan technological university 2016 V.19 №10 p.152-153.

[10] Zainutdinova E.M. Yagafarova G.G. Wastewater treatment from heavy metal ions using aquatic plants. Bashkur chemical journal. 2013 V.20 №3 p.150-152.

[11] Yagafarova G.G. Akchurina L.R. Fedorova I.A. Leontieva S.V. Yagafarova D.I. Fatikhova N.I. Safarov A.K. The method of cleaning water objects from film and emulsified petroleum products. Patent Russian Federation №26240238 from 30.03.2016 Published 30.06.2017 Bulletin №19.

On the effect of a weak magnetic field on corrosion wearing of steel plates

S.N. Yakupov[1], **G.G. Gumarov**[1,2], **N.M. Yakupov**[1]

[1] Institute of Mechanics and Engineering - Subdivision of the Federal State Budgetary Institution of Science "Kazan Scientific Center of the Russian Academy of Sciences" 2/31, Lobachevsky str., Kazan 420111 Russia

[2] Kazan E. K. Zavoisky Physical -Technical Institute (KPhTI), Sibirsky tract, 10/7, 420029 Kazan, Russia

gumarov@kfti.knc.ru

Abstract. The previously discovered effect of the corrosive wear increasing of thin-walled St3 steel plates oriented parallel to the magnetic field lines of the Earth's magnetic field was investigated. The remanent magnetization of the samples was measured on a rectangular grid. A relatively large value of the normal components of the residual magnetization was established for parallel orientation of the samples. The slowing down of corrosion is associated with the action of the magnetic field gradient on paramagnetic ions in the electrolyte.

Introduction

Corrosion is a spontaneous destruction of metallic materials due to chemical or electrochemical interaction with the environment [1]. At the same time, electrochemical corrosion differs in that the ionization of metal atoms and the reduction of the oxidizing component of the corrosive medium do not occur in one act and their rates depend on the electrode potential of the metal. This type of corrosion is most common and occurs when metals interact with liquid electrolytes. Thus, corrosion is directly related to the motion of charged particles, which causes its dependence on magnetic fields of different nature. The source of magnetic fields can be external magnetic fields, residual magnetization of steel structures, external currents, terrestrial magnetic field, etc. Magnetic fields can affect the stability (integrity) of the passivating layer, which, as is known [2, 3], is formed on the metal surface, which is in an aggressive environment.

The magnetic field can influence on metal corrosion by acting on the ionic electrode kinetics [4], on mass transfer [5], on the formation of an oxide / hydroxide layer [6] on the interface, or on the potential difference at the metal-solution interface [7]. To describe the effect of the magnetic field on the transport of ions, a magnetohydrodynamic theory is used, to which, in particular, reviews [8-10] are devoted.

Mechanisms of the effect of magnetic fields on corrosion are not completely established. Often the results of researches are contradictory [11]. It is reported that the magnetic field reduces the susceptibility to corrosion of stainless steel [12]. On the other hand, a magnetic field applied perpendicularly to the surface of stainless steel reduces the repassivation potential [13]. Under the action of a constant magnetic field, the corrosion-fatigue life of 17G1C steel increases by 1.37 times in air and by 1.2 when exposed to a corrosive medium (3% NaCl) [14]. In a NaCl solution, the magnetic field increases the corrosion resistance of nickel-zinc alloys with a low nickel content [15]. A magnetic field applied to iron samples in a solution of sulfuric acid makes it possible to reduce the corrosion of an iron sample by 80% [16]. The effect of benzene compounds on the corrosion of steel under the action of a magnetic field is considered in article [17].

The negative effect of the influence of the residual magnetic field on the process of corrosion wear was obtained by the author of [18]. It is noted that the enhancement of corrosion contributes to five factors (complex), among which residual magnetization.

A number of studies have shown the effect of reducing the corrosive wear of steel samples in an aggressive liquid in the presence of a magnetic field [19-201], a method and a device [21], developed on the basis of this effect, are proposed. The effect of the orientation of steel samples on the direction of the Earth's magnetic field lines on corrosion wear was observed [22]. It has been established that samples with surfaces that are parallel to the lines of force of the Earth's magnetic field are subject to greater corrosion wear. It was suggested that the magnetic field appearing on the surface of the sample creates an additional force of attraction of the passivating layer to the base of the sample.

The proposed work investigates the mechanism of the influence of a weak magnetic field on the corrosion of steel samples. For this purpose, the remanent magnetization along the surface of both the original samples and those that have passed the corrosion test cycle was measured. Also forces that can be responsible for the effect of increasing the corrosion of thin-walled steel plates oriented parallel to the earth's magnetic field lines were analyzed.

Materials and Methods

To study the corrosion wear of the samples [20], circular plates of St3 with a diameter of 10 cm and a thickness of 0.6 mm were used. Two groups of samples (ten in each) were placed in one container with 16% sodium hypochlorite solution in water. The mutual orientation of the Earth's magnetic field and samples of the first group (W-E) is shown in Fig. 1. The lines of force of the magnetic field lie in the plane of the plate. Components of the magnetic field [23]: horizontal H_{HOR} = 0.16 G, vertical H_{VERT} = 0.51 G, the resulting field H_E = 0.53 G. In this case, the lines of force of the Earth's magnetic field penetrated the second group (N-S) of samples at an angle of 73° to the normal to the surface of the plates. Samples were being held in the medium under study for 30 days.

Results

To evaluate the degree of corrosion of the samples, an experimental-theoretical method was used, which is a combination of experimental study of samples and theoretical processing of experimental data using the ratio of the nonlinear theory of shells [24]. According to [24], the samples are fixed along the contour in a special installation and are loaded with a uniform pressure p. Knowing the deflection of the sample H at a given pressure p, proceeding from the relations of the nonlinear theory of shells, one can determine their reduced (integral) modulus of elasticity.

The dependences of the mean deflection on pressure for two groups of samples are shown in Fig. 2. Curves are the result of averaging over 10 measurements. The measurement error was about 5% (shown in the figure). As can be seen from Fig. 2, the deflection of samples from the group (N-S) is less than for samples from the group (W-E). Those samples from group (N-S) undergo less corrosive wear than samples from group (W-E).

Normal components of the residual magnetization M_{RN} of the samples were measured using a GM2 magnetometer from AlphaLab Inc (accuracy of ± 0.01 G). The measurements were carried out on a rectangular grid with a step of 1 cm. The distribution of the remanent magnetization of the initial samples and after holding for 7 days in different plate orientations relative to the earth's magnetic field are shown in Figures 3 and 4, respectively.

The central (working) part of the sample with a diameter of 8 cm is included in the stage of determination of the stiffness of the samples under study [23]. Measurements of the normal component of the remanent magnetization in the central part of the plates with averaging over all measured points have shown that for the initial plates the average value of M_{RN} is of the order of 0.71 Gauss, and for samples in the WE orientation 0.2 G, in the NS orientation 0.25 G.

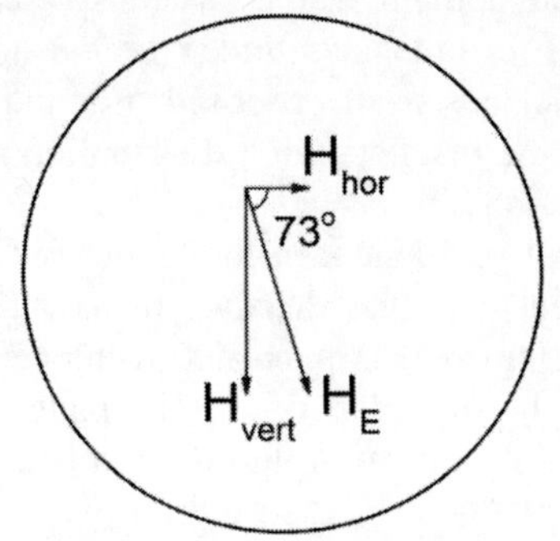

Fig. 1. Mutual orientation of the Earth's magnetic field and group samples (W-E). All the components of the field lie in the plane of the plate.

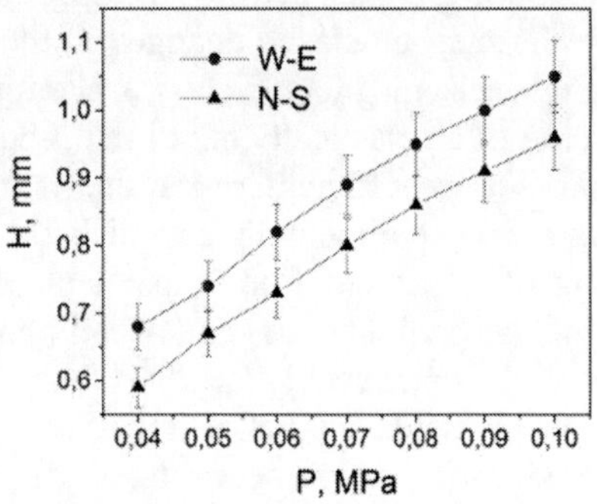

Fig.2. Dependence of the average deflection H on the pressure p for two groups of samples.

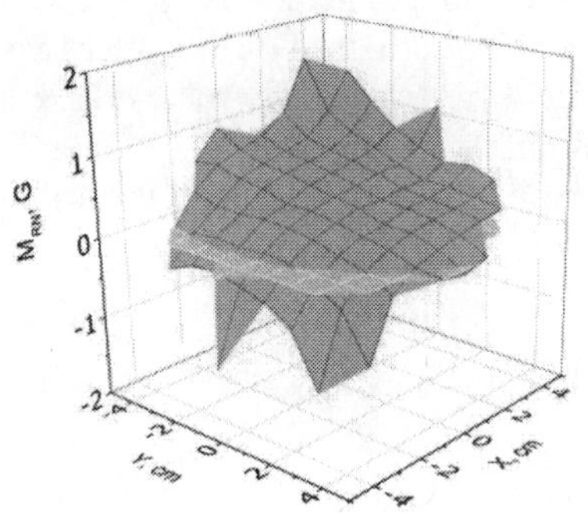

Fig. 3. Characteristic distribution of the normal component of the residual magnetization MRN of the original plates.

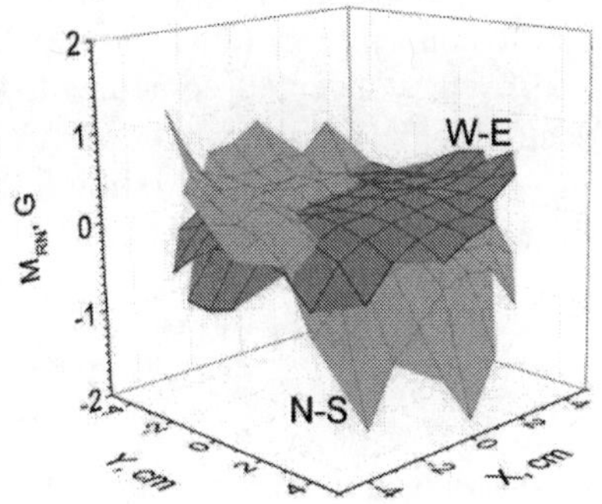

Fig. 4. Typical distribution of the normal component of the residual magnetization of MRN plates with different orientations.

The change in the magnetization of steel parts and structures with long exposure in weak magnetic fields can be related to the effects of vibrations and alternating loads, which are always present in one form or another. In vibrations, the frictional forces between domains are relaxed and their orientation is facilitated in the direction of the external field, i.e. magnetization is facilitated [25]. Variable mechanical stresses lead to an increase in the magnetoelastic energy [26], which in turn causes a change in the magnetization. The increase in the magnetization intensity of a sample in a magnetic field also occurs under the action of ultrasound [27]. Ultrasound causes in the sample variable mechanical stresses, which as a result leads to an increase in the magnetization. The residual magnetization of the sample with the field switched on is reduced by ultrasound.

It should be noted that the topography of the residual magnetization distribution of the initial plates of St3 is largely determined by their prehistory - the methods of rolling, cutting, deformation, storage, etc. The relatively large integrated M_{RN} is associated with the storage conditions in the form of a vertical stack. This changes the demagnetizing factor, which causes the growth of the normal components of the M_{RN}.

In the theoretical description of the change in the magnetization of a ferromagnet under mechanical loading, it is necessary to take into account factors such as the level of internal stresses and internal scattering fields that determine the magnitude of the potential barriers of 90 and 180-degree domain walls, texture, grain anisotropy, the pressure mechanism on domain walls, structure, etc. [28]. To describe the magnetoelastic changes in the magnetization of ferromagnets under the tensor nature of loading in weak magnetic fields, a phenomenological model has been proposed. For the relatively simple case of an elastically bent long tube, it is shown that the magnetization distribution will, albeit with a nonlinear coefficient, repeat the flexural stress distribution [29].

It has been experimentally established [30] that the normal component of the magnetic field of the scattering of a channel from Cm3 depends practically linearly on the distance measured from the middle of the part. The tangential component of the scattering field is maximal at the center of the channel and is noticeably smaller than the normal one. In the case of a thin plate and small magnetization fields, as well as small stresses, the residual magnetization distribution is complicated and unpredictable, which necessitates experimental measurements of its distribution over the surface. The magnetization of the surface of the ferromagnetic component leads to a strongly inhomogeneous scattering field above its surface. Different forces of magnetic origin, which can affect the electrochemical processes, which is the process of corrosion, are considered. This is the force of the paramagnetic gradient, caused by the cation concentration gradient; the force associated with the gradient of the magnetic field; The Lorentz force and the electrokinetic force [31]. The force of the paramagnetic gradient acts in the direction of the concentration gradient, i.e. in the direction of ordinary diffusion, and its influence can usually be neglected. The influence of the latter two forces is determined by the tangential component of the magnetic field [31], which is relatively small in the case of scattering fields associated with remanent magnetization.

The effect of residual magnetization on the pitting corrosion of low-carbon steel is attributed to the effect of the magnetic-field gradient force [32]:

$$\boldsymbol{F}_B = \frac{\chi_m c B \nabla B}{\mu_0}, \tag{1}$$

Here B is the induction of the magnetic field, c is the concentration of the electrolyte, and χ_m is the molar susceptibility of the electrolyte. On the basis of studies using optical and scanning electron microscopy, it was shown that the maximum depth of the corrosion pits, as well as their average depth, is much smaller in magnetized samples. Finite element calculations using the FEMM program confirm the model proposed by the authors for the effect of the magnetic field on corrosion, indicating the maximum gradient of the magnetic field at the edge of the wells [32].

The gradient of the magnetic field can lead to the outflow of metal chlorides that prevent the formation of the passivating layer from the bottom of the pit towards the edges. As a result, the growth of the pits slows down, and the average mass loss due to corrosion in the unmagnetized sample is several times greater [32]. The maximum values of the residual magnetic induction, according to our measurements for the samples of plates from St3, amounted to about 1-2 Gauss, i.e. close to the values used in [32]. As a result, taking the value of the magnetic field gradient of 150 T/m [31], c=10^3 mol/m^3, $\chi_m = 10^{-8}$ m^3/mol according to [31], we obtain the value $\boldsymbol{F}_B \approx 0.5$ N/m^3 for the force. Such a force, acting at the beginning of the process of formation of a corrosive pit, can have many large values in the process of its growth.

The strength of the magnetic field gradient is also explained by the weakening of corrosion on high-purity iron samples (99.9%) and single-crystal iron in a strong external magnetic field (7.5 kOe) [33]. The effect is explained by the attraction of iron ions to regions with a high flux density of the magnetic field, that is, to the edge regions. Hydrogen ions are displaced from these regions in connection with the law of electroneutrality. As a result, the dissolution rate of iron in the presence of a magnetic field is reduced by approximately 80%.

Conclusion

In weak magnetic fields, including in the Earth's magnetic field, due to the presence of background mechanical influences and vibrations, the magnetization of the ferromagnetic elements of the structures changes. To study the effect of magnetic fields on corrosion processes, information on the topography of the remanent magnetization, on the regions of the maximum gradients of the magnetic field, is of particular importance. The gradient of the magnetic field, which is formed, including in the areas of corrosive depressions, can to some extent influence the corrosion process. In particular, this may be due to the outflow of metal chlorides that prevent the formation of a passivation layer from the bottom of the pit towards its edges [32]. An additional force of attraction of the passivating layer to the base of the sample also contributes to this role [22] due to the gradient of the magnetic field.

The nature and location of regions with the maximum gradient of the magnetic field depends not only on the presence of various inhomogeneities in the structure of the material and the concentration of mechanical stresses, but also on the orientation of the samples in a magnetic field. The performed measurements of the normal components of the residual magnetization of the circular plates of St3 showed its relatively large values for the N-S orientation of the samples. Correlation of the magnetic field gradient with its component normal to the surface indicates the influence of the corresponding force on the corrosion process.

It is necessary to further investigate the relationship between weak magnetic fields, residual magnetization, mechanical stresses and corrosion processes.

This work was carried out within the framework of the research of the FIC KazNC RAS on the theme "Development of the mechanics of multiphase media, aerohydroelastic systems and nonlinear mechanics of a deformable solid with applications in machine building and power engineering", No.AAAA-A18-118032690289-5.

References

[1] Zhuk N.P. *Kurs teorii korrozii i zashchity metallov [Course of the theory of corrosion and protection of metals].* M.: Metallurgiia, 1976. 472 p.

[2] Frumkin A.N., Bagotckii V.S., Iofa Z.A., Kabanov B.N. *Kinetika elektrodnykh protcessov [Kinetics of electrode processes].* M.: Moscow University. 1952. 319 p.

[3] Kolotyrkin Y.M. *Metall i korroziia [Metal and Corrosion].* M.: Metallurgiia, 1985. 88 p.

[4] Hinds G., Coey J.M.D., Lyons M.E.G. Magnetoelectrolysis of copper 1998 *Journ. Appl. Phys.* **V.83.** P. 6447-6449.

[5] Novinski V. Magnetic field effect on copper electrodeposition in the Tafel potential region 1997 *Electrochimica Acta.* **V.42.** P. 251-254.

[6] Chiba A., Kawazu K., Nakano O., Tamura T., Yoshihara S., Sato E. The effects of magnetic fields on the corrosion of aluminum foil in sodium chloride solutions 1994 *Corrosion Science.* **V.36.** P. 539-543.

[7] O'Brien R.N., Santhanam K.S.V. Magnetic field assisted convection in an electrolyte of nonuniform magnetic susceptibility 1997 *Jour. of Appl. Electrochem.* **V.27.** P.573-578.

[8] Fahidy T. Z. Magnetoelectrolysis 1983 *Jour. of Appl. Electrochem.* **V.13.** P.553-563.

[9] Aaboubi O., Chopart J.P., Douglade J., Oliver A., Gabrielli C., Tribolet B. Magnetic Field Effects on Mass Transport 1990 *Jour. Electrochem. Soc.* **V. 137**, №6. P.1796-1804.

[10] Tacken R.A., Janssen L.J.J. Applications of magnetoelecrolysis 1995 *Jour. Applied Electrochemistry.* **V.25.** P.1-5.

[11] Tang Y.C., Davenport A.J. Magnetic Field Effects on the Corrosion of Artificial Pit Electrodes and Pits in Thin Films 2007 *Jour. Electrochem. Soc.* **V.154.** P.362-370.

[12] Rucinskiene A., Bikulcius G., Gudaviciute L., and Juzeliunas E. Magnetic field effect on stainless steel corrosion in FeCl3 solution 2002 *Electrochem. Commun.* **V.4,** No.1. P.86-91.

[13] Linhardt P., Ball G., and Schlemmer E. Electrochemical investigation of chloride induced pitting of stainless steel under the influence of a magnetic field 2005 *Corros. Sci.* **V.47.** P.1599-1603.

[14] Hudiakov M.A., Altynova R.R. Effect of a permanent magnetic field on the cyclic crack resistance and corrosion resistance of steel 17Г1С 2006 *Oil and gas business.* **T.4,** №1.

P.286.

[15] Chouchanea S., Levesqueb A., Zabinskic P., Rehamniaa R., Chopart J.-P. Electrochemical corrosion behavior in NaCl medium of zinc–nickel alloys electrodeposited under applied magnetic field 2010 *Journ. of Alloys and Compounds*. №506. P.575-580.

[16] Sueptitz R, Tschulik K., Uhlemann M., Gebert A., Schultz L. Impact of magnetic field gradients on the free corrosion of iron 2010 *Electrochimica Acta*. №55. P.5200-5203.

[17] Maayta A. K., Mohammad M. Fares, Al-Shawabkeh Ali F. Influence of Linear Alkyl Benzene Sulphonate on Corrosion of Iron in Presence of Magnetic Field: Kinetic and Thermodynamic Parameters 2010 *Int. Journal of Corrosion*. Article ID 156194. P.1-9.

[18] Kuzmin M.I. *Increase of efficiency of oil wells exploitation in conditions of magnetization of underground equipment*. The dissertation author's abstract on competition of the uch. ct. Ph.D. S.-P.2013. 20 p.

[19] Yakupov N.M., Giniyatullin R.R., Yakupov S.N. Effect of a Magnetic Field on Corrosive Wear 2012 *ISSN 1028_3358, Doklady Physics*. **V.57**, No.3. P.104-106.

[20] Yakupov N.M., Giniyatullin R.R., Yakupov S.N. 2012 *Influence of the magnetic field on corrosive wear*. Proceedings 19th European Conference on Fracture: Fracture Mechanics for Durability, Reliability and Safety. Kazan, Russia, 26-31 August,. P.6.

[21] Yakupov N.M., Veliiulin I.I., Yakupov S.N., Nurullin R.G., Giniiatullin R.R. Device for corrosion prevention 2015 *Patent №2547067 Russian Federation. Bull. №10.*

[22] Yakupov N.M., Giniyatullin R.R., Yakupov S.N. Influence of the direction of the lines of force of the earth's magnetic field on the corrosive wear 2015 *Doklady Academii Nauk*, **V.463**, No. 6. P.684-686.

[23] http://www.geomag.bgs.ac.uk/data_service/models_compass/wmm_calc.html

[24] Yakupov N.M., Nurgaliev A.R., Yakupov S.N. Metodika ispytaniia plenok i membran v usloviiakh ravnomernogo raspredelennogo poverkhnostnogo davleniia [Method for testing films and membranes under conditions of uniform distributed surface pressure] 2008 *Zavodskaia laboratoriia. Diagnostika materialov - Factory laboratory. Diagnostics of materials*, **V.74.** №11. P.54-56.

[25] Shelihov G.S. *Magnitoporoshkovaia defektoskopiia detalei i uzlov [Magnetic particle inspection of parts and assemblies]*. M.: Metallurgiia, 2002. - 224 p.

[26] Belov K.P. *Magnitostriktcionnye iavleniia i ikh tekhnicheskie primeneniia [Magnetostrictive phenomena and their technical applications]*. M.: Nauka. 1987. 160 p.

[27] Kirenskii L.V., Drokin A.I., Cherkashin V.S. Vliianie ul`trazvuka na magnitnye svoistva ferromagnetikov pri razlichnykh temperaturakh [Effect of ultrasound on the magnetic properties of ferromagnets at various temperatures] *1960 Magnetic structure of ferromagnets. Novosibirsk, P.165-173.*

[28] Kuleev VG, Rigmant MB Phenomenological model of magnetoelastic changes in the magnetization of ferromagnets under the tensor character of their loading in weak magnetic fields 1994 *Defectoscopy*. № 9. P.79-91.

[29] Kuleev V.G., Lopatin V.V. Magnetic Leakage Field of an Elastically Bent Ferromagnetic Steel Pipe 2003 *Russian Journal of Nondestructive Testing*. **V.39,** №5. P. 381-386.

[30] Arkulis M. B., Misheneva N. I., and Savchenko Yu. I. Studying magnetic stray field at the surface of elastostrained extensive object 2016 *Russian Journ. Nondestruct. Testing.* **V. 52,** No. 11. P.623–626.

[31] Hinds G., Coey J.M.D., Lyons M.E.G. Influence of magnetic forces on electrochemical mass transport 2001 *Electrochem. Communications.* **V.3.** P.215-218.

[32] Espina-Hernández J.H., Caleyo F., Venegas V., Hallen J.M. Pitting corrosion in low carbon steel influenced by remanent magnetization 2011 *Corrosion Science.* **V.53.** P. 3100-3107.

[33] Sueptitz R., Tschulik K., Uhlemann M., Schultz L., Gebert A. Effect of high gradient magnetic fields on the anodic behaviour and localized corrosion of iron in sulphuric acid solutions 2011 *Corrosion Science.* **V.53.** P.3222-3230.

IOP Conf. Series: Earth and Environmental Science **288** (2019) 012035 doi:10.1088/1755-1315/288/1/012035

On-line express-analyzer and method of water monitoring on the base of proton magnetic resonance

Kashaev R.S., Kozelkov O.V.

Kazan State Power Engineering University, 420066, Kazan, str. Krasnoselskaya, 51

kashaev2007@yandex.ru, ok.1972@list.ru

Annotation. Described flow analyzer of water on the presence of impurities, based on proton magnetic resonance - PMRA. Presented common appearance and construction, mode of operation PMR-analyzer. Presented equations of received dependences for determination of salts and oil impurities in water. Proposed structure and technologic schemes for removal of impurities from water in the installation with rotating magnetic and nonuniform electric field with continual on- line control of water purity bu flow PMR-analyzer.

Introduction

Last years observed increasing mining of high water containing (up to 98%) oils in the form of high disperse emulsions, separation of which is a sufficient problem. It is connected with their high stability due to great asphaltenes concentration. After separation of such emulsions water inevitably contain impurities in the form of oil droplets and diamagnetic/paramagnetic salts of oil stratum waters. In all the case it is poured in surface waters, which are used for heat stations and localities water supply.

The quality of drainage water is regulated and must satisfy resolution of International committee on the protection of sea waters MEPC.107(49) by 18.07.2003 (not more, than 15 mg/l of oil).

At present time concentration of oil in water can be measured by laboratory analyzer KN-1 (Ecros) in the range of 0,05-50 mg/l as well as of 0-1000 mg/l by dissolving of the sample in CCl_4. In flow mode it can be controlled by signalyzer Fluorat *AE*2 (analyzer is not present), which signalize the oil presence in water in the range of 0,01-10 mg/l with accuracy □ 10%. In the measurement process is used extract of oil from water byheksane (10 mg of heksane on 100 ml of sample). Measurement time is 15-30 minutes.

Among foreign analyzers must be mentioned OCMA-300 HORIBA (Bella-Diffusion, France) and ELE International with measurement range of 0,01-200 mg/l (ppm) and accuracy □ 10%, based on the near infrared (NIR)-spectroscopy. For sample preparation is used dissolver Horiba S-316. Measurement time – 15-30 minute.

Flow (on-line) PMR express – analyzer for water quality monitoring

For monitoring of water quality, used in the heat stations and for industrial purposes we proposed elaborated and having no analogues flow express-analyzer on the base of proton magnetic resonance - PMRA. On the fig.1, 2 presented common view and functional scheme of the first generation PMRA-I [1].

Fig.1. Flow (on-line) PMR-analyzer

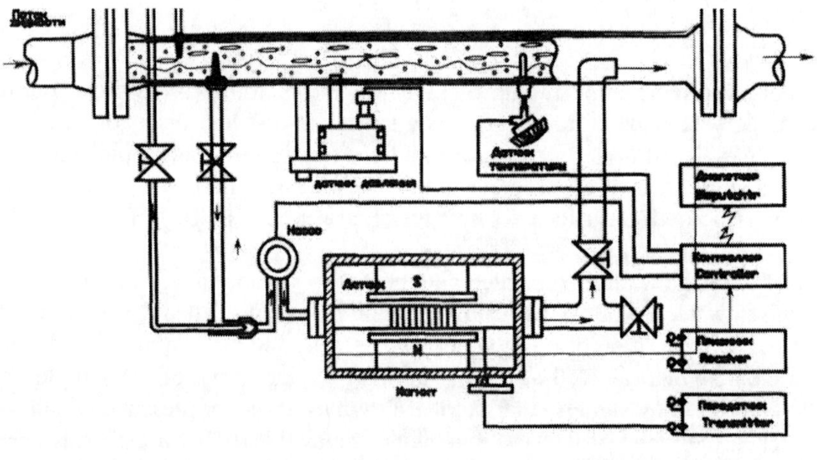

Fig.2. Functional scheme of flow (on-line) PMR-analyzer

Express mode of analysis is the foundation of the PMR method, which is attributed to methods with inner standard, noncontact and not destroing, does not need sample preparation. Time of measurement depend from number of accumulations *n*, increasing measurement presicion in $\sqrt{n}$ times and in mean time of measurement equals 2-3 minute.

Controlled media – water-oil and gas mixtures; temperature of measured mixture is +5 - +50^{0}C, temperature of environment -60^{0}C - + 50^{0}C, pressure 0.1 - 6.4 MPa, permissible pressure drop on the analyzer is 0,25 – 0,5 MPa, density of the media 400 - 1100 кг/м^3. Working mode is automatic, continuous-cyclic. Length of the communication line of analyzer with controller of computer - $\lambda/4$ (near to 15 m on the resonance frequency 3 MHz). Measurement results through radiochannel is translated on the central control panel of measurement installations.

Method of salts and oil in water determination by the data of pulse nuclear (proton) magnetic resonance method [2]. Concentration of diamagnetic salts in water

For determination of NaCl, $CaCl_2$, $MgCl_2$ salts C_B (mol/1000 g H_2O) concentration in water, may be used dependences of proton relaxation rates $(T_{1B})^{-1}$ from salts concentration.

Especially, for NaCl dependence for C_B of salt in water from relaxation rate $(T_{1B})^{-1}{}_{NaCl}$ (in solutions $T_{1B} \equiv T_{2B}$), can be approximated with regression coefficient $R^2 = 0{,}9962$ by equation:

$$C_B = -42{,}765 + 220{,}9\,(T_{1B})^{-1} + 247{,}8(T_{1B})^{-2} \quad (1)$$

For $CaCl_2$ – with regression coefficient $R^2 = 0{,}9996$ by equation:

$$C_B = -4{,}55 + 20(T_{1B})^{-1} + 9{,}7(T_{1B})^{-2} \quad (2)$$

For $MgCl_2$ – with regression coefficient $R^2 = 0{,}9924$ by equation:

$$C_B = -2{,}23 + 9{,}8(T_{1B})^{-1} + 3{,}99(T_{1B})^{-2} \quad (3)$$

Every diamagnetic ion, contained in solution, has its own coefficient of relaxation effectiveness. In stratum water, remained after its separation from oil-water emulsion, dominantly presented salts of NaCl, $CaCl_2$ and $MgCl_2$ in the proportion 73:20:7. That's why for mixtures of this salts we must use equation, considering contribution of all of them:

$$C_B = k_{C1} + k_{C2}\,\{P_1(T_{1B})_{-1NaCl} + P_2\,(T_{1B})_{-1CaCl2} + P_3\,(T_{1B})_{-1MgCl2}\} = \\ = -32{,}284 + 165{,}94(T_{1B})_{-1} + 183{,}1(T_{1B})_{-2} \quad (4)$$

Here for stratum water $P_1 = 0{,}73$, $P_2 = 0{,}2$, $P_3 = 0{,}07$.

Concentration of oil in water

Determined by us dependences between PMR relaxation parameters and summarized signal amplitude $A_{\Sigma 1}(C_{H1})$ for the range 0-300 мg/l and $A_{\Sigma 2}$ (C_{H2}) for the range 0-30 000 mg/l are described with regression coefficients $R_1{}^2 = 0{,}9976$ и $R_2{}^2 = 0{,}9812$ by equations:

$$C_H\,(\text{мг/дм}^3) = -58{,}42 - 0{,}23A_{\Sigma 1} + 0{,}006\,A_{\Sigma 1}{}^2 \quad (5)$$

$$C_{H2}\,(\text{г/дм}^3) = -102{,}4 + 0{,}877A_{\Sigma 2}, \quad (6)$$

Concentration of paramagnetic salts in water

Determination of soiling of pollution waters by heavy (paramagnetic) metals salts (integral index) and elemental composition by using method of addition or comlex-formation is performed in the range from

0.05 mg/l up to saturated solution with relative precision ± 4% by equation:

$$N = K_{1(2)\text{-}1}\,[(T_{1(2)\text{-}1})_{\text{изм}} - (T_{1(2)\text{-}1})_0] \quad (7)$$

where $(T_{1(2)\text{-}1})_{\text{изм}}$ and $(T_{1(2)\text{-}1})_0$ - relaxation rates, measured in the sample of polluted water and in distilled water without paramagnetic ions, $K_{1(2)}$ – coefficient of relaxation effectiveness.

Minimal measured paramagnetic ions concentration N_{min} is determined by the difference between $(T_{1(2)\text{-}1})_{\text{изм}}$ and $(T_{1(2)\text{-}1})_0$, i.e. by apparature precision, which can be characterized by coefficient f. Then :

$$N_{min} = (T_{1(2)})_{-1}{}_{0}(f - 1)/K_{1(2)} \quad (8)$$

Increasing of the measurements precision decrease lower limit of measured concentrations.

For the paramagnetic ions mixture with different concentrations can be used method of addition. Initially is prepared solution with known ion concentration N2 and in doses it is added in measured sample with N1 ion concentration. Then is measured relaxation rate (T1(2)-1)изм and is drawn up graph of (T1(2)-1)изм /N1 dependence from N1-1. Tangent of slope angle of this line is equal to N1/N2.

Cleaning of water from impurities (ions and oil droplets) in the rotating magnetic and nonuniform electric fields with permanent on-line control of water quality by flow PMR express-analyzer [3].

For purifying of water in the form of fine disperse distributed oil emulsion and diamagnetic/paramagnetic salts the use of standard industrial installations, based on thermal affect is embarrassing due to their resource- and energy consumption. Industry need compact installations with low energy consumption and which use methods of soft affect by physical fields, which does not cause additional droplets dispergation and does not need reagents to use.

For this aim we propose installation [3], presented at fig.3.

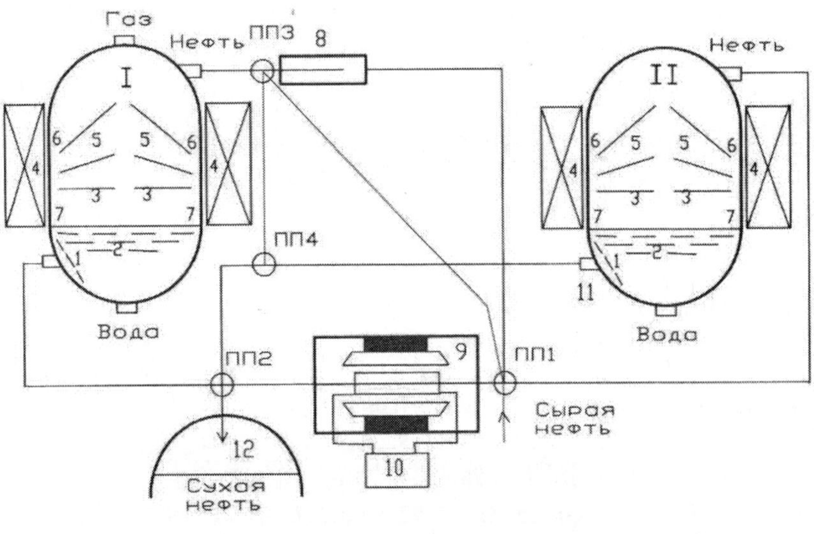

Fig.3. Block for emulsions destruction in the rotating magnetic and nonuniform electric field and installation for process control by flow PMR-analyzer.

Method of water purifying from impurities include the following stages:

1. Water emulsion through three way flow switcher ПП1 feed to the input control in PMR-analyzer 9 for the determination of emulsion parameters to choose process mode – electric fields tension, magnetic field induction, flow velocity, charge on the net 1 of block I for water separation from oil. Then through flow switcher ПП2 emulsion is feed in block I and flow through charged net 1, where droplets and impurities get charges determined by equation:

$$q \approx 1{,}64 E_o r_o^2 \quad (9)$$

Charged impurity (oil droplet, salt ion) is pushed from the net 1 with the force:

$$F \approx 1{,}37 E_o^2\, r_o^2\, \varepsilon_{cp} = 0{,}835\, qE_o. \tag{10}$$

The flow feed can be performed not on radius direction but by perimeter side of trunk in the direction coinciding with the electrodes rotation. This leads to the turning of emulsion flow and additional centrifugal force, throwing impurities off to the outer perimeter of block I.

2. Beyond of precipitated water layer is settled intermediate layer with petty droplets, where they coalescence, subjected by treatment of electromagnetic (EM) fields, rotating magnetic field, created by windings 4 and simultaneously nonuniform DC and AC electric field, formed between water surface and several electrodes 5 with AC/DC tension, while field quantity between electrodes 5 is changed (increased) with the electrode higher position on the rotor in block I. Such demand of specific electric field E_{sp} growth is requested with the circumstance that at low droplets concentrations (for instance < 2,9 %), distance between droplets-dipoles become more, than 6 radius, i.e. at low concentrations forces of dipole-dipole attractions become neglecting small and separation in electric field does not work. Increasing of the E_{sp} between electrodes 5 create conditions for increasing forces $\boldsymbol{F}_D$ between small droplets at low concentrations.

So in the electrodes 5 and magnetic field area formed by winds 4, on the droplets affect the forces: a) attraction forces, determined by dipole-dipole interactions between droplets by equation:

$$\boldsymbol{F}_D = 6(|\alpha|)^2 \boldsymbol{E}_o^2 \varepsilon_{cp}/r^4, \tag{11}$$

As electrodes are settled on ferromagnetic rotor, they also rotate with it by asynchronous drive principle decreasing opportunity for chains of droplets formation between electrodes;

б) forces $\boldsymbol{F}_G$ of electric field E_o gradient grad$E$ by equation:

$$\boldsymbol{F}_G = 2\pi\varepsilon_o(\varepsilon - 1)E_o R^3 \mathrm{grad}\boldsymbol{E}^2/(\varepsilon+2), \tag{12}$$

where R – outer radius of electrode. Under this forces droplets and impurities will move along grad$\boldsymbol{E}$ direction in the area of maximal field $\boldsymbol{E}_o$.

в) forces $\boldsymbol{F}_L$ of Lorentz by equation:

$$\boldsymbol{F}_L = q[\boldsymbol{v}\, \boldsymbol{B}] \tag{13}$$

where $\boldsymbol{B}$ – magnetic induction of alternating (rotating) magnetic field; $\boldsymbol{v}$ – linear velocity of magnetic field rotation. Here we observe move not of the droplets, but of the rotating magnetic field with velocity $\boldsymbol{v}$. Under effect of force $\boldsymbol{F}_L$ charged small droplets move down where their concentration is greater. Due to charge and dipole moment the droplets also aspire to move in the direction of magnetic field gradient grad$\boldsymbol{B}$, i.e. to the sources of magnetic field 4 - to the perimeter of the block I, where droplets coalescence and precipitate.

3. Then purified water and oil through flow switchers ПП3 and ПП1 feed on the output (after block I) control in PMR-analyzer 9 for determination of water parameters.

Conclusions

With the object of operative control in the sole device, measurement range increasing and permanent flow analysis of all soiling of water by impurities we proposed the method of salts and oil determination on the base of pulse proton magnetic resonance, including excitation of protons spins in the sample, sited in constant magnetic field the spin-echo proton magnetic resonance signals by the series of radiofrequency pulses, registration of their amplitudes A_i and relaxation times, using which the concentrations of salts and oil in water is determined.

References

[1] Kashaev R. Apparature and methods of NMR-analysis of oil disperce systems 2012. Lambert Academic Publishing (Saarbruken, Germany) 91

[2] Patent of RF № 2519496 C1 2012 Kashaev R., Idiyatullin Z., Temnikov A. Bull. 2014. №16.

[3] Kashaev R., Faskchiev N. Study by the method of nuclear (proton) magnetic resonance relaxometry effect of influence of the rotating magnetic field on the emulsions structure 2011 Applied Magnetic Resonance V.41, i.1 pp. 31-43

IOP Conf. Series: Earth and Environmental Science **288** (2019) 012036 doi:10.1088/1755-1315/288/1/012036

Innovative products and environmental aspects of modern fodder production for aquaculture objects

L T Akhmetova[1], M L Kalayda[2], R R Safiullin[3], G G Sibgatullin[3], N N Khazipov[4], A A Kalayda[2], D S Dementiev[2]

[1]LLC "Scientific and Production Association "Bitime", Russia, 420138 Kazan, Garifyanova Str., 44

[2]"Water bioresources and aquaculture" department, Kazan state power engineering university, Russia, 420066 Kazan, Krasnoselskaya, 51

[3]Management of aquaculture and beekeeping of the Republic of Tatarstan, Russia, 420054 Kazan, Teachers Str., 7

[4]Ministry of Agriculture and Food of the Republic of Tatarstan, Russia, 420111 Kazan, Fedoseevskaya Str., 36

lilia_015@mail.ru, kalayda4@mail.ru

Abstract. The quality of water and fodder components in conditions of industrially developed regions with a high degree of anthropogenic pressure on the territory become especially significant factors. The increasing demand for hydrobionts as a source of food continues to grow annually. This fact is connected not only with the increase in population, but also with the fact that preference is increasingly given to healthy food. To improve the rates of growing aquaculture and expand its species diversity, it is necessary to increase the production of feed, expand their range and develop new technologies for obtaining highly effective feed additives. A feed additive for fish feeds based on secondary raw beekeeping materials is characterized by a complex chemical composition and is a universal source of natural biologically active complexes. The practical absence of heavy metals, radionuclides, pesticides and other harmful substances in the feed additive based on bee products does not pollute the aquaculture of water bodies, and its properties due to a unique composition contribute to the accelerated elimination of toxic substances from the body of fish. The combination of ecological approaches and the introduction of innovative technologies in the production of aquaculture facilities allows to solve such key tasks of social development as ensuring food security and maintaining public health.

Introduction

In the modern world, the production of fish and fish products is considered an important economic indicator in many countries. The increasing demand for hydrobionts as a source of food continues to grow annually. This fact is connected not only with the increase in population, but also with the fact that preference is increasingly given to healthy food [1], because the fish in its composition contains high-quality proteins, unsaturated fatty acids and other important nutrients necessary for maintaining human health. It is interesting that the increase in consumption of fish products coincides with the increase in the life expectancy of both men and women [2].

According to FAO data, mankind annually consumes 90 million tons of products produced by aquatic ecosystems, including 13 million tons of aquaculture. In the 21st century, 130 million tons will be needed to meet the nutritional needs of the growing population [3]. The contribution of fish to the diet is more significant from the point of view of proteins of animal origin, since a portion of fish in 150 grams covers about 50-60% of the daily need of an adult in proteins. In 2009, fish accounted for 16.6% of the total content of protein of animal origin consumed by the world's population, and 6.5% of total protein intake [4]. It is also important to have a relatively low cost of growing fish (for comparison, the cost of 1 kg of pond fish is 4-5 times lower than the cost of 1 kg of meat).

In general, the situation is as follows: in the world over the past half century there has been a steady increase in fish consumption - from about 9 kg in 1961 to 18.8 kg in 2011 (per person per year). However, fishing can not continue to meet the growing demand for fish, as its catch has reached a relatively stable limit since the mid-nineties and amounts to approximately 85.0 -90.4 million tons, but not more than 100 million tons per year [3, 5].

To increase the volume of production of commercial fish, it is necessary to ensure the maximum realization of the genetic potential of fish productivity and the growth of its safety, which is possible only with full-fledged feeding, including the use of feed containing a full range of vitally important nutrients - amino acids (irreplaceable and replaceable) - and macronutrients, vitamins, etc. [6].

Along with this, it should be noted that the requirements for the quality of food products are increasing all over the world, including quality of fish. High quality of fish products can be provided only if ecological requirements are met for ecosystems.

On the other hand, there has been a recent increase in interest in the use of beekeeping products, characterized by the absence of harmful side effects on the environment, which is typical for all paleo products [7]. In this regard, they are of interest both in the production of innovative feed additives for mixed fodders for growing fish, and for solving a number of environmental problems in the industry.

Results and discussion

Ecological aspects of modern fodder production for objects of aquaculture

The Middle Volga region is one of the most industrially developed regions with a high degree of anthropogenic pressure on the territory. The quality of water and feed components under these conditions becomes especially significant factors.

A significant characteristic of feed components is the presence of heavy metals in them. In the muscular tissue of fish, zinc and copper accumulate at the maximum concentrations, while the permissible residual amounts of zinc and copper in the fishery products are 40 and 10 mg / kg for wet weight, respectively [8]. For lead, cadmium and mercury, the permissible residual quantities are 1.0, 0.1 and 0.5 mg / kg of wet weight, respectively, in predatory fish, and 1.0, 0.1 and 0.3 mg / kg of wet weight in non-predatory fish [9]. The permissible standards of toxic metals in fish were developed taking into account the analysis of the current legislation on hygienic regulation in Russia and foreign countries, the international recommendations of FAO / WHO. This took into account the natural content of elements in food products.

The peculiarities of the accumulation of heavy metals in fish from different habitats can be related to the age-specific features, since with the increase in the age of fish the concentration of heavy metals in their organs is influenced by two interrelated trends: accumulation due to an increase in the volume of consumed food and a decrease in the specific content due to an increase in the total body weight of fish. With increasing age of fish, the content of heavy metals in their bodies increases, especially in fish with a lower growth rate. Accumulation of heavy metals in fish is also associated with their entry into the body of fish by chemisorption of ions with mucous membranes; mechanical capture of suspended particles containing heavy metals; absorption by gills during breathing and, of course, with food. In this regard, the use of fish feeds that do not have a negative effect on the biocenosis of aquatic systems becomes important.

IOP Conf. Series: Earth and Environmental Science **288** (2019) 012036 doi:10.1088/1755-1315/288/1/012036

In the last period there has been an increase in interest in the use of beekeeping products both in environmental problems [10] and as a feed additive [11].

New raw materials for the production of feed additives for aquaculture objects

Bee products - perga (bee bread) and merva are potential raw materials for obtaining feed additives and their use in animal husbandry because they contain a vast complex of vitamins, microelements and amino acids [12]. However, under the influence of external factors - humidity, temperature, ultraviolet irradiation, seeding by various microorganisms - the useful properties of raw beekeeping resources are quickly lost. In order to preserve the composition and properties of these valuable products, technologies have been developed for their processing and obtaining purified substances and feed additives based on them [13,14].

Since the quality of commercial fish depends on the initial components of feeds and their composition, study was conducted on the amino acid, vitamin and macro- and microelement composition of the products obtained - the substance "Perga refined", the substance "Merva" and the fodder additive for fish food "Vinivet F" based on them. The results of the study are shown in Tables 1-3.

A comparative analysis of the chemical composition of bee products and biocomplexes on their basis showed that during the processing of bee perga and merva their original composition is completely preserved.

Table 1. The averaged amino acid composition of biologically active substances and fodder additive "Vinivet F".

Index	Substance "Perga refined"	Substance "Merva"	Fodder additive "Vinivet F"
		%	
Alanin	1,84	0,4	0,137
Arginine	0,89	9,3	3,078
Aspartic acid	2,10	16	5,326
Valine	1,10	1,2	0,213
Histidine	0,93	9,1	3,0
Glycine	0,84	18,4	6,073
Glutamine	-	1,4	0,461
Glutamic acid	2,24	5,4	1,80
Isoleucine	0,92	3,6	1,191
Leucine	1,50	2,3	0,80
Lysine	1,00	3,8	1,27
Methionine	0,62		
O-phospho-L-serine	Not detected	12	3,956
Tryptophan	2,11	2,28	2,01
Proline	2,05		
Serin	1,05	9,3	3,07
Tyrosine	0,65		
Threonine	0,97	1,3	0,429
Phenylalanine	0,88	1,2	0,41
Cysteine	2,44	-	0,0008

Table 2. The averaged vitamin composition of biologically active substances and fodder additive "Vinivet F".

Index	Substance "Perga refined"	Substance "Merva"	Fodder additive "Vinivet F"
		mg / kg	
A	0,7	Not detected	0,007
B_1	8	1,5	1,57
B_2	5,5	3,5	3,5
B_3	36	Not detected	0,42
B_5(PP)	51	Not detected	0,51
B_6	21	Not detected	0,21
B_{12}	2	Not detected	0,02
B_c	1,2	0,8	0,08
C	1080	Not detected	11,2
D_3	4	Not detected	0,09
E	12	8,7	8,64
H	0,8	Not detected	0,008
K_3	39,4	7,1	19

Table 3. The average chemical composition of biologically active substances and fodder additive "Vinivet F".

Index	Substance "Perga refined"	Substance "Merva"	Fodder additive "Vinivet F"
		%	
Moisture	6,05	8,2	8,2
Crude protein	22,50	33,3	30,1
Raw fiber	2,62	6,1	5,0
M. F. of fat	1,04	2,1	1,9
	Carbohydrates, %		
Glucose	23,7	1,78	2
Fructose	10,52	3,64	3,71
Sucrose	-	-	-
	Macro and microelements, mg / kg		
Iron	47,59	102,83	102,3
Zinc	18,53	59,54	59,1
Manganese	19,84	31,77	31,6
Copper	0,083	0,243	0,241
Cobalt			1,2
Molybdenum			1,2
Phosphorus			4,7 g / kg
Potassium			0,7 g / kg
Calcium			3,2 g / kg

As shown in the composition of the feed additive "Vinivet F", all vitamins are present, both fat and water-soluble; macro- and microelements are widely represented; and the revealed composition of irreplaceable and substitutable amino acids in the supplement surpasses in its balance any artificially created product. The value of the fodder additive "Vinivet F" based on perga and merva is exceptionally high not only in composition, but also in the level of each essential nutritional factor.

Concentrations of micronutrients in the composition of the fodder additive being studied are insignificant, but quite sufficient for the normalization of metabolism and optimization of the physiological and biochemical status of the fish organism. Due to the natural balance of the composition, the risk of overdosing both individual components and the whole complex of nutrients as a whole is eliminated, but also there is an alignment of antagonistic relations between the individual elements. As a consequence, bioavailability and nutrient uptake in the body is increased, which is also an important factor in optimizing nutrition when feeding fish.

So, the feed additive for fish feeds based on secondary sources of raw beekeeping is characterized by a complex chemical composition and is a universal source of natural biologically active complexes. Practical absence of heavy metals, radionuclides, pesticides and other harmful substances does not pollute the aquaculture of water bodies. Moreover, the properties of the Vinivet fodder supplement, due to its unique composition, contribute to the accelerated elimination of toxic substances from the body of fish. In this regard, the quality of commercial fish is improving, what improves the health of people.

Thus, the combination of the use of waste water of energy facilities, ecological approaches and the introduction of innovative technologies in the production of aquaculture objects allows solving such key tasks of social development as ensuring food security and maintaining public health.

Conclusions

Feed additives based on beekeeping products meet the criteria of ecological safety and improve the quantitative and qualitative characteristics of the composition of fish feeds for amino acids, microelements and vitamins in conditions of highly industrial fish farms at high planting densities and especially in warm water conditions of energy facilities that allow to grow aquaculture objects all year round.

References

[1] Abimorad E G and Carneiro D J 2007 Digestibility and performance of pacu (Piaractusmesopotamicus) juveniles – fed diets containing different protein, lipid and carbohydrate levels *Aqua.Nutr* **13** 1–9

[2] Kalayda M L and Islyamova A A 2016 Роль The role of fish in ensuring public health *Ecological problems and public health. Digest of articles All-Russian Scientific and Practical Conference* (Penza: RPO PSAMA) 38-42

[3] Bartley D M , Brugere C, Soto D and Gerber P&Harvey B (eds) 2007 Comparative assessment of the environmental costs of aquaculture and jther food production sectors: methods for meaningful comparisons. *FAO/WFT Expert Workshop. 24-28 April, 2006, Vancouver, Canada. FAO Fisheries Proceedings* (Rome, FAO) **10** 241

[4] Bogeruk A C, Lukanova I A 2010 *World aquaculture: experience for Russia* (Moskow:FGSO Rosinformagrotekh) p 364

[5] Recommended citation: FAO 2016 The State of World Fisheries and Aquaculture *Contributing to food security and nutrition for all* (Rome) 200

[6] Kalayda M L, Khazipov N N, Safiullin R R, Nabiullin R G, Akhmetova L T, Kalayda A A and Dementiev D S 2017 Development of fodder production is the most important task of development of aquaculture in the Republic of Tatarstan at the present stage *Status and development of aquaculture in the Russian Federation in the light of import substitution and ensuring the country's food security: materials of the II national scientific-practical conference, St. Petersburg, 13-15 of September 2017* (Saratov: LLC TseSain) pp 48-55

[7] Akhmetova L T, Harmonov S U, Zevakov I V, Maksimova T V and Sibgatullin G G 2013 Biologically active substance based on perga *Bulletin of the Russian University of Friendship of Peoples* **3** 73-7

[8] *Methodological instructions for cleaning with artificial sorbents of fish raw material contaminated with heavy metals* 1997 (St. Petersburg: GosNIORH) p 26

[9] *Medical and biological requirements and sanitary standards of quality of food raw materials and food products* 1990 (Moskow)
[10] Skrebneva L A 2016 Evaluation of the patterns of accumulation of heavy metals in honey bees and beekeeping products for the development of bioindicative criteria for atmospheric air pollution *abstract of the dissertation of the candidate of biological sciences* (Tolyatti) p 22
[11] Akhmetova L T 2015 Scientific substantiation and assessment of the effectiveness of the use of fodder additives in poultry farming, developed on the basis of raw sources of beekeeping: *abstract of the dissertation of the doctor of biological sciences* (Kazan) p 45
[12] Akhmetova L T, Harmonov S U, Sibgatullin G G, Salakhov I A, Zevakov I V and Irtuganova E A 2012 Biochemical composition and properties of a substance based on perga *Questions of biological, medical and pharmaceutical chemistry* **9** 27-31
[13] Akhmetova L T, Sibgatullin G G, Garmonov S Y and Akhmetova R T 2012 Technology for extraction of bee-bread from the Honeycomb *Procedia Engineering* **42** 1997-2000
[14] Akhmetova L T, Yusupova A A, Akhmetova R T and Sibgatullin G G 2011 New technologies of bee-keeping products *Materials of Ith European Congress of Applied Biotechnology ECAB* (Germany, Berlin) p 286

IOP Conf. Series: Earth and Environmental Science **288** (2019) 012037 doi:10.1088/1755-1315/288/1/012037

Aquabiotechnology of directed cultivation of microalgae Chlorella sorokiniana biomass

Bazarnova Yu.G., Kuznetsova, T, Trukhina E

Saint Petersburg State Polytechnical University named after Peter the Great

j.bazarnova@spbstu.ru, kuznetsova.ta1@spbstu.ru, Trukhina_ev@spbstu.ru

Abstract. We carried out cultivation of *Chlorella sorokiniana* microalga using various irradiation sources: daylight lamp with spectral range 530-610 nm and filament lamp with spectral range 3.5-5.0 μm. Specific growth rates were evaluated for these conditions at exponential phase. From the obtained biomass we extracted lipids and determined its' fatty-acid content. Also we analyzed triacylglycerol (TAG) content and unsaturated fatty acids, as well as oleinic acid content. Biomass irradiation by infrared source causes enlarging of total lipid content in alga biomass, as well as saturated fatty acids and oleinic acid portions. It is viable to use infrared irradiation mode for accumulation of non-polar lipids (TAG) at stationary growth phase, which might be used for biofuel obtaining after fractionating. Polar lipid fraction, extracted from biomass at linear growth stage, might be used for obtaining essential polyunsaturated fatty acids.

Introduction

Chlorella sorokiniana microalga is able to rapidly increase its biomass during cultivation and efficiently transform solar energy into chemical one. Biomass of *C. sorokiniana* might be considered as a renewable biofuel source of the third-generation (biodiesel, bioethanol, biohydrogen, biobutanol) [1–6]. Despite this, there is still no technology for biofuel obtaining from microalgae of adequate competitiveness price [7].

Biomass of this microalga type is also a perspective raw material for production of feed and high value-added products: essential lipids, pigments, carotenoids, proteins, carbohydrates [8].

Biomass of Chlorella family microalga is characterized by high specific growth rate, and for alga cultivation minimum requirements are needed for creation of optimum cultivation conditions [9]. For Chlorella cultivation wastewater may be used [10–12]

Development of aquabiotechnologies for directed cultivation and cascade processing of Chlorella microalga biomass are of immediate interest. Impact of various factors allows one to perform directed synthesis of lipids and phytochemical substances of Chlorella microalgae [13]. It is known that stress factors connected with shortage of nutrient broth components (for example, nitrogen)[7,13], result in lipid accumulation in biomass. Also chemical stressors of lipid accumulation may be used [7,14]. However, directed usage of stress factors for lipid accumulation must not be accompanied with reduction of biomass output [15]. It is known, that lipid accumulation generally takes place in alga cover [16].

The aim of the present investigation is to study influence of light spectral characteristics during cultivation on lipid content in biomass of single-cell alga *C. sorokiniana* and its fatty-acid content.

IOP Conf. Series: Earth and Environmental Science **288** (2019) 012037 doi:10.1088/1755-1315/288/1/012037

Materials and methods

The objects under investigation were *Chlorella sorokiniana* preculture (strain 211-8k) from Culture Collection of Algae at Göttingen University, international acronym SAG.

Cultivation of microalgae Chlorella sorokiniana biomass was performed in the laboratory photobioreactor [17], its dimensions are the following: height 380 mm, diameter 50 mm, volume 500 ml.

For microalga illumination we used daylight fluorescent lamp (DFL) with intensity 2500±300 lx, T=400 K, spectral range 530-610 nm; and filament lamp of 14100 lx intensity, spectral range 3.5-5.0 μm (IR).

Cultivation temperature was (23±1) °C; mixture aeration intensity was 1.5 l/min; mixing mode was periodical (15 minutes once a day); mixing speed was 500 rpm.

For microalga cultivation we used cultural medium, which was well-balanced in macro- and microelements content [18]. Duration of cultivation before reaching maximum cell concentration in the suspension was 9-10 days in the laboratory conditions.

Microalga cell shapes were studied by microphotographs from digital camera IS-500 (AO "LOMO", Saint-Petersburg). Microphotograph processing was implemented using Microanalysis FOTO software (AO "LOMO", Saint-Petersburg) and Levenguk ("Levenhuk LabZZ").

Microalgae samples were preliminary lyophilized at 1 mbar at temperature -55 °C in lyophilic drying installation Alpha 1-2 LDplus (Martin Christ Gefriertrocknungsanlagen GmbH). At cell wall disintegration the sample of 3 g weight was mixed with 10 ml of n-hexane and homogenized using high-speed homogenizer (IKA® Werke, T25 Basic) at 10 000 rpm for 5 min. Lipid extraction was carried out at Buchi Extraction Unit e-812 SOX installation. Mixture of 96%-ethanol and n-hexane 1:9 was used as an extracting agent. Composition of higher fatty acids in lipid extract of *Chlorella sorokiniana* microalga was determined using gas chromatograph with flame-ionization detector Agilent Technologies Sales & Services GmbH & Co.KG, at column BPX70 (60m×0.25mm×0.25 μm), SGE Analytical Science, VWR International GmbH; carrier gas is nitrogen.

Results and discussion

Research of influence of various irradiation conditions on C. sorokiniana biomass accumulation.

Analysis of cultivation curves (Fig.1) shows that the most favorable conditions for biomass growth are at DFL illumination mode.

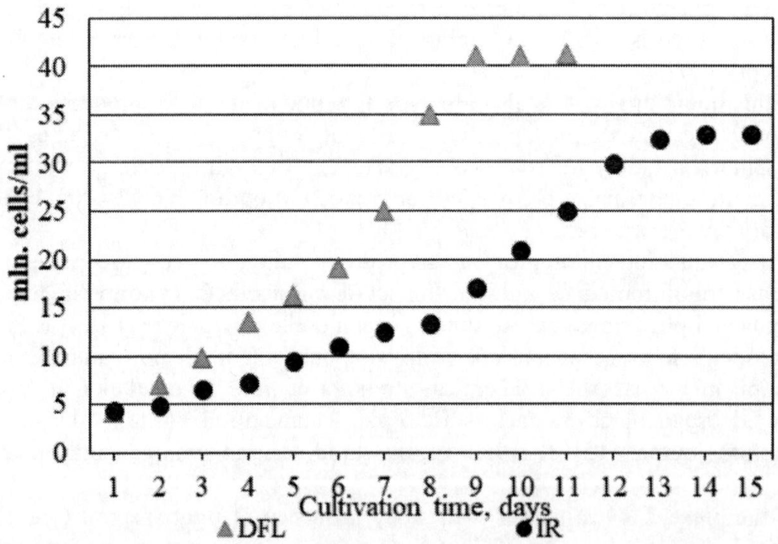

Fig.1 Cultivation curves for *C. sorokiniana* microalgae for various illumination modes.

It was determined, that in accumulation culture of single-cell alga biomass its growth is well described by S-shape curve [9], at which lag-phase, exponential growth phase, stationary phase are clearly observed. It was stated that maximum concentration of cells in biomass equal to 41×10^6 cells/ml was achieved at the 9-th day of cultivation. Specific rate of linear phase of biomass growth was $\mu = 0.26$ days^{-1}. At IR irradiation mode we observed less intensive population growth. Maximum growth of population was achieved at the 14th day, whereas specific biomass growth rate was $\mu = 0.15$ days^{-1}.

It was revealed, that linear phase of microalga biomass growth at IR mode was accompanied by cell flocculation, which is driven by pH shift to alkali region. Increase of alkalinity is definitely a stress factor. Flocculation might be considered as a response cell reaction to alkalinity increase, which reduces influence of this factor on single cells (Fig. 2b). At DFL mode increase of flocculation cell amount is observed only at the 9th day of cultivation, which corresponds to stationary phase.

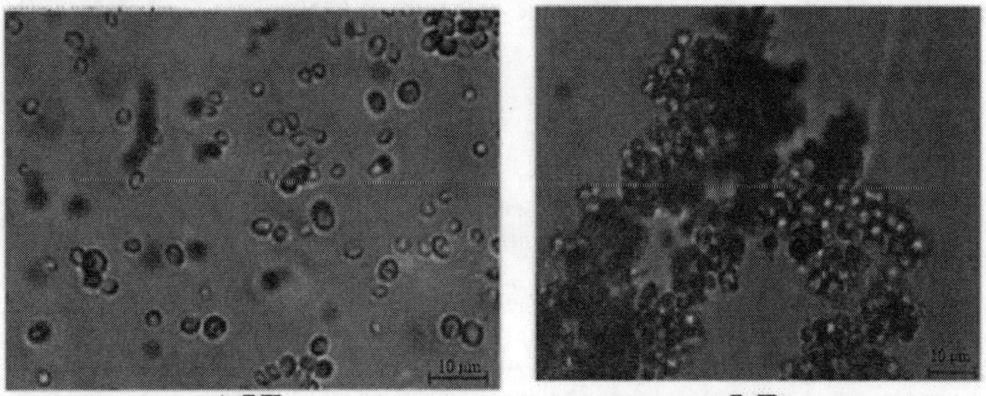

A. DFL B. IR

Fig.2. Microscopic image of *C. Sorokiniana* population while cultivation at various illumination conditions

Investigation of influence of illumination spectral characteristics on accumulation of lipids and higher fatty acids in biomass.

Cultivation of *C. Sorokiniana* at IR-irradiation results in increase of lipid content in biomass as compared to DFL irradiation mode for 5% in average (Fig.3).

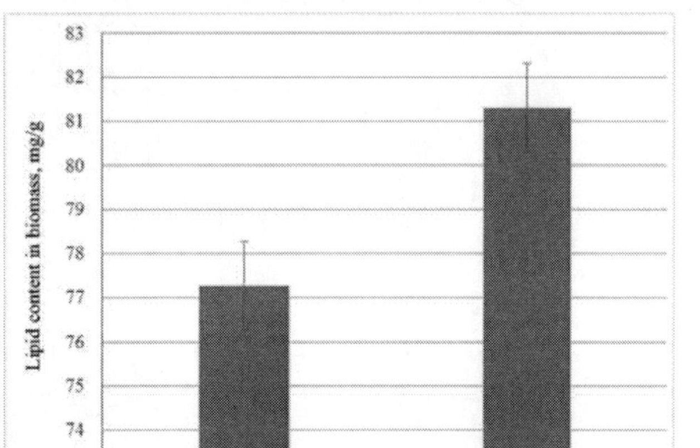

Fig. 3 Influence of illumination on lipid content in C. Sorokiniana biomass

IOP Conf. Series: Earth and Environmental Science **288** (2019) 012037 doi:10.1088/1755-1315/288/1/012037

At IR irradiation accumulation of saturated higher fatty acids takes place (Fig.4a) while DFL irradiation mode activates synthesis of unsaturated fatty acids (FA) (Fig.4b)

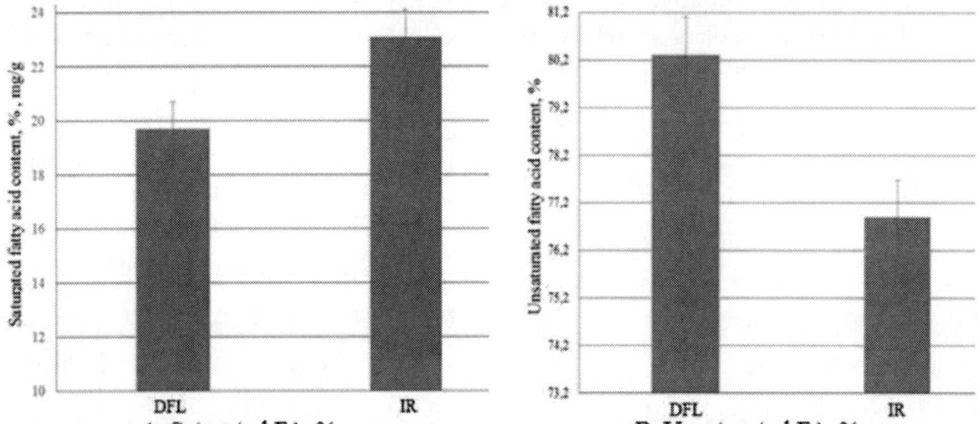

A. Saturated FA, % B. Unsaturated FA, %

Fig. 4. Influence of irradiation conditions on saturated and unsaturated FA content in lipids of C. Sorokiniana biomass

In literature there are some data allowing one to distinguish three aspects of adaptive function of triacylglycerol (TAG) of microalgae [19]. Neutral lipids have a function to be a source of long-chain fatty acids, which are construction blocks for membranes. Simultaneously with that, TAG biosynthesis provides runoff for photoassimilants, which prevent development of photooxidative damages during stresses, which reduce cell ability to utilize photosynthesis products. And, finally, TAGs which are deposited in the form of lipid globules, form storage area for secondary carotenoids, which provide protection from photodamage. After interruption of stress conditions the biochemical composition of cells restores[20].

Conditions for microalgae biomass illumination significantly affect oleinic acid content (C18:1) in total lipid fraction (Fig.5). At thermal exposure metabolization of polyunsaturated fatty acids takes place and accumulation of monounsaturated ones.

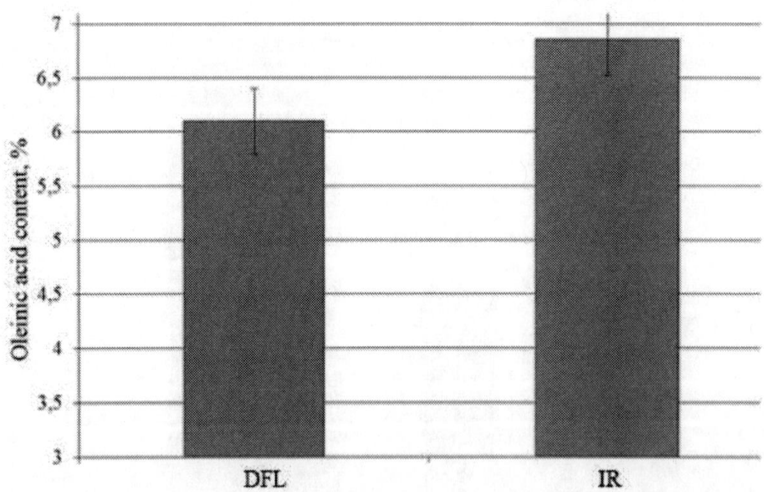

Fig.5. Oleinic acid content in a lyophilized biomass of *C. Sorokiniana.*

According to one of the hypothesis [21], reduction of light intensity in terms of one microalga cell is considered as a signal for termination of exponential growth phase. This leads to redirection of usage of surplus of absorbed light energy predominantly to synthesis of reserved TAG. TAG composition of these types is characterized by prevalence of saturated and unsaturated FA in them.

Thus, one of the important conditions for directed cultivation is usage of stimulators of lipid synthesis at preservation of high biomass output. It is valuable to use IR-radiation sources for illumination of the accumulated biomass in stationary growth phase. This leads to increase of nonpolar lipids (TAG) content, which will be used for biofuel obtaining after fractioning. Polar lipid fraction, extracted from microalga biomass at linear growth phase is aimed to be used for obtaining essential unsaturated fatty acids.

Acknowledgements

This work was performed within the implementation of Federal Targeted Programme for Research and Development in Priority Areas of Development of the Russian Scientific and Technological Complex for 2014-2020, the project «Development and implementation of innovative biotechnologies for treatment of microalgae Chlorella Sorokiniana and duckweed Lemna minor» (Agreement No. 14.587.21.0038), the unique project identifier is RFMEFI58717X0038.

References

[1] Lal A and Das D 2016 Biomass production and identification of suitable harvesting technique for Chlorella sp. MJ 11/11 and Synechocystis PCC 6803 *3 Biotech* **6** 1–10

[2] Kamyab H, Lee C T, Din M F M, Mohamad S E, Mohanadoss P, Khudhair A B and Roudi A M 2014 Biodiesel production from microalgae-Chlorella sorokoniana. *Aust. J. Basic Appl. Sci.* **86**,140-145

[3] Chader S, Mahmah B and Al E 2011 Biodiesel production using Chlorella sorokiniana a green microalga *Rev. des Energies Renouvelables* **14** 21–6

[4] Vishnyakova M A, Seferova I V and Samsonova M G 2017 Genetic sources required for soybean breeding in the context of new biotechnologies: (Review) *Sel'skokhozyaistvennaya Biol.* **52** 905–16

[5] Politaeva N A, Kuznetsova T A, Smyatskaya Y A, Trukhina E V and Atamanyuk I 2018 Energy Production from Chlorella Algae Biomass Under St. Petersburg Climatic Conditions *Chem. Pet. Eng.* 1–5

[6] Politaeva N, Smyatskaya Y, Slugin V, Toumi A and Bouabdelli M 2018 Effect of laser radiation on the cultivation rate of the microalga Chlorella sorokiniana as a source of biofuel *IOP Conf. Ser. Earth Environ. Sci.* **115**

[7] Dvoreckij D S, Dvoreckij S I, Temnov E I and Peshkov E V. 2015 *Tekhnologiya polucheniya lipidov iz mikrovodoroslej [Technology of lipids obtaining from microalga]* (Tambov: Izdatel'stvo FGBOU VPO «TGTU»)

[8] Skjånes K, Lindblad P and Muller J 2007 BioCO 2 – A multidisciplinary , biological approach using solar energy to capture CO 2 while producing H 2 and high value products *Biomol. Eng.* **24** 405–13

[9] Vladimirova M G and Semenenko V E 1977 Mass microalgae cultivation [Massovoe kul'tivirovanie mikroskopicheskih vodoroslej] *Plant life [Zhizn' rastenij]* (Moscow: Izd-vo «Prosveshchenie») pp 367–76

[10] Yao L, Shi J and Miao X 2015 Mixed wastewater coupled with CO2 for microalgae culturing and nutrient removal *PLoS One* **10**

[11] Franco M C 2014 Batch cultivation of microalgae in the Labfors 5 Lux Photobioreactor with LED Flat Panel Option *InforsHT*

[12] Alexeev D V, Shagivaleev A A, Urazbakhtina L R and Napoykina E A 2015 Improving the efficiency of water recycling systems at wastewater treatment power plants from insoluble suspensions *Proc. High. Educ. institutions. ENERGY Sect. Probl.* **1–2** 11–7

[13] Rai V, Muthuraj M, Gandhi M N, Das D and Srivastava S 2017 Real-time iTRAQ-based proteome profiling revealed the central metabolism involved in nitrogen starvation induced lipid accumulation in microalgae *Sci. Rep.* **7**

[14] Wase N, Tu B, Allen J W, Black P N and DiRusso C C 2017 Identification and metabolite profiling of chemical activators of lipid accumulation in green algae *Plant Physiol.* pp.00433.2017

[15] Ngangkham M, Ratha S K, Prasanna R, Saxena A K, Dhar D W, Sarika C and Prasad R B N 2012 Biochemical modulation of growth, lipid quality and productivity in mixotrophic cultures of chlorella sorokiniana *Springerplus* **1** 1–13

[16] Watanabe K, Imase M, Sasaki K, Ohmura N, Saiki H and Tanaka H 2006 Composition of the sheath produced by the green alga Chlorella sorokiniana *Lett. Appl. Microbiol.* **42** 538–43

[17] Politaeva N, Kuznetsova T, Smyatskaya Y, Elena T and Ovchinnikov F 2017 Impact of various physical exposures on Chlorella Sorokiniana microalgae cultivation *Int. J. Appl. Eng. Res.* **12** 11488–92

[18] Czarena L Crofcheck, Michael Monstross, Xinyi E, Aubrey P Shea, Mark Crocker and Rodney Andrews 2012 Influence of media composition on the growth rate of Chlorella vulgaris and Scenedesmus acutus utilized for CO2 mitigation *J Biochem Tech* **4** 589–94

[19] Solovchenko A E 2012 Physiological role of accumulation of neutral lipids by eucariotic (in Russian) *Physiol. plants* **59** 192–202

[20] Klyachko-Gurvich G L and Rudova T S 1973 Influence of imidazol on fatty acid exchange at reconstruction of Chlorella cells after nitrogen depletion *Physiol. plants* **20** 326–31

[21] Cohen Z and Khozin-Goldberg I 2010 Searching for Polyunsaturated Fatty Acid-Rich Photosynthetic Microalgae *Single Cell Oils: Microbial and Algal Oils: Second Edition* pp 201–24

Year-round fish breeding on energy saving principles of industrial aquaculture

P.E. Garlov, N.B. Rybalova, E.D. Shinkarevich, T.A. Nechayeva, S.U. Temirova, B.S. Bugrimov

Saint-Petersburg State Agrarian University, Russia, 196601, Pushkin Peterburgskoe Highway, 2

wba@mail.ru

Abstract. With a view to year-round valuable fish farming a control system terming their breeding and growth, protected by USSR inventor's certificates and Russian inventor's patents, was developed. The management principle is fish contents in "critical" salinity along with the simultaneous thresholds "signal" factors (temperature and light) and subsequently obtaining offspring and juveniles in complex of optimal ecological conditions.
For the implementation of the proposed industrial biotechnology and year-round fish farming a new installation of water recycling system (WRS) hatcheries and farms was developed by means of the underground conditioning of fish cultivation medium. Their essence is that fish farm water supply systems are additionally provided by underground, or in a half in ground large volume tank-sump tanks. Such WRC, being isolated from climate conditions, allow to conserve and use natural seasonal water resources and first consensually resolve earlier alternative volume-dependent energy problem (requiring lower volumes of water) and water purification (requiring its increasing) in septic tank-sump tanks. The basic principle of their operation is filling one tank-conditioner by “cold” water and the other by “warm” at the appropriate seasons and by additional water supply fish farm systems water circulation. Techno-economic calculation shows, that with the increasing volume of such water air conditioning heat transfer will be proportionally reduced, but productivity increases and reduces its system specific cost while maximizing reliability.

Introduction

On the basis of constructive working schemes of fish neurosecretory brain centres participation in reproduction by means of their maintenance in material-energic body homeostasis the control system terming breeding and growth of valuable fish was developed with a view to year-round fish aquaculture [3]. Specifically to stimulate breeders maturity preparations of isolated anterior and posterior lobes of the pituitary gland were developed and introduced in sturgeon culture. It allows to enhance breeders industrial usage to an average of 15 % (inventor's certificates ## 719671, 1163817). To delay breeders maturity, to reserve their brood stocks and to accelerate growth of juveniles the method of their industrial reservation and growing in critical salinity medium 4-8‰ was produced (inventor's certificates ## 965409, 682197, Russian inventor's patent # 2582347). This external nature medium, close to the inner body medium, provides energy-efficient osmotic equilibrium between them and therefore optimizes fish reproduction, growth and survival [3]. Ecological-physiological principle of management is The management principle is fish contents in "critical" salinity along with the

IOP Conf. Series: Earth and Environmental Science **288** (2019) 012038 doi:10.1088/1755-1315/288/1/012038

simultaneous thresholds "signal" factors (temperature and light) and subsequently receipt offspring and cultivation juveniles in complex of optimal ecological conditions (figure 1).

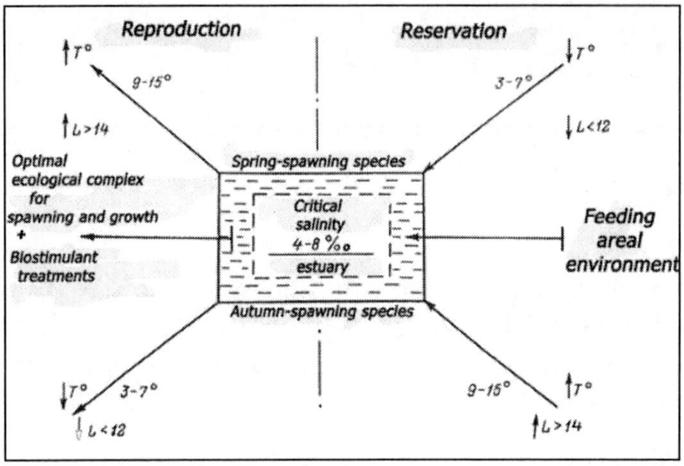

Figure 1 Principle of fish breeding and accelerate growth management by the triad of leading environmental factors: signal (T°, L) and phylogenetic (‰) values, providing metabolic (material and energic) body homeostasis on the example of fish migrations basic mechanism [by: 3]

Results

For the implementation of the proposed industrial biotechnology and year-round fish farming a new installation of water recycling system (WRS) hatcheries and farms was developed by means of the underground conditioning of fish cultivation medium (inventor's certificates #982614, figure 2A.; Russian inventor's patent # 2400975, figure 2**B**.).

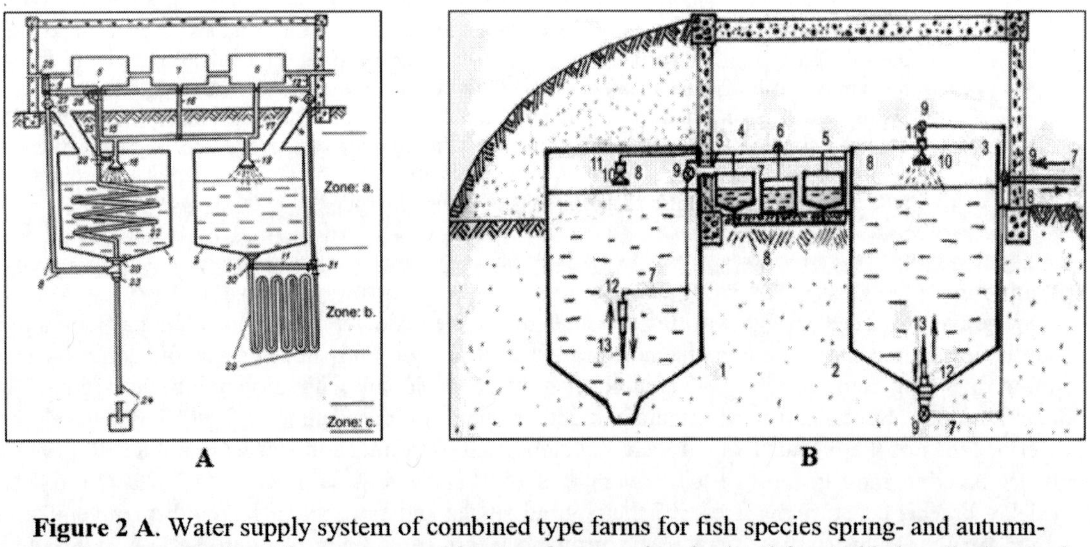

A **B**

Figure 2 A. Water supply system of combined type farms for fish species spring- and autumn-spawning reproduction. The system contains 2 underground tanks (1, 2) located below the layer of seasonal frost penetration (zone a.), each of which is connected with fish pools (5, 6), with means of aeration and water purification (7).
B. Water supply system of the fish hatcheries, which includes sediment tanks, partially buried in the ground (1, 2), fish ponds (4, 5), auxiliary water treatment (6).

Systems are based on natural and industrial principles of environmental engineering [1] and the new energy-saving principle of subterranean hydro-conditioning of fully managed medium for fish cultivation [3].

They also are operated on the basis of a new principle of fish cultivation management biotech by the triad of environmental factors (see figure 1) and therefore, are fundamentally different from all the WRS developed, that now using in fish farms and considered the most perspective [2, 4, 5].

The essence of new technological solutions is that fish hatcheries and farms water supply systems systems are provided additional by large volume underground tanks buried either in the ground or in a half undergrounds (figure 2). Such in fact fundamentally new motherland WRS, isolated from climate conditions, allows to maintain and use the natural seasonal water-energic resourses (instead of reproducing the seasonaly lost) and firstly to resolve consensually earlier alternative volume-dependent energy problem (requiring lower volumes of water) and water purification (requiring increasing amounts of water) in tank-sump tanks.

Storage tanks (figure 2a) may be equipped with heat exchangers that provide long-term seasonal water term stabilization in the system, which consists of 2 different tanks placed below ground layer below seasonal frost soils (zone a: 1, 2) and communicating with the ground mines (3, 4). 1 reservoir associated with fish pools (5, 6), with means of aeration and water purification (7) through a system of pipelines (8, 9) and pumps (10), and the tank 2-through piping system (11-13)-with pumps (14). Return water from fish basins, aeration and cleanup tools is realized through pipelines (15-17) running spray aeration nozzle (18, 19). At the bottom of the tanks placed water intake tube (20, 21), connected to the piping. 1 tank is equipped with an internal heat exchanger (22), United with artesian water supply system (zone c: 24). Heat exchanger (22) connected through the control valve with a nebulizer (18). The outside of the tank 2 installed heat exchanger (29), housed in the groundwater zone (zone b) and connected in parallel to the reservoir and pipeline 2 13. Regulating valves (30, 31) allow you to skip the water from the tank through a heat exchanger 2 (29), or bypassing it. The basic principle of operation systems is filling one tank-air conditioners «cold» water (e.g. with $3\text{-}7^0$) and the other "warm" ($9\text{-}15^0$) in the appropriate seasons and water them further land-based hatcheries pools closed water circulation systems. Reviewed and management options composition of water and prolonged heat sunk its system of interseasonal heat exchangers in the relevant ground zones (figure 2A).

For breeding spring-spawning fish and reservation autumn-spawning water supply is carried out from the tank 2 ($9\text{-}15^0$); for breeding autumn-spawning and reservation spring-spawning fish – 1 tank ($3\text{-}7^0$). In summer, the heat setting of cold water in the water tank 1 is carried out by artesian water through a heat exchanger (22) and heat setting (additional insulation) "warm" water through the heat exchanger in winter (29) in the zone all year round warm (with $10\text{-}15^0$) groundwater. There are a variety of options for use of tanks-in isolation, with a mix of warm, cold river and artesian waters, use of geothermal waters in the lines, etc.

Technical-economic calculations shows that already in the volume of water in the tank from 10.000. m^3 speed heat transfer in soil decreases to 0, 1°C per month and below, and the degree of water purification progressively increases due to the sumption effect. With the increasing volume of reservoirs-hydro-conditioners proportional increases productivity and reduces its system specific cost while maximizing reliability, available for any culture.

Principle of this system operation is based on the following common biological and bio-geo-coenological patterns, proven us reference and estimates (if protection of inventions, scientific discussions, research reports):

1) match temperature ranges spring-spawning fish breeding with temperatures reservation autumn-spawning ($9\text{-}15^0$C) and vice versa (with $3\text{-}7^0$C), that allows to use 2 temperature modes in two autonomous circulating systems;

2) groundwater and soil temperatures (below freezing) coincide with the seasonal spawning fish bred for local climatic zone (South: sturgeon, carp – $9\text{-}15^0$C; North: whitefish, salmon – $2\text{-}7^0$C with year-round);

3) most common of artesian water temperature (with 5-9^{0}C) close to the essential (3-7^{0}C) tank system 1,

4) any geothermal waters are suitable for additional insulation (with 9-15^{0}C), as well as for different use cases in any system of tanks. ROI application system must grow in regions with short or generally unfavorable growing season.

However, the disadvantage of such a system are the high costs for the construction and operation of underground tanks with mines and heat exchangers, reducing its reliability, as well as large energy consumption for water supply in land-breeding facilities that reduces the effectiveness of the work as a whole.

Therefore, to reduce the cost of building and operating the system, increasing the reliability and efficiency of its work, we developed a new simplified (modernized) device (figure 2B). This task is solved by the construction of a special device the water system tanks households of which fish are recessed into the ground and have calculated the amount of taper and providing heat gradient from inner medium, unlike the previous one, no more than 0,1°C per month. It differs from the previous one so that the water tanks buried in the ground in the upper part and its part, being above the earth's surface, isolated from climatic influences various insulating materials. Their filling with water heat gradient not exceeding0,1°C per month, they have a conical bottom corner not less than the 45^0 that provides maximum water circulation for the spillway. On pipelines for water are a means of its aeration and physico-chemical treatment, but for the first time their water pipes are equipped with devices for automatic vertical movement, managed water quality sensors. The latter are designed for water intake of pure water higher than defending the beleaguered mist and its complete removal from the bottom when water outlet. It is important to emphasize that if the gradient of heat transfer to the surrounding (outer) medium 0,1°C per month, the tank volume provided securely 10,000 m^3 increased, this ratio decreases even more.

The cost of iron-concrete-underground tanks, thousands rub. in prices 2007 year	Volume of tanks m^3	#the model draft (SNIP)	Type of plant farming main indicators date design	Regulatory and estimates					
				Techno-economic indicators		Indicators of water consumption			Additional indicators for the calculation
				Standards unit costs per unit of output (in LCY for 1 thousand pieces. FRY)	Supplementary products (% increase in these stages	Water supply for tank, m^3 water speed recycling	Specific water supply for tank, m^3/unit. products, the estimated calculated under the new scheme	Minimal reduction of specific water consumption, %	
2.700	1000	4-18-850	Sturgeon farm (project No. 1577) power: 30 million. PCs. the main production of fingerlings governmental funds:-11.9 million. rub. Water consumption: 33 million. m3/year of energy consumption-153.2 thous. USD	1673	Increase: 1) ROE: 10-15, 2) larvae and juveniles:5-10 3) total products: 20-30 due to the development and expansion of the minimum of the I	2000 1,5 6000 4 10000 7	1,1 0,88	20	Calculations are made without taking into account additional underground tanks heat exchanger, comprising one hundred-imost (at 1985 g.) artesian wells and water ponds, 10-12, 5 thousand. rub. Gradient of heat transfer tank volume of 10000 m^3 : 0.1 - 0.79^{0}C/month. Optimal volume of tanks for fish farms of various types-6000-10000 m^3 (the ratio of the volume of tank-sump and fish basins for carp farms-5-10:1; for trout farms-100-200:1. Coefficient of water utilization management in at operation of reservoirs in water-recycling increases from 0 to a value close to 1. Technical readiness factor of concrete underground tanks on building standards (SNIP) is close to 1.
3.500	2000	4-18-851	Salmon farm (project No. 1428) power-0.5 million. PCs. juveniles per year principal production funds-3.3 million. rub. Water consumption by 50 million. m3/year Energy-71 thousand. USD/year	27.000	Increase: 1) caviar incubation:-20, 2) larvae at growing: 5-10, 3) juveniles due to optimization of hatching: 20-30	2000 1,3 6000 3,9 10000 6,4	60 40	33,4	
4.700	6000	4-18-853							
5.800	10000	-	Whitefish farm (object no. 1316) power-1.5 million. PCs. yearlings. The main production funds 344.2-thousand. rub. Water consumption:-3.73 million. m3/year	5.092	Outputs: 1) caviar incubation:-20, 2) larvae at growing-5-10, 3) juveniles due to optimization of hatching-20-30	2000 1,7 6000 5,1 10000 8,5	3,4 2,83	17	
8.000-10.000	20000								

Figure3. Table estimated and estimates for construction and operation of underground tanks for fish farms

IOP Conf. Series: Earth and Environmental Science **288** (2019) 012038 doi:10.1088/1755-1315/288/1/012038

This solution provides lower costs to build and operate the system, increasing its reliability and efficiency through optimizing placement, design and volume tanks. This means the availability of equipment for maintenance and repair, as well as a single level of water filling throughout the system. You do not need the buried heat exchangers and mines and with the increasing volume of tanks and additional means of thermoregulation and water purification. Main technical and economic indices calculation possible exploitation of the system are given in the table. For the effective valuable fish species growth in such WRS it appears appropriate to continue the search, development and testing of such energy-saving critical salinity medium, but already artificial (retrofit) similar tested for effective breeders reserve migratory and local forms of sturgeon and bony fish (inventor's certificate # 965409).

References

[1] Astvacaturov A.e. Environmental Engineering: Stud. allowance. -Rostov n/w: DSTU publishing center. 2006-250 with.

[2] Brajnballe I guidelines for aquaculture in closed water installations. Introduction to new environmental and highly productive fish closed system. -Copenhagen: Eurofish, FAO subregional office for Central and Eastern Europe, 2010. -71 c.

[3] Garlov, P.e. Nechaeva t.a., Mosjagina M.v. "mechanisms of neuroendocrine regulation of breeding fish and prospects of artificial reproduction of their populations. UPB: IZD-vo "Science Avenue", 2018. 335 s.

[4] Summerfelt, S.T. Hankins J.A. Weber, A.L., Durant M.D. Ozonation of a recirculating rainbow trout culture system. 2. Effects on microscreen filtration and water quality//Aquaculture aims. -1997. -Vol. 158. -N 1-2. -R. 57-67.

[5] Van Rijn (J). The potential for integrated biological treatment systems in recirculating fish culture//Aquaculture aims. -1996. -Vol. 139. -N 3-4. -R. 181-201.

IOP Conf. Series: Earth and Environmental Science **288** (2019) 012039 doi:10.1088/1755-1315/288/1/012039

Using Redox potential in water quality assessment of energy facilities

M E Gordeeva and M L Kalayda

"Water bioresources and aquaculture" department, Kazan state power engineering university, Russia, 420066 Kazan, Krasnoselskaya, 51

Maria.Galeeva@gmail.com

Abstract. The question of natural water quality for a long time has wondered researchers and ordinary citizens. Cities expanded, population increases, it leads to a deterioration in natural waters by sanitary-chemical and microbiological indicators. In this regard, this article concerns the question of the monitoring of physicochemical indicators and introduction the new indicator – Redox potential (Eh). Dynamics of Eh changes in the rivers of the Republic of Tatarstan and surrounding areas as well as natural and artificial water reservoirs, including basin-coolers of energy facilities are researched in this article. Conducted full-size data and experiments made it possible to establish the Eh sensitivity in relation to the pH. In the article has showed the possibility of introducing Eh indicator into the assessment of water quality on the assumption of further consideration and detail.

Introduction

Currently, the expansion of urban areas a matter of quality drinking and natural water is really an issue. In this regard the relevance of monitoring water facilities no doubt.

Public monitoring of water reservoirs is a system of observation, assessment and prediction the changes of water objects within federal property, property of subjects of the Russian Federation, municipal property, property of natural and legal entities [1]. It is a part of the Public environmental monitoring.

The object of water and environmental monitoring is assessment of there quality and level of pollution as necessary condition for the adoption of science-based decisions about the effectiveness of environmental protection measures. Conducted monitoring allows to determine the practical use of the water body: drinking, recreation, fisheries, irrigation, technical. The greatest interest is the reservoirs, related to multiple uses. An example of such reservoir is the Lake Sredniy Kaban in the city of Kazan, which since 1931 year has been the basin-cooler of the Kazan CHPP-1, has experienced anthropogenic load from sewage of industrial enterprises and since 2013 has been the sport object on which international competitions for rowing are held [2].

Currently formed a lot of parameters for controlling the water quality. A complete list of monitored parameters is provided in the normative and technical documents of water-sanitary legislation. However, this list is missing such indicator as redox potential. Redox potential is a measure of the chemical activity of elements or their compounds in reversible chemical processes associated with the change in the charge of ions in solution [3].

To answer the question about the possibility of using Redox potential in water quality assessment of energy facilities is required:

- analysis of the possible range of Eh values in natural waters;
- general understanding of changing processes of water Eh values.

Redox potential of natural waters

To analyze the possible range of Eh values changes in natural waters we researched Eh values in rivers and in natural and artificial reservoirs. To determine the Eh values used ion meter I-160 MI.

Focusing on major rivers, the territory of the Tatar ASSR divided into 3 parts on the orography – Predvolzhie, Predkamie and Zakamie. Predkamie on the river Vyatka is divided into West and East. Zakamie on the rivers Verhnii Chereshman and Sheshma River is divided into Western Zakamie (low) and Eastern Zakamie (high). In accordance with this division all collected data of Eh-value in rivers (total amount of investigated rivers is 30) are shown in Figure 1.

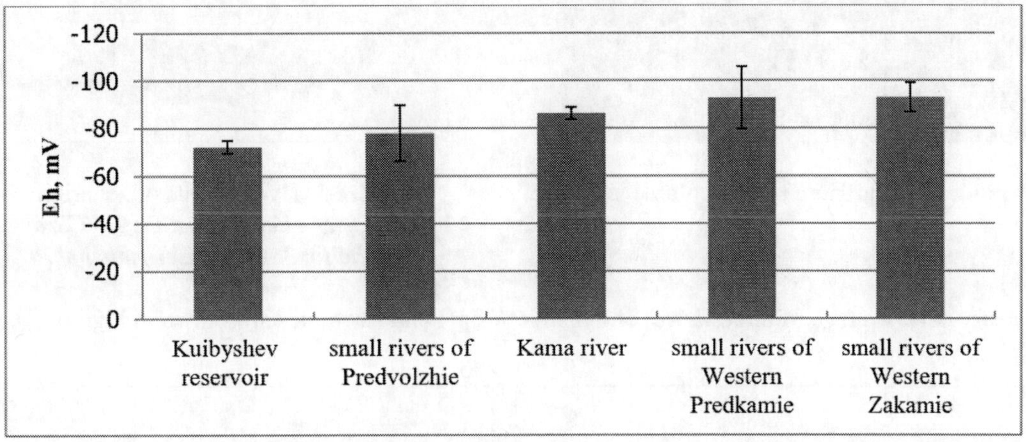

Figure 1. The mean values of Eh (mV) with the standard error of average in Kuibyshev reservoir and the rivers of the Republic of Tatarstan.

The analysis of the collected data showed that the range of Eh changes from -21,1 mV to -166,1 mV [4]. Minimum and maximum values are in the small rivers of Western Predkamie. The amplitude of variational series is 145 mV. The great range of values is confirmed by the standard error of the mean (Figure 1). Minimal variations observed in the Kuibyshev reservoir and on the Kama River.

The range of Eh values changes in lakes and natural ponds more wide compared to Eh values in rivers. For example, only in the Republic of Tatarstan the range of Eh values changes in lakes from 7,5 mV to -226,4 mV [5]. The amplitude of variational series is 233,9 mV. It is in 1,6 times higher the Eh amplitude in rivers. For reservoirs, experiencing the action of energy facilities (for example, the Sredniy Kaban lake of Kazan) the average value of Eh in zone, not suffering from action of energy facility (Kazan CHPP-1), is -53±8,5 mV. In the area of direct impact of energy facility Eh value drops to -80 mV, average value is -70±8,5 mV [2]. The value of -80 mV, characterized for zone of confluence heated water of offtake into the lake Sredniy Kaban, is according to Eh values in small rivers of Predvolzhie and the Kama River.

Redox potential in artificial systems

For the general understanding of changing processes of water Eh values at the "Water bioresources and aquaculture" Department of Kazan state power engineering university was devised an experiment. Artificial ecosystem was created by using aquarium (water volume is 600 l) (Figure 2) with the continual aeration of water. At the time of the experiment in the aquarium contained 2 *Acipenser gueldenstaedtii*, 12 *Carassius gibelio* and 8 *Acipenser ruthenus* (Figure 3). For control was used tap water by which filled the aquarium. The experiment was conducted without using biofilter and with water changing (1 time in 3 days) for estimation of fish metabolites. The number of series of the experiment is 3. Measurable indicators: Eh (mV), water temperature (°C), pH, concentration of ammonium nitrogen (mg/l) and concentration of ammonia (mg/l).

Figure 2. Aquarium (water volume is 600 l).

Figure 3. Hydrobionts of aquarium: 1 –*Acipenser gueldenstaedtii*; 2 – *Carassius gibelio*; 3 –*Acipenser ruthenus*.

Summary data (average of three series) of dynamics of pH and Eh values are shown in Figure 4.

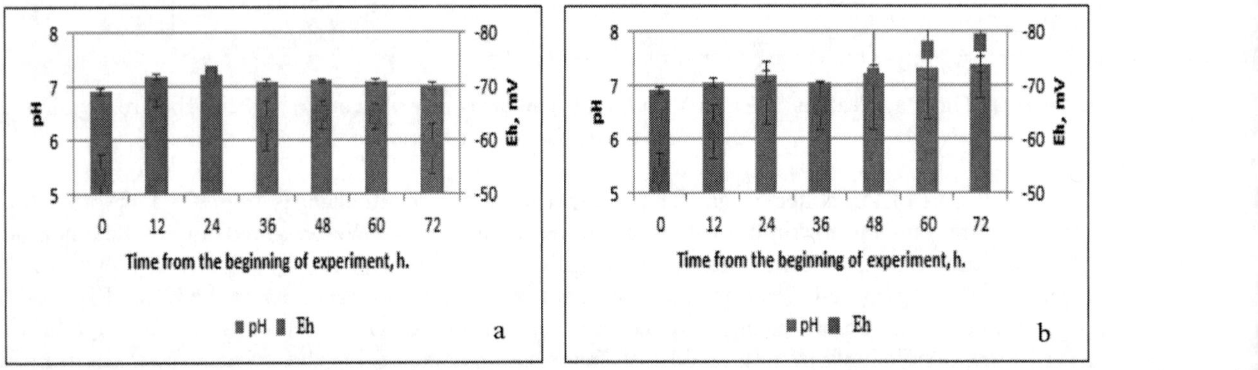

Figure 4. Dynamics of pH and Eh (mV) values in the water with hydrobionts (a) and in water without hydrobionts (b).

From data provided in Figure 4 you can see that on the first day of the experiment, both in control and in an aquarium with fish, the Eh values in water have been dropping, and pH values have been increasing. On the second and third day there has been a gradual increasing of Eh values in aquarium water (oxidation processes) associated with the accumulation of metabolites and lowering of Eh values in the control sample (reduction processes). Thus, redox potential allows you to keep track of time when it's need to improve the quality of water.

Considering that the initial water for the aquarium and control are identical, it can be assumed that the difference of Eh values in aquarium water and control sample is the value characterized the highlight activity of hydrobionts and mostly concerned with nitrogen substances. Monitoring and analysis of the allocation of nitrogenous substances (Figure 5) by fishes in the context of the experimental aquarium is make possible to find relation between the concentration of ammonia nitrogen and redox potential which corresponds to the vital functions of hydrobionts. The coefficient of determination is 92.73%. The regression equation is presented in Figure 5.

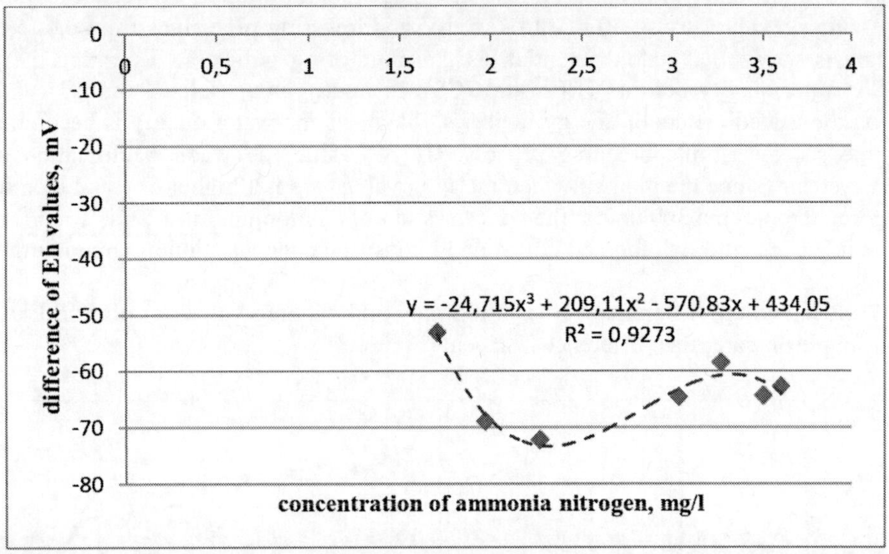

Figure 5. Relation between the concentration of ammonia nitrogen (mg/l) and redox potential (mV).

A similar experiment was done in closed-containment aquaculter system (CCS) with using the biofilter. In the CSS contained *Cyprinus carpio var. koi* (Figure 6). The experiment lasted from the moment when the process was characterized by dynamic and swinging until the sustainable up trend of Eh. This time amounted to 17 days. The experiment allowed to take into account not only the influence of fish metabolites on changing of the hydrochemical indicators, but also see the influence of biofilter organisms to changes of characteristics of water ecosystem.

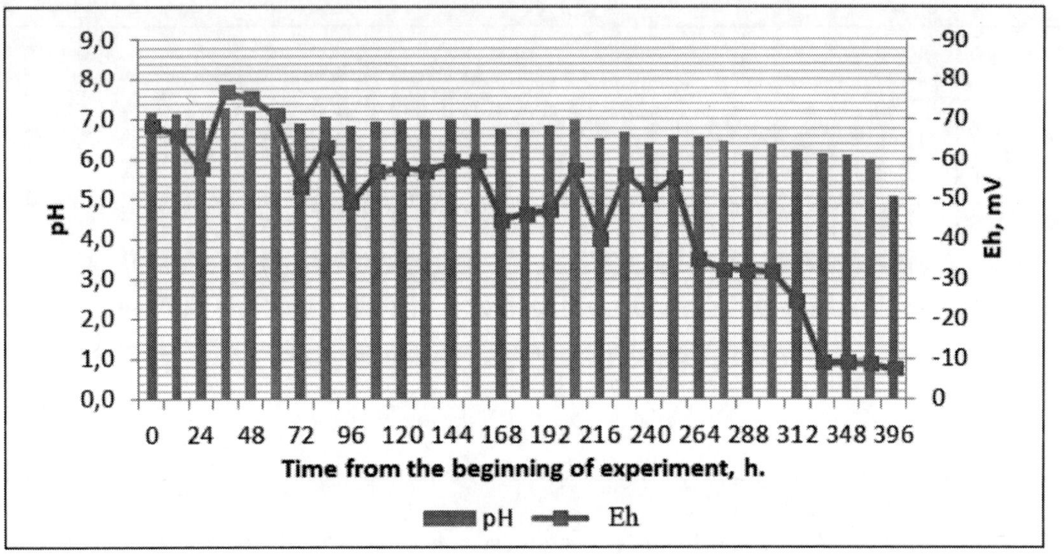

Figure 6. Dynamics of Eh and pH values in CCS (orange marked morning measurements, before fish feeding).

Analysis of the obtained data allowed to find a prosperous phase of water ecosystem, which is characterized by fluctuations of pH and Eh values ranging from 6,4 to 7,3 and from -40 mV to -77 mV respectively (the first 11 days) (Figure 7). The nest stage is deterioration of water ecosystem. In this phase increasing Eh values from -40 mV to -7,5 mV and lowering pH values from 6,4 to 5,1 (Figure 7). The analysis shows that redox potential, as the monitoring indicator, determine the process of declining the water quality before pH (in the CCS). In highly industrialized forms of fish farming, especially on the heated waters of energy facilities, diagnosis the water quality is becoming a critical factor for the success of the breeding process. By pH value the water acidification and out of equilibrium systems (since the moment when morning values are not higher or equal then the evening values) is going through the 300 hours (the concentration of ammonium nitrogen is 1 mg/l) or 12 days. By Eh value it happens through the 264 hours, or 11 days (the concentration of ammonium nitrogen is 0,9 mg/l).

Figure 7 illustrated the difference between morning and evening values of pH and Eh in CCS. In the control sample of water this difference is absent (Figure 8).

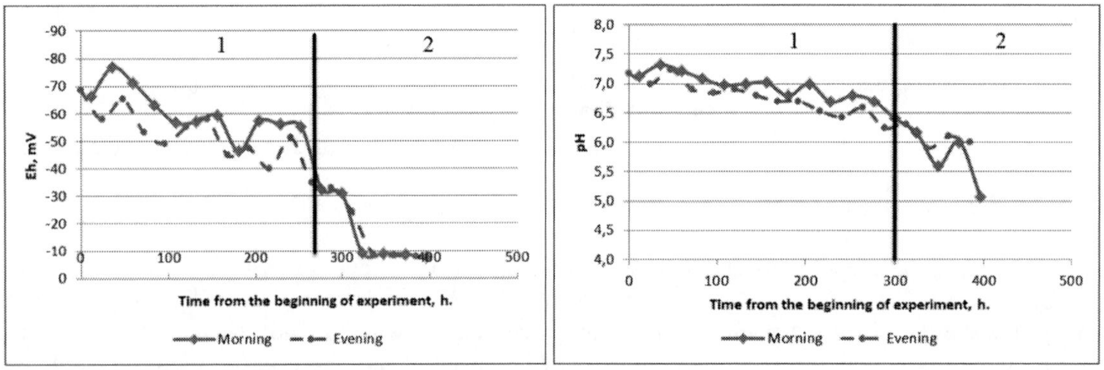

Figure 7. Morning and evening pH and Eh values in CCS: 1 – prosperous phase of water ecosystem; 2 – deterioration phase of water ecosystem.

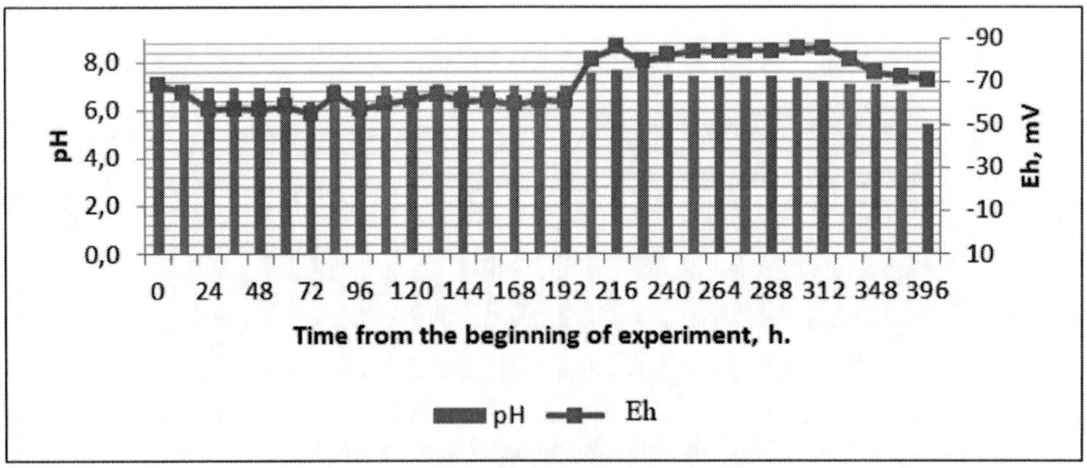

Figure 8. Dynamics of Eh and pH values in control sample (orange marked morning measurements, before fish feeding).

The amplitude range of pH is from 5,1 to 7,3; Eh from -7,5 mV to -76,9 mV. The amplitude of variation series of pH is 2,2 units; of Eh 69,4 units. The amplitude of variation series of Eh in 31 times higher compared to pH.

Thus, redox potential is more sensitive indicator compared to pH. According to our researches we revealed that 1 unit of pH corresponds to 58,873 mV of Eh. Experiments have confirmed the sensitivity of Eh indicator to lower water quality (for example, in CCS). Redox potential can be used in water quality evaluation of water ecosystems and basin-coolers of energy facilities after detailed elaboration.

References

[1] Water codex of Russia Federation 2006 (red. 29.07.2017) **74-FS**

[2] Kalaida M L, Gordeeva M E 2015 Complex assessment of status of lakes ecosystem. Urbanized territories *LAP LAMBERT Academic Publishing* 228

[3] Tokarenko O G 2014 *General hydrogeology: chemical composition and properties of natural waters* (Tomsk: TPU) p 57

[4] Gordeeva M E, Kalaida M L 2018 Redox potential as an indicator of the water quality in the monitoring of water ecosystems *All-Russian scientific and practical Conference «Water bioresources and aquaculture of southern Russia»* 40-4

[5] Kalaida M L, Gordeeva M E 2017 Redox potential as an indicator of the quality of natural and artificial reservoirs in the structure of fisheries monitoring *The role of education and science in the development of Russian society: collection of articles of the International scientifically-practical Conference* (Electronic resource) (Moscow: Impuls) pp 279-91

IOP Conf. Series: Earth and Environmental Science **288** (2019) 012040 doi:10.1088/1755-1315/288/1/012040

Modification of the temperature condition of the basin-cooler in the energy object coverage

M E Gordeeva and R V Zanozeev

"Water bioresources and aquaculture" department, Kazan state power engineering university, Russia, 420066 Kazan, Krasnoselskaya, 51

Maria.Galeeva@gmail.com

Abstract. In the context of the growing interest to urban ecology, in particular the forces of human impacts on aquatic ecosystems, and the inevitable need for electricity generation for the life of the townspeople, becoming especially topical study thermal effects on basin-coolers. In article is described modification of the temperature condition of the basin-cooler within range of the Kazan CHPP-1, namely the process of cooling water during open offtake (in the dynamics of years). Demonstrated the lowering of water temperature (average of the season) in open offtake and in the basin-cooler – the Sredniy Kaban lake.

Introduction

Currently, aquatic ecosystems are objects of multi-purpose industrial and/or agricultural uses. Particularly relevant for research become reservoirs located within major cities experiencing increased anthropogenic impact. In Kazan, a city with over a million dwellers, is located the system of Kaban lakes (from 1902 onwards), attracting the attention of researchers. Special interest is the Srednyi Kaban lake which is the basin-cooler of the Kazan CHPP-1 from the year 1931.

Characteristics of the object of research

Total water surface of the Srednyi Kaban lake, taking into account open offtake of the Kazan CHPP-1, is 131.8 hectare. Since the use of the lake as a basin-cooler the 23% of its water area not covered by ice during winter time, which opens the possibility of using a reservoir for aquaculture. Length of the open offtake and the object of our detailed research are 760 meters. From 1931 year warmed water waste were cooling directly on the CHPP and next on open offtake and entered to the Srednyi Kaban lake. According to 2010/2011 for the full length of open offtake the water cooled to 3.5 °C [1, 2]. In March 2011 the dam was built to reduce the heat warmed waters of CHPP on reservoir. This dam bearing open offtake from the Srednyi Kaban lake (Figure 1). Circulation in the lake was maintained by pipes running under the dam.

The results of the research

Based on the results of data which collected from 2010 to 2017 were built charts of temperature modification (average of the season) of water surface layer in the process of moving from the beginning of the open offtake to the open plot of basin-cooler (100 meters from the dam) (Figure 2).

Figure 1. The location of the dam, separating the open offtake of the Kazan CHPP-1 from the basin-cooler the Srednyi Kaban lake.

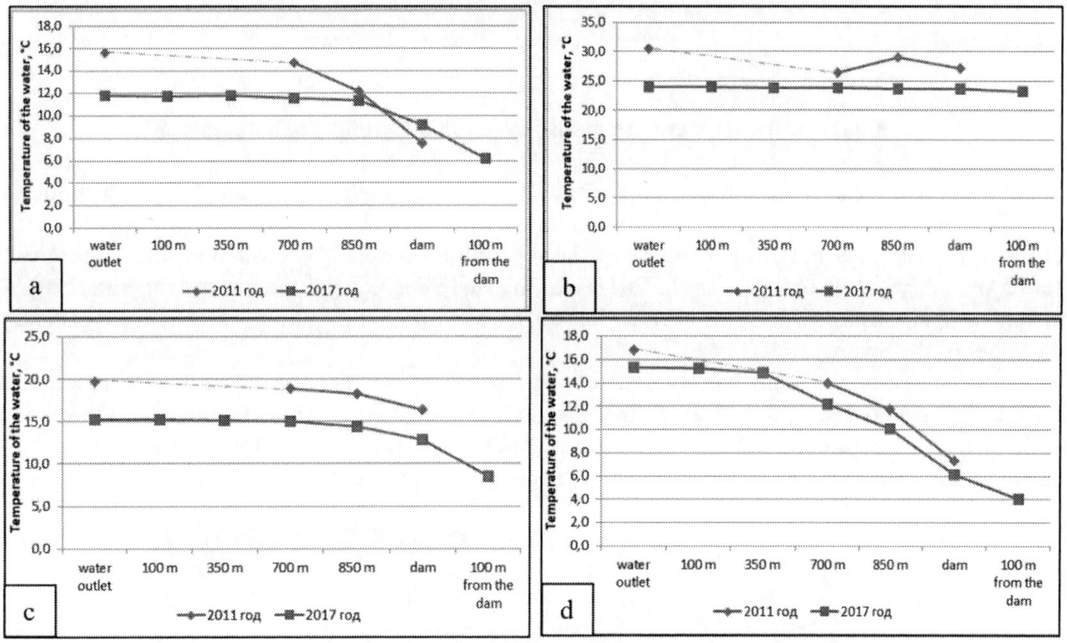

Figure 2. The temperature modification of the water surface from the beginning of the open offtake to the open plot of basin-cooler: a – spring; b – summer; c – autumn; d – winter.

Figure 2 shows that in 2017 year the water temperature in the beginning of the open offtake lower an average at 4.1 ± 1.02 °C: in spring at 3.8 °C, in summer at 6.5 °C, in autumn at 4.4 °C, in winter at 1.5 °C. Perhaps this is due to the improvement of the cooling system of the Kazan CHPP-1.

According to the data on Figure 2 we can note that the water, moving through the territory of the open offtake (760 meters), cooled in spring at 0.2 °C, in summer at 0.1 °C, in autumn at 0.3 °C and in winter at 3.2 °C in 2017 (in 2011: at 0.9 °C; at 4°C; at 0.8 °C; at 2.9 °C accordingly). Thus, in 2017 the maximum is observed in winter period, which is associated with low air temperature, and the minimum in summer, when the temperature of the waste water and the temperature of the reservoir varies by 0.7 °C.

It should take into account that the air temperature in the summer time in the year 2017 is lower than in the year 2011: the temperature of the warmest month – July – is declined by 4.2 °C. If in the 2011 year the maximum temperature was observed only in July, in 2017 year the average monthly air temperature for 23° C continues 2 months – July and August (Figure 3).

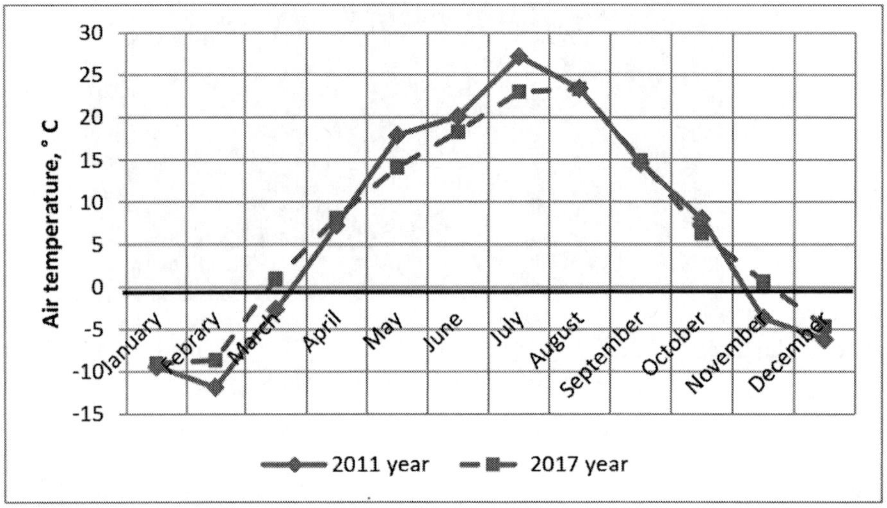

Figure 3. Air temperature modification in different time intervals [3].

In General, we can point out that built in 2011 year dam changes the water temperature to a level similar to its construction (Figure 3). Maximum cooling of water is carried out by mixing with the water of the lake, but not at the expense of the separation barrier. It is clearly visible in Figure 2 a – spring period, where is observed in the 2011 year the sharp decrease of the water temperature after 700 meters of open offtake, where has been an increasing of offtake and mixing water of open offtake with the water of the Srednyi Kaban lake.

Construction of the dam has a reverse side. Construction has led to slow circulation and as a consequence to the eutrophication the plot which fenced by dam (Figure 4): obliteration of higher aquatic vegetation, increased algal mass, flock of birds in the territory. Besides, near the dam began to accumulate trash polluting the aquatic ecosystem (Figure 4).

Figure 4. The plot of the Srednyi Kaban lake fensed by dam.

As a result of heavy precipitation in summer of 2017 the dam was almost completely flooded, causing the water of the open offtake mixed with the water of the lake by the underground pipes and ground mixing (Figure 5).

Figure 5. Flooding the dam as a result of the heavy rainfall in the summer time.

Thus, it is need to do the design revision of the dam as a way to reduce the heat influence to the basin-cooler in order to improve the environmental condition of the plot which flooded by the dam.

References

[1] Kalayda M L, Galeeva M E 2011 Temperature condition of not deeps lakes of temperate zone of XXI century *Ecologically safe and resource-saving technologies and materials* (Ulan-Ude: Buryat State University) pp 177-79

[2] Kalayda M L, Galeeva M E 2011 Temperature "portraiture" of not deep lakes of temperate zone in historical aspect Electromechanical and integral processes in power plants, blast acoustics and diagnostics, devices and methods for controlling of environment, substances, materials and products (Kazan: Kazan higher military command school) **2** 156-8

[3] Journal of weather [Electronic resources]: https://www.gismeteo.ru/

IOP Conf. Series: Earth and Environmental Science **288** (2019) 012041 doi:10.1088/1755-1315/288/1/012041

Integrated use of lake-type basin-coolers for recreational purposes

M E Gordeeva

"Water bioresources and aquaculture" department, Kazan state power engineering university, Russia, 420066 Kazan, Krasnoselskaya, 51

Maria.Galeeva@gmail.com

Abstract. At present the bulk of all water bodies in the Russian Federation and the Republic of Tatarstan in particular is experiencing significant anthropogenic impact. One of the types of impact is thermal. The article shows the positive effects of thermal contamination on lake-type basin-cooler such as increasing of the concentration of dissolved oxygen in the water in winter, improving the state of bottom sediments as a result of the water circulation, and expanding the possibilities of aquaculture. The possibilities of integrated use of lake-type coolers for recreational purposes are shown (combining the technical use of lake water with recreational purposes through the implementation of sporting events and the creation of a park and observation zone on the lake).

Introduction

Since ancient times people have settled near water bodies that played an important role in their life. In this regard, water resources should be given special attention as one of the most important elements of economic development, taking into account its social and environmental advantages [1]. The rapid growth of urbanization has a negative impact on intra-urban water bodies. Rivers, water passages and reservoirs are basin-coolers of thermal power stations, sewage receivers, these negatively affect the quality of water and bottom sediments, the life of hydrobionts, aquatic vegetation and the coastal zone. However, if rivers have a flow and are capable of carrying pollutants, the lakes accumulate them and as a result they are most susceptible to various kinds of pollution. Due to this, it becomes urgent to assess the ecological status of water bodies especially those that serve multi-purpose use.

One of the main physical and chemical indicators for determining the state of water bodies is temperature. The thermal regime of water bodies determines the rates of organic substances' circulation and their involvement in food chains. Contemporary research gives much data on the negative impact of thermal pollution on the lake's aquatic ecosystem: the oppression of most species of hydrobionts, prerequisites for eutrophication and silting of the bottom, increased evaporation and, as a result, increased water hardness, incremented pathogenic microflora and macrophyte production. However, most of the described consequences can be avoided with the rational management of the aquatic ecosystem and the selection of ways to improve the condition of water bodies.

In addition, as shown in the current article the thermal pollution does not always lead to the degradation of aquatic ecosystems and may prove useful to humans. For example, prospects for the development of aquaculture are expanding. Spawning season of fish is shifted to earlier periods, the period of feeding of young fish is extended. A new warm-water biotope with a longer vegetation period is created, which has a significant impact on both the composition of the fauna of fish and the change in the biological indices of individual species.

The identification of the features of the aquatic ecosystems formation in the urban environment allows to combine the interests of different economy sectors with the integrated use of lakes in the urbanized area. It also provides the necessary quality of the ecosystems state for completing the main task.

Kazan, is an industrialized city with a million population, has a system of Kaban lakes (Figure 1). One of its lakes – the Lake Sredniy Kaban is a basin-cooler of Kazan CHPP-1 and, therefore, its main task is to provide the uninterrupted supply of water to cool the turbogenerators. In addition to using the Lake Sredniy Kaban as a cooling pond, it receives sewage from 7 industrial enterprises, stormwater sewage and surface runoff from municipal lands, private households and urban garden associations. Since 2013 the Lake Sredniy Kaban is also a sports venue of international level which annually hosts competitions in rowing sports of different levels.

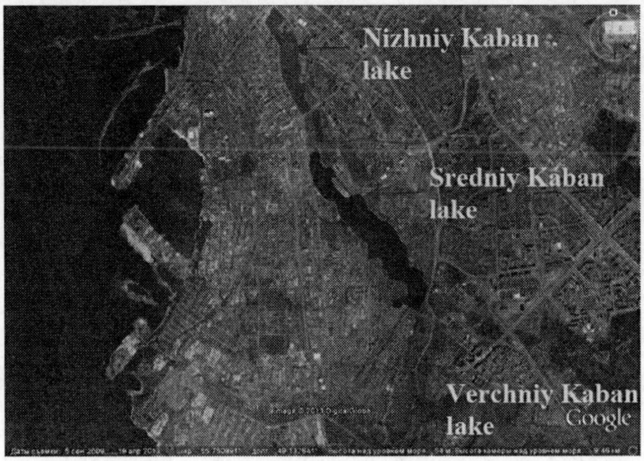

Figure 1. The Kaban Lake system (view from the space).

Figure 2. Zoning of the Lake Sredniy Kaban.

To create a rational approach to the integrated use of the natural cooler for recreational purposes, it is necessary to obtain a comprehensive assessment of the ecological state of the lake.

Features of the Lake Sredniy Kaban as a result of the activities of Kazan CHPP-1

The long-term studies (from 2010 to 2018) of abiotic factors of the environment made it possible to identify 3 zones [2] of the lake with industrial, thermal pollution (Figure 2).

Zone of the I type shows industrial pollution. This zone is characterized by a lack of thermal impact on the reservoir, resulting in 6 periods in the temperature regime: summer and winter stagnation; a partial summer and winter, full autumn and spring circulation. The lake temperature in the surface layer of water varies from 0.3 to 31.1 °C during the year, from 1.9 to 15.4 °C in the bottom. The average annual concentration of oxygen in water is 6.3±0.9 mg/l. The presence of open water patch, as a result of thermal action, leads to an increase in the oxygen concentration in the winter period to 6.2±0.9 mg/l in the neighboring zone (zone of III type), while in a lake with a minimum anthropogenic impact the concentration is 4.4±0.9 mg/l. Ph reaction of the environment during the year with the exception of July is alkalinous (10.03), it is caused by the massive development of phytoplankton.

Zone of the II type is the upstream of the lake, here water circulation is observed as a result of Kazan CHPP-1 work. The zone shows higher average seasonal water temperature in spring and autumn in comparison with the zone of I type: an increase of 65.7% and 8.1%, respectively; the number of temperature periods is 6. Reverse stratification and ice cover are observed in the temperature regime in winter. Higher concentrations of dissolved oxygen in water in spring (in the surface layer of water by 39%, in the bottom layer by 13.3%) are noted in comparison with the zone of I type. The pH values shift to a weakly alkaline but do not exceed the limits of the established standards of 6.5-8.5 with the exception of summer (11.1).

Zone of the III type is specified by industrial and thermal pollution. This zone is characterized by the absence of periods of complete autumn circulation and partial winter circulation, winter stagnation. The absence of the ice cover leads to a direct stratification of the water temperature. The average seasonal values of water temperature are maximum (in winter – 6.1±1.2 °C, in spring - 7±1.2 ° C, in summer – 26.8±1,2 °C, in autumn – 15.1±1.2 °C). Despite a significant temperature increase in the summer, the values of dissolved oxygen in the surface are similar to a lake with a natural temperature regime, in winter it increases by 81% as a result of the intake of oxygenated waters from the open discharge channel of the CHPP and the absence of ice cover.

A detailed analysis of the physical and chemical parameters of the Lake Sredniy Kaban allowed to develop a water quality index (I_{QW}) [3] which makes it possible to assess the quality of water by the contamination degree based on the use of a prospective complex indicator-total antioxidant activity [4]. This index is derived from the balance equation of the values of antioxidant activity (mg.rutin/l) and abiotic factors of the environment: surface water temperature (°C), dissolved oxygen (mg/l), pH, Eh (mV) (formula 1).

$$I_{WQ} = \frac{-18{,}96 - 0{,}32*\mathrm{Ts} - 0{,}33*\mathrm{Co_2} + 4{,}26*\mathrm{pH} + 0{,}19*\mathrm{Eh}}{\mathrm{TAOA}}, \quad (1)$$

where TAOA - total antioxidant activity (mg.rutin/l); t - temperature of the surface water layer (°C); Co_2 is the concentration of dissolved oxygen in water (mg/l); pH is the pH of the environment; Eh is the redox potential of the environment (mV).

The reliability of the predicted model is confirmed by a standard error of 0.09 and a determination index of 99.32%. If the index values are in the range from -1 to +1, the waters are classified as clean (oligosaprobic according to the data of the parallel microbiological analysis). If it is in the intervals from +1 to +2 and from -1 to -2, then it is α-mesosaprobic zone, if from +2 to +3 and -2 to -3, then it is β-mesosaprobic zone. At values of the index in the range -3 < I_{QW} <3, contaminated water is polysaprobic. The water quality in zone II is characterized as the lowest (in the summer, the I_{QW} reaches -3.2). Zone III is characterized by the maximum variability of the I_{QW} (from -2.5 to +2.4, V = 909.4%). At different times water quality is characterized as oligo-, β-mesosaprobic

Integrated assessment of the Lake Sredniy Kaban

In order to map out the ways of rehabilitation of the lake in conditions of high anthropogenic load when the lakes are located within the industrial developed urban environment and are used for multi-diversity purposes, it is necessary to identify the most important parameters that can be used to upgrading the condition of the lakes. As a result of the realized investigations [10] the integrated assessment (Figure 3) was carried out on 14 indicators (dissolved oxygen, pH, TAOA, water temperature, Eh, saprophytic microorganisms quantity, number and biomass of phyto -, zooplankton, zoobentos, ikhtyomass), described in 6 methods of water quality evaluation [2, 5-7]. Since in the methods considered, the degree of water pollution and trophicity of lakes is represented by a different number of water quality categories, we have made the translation of these methods into a scoring assessment, where each category of water quality was assigned 1 point [3]. The water quality decreases as the number of points increase. The minimum score is 33, the maximum is 208.

The results of the integrated assessment (Figure 3) indicate that during the summer and winter periods, the lowest water quality is observed compared to the spring and autumn periods. Herewith, in the summer period the maximum proportion of pollution of type zones II and III is contributed by the physicochemical indicators (50% and 62% of the maximum number of scores respectively), and in I type zone is contributed by the microbiological indicators (52% and 53% respectively). In winter period physicochemical indices make the maximum contribution to the contamination (form 56% to 60% of the maximum number of points).

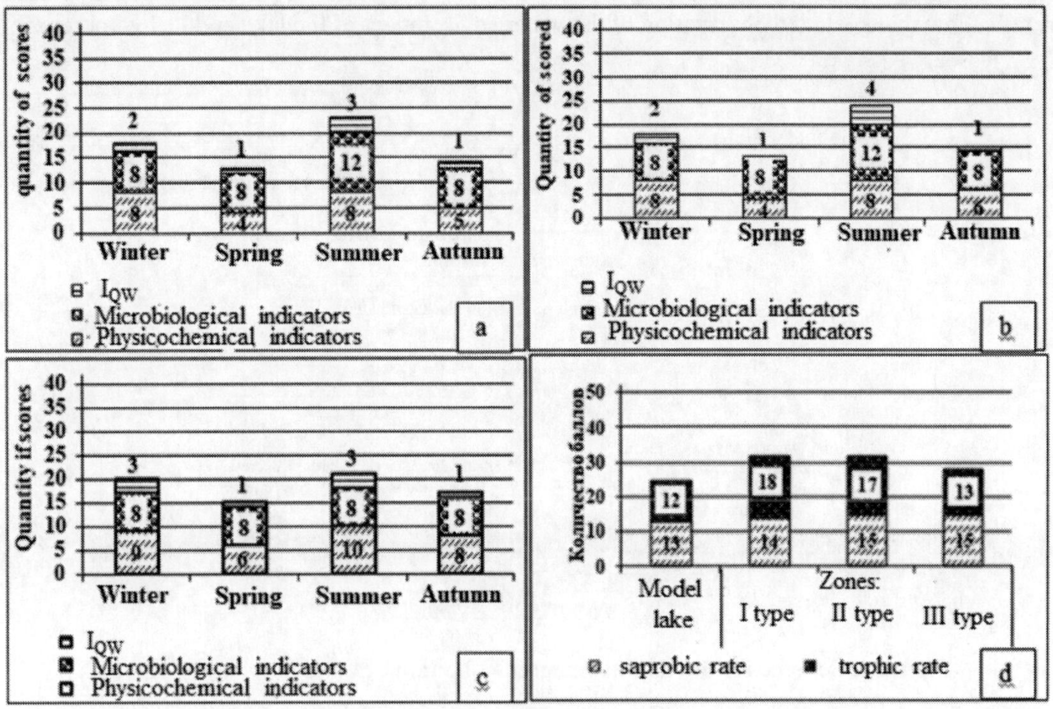

Figure 3. Integrated assessment of the pollution degree of Lake Sredniy Kaban: a – I type zone; b – II type zone; c – III type zone; d – integrated assessment of biota.

A water quality assessment based on biota characteristics (Figure 3 d) revealed the worst state of the aquatic ecosystem in I and II type zones (32 points each), which are characterized by a complex anthropogenic impact. In general, an integrated scoring assessment of the pollution degree and trophicity indicates the average level of pollution (mesotrophy) of the lake with industrial pollution .

Thus, the complex analysis of the state of the lake ecosystem allows us to rationally select the integrated use of the basin-cooler for recreational purposes.

Recreational zone design on the base of basin-cooler of energy facility

The realized integrated assessment of the ecosystem state of the Lake Sredniy Kaban allowed composing the scheme for the improvement of condition of the lake (figure 4).

In connection with the increasing anthropogenic pressure on the lake and the availability of a center for rowing sports on the shore of the lake, the realization of recreational potential is proposed. This will solve the issue of pollution of the lake by storm sewage in the I type zone (18 out of 21), which cause maximum damage. This event is possible through the formation of a rational and promising system of recreational facilities (a zoo and a well-equipped embankment) and engineering infrastructure (the creation of a drainage system for storm water drainage). The general layout of the territory after the construction of the recreational zone is shown on Figure 5.

Due to the fact that the modern zoo is under construction at the moment, the center of rowing sports was built for the Universiade 2013, we offer to landscape the territory creating:

- Forest park zone (№ 5 on the Figure 5, green color);
- Upgrading in small forms (№ 4 on the Figure 5, sand color). Creation of comfortable recreation areas on the water body and the possibility of monitoring the competitions in rowing sports

from any angle and distance. Achieved by equipping the area with benches, pedestrian and bicycle paths, flowerbeds, pergolas, retaining walls;

- Hanging terraces above the water area (№ 6 on the Figure 5, light blue color). The hanging terraces allow more detailed observation of the water environment of the lake and its hydrobionts.

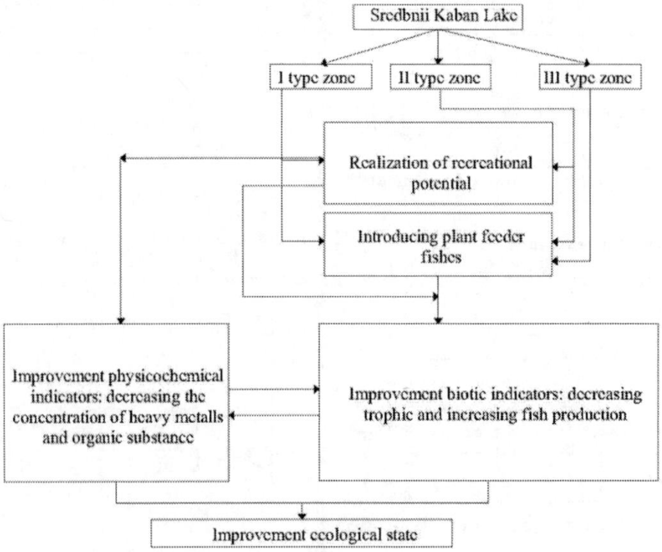

Figure 4. Scheme for the improvement of condition of the Lake Sredniy Kaban.

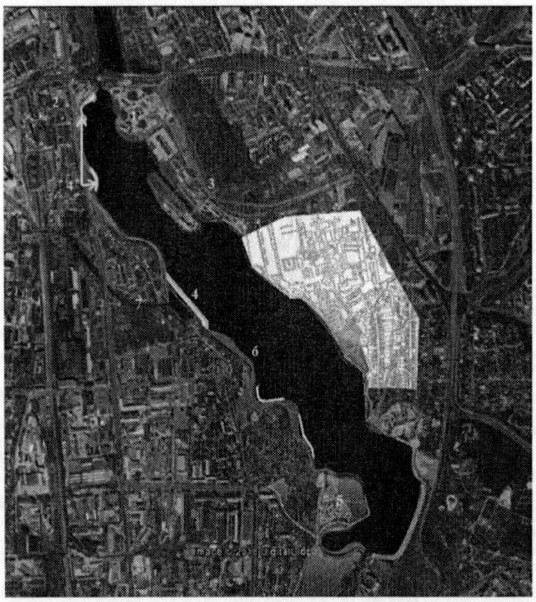

Figure 5. General layout after the creation of recreational zone: 1 – zoo; 2 – Kazan CHPP-1; 3 – center of rowing sports; 4 – Upgrading in small forms; 5 – forest park zone; 6 – hanging terraces above the water area; 7 – open offtake.

Due to the fact that recently the number of floating higher aquatic plants in the Lake Sredniy Kaban increases, it is advisable to stock the reservoir with herbivorous fish. This will also reduce the level of overgrowth of the lake. From the calculations carried out for the fodder base for the Lake Sredniy Kaban, it is generally recommended that *Ctenopharyngodon idella Val* introduced with a density of planting of 705 pcs/ha, which will ensure an annual seizure of the highest aquatic vegetation with an area of 5.2 hectares and the possibility of catching fish products in the volume of 1505 kg with an annual stocking by yearlings with a density of planting of 705 pcs/ha.

All planned events will reap the appearance of the cooling pond located in the city boundary, make the lake more attractive for the residents of Kazan in connection with the possibility of using it as a park and sports zone, and improve the physicochemical and biotic indicators of the lake.

References

[1] 6th World Water Forum 2012 *Ministerial declaration* (Marseilles) 7

[2] Kalaiyda M L, Gordeeva M E 2015 Integrated scoring assessment of the lake ecosystem state. Urbanized areas *LAP LAMBERT Academic Publishing* 228

[3] Galeeva M E, Lapin A A, Chugunov Y V, Kalaiyda M L 2012 Antioxidant activity – a promising indicator for determining the water quality index *Butlerov communications* **29(3)** 110-19

[4] Lapin A A, Kalaiyda M L, Galeeva M E Zelenkov V N Investigation of antioxidant activity of water. Non – traditional natural resourses, innovative technologies and products *Collection of scientific papers* (Moscow: RANS) **19** 54-60

[5] Kitaev S P 1984 *Ecological bases of bioproductivity of lakes of different natural zones* (Moscow: Science) p 208

[6] Romanenko V D 1990 *Ecological assessments of the impact of hydraulic structures on water bodies* (Kiev: Naukova Dumka) p 256

[7] Galeeva M E, Kalayida M L 2011 Abiotic factors of the environment of a fishery body. *Herald of the state polar academy* **1(12)** 42-3

IOP Conf. Series: Earth and Environmental Science **288** (2019) 012042 doi:10.1088/1755-1315/288/1/012042

Problems and perspectives of using warm waters of the state district power station in solving the provision of the population with fish

L K Govorkova and M L Kalayda

"Water bioresources and aquaculture" department, Kazan state power engineering university, Russia, 420066 Kazan, Krasnoselskaya, 51

govorkovagoncharenco@mail.ru

Abstract. One of the sources of increasing fish production in Russia is intensive fisheries development of cooling reservoirs of thermal power plants and nuclear power plants. Cooling reservoirs have a unique temperature regime, which differs from natural water bodies by a higher temperature throughout the year. The increased water temperature causes changes in the quantitative and qualitative composition of the forage reserve, the ichthyofauna, the habitat of the hydrobiont and creates favorable opportunities for the introduction of a complex of thermophilic fishes and enhance the fish productivity of water bodies. The rise in water temperature also accelerates chemical and biochemical processes, promotes the intensive decomposition of organic substances, influences the gas regime of the reservoir, as a result, there may be overseas phenomena and the death of fish in cooling reservoirs. Therefore, the growing of aquaculture objects, the most resistant to high temperatures and temperature and gas changes, is becoming topical.

Introduction

The most large-scale single use of water is the production of electricity, where it is used mainly for cooling and condensing steam produced by turbines of thermal power plants. At the same time, water is heated by an average of 7 ° C, then dumped directly into rivers and lakes, being the main source of additional heat, called "thermal pollution". The temperature of water used in thermal power plants for steam cooling increases by 3-10 °C, and sometimes up to 20 °C. Powerful power plants noticeably heat up the waters in the rivers and bays on which they are located. In the summer, when the demand for electrical energy for air conditioning is very high and its production increases, these waters often overheat. The term "thermal pollution" refers specifically to such cases, because excess heat reduces the solubility of oxygen in water, accelerates the rate of chemical reactions and, consequently, affects the life of animals and plants in water intake basins [1]. Under such conditions, a change in the fish-breeding zone may occur.

The effect of thermal pollution on water bodies

The protection of aquatic biological resources, the improvement of the ecological status of water bodies at the current stage should be based on the reproduction of aquaculture facilities. It is the reproduction of aquatic biological resources that is the basis for the successful development of aquaculture in the reservoirs of the Volga region [2].

The main problem is the high temperature background in the summer. Against the backdrop of climate change in the Volga region of Russia, when the natural temperature background in some years reaches more than 30 °C, thermal contamination of power plants becomes a critical factor due to oxygen deficiency and overseas phenomena. With the change in the water temperature in the water body, the gas exchange in fish changes, and the concentration of free oxygen in water changes [1].

The deterioration of the oxygen regime of the reservoir-cooler leads to a reduction in the number of herds of fish, the replacement of valuable commercial species by weed, and the occurrence of parasitic diseases of the ichthyofauna. The impact of numerous and powerful sources of anthropogenic chemical, biological and thermal pollution in conditions of regulated flow contributes to a significant extension of the initial period of formation of the oxygen regime of the reservoir. The accumulation of the biomass of algae and its decomposition at the end of the growing season in a limited part of the water area of the reservoir leads to the accumulation of nutrients. This, in turn, reduces the pH of the aqueous medium, which increases the solubility and mobility of heavy metal ions, their desorption from the bottom sediments of the reservoir-cooler, the transition of toxic heavy metal salts into water. From the bottom sediments, phosphorus enters the aquatic environment, and as a result, the eutrophication process accelerates [3; 5].

Their important feature is a seasonal thermocline. As the surface waters warm up in the summer, a temperature gradient is established with denser and colder water at the bottom and a warmer one at the surface. This is not at all surprising, except that the change in temperature, with increasing depth, is not gradual. Instead, a sharp break occurs a few meters down, while the cold water is locked down. The fission line is thermocline. This delineation of two large masses of water is very important for fish. Typically, the wind can mix water, and raise the cold, oxygen-rich water from the depths. However, thermocline blocks the mixing of water at different temperatures, resulting in a lack of oxygen and massive fish loss [4].

In summer, when the surface waters are warm, algae multiply intensively, which promotes the spread of microscopic life. In itself, this is fine, until the algae switch from photosynthesis (the process of producing oxygen) to respiration (oxygen uptake), in low light, for example, at night or during prolonged cloudy periods. This process significantly reduces the concentration of oxygen in the water, and the fish begins to die from suffocation. However, the situation may worsen even more when algae begin to die off at a rapid pace. In the biological decomposition caused by bacteria, a considerable amount of oxygen is used, which reduces its concentration in the water even more. People can also significantly aggravate the problem of lack of oxygen in the water. This occurs when water is contaminated with certain types of nutrients, including manure drainage or fertilizer from farms. Phosphorus and nitrogen from these effluents fall into water bodies and are fixed in algae, because of what they consume more oxygen [5].

So, for example, in August 2016 according to the materials of the press service of BUSINESS Online in Zainskoe reservoir, 170 tons of fish were killed. Zainsk fish farm is the only farm in Tatarstan, located in a warm pond. It is located on the river Steppe Zai, on which the reservoir is organized. It is a "warm body of water", where, as a rule, it discharges hot water. Zainskaya GRES uses it as a cooler.

The reservoir under ordinary conditions cools the water discharged by Zainskaya GRES, but in 2016 this did not happen. The reason is an abnormal heat and, in addition, a 100% load of the Zainskaya GRES. The water temperature reached critical levels for 40 degrees Celsius, there was an oxygen starvation of the fish, there was a loss of her literally in two hours. In the fish farm, carps and carnivores were killed. So, for 2016 the production of commercial fish by specialized fish farms was 212 tons, while in 2015 the figure was twice as high-429 tons. The damage to the fish farm, according to its director, was 35-40 million rubles [6].

Perspective objects of industrial cultivation on elevated warm waters

A particularly attractive object of high-grade nutrition is the fish, which gives easily digestible protein. The optimal level of consumption of fish products should be at least 20 kg/year per capita (with the

Institute of Nutrition of the Academy of Medical Sciences of Russia recommended physiological norm of 23.7 kg/year). Currently, the level of consumption of fish and seafood in Russia is 18 kg/year, which completely does not meet the needs of the population (Figure 1) [7].

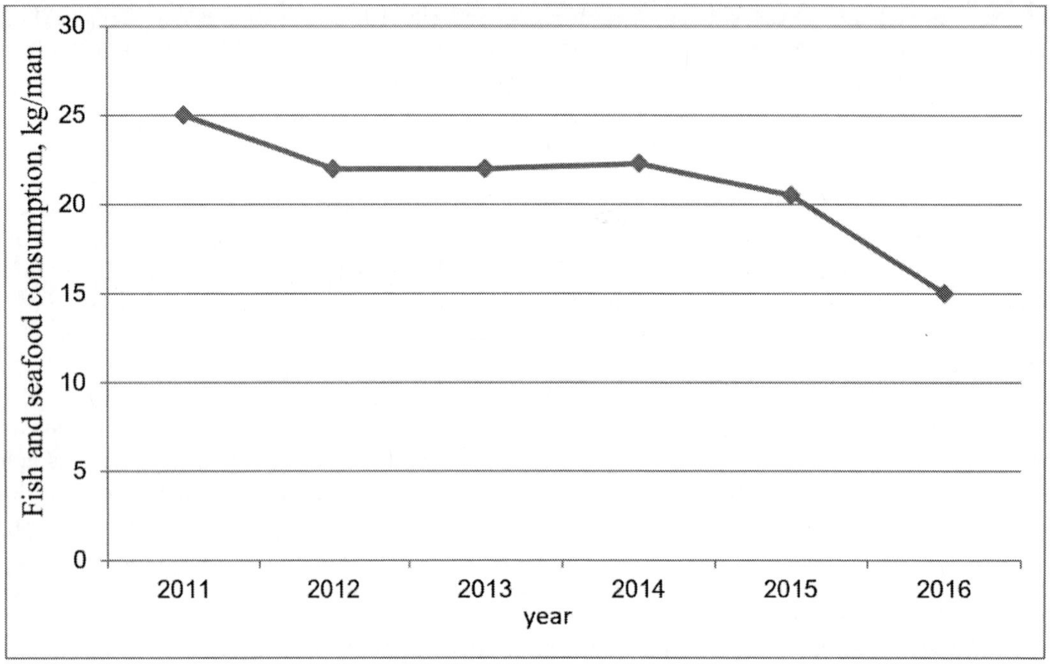

Figure 1. Average annual fish consumption in Russia.

To meet the population's demand for fish products, it is important to grow fish in fish farms. One of the promising forms of fish farming has been the use of warm waters of reservoirs-coolers. Its integrated development allows not only to create by the method of pasture aquaculture the uterine livestock of thermophilic species of fish and to establish their reproduction, but also to use the considerable possibilities of cage farming on the cooling pond for growing viable planting material. Effective use of cage farming for the cultivation of commercial fish.

If earlier in the summer temperatures the recommended object of industrial cage fish culture was carp, and herbivorous fish were introduced into the body of water as biomeliorators, in modern conditions it is necessary to change the recommended objects of industrial fish farming.

Under these conditions, one of the most interesting objects of growing on warm waters is the Clareeum catfish, which prefers higher water temperatures than carp, and is practically insensitive to oxygen because of its biological characteristics. The value of the clarium catfish determines good growth, effective feed intake, the ability to adapt to different growing conditions, including high planting density, and excellent taste qualities. This thermophilic fish: the temperature optimum lies in the range 25 - 30 °C, at a temperature of 17-19 °C it ceases to feed, perishes during a prolonged stay in water with a temperature of 14-15 °C, but can withstand a short decrease to 5 °C. Clarias withstand a high percentage of water in the nitrogen compound. The Claire's catfish provides buoyancy for itself with the help of air coming from the nadzhiburial cavity. This is due to the fact that it is in the plane that the ejaculatory organ of respiration is located (Figure 2). On the second and fourth bronchial arc, and it is located, abundantly covered with vascularized tissue, has branched formations. With its help, clarias separate oxygen from the air. Connects the nadestial cavity with the pharynx. If the oxygen content in the water is low, then to breathe in air the catfish cat rises to the surface of the water.

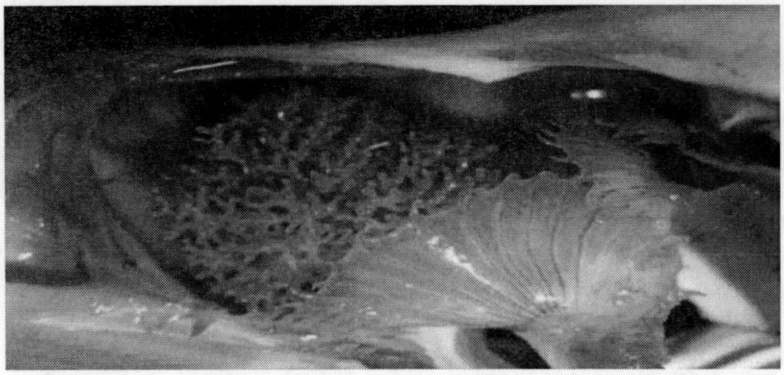

Figure 2. Appearance of the supra-lateral organ of respiration of the clary catfish.

Breathing with the help of atmospheric air allows the fish to live out of the water for a long time, move along the ground and even be in muddy water. Clary feels best when the concentration of dissolved oxygen in the water exceeds 4.3 mg / l and access to the surface is unlimited. As soon as the reservoir ceases to be comfortable for the clareef catfish, it passes into another pond. A comfortable environment for the clareef catfish is water with pH 6.5-8, it is resistant to temperature changes, withstands the salt level in water to 10 ppm [8].

Claristic catfish do not represent sanitary - epidemiological and ecological danger. And precisely because, this species has been reproduced in hatcheries for many generations without contact with other hydrobionts, which could be intermediate hosts of parasites, including those that pose a danger to humans, so the probability of accidental infection of such parasites is virtually excluded. Clariium catfish under variable and constant temperature jumps, has a high tolerance to diseases, in particular immunity to parasites of Dactylogyrus [9].

In conditions of unstable thermal regime, the fish of the North American complex are also promising for growing in the pond on the non-soft waters of the state district power station: large, small and black buffalo. The first species is especially valuable. Bigger Buffalo - a fast-growing and large fish, not inferior to the growth rate of carp, reaches a weight of 45 kg. It well tolerates turbid water, it is not exacting to the hydrochemical regime. In contrast to the carnivores in reservoirs-coolers, natural spawning is possible. Valuable fishing quality is the ability of buffalo to gather in flocks in the bottom layers and its accessibility for catching by active fishing gear.

The rise in water temperature will intensify the development of shallow water in the shallows, which will create the possibility of introducing a white cupid, a consumer of macrophytes, into the pond.

Another promising area of use of warm waters of the state district power station is the growing of rainbow trout in the cold season. In Russia, rainbow trout and its forms are practically the only object of aquaculture of salmon breeding and so far constitute an insignificant part of the total volume of fish products produced [10]. Trout has high economic-useful attributes. Data characterizing the female: the total survival rate during the period of growing the two-year-olds (including the incubation of caviar) is 60%. The waste of three-year-olds is equal to 5% of the number of two-year-olds. The average weight of commercial two-year-olds is 250 g, three-year-olds - 1.5 kg. The yield of commercial fish per female during the cultivation of two-year-olds will be 270, three-year-olds - 3420 kg.

Along with the rainbow trout, its passage form - steelhead salmon and closely related species - lake trout, belong to the most common objects of aquaculture of salmonids. Steelhead salmon transfers the temperature rise to 28 ° C. According to the pattern of growing is identical to rainbow trout. Due to late maturation, the planting material of steelhead salmon can be successfully used for cage fish culture in the northern regions of our country, where the growing season starts in April-May.

Thus, the cultivation of fish species more resistant to fluctuations in the temperature and gas regimes of reservoirs will make it possible to use the potential of warm waters of energy facilities much more efficiently.

References

[1] Safronov E N and Varyukhin A V 2003 Temperature of water in pond fish farming Zoo industry **11** 12-5
[2] Kalaida M L 2016 Current state and tasks of aquaculture development in the Republic of Tatarstan (Saratov: Scientific Book) p 152
[3] Danilov-Danilyan V I and Losev K S 2008 Water consumption and its deficit: the ecological aspect Global environmental problems of Russia (Moscow: Nauka) p 35-9
[4] Pavlov D S and Mochek A D 2006 Ecology of fish in the Ob-Irtysh basin (Moscow: Fellowship of Scientific Publications of KMK) p 596
[5] Morozova O G, Morozov S V, Pen R Z, Repyah S M 2002 Formation of the oxygen regime of the reservoir-cooler BGRES-1 *Izv. universities. Chemistry and Chemical Technology* **45**(6) 185-8
[6] BUSINESS Online: https: //www.business-gazeta.ru/article/339972
[7] Extreme A A 2012 The power of the word [legal regulation issues of the development of the fishery complex of the Russian Federation] *Fish protection of Russia* **3** 20-7
[8] Nikiforov A I and Mailkova A I 2005 Morphological features of Soma Clarias gariepinus *Freshwater aquaculture: state, trends and development prospects Collection of scientific articles devoted to the 60th anniversary of the Scientific Research Fishery Station* 56-8
[9] Pillow S B 2006 Clarion catfish and its use in aquaculture *Status and prospects of development of farmer fish culture in the arid zone* 71-4
[10] Esavkin Y I 2012 Intensive technology of freshwater trout breeding *Dissertations of Doct. s.-. n.* p 299

Assessment of the quality of fish habitat in the area of Ulyanovsk CHHP-1

L K Govorkova

"Water bioresources and aquaculture" department, Kazan state power engineering university, Russia, 420066 Kazan, Krasnoselskaya, 51

govorkovagoncharenco@mail.ru

Abstract. Water - the most important mineral on Earth, which can not be replaced by any other substance. It makes up most of any organisms, both plant and animal. The problem of preserving water quality is at the moment the most urgent. Science knows a lot of pollutants of natural waters, which adversely affects the health of the population, leads to the death of fish, waterfowl and other animals, as well as the death of the plant world of reservoirs. At the same time, not only toxic chemical and oil pollution, but also the excess of organic and mineral substances coming with flushing of fertilizers from fields, are dangerous for aquatic ecosystems. A very important aspect of pollution of the Earth's water basin is thermal pollution, which is the discharge of heated water from industrial plants and thermal power plants into rivers and lakes.

Introduction

Among the natural resources of great importance are the water resources, which are involved in social production in ever-increasing scale. Intensive development of industry, agriculture, population growth are the cause of pollution of water bodies, industrial, agricultural and domestic wastewater are discharged into them [1]. Pollution of reservoirs by sewage can cause sustainable water pollution and adversely affect the fauna and flora of water sources. At the same time, the commercial qualities of fish as a food product deteriorate, young fish and food organisms die, the productivity of water bodies decreases. Pollution of water bodies can affect people's health in the form of infectious diseases and poisonings. Especially dangerous are discharges to the reservoir of untreated or under-treated sewage containing toxic substances [2].

One of the most acute and urgent environmental problems of water reservoirs-coolers of energy facilities is their thermal pollution, which is observed as a result of the receipt of thermally heated water discharged after the cooling of power equipment. There is a so-called "thermal pollution". The rise in water temperature in the reservoir leads to the growth of blue-green algae, as a result of which the concentration of dissolved oxygen in the water decreases. This is the cause of death of the inhabitants of water bodies that do not tolerate high temperatures. An increase in water temperature can lead to the inhibition of aquatic organisms, and when the food base is limited or the threshold (lethal) temperatures are exceeded, they die [3].

Quality of the aquatic environment by chemical indices

Chemical pollution is the main, both in terms of scale and impact, and causes very great damage to water bodies. The most dangerous chemical pollutants of the environment are heavy metal ions. This is due, above all, to their high toxicity in relation to hydrobionts and humans. The danger is aggravated by the fact that metals are not subjected to chemical and biological degradation, as is characteristic of organic toxicants, so once they fall into the surrounding environment, they are only redistributed between biotic and abiotic links, changing the form of finding and interacting with different categories of living organisms [4]. The special danger of the accumulation of heavy metals in water bodies is that they are stored permanently under any conditions.

Deterioration of water quality in polluted water bodies has a negative impact on the conditions of feeding, feeding, spawning of fish, reduces the effectiveness of their natural reproduction and, consequently, the population as a whole [5]. There is also a decrease in the quality of processed fish products due to the accumulation of various toxicants in their organisms. Petroleum products, along with heavy metals, are dangerous contaminants in water bodies. They are distinguished by great resistance, due to which they remain in the reservoir for years, without losing their toxic properties [6].

The ability of reservoirs to self-purification is not unlimited, and their protection from pollution is impossible without the maximum neutralization of sewage. If the wastewater were not treated, in a few months the ecosystems of inland water bodies would be completely degraded. Therefore, the discharge of untreated sewage is universally prohibited, the required degree of purification is regulated by law. With the help of physical, chemical and biological methods, waste water at treatment plants is brought to a state that meets certain hygienic and biological requirements. The various substances remaining in the waste water should not be in an amount exceeding the maximum permissible concentration (MPC). Establishment of MPC is conducted taking into account hygienic requirements and interests of ecosystems protection of water bodies [1].

The purpose of the studies, the results of which are outlined in this paper, was to determine the effect of sewage from Ulyanovskaya CHPP-1 on the quality of the river Sviyaga. Water quality was assessed by chemical and hydrobiological indicators. In connection with this, water and soil samples were taken in the river. Sviyaga, as well as hydrobiological tests in the spring-autumn period 2016-2017. Samples were collected at different stations above and below the sewage discharge of Ulyanovskaya CHHP-1 (UCHHP-1).

At the beginning of December 1946, the first plant in Ulyanovsk was built at an automobile plant. Over the years, constantly expanding and improving in accordance with the growing needs of the city in heat and electricity, UCHHP-1 has turned into a powerful enterprise. In 1973, UCHHP-1 switched to gas as the main fuel. At present, the installed electric power of UCHHP-1 is 435 MW, the installed heat capacity is 1539 Gcal/h.

In the composition of waste water, a large number of various pollutants are discharged into reservoirs, which are included in the trophic chains and the circulation of substances. Therefore, in order to preserve the environment, sewage treatment plants were put into operation at UCHHP-1. Formation of the hydrochemical regime of the river. Sviyaga occurs under the influence of surface and ground runoff. The oxygen content in the surface and deep horizons of the river. Sviyaga did not fall below 4.5-5.5 mgO/l, i.e. corresponded to normative values. Sviyaga refers to rivers with high mineralization (hydrocarbonate character with predominance of calcium ions in the salt composition). Mineralization of the river. Sviyaga ranges from 400-1100 mg/l. The heterogeneity of the chemical composition of water is due to the underground feeding of the river, as well as the influence of tributaries. The average long-term content of organic and biogenic substances in the waters of the Sviyaga River is presented in Table 1.

Table 1. Average long-term content of organic and nutrient substances p. Sviyagi, (mg / l).

Stations	pH	Mineralization	BOD_5	Chlorides	Sulphates	Nitrogen ammonium	Nitrites	Nitrates	Phosphorus	Iron
1	8,9	627	6,0	48,66	399,2	0,20	0,043	0,04	0,015	0,21
2	8,2	548	3,2	24,93	60,15	0,26	0,145	3,97	0,199	0,51
3	8,2	577	4,8	17,00	40,10	0,21	0,112	4,21	0,165	0,04

Note: 1 - the place of production of waste water; 2 - above the output of 100 m; 3 - below the output of 200 m.

In wastewater of various industrial enterprises, including UCHHP-1, a certain amount of pollutants such as heavy metals is contained. In this case, they can accumulate in water, in soils, in hydrobionts, in fish. Average annual concentrations of heavy metals in the water and in the soils of the river. Sviyagi are represented in Figures 1 and 2.

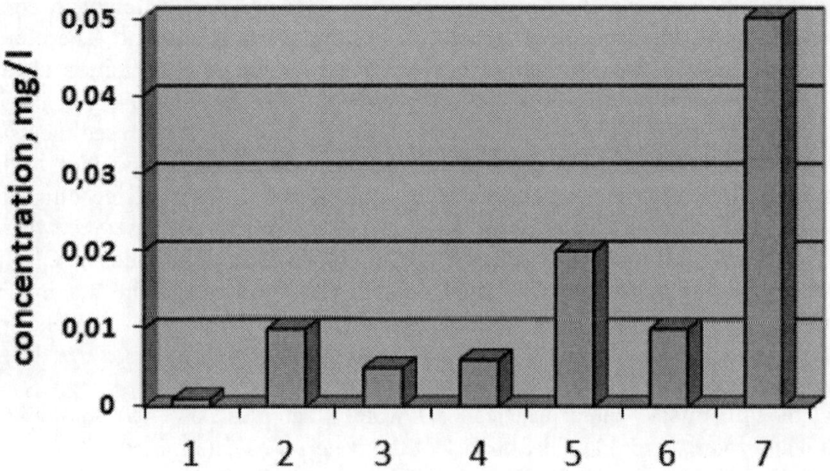

Figure 1. The average annual concentration of heavy metals in the water p. Sviyagi: 1 - Copper, 2 - Zinc, 3 - Cadmium, 4 - Lead, 5 - Chrome, 6 - Manganese, 7 - Petroleum products.

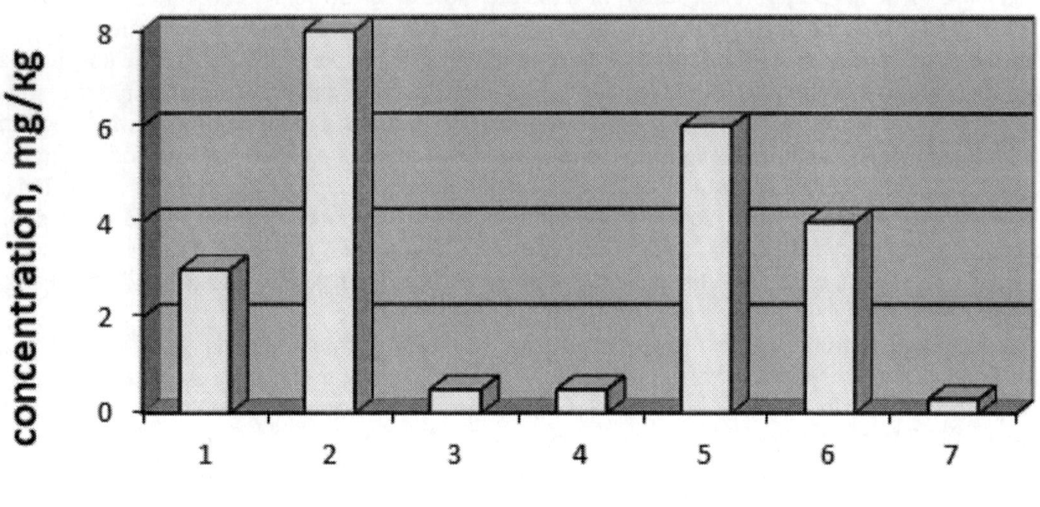

Figure 2. The average annual concentration of heavy metals in the ground r. Sviyagi: 1 - Copper, 2 - Zinc, 3 - Cadmium, 4 - Lead, 5 - Chrome, 6 - Manganese, 7 - Petroleum products.

The analysis of the obtained results showed that in water samples the cadmium content is less than 0.005, and the chromium is less than 0.02, which corresponds to the standards, since the maximum permissible concentration of cadmium in water is 0.005 mg / l, and the MPC of chromium is 0.02 mg / l. Concentration of lead is also within normal limits. The copper content did not exceed the maximum allowable concentrations of 0.001 mg / l. The content of zinc in water samples was in the range from 0.01 mg / l to 0.017 mg / l, manganese - in the range from 0.01 mg / l to 0.013. Given that the maximum permissible concentration for zinc is 0.01 mg / l, and that of manganese 0.01 mg / l. Concentration of oil products in the water samples of the Sviyaga River and the discharged waters of the CHP is within the permissible concentrations of 0.04-0.05 mg/l.

There are a large number of standards for the content of pollutants in the bottom sediments of the reservoir. These are background concentrations used in the USA, Belgium, Holland, the Netherlands. There are also such standards, which are offered in Russia on the example of different reservoirs in different regions of the country. But the fact is that the soils are very different in composition. On different reservoirs try to determine their standards of pollutants in the ground. Therefore, it is best to use the norms of those freshwater systems whose soils are similar in composition to the soils of the particular water body under investigation. For several years the Kuibyshev reservoir has been carrying out soil research and establishing certain standards of pollutants in the bottom sediments [7]. The soils of the Kuibyshev Reservoir of the Ulyanovsk Reach are very similar in composition to the soils of the site under study. Therefore, we relied on these proposed standards. Exceeding the maximum permissible concentrations of the metals under investigation in soils is not observed.

Thus, in the samples of water and soil r. Sviyaga in the area of Ulyanovsk CHHP-1 there are certainly certain polluting substances, but their content does not exceed the maximum permissible concentration.

Quality of aquatic environment by hydrobiological indicators

Along with the chemical characteristics of water, samples of hydrobiological indicators (phytoplankton and zooplankton) and the quality of the waters was determined.

As you know, water is an essential component of human life. The impact of sewage discharged into reservoirs adversely affects both the inhabitants of reservoirs and humans. Pollution of water bodies strongly affects the species composition of living organisms. Disappearing ambitious species, and there are organisms that are characteristic of contaminated sites. As a result, the productivity of the fodder base of fish is undermined, which in turn leads to further profound consequences. At present, the quality control of surface waters by hydrobiological indicators has been widely applied [8].

In the practice of ranking the level of water pollution, an integrated biological indicator of species diversity (IVR) is widely used - the Shannon index. Therefore, the quality of water in the investigated plots by phytoplankton was determined with the help of this index. The degree of water pollution, characterized by the index of saprobity, was also determined. For this, the method of indicator zooplankton organisms according to Pantle and Buk in the modification of Sladechek was applied [8].

The species diversity of the zooplankton community in September is small in the areas studied. Indices of species diversity vary at different stations from 0 to 2 (in number) and from 0 to 2.8 (biomass). The average values of the indices in the sector are 1.05 ± 0.18 in number and 0.75 ± 0.22 in terms of biomass.

The average saprobity index (by number) in the studied water area is 1.53 ± 0.06, which corresponds to the mesosaprobic zone, moderately polluted degree, to the III class of water quality. In general, the index values at different stations vary from 1.3 (oligosaprobic zone) to 2.1 (mesosaprobic zone).

The species diversity of the zooplankton community in October is very low. Indices of species diversity vary at different stations from 0 to 2.32 (in number) and from 0 to 1.62 (by biomass). The average values of the indices in the sector are 1.05 ± 0.041 in number and 0.81 ± 0.35 in terms of biomass. According to the study area, the saprobity index (by number) is 1.75 ± 0.08, which corresponds to the -mesosaprobic zone, the moderately polluted degree, the III class of water quality (Table 2).

Table 2. Integrated indicators for determining water quality.

№/№	Index of species. diversity in Shannon	Index of saprobity for Pantle and Buk	Zone saprobity	Characteristic quality water
1	2,69	2,21	β-α- mesosaprobic	Moderately polluted - contaminated
2	2,30	1,57	β- mesosaprobic	Moderately polluted
3	1,32	1,55	β- mesosaprobic	Moderately polluted
4	2,18	1,55	β- mesosaprobic	Moderately polluted
5	2,37	1,55	β- mesosaprobic	Moderately polluted
6	1,93	1,50	oligosaprobic	Pure
7	1,62	1,55	β- mesosaprobic	Moderately polluted
8	1,90	1,37	oligosaprobic	Pure
9	1,94	1,64	β- mesosaprobic	Moderately polluted

Thus, the values of the IWR at different sampling stations in the fall period are in the range 0.18-1.62, and the index of saprobity varied from 1.3 to 2.1. The waters are characterized as "moderately polluted" and belong to the third class of water quality.

References

[1] Valiev V S, Zakirov A G and Frolova L L 2000 Methodological aspects of creating a database on the content of heavy metals in aquatic ecosystems *Ecological and ichthyotoxicological aspects of monitoring freshwater objects* **326** 66-9
[2] Wundzettel M F 2002 Sanitary Hydrobiology (Fish: VNIIPRKh) p 62
[3] Melekhova O P, Egorova E I and Evseeva T I 2007 Biological control of the environment Bioindication and biotesting (Moscow: Academy) p 288
[4] Demina T A 1995 Ecology, Nature Management, Environmental Protection (Moscow: Aspect Press) p 86
[5] Komarovskiy F Y and Polishchuk L R 1981 Mercury and other heavy metals in the aquatic environment: migration, accumulation, toxicity for hydrobionts Hydrobiology (Moscow: Academy) **17(5)** 71 - 83
[6] Komov V T, Komarovskii F Y and Polischuk L R 2004 Content of mercury in fish muscles from the reservoirs of the Northwest of Russia: causes of intensive accumulation and evaluation of the negative effect on human health *Current problems in aquatic toxicology* 99-123
[7] Stepanova N Y, Latypova V Z and Yakovlev V A 2004 Ecology of the Kuibyshev Reservoir: bottom sediments, benthos and benthic fishes (Kazan: Academy of Sciences of the Republic of Tatarstan) p 268
[8] Sadchikov A P 2003 Methods of studying freshwater phytoplankton (Moscow: University and School) p 200

IOP Conf. Series: Earth and Environmental Science **288** (2019) 012044 doi:10.1088/1755-1315/288/1/012044

Release of sterlet (lat. Acipenser ruthenus) in the Kuibyshev reservoir is an important task for the development of aquaculture

M L Kalayda, I K Abdrakhmanov, M F Khamitova and A A Kalayda

"Water bioresources and aquaculture" department, Kazan state power engineering university, Russia, 420066 Kazan, Krasnoselskaya, 51

Kalayda4@mail.ru

Abstract. The development of hydropower and the creation of reservoirs of the Volga cascade had an impact on the state of populations of sturgeon species of fish, while creating incentives for improving the biotechnology of their reproduction and cultivation. This period is characterized by the desire and technical and technological capabilities to restore the number of sterlet in the Kuibyshev reservoir. Release of sterlet can be conducted not only in the vicinity of potential spawning grounds, but also in the feeding areas. Taking into account the possibilities of long-term development of the Kuibyshev reservoir with the directed formation of ichthyofauna and planning of sterlet catches in the volume of 480 tons, annual releases of juveniles with a mass of 3 g in the amount of 57 million pieces are required. The analysis shows the need to organize significant production of sterlet juveniles for stocking the Kuibyshev Reservoir.

Introduction

The development of aquaculture in the world is one of the most important priorities of our time. The growth of the population, the increasing press on the environment, the requirements for quality of life, and, consequently, the consumed protein products make the trend of growth of aquaculture production in the world steady. The prerequisites for the successful implementation of aquaculture tasks in Russia are a significant number of inland waters and their quality. Since the beginning of the twentieth century, the problems of the development of fish industry have been closely linked with the development of energy [1, 2].

In 1931, the working hypothesis of a comprehensive scheme for the use of the Volga for energy and transport purposes was developed, various projects for the construction of hydroelectric power stations were put forward, however, among them there were no projects that took into account the whole complex of problems that were considered in the complex scheme for the use of the Volga. In the subsequent period, the problem of the "Great Volga" was solved by creating a cascade of reservoirs: the Upper Volga (1944), Ivankovsky (1937), Uglich (1939-1943), Rybinsk (1940-1949), Gorky (1955-1957), Cheboksary (1981), Kuibyshev (1955-1957), Saratov (1967-1968), Volgograd (1958-1960 gg.), Kamsky (1954-1956 gg.), Votkinsky (1961 -1964 gg.), Nizhnekamsk (1978) [1]. At the same time, with the solution of energy problems, were the foundations of nowadays problems of aquaculture, connected both with changes in the hydrological and hydrochemical characteristics of the reservoir, and with the problems of directed formation of ichthyocenosis and reproductive problems, primarily for sturgeons.

The catches in the largest reservoirs of the Volga cascade at the beginning of the 21st century are presented in the Table 1 from [3].

Table 1. Catches of fish in the largest reservoirs of the Volga cascade, thous. tons.

Reservoir	1995	2000	2001	2002	2003	2004	2005	2006	2007
Rybinsk	1.4	1.5	1.6	1.7	1.0	1.3	1.04	1.0	1.1
Kuibyshev	3.2	2.8	2.7	2.7	2.0	1.94	2.11	2.62	2.7
Saratov	0.8	0.5	0.7	0.8	0.6	0.67	0.6	0.69	0.7
Volgograd	1.0	1.0	1.3	1.3	1.5	1.69	1.72	2.17	2.05

The construction of reservoirs assumed their fishery use by methods of pasture aquaculture. Historically, the formation of ichthyocenoses was based on the fish resources of the Volga River. It was also planned to introduce valuable species of fish, the reproduction of which was to be handled by regional fish breeding and spawning-growing economies.

Results and discussion

The current state of pasture aquaculture at the hydropower object - the Kuibyshev Reservoir

When planning fish catches in the Kubyshevsky reservoir, the main attention was paid to bream, carp, pike perch, pike and sturgeon fishes (Table 2) [4].

Table 2. Catches (%) in the Kuibyshev Reservoir.

Species	In 1950 it was planned	2014	Quotas for the catch in 2017, tons	%
bream	35	23.7	785.6	22.85
carp	15	1.2	29.2	0.84
pike perch	10	5.4	224	6.52
pike	8	0.1	19.2	0.56
sturgeons	2	0.02	5.9	0.17
other	30	69.58	2374.2	69.06
Catch	*20-24 thous. t.*	*4196.8 t.*	*3438.1 t.*	*100*

Other species had to provide 30% of catches. The current state of catches in comparison with the planned indicators is presented in the Table 3. Data on catches indicate an increase in the proportion of low-value fish in the recent period. If in 2010 low-value species accounted for 53.4% in catches, in 2014 for 64.8%. Analyzing the planned indicators for catches in the Kuibyshev reservoir, it can be noted that the planned harvest of sterlet was 480 tons. Actual catches of sterlet in the beginning of the XXI century varied about 0.850 tons.

Table 3. Catches (%) of valuable and low-value fish species in the Kuibyshev Reservoir.

Species	2010	2014	Quotas 2017	Species (lat.)	2010	2014	Quotas 2017
	% in catches				% in catches		
	Valuable species			*Low-value species*			
sterlet	0.04	0.02	0.17	Blicca bjoerkna	16.7	17.2	22.31
carp	1.59	1.23	0.84	Ballerus ballerus	11.3	14.4	16.99
catfish	0.57	0.19	0.26	Rutilus	9.1	12.0	7.50
pike perch	5.82	5.38	6.52	Pelecus cultratus	4.8	5.1	3.40

Ending of table 3

pike	0.80	0.11	0.56	Sander volgensis	2,9	3.1	3.32
burbot	0.24	0.31	0.46	Perca fluviatilis	2.5	5.1	4.13
bream	31.06	23.7	22.85	Carassius	3.0	4.5	2.33
silver carp	0.08	0.07	0.06	Abramis alburnus	3.1	3.4	5.09
				Clupeonella		2.93	1.54
				Aspius aspius		0.60	0.73
				Leuciscus idus		0.36	0.47
				Cyprinidae		0.45	0.47

The tendency of decline in the number of valuable fish species in the Kuibyshev reservoir was clearly revealed by the 2000s [2, 4] (Table 3). In the catches it was noted not only an extremely small number of sterlets (Table 4), but also pike, catfish and others (Table 3), at the same time the number of low-value and weedy species of fish has grown.

Table 4. The fishing stocks and the total allowable catches (TAC) of sterlet in the Kuibyshev reservoir [5].

Species	Catch, t 2014	Fishing stock, t						TAC, t 2016
		Years						
		2009	2010	2011	2012	2013	2014	
Sterlet	0.9	79	82	89	89	83	83	8
Total fish	1285.4	16016	16618	14459	16648	16671	15894	2623.0

As can be seen from the data given (Table 3, 4), in connection with the reduction of the stock of sterlet, it is necessary to carry out works to reproduce it and obtain juveniles for release into the reservoir.

Sterlet - as an important object of reproduction for pasture aquaculture on the Kuibyshev reservoir.

Special place belongs to sterlet in the composition of the fish population of the Volga and Kama rivers and, respectively, of the Kuibyshev and Nizhnekamsk reservoirs. It is the most valuable species in the ichthyofauna of the region. Prefers areas with a rapid current, does not make large migrations, benthophagus. Sterlet spawns at the end of May - early June on sites with a rapid current.

Works on the study of sterlet in the Middle Volga region were carried out as far back as 1913-1916 [6, 7]. Caviar was taken from mature fluid females, it was incubated in apparatuses at the demonstration plant at the Kazan City Museum. The yield of artificially produced fry was 66.9%. Thus, at the beginning of the 20th century, the basis for the artificial reproduction of sterlet was created for the first time in Russia in the territory of modern Tatarstan.

Since sterlet is included in the Red Book of the Republic of Tatarstan, and the characteristics of the producers of this fish in recent years are close to those of the previous period (Table 5), there are objective prerequisites for the organization of work on its reproduction. For the organization of release of juvenile sterlet, it is necessary to estimate the total amount of need for juvenile sterlet in the Kuibyshev Reservoir

Table 5. Average sizes (cm) of mature sterlet.

Gender	Volga river [8]	Kuibyshev Reservoir				
		[9]		[5]		
		1966-1969	1973-1974	1990-1993	1998-2009	2010
Females	47.6	47.4	52.4	47.2	47.1	47.4
Males	42.4	42.1	47.6	43.9	44.8	44.9

Coefficients of replenishment of the fishing stock (commercial return) from caviar, larvae, young sterlet in the reservoirs of the Volga cascade are given in Table 6 [10].

Table 6. Coefficients of replenishment of the fishing stock (commercial return) of sterlet for the reservoirs of the Volga cascade [10].

Index	Juveniles sterlets with weight (g)							
	1.5	3.0	5.0	10.0	11-20	21-30	31-40	41-50
Coefficient of commercial return	3.0	5.5	5.9	6.7	7.5	9.1	10.7	12.8

Currently, the harvest of sterlet in the Kuibyshev reservoir is prohibited.

With the organization of juvenile releases, the restoration of the number of sterlet to a scale permitting the organization of the fishery, its probable fishing length will be 42 cm with an absolute length of 48 cm. It should be noted that in studies before the regulation of Volga river [8] mature sterlet was encountered with an absolute body length of more than 33 cm. The majority of mature individuals were characterized by an absolute length of 42-53 cm. The size composition of the sterlet nowadays presented in Table 7.

Table 7. The size composition of sterlet in scientific research catches in the Kuibyshev reservoir [5], %.

Catch year	Size of fish, sm 15 – 20 – 25 – 30 – 35 – 40 – 45 –50 –55 – 60 – 65–70 - 75 - 80													N, ex.	Average length, sm
2009	1.1	16.9	21.4	19.4	22.7	12.7	2.8	1.5	1.0	0.2	0.1	0.1	0.1	320	36.6
2010	1.2	16.0	20.7	19.3	22.1	12.3	3.5	2.6	1.2	0.3	0.2	0.1	0.1	210	36.9
2011	1.6	16.4	20.9	19.9	21.8	12.6	3.7	2.6	1.1	0.2	0.2	0.1	0.1	225	37.3
2012	2.0	16.0	21.6	19.2	22.4	12.0	3.3	1.5	1.0	0.5	0.3	0.1	0.1	310	37.5
2013	1.9	16.1	20.8	19.6	22.1	12.5	3.1	1.8	1.2	0.4	0.3	0.1	0.1	240	37.4
2014	1.6	16.3	20.6	21.7	19.4	12.5	3.4	2.7	1.2	0.1	0.3	0.1	0.1	220	37.2

The conducted studies of the size and age composition of sterlet have revealed that sterlet has a length of more than 48 cm at the age of over 9 years. These are mature individuals. In the catches they make up 5.8 - 8.0%.

Taking into account that the actual fish catches in the Kuibyshev reservoir in the last period did not exceed 4 thousand tons with the ratio in the sterlet catch of 2%, the desired catch could be 80 tons. Taking into account that the actual fish catches in the Kuibyshev reservoir in the last period did not exceed 4 thousand tons with the ratio in the sterlet catch of 2%, the desired catch could be 80 tons.

Fishery calculations show that it is necessary to provide 72 tons of commercial sterlet by the means of artificial reproduction. With an average weight of 0.5 kg, its number is 144 thousand exemplars. Taking into account the annual natural mortality calculated according to P.V. Turin [11] and the coefficient of fishing return of 5.5% (Table 6), it will be necessary to issue 8587.3 thousand units. Thus, the annual requirement for the production of sterlet juveniles with a weight of 3 g in an amount of 8.6 million exemplars can provide for the catch of a commercial sterlet about 80 tons in 10 years. Release of sterlet can be conducted not only in the vicinity of potential spawning grounds, but also in the feeding areas. Taking into account the possibilities of long-term development of the Kuibyshev reservoir with the directed formation of ichthyofauna and planning of sterlet catches in the volume of 480 tons, annual releases of juveniles with a mass of 3 g in the amount of 57 million exemplars.

Conclusions

The development of hydropower and the creation of reservoirs of the Volga cascade had an impact on the state of populations of sturgeon species of fish, while creating incentives for improving the biotechnology of their reproduction and cultivation. This period is characterized by the desire and technical and technological capabilities to restore the number of sterlet in the Kuibyshev reservoir. The annual need for the release of sterlet juveniles with a weight of 3 g in the amount of 8.6 million exemplars can provide catch of the commodity sterlet about 80 tons in 10 years. Выпуски стерляди могут проводиться не только в районе потенциальных нерестилищ, но и в местах нагула рыб. Taking into account the possibilities of long-term development of the Kuibyshev reservoir with the directed formation of ichthyofauna and planning of sterlet catches in the volume of 480 tons, annual releases of juveniles with a mass of 3 g in the amount of 57 million pieces are required. The analysis shows the need to organize significant production of sterlet juveniles for stocking the Kuibyshev Reservoir.

References

[1] Avakyan A B 1998 *Volga in the past, present and future* (Moskow: «Экопресс-3М») p 20
[2] Kalayda M L 2001 *History and prospects of development of the fish industry of Tatarstan* (Kazan: Matbugat Yorty) p 96
[3] Dyomin A P 2010 Water management complex in Russia: concept, state, problems *Water resources* **37(5)** 617-32
[4] Kalayda M L 2017 The tasks of aquaculture development in the Republic of Tatarstan at the present stage. *Fish farming and fish industry* **8(139)** 7-16
[5] The materials of the total allowable catch of aquatic biological resources in the Kuibyshev and Nizhnekamsk reservoirs for 2016 with an environmental impact assessment (Area of work - Samara, Ulyanovsk, Tatarstan, Bashkortostan, Udmurtia, Mari El, Chuvashia) in the area of responsibility of the Tatar Division FGBSI «GosNIORH» **2015** (Kazan) 154
[6] Report of the Board of the Kazan Department of the Imperial Russian Fish Farming and Fisheries Society on the activities of the department for 1913 **2014** (Kazan: The printing house of the provincial government) 47
[7] Report of the Board of the Kazan Division of the Russian Society of Fish Culture and Fisheries on the activities of the Division for 1916 **2017** (Kazan: Print-litography of Provincial Board) p 62
[8] Lukin A V 1947 Main features of ecology of sturgeon in the Middle Volga *Proceedings of the Society of Naturalists at the Kazan University* **LVII(3-4)** (Kazan: Pub. h. of Kazan State university) pp 39-143
[9] Tsyplakov E P 1978 Migration and allocation of sterlet Acipenser ruthenus L. in the Kuibyshev Reservoir *Issues of ichthyology* **18(6)** 1020-28
[10] Methodology for calculating the amount of damage caused to water bodies as a result of violation of water legislation 2006 n 639 (collection of legislation of the Russian Federation, n 46, art. 4791)
[11] Turin P V 1972 "Normal" curves of experience and rates of natural mortality of fish as a theoretical basis for fishery regulation *News of GosNIORH* **71** 269

IOP Conf. Series: Earth and Environmental Science **288** (2019) 012045 doi:10.1088/1755-1315/288/1/012045

Crayfish in the reservoirs of the Republic of Tatarstan

M L Kalayda and I A Bogatyrev

"Water bioresources and aquaculture" department, Kazan state power engineering university, Russia, 420066 Kazan, Krasnoselskaya, 51

kalayda4@mail.ru

Abstract. Among the few commercial invertebrates in Europe, the most valuable resource species of crustaceans are representatives of the genus Astacus, primarily broad-fingered crayfish *(Astacus astacus* L.), which lives in inland waters - lakes, rivers, reservoirs. Crayfish are an integral part of the trophic chain of freshwater ecosystems, performing sanitary functions, utilizing animal and plant remains, thereby improving the epizootic situation and hydrochemical regime of water bodies. In the Republic of Tatarstan, before the regulation of the Volga River, there were 2 species of the genus *Astacus* - narrow-fingered crayfish (*A. leptodactylus* Esch.) and broad-fingered crayfish (*A. astacus* L.), and in the 30s of the 20th century, only the narrow-fingered crayfish in consequence of epizootics plague of crayfish. Presently, both species of crayfish are represented in crayfish populations, and in the Kuibyshev reservoir zone the ratio of narrow-fingered to broad-fingered crayfish is 3: 2, in the zone of the Nizhnekamsk reservoir - 1: 1. In the catches there are individuals from 8 cm to 17 cm with a body weight of 20 to 100 g. The females of narrow-fingered crayfish were 27.8%, and females of broad-fingered crayfish - 20% of the total number. In connection with the worldwide trend in the growth of the popularity of crayfish in food use, the importance of research of natural populations as a potential breeding stock for the tasks of industrial crayfish farming grows.

Introduction

Aquaculture rapidly developing around the world, providing for the development and cultivation of commercial species of hydrobionts under controlled conditions, is a promising direction of applied hydrobiology. Among the few commercial invertebrates in European countries, the most valuable resource species of crustaceans are representatives of the genus *Astacus*, primarily broad-fingered crayfish (*Astacus astacus* L.), inhabiting inland water bodies - lakes, rivers, reservoirs. Crayfish are an integral part of the trophic chain of freshwater ecosystems, performing sanitary functions, utilizing animal and plant remains, thereby improving the epizootic situation and hydrochemical regime of water bodies. Broad-fingered crayfish are recognized as bioindicators of environmental cleanliness and are currently under the scrutiny of international environmental organizations. Broad-fingered crayfish is included in the lists of protected species of the Berne Convention and the European Community Directive on species and habitats. Having the status of a protected species in European countries, broad-fingered crayfish are of great commercial value. Not only meat of crayfish used for food, but also a shell has applications in medicine and cosmetology [1, 2, 3].

IOP Conf. Series: Earth and Environmental Science **288** (2019) 012045 doi:10.1088/1755-1315/288/1/012045

Results and discussion

River crayfish as components of water ecosystems of the Republic of Tatarstan

Information on the status of the population of crayfish in the territory of the Republic of Tatarstan before 1932 is absent [5]. Prior to the regulation of the Volga River, practically on all small rivers, there were mill ponds in which river crayfish could be found. By the 30th of the 20th century, the populations of crayfish began to recover after the epidemic of crayfish plague that exterminated them in Volga and Kama [5]. Most crayfish survived in mill ponds on small rivers.

In the European part of Russia, there are 15 species of crayfish belonging to 3 genera: *Astacus*, *Pontoastacus*, *Caspiastacus* [1]. In the Republic of Tatarstan there were 2 species of the genus *Astacus* - narrow-fingered crayfish (*A. leptodactylus* Esch.) and broad- fingered crayfish (*A. astacus* L.) [6].

By the time of regulation of the Volga River, a study on the biology of crayfish revealed that the ratio in the catches of females to males was 30%: 70% in summer catches and 44% 62% in autumn. Dimensions of crayfish according to [5] in spring catches are shown in Figure 1.

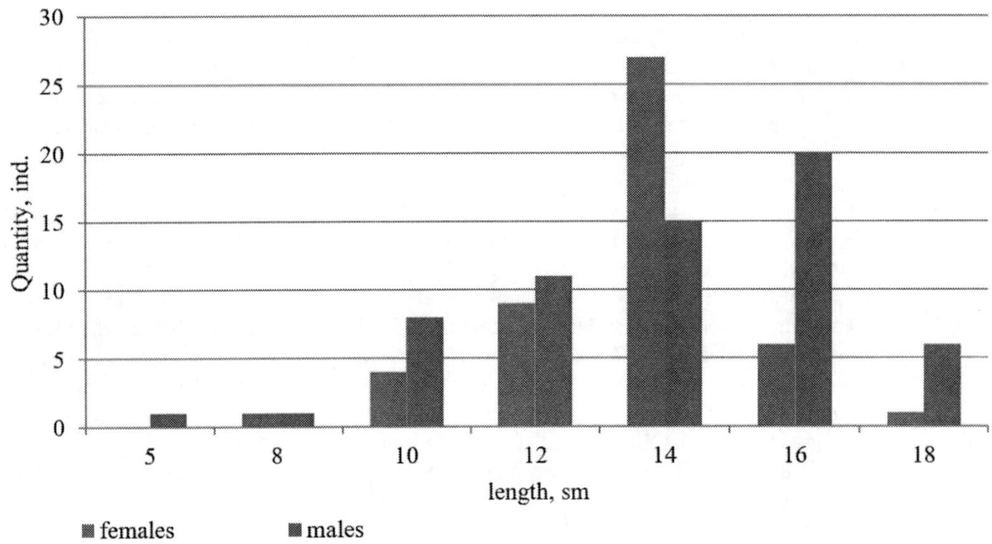

Figure 1. The size of broad- fingered crayfish in the reservoirs of the Republic of Tatarstan prior to regulation the Volga River according to [5].

In the studies of M S. Stroikova [5] it was shown that in the reservoirs of the Republic of Tatarstan in the 30s of the XX century there was only a narrow-fingered crayfish – *Astacus leptodactylus*. It is noted that at the beginning of life, females and males of crayfish grow at the same rate, and then at puberty the rate of growth of females decreases. One-year-old crayfish had a length of about 4 cm, two-year females and males on average - 10.5 cm, three-year-old males - 14.5 cm, females - 13.5 cm. Four-year-olds were met about 17.5 cm in size and all male. The rate of growth of crayfish in floodplain lakes of the Volga River was higher than in small rivers (Table 1).

Relation between the weight of a narrow-fingered crayfish and the length under conditions of the river Volga taken from the net catches is presented in Table 2.

Sexual maturity occurred for female crayfish in the third year of life and is related to the growth rate. In organizing the fishery, the fishing length of crayfish for small rivers was not less than 8.0-8.5 cm, for productive lakes - more than 10.5-12.5 cm [5]. Crayfish over 4 years old have not occurred.

Table 1. Growth rate of narrow-fingered crayfish in reservoirs of the Republic of Tatarstan according to [5].

Water body	Age, years			
	1	2	3	4
Floodplain lake Long	5,5	10,5	14,0	17,5
River Nurma	4,21	6,48	8,25	11,93
River Ik			11,5	16,0

Table 2. Size and weight composition of narrow-fingered crayfish from net catches in the Volga River [5].

Mass, g	Fish size, sm									N, ind.
	8 - 9	- 10	– 11	– 12	– 13	– 14	–15	– 16	– 17	
10-20	4	1	-	-	-	-	-	-	-	5
20-30	1	15	9	1	-	-	-	-	-	26
30-40	-	-	10	1	-	-	-	-	-	11
40-50	-	-	-	7	8	-	-	-	-	15
50-60	-	-	-	2	10	3	-	-	-	15
60-70	-	-	-	-	7	10	1	-	-	18
70-80	-	-	-	-	1	10	5	-	-	16
80-90	-	-	-	-	-	6	4	1	-	11
90-100	-	-	-	-	-	2	2	2	-	6
100-110	-	-	-	-	-	-	2	1	1	4
110-120	-	-	-	-	-	-	1	2	-	3
120-130	-	-	-	-	-	-	1	3	2	6
n	5	16	19	11	26	31	16	9	3	136

In studies of the first half of the 20th century, there was a high degree of infection of crayfish with trematodes. Crayfish larger than 43 cm were infected in 82.4% of cases. The number of parasites varied in different individuals from 4 specimens to several hundred. Places of localization of parasites – common liver fluke were: sexual glands, abdominal muscles, gill covers, cephalothorax cavity, muscles and side walls, liver, intestinal walls, genital ducts, heart [7]. With the presence of parasitic frowth *Astacotrema tuberculatum* (A.cirrigerum) and *Distomum isostomum* the caryfish plague was associated, which destroyed crayfish in many parts of Europe. It was suggested that crayfish are the second intermediate hosts for the genus Astacotrema.

While regulating the Volga River in the Republic of Tatarstan, crayfish were not considered as an essential objects of pasture aquaculture. Currently, there is no commercial catch of crayfish in the Republic of Tatarstan. At the same time, in the European part of Russia at the present time, the Lower Volga region, the Azov-Black Sea basin, the reservoirs of the Middle Volga, the Northwest (Pskov and Leningrad oblasts) are the main crayfish fishing areas. The distribution of crayfish stocks by regions of the European part of Russia and the volumes of possible catch of crayfish in water bodies are presented in the Table 3.

In connection with the worldwide trend in the growth of the popularity of crayfish in food use, the importance of research of natural populations as a potential breeding stock for the tasks of industrial breeding grows. To date, aquabiotechnologies for the artificial reproduction of broad-fingered cancers have been developed [8], for which there is an uneven, sporadic nature of distribution within the range of its area, caused mainly by epizootics of crayfish plague, pollution of reservoirs and destruction of natural biotopes [3].

Table 3. Reserves and total allowable catches of crayfish in the crayfish fishing reservoirs of the European part of the Russian Federation по [4].

Region	Area, ha (useful for crayfish)	Stock, t	Total allowable catches ,t
	Reservoirs of the North-West:		
Pskov region	46800	107	41
Leningrad region	26100	71	17
Tver region	6960	96	24
	Reservoirs of the Middle Volga:		
Gorky Reservoir	36800	13,5	3
Cheboksary Reservoir	53600	19,7	5
Kuibyshev Reservoir	170000	100	25
	Reservoirs of the Lower Volga:		
Saratov Reservoir	40900	58	15
Volgograd Reservoir	126664	176,5	40
Reservoirs of the Volga-Akhtyubinsk floodplain and the Volga delta	8880	167	44
Reservoirs of the Azov-Black Sea basin	5700	1200	210
Tsimlyansk Reservoir	2700	740-760	140-150
Reservoirs of the Ural	28050	380	50
Всего		2390	475

Current status of crayfish populations in the Kuibyshev and Nizhnekamsk reservoirs

Currently in the Republic of Tatarstan there are two types of crayfish: broad-fingered and narrow-fingered (Figure 2). In the spring catches in the Kuibyshev reservoir, the ratio of narrow-fingered and broad-fingered crayfish was 60%: 40%. At a number of sites in the Nizhnekamsk reservoir, the ratio was 50%: 50%. In this case, females of narrow-fingered crayfish were 27.8%, and females of broad-fingered crayfish - 20% of the total number of crayfish of this species.

Figure 2. Crayfish from the catches on the Volga, Kama and Belaya rivers in 2017.

Broad-fingered crayfish in the lifestyles are classed as "K-strategists", which are characterized by relatively low fertility, slow growth and long (up to 20 years) life expectancy [8].

In the conditions of the Middle Volga, crayfish were found in catches at different depths: from the beginning of February - from 2 to 3 m, at the end of February they were not caught at any depths, in March they were caught at depths of more than 3 m and from the middle of March even deeper - up to 6 m.

The distribution of crayfish (%) in the reservoirs and rivers of the Republic of Tatarstan is shown in the Figure 3.

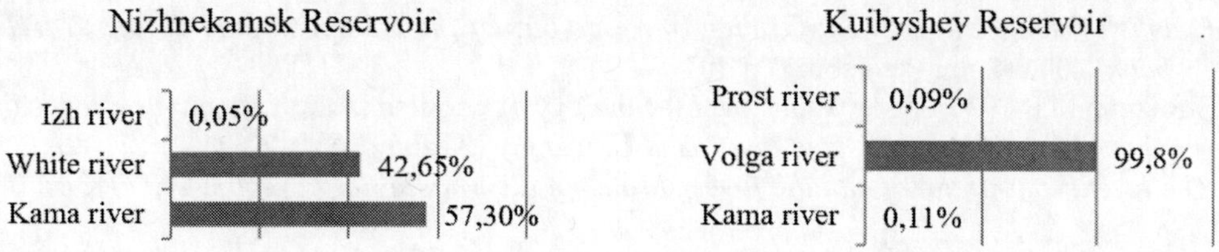

Figure 3. Distribution (%) of crayfish from catches in 2017 in the Republic of Tatarstan.

In Europe, the maturity of females occurs at the end of the 4th summer of life, at the age of 3+, with a body length of about 8 cm [3]. In the studied materials, female crayfish with caviar were found in catches only from 10 to 20 May. The copulation of broad-fingered crayfish usually occurs in late autumn (October-November) at a water temperature of less than 10 ° C and lasts for 2-3 weeks. Sexually mature males mate each year and are able to fertilize several (more than 3) females. The period of embryogenesis of broad-fingered crayfish occurs within 8 months (November-June) in a wide range of temperatures and is about 1300 degree-days. The timing of hatching depends on the thermal regime at the end of the incubation period. The working fecundity of females of broad-fingered crayfish increases with sizes and ranges from 90 to 260 eggs [8]. Crayfish belong to omnivorous animals, they feed on detritus, aquatic vegetation, animal food: crustaceans, insect larvae and mollusks. Their favorite food is fish. The nutritional components of juveniles of broad-fingered crayfish are detritus, filamentous algae, protozoa, rotifers, crustaceans, larvae of chironomids and caddisflies. The daily diet for mature broad-fingered crayfish varies between 0.3-1%, for young - 1-4% of their body weight [3]. The length of the body for juveniles of broad-fingered crayfish varies depending on the season and the water content of the pond in the range from 15 to 23 mm.

In the Volga river in the catch in February-March were caught individuals aged 2 to 4 years. The maximum mass of crayfish in February was 60 g with a body length of about 13 cm, in March, there were individuals up to 100 g with a body length of 17 cm. Taking into account that large males of broad-fingered crayfish reach a body length of 15 cm from the rostrum to the end of the abdomen and sometimes there can be met larger specimens, 17-25 cm in size and weighing 270-300 g [8], can be noted that adult crayfish in the Kuibyshev and Nizhnekamsk reservoirs smaller and at a length of 17 cm have a mass of no more than 100 g.

Conclusions

Crayfish farming can become an important component in the structure of aquaculture in the Middle Volga region. At present, natural populations of crayfish in the reservoirs and rivers of the Republic of Tatarstan include 2 species of the genus *Astacus* - narrow-fingered crayfish (*A. leptodactylus* Esch.) and broad-fingered crayfish (*A. astacus* L.) [6]. The ratio of narrow and broad-fingered crayfish in the zone of the Kuibyshev reservoir was 3:2, and in the Nizhnekamsk reservoir - 1:1. Adult individuals reach 17 cm at a mass of up to 100 g. Sexually mature females with caviar at the age of 4 years were met in catches in May. The most significant crayfish populations are found in the Volga, Kama, Belaya, Izh, Prost rivers.

The development of aquabiotechnologies makes it possible to use natural crayfish populations for the tasks of creating industrial breeding farms.

References

[1] Fyodorova Z V Cultivation of crayfish abroad 1991 *Fish management. Aquculture* (VNIIERKh) **1** 1-16.

[2] Huner J V, Romaire R P 1990 Crayfish culture in the southeastern USA *World Aquacult* **21(4)** 14-7

[3] Zukersis Y M 1989 *Crayfish* (Vilnius: Mokslas) p 143

[4] *Crayfish fishing and crayfish farming on reservoirs of the European part of Russia (reference book)* 2006 (Saint Petersburg) p 207

[5] Stroikova M S 1937 Observations over the biology of crayfish in the Tatrepublic *Proceedings of the Society of Naturalists at the Kazan University* (Kazan: KSU) **LV (1-2)** 171-80

[6] *The brief determinant of aquatic invertebrates of the Middle Volga region* 1977 (Kazan: KSU) p 156

[7] Sosnina M F 1947 To the biology of the parasite of long-fisted crayfish Astacotrema tuberculatum Zaw. *Proceedings of the Society of Naturalists at the Kazan University* (Kazan: KSU) **LVII (3-4)** 165-71

[8] Mackevichene G, Mickenene L, Plyurayte V and Joninene B 2008 Artificial reproduction of a broad-fingerd crayfish Astacus astacus L *Biotechnics of artificial reproduction of fish, crayfish and preservation of stocks of commercial fish* (Vilnius) p 178-95

[9] Souty-Grosset C., Holdich D M, Noel P Y, Reynolds J D and Haffner P (eds.) 2006 *Atlas of crayfish in Europe. Museum national d'Histoirenaturelle. Paris* (Patrimoinesnaturels, 64) p 187

Prospects for the development of biogas technologies at reservoirs of energy facilities

M L Kalayda and S D Borisova

"Water bioresources and aquaculture" department, Kazan state power engineering university, Russia, 420066 Kazan, Krasnoselskaya, 51

Svetlana-zag@bk.ru

Abstract. In connection with the increase in energy prices and the deterioration of the environmental situation, the problem of introducing biogas plants is becoming increasingly important. In this case, the selection of promising raw materials for the production of biogas becomes actual. The organization of energy-biological complexes on coolering reservoirs will allow not only to solve the problem of thermal emissions of power plants, but also to obtain biogas. The article deals in detail with the creation of energy-biological complexes at reservoirs of power engineering objects. Special attention is paid to the use of aquatic vegetation for the production of biogas, as a stage of utilization of excesses of the higher aquatic vegetation of the cooling reservoir of the power station.

Introduction

One of the promising areas of alternative energy is bioenergy. This direction unites ways of reception of energy from a waste, biofuel from plants, use of nonconventional kinds of fuel.

To date, around 60 varieties of biogas technologies are used or are developing in the world, which differ in the composition and preparation of raw materials, the conditions and technological parameters of methanogenesis and the designs of biogas plants. In Europe, the use of biogas systems has found wide application, primarily as one of the ways to solve environmental problems associated with the utilization of agricultural waste, as well as a way of obtaining additional energy and organic fertilizers.

In this case, the use of biogas technologies in the organization of energy-biological complexes on reservoirs of energy facilities, where promising raw material for obtaining biogas is the higher aquatic vegetation.

The relevance of studies of the vegetation of the cooling reservoirs is attached to the importance of their results for solving a number of problems related to ensuring the safety of the operation of power plants. Excessive development of vegetation makes it difficult to operate the cooling reservoir. Floating plants and weakly attached forms, clogging the grids of water intake facilities, can even create an emergency situation in the operation of the technical water supply system of a power plant. Thus, preventing the overgrowing of cooling reservoirs macrophytes allows both to ensure the safety of the operation of power plants, and to obtain a promising fuel for biogas plants.

Energobiological complex on reservoirs of energy facilities

One of the most effective ways of utilizing low-potential waste water is the organization of energy biological complexes (EBC).

The EBC includes a number of mini-enterprises: basin aquaculture enterprises, cooling pond, microbiological production, refrigeration, heating ground, mushroom growing, greenhouse production. Separately, they are self-sufficient, modern, at high level of technology of production. In EBC, the technology of each of them is associated with the use of warm water and with each other in such a way that the final cycle of production of one of them becomes the beginning of the production of the other, thereby achieving an almost complete waste-free and intensified production.

Basin aquaculture - cultivation of aquatic organisms. Aquaculture is currently progressing rapidly in its development - an increase in production of 10% per year for 20 years. The use of warm water for fish breeding is, at present, the most developed way of utilizing the heat taken off, both in our country and abroad [1].

Temperature plays an extremely important role in the life of fish and other aquatic organisms that are related to cold-blooded animals. When growing fish in warm water, the terms of ripening of valuable commercial fish are shortened: sturgeon eggs can be obtained at the age of 4-8 years instead of 12-18 years in natural conditions, and on average the growth process is accelerated 3-4 times or more (up to 10 times). The diverse requirements of fish for optimal temperature regimes make it possible to use practically all the waste cooling water: heat-loving fish (carp, eel, buffalo, etc.) - + 20-30 °C (up to + 35 °C), cold-loving (salmon, trout) 16 °C. The yield of commercial fish from 1 m^3 of fish-breeding capacity can reach 40-300 kg depending on the fish species and biotechnology used. Many species of higher aquatic plants and algae, rich in iodine and other useful substances, can also be grown in these conditions.

The cooling pond ensures the reliability of the power plant operation, it can grow numerous highly productive thermophilic fish species - meliorators feeding on zoo- and phytoplankton, detritus (white, bighead, silver carp, grass carp), a number of mollusks for water purification, directional shaping of fodder base allows to increase its biological productivity several times.

Microbiological production is characterized by an exceptionally high rate of accumulation of bioproducts - in single-celled organisms protein synthesis is 1000 times higher than in higher plants, efficiency of use of solar energy for photosynthetic microalgae is up to 5%, and in special installations up to 15-20%, while for higher plants - no higher than 1-1.2%. Among the important, already proven food products are 119 different microalgae from the known more than 34,000 species. Feed yeast, containing up to 30-40% protein with a full set of micro- and macroelements, can be up to 30-40% in the carp's diet, for other fish - slightly less.

Schemes for processing biowaste in biogas were mastered [2]. The main equipment of the biogas plant is a hermetically sealed container with a heat exchanger (Figure 1) (coolant - water, heated to 50-60 °C), devices for input and output of biomass and for gas removal.

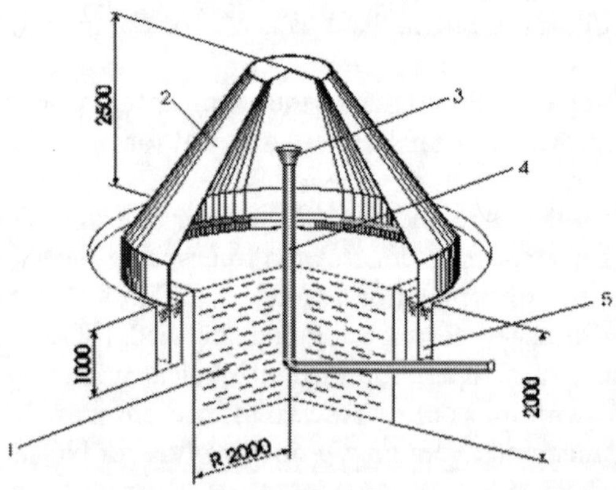

Figure 1. Scheme of tank for producing biogas with a conical bell, where 1 - pit with plant raw materials; 2 - the bell; 3 - outlet connection; 4 - a pipeline (hose) of biogas feed; 5 - water sealed channel with water.

The biogas plant is intended for processing of various production wastes for obtaining environmentally friendly organic fertilizers of natural type, energy resources, feed additives, waste utilization. The biogas plant uses the process of methane fermentation. A high degree of conversion of organic matter in waste allows obtaining an increased yield of biogas and liquid sludge. Depending on the raw material, the sludge can be used as ready-to-use fertilizer (manure processing, litter) or high-efficiency feed additives (food industry waste, beer industry) [1, 3, 8-10].

The refrigeration facilities of the energy-biological complex are based on the use of warm water (for obtaining cold) by the adsorption method. Refrigeration plants are used not only to store products, but also in the production process for a short-term change in temperature at a certain point in the cultivation of organisms, creating an optimal temperature regime in a given period, in a particular installation. Heated soil intensifies all soil processes, allows to increase the yield of plants several times.

Mushroom production needs for its purposes heat of a relatively low potential - + 16-17 °C, and only in the production of mycelium - +29 °C. For substrates can be used waste from many industries, yield of protein with mushrooms is 25-30 times higher than with 1 hectare of the most used arable land. About 40 species of fungi are artificially cultivated. In greenhouse production, the productivity of 1 hectare is 20-30 times higher than the productivity of 1 ha of open ground, in high-rise greenhouses it is hundreds of times higher. The heating of 1 hectare of greenhouse area consumes 1.5-1.8 million m^3 of natural gas. At the same time, the optimal temperature for growing plants in greenhouses is + 18-35 °C and even lower (cucumbers and tomatoes + 20-21 °C, flowers + 14-16 °C). To create such a temperature, coolant with a temperature of +90-70 °C is used [2].

The constant growth of energy prices places the industry on the brink of existence and encourages to look for new ways of heating greenhouses, layout solutions that would suit both greenhouse production and energetics. Heating high-rise greenhouses can be carried out with the help of dry cooling towers, cascade greenhouse-cooling towers can potentially perform a dual role: serve as a cooling system for waste water from a power plant and act as a greenhouse for growing plants in it.

The biomass of greenhouse plants is a potential reserve, both for microbiological production and for processing by earthworms in fish feeds [1-3, 11]. One ton of organic material is potentially capable of yielding 70 to 100 kg of this protein product.

Biogas plants

The leader in the application of biogas technologies is China, where 40 million of small units have been built, with reactor volumes of no more than 10 m^3, up to 2020 it is planned to increase their number to 80 million, in India - 4.4 million units, Nepal and South Asian countries - 355 thousand installations. In China, India there are state programs for the introduction of small biogas plants. In Europe, more than 10,000 large biogas plants have been built, Germany leads with 5000 large agricultural biogas plants. In the USA, there are more than 170 large plants with a reactor volume of more than 2000 m^3, of which 153 produce electric and thermal energy.

A biogas plant is a closed container in which raw materials are processed (Figure 2). Any organic waste can be used as raw material. The principle of the biogas plant is that after preparation, the raw material is fed to the reactor, where it is processed while maintaining certain conditions. During fermention, biogas is released, which can be used in various gas appliances. Fermented mass is a highly efficient and environmentally friendly fertilizer [4].

Biogas plants differ in the way of loading raw materials. Batch loading plants are fully loaded with raw materials, and then completely released after a certain processing time. For this type of loading, installations of any design and any type of raw materials are suitable, but such installations are characterized by unstable production of biogas. Continuous loading plants are loaded daily with small amounts of raw materials. When loading new raw materials, an equal portion of the processed sludge is unloaded. Raw materials processed in such plants must be liquid and homogeneous. Gas production is stable and quantitatively exceeds the volume of biogas produced in batch plants. Virtually all installations under construction in developed countries now operate as continuous load installations [5, 9].

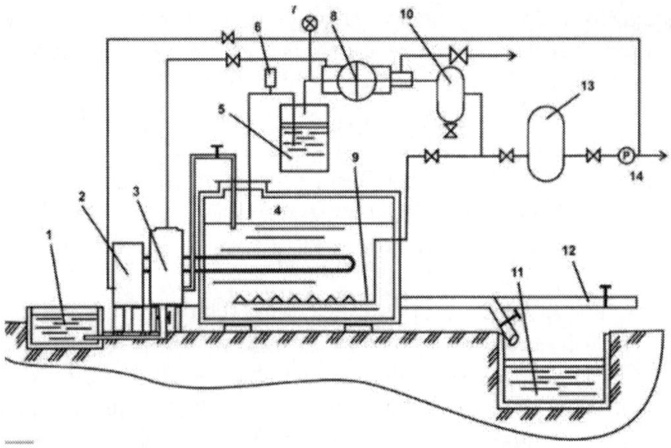

Figure 2. Biogas plant: 1 – raw material receiver; 2 – water heating boiler; 3 – loading hopper; 4 – reactor; 5 – water seal; 6 – safety valve; 7 – electrocontact manometer; 8 – compressor; 9 – gas stirrer; 10 – receiver; 11 – biofertilizer storage; 12 – branch pipe for loading in transport; 13 – gasholder; 14 – gas reducer.

The amount of biogas produced depends on the type of feed and temperature in the reactor. On average, the output of biogas is 1 - 2 m^3/day from 1 m^3 of the reactor - 30 m^3 from 1 ton of raw materials. The composition of biogas: 60-70% of methane, 30-40% of carbon dioxide. Calorific value: 20-25 MJ/m^3, which is equivalent to combustion of: 0.6 liters of gasoline, 0.85 liters of alcohol, 1.7 kg of firewood, or use - 1.4 kW of electricity [3,8].

Water vegetation of reservoirs of energy facilities as a promising raw material for biogas plants

In cooling reservoirs of electric stations, higher aquatic plants are actively developing, which disturbs the ecological balance of the reservoir, causes its siltation and shallowing, and can disrupt the operating of power plants. It is advisable to apply the clearing of reservoirs of power engineering objects from higher aquatic vegetation and use this vegetation and bioorganic waste of the blocks of the energy-biological complex as a raw material for producing biogas.

Such higher aquatic plants as *Elodeya Canadian*, hornwort and eichornia can be used as the main element for biofuel production in combination with purification of water of cooling reservoirs [6]. The macrophytes under consideration have high growth rates.

The amount of green biomass per day, ensuring the uninterrupted operation of the biogas reactor, is determined by the production capacity of a particular complex. For intensive plant growth, certain conditions are necessary: water temperature + 24-28 °C, air temperature + 24-28 °C, humidity above 90%, intensive lighting, balanced nutrition and a number of other parameters. In open water bodies, the amount of biomass produced depends on climatic conditions and nutrient medium. In the conditions of the middle and southern strip from one hectare of the water surface per season (four months), several hundred tons of fresh green mass can be obtained. Elodea, hornwort and eichornia are well kept and do not lose their nutritional qualities in dry form, and also in fresh condition in the form of dense rolls wrapped in polyethylene film. This allows growing and storing surpluses of grown green biomass, compensating for the decline in the intensity of growth during the winter season, and also growing them in artificial and natural water bodies in the spring-summer-autumn period in an amount sufficient to ensure year-round operation of biogas complexes [6].

Collected in the pond macrophytes are placed in the chopper, and from there the ground biomass falls into the methantanks. In methane tanks, under the influence of a special preparation, methane fermentation takes place with the formation of biogas, which is used by the plant for burning and generating electricity. The cost of such electricity is very low, since no special expenses are required to reproduce higher aquatic plants. The main investments are necessary for the construction of the installation, and the costs of its maintenance become current expenses.

The introduction of microorganisms into the reactor occurs only once during the first run. In the future, the addition of microorganisms and additional costs are not needed. At the output there are two main products: biogas and a series of biofertilizers in the form of composted and liquid substrates. For storing biogas, a special tank is used - a gasholder, and from it biogas is continuously supplied to a gas power generator that produces electricity and heat. The fermented mass is actually ready for use biofertilizer.

It is estimated that from 1 ton of biomass of higher aquatic plants, about 500 cubic meters of biogas is produced, with the combustion of 1 cubic meter of which 2 kW of electricity and 2 kW of heat are generated [5].

On the basis of experimental data, we developed a computer program for modeling the operation of a water treatment plant using the higher aquatic vegetation "CLEANING", which not only makes it possible to evaluate the efficiency of cleaning ("self-cleaning") of aquatic plants in natural flowing water bodies, but also to simulate the volumes of aquatic vegetation for subsequent disposal [7].

Thus, at present, the use of higher water plants in reservoirs of energy facilities as raw material for biogas plants is a promising direction in bioenergy.

References

[1] Rostuntsova I A and Shevchenko N U 2015 Efficiency of using low-grade heat of TPPs as secondary energy resources *International Journal of Applied and Fundamental Research* **9-4** 618-21

[2] Vasiliev U V and Remizov N N 2004 Energobiological complexes as a way of utilizing the waste heat of large energy facilities and creating a high-intensity waste-free production *Questions of Atomic Science and Technology. Ser. Thermonuclear fusion* **1** 57-60

[3] Kovalev A A 2001 Efficiency of biogas production on livestock farms *Technology in agriculture* **3** 30-3

[4] Kalayda M L and Zagustina (Borisova) S D 2007 Analysis of existing biogas plants for production from biomass of aquatic plants *Materials of the II International Youth Conference "Tinchurin Readings"* (Kazan: KSPEU) pp 67-9

[5] Masaev I V 2000 The use of biowaste in agriculture as a fuel and rational combustion technologies *News of the Academy of Industrial Ecology* **4** 84-6

[6] Kalayda M L and Borisova S D 2010 Accumulation of pollutants by aquatic plants and the possibility of utilizing plant matter *Butlerov Communications* **9** 33-9

[7] Kalayda M L, Zagustina (Borisova) S D and Petrov A V 2009 Computer program for modeling the operation of a water treatment plant using higher aquatic vegetation «CLEANING» *Certificate of the Russian Federation on the industrial registration of an electronic resource №00176 from 19.08.2009*

[8] Zuyeva O S, Kalayda M L, Zuyev U F 2014 Biofuel from algae: directions of development and nanotechnology *Nanomaterials and nanotechnologies in power engineering. Monograph* (Kazan:KSPEU) **II** 81-99

[9] Zuyeva O, Kalayda M and Chichirov A 2014 Biofuel and Bionanotechnology (LAP LAMBERT Academic Publishing (OmniScriptumGmbH&Co.KG), Saarbrücken, Germany) 112

[10] Kalayda M L, Zuyeva O S, Zuyev U F and Galimova R Sh 2014 Use of algae for biofuel production as a promising area for alternative energy *Proceedings of higher educational institutions Energy problems* **7-8** 96-103

[11] Kalayda M L and Abdullin R R 2015 Biogeochemical bases of inclusion in feed of fishes of worms Eisenia foetida as a dietary supplement *Butlerov Communications* **44(11)** 130-5

IOP Conf. Series: Earth and Environmental Science **288** (2019) 012047 doi:10.1088/1755-1315/288/1/012047

Creation of trout farms on the basis of power plant water cooler reservoirs

M L Kalayda and D S Dementiev

"Water Bioresources and Aquaculture" department, Kazan State Power Engineering University, Russia, 420066 Kazan, Krasnoselskaya, 51

kalayda4@mail.ru

Abstract. Increase of production of fish by traditional methods, based primarily on the extensive use of natural resources, has certain natural limitations.The limiting factors are land, water and the environment. Considering this fact, the perspective expansion of industrial farms, provided with superintensive technologies, is actual. The role of abiotic environmental factors in the life of fish is enormous. For example, the water temperature determines the intensity of metabolism and is a natural stimulus determining the begining of spawning migration of fish. Water temperature has a very important influence on the life of the organism, particularly on the metabolic processes, the behavior and dispersion of organisms. Water quality is one of the most important factors stimulating the development of water biotechnology. In this regard, we explored the physical and-chemical features of the Volga water from the viewpoint of the trout-breeding. The analysis revealed that the water does not fully meet the required water quality for trout ponds. Characteristics as pH, absence of hydrogen sulfide meet the requirements; such indicators as permanganate and bichromate oxidation, BPK5, ammonium ion, nitrates and nitrites, oxygen content are noted exceeding the standard values for trout farms. Power engineering and fish farming are connected closely – power plants contribute to the creation of favorable conditions for the cultivation of commercial fish, including valuable and fish-breeding complexes based on water coolers help to improve their environmental condition. At the same time in Russia in recent years, this industry is beginning to receive attention it deserves. The experience of Western countries shows a new direction of the industry development: in the previous period the favorable conditions for growing and breeding fish were created on the basis of the warm water for increasing production capacity of energy enterprises due to energy production and fish production, nowadays the power plants are built at fish-breeding enterprises in order to optimize costs and increase profitability of fish production.

Introduction

In the context of the ever-growing global demand for food, fish farming comes to the fore in the field of providing the population with animal protein. In the ecological sense, fish farming is a more productive way of getting animal protein, as fish use feed for growth more effectively than farm animals. The annual growth in the fishing industry for about two decades is caused by aquaculture, as the volume of fish caught only decreases.

Increasing the production of fish by traditional methods, based primarily on the extensive use of natural resources, has certain natural limitations. The limiting factors are land, water and the environment. Considering this fact, the perspective expansion of industrial farms, provided with

IOP Conf. Series: Earth and Environmental Science **288** (2019) 012047 doi:10.1088/1755-1315/288/1/012047

superintensive technologies, is actual. These is especially true for fish-breeding systems with a closed cycle of water supply, allowing for year-round cultivation of all types of aquaculture, regardless of climatic conditions, achieving maximum growth and productivity rate while conserving resources and keeping environmental cleanliness of the production process [1].

The role of abiotic environmental factors in the life of fish is enormous. For example, the water temperature determines the intensity of metabolism and is a natural stimulus determining the beginning of spawning migration of fish.Water temperature has a very important influence on the life of the organism, particularly on the metabolic processes, the behavior and dispersion of organisms.The highest concentration of fish is observed in the areas of contact between cold and warm flows - where frontal zones are formed.

Fish is highly dependent on the ambient temperature. For most fish body temperature is only 0.5-1°C different from the temperature of the surrounding water. The water temperature affects the change in the intensity of metabolism and is the main reason determining the beginning of such processes as spawning, migration, wintering, etc. In addition, the water temperature changes have big impact on the growth and development of fish.

Rainbow trout is one of the most promising aquaculture objects, but in Russia the growth rate of commercial trout production is significantly behind foreign ones. Including for this reason, about 90% of the red fish consumed in Russia is imported.

The development of trout farming in the Republic of Tatarstan is particularly relevant from the point of solving the problems of import substitution and providing the region population with high-quality fish products. There are some objective preconditions for it. Back in 1915, in the summary of M. Somov, the existence of two trout farms with a total area of 9.8 hectares located on the territory of the Kazan province was indicated. During this period, in the region were cultivated: American Palia; river trout; rainbow trout; whitefish and salmon[2].

Water quality is one of the most important factors stimulating the development of water biotechnology. In this regard, let's explore the physical-chemical features of the Volga water from the viewpoint of the trout-breeding.

Results and discussions

The quality of water as an important section of trout fish farming development in Middle Volga region.

Geographical zoning in the Volga cascade reveals itself in the change of a number of abiotic characteristics: increase in transparency and general mineralization from North to South, decrease in color and suspended solids.These patterns are broken on the area of the Middle Volga, which is characterized by high anthropogenic impact [3]. The Volga basin accounts for more than a third of Russia's wastewater discharge.

Volga is usually divided into three main parts – Upper, Middle and Lower. The border between the Upper and Middle Volga is the confluence of the river Sheksna in Rybinsk city. The border between the Middle and Lower Volga river passes through the Volga hydroelectric dam [4].

The Middle Volga region is rich in water resources. There are such large rivers as Volga, Kama, and their tributaries: Vyatka, Sviyaga, Mesha, Sesma, Ik, Toyma, Izh, Stepnoy Zay, Kuibyshev and Nizhnekamsk reservoirs, lakes and complex purpose irrigation reservoirs. Kuibyshev reservoir – lake-reservoir - is the largest in Europe - 6450 km^2 [4,5]. In recent years, the waters of the Kuibyshev reservoir are estimated as «moderately polluted». The main pollutants are copper compounds, petroleum products, phenols and others [6]. The water quality is influenced by the transfer of polluted water from the upper reaches of rivers flowing into reservoirs, as well as the discharge of insufficiently treated wastewater from industrial enterprises. Because of the surface water quality deterioration, there are problems not only of its purification by physical and chemical methods, but also by biological methods [7-9].

Requirements for the quality of water entering the trout farms ponds are presented in the Table 1.

Table 1 presents the requirements to the quality of water entering the trout ponds. It shows the data of the chemical quality of water at the test sites of the Volga reach of Kuybyshev water reservoir, that has the highest preserved fluvial regime of the Volga river and chemical quality of water of small rivers of the Kama basin. The authors studying the small rivers water quality are noting that the maximum permissible concentration (MPC) of harmful substances in this water is almost never seeing, a high total iron content is typical for natural waters of the region [14]. Water quality in the Volga river, according to different authors, is given in Table 2.

As can be seen from the data above (Table 1, 2) the water of Volga river in the region of the Middle Volga and small rivers of the Kama basin does not meet required water quality for the trout ponds. Such characteristics as pH, absence of hydrogen sulphide are meeting the requirments; such indicators as permanganate and dichromate oxidability, BOD5, ammonium ion, nitrates and nitrites recognised as regulatory for trout farms.

Table 1. Requirements for the quality of water entering the trout farms in comparison with the quality of natural surface waters.

Characteristic	Regulatory values for trout ponds	Control plot Volga reach Kuibyshev reservoir [11]	Kama pool small rivers [14]
Transparency, m	1.5	-	-
pH	7.0-8.0	-	7.45-8.1
Suspended solids, mg/dm^3	Less than 10	-	0.129-0.263
Oxygen dissolved, mg/ dm^3	Not lessthan 9.0	7.53±0.75	-
Free carbon dioxide, mg/ dm^3	10.0	-	-
Hydrogen sulfide, mg/L	absense	abs.	-
The permanganate oxidability, mg O/dm^3	10.0	8.9±0.89	-
Dichromate oxidability, mg O/dm^3	30.0	50±12	-
Indicator biological oxygen (BOD5), mg O_2/dm^3	2.0	1.08±0.28	1.5-2.5
Ammonium ion, mg/dm^3	0.5	0.2±0.07	0.05-0.07
Nitrates, mg/dm^3	1.0	0.5±0.0	3.9-4.5
Nitrites, mg/dm^3	0.02	0.02±0.0	0.43-0.59
Phosphates, mg/dm^3	0.3	0.05±0.01	0.68-0.80
Total iron, mg/dm^3	0.1		0.42-0.61

Table 2. Water quality in Volga river.

Characteristic	Volga river [4]	Volga river above Kazan [6, 15]	Volga river 2 km below Ulyanovsk [12]	Volga river near Astrakhan [13]
Transparency, m	0.90-2.10			29-30
pH	7.87-8.96	6.9-7.0	6.78-8.0	7.83-7.87
Suspended solids, mg/dm^3		9.76-12.0	2.3-12.0	20.5-23.5
Oxygen dissolved, mg/dm^3	7.1-13.6	7.6-9.8	9.2-10.3	7.9-10.2
Hydrogen sulfide, mg/L	-	-	-	-
Perm. the oxidability, mg O/dm^3	7.1-10.1	8.8-11.4		4.4-7.89
Dichromate oxidability, mg O/dm^3		25.5-31.4		34.5-43.74
BOD5, mg O2/dm^3	1.2-7.8	1.1-1.4	3.22-3.50	1.17-1.55
Ammonium ion, mg/dm^3	0.17-1.50	0.32-0.45	0.198-0.354	0.395-0.401
Nitrates, mg/dm^3	0.3-2.9	0.3-0.48	0.285-0.382	1.24-3.795
Nitrites, mg/dm^3	0.03-0.15	0.01-0.6	0.047-0.054	0.044-0.051
Phosphates, mg/dm^3	0.016-0.28	0.07-0.10	0.050-0.110	-
Total iron, mg/dm^3	0.03-0.05	0.04-0.09	0.12-0.21	0.255-0.315

IOP Conf. Series: Earth and Environmental Science **288** (2019) 012047 doi:10.1088/1755-1315/288/1/012047

A similar situation is observed in water quality of small rivers of the Kama basin (Table 1) indicators: such characteristics as pH, absence of hydrogen sulphide meet the requirements; for trout farms of the values indicators such as BOD5, nitrates and nitrites, phosphates and iron marked with regulatory.

Heat-water fish farming experience

The fish farming use of warm water is one of the fish farming types closely related to the energy complex. It includes a set of activities to improve the efficiency of hydrobionts cultivation by optimizing the conditions of cultivation and growing through using warm water waste of heat power plants (HPP) and nuclear power plants (NPP). This method of fish farming is actively practiced in Germany, Poland, Japan, USA and in a number of other countries. Fish farms on warm waters do not depend on natural and climatic conditions, the growing season lasts all year round. Thus, the efficiency of fish farming significantly increases, including through the use of cooling water bodies it is possible to supply fish farms with larger and more resilient young fish.

The development of fish farming in warm waters refers to the end of the 1960s, when fish began to grow in cages and pools at 8 condenser and heat power plants in Russian Federation, two condenser power plants in Kazakhstan and Belarus, 5 energy facilities in Ukraine, as well as in Lithuania, Moldova and Uzbekistan. In total at 21 condenser and heat power plants [15]. Rainbow trout is among the main objects of cultivation. Back in the USSR, there were more than 200 thermal power plants with an area of cooler reservoirs of 140 thousand hectares, by 1990 it was 1 million hectares, with more than 4 billion gigacalories of heat per year discharged into reservoirs [16].

Thereafter, fish farming on warm waters arose and formed into an independent direction of modern aquaculture in Russia, going from the first experiments to the development of scientific foundations and the creation of fish farms with controlled temperature regime. The relevance of this direction is primarily due to the lack of heat in most of the territory of our country, which prevents the effective reproduction and cultivation of the main objects of domestic fish farming [17].

Nowadays, the biological melioration of cooling water from thermal power plants and nuclear power plants has become important for both power engineers and fish farmers. Heat loads and eutrophication of such water increase, as a result of which the amount of oxygen in the area of farms decreases, the intensity of zooplankton development increases, which leads to disruption of the power plants cooling systems. Directly in the water cooling reservoirs herbivorous fish can be bred: white and black Amur, white and motley carp, which not only have valuable biological qualities, but also are biological «meliorators». They purify the water from water plants and biofouling.

Power engineering and fish farming are connected closely – power plants contribute to the creation of favorable conditions for the cultivation of commercial fish, including valuable, and fish-breeding complexes based on water coolers help to improve their environmental condition.

It is necessary to take into account the experience of other countries for the development of this technique of fish farming. The experience of some European countries is interesting in this regard. For example, in Turkey, in a country with a mild climate and, as a result, the absence of problems with maintaining stable conditions of the water environment, there is also a practice of growing fish in warm waters of cooling ponds. KonyaSeker Corporation, which owns a giant food production (animal products, sweets, vegetable, plant) uses its own power plants in order to optimize its production [18]. They decided not to cool the condensed water, but to fill the pools for the cultivation of valuable fish - Tilapia with it. Such production is waste-free, as the feces from these pools are processed and used as fertilizer on the KonyaSeker plantations for vegetable and crop production. Since the Corporation is also engaged in the production of its own combined fodder, the profitability of such enterprise should be impressive. This production can be attributed to the energy-biological complexes, where priority is given to the production of food products.

Species and breeds of rainbow trout are promising among the objects of cultivation, for which temperatures of about 25°C are suitable, the possibility of active growth of fish in winter is especially important.

In Western Europe, where, unlike Russia, there are not so many power plants in absolute numbers, the relationship between aquaculture and electricity is carried out in a different way. For example, in Mecklenburg (Germany) for the supply of recirculation system for the cultivation of African catfish, solar thermal power plant has been assembled, which was developed by a unique technology Solarlite [19]. The optimum temperature for African catfish cultivation is considered 26°C [20]. At this fish farm, such temperature is maintained around the clock throughout the year, which makes it possible to obtain the maximum possible volume of production of the African catfish.

Conclusions

Power engineering and fish farming are not only mutually beneficial in their interaction, but also constantly develop, finding more and more points of contact. In the previous period the favorable conditions for growing and breeding fish were created on the basis of the warm water for increasing production capacity of energy enterprises due to energy production and fish production, nowadays the power plants are built at fish-breeding enterprises in order to optimize costs and increase profitability of fish production

Reference

[1] Kozlov V I, Nikiforov-Nikishin I A and Borodin A L 2006 Aquaculture (Moskow: Coloss) p 445

[2] Kalayda M L 2001 *History and prospect of fish farming development in Republic of Tatarstan* (Kazan: Matbugat Yorty) 96

[3] Debolskij V K, Grigorjeva I L, Komissarov A B, Korchagina Y P, Khrustaleva L I and Chekmareva E A 2010 Current hydrochemical descripthion of the Volga river and its reservoirs *Water: Chemistry and ecology* **11** 2-12

[4] *Volga and its life* 1978 (Leningrad: Science) 350

[5] Kalayda M L 2003 *Environmental assessment of the Kuibyshev reservoir under conditions of anthropogenic impact* 135

[6] Kalayda M L, Uryadova L F and Ashadullina A R 2010 Results of the study of aquatic organisms for heavy metal content in the conditions of varying degrees of anthropogenic stress *Butlerov Communications* **22(12)** 61-6 ROI: jbc-02/10-22-12-61

[7] Kalayda M L, Khamitova M F and Novotochinov S I 2016 Results of algolization of wastewater contaminated with organic substances, unicellular algae Chlorella vulgaris. Part 1. Change of chemical oxygen density due to algolizing water by chlorella *Butlerov Communications* **48(10)** 143-9 ROI: jbc-02/16-48-10-143

[8] Kalayda M L, and Khamitova M F 2015 The possibilities of application of eichornia in the purification of the waters of pulp and paper mill. Part 1. The specificity of chemical composition of the water in the Volga reach of the Kuibyshev reservoir *ButlerovCommunications* **44(11)** 97-103. ROI: jbc-02/15-44-11-97

[9] Kalayda M L and Zagustina S D 2010 Accumulation of pollutants by hydrophytes and utilization capabilities of vegetation pulp *Butlerov Communications* 21**(9)** 33-39 ROI: jbc-02/10-21-9-33

[10] Kalayda M L and Khamitova M F 2014 Peculiarities of chemical composition of juvenile perch in the Volga reach of the Kuibyshev reservoir *Butlerov Communications* **40(12)** 37-41 ROI: jbc-02/14-40-12-37

[11] Zavaltseva O A, Konovalova L V and Svetukhin V V 2011 Modern eco-hydrochemical state of Kuibyshev reservoir within Ulyanovsk region *Water: Chemistry and ecology* **9** 17-22

[12] Evseeva S S 2011 The analysis of hydrochemical parameters of water of the Volga river as of December 2010. *The regions' scientific potential at the service of modernization* 94-97

[13] Kuramshin E M, Imashev U B, Kuramshina N G and Nurtdinova E E 2014 Hydrochemistry of a surface water of the small rivers of Kamsky pool in a zone of influence of objects of oil production *Bashkir chemical journal* **21(2)** 88-93

[14] Kalayda M L and Al-Bachry Waleed Sami Jawad 2016 Comparative characteristics of the chemical composition of water in the Tigris River in Iraq and the Volga River in the Republic of Tatarstan *ButlerovCommunications* **46(4)** 47-53 ROI: jbc-02/16-46-4-47

[15] Privezentsev J A 1985 *The Use of warm water for fish farming* (Moscow: Agropromizdat) 176

[16] Nikolsky G V and other 1979 Modern state and prospects of fish farming on warm waters *Biological resources domestic water bodies of the USSR* (Moscow: Science) 125-138

[17] Korneev A N 1990 *Biological basis of industrial fish farming on the basis of thermal waters of energy facilities* (Moskow) 65

[18] Fish Farming Plant [Electronic resource]: http://www.konyaseker.com.tr/en/icerik/detay/6502/fish-farming-plant

[19] Solarlite [Electronic resource]: http://www.solarlite.de/en/project_woltow.cfm

[20] Sevryukov V N, Semenikhin V V and Labunets A V 1999 The first experience of industrial cultivation of African catfish *Resource-Saving technologies in aquaculture. The second international Symposium. Materials of reports* (Krasnodar) 92-3

IOP Conf. Series: Earth and Environmental Science **288** (2019) 012048 doi:10.1088/1755-1315/288/1/012048

Prospects for the cultivation of Australian red claw crayfish Cherax quadricarinatus on warm waters of energy facilities

M L Kalayda and L N Sadykova

"Water bioresources and aquaculture" department, Kazan state power engineering university, Russia, 420066 Kazan, Krasnoselskaya, 51

kalayda4@mail.ru

Abstract. River crayfish - a delicious food product. On the caloric content (72kcal/100g), the content of fats (2.83%), protein (17.13%) and vitamins of group B (thiamine, riboflavin, etc.), crayfish are not inferior to freshwater fish, and by a number of parameters exceed them. Meat of crayfish is rich in sulfur, phosphorus, iron, cobalt, magnesium, fluorine, potassium, selenium, chromium, calcium and vitamins of group B (B12, B6, B2, B1, B9,), C, E, K, PP and D. There is practically no cholesterol in it, but there are a lot of organic acids. A promising target of cultivation, both in Russia and abroad, is currently the Australian red-cheeked crayfish *Cherax quadricarinatus*. It is cultivated in such countries as: China, Brazil, Ecuador, America, and others. Currently, in spite of the presence of active crayfish farms in Russia, the biotechnology of growing *Cherax quadricarinatus* is not sufficiently developed. The main advantages of *Cherax quadricarinatus* over crayfish that live in our region (*Astacus leptodactylus*, *Astacus astacus*, etc.) are: high growth rate, tolerance to water quality, lack of pelagic larva stage. It is advisable to use cooler reservoirs of energy objects for growing valuable warm water aquaculture objects, since the water temperature of these reservoirs is higher than the natural. Energy can contribute to the development of warm water crayfish farming both in the Republic of Tatarstan and in Russia as a whole. Due to the excessive heat that enters the natural reservoirs from energy facilities, an excellent opportunity appears in the absence of the cost of expensive equipment, to grow valuable products. According to expert estimates, the cultivation of crayfish is 5-6 times more profitable than fish.

Introduction

River crayfish – a delicious food product. On the caloric content (72 kcal/100g), the content of fats (2.83%), protein (17.13%) and vitamins of group B (thiamine, riboflavin, etc.), crayfish are not inferior to freshwater fish, and by a number of parameters exceed them. Meat of crayfish is rich in sulfur, phosphorus, iron, cobalt, magnesium, fluorine, potassium, selenium, chromium, calcium and vitamins of group B (B12, B6, B2, B1, B9,), C, E, K, PP and D. There is practically no cholesterol in it, but there are a lot of organic acids [10, 12]

Meat of crayfish is used as food as well, as their skeletal parts. Hard outer cover of crayfish contains a large amount of calcium and phosphorus, so it is used to make medicines, for example, chitosan is widely used in medicine and in cosmetology.

Due to the high nutritional value, crayfish is traditionally in high demand in many countries. So, in Sweden annual consumption is 2000-3000 tons of crayfish, in France - more than 2000 tons, the demand for crayfish in Spain reaches 12,500 tons [2]. For highly developed countries, crayfish is an exquisite delicacy food product, for countries that are economically less developed - the object of exports and additional income. In the markets of Scandinavian countries, France, Spain, the purchase price of 1 kg of crayfish reaches 35 US dollars and above (*Astacus astacus* and some local crayfish) [3]. The high value of crayfish as objects of farming also determines the high interest in the development of aquabiotechnology for their cultivation.

When choosing the objects of crayfish cultivation, the limiting factors are temperature, water quality, the presence of shelters. Among the promising objects of cultivation on warm waters of energy objects - Australian red claw crayfish *Cherax quadricarinatus*. It is cultivated in such countries as: China, Brazil, Ecuador, America, and others. [4].

The shell is blue, and there may also be colored spots. A distinctive feature of the males is a bright orange spot on the outside of the claw. It is assumed that stains are used in communication between individuals of the population, including information about the physiological state of the individual [5]. In the natural environment, the length of the red claw Australian crayfish can reach 20-25 cm. Weight up to 500 g for males, for females - up to 400 g [6].

Figure 1 shows the layout of the structure *Cherax quadricarinatus*.

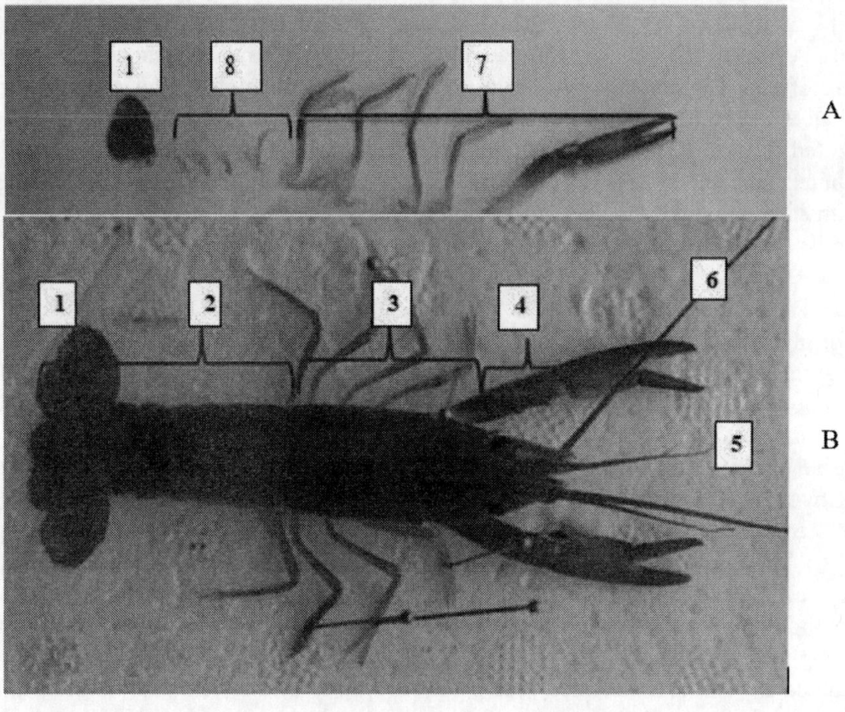

Figure 1. A, B – structure *Cherax quadricarinatus* (von Martens, 1868): 1 – telson and uropods; 2 – abdomen; 3 – carapace; 4 – rostrum; 5 – antennules; 6 - antennas 7 – pereiopods (the first 3 pairs in the form of claws); 8 – pleopods.

Results and discussion

The quality of water required for cultivation Cherax quadricarinatus

The temperature of effective intensive cultivation of Australian crayfish is 25-30 °C [13, 14]. Lethal are temperatures below 10 °C and above 36 °C [6]. At a temperature below 20°C there is a decrease in activity, growth rate, resistance of individuals to diseases. For juveniles, the critical temperature is below 20 °C and above 32-34 °C [7].

Possibilities of using the potential of natural reservoirs of the Volga region are extremely limited, since the suitable water temperature for the cultivation of *Cherax quadricarinatus* lasts about 2-3 months. This time period is not enough for the young *Cherax quadricarinatus* to reach commercial size. This makes it especially promising to use warm waters of thermal power facilities for the cultivation of red claw crayfish. In accordance with environmental requirements, the water temperature in reservoirs with discharges of warm waters of energy objects may rise in winter by 5 ° C, and in summer - by 3 ° C above the natural maximum temperature [1]. Thanks to the dumping of warm waters by thermal power stations into cooler reservoirs, it becomes possible and promising to locate industrial breeding farms for the cultivation of valuable aquaculture objects of Australian red claw crayfish (*Cherax quadricarinatus*).

Features of cultivation Cherax quadricarinatus

The main advantages of Australian red claw crayfish in comparison with the crayfish living in the middle reions of Russia (*Astacus leptodactylus*, *Astacus astacus*, etc.) are: high growth rate, tolerance to water quality, absence of pelagic larva stage. Under optimal growing conditions, the Australian red claw crayfish can reach a commodity weight of 100-120 g in 1 year, while a *Astacus leptodactylus* crayfish reaches these sizes in more than 6 years [11]. Preliminary calculations show that when creating a crayfish farm on the basis of a reservoir-cooler of a power plant for one season (May-September), it will be possible to grow commodity crayfish with a weight of 70 to 150 g.

Since the red claw crayfish is a new aquaculture object for Russia, at present the biotechnology of the cultivation of *Cherax quadricarinatus* is not sufficiently developed and requires research. The factor that reduces the results of commodity cultivation of crayfish is cannibalism. It is believed that for Australian red claw crayfish cannibalism is less pronounced [8].

We conducted studies on establishing the values, causes, and methods of reducing mortality associated with cannibalism of red claw crayfish.

At the first stage of the experiment, young Australian crayfish with linear sizes from 1 to 1.5 cm were grown for 30 days with a density of 93.75 exemplars/m^2 at different feeding rates. In biotechnological standards for the cultivation of these crayfish, the recommended density of planting is 250 g/m^2 for juveniles up to 5 grams [2].

Since in the industrial cultivation of aquaculture objects the main component of the cost is the cost of feed, it is important to minimize its quantity. At the same time, the lack of food can lead to an increase in the manifestation of cannibalism, which refers to the biological characteristics of red claw crayfish. The results of cannibalism for juvenile crayfish at different feeding rates (in % of body weight) are shown on the Figure 2.

As can be seen from the data shown in Figure 2, when the recommended rate of feeding is reduced from 2% [2] to 1% of the mass, manifestation of cannibalism begin from the first week. With an increase in the feeding norm by 2%, the amount of elimination in the first stages decreases, but by the end of the second month of cultivation the survival rate of crayfish at the feeding norms of 2 and 4% of the body weight becomes the same.

The conducted experiments revealed that if the feeding is insufficient, the amount of elimination increases. With an increase in the norm of feeding, the level of cannibalism does not decrease. Experiments have shown that when growing juveniles of Australian red claw crayfish in the presence of shelters and providing feed, a decrease in the numbers due to cannibalism over a period of 30 days is 20%.

At the second stage of the experiment Australian crayfish with initial linear sizes from 3 to 4.9 cm at the age of 4 months were grown for 50 days at a planting density of 0.16 ind./m^2 at varied feeding rates. The value of elimination as a result of cannibalism was - 30.3%.

As a result of the experiment it was shown that cannibalism is a significant factor in the cultivation of Australian red claw crayfish. In the presence of shelter, optimal temperature, manifestations of cannibalism are associated with the norm of feeding and age of crayfish. Average minimum elimination values for juveniles of red claw crayfish were 20%, for crayfish at the age of 0.5 years - 33%. Despite the manifestations of cannibalism and the observed elimination, according to expert estimates, the cultivation of crayfish is 5-6 times more profitable than fish [9].

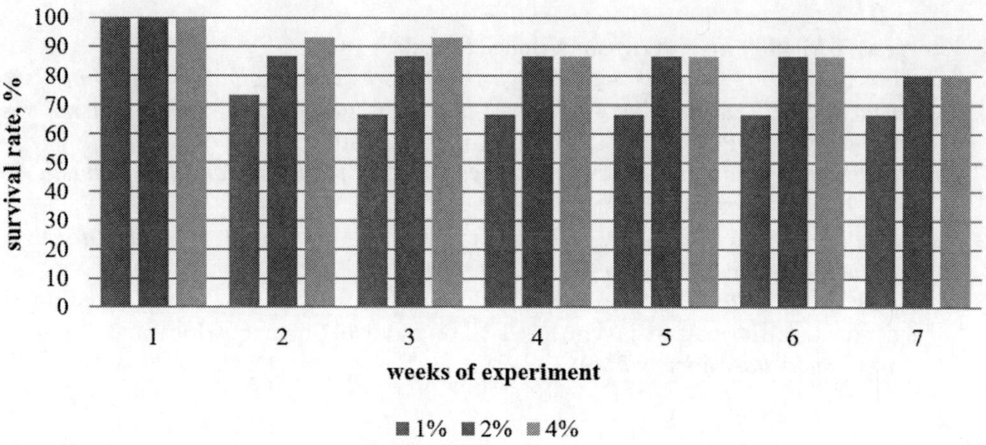

Figure 2. Survival rate (%) associated with cannibalism of Australian red claw crayfish at different rates of feeding (1%, 2%, 4%).

Conclusions

Efficiency of secondary use of warm waters of power objects can be increased due to the development of a breeding with the use of acclimated Australian red claw crayfish both in the Republic of Tatarstan and in Russia as a whole. Due to the excessive heat that enters the natural reservoirs from energy objects, it becomes possible, in the absence of the cost of expensive equipment for heating water, to grow freshwater crayfish. According to expert estimates, the cultivation of crayfish is 5-6 times more profitable than fish. When planning commodity cultivation of crayfish, cannibalism must be considered. Value of the elimination for juveniles in conditions of complete provision with feed is 20%, for crayfish of the older age group - 33%.

References

[1] Topachevsky A V and Pidgayko M L 1971 Goals and objectives of hydrobiological research of reservoirs-coolers of thermal power plants *Coll. hydrochemistry and hydrobiology of reservoirs-coolers of thermal power stations of the USSR* (Kiev: Naukova Dumka) **15** 6-10

[2] Fyodorova Z V 1991 Cultivation of crayfish abroad *Fish management. Ser. Aquaculture* (VNIERKh) **1** 1-16

[3] Huner J V and Romaire R P 1990 Crayfish culture in the southeastern USA *World Aquacult.* **21(4)** 14-7

[4] Holthuis L B 1986 The freshwater crayfish of Newvbfd Guinea *Freshwater Crayfish* **6** 48-58

[5] Karplus I 2003 The soft red patch of the Australian freshwater crayfish (Cheraxquadricarinatus (von Martens)): a review and prospects for future research *J. Zool.* (London) **259** 375-9

[6] Lawrence C and Jones C 2002 *Cherax. In: Biology of Freshwater Crayfish. Holdich D.M. (Ed.)* (UK, Oxford: Blackwell Science) **17** 635-70

[7] King C R 1994 Growth and survival of redclaw hatchlings (Cheraxquadricarinatus (von Martens)) in relation to temperature, with comments on the relative suitability of Cheraxquadricarinatus and Cherax destructor for culture in Queensland *Aquaculture* **122** 75-80

[8] Gigin A V, Arystangaliyeva V A, Tyrin D V and Kovacheva N P 2017 Determination of the optimal temperature for growing juveniles of Australian red claw crayfish *Environmental Engineering* **3** 121-8

[9] Kolmykov E V 1997 Problems and prospects of commodity cultivation of crayfish in delta of Volga *Problems of protection, rational use and reproduction of crayfish* (Moskow: Medinor)

pp 116-8
[10] Zuckersis Y M 1989 *River crayfish* (Vilnius: Mokslas) 143
[11] Lebedev N A 2008 *Astacus leptodactylus – economically valuable object of fishery and aquaculture: a workshop for students of the specialty "Biology" with additional specialties* (Mozyr: UO "MSPU named after I. P. Shamyakin") 50
[12] *Crayfish catching and crayfish farming on reservoirs of the European part of Russia (reference book)* 2006 (St. Petersburg) 207
[13] Xiaoxuan C 1995 Effects of Water Temperature on Ingestion and Growth of Cheraxquadricarinatus *Journal of Huazhong Agricultural*
[14] Meade M E 2002 Effects of temperature and salinity on weight gain, oxygen consumption rate, and growth efficiency in juvenile red claw crayfish Cheraxquadricarinatu *Journal of the World Aquaculture Society* **33(2)** 188–98

Fish protection structuers on reservoirs of energy facilities as an important measure for the conservation of the herd of fish

M L Kalayda and A R Saetov

"Water bioresources and aquaculture" department, Kazan state power engineering university, Russia, 420066 Kazan, Krasnoselskaya, 51

kalayda4@mail.ru

Abstract. Directed formation of ichthyocenoses on the basis of reservoirs - energy objects - an important task for pasture aquaculture. At the energy facilities to ensure the operation of the main equipment, the main issue is water supply. The water intake structure, as the main technological element of the water supply system, must uninterruptedly take out water, at the same time functioning as a fish protection facility. According to the requirements of the Russian legislation, the effectiveness of preventing fish and other aquatic biological resources from entering water intake facilities should be at least 70%. Fish protecting device (FPD), along with high fish protection characteristics, should have a reliability that allows to maintain their functions during the entire period of operation. There are three main principles for preventing the ingress of juvenile fish into water intake facilities using FPD: ecological, behavioral and physical. At the present time, the most common at water intakes are net FPD, morally obsolete, according to many practitioners. They account for more than 60% of all FPDs. Among the promising types of fish protection structures are air-bubble type structures that operate on the principle of an ascending water-air curtain formed by a system of bottom modules with aerating nozzles. Among the advantages of this type of FPD is the positive effect on the hydrochemical state of water bodies. Aeration promotes the oxidation of organic substances in water, contributing to a decrease in the level of eutrophication of the aquatic ecosystem. Further improvement of the FPD at power facilities is important for the work on the release of valuable fish species, especially sterlet juveniles when stocking the Kuibyshev and Nizhnekamsk reservoirs.

Introduction

In the developed "Energy Strategy of Russia for the period until 2020", the development of the Russian electric power industry is oriented towards the scenario of the country's economic development. Achievement of the planned scale of electric power consumption takes into account active energy saving - both due to structural reorganization of the economy, and through organizational and technical measures in the industry. In our country, a huge amount of electricity is produced and consumed. It is almost entirely produced by three main types of power plants: thermal, nuclear and hydroelectric power plants. Virtually all energy facilities are closely connected with water bodies and exert a certain impact on water ecosystems.

At the same time, all water bodies at the present stage of the development of society are involved in the production of fish products in varying degrees of intensification. The desire to improve the condition of inland water bodies in the Republic of Tatarstan is evidenced by the adopted resolutions of the Cabinet of Ministers of the Republic of Tatarstan dated December 28, 2015 № 991 "On

IOP Conf. Series: Earth and Environmental Science **288** (2019) 012049 doi:10.1088/1755-1315/288/1/012049

approval of the State Program for the Development of Commodity Aquaculture (Commodity Fish Culture) in the Republic of Tatarstan for 2016-2020", Concept for the Creation of Aquabioculture Technopolis in Republic of Tatarstan, which was approved by the President of the Republic of Tatarstan at a meeting on December 18, 2016. As part of the development of aquaculture in the region, it is proposed to reduce wastewater discharges into water bodies, carry out measures for the rehabilitation of water bodies. These same tasks are important for the implementation of the Food Security Doctrine of the Russian Federation, approved by Presidential Decree No. 120 of January 30, 2010, "On Approving the Doctrine of Food Security of the Russian Federation" and implementing the Decree of the Government of the Russian Federation of April 15, 2014 No. 314 "On approval of the state program of the Russian Federation "Development of the fishery complex" (subprogram "Development of aquaculture").

In accordance with the "Environmental Requirements for the Placement of Buildings, Structures, Structures and Other Objects" of the Federal Law No. 7-FL of January 10, 2002 "On Environmental Protection" when placing buildings, structures, structures and other objects, ensuring compliance with environmental protection requirements, restoring the natural environment, rational use and reproduction of natural resources, ensuring environmental safety, taking into account the nearest and remote ecological, economic, demographical and other consequences of the operation of these facilities and compliance with the priority of preserving favorable environment, biological diversity, rational use and reproduction of natural resources".

At the energy facilities to ensure the operation of the main equipment, the issues of ensuring water supply are the main concern. A water intake facility (water intake) is a hydraulic structure for the extraction of water from a reservoir [1, 3]. Any water intake structure, as the main technological element of the water supply system, must continuously take out water, at the same time functioning as a fish protection facility.

In accordance with the Water Code of the Russian Federation, Resolution of the Government of the Russian Federation No. 380 of 29.04.2013 "On Approval of the Regulation on Measures for the Conservation of Aquatic Biological Resources and their Environment" - the use of water intake facilities not equipped with fish protection devices or structures is forbidden.

Fish protection facilities (hereinafter referred to as "FPF") installed on water intake facilities, serve to prevent fish from entering working equipment. Modern fish protection devices are a composite, technically complex, part of water intake structures and influence the conditions of its design and operation. The main document for the design of water intake facilities is "SNiP 2.06.07-87 Retaining walls, navigable sluices, fish-passing and fish-protecting structures", approved by Decree N76 of the State Construction Committee of the USSR of April 14, 1987.

Directed formation of ichthyocenoses on the basis of reservoirs - energy objects - an important task for pasture aquaculture. The increase in fish production by traditional methods based mainly on the extensive use of natural resources, nevertheless implies a significant amount of high-tech fishery operations in terms of the reproduction of valuable fish species for release tasks in water bodies in the directional formation of herds. Under these conditions, fish protection structures become especially significant factors of possible negative impact on the aquatic ecosystem. In connection with this, the improvement of the systems of FPF and FPD, which minimize the negative impact on ichthyocenosis during water intake, is topical [3, 7, 8].

Results and discussion

Modern applied models of FPD

FPD, along with high fish protection characteristics, should have a reliability that allows to maintain their functions during the entire period of operation. According to the requirements of the Russian legislation, the effectiveness of preventing fish and other aquatic biological resources from entering water intake facilities should be at least 70% [4, 5]. Modern efficient designs of FPD, in contrast to old ones, fully meet these requirements [6].

There are three main principles for preventing the ingress of juvenile fish into water intake facilities using FPD: ecological, behavioral and physical. Based on these principles, protection methods have been developed that determine the nature of the impact on the protection object. The main reason for the entry of fish into water intake facilities is the passive water transfer. The ingress of fish into water intake facilities is connected, on the one hand, with the physical impossibility of resisting the current, and on the other hand, with the absence of conditions for the orientation of fish in the stream, primarily with the lack of conditions for the visual orientation.

At the present time, the most common at water intakes are net FPD, morally obsolete in the opinion of many practitioners [9, 10]. They account for more than 60% of all fish protection devices. It is natural that in recent years more and more widely used are more technological and modern types of FPD [11]. The principle of operation of which is based on different methods. For example, the device of electric barriers is based on the irritating effect of electric current on the fish organism and the specific behavioral reaction of fish. This reaction is manifested in spatial orientation, involuntary migration of fish along the gradient of the field from the negative pole to the positive and sedative effect of the electric field on the central nervous system of fish. The effect of the electric field on the reaction of the fish is due to the presence in fish, as specific organs for the perception of electrical signals in the aquatic environment, the so-called electroreceptors (in some orders of fish it is Ampoules Lorentzini), and in the electrical nature of the processes in the nerve structures that control the muscle tissue of fish. Electroceptors are specialized sensory organs for the perception of weak electric fields. That is, an electric field of the appropriate voltage is created in the way of fish, getting into which the fish feels irritated and tends to go beyond it.

Another modern type of fish protection structures are air-bubble type constructions (“air curtain”) [2, 12]. The FPD of this type operates on the principle of an ascending air curtain formed by a system of bottom modules with aerating nozzles. The principle of operation of this type of structure is to pump water to the FPD modules while simultaneously taking air through a separate pipeline-duct. The fish-protective effect of FPD using the principle of a water-air curtain is created by feeding a finely dispersed water-air suspension into the perforated pipeline. Experience in the application of hydrocavitation technology has made it possible to create a hydrokavitational aerator for obtaining a water-air suspension with air ejection from the atmosphere [4]. The outflow of the rolls of water-air curtains creates a stable surface fish-diverting flow, ensuring the removal of young fish to the area of the transit flow of the river. As a result, on the selected part of the reservoir, from the bottom to the water surface, a fish protection front is formed, which does not allow the fish to enter the water intake devices.

The use of the V-shaped beam of the barrier of the “Water-air curtain” type FPD improves the efficiency of work not only due to the presence of a double barrier front, but also due to the emerging outflow between divergent rays on the surface of the reservoir, this ensures the removal of juvenile fish to the edge of the fence.

The main types of impact of FPD based on a water-air curtain on fish juveniles are: tactile, visual, acoustic and mechanical. FPD based on a water-air curtain is distinguished by high reliability, simplicity of construction, absence of environmental pollution. An additional advantage of a water-air curtain is the protection of water intake from suspended and floating debris, which rises upward to the surface and is diverted from the water intake area by the surface current. Also, in winter, in the area of the water-air curtain, an airhole is formed, which, thanks to surface currents, protects the water intake from the ingress of floating ice and sludge. This FPD has a positive effect on the hydrochemical state, increasing the oxygen content both in summer and in winter due to the formation of airhole. Aeration promotes the oxidation of organic substances in water, thereby contributing to a decrease in the level of eutrophication of the aquatic ecosystem.

Prospects for the practical implementation of the development of FPD on the example of the Republic of Tatarstan

To reduce the ingress of fish into the water intake facilities at the shore pumping station No. 1 of the branch of JSC "Tatenergo", Zainskaya GRES in 2015, FPD of the type "Water-air curtain" was installed, based on the reaction of avoidance by fish of low-frequency oscillations and intense ascending water-air jets. The indicator of fish protection effectiveness (Kef) of FPD for the study period for all seasons in 2016 averaged 72.4%, which is quite consistent with the normative indicators. Having received and analyzed the obtained data with positive results of the implementation of this investment project with high energy-efficient and environmental potential, the management of JSC "Tatenergo" decided to further install "Water-air curtain" type FPD at shore pumping stations No. 2 and No. 3 of the branch of JSC "Tatenergo" Zainskaya GRES. The works on the implementation of the 2 launch complex at the BNS No. 2 were implemented in 2017. In 2018, work is carried out to determine the effectiveness of the FPD installed on the water intake structure of the coastal pumping station No. 2. According to preliminary data, the indicator of fish protection effectiveness (Kef) FPD-2 will be higher than the normative indicators, which will positively affect the ecological situation in the reservoir.

In 2018, also at Zainskaya GRES, work is underway to install a water-air curtain-type FPD at the coastal pumping station No. 3. The implementation of the launch complexes of the FPD is accompanied by the development of not only the design characteristics of the FPD, taking into account the actual tasks, but also the operational and technical maintenance of the fish protective devices. For example, for a containment filter located above the water surface, a special double-purpose metal head was made: it closes the access to the equipment and additionally filters foliage, down, dust and other debris carried by the air masses from the air to the vestibule. Another example of improving the operation of the FPD equipment such as the "air curtain" is the introduction of a revision and subsequent mechanical cleaning of the aerating collectors and nozzles, since the efficiency of their operation in connection with the biofouling can be reduced. The problems of biofouling of hydraulic structures are acute at energy facilities [1, 14]. Exit from the working condition of aerators for the period of 2 years of operation was 2.8%. To ensure smooth and trouble-free operation of fish protection devices, it is recommended that in the complete set with the main equipment delivery of additional aerators and maintenance work [13].

Conclusions

The technology of "air curtain" has a number of advantages in comparison with other FPD, including: air intake from the atmosphere to create a water-air mixture in the nozzle aerator modules aeration in the absence of rubbing surfaces does not lead to contamination of the reservoir with oil products; the presence of air in the composition of the air-water mixture allows to significantly increase the diameter of the holes of the perforation of the aerating collector (up to 12 mm), which reduces the likelihood of their biofouling and increases the reliability of the FPD. The air supply to the distribution pipeline as a part of the water-air mixture reduces the dependence of the conditions for the formation of the air-bubble curtain on the depth of the distributor. Water-air modules allow for routine diving operations to clean up almost any of the clogged sections of the FPD circuit without dismantling them and raising them to the surface.

It is important that the air curtain does not injure fish of any size. The design of the "water-air curtain" type FPD is distinguished by high reliability, it is easy to maintain, besides, the curtain physically repels weighted and floating debris from the water intake, and in winter protects the water intake from floating ice and sludge. At the same time, the saturation of water with oxygen improves the oxidative capacity of water in the places of placement of FPD.

Further improvement of the FPD at power facilities is important in the work on the production of valuable fish species, especially sterlet juveniles when stocking the Kuibyshev and Nizhnekamsk reservoirs.

References

[1] Bespalov A G, Salnikov A L, Bespalova O N and Davydova S A 2016 *Hydraulic engineering structures* (Moskow: KNORUS) p 94

[2] Bulgakov A B, Romancev V P, Blancevich Z L and others 2000 Air-bubble fish protective devices (FPD) with the use of hydraulic cavitation apparatus (HCA) *News of the Academy of Industrial Ecology* **43** 60-63

[3] Dyomin A P 2010 Water management complex in Russia: concept, state, problems *Water resourses* **37(5)** 617-32

[4] Kaveshnikov N T 1989 *Operation and repair of hydraulic structures* (Moskow: Agropromizdat) p 272

[5] Kuzovlov G M 1966 On the design of reservoir and sea water intakes *Water supply and sanitary engineering* **1** 18-23

[6] Kuzmin U I 1972 *Methodical recommendations for the design of fish protection devices for water intake facilities* (Leningrad: GosNIORH) p 56

[7] Malevanchik B S and Nikonorov I V 1984 *Fish-passing and fish-protecting constructions* (Moskow: Light and food industry) p 256

[8] Manteuffel B P, Pavlov D S and Pakhorukov A M 1978 Biological bases of the device of fish-protecting and fish-passing constructions *Biological bases of application of fish-protective and fish-passing structures* (Moskow: Science) p 178

[9] Motinov A M and Kolesnikova T V 1983 Fish-passing structures and fish protective devices *Hydraulic engineering structures* (Moskow: STROIZDAT) p 543

[10] Mikheev P A 2000 *Fish protection constructions and devices* (Moskow) 405

[11] 1987 Building codes and regulations: Retaining walls, navigable sluices, fish-passing and fish-protecting structures: SniP 2.06.07-87 (Moskow: STROIZDAT) p 35

[12] Sturane R Y and Udin V C 1988 Constructions of filter type fish protection devices with air blowing *Express information. Melioration and water management. Series 5. Water management* (Moskow: TsBNTI of the Ministry of Water Resources of the USSR) **11** 2-6

[13] Ersler A A and Sheramet N G 2002 Express-method for determining the functional efficiency of fish protection structures at water intakes (Moskow: VNIERKH) p 43

[14] Kalayda M L and Mugantseva T P 2012 Increasing the efficiency of the technical water supply system of TPPs *University news: Problems of Power Engineering* **7(8)** 128-131

Pasture and industrial aquaculture as actual elements of the use of hydro and heat power facilities

M L Kalayda, M F Khamitova and A A Kalayda

"Water bioresources and aquaculture" department. Kazan state power engineering university, Russia, 420066 Kazan, Krasnoselskaya, 51

kalayda4@mail.ru

Abstract. Aquaculture products provide about 58% of the total production of aquatic ecosystems. The global trend of population growth, increasing the press on the environment, the requirements for improving the quality of life and consuming products make the trend of growth in aquaculture products sustainable both in the world and in Russia. The role of abiotic environmental factors in the life of fish is enormous. Thus, the water temperature determines the intensity of metabolism and is a natural stimulus determining the onset of spawning migration of fish. The water temperature has a very important effect on the vital activity of the organism, in particular, on metabolic processes, on the behavior and distribution of organisms.Since the beginning of the 20th century, aquaculture has been closely linked to the development of energy. At the first stage - with hydropower, and then with thermal power. If the use of water reservoirs - hydropower facilities is based on extensive technologies that are related to pasture aquaculture, aquaculture farms - live fish farms - highly intensive industrial farms. These areas of aquaculture development coexist and complement each other. Modern pasture fish farming today is not possible without the release of valuable species of fish to form a species composition of ichthyocenoses. At the same time, the development of highly industrial commodity farms implies the use as a brood of individuals of valuable species of fish seized from the natural environment to improve the genetic characteristics of juveniles. At the present time, there is an increase in the range of fish grown due to the use of warm waste waters of energy facilities.

Introduction

Aquaculture currently belongs to globally significant world production [1, 2]. Currently, more than 600 species of animals and plants are grown both in monoculture farms and in multicultural production with a very high degree of intensification. Aquaculture products provide about 58% of the total production of aquatic ecosystems. The global trend of population growth, increasing the press on the environment, the requirements for improving the quality of life, and consuming products make the trend of growth in aquaculture products sustainable both in the world and in Russia. Since the beginning of the 20th century, aquaculture has been closely linked to the development of energy [3, 4].

Results and discussions

Pasture aquaculture at hydropower facilities

In 1931, the working hypothesis of a comprehensive scheme for the use of the Volga for energy and transport purposes was developed, various projects for the construction of hydroelectric power stations were put forward, however, among them there were no projects that took into account the whole complex of problems that were considered in the complex scheme for the use of the Volga. In the subsequent period, the problem of the "Great Volga" was solved by creating a cascade of reservoirs: the Upper Volga (1944), Ivankovsky (1937), Uglich (1939-1943), Rybinsk (1940-1949), Gorky (1955-1957), Cheboksary (1981), Kuibyshev (1955-1957), Saratov (1967-1968), Volgograd (1958-1960 gg.), Kamsky (1954-1956 gg.), Votkinsky (1961 -1964 gg.), Nizhnekamsk (1978) [1]. The catches in the largest reservoirs of the Volga cascade in the early 21st century are shown in Table 1 from [5].

Table 1. Fish catches in the largest reservoirs of the Volga cascade, ths. tons.

Reservoir	1995	2000	2001	2002	2003	2004	2005	2006	2007
Rybinsk	1.4	1.5	1.6	1.7	1.0	1.3	1.04	1.0	1.1
Kuibyshev	3.2	2.8	2.7	2.7	2.0	1.94	2.11	2.62	2.7
Saratov	0.8	0.5	0.7	0.8	0.6	0.67	0.6	0.69	0.7
Volgograd	1.0	1.0	1.3	1.3	1.5	1.69	1.72	2.17	2.05

At the stage of planning the construction of reservoirs, they were expected to be used by pasture aquaculture methods (Table 1, 2). For this purpose, hatcheries and spawning-vigorous farms were built, which were to ensure the directional introduction of valuable fish species. When planning fish catches in the Kubyshevsky reservoir, the main attention was paid to bream, carp, pike perch, pike and sturgeon fishes (Table 2) [4]. The remaining species had to provide 30% of catches.

The planned catch volumes in the Kuibyshev Reservoir and the actual state of the fish herd structure are shown in Table 2.

Table 2. The planned catches of fish in the Kuibyshev reservoir in connection with hydro-construction in the 1950s and the current state of the catch.

Period	Catch, ths. tons	Fish species (% in catch)					
		Bream	Carp	Pikeperch	Pike	Sturgeon	Other
1950s	20-24	35	15	10	8	2	30
2017	3.438	22.85	0.84	6.52	0.56	0.17	69.06

In the construction of hydroelectric power plants, simultaneously with the solution of the tasks of supplying electricity to the production sector of the economy, the foundations of the ecological problems of the day, connected both with changes in the hydrological and hydrochemical characteristics of the reservoir. These problems can be solved by aquaculture methods.

The current state of catches in comparison with the planned indicators is presented in Table 3. Data on catches indicate an increase in the proportion of low-value fish in the recent period. If in 2010 low-value species accounted for 53.4% in catches, in 2014 for 64.8%. Analyzing the planned indicators for catches in the Kuibyshev reservoir, it can be noted that the planned harvest of sterlet was 480 tons. Actual catches of sterlet in the beginning of the XXI century varied about 0.850 tons.

Table 3. Catches (%) of valuable and low-value fish species in the Kuibyshev Reservoir.

Species	2010	2014	Quotas 2017 г.	Species	2010	2014	Quotas 2017
	% in catches				% in catches		
	Valuable species			*Low-value species*			
sterlet	0.04	0.02	0.17	Blicca bjoerkna	16.7	17.2	22.31
carp	1.59	1.23	0.84	Ballerus ballerus	11.3	14.4	16.99

Ending of Table 3

catfish	0.57	0.19	0.26	Rutilus	9.1	12.0	7.50
pike perch	5.82	5.38	6.52	Pelecus cultratus	4.8	5.1	3.40
pike	0.80	0.11	0.56	Sander volgensis	2,9	3.1	3.32
burbot	0.24	0.31	0.46	Perca fluviatilis	2.5	5.1	4.13
bream	31.06	23.7	22.85	Carassius	3.0	4.5	2.33
silver carp	0.08	0.07	0.06	Abramis alburnus	3.1	3.4	5.09
				Clupeonella		2.93	1.54
				Aspius aspius		0.60	0.73
				Leuciscus idus		0.36	0.47
				Cyprinidae		0.45	0.47

The tendency of decline in the number of valuable fish species in the Kuibyshev reservoir was clearly revealed by the 2000s [2, 4, 6] (Table 3). In the catches it was noted not only an extremely small number of sterlets, but also pike, catfish and others (Table 3), at the same time the number of low-value and weedy species of fish has grown. If the statistics of catches revealed only a decrease in the number of valuable commercial fish species, the 1999 survey of amateur fishermen revealed an increase in the share in catches of silver crucian, rotan, gobies. For example, in the Mesha river the number of silver crucian was estimated as 6% in catches, and as much zope (Ballerus ballerus) and sabrefish (Pelecus cultratus) [6]. The average annual catch in the Kuibyshev Reservoir within the Republic of Tatarstan for amateur fishermen was 269 ± 60 tonnes, the average annual catch of poachers was 2504 ± 93 tons [4]. It was shown that the catch of fish from the reservoirs of the Republic by amateur fishermen and poachers is 1.8 times higher than its catch by fishing organizations. In 1989, the catch was 5981.1 tons, in 1994 - 2619.7 tons, in 2005 - 2114.2 tons, in 2008 - 3140.2 tons. In the last period, the total catch of fish in the Kuibyshev Reservoir is 3,093.5 tons (2010) - 4196.8 tons (2014). Analysis of data on catches clearly demonstrates a decrease in the catches of valuable fish species, including sterlet, which is currently included in the Red Book of the Republic of Tatarstan. There is little carp, catfish, pike, burbot and pike-perch in the catches (Table 3). The share of bream in catches also decreases. Especially it should be noted an extremely small amount of silver carp. As biomelators, it is most beneficial to improve the state of the ecosystem of the reservoirs of the central Russia, by releasing silver and bighead carps. Evaluation of the production potential of the Kuibyshev reservoir confirms the possibility of releasing silver carp in the reservoir with a density of planting 100 pc / ha in high productivity zones. Directed ichthyofauna formation is the most effective in the long term to maintain high water quality while simultaneously addressing the problems of maintaining a high number of valuable fish species. Since the share of benthophagous fish accounts for 12.7% of potential fish products, carp is no longer the best target for the work on the directed formation of the ichthyofauna. Sturgeon producing factories, established in the 1950s to compensate for the loss of natural spawning grounds of sturgeon on the Volga in connection with hydroelectric construction, over the years of its existence, released into the river more than 2.7 billion exemplars of young sturgeon. This indicates that a powerful sturgeon industry has been created in Russia. Since 1990, the production of sturgeon juveniles has declined and has now reached a minimum.

The crisis of sturgeon farming at the end of the 20th century, the reduction of broodstock in the Volga River in connection with the reduction of spawning grounds and the demand for high costs for the reproduction of valuable species, while regulating their numbers in reservoirs, stimulated the development of aquabiotechnologies. These problems served as an impetus to the solution of scientific and technological problems: the development of methods for pituitary injection with the replacement of the pituitary hormone for a wide range of gonadotropic drugs, the lifetime determination of stages of puberty of gonads [7], ultrasound - diagnosis, lifetime caviar receiving. In recent years, the most effective method of selecting ovulated caviar from sturgeon is the method of incision of oviducts with subsequent pouring of eggs [8]. This least invasive microsurgical method has been used for more than 20 years, and many females of different sturgeon species have been subjected to a pouring procedure more than seven times.

The development of aquabiotechnologies and the reduction in the catches of valuable fish species in virtually all water bodies stimulated the development of industrial commercial aquaculture, which is interfaced with energy facilities.

Industrial aquaculture on warm waters of energy facilities

In Russia, the dominant variants of commercial breeding of fish, including sturgeon, are pond, cage and breeding in closed water circulation installation (CWCI). In the 1970s-1990s, the country's industrial fish farming was created on the basis of cage farms on cooler reservoirs of power facilities in the structure of thermal power plants, nuclear power plants, and hydroelectric power stations, which had warm water in excess. Such farms functioned practically in all regions. Output of commercial products from one square meter of basins and cages reached 100-150 kg per year. The total production of commercial fish exceeded 30 thousand tons per year [9].

In the middle zone of Russia at the end of the 20th century, industrial fish farming: trout breeding and sturgeon breeding was based on the use of floating cages in reservoirs, which were stocked by whitebait weighing 3-5 g [10]. The most successful of sturgeon was a bester on the basis of cooler reservoirs of the state district power stations, which reached in the third year a mass of 0.85 kg. For example, it was successful to grow a bester in cages on the discharge channel of Novocherkasskaya GRES.

Methods of cage culture by the beginning of the 21st century were replaced by the cultivation methods in the CWCI. This was due to the possibility of complete control of habitat, mechanization of fish-breeding processes, reducing the impact on the environment [10, 11].

Commercial living fish factories of open or closed type of various capacities were built on the exhausted waters of the state district power stations, for example at Konakovskaya GRES - a closed type with a capacity of 360 tons of commercial fish per year (Table 4). At present, the term of use of warm waters of thermal power stations for fish farming can range from four to twelve months (live fish farms, cage farms, cooling ponds). Such enterprises allow the most efficient use of warm water from TPPs in any water supply system of a power plant irrespective of its geographic location. In the pools, commercial fish are grown with high density, strong flow of warm water and intensive feeding with artificial feeds.

Table 4. Fish farms at energy facilities.

Power station	Fish farm	Farm type / tonnage	Grown objects
Konakovskaya GRES	Konakovsky factory of commodity sturgeon breeding	Basins / 360 tons	Sturgeon; sterlet, beluga, bester
Karmanovskaya GRES	Karmanovskiy fish farm	Cages / 1200 t carp, 150 t sturgeon, 150 t trout.	Carp, white amur, silver carp; sturgeon, bester, sterlet; rainbow and gold trout; channel catfish.
Belovskaya GRES	Belov fish industry	Cages / 1000 t	Carp, white amur, silver carp, carp koi, trout, sturgeon, channel catfish
Cherepovets GRES	LLC “Fish Company “Diana”	Cages, basins / 1000 t carp, 400 t sturgeon, 20 t black caviar	Beluga, sturgeon, sterlet, stellate sturgeon

Ending of Table 4

Cherepovets GRES	LLC "Fish Company "Diana"	Cages, basins / 1000 t carp, 400 t sturgeon, 20 t black caviar	Beluga, sturgeon, sterlet, stellate sturgeon
Nazarovskaya GRES	LLC "Nazarovo Fish Farm"	Cages, incubation shop / 170 t	Carp, carp koi, channel catfish, sturgeon, trout
Permskaya GRES	LLC "Dobryansk Fish Center"	Basins, incubation shop / 1650 t	Carp, sterlet, sturgeon, beluga, stellate sturgeon, trout
Pechora GRES	LLC "Aquacomplex"	Cages, incubation shop	Lena sturgeon, sterlet, rainbow trout
Sredneuralskaya GRES	LLC ACE "Sredneuralsky fish-breeding complex"	Cages (winter), basins (summer), incubation shop / 200 t	Carp, white amur, silver carp, sturgeon, trout
Mainskaya GRES	LLC "Sayan trout"	Cages / 300 t	Trout
Kursk NPP	Kurchatov	Cages	Carp, carp koi, black and white amur, silver carp, shrimp
Ust-Ilim CHPP	"Fishery of Ust-Ilimsk CHPP"	Basins, incubation shop / 600 t	Carp, silver carp, sturgeon, тиляпия, trout, omul, peled.
Krasnodar CHP	LLC "Kuban Institute of Sturgeon"	Cages	Sturgeon, sterlet, beluga, stellate sturgeon

Thus, the effective use of waste warmed-up waters, the utilization of secondary energy resources by aquaculture methods in the context of the need to ensure food security and human health are of particular relevance.

Conclusions

The development of energy and the creation of reservoirs led to the need for work on the formation of ichthyocenoses and the development of pasture aquaculture. At the same time, incentives were laid for the improvement of aqua-biotechnologies for the artificial production of eggs and juveniles of various fish species, which served as an impetus for the development of industrial forms of commercial fish farming: cage, basin, the use of installations with a closed cycle of water supply. These areas of aquaculture development coexist and complement each other. Modern pasture fish farming today is not possible without the release of valuable species of fish to form a species composition. At the same time, the development of highly industrial commodity farms implies the use as a brood of individuals of valuable species of fish seized from the natural environment to improve the genetic characteristics of juveniles. At the present time, there is an increase in the range of fish grown due to the use of warm waste waters of energy facilities.

References

[1] FAO 2014 The State of World Fisheries and Aquaculture. Opportunities and challenges

[2] Bartley D M, Brugere C, Soto D, Gerber P and Harvey B 2007 Comparative assessment of the environmental costs of aquaculture and other food production sectors: methods for meaningful comparisons *FAO Fisheries Proceedings* **10** 241

[3] Kalayda M L 2001 *History and prospects of development of the fish industry of Tatarstan* (Kazan: Matbugat Yorty) p 96

[4] Kalayda M L 2017 the tasks of aquaculture development in the republic of tatarstan at the present stage *Fish farming and fisheries* **8(139)** 7-16

[5] Dyomin A P 2010 Water management complex in Russia: concept, state, problems *Water*

resources **37(5)** 617-32

[6] Kalayda M L 2002 The need to take into account amateur and poaching fisheries in the aquaculture of reservoirs *Aquaculture of the beginning of the XXI century: sources, condition, development strategy. Materials of the International Scientific and Practical Conference* (Moskow:VNIRO) pp 88-91

[7] Chebanov M S and Galich E V 2013 *Manual on artificial reproduction of sturgeon* (Ankara:The Food and Agriculture Organization of the United Nations) p 370

[8] Podushka S B 1986 Method for obtaining caviar from female sturgeon *Certificate of authorship USSR* 1412035

[9] Raspopov V M, Chipinov V G and Sergeeva U V 2007 Formation of fish stocks in market conditions *Thermal aquaculture and biological productivity of arid climate basins* (Astrakhan:AGTU) pp 85-6

[10] Mikheev V P 1982 *Cage culture of marketable fish* (Moskow: Light and food industry) 216

[11] Filatov V I, Dokukina K N and Petrov F A 2002 *Fish farming in closed systems* (Dmitrov: The North of Moscow region) **2(III-IV)** 100-2

IOP Conf. Series: Earth and Environmental Science **288** (2019) 012051 doi:10.1088/1755-1315/288/1/012051

Features of the physicochemical water state of reservoirs of energy facilities

M L Kalayda and M E Gordeeva

"Water bioresources and aquaculture" department, Kazan state power engineering university, Russia, 420066 Kazan, Krasnoselskaya, 51

Maria.Galeeva@gmail.com

Abstract. In a time when increasing anthropogenic impact on the environment, particularly on the water ecosystems, determine the water status of reservoirs (by physicochemical and biotic indicators) in order to highlight priorities at the field of water protection and management is the urgent question. In this case, the particular significance has lakes located within megalopolis and which is multiple-use objects. The article discusses in detail how to change temperature condition of basin-cooler of heating plant: vertical and horizontal. As a result of analysis of physicochemical indicators the water quality index is displayed. Index allows you to estimate the water quality by the rate of pollution through the use of perspective an integrated indicator-total antioxidant activity. Special attention is given to such physicochemical indicator as redox potential.

Introduction

Thermal power plants (TPP) use a significant volume of water to cool the turbine generators. In accordance with the Rules of technical operation of power plants and networks of Russia, during the operation of technical water supply systems should be ensured uninterrupted supply of cooling water to normative temperature in necessary quantity and the required quality. As a result, formed the overflow heated water. Utilization of secondary energy resources is becoming an urgent problem on thermal power plants. Modern power plants convert 30-40% of fuel heat into useful electrical energy. The rest 60-70% is dispersed into the environment. Temperature of the overflow water at TPP in summer rises to +30-35°C, in winter it is +8-11°C. The temperature potential from +8°C to +35°C is comfortable for almost all biological creatures. At the nuclear power plant the temperature of the overflow water in summer rises to + 35-40°C, in the winter it does not drop below + 18-20°C [1].

An important aspect of the connection the status of aquatic ecosystems and the energy is the thermal impact on reservoirs of heating plants. The problem becomes particularly acute in cases when lakes are using within the city limits as basin-cooler of combined Heat and Power plant (CHPP).

Water temperature condition of reservoirs of energy facilities

In Kazan (the Tatarstan Republic) the lake Sredniy Kaban is basin-cooler of the Kazan CHPP-1. When choosing the place for power plant the crucial role has played the fact that water in the lake Sredniy Kaban was not only clean, but also soft. In 1866 the famous Russian chemist A.M. Butlerov recommended water of the lake Sredniy Kaban for its high natural qualities for the central urban water supply. As a result of the construction of CHPP-1 and continuing process of water abstraction and discharge, artificial water circulation in the north-western part of the lake is appeared (Figure 1).

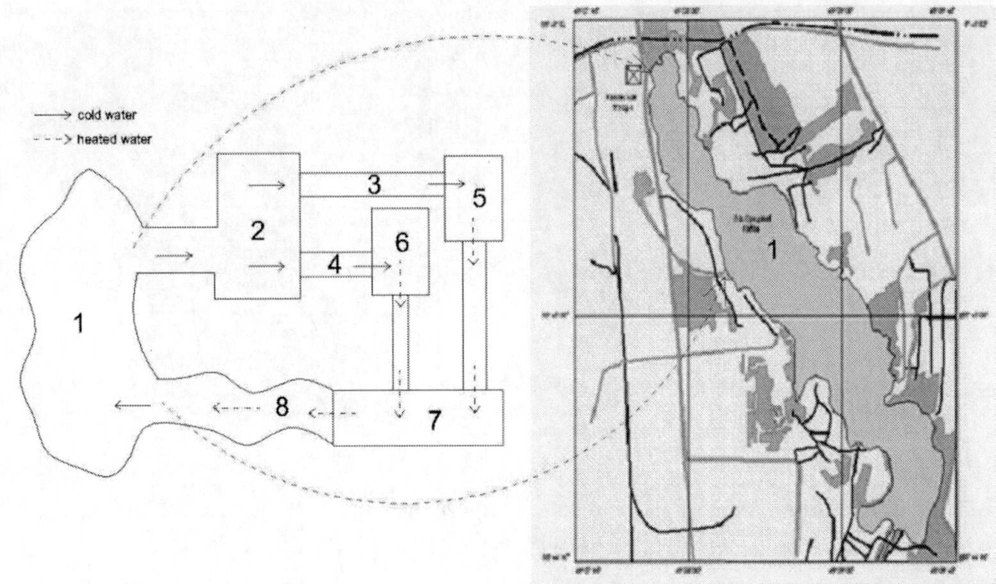

Figure 1. Plan of water using of the lake Sredniy Kaban in technical water supply system of CHPP-1: 1 – Sredniy Kaban lake; 2 – outdoor gravity feed; 3 – forebay of Kazan CHPP-1; 3, 4 – underground water; 5-6 – turbine condensers; 7 – underground circulation channel; 8 – open offtake.

After the construction of a dam at the confluence of open offtake in the lake circulation ceased and water masses are moving only due to wind processes. Based on the speed of withdrawal and the volume of water in the lake, it has been calculated that the total water cycle in the lake is going for 96 days. Our researches [2, 3] pointed out that the lake of Sredniy Kaban – lake with artificial thermal regime – belongs to the "very warm" (4074°) according to the number of the heating degree days on the classification of S. P. Kitaev. Modern temperature condition of the lake Sredniy Kaban is characterized by 5 periods [2]: summer stagnation; partial summer and winter, full autumn and spring circulation. When compared to the lake with a natural temperature regime we can see the reduction of one temperature period – fading winter stagnation. During the year the temperature of the water surface layer varies from +0,3°C to +34,5°C and benthic – from +1.9°C to +28.2°C. Significant contributions, especially in winter time, make heated water of Kazan CHPP-1, which does not allow an average temperature drop below +4°C. The large part of the lake (77%) is covered by ice, which lasts 4 months. The temperature of water surface layer in a horizontal direction in the warmest month (July) at the lake of Sredniy Kaban varies within 9°C (from +21°C to +30°C), while at the lake Verhnii Kaban (lake with natural temperature regime) –within 2°C (from +26°C to +28°C) (Figure 2).

If at the lake Sredniy Kaban at the place of the heated water influence of CHPP is observed direct stratification in the winter time (Figure 3a), at the lake Verhniy Kaban (without thermal pollution) – backward stratification (Figure 3b). At the lake Sredniy Kaban thermocline continues 2 months: July and August (Figure 3c), at the lake Verhniy Kaban – 1 month (July) (Figure 3d). At the bottom layer of lakes we can see rising temperature at 0,2-0,3°C associated with the decomposition of organic matter in the subsurface layer of soil.

Thus, thermal anthropogenic impact on the lake is lead to change the temperature condition of the basin-cooler. Coefficient of variation (V%) of water temperature at the lake without thermal pollution is 4,6% and at the lake with thermal pollution is 16,4% on conditions that the air temperature is similar.

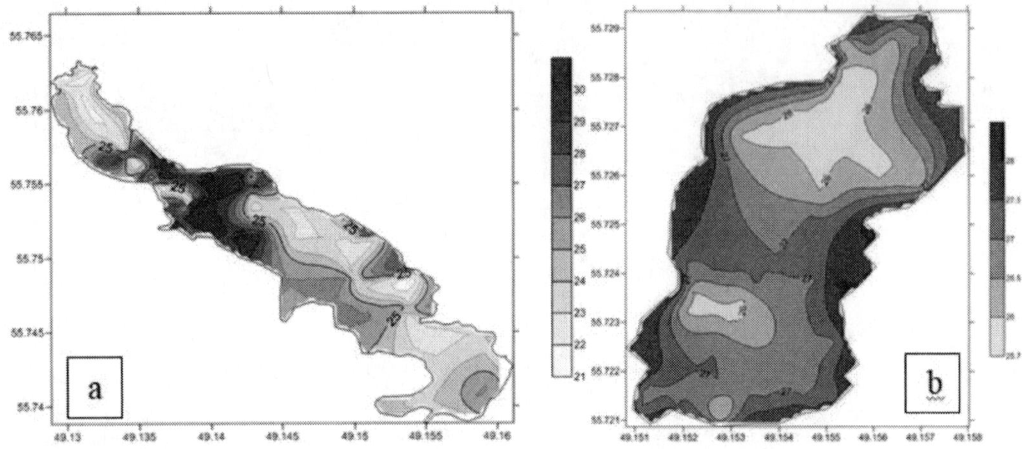

Figure 2. 2D distribution model of surface water temperature in July at the lake Sredniy Kaban (a) and Verhniy Kaban (b).

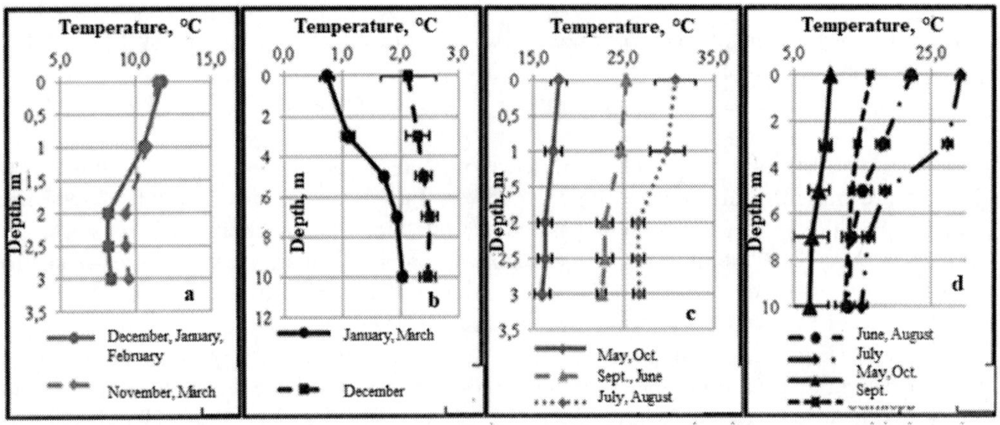

Figure 3. Temperature distribution in depth at the lake Sredniy Kaban (a, c) and Verhniy Kaban (b, d).

Water quality index

For the first time to evaluate the quality of the water has been used new complex index – total antioxidant activity (TAOA). The range of water TAOA at the lake Sredniy Kaban from 2,61 to 6,09 mg.rutin/l [4, 5]. Values of TAOA on the territory of withdrawal of the Kazan CHPP-1 have been higher than on the territory where heated waters inflow to the lake: 4,64 and 3,19 mg.rutin/l respectively. To compare data has done analysis of TAOA of distilled water – 2,61 mg.rutin/l; different brands of drinking water – 3,19 mg.rutin/l; melt water – 3,19 mg.rutin/l; artesian water – 5,37 mg.rutin/l.

In addition to the listed indicators were researched other physicochemical parameters: pH-value, concentration of dissolved oxygen, redox potential (Eh), which entered to the water quality index (I_{WQ}) developed by us. I_{WQ} evaluate the water quality on the degree of contamination through the use of perspective and complex indicator – total antioxidant activity.

This index is obtained from the equation balance between values of antioxidant activity (mg.rutin/l) and abiotic water factors: water surface temperature (°C), dissolved oxygen (mg/l) pH, Eh (mV) (formula 1).

$$I_{WQ} = \frac{-18{,}96 - 0{,}32*Ts - 0{,}33*Co_2 + 4{,}26*pH + 0{,}19*Eh}{TAOA}, \quad (1)$$

TAOA – total antioxidant activity (mg.rutin/l); Ts – water surface temperature (°C); Co_2 – concentration of dissolved oxygen in the water (mg/l); pH – water pH-value; Eh – redox potential (mV).

Forecast model fidelity confirms by the standard error 0,09 and the coefficient of determination 99,32%.

If the values of I_{WQ} in the range from -1 to +1 it means that water is pure (oligosaprobic in accordance with the data of parallel microbiological analysis). If the values of I_{WQ} in the range from +1 to +2 and from -1 to -2 that is α-mesosaprobic zone. If the values of I_{WQ} in the range from +2 to +3 and from -2 to -3 that is β-mesosaprobic zone. When the values of the coefficient in the range $-3 < I_{WQ} < 3$ water is contaminated – polisaprobic. Water quality in the area of withdrawal of Kazan CHPP-1 is regarded as the lowest (in the summer period I_{WQ} reaches -3,2 – polisaprobic water quality). The area of heated pollution (confluence of the heated waters of the Kazan CHPP-1 into the lake) is characterized by maximum variability of I_{WQ} (from -2.5 to + 2.4; V = 909,4%). In different periods of time the water quality in this zone is characterized as oligo-, β-mesosaprobic.

Redox potential as a promising monitoring indicator

Eh is a complex physicochemical indicator that is able to indicate water quality [3]. Eh is a measure of the chemical activity of elements or their compounds in reversible chemical processes associated with the change in the charge of ions in solution. Positive values of Eh indicate the oxidizing conditions, negative – on restoration. Eh value is defined by the presence in solution potentially specify ions [6]:

- elements-oxidants, which are capable to accept electrons (Fe^{3+}, Mn^{4+}, S^{6+}, Cr^{6+});
- elements-reductors, which are capable to give electrons (Fe^{2+}, Mn^{2+}, S).

In addition, the most important oxidant is oxygen, and reductor – organic matter and certain types of bacteria.

Eh researches have their roots in the middle of 20th century (Figure 4). In conditions of active water cycle usually present free oxygen and geochemical environment is oxidative. The Eh value more than 0,15 V, sometimes more than 0,4 V [7].

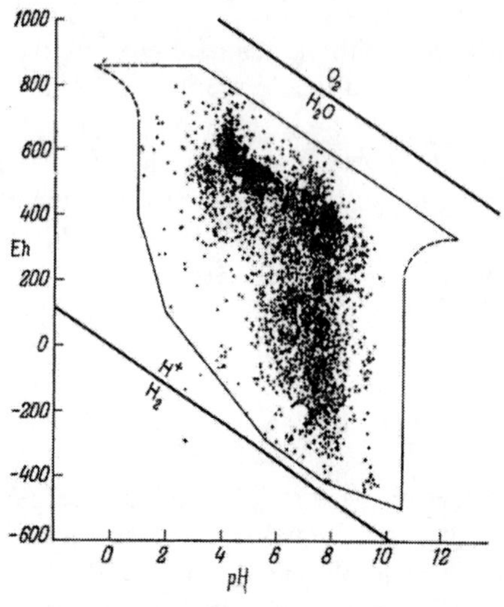

Figure 4. The results of measurements of pH and Eh values in natural waters by Garrels, 1968 [7].

At a time of slow water cycle oxygen spend on oxidation of organic substances and elements with variable valency (Fe, Mn, As, etc.) and geochemical environment is Gley (oxygen-free). The Eh value usually less than 0,4 V, sometimes less than 0 V (bogs of tundra, taiga and steppe zones, lower horizons of groundwater) [7].

In a very slow water cycle in deep horizons of underground waters is formed reductive atmosphere with hydrogen sulfide, Eh-value < 0, often up to 0,5 V due to extremely weak inflow of oxygen and active work of sulfate-reducing bacteria [6].

Interest to Eh researches doesn't extinguished in the XXI century. Research of physicochemical indicators of rivers in the Republic of Tatarstan has shown that the range of variation of Eh from -21,1 mV to -166,1 mV [60]. The amplitude of variational series is 145 mV. Focusing on major rivers, the territory of the Tatar ASSR divided into 3 parts on the orography – Predvolzhie, Predkamie and Zakamie. Predkamie on the river Vyatka is divided into West and East. Zakamie on the rivers Verhnii Chereshman and Sheshma River is divided into Western Zakamie (low) and Eastern Zakamie (high). In accordance with this division all collected data of Eh-value in rivers are shown in Figure 5.

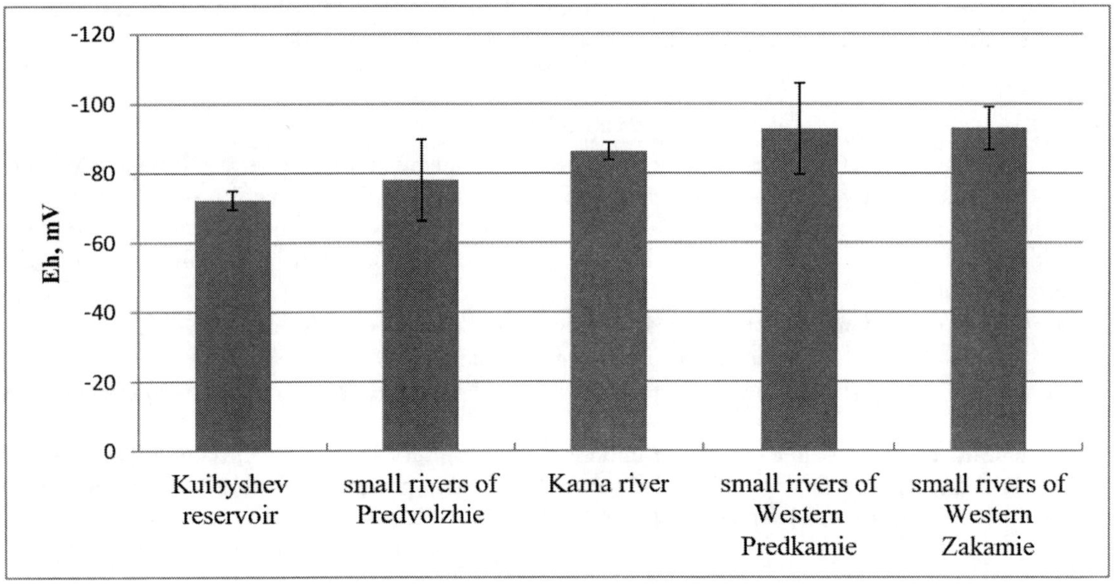

Figure 5. The mean values of Eh (mV) with the standard error of average in Kuibyshev reservoir and the rivers of the Republic of Tatarstan.

From data represented in Figure 5 shows that the minimum variations is in the Kuibyshev reservoir and on the Kama river. Water sampling in the Kuibyshev reservoir was in 3 locations around Kazan (next to the urban-type settlement Vasilevo, Oktyabrskoe rural settlement and in the Kirovskyi district of Kazan) and 2 locations near the Tolyatti city (in Primorskii district of Tolyatti and near of Fedorovskie luga of Tolyatti). The value of Eh in Kuybyshev reservoir increased from the urban-type settlement Vasilevo to Fedorovskie luga of Tolyatti (from -78,5 mV to -63 mV).

Maximum variation of Eh values and maximum the standard error of average observed in the rivers of Western Predkamie (Eh values range from -21,1 mV to -166,1 mV) and Predvolzhie (Eh values range from -38,8 mV to -104,7 mV).

In general, the range of Eh values changes in lakes and natural ponds more wide compared to Eh values in rivers. For example, only in the Republic of Tatarstan the range of Eh values changes in lakes from 7,5 mV to -226,4 mV. The amplitude of variational series is 233,9 mV. It is in 1,6 times higher the Eh amplitude in rivers. Our researches corroborate direct dependence of Eh from pH [8], therefore the range of Eh values conforms to the values of pH from 5,87 to 9,85. For reservoirs, experiencing the action of energy facilities (for example, the Sredniy Kaban lake of Kazan) the average value of Eh in zone, not suffering from action of energy facility (Kazan CHPP-1), is -53±8,5 mV. In the area of direct impact of energy facility Eh value drops to -80 mV, average value is -70±8,5 mV [2]. Thus, Eh is more sensitive indicator to determine changes of physicochemical condition of the water compared to pH.

One of the most effective ways of recycling low-potential overflow water is creating energobiological complexes.

References

[1] Rostuncova I A, Shevchenko N Y 2015 The efficiency of using low-potential heat of TPP as secondary energy resources *International Journal of applied and fundamental research* **9-4** 618-21

[2] Kalaida M L, Gordeeva M E 2015 Complex assessment of status of lakes ecosystem. Urbanized territories *LAP LAMBERT Academic Publishing* 228

[3] Kalaida M L, Gordeeva M E 2017 Redox potential as an indicator of the quality of natural and artificial reservoirs in the structure of fisheries monitoring *The role of education and science in the development of Russian society: collection of articles of the International scientifically-practical Conference* (Electronic resource) (Moscow: Impuls) pp 279-91

[4] Galeeva M E, Lapin A A, Chugunov Y V, Kalaida M L 2012 Antioxidant activity is a promising indicator for determining water quality index *Butlerov communications* **29(3)** 110-19

[5] Lapin A A, Kalaida M L, Galeeva M E, Zelenkov V N 2011 Study on water antioxidant activity. *Non-traditional natural resources, innovative technologies and products. Collection of scientific papers* (Moscow: RANS) **19** 54-60

[6] Tokarenko O G 2014 *General hydrogeology: chemical composition and properties of natural waters* (Tomsk: TPU) p 57

[7] Garrels R M, Kraist Ch L 1968 *Solutions, minerals, balance* (Moscow: Mir) p 368

[8] Gordeeva M E, Kalaida M L 2018 Redox potential as an indicator of the water quality in the monitoring of water ecosystems *All-Russian scientific and practical Conference «Water bioresources and aquaculture of southern Russia»* 40-4

IOP Conf. Series: Earth and Environmental Science **288** (2019) 012052 doi:10.1088/1755-1315/288/1/012052

Features of the hydrobiocenosis of the Kuibyshev reservoir in conditions of local pollution

M F Khamitova and M L Kalayda

"Water bioresources and aquaculture" department. Kazan state power engineering university, Russia, 420066 Kazan, Krasnoselskaya, 51

it-sk@bk.ru

Abstract. A complex ecological situation develops in the area of the largest reservoir of the Volga cascade – Kuibyshev, especially in the areas of concentration of population and production. In connection with the development of aquaculture, the problems of conservation and reproduction of sterlet in the Kuibyshev reservoir become actual. Taking into account the tasks of preserving water quality in the upper part of the Volga reach of the Kuibyshev Reservoir as a spawning site of sturgeon in the early 20th century, special attention was paid to the impact of waste water discharges from the Mari Pulp and Paper Mill (JSC "MTSBK"). The hydrochemical and hydrobiological characteristics of section of the Volga reach of the Kuibyshev reservoir and secondary settler of JSC "MTSBK" were studied. The change in the qualitative and quantitative characteristics of the hydrobiocenosis in the conditions of functioning water treatment plant was shown.

Introduction

A complex ecological situation develops in the area of the largest reservoir of the Volga cascade – Kuibyshev, especially in the areas of concentration of population and production. In connection with the development of aquaculture, the problems of conservation and reproduction of sterlet in the Kuibyshev reservoir are actual. Taking into account the tasks of preserving the quality of waters in the upper part of the Volga reach of the Kuibyshev reservoir as a spawning site of sturgeon fish in the early 20th century [1], particular attention was paid to the impact zone of wastewater discharges (JSC "MTSBK"). Investigation of the local contamination site of the Volga Reach in the area of wastewater discharges of JSC "MTSBK" as a model, with a high content of organic substances, makes it possible to trace the process of changing the characteristics of the biocenosis as the content of organic substances decreases: from the maximum contamination zone in the secondary settler, to the section of the Volga reach – less exposed to sewage.

Monitoring of hydrochemical and biotic indicators of the state of water ecosystems of energy objects becomes extremely important. It allows assessing the limits of stability of hydrobiocenoses, outline measures to reduce the negative anthropogenic impact on freshwater bodies.

Results and discussion

Characteristics of the study area, material and methods of research

The model site of the study is located in the Volga reaches of the Kuibyshev reservoir in the Lopatinskaya Volozhka area. JSC "MTSBK" commissioned in 1938. The secondary settler is a system of inter-island closed gulfs, from which the sewage flows through the canal into the channel part of the

Kuibyshev reservoir where the waters are further diluted by the main stream of the reservoir and the waters of the Ilet River that confluence upstream.

Hydrochemical and hydrobiological tests in the area of the Lopatinskaya Volozhka were conducted every two weeks during the growing seasons of 2011-2014 at the stations indicated in Figure 1. Hydrochemical, hydrobiological studies were carried out by standard methods.

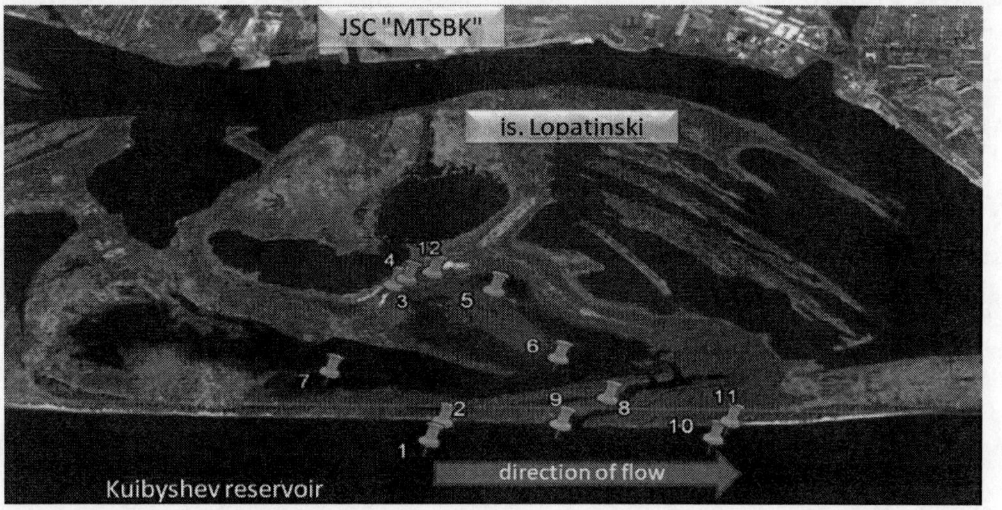

Figure 1. Scheme of sampling stations in the area of wastewater discharge of JSC “MTSBK”.

Investigation of the state of water according to physicochemical factors of the environment
In the zone of maximum organic pollution - wastewater discharges into the secondary settler, the COD varied from 69.4 to 270 mgO_2 / dm^3. The control stations are characterized by COD values from 19.1 to 27.7 mgO_2/dm^3, what indicates a low concentration of organic substances. If the values of BOD_5 are more than 10 mg / l for "dirty" waters, then in the most contaminated sections of the settler the BOD_5 values are on average twenty times higher. At the control points of the Volga reach, the BOD_5 index varied from 1.4 to 2.6, in the zones of self-cleaning processes – 4.02 mgO_2/l. During the operation of the treatment plant for about 60 years, on the control sites the BOD_5 indicator corresponds to clean and moderately polluted waters. The conducted study of the concentration of biogenic and organic substances in the control area showed an increase (within the limits of the MPC) of the concentrations of ammonium-ion, nitrate ion and phosphate ion as compared to the 2000s [2].

Analysis of the chemical composition of the water of the Volga reaches on the control site, carried out by the X-ray fluorescent method, revealed an excess of the MPC [3]: for strontium by 1.58, for iron by 2, for zinc by 6, and for copper by 40 times. Of the heavy metals in the water of the secondary settler there were Sr>Mn>Fe>Co>Zn>Cu>Rb. In the most polluted section of the secondary settler were noted Mn, Co, Rb, which were not observed in the water of the control section (Table 1).

Table 1. Concentrations of heavy metals (in fractions of MPC) in the water of the secondary settler and the Volga Reach.

Site of the study /Index	Elements								
	Sr	Fe	Zn	Mn	Cu	Rb	Co	Ni	Pb
	Secondary settler of JSC "MTSBK"								
At wastewater discharge from the primary settler	1,60	3,0	4,0	60,0	30,0	0,05	4,0	-	-
Gulf of the secondary settler	2,2	4,0	355	40,0	-	-	41,0	-	-

Ending of Table 1

	The Volga reach								
The control section (500 m above the mixing zone)	1,58	1,2-2,1	6,0	-	0,75-40,0	-	-	-	-
Water mixing zone	1,68	2,0	48,0	20,0	-	-	-	-	-
Downstream (500 m from the water mixing zone)	1,65	6,0	14,0	10,0	50,0	-	-	3,0	13,3
MPC (mg/l)	0,4	0,1	0,01	0,01	0,001	0,1	0,01	0,01	0,006

Assessment of changes in the state of water by hydrobiological characteristics under conditions of local pollution.

The evaluation of the microbial component of the biocenosis shows that the section of the Volga reach (including the secondary settler) refers to the mesosaprobic zone. In different sections of the settler the total microbial number varied from 9813 to 49000 CFU/ml. Wastewater discharges showed a high number of coliform bacteria: up to 773 070 CFU/100ml. In all soil samples bacteria of these genera were found: *Bakterium, Pseudomonas* and *Clostridium*, which are ammonifying, and some members of the genus *Clostridium* – cellulose-destroying bacteria.

In the composition of phytoplankton, 70 species and taxa of algae were found from 6 divisions: in the control section of the Volga reach – 49, in the secondary settling basin – 59 (Figure 2).

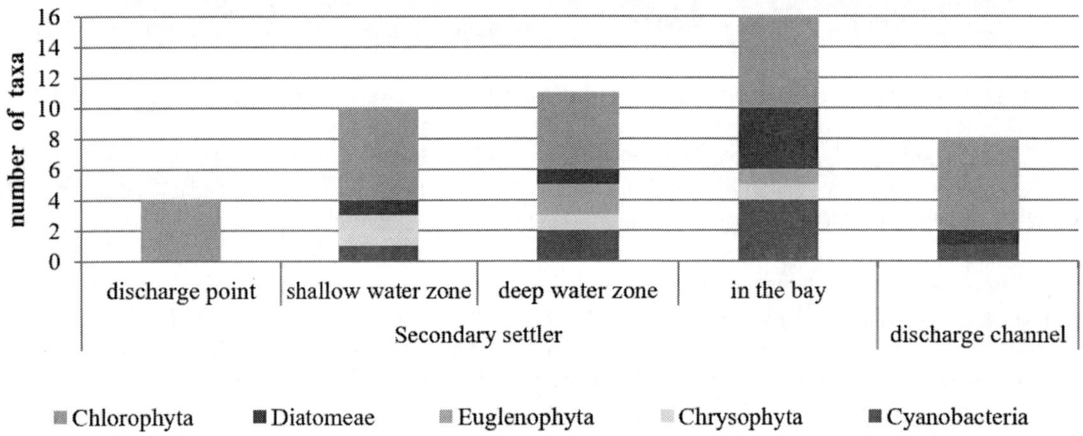

Figure 2. Taxonomic diversity of phytoplankton in the secondary settling basin.

In the secondary sedimentation basin, exept the wastewater discharge site, cyanobacteria were the most numerous. The greatest abundance and biomass of phytoplankton was noted in the shallow section of the sedimentation basin – up to 708,879 cells/l and 325.78 mg/l, respectively. The basis of abundance was *Aphanothece globosa* Elenkin (up to 57%), 42% of the biomass formed large golden algae *Dinobryon divergens* Imhof. The density indices in the settler basin were dominated by green algae *Chlamydomonas* sp. (12.98), cyanobacteria *Microcystis aeruginosa* f. *flos-aquae* (Wittr.) Elenk. belonged to subdominants (9.99).

In the composition of zooplankton, 110 species and forms were encountered, in the control section of the Volga reach – 63, in the secondary settling basin – 71. The current state of the structure of zooplankton in the control areas of the Volga reach revealed ratio similar to the previous period [4].

In the control plot were well represented predatory species of branchy crustaceans – *Polyphemus pediculus* (Linne, 1778), *Bythotrephes longimanus* (Leydig, 1860), *Leptodora kindtii* (Focke, 1844), which were not observed in areas of wastewater discharges in Volga reach in previous studies [4]. At all stations of the Volga reaches, the Dreissena veligers and the bryozoa resting eggs were noted.

The study of zooplankton of the secondary sedimentation basin revealed a reduction in the qualitative (Figure 3) and quantitative indices of its development, there were no rotifers and crustaceans among the dominant species. There was a greater development of the infusoria of *Paramecium caudatum* (Ehrenberg, 1836) and representatives of class *Peritricha* sp., in the area of discharge of water into the secondary settler they amounted to 89.5% of the zooplankton population, small roundworms, annelids and various plankton forms of insect larvae were met.

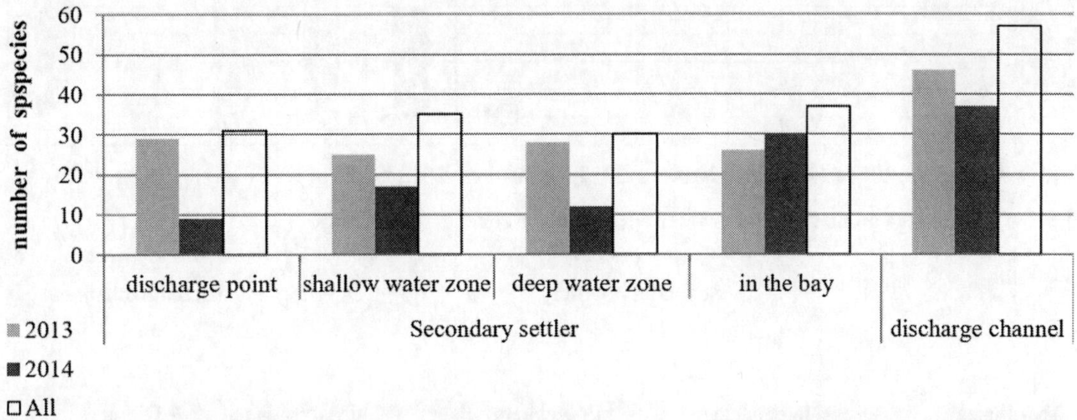

Figure 3. Taxonomic diversity of zooplankton in the secondary settler.

Regularities in the change of zooplankton communities have been revealed: in the most polluted area (polysaprobic zone) the infusoria *Paramecium caudatum* and representatives of the *Peritricha* group predominate, the variability of the species diversity indicators was more than 40%; at the site of reduction of organic contamination (α-mesosaprobic zone), maximum development of phytoplankton and low development of zooplankton are noted first (the maximum quantity of zooplankton varied from 0.26 to 7.18 thousand specimens/m^3). With a further decrease in organic contamination (β-mesosaprobic zone), the maximum biomass of zooplankton (up to 76.7 g/m^3) was noted, first due to the development of *Daphnia pulex*, followed by *Moina macrocopa* and *Ceriodaphnia laticaudata*. In the areas of restoration of water quality from the branching crustaceans can be related to typical forms: *Bosmina longirostris, Chydorus sphaericus, Daphnia cucullata* (Sars, 1862); among copepods: *Eudiaptomus graciloides* (Lilljeborg, 1888), *Cyclops kolebsis* and *Metacyclops gracilis*, copepodite and nauplial stages of copepods. The coefficients of variation of Shannon species diversity indices [5] at the restoration sites – 13-22%; in the zone of water mixing the variability of species diversity is higher – up to 37%; on these sites are widely represented predatory species of branching crustaceans – *Polyphemus pediculus, Bythotrephes longimanus, Leptodora kindtii.*

According to the "trophy scale" [6], the entire stretch of the Volga reach can be related to the low class, and the shallow section within the secondary sediment basin - to the elevated class of trophia. The density indices in the secondary sedimentation basin showed an increase in the dominance of *Asplanchna priodonta* (Gosse, 1850) and the change of dominance from *Bosmina longirostris* to *Daphnia pulex* and *Moina macrocopa*.

As a part of the zoobenthos, 160 species and forms were encountered, of which 49 in the control section of the Volga reach. The most commonly noted were mollusks Lithoglyphus naticoides, from 2013 their frequency of occurrence - 100% and *Dreissena polymorpha* (Pallas, 1771).

The basis for the abundance of soft benthos (up to 60%) in summer was larvae of chironomids, and in the autumn - oligochaetes (62%). In the coastal zone the zoobenthos population varied up to 0.62-1.12 thousand specimens / m^2. In 2014 there was a significant increase in the number of polychaetes *Hypania invalida* (Grube, 1860), the maximum of their development occurred at the beginning of August (0.22 thousand specimens/m^2).

40 species and forms of zoobenthos were encountered in the secondary settler of JSC "MTSBK" (Figure 4), of which the most frequent were *Culex* sp., *Eristalis* sp., *Chironomus* гр. *plumosus* (Linne, 1758), *Glyptotendipes* гр. *gripekoveni* (Kieffer, 1913).

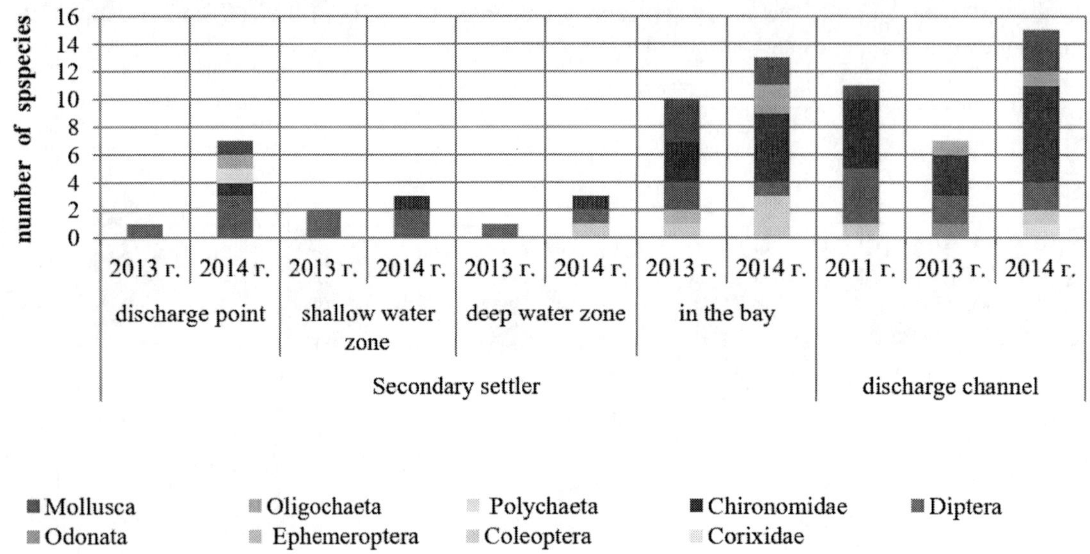

Figure 4. Taxonomic diversity of zoobenthos in the secondary settler.

The structure of the main complex of zoobenthos species on the sites of the greatest organic pollution is represented by the forms of insects capable of atmospheric respiration. With the improvement of the state in the ecotope, the diversity and representation of the secondary-water insects breathing dissolved oxygen increased, the control areas were dominated by primary water animals.

In the secondary settling tank, the zoobenthos population varied from 20 to 420 specimens/m^2, the largest biomass in 2013 in the Gulf (up to 51.02 g/m^2) and in the channel in 2014 (up to 64.5 g/m^2). At the remaining stations of the settling basin, the zoobenthos biomass did not exceed 1.76 g/m^2. According to the quantitative development of the zoobenthos, the class of waters [6] in the secondary sediment basin is estimated as ß-oligotrophic and, only in the gulf, as hypertrophic.

The study of the components of hydrobiocenoses and hydrobionts in the areas of wastewater discharges revealed on all sites from 4 to 7 pollutants of heavy metals. Of the highly hazardous heavy metals in the bottom sediments and water, lead was present, from moderately hazardous - Cu, Cr, Zn, Ni. At the wastewater discharge sites, the main pollutants that enter the food chains into the fish are currently copper and zinc. In the Volga reach in the perch, there is an excess of the permissible residual amount for zinc and copper 3.5 and 4.7 times respectively. Lead is also a potential threat, as it accumulates in hydrobionts. The conducted studies revealed a decrease in the presence of copper, mercury and cadmium in hydrobionts in comparison with the end of the twentieth century [7, 8].

Conclusions

At the wastewater discharge sites in the Volga reach of the Kuibyshev Reservoir, the main pollutants that enter the food chains into the fish are currently Cu and Zn. Pb, Hg and Cd – the characteristic pollutants of hydrobionts at the end of the 20th century are not currently noted in the hydrobionts of the Volga Reach.

The main pollutants in wastewater discharge in the secondary settler of the pulp and paper mill are organic substances. In the secondary settling tank there is a reduction in COD of water from 69.4-265.0 mgO_2 / dm^3 to a level characteristic for the control section of the river (in 2.9-4.4 times). Despite the 60-year service life of the treatment plant, the studies showed no significant changes in the state of the hydrobiocenosis of the Kuibyshev reservoir downstream of the discharge compared to the control site.

49 species and intraspecific taxa of algae, 88 species and forms of zooplankers, 72 – zoobenthonts were encountered in the area of the Kuibyshev reservoir in the area of wastewater discharge of JSC "MTSBK". The whole studied area of the Volga reach refers to the low trophy class and is estimated as moderately polluted.

In the secondary sedimentation basin, 50 species and intraspecific taxa of algae were encountered, 71 species and forms of zooplacters, 40 – zoobenthonts. As the concentration of organic substances decreases in local contamination sites, the change in the prevailing groups of internal communities of hydrobionts proceeds according to the following scheme: bacterioplankton → phytoplankton (in the direction of green → blue-green algae) → zooplankton (due to branchy crustaceans). The structure of the main complex of macro-invertebrate species on the sites of the greatest organic pollution is mainly represented by insect species capable of atmospheric respiration; when the state in the ecotope improves, the diversity and representation of the secondary-water insects increases: forms with breathing dissolved oxygen, replace forms consuming atmospheric air. Primary-water animals prevailed in the control areas and in the restoration zones.

References

[1] Lukin A V 1947 Main features of ecology of sturgeon in the Middle Volga *Proceedings of the Society of Naturalists at the Kazan University* (Kazan: GSU) **LVII(3-4)** 39-143
[2] Kalayda M L, Khamitova M F and Novotochinov S I 2016 Change in the chemical index of oxygen during algolization of chlorella *Butlerov Communications* **48(10)** 143-9
[3] Kalayda M L and Khamitova M F 2015 Possibilities of applying eichornia in post-treatment of waters of pulp and paper mill. Part 1. Features of the chemical composition of water in the Volga reach of the Kuibyshev reservoir *Butlerov Communications* **44(11)** 97-103
[4] Kalayda M L 2003 *Ecological assessment of the Kuibyshev reservoir under anthropogenic impact* (Kazan: Kazan State Energ. Un.) p 135
[5] Wilhm J L and Dorris T C 1968 Biological parametris for water quality enteric *Bioscienѕe* **18(6)** 477-80
[6] Kitaev S P 1984 *Ecological bases of bioproductivity of lakes of different natural zones* (Moskow: Science) 207
[7] Kalayda M L, Uryadova L F and Askhadullina A R 2010 Results of the study of aquatic organisms on the content of heavy metals in conditions of different anthropogenic load *Butlerov Communications* **22(12)** 61-6
[8] Kalayda M L and Khamitova M F 2014 Peculiarities of the chemical composition of perch juveniles in the Volga reach of the Kuibyshev reservoir *Butlerov Communications* **40(12)** 37-41

Assessment of an ecological state of the Kuibyshev reservoir based on zoobenthos indicators

T A Kondrateva[1], N Yu Stepanova[2]

[1]Federal State Budget Institution «Management of Hydrometeorology and Environmental monitoring of the Republic of Tatarstan, Zavodskaja str., 3, Kazan city, 420021, Russia

[2]Institute of Environmental Sciences, Kazan Federal University, Kremlevskay str., 18, 420008, Russia

tatjana_kondrate@mail.ru

Abstract. On the basis of the analysis of long-term data of chemical and biological monitoring, the ecological state of the Kuibyshev reservoir (the Republic of Tatarstan, Russia) was defined. There were used the dissolved oxygen, BOD, ammonium content and pollutants with excess of maximum allowable concentration (15 ingredients in total) as markers of chemical pollution. Total number of a zoobenthos and relative number of oligochaetes were chosen as biological indicators. Based on chemical and biological indicators it was noted that the lower sites of the Kuibyshev reservoir are less polluted in comparison with top sites which ecosystem now is in transition state from equilibrium to crisis.

Introduction

By the end of the 20th century more than 45 000 large dams were constructed in more than 140 countries. It is expected that the construction of reservoirs will be a continuous increase in the future. These artificial reservoirs meet needs of the human for providing with drinking water, electricity generation, agricultural irrigation, industry, fishing trade and a recreation [2]. All this belongs also to the Kuibyshev reservoir which is located in the central part of the Middle Volga area.

The enterprises of housing and communal services, the power and petrochemical industry, agriculture make negative impact on a water quality of the Kuibyshev reservoir. The largest volumes of the polluted waste waters come to the reservoir from the enterprises of the cities of Kazan and Naberezhnye Chelny. Poor quality of water is the result of such influence. In the same time the quality of the water is the prime factor which defines the existence and a possibility of water biological resources using [3].

Significance/Relevance of the study

Two main approaches, chemical and biological, are used in surface waters monitoring system. The first one is based on the determination of pollutants content and their comparison with the maximum permissible concentrations (MPC), calculation of water pollution indexes. Despite the fact that this system does not take into account the state of the biotic component of the ecosystem, at the same time, formalized indicators, for example, the specific combinatory water pollution index are very useful and informative for characterizing the dynamics of the anthropogenic component of the chemical pollution of the hydrosphere. As an alternative to MPC, recently it has been proposed to use the so-called ecologically permissible levels, developed by mathematical modeling methods for individual regions with taking into account the biotic component by in situ methods [4, 5]. In this case the limits of the environmental factors (chemical substances content) are introduced as levels that do not violate the norm of the ecological state, established by biological indicators.

The second approach takes into account the state of the biotic component of the ecosystem. To implement the biotic approach, a set of methods is necessary for obtaining assessments of the state of biotic communities, with the help of which it would be possible to distinguish an ecologically safe ecosystem from an ecosystem in which there were significant changes caused by external (mainly anthropogenic) impacts [6]. The most common methods of indication are based on expert estimates for specific groups of hydrobionts. Most scientists agree that the most objective assessment of the quality of the aquatic environment by the biotic component can be obtained by using zoobenthos organisms as bioindicators [7, 8, 9, etc.].

The objective of the study

The objective of this study is the determination of the ecological state of the Kuibyshev reservoir based on characteristics of abiotic and biotic components.

Materials and methods

Sampling zoobenthos and water for the chemical analysis was performed at the following locations: the Zelenodolsk and Kazan cities, the settlement Tenishevo, the settlement Zaovrazhnye Karatai, the Laishevo town, the Tetyushi town. Collection and processing of hydrobiological and hydrochemical samples was carried out according to standard methods [10].

Water quality by chemical indicators was evaluated using a specific combinatorial index [11].

To assess the state of ecosystems according to the abiotic component, statistical analysis of long-term information on the variation in the concentration of dissolved oxygen, easily oxidized organic substances according to BOD (Biological Oxygen Demand), ammonium nitrogen, and modal intervals MI (intervals including the most frequent values in this variation series) was carried out. In addition, calculated indicators, such as the proportion and degree of anthropogenic impact, were used. The share of anthropogenic impact (S) was calculated by the formula 1:

$$S = N_1/N*100 \qquad (1),$$

where N_1 – the number of ingredients exceeding the MPC;
N – total number of rated priority pollutants.

The degree of anthropogenic impact was calculated by the formula (2):

$$C = N_2/N_1*100 \qquad (2),$$

where N_2 – the number of ingredients in excess of 10 MPC.

In total, 15 ingredients were used for calculations: oxygen dissolved, BOD, COD (Chemical Oxygen Demand), nitrates, nitrites, ammonium, chlorides, sulfates, total iron, phenols, total petroleum hydrocarbons (TPH), Cu^{2+}, Zn^{2+}, Cr total, Mn^{2+}.

The following indicators were used to assess the status of ecosystems using the biotic component:

- the total number of zoobenthos, N_{zb}, mill ind.M^{-2};
- the relative abundance of oligochaetes in zoobenthos, N_{ol},%.
- Woodiwiss biotic index.

Based on the results an invariant condition of an ecosystem was defined according to a procedure of ecological modifications [12, 13].

Result and Discussion

Generalization and analysis of long-term information of hydrochemical and hydrobiological indicators allowed to determine the ecological status of the Kuibyshev reservoir.

Assessment of the abiotic component

Analysis of hydrochemical information showed that the exceed of MPC was for 12 ingredients. The main contribution to the pollution of surface waters of the Kuibyshev reservoir introduced nitrogen nitrite, nitrogen ammonium, TPH, copper compounds, BOD, COD and water quality was characterized as “characteristic”. At the same time the oxygen regime was satisfactory in all investigation points. Water quality on the upper parts of the reservoir was characterized as “dirty”, the other as “very polluted”.

Table 1. The state of the Kuibyshev reservoir based on abiotic parameters

№[a] Water quality	MI[b] BOD_5 mgO_2/L (moda)	MI NH_4^-, mg/L (moda)	The share of anthropogenic impact	
			range values, %	state of the ecosystem
1) dirty	1.54–3.25 (2.29) crisis	0-0.38 (0.21) equilibrium	0-80[c] 6.7-53.3[d] (33.3)[e]	transition from equilibrium to crisis
2) dirty	0.84–2.89 (1.96) transition from equilibrium to crisis	0-0.62 (0.32) transition from equilibrium to crisis	0–80 6.7-66.7 (33.3)	transition from equilibrium to crisis
3) very polluted	1.02-1.67 (1.59) equilibrium	0.14-0.25 (0.17) equilibrium	0-40 20-27 (20)	equilibrium
4) very polluted	1.04-1.82 (1.38) equilibrium	0-0.19 (0.1) equilibrium	0-33 20-30 (20)	equilibrium
5) very polluted	0.98-1.45 (1.33) equilibrium	0-0.09 (0.06) equilibrium	0-33.3 13.3-20 (11.7)	natural
6) very polluted	0.71-2.45 (1.64) transition from equilibrium to crisis	0-0.22 (0.12) equilibrium	0–80 6.7–46.7 (26.7)	equilibrium

a – points of investigation: 1) the Zelenodolsk city. 2) the Kazan city, 3) the settlement Tenishevo, 4) the Laishevo city, 5) the settlement Zaovrazhnye Karatay, 6) the Tetyushi city;
b – modal interval; c – total range; d – MI; e - long-term value.

According the share of anthropogenic loading the state of the water ecosystems on the upper sites of the reservoir was assessed as transitional from equilibrium to crisis, the lower sites as equilibrium, and near the Kama Mouth as natural. Based on the extent of anthropogenic influence the state of the ecosystem of the Kuibyshev reservoir was characterized as equilibrium (cases of excess of MPC more than 10 times were not fixed). Such differences in ecological state assessment of the top and lower sites of the reservoir are caused, first of all, by impact of large industrial centers like Zelenodolsk and Kazan cities.

Assessment of the biotic component.

Typical groups of zoobenthos for reservoirs were marked out. These are, in general, Mollusca, Olygochaeta, Chironomidae, Gammaridae (Crustacea). There were also Hirudinea, Hydracarina, Polychaeta, Cumacea (Crustacea), Misidacea (Crustacea), Nematoda. There were fewer larvae Trichoptera, Ephemeroptera and others. The greatest group and specific variety of a zoobenthos was observed on the lower sites (3-14 taxons by a rank higher than a genus), the smallest on the top (1-10 taxons by a rank higher than a genus). The number of a zoobenthos varied from 0.02 to 9.48 mill ind.M^{-2} (Table.2).

Table.2. Statistic characteristics of a zoobenthos

№, Water quality	BI[a] Min-max M±Δ	N_{zb}, (mill ind.M^{-2}) Min-max M±Δ	MI (w, %)	N_{ol}, % Min-max M±Δ	MI (w, %)	Invariant state
1) polluted	1.0-6.0 3.6±0.1	0.1–8.32 0.96±0.14	0.1–1.22 (77)	0–100 50.8±3.8	31–87 (55)	Elements of ecological regress
2) polluted	2.0-6.0 3.4±0.1	0.04–22.4 2.67±0.32	0.04-3.88 (83)	0–93 33.6±2.3	27-93 (52)	Elements of ecological regress
3) polluted	2.0-8.0 3.9±0.2	0.28-4.16 1.86±0.13	0.28-1.28 (55)	1.3–80.6 25.9±2.2	6-51 (61)	Anthropogenic stress with elements of ecological regress
4) polluted	2.0-7.0 4.3±0.2	0.18-1.68 3.26±0.4	0.18-5.16 (81)	1.2-92.9 23.4±2.1	30-82 (60)	Elements of ecological regress
5) poorly polluted	4.0-5.0 4.6±0.1	0.02-9.08 1.7±0.3	0.04-1.64 (74)	2-100 19.6±2.1	1-25 (61)	Background
6) poorly polluted	4.0-6.0 4.6±0.1	0.34-9.48 5.0±0.9	0.34-4.64 (68)	0.9-66.7 22.8±2.4	1-17 (64)	Background

a – Woodiwiss biotic index

The largest average values of the number are recorded in the lower parts, the smallest in the upper and middle one. Mollusca (Dreissena spp.), Olygochaeta, Chironomidae dominated in the benthic community. The total number of zoobenthos remained at the same level during last two decades (Figure 1), while slight increase of the number of Mollusca was noted (Figure 2).

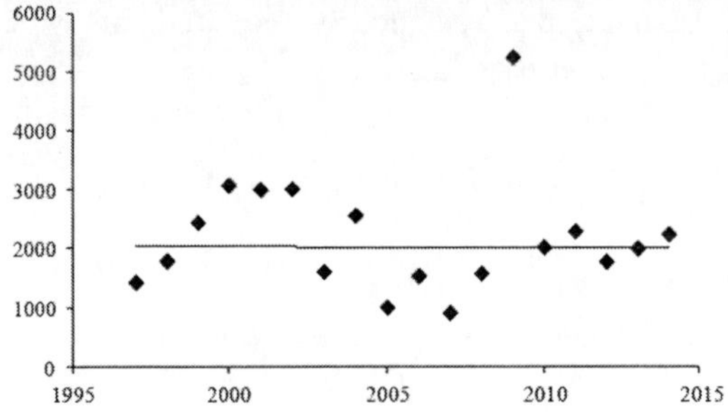

Fig. 1. Dynamics of the total number of zoobenthos, ind.M^{-2}

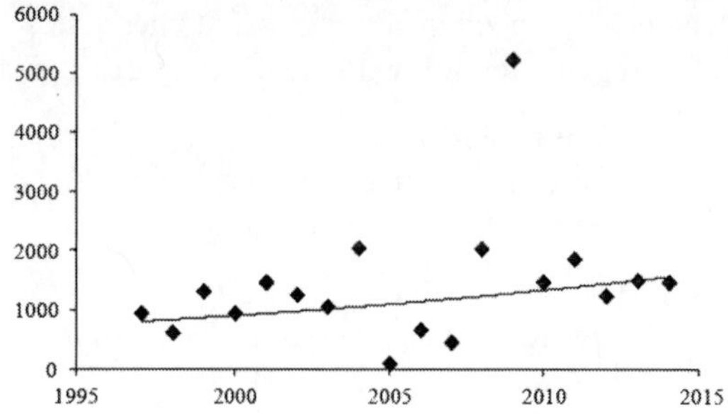

Fig. 2. Dynamics of the number of Mollusca,ind.M^{-2}

The relative number of oligochaetes changed from 0 to 100%. The greatest average values were noted also on the top sites of the reservoir, the smallest on the lower. The analysis of long-term data of the number of oligochaetes revealed a tendency of it decrease (Figure 3). This fact was already noted earlier [14, 15]. At the same time small, but steady increase of the number of polychaetes - the main competitors of oligochaetes were observed since 2005 (Figure 4).

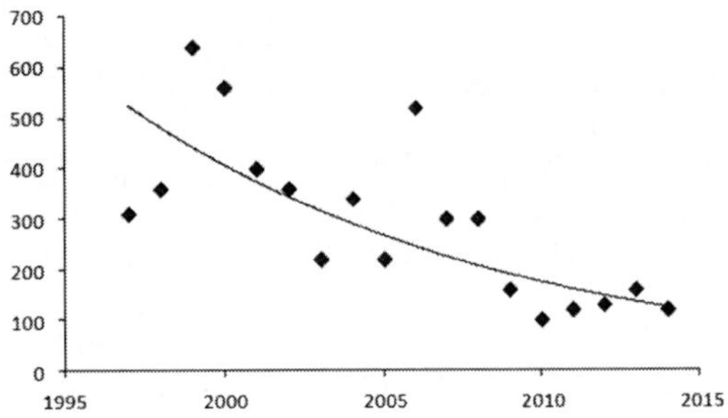

Fig. 3. Dynamics of number of Olygochaeta, ind.M^{-2}

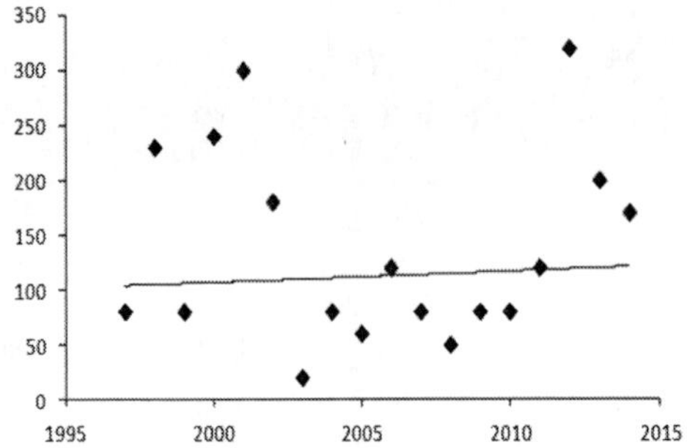

Fig. 4. Dynamics of number of Polychaeta, ind.M^{-2}

The quality of benthonic water determined by the biotic Woodiwiss index was characterized as polluted on the majority of investigated sites of the reservoir and only in area of the settlement Zaovrazhnye Karatay and the Tetyusha city as poorly polluted. The obtained in this study water quality assessment for the upper sites of the reservoir agree with the previously reported [16, 17].

The invariant condition of ground ecosystems on the majority of sites of the reservoir was characterized by elements of ecological regress. And only the lower sites of the Kuibyshev reservoir can be characterized as background (natural) on which ecological modulations are observed.

In case of increased level pollution, when the ecosystem moves from an equilibrium state to a crisis, the benthic communities simplify their structure by reducing the species diversity with a simultaneous increase in abundance with dominating of 1-2 species.

Conclusion

Based on results of chemical analysis the Kuibyshev reservoir in investigated area was characterized as dirty and very polluted. The reservoir ecosystem on a share of anthropogenic influence an ecosystem on the majority of sites was in an equilibrium state and only in a zone of influence of Zelenodolsk and Kazan cities – in transitional of equilibrium in crisis. The quality of benthonic water based on zoobenthos indicators on the same sites was estimated as polluted with elements of ecological regress. Other sites were characterized as poorly polluted in which anthropogenic tension with elements of ecological regress was observed.

Acknowledgement

This work was supported by a subsidy allocated to Kazan Federal University from Russian Foundation for Basic Research, project No 18-44-160027

References

[1] Seitzinger, SP, Mayorga, E., Bouwman, AF et al. 2010. Global Biogeochemical Cycles **24** 1-16.

[2] Jørgensen, SE, Löffler, H., Rast, W., Straškraba, M. 2005. Lake and Reservoir Management. 43-106.

[3] Rosenberg GS, Evlanov IA, Seleznev VA, Mineev AK, Selezneva AV, Shitikov VK 2011 Issues of ecological rationing and the development of a system for assessing the state of water bodies. *Proc. of the Joint Plenum*. (Moscow) 5-29.

[4] Levich AP, Bulgakov NG, Maksimov VN 2004 Theoretical and methodological basis of the technology of regional control of the environment according to environmental monitoring (Moscow: NIA-Nature). p. 271.

[5] Levich A 1994. *J. Reports of RAS*. (Moscow, Science) **337**(2) 280-282.

[6] Bulgakov, NG. 2002. J. Usp. Sovrem. Biol. **122**(2) 115-135

[7] Popchenko VI., Rosenberg GS, Shitikov V.K. 2015. Izvestiya of the Samara Scientific Center of the Russian Academy of Sciences **17** (4) 179-188.

[8] Bezmaternykh D.M. 2007. Zoobenthos as an indicator of the ecological state of aquatic ecosystems in Western Siberia Anal. Overview. Ecology series (Novosibirsk). **8** 87 p.
[9] Zinchenko T., Vykhristyuk LA, Shitikov V.K. 2000. Izvestiya of the Samara Scientific Center of the Russian Academy of Sciences. **2** (2) 223-233.
[10] Guide to hydrobiological monitoring of freshwater ecosystems. 1992. (St. Petersburg, Hydrometeoizdat). 318 p.
[11] Method of complex assessment of the degree of surface contamination by hydrochemical indicators. 2002 (Rostov-on-Don, Hydrometeoizdat) 55 p.
[12] Abakumov VA. 1991. Proc. Int. Symp. (Leningrad, Hydrometeoizdat). 18-40.
[13] Recommendations. Assessment of the risk of anthropogenic impact of priority pollutants on the surface waters of the land. 2006 (Moscow Meteoagency of Roshydromet) 25 p.
[14] Kuibyshev Reservoir: Scientific and Information Handbook 2008 (Togliatti, IEEB RAS) 123 p.
[15] Stepanova N.Yu., Kondratieva TA, Latypova V.Z. 2009. Bulletin of the Peoples' Friendship University of Russia. Series "Ecology and life safety" **4** 22 – 27.
[16] Kalaida M.L. 2003. Ecological assessment of the Kuibyshev reservoir in conditions of anthropogenic impact. (Kazan) 135 p.
[17] Melnikova A.V., Ilyasova A.R. 2017. Monitoring of the quality of the Upper Kuybyshev Reservoir water in terms of zoobenthos. Materials of the International Scientific and Practical Conference "Complex Problems of Technospheric Security" **4** 10-11.

IOP Conf. Series: Earth and Environmental Science **288** (2019) 012054 doi:10.1088/1755-1315/288/1/012054

Biochemical effects of molecular hydrogen in aqueous systems

A A Lapin [1], A A Kalayda[1], S D Filippov[2], V N Zelenkov[3]

[1]"Water bioresources and aquaculture" department, Kazan state power engineering university, Russia, 420066 Kazan, Krasnoselskaya, 51

[2]JSC AKVALON, Russia, 454012 Chelyabinsk, Kopeiskoe shosse, 5

[3]All-Russian Research Institute of Vegetable Growing – branch of the Federal State Budget Scientific Institution "Federal Scientific Center for Vegetable Growing", Russia, 140153 Moscow region, Ramensky district, v. Vereya, building 500

Andrey-8004@mail.ru

Abstract. Activated water - such water, in which the structural network of hydrogen bonds loosens, water molecules acquire additional degrees of freedom, which facilitates the assimilation of such activated water by the cells of living organisms and accelerates the removal of biological slags. The essence of water activation is the destruction of cluster structures to saturate water with monomolecules. The cluster of activated water contains 5-6 molecules (ordinary water 13-16 molecules). Such water is considered more active in terms of biophysical and biological indicators. Water, activated in any way, has a high fluidity (has a low surface tension) and a dissolving power. Water with micronized clusters has higher reactive properties, better penetrates through biological membranes, is quickly excreted from the body. Effectively replaces and supplements all kinds of cleansing food supplements and physiotherapeutic procedures. Used for complex purification of the body: neutralization and excretion of slags, toxins, carcinogens and radionuclides. The paper considers experimental data on the determination of the total antioxidant activity of water systems saturated with hydrogen using the coulometric method of analysis, with the water having the highest increase in activity up to 20 times. We studied the functional drink "Arkhyz + Antioxidant = Zhivitsa", all the ingredients that make up it have a natural origin. Saturation of the drink with hydrogen leads to an insignificant increase in the total antioxidant activity (by 16.5% rel.). The presence of active organic compounds in water leads to a decrease in the effect of hydrogen on the total antioxidant activity, probably due to a change in the structure of its aggregates consisting of a different number of molecules.

The first problem is the activation of water and water systems

Hydrogen can serve as a replenished source of energy, but the industrial methods for obtaining it are not environmentally friendly. They are replaced by biological methods with the help of hydrogenases - enzymes, which are responsible for the restoration of protons and the generation of molecular hydrogen in aqueous systems without the participation of platinum, palladium and other metals [1].

Water with negative oxidation-reduction potential (ORP) has reducing properties, and with positive ORP oxidizing properties. Both varieties of water are usually called "activated". This has a profound biological meaning, reflecting the role of changes in the degree of structuring of water, both within

IOP Conf. Series: Earth and Environmental Science **288** (2019) 012054 doi:10.1088/1755-1315/288/1/012054

living organisms and the water of the external environment, in the processes of organization and vital activity of living systems, besides, water creates many problems in industry, energy, ecology, production and storage of food products and other areas of human life [2]. The main technology of water activation is electrolysis. The catholyte and anolyte have a negative and a large positive ORP, respectively. It is generally believed that the processes leading to both a decrease and an increase in ORP occur in the region of the boundary with the surface of the electrodes, where high electric field intensities. A number of studies have been devoted to the investigation of mechanisms of water activation [3].

In the process of electrolysis, molecular hydrogen is formed at the cathode, which is a reducing agent. Therefore, it is natural to assume that the negative value of ORP of catholyte is due to the dissolution of hydrogen in water. It was noted in [4] that saturation of water with hydrogen can not lead to the experimentally observed value of ORP of about minus 600 mV. In the analysis of the process, the reaction 2H + + 2e-H_2 was considered as the mechanism of action of hydrogen, the standard potential of which with respect to the normal hydrogen element (NE) at pH = 7 is minus 414 mV [3]. In this paper it is shown that when the water is saturated with hydrogen (concentration ~ 8 10-4 mol/l), the ORP values can be minus (500 ÷ 700) mV. The reaction that will provide such a potential, we can consider the process: $2H_2O + 2e^- \leftrightarrow H_2 + 2OH^-$ for which the standard potential is relatively n.h.e. is minus 828 mV. Thus, the ORP value achievable by electrolysis can be explained by dissolving hydrogen and diffusing it throughout the entire volume of the liquid.

Hydrogen (food additive E949), finds application in the food industry as a propellant, which helps to prevent oxidation and spoilage of products. As a food additive, it is allowed in the Russian Federation and the EU, but it is banned in the US, Australia and New Zealand. Hydrogen is non-toxic and its use in food production is considered safe. Today in science and medicine it is known that one of the most important parameters of water is its ORP. When ordinary drinking water penetrates the tissues of the human body, it takes electrons away from cells and tissues that consist of water by 80-90%. As a result, the biological structures of the body undergo oxidative destruction. So the body wears out, grows old, vital organs lose their function. These negative processes can be slowed down if in the body with a drink and food will receive a fluid that has the properties of the internal environment of the body, that is, has protective restorative properties. This is confirmed by numerous studies in specialized scientific centers in Russia and abroad. If the liquid that enters the body, in particular, drinking water has an ORP close to the ORP value of the internal environment of the human body, then the "vital energy of the organism" is assimilated, since it has the best biological compatibility with the human body in this parameter. For the treatment of water, aqueous solutions of salts, juices and other beverages, various devices are used to change their oxidation and reduction properties, including by saturation with hydrogen to produce a liquid medium with a negative oxidation-reduction potential [5].

The second problem is the analysis of aqueous systems saturated with hydrogen

To study aqueous systems saturated with hydrogen was used technique of determination of the total antioxidant capacity (DTAC) [4] of the samples in mg Routine (Ru) per 1 dm^3. The measurement was carried out at rest and with stirring by means of a magnetic stirrer or by passing air. We estimated the potential populations of clusters according to their DTAC in terms of the standard sample in mg of routine (Ru) per 1 dm^3 through the modal value (mode Mo) [6].

For the research, drinking cold water was used, which was pre-treated by municipal services and met the requirements of SanPiN 2.1.4.1074-01 "Drinking water". The water was purified through a household reverse osmosis filter Atoll, reverse osmosis systems manufactured in Russia (according to TU 3697-009-18261557-03) and the USA (NSF / ANSI Standard 058). Purified water has the following parameters: oxidation-reduction potential from +200 to +343 mV, total mineralization from 25 to 130 mg/l, pH from 6.9 to 8.3. Saturation of the samples with hydrogen was carried out under pressure in standard polyethylene terephthalate bottles with a volume of 0.5 dm^3 using a device for aerating and sealing liquid contained in a container [7].

The hydrogen saturated samples of the functional beverage "Arkhyz + Antioxidant = Zhivitsa" were studied, which is produced by CJSC Visma-Arkhyz.

To test the technique, we conducted studies of the effect of hydrogen on the antioxidant activity of water and a food drink

The content of dissolved hydrogen in water can be from 1.6 to 10 mg/l and depends on the pressure. The ORP of water in this case is from minus 600 to minus 50 mV, but the total mineralization of water and pH does not change. The solution does not lose substantially its properties for 30 days, which allows it to be transported and stored. When the solution is sealed in glass containers, the storage life is up to one year [8].

Figure 1 shows that the fall of the DTAC of water activated by hydrogen when the sample is standing in a 100 dm^3 polyethylene measuring cup in open air is described by the linear equation $Y= -1{,}697x + 86{,}805$, for $R^2 = 0{,}970$. According to the above equation, DTAC decreases by 47% of rel. in 25 minutes. In the control sample of the initial water, the DTAC of the original water sample twists within the measurement error and averages 4.28 ± 0.36 mg Ru per 1 dm^3, a similar effect was observed by us when studying the effect of temperature on the DTAC water [9].

While stirring the activated water with a magnetic stirrer, the DTAC falls to the maximum value of the control sample (4.28 ± 0.36 mg Ru per 1 dm^3) in 5.7 minutes and is described by equation $Y= -8{,}214X + 49{,}948$, for $R^2 = 0{,}953$, and when air is passed through it for 9.6 minutes ($Y= -5{,}084X + 52{,}53$, for $R^2 = 0{,}965$). We studied the functional drink "Arkhyz + Antioxidant = Zhivitsa", all the ingredients that make up it have a natural origin. And most importantly - this useful energy drink does not contain caffeine and taurine, which, if over-consumed, have a negative effect on health. The basis of the drink is the water "Arkhyz" with the addition of the biologically active additive "Flavomix-R" (dihydroquercetin of high purity), the natural flavoring (Limon-Lime), the natural color "Sugar Colors" E 151a, vitamin C (ascorbic acid). Due to the BAA, the beverage exhibits a DTAC, twisting between values of 268.50 and 304.78, an average value of 286.64 ± 18.14 mg Ru per 1 dm^3. Saturation of the drink with hydrogen leads to an insignificant increase (by 16.5% rel.) of the DTAC to 333.81 mg Ru per 1 dm^3.

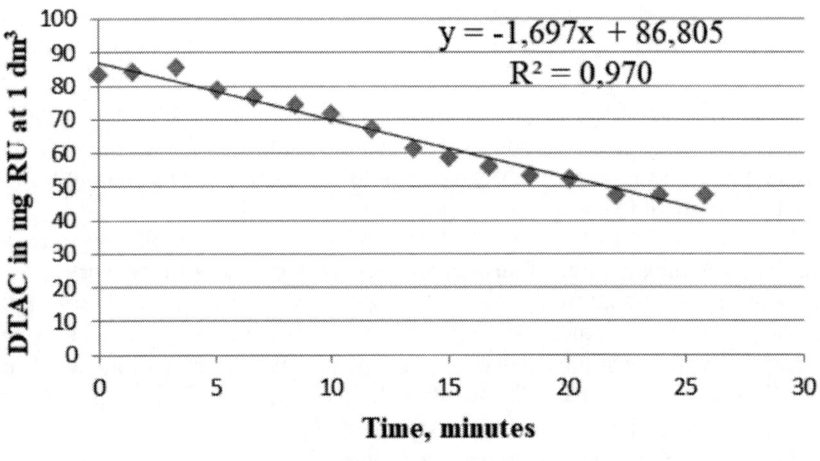

Figure 1. The fall of the total antioxidant capacity of hydrogen-activated water in a 100 dm^3 polyethylene measuring cup in air in open form without stirring.

References

[1] Wu C H, McTerman P M, Walter M E and Adams M W 2015 Production and application of soluble hydrogenase from *Pyrococcus furiosus*. *Archaea*. Art. ID912582

[2] Lapin A A and Chugunov U V 2015 Influence of the nature of water nanoclusters on antioxidant activity *V International "From nanostructures, nanomaterials and nanotechnologies to the nanoindustry" Theses of reports* (Izhevsk: Pub. house IzhSTU named after M T Kalashnikov) pp 109-10

[3] Aristova N A, Berkutov N A and Piskarev I M 2012 Storage of water with negative oxidation-reduction potential Water: chemistry and ecology **7** 81-3

[4] Petrushanko I U and Lobyshev V I 2004 Physico-chemical properties of aqueous solutions obtained in a membrane electrolyzer *Biophysics* **49 (1)** 22-31

[5] Piskarev I M, Tugolukov S N, Milyavsky M A and Volkov L S 2007 Device for obtaining a liquid medium with a negative redox potential by saturation with hydrogen *Pat. RF 71332 C02F*

[6] Lapin A A, Chugunov U V and Filippov S D 2015 The total antioxidant capacity of water systems saturated with hydrogen *Butlerov Communications* **44 (12)** 61-6. ROI: jbc-01/15-44-12-61.

[7] Filippov S D 2001 Device for aerating and sealing liquid contained in a container *Pat. RF 2167560. A23L2/00*

[8] Filippov S D 2015 Hydrogen cocktail - a means to stimulate lymphatic drainage, the way it is produced *Pat. RF 2557974. A61P43, A61K33*

[9] Lapin A A and Chugunov U V 2012 Effect of temperature on the antioxidant activity of water *Butlerov Communications* **30 (6)** 113-9

Sustainable water use of the Kuibyshev reservoir as an hydropower facility

E A Minakova[1], A P Shlichkov [2], V Z Latipova[1]

[1]Kazan Federal University, Institute of Fundamental Medicine and Biology, Kazan 420008, Kremlyovskaya street 18

[2]Institute of Ecology and Mineral Sciences of the Republic of Tatarstan, Kazan 420087, Daurskaya street 28, Russian Federation

ekologyhel@mail.ru

Abstract. Sustainable water use in a broad sense is a combination of all forms and uses of water resources that directly and indirectly affect the formation of the water fund. When operating hydropower facilities, it is necessary to comply with the requirements for the rational use and protection of water bodies. The article presents the results of the assessment of the specificity of water consumption and water disposal in the Kuibyshev reservoir within the Republic of Tajikistan, as well as the calculation and assessment of the main trends in water reservoir quality dynamics. Calculation of the index of the combinatory water pollution index during the period 2009-2017 yrs allowed to reveal a moderate tendency of its decrease within the limits of the 4th class of quality from the upper reaches to the dam of the Kuibyshev reservoir.

Introduction

In Russia, the concept of sustainable water use began to form in 1997. Гn accordance with the Concept of the transition of the Russian Federation to sustainable development, legally fixed by the Presidential Decree of 1.01.1996. № 440. During these years the scientific literature actively discusses the issues of sustainable water use, the draft Concept of the State Policy of Sustainable Water Use in the Russian Federation (RF). Ensuring sustainable water use is the highest priority in the field of environmental protection and rational nature management, as reflected in the Law of the RF On Environmental Protection (2002), the Water Code of the RF (2006), and the Environmental Doctrine of the RF. The Russian Government adopted the Water Strategy of the Russian Federation for the period up to 2020 adopted by the Government of the RF dated August 27, 2009 N 1235-r. Within the framework of this program, work is also being done to improve the regulatory legal regulation of rational use of water resources, reduce the discharge of contaminants in sewage, economic mechanisms.

In general, the structure of water consumption in the Russian Federation is as follows: production needs are 58%, domestic and drinking needs 20%, irrigation 14%, agricultural water supply 2%, other needs 6%. Thus, industry is the main water consumer. Within this sector, the main water-consuming industry is energy (75% of total industrial water consumption), followed by ferrous and non-ferrous metallurgy (6%, pulp and paper industry (5%), chemical and food industries (5%), engineering (3%) [1].

On the territory of the RF, about 3% of water reserves, or approximately 90 km3, are annually used for water abstraction, extraction and transportation to places of use. However, for individual basins, the ratio of the fence to the reserves is significantly differentiated. A third of all water abstraction and more than 40% of the total wastewater discharge in the RF is carried out in the basin of the river Volga. Virtually all the watercourses of the basin of the river Volga is subject to anthropogenic impact, the average annual concentrations of many pollutants exceed the maximum permissible levels, and the water quality of most of them does not meet regulatory requirements. In addition, the contribution of the anthropogenic load to the formation of the quality of the water resources is already commensurate with the natural factors [2].

The Kuibyshev Reservoir is the largest in Eurasia and the third largest in the world after the African Lake Volta (Ghana) and the North American Reservoir Smallwood (Canada), was established in 1955-1957 yrs dam of the Zhiguli hydroelectric power station (HPP) and is part of the Unified Energy System of the European part of the Russian Federation. The idea of energy use of the Volga on the site of the Samara Luke proposed by G.M. Krzhizhanovsky and developed by engineer K.V. Bogoyavlensky and Professor A.V. Chaplygin.

The Kuibyshev reservoir is a seasonal regulation water basin and is used in the interests of industry, energy, drinking and domestic water supply, health, agriculture and forestry, mining, transport, recreation, construction, fire safety, etc. [3].

The main contribution to the pollution of the waters of the Kuibyshev reservoir is made by the transit transfer of pollutants from the upper regions of the RF, which is mainly carried out along its bed part. A significant contribution to the pollution of the Kuibyshev reservoir waters is caused by discharges of insufficiently treated sewage from industrial and communal enterprises adjacent to the water area of the reservoir [4, 5] and diffuse runoff. In addition, a significant contribution to the pollution of water bodies is made by the geological environment and aerotechnogenic pollution. And this is far from a complete list of the most significant factors that cause the pollution of the water resources of the Kuibyshev reservoir and its tributaries. Studies conducted by the Institute of Ecology of the Volga Basin of the Russian Academy of Sciences in 1997-2010. on the Middle and Lower Volga [6], showed that, despite the decrease in anthropogenic pressure over the past decade, the quality of the Volga water remains unsatisfactory. For a number of hydrochemical indicators, the water of the Kuibyshev reservoir does not meet the regulatory requirements for fishery water reservoirs.

In connection with the foregoing, the purpose of this work was to assess the specificity of water consumption and wastewater in the Kuibyshev reservoir within the Republic of Tatarstan (RT), as well as the calculation and assessment of dynamics of water quality in the reservoir.

Materials and Methods

The work uses data from observations of the quality of the waters of the Kuibyshev Reservoir, carried out by FGBU "Sredvolgavodkhoz" since 2009. The work on the sampling of water and hydrochemical analysis is carried out by the hydrochemical laboratory of the Svervolgavodkhoz FGBU in the main phases of the water regime: winter low water, high water and recession), summer low water, in the autumn before freezing, as well as during the passage of rainwater. In 2009, sampling was carried out at 8 points of observation, and the analysis of the selected samples was performed on 25 ingredients and water quality indicators. Starting in 2015, sampling was conducted at 12 points of observation, and the analysis of samples taken was performed on 45 ingredients and water quality indicators.

Calculation of the combinatory water pollution index (CWPI) was carried out to assess the quality of surface waters of the Kuibyshev Reservoir for the period from 2009 to 2017.

Indicators of water consumption and sanitation in the Republic of Tatarstan (RT) in the retrospect of years 1998 - 2014. performed according to the materials of state statistical reporting with the purpose of assessing the impact of indicators on the level of water pollution.

IOP Conf. Series: Earth and Environmental Science **288** (2019) 012055 doi:10.1088/1755-1315/288/1/012055

In order to obtain comparable information, the data processing of water pollution observations of the Kuibyshev reservoir was carried out using physical and statistical methods of analysis.

Results and Discussion

The dynamics of consumption of water from natural sources in RT is shown in Fig. 1. Analysis of Fig. 1. shows that the amount of water consumption from natural water bodies for the considered period decreased by 202,47 million m^3 or by 18,7%. During the period 1998-2009 yrs the decrease in water consumption from natural water bodies has been registered, and since 2010 there has been an increase. The maximum water consumption from natural water bodies was 1081,88 million m^3 in 1998 and the minimum in 2009 was 736,74 million m^3. On the one hand, the decrease in water withdrawal from natural water bodies was caused by a decrease in production volumes, and on the other hand introduction of technologies aimed at reducing water consumption at economic facilities.

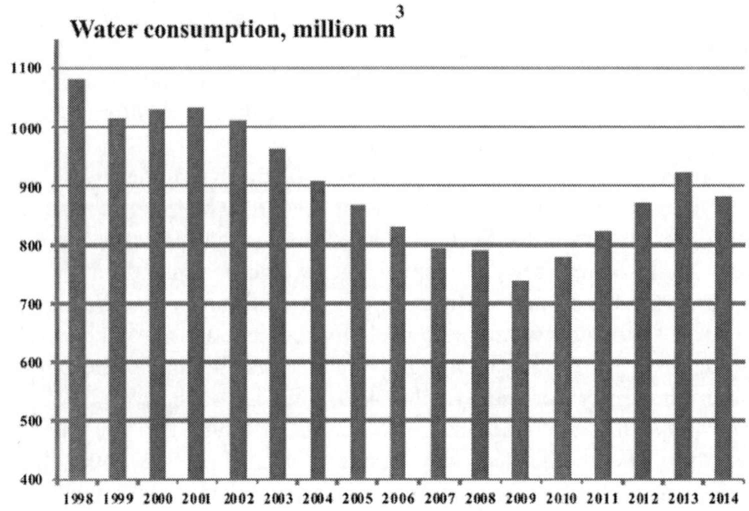

Fig. 1. Water consumption from natural water bodies, million m^3

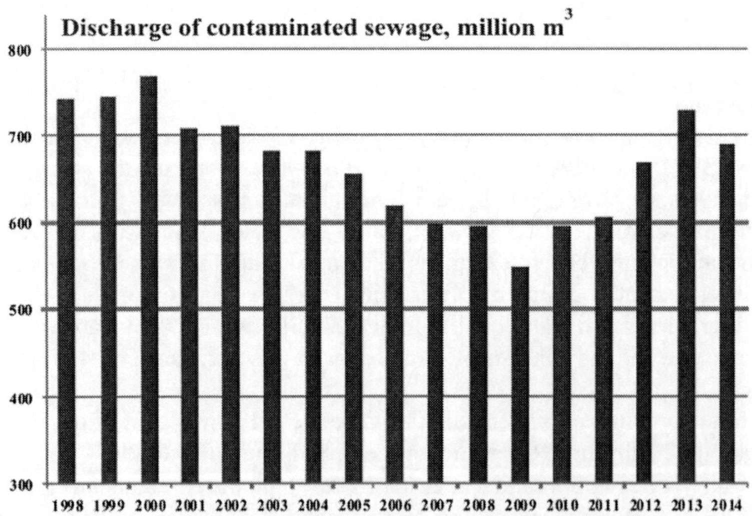

Fig. 2. Discharge of contaminated sewage into surface water bodies of the RT, million m^3
Discharge of contaminated sewage, million m^3

The dynamics of discharge of contaminated sewage into surface water bodies of RT is shown in Fig. 2. Analysis of Fig. 2. shows that the discharge of contaminated sewage into surface water bodies of the RT for the considered period decreased by 54,131 million m^3 or by 7,3%. In the period 2000 - 2009 yrs there is a decrease in discharges of polluted sewage into surface water bodies, and from 2010-2014 yrs there is an increase. Maximum discharges of contaminated sewage into surface water bodies were registered in 2000 and amounted to 768,262 million m^3, while the minimum in 2009 was 548,16 million m^3.

We obtained the regression equations between a water intake from natural water objects and dumping of the polluted sewage in superficial water objects of RT describe by the equation:

$$V_i = 0{,}57 \cdot V_d + 154{,}28, \tag{1}$$

where – V_i water intake from natural water bodies, V_d - discharge of contaminated sewage into surface water bodies of RT.

In the regression equation, the correlation coefficient between the values of V_i and V_d is r = 0.93. The correlation coefficient is significant at the level α = 0.05. The factor V_i determines 87% of the variance Vd. For the analyzed case R^2> 0.7, which indicates a high binding strength between the values of V_i and V_d.

Values of actual (V_a) and calculated (V_c) values of discharges of contaminated sewage into surface water bodies of RT using the regression equation (1) are shown in Fig. 3. As the analysis of Fig. 3, the actual and calculated values of discharges of contaminated sewage into the surface water bodies of the RT are well connected.

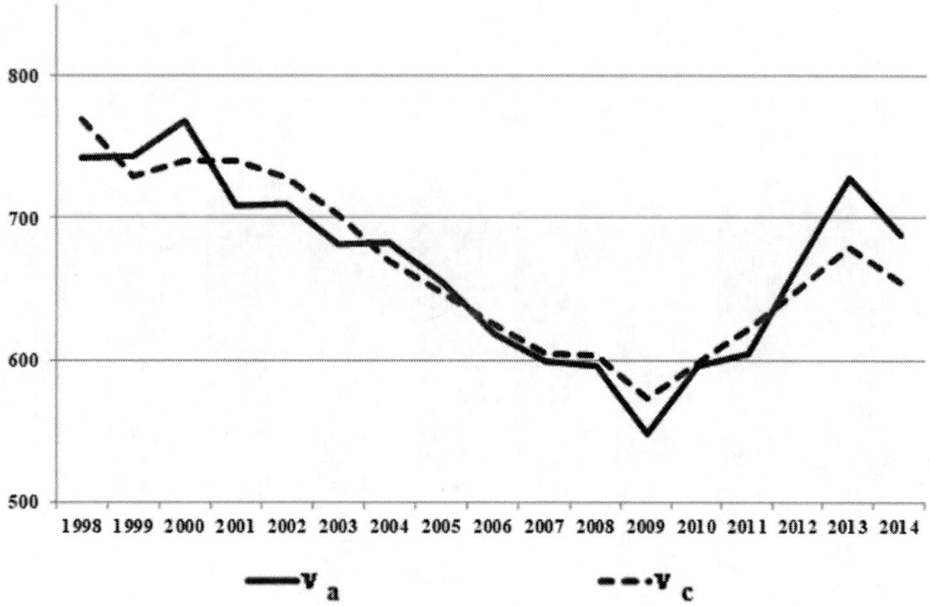

Fig. 3. Variability of actual (V_a) and calculated (V_c) values of discharges of contaminated sewage into surface water bodies of RT

The resulting regression equation can be used to predict the discharge of contaminated sewage into surface water bodies of RT.

Assessment of the quality of surface waters of the Kuibyshev Reservoir for the period from 2009 to 2017 yrs. is performed using the CWPI [7, 8]. It was revealed that the period under consideration of the value of the CWPI varied within the limits from 3,9 to 5,5 (Fig. 4). The maximum value CWPI of the – 5,5 was observed in 2010, and the minimum value – 3,9 in 2011. Analysis of Fig. 4 shows that in the period 2012-2017 yrs. there was a weak tendency to reduce the CWPI within one quality class. Substantial improvement (deterioration) in the quality of the waters of the Kuibyshev reservoir during the considered period was not revealed.

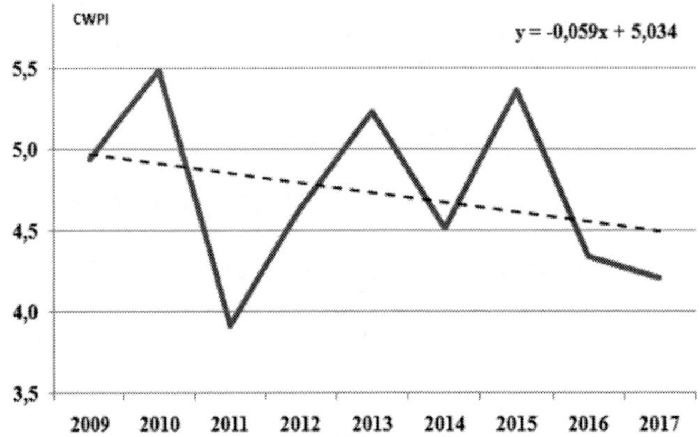

Fig. 4. Annual average dynamics and trend of surface water quality in the Kuibyshev reservoir (2009 – 2017 yrs)

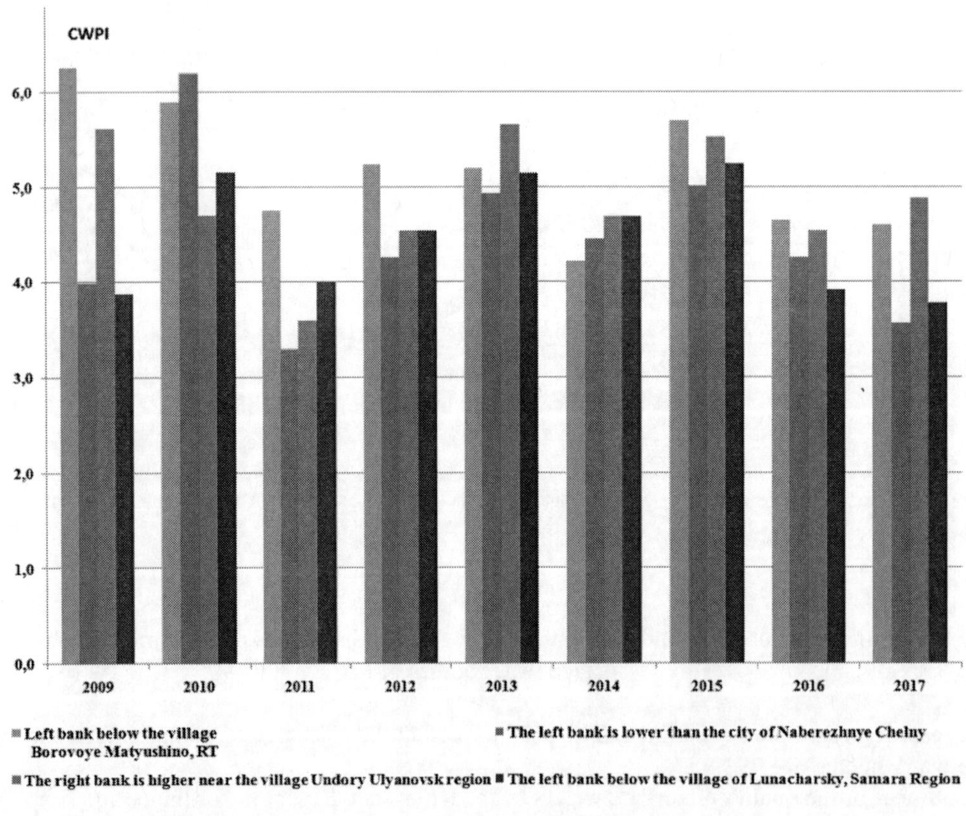

Fig. 5. Dynamics and trend of CWPI of waters along the water area of the Kuibyshev reservoir

Dynamics and trend of changes in the quality of surface waters over the water area of the Kuibyshev Reservoir for the period from 2012 to 2017 yrs. are shown in Fig. 5.

The analysis of Fig. 5 shows that in the period 2009-2017 yrs. there was a moderate decline in the CWPI within the 4 th grade class from the upper reaches to the dam of the Kuibyshev Reservoir.

During the period under review, the values of the CWPI varied between 4,4 and 5,2. The maximum value of the CWPI equal to 5,2 was observed on the left bank of the Kuibyshev reservoir below the settlement Borovoye Matyushino RT. The minimum value of the CWPI equal to 4,4 is registered on the left bank of the Kuibyshev reservoir below the city of Nabereznie Chelny.

Conclusions

The analysis of the received materials shows that there has been no significant improvement (deterioration) in the quality of the waters of the Kuibyshev Reservoir during the period under review. All changes in water quality, both in time and in the water area of the reservoir, were observed within the limits of grade 4, and the waters were still characterized as "dirty".

In order to improve the quality of water in the Kuibyshev reservoir and its tributaries, it is necessary to reduce the diffuse discharge through the purification of sewage and meltwater from the territory of large industrial complexes located in the catchment area, as well as the conduction and reforestation of water protection zones. In addition, it is necessary to continue work on further improvement of wastewater treatment of industrial and municipal enterprises that discharge to the Kuibyshev Reservoir and its tributaries.

References

[1] Bobylev S N and Khodzhaev A Sh 2003 *Economics of Nature Management Textbook* (Moscow) p 567

[2] *Quality of surface waters of the Russian Federation in 2010 Yearbook* 2011 (Rostov-on-Don: Hydrochemical Institute) p 552

[3] *Kuibyshev Reservoir: Environmental Aspects of Water Management* 2007 ed V Z Latypova and O P Ermolaev (Kazan: Publishing house Foliant) p 320

[4] Khublaryan M G and Moiseenko T I *Bulletin of the Russian Academy of Sci*, 2009 **79(5)** p. 403-410

[5] Pavlov D S, Rosenberg G S and Shatunovsky M I 2011 *Issues of ecological rationing and development of a system for assessing the state of water bodies* (Moscow: Association of Scientific Publications) p 196

[6] The Volga basin. Sustainable development: experience, problems, prospects 2011 ed G.S. Rosenberg (Moscow: Institute for Sustainable Development of the Public Chamber of the Russian Federation) p 104

[7] Latypova V Z, Yakovleva O.G and Semanov D A 2001 *Ecol chemistry* **10** 115 – 120

[8] Minakova E A, Mukhametshina E G, Shlychkov A P and Sushkova A V 2013 *Ecology and industry of Russia* **6** 39- 41

IOP Conf. Series: Earth and Environmental Science **288** (2019) 012056 doi:10.1088/1755-1315/288/1/012056

The role of the silver carp in increase fish productivity and the reducing the level of eutrophication of eutrophication of reservoirs of Kuban basin

G A Moskul, A V Abramchuk and N G Pashinova

Kuban state university, Russia, 350058 Krasnodar, Stavropolskaya 149

apilab@yandex.ru

Annotation. The data on the nutrition of silver carp and its role in reducing the level of eutrophication and increasing the fish productivity of the reservoirs of the Kuban basin are given.

Introduction

Silver carp (*Hypophthalmichthys molitrix* (Valenciennes, 1844) is the only representative of the ichthyofauna in reservoirs of Kuban, feeding on primary products (phytoplankton) and transforming it into valuable high-quality fish products.

Studies on the role of a silver carp – fitoplanktona in reservoirs of Kuban basin (Krasnodarskoe, Kryukovskoe and Varnavinskoe) ware carried out in 2000-2013. Until that time such studies on the role of the silver carp in reservoirs of Kuban basin have not been conducted. Therefore, our studies have great scientific and practical interest.

It should be noted that studies on the nutrition of fish has not been done recently. There are very few works devoted to this important issue in the literature.

Knowledge of fish nutrition is necessary to solve a lot of fisheries issues, namely: analysing the causes of fluctuations in the number and rate of growth, the establishment of the optimal fishing size, the development of measures for the most rational use of feed resources, increasing fish production and reducing the level of eutrophication in reservoirs, etc.

Materials and methods of studies

The study of fish nutrition was given according to the «Manual for studying fish nutrition in natural environment» [14] and «Manual for studying nutrition and food relations of fish in natural environment»[9]. Collecting and processing of hydrobiological samples (phytoplankton, zooplankton) was carried out according to generally accepted methods. 494 samples of phytoplankton and 84 zooplankton were collected and processed, as well as the content of the intestines of 357 silver carp of different ages was studied.

Results and discussions

It is known that silver carp in the diet is fitoplankton, but in case of lack of phytoplankton, it actively consumes detritus [3,4]. According to E. V. Borutsky [3], in spring and early summer the main component the silver carp food is detritus (over 90 %). The same data were obtained from fish living in Kuban estuaries [10]. At the same time, Omarov O. M. [12] notes that in the nutrition spectrum of the silver carp phytoplankton was occupied 95 %, the zooplankton - 3 %, and detritus – 2% from the total food.

The content of phytoplankton in the intestinal tract of silver carp ranges from 45,9 to 59,0%, on average 52,43±0,82 % of the mass of the food lump. The maximum content of phytoplankton in the food lump of the silver carp is observed in July and August, which coincides with the maximum biomass of phytoplankton in this period.

Detritus is an essential component of nutrition of different age groups of silver carp.

Detritus is found in the intestines of all the studied fish from 40,2% to 53,8%, on average 46,95±0,86 % of the mass of the food lump, including mineral particles, in the intestines of the fish was noted in spring, when the phytoplankton biomass was minimal (table 1).

Table 1-Seasonal changes in the abundance and biomass of phytoplankton in reservoirs of Kuban basin

Reservoir	Spring	Summer	Autumn	Average for season	n
	M ± m	M ± m	M ± m	M ± m	
Krasnodarskoe	3,20 ± 0,42	26,63 ± 1,34	17,63 ± 0, 76	15,82 ± 0,76	265
	3,64 ± 0,36	23,84 ± 0,98	12,21 ± 1,15	13,23 ± 0,48	
Krukovskoye	10,43 ± 0,76	29,36 ± 0,67	9,54 ± 1,23	19,77 ± 0,56	112
	10,10 ± 0,98	27,70±1,26	13,80 ± 1,76	17,20 ± 0,21	
Varnavenskoe	12,75 ± 1,21	24,99 ±1,43	15,78 ±1,17	17,84 ± 0,36	117
	9,30 ± 0,76	26,80 ±1,54	13,00 ± 1,98	16,36 ± 0,78	

Note: the top row is numbers, thousand cells/l; bottom line-biomass, g / m3

Regarding the selective characteristics of the silver carp to food organisms, there are conflicting opinions: some authors [5,6] note that he has no selective ability to feed, that in his intestine are dominated by those forms of phytoplankton, which have a mass distribution in the reservoir; other authors [1,15] note the obvious selection of certain species and groups of phytoplankton by the silver carp.

Our research has shown that silver carp prefers euglenoid (Euqlena acus Ehr., E. viridis Ehr., E. oxyuris Schmarda, Phacus acuminatus Stokes, Ph. longicauda (Ehr) Duj, Trachelomonas intermedia Dang, T. nigra Smir, etc.), protococcus (Actinastrum hantzchii Lagerh, Actinastrum sp. Laqerh, Ankistrodesmus Scenedesmus acominatus Chod. S. apiculatus Chod, Tetrastum glabrum (Roll) Ahlstr, etc.), volvox (Chlamydomonas ehrenbergii Gorosch, Ch. Gleucystoformis Dill. Pondorina morum(Mull) Bory, Pediastrum boryanum Meneg, P. duplex Meyen et al.), diatoms (Asterionella Formosa Hass, Bacillaria paradox Gmelin, Cocconeis pediculus Ehr, Cyclotella comta (Ehr.) Kutz, Navicula sp. Bory, Diatoma anceps (Ehr.) Kirchn, Stephanodiscus binderanus Kutz.) and other algae (table 2).

Perogitive, blue-green, golden, yellow-golden, disminuye, apparently, are junk food for the silver carp. Blue-green and ductal algae in the plankton are almost at the same level (26,65±0,42% and 26,82±0,72%), however, the index of selection of ductal is 1,33, and blue-green algae is 0,54. In the food lump of the silver carp, the ductus is 2,46 times higher than the blue-green.

According to the index of selection, blue-green algae are in the fifth place (0.54%). But in general, in the food lump of the silver carp all groups of algae which are part of the phytoplankton of the reservoirs are found.

The intensity of silver carp nutrition in all age groups during the growing season (April-October) is at a fairly high level (table.1). The average index of intestinal filling of silver carp is 284±0,93%. High index of intestinal tract filling indicate high feeding of silver carp in the reservoirs of Kuban basin. This ensures the intensive growth of fish of all age groups (table 3).

IOP Conf. Series: Earth and Environmental Science 288 (2019) 012056 doi:10.1088/1755-1315/288/1/012056

Table 2 – The balance of groups of algae in the plankton and in the diet of the silver carp, %

The divisions of algae	In plankton	In the food	The index of selection
Green, including			
Ductal	26,82±0,72	35,77±0,61	1,33
Volvox	9,84±0,34	11,27±0,34	1,14
Desmidium	1,25±0,24	0,39±0,02	0,31
Diatom	16,32±0,61	17,04±0,92	1,04
Euglenid	14,81±0,73	20,67±0,98	1,40
Perogitive	1,56±0,24	0,16±0,05	0,10
Blue-green	26,65±0,42	14,55±0,32	0,54
Golden	1,92±0,01	0,12±0,01	0,06
Yellow-green	0,83±0,01	0,03±0,01	0,04

Table 3 - Linear and weight growth of silver carp in the reservoirs of Kuban basin

Age	Reservoir						Average index		n
	Krasnodarskoe		Krukovskoe		Varnavskoe				
	cm	*gr*	*cm*	*gr*	*cm*	*gr*	*cm*	*gr*	
1+	35,4	680	30,5	520	31,8	495	32,5	565,0	250
2+	52,2	2240	44,2	1950	40,5	1650	45,6	1946,0	220
3+	63,5	3670	52,5	1960	48,4	2670	54,8	2766,6	180
4+	65,4	4500	56,8	3280	57,6	4100	59,9	3960,0	148
5+	69,3	5900	62,5	3850	61,5	5250	64,4	5000,0	154
6+	72,6	7600	66,4	5140	69,8	7480	69,6	7843,3	167
7+	75,5	8900	69,5	7150	74,5	8950	73,2	8333,3	123
8+	79,4	10100	71,6	8850	-	-	75,5	9475,0	80
9+	82,2	11300	-	-	-	-	82,2	11300,0	36
10+	84,8	12450	-	-	-	-	84,8	12450,0	28

As it can be seen from the data in table 3, the annual increase of weight of silver carp ranges from 490,0 to 2843,3 gr. The intensive growth of silver carp in the reservoirs of Kuban basin is due to good conditions of feeding (high food supply, long growing season, more than 200 days with water temperature of 20-25 °C and the lack of competitors in the nutrition).

However, despite the high rates of fish forage and growth, the fish productivity of the reservoirs is low and in most cases does not meet their potential opportunities.

Over the past 10 years, the average fish productivity of the reservoirs of Kuban basin is on average 1.6 kg/ha, with a fluctuation of 0.6 to 2,5 kg/ha, and for herbivores (white and motley carp) – 0,2 kg/ha. This fish production is very low and can not serve as a reference point for south reservoirs of the country. In this regard, having long-term data on the number and biomass of phytoplankton, we have attempted to determine the potential fish production of reservoirs for feed resources.

As it is known, the yield of fish products is determined not by the residual biomass of forage organisms, but by the value of their annual (seasonal) production. Therefore, it is important to study defined production values of individual species and groups of aquatic organisms that play a major role in the food of fish.

To determine the production of phytoplankton in the reservoirs of the Kuban basin, we used the available in the literature P/B coefficients, which range from 40-71 for the Khakhovskoe reservoir [16], to 350-400 generations per year for the world ocean [2]. For the Azov sea of V. G. Datsko [7] uses P/B-coeficient 340. In the Tsimlyanskoe reservoir [13] it was found that the biomass of algae is renewed during the growing season 240 times. This P/B - coefficient is used by I. Lapitsky [8] for calculation of phytoplankton production in the Tsimlyanskoe reservoir. To determine the production of

IOP Conf. Series: Earth and Environmental Science **288** (2019) 012056 doi:10.1088/1755-1315/288/1/012056

phytoplankton in reservoirs of Kuban basin, we used the minimum P/B-coefficient – 40, defined for the Khakovskoe reservoir for the summer period [16]. With average for the vegetation period of phytoplankton biomass and taking P/B-coefficient equal to 40, we obtained the value of phytoplankton production for the reservoirs of the Kuban basin (table 4).

Table 4 – Average annual biomass and production of phytoplankton in the reservoirs of Kuban basin

Reservoir	Biomass			Production P/B – 40	
	gr/m³	*kg/ha*	total, *m*	*kg/ha*	total, *m*
Krasnodarskoe	13,23±0,48	463,0	13891	18520	555660
Krukovskoe	17,20±0,21	481,6	1686	19264	67424
Varnavskoe	16,36±0,78	458,1	1603	18324	64120

The obtained data on the production of phytoplankton are indicative, but on their basis it is possible to approach the direct determination of potential fish production (table 5).

Table 5- Potential and commercial fish productivity in reservoirs of Kuban basin

Indicators	Reservoir		
	Krasnodarskoe	Krukovskoe	Varnavskoe
Phytoplankton production, kg / ha	18520	19264	18324
Product utilization rate (25%), kg / ha	4630	4816	4581
Feed ratio, u	30	30	30
Potential fish production, kg / ha	154	161	153
The fish return rate, %	15	20	20
Commercial fish production, kg / ha	23	32	31

The data of table 5 show that due to the rational use of natural resources of the reservoir (phytoplankton) can be obtained in average per hectare water areas of the Krasnodarskoe reservoir 23 kg/ha, Kryukovskoe – 32 kg/ha, Varnavinskoe – 31 kg/ha of fish products.

To obtain the above-mentioned fish products, it is necessary to carry out the annual stocking of the reservoirs by fingerlings or yearlings of silver carp from the calculation: Krasnodarskoe reservoir-85 copies/ha, Kryukovskoe reservoir – 60 specimens/ha, Varnavinskoe reservoir – 55 specimens / ha.

Fishing for carp should be start on the second or third year after stocking, when the individual weight of fish will reach more than 2 kg. For 5-7 years of fishing the fish population of the first stocking will be almost completely removed. The total potential catch of silver carp in Krasnodarskoe reservoir will be more than 700 tons, in Kryukovskoe – 84 tons, in Varnavinskoe – 77 tons.

In addition, annually the silver carp, utilizing a significant part of the production of phytoplankton (21,000 tons in Krasnodarskoe reservoir, 2520 tons in Kryukovskoe reservoir, 2310 tons in Varnavinskoe reservoir), significantly reduce the level of eutrophication of reservoirs.

Thus, the calculations carried out by us in relation to the reservoirs of Kuban basin are very approximate, and, nevertheless, they show quite real values of the planting material of the silver carp, which can ensure the removal of a significant part of the seston, including phytoplankton production and its transformation into valuable fish products.

Whereas there are no appropriate conditions for the natural reproduction of silver carp in the reservoirs of Kuban basin [11], their stocking is possible only on the basis of artificial breeding and cultivation of fish planting material in pond farms.

IOP Conf. Series: Earth and Environmental Science **288** (2019) 012056 doi:10.1088/1755-1315/288/1/012056

References

[1] Baltagi R A 1972 The diet of herbivorous fish in ponds constructed on the basis of the warm waters of the Mironovskaya TPP *Acclimatization of herbivorous fish in the waters of the USSR* 210 p.

[2] Bogorov V G 1974 Plankton of the World ocean 210 p.

[3] Borutsky E.V 1950 Materials on nutrition of Amur silver carp (*Hypophthalmichthys molitrix* Val.) *Proceedings of the Amur ichthyological expedition 1945-1949* P. 267–302.

[4] Borutsky E V 1973 Materials on nutrition of Amur silver carp (*Hypophthalmichthys molitrix* Val.) *Trophology of aquatic animals* P. 299–322.

[5] Vovk P S 1974 The possibility of using the silver carp *Hypophthalmichthys molitrix* (Val.) to increase fish productivity and reduce the level of eutrophication of the Dniepr reservoirs *The questions of ichthyology* P. 406–414.

[6] Vovk P S, Stetsenko L A, 1985 Phytophagous Fish in the ecosystem of reservoirs 135 p.

[7] Datsko V G 1959 Organic material in the waters of southern seas of the USSR 271 p.

[8] Lapitsky I I 1970 Directed formation of fish fauna and management of fish populations in the Tsimlyanskoe reservoir 279 p.

[9] Methodological guide for the study of food and food relations of fish under natural conditions 1974 250 p.

[10] Moskul G A 1977 Feeding characteristics of two-year-olds of white and motley silver carp in feeding estuaries of Krasnodar region *Hydrobiological journal* P. 45–50.

[11] Moskul G A 1994 Fishery development of the Krasnodarskoe reservoir 136 p.

[12] Omarov M O 1970 Daily diet of silver carp (*Hypophthalmichthys molitrix* Val.) *Questions of Ichthyology* P. 580 – 582.

[13] Pototskaya I V 1965 Intensity of development distribution and seasonal dynamics of phytoplankton in the Tsimlyanskoe reservoir *Tr. Volgograd branch of FSBSI"GosNIORH"* P. 58–74.

[14] Manual on the study of fish nutrition in natural conditions 1961 262 p.

[15] Savino R A 1968 Nutrition of silver carp (*Hypophthalmichthys molitrix* Val.) in ponds 24 p.

[16] Ceeb Y Y 1996 Feed resources Khakovkskoe reservoir *Questions of ichthyology* P. 319–335.

Influence of IR radiation on the process of phytoremediation of copper by eichhornia

Politaeva N.A.[1], Olshanskaya L.N.[2], Ovchinnikov F.A.[3], Sverguzova S.V.[1]

[1]Higher School of Biotechnology and Food Technology, Peter the Great St. Petersburg Polytechnic University, Saint Petersburg, Russia, 195251, Polytechnicheskaya Str., 29

[2]Engels Technological Institute (branch) of Saratov State Technical University named after Yuri Gagarin, Engels, Saratov region, Russia, 413100, Pl. Liberty, 17

[3]Belgorod State Technological University named after V.G.Shoukhov, Kostyukov str., 46., Belgorod, Russian Federation

politaevana1971@gmail.com

Abstract. The effect of infrared irradiation on the process of higher water plants phytoremediation of heavy metal ions is in the focus of this study. The influence of various factors (initial concentration of the solution, duration of the irradiation) on the extraction of copper by eichhornia is discussed.

Introduction

The ability of higher water plants (HWPs) to remove pollutants from water allows them to be used in the practice of cleaning industrial, domestic, sewage and natural waters, both in Russia and throughout the world. In many countries of the Americas, the mine water purification systems on the plantations of eichhornia and reeds are widely used. Facilities with reed vegetation for cleaning domestic sewage in the Netherlands, Japan, Norway, Australia and other countries are described in a number of works [1-3].

In Benton (USA), with a population of 4,700 people, since 1985, domestic wastewater has been treated in ponds with thickets of reeds and water hyacinth (eichhornia). In Ireland (Viljamestown), a system of joint cleaning of domestic water (72%) and surface runoff (28%) is successfully operated, designed in the form of three shallow lagoons, two of which are planted with reeds and cattail, and the third is a bioproduct with floating aquatic plants - eichhornia, lily and duckweed. In China, water hyacinth is used to purify film factory sewage from silver. It was found that the efficiency of water purification from silver, suspended solids, phosphorus and nitrogen compounds, respectively, was 100%, 91%, 53.9%, and 92.9%, with BOD and COD decreasing by 98.6% and 91 %. Australian scientists have developed a method for cleaning surface runoff from highways. The roads are not equipped with curbs; the flow is collected by filtration trenches filled with a depth of 0.8 m of gravel. At the bottom of the trench, prefabricated pipelines with a diameter of 150 mm are laid, which transport the drain for further purification into the bioplatto [1-3].

In Russia the Institute of Cytology and Genetics developed a technology for wastewater treatment using water hyacinth. This technology is widely used in Krasnodar region (a region with a higher average annual temperature). In spring and summer period this method is also used for sewage treatment plants in the city of Engels, Saratov region.

The main disadvantages of the method are seasonality and low cleaning efficiency. To increase efficiency the processes is stimulated by physical influences. In works [4-6], the effect of a magnetic field (MF) of various intensity and direction on the processes of phytoremediation was studied. It is shown [5,6] that a magnetic field strength of 2 kA/m exerts a stimulating effect on copper recovery processes, as a result of which the cleaning efficiency increases by ~8-12% and reaches 60% in a day in comparison with purification without the influence of MF.

It is also interesting to study other physical effects, for example influence of infrared (IR) radiation on phytoremediation processes.

The purpose of this work was to study the effect of the duration of IR irradiation on the extraction of copper by HWP by eichhornia plant.

Materials and Methods

To study the effect of IR on the process of phytoremediation, model solutions based on copper sulfate were prepared. For this purpose, 5 mg/l of $CuSO_4$ was added to conditionally clean sewage (after biological treatment in sewage treatment plants in the city of Engels, Saratov region). The processes of accumulation and extraction of copper cations from a solution of one liter were investigated. Higher aquatic plant eichhornia of the same mass (20 g) and maturation periods was placed into vessels. The plant was exposed to IR radiation for 1, 5 and 168 hours. Samples were analyzed after 1, 5, 24, 48, 72, 120 and 168 hours. For IR irradiation an artificial source of thermal radiation-an "infrared radiator (a "Minina" lamp with λ = 780-1400 nm) was used. The solution with eichhornia was located at a distance of 1 m from the source of IR irradiation. The scheme of the phytoremediation process under the action of IR irradiation is presented in Fig.1.

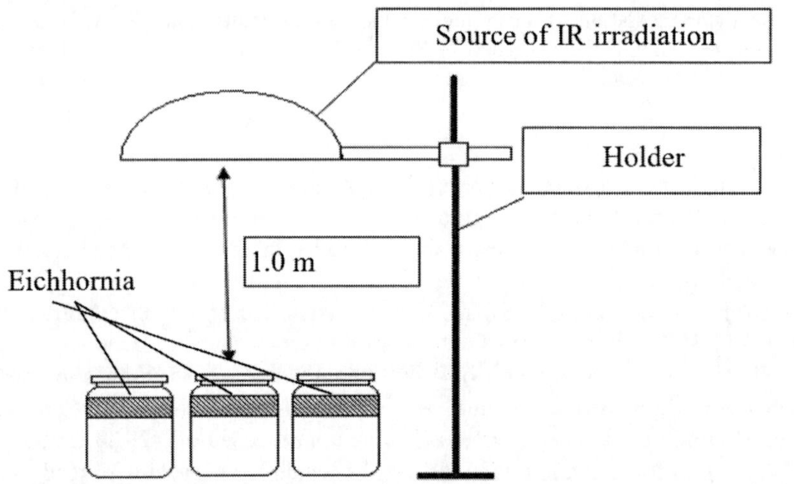

Figure 1: The scheme of the effect of IR radiation on the phytoremediation process

Analysis of water for the residual content of heavy metal ions (HMI) was carried out using photoelectrocolorimetric method (device KFK 3-01) in accordance with Russian State Standard "Drinking water. Methods for determining the mass concentration of copper." To determine the final concentration, 5 ml of the stock solution were placed in a 50 ml flask and adjusted to 50 ml with distilled water. 1-2 drops of aqueous hydrochloric acid HCl (1: 1) was added to the solution, then 1 ml of Rochelle salt (1:1), 5 ml of NH_3 ammonia (25%), 1 ml of starch (0.5%) and 5 ml of Sodium diethylenedithiocarbamate (0.1%). After the performed operations a yellow-brown solution was

IOP Conf. Series: Earth and Environmental Science **288** (2019) 012057 doi:10.1088/1755-1315/288/1/012057

obtained. A blue light filter with a wavelength λ = 430 nm and a cuvette with a working layer thickness of 50 ml was used in the photocolorimetric analysis.

Alongside with the analysis of changes in the concentration of copper in solutions of wastewater, the temperature and pH were fixed by means of an ionometric I-500 transducer of the firm "Aqualon".

Results and Discussion

Obtained results of the influence of IR radiation (Fig. 2) made it possible to establish that the processes of copper phytoremediation differ from those without IR. During the first hour, extraction of copper by eichhornia under the influence of IR occurs faster by 25%. Most likely this is due to the acceleration of bioelectrochemical processes in plant cells, and this is reflected in their growth, development and reproduction, with thermal exposure to IR irradiation. According to the rule of Vant-Hoff, the rate of chemical reactions increases by 2-3 times with an increase in temperature by 10 °C, and upon reaching the optimum, it begins to decrease. It is known that physiological processes in plant tissues begin at a temperature of +5 °C and are activated at + 10 °C and higher [7]. This fact is confirmed by data on the measurement of temperature. After the first hour, the temperature (T) of the solution in all three cases increased from Tini = 20 °C (ambient temperature) to ~ 1.5-1.8 °C, after 5 hours - by ~ 4 °C for samples of IR-1 and IR- 2, for which IR irradiation was continued. In the future, a slight increase in temperature was observed only for the IR-1 sample (Table 1).

Table 1. Dependence of T and pH from the time of IR exposure

Hours	T, temperature ° C			pH		
	IR-1	IR-2	IR-3	IR-1	IR-2	IR-3
1	21.8	21.9	21.8	6.94	6.94	6.94
5	24.3	24.5	22.0	6.83	6.83	6.85
12	24.7	25.0	22.3	6.97	6.73	6.93
24	25.0	22.4	22.4	6.8	6.98	6.88
72	24.8	20.6	20.7	6.96	7.09	7.11
120	23.8	20.1	20.2	6.98	7.32	7.14
168	24.2	22.0	21.6	7.05	7.30	7.07

In the Table IR-1 means exposure to the IR throughout the phytoremediation process, IR-2 means exposure to the IR for 5 hours, then the phytoremediation process proceeds without IR, IR-3 means exposure to UV for 1 hour, then the phytoremediation process proceeds without IR exposure.

After 5 hours, the excess copper is discarded by the plant into the solution. On the curves of the concentration change in time, a large peak appears for the samples under IR exposure (Fig. 2), and then there is a gradual decrease in concentration. This fact indicates that during this time, the maximum penetration of infrared radiation into the tissue occurs. It is known that infrared radiation penetrates tissues to a depth of 20 to 50 nm [7], therefore, surface layers are heated to a greater extent. The plant cell contains a large amount of water. The water in the cell is heated evenly, if necessary, the cell evaporates water, protecting plants from overheating by infrared rays [8].

According to the data obtained, penetration of infrared rays and heating of the plant occurs in the first hours of exposure (up to 5 hours). At the same time, the processes of photosynthesis and formation of chlorophyll are accelerated. Bright green color of the leaves of the eichornia is the evidence of it. Therefore, after the first hour, the plant cell, being adapted to conditions after IR exposure, is released of excessively absorbed copper ions (Fig. 2a). In the future, the effect of IR radiation inhibits the processes of phytoremediation and the absorption of copper by a plant without physical effects occurs more quickly (Fig. 2b).

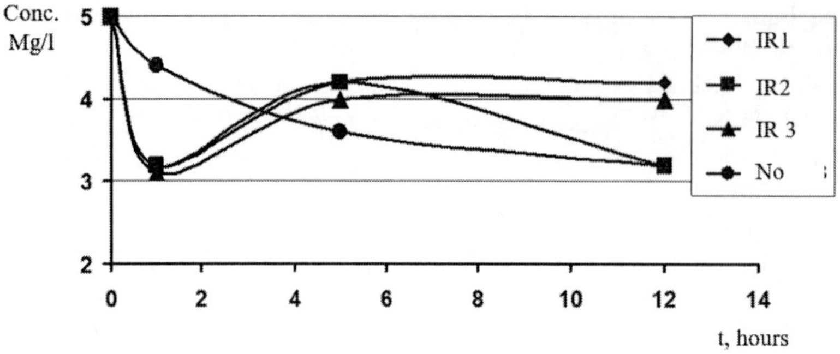

a)

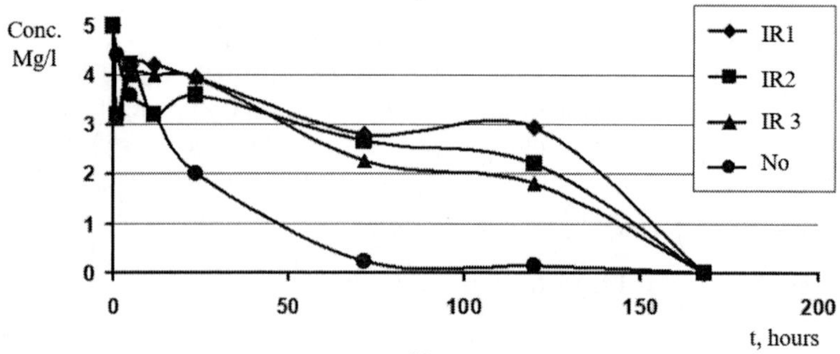

b)

Figure 2: Change in the concentration of Cu^{2+} in solution in the process of its extraction by eichhornia without and under the influence of IR radiation: a) at the initial time; b) after 168 hours

Measurement of the pH value of the solution in the process of phytoremediation without and under the influence of IR radiation showed that under the influence of infrared radiation a small increase in the pH of the medium occurred (Table 1). Most likely this may be due to the absorption of hydrogen ions by the H+ plant cell, accompanied by the displacement of copper ions from near the membrane space at the cell surface, and, as a consequence, by a decrease in the rate of extraction of Cu2+ cations. Hydrogen ions have a much smaller size of the ionic radius (r = 0.046 nm) as compared with copper ions (r = 0.128 nm) [9]. After the first hour, a DES is formed on the surface of the plant membrane, which is filled with hydrogen ions. The amount of hydroxyl ions (OH-) increases in the solution and the pH value increases (Fig. 3).

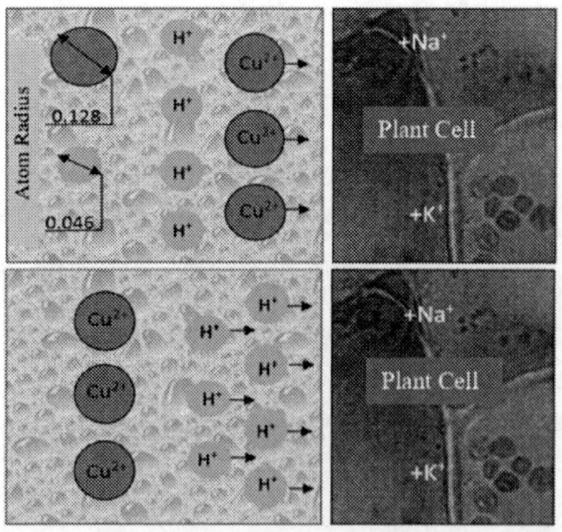

Figure 3: Arrangement of H^+ and Cu^{2+} ions in an aqueous solution of $CuSO_4$ at the border of the plant cell / solution

Conclusions

Conducted studies of the effect of IR irradiation on the process of phytoremediation of eichhornia of copper cations allowed us to establish that:

1. After IR irradiation of the HWP of eichhornia within 1 hour, the acceleration of the phytoremediation process is achieved by ~ 25%. Subsequent action leads to a retardation of the purification of effluents from copper ions. The effect obtained is most likely associated with the thermal effect at which the rate of the course of the physico-chemical reactions in the cells varies, since according to the Van't Hoff rule the rate of chemical reactions increases by 2-3 times with an increase in temperature by 10 °C, and decreases upon reaching the optimal state.
2. Influence of infrared radiation leads to a change in the physicochemical properties of water: an increase in temperature and a decrease in the pH of the solution, which in turn affects the growth and development of the plant cell and the processes of bioelectrochemical phytoremediation.

References

[1] Kravets V.V., Accountant L.B., Akolzin A.P., Higher aquatic vegetation as an element of wastewater treatment. *Ecology and industry of Russia*, **8**, 20-23, **(1999)**.

[2] Direnko A.A., Kotsar' E.M., Use of higher aquatic plants in the practice of wastewater treatment and surface runoff. *C.O.K.*, **5**, 5-8, **(2006)**.

[3] Bulat T.P., Petrushenko, V.V., Tovstukh N.I., Przhevalskaya L.D., The role of higher aquatic plants in the purification of reservoirs from industrial pollution, *Environment ecology of CIS countries*, **(2008)**.

[4] Olshanskaya L.N., Sobgaida N.A., Kolesnikova O.N., Effect of the force and direction of the magnetic field on the efficiency of wastewater treatment from copper and cadmium by phytoremediation, *Scientific research, nanosystems and resource-saving technologies: materials of the International Scientific and Practical Conf.* Belgorod, September 18-19, 2007 Belgorod: BSTU, 135-138, **(2007)**.

[5] Olshanskaya L.N., Sobgaida N.A., Tarushkina Yu.A., Stoyanov A.V., Influence of the magnetic field on the extraction of heavy metals from wastewater drains, *Chemical, Oil and Gas Engineering*, **8**, 41-44, **(2008)**.

[6] Olshanskaya L.N., Sobgaida N.A., Stoyanov A.V., Aleksandrova A.E., Influence of the force and direction of the magnetic field on the process of bioelectrochemical sorption of heavy metal ions, *Tatishchev Readings: Actual Problems of Science and Practice: Materials V1 Intern. Scientific and practical conference*, Togliatti: The Volga University V.N. Tatischeva (WUIT), 66-68, **(2009)**.

[7] Kuklev Yu. I. Physical ecology, Moscow: Higher School, 357p, **(2001)**.

[8] Kozhakaru A.F., The mechanism of energy-information impact of EMP of low intensity, *Problems of human security. Foundation and applied research: theses of the report 1 of the Russian Conference* Moscow, MGU, 21-22, **(1996)**.

[9] Rabinovich V.A., Khavin Z.Ya., Brief Chemical Handbook, Leningrad: Chemistry, 392 p, **(1978)**.

The role of reservoirs in changing the species composition of the ichthyofauna (on the example of the Kuibyshev reservoir)

F M Shakirova, V Z Latypova*, N Yu. Stepanova*, V G Tereshchenko, O K Anokhina, Yu.A Severov**

Tatar branch of Federal State Budgetary Scientific Establishment "Berg State Researh Institute on Lake and River Fisheries" (FSBSI "GosNIORH"); Tazi Gizzata str, 4, Kazan, 420111 Russia.

*Kazan Federal University, Kremlovskaia str, 18, 420008 Russia.

**Papanin Institute for Biology of Inland Waters Russian Academy of Sciences, Yaroslavl region, Borok, 152742 Russia.

shakirovafm@gmail.com

Abstract. The article presents data on the composition of the ichthyofauna of the Kuibyshev reservoir for the period (1957-2017). A list of fish in the reservoir is presented, compiled from recent studies, taking into account new taxonomic revisions and reports (13 units, 19 families, 47 genera and 59 species). During the existence of the reservoir, its ichthyofauna was replenished with 18 new species, 9 of which formed self-reproducing populations.

Introduction

Human economic activity, including the construction of canals and reservoirs, the activation of navigation, artificial introduction, cage culture of fish and many others lead to a change in the species composition of the ichthyofauna of water bodies and often to the expansion of the natural area of species. The Kuibyshev Reservoir is located in the industrial and densely populated area of the Middle Volga, belongs to multi-purpose reservoirs. The Kuibyshev Reservoir is the largest in Europe and in the Volga-Kama cascade, is regulating more than 90% of the water resources of the Volga basin [1]. The reservoir arose due to the regulation of the Volga River stream (1955-1957) by hydro-engineering structures of the Kuibyshev hydroelectric complex in the Zhiguli Mountains, which led to a significant change of the composition of the ichthyofauna.

In the first years after the Volga River damming, almost all the migratory fish species fell out the ichthyofauna composition [2]. During the sixty-year period of the existence of the reservoir, registered catches of these species occurred only in the first years after the creation of the reservoir, except of some cases [3]. Subsequent changes of the composition of the ichthyofauna took place due to acclimatization and fish-breeding activities conducted on the reservoir and accidental importation of species (incidental acclimatization), as well as penetration and dispersal of alien species both from the north and from the south, and still go on in present time [4].

The purpose of this work was to analyze the changes of the species composition of the ichthyofauna of the Kuibyshev Reservoir during period of its existence (1957-2017), the ways of alien species introducing and to compile a list of fish inhabiting in the reservoir, taking into account new taxonomic revisions and summaries.

Table 1. Fish and fish composition of the Middle Volga River and Kuibyshev Reservoir

Family	River Volga	Reservoir	
		1957-1969	1970-2017
Fam. Petromyzontidae Bonaparte, 1832 – Minnow	+	-	-
Number of species	2	-	-
Fam. Acipenseridae Bonaparte, 1831 – sturgeon	+	+	+
Number of species	5	3	2
Fam. Anguillidae Rafinesque, 1815 – river eels	-	+	+
Number of species	-	1	1
Fam. Clupeidae Cuvier, 1816 - herrings	+	+	+
Number of species	3	1	1
Fam. Cyprinidae Fleming, 1822 - carp	+	+	+
Number of species	25	24	27
1.1. Fam. Balitoridae Swainson, 1839 – baloric	+	+	+
1.2. Number of species	1	1	1
1.3. Fam. Cobitidae Swainson, 1838 – void	+	+	+
1.4. Number of species	2	2	3
1.5. Fam. Siluridae Cuvier, 1816 – catfish	+	+	+
1.6. Number of species	1	1	1
Fam. Esocidae Cuvier, 1816 – pike	+	+	+
Number of species	1	1	1
Fam. Osmeridae Regan, 1913 – smelt	-	+	+
Number of species	-	1	1
Fam. Coregonidae Cope, 1872 – whitefish	+	+	+
Number of species	1	2	2
Fam. Thymallidae Gill, 1884 – grayling	+	-	+
Number of species	1	-	1
Fam. Salmonidae Rafinesque, 1815 - salmonids	+	+	+
Number of species	3	2	2
1.7. Fam. Lotidae Bonaparte, 1837 – billets	+	+	+
1.8. Number of species	1	1	1
Fam. Gasterosteidae Bonaparte, 1831 – sticky	-	-	+
Number of species	-	-	3
1.9. Fam. Syngnathidae Rafinesque, 1810 – needle	-	+	+
1.10. Number of species	-	1	1
Fam. Cottidae Bonaparte, 1832 - kerchak, slingshot	+	+	+
Number of species	1	1	1
Fam. Percidae Cuvier, 1816 – percids	+	+	+
Number of species	4	4	4
Fam. Odontobutidae Hoese et Gill, 1993 - Headed	-	-	+
Number of species	-	-	1
Fam. Gobiidae Fleming, 1822 – goby	-	-	+
Number of species	-	2	5
Total species:	51	48	59

Note: + - species of the family are marked in the reservoir; - family species in the reservoir are not found.

In the Middle Volga area, on the area of the Kuibyshev reservoir, before the regulation of the runoff (1955-1957) 51 species of fish were found. 36 species were the permanent inhabitants of the river, among which the most numerous were bream, pike, bluefish, roach, ide, perch, bleak, etc. Among the valuable commercial species, a lot of sterlets were found. Zunder was observed in commercial value, although the herd was not so big. The commercial herd of river carp was insignificant, and the total commercial river catches during this period ranged from 1.35 to 1.91 thousand tons per year [5, 3].

Dams construction of the Zhiguli Hydroelectric Power Station on the Volga River radically changed the habitat of fish and adversely affected on the number of migratory species, significantly changing fish composition of the reservoir (Table 1).

The spontaneous formation of the ichthyofauna of the Kuibyshev reservoir during its formation, often unfavorable conditions for the reproduction of fish, prevented the creation of large commercial stocks of valuable species in the reservoir. While secondary and low-value fish, having high ecological flexibility, increased their numbers sharply. This was also promoted by the absence of large-scale works of the fish population reconstructing by increasing the number of economically important high-value species [5]. In the initial period of the reservoir formation, a number of fish-protecting measures, for example, acclimatization and fish-breeding operations, including a prohibition of fishing for sturgeon, bream, carp and pike-perch, but allowing the capture of low-value fish and pike were carried out. At the same time during this period, in order to increase the carp population, 1 million young carps and 33 thousand of its producers, caught in the lower reaches of the Volga, were let out the flooding zone. In addition, about 1000 specimens of sturgeons were transplanted from the lower reaches of the reservoir and the output of pike-perch producers was planned [6].

In 1961, a fish lift was installed at the hydroelectric station dam from the lower tail of the Volga hydroelectric power station to the upper for passing fish (sturgeon, herring, etc.). For the period of fish lift functioning (from 1961 to 1988), about 13 million valuable species of fish were led to pass. After stop of fish lift functioning at the Volgograd hydroelectric power station in 1988 a single penetration of sturgeon, stellate sturgeon and whitefish continued through the shipping lock of the Saratov hydro system up to 1993, but the beluga was no longer marked [7].

According to Lukin [8] and Kuznetsov [3], after the construction of the dams on the Volga River, large spawning migrations of beluga to the Middle Volga and the Kama River ceased, and juvenile beluga did not occur from 1957 to 1962 in the reservoir. According to the authors, the remaining producers matured and spawned, and their young fishes were caught for several years.

The last recorded capture of the beluga was noted in the Kuibyshev Reservoir at the end of October 2000 on the border of the Volga-Kamsky and Tetyushsky reaches above the Tetyusha town. This was the female beluga in the fourth stage of maturity, weight 255 kg, and length 270 cm which was caught in the trawl. The age of the beluga was 48-54 years (51.75 ± 0.75). Probably, this fish appeared in the first years after the dam construction from the remaining producers. This is indicated by the ratio of the age of the reservoir and the age of caught fish.

Changes of the conditions of fish habitat in the reservoir affected the number of native species. Ecologically flexible species, capable of adaption to new conditions, not only remained numerous, but their numbers increased significantly. However, the number of carp and some rheophilic species such as the Volga sub- stance, sterlet, etc. decreased as a result of the instability of the level regime in the spring spawning period. However, in new conditions, the number of pike-perch, borsch and perch began to increase.

According to our and literary data [9, 10], ommon and lake minnow, white gudgeon, moustached char, Siberian plucking, ordinary plucking, common loach, European grayling, brook trout, ordinary stalking occur in the mouths of rivers flowing into the reservoir. The absence of such species as the white-fronted gudgeon and the Siberian shchipovka [10] in some of the earliest fish lists of the Volga River and the Kuibyshev reservoir [11, 12] was explained by not sufficiently developed of the taxonomy at that time. But in our opinion, clear species criteria were not developed at that period. This is confirmed by the description of a white minnow in the identification guide of Berg [13], who was noted in the Volga River near the Kazan city.

Thus, at present time 59 species of fish belonging to 13 orders, 19 families and 47 genera are found in the Kuibyshev reservoir. More than half of them - 30 species or 51.7% - commercial, 17 or 29.3 - invaders, 9 or 15.5% - rare species included in the Red Book of the Republic of Tatarstan.

Conclusion

At the present time, migratory fish species fall out from the fish population of the Kuibyshev reservoir, but species appeared that not inhabited previously, but was purposefully or independently introduced. Some of them have successfully naturalized, reached a sufficient number and are used in fishery. Others multiply and become common species with a local distribution. The Far Eastern pelagophilic herbivorous species did not naturalize, the number of which is supported by production volumes and controlled. These species, being valuable objects of pond cultivation, with optimum volumes of their production could be effectively used to grow in the natural conditions of the reservoir, which would contribute to improving fish productivity of the reservoir.

The ecological relationships of fish in the Kuibyshev reservoir formed. But due to the poor knowledge of the way of life of the majority of invaders and the lack of information of the dynamics of the number of their populations, it is difficult to assess their impact on the ecosystem of the reservoir. However, our own monitoring investigation demonstrates that the process of forming the ichthyofauna of the Kuibyshev reservoir is going on.

Acknowledgments

The work was supported by a grant presented by the Russian Foundation for Basic Research (FGBU), No. 18-44-160023 / 18

References

[1] Kuibyshev reservoir. 1983. L .: Science. 213 p.
[2] Tsyplakov E.P. 1974. The expansion of the range of some fish species in connection with hydro construction on the Volga and acclimatization works, Vopr. ichthyology. T. 14. Issue. 3. P. 396-405.
[3] Kuznetsov V.A. 2005. Pisces of the Volga-Kama region. Kazan: Kazan-Kazan. 208 p.
[4] Shakirova FM, Severov Yu.A. Species composition of the ichthyofauna of the Kuibyshev reservoir // Vopr. ichthyology. 2014. T. 54. № 5. P.520-532
[5] Tsyplakov E.P. 1980. Fish production opportunities of Kuibyshev reservoir // Biology of inland waters: Inf. BULL. - L. № 47. P. 46-49.
[6] Lukin A.V. 1961. Kuybyshev Reservoir, Izv. GosNIORH. T.50. Pp. 62-76.
[7] Ermolin V.P. 2010. The composition of the ichthyofauna of the Saratov reservoir, Vopr. ichthyology. T. 50. № 2. P. 280-284.
[8] Lukin A.V. 1964. The main regularities of the formation of fish reserves of the Kuibyshev reservoir and the way to their rational use // Proc. The Tatars. ord. GosNIORH. Issue. 10. P.3-26.
[9] Mikheev VA, Aleev FT, Nazarenko V.A. 2004. A Brief Overview of the ichthyofauna of the Ulyanovsk Region // Nature of the Simbirsk Povolzhie. Issue. 5. Ulyanovsk: Corp. technol. promotion. Pp. 97-102.
[10] Semenov D.Yu. 2010. Dynamics of the species diversity of the cyclostomes and fishes of the Kuibyshev reservoir // Vopr. Ichthyology T. 50, No. 6. P. 790-795.
[11] Lukin AV, Smirnov GM, Platonova OP 1971. Pisces of the Middle Volga region. Kazan: Publishing house of Kazan. state. University. 85 p.
[12] Kuznetsov V.A. 1978. Features of reproduction of fish in conditions of regulated river flow. Kazan: Publishing house of Kazan. state. University. 160 p.
[13] Berg L.S. 1948-1949. Freshwater fish of the USSR and neighboring countries. Ch. 1-3. M.-L .: Publishing House of the Academy of Sciences of the USSR. 1381 sec.

Mathematical modeling of surfactant self-organization in water solution in the presence of carbon nanotubes

O S Zueva, V S Rukhlov, E V Gazeeva, Y K Mongush

Kazan State Power Engineering University, 51 Krasnoselskaya str., Kazan 420066, Russia

ostefzueva@mail.ru

Abstract Based on the experimental results mathematical modeling of surfactant molecules organization in aqueous solutions into nanostructured formations was performed. The effect of the addition of carbon nanotubes on the processes was investigated. Numerical calculations of CTAB micelles dimensionless electrostatic potential were carried out in the framework of two most commonly used approximations: the Poisson-Boltzmann model and the jellium-approximation model for various surfactant concentrations in the presence and without of carbon nanotubes. A calculation procedure was developed that made it possible to determine the region of the most active counterions redistribution. The possibility of the interaction between micelles was noted, as a result of which a common self-consistent electric field appears, leading to spatial ordering in the bulk of solution. It was shown that there is a more rapid decay of the potential for the case of CTAB micelles in the presence of carbon nanotubes, which may indicate a change in the micelles morphology and their transition to a more compact form. It has been suggested that because increasing in the local surfactant concentration is observed near the carbon surface, carbon nanotubes can act as stimulators for the formation and growth of supramolecular structural complexes.

Introduction

Surfactants are widely used in a large variety of technological applications. The versatility of practical use of surfactant dispersions both in pure form and in various mixtures gives a basis for their comprehensive including theoretical study [1-4]. One of the modern applications of surfactants is their use for dispersing and functionalizing carbon nanotubes (CNTs) [5-9]. Carbon nanotubes are employed for many practical purposes, including for creation of advanced polymer, construction and rubber composite materials [10,11] and even in the additives for motor [12-14] and boiler fuel [15,16].

In many cases for a uniform distribution over the bulk carbon nanotubes are previously dispersed in aqueous surfactant solutions under the ultrasound influence. In different solvents through the features of their structure surfactant molecules have the ability to adsorb on surfaces and to spontaneous aggregation into various nanostructures. Due to adsorption and appearance of supramolecular formations, surfactant molecules functionalize carbon nanotubes, preventing them from re-adhering. Theoretical and experimental study of the processes occurring under the interaction of carbon nanotubes and the surrounding supramolecular complexes with the dispersion medium are important and relevant for all areas of technical use of CNTs. Since in the organized solutions the changes in the spatial distribution of charges occurring in any part of its volume inevitably must affect other parts of it, in this paper we attempted to theoretically analyze the processes occurring in a surfactant solution in the presence of carbon nanotubes. The aim of this work was the mathematical modeling of the character of the surfactant molecules organization in aqueous solutions into nanostructured formations and the effect of the addition of carbon nanotubes on the occurring processes.

Experimental results

Experimental study of the processes occurring in pure anionic surfactant water solutions [17] and at the dispersing of multiwalled carbon nanotubes of the carbon nanomaterial "Taunit" produced by "NanoTechCenter" (Tambov, http://www.nanotc.ru), including the determination of the type of dominant mechanisms of interaction of surfactants with a carbon surface, was carried out by us in [9,18]. However, many authors believe that the most successful is dispersing of carbon nanotubes in cationic surfactant solutions [19], such as cetyltrimethylammonium bromide (CTAB) [8], dodecylpyridinium bromide [20,21], etc. This is due to the fact that the carbon nanotubes surface in an aqueous solutions has a negative charge and the positively charged hydrophobic hydrocarbon tails of cationic surfactants can bond more firmly to the surface of the CNT, stabilizing it in solution. Typically, CTAB with the chemical formula $C_{19}H_{42}BrN$ and a molar mass of 364.46 g/mol are used to disperse the CNTs. High critical micelle temperature (Kraft point is near 25-30 °C) and low critical micelle concentration (CMC), not exceeding 1 mM, are characteristic for CTAB dispersions. The rest of the literature data, describing the dispersions of CTAB, are highly contradictory. Thus for example at 30 °C the micelle aggregation number is 70-120, the degree of micelle ionization is 0.1-0.24, and the micellar radius is 2.56-3 nm. And only when determining the critical micelle concentration the values obtained by different authors are fairly uniform and fall within the interval 0.9-1 mM. The relatively low value of the CMC may have a significant effect on the spontaneous self-organization of CTAB molecules into the supramolecular structural complexes of the nanosheets type and the nanoflowers type described by us in [8,22].

CTAB dispersions and suspensions of carbon nanotubes in CTAB dispersions at 30 °C were investigated in our work [23] by two methods: conductometry and tensiometry. Comparison of the behavior of SDS and CTAB dispersions in the presence of carbon nanotubes and without them, together with available data on the behavior of SDS molecules on carbon surface, led to the conclusion that the carbon nanotubes strongly influence the aggregation properties of CTAB micelles. The obtained by us experimental data do not contradict the fact of the simultaneous existence of micelles with different aggregation numbers and different ionization degrees. The fact that part of CTAB ions are adsorbed on carbon nanotubes surfaces leads to an apparent change in the CMC with 1 mM for pure CTAB dispersion to 1.3-1.4 mM for CNT suspensions in CTAB dispersions.

In addition, data on specific electrical conductivity in the presence of CNTs and without them indicate a possible change in the micelles conformation toward to small micellar formations that have a larger mobility and a larger ionization degree. Earlier we showed that the dominant mechanism of adsorption of some anionic surfactants can be micellar [9,18]. Adsorption of surfactant molecules on carbon surface leads to an increase in their local concentration near the nanotubes. Most likely, rather uniform configurations of surfactant associates having a micellar or hemimicellar structure appear on the surface of CNTs. The contact and interpenetration of remote parts of diffusion shells of micelles leads to interaction of micelles on the surface and in the bulk. As a result, a common self-consistent electric field appears in the solution, leading to spatial ordering throughout the solution. Thus in organized media the orderliness of the conformational properties of micelles arising near the CNT surface due to interaction between micelles is transmitted throughout the bulk. This is facilitated by a short lifetime of a single micelle ($10^{-5} - 10^{-3}$ s) in an equilibrium disperse system.

Theory and calculations

For theoretical description of the phenomena occurring in organized solutions, an approximate statistical method – the self-consistent field method is usually used. The possibility of independent ion motion is allowed in this field. In the framework of the self-consistent field method, the electrostatic potential $\Psi(r)$ created by the micelle nucleus of radius a at the points of the diffuse part of the micelle is associated with the spatial distribution of the charge density around the micelle by the Poisson

IOP Conf. Series: Earth and Environmental Science **288** (2019) 012059 doi:10.1088/1755-1315/288/1/012059

equation [24, 25]

$$\frac{1}{r^2}\frac{d}{dr}\left(r^2\frac{d\Psi(\vec{r})}{dr}\right)=-\frac{\rho_c(\vec{r})}{\varepsilon_0\varepsilon} \quad (1)$$

Here r is the radial coordinate with the origin at the center of the particle, $\rho_c(\vec{r})$ is the charge density of the diffuse part of the micelle, ε_0 and ε are the dielectric constant and the relative permittivity of the medium. As boundary conditions, the potential value $\Psi(a)$ on the surface of the micelle is used, as well as the fact that at large distances from the particle ($r=\infty$) the electrostatic potential tends to zero. The solutions of the Poisson equation allow us to determine the potential value $\Psi(\vec{r})$, electric field strength $E=-d\Psi(\vec{r})/dr$ and counterion concentration $\rho_c(\vec{r})$ at any point near the micelle core.

In the relation (1) one can proceed to the dimensionless potential defined as $\Phi=\pm e\,\Psi(\vec{r})/(kT)$, where k is the Boltzmann constant, and T is the absolute temperature. The choice of signs on the right-hand side of this relation corresponds to the fact that the dimensionless potential Φ must be essentially positive.

In the framework of the theoretical description of phenomena, we have studied numerical solutions of the Poisson equation that give the dimensionless electrostatic micelle potential $\Phi(r/a)$ as a function of the relative distance from their center r/a for various concentrations of dissolved surfactant (2, 4, 8, 16, 32, 64, 128 mM). Similar dependencies were calculated for the CTAB dispersions for the Poisson-Boltzmann (PB) model and the jellium-approximation (JA) model [24,25] which differ in the spatial micelles distribution in solution. The Poisson-Boltzmann model is based on the assumption that the spatial distribution of all particles, including micelles, is described by the Boltzmann distribution. In the jellium-approximation model, it is believed that micelles are distributed uniformly and surrounded by an ionic atmosphere in which counterions are subject to the Boltzmann distribution. For these models, the Poisson equation has a simpler species, but still requiring a numerical solution. In the course of numerical calculations, the following values of the main parameters of the CTAB micelles were used: the critical micelle concentration by our data [23] at $T = 30\ ^\circ$C was taken equal to $C_{KKM} = 1$ mM, the degree of micelle ionization $\alpha = 0.24$, the average aggregation number of ions in micelle $N_{agg} = 70$, micellar radius $a = 2.56$ nm, micelle surface potential $\Psi(a) = 101$ mV [26-28].

The addition of carbon nanotubes leads to redistribution of surfactant ions. In solution the number of surfactant ions decreases, since the carbon surface is rather attractive for the adsorption of these ions. However, even those ions that are adsorbed on the surface of CNTs can be organized into micelles or hemimicelles [9,18]. In the presence of carbon nanotubes, the CTAB dispersions have changes associated with alteration in the critical micelle concentration to 1.3 mM due to the adsorption of a certain number of surfactant ions on their surface. Therefore, in the first approximation, a theoretical analysis of the effect of carbon nanotubes can be carried out taking into account the change in this parameter.

We calculated the dimensionless electrostatic potential $\Phi_{CNT}(r/a)$ of CTAB micelles as a function of the relative distance from their center in the presence of CNTs at various concentrations of the dissolved surfactant (2, 4, 8, 16, 32, 64, 128 mM). Comparison of the corresponding values in the presence of CNTs and without them made it possible to calculate the difference curves

$$\Delta\Phi(r/a)=\Phi(r/a)-\Phi_{CNT}(r/a), \quad (2)$$

characterizing the potentials of micelles in CTAB dispersions in the absence and presence of carbon nanotubes. Calculations of the change in the dimensionless potential $\Delta\Phi(r/a)$ as a function of the relative distance from the center of the micelles were performed within the framework of the Poisson-Boltzmann model and the jellium-approximation model. The results are shown in Figure 1.

Results and discussion

Within the framework of the considered models taking into account the experimentally obtained parameters of CTAB micelles, we obtained numerical solutions of the Poisson equation that allow us to determine the electric field potential at any point of the diffuse layer of CTAB micelles. Accounting the change in the CMC in the presence of carbon nanotubes made it possible to calculate similar data for CNTs suspensions in surfactant dispersions in the first approximation. Comparison of the corresponding values of the potentials in the presence of CNTs and without them made it possible to find the difference curves $\Delta\Phi(r/a)$ (Fig. 1).

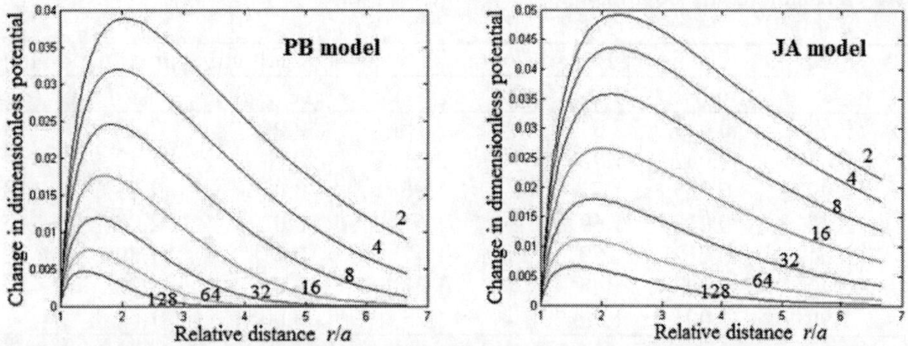

Figure 1. Calculations of the change in the dimensionless potential $\Delta\Phi(r/a)$ of micelles as a function of the relative distance from their center, performed in the framework of the Poisson-Boltzmann model (PB model, left) and the jellium-approximation model (JA model, right) for different concentrations of CTAB (2 , 4, 8, 16, 32, 64, 128 mM)

The obtained curves have a clearly pronounced maximum. The presence of maximum indicates the existence of a region of the most active redistribution of counterions. Calculations show that at small distances from the micelle core (of the order of several internuclear distances) there is almost no change in the potential distribution, i.e. the microenvironment of the micelle core is almost constant. The next part of the micelle diffuse layer corresponds to the region of the most active redistribution of counterions. The maximum value of the change in the dimensionless potential $\Delta\Phi(r/a)$ and the value of the relative distance r_{max}/a corresponding to the maximum change depend on the concentration C of the dissolved surfactant and differ slightly depending on the used model. The values of these quantities (C, $\Delta\Phi(r/a)$, r_{max}/a) are given in Table 1. Also in Table 1 are presented the results of calculations of the Debye wavelength [29] for particles with the parameters of CTAB micelles for both models. The Debye wavelength λ is introduced to estimate the thickness of the ionic atmosphere created around a weakly charged spherical particle, i.e. to estimate the distance $a+\lambda$ (from the center), which is affected by the electrostatic field of the particle. In addition, Table 1 gives the values of half the relative distance between the centers of the micelles $D/2a$ ($D/2$ is the radius of the sphere per micelle), determined by the volume per micelle

$$\frac{D}{2a} = \frac{1}{2a}\sqrt[3]{\frac{N_{agg}}{N_A(C - C_{CMC})}} \quad (3)$$

In the course of calculations, it was found that the comparison of the Debye length with the distance which is affected by electrostatic field of micelle is not very correct, since the potential at these points is not small. This is also indicated by the fact that the region of the most active redistribution of counterions r_{max}/a is located close enough to the relative Debye length $(a+\lambda)/a$, and even goes beyond this length for Poisson-Boltzmann model at concentration more than 16 mM. Moreover, practically for all considered concentrations the potential of the micelles does not vanish even at the boundaries between neighbouring micelles (at distances $D/2$). This means the presence of interaction between the micelles caused by the interpenetration of the remote parts of micelle diffuse layers, as a result of which a general self-consistent electric field appears in the bulk, leading to spatial ordering in the entire solution and to appearance of supramolecular structural complexes [8,22].

Table 1. The concentration C of CTAB in solution, the maximum values of the change in the dimensionless potential $\Delta\Phi(r/a)$, the corresponding relative distances r_{max}/a, and the relative Debye lengths $(a+\lambda)/a$ calculated for both models.

C (mM)	$R/2a$	Poisson-Boltzmann model			Jellium-approximation model		
		$\Delta\Phi(r/a)$	r_{max}/a	$(a+\lambda_{PB})/a$	$\Delta\Phi(r/a)$	r_{max}/a	$(a+\lambda_{JA})/a$
2	9.54	0.039	2.04	3.16	0.049	2.27	4.63
4	6.62	0.032	1.92	2.41	0.043	2.22	4.29
8	4.99	0.025	1.82	1.96	0.036	2.13	3.44
16	3.87	0.018	1.69	1.67	0.027	2.00	3.29
32	3.04	0.012	1.59	1.47	0.018	1.89	2.77
64	2.40	0.008	1.48	1.33	0.012	1.72	2.31
128	1.90	0.005	1.39	1.23	0.005	1.61	1.95

In the presence of CNTs a faster potential drop of CTAB micelles can speak about micelles of a more compact form having a smaller diffuse layer. This assumption allows to explain the observed by us increase in the specific electrical conductivity of CTAB dispersions in the presence of carbon nanotubes. In addition, the adsorption of surfactant molecules on the carbon surface leads to an increase in their local concentration near the nanotubes. This fact can enhance the interaction between micelles and contribute the appearance of centers of spatial ordering of surfactant molecules in aqueous solutions. Thus, carbon nanotubes can act as stimulators of the formation and growth of supramolecular structural complexes. It has been suggested that because increasing in the local surfactant concentration near the carbon surface is observed, carbon nanotubes can act as stimulators for the formation and growth of supramolecular structural complexes.

Conclusions

Based on the experimental results mathematical modeling of surfactant molecules organization in aqueous solutions into nanostructured formations was performed by us. The effect of the addition of carbon nanotubes on the processes was investigated. It was assumed that, in the first approximation, a theoretical analysis of the effect of carbon nanotubes can be performed taking into account the change in the value of the critical micelle concentration. Numerical calculations of CTAB micelles dimensionless electrostatic potential were carried out in the framework of two most commonly used approximations: the Poisson-Boltzmann model and the jellium-approximation model for various surfactant concentrations in the presence and without of carbon nanotubes. A calculation procedure was developed that made it possible to determine the region of the most active counterions redistribution. The possibility of the interaction between micelles was noted, as a result of which a common self-consistent electric field appears, leading to spatial ordering in the bulk of solution. It was shown that there is a more rapid decay of the potential for the case of CTAB micelles in the presence of carbon nanotubes, which may indicate a change in the micelles morphology and their transition to a more compact form. It has been suggested that because increasing in the local surfactant concentration is observed near the carbon surface, carbon nanotubes can act as stimulators for the formation and growth of supramolecular structural complexes.

References

[1] Myers D 2006 *Surfactant science and technology* (New Jersey: John Wiley & Sons)

[2] Rosen M J and Kunjappu J T 2012 *Surfactants and interfacial phenomena* (New Jersey: John Wiley & Sons)
[3] Pashley R and Karaman M 2005 *Applied Colloid and Surface Chemistry* (New Jersey: John Wiley & Sons)
[4] Bourrel M and Schechter R S 2010 *Microemulsions and related systems: Formulation, solvency, and physical properties* (Paris: Edition Technip)
[5] Jiang L, Gao L and Sun J 2003 *J. Colloid Interface Sci.* **260** 89
[6] Vaisman L, Wagner H D and Marom G 2006 *Adv. Colloid Interface Sci.* **128-130** 37
[7] Tummala N R and Striolo A 2009 *ACS Nano* **3** 595
[8] Zueva O S, Osin Y N, Salnikov V V and Zuev Yu F 2014 *Fundamental research, Russian* **11** 1021
[9] Zueva O S, et al. 2016 *Russ. Chem. Bull.* **65** 1208
[10] Mai Y-W and Yu Z-Z 2006 *Polymer Nanocomposites* (Cambridge: Woodhead Publishing Limited)
[11] Dasari A, Mai Y-W and Yu Z-Z 2016 *Polymer Nanocomposites: Towards Multi-Functionality* (London: Springer-Verlag)
[12] Shaafi T, Sairam K, Gopinath A, Kumaresan G and Velra R 2015 *Renew. Sust. Energ. Rev.* **49** 563
[13] EL-Seesy A I, Abdel-Rahman A K, Bady M and Ookawara S 2016 *Energy Procedia* **100** 166 Basha J S and Anand R B 2014 *Alexandria Engineering Journal* **53** 259
[14] Zvereva E R, Khabibullina R V and Zueva O S 2017 *Solid State Phenomena* **265** 374
[15] Zvereva E R, et al. 2016 *Chem. Technol. Fuels Oils* **52** 488
[16] Gubaidullin A T, et al. 2016 *Russ. Chem. Bull* **65** 158
[17] Borovskaya A O, Idiatullin B Z and Zueva O S 2016 *J. Phys. Conf. Ser.* **690** 012030.
[18] Clark M D, Subramanian S and Krishnamoorti R. 2011 *J Colloid Interface Sci.* **354** 144
[19] Zakharova L Y, et al. 1999 *Mendeleev Comm.* **6** 245
[20] Zakharova L Y, et al. 2002 *Russ. J. Gen. Chem.* **72** 426.
[21] Zueva O S, Salnikov V V, Osin Yu N and Zuev Yu F 2016 *Liq. Cryst. and their Appl., Russian* **16** 90
[22] Zueva O S, Mongush Y K and Makarova A O 2017 *Solid State Phenomena* (in press)
[23] Danov K D, Kralchevsky P A and Ananthapadmanabhan K P 2014 *Adv Colloid Interface Sci.* **206** 17
[24] Danov K D, Basheva E S, Kralchevsky P A, Ananthapadmanabhan K P and Lips A 2011 *Adv. Colloid Interface Sci.* **168** 50
[25] Bunton C A, Nome F, Quina F H and Romsted L S 1991 *Acc. Chem. Res.* **24(12)** 357
[26] Robinson B H, White N C and Mateo C. 1975 *Adv. Mol. Relax. Proc.* **7** 321
[27] Pisárčik M, Devínsky F and Pupák M 2015 *Open Chem.***13** 922
[28] Ohshima H. *Electrical Phenomena at Interfaces and Biointerfaces: Fundamentals and Applications in Nano-, Bio-, and Environmental Sciences* 2012 (New Jersey: John Wiley & Sons)

Application of fuzzy neural network for estimating the intensity of insulation aging of cable lines

V.I. Biryulin[1, a], **A.N. Gorlov**[2, b], **O.M. Larin**[3, c], **D.V. Kudelina**[4, d]

[1]Soyuznaya, 69b, 36, Kursk, Russia, 305009

[2]Teatralnaya, 78 A, 10, Kursk, Russia, 305009

[3]Profsoyuznaya, 3, 1, Kursk, Russia, 305009

[4]Schepkina 11, 141, Kursk, Russia, 305016

[a]bir1956@mail.ru,[b]alexjulia17@yandex.ru, [c]larin77@mail.ru, [d]mary_joy@mail.ru

Abstract. The article considers the issues of forecasting the state of insulation of cable lines. It is shown that the insulation state is influenced by a large number of factors, some of which are indeterminate and can not be expressed in the quantitative form. It was proposed to use a fuzzy neural network for predicting the state of insulation, which ensures processing both quantitative and qualitative information, which makes it possible to obtain more valid results, in comparison with the use of only quantitative information.

Problem definition

In the initial state, the value of the breakdown voltage, which the insulation elements of the electrical equipment have, including cable lines, is quite high. The value of the probability of the insulation work without failures in this period of time is usually close to unity. As a result of insulation materials aging during the operation of equipment, the value of the breakdown voltage decreases, and the probability of the insulation work without failures also decreases. An important task to be solved in the operation process is to assess the intensity of the aging process of insulation.

Theory

In the insulation of the electrical equipment the aging processes are developing under the influence of the operating voltage. With the short lifetime of electrical installations (from units to 10^4 hours), the dependence [1] was experimentally established:

$$\tau = \frac{A}{U^n}, \tag{1}$$

where A – a constant, its value depends on the properties of the insulation; n– the exponent which is affected by the design features of the insulation and the type of voltage of the electrical equipment.

In case of the long service lives ($\tau$is more than 10^4 hours), the amount of experimental data is rather small, which is explained by the considerable duration of the experiments; therefore, the course of the time dependence $\tau = f(U)$ is determined with the lower reliability.

IOP Conf. Series: Earth and Environmental Science **288** (2019) 012060 doi:10.1088/1755-1315/288/1/012060

Operating the electrical equipment, the following dependence is revealed: when the voltage U decreases, the service life τ increases faster than it follows from the expression (1), and below the certain voltage it becomes unlimitedly long.

The expression (2) corresponds to this course of the dependence $\tau = f(U)$:

$$\tau = \frac{A_1}{\left(U - U_{Ч.Р.}\right)^{n-1}}, \tag{2}$$

where U_{PD}– the voltage of partial discharges in the insulation, which is one of the main reasons for the electrical aging of electrical equipment insulation.

The intensity of the process of electrical aging of the insulating structure depends not only on the value of the operating voltage, but also on the number of factors whichcan not be quantified accurately [2-4]. Therefore, the evaluation of the intensity of the aging process of insulation and changes in its electrical properties is the task for the solution of which both quantitative and qualitative data should be used, some of which will inevitably have an indistinct or indeterminate character.

At present time the mathematical apparatus of fuzzy neural or hybrid networks is considered to be one of the most sought after solutions for such poorly or poorly structured problems [5, 6]. In general case a hybrid network is a multilayer neural network containing special structures which do not have feedbacks [7].

The usual (not fuzzy values) data, weights and functions of their activation are used for the operation of such network. The network performs the summation operation using an unchanged T-norm or some other continuous operation, for example, a T-conorm. In such network, all inputs, outputs and weights of the hybrid or fuzzy neural network are real numbers from the interval [0, 1].

The basic idea used for the operation of hybrid networks is the following: to use the existing sample of the initial data to determine the constituents of the membership function which are the most suitable for the operation of the fuzzy inference system.

There is a class of hybrid networks whichis functionally equivalent to the systems of fuzzy reasoning. This architecture is called ANFIS (Adaptive Network Based Fuzzy Inference System). ANFIS represents one of the first versions of the hybrid neural-fuzzy networks – a neural network of direct signal propagation of the special type. The architecture of the neural-fuzzy network is isomorphic to the fuzzy knowledge base.

The hybrid networks use the differentiable implementations of triangular norms (multiplication and probabilistic OR), as well as the smooth membership functions. It allows to usewell-proven fast algorithms for neural networks training which use the method of the back propagation of the error.

The data about the variable for the certain time interval make up an image (time series), which is determined by its value at some time point outside the given interval, i.e. the value of the variable through the prediction interval. Let's assume that the current value of the time series is the dependent on the previous values or the process of changing of the insulation properties can be expressed as a regression model.

All the variables used to create such regression model can be expressed through the fuzzy sets $\{A\}$ and it is possible to specify the membership functions of these variables to these fuzzy sets. Therefore, it is necessary to find the set of fuzzy sets $\{A\}$ and determine the parameters of the membership functionsto evaluate the intensity of the aging process.

Choosing the type of the membership functions, it is necessary to take into account that the degrees of the membership of the boundary values of the interval of variation of the variables to the extreme terms should be approximately equal to unity. In this casethe visual rule will generally be carried out: the smaller or larger the value of the variable being considered, the more it will correspond to the extreme terms "very low" or "very high". Thus,for the membership function of the extreme terms, the trapezoidal forms were chosen.

IOP Conf. Series: Earth and Environmental Science **288** (2019) 012060 doi:10.1088/1755-1315/288/1/012060

Experimental results

The neural network which is being developed should have only one output, on which the estimate is formed, and as many inputs as the previous values we want to use for evaluation – for example, the last 5 values.

The training sample for the created network is compiled as follows: the input values for the network operation will be values for 5 selected time points, and the output value will be the known value at the next time point for the selected input quantities. To compose the training sample, it is necessary to copy the data into 4 columns and move the second column 1 cell up, the third column – 2 cells up, etc.

Hybrid or fuzzy neural networks have the advantages of the neural networks and the fuzzy inference systems. These networks provide the development and presentation of various research systems in the form of sets of rules for fuzzy products. Today, this mathematical apparatus is considered promising for obtaining solutions to weakly or poorly structured problems.

A hybrid network can be represented as a multilayer neural network which have a special structure without feedbacks. These networks use conventional (non-fuzzy) signals, weights and activation functions. The execution of the summation operations is based on the use of a fixed T-norm or some other similar continuous operation. Using these networks, it should be taken into account that all values of the inputs, outputs and weights of the hybrid network are expressed by real numbers from the interval [0, 1].

The work of the hybrid networks is based on the use of existing data sample to perform the operation of finding the parameters of the membership functions used in this network. These operations are performed on the basis of the known procedures for neural networks training. It should be noted that the use of hybrid networks ensures mutual strengthening of the positive sides and compensation of thedisadvantages of fuzzy systems and neural networks.

The work of the hybrid network in performing of prediction tasks is to predict future values at the output of the system, based on its previous behavior. Having initial information on the values of the studied variable x at the certain time moments preceding the forecasting process, this network forms the most probable value of the predicted variable at thenecessary time moment t.

Practical implementation of the fuzzy neural network for solving the problems of forecasting the state of isolation of the cable lines was carried out using the MATLAB Fuzzy Logic Toolbox expansion system. In this package fuzzy neural networks are represented in the form of adaptive systems of neuron-fuzzy output ANFIS.There are two the most commonly used algorithms for fuzzy inference which can be used for forecasting purposes: the Mamdani and the Sugeno algorithms. To develop the hybrid network the Sugeno algorithm is chosen, because the fuzzy neural network based on this Sugeno algorithm has the less error in both learning and predicting processes [7].

Figure 1 shows the structure of the ANFIS network, which uses two rules for its work [8]:

$$\text{If} x_1 = A_1 \text{ И } x_2 = B_1, \text{ then} y_1 = c_{11}x_1 + c_{12}x_2;$$
$$\text{If} x_1 = A_2, \text{ И } x_2 = B_2, \text{ then } y_2 = c_{21}x_1 + c_{22}x_2.$$

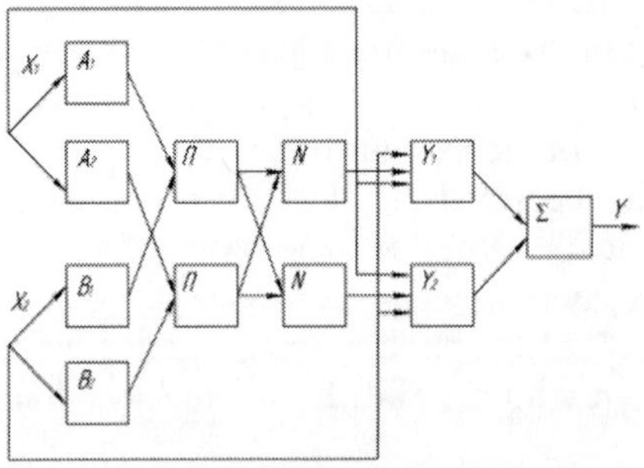

Figure 1. The ANFIS network structure

The output value of the network is determined by the formula:

$$y = \frac{\omega_1 y_1 + \omega_2 y_2}{\omega_1 + \omega_2}. \qquad (3)$$

On the one hand, the ANFIS hybrid network can be represented as the neural network with one output and several inputs which work with the fuzzy linguistic variables. The terms of the input linguistic variables are formed by the standard membership functions, and the terms of the output variable are represented by the linear or the constant membership function.On the other hand, the ANFIS hybrid network is essentially an inference FIS system of the Sugeno type, in which each of the used fuzzy product rules has a constant weight equal to 1. Creating the network, the user has the ability to edit and configure the hybrid ANFIS network [9].

Consider the creating process of the fuzzy model of the hybrid network to solve the problem of assessing the insulation state of cable lines *Q*. The meaning of this problem is that, knowing the dynamics of *Q* changing over the certain time interval, to evaluate the value of this parameter at the certain time point in in the future.As initial data we use information about the dynamics of the changes in *Q* over the certain time interval. For the convenience of the further work, we will present this information in the tabular form (Table 1). For the origin we take the time $t = 0$.

Table 1. The dynamics of the changesin *Q*

T, month	1	2	3	4	5	6	7	8	9	10	11	12
Q	90	89,8	89,8	89,7	89,6	89,4	89,3	89,2	89,1	89	88,9	88,8
T, month	13	14	15	16	17	18	19	20	21	22	23	24
Q	88,7	88,5	88,4	88,3	88,2	88,1	88,9	88,8	88,8	88,6	88,5	88,4

Suppose that the fuzzy model of the hybrid network will contain 4 input variables. In this case the first input variable will correspond to the *Q* for the current time, the second –*Q* to the previous time, i.e., for the time (*i*– 1), where *i* denotes the current time. Then the third input variable will correspond to the *Q* for the time (*i*– 2), and the fourth one tothe *Q* for the time (*i*– 3). The relevant training data is summarized in the separate table of the training sample (Table 2).

Save the training sample in the external file named proba_Q.dat. After that, open the ANFIS editor, in which we will load this file with the training data. The appearance of the ANFIS editor with the downloaded training data is shown in Figure 2.

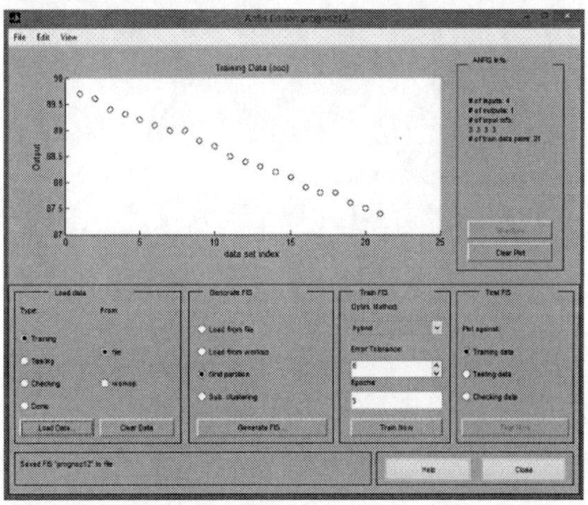

Figure 2. The graphical interface of the ANFIS editor after downloading the training sample

Table 2. The training sample

The values of the input variables				The value of the output variable
1	2	3	4	
89,7	89,8	89,8	90,0	89,7
89,6	89,7	89,8	89,8	89,6
89,4	89,6	89,7	89,8	89,4
89,3	89,4	89,6	89,7	89,3
89,2	89,3	89,4	89,6	89,2
89,1	89,2	89,3	89,4	89,1
89,0	89,1	89,2	89,3	89,0
88,9	89,0	89,1	89,2	89,0
88,8	88,9	89,0	89,1	88,8
88,7	88,8	88,9	89,0	88,7
88,5	88,7	88,8	88,9	88,5
88,4	88,5	88,7	88,8	88,4
88,3	88,4	88,5	88,7	88,3
88,2	88,3	88,4	88,5	88,2
88,1	88,2	88,3	88,4	88,1
87,9	88,1	88,2	88,3	87,9
87,8	87,9	88,1	88,2	87,8
87,8	87,8	87,9	88,1	87,8
87,6	87,8	87,8	87,9	87,6
87,5	87,6	87,8	87,8	87,5
87,4	87,5	87,6	87,8	87,4

Before generating the structure of the Sugeno type fuzzy inference system, after setting the properties of the dialog box we set three linguistic termsfor each of the input variables.We select the triangular functions as the type of their membership functions which are set by the MATLAB system by default. We define the linear function as the type of the membership function of the output variable.

We use the hybrid training method with the error level 0 to train the hybrid network.The number of training cycles is set at 100. After learning the hybrid network, the structure of the constructed fuzzy model can visually evaluated (Figure 3).

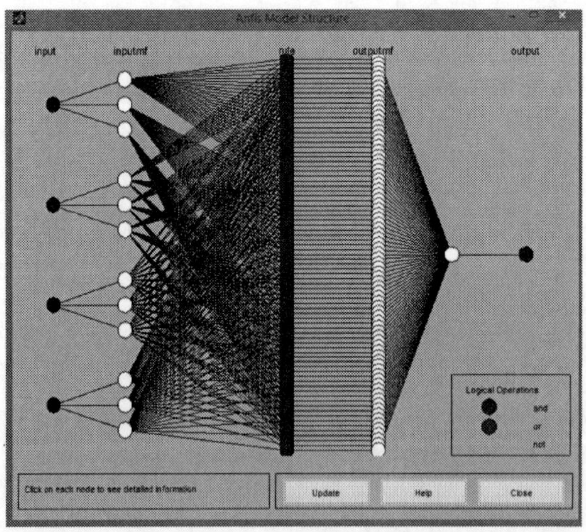

Figure 3.The structure of the created fuzzy model

Conclusion

Perform the test of functioning of the constructed fuzzy hybrid network model. For this purpose we will make the retrospective forecast of the value of Q for the future, for example, for 25 months, counting for this case the current 24 month.

We use the function in the command line for this purpose: evalf. We indicate the vector of the values $Q$for the current time and 3 preceding ones as the arguments of this function. The full format for calling this function is:

q1 = evalfis([87.4 87.5 87.6 87.8], prognoz12).

where q1 – the name of the output variable; 87.4 – the value of Q for 24 months; 87.5 – the value of Q for 23 months; 87.6 – the value of Q for 22 months; 87.8 – the value of Q for 21 months; prognoz12 - the name of the FIS structure preloaded into the working area of the MATLAB system.

The value of the output variable for the future time, equal to 87.2 will be obtained after executing this command with the help of the developed fuzzy model.

References

[1] Gutov IA 1997. *Predicting the state of electric motors based on the use of multifactorial models of insulation aging* (Barnaul:Barnaul State University)p 265

[2] Kudelina D V Fuzzy control system for insulation state 2017 *Scientific bulletins of Belgorod State University***2** pp102–107

[3] Korzhov A V Method for assessing the significance of design and operational factors influence on the service life of the power cables 6 (10) kV of the city electric networks2014 *Vestnik SUSU***1** pp 31–34

[4] Yurchenko E Yu 2009.*Evaluation of the insulation state of the city cable lines with the voltage 6–10 kV with the development of recommendations for improving the electrical safety conditions*(Chelyabinsk: South Ural State University).

[5] Galperova E V, Kononov Yu D and Mazurova O V2008 Forecasting the demand for energy resources in the region taking into account their cost *Region***3** pp 207–219

[6] Uspenskaya I G 2009 Modern problems of energy consumption forecasting in the region (on the example of the Komi Republic)*Problems of forecasting***5** pp 120–133

[7] Rutkovskaya D, Pilinsky M and Rutkowski L 2004. *Neural networks, genetic algorithms and fuzzy systems*(Moscow: Goryichaya liniya-Telecom)p 452

[8] Manusov V Z and Biryukov E V Short-term forecasting of electric load on the basis of fuzzy neural network and its comparison with the other methods 2006*Bulletin of Tomsk Polytechnic University***6**pp 153–158.

[9] DyakonovVP, KruglovVV 2006*Matlab 6.5 SP1 / 7/7 SP1 / 7 SP2 + Simulink 5/6. Instruments of artificial intelligence and bioinformatics*(Moscow: SOLON-PRESS)p 456.

IOP Conf. Series: Earth and Environmental Science **288** (2019) 012061 doi:10.1088/1755-1315/288/1/012061

Energy-saving management of liquid's transportation in pipelines

K.H. Gilfanov[1], N.V. Bogdanova[2], V.V. Plotnikov[3]

[1] Flat 71, 9 Daurskaya St. Kazan, RU 420087

[2] Flat 59, 59-b Musina St. Kazan, RU 420103

[3] Flat 111, 4 Przhevalsky St. Kazan, RU 420037

kamil.gilfanov@yandex.ru

Annotation. Transportation of coolants, oil and oil products are associated with the flow of liquid medium through pipelines of circular cross-section. When the fluids move through the pipes, pressure losses arise on the friction of the flow against the pipe wall. Reduction of wall friction, and accordingly energy loss for pumping liquid can be achieved by pulsating flow in the pipeline. The pulsating mode has acceleration phases, which corresponds to a more complete velocity and deceleration profile, which is characterized by less filled velocity profile, in comparison with a stationary flow. The mathematical model is based on the theory of the boundary layer, using the Prandtl hypothesis on the length of the mixing path. As a result of the numerical experiment, qualitative and quantitative information on the effect of harmonic oscillations of the fluid flow on the hydrodynamic flow characteristics was obtained. The influence of frequency factors and the amplitude of flow oscillations is shown, as well as one of the main characteristics determining the economic effect of the process that is pressure loss on friction along the length of the channel. A method for creating harmonic and nonharmonic oscillations is proposed.

Introduction

The territory of the Russian Federation has a huge length of heating mains and oil product pipelines. Pipeline transportation is characterized by the lowest costs, high speed of delivery and throughput and it is the most environmentally friendly transportation mode. Losses of the transported product in pipeline transport can be minimized [1].

Relevance

To move the fluid through the pipeline of any section and any configuration, a pressure drop is necessary. The flow of liquids is associated with a change in a large number of hydrodynamic and thermal parameters. These include the properties of the fluid being transported, the characteristics of the pipeline and the technological parameters of the equipment that creates the driving force. All these parameters mutually affect to each other and they are interconnected by complex dependencies.

The friction of the fluid on the streamlined surface has a major effect on the pressure loss in the pipelines, which in turn is determined by the velocity gradient in the near-wall region. The velocity gradient depends on the velocity profile, i.e. on the flow regime and the influencing factors, such as the longitudinal pressure gradient, the injection or suction of the boundary layer, heat and mass transfer in the near-wall region, and the nonstationarity of the hydrodynamic and thermal flow parameters.

The choice and justification of the mathematical model

As a rule, the flow of liquid through main pipelines occurs in a turbulent mode. Two flow sections can be distinguished in the pipe. In the first one, the boundary layer originates and develops, which is formed due to the fact that the fluid particles flow "glued to" on the pipe wall. The boundary layer increases in thickness along the pipe. Gradually, the boundary layers in the pipe are closed, and the second section occurs that is a developed turbulent flow. Moreover, under the influence of the longitudinal pressure gradient and the nonstationarity of the hydrodynamic parameters, the developed flow may collapsed, recurrent flows may arise and the boundary layers may develop again.

The mathematical model is based on the theory of the boundary layer. A two-layer model, consisting of a viscous sublayer and a turbulent core, is simple, but insufficiently complete, describing complex hydrodynamics in channels of circular cross-section.

In this paper, the intensification [4] of the process of turbulent flow along the circular channel is carried out in order to reduce pressure losses due to friction along the length of the channel. Effect of flow nonstationarity is created by realization of flow's liquid fluctuations at the entrance to the channel.

The hydrodynamic parameters of the boundary layer [2], [3] are calculated on the basis of a two-dimensional flow model. The system of differential equations is written with respect to the integral parameters of the boundary layer. Such mathematical model includes a greater number of empirical relations for closing the system of differential equations. The model is relatively simple in calculation, which is important for the possibility of application in engineering practice.

A nonstationary turbulent flow of an incompressible single-phase fluid flow in a circular channel is studied. The flow in the boundary layer is assumed to be turbulent directly from the inlet section of the tube. The velocity and temperature profiles in the inlet section are uniform (Fig. 1).

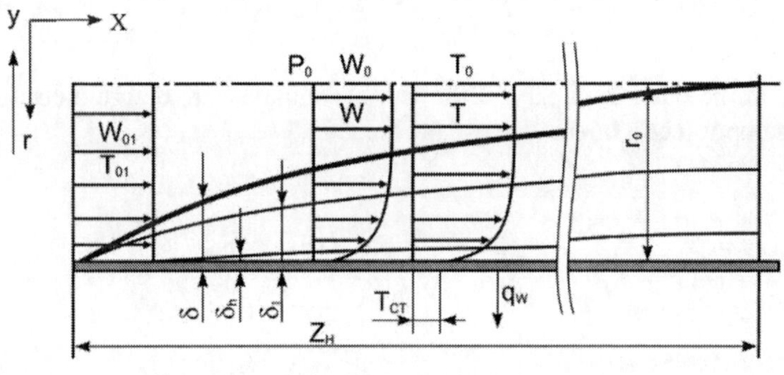

Fig.1. The nonstationary flow of a single-phase flow of a Newtonian fluid in a circular channel: $X = x/2r_0$ is dimensionless longitudinal coordinate; $W = w/w_{01}$ is the relative velocity; P is pressure, N/m2; δ is the thickness of the boundary layer.

The influence of the amplitude and frequency of flow oscillations on hydrodynamic flow parameters are determined by means of mathematical modeling.

The analys of joint influence of factors is carried out using parametric methods developed by S.S. Kutateladze and A.I. Leont'ev [5]. They allow quantitative expression of the investigated characteristics of the fluid flow and establish the interrelations between the factors.

Nonstationary differential equations [6] of motion and continuity:

$$\rho\frac{\partial w_X}{\partial t}+\rho w_r\frac{\partial w_X}{\partial r}+\rho w_X\frac{\partial w_X}{\partial x}=\frac{\partial P}{\partial x}+\frac{1}{r}\frac{\partial(r\tau_X)}{\partial r}\quad, \tag{1}$$

$$\frac{\partial \rho}{\partial t}+\frac{\partial \rho w_X}{\partial x}+\frac{\partial \rho w_r}{\partial r}+\frac{\rho w_r}{r}=0 , \tag{2}$$

are written in an axisymmetric formulation and are transformed into integral:

$$\frac{dW_0}{dX}=\frac{Re_{01}\left(C_g+\frac{C_f}{2}\right)+P}{\frac{Re_{01}}{4H}+\frac{Re^{**}}{W_0}(2+H)} , \tag{3}$$

$$4HRe^{**}=Re_{01}(W_0-1), \tag{4}$$

where ρ is density, kg/m³; w_0 is main flow velocity; $W = w/w_{01}$ is relative velocity; Re_{01} is Reynolds number from the input parameters of the channel; Re^{**} is specific Reynolds number in terms of the thickness of the momentum loss; P is pressure, N/m²; t is the time, s; $C_f/2 = \tau_w/\rho_0 w^2$ is dimensionless coefficient of friction; τ is wall shear stress, N/m²; δ is thickness of the boundary layer; $\xi = x/\delta$ is dimensionless lateral coordinate; η is stability parameter of the viscous sublayer; indexes: $_0$ is parameters of the main (external with respect to boundary layer) flow; $_1$ is parameters at the boundary of the viscous sublayer; $_w$ means parameters on the wall; * is parameter of braking; $_{01}$ is parameters at the entrance to the channel.

These integral equations can be solved numerically with help of the momentum transfer law obtained according to the Prandtl hypothesis and the tangential stresses approximation

$$\sqrt{\frac{C_f}{2}}=(1-w_1)\Big/\int_{\xi_1}^{1}\sqrt{\frac{\bar{\tau}}{\tau_0}}\frac{d\xi}{\chi\xi} , \tag{5}$$

The parameters at the outer boundary of the viscous sublayer are determined from the equation of motion and the assumption of a linear velocity profile:

$$w_1=Re^{**}\frac{\delta}{\delta^{**}}\frac{C_f}{2}\xi_1\left(1+\frac{\bar{\tau}'_w\xi_1}{2}\right), \tag{6}$$

$$\xi_1=\eta_1\left[Re^{**}\frac{\delta}{\delta^{**}}\sqrt{\frac{C_f}{2}\left(1+\frac{\bar{\tau}'_w\xi_1}{2}\right)}\right]^{-1} , \tag{7}$$

The loss of pressure on friction is calculated by formula (9).

$$\Delta P_{C_f}=4\frac{C_f}{2}\rho_0 w_0^2 , \tag{8}$$

The system of equations (3-7) is numerically integrated by the Runge-Kutta method for the initial and boundary conditions presented in Fig. 1

Results of calculations

Figure 2 shows the results of calculating the frictional pressure loss in a non-stationary pulsating flow in a pipeline. The calculation was carried out at different flow rates.

As can be seen from the graph, the pressure loss is minimal if the flow pulsations occur in the frequency range from 0.1 to 0.5 Hz [7]. As the frequency and amplitude of the oscillations increase, the pressure loss also increases, which agrees with the data [8].

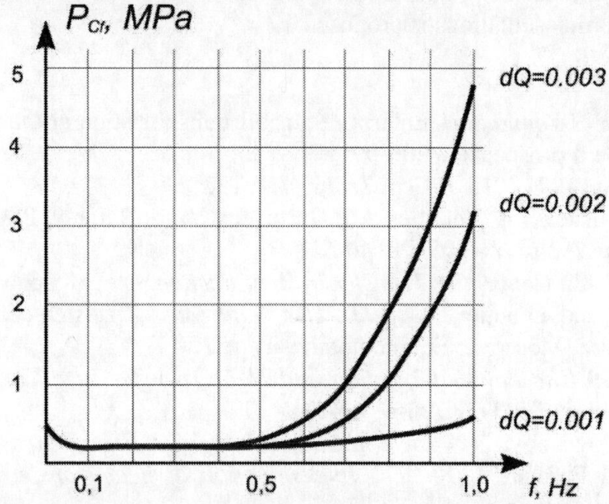

Fig.2. Loss of pressure on friction with harmonic fluctuations in fluid flow

The magnitude of the frictional pressure loss increases during the acceleration phase of the fluid flow. This effect is observed because of the increase in the values of tangential stresses due to the contraction of the boundary layer by the turbulent core [9]. Accordingly, the frictional pressure loss decreases in the phase of the flow deceleration. Creation of the necessary oscillations is carried out by an electric drive with frequency control.

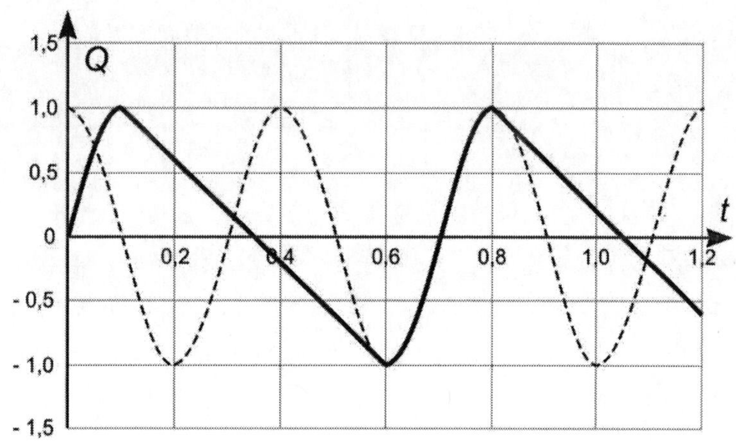

Fig. 3. Change in fluid velocity, where Q, m^3/s is flow rate of the liquid in the pipeline; *t*, s is time

Using of non-harmonic flow oscillations to influence the flow of liquid can be carried out as shown in Fig. 3. Acceleration of the flow occurs according to a linear law, and the deceleration - according to the sinusoidal. By increasing the deceleration phase in time and decreasing the acceleration phase of the flow, it is possible to achieve a reduction in pressure loss due to friction.

Conclusions

The numerical experiment on the effect on wall friction and pressure loss on pumping liquid in a pipeline of finite length of the frequency of flow oscillations 0,1 -1,0 Hz and amplitude of 16%, 33% and 50% of the nominal flow is carried out. It is shown that pressure losses in a pipeline of 23,000 diameters in length using a pulsating harmonic regime can be reduced by 64%. A method for creating harmonic and nonharmonic oscillations is proposed.

References

[1] Pluhina N A 2015 *J Young Sci.*. Enterprises of oil transportation at the present stage: problems and development prospects. 8 606-09
[2] Rao Y, Feng Y and Li B 2015 *J. Heat Trans.* Vol. 137
[3] Schelchkov A V, Isaev S A, Leontiev A I, Gortyshov Yu F, Baranov P A, Popov I A 2017 *Int. J. Heat and Mass Trans.* Vol. 95. Pp. 40-62
[4] Gukauskas A A 1982 *Convective Transfer in Heat Exchangers* (Moscow: Nauka) p 472
[5] Kutateladze S S and Leontiev A I 1972 *Heat and Mass Transfer and Friction in Turbulent Boundary Layer* (Moscow.: Energoatomizdat) p 320
[6] Fafurin A V 1979 *The Laws of Friction and Heat Transfer in a Turbulent Boundary Layer* (Kazan: Sat. scientific. Tr./KAI) pp 62-69
[7] Plotnikov V V, Bogdanova N V and Samoilov A V 2014 *J. Bul. of Kaz. Technol.Un.* 15 p 81
[8] Fedotkin I M nd Firisyuk V R 1971 *Intensification of heat transfer in devices of chemical manufactures* (Harkov: Tehnika) p 216
[9] Gilfanov K H, Podymov V N, Minvaleev N Yu, Sibgatullin I F and Gainullin R N 2014 *J. Izv. Uni. Prob. of Pow. Eng.* 11-12 pp 81-88

Lean production instruments as basis for resource-saving management

Ketoeva N.L.[1], Soldatova N.F.[1], Solovev I.V.

[1]National Research University "Moscow Power Engineering Institute", Krasnokazarmennaya 14, Moscow, 111250 Russia

nhl76@mail.ru

Abstract. Russian power companies and those partnering with the power industry generate a strategic objective to optimize production and management costs while maintaining high quality of the products and services conforming to the international and domestic standards. This inevitably requires reorganization of business processes at all management levels.
Sustainable development and energy efficiency of the Russian economy depends on a number of factors; one of the essential factors is effective management of production enterprises.
The key method for improving the management efficiency is an array of instruments applied in leading industrialized countries worldwide and known as the "lean" production ("lean manufacturing") concept.
In the last 20 years, Russian production enterprises have tried to introduce both lean production methods, and separate tools in practical management. Large power corporations, such as Gazpromenergoholding, Rosatom arrived at meaningful positive results. Lean production practices in Russia are topical for research into methods and primary ways of adopting the lean production (LP) instruments by the domestic industry.
The objective of the research is to classify and provide quantitative assessments of expected positive economic effect through adoption of the lean production instruments at nuclear industry enterprises.
The methodological framework for the research included lean production state standards (GOST) and research reports by foreign and domestic authors doing research into the LP concept, models and instruments.
The information base of the research consists of data from foreign and domestic scientific publications, Rosatom State Corporation's official data and open data from professional communities' websites.
Keywords: lean production concepts, business processes, lean production instruments, sustainable development of a power company, resource-saving management, quality management.

Practical application of the lean production instruments by Russian nuclear power enterprises is conditioned by a range of specific features of the Russian industry. They are:

- specifics of the Russian ISO 9001:2015 standards;
- market orientation of the power companies and those partnering with the power industry;
- rise in electricity, gas and water utility tariffs;
- lower level of labor productivity as compared to that in the leading industrialized countries and BRICS counties;
- modest corporate culture and loyalty to the company;

• material incentives/discouragement orientation of employees, etc.

The practical application of the LP instruments at Russian nuclear power enterprises is aimed at achieving top economic results. Table 1 summarizes improvements to the company's internal processes turned into economic results, which are equally important to the partnering enterprises.

Table 1. Effect of the LP instruments on the company's internal processes and the results achieved

Internal processes of the company using the LP instruments	Achieved results: economic, information	General result
Timely execution of orders	Shorter cumulative lead time	RESOURCE SAVING
Increase in labor productivity	Higher operational efficiency	
Reduction of changeover time	Longer in-service time	
Operating and management cost saving	Cost reduction	
Optimization of raw material stocks	Reduced requirement for current assets	
Minimization of losses due to spoilage	Decreased costs	
"Transparency" of production processes and management	High-quality products	
Full compliance with occupational safety rules	Zero rate of accidents	

Implementation of the lean production management system (LPMS) as a resource-saving management tool is aimed at increasing the energy efficiency; research, development and production companies of the nuclear industry are chosen as a platform for studying the experience in application of the "lean production" concept.

According to the research at the Russian nuclear power enterprises the practical application of the LP instruments is positive and exerts an impact on the labor productivity, quality of products (services rendered), contributes to resource saving and increases the company's economic activity.

The external factor that encouraged the Russian enterprises to apply and adapt the LP instruments is the necessity to integrate partners into a single system, create partner alliances, increase effectiveness of supply chains management, and ensure high quality of products and services. The internal factors include the tough necessity to optimize production management costs, increase management efficiency and raise labor productivity.

The application of the LP instruments at the R&D and production enterprises under the Rosatom State Corporation involved the use of such tools as the Value Stream Mapping (VSM), the Workplace Organization (5S), the Total Production Management (TPM), the Single Minute Exchange of Die, (SMED), statistic analysis and process improvement (Six Sigma), and the Just-in-time system.

The energy efficiency buildup program based on the LP instruments in the electric power division of the Rosatom State Corporation (AO Rosenergoatom Concern) resulted in an increase in own power production in 2017 (by 3,3% in contrast to 2016) and an increase in labor productivity from 7,04 million rubles per person in 2016 up to 8,65 million rubles per person in 2017 [1, p. 24, 42].

The number of the LP instrument used by Russian companies is steadily increasing (the rate of increment of the companies using three – four tools amounted to 5% in recent years as compared to 2008, and the number of the enterprises applying five – six tools increased by 6% for the same period) [1, p. 22].

A little over one third of Russian enterprises use the LP instruments. Russian R&D and production enterprises of the nuclear industry make use of foreign experience in applying the LP instruments through their adaptation to Russian specificities.

A peculiarity of the use of the LP instruments at R&D and production companies of the nuclear industry is setting up an LP management system based on a combination of six tools. Firstly, it is Kaizen approach including the TPM and Kanban systems.

The other aspect of development of the LP management system is the application of the Lean system/lean production system by the Russian R&D and development companies targeted to optimization of the company's expenditures and elimination of waste, and correspondingly, growth of the consumer value.

Table 2. Cost structure for application of LP instruments by Rosatom organizations

Cost of LP instruments	Structure, as percentage of total		
	2015	2016	2017
Lean system/lean production	23,0	23,5	23,4
Business system	19,8	19,4	19,7
TOC	10,1	10,0	10,1
Six Sigma	12,8	12,9	8,9
kaizen	21,3	20,6	23,0
kanban	13,0	13,6	14,9
Total	100,0	100,0	100,0

The table is made up on the basis of Rosatom experts' opinions.

The Business system is the platform for management of the corporation's production systems. It is integrated with the Lean system/lean production, Six Sigma and TOC. Thus, the combined use of several LP instruments based on several approaches takes place at a practical level.

The application of these LP instruments yields favorable results in the company's activities. We see the following exponents as criteria of economic results: rate of wage-cost creep, rate of change of the product cost, rate of change of spoilage in actual output, rate of change of labor productivity.

Depending on the structure of the LP instruments applied, the management system requires a specific business mechanism. New controlling mechanisms influence internal transformation processes at all company's management levels.

Focus on continuing growth in energy efficiency of an R&D company in the nuclear industry is conditioned not only by the external factors, but also the company's internal processes. The market orientation of the R&D company stipulates compliance of the internal processes to market requirements; in case of deviations the company takes corrective actions using the LP instruments. The continuous change of the market processes and the company's strive for compliance with the market requirements necessitate implementation of an algorithm of continuous improvement of the organizational mechanism and development of the LP instruments by R&D companies and production associations of the domestic industry.

Development and buildup of the LP instruments by the companies of this type implies implementation of a step-by-step improvement model for the research and production process taking into account the influence of the external factors, assessment and management of the internal processes from the perspective of the market requirements.

The step-by-step improvement model for the research and production process implies setting up a cross-functional group that assesses and chooses the LP instruments to achieve certain economic results.

Many Russian companies have been developing a production system, in our case a research, development and production system, based on the LP concept and instruments. The external factors and the company's economic capacity contribute to recognizing the importance of implementation of the LP concept, the organizational and economic mechanism of application of the LP instruments, and the level of internal management reforms.

References

[1] Annual report 2017, AO Concern Rosenergoatom.
[2] Safronova K.O. Lean production adaptation under conditions of economic instability. Thesis for PhD in Economis.
[3] http://www.rosatom.ru portal.

The development and energy efficiency of LNG in Russia

N.L. Ketoeva

Ph.D., FSEI HE «NRU «MPEI»
Russia, Moscow, Krasnokazarmennaya Str., 14.

e-mail: KetoyevaNL@mpei.ru

Y.S. Orlova

FSEI HE «NRU «MPEI»
Russia, MA, Podolsk, ul. Kirova 5/38

e-mail: OrlovaYS@mpei.ru

Abstract. The state data of the market structure and the development of liquefied natural gas (LNG). The major LNG projects being implemented or planned for implementation on the territory of Russia. Characterized prospects for the development of LNG production in Russia up to 2023. The main problem of LNG projects implementation. Marked instruments contributing to the development and the energy efficiency industry improvement.

Keywords: energy efficiency, market liquefied natural gas, LNG projects in Russia, the prospects of LNG market development in Russia..

In the last decade, speaking about energy resources and their use is increasingly discussed issue of energy efficiency. With the ultimate resources used also plays a significant role and makes an active search and exploration of alternative energy sources. So share the use of gas, oil and coal in the country significantly activities has changed (Figure 1) [1].

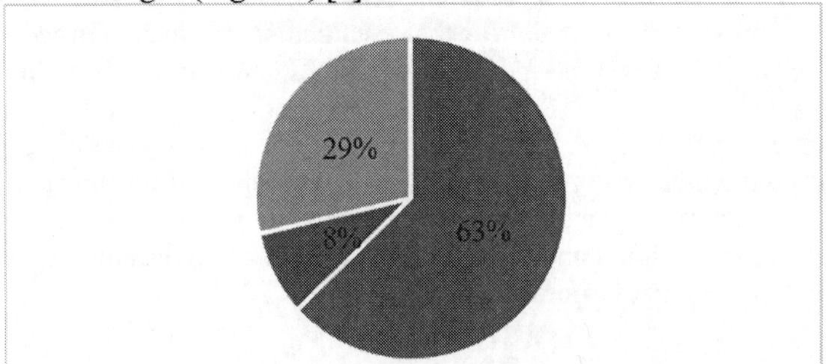

Figure 1. Structure of the use of energy in 2017.

IOP Conf. Series: Earth and Environmental Science **288** (2019) 012063 doi:10.1088/1755-1315/288/1/012063

The threat of global warming, caused by increased emissions of greenhouse gases into the atmosphere, also led the heads of state to think about improving the environment and energy efficiency of production. So on December 12, 2015, Russia signed the Paris Agreement, which will enter into force in 2020, by which the member countries pledged to reduce emissions of hydrocarbons to the atmosphere (currently 68%), by restructuring the energy structure and transferring production to more efficient and environmentally friendly technologies.

At the same time, alternative generation and non-traditional hydrocarbons are introduced as new sources of energy. The given process proceeds impermissibly slowly, in comparison with a rate of growth of volumes of consumption of energy by the population. However, despite the existing barriers and specifics of implementation, significant results over the past 5 years are still achieved. So the percentage of energy received from renewable energy sources in Russia's total energy production increased by an average of 2%. The non-traditional hydrocarbons production in Russia is about 5% in the total world production.

At the same time, one of the most promising areas for the development of Russia's energy industry is the production and liquefied natural gas (LNG) supply.

The LNG production in the world is stimulated, first of all, by the need to transport fuel to countries that are disadvantaged in geographical terms from the perspective of the exporter. As the range of transportation increases, it becomes cheaper to transport gas by tankers, rather than to pump it through the main pipeline. Liquefying natural gas increases its density by hundreds of times, which reduces the volume during transportation and storage. In addition, the delivery of natural gas in the form of LNG allows you to avoid problems associated with transit and negotiations on transport routes. Over the past decade, the volume of the liquefied natural gas market has more than doubled - up to 250 million tons per year. About 30 countries are engaged in importing LNG, producing and exporting about 20 ones.

Currently, according to various experts, domestic producers occupy about 4-5% of the world LNG market. At the end of 2016, the volume of production of liquefied natural gas in Russia amounted to 10,928.7 thousand tons. In 2017, the volume of production has worsened by an average of 19%. At the same time, the share of exported liquefied natural gas occupies 88.5% of the total volume of production.

In 2016, the volume of export supplies of liquefied natural gas amounted to 10,867.9 thousand tons, which is 13.2% more than the same indicator in 2015, which was 9601.7 thousand tons. In 2017, Russia exported 21.4% more energy resources.

According to forecasts, by 2020 the LNG production volume will reach 15109 thousand tons (an increase of 38.2% relative to 2016) [2].

As part of the Energy Strategy project, Russia's share as an LNG supplier on the world market should grow to 12% (against today's 5%) [3]. This is possible due to the introduction of export-oriented LNG projects. The largest of them are: Sakhalin-2, Far East LNG, Vladivostok LNG, Yamal LNG, Pechora LNG, Baltic LNG, Vysotsk and Arctic LNG.

As of 2017, the LNG plant, launched under the Sakhalin-2 project in 2010, is being operated. However, the project plans to increase the capacity and productivity of the plant by 56% by 2021. Another project - Yamal LNG is in the process of implementation. To date, it is planned to launch the first of the three lines of the plant, the production capacity of which will be 5.5 million tons / year. The output to full design capacity is planned in 2019. The remaining projects are in a state of freezing for a number of reasons. First of all, the economic sanctions of the US and the EU led to problems with attracting investments to projects. The fall in oil prices entails a fall in prices for LNG, which also has a negative impact on the return on investment. Finally, the question of import substitution of the equipment used [4].

The analysis shows that foreign equipment is used for running and planned to launch projects. This is mainly due to the lack of domestic analogues with the required characteristics. At the same time, Russian companies have all the necessary resources (technological, legal, personnel, financial) to create competitive equipment. However, if we look at the issue in more detail, it becomes clear that the main brake on decision-making is the managerial resource, namely, it is not always clear how to assess in advance the degree of competitiveness of the projected product.

IOP Conf. Series: Earth and Environmental Science **288** (2019) 012063 doi:10.1088/1755-1315/288/1/012063

The solution to these problems can be the production control algorithm, built on the basis of the cost model. Obviously, 80% of the cost of technological equipment is made up of its materials, while the overall dimensions directly depend on the characteristics that the equipment must meet. The calculation is based on the design methodology and design features of the equipment, but the mass-size indicator in the general case can be determined by the formula (1).

$$M = \sum_i \rho_i \iiint Q_i\,(V)\,dV \equiv \sum_i \rho_i \iiint f(P, v, p, t)\,dV \qquad (1)$$

here M - is the mass of the article, $Qi\,(V)$ - is the mathematical description of the unit of the surface of the body, f - is the function of the mass-dimensional parameters of the product in question, P - is the product power, v - is the productivity, p - is the working pressure, and t - is the temperature of the working medium.

Then the model of calculation of the cost price will have the form (2).

$$S = f(M, s) \equiv f(P, v, p, t, s) \qquad (2)$$

here S - product cost, s - production seriality

Thus, the cost price can be represented as a function of the consumed materials and the parameter that affects the technological costs due to the production volumes of the goods.

Management of a manufacturing enterprise always involves preliminary analysis and identification of key parameters characterizing the economic and economic activity of the enterprise, the so-called significant factors, the indicators for which must be fixed or optimized. Identify these indicators is possible by analyzing the strengths and weaknesses of the enterprise, using a balanced system of indicators, expert method, etc. Using a balanced system of indicators, it is possible to present a strategic map of the LNG industry in Russia in the form of Fig.2.

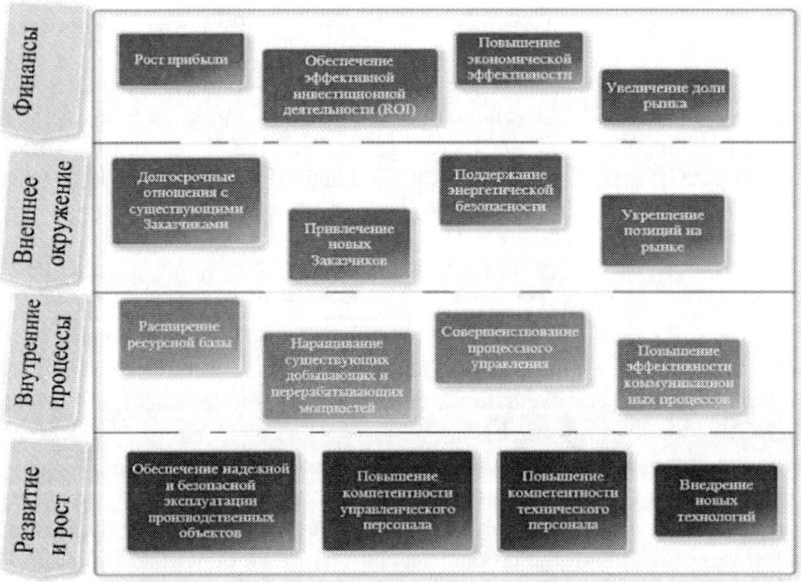

Figure 2. Strategic map of the Russian LNG industry

By grouping the selected activities, seven main factors can be formulated for the enterprises of the oil and gas engineering industry:

- production energy efficiency,
- production environmental friendliness,
- innovative production,
- production volumes,
- market share,
- qualified personnel,
- production costs optimization

The management algorithm is a systematic approach to managing key indicators, according to nine basic management functions. These functions: forecasting, planning, organization, rationing, accounting, control, analysis, regulation and coordination allow each of the significant factors to be considered in detail and to compile the process graph using the matrix data of the management subprocesses (Table 1) [5].

Table 1. The matrix of control subprocesses

Objects	Control function								
	standardi-zation	prediction	planning	organiza-tion	accounting	control	analysis	regulation	coordination
production energy efficiency	-	-	+	+	-	+	+	+	0
production environmental friendliness	+	+	+	-	+	-	+	-	0
production innovation,	-	-	+	+	-	+	+	+	+
production volume,	+	+	+	+	+*	+	+	+	+
market share,	+	+	-	-	+	-	+	-	0
qualified personnel,	0	-	+	0	+	+	+	+	+
cost optimization	-	+*	+	+	+	+	+	+	-

In the matrix, the sign "+" marks real problems, "0" means the absence of the physical meaning of this operation, "-" - the existence of the problem is possible under certain conditions, "+ *" is a task related to the task of another object.

The system of graphs obtained from the analysis of tasks of the matrix of control of subprocesses and will make up an algorithm of production management, based on the model for calculating the prime cost (object "optimization of cost"). This algorithm will not only assess the competitiveness and rationality of production of a new product, but also optimize the production process of the existing one, while increasing the energy efficiency of both the enterprise and the entire industry.

References

[1] Analytical Center under the Government of the Russian Federation. Electronic resource. Access mode: http://ac.gov.ru;

[2] Russia's energy strategy for the period up to 2035;

[3] Forecast of the development of energy in Russia and the world until 2040. Institute of Energy Research of the Russian Academy of Sciences and Analytical Center under the Government of the Russian Federation;

[4] Description of the Sakhalin-2 project. Electronic resource. Access mode: http://www.sakhalinenergy.ru;

[5] Bogatyrev. Fundamentals of the theory of management of economic systems: Textbook. Allowance. V.D. Bogatyrev, B.N. Gerasimov. - Samara: Samara Publishing House. state. aerospace. University. 2008. -260c

International Scientific and Practical Conference: Water Power Energy Forum 2018 IOP Publishing
IOP Conf. Series: Earth and Environmental Science **288** (2019) 012064 doi:10.1088/1755-1315/288/1/012064

Efficiency of an alternative LED-based grow light system

E.G. Kulikova[1], S.Yu. Efremova[2], Natalia Politaeva[3], Y. Smyatskaya[3]

[1]Penza State Agrarian University (440014, Russia, Penza city, Botanicheskaya st., 30)

[2]Penza State Technological University (440039, Russia, Penza city, 1a/11 Baidukova psg./Gagarina st.)

[3]Higher School of Biotechnology and Food Technology, Peter the Great St. Petersburg Polytechnic University, 194021, Russia, St.Petersburg, Novorossiyskaya, 48-50

kuleg@mail.ru

Abstract. This article is dedicated to the assessment of the usage influence of LED lights of varying color spectrum on plant growth and development, and determining their usage impact. The research interest is focused on the usage of LED-based light system inside greenhouses for optimization of lighting conditions.

Research methods. The current research focused on the following Green-LP™ LED-based grow lights: GLP-FH-56-B (30/70 % red(660 nm)/blue(445 nm) spectral composition), GLP-FH-56-R (70/30 % red/blue spectral composition) and GLP-FH-56-RB (50/50 % red/blue spectral composition). Examination of the influence of plant lamps on the growth processes was carried out at the laboratory conditions (23 °C during the day and 20 °C at night, humidity from 75 % to 80%, with the varying PAR intensity and the daylight of 16 h (supplementary illumination was provided for 6 to 8 hours)) with lettuce as a subject. Phenological observations and biometric registrations were conducted according to generally accepted techniques. Chlorophyll content was measured using the express method with atLEAF+ device.

Results. Based on the results of the conducted experiments the highest germination rate was observed under the lamps with blue and blue-red spectral peaks, salad seedlings had largest shoot fresh weight under grow lights with red and blue-red spectrum peaks (0.235 and 0.250 g, respectively). The shoot weight exceeded that of control group of plants 1.8 to 2.1 times. Additionally, plants grown under the lamp with a blue spectral peak had 17 to 29 % more powerful root system than those grown under red one and 43 to 73 % higher than control group. Highest chlorophyll content was observed in plants grown under the lamps with blue spectral peak (440-450 nm), which exceeded by two times that in plants grown without supplementary illumination and was 1.8 times higher than that of plants grown under a high-pressure sodium arc lamp.

Conclusion. LED lamps with different illumination spectrum can serve as an additional light source for plants grown inside greenhouses, positively affecting both growth and development of plants, and economical efficiency.

Introduction

Currently, energy efficiency and energy saving are among the five strategic directions of prioritized technological development, with decreasing the energy consumption of domestic industry being one of the strategic goals [1-3]. In order to recognize and evaluate the energy saving potential of the engineering solutions in question it is necessary to possess deep knowledge of the variety of the LED lights options, as their properties and performance characteristics are quite diverse [4-6]. Investments into the implementation of new specifically developed greenhouse plant growing technologies for healthy plant growth at maximum speed can result in significant decrease of power costs during the operation time. Thus, we should consider the positive experience of both foreign researchers and those in the regions of Russia where thanks to use of new modern electric technologies even with reduction in cultivated area a significant increase of productivity and energy economy was achieved. Considering the unceasing cost growth of energy carriers, scientific justification of the technical solutions for the introduction of electrical technologies is undoubtedly relevant [7].

In order to ensure the quality photosynthesis process in the plant cells a full light spectrum mimicking that of the sunlight is needed, at 440-450 nm and 650-660 nm wavelengths in particular. Thus, artificial illumination has to be as identical to the sunlight as possible, containing light of various wavelength and color (except for green). Most interesting studies of efficient spectral compositions for plant growth are presented in [8-13].

The artificial influence on plants can significantly decrease the costs for thermal and electric energy used for growing plants in greenhouses. Thus, it is expedient to find a scientific basis for determining the optimal PAR spectral composition for each plant species cultivated in the greenhouses. Thereby, the goal of present work was the study of influence of the 440-450 nm and 650-660 nm peaks in the spectral composition of the experimental LED lamps on the phases of crop development.

Research methods

The current research focused on the following Green-LP™ LED-based grow lights: GLP-FH-56-B (30/70 % red(660 nm)/blue(445 nm) spectral composition), GLP-FH-56-R (70/30 % red/blue spectral composition) and GLP-FH-56-RB (50/50 % red/blue spectral composition). Two control groups used a sodium arc discharge lamp or no supplementary illumination.

Examination of the influence of plant lamps on the growth processes was carried out at the laboratory conditions with cruciferae family vegetable crop (lettuce) as a subject.The plants were grown at 23 °C during the day and 20 °C at night, humidity from 75 % to 80%, with the varying PAR intensity and the daylight of 16 h inside Petri dishes. Supplementary illumination was provided for 6 to 8 hours. There were four replications of the experiment. Phenological observations and biometric registrations were conducted according to generally accepted techniques. Chlorophyll content was measured using the express method with atLEAF+ device.

Results and comprehensive assessment

Usually, sowing of the early maturation vegetables is done during the period of low solar activity and short daytime. The daylight length required for production of strong and healthy planting stock is at least 10 to 15 hours, but natural conditions in our latitude cannot ensure such lighting mode. Thus, the primary task of any vegetable grower is the organization of supplementary illumination.

Currently, the primary means of supplementary and full artificial lighting are obsolete lamps, such as incandescent lamps and gas-discharge lamps. Those, like all the luminescent ones, are subject to degradation – exhaustion of the luminous flux. After 6 months in action, luminous flux strength of such lamp drops by 30%, and after one or two years it decreases twofold. Besides, there are serious deviations in the spectral composition of the emitted light. Because of this, it is necessary to replace such lamps every half a year to a year, and their reliability suffers. The main downside of incandescent lamps is lack of blue light radiation in their spectrum and very low luminous efficiency of 17 to 25 Lm/Wt. Additionally, they create excess heat: when placed at a height below 1 m they may cause a burn of plants, and when placed at a height above 1 m they are not able to provide efficient illumination. In order to ensure the balanced spectral composition it is recommended to alternate gas-discharge sodium lamps with metal halide lamps, which is a labor-intense procedure.

The material used in the semiconducting element of an LED determines its color, and the intensity of the emitted light can be regulated by varying the electric current. A complex lighting fixture consisting of several LEDs, each emitting the light of its own part of the spectrum, can be controlled by changing current through each of the components [14] to emit light of any color.

It is traditionally assumed that the cruciferae family plants are the most sensitive to external influences [15]. A study of influence of light of specific wavelengths on growth process has shown that the highest germination rate was under the lamps with blue and blue-red peaks in the spectral composition (89 and 88 %), but the highest mass was obtained by salad sprouts under the lamps with red and blue-red peaks in the spectral composition (0.235 and 0.250 g per 20 specimen respectively).

Table 1 – effect of light of specific wavelengths on growth process of lettuce.

No.	Options	Germination energy, %	Germination rate, %	20 seedlings' total weight, g
1	Control (no supplementary illumination)	45	82	0.165
2	Control (high-pressure sodium arc lamp)	60	83	0.184
3	LED lamp (440-450 nm) (660nm-30%: 445nm-70%)	47	89	0.212
4	LED lamp (650-660 nm) (660nm-70%: 445nm-30%)	42	81	0.235
5	LED lamp (440-660 nm) (660nm-50%: 445nm-50%)	57	88	0.250

The influence of light of specific wavelengths on lettuce mass gain dynamics is shown in Figure 1:

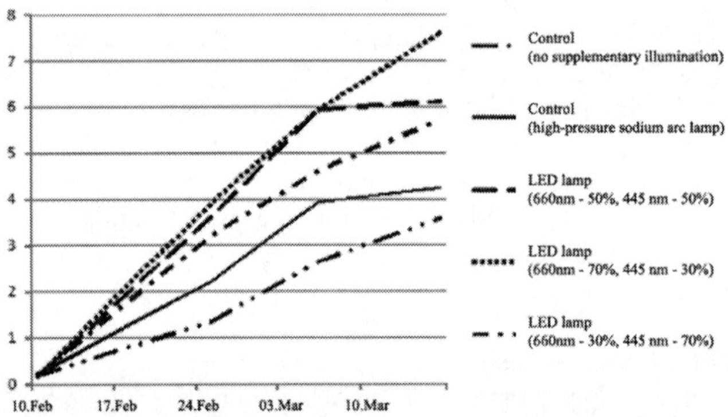

Figure 1 – Lettuce weight gain dynamics for different lighting modes.

Analysis of the results demonstrated that lamps with blue and red spectral peaks have had the greatest influence on the growth process of lettuce, and until a certain point there was no observable difference between the two. Later, plants grown under the lamp with the red spectral peak began to gain biomass more intensively. The difference when compared to control group specimen was 1.8 to 2.1 times, and the lettuce control specimens were even visually observed having some light deficiency. However, under the lamp with a peak in the blue part of the spectrum, a more powerful formation of the root system was observed, 17-29% higher compared with those grown under the lamp with a peak in the red part of the spectrum and 43-73% higher compared with the control group [16].

The measurement of chlorophyll content in plants' leaves has demonstrated that the highest amount of chlorophyll was found under the lamps with the blue spectrum peak (440-450 nm), which was two times higher compared than that without the supplementary illumination and 1.8 times higher than under the high-pressure sodium arc lamp (Table 2).

Table 2 – effect of light of specific wavelengths on chlorophyll content (scores).

No.	Options	Chlorophyll content (scores)
1	Control (no supplementary illumination)	11.3
2	Control (high-pressure sodium arc lamp)	12.5
3	LED lamp (440-450 nm) (660nm-30%: 445nm-70%)	22.8
4	LED lamp (650-660 nm) (660nm-70%: 445nm-30%)	19.5
5	LED lamp (440-660 nm) (660nm-50%: 445nm-50%)	20.7

Analysis of LED lamps economic efficiency has shown that low electric power consumption was one of the advantages of the tested lamps. And most importantly, these lamps could be placed close to plants without any harmful consequences, as the lamps did not overheat. Additionally, the number of watering could be reduced, since the soil was not drying out too quickly.

Economic and energy characteristics of the instruments under study indicate their high reliability and economic feasibility [16, 17]. Despite the fact that the cost of LED lamps is 1.7 times higher than that of the discharge lamps, they consume 2.7 times less electricity and possess a 5 times longer guaranteed lifetime (Table 3).

Table 3. Economic and energy characteristics of the tested lamps.

Lamp	Manufacturer	Price with VAT, rub	Nominal power consumption, W/hour	Warranty, years	Expected service life, years	Expected service life, hours
GLP-FH-56-R	Green-LP	5,800	56	5	10	At least 60,000
GLP-FH-56-B	Green-LP	5,700	56	5	10	At least 60,000
GLP-FH-56-RB	Green-LP	5,800	56	5	10	At least 60,000
High-pressure sodium arc lamp ДНаТ 150	Russia (entire procurement set)	3,500	150	1		

Conclusions

Studies conducted demonstrated that LED lamps with different illumination spectrum can serve as an additional light source for plants grown inside greenhouses, positively affecting both growth and development of plants, and economical efficiency.

References

[1] Kasyuk S.T. Energy saving and energy efficiency policies in the European union. *Proceedings of the higher educational institutions. ENERGY SECTOR PROBLEMS.* 2015;(9-10):95-98. https://doi.org/10.30724/1998-9903-2015-0-9-10-95-98

[2] Gorshkov A S, Vatin N I, Rymkevich P P and Kydrevich O O 2018 Payback period of investments in energy saving *Mag. Civ. Eng.* **78** 65–75

[3] Savina N.V., Artjushevskaja E.J. Actual problems of realization of the federal law № 261 from 23.11.2009 in the Russian Federation in the part of heat supply. Proceedings of the higher educational institutions. ENERGY SECTOR PROBLEMS. 2017;19(3-4):31-40. https://doi.org/10.30724/1998-9903-2017-19-3-4-31-40

[4] Zubkov V I, Evseenkov A S, Orlova T A and Zubkova A V 2015 Monte Carlo Simulation of the Radiation Output from a Led Structure with Textured Interfaces *Russ. Phys. J.* **58** 1172–80.

[5] Yisihakefu A. Development of measuring complex for research of quality of led lighting

devices. *Proceedings of the higher educational institutions. ENERGY SECTOR PROBLEMS.* 2017;19(7-8):144-148. https://doi.org/10.30724/1998-9903-2017-19-7-8-144-148.

[6] Yisihakefu A., Tukshaitov R.H. Control the body temperature of LED lamps at work in different lighting devices. Proceedings of the higher educational institutions. ENERGY SECTOR PROBLEMS. 2017;19(9-10):146-150. https://doi.org/10.30724/1998-9903-2017-19-9-10-146-150.

[7] R.G. Bolshin. Increasing the efficiency of the potato meristem plants irradiation by LED phyto installations // Dissertation for a degree of Candidate in Technical Sciences/ M.: ARIAE SSE, 2016, page 148.

[8] N.N. Protasova. Importance of individual spectrum sections for plants photosynthesis, growth and productivity (under irradiation performed with energy or number of quanta). Greenhouse Service information bulletin / N.N. Protasova // 1995, No. 6-7, pages 24 -25.

[9] D.A Aliev. Photosynthetic activity, mineral nutrition and plants productivity // D.A. Aliev. Baku: Elm, 1974, page 344.

[10] V.V. Malyshev. On the possibility of assessing quantitative criteria for the different spectrum lamps in plant growing according to light parameters // Association of Greenhouses of Russia information collection. 1999. No. 2, pages 16-19.

[11] N.P. Kondratyeva. Increasing efficiency of electric irradiation installations in sheltered ground // Dissertation for a degree of Doctor in Technical Sciences/ M.: ARIAE SSE, 2003, page 251.

[12] A.N. Vinokurov. Thermal regimes of DORADO high-power LEDs / A. Vinokurov // Components and Technologies, No. 5, 2006.

[13] A.A. Tikhomirov. Light spectral composition and plant productivity. A.A. Tikhomirov, G.M. Lisovskiy, F.Ya. Sidko / M.: Novosibirsk, 2001, page 163.

[14] L.P. Varfolomeyev. LEDs and application thereof / L.P. Varfolomeyev - M .: News of lighting engineering, issue 3, 2000, page 131.

[15] E.G. Kulikova. Study of the urban ecosystems soil phytotoxicity // XXI Century: Results of the Past and Problems of the Present Plus, Periodical Scientific Publication, Penza State Technological Academy Publishing House, 2011, pages 88-91.

[16] E.G. Kulikova. Study of the special spectrum lamps influence on the Solanaceae family plants growth processes / In the collection: Agro-industrial complex: state, problems, prospects. Penza: PSAU ISSRC, 2017, pages 44-46.

[17] http://green-lp.ru/

Analysis of the change impact in the share of generation from HPPs in the structure of the unified power system energy balance to residential tariff

[1]O V Novikova, [1]A N Grushkin, [1]I S Khrebtenko, [2]N A Yudina

[1]Peter the Great St. Petersburg Polytechnic University, St. Petersburg, Russian Federation
[2]Kazan State Power Engineering University, city of Kazan, Russia Federation

novikova-olga1970@yandex.ru

Annotation. The paper analyzes the structure of the energy balance of the Russian Federation for power systems, draws conclusions on the impact of HPPs in the structure of electricity generation on the price of electricity for ultimate customers. The influence of the construction of new hydropower stations on the development of waterways is considered.

1. Introduction

The structure of the energy balance of any country can be viewed both from the point of commercial interests view of specific energy producers, the state, and from the point of view of the final consumers interests [1]. The development of the economy, especially in the part of energy-intensive industries, is directly related to the cost of energy resources [2]. It becomes important to understand the influence of the generation structure on the formation of the electricity price. The strategy of power capacities development should take into account the interests of all groups of consumers, including the population.

2. Assessment of the hydropower share in the energy balance

The structure of the installed capacity of the Unified Power Systems (UPS) power plants and UPS of Russia as of 01.01.2018, as well as the generation of electricity in the UPS of Russia in 2017 are shown in Tables 1 and 2 (information was obtained from an open source [3]).

Table 1. Balance of IPS power plants for installed capacity and for electricity generation (%)

Power system	Installed capacity, MW	Percentage of capacity / percentage of generation				
	Generation, mln kWh	HEP	CHP	NPS	WPP	SPP
UPS of Russia	239812,20	20,20	67,88	11,64	0,06	0,22
	1053862,03	16,98	63,70	19,25	0,01	0,05
IPS of Center	53077,07	3,37	71,01	25,62	-	-
	237546,50	1,84	55,91	42,25	-	-
	27203,76	25,60	59,23	14,97	0,13	0,07

IPS of mid-Volga Region	107782,40	23,99	46,07	29,94	0,00	0,00
IPS of Ural	52714,90	3,52	93,40	2,82	0,01	0,25
	260659,93	2,50	93,54	3,91	0,00	0,05
IPS of Northwestern	23865,23	12,36	63,48	24,14	0,02	-
	108354,80	13,07	52,87	34,06	0,00	-
IPS of Southern	21538,55	27,59	56,55	13,93	0,43	1,50
	100006,40	21,77	54,52	23,18	0,13	0,40
IPS of Siberia	51911,19	48,71	51,18	-	-	0,11
	202657,80	46,36	53,63	-	-	0,01
IPS of East	9501,50	38,52	61,48	-	-	-
	36854,20	33,30	66,70	-	-	-

Analysis of the table data shows that the potential for hydro resources of the regions, taking into account the already commissioned capacities, is not fully utilized. The best indicators for the power and generation balance were revealed in the North-West, where 12.03% of the hydroelectric power accounted for 13.03% of the power system. The worst balance ratio in the IPS of the South. And at the same time, there is a shortage of capacities, which can be realized precisely due to the energy of water resources.

Taking into account the different potentials for hydropower generation in the Russia regions and referring to separate integrated power systems, a study was made of the variable component of generation facilities separate groups in the IPS, where hydropower is present.

Table 2 shows data on variable costs in the generation structure for 2017. Variable costs depend on the amount of energy production. They are taken into account when forming a tariff for consumed electricity. Unlike them, fixed costs form a fee for capacity.

Table 2. Variable costs in the generation structure for 2017, RUR / MWh

Power system	HEP	CHP	NPS	SDPP
IPS of Center	21,75	1008,33	250,5	1087,94
IPS of Northwestern	21,10	938,78	250,3	1230,25
IPS of Southern	18,87	1020,97	226,9	1181,74
IPS of mid-Volga Region	16,56	1081,19	n/a	1411,9
IPS of East	n/a	n/a	-	-
IPS of Siberia	21,65	637,30	-	613,01

The results of the variable cost analysis unambiguously demonstrate the advantage of HPPs and NPPs, in the structure of circulating assets of which there are no (HPPs) or minimal (NPP) costs for fuel resources. An analysis was made of the connection of the share of HPPs in the IPS with the electricity tariff for the population in large settlements provided by electricity in the framework of the IPS considered (Table 3, Figures 1, 2).

Table 3. Dependence of the residential tariff on the HPPs share

Power system	HEP, %		Residential tariff for 2017 (averages taking into account changes from 1.07.2017), rubles / kWh
	capacity	generation	
IPS of Center	3,37	1,84	from 4,43 to 5,38
IPS of Ural	3,52	2,50	from 3,77 to 4,85
IPS of Northwestern	12,36	13,07	from 3,88 to 4,59
IPS of Southern	27,59	21,77	from 3,66 to 4,28
IPS of mid-Volga Region	25,60	23,99	from 3,50 to 3,64
IPS of East	38,52	33,30	from 2,77 to 4,27
IPS of Siberia	48,71	46,36	from 0,99 to 3,07

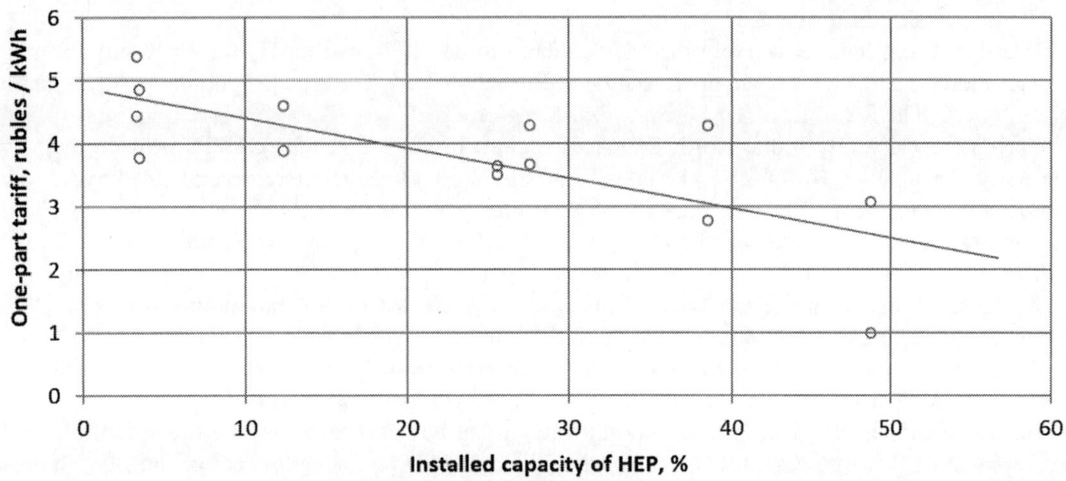

Figure 1. Dependence of the residential tariff on the HPPs share in the IPS at the installed capacity

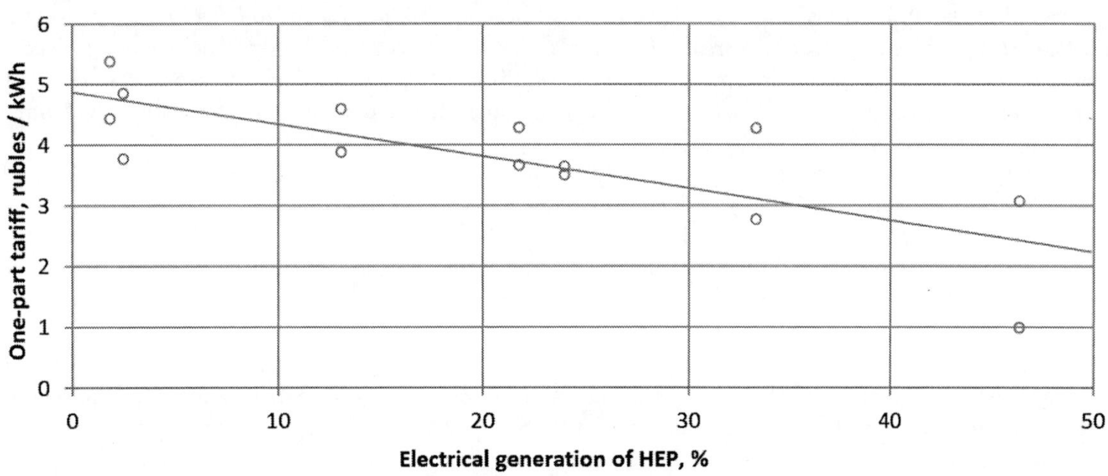

Figure 2. Dependence of the residential tariff on the HPPs share in the IPS on the electricity generated in 2017

IOP Conf. Series: Earth and Environmental Science **288** (2019) 012065 doi:10.1088/1755-1315/288/1/012065

The method of forecasting changes in tariffs with a change in the share of electricity generation with the use of hydro resources, allows to investigate graphical dependencies. Practically linear dependence is revealed. Although these schedules are somewhat arbitrary, since a number of factors influence the tariff, nevertheless, a clear dependence of the tariff reduction on the population on the share of electricity generation at hydroelectric power plants in the IPS is traced, which justifies the economic and social significance of increasing the generation share at hydroelectric facilities.

The confirmed connection allows making preliminary forecasts for the tariff taking into account possible changes in the HPPs share in the energy balance structure. The revealed parameters and relationships confirm the presence of both social significance of increasing the generation share at hydroelectric facilities and, undoubtedly, have a significant effect on the cost of industrial enterprises production, especially energy-intensive ones [4-5].

An analysis of regional electricity tariffs shows that they are significantly lower in regions with a high proportion of hydroelectric power stations in the structure of generating capacities. For example, in the Siberian regions [8], on the territory of three of which (the Krasnoyarsk Territory, the Irkutsk Region and the Republic of Khakassia) there are large hydroelectric power stations, the one-part tariffs for the population in 2018 are: RUR / kWh: Irkutsk Region - 1, 01; The Republic of Khakassia - 1,99; The Krasnoyarsk Territory is 2.37.

In neighboring regions where, practically, there are no HPPs, the tariffs for electricity are much higher, amounting to RUR / kWh: the Republic of Buryatia - 4.24; The Altai Republic - 4.84; Kemerovo region. - 3.15; The Republic of Tuva - 3,3; Tomsk region - 3.25.

Tariffs for electricity in the North Caucasus region: in Dagestan, where HPPs provide their own electricity consumption by 90%, -2.41 rubles / kWh; in North Ossetia, where several small hydropower stations generating about 20% of electricity, -3.86 rubles / kWh; in the Republic of Kabardino-Balkaria, in which hydroelectric power plants produce less than half of the power consumption - 3.58 rubles / kWh.

At the same time, in Kalmykia, where there is no hydroelectric power station, the tariff is one of the highest in the region and is 4.38 rubles / kWh.

Thus, hydropower facilities create the conditions (prerequisites) for a possible reduction of tariffs or containment of their growth while increasing the cost of hydrocarbon fuel for TPPs.

Based on the results of the performed analysis, it can be concluded that the transition to a wider development of fuel-free power is the most rational strategy for the development of the electric power industry in the conditions of the hydrocarbon resources depletion [6] and the predictable increasing redistribution in favor of the chemical industry. In these conditions, the role of hydropower, using renewable water resources, increases [7]. The substitution of thermal generation due to the development of nuclear and hydropower will make it possible to build a balanced highly efficient energy sector, reduce the burden on the environment [9], reduce the final cost of electricity, release hydrocarbon raw materials (for export, chemical industry), provide a comprehensive solution to the water development problems of the country.

3. Complex influence of hydropower engineering

The period 1960-1980. was a period of active hydropower construction, when 36 GW of installed capacity was commissioned in the Russian Federation. Thus, a little less than 2.0 GW was introduced per year. In these years, the creation of a transport-water-energy system in Europe - the Volga-Kama cascade of hydroelectric power stations - was completed.

Today, the Volga-Kama Cascade of HPPs provides the opportunity to operate the largest in Europe water transport system of Russia - the United Deep Sea System of the European part of the Russian Federation that connects the White Sea, the Baltic Sea, the Volga, Moscow, the Caspian Sea and the Azov Sea with the Black Sea [10].

In the regions of the Russian Federation, where there are operated inland waterways, about 90% of the gross domestic product is created and 80% of the population lives.

IOP Conf. Series: Earth and Environmental Science **288** (2019) 012065 doi:10.1088/1755-1315/288/1/012065

At the same time, inland waterway transport accounts for less than 2% of the total volume of freight transport and turnover of all types of transport in the Russian Federation, while in Germany 11%, the Netherlands 34%, France 10% trends in river transport growth.

The cost of transporting goods along rivers is 10 times lower compared to road transport and almost 2-fold - with rail. Cascade construction of hydropower facilities on the northern rivers of the Russian Federation, taking into account the need for development of waterways, will ensure transport access to explored and new deposits of natural resources, create conditions for the construction of enterprises for their processing and enrichment, will provide the opportunity to export the products. All this is extremely important for Siberia and the Far East, will contribute to the development and additional loading of the Northern Sea Route.

Thus, the construction of hydroelectric power stations can solve the problems of both electricity supply and transport accessibility of the newly developed regions of Russia.

The construction of hydroelectric and hydraulic structures not only provides an opportunity for the development of water transport, but also solves the problems of technical and drinking water supply, irrigation, irrigation of rivers for sanitary purposes, reduces damages from flooding of territories during the passage of floods. It should be noted that at present the expenses of our state for the liquidation of the consequences of floods, flooding and other natural disasters on water bodies are many times higher than the amounts allocated for the prevention of these phenomena and disasters.

Today, taking into account the need for a breakthrough development of the Russian Federation economy in conditions of limited financial resources, it is advisable to focus on projects that can have a cumulative effect on the entire economy, for the entire industrial development of Russia, and such a project can be modern energy in general and hydropower in particular.

In addition, given the existing redundancy of water resources in some regions of Russia, the creation of integrated use large reservoirs makes it possible to provide for the long-term prospect the possibility of exporting another strategic resource, in the short term, water. Currently, it is difficult to assess the economic effect of the such a resource sale, but this can already be accurately assessed as a possible strategic project of cooperation with neighboring countries that is not inferior to the project of China, the New Silk Road, in political and economic aspects.

Unfortunately, up to now there has not been developed a guideline document that would allow to link the construction of hydropower facilities with the development of electric power engineering, water transport routes, with the decision to reduce the costs of flood control, integrated water supply of territorial conglomerations. And this problem requires an early decision in order to increase the efficiency of each ruble invested in the development of energy.

4. Conclusions

Hydropower should be a key link in the development of energy in Russia. The construction of hydroelectric power stations not only creates the conditions for ensuring the possibility of regulating the tariff component for ultimate customers, but also provides an opportunity for more efficient operation of nuclear and thermal power plants. Unlike other sources of generation, it creates all the conditions for the integrated development of territories, one of the most important - the development of water transport routes, which our North and the Far East are in urgent need. Only this factor, namely the creation of water transport routes and the loading of the Northern Sea Route, can block the resources invested in the construction of hydroelectric power stations. This branch of the electric power industry has a number of advantages - reproducibility of resources, the absence of production waste, low operating costs, the possibility of multi-purpose use of reservoirs.

The revealed tendency of decrease in the price for the electric power at increase in a share of generation from hydroelectric power station in modern conditions of formation of the tariff and earlier at planned economy is the indisputable fact. At present, there is a task to minimize the burden on the regional budget, which is forced to cover the gap between the real cost of energy and the tariff that is set for the population. And the influence of the change in the structure of generating capacities in the direction of increasing the share of HPPs contributes to the solution of this problem.

References

[1] Erastov AE, Novikova OV. Rating of the attractiveness of regions for the implementation of energy saving projects. In the book: Innovative Economy and Industrial Policy of the Region (ECOPROM-2016), the proceedings of an international scientific and practical conference. Edited by A.V. Babkin. 2016. P. 103-111.

[2] KamenikL.L. An innovative model of the resource base for the reproduction of the economy is the strategic direction of the renewal of Russia's economic policy. In the collection: Forsyte "Russia": the design of a new industrial policy. Collection of materials of the St. Petersburg International Economic Congress (SPEK-2016). Under the Society. Ed. S.D.Bodrunova. M.: Cultural Revolution.

[3] Website: JSC "SO UES", "Report on the functioning of the UES of Russia in 2017" [Electronic resource] -: www.so-ups.ru.

[4] Erastov AE, Novikova OV Innovative energy saving: an integral method for assessing the motivational environment. Bulletin of the Ivanovo State Power University. 2017. № 2. P. 75-86.

[5] Erastov AE, Novikova OV Federal subsidization as a driver of innovative development of regions in the field of energy saving and energy efficiency. In the book: Restructuring the Russian economy and industrial policy Proceedings of the scientific and practical conference with foreign participation. Edited by A.V. Babkin. 2015. P. 102-106.

[6] KamenikLL. Increase in resource availability - new opportunities for Eurasian integration / analytical materials / Analytical materials of the First Congress of the Assembly of Peoples of the World. Moscow, May 2017 (Electronic resource) -www, rosprodunion.ru.

[7] Makarov V.M., Novikova O.V., Tabakova A.S. Energy efficiency in "green construction": experience, issues, trends. Reliability, Infocom Technologies and Optimization (Trends and Future Directions) 6th International Conference ICRITO. 2017. C. 732-737.

[8] Grushkin AN, Khrebtenko IS, Novikova OV Approaches to the development of a methodology for assessing the efficiency of the use of fuel and energy resources in certain territories. Economics of Energy and Energy Saving Materials of the International Scientific Conference.

[9] Kamenik LL Ecological and economic balance - a strategy for managing the innovative development of society XX! century / / Issues of innovative economy - 2018-Volume 8.-№1.

[10] Sergey Glazyev. The economy of the future. Does Russia have a chance? // The Economist. No. 3. 2015. M: With. 32-42

IOP Conf. Series: Earth and Environmental Science **288** (2019) 012066 doi:10.1088/1755-1315/288/1/012066

Prospects for the rational use of the resource potential of the Russian hydropower engineering

[1]O.V. Novikova, [1]L.L.Kamenik, [1]A.S. Furtatova, [1]P.I. Denisova and [2]T.Y.Dunaeva

[1]Peter the Great St. Petersburg Polytechnic University St. Petersburg, Russia
[2]Kazan State Power Engineering University, city of Kazan, Russia Federation

novikova-olga1970@yandex.ru

Abstract. The following aspects of development have been identified based on the introspective approach: conditions for the sustainable development of hydropower, hydropower and ecology, hydropower and scientific and technical development, development of the social environment and human capital. On this basis, the conditions and principles of sustainable long-term development of the industry are determined. The analysis made it possible to formulate recommendations on targeted long-term proposals for the development of the industry in the areas analyzed.

In 2017 in Russia, the Parliamentary Assembly of the Collective Security Treaty Organization formulates a model law on energy security. Energy security (hereinafter EB) assumes the desire to protect citizens, society, the state, the economy from the threat of a deficit in providing their energy needs with affordable energy resources of acceptable quality, from threats of disruptions in the continuity of energy supply. It is becoming urgent to study the role of hydropower in the provision of these conditions. The study of the basic principles of the provision of energy security makes it possible to highlight the consideration of environmental safety requirements, which implies the development of the fuel and energy complex (FEC) in accordance with increasing environmental requirements, the prevention of irrational use of energy resources (the relationship with the energy efficiency policy), the continuity of the process of ensuring energy security[1].

Conditions for sustainable development of hydropower

The main threats to energy security include the deterioration of the state of the fuel and energy base. This parameter is positively influenced by the increase in the share of non-carbon generating capacities, which include hydropower. The regions of Siberia and the Far East should now occupy a special place in the development of the hydropower industry of the Russian Federation. This was said by the President of the Russian Federation V.V. Putin in his messages to the Federal Assembly in 2012: "In the 21st century, the development vector of Russia is the development in the East of Siberia and the Far East. This is our colossal potential. " The President's address to the Federal Assembly on December 12, 2013 contains a direct instruction: "Russia, with its huge hydro resources, uses its potential for less than 20%. It is necessary to begin the construction of large hydroelectric power stations, primarily in Siberia and the Far East[2]. "

Among the threats to energy security is the lack of competitiveness of certain types of fuel resources and, as a consequence, the lack of the necessary diversification of the structure of the fuel and energy balance of the regions and the state as a whole.

Sustainable energy development is determined by the balance of production (availability) and consumption of energy resources. Sustainable development, with regard to hydropower, is predetermined by the fact that reproduction of the mineral and raw materials base is in a stable state, which ensures a sustainable long-term development of the industry. This is the fundamental difference and advantage of hydropower from all other types of power in the factor of subsoil use .

In the context of assessing the future development of the electric power industry, this fact should be paid special attention. The document "Forecast of the scientific and technological development of the fuel and energy sector up to 2035" states: "According to the available forecasts, for the entire period until 2035 the main areas for the growth of continental oil and gas reserves will remain the West Siberian , Leno-Tunguska , Timano-Pecherskaya and Volga -Ural oil and gas bearing provinces[3]. However, in the main areas of oil and gas production, the available reserves and resources of a distributed subsoil fund will be able to ensure the reproduction of the mineral resource base in the medium term by about half. "

Under the new conditions of development, the existing balance of hydropower and subsoil use is a balance that is of strategic importance for the electric power industry and for the country's economy as a whole. Therefore, today the task should be to increase the efficiency of using the available resource potential of hydropower.

Positive conditions for the development of generation in hydropower include the existence of a stable and significant reserve of water and energy potential, reserves for energy saving and energy efficiency of the industry , development of small hydropower, which will provide an opportunity to increase hydrocarbon exports, and through this, indirectly, an increase in the country's budget. Hydropower development and construction of pumped storage stations will significantly expand the market for renewable energy, based on the wind - and the photogeneration, remove their negative impact on the unified energy system of Russia.

The shortcomings include: low innovative potential, significant development of the park resource of generating equipment and, as a consequence, the technical backwardness of the industry development, the risk of making the wrong decision about the prospects of hydropower development.

The role of hydropower in the future of Russia's energy industry can not be underestimated. So, for example, one of the promising areas of energy development that allows solving ecology problems and partial replacement of natural hydrocarbon fuel at the same time in the future is the production of hydrogen on the basis of large hydrogenerating objects.

All of the above dictates the need to develop a long-term comprehensive program for the rational use of Russia's existing resource potential with the active development of carbon-free energy, including through the development of hydropower as its main part. This program should be built on the basis of the unity of energy production with the resource potential of Russia in the future for a period up to 2050. Such a temporary landmark is dictated by the stability of the resource potential of hydropower and the forecast of the development of the energy resource base. After 2050, a profound transformation is possible, both in the economy and in the direction of energy development. By this time, the resource potential of hydropower as a stable and stabilizing branch of energy supply should be realized. This strategy should not only become the general direction of hydropower development as an industry, but it should also be implemented in the interests of developing the energy sector in general and the country's economy in general.

Hydropower and ecology

The domestic experience of hydropower construction testifies to the possibility of solving acute environmental problems that periodically occurred at various stages of the use of hydropower resources. Until now, a one-sided approach to environmental impact assessment, which is provided by the construction of hydrotechnical facilities on the environment, prevails. This view requires rethinking.

The overall ecological effect to date has not been fully appreciated by the scientific community and deserves a separate study.

Environmental problems associated with the development of hydropower in recent decades are constantly attracting the attention of various specialists and are the subject of discussions at numerous international and domestic sites, which marked the ongoing massive construction of hydroelectric power stations in the world and an increase in the scale of the impact on nature[4]. The need was recognized for using hydropower resources, shifting priorities towards environmental protection, in analyzing environmental problems, in recognizing the urgent adoption of environmental measures, in finding the most accessible ways to implement them.

By the end of the eighties of the last century, a set of normative and methodological documents on environmental protection for the ecological and economic justification of hydropower and water management facilities was prepared in our country[5]. Despite a sharp reduction in the rate of hydropower construction in modern Russia, environmental studies are continuing and the necessary sectoral documents are being prepared in the following key areas:

- analysis and assessment of the impact of hydropower facilities on the environment;
- modeling of impacts;
- forecast of environmental consequences;
- management of the impact of hydropower facilities on the environment and the rationale for environmental protection measures.

This contributes to a qualitative consideration of environmental issues in the design of new and reconstruction of existing hydropower facilities, justification of the implementation of affordable measures to ensure environmental safety of hydropower facilities.

Currently, in connection with global climate change, the issue of reducing greenhouse gas emissions becomes acute. In view of the contradictory nature of the assessments, it has become necessary to conduct large-scale studies at reservoirs in order to determine their impact on the ecological environment. There is a place for the existence of opposing points of view, justifying the contribution of reservoirs in creating or improving the ecological situation in the developed territory.

It should be borne in mind that the hydropower industry of Russia annually saves about 50 million tons of equivalent fuel. It should be noted and the life-saving function of hydroelectric power stations. Thus, the development of each bln. KW • h hydro instead of coal-fired CHP leads to a reduction in mortality at 110-220 people. in year. In general, the operation of the HPP allows avoiding damage to Russia by reducing the environmental impact of industry on a person . The recreational effect of HPP reservoirs is estimated by reducing the incidence rate for 10 days per year per person as a result of a month's recreational recreation near water bodies [6].

Obviously, in order to preserve and develop hydropower, it is necessary to develop a concept aimed at preventing negative phenomena in the natural environment during the life cycle of the facility[7]. It can be based on the idea of Academician M.P. Fedorov providing rational use of natural resources in hydropower through the creation of natural-technical systems, which served as the beginning of the formation of an appropriate scientific direction. The increase of the ecological and economic effect in the development of hydropower, carried out on the basis of innovative and resource potential, is critically important.

The existing impact of hydropower, as well as all areas of energy, on the environment and the livelihoods of the population is assessed negatively by society. It should be noted that with the unique utility of each of the energy sectors for society, the manifestation of negative consequences for each of them has its own specifics, and therefore each of them has its own tasks and ways to improve environmental and economic efficiency.

The overall goal for energy is formulated in the "Report on the Sustainable Development Goals, 2017" by the United Nations, which identifies 17 goals as an independent goal - Goal 7: Low- cost and clean energy.

According to the Environmental Policy of JSC RusHydro , the main challenges associated with the technical and technological features of the Company's production activities are the operation of existing ones, the design and construction of new hydroelectric power stations in the field of environmental protection are as follows:

1. Insufficient orientation of the existing and accessible design (technical and technological) solutions for the environmental aspects of hydrogeneration limits the possibilities of increasing

environmental efficiency and minimizing environmental risks in the design, construction and operation of HPPs. There are technical and economic limitations in the implementation of modern requirements in the direction of ensuring environmental protection at existing HPPs built and commissioned in previous periods.

2. The most promising sections of the possible (expedient) deployment of new hydroelectric power plants are located in the least developed areas that have preserved their natural natural wealth. The construction of new hydroelectric power plants under such conditions inevitably affects the habitat of rare species of animals and plants, including those listed in the Red Books of different levels.

3. Construction of hydropower plants and reservoirs inevitably leads to the formation of a new water ecosystem, which is primarily determined by the use of water resources in the technological cycle of electricity production.

Ways of increasing the ecological and economic effect in the development of hydropower are as follows:

1. Necessity of confirming the status of hydro generation as the most environmentally friendly form of electricity generation.

For this, it is necessary to conduct an objective socio-ecological and economic assessment of the projected and economic activity of all types of energy generation based on the foresight methodology : a single end-to-end system of indicators that comprehensively assesses the performance of all activities. Based on the results of this foresight methodology , a comparative assessment of the consequences - positive and negative characteristics of the analyzed energy generation trends - should be made. This will help to get away from the magic of calculations in one direction or another. Using the opportunities that opens up the work with data , dast the possibility of this is really done. An expert evaluation of the foresight methodology should be carried out throughout the life cycle of energy generation: the invasion of nature-production-waste.

2. It should take a strategic decision on the intensive transfer of hydropower from traditional large- scale facilities to small hydropower. Small hydropower is one of the most promising sources of energy in the future, economically feasible and environmentally justifiable. Feasibility is determined by a wide range of indicators based on the foresight methodology .

3. T he best world practices should be disseminated and implemented in the field of hydropower.

4. According to the results of the expert evaluation of the foresight methodology , an order should be made for the development of innovative technologies of domestic production to solve existing problems in hydropower.

Hydropower projects, taking into account the innovative and resource potential, should become a model of ecological security: how resources are used -such is the ecology. The increase in environmental safety is measured in economic terms (RUR , $) and directly leads to an increase in the economic effect in hydropower.

Hydropower and scientific and technological development

The Russian Federation, having the largest resource base capable of meeting the needs of the country and foreign markets in energy resources, traditional and promising fuels for the coming decades, should actively participate in the process of energy conservation and the formation of a future energy sector that meets the criteria of energy and economic efficiency and environmental safety. It is necessary now to create all the conditions and prerequisites for building such energy, which in the short term could provide competitive advantages for our industry and create the prerequisites for energy expansion into international markets. A key role in this should be played by hydropower.

Hydropower is a branch of industry and energy that unites the possibility of developing the branches of the national economy that are under the jurisdiction of the Ministry of Energy, the Ministry of Transport, the Ministry of Natural Resources and the Ministry of Industry and Trade [8]. All this proves the need for a comprehensive approach and consideration of all the facts that affect the development of the state when developing the hydropower potential of Russia.

Taking into account the necessity to build horizontal inter-branch relations, the scientific and technological and resource capacity of hydropower construction, we can talk about the advisability of creating a digital polygon based on this energy sector. The transition to a new technological level of planning, construction, operation of energy facilities will not only improve the reliability and safety of generating facilities, but also have a positive impact on their investment attractiveness, increase financial transparency for the entire life cycle of the facility. It is possible to realize this by creating a single information space that links both the resource flow and the entire technological cycle of the creation and operation of a hydroelectric power facility based on domestic software and technological developments. The expediency of initiating the construction of a new technological digital system based on hydropower, among other things, is due to the fact that hydropower still retains full technological independence from imported equipment and technologies. While the scientific-design and laboratory-experimental base is preserved, the experience and qualification of the personnel are preserved. We believe that this will contribute to the achievement of the set goals for the development of not only the energy industry, but also the state.

The industry has the potential and understanding of the need to build close cooperation between related industry companies, science and higher education, and certain steps are taken independently in this direction. The task is to preserve the human potential, to involve young people in active scientific and practical work, including, within the framework of the program "From individual abilities of man to an efficient economy"[9].

However , in the hydropower industry today, there are certain negative trends in reducing the activity of construction of new generating facilities, which has already begun to negatively affect the training of personnel for the industry, the degradation of education, the loss of qualified personnel, retirement or redirection of industry organizations. This process must be stopped already in the short term, otherwise this technological key direction of the energy sector will be irretrievably lost for Russia and will become completely dependent on Western companies and, first of all, China[10].

The problem is that when assessing the investment attractiveness of hydropower facilities, only the direct economic effect from the sale of electricity is taken into account, without assessing the complex impact of hydrotechnical facilities on the national economy. With a direct account because of the technical peculiarities, the resource capacity of hydropower facilities they will always lose thermal generation at the initial payback periods. Here, of course, it is necessary to change the methodological basis for calculating the economic efficiency of hydroelectronics and take into account the need for state participation.

The design of new facilities should be carried out on new principles:

- increase the role of science in making decisions to justify constructive and layout decisions, the main task for which should be to ensure the introduction of the most modern technical and technological solutions that are not inferior to, and superior to foreign ones, and for this as long as there are all conditions;
- ensuring the implementation of a new project in the format of a single information space created for all key stakeholders of the process;
- ensuring the possibility of a comprehensive solution of the issue of development of territories and industry;
- eduction of environmental impact and more rational use of non-renewable resources.

According lnaya or partial implementation of the HS energy security roses inevitably leads to destabilization of functioning fuel systems - and the power that can slow down economic development, as well as the aggravation of problems of social and environmental protection of the state.

Development of the social environment and human capital.

The development of the hydropower sector can not be achieved without the accelerated development and effective use of human capital in hydropower.

To ensure development and effective use of human capital in hydropower, the following tasks must be solved:

IOP Conf. Series: Earth and Environmental Science **288** (2019) 012066 doi:10.1088/1755-1315/288/1/012066

1. Development and wide discussion of the long-term strategy of hydropower development for 50-100 years as a sustainable perspective direction of the present and future development of the country's energy and economy. Formation of the image of the future hydropower.
2. Development and implementation of long-term strategies in the field of human capital management by hydropower structures, ensuring an effective level of investment in human capital from hydropower structures and creating attractive high-performance jobs.
3. Creation of tracking and anticipating branch trends in the system of vocational education and training of specialists and workers, ensuring systematic development and implementation of innovations and development of breakthrough technologies in hydropower.
4. Development and distribution of new forms and programs of state and corporate training, training, retraining and advanced training based on the integration of production, science and education, including the creation of industry competence centers for hydropower.

Effective creation of human capital should be supported by appropriate social measures and social programs, both public and corporate, ensuring:

- Creation of a system of retaining competences (accumulated experience and special knowledge) in the case of a change of generations;
- worthy social benefits, guarantees and compensation to employees of hydropower based on social standards, enshrined in sectoral agreements between employers of the industry and trade unions of workers[11].

The solution of the set tasks will contribute to the increase of labor productivity and the qualification of personnel in the traditional and "breakthrough" technological areas.

As a result, a new generation of future-oriented workers is formed, with sufficient competence to work on high-performance and high-tech jobs in hydropower, capable of developing, mastering and exploiting new technology and technologies that are responsibly related to the future of future generations.

Thus, the development of hydropower in the context of human capital acts as:

- the source of investment demand for new innovative knowledge and technology,
- Environments for the introduction, use and dissemination of new scientific knowledge and technologies

Both these aspects are realized on the basis of developed human capital.

Literature:

[1] Erastov A.E., Novikova O.V. Innovative energy saving: integrated method of estimation of motivation enviroment, Bulletin of the Ivanovo State Power University. 2017. № 2. P. 75-86.

[2] Erastov A.E., Novikova O.V. Federal subsidisation as a driver of innovative development of regions in the field of energy saving and energy efficiency, In the book: Restructuring the Russian economy and industrial policy Proceedings of the scientific and practical conference with foreign participation. Edited by A.V. Babkin. 2015. P. 102-106.

[3] Erastov A.E., Novikova O.V. Rating of attractiveness of regions for implementation of projects in the field of energy saving In the book: Innovative Economy and Industrial Policy of the Region (ECOPROM-2016), the proceedings of an international scientific and practical conference. edited by A.V. Babkin. 2016. P. 103-111.

[4] Makarov V.M., Novikova O.V., Tabakova A.S. Energy efficiency in "green construction":experience, issues, trends In the collection: Reliability, Infocom Technologies and Optimization (Trends and Future Directions) 6th International Conference ICRITO. 2017. pp. 732-737.

[5] Kamenik L.L. Ecological and economic balance - a strategy for managing the innovative development of society XX! century / / Issues of innovative economy - 2018-Volume 8.-№1.

[6] Kamenik L.L., Furtatova A.S. Modern problems of modernization of the water supply system in the conditions of innovative development (on the example of the city of St. Petersburg). Economics and entrepreneurship. 2017. №4-2 (81-2).

[7] KamenikLL. Increase in resource availability - new opportunities for Eurasian integration / analytical materials / Analytical materials of the First Congress of the Assembly of Peoples of the World. Moscow, May 2017 (Electronic resource) -www, rosprodunion.ru
[8] Kamenik LL The past and future of Russia in the global raw materials market: a new "super-reality", // Competitiveness in the global world: science, technology. Economy. № 5 2017
[9] Sustainable development: new challenges: Textbook for high schools / Under total. Ed. VI Danilov-Danilyan, N.A. Pikulova. _M .: Aspect-Press Publishing House, 2015
[10] Glazyev Sergey. The economy of the future. Does Russia have a chance? // The Economist. No. 3. 2015. M: With. 32-42
[11] CamenikL.L. An innovative model of the resource base for the reproduction of the economy is the strategic direction of the renewal of Russia's economic policy. In the collection: Forsyte "Russia": the design of a new industrial policy. Collection of materials of the St. Petersburg International Economic Congress (SPEK-2016). Under the Society. Ed. S.D.Bodrunov. M.: Cultural Revolution.

Substantiation of the need for long-term planning of hydropower development as an element of energy management in Russia

[1]O.V. Novikova, [1]L.L. Kamenik, [1]A.S. Furtatova, [1]P.I. Denisova and [2]S.A. Livshits

[1]Peter the Great St. Petersburg Polytechnic University St. Petersburg, Russia
[2]Kazan State Power Engineering University, city of Kazan, RussiaFederation

novikova-olga1970@yandex.ru

Abstract. At the present stage of social development, the need to assess the prospects for the rational use of the hydropower resource potential in Russia had been manifested as an independent unresolved problem, which urgently requires the study of its essence, regularities, the specifics of manifestation and the mechanisms for its solution. This testifies to the urgency of the conducted research, whose purpose was to conduct analytical work on the evaluation of current priorities and directions in energy development and the study of new approaches to long-term planning in hydropower development. The solution of the task was accomplished on the basis of a systematic approach to forecasting the resource potential, energy, hydropower and economic development of the country, which is the novelty of the work. The conclusion is the increasing role of hydropower. The article is interest to scientists, government employees, professors, students and all those who are interested in sustainable development of the society.

The need to build a stable foundation for the modern economy, create conditions for the development of the internal market, concentrate resources and capital to solve the strategic tasks of the energy sector is becoming especially necessary nowadays.

The previously developed scenarios for the socioeconomic development in Russia have lost their relevance and require revision by far, taking into account the newly discovered external and internal factors. For example, in the normative document "Scheme and program for the development of the unified energy system of Russia for 2017-2023", approved by the Order of the Ministry of Energy of 01.03.2017, the average annual increase in demand for electricity in 1.0%. These days, the forecast demand for electricity is formed on the basis of the medium-term forecast about the main macroeconomic parameters of the basic scenario in Russia's socio-economic development forecast. According to this scenario, prepared by the Ministry of Economic Development, until 2023, the annual average growth in Russia's GDP is expected to increase over 2.0%.

Thus, today, the development of the energy industry (including the power grid infrastructure) is directly related to the forecast of the state's economic development, and the pace of this development, with objective reasons, for the accepted scenarios essentially lag behind the pace of development of the economy itself. [1] How true is this approach? Moreover, Russia today is one of the few non-volatile states, and in the future the country has every reason to claim the role of an energy leader. Such a statement of the question requires serious consideration.

1. Analysis of the current development trends in energy

Currently, the leading role in the energy sector of Russia is played by thermal stations that represent more than 65% of the electric power production structure, which is about 160 GW of capacity and more than 670 billion kWh of electricity generation (230 million tce). More than 70% of the electricity produced by TPPs is provided by gas stations. It can be assumed that, for objective reasons, during the planning horizon of 25-30 years, there will be a decrease in the share of thermal generation in the structure of electricity generation. The main replacement of this will be because of nuclear energy. [2] This trend has already evidence with the approval from the Ministry of Energy "Scheme and program for the development of the Unified Energy System of Russia for 2016-2022".

Today, there is a strategy for the development of the electric power industry in the conditions of the depletion of hydrocarbon resources and the predicted ever-increasing redistribution in favor of the chemical industry. At the same time, one should not discard the energy sector based on traditional renewable resources. These resources doesn't comtemplate a lack of fuel component and is highly maneuvering, providing system services in improving the quality of electrical energy and creating conditions for more efficient operation of nuclear power plants. [3]

Hydroenergy potential of water resources in our country is used only by 20% of its capacity, and in the Far East - only by 6%. Replacement of thermal generation due to the development of nuclear and hydropower will allow to build a balanced high-efficiency power system, reduce the load on the environment, reduce the final cost of electricity, release hydrocarbon raw materials (for export, chemical industry), solve flood protection problems and flood river transport infrastructure. [4]

In the conditions of limited financial resources, one of the main tasks for the state and energy companies is to reduce the capital costs per 1 kW of installed capacity, then, under the existing conditions of the economy, the payback period. [5] At the same time, there is an unrealized potential in the hydropower industry to increase the investment attractiveness of hydropower facilities due to the introduction of modern technical and technological solutions, the use of modern materials, a new, more efficient approach to the organization of interaction between scientific and design organizations, the introduction of a unified information space for design, construction and operation of hydropower facilities.

The risk analysis of the materials of the Energy Strategy of Russia from the point of view of the development of the energy management system begins with the analysis of energy policy. Even in the case of realization with possible risks they are not able to change the purpose, tasks, priorities, directions of development and basic elements of the state energy policy. Observed shifts in the timing to achieve the specified values of some of the target indicators in 3-5 years. Selected perspective directions are being revised. The target scenario is taken as a basis and the backlog from it was solved with relatively low costs. It was corrected by a slowdown in the implementation of energy programs and projects. The initial orientation in the forecast for the slow development of the economy and the fuel and energy sector is fraught with excessive damage - down to the reduction of export niches and containment of the country's economic growth lack of energy capacity. [2]

The authors of this article agree with this forecast and consider its implementation a necessary condition for an intensive, breakthrough development of the Russian economy.

At the same time, it should be noted that the draft energy strategy envisages an increase in electric power generation by 2020 by 6%, and by 2035 - by 27-43% (from 1,062 to 1,352-1514 billion kWh) with an increase in the installed capacity of power plants at 13- 25% (from 250 to 282-312 million kW). [2]

Taking into account the need to modernize the obsolete and exhausted equipment of power plants, which now represent more than 60%, this forecast looks more than ambitious. In fact, if we take into account the main power equipment, the production capacity will be the double.

Moreover, there are already serious concerns about the possibility of realizing the target scenario of development, primarily for the electric power industry. This is due to:

- the absence of well-developed mechanisms and plans aimed at implementing the tasks outlined in the Strategy to achieve the set goals;

- the presence of serious limitations in the possibility of attracting long-term investment resources, which is not enough even for a comprehensive modernization of existing energy capacities;
- direct linking the pace of development of the industry with forecasts of the growth rates of the Russian economy;
- the dependence of the Russian energy sector on foreign technologies, equipment, materials, software and services, which in a number of areas has reached a critical point and created threats to Russia's energy security. First of all, this relates to TPPs and oil production;
- shifting the priority of development in favor of carbon-free energy (wind and photo energy), which has a low economic return and requires the creation of capacity reserves.

These constraints must be lifted in the shortest time possible, otherwise the declared goals cannot be achieved. A starting momentum is needed, which would be the starting point in the implementation of the Strategy and in the breakthrough development of the economy of the Russian Federation. This starting point is definitely to be searched inside the fuel and energy complex. [6]

Today we must use the existing foreign policy and economic situation to create and develop internal conditions for increasing with efficiency the development of our industry, using the existing potential and developing a promising one. The priority here must be the energy, whose main tasks should be a significant increase in energy efficiency in a short term or in a medium term - creating the conditions for reducing the costs of industry. [7]

Separately, it should be noted that the implementation of the target scenario for the development of energy in Russia requires the availability of appropriate scientific and technical competencies, production capacities and profile contracting organizations. To solve the stated ambitious tasks, this entire complex requires the rapid and serious qualitative changes.

2. A new approach to long-term energy development planning

A new approach to long-term energy development planning is required in connection with the following needs:
- accounting and linking the development of the country's energy industry with the development of industrial production;
- the creation of competitive advantages in the near future for the domestic industry;
- stimulating the development of the domestic market due to the current profound transformations in the global economy;
- opening of "bottlenecks" and expanding the boundaries of the area of decision-making it seems advisable to shift the time horizon for the planning of the new Energy Strategy of Russia for 2050.

Until the end of the 2030s, it can be assumed that the development of the world and Russian energy will be determined mainly by the usual fuel energy and its problems. But, since 2050, a profound transformation in the direction of renewable energy sources, a new way of energy, a fundamentally new structure of world energy services and technologies is possible.

Simultaneously, the matrix of long-term planning for the development of the electric power industry should be based on a new approach.

Earlier, the forecasts were based on the model: "need - provision", where new available deposits were opened and put into operation to meet the demand, new generation facilities were built without taking into account the limited initial resource base, the environmental consequences, the economic effectiveness of the decisions made in the long term . [8]

In modern conditions, the situation has changed significantly. First, there is a reduction in the initial resource base for the A + B + C field groups, which leads to an increase in production costs and, consequently, a constant increase in prices for natural raw materials (gas, coal, oil). Market prices no longer cover the costs of their production. Secondly, there is an extremely slow decline in the energy intensity and electrical capacity of GDP, which, of course, affects the cost of production. Thirdly, there were strict requirements to minimize the environmental impact. All this dictates the need to create conditions for sustainable effective development in the energy sector, with support in the medium term for own carbon-free energy, domestic producers, and domestic scientific and technical potential. [9]

In the context of the new conditions for the development of the Russian economy, a new approach to building long-term development of the electric power industry is needed. The new approach should be based on a closer linkage of energy production with resource potential and environmental consequences according to the model: "capacity - demand - provision".

An integrated approach based on the unity of production of energy resources, the forecast of electric generation, the forecast and implementation of conditions for the integrated development of productive forces, should become the basis for long-term planning of the development of the electric power industry. The principle of this unity is that a comparative analysis of options for long-term development of the electric power industry should be carried out. It is necessary to avoid the "situational" solution of emerging problems, taking into account the conditions of the industry development on the basis of modeling the entire life cycle of both a single object and the entire system – in conclusión, the whole industry.

The target vision of the development of Russia's energy industry for 50 years ahead is to determine the long-term guidelines for the development of the industry with the basis of a realistic approach to the innovative development of the Russian economy, the expected structural shifts in the world economy, with an assessment of the positions of their influence on Russian energy and its place industrial economy of the world. [10]

The energy strategy should include the following tasks:

• advanced development of the electric power industry in comparison with the development of the economy on the basis of the use of the most efficient production complexes and scientific developments;

• creating conditions for increasing economic growth rates by reducing production costs both fuel and energy complex and the economy as a whole;

• expanding the resource base of fuel industries and product markets;

• develop with the support of the national scientific and technological base and branch enterprises;

• take into account the complex effect of the construction of certain types of generation on the industrial and technological development;

• ensure the creation of conditions and mechanisms for increasing the investment attractiveness of new construction projects;

• ensure the construction of a self-sufficient efficient industry with the prospect of explosive expansion to foreign markets.

At present, the most objective and complete are assessments of the life cycles of energy facilities, including nuclear power, traditional carbon energy and renewable energy. The important characteristics of energy technologies are estimated by their payback periods, as well as specific environmental assessments of their impact on it. This approach, taking into account the complex influence of the generating facilities under construction on the development of the economic complex, will allow building the most effective energy in Russia and solving the strategic tasks of the state development.

The development of the energy complex will provide an opportunity to develop natural resources in the Far East and Eastern Siberia and will ensure the conditions for the creation of large industrial power-intensive complexes oriented there, including for export.

3. Strengthening the role of hydropower in the development of the energy complex

Under the new conditions, the role and importance of various sectors of the Russian energy sector is changing: carbon (coal, gas, etc.), hydropower, nuclear and renewable.

The main goal of the Energy Strategy of Russia for the long-term perspective is "sustainable, safe and efficient functioning of the energy sector for the development of the economy and improving the welfare of the population of the country."This long-term goal has not lost its relevance nowadays. What is the role of hydropower in the future development of the Russian energy sector? What rates should hydropower develop under the new conditions and challenges? What resources are available for this? [11]

IOP Conf. Series: Earth and Environmental Science **288** (2019) 012067 doi:10.1088/1755-1315/288/1/012067

The general scheme for the development of the electric power industry until 2035 provides the construction of only two new hydraulic stations: Nizhne-Zeyskaya and Motyginskaya, with a final capacity of 1,482 MW. [2]

In the current conditions of planning for the near future, the share of HPPs in the structure of electricity generation, as much and taking into account the ongoing modernization program, will remain at the achieved level of 19% of its capacity.

At the same time, hydropower is a powerful complex-forming center in its regions. On the basis of cheap electricity, modern industrial and social infrastructure is formed, contributing to sustainable regional development, raising the standard of living of the population. After commissioning the HPP, the cost of electricity does not depend on the level of fuel prices and the situation on the world market. Hydroelectric power plants perform a number of system and other functions to ensure the reliability of the functioning of the power systems. Many of Russia's hydroelectric power plants are objects of integrated importance: they regulate river runoff, ensure the requirements of navigation, irrigation, water supply or flood protection, among others. [12]

The main disadvantage of HPPs, in comparison with other generation facilities, is the high unit cost of capital investments per 1 kW of installed capacity. In part, this is a constructive feature of this source of electricity. But this gap can and should be shortened. [13] It can be stated that there is an unrealized potential in the hydropower industry to increase the investment attractiveness of hydropower facilities due to the introduction of modern technical and technological solutions, the use of modern materials, and a new, more effective approach in organizing interaction between scientific and design organizations.

The positive contribution of hydropower stations to the development of energy was assessed by the international community. For example, the hydropower potential in the US and Canada was used by about 58-75%, in Brazil by 53%, in Western Europe by 80%, and in Norway by 96% . China today is actively building hydroelectric facilities, outstripping the rest of the world in commissioning capacities. For comparison, the indicators of the use of hydropotential in Russia: about 40% - in the European part of the country, 23% - in Siberia and less than 6% - in the Far East.

However, it should be noted that the indicators of Russia are from 1967. Currently, due to the socio-economic changes in the country, the economic hydropower potential needs to be adjusted (due to the strengthening of environmental restrictions, the emergence of alternative opportunities for the use of natural resources aligned in the construction of hydropower, changes in the economic situation and the system of market assessments used in the analysis of efficiency implementation of hydropower projects). Therefore, the definition of the current economic hydropower potential today is the most important and one of the priority tasks. [14]

4. Conclusions

1. The maximum effect for the development of our economy, the development of industrial production can be obtained with the rational use of local resources, primarily renewable. Hydropower should play an important role in this process in the long-term planning of its development.

2. In the face of great uncertainty and disproportions in the development of energy systems, there is a need for long-term planning. Based on the definition of energy management, any actions within the framework of general managerial functions that contribute to the realization of the goal and objectives of the state energy policy, it is important to understand the target parameters. It is the long-term hydropower planning that acts as an element of energy management that meets the target task of minimizing the costs of fossil fuels, by developing renewable energy sources.

3. A new approach to long-term planning is necessary, first of all, in connection with the need to take into account the limitations of the initial resource base.

References

[1] Kamenik L L 2016 Innovatsionnaya model' resursnoy bazy vosproizvodstva ekonomiki - strategicheskoe napravlenie obnovleniya ekonomicheskoy politiki Rossii. / Kamenik L L.

Forsayt "Rossiya": dizayn novoy promyshlennoy politiki / sbornik materialov Sankt-Peterburgskogo mezhdunarodnogo ekonomicheskogo kongressa (SPEK-2016) / pod obshch. red. Bodrunova S D. — M.: INIR, Kul'turnaya revolyutsiya, pp 375-380

[2] Prognoz nauchno-tekhnicheskogo razvitiya otrasley toplivno - energeticheskogo kompleksa Rossii na period do 2035 goda. Moskva, 2016

[3] Kamenik L L 2017 Proshloe i budushchee Rossii na global'nom rynke syr'ya: novaya "sverkhreal'nost'" *Competitiveness in a global world: economics, science, technology*, №**5**, pp 69-73

[4] Erastov A E, Novikova O V 2015 Federal'noe subsidirovanie kak drayver innovatsionnogo razvitiya regionov v oblasti energosberezheniya i energoeffektivnosti. Restrukturizatsiya ekonomiki Rossii i promyshlennaya politika / Trudy nauchno-prakticheskoy konferentsii s zarubezhnym uchastiem / pod red. Babkina A V, pp 102-106

[5] Kamenik L L 2017 Povyshenie resursoobespechennosti - novye vozmozhnosti evraziyskoy integratsii / analiticheskie materialy / Analiticheskie materialy I s"ezda assamblei narodov mira. Moskva, [Electronic resource] – www.rosprodunion.ru

[6] Danilov-Danil'yan V I **Ustoychivoe razvitie: Novye vyzovy**. Uchebnik dlya VUZov / pod obshch. red. Danilova-Danil'yana V I, Piskulovoy V I. — Aspekt-Press Moskva, p 336 (2015)

[7] Erastov A E, Novikova O V 2017 Innovative energy saving: the integrated method of estimating motivational environment *Vestnik of Ivanovo State Power Engineering University,* №**2**, pp 75-86

[8] Kamenik L L **Resursosberegayushchaya politika i mekhanizm ee realizatsii v formate evolyutsionnogo razvitiya**. Monografiya / Kamenik L L. Sankt-Petersburg. (Izd. 2-e, dop.) pp 476 (2012)

[9] Kamenik L L 2018 Ecological and economic balance - a strategy for managing innovative development of the society of the XXI century *Innovative Economy Issues* Vol. **8**, №1, pp 25-38

[10] Glazev S Yu, Chistilin D K 2017 Where goes Russia? (Analysis of the proposed programs for socio-economic development of the country) *Russian Economic Journal*, №**4**, pp 36-48

[11] The Sustainable Development Goals Report. United Nations, New York, 2017, pp 32-33

[12] Furtatova A S, Kamenik L L 2017 Contemporary problems of modernization of a water supply system in the reality of innovative development (by the example of the city of St. Petersburg) *Journal of Economy and enterepreneurship*, № 4-2 (81-2) Vol.**11**, Nom. 4-2, pp 868-873

[13] Erastov A E, Novikova O V 2016 The rating of the regions attractiveness for implementation of energy-efficiency projects / Innovation economy and industrial policy of region (ECOPROM – 2016) / Works with the research-to-practice conf. with foreign participation 22-24 September 2016 / pod red. Babkina A V, pp 103-111

[14] Makarov V M, Novikova O V, Tabakova A S 2017 Energy efficiency in “green construction”: experience, issues, trends / Reliability. Infocom Technologies and Optimization (Trends and Future Directions) 6th International Conf. ICRITO, pp 732-737

Energy-saving technologies for soil purification from chromium (III) ions using higher plants

Lubov' Ol'shanskaya[1], Natalia Politaeva[2], E Tatarinceva [1], E Danilova[1], A Tareeva[1]

[1]Saratov State Technical University named after Yu. Gagarin Russian Federation, Saratov, Politekhnicheskaya, 77

[2]Higher School of Biotechnology and Food Technology, Peter the Great St. Petersburg Polytechnic University, 194021, Russia, St.Petersburg, Novorossiyskaya, 48-50

ecos123@mail.ru

Abstract. This work presents investigation of IR-irradiation influence on seeds of beans (Phaseolus), its' growth and development affection during soil purification from chromium (III) ions by energy-saving reagentless phytoremediation technique.

Introduction

The territory of the Russian Federation, which is polluted by heavy metals (HM) is now more than 70 mln. hectares, 1 mln hectares of which are of extremely dangerous level of pollution. Soil and water pollution is a global problem and might lead to irreparable consequences [1–8]. Damage of the fertile layer inexorably leads to destruction of natural balance, metabolic processes in nature, which might result in destruction of other ecosystems [9–11]. In our country there arises a strong demand for urgent actions aimed at reduction of soil pollution by heavy metals. In recent years in many countries intensive works for polluted soil purification by higher plants are carried out. The advantages of this method, known as phytoremediation (from Ancient Greek φυτό (phyto), meaning 'plant', and Latin remedium, meaning 'restoring balance') are possibilities of soil purification from heavy metals, organic compounds, oils and petroleum products, etc. [12]. At this, harmful bacteria are destroyed by this phytomass, soil is enriched by oxygen, obtained during phytosynthesis process. At this the major part of toxic substances is decomposed by components. In plants there is no excessive accumulation of harmful substances and the grown green mass after harvest might be used as a feed for livestock, birds, for production of paper and biofertilizers, recycling for gas and liquid fuel [13,14]. Soil purification by phytoremediation may be considered as the most efficient and cheap method, as special estimations show that today's expenses for it usually do not exceed 20% from alternative technologies expenses [15]. For implementation of this method one need only solar energy.

In the recent decades numerous facts have been discovered, which stated for high sensitivity of plants to impact of external physical fields (EPF) of various nature: ultraviolet (UV), infrared (IR) radiations, which create additional electrical currents at biological objects. These exposures vary the membrane potential of the plant cell and can govern the organism growth and evolution processes having both stimulative and inhibitory impacts. This impact depends on external exposure characteristics: wavelength, oscillation frequency of electromagnetic radiation, intensity and time [16]. Such seed treatment is a progressive way of its preprocessing to planting, which allows one not only to remove them from dormant state, but to activate functioning of various biological catalytic agents - ferments, which provide rapid growth and development of plants, and, consequently, speed-up of phytoremediation processes. The cell wall contains proteins, pectins, phospholipides and others containing fixed charged groups (primarily carboxyl ones). They denote cation-exchange ability and affect HM cation accumulation in a cell from soil solution by higher plants.

This work presents investigation of IR-irradiation influence on seeds of beans (Phaseolus) higher plants, its' growth and development affection during soil purification from chromium (III) ions by phytoremediation technique.

Materials and Methods

Test culture grain bean (*Phaseolus vulgaris*), breed Ruby was selected as a highly productive culture, which might be used for phytoremediation purposes. As pollutants we have chosen model solutions of $Cr_2(SO_4)_3$ of 1, 5 and 10 maximum permissible concentrations (MPC) [17] (general sanitary maximum permissible concentration for Cr^{3+} ions is 6 mg per 1 kg of soil). We took moving form of metal in soil. As a source of IR-irradiation for bean seeds we used blue "Minin" lamp with λ=780-1400 nm. Seeds treatment was performed at a selected time for 6 hours [18].

Results and discussion

The conducted investigations on influence of nature and concentration of Cr^{3+} ions on plant height and development of Phaseolus lamina (Table 1, Figs.1-3) have shown that after 14-21 days the plants haven't shown any evidence of growth depression and leaves dying away.

It is known, that chromium presence positively affects plant growth. Concentration of this element in seed nucleotides is approximately 100 times higher than in total mass of the plant cell [19], which might be caused by its functional role. Chromium increases chlorophyll content and productivity of photosynthesis in leaves. Positive impact of this element on corn photosynthesis productivity was shown by A.T. Scheglov. The productivity of grown plants has increased for 24-40%, chlorophyll concentration for 16-29%, weights of green and dry mass for 34-65% [20].

Table 1 - Influence of IR and Cr^{3+} ions concentration on growth and development of Phaseolus

Phaseolus Plant height, mm					Phaseolus Leaf length, mm				Phaseolus Leaf width, mm			
Days	K*	1 MPC	5 MPC	10 MPC	K*	1 MPC	5 MPC	10 MPC	K*	1 MPC	5 MPC	10 MPC
7	130	0	0	20	0	0	0	0	0	0	0	0
14	210	80	80	210	70	40	10	53	65	20	10	60
21	290	260	260	265	70	60	60	54	65	60	60	57
28	300	270	280	284	60	63	50	55	60	63	52	60

Note: K* is control without IR-irradiation and HM

Chemical properties make it possible to suppose that chromium, being in the organisms, is not an indifferent metal, but plays an important role in plant lifecycle[21]. It is referred to metals with alternating valence, which are especially active at complex-formation processes.

In 28 days after adding of Cr^{3+} in soil the Phaseolus plants were took out of soil, then dried out and weighted. Roots were separated, as the previous histochemical investigations had shown, that metal accumulation generally took place in roots of plants-excluders, to which the Phaseolus belongs [22,23]. Complexes are formed in roots which are partially immobilized with donor-ligands. According to [24] a significant amount of chromium also moves to leaves and stems.

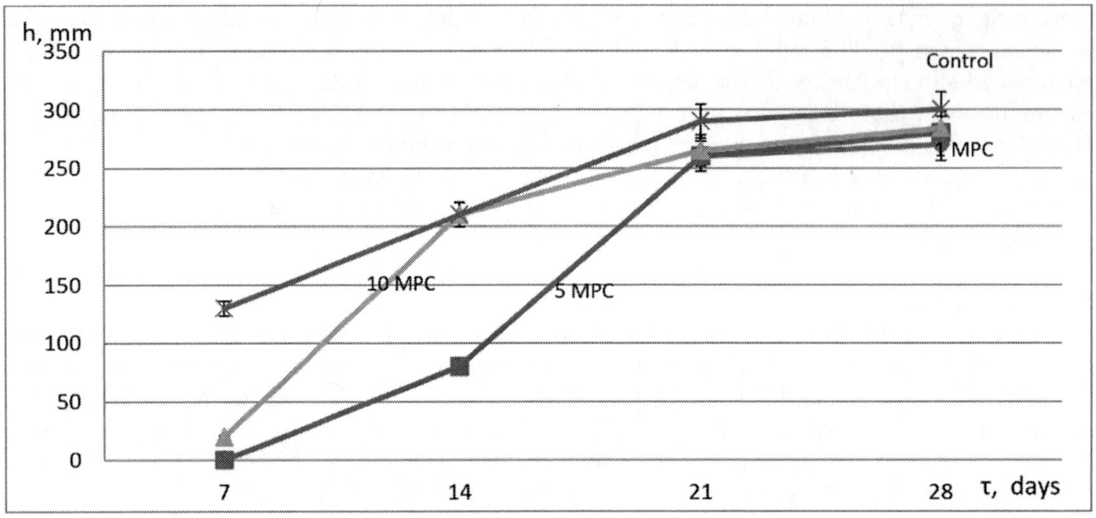

Fig. 1 Combined influence of Cr^{3+} ion concentration and IR-irradiation of seeds (τ_{irrad}= 6 h) on height of Phaseolus plants

In order to estimate influence of chromium ion concentration and its combined impact with IR-irradiation on soil phytoremediation processes we carried out determination of residual concentration of Cr^{3+} ions in soils according to Russian State Standard GOST 17.4.3.01-83 "Protection of soil nature. Basic requirements for sampling".

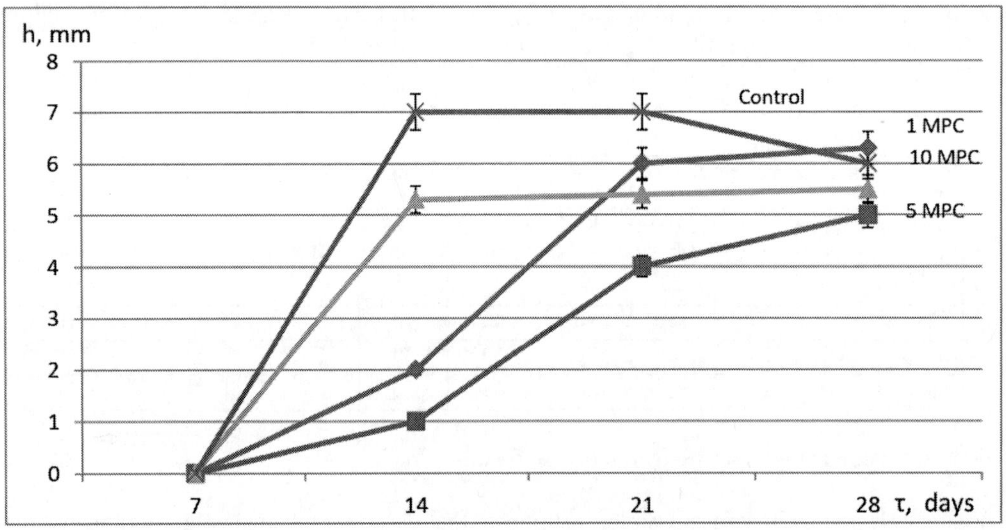

Fig. 2 Phaseolus leaf length change in time depending on Cr3+ ion concentration and IR-irradiation of seeds (τirrad.= 6 h)

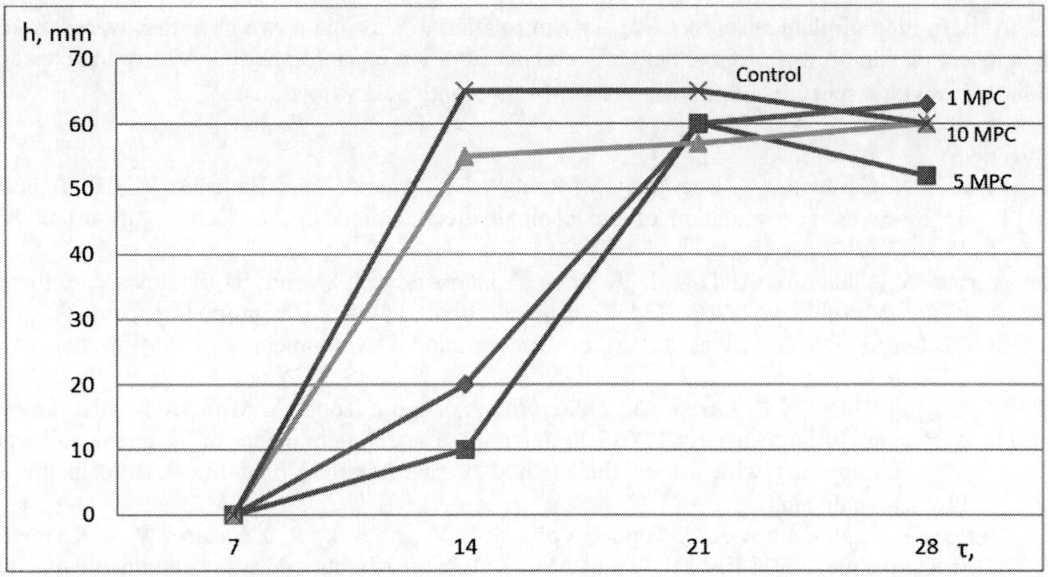

Fig. 3 Phaseolus leaf width change in time depending on Cr3+ ion concentration and IR-irradiation of seeds ($\tau_{irrad.}$= 6 h)

We prepared soil extractions and analyzed content of residual metal concentration by photometric technique.

Data on efficiency (Epur, %) of soil purification from chromium ions by phytomeliorant plants seeds of which were treated by IR-radiation and without treatment are presented in Table 2.

It was established that at the lowest Cr^{3+} concentrations (1 MPC) the efficiency of soil purification by Phaseolus plants decreased. Phaseolus showed higher purification efficiency at chromium concentration in soil of 10 MPC and IR-irradiation.

Table 2 - Impact of Cr3+ ion concentration with and without IR-exposure on efficiency (E, %) of soil purification by Phaseolus and adsorption capacity of plants (Ai, μkg/kg) after 28-days incubation.

	E, %			Ai, μkg/kg		
Cr^{3+} concentration	1 MPC	5 MPC	10 MPC	1 MPC	5 MPC	10 MPC
Without IR	71.1	92.7	90.6	70.6	91.2	79.1
IR	73.4	94.9	97.3	74.1	99.3	86.4

Adsorption capacity (Ai, μkg/kg of soil) of plants (Table 2) was established according to the amount of metal absorbed by Phaseolus after drying-out and wet-ashing of phytomass (in terms of absolutely dry weight of plants). The highest adsorption capacity of Phaseolus appeared to be at chromium concentration of 5 MPC and at IR-exposure of seeds.

Conclusions

1. We have studied impact of chromium (III) nature and content on processes of growth and development of Phaseolus plants, as well as its properties for soil phytoremediation. The highest toxical impact on the plants is seen at increase of HM concentration in the soil.

2. It was shown that efficiency for purification of soils polluted by Cr^{3+} ions was 73-97% for Phaseolus and increased with metal concentration growth. This might be caused by functional role of metal. According to literature data, chromium increases chloroplhyll content and photosynthesis productivity in leaves which, as a consequence results in completeness of pollutant removal.

3. At definition of plant adsorption capacity in relation to chromium, we have discovered extreme dependence: maximum was observed at Cr^{3+} ion concentration equal to 5 MPC. Adsorption capacities of Phaseolus with respect to chromium were ≈ 74 - 99 μm/kg of dry phytomass.

References

[1] Arefiev N, Garmanov V, Bogdanov V, Ryabov Y, Terleev V and Badenko V 2015 A market approach to the evaluation of the ecological-economic damage dealt to the urban lands *Procedia Engineering* vol 117pp 26–31

[2] Terleev V, Nikonorov A, Togo I, Volkova Y, Garmanov V, Shishov D, Pavlova V, Semenova N and Mirschel W 2016 Modelling the Hysteretic Water Retention Capacity of Soil for Reclamation Research as a Part of Underground Development vol 165(Elsevier Ltd)pp 1776–83

[3] Terleev V, Ginevsky R, Lazarev V, Nikonorov A, Togo I, Topaj A, Moiseev K, Abakumov E, Melnichuk A and Dunaieva I 2017 Predicting the scanning branches of hysteretic soil water-retention capacity with use of the method of mathematical modeling vol 90(Institute of Physics Publishing)

[4] Terleev V V, Nikonorov A O, Togo I, Volkova Y V, Ginevsky R S, Lazarev V A, Khamzin E R, Garmanov V V, Mirschel W and Akimov L I 2017 Hysteretic water-retention capacity of sandy soil *Mag. Civ. Eng.* **70** 84–92

[5] Orlova T, Melnichuk A, Klimenko K, Vitvitskaya V, Popovych V, Dunaieva I, Terleev V, Nikonorov A, Togo I, Volkova Y, Mirschel W and Garmanov V 2017 Reclamation of landfills and dumps of municipal solid waste in a energy efficient waste management system: Methodology and practice *IOP Conference Series: Earth and Environmental Science* vol 90

[6] Svatovskaya L, Kabanov A and Sychov M 2017 Soling, Aerating and Phosphating for Soil Strengthening and Detoxication vol 189(Elsevier Ltd)pp 398–403

[7] Alexeev D V, Shagivaleev A A, Urazbakhtina L R and Napoykina E A 2015 Improving the efficiency of water recycling systems at wastewater treatment power plants from insoluble suspensions *Proc. High. Educ. institutions. ENERGY Sect. Probl.* **1–2** 11–7

[8] Sandakov V D, Podrezov A N and Valeev I M 2015 Improving performance purification plants when exposed to ultraviolet radiation of the atmosphere *Proc. High. Educ. institutions. ENERGY Sect. Probl.* **3–4** 96–100

[9] Ilin I, Kalinina O, Iliashenko O and Levina A 2016 Sustainable Urban Development as a Driver of Safety System Development of the Urban Underground vol 165(Elsevier Ltd)pp 1673–82

[10] Didenko N, Skripnuk D and Mirolyubova O 2017 Modeling the changes in global temperature due to pollution vol 17(International Multidisciplinary Scientific Geoconference)pp 577–86

[11] Usoltceva M, Volkova Y, Radchenko A, Guseva I, Makarova T, Garmanov V and Wenkel K-O 2016 Assessment of Natural Resource Potential of a Territory for Planning of Investment Development and Construction in Suburban Areas vol 73(EDP Sciences)

[12] Nikitin D P and Novikov Y V 1986 *Environment and man (in Russian)* (Moscow: Higher school)

[13] Ol'shanskaya L N and Sobgayda N A 2012 *Phytoremediation of metals from water. Influence of external physical fields on phytoremediation processes acceleration. Monography (in Russian)* (Berlin: LAP LAMBERT Academic Pablishing)

[14] Fernandes M C, Ferro M D, Paulino A F C, Chaves H T, Evtuguin D V and Xavier A M R B 2018 Comparative study on hydrolysis and bioethanol production from cardoon and rockrose pretreated by dilute acid hydrolysis *Ind. Crops Prod.* **111** 633–41

[15] Protasov V F 2000 Ecology, health and environment protection in Russia *Moscow Financ. Stat.* 116–9

[16] Altukhov I V, Fedotov V A and Ochirov V D 2010 Change of basic quality characteristics of wheat seeds after various irradiation (in Russian) *Bull. IrSAgA* **40** 107–15

[17] Kolesnikov S I, Zharkova M G, Kazeyev K S and Val'kov V F 2009 Modeling of Ordinary Chernozem Pollution by Lead with the Purpose of Determination of Ecologically Safe Concentration *Ecol. Ind. Russ.* **8** 34–6

[18] Ol'shanskaya L N, Titorenko O V and Eremeeva Y A 2015 Influence of constant magnetic field and ultraviolet radiation on growth of higher plants and phytoremediation of soil from oil products *Chem. Pet. Eng.* **5** 43–5

[19] Mysienko N N and Ternavskiy A I 1989 *Root nutrition of plants* (Kiev: Higher school)

[20] Scheglov A T 1981 Influence of chromium on some physiological characteristics of corn *Usage Fertil. microelements growth Regul. Agric. Proc. Stavropol Agric. Univ.* **44** 35–40

[21] Alekseev Y V 1987 *Heavy metals in soil and plants* (Leningrad: Agropromizdat)

[22] Ol'shanskaya L N, Khalieva A S and Titorenko O V 2015 Lokalization of heavy metals (Pb, Ni) in tissues and organs of beans and soya beans during the process of its removal from soil with and without magnetic field action and UV-irradiation on plant seeds *Izv. Vyss. uchebnyh Zaved. Him. i Him. tekhnologiya* **58** 89–94

[23] Micera G and Dessì A 1988 Chromium adsorption by plant roots and formation of long-lived Cr(V) species: An ecological hazard? *J. Inorg. Biochem.* **34** 157–66

[24] Wallace A, Soufi S M, Cha J W and Romney E M 1976 Some effects of chromium toxicity on bush bean plants grown in soil *Plant Soil* **44** 421–73

IOP Conf. Series: Earth and Environmental Science **288** (2019) 012069 doi:10.1088/1755-1315/288/1/012069

Perfection of the methodology for developing industrial secondary energy generation systems

L V Plotnikova, R R Giniyatov, S Y Sitnikov, M A Fedorov and R S Zaripova

Kazan State Power Engineering University, Russia, 420066 Kazan, Krasnoselskaya, 51

mikhailovalv@mail.ru

Abstract. The algorithm for selecting an energy-efficient secondary energy generation system for large industrial enterprises is proposed. The updated coefficients of the system use of heat and exergy and the Sankey diagram are given. The software implementation of thermodynamic calculations of complex industrial schemes with secondary energy generation devices is presented.

Introduction

Industrial enterprises are complex combinations of tens and hundreds of elements of equipment that are a consumer or generator of thermal energy. Thermal energy is directed not only to useful needs, but also is lost in significant quantities with waste energy (secondary energy resources (SER)). The issues of SER recycling were considered at the end of the XX century, but the main attention was paid to the utilization of high-temperature and medium-temperature SER [1]. The low-temperature SER for a long time remained unreached [2] due to the lack of appropriate utilization equipment for flows of this potential. The appearance of heat transformers, currently used mainly in the heat supply of buildings [3], opened the possibility of using low-capacity resources in industry [4-6]. Consequently, the issue of organizing secondary energy generation systems at industrial enterprises remains topical. Therefore, the subject of research is the organization of the optimal structure of hardware systems for the recovery of secondary energy flows.

In addition, the devices in the technological scheme of production are interrelated, each connection carries information about pressure, temperature, enthalpy, flow rate of the substance. On large objects, many connections are reversed, that is, they are sent to one of the previous elements [7, 8]. That is, the operation of each device of the scheme can affect the operation of the previous device, as well as the graphics of energy consumption and the output of secondary resources in the technological line in which this connection is included. As a result, the process of calculating such a system with feedbacks can be in the nature of numerous iterations. Therefore, an important task is the choice of methods and means of calculation of complex structured heat power systems. The solution of this problem will allow to find the optimal structure of the energy system of the industrial enterprise, in particular, with the inclusion of secondary energy generation units.

The following method is usually used to select the variant of secondary energy generation system [9]:

- analysis of the structural organization of the initial production, with the waste energy, and organized energy systems of secondary energy generation (1st stage of system analysis);
- analysis of the thermodynamic efficiency of the system (the second stage of system analysis);

- selection of the SER recycling system with the highest potential and their redirection to the devices with the lowest efficiency.

It is proposed to correct the algorithm for selecting a secondary energy generation system; coefficients of effectiveness evaluation; carry out structural and thermodynamic analysis of the system in a single software product, which will confirm the correctness of improving the methodology for developing industrial secondary energy generation systems.

Evaluation of the efficiency of the secondary energy generation system

As a result of the implementation of this technique, secondary energy generation systems were selected that do not consume all of the waste energy streams of the heat technological scheme of production. For example, for a pyrolysis scheme in the production of ethylene, the streams not found in the secondary energy generation system are presented in Table 1 [10].

Table 1. Flows without potential for use in the scheme

Discharge energy	Flow rate G	Temperature t	Heat flow Q	Exergy of flow E	Heat value factor K=E/Q
	ton/h	°C	KW	KW	
Recycled water of scrubbers	91,8	30	31909	4110,4	0,13
Smoke gases from the recovery boiler unit	2,33	150	1371	505,9	0,37
Total:			33280	4616,3	0,14

The coefficient of system use of heat [9] in such a system of $KCИ_Q$ was 83 %, the coefficient of the systematic use of the exergy $KCИ_Q$ – 72 %.

The flows presented in Table 1 have an impact on the reduction of the final efficiency coefficients; however, all possible to the application of waste energy flows in the production of ethylene are used in the generation system completely [6, 8].

It is proposed to adjust the coefficient of system use of heat $KCИ_Q$ (formula 1), which will determine the true share of the useful use of waste energy in the secondary energy generation system:

$$KCИ_Q = \frac{\Sigma Q^{**}}{\Sigma Q^{*}} = \frac{\Sigma Q^{*} - \Sigma D}{\Sigma Q^{*}} = \frac{\Sigma Q^{*} - (\Sigma Qn + \Sigma Q' сбр1 + \Sigma Q' сбр2 + \Sigma Q'' сбр)}{\Sigma Q^{*}} \quad (1)$$

where $\Sigma\ Q$ *- heat supplied to the object; $\Sigma\ Q$ ** - useful heat used; $\Sigma Qсбр$ - the heat abstracted from the object flows of the substance (particularly with secondary energy); ΣQn - heat loss; ΣD - total losses; the sign «'» means heat with the streams taken away from the object, which can be useful used on a particular object under consideration; the sign «"» means heat with the streams taken from the object, which can not be used for a particular object under consideration in connection with its infrastructure; "1" - the waste energy flow used (in case of use, it is removed from the numerator); "2" is an unused flow of waste energy.

It is suggested in the choice of the system of generation of secondary energy to highlight and delete the formula component $Q''сбр$. The calculation results for the production of ethylene in this case: $KCИ_Q$ = 18 % - if none of the waste energy streams in the production of ethylene is used (useful energy is the energy of production, and all other streams will be considered losses, for example, heat loss with flue gases, condensate etc.); $KCИ_Q$ = 83 % - without exception $Q''сбр$; 88 % - with the exception of $Q''сбр$. In the latter case, the coefficient of system use will approach its maximum, that is, the efficiency, if all possible secondary energy flows are recycled in the system, which will objectively reflect the situation. The exergy coefficient of the system use is calculated similarly (*КСИЕ* "grown" from 65 % to 72 %).

The correction of the coefficient of system use is relevant when production has surplus of waste energy, or does not take into account the impossibility of using the potential of SER. Underestimation of the coefficient of system use in connection with this fact is eliminated.

It is also proposed to take into account the ratio of planned to recycle flows with the possibility of their "acceptance" in the "weak spot" in the available quantity *G*. Previously, the choice of the flow for the production of additional amounts of secondary thermal energy was made on the basis of data on the flow temperature without regard to its flow, hence the value of heat and exergy did not indicate the most energy-intensive flow.

Further, the potential of flow exergy K=E/Q was evaluated. For a visual assessment, Sankey diagrams were used (Fig. 1a). The ratio of the planned recycling streams and their possible "acceptance" in the "weak spot" in the available quantity *G* it is proposed to take into account the modification of the diagram (Fig. 1b).

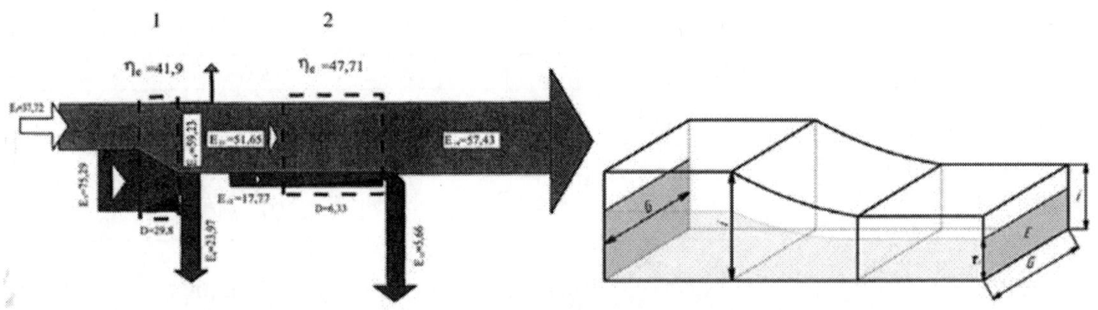

a) b)

Fig. 1. Flow diagram of exergy for the elements in the schema production: a) Diagram of Sankey for the system of "1" and " 2»; b) addition to the Sankey diagram

So, the sequence of selection of hardware systems of secondary energy generation from the discharge is as follows (after structural and thermodynamic analysis):

1. in order to avoid mistakes when choosing the flow of waste energy and the "place" of its use, it is necessary to initially choose the flows with the greatest exergy $E = f(G, T)$;
2. compare the selected flows by the availability of consumers for them in production, taking into account the value of G in order to reduce the unused "surplus" of secondary energy and to discard flows that do not satisfy any consumer. The problem of "surplus" secondary energy affects the artificial understatement of efficiency;
3. to evaluate the secondary energy generation system in accordance with the above-described criteria.

Software implementation

To calculate the schemes with feedbacks, which are the schemes of industrial enterprises with secondary energy generation systems, to exclude numerous iterations, the software is developed, which carries out the above thermodynamic calculations and analysis of the structural organization of the system in a single product [6, 11]. Example – a section of the ethylene production scheme (Fig. 2), for which the Sankey diagram was constructed (Fig. 1a). As can be seen from Fig. 2, flow rates and enthalpies of flows are associated with each flow of the scheme.

The developed software works as follows (Figure 3):

1. filling in the flow matrix (fig. 3a). It is proposed to denote the flows by specific numbers, so

that the program binds thermodynamic parameters to each flow;

2. finding connections matrix multiplication, reflecting the connection between the devices (Fig. 3b is formulated as "перемножение матриц") [6, 7, 11]);
3. search for contours (cycles) - the connection is closed to itself (in Fig. 3c is defined as "поиск циклов");
4. it is proposed to add search for sequences of unclosed elements (Fig. 3d formulated as "поиск хвостов");
5. determination of the places of contour rupture (Fig. 3i is formulated as "разбиение циклов");
6. after the break threads are extracted from the MS Excel file, the specified thread parameters are extracted (Fig. 3f; here "№ потока" – No flow's);
7. When the flow data is set, the calculation of thermodynamic parameters is started (Fig. 3g; here "Расчет параметров контура" - this is "Contour parameters calculation", "Расчет закончен, т.к. поток с № 3 разорван" - "The calculation is finished, because the flow number 3 is broken").

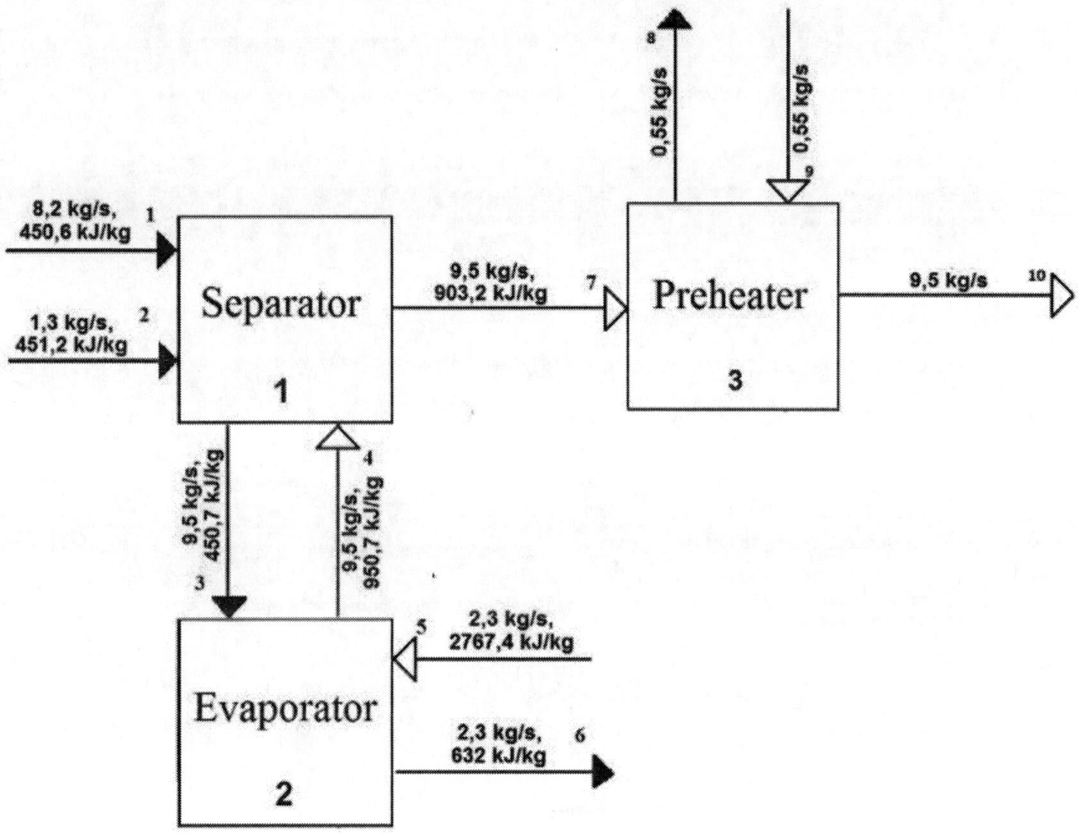

Fig. 2. Heat power system section: 1 - separator; 2 - evaporator; 3 – heater

In Fig. 3f it can be seen that the raw data is taken from Fig. 2

In Fig. 3g it can be seen that the enthalpies of the flux are calculated for the circuit of Fig. 2.

In the same block (Fig. 3g) also incorporated the calculation of the calculated effectiveness ratios of elements and system as a whole.

Summary

The scientific significance of the work lies in the fact that it offers an improved method of finding energy-efficient options for circuit solutions for the organization of systems of complex use of secondary energy. New criteria for the choice of options for the inclusion of secondary energy generation units are substantiated.

The method of system modeling and its software implementation in the C# programming language in Microsoft Visual Studio to assess the energy efficiency of industrial thermal technological schemes is developed.

On the basis of the materials of this article, the optimal complexes of recovery of secondary energy for petrochemical production, gas separation processes, chemical production, pulp and paper production are developed [5, 6, 8, 12].

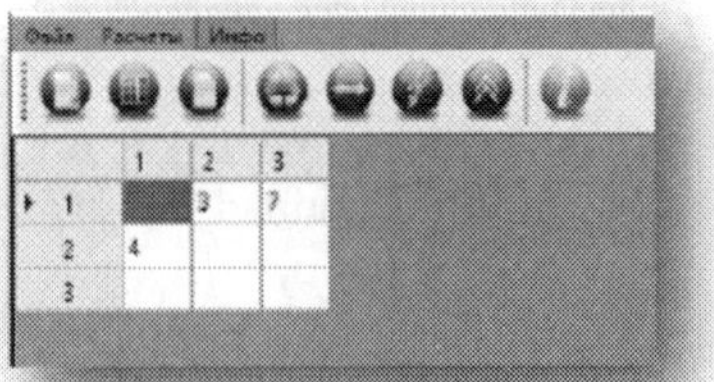

a) The matrix presentation of flows

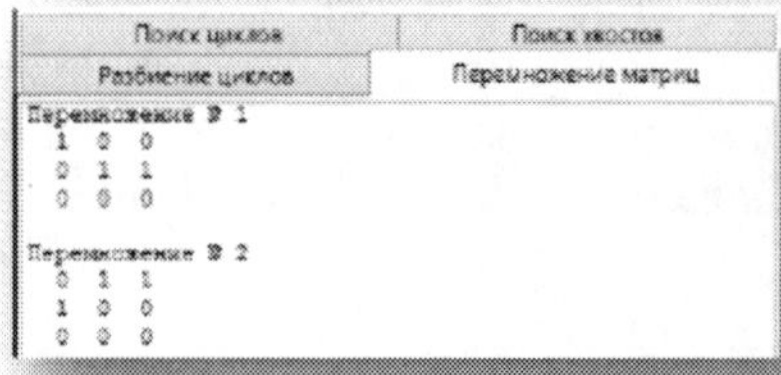

b) Results of multiplication of the adjacency matrix

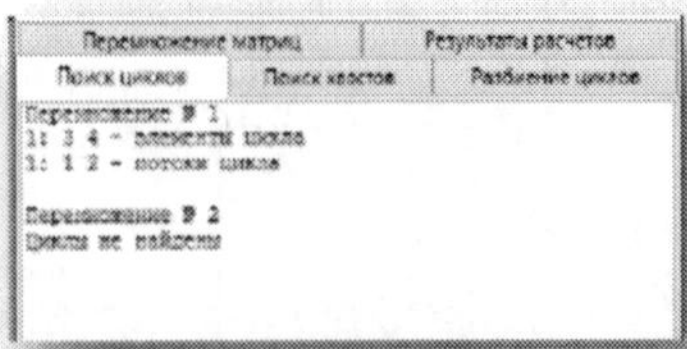

c) Search results for paths (cycles)

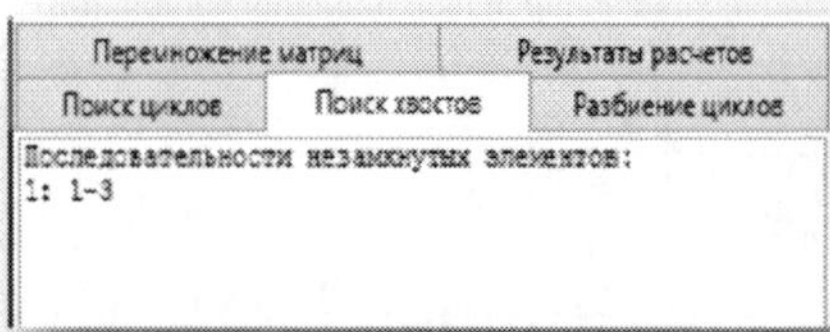

d) Search results for open sequences

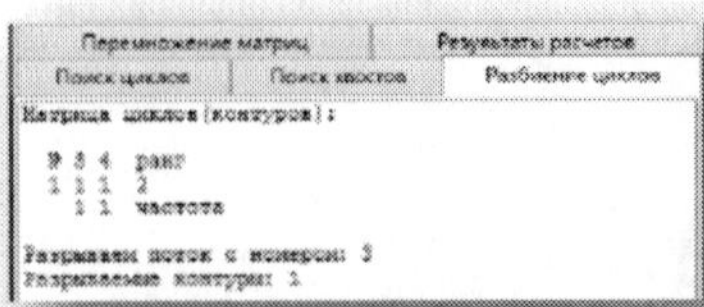

i) The matrix circuits and the results of the search for optimal tear streams

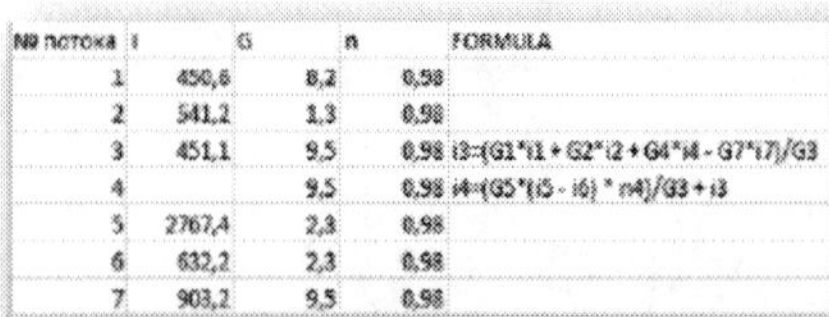

f) The flow variables defined in the MS Excel file

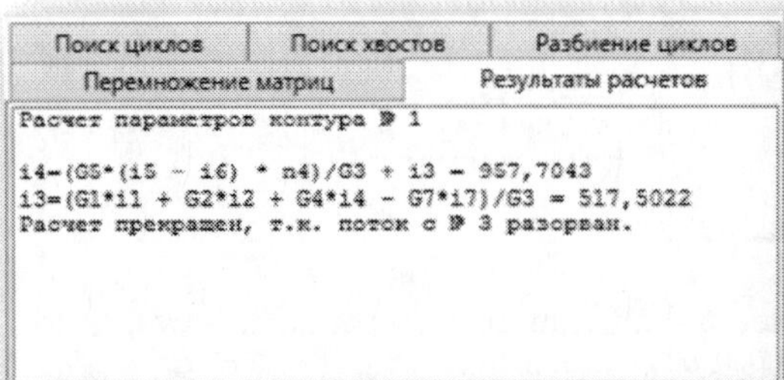

g) Calculation of flow parameters

Fig. 3 Software for structural and thermodynamic analysis

References

[1] Mukhametshina E I and Shamsutdinov E V 2009 Development of a system for utilization of high-potential secondary energy resources *Proceedings of Akademenergo* **3** pp 58-67

[2] Yantovsky E I and Levin L A 1989 Industrial heat pumps (Moscow: Energoatomizdat) 128 p.
[3] Kladov I V, Sedlov A S, Shelginsky A Ya and Galaktionov V V 2012 Analysis of the efficiency of using open-type electric pump systems with electric drive in the production of phosphoric acid extraction *Electrical equipment: operation and repair* **6** pp 13-17
[4] Shelginsky A Ya and Malenkov A S 2014 Cogeneration systems based on secondary energy resources vare in the production of mineral fertilizers *Industrial energy* **7** pp 32-37
[5] Krasavina E O and Plotnikova L V 2016 Energy-saving heat pump in industrial separation systems *Bulletin of Kazan State Power Engineering University* **4** (32) pp 95-105
[6] Plotnikova L V, Chilikova I I, Sitnikov S Y and Efremov G I 2016 Systematic approach to the assessment of energy complex efficiency for thermal energy production with heat power saving transformer turning on *International Journal of Pharmacy and Technology* **8**(4) pp 26727-37
[7] Popyrin L S 1978 Mathematical modeling and optimization of heat power plants (Moscow: Energy) 416 p.
[8] Plotnikova L V, Zvegintsev A A, Kashipova L A, Ishmuratov R A and Nureyev N K 2015 Program implementation of the system analysis of the chemical and technological scheme complex structures of petrochemical productionn *Messenger of the Kazan techological university* **18** (8) pp 198-202
[9] Nazmeev Yu G and Konakhina I A 2001 Organization of power technological complexes in the petrochemical industry (Moscow: MEI publishing house) 364 p.
[10] Plotnikova L V Andreeva S A and Efremov G I 2009 Organization of an energy-saving recycling system for secondary pyrolysis stage in the production of ethylene *Energy saving and water treatment* **2** pp. 9-12
[11] Plotnikova L V and Kashipova L A 2016 Development of software for analysis of the structure of industrial heat and power systems *2016 2nd International Conference on Industrial Engineering, Applications and Manufacturing, ICIEAM 2016 - Proceedings 2* pp 7911675
[12] Plotnikova L V, Chilikova I I and Dodonov V A 2016 Structural analysis of the heat technological complex for the production of paper OAO Polygraphkarton *Engineering journal of Don* **4** (43) pp 29

The use of continuous monitoring systems for simultaneous-separate well operation in order to increase the rate of oil extraction

Safin M. and Semenova K.

Federal State-Funded Educational Institution of Higher Professional Education "Kazan State Power Engineering University", Krasnoselskaya street, 51.

cmvorkut@mail.ru, semenova.kseniya.95@mail.ru

Abstract. The article contains theoretical and methodological materials based on the analysis of literature data and research in the field and in laboratories using actual methods of processing initial information and analyzing them.

Introduction to the technology of simultaneous-separate operation (SSO)

Fuel resources provide energy not only the entire industry of any country in the world, but virtually all spheres of human life. The most important part of the fuel and energy complex in Russia is the oil and gas sector.

According to experts, more than 70% of the oil reserves in Russia are among the hard to recover, among them a significant amount of reserves is concentrated in reservoirs with low filtration and capacitive properties. At the same time, the task of increasing the rate of oil production at existing and new fields remains topical and today, most of the oil and gas fields in the Russian Federation have a multi-layer structure. [1]

At present, the technology of separate exploitation is finding increasing use in the development of a multi-layer deposit. This allows, using a single well grid, to select oil from several layers at the same time. This technology accelerates the production of reserves and allows increasing the volume of extracted products. The problem of accounting and the current state of development of reserves from a multi-layer deposit has not been resolved to date. Particular attention is paid to the tasks of assessing the efficiency of oil extraction. From a multi-layer reservoir with the use of effective technologies for reservoir energy maintenance, their impact on the modes and operation of wells equipped for simultaneous-separate operation (SSO) in conditions of water-gas impact on the reservoir, stationary and non-stationary waterflooding. Also pay attention to the problems that are associated with determining the criteria for selecting wells for the technology of simultaneous-separate operation based on physicochemical and geological characteristics, the reservoir's reservoir properties. Important issues are the creation of a technology for the separation of recoverable products with the simultaneous development of seams.

Analysis of the effectiveness of the SSO

In the industrial development of NGDU Yamashneft there are 9 deposits, which are characterized by small size, difficult to extract reserves, low productivity, zonal and layered heterogeneity of the seams.

The dynamics of oil production and watering for the period of formation and development of NGDU Yamashneft can be divided into three stages. The first and second stages are characterized by active input into the development of new deposits. The third stage is characterized by additional development of existing deposits, optimization of the carbonate deposit development system. Since the discovery of new deposits was not foreseen, an increase in oil production was possible only if the geological section was involved in the development of reserves throughout the vertical. All of the above prompted NGDU Yamashneft to test the WEM technology in 2003 at well # 2046 of the Berezovskoye field. The number of installed units of simultaneous separate oil production today is 238 wells (18% of the total for OAO Tatneft, 19% of total production).

The large-scale application of simultaneous separate operation facilities at OAO Tatneft to date has reached 1,269 wells, but the question of finding ways to improve the profitability of the well has remained topical. As a result of the introduction of simultaneous separate production facilities, 33 wells (or 14% of the total SSO fund) withdrew from the unprofitable category. 25 wells from them have already paid for the incurred costs.

Using SSO technology, they introduced single-shaft plants at 7 sites. The total increase for 7 wells was 37.1 tons per day. on oil.

At the Yersubaykinskoye field, using SSO technology, they successfully installed one-lift units in wells No. 4871 and 4873 and received a total increase of 14.6 tons per day.

Thanks to the combination of simultaneous separate extraction and injection technologies at the Yamashinskoye field, a classical scheme for the development of the site was implemented, which included both the injection zone and the sampling zone. In the wells No. 2524, 2526, 2528, 2531, exploiting the deposits of the Turney stage, the Tula horizon was added. The daily increment in production by the site was 11 tons for oil.

In order to maintain the reservoir pressure and to compensate for the selection during the same period, the injection in the injection well No. 2527n, through which water was pumped into the formation, was injected using the ORZ technology. Accumulated additional production at this site amounted to more than 25 thousand tons of oil.

Using the technology of separate operation at the next section of the Shegurcha field, they incorporated the Tula layer in Bobrikov wells No. 1285, 4772, and introduced single-shaft units, thereby significantly increasing the coverage of the area produced by injection and, therefore, the efficiency of the injection itself. Accumulated additional production in this area along the Tula horizon amounted to 14 thousand tons, the total increase - 10 tons / day. on oil.

Since February 2013. At the Arkhangelskoye field, a zero rate of mineral extraction tax (MET) was applied for high-viscosity oil from the Bashkirian stage. An obligatory condition for such production is the separate accounting of extracted preferential oil. And here the installation of simultaneously separate operation has become an indispensable tool for solving this problem. Preserving reserves for existing facilities, the Bashkirian stage was introduced into the development by introducing two-lift units of separate extraction and pumping (ORDiZ) at 13 wells with a total increase of 40 tons per day.

The technology of ORDiZ can be called an extraordinary solution, and earlier it could not be imagined about the combination of two assignments by one well. New capabilities of the WEM enabled the introduction of the ORDiZ technology in production well No. 10969 in 2010.

Yersubaykinskoye field for the organization of additional waterflooding sites without additional capital costs for development. At the injection well No. 7297, to connect the Bashkirian Stage, thereby increasing the production of reserves for this facility.

All new specialists of the NGDU Yamashneft have always been welcomed. The first installation of a separate small-diameter production at TATNEFT was simultaneously implemented at well No. 16502 of the Berezovskoye field in December 2012, which allowed the development of the western part of the Tula horizon, the central part of which was already exploited by simultaneous separate production facilities. The increase was 4.9 tons per day.

In 2012, the first installation for the simultaneous separate operation of three facilities was tested at well # 4752. The gain was 3,6 tons / day.

IOP Conf. Series: Earth and Environmental Science **288** (2019) 012070 doi:10.1088/1755-1315/288/1/012070

The above-mentioned technologies are very relevant for today, since there is a fund of wells for the introduction of these facilities. Potential fund for the implementation of the simultaneously separate extraction of 3 layers is 35 wells. [2]

Analysis of the operation of wells at the Alekseevsky deposit

The development of reserves with simultaneous separate utilization of seams is determined by the adjustment of the pumping equipment based on the potential capacity of the formation determined by the geological service of the oil enterprise.

At the Alekseevskoye field oil reservoir capacity allows the development of a number of facilities with a single well grid. Exploitation of reservoirs by a common filter over the entire history of field development was not carried out by Rostechnadzor bodies and the use of separate exploitation was simultaneously approved to develop reserves of Kizel, Zavolzhsky and Dankovo-Lebedyan productive horizons.

As of October 2016, a significant part of the wells at the same time of separate operation (14 wells) simultaneously conducts separate operation of the Kizel and Zavolzhskiy horizons and 8 wells, which simultaneously operates a separate operation of the Kizelov and Dankovo-Lebedyan horizons. The histogram of the distribution of liquid flow rates is shown in Figure 1. Most wells at the same time separate operation of the Kizel and Zavolzhskiy horizons (Fig. 1, a) operate with a rate of 5.1 t / day (9 wells) along the Kizel horizon, with a daily rate of 5.3 t / d (6 wells). A similar distribution of wells along the selected intervals of fluid flow rates is associated with similar geological and physical characteristics of the seams.

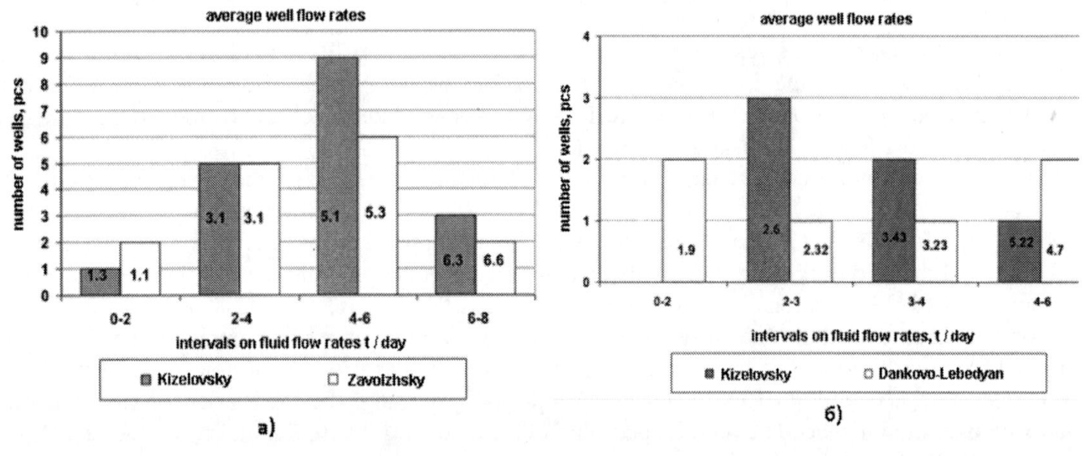

Figure 1 - Histograms of the distribution of liquid flow rates along the wells at the same time as separate exploitation:
a) Kizel and Zavolzhskiy horizons; b) Kizel and Dankovo-Lebedyan horizons

Most of the wells in the Kuzelovsky and Dankovo-Lebedian horizons (Fig. 1, b) operate with an average flow rate of 2.6 t / d (3 wells) along the Kizel horizon. According to the Dankovo-Lebedyan horizon, two wells operate at an average flow rate of 1.9 tons per day and two wells take an average of 4.7 tons of fluid per day. We note a different character of the distribution of wells in terms of the selection of the Kizel and Danko-Lebedian horizons within the allocated groups.

Figure 2 shows a comparative estimate of oil production. Given horizons with reference to the development using the WEM technology and the exploitation of only the Zavolzhskiy horizon and only Kizelov horizons.

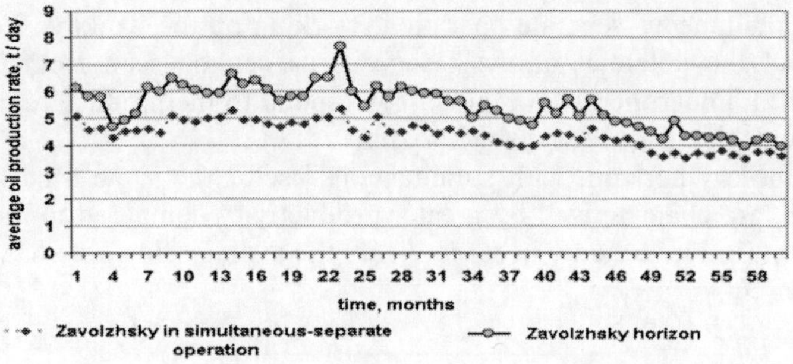

Figure 2 - Dynamics of average oil production through the Volga horizon in the wells at the same time-separate operation and the leading separate operation of the Transvolga horizon

Figure 2 shows that for the wells with simultaneous-separate operation and the wells operating only the Zavolzhsky horizon, there are insignificant differences in operating modes. This indicates the correct selection of the parameters of the installed pumping equipment.

Wells that extract oil only from the Kizel horizon are characterized by lower rates (Figure 3). Perhaps this is due to the fact that wells operating only on the Kizel horizon are in less productive areas (Figure 4).

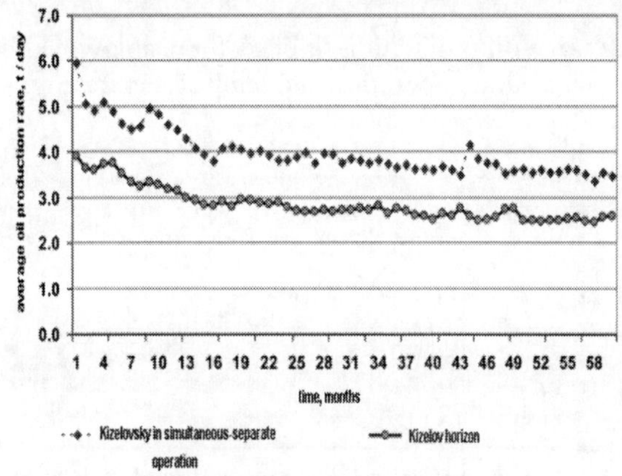

Figure 3 - Dynamics of average oil production through the Kizel horizon in wells of simultaneous-separate operation and leading separate operation of the Kizel horizon

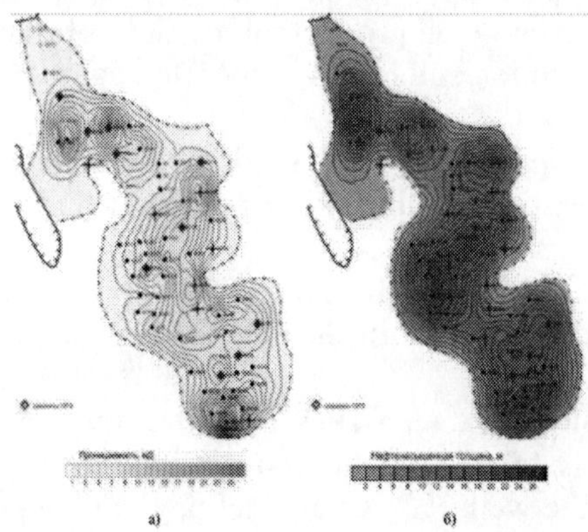

Figure 4 - Permeability maps (a) oil-saturated thicknesses (b) Kizelovskoe horizon

IOP Conf. Series: Earth and Environmental Science **288** (2019) 012070 doi:10.1088/1755-1315/288/1/012070

On wells with simultaneous separate operation, working on the dankovo-Lebedyansky horizon, it can be noted that the potential of the reservoir, as can be seen from the dynamics of oil production. (Figure 5), under the given regime, correspond to the productive possibilities for wells working only on the dankovo-Lebedyansky horizon.

Wells of the Bobrikov horizon, with simultaneous use at the same time for almost the entire development period, are characterized by greater productivity compared to the wells that separate the Bobrikov horizon (Figure 6), only in recent years the average flow rate has been reduced due to the increase in water cut.

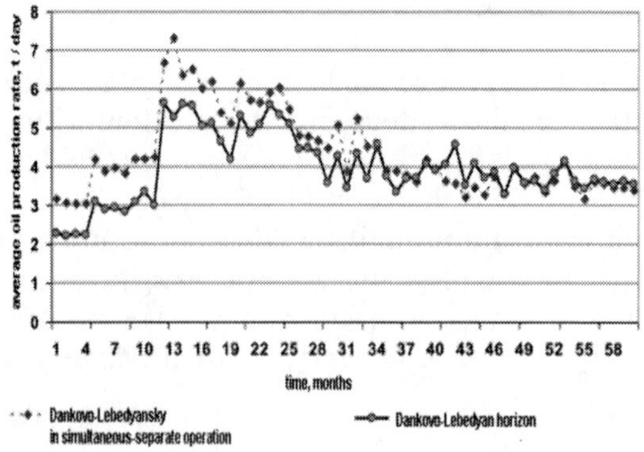

Figure 5 - Dynamics of average oil production through the Dankovo-Lebedyan horizon in wells of separate-separate operation and leading separate operation

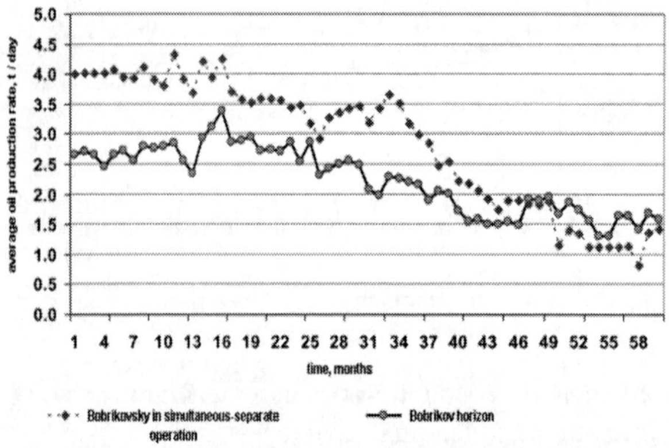

Figure 6 - Dynamics of average oil production. on the Bobrikov horizon in wells with simultaneous-separate operation and leading separate operation of the Bobrikov horizon

Unlike reservoir productivity at various ways oil of the same productive horizon due to the selection of predetermined operating parameters of pump set that optimizes individual wells dual completion formations.

Alekseevskoye field. For this purpose, the geological and physical parameters of the wells are comparable with accumulated oil withdrawals. for the entire period of simultaneous-separate operation.

Basic geological and physical characteristics of productive formations Alekseevskogo field, which is applied simultaneously-lobed operation is: Average depth, reservoir type, reservoir type, the average total thickness of the well, the average net pay thickness, porosity, permeability, gross ratio, dissected coefficient initial reservoir temperature, the initial reservoir pressure, oil viscosity at reservoir conditions, the density of the oil under surface conditions, in absolute mark K, the volume ratio of oil, sulfur and paraffin oil pressure of oil saturation and gas, GOR and water density at surface conditions. Since oil horizons are considered differences in physicochemical properties, the reference parameter to identify patterns for wells with dual completion is received transmissibility layers.

The ratio of accumulated oil abstraction to formations developed simultaneously and separately is directly proportional to the ratio of the hydroconductivity of the corresponding strata:

$$\frac{Q_{нак}^{1}}{Q_{нак}^{2}} = f\frac{\xi_1}{\xi_2} \tag{1}$$

Figure 7 shows the distribution of wells at the same time-separate operation, depending on the ratio of accumulated oil sampling and the difference in geological characteristics of the seams. According to the presented distribution of wells with simultaneous-separate operation, four quadrants have been identified, with respect to which three groups of rational application have been singled out simultaneously for separate operation:

Group I (rational simultaneous-separate operation) – the specified operating modes of each of the strata correspond to the potential capabilities of the formation based on the geological and physical characteristics of the reservoirs, the installed pumping equipment operates in the optimal mode. Group II (irrational simultaneous-separate operation for the first layer) – the reservoir properties of the first layer indicate a possible increase in the selection by technological optimization of the pumping equipment, which will allow the well to be transferred from Group II to Group I. Group III (irrational simultaneous-separate operation for the second formation) – the reservoir properties of the second layer indicate a possible increase in the selection by means of technological optimization of pumping equipment, which will allow the transfer of the well well from a group III to group I.

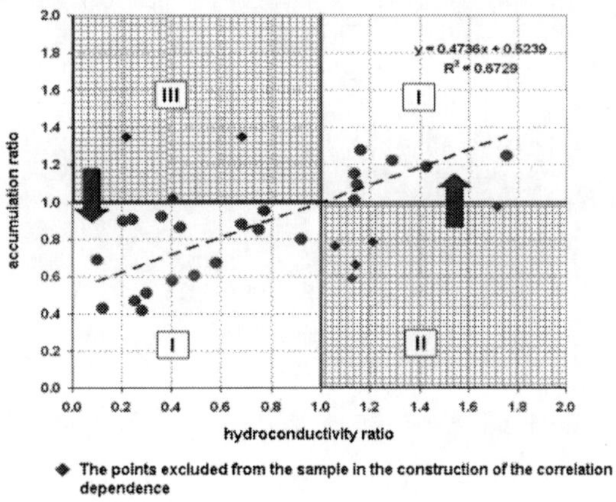

Figure 7 – Distribution of wells at the same time as a separate operation depending on the ratio of accumulated oil abstraction and the difference in the geological characteristics of the seams

Figure 7 shows the main part of the wells (75%), which operates in the optimal mode, that is, simultaneous-separate operation is rational. For 25% of the wells, technological optimization is recommended, aimed at increasing the productivity of pumping equipment. [3]

Thus, it can be concluded that, in general, the implemented technology of simultaneous-separate operation at Alekseevskoye field "Aloyl" is characterized by the coordinated work of carefully selected pumping equipment and productive layers combined for simultaneous separate oil extraction. Such a picture indicates a uniform production of reserves, excluding inter-river flows and other problems of joint development.

Only the expansion of the field of application of the technology, simultaneous and separate production, will allow us to conduct cost-effective operation of wells and deposits, ensuring the maintenance of reservoir pressure, early involvement in the development of reserves while maintaining the existing development system, sealing wells without additional drilling, reducing the project fund, creating water-costs.

References

[1] Nurgaliyev A A, Khabibullin L T 2015 *Analysis of simultaneous-separate operation efficiency at small reservoirs in the south-east Tatarstan* (Almetyevsk)
[2] Gulyaev D N, Melnikov S I, Kokurina V V, Kremenetsky M I, Krichevsky V M 2012 *Analysis of the interference of wells based on the results of in-depth depth monitoring based on sectoral modeling*: oil Industry Journal (Moscow)
[3] Faysullin N N *Automation of wells based on continuous monitoring systems of reservoir meters SPM. AS*: scientific and technical journal "Automation, telemechanization and communication in the oil industry" (Moscow)

Development strategy of the energy supply system for MPEI campus blocks based on green building

I.A. Sultanguzin[1], E.V. Zhigulina[1], Y.V. Yavorovsky[1], I.D. Kalyakin[1], A.V. Govorin[1], A.V. Fedyukhin[1], A.A. Krolin[1], S. V. Guzhov[1], O. Derevianko[2], L. Mukhametova[3]

[1]National Research University "Moscow Power Engineering Institute", Krasnokazarmennaya 14, Moscow, 111250 Russia

[2]Peter the Great St. Petersburg Polytechnic University, Polytechnicheskaya 29, Saint Petersburg, 195251 Russia

[3]Federal State Budgetary Educational Institution of Higer Education "Kazan State Power Engineering University", Kazan, Krasnoselskaya street, 51

FedyukhinAV@yandex.ru

Abstract. The article presents the development strategy of the energy supply system for National Research University "Moscow Power Engineering Institute" (MPEI) buildings based on green building. The designing mathematical model of one of MPEI academic buildings connected with the Scientific and technical library of the University has been created in Passive House Planning Package and designPH software. The parameters of the building energy supply system with the lowest energy consumption are presented.

Energy-efficient buildings construction with minimum energy consumption in Russia is one of the basic tasks in solving of energy saving problems [1–14] . Here optimum solutions are to be estimated for north, moderate and south climate zones taking into account their specific characteristics [15–19].

The solution for the optimization task through a mathematical model allows one to determine optimum parameters of energy supply system at which power independence of the considered object would be reached. It will allow to use the renewable energy and to preserve fossil fuels that is urgent because of existing reserves exhaustion [20–24].

In accordance with Federal Law No. 261-FZ from 23.11.2009 "On energy saving and increasing energy efficiency and on amendments to certain legislative acts of the Russian Federation" starting from 01.01.2010 the state (municipal) institutions should ensure the reduction under comparable conditions the uptake of water, diesel and other fuels, natural gas, heat and electricity, coal for five years, not less than 15% of the volume actually consumed by them in 2009 each of these resources with the annual decrease this amount by at least 3%.

National Research University "Moscow Power Engineering Institute" is one of the largest technical universities of Russia. MPEI infrastructure consists of nine educational buildings that contain hundreds of academic and scientific laboratories, a training-experimental heat and power plant, five buildings of students hostel, pilot-production plant, health center, culture house, dining-hall, sports complex "Energia", swimming pool.

Being the main power engineering university in Russia National Research University of MPEI realizes its responsibility for students and for society to represent itself an example of the most energy efficient organization and to develop the energy-saving technologies. Despite the fact that MPEI regularly hosts energy survey (Fig. 1) [25,26] and there were implemented many of the energy efficiency measures, the conditions of all buildings of the campus are quite far from passive house or nearly zero energy building standards [1,12,18,27] .

The author collective has developed and suggested the following strategy of the energy supply system for MPEI buildings based on green building.

Similar projects are being realized in Europe in the recent years. For example, the administrative building of Vienna Technical University [28,29], renovated in 2015 is the first high-rise building in Europe, which produces more energy than it consumes. This building operates as a public building which is visited by approximately 800 employees and up to 1800 students every day. In October 2015, Vienna Technical University, in whose ownership the building is, was awarded the Austrian state prize for outstanding achievements in the field of technologies of environmental protection and energy efficiency (category "Research and innovation"). Also there are similar examples among multi-store residential buildings, such as Aktiv-StadtHaus at Frankfurt am Main, with total living area of 10714 square meters, which was constructed in 2015 [30].

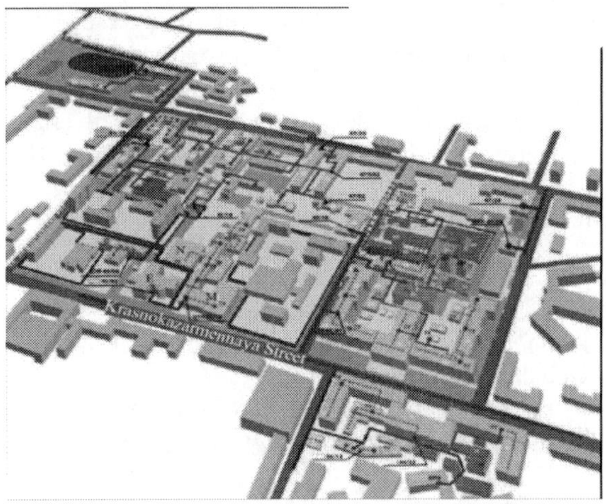

Figure 1. Energy distribution and consumption of MPEI visualization subsystem

The main stages of the forthcoming work at MPEI are the following:

I. Reconstruction of the following MPEI blocks: E, M and Scientific and Technical Library (STL) of MPEI on the basis of implementation of green building strategy Passive House Planning Package (PHPP)

1. Determination of the existing energy consumption and setup of mathematical models for buildings E, M and STL in the three-dimensional modeling software package, design PH and passive house design PHPP based on energy audit results.

2. Development and implementation of the buildings thermal protection program through application of modern and advanced building materials and technologies

- Thermal protection of external walls
- Energy-saving windows and doors installation
- Insulation of the roof and technical floor
- Insulation of basements
- The use of phase change materials (PCM)

3. Development of the energy supply system for buildings based on modern and advanced energy-saving technologies and materials [28–35]:

- Heating system
- Hot water system
- Ventilation system
- Air Conditioning system
- Lightning system
- Power supply system
- The automation system of the individual subsystems
- The system of monitoring and integrated management of the entire energy supply system

4. Implementation strategy of green technology through renewable energy sources usage:

- Solar collectors
- Solar panels
- The combined production of electric and thermal energy technology based on BIPV and BIPVT (Building Integrated PhotoVoltaic Thermal) solar batteries and collectors integration in the roof and the facades
- Trigeneration of electricity, heat and cold with the use of PVT technology and absorption chillers to provide air conditioning systems in summer
- Wind turbines
- Inverters and electric accumulators
- System daily, weekly and seasonal thermal energy storage

5. Economic evaluation of green building projects implementation

6. Analysis and choice of energy saving projects with minimal funding

7. Analysis and selection of green building projects with the necessary funding leading to nearly zero energy building

II. Reconstruction of other blocks of MPEI campus based on the implementation strategy for green building:

1. Academic and administrative buildings;
2. Residential buildings and dormitories;
3. Institutions of culture and medicine, buildings of public catering and shops.

By now the first point is being done. The mathematical models of buildings M and STL have been developed in the three-dimensional modeling software package, designPH and passive house design PHPP [31,32] based on energy audit results [25,26].

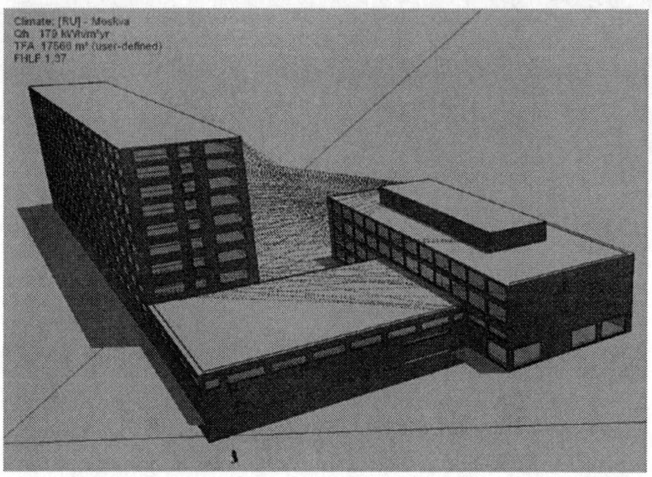

Figure 2. Buildings of M -block and the Scientific and Technical Library

The model of M-block and of the Scientific and technical library is presented in Fig.2. The results of modeling the buildings before the reconstruction are shown in table 1.

Table 1. The results of modeling the buildings at present

▾ Annual Heat Demand

Total heat losses (kWh/a)	Total free heat gains (kWh/a)	Utilisation factor	Treated Floor Area (m²)	Ann. Heat Demand (kWh/a)	Specific Ann. Heat Demand, Q_h (kWh/m²a)
3663513.34	527381.10	1.00	17566.41	3136160.15	178.53

▾ Transmission heat losses

Total Heat Loss Area (m²)	Area Weighted U-value (W/m²K)	Av. Temp. Factor	Ann. Htg. Degree Hours (kKh)	Transmission Heat Loss (kWh/a)	Q_t (kWh/m²a)
24111.59	1.34	0.89	116.00	3431520.82	195.35

▾ Ventilation heat loss

Treated Floor Area (m²)	Ventilation volume (m³)	Eff. air exchange rate	Heat capacity of air	Ann. Htg. Degree Hours (kKh)	Ventilation heat loss (kWh/a)	Q_v (kWh/m²a)
17566.41	43916.02	0.14	0.33	116.00	231992.52	13.21

▾ Solar heat gains

Group nr.	Area Group	Win. area (m²)	Glazing area (m²)	g-value	Reduction factor	Radiation, G_s	Solar heat gain (kWh/a)	Q_s (kWh/m²a)
2	North Windows	219.10	176.06	0.50	0.49	148.15	7898.35	0.45
3	East Windows	2108.01	1734.09	0.50	0.50	322.18	169187.71	9.63
4	South Windows	338.83	277.24	0.50	0.50	448.60	37659.61	2.14
5	West Windows	2116.02	1737.66	0.50	0.50	219.29	115380.12	6.57
6	Horizontal Windows	0.00	0.00				0.00	0.00
		4781.95	**3925.05**				**330125.80**	**18.79**

▾ Internal heat gain

Treated Floor Area (m²)	Internal heat gain rate (W/m²)	Heating period (days/a)	Heating period (kh/a)	Internal heat gain (kWh/a)	Q_i (kWh/m²a)
17566.41	2.10	222.80	5.35	197255.30	11.23

Energy audit showed that the buildings M and STL have high heat losses. Thermovision inspection has been carried out outside of building. The temperature defect of the building guarding construction (walls and windows) is watched in Fig. 3. It means that it is necessary to increase wall insulation and change some windows to energy-saving windows.

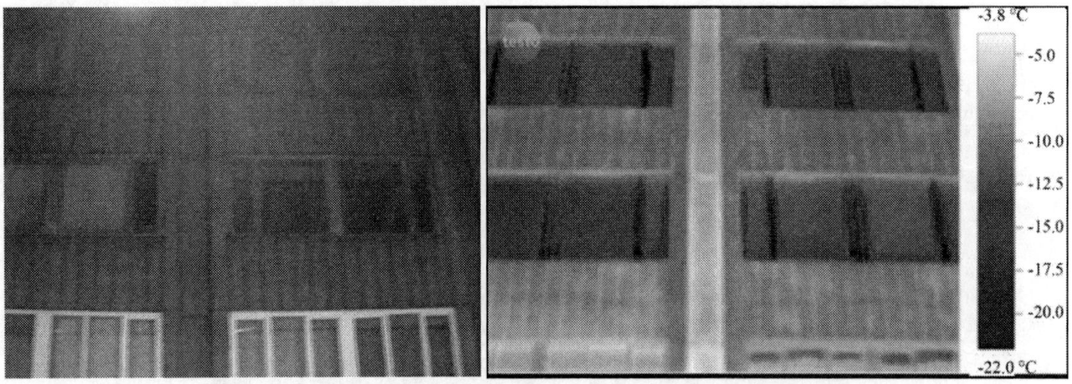

Figure 3. Thermovision inspection of outside of building M

Also, energy audit showed that inside temperature in building M was 22.5 – 25.2 °C in December 2016 – January 2017 when outside temperature was in the range from +1.5 to -26 °C. Indoor temperature is significantly higher than normal temperature 20 - 22 °C. It means that it is necessary to develop the heating automation system, that will allow one to get the essential energy saving for heating.

Just time ventilation system for building M and STL is used without recuperation of waste air.

IOP Conf. Series: Earth and Environmental Science **288** (2019) 012071 doi:10.1088/1755-1315/288/1/012071

Installation of ventilation units with 80–90% heat recovery efficiency and 50–80% humidity recovery efficiency allows decreasing loss of heat with exhaust air. According to Passive House Institute recommendations [27] the building envelope must be built in a very airtight way, $n_{50} < 0.6\ h^{-1}$ is required and < 0.3 is recommended. Also, in the interest of justifiable operational costs, the ventilation systems in University buildings must be operated periodically or according to demand. Regulation of the air quantities according to demand should be strived for the CO_2 content of the air and depend on the occupancy density.

For decision problem of air conditioning system in summer it is necessary to select energy effective equipment for specific conditions. Here different schemes, including heat pumps could be analyzed and chosen.

Realization of energy saving program on building M and STL reconstruction allows one to decrease specific energy consumption for heating from 179 kW·h/(m^2·year) to 25 kW·h/(m^2·year) approximately. The results of preliminary modeling the buildings after reconstruction are shown in table 2.

Table 2. The results of modeling the buildings after reconstruction

▾ Annual Heat Demand

Total heat losses (kWh/a)	Total free heat gains (kWh/a)	Utilisation factor	Treated Floor Area (m²)	Ann. Heat Demand (kWh/a)	Specific Ann. Heat Demand, Q_h (kWh/m²a)
954294.72	533305.57	0.98	17566.41	434216.34	24.72

▾ Transmission heat losses

Total Heat Loss Area (m²)	Area Weighted U-value (W/m²K)	Av. Temp. Factor	Ann. Htg. Degree Hours (kKh)	Transmission Heat Loss (kWh/a)	Q_t (kWh/m²a)
24111.59	0.27	0.89	116.00	722302.20	41.12

▾ Ventilation heat loss

Treated Floor Area (m²)	Ventilation volume (m³)	Eff. air exchange rate	Heat capacity of air	Ann. Htg. Degree Hours (kKh)	Ventilation heat loss (kWh/a)	Q_v (kWh/m²a)
17566.41	43916.02	0.14	0.33	116.00	231992.52	13.21

▾ Solar heat gains

Group nr.	Area Group	Win. area (m²)	Glazing area (m²)	g-value	Reduction factor	Radiation, G_s	Solar heat gain (kWh/a)	Q_s (kWh/m²a)
2	North Windows	219.10	179.58	0.50	0.50	148.15	8056.39	0.46
3	East Windows	2108.01	1764.93	0.50	0.51	322.18	172194.94	9.80
4	South Windows	338.83	282.30	0.50	0.50	448.60	38346.77	2.18
5	West Windows	2116.02	1768.85	0.50	0.51	219.29	117452.17	6.69
6	Horizontal Windows	0.00	0.00				0.00	0.00
		4781.95	**3995.66**				**336050.27**	**19.13**

▾ Internal heat gain

Treated Floor Area (m²)	Internal heat gain rate (W/m²)	Heating period (days/a)	Heating period (kh/a)	Internal heat gain (kWh/a)	Q_i (kWh/m²a)
17566.41	2.10	222.80	5.35	197255.30	11.23

Building M and STL are located at Krasnokazarmennaya street, and from south side there are no buildings which cover facades and roofs for solar radiation. It means that it is possible to install building integrated photovoltaic panels on ventilated facades and BIPVT panels [33–35] on roofs of building M and STL.

Calculation with PHPP software [27,31] showed that it is possible to rebuild M block and STL with nearly zero energy balance like buildings in [28–30] cases with using BIPV and BIPVT technologies for good insulating energy saving buildings.

We hope that this will be a demonstrative project of energy saving green building in MPEI.

All other buildings of MPEI Campus could be reconstructed as green nearly zero energy buildings realized on the base of M block building and STL experience.

Conclusion

The present work proposes the development strategy of the energy supply system for MPEI campus blocks based on green building.

Realization of this strategy is considered on the example of M block and Scientific and Technical Library modeling and further reconstruction.

References

[1] Perlova E. et al. Concept project of zero energy building // Procedia Engineering. 2015. Vol. 100, № January. P. 1505–1514.

[2] Zaborova D., Petrochenko M., Chernenkaya L. Thermal Stability Influence of the Enclosure Structure on the Building's Energy Efficiency // MATEC Web of Conferences. 2016. Vol. 73.

[3] Murgul V. Methodology to Improve Energy Efficiency of Residential Historic Buildings in St. Petersburg // MATEC Web of Conferences. 2016. Vol. 53.

[4] Zadvinskaya T.O., Gorshkov A.S. Comprehensive method of energy efficiency of residential house // Advanced Materials Research. 2014. Vol. 953–954. P. 1570–1577.

[5] Nemova D. V, Bogomolova A.K., Kopylova A.I. The impact of green roofs on thermal protection and the energy efficiency of buildings // Advances and Trends in Engineering Sciences and Technologies II - Proceedings of the 2nd International Conference on Engineering Sciences and Technologies, ESaT 2016. 2017. P. 579–584.

[6] Vatin N.I. et al. The energy-efficient heat insulation thickness for systems of hinged ventilated facades // Advanced Materials Research. 2014. Vol. 941–944. P. 905–920.

[7] Starkov V., Ovchinnikov P., Dzampaev T. Optimization of Energy Performance of Windows by Applying Self-Adjustable Shadings // MATEC Web of Conferences. 2016. Vol. 73.

[8] Staritcyna A. et al. Energy efficiency in multi-story buildings // MATEC Web of Conferences. 2016. Vol. 73.

[9] Harmati N., Jakšić Z., Vatin N. Energy consumption modelling via heat balance method for energy performance of a building // Procedia Engineering. 2015. Vol. 117, № 1. P. 791–799.

[10] Penić M., Vatin N., Murgul V. Double skin facades in energy efficient design // Applied Mechanics and Materials. 2014. Vol. 680. P. 534–538.

[11] Zeb A. et al. LED Lightbulbs as a Source of Electricity Saving in Buildings // MATEC Web of Conferences. 2016. Vol. 73.

[12] Statsenko E. et al. Thermal Properties of the Building with Low Energy Consumption (LEB) // Advances in Intelligent Systems and Computing. 2018. Vol. 692. P. 417–431.

[13] Vatin N., Gamayunova O. Energy saving at home // Applied Mechanics and Materials. 2014. Vol. 672–674. P. 550–553.

[14] Mukhametova L.R., Akhmetova I.G., Akhmetov T.R. Evaluating the effectiveness of energy efficiency programs. Problems of energy saving in heating. Proceedings of the higher educational institutions. ENERGY SECTOR PROBLEMS. 2015;(9-10):12-21. https://doi.org/10.30724/1998-9903-2015-0-9-10-12-21.

[15] Baranova D. et al. Correlation of energy efficiency and thermal comfort depending on the ventilation strategy // Procedia Engineering. 2017. Vol. 205. P. 503–510.

[16] Zaychenko I., Gutman S., Kalinina O. Adjustment of Energy Strategy of Russia to Specific Nature of Far North: Analytic Hierarchy Process // Advances in Intelligent Systems and Computing. 2018. Vol. 692. P. 453–462.

[17] Pukhkal V., Bieliatynskyi A., Murgul V. Designing energy efficiency glazed structures with comfortable microclimate in northern region // J. Appl. Eng. Sci. 2016. Vol. 14, № 1. P. 93–101.

[18] Pukhkal V. et al. Studying humidity conditions in the design of building envelopes of "passive house" (in the case of Serbia). Elsevier Ltd, 2015. Vol. 117, № 1. P. 864–869.

[19] Aronova E., Vatin N., Murgul V. Design energy-plus-house for the climatic conditions of Macedonia. Elsevier Ltd, 2015. Vol. 117, № 1. P. 771–779.

[20] Arseniev D.G. et al. The model of intelligent Autonomous Hybrid Renewable Energy System based on Bayesian Network // 2016 IEEE 8th International Conference on Intelligent Systems, IS 2016 - Proceedings. 2016. P. 758–763.

[21] Malinina T., Shklyaruk M. Comparative evaluation of supporting systems for the use of renewable energy sources in the European Union and Russia. Technical University of Kosice, 2015. P. 420–421.

[22] Zubkova M. et al. Technological Decision to Renewable Energy Usage Biogas for Off-grid Systems Consumption // MATEC Web of Conferences. 2016. Vol. 73.

[23] Timerbaev N.F., Tuntsev D.V., Khairullina M.R., Kitaev S.V. Technologies and equipment for the production of electricity from wood waste. Proceedings of the higher educational institutions. ENERGY SECTOR PROBLEMS. 2017;19(1-2):102-107. https://doi.org/10.30724/1998-9903-2017-19-1-2-102-107

[24] Politaeva N A, Kuznetsova T A, Smyatskaya Y A, Trukhina E V and Atamanyuk I 2018 Energy Production from Chlorella Algae Biomass Under St. Petersburg Climatic Conditions *Chem. Pet. Eng.* 1–5

[25] Dragunov V.K. et al. The energy consumption, monitoring and energy saving measures in the University with the use of information systems // Sci. Methodol. Probl. new Educ. Technol. 2014. Vol. 6. P. 32–38.

[26] Krolin A.A., Guzhov S. V. MPEI — modern platform for the promotion of energy-saving technologies // Reg. energy energy Effic. 2015. Vol. special is. P. 4–5.

[27] Passive House schools – Requirements [Electronic resource]. URL: https://passipedia.org/planning/non-residential_passive_house_buildings/passive_house_schools/passive_house_schools_requirements.

[28] Bismarck M. Reconstruction of the Vienna technical University // Build. high Technol. 2016. Vol. 4. P. 18–21.

[29] Prize-winning energy-plus building in Vienna TU // SAUTER Facts. 2015. Vol. 33. P. 22–23.

[30] Elokhov A.E. An example implementation in Germany residential buildings with positive energy balance // EuroStroyProfi. 2016. № special issue on “Energy efficiency. Energy saving. Environment.” P. 28–33.

[31] Strogonov K. et al. Estimation of Practical Significance for Application of Composite Pipes in Comparison with Metal and Polymer Materials. Springer Verlag, 2018. Vol. 692. P. 1024–1035.

[32] Kharkov N.S. Nonstationary heat and mass transfer in the multilayer building construction with ventilation channels. Institute of Physics Publishing, 2017. Vol. 891, № 1.

[33] Shatornaya A. et al. Ventilated Facades: Insulation Materials of Different Manufacturers. EDP Sciences, 2016. Vol. 53.

[34] Olshevskyi V. et al. Moisture Transfer in Ventilated Facade Structures. EDP Sciences, 2016. Vol. 53.

[35] Petrichenko M. et al. Functionality of Ventilated Facades: Protection of Insulation. EDP Sciences, 2016. Vol. 53.

[36] Gorshkov A.S., Rymkevich P.P. A diagram method of describing the process of non-stationary heat transfer // Mag. Civ. Eng. St-Petersburg State Polytechnical University, 2015. Vol. 60, № 8. P. 68–82.

[37] Rudskoy A.I. et al. Advanced metallic materials and processes // Mater. Phys. Mech. Institute of Problems of Mechanical Engineering, 2016. Vol. 25, № 1. P. 1–8.

[38] Rudskoy A.I. et al. Key Engineering Materials: Preface // Key Engineering Materials. 2005. Vol. 291–292.

[39] Feist V. Basic provisions on design of passive houses. Moscow: KONTIPRINT, 2015. 144 p.

[40] Sultanguzin I. et al. No Title // INGENIUERTAG 2016. Der Fakultat Maschinenbau, Electround Energiesysteme Gus- und Osteuropatag. NESEFF-NETZWERKTREFFEN 2016.

Tagunsband. Brandenburgische Technische Universitat. Cottbus-Senftenberg. 14-15 November 2016. 2016. P. 8–12.

[41] Debbarma M., Sudhakar K., Baredar P. Thermal modeling, exergy analysis, performance of BIPV and BIPVT: A review // Renew. Sustain. Energy Rev. 2017. Vol. 73. P. 1276–1288.

[42] Biyik E. et al. A key review of building integrated photovoltaic (BIPV) systems // Eng. Sci. Technol. an Int. J. 2017. Vol. 20. P. 833–856.

[43] Lamnatou C. et al. Modelling and simulation of Building-Integrated solar thermal systems: Behavior of the system // Renew. Sustain. Energy Rev. 2015. Vol. 45. P. 36–51.

IOP Conf. Series: Earth and Environmental Science **288** (2019) 012072 doi:10.1088/1755-1315/288/1/012072

On directions of improving the energy efficiency in Russia

[1]N.A. Yudina, [1]T.U. Dunaeva, [1]S.A. Livshits, [2]D.B. Katz, [3]S.R. Enikeeva, [1]E.K. Nikolaeva

[1]Kazan State Energy University, Kazan, Russian Federation

[2]Kazan Federal University, Kazan, Russian Federation

[3]Kazan Research Technological University, Kazan, Russian Federation

semen19772004@mail.ru

Annotation. The paper analyzes the structure of the energy balance of the Russian Federation for power systems, draws conclusions on the impact of HPPs in the structure of electricity generation on the price of electricity for ultimate customers. The influence of the construction of new hydropower stations on the development of waterways is considered.

Introduction

Over the last few decades energy efficiency has become more often seen as a central factor in ensuring the competitiveness of individual companies, the industry as a whole, and the national economy. Energy efficiency can be identified as a unique potential [1] that promotes economic growth, long-term energy security and even better people's well-being and health. The reduction in energy consumption was ensured by investments in energy efficiency, which exceeded the production of another energy resource in many countries. However, experts note a significant lag in Russian energy efficiency compared with other countries. Russia ranks fourth in the world in terms of energy consumption. However, it spends more energy per unit of GDP than the countries that are among the ten largest energy consumers. We can conclude that the rating of Russia raise from the data of the international energy agency. According to this data, in 2009 Russia was on the 12th place in the list of 121 countries in terms of energy intensity of the economy. In 2017 there was a move to the 15th place. The World Bank, having conducted the study, showed that implementation of measures aimed at increasing energy efficiency can save about 45% of the total primary energy consumption every year in Russia. Investments of private and public organizations in the amount of 320 billion US dollars are needed to actualize the potential for increasing energy efficiency. Investments in energy efficiency at the level of the national economy can pay off in three years. After actualizing the potential for increasing energy efficiency Russia will be able to save 340 billion kWh of electricity, 240 billion cubic meters. m. of natural gas, 89 million tons of coal, 43 million tons of oil products in terms of crude oil. As practice shows, energy efficiency is three times cheaper than increasing energy production [2].

Improving energy efficiency in Russia

The fearsome argument that influences the urgency of the task of improving energy efficiency is the climate change. The International Energy Agency (IEA) and the Organization for Economic

Cooperation and Development (OECD) identified challenges in the structure and energy efficiency of energy: "At the global level, energy intensity of GDP and the carbon intensity of primary energy consumption should be reduced by about 60% by 2050 compared to today's level ". The International Energy Agency proposed three possible options for further development of energy for the period until 2050:

1. A scenario that leads to disastrous consequences. Unfortunately, at present the world is moving precisely toward it. According to forecasts, the ambient temperature can rise by 6° C (6DS).

2. A scenario that reflects the initiatives that many countries have announced with regard to reducing emissions and improving energy efficiency. In this scenario the outlook is not so sad and presumably the ambient temperature will rise by 4° C (4DS).

3. A scenario that offers the latest concept of developing a balanced energy system with a reduction of greenhouse gas emissions, including carbon dioxide (CO2). As a result, the ambient temperature will rise by 2° C (2DS) [2].

The goal of many countries in the world is to reduce the absolute consumption of energy that is obtained from fossil carbon fuels. This goal is related both to ensuring energy security (exhaustion of cheap hydrocarbon reserves and uneven distribution of those reserves) and the consequences of burning carbon fuels for the climate.

The Russian economy is characterized by a high level of energy intensity and is the fourth largest in the world in terms of total energy consumption. Inside Russia the manufacturing industry holds the first place and the second place is occupied by the housing sector in terms of energy consumption.

At one of the annual meetings of the commission on the modernization and technological development of the Russian economy special attention was paid to five strategic areas, among them there are energy efficiency and energy saving which have priority technological development.

An important point in the development of the entire Russian economy and increasing competitiveness in the world, according to the Minister of Energetics A. Novak, is the development of energy efficiency and energy conservation. Increase in energy efficiency can be compared figuratively with the fifth type of fuel. In 2008 the president of Russia set a goal by his decree - to reduce the energy intensity of the gross domestic product by at least 40% of the 2007 level by 2020. In recent years experts have been observing the rate of decrease of energy intensity, but in general, for period from 2008 to the present, the decrease is about 13%, which is much lower than planned. What could affect this significantly? Let us note several objective reasons:

1. The understated growth of gross domestic product over the period under review.
2. Structural changes in the economy which have not finally reached their goals.

We can say that the decree of the designation of the goal and the beginning of the system work was laid [3].

In 2009, the federal law on energy conservation and energy efficiency was enacted (Federal Law No. 261-FZ of November 23, 2009 "On energy conservation and on improving energy efficiency and on amending certain legislative acts of the Russian Federation") [4]. Virtually all sectors of the Russian economy have the potential to increase energy efficiency, but the fuel and energy complex, industry, housing and communal services, transport and the public sector have the best. By now, 55 regional energy supply centers were created and 78 regional energy laws were adopted by the subjects of the Russian Federation. To comply with the world practice in the field of energy efficiency, a regulatory and legal framework has been created, which has proved quite successful in recent years. This is confirmed by the 17th place of Russia according to the rating of state regulation in the field of energy efficiency of the World Bank. Russia was able to enter the list of leaders which is a breakthrough and gives prerequisites for further development.

Nowadays there are several main areas for the implementation of the state policy in the field of energy efficiency and energy saving.

The first direction is the introduction of energy management at the state level, which will allow to form a public energy management system and which will contribute to an increase in energy efficiency of up to 5%. Key points of this direction are energy efficiency indicators and sectoral responsibility. There is a need to add indicators to government programs. Despite the fact that not all changes have been accepted through and through, most state programs do have them. As for the fuel and energy complex, it already includes 7 sectoral indicators of energy efficiency. For example, the specific consumption of electricity for oil transportation and gas transportation; specific consumption of fuel equivalent for the production of electricity (five-year observations showed a 9% decrease in this indicator); the useful use of associated petroleum gas. Here we can draw an unambiguous conclusion about the positive dynamics.

Similar work can be observed in other large industries and companies. According to statistics, about 2/3 of the largest companies have implemented energy management long ago [5]. In accordance with the federal law, state and regulated companies implement special programs for energy conservation and energy efficiency necessarily. The boards of directors periodically review these programs. The significant effect of the program realization has the form of saving heat and electric energy, as well as from the introduction of modern technologies.

The second direction can be considered as the issue of the policy inclusion of information on the achievement of energy efficiency indicators in the public corporate reports of the largest state-owned companies. In autumn of 2017 the government supported and submitted a federal law project to the State Duma, which provides the transition to the sectoral principle of forming requirements for energy efficiency programs. This federal law was introduced to improve work in this area and improve energy efficiency.

Thus, all sectoral departments form their own requirements for energy saving programs of state companies in their industries. The state program of the regions also includes the necessary indicators. At the moment, energy efficiency indicators are included in 63% of regional sectoral government programs. To this end, a tremendous effort was made to train 40,000 municipal and state employees.

There is an annual federal state report confirming the energy efficiency of enterprises and regions with public ratings. This technique allows benchmarking and highlighting best practices. For example, in several regions LEDs were introduced into the public sector. In the city of Vladimir over the past 3 years the traditional street lighting has been replaced with LED lighting, it is about 44% if compared to a percentage. This resulted in economy of more than 30% of power. The illumination of Moscow was increased by 40%. In Kazan a project for the modernization of heating was implemented, according to which 1.4 thousand individual heat points with automatic control was installed. This project in turn allowed to lower the total payments of the population for heating. The work on the preparation and implementation of standards for the openness of government bodies (which necessarily includes energy saving [3]) is in full progress within the framework of the open government in 60 regions of Russia.

The functioning public sector monitoring information system is aimed to increase the number of buildings that have reached the permissible energy efficiency class in the process of overhaul. By means of such a system we can identify the potential savings. The Moscow region was the first to implement such practice of control and planning. 80% of the total number of state and municipal institutions in the country are already included in the system. The main goal is to provide annual information to the system, and the provision of information is planned to be made mandatory. This bill passed the second reading and was submitted to the State Duma.

The third direction of state policy is to introduce technological and environmental regulation concerning energy saving issues. The main points in this direction are the use of the latest available technologies, energy-efficient requirements and standards. The Ministry of Energy has prepared projects of guidelines for the best available technology (BAT) with respect to energy efficiency for the fuel and energy complex. Innovative technologies occupy the main positions here. They can claim the role of a new industry standard. We can take "smart" electrical networks and digital substations as an example. After introduction of such technologies in the Kaliningrad region a reduction in energy loss of 45% was recorded. In technical regulation the use of LEDs for lighting gives us a good example.

Since 2017 there is a ban on the purchase of non-energy-efficient light sources for state and municipal institutions. The goal by 2020 was to demand bringing the share of LED lighting to 75% as an additional responsibility of regulated organizations. Compiled and approved set of rules obliges the use of LED light sources when designing street lighting. The use of LEDs in the public sector has made significant savings in cash per year. If we roughly calculate the size and potential savings then according to the most conservative estimates it can be about 200 billion rubles a year while consumption is at least 12% of the total electricity consumption. Obviously it is quite real to save 40% of consumption. If we study the international practice, every month there is new measures aimed to ban incandescent lamps. More than 100 countries have taken such measures. In Russia measures to reduce the permitted threshold for incandescent lamps from 100 to 75 or 50 W is planned to introduce in the next two or three years. An obligatory condition for taking such measures should be the introduction of a minimum warranty period for the product and the introduction of certification. Our industry's readiness to replace the most sold lamps with energy-efficient lamps [4] is high.

The formation of an economic stimulus can be highlighted as the fourth direction of state policy. The funding will accelerate the growth of energy efficiency projects. As international practice has shown, the most significant points are the set prices of energy resources in our region, investment surcharges, excises and fees for harmful emissions.

We have all the socio-economic opportunities for carrying out and implementing such events [4]. For example, a new model of the heat market, fines for associated petroleum gas, stimulating the withdrawal of inefficient generation and its modernization. Moreover, there are promising proposals for new modernization projects.

Since 2016 a mechanism of tax incentives for energy efficient equipment has been developed and successfully put into operation together with the Ministry of Industry and Trade. The list is approved by a government decree and gives the right to use property tax concessions and accelerated depreciation. This is the essence of economic recoupment and economic stimulus, which can become the driver of investment. There is a positive experience with the extension of subsidies distributed from 2011 to 2018 (about 10 billion) [6]. The projects in the subjects of the Russian Federation were co-financed and extrabudgetary funds and subject funds were raised thanks to them. The attraction of additional investments is in the range of 10 to 20 rubles per each ruble. As a result of the calculations the Ministry of Economic Development plans to consider extending and improving such an incentive system [3].

The fifth direction is the promotion of energy saving. Our regions have been actively involved in the implementation of this direction [7] and have done a great job. Undoubtedly, this is a step forward. Of course, we have considerable potential and much remains to be done.

The energy efficiency and energy saving is an interdisciplinary task which affects all areas of activity. In order to intensify the work and increase efficiency in the field of energetics, the Ministry of Economic Development has developed and approved a comprehensive plan to increase the economy, which in turn affects all sectors of the economics [8]. For today the project is undergoing an approval procedure. The state policy of Russia is intersectoral, the indicator of energy efficiency and energy intensity of the economy is included in the state program "Economic development and innovative economy" [9].

Also, work on inclusion of our federal sectoral ministries in all government programs and assigning an annual report on the state of energy conservation on the Ministry of Economic Development continues and improves.

Conclusions

The structure of the energy balance of any country can be viewed both from the point of view of commercial interests of specific energy producers, the state, and from the point of view of the interests of final consumers [1]. The development of the economy, especially in the part of energy-intensive industries, is directly related to the cost of energy resources [2]. It becomes important to understand the influence of the generation structure on the formation of the price of electricity. The strategy for the development of energy capacities should take into account the interests of all consumer groups, including the population.

IOP Conf. Series: Earth and Environmental Science **288** (2019) 012072 doi:10.1088/1755-1315/288/1/012072

References

[1] On the issue of a single terminology in the system of regional energy management. Erastov AE, Novikova OV Scientific and technical lists of the St. Petersburg State Polytechnic University. Economic sciences. 2015. No. 2 (216). Pp. 68-75.

[2] Alternative Energy Portal https://altenergiya.ru/energosberezhenie/energoeffektivnost-sovremennoj-ekonomiki.html

[3] Business magazine "Neftegaz.RU" https://neftegaz.ru/analisis/view/8597-Doklad-A.-Novaka-o-razvitii-energoeffektivnosti-i-energosberezheniya-na-zasedanii-Pravitelstva-RF-19 - oktyabrya-2017-g.

[4] Ministry of Energy of the Russian Federation https://minenergo.gov.ru/

[5] Methodical approach to self-test for compliance of the energy management system with the requirements of the standard. Tolstova LV, Novikova OV In the collection: Week of Science of SPbPU materials of the scientific-practical conference. Engineering and Economics Institute of SPbPU. S.V. Shirokova (Ed.), A.A. Kovalenko (editor's note). 2015. P. 39-43.

[6] Federal subsidization as driver of innovative development of regions in the field of energy saving and energy efficiency. Erastov AE, Novikova OV In the book: Restructuring the Russian economy and industrial policy Proceedings of the scientific and practical conference with foreign participation. Edited by A.V. Babkin. 2015. P. 102-106

[7] Methodology for calculating the integrated indicator of the effectiveness of regional energy saving programs. Erastov AE, Novikova OV Bulletin of the Ivanovo State University. Series: The humanities. 2015. № 3. P. 73.

[8] Mukhamedova LR, Akhmetova IG Assessment of the regional level of diffusion of innovative solutions in the field of energy conservation and energy efficiency / Kazan economic bulletin. №3 (23): - Publishing house of the Kazan University, 2016 ISSN 2305-4212.

[9] Yudina NA, Livshits SA, Farrakhova AF, Abdullina ER Preliminary results of the federal law of November 23, 2009 "On energy saving and improving energy efficiency and on introducing amendments to certain legislative acts of the Russian Federation" and the problem of energy service contracts "

IOP Conf. Series: Earth and Environmental Science **288** (2019) 012073 doi:10.1088/1755-1315/288/1/012073

IMPROVING ENERGY AND ENVIRONMENTAL EFFICIENCY OF FLUE GAS CLEANING AT THERMAL POWER PLANTS

A.T. Zamalieva [1], M.G. Ziganshin [2]

[1] LLC Gazprom transgaz Kazan, Russia, Mayakovskogo str., 3 – 422000 Arsk, Tatarstan, Russia. E-mail: Albina-0587@rambler.ru

[2] Kazan State University of Architecture and Engineering, Russia, Zelenaya Str., 1 - 420043 Kazan, Tatarstan, Russia. E-mail: mjihan@mail.ru

Abstract. The aim of the work is to improve cyclone and filter dust collectors by increasing the efficiency of fine particle deposition classes PM_{10}, $PM_{2,5}$ as well as reducing energy and material costs. The analysis of existing methods of cyclonic filtration of industrial dispersed emissions is carried out. The laboratory tests of the cyclone filter, made by modernization of the cyclone CN-11-200, were performed, the aerodynamic characteristics of its operation were found. The calculated determination of the efficiency of suspension deposition from multiphase flow in the separator using a dimensionless Re_r criterion is performed, derived from the Navier-Stokes single-phase flow equations of motion and the particle motion equation based on Newton's law. A numerically studied the aerodynamic parameters of the cyclone on the basis of computational fluid dynamics methods. For mathematical modeling of turbulence we used the DES model

1. Analysis of ways to improve the efficiency of separation in devices that implement the principle of cyclon filtratione

In present time the problem of increase in power and environmental efficiency of cleaning of flue gases and emissions of coal dust preparation systems of power plants and the systems of preparation of gas on thermal power plants with gas generation. is very relevant, and is regulated in Russia by requirements of the Federal law No. 261-FZ from 11/23/2009 (an edition from 7/29/2017) "About energy saving and about increase in power efficiency and about introduction of amendments to separate acts of the Russian Federation" and also requirements for control of emission of greenhouse gases (e.g., the order of the Ministry of Natural Resources and Environmental Protection of the Russian Federation No. 330 from 6/29/2017 "About the approval of methodical instructions on quantitative scoping of indirect power emissions of greenhouse gases", etc.).

Now at the production facilities for dust deposition is widely used cyclone dust collectors, which is explained by the advantage of a number of their technical and economic indicators in comparison with other dust separators. It is the simplicity of design and manufacture, high capacity per unit mass of equipment, small operating costs, the possibility of dust capture in dry form, maintaining the required fractional cleaning efficiency in case the flow of the solid phase increasing, the reliability of work at a sufficiently high temperature. In the world, the search for ways to improve cyclonic devices in the direction of increasing the degree of deposition of suspended particles and reducing the energy intensity of cyclonic processing is intensively conducted. The authors of [1] presented HL-cyclone of high performance (Hochleistungszyklon). It differs from the standard classic cyclones that it has the cross-cutting Central (axial) "guiding" pipe, strongly narrowed inlet pipe of dusty gas, significantly shortened conical part of the cyclone, connect with into the hopper. According to the authors of the development, the structural addition in the form of a Central pipe located along the HL-cyclone axis (Fig.1), provides high-performance operation of the device of this design. According to the authors [2], in the classical standard cyclones without a Central guide tube, the transition from laminar to turbulent flow occurs inside the exhaust pipe of the device. As a result, the core of the vortex flow in the apparatus becomes unstable, which causes great losses of pressure and energy, apparently, to rebuild the structure of the flow, which consists in the destruction of the organized rotational motion of the flow with the formation of small vortices of stochastic nature. According to [3], this in turn causes high losses (up to 90% of the total pressure loss) due to energy dissipation in the flow. The authors [4] conclude that the role of the Central ("guiding") pipe on the axis of the cyclone is to stabilize the rotating flow, and to shift the transition from laminar to turbulent regime in the region of higher Reynolds numbers. In this case, the static pressure loss is reduced, and the degree of separation is improved due to the increase of the dynamic pressure.

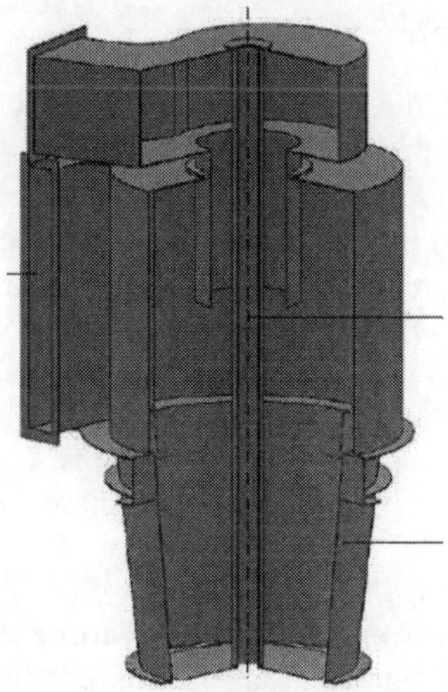

Fig.1. HL-cyclone design [3]

At the same time, the data of the authors [1] show that the tests of this cyclone design were carried out at loads 1.5...2 times lower than the optimal for the classical standard cyclones of this size. Therefore, the tested modes may not correspond to the conditions of industrial operation of the devices, and the efficiency achieved in the experiments may not be confirmed in real conditions.

A cyclone filter [5] can be used to clean up industrial emissions, including those contaminated with particularly hazardous suspensions, and if necessary, for fine cleaning of emissions (Fig.2).

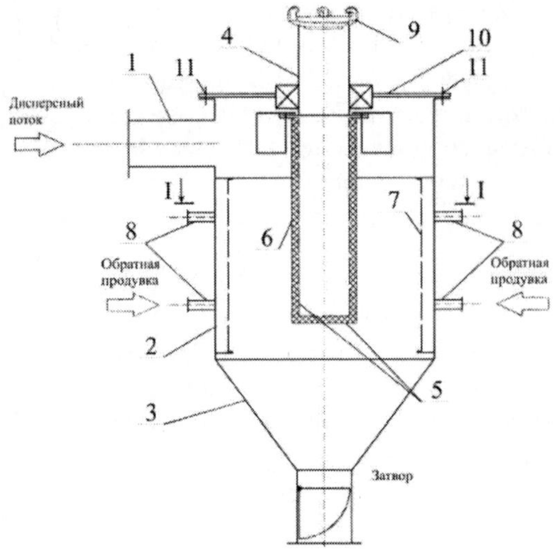

Fig. 2. The cyclone filter (Patent No. 2361678 RF) [5]: 1-inlet, 2-body, 3-conical bottom, 4-exhaust tube, 5-cylindricity frame, 6-dense filtering material, 7-thin filter fabric, 8-purge fittings, 9-pin element 10 to the upper end, 11-retaining elements of the connector

The technical problem of intensification of cyclonic processing and increasing the degree of deposition of fine particles is solved here by increasing the velocity of the dispersed flow while taking measures to prevent abrasive wear internal surfaces of the apparatus. The device allows to reduce the resource intensity, but energy costs increase in proportion to the speed increase .

Analysis of many other existing cyclone filtration devices for industrial particulate emissions (e.g., [6, 7, 8, 9]) it also showed that the devices with a high degree of deposition of PM_{10} and $PM_{2,5}$ particles are both material- and energy-intensive. Devices with low resistance, and therefore not energy-intensive, have a low degree of deposition of particles of these classes.

2. Cyclone filter unit to improve the energy and environmental efficiency of cleaning

For the purpose of good cleanup of gases from the suspended particles of the classes PM_{10}, $PM_{2,5}$ without increase in energy consumption at sedimentation of a suspended matter the advanced design of a cyclone filter is offered. The achievement of high extent of cleaning of a fine suspension is necessary at gas power plant for ensuring reliable operation of units of gas-turbine installations (GT) of steam-gas cycles, and at coal power plant – for increase in energy efficiency of operation of gas-cleaning devices. So, for example, increase in efficiency of trapping of the weighed part from a stream of natural gas in the point of gas preparation (PPG) of power plant leads to decrease in wear of cylinder walls of piston compressors or the surfaces of screws and cases of a complex configuration (which accuracy of processing is 10 microns) the screw compressors of the booster compressor stations (BCS). And placing a highly effective precipitators at the gas pipeline before GT (after the safety locking valve) for increase in extent of sedimentation of products of internal corrosion of pipelines and prevention of their penetration in GT for goal of reduce abrasive wear of turbines is provided by special articles of federal norms and rules of Russia of industrial safety "Safety rule of networks of gas distribution and gas consumption" (items 93, 94).

The proposed design of the cyclone filter can also be used in coal-fired power plants to improve the degree of cleanup of atmospheric emissions of dust preparation systems from coal dust and flue gas fromPM_{10}, $PM_{2,5}$ fine particle ash. By data [10], in flying ashes of power plants the average content of such particles about 30%, and in emissions of coal and slate dust can reach 35%. The offered cyclone filter (fig. 3) with a tangential inlet combines two steps of cleaning. Tests have shown that it provides rather high efficiency of cleaning of the emissions containing particles of the classes PM_{10} and $PM_{2,5}$

at acceptable power expenses and low abrasive wear. To increase the efficiency of sedimentation of fine particles with a minimum of energy and material costs cyclone has an additional device –the insertion of a fabric filter.

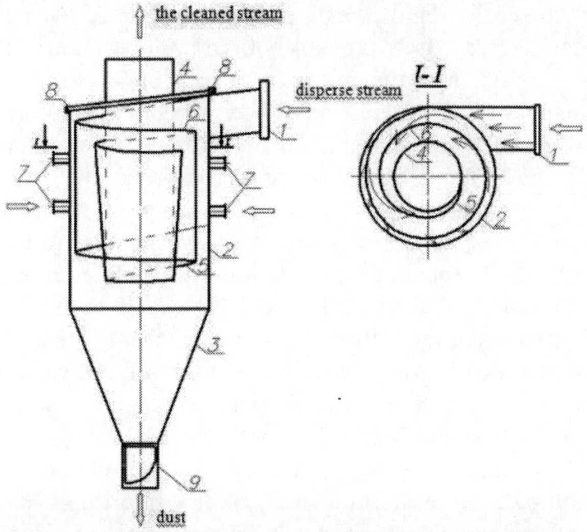

Fig. 3. Scheme of a design of a cyclone filter: 1 - entrance branch pipe; 2 - case; 3 - bunker; 4 - exhaust pipe; 5 - filtering material; 6 - metal framework; 7 - a purge fitting; 8 - details of fixture of a removable end face; 9 – bunker lock

Laboratory tests of a cyclone filter are carried out on the basis of the serial CN-11 cyclone with a diameter of 200 mm. Tests were carried out on specially prepared disperse material – aluminous (chamotte) dust with a size of particles up to 63 microns. Test of material for tests selected according to good laboratory practice and classified on a sieve in accordance with GOST P 51568-99 (ISO 3310-1-90). Radiuses of particles by results of the sedimentation analysis of fractional structure have turned out from 1,23 microns to 31,51 microns; therefore, the largest diameter of particles of dust by results of the sedimentation analysis corresponds to the sieve cell size. The disperse composition of dust in emissions after a cyclone filter was defined using an impactor. Based on the analysis of the dispersed composition of the flow, the median particle size d_{50} was 1,6 μm.

The carried out experimental studies confirmed the expediency of constructive addition of the return-flow cyclones with a filter insert in the zone where inertial dust deposition occurs. At the same time, in order to refine the design, a number of studies were performed using the computational methods. To determine the efficiency of separation of the suspended part of the flow in this cyclone filter, the dimensionless criterion Re [11] was used:

$$\mathrm{Re}_r \quad \frac{U_o\ \rho_p^2\ D_p^4}{c\ \rho_g \cdot R_2^3\ \eta} \tag{1}$$

where: U_0 – initial velocity, m/s, ρ_p – density of particles, kg/m^3, D_p – diameter of particles, m, c– the coefficient depending on the swirling device, ρ_g – density of gas, kg/m^3, R_2 – the radius of a cyclone, m, η – coefficient of dynamic viscosity, Pa·s.

The Re_r parameter is received as a result of use of two standard methods of the theory of similarity (the analysis of dimensions and scaling), with reduction of system of the equations consisting of Navier-Stokes equations for flow movement and Newton's law for the movement of a particle to a dimensionless form. Calculations have shown that to degrees of sedimentation of particles in a cyclone there correspond strictly definite values of numbers Re_r. So, at a speed of 5,0 m/s the number $Re_r=2,4\cdot10^{-3}$ to be compared to degree of sedimentation of 99%, and degree of sedimentation of 50% corresponded to $Re_r=4,65\cdot10^{-8}$. The carried out researches have shown that calculations on the basis of Re_r lead to the results close to the experience values of the sedimentations degree. At the same time the Re_r parameter allows to set the regimes of inertial separation of various sizes particles from any flow with curvilinear streamlines, i.e. from streams both in cyclones, and in the porous media.

Also numerical modeling of aerodynamic parameters of cyclone filter operation using the Computational Fluid Dynamics (CFD) methods has been performed. Today there is a set of works on CFD modeling of streams in cyclones, e.g., [12,13,14]. However only individual characteristics of a stream in concrete modifications of cyclones were, as a rule, observed. Therefore numerical researches of a 3D model of a stream in the developed cyclone filter design have been carried out. The accepted Detached Eddy Simulation (DES) model is "hybrid", based on sharing Reynolds Averaged Navier-Stokes (RANS) equations and Large Eddy Simulation (LES). It combines strengths of these two methods, namely, high precision of statistical models in the field of an interface and acceptable computing expenses of RANS and universality of LES in detachable areas of a stream.

The additional reserve of energy saving when using of the offered way of dust removal consists in optimization of details in communications. It is known that degree of sedimentation of the weighted particles from a heterogeneous stream correlates with the size of the power expenses which are directly related with sedimentation processes. In a construction of the presented cyclone filter those is processes of sedimentation of a suspension on surfaces of a filtering insert and an internal wall of a cyclone. Energy losses on local resistances arising at the movement of a dusty stream in shaped details before a cyclone and at entrance to a cyclone those aren't, but they are inevitable in use the device. Numerical and experimental studies of authors of works [15,16,17,18] have shown that losses may decrease by inserts which outlines can be also found using computer modeling of flows in connecting elements by the CFD methods.

Thus, the offered design of a cyclone filter allows to reach increase capacity of clearing devices with improvement of quality of clearing of gas which is expressed in reduction of the size of the particles caught for 50% (cutoff diameter) from averages for usual cyclones of 5-10 microns up to 1,6 microns.

The specified improvement of quality of cleaning doesn't demand essential increase in expense of energy. It is one of the main advantages over the usual cyclones at which for decrease cutoff diameter to 0,1 microns (since 2-3 microns) it is required to increase energy expenses to 15%.

References

[1] Brunnmair E 2015 Entwicklung und Modellierung eines neuen Hochleistungszyklons zur Trennung von Feststoff *Gas-Gemischen: Dissertation zur Erlangung des akademischen Gradeseines Doktors der montanistischen Wissenschaften* (Leoben: Montanuniversität) p 168

[2] Bielefeldt E A 2013 Wirbelkammern neurer Technologie BE-Bericht *Neue Patentanmeldung Wirbeltrennverfahren* **1/86** p 286

[3] Muschelknautz E 1980 Theorie der Fliehkraftabscheider mit besonderer Berücksichtigung hoher Temperaturen und Drücke *VDI-Bericht Nr.* **363** 49–60

[4] Staudinger G 1992 Zyklone mit Zentralkörpern *Veröffentlichung in Chem.Ing.-Tech.* **64** 769–888

[5] Ziganshin M G 2009 Cyclone-filter *Patent RUS* 2361678

[6] Balestrin E, Decker R K, Noriler D, Bastos J C S C and Meier H F 2017 An alternative for the collection of small particles in cyclones: experimental analysis and CFD modeling *Separation and Purification Technology* **184** 54–65

[7] Mazyan W I Ahmadi A Ahmed H and Hoorfar M 2017 Increasing efficiency of natural gas cyclones through addition of tangential chambers *Journal of Aerosol Science* **110** 36–42

[8] Temnikova E.Yu., Bogomolov A R and Petrik P T 2009 Research of characteristics of a

cyclone with internal elements *News Kuzbassky of the state technical university* **2** 140–4

[9] Serebryansky D A, Zakharov A A and Plashikhin S V 2013 Cyclonic dust collectors. Low-cost modernization *Chemical industry of Ukraine* **3** 70–4

[10] Scriabina L Y 1980 *Atlas industrial dusts. Part 1. Fly ash from industrial power plants.. Containertypes* p 48

[11] Ziganshin M G and Ziganshin A M 2012 Laboratory course of a dusting. Part 1. methodology and instructions on educational and research for students *KGASU publishing house* p 188

[12] Fassani F L and Goldstein L 2000 A study of the effect of high inlet solids loading on a cyclone separator pressure drop and collection efficiency *Powder Technology* **107** 60–5

[13] Hoffmann C, Vansanten A, Allen R W K and Clift R 1992 Effects of geometry and solid loading on the performance of gas cyclones *Powder Technology* **70** 83–91.

[14] Chu K W, Wang B, Yu A B and Vince A 2009 CFD-DEM modeling of multiphase flow in dense medium cyclones *Powder Technology* **193** 235–247.

[15] Ziganshin A M 2015 Reduction of energy consumption of flow by profiling of fittings in power plants communications *Safety & Reliability of Power Industry* 63–8.

[16] Ziganshin A M, Aleshchenko I S and Ziganshin M G 2016 Connection fitting with profiling inserts *Patent RUS* 2604264

[17] Ziganshin A M, Gimadieva A G and Batova K E 2017 A pressure drop and characteristics of the jet exiting through the secondary side opening *Proceedings of the University* **4 (42)** 257-265

[18] Logachev K I, Ziganshin A M, Averkova O A and Logachev A K 2018 A survey of separated airflow patterns at inlet of circular exhaust hoods *Energy Build.* **173** 58–70

IOP Conf. Series: Earth and Environmental Science **288** (2019) 012074 doi:10.1088/1755-1315/288/1/012074

Development of the forest waste gasification system with electricity production

A Aiyzan[1], A Fedukhin[1], K Strogonov[1], O. Derevianko[2], S. Makoev[2], M. Egorov[2]

[1]National Research University «Moscow Power Engineering Institute», 111250 Russia, Moscow, Krasnokazarmennaya 14.

[2]Peter the Great St. Petersburg Polytechnic University, 194021, Russia, St.Petersburg, Polytechnicheskaya 29

fedyukhinav@yandex.ru

Abstract. The aim of this work is development of a technological scheme of gasification of biomass to generate electricity, as well as development of efficient circulating fluidized bed gasifier with heat exchanger. The paper deals with the existing technology of biomass gasification and the influence of thermal parameters on the composition of gasification products.

Introduction

The total wood stock in Russia is more than 80 billion m^3, our country is the leader in the area of forest resources and its amount per an inhabitant. The potential of biomass usage in Russia is the following: gross (467 million TOE/year), technical (129 million TOE/year), economic (69 million TOE/year), including agricultural waste, which is up to 80 million TOE/year; forest waste is more than 30 million m3 [1–7].

The volume of wood waste generation in the timber industry is 32% of the raw material. The amount of waste, for example, in a furniture factory reaches about 62% of all used wood.

In Russia at the enterprises, plants and factories wood waste is burned out or thrown into dumps and landfills. At present, energy-saving programs are being introduced, as well as that for reduction of harmful emissions into the atmosphere and optimization of primary resources usage [8–21].

Developed technological scheme

The area of application of generator gas is very wide. From the generator gas, electricity can be obtained by burning in a steam boiler, in gas engines or in steam and gas installations. We have investigated the scheme of the steam-gas installation (Fig.1), because the efficiency of a steam-gas installation is greater than for other schemes [22–24].

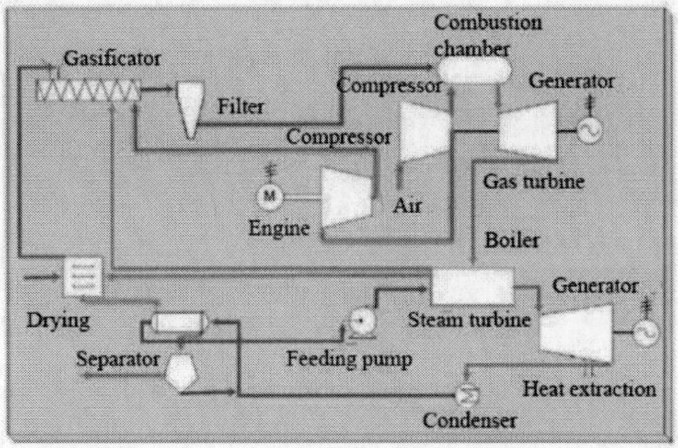

Fig. 1. The scheme of the steam-gas installation based on biomass gasification

The feedstock after drying is fed to the gas generator, where steam and air are pumped under pressure of 20 atmospheres for thermal conversion process. As a result of heat treatment, a generator gas is formed, which is purified from dust and ash in the filter. The purified gas is sent to a gas turbine that produces electricity. The exhaust gases after the gas turbine are sent to the waste heat boiler. In the boiler, the heat exchanger produces overheated steam. Part of the steam goes to the gas generator, and another part of the steam goes to a steam turbine, which generates electricity.

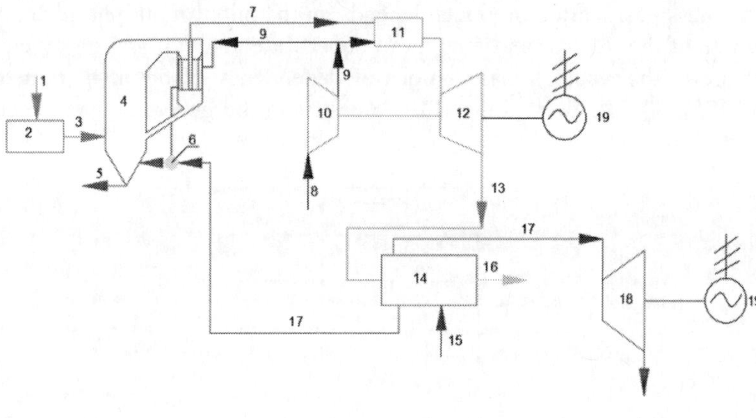

Fig. 2. Developed technological scheme of biomass gasification from steam-gas installation:
1 – biomass; 2 – crusher; 3 – crushed biomass; 4 – circulating gas generator of boiled layer with heat exchanger; 5 – ash; 6 – ejector; 7 – gas of generator; 8 – cold air; 9 – compressed air; 10 – compressor; 11 – combustion chamber; 12 – gas turbine; 13 – wasted gases; 14 – utilizing boiler; 15 – feed water; 16 – output gases; 17 – overheated steam; 18 – steam turbine; 19 – generator.

In Fig. 2 the developed technological scheme of the steam-gas installation based on wooden wastes is presented. The principle is the same as in Fig. 1, but there are some distinctive features:

- Absence of drying installation for biomass;
- Usage of physical heat of the generator gas for heating the compressed air in the heat exchanger;
- No additional air compressor under the engine;
- Usage of an ejector to disperse compressed air using overheated steam.

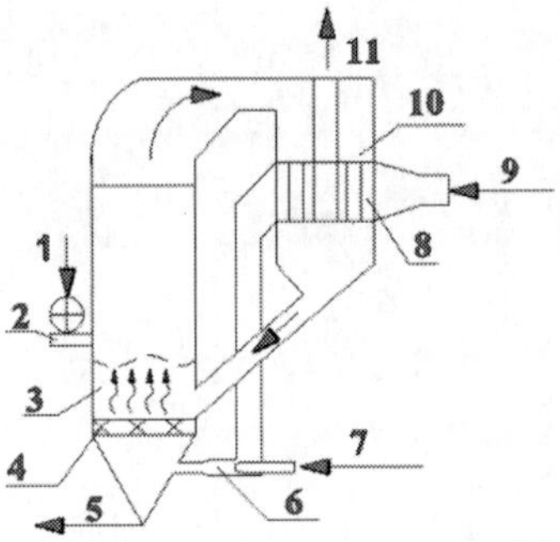

Fig. 3. The developed scheme of the circulating gas generator of boiling layer with heat exchanger. 1 – biomass; 2 – feeder; 3 — gasification zone; 4 — fuel layer; 5 – ash; 6 – ejector; 7 – overheated steam; 8 – heat exchanger; 9 – compressed air; 10 – cyclone dust collector; 11 – gas of generator

In the paper we present a circulating gas generator of boiled layer with heat exchanger, which is shown in Fig. 3. The heat exchanger 8 can increase the temperature of the compressed air 9 and raise the efficiency of gasification.

The main feature of the developed gas generator is the supply to the gasification zone 3 of the steam-air mixture from the ejector 6, dust purification in the vertical cyclone dust collector 10 and heating of the compressed air 9 in the heat exchanger 8.

The research of the dependency of the output parameters of the biomass gasification products

An output of biomass gasification products depends on the initial input parameters to the generator (ratio of the flow rate of the oxidant and amount of superheated steam) and on given parameters (gas temperature, pressure in the reactor). Calculations of dependency of output parameters were carried out in MathCad15 [25]. The air flow rate and the pressure in the generator were accepted as 0.3 and 2.5 MPa respectively [26].

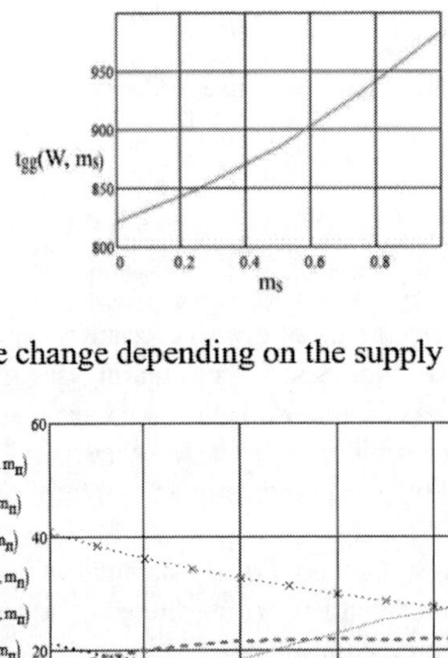

Fig. 4. Temperature change depending on the supply of superheated steam

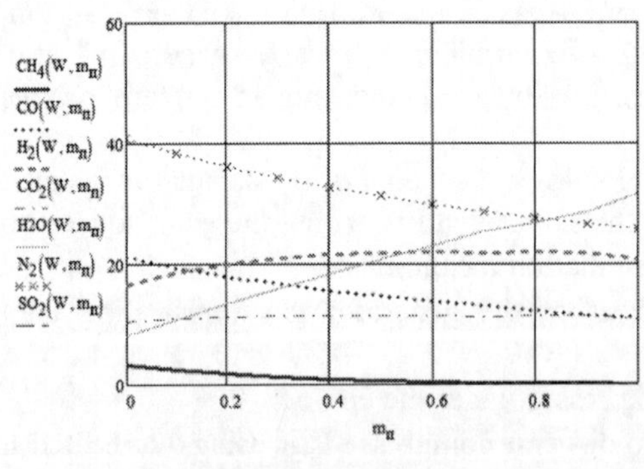

Fig. 5. Relation between the output of gasification biomass products and the steam amount

With an increase of the amount of overheated steam supplied to the ms, the temperature of the generator gas linearly increases (Fig. 4).

In general, the amount of overheated steam in the gas generator mostly influences on the output of H2 and H2O (Fig. 5). The maximum value of H2 at the ms boundary varies from 0.2 to 0.8, but with increasing of ms the CO output decreases.

From Fig. 6 it can be seen that with the increase in the amount of overheated steam, the chemical heat of combustion of the generator gas decreases, but the physical heat increases. So, it is not effective to supply more steam to the gas generator, since an important role in thermal installations is played by the heat of combustion, and the use of physical heat can lead to pressure loss.

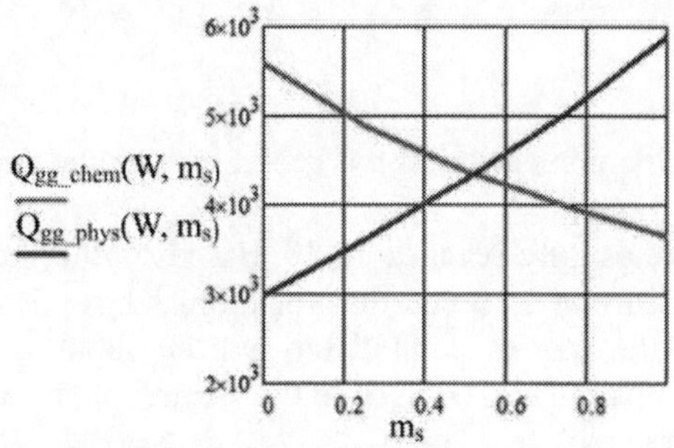

Fig. 6. Change in the heat of combustion and the physical heat of the generator gas depending on the amount of steam

Based on the research of the dependence of the heat of the gas on the amount of steam, the amount of overheated steam can be taken ms = 0.2 kg/kg (biomass) to obtain a certain pressure in the gas generator.

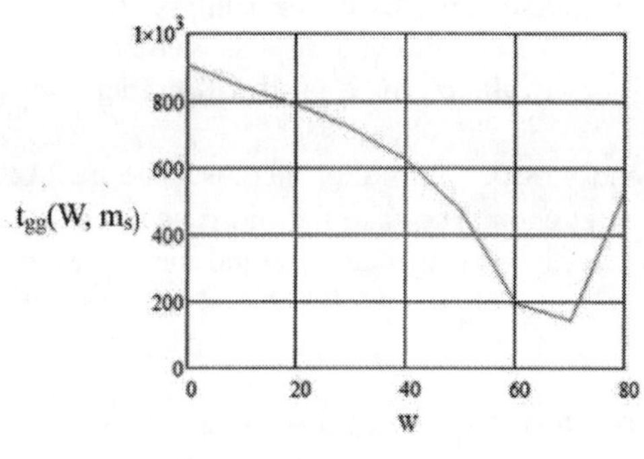

Fig. 7. Change of gas temperature in the dependency on biomass humidity.

Raw materials humidity W greatly affects the process of biomass gasification. So in fig. 7 it is displayed that the temperature of the gas of generator is reduced to 70 % with an increase in the humidity of the raw material, then increases sharply from 70 %. This is due to the fact that the heat is lost on the evaporation of moisture and the thermal conversion reaction of the evaporated moisture.

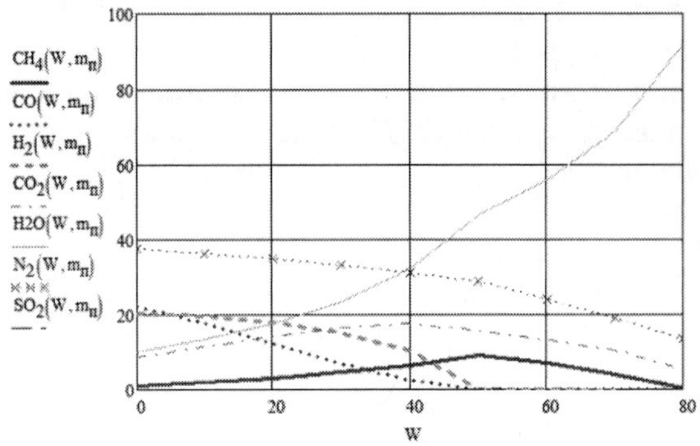

Fig. 8. Relation between composition of the generator gas and raw materials humidity.

The humidity of wood waste usually reaches 15 %. Based on the graph of the dependence of the amount of heat on humidity, it can be assumed that for biomass gasification it is enough to take raw materials with W=10 %, since the heat of combustion will be about 5 MJ/m3. The gas of generator with such chemical heat of combustion can be used in the steam-gas installation.

Conclusions

The new technological scheme of the steam-gas installation based on wood wastes was developed. It has a number of features in the comparison with a scheme in Fig. 1:

- Absence of drying installation for biomass;
- Usage of physical heat of the generator gas for heating the compressed air in the heat exchanger;
- No additional air compressor under the engine;
- Usage of an ejector to disperse compressed air using overheated steam.

Also we developed a new effective circulating gas generator of the boiling layer with heat exchanger, in which the efficiency of the biomass gasification increases due to heating of compressed air.

As the overheated steam and biomass humidity increase, the heat of combustion of the generator gas decreases. Based on this, it is enough to take the biomass moisture W = 10%, and the amount of steam is 0.2 kg/kg (biomass) to disperse the compressed air to reach a certain pressure in the gas generator.

References

[1] Kirpichnikov I M and Fil' N S 2012 wood waste as a renewable source for obyaining thermal and electrical energies *Vestn. Yuzhno-Ural'skogo Gos. Univ. Seriya Energ.* **16** 18–21

[2] Saari J, Sermyagina E, Kaikko J, Vakkilainen E and Sergeev V 2016 Integration of hydrothermal carbonization and a CHP plant: Part 2 –operational and economic analysis *Energy* **113** 574–85

[3] Zhao Y, Wang L, Huang L, Maximov M Y, Jin M, Zhang Y, Wang X and Zhou G 2017 Biomass-derived oxygen and nitrogen co-doped porous carbon with hierarchical architecture as sulfur hosts for high-performance lithium/sulfur batteries *Nanomaterials* **7**

[4] Politaeva N, Smyatskaya Y, Slugin V, Toumi A and Bouabdelli M 2018 Effect of laser radiation on the cultivation rate of the microalga Chlorella sorokiniana as a source of biofuel *IOP Conf. Ser. Earth Environ. Sci.* **115**

[5] Politaeva N, Kuznetsova T, Smyatskaya Y, Atamaniuk I and Trukhina E 2018 Chlorella Microalga Biomass Cultivation for Obtaining Energy in Climatic Conditions of St. Petersburg *Adv. Intell. Syst. Comput.* **692** 555–62

[6] Fedyukhin A, Sultanguzin I, Gyul'Maliev A and Sergeev V 2017 Biomass pyrolysis and gasification comprehensive modeling for effective power generation at combined cycle power plant *Eurasian Chem. J.* **19** 245–53

[7] Timerbaev N F, Tuntsev D V, Khairullina M R and Kitaev S V 2017 Technologies and equipment for the production of electricity from wood waste *Proc. High. Educ. institutions. ENERGY Sect. Probl.* **19** 102–7

[8] Islamova S I and Vachagina E K 2015 No Title *Proc. High. Educ. institutions. ENERGY Sect. Probl.* **9–10** 3–11

[9] Klochkov Y S, Didenko N I, Makov K M, Zapivahin I O, Ostapenko M S and Volgina A D 2018 Employment and professional adaptation of specialists vol 2018–Janua (Institute of Electrical and Electronics Engineers Inc.) pp 162–4

[10] Gorshkov A, Murgul V and Oliynyk O 2016 Forecasted Payback Period in the Case of Energy-Efficient Activities vol 53, ed M V. (St. Petersburg State Polytechnical University, Polytechnicheskaya st. 29, St.-Petersburg, Russian Federation: EDP Sciences)

[11] Tanic M, Stankovic D, Nikolic V, Nikolic M, Kostic D, Milojkovic A, Spasic S and Vatin N 2015 Reducing energy consumption by optimizing thermal losses and measures of energy recovery in preschools *Procedia Eng.* **117** 924–37

[12] Perlova E, Platonova M, Gorshkov A and Rakova X 2015 Concept project of zero energy building *Procedia Eng.* **100** 1505–14

[13] Strelets K, Perlova E, Platonova M, Pankova A, Romero M and Al-Shabab M S 2016 Post Occupancy Evaluation (POE) and Energy Conservation Opportunities (ECOs) Study for Three Facilities in SPbPU in Saint Petersburg vol 165(Elsevier Ltd)pp 1568–78

[14] Polosin A N and Chistyakova T B 2018 Mathematical modeling for resource and energy saving control of extruders in multi-assortment productions of polymeric films vol 1015(Institute of Physics Publishing)

[15] Gorshkov A S, Vatin N I, Rymkevich P P and Kydrevich O O 2018 Payback period of investments in energy saving *Mag. Civ. Eng.* **78** 65–75

[16] Ilin I, Levina A and Iliashenko O 2018 Enterprise Architecture Analysis for Energy Efficiency of Saint-Petersburg Underground *Adv. Intell. Syst. Comput.* **692** 1214–23

[17] Akhmetova I, Chichirova N and Derevianko O 2017 Revisiting heat losses calculation at district heating network *Int. J. Civ. Eng. Technol.* **8** 694–702

[18] Brunman V E, Vataev A S, Volkov A N, Petkova A P and Plotnikov D G 2017 Optimizing pump-drive operation to improve the energy-efficiency of oil extraction *Russ. Eng. Res.* **37** 479–84

[19] Brunman V E, Vataev A S, Volkov A N, Volkov E A, Petkova A P, Kochanzhi F I and Plotnikov D G 2017 Energy-efficient oil extraction by sucker rod borehole pumps in low-yield fields *Russ. Eng. Res.* **37** 378–82

[20] Korotkov A and Frolov V 2015 Experimental studies of electrical load schedules of residential electricity consumers in urban distributive power grids *Acta Tech. CSAV (Ceskoslovensk Akad. Ved)* **60** 337–46

[21] Politaeva N A, Kuznetsova T A, Smyatskaya Y A, Trukhina E V and Atamanyuk I 2018 Energy Production from Chlorella Algae Biomass Under St. Petersburg Climatic Conditions *Chem. Pet. Eng.* 1–5

[22] Fedyuhin A V 2014 *Development of systems of combined generation of thermal and electrical power basing on investigation of biomass pyrolysis and gazification processes dis. ... kand. tekhn. nauk* (Ivanovo: IGEHU)

[23] Galaev S A, Ris V V, Smirnov E M and Babiev A N 2018 Experience Gained from Designing Exhaust Hoods of Large Steam Turbines Using Computational Fluid Dynamics Techniques *Therm. Eng.* **65** 352–61

[24] Galaev S A, Ris V V, Simoyu L L and Smirnov E M 2017 Experience in numerical simulation of turbulent wet-steam flow in the last stage of a high-power condensing turbine under

conditions defined by full-scale experiments at a power plant *12th European Conference on Turbomachinery Fluid Dynamics and Thermodynamics, ETC 2017*

[25] Popov S K and Ippolitov V A 2009 *Solving of problems of high-temperature heat technology using MathCad: textbook* (Moscow: MPEI Publishing House)

[26] Klimenko A V and Zorin V M 2007 *Heat-and-power engineering anf Thermal engineering: reference books* ed A V Klimenko (Moscow: MPEI Publishing House)

IOP Conf. Series: Earth and Environmental Science **288** (2019) 012075 doi:10.1088/1755-1315/288/1/012075

Application of methane-hydrogen fraction as fuel in mix with fuel oil for coppers of thermal power plant

R.V. Akhmetova, M.A. Taymarov, E.A. Akhmetov

Russia, Kazan, Kazan state power engineering university

ahmetova_rv@bk.ru

Summary. In this article the heat exchange processes happening at reactions of burning of methane-hydrogen fraction in power coppers of TGM-84A and TGM-84B are investigated. Questions of distribution of thermal streams on fire chamber height are considered. Results of calculations of operating conditions on burning of methane-hydrogen fraction are compared with operating conditions of work of coppers at combustion of natural gas.

Description of an object and technique of a research

Properties of methane-hydrogen fraction were experimentally studied on the operating oven equipment ON-1000/1 and ON-1000/2 of installation of hydrotreating of vacuum gasoil of the shop No. 2 in JSC Ryazan Oil Refining Company. To research objects on which application of methane-hydrogen fraction as fuel instead of natural gas is planned coppers of TGM-84A and TGM-84B are taken. During the experiments at measurement of thermal streams radiometers of full radiation TEPA-50 were used, and the chromel-alumel doubled thermocouples and the ThermaCAM R-50F thermal imager were used to determination of temperature of a torch. The analysis of structure of tests of methane-hydrogen fraction was carried out in the central factory laboratory. The method of experiments of measurement of parameters of operation of furnaces and a copper is stated in works [1]. Calculation of parameters of work of coppers of TGM-84A and TGM-84B when burning methane-hydrogen fraction is executed with use of materials [2-7]. The scheme of ON-1000/1 and ON-1000/2 furnaces on which operating conditions of process of methane-hydrogen fraction burning were studied is provided on fig. 1.

The sizes of ON-1000/1 and ON-1000/2 furnaces on fire chambers: length is 8,6 m, width is 3,3 m, height is 5,5 m. Hearth gas burners 1 (see fig. 1) in number of 40 pieces of the LE-CFSG-2W brand are executed in the form of single pipes with an output diameter of gas supply openings of 6,3 mm and provide an expense of methane-hydrogen fraction on one furnace in quantity from 1000 m^3/h to 1500 m^3/h.

Measurements of the thermal streams falling from a torch and temperature were carried out through side hatches. Heat of combustion of methane-hydrogen fraction is perceived in a fire chamber by radiant coils 3 (see fig. 1), providing heating of a product – vacuum gasoil in number of 177 t/h in mix with hydrogenous gas in number of 97,4 thousand m^3/h and convective coils 4 in the taking- away chimney.

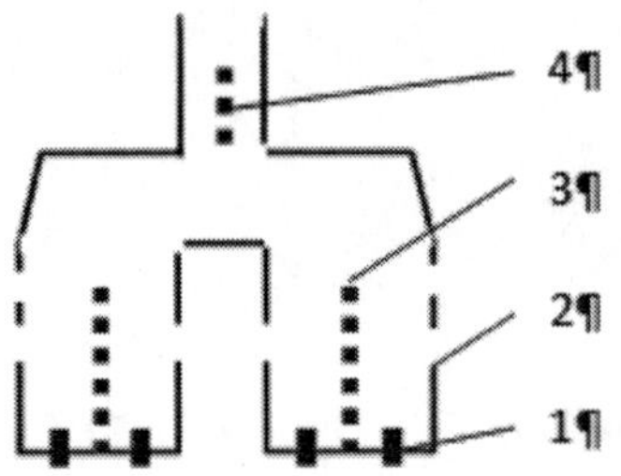

Fig. 1. Cross-section of ON-1000/1 and ON-1000/2 furnaces: 1 – torches, 2 – hatches, 3 – radiant coil, 4 – convective coil

Heating of a product is made from temperature 322 °C to 347 °C. Temperature of an external surface of radiant coils on measurements by the ThermaCAM R-50F thermal imager is equal to 530 °C
The scheme of a copper TGM-84A, for which the possibility of application of methane-hydrogen fraction as fuel instead of natural gas is considered, is provided on fig. 2.

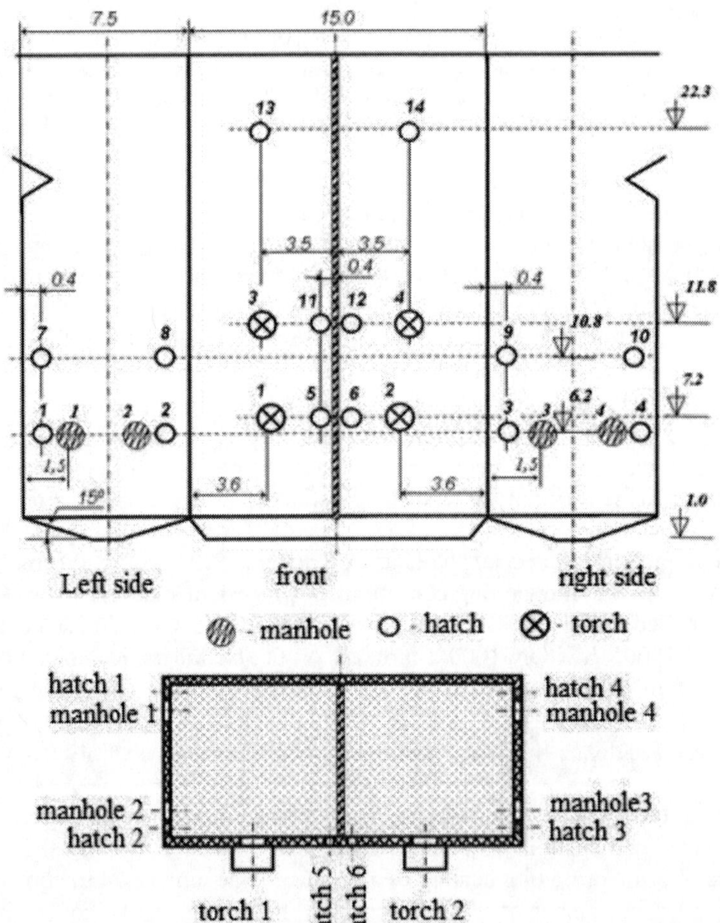

Fig. 2. Scheme of a copper TGM-84A

TGM-84A copper oil-gas drum with the steam generating capacity of 420 t/h and steam parameters

IOP Conf. Series: Earth and Environmental Science **288** (2019) 012075 doi:10.1088/1755-1315/288/1/012075

560 °C and with pressure of 14 MPas. The fire chamber on height is divided into two half by the two-light screen. Fire chamber sizes: width is 15 m, depth is 7,5 m, height is 27 m. Torches on 2 pieces are located on marks of heights of 7,2 m and 11,8 m. The consumption of natural gas on TGM-84A copper with a nominal productivity of 420 t/h is equal to 32,5 thousand m^3/h at the lowest heat of combustion of $Q_н^р$=34,24 MJ/m^3 and composition of gas: methane CH_4=98,89 % (on weight), ethane C_2H_6= 0,47 %, propane C_3H_8= 0,18%, nitrogen N_2=0,76 %. On a copper of TGM-84A vortex torches of HF-TsBK-VTI-TKZ with a single power are established during the work on gas of 76,7 MW with peripheral tubular gas-distribution: ∅23 - 12 pieces and ∅33 - 12 pieces. Air turning two-line on hot air also consists of two sections of an axial (central) swirler and one section of a tangential (peripheral) swirler.

Results of a research and their discussion

The structures of the methane-hydrogen fraction used as fuel of ON-1000/1, ON-1000/2 furnaces of vacuum gasoil heating in JSC Ryazan Oil Refining Company are brought in the tab. 1.

Table 1. Structures of tests of the methane-hydrogen fraction used as fuel of ON-1000/1 and ON-1000/2 furnaces of vacuum gasoil heating on hydrotreating installation

Test No.	Density ρ at 760 mm Hg and 0 °C, $Q^р_н$	Hydro gen, % on weight	Methane, % on weight	Ethylene, % on weight	Ethane, % on weight	Propene, % on weight	Propane, % on weight	I-butane, % on weight	N-butane, % on weight	The sum of butylene, % on weight
	0,440	14,75	18,53	2,06	12,45	3,34	13,26	5,40	9,95	0,71
	0,380	18,36	18,28	0,86	10,74	1,00	13,34	4,55	12,94	0,20
	0,444	14,33	24,41	0,54	10,50	0,73	14,45	4,87	13,39	0,35
	0,431	13,76	36,71	0,55	9,86	0,66	12,48	2,88	7,03	0,36
	0,481	11,04	44,18	0,50	8,08	0,36	10,56	3,15	8,44	0,00
	0,400	15,42	31,51	1,07	8,43	0,57	8,04	2,17	4,67	0,06
	0,445	15,99	3,64	0,03	7,26	0,04	16,54	6,13	18,46	0,00
	0,434	14,59	24,82	0,73	10,85	1,72	14,71	5,05	8,85	0,24

Test No.	I-pentane, % on weight	N-pentane, % on weight	Hexane and above, % on weight	Oxygen, % on weight	Nitrogen, % on weight	Carbon monoxide, % on weight	Carbon dioxide, % on weight	Hydro gen sulfide, % on weight	Heat of combustion $Q^р_н$, MJ/m^3
	4,21	2,39	0,00	2,95	9,27	0,11	0,58	0,03	20,91
	4,96	4,78	0,23	1,91	7,25	0,03	0,36	0,20	22,83
	3,94	4,10	0,19	1,35	5,98	0,03	0,36	0,48	21,27
	2,39	2,76	0,60	1,79	7,65	0,06	0,18	0,28	23,98
	3,50	3,06	0,27	1,57	4,93	0,03	0,25	0,10	22,96
	1,20	1,14	0,00	6,18	19,19	0,03	0,20	0,12	25,45
	7,32	5,03	0,39	5,06	13,93	0,00	0,18	0,00	18,80
	3,40	2,58	0,20	3,03	8,29	0,06	0,23	0,67	21,88

Apparently from tab. 1, heat of methane-hydrogen fraction combustion makes from 55,5 % to 75,1 % of heat of the natural gas combustion used as fuel for coppers of thermal power plant. Experimentally received distribution of temperature of a torch and the thermal streams q_o falling from an internal brickwork envelope on height h of fire chambers when burning methane-hydrogen fraction with structure of tests No. 6 and 7 at coefficient of excess of air 1,3 is given in ON-1000/1 and ON-1000/2 furnaces in fig. 3.

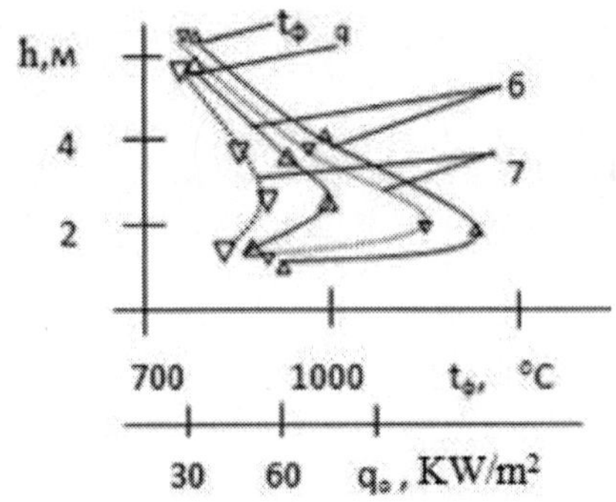

Fig. 3. Distribution of temperature of a torch $t_ф$ and thermal streams from an internal brickwork envelope q_o on height h of ON-1000/1 and ON-1000/2 furnaces when burning methane-hydrogen fraction with structure of tests No. 6 and 7 at coefficient of excess of air 1,3.

Apparently from fig. 3, the transfer of OH-1000/1 and ON-1000/2 furnaces from burning of methane-hydrogen fraction with heat of combustion of 25,45 MJ/m³ (test 6) on burning of structure with heat of combustion of 18,8 MJ/m³ (test 7) leads to decrease in temperature in a torch kernel on average on 100 °C. Thermal streams q_o also significantly decrease. Therefore operation of furnaces when burning methane-hydrogen fraction with low heat of combustion on installation of hydrotreating of gasoil is carried out only at the fresh catalyst which allows lower temperatures of a torch in a fire chamber.
Results of experiments on definition of concentration of NO_x nitrogen oxides and speed of burning w of methane-hydrogen fraction in the ON-1000/1 furnace and natural gas in TGM-84A copper depending on heat of combustion $Q_н^р$ are given in tab. 2.

Table 2. Results of experiments on determination of speed of methane-hydrogen fraction combustion w in the ON-1000/1 furnace and natural gas in TGM-84A copper depending on heat of combustion $Q_н^р$

Heat of combusti on $Q_н^р$, MJ/m³	Density ρ, kg/m³	The maintenance of H_2 on weight	The maintenance of CH_4, % on weight	Coefficient of air excess α	Fuel usage, m³/h	Temperature at the exit from a fire chamber $t_п$, °C	Concentration NO_x, mg/m³ (in terms of α=1.4)	Burning speed w, m/s
25,43	0,540	10,1	28,3	1,3	1115	802	84,8	2,02
26,68	0,536	10,05	28,27	1,1	1375	881	63,6	2,11
22,83	0,38	18,36	18,28	1,1	1458	884	56,1	3,07
33,94	0,69	-	98,89	1,1	32500	1150	136	0,84

Apparently from tab. 2, increase in content of H_2 hydrogen from 10,05 % (on weight) up to 18,36 % leads to growth of speed of burning w by 45 %. Speed of natural gas burning with the content of methane CH_4 =98,89 % in TGM-84A copper with leaves 0,84 m/s, that is speeds of burning of methane-hydrogen fraction with the maintenance of H_2=10,05 % are 2,5 times lower. Owing to low heat of methane-hydrogen fraction combustion concentration of NO_x nitrogen oxides is on average twice lower in comparison with natural gas combustion.

Distribution of thermal streams from $q_ф$ torch on fire chamber height h in TGM-84A copper at natural gas combustion is given in fig. 4 in comparison with settlement data when burning methane-hydrogen fraction at the content of hydrogen of 10,05 % and methane of 28,27%.

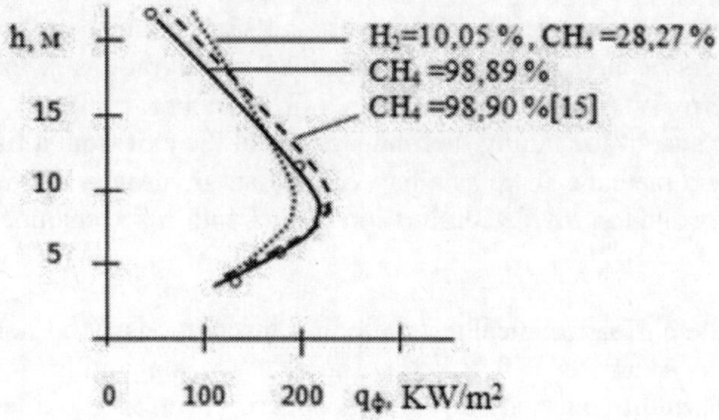

Fig. 4. Thermal streams from $q_ф$ torch on fire chamber height h in TGM-84A copper at natural gas combustion (CH_4 =98,89 %) in comparison with settlement data when burning methane-hydrogen fraction (H_2=10,05 %, CH_4 =28,27 %) at steam loading of 255 t/h and with data [8] for PK-41 copper

Apparently from fig. 4., burning of methane-hydrogen fraction (H_2=10,05 %, CH_4 =28,27 %) causes growth of heat fluxes of $q_ф$ on an output from a fire chamber that increase in heat loads on the boiler superheater of the boiler TGM-84A is followed and leads to lowering of efficiency in comparison with operation of this boiler on natural gas. However distribution of heat fluxes of a torch on height h when burning of methane-hydrogen fraction most is suitable for the direct-flow boiler PK-41[8] of the supercritical pressure having an over clamping of the furnace chamber.

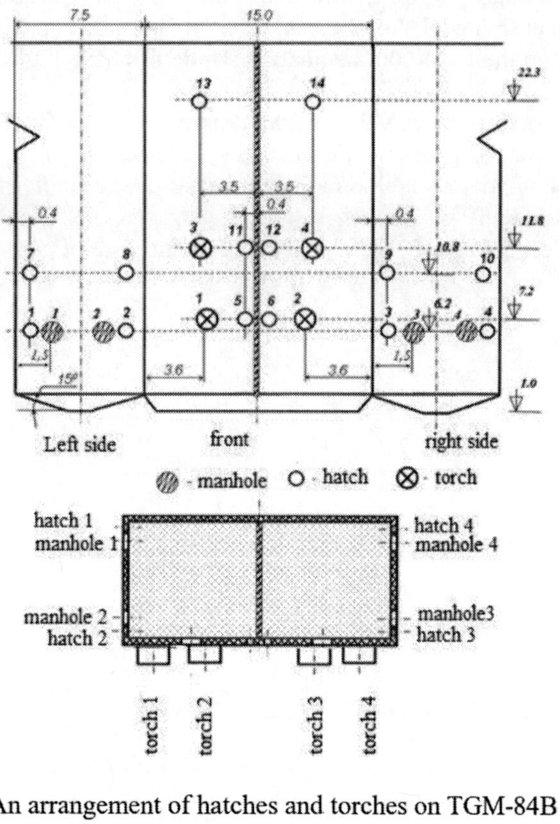

Fig. 5. An arrangement of hatches and torches on TGM-84B copper

Conclusions

1. Burning of methane-hydrogen fraction with the high content of hydrogen over 10 % (on weight) in power coppers of thermal power plant instead of natural gas is followed by the increased values of the falling thermal streams at the exit from a fire chamber.
2. For decrease in size of the falling thermal streams at the exit from a fire chamber of coppers it is necessary to increase a methane share as a part of methane-hydrogen fraction that allows to reduce the speed of fuel burning and to provide combustion of fuel within fire chamber volume.

References

[1] Trembovlya V.I., etc. Heattechnical tests of boiler installations. M.: Energy, 1977, 297 pages.
[2] Taymarov M.A., Akhmetova R.V., Lavirko Yu.V. Sungatullin R.G., Zheltukhina Ye.S. Decrease in harmful emissions in the atmosphere of nitrogen oxides by coppers of thermal power plant. KSUAE news. 2017. No. 1(39). p. 180-188.
[3] Rawdon A.H, Sadowski. R.S. An exsperimental correlationof oxide of nitrogen emissions from of ASME. 1973. Vol. 95, N 3, p.32-39
[4] Zeldovich Ya.B., Sadovnikov P.Ya., Frank-Kamenetsky D.A. Oxidation of nitrogen during combustion. M.-L.: Academy of Sciences of the USSR publishing house, 1947.-147 pages.
[5] Fenimore C.P. Formation of nitric oxide in premixed hydrocarbon flames. 13 International Symposium on Combustion Pittsburg. 1971. p.374-384.
[6] Taymarov M.A. Definition of concentration of nitrogen oxides formation in PK-47 copper fire chamber. Certificate on the state registration of the computer program No. 2012660912 of November 30, 2012.
[7] Knorre G.F., Arefyev K.M., Bloch A.G., Nakhapetyan Ye.A., Paleyev I.I., Steinberg V.B. Theory of furnace processes. M.-L., Energy. 1966, 492 pages.
[8] Bloch A. G. Heat exchange in fire chambers of boilers. L., Energoatomizdat, 1984. 240pages.
[9] Taymarov M.A., Kuvshinov N.Ye., Stepanova T. O. Optimization of operation of P-101 and P-101A furnaces on the L-24-600 installation. Bulletin of the Kazan technological university. 2015, b.18, No.22, p. 50-53.
[10] Tanatarov M.A., Akhmetshina M.N., Faskhutdinov R.A. Technological calculations of oil refining installations, Chemistry, Moscow 1987. 352 pages.
[11] Sardanashvili A.G., Lvova A.I., Examples and tasks of technology of oil refining and gas, Chemistry, Moscow, 1980. 256 pages.
[12] Manovyan A.K. Technology of primary oil refining and natural gas, Chemistry, Moscow 2001. 568 pages.

Determination of additional factors in assessing the reliability of heat supply systems

I G Akhmetova[1], A A Kalyutik[2], A V Fedukhin[3], O V Derevianko[2], L R Mukhametova[1]

[1]Federal State Budgetary Educational Institution of Higer Education "Kazan State Power Engineering University", Kazan, Krasnoselskaya street, 51

[2]Peter the Great St. Petersburg Polytechnic University, Polytechnicheskaya 29, Saint Petersburg, 195251 Russia

[3]National Research University «Moscow Power Engineering Institute» Krasnokazarmennaya 14, Moscow, 111250 Russia

Irina_akhmetova@mail.ru

Annotation. As a result of the research, additional factors that significantly influenced the reliability of the consumers' heat supply were assessed, an assessment of the influence of these factors on the reliability of the operation of the heat network was made. The influence of additional factors is taken into account when developing a new methodology and algorithm for calculating the reliability of consumers' heat supply; a program for calculating the reliability of consumers' heat supply was developed.

Determination of factors unaccounted for in the existing methodology

To determine the reliability of heat supply to consumers, the initial data: failure rate and average recovery time of heat pipelines and equipment should be taken into account [1–8].

The actual level of reliability of the heating system should be assessed on the basis of statistical data on failures of the system components.

To assess the reliability of existing and newly developed schemes of heat networks, a technique described in [9,10] is developed, which allows to determine the availability and probability of failure of the heat network on the basis of such initial data as the length, diameter and duration of operation of the pipeline system sections.

The disadvantage of this approach is that the calculations do not take into account a number of factors that directly affect the reliability of the heating system.

The aim of the study was to identify and take into account additional factors, as well as the development of a new method of calculating reliability indicators.

Research problems are the following:

- Determination of additional factors affecting the reliability of heat supply to consumers;
- Assessment of the impact of these factors on the reliability of the heat network;
- Taking into account the influence of these factors in the development of a new method of calculating the reliability of heat supply to consumers;
- Development of the program of reliability indicators calculation for consumer heat supply on

the basis of the new methodology.

It was assumed that the failure rate of the elements of the heat network (in addition to the service life) is influenced by the following factors: the residual thickness of the metal of the pipeline, the previous gusts, the corrosion activity of the soil, the presence of the channel flooding, the material of the pipeline, the presence of intersections with communications, the percentage of failure of the pipeline [7,11–21]

In order to solve the issue of including factors in the new methodology, the analysis of statistical data on gusts in different parts of the heat network of the city of Kazan was carried out [10].

Information on gusts was grouped by a set of factors. The influence on the failure rate of the elements of the heat network was revealed for the following factors: the residual thickness of the metal walls of the pipeline (K1), the presence of previous gusts (K2), the corrosion activity of the soil (K3), the presence of the channel flooding (K4), the presence of intersections with communications (K5). The influence of these factors on the reliability of heat supply of consumers was evaluated [10].

To take into account the factors identified in the existing methodology was introduced the coefficient of additional factors:

$$\mathrm{K}i = f(\mathrm{K1;K2;K3;K4;K5}) \tag{1}$$

With the calculated values of the coefficient Ki it is possible to estimate the effect of additional factors on the failure rate of the heat network sections.

The linear dependence of the coefficient of additional factors Ki on the residual wall thickness of the pipeline metal was revealed during the comparison of the parameters (figure 1):

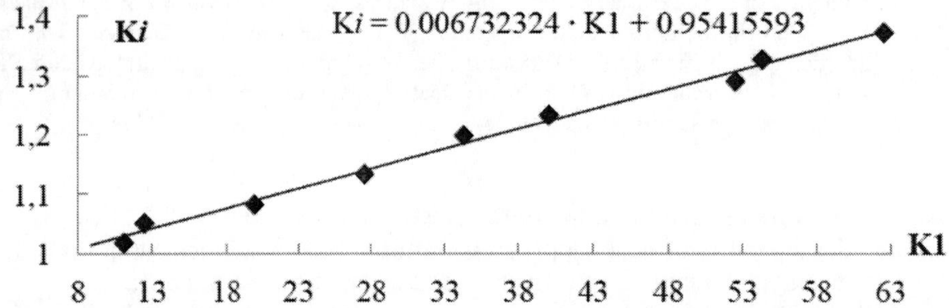

Figure 1. Influence of residual wall thickness of the pipeline metal on the coefficient of additional factors Ki

The next stage was the failure analysis, which includes two additional factors-the thickness of the pipeline metal and the corrosion activity of the soil (figure 2).

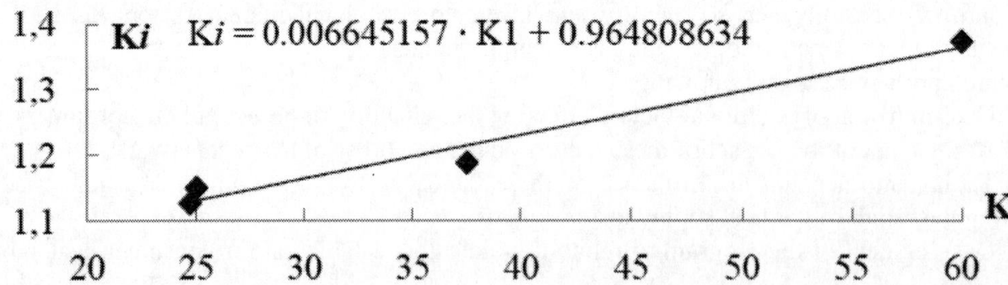

Figure 2. The influence of parameters K1 and K3 on a coefficient by accounting additional factors Ki

Next analysis was performed, which includes two additional factors-the thickness of the pipeline metal and the presence of the channel flooding (figure 3).

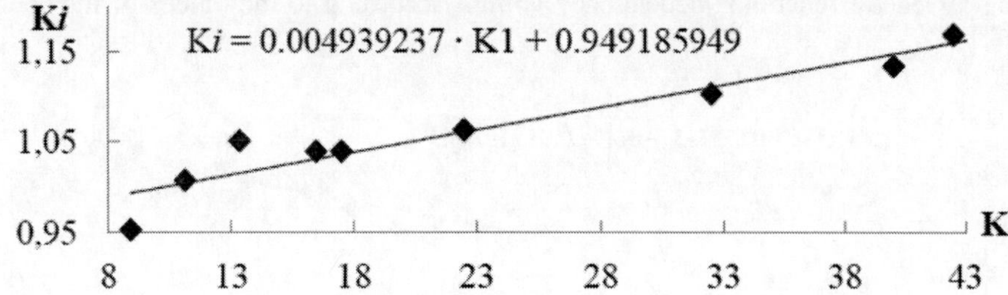

Figure 3. The influence of parameters K1 and K4 on the coefficient by accounting additional factors Ki

The next sample included parameters-the thickness of the pipeline metal and the presence of intersections with communications.

The functional dependence of the coefficient Ki on the parameters K1 and K5 is shown in figure 4.

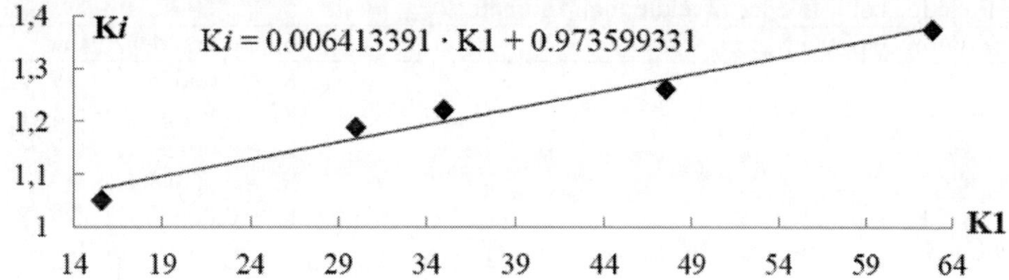

Figure 4. The influence of parameters K1 and K5 on the coefficient by accounting additional factors Ki

The final analysis of failures was combined with five parameters – the thickness of the pipeline metal, the presence of previous gusts at the site, the corrosion activity of the soil, the presence of the channel flooding and the presence of intersections with communications.

The functional dependence of the coefficient Ki on the parameters K1, K2, K3, K4 and K5 is shown in figure 5.

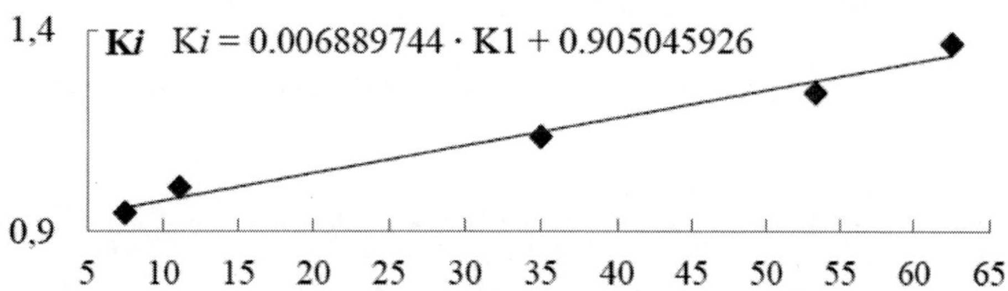

Figure 5. The influence of parameters K1, K2, K3, K4 and K5 on the coefficient by accounting additional factors Ki

Consideration of the influence of parameters not included in the analysis

The remaining isolated cases of refusal of sections of heating networks (for example, the combination of K1-K2, K1-K3,-K5, etc.) that were not included in the analysis were taken into account using a General functional dependency is built according to the values of the coefficient Ki (figure 6).

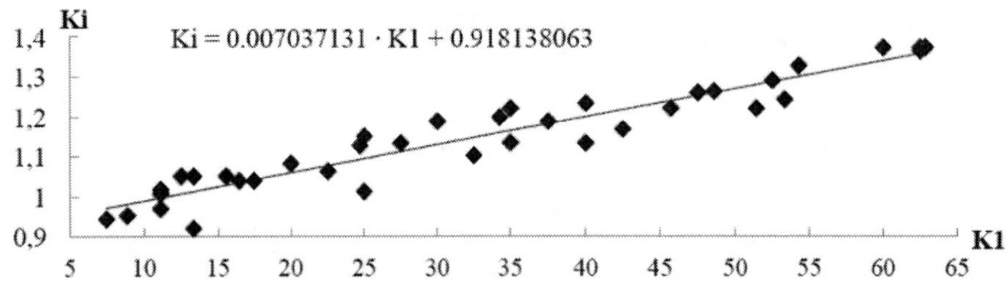

Figure 6. General functional dependence in all cases bounce sections of a heating system

Based on the functional dependencies presented in table 1, the theoretical values of the coefficient Ki were calculated.

Table 1. Functions of influence of additional parameters

№	The presence of the factors	Functional dependence
1	K1	$Ki = 0.00673 \cdot K1+0.954$
2	K1, K3	$Ki = 0.00664 \cdot K1+0.964$
3	K1, K4	$Ki = 0.00494 \cdot K1+0.949$
4	K1, K5	$Ki = 0.00641 \cdot K1+0.973$
5	K1, K2, K3, K4, K5	$Ki = 0.00689 \cdot K1+0.905$
6	A combination of factors different from options 1-5	$Ki = 0.00704 \cdot K1+0.918$

Figure 7 compares the theoretical and experimental values of the coefficient of additional factors Ki.

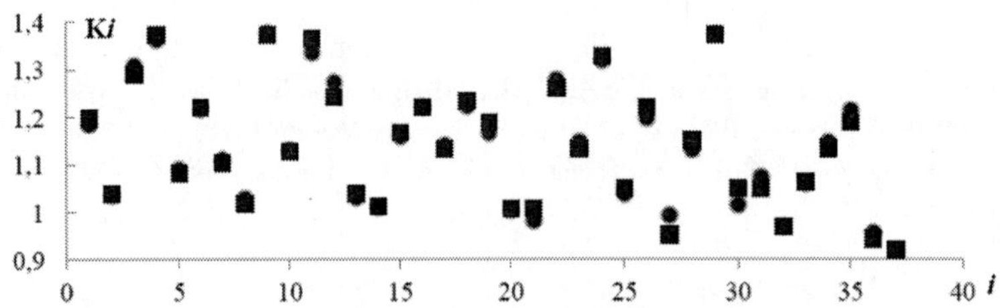

Figure 7. Theoretical and experimental values of the coefficients taking into account additional factors Ki

Conclusion

As can be seen from the graph, the deviation of theoretical values from the experimental data is not so significant – the average error does not exceed 1.13%. Therefore, the obtained functional dependences can be used to calculate the reliability of heat supply to consumers.

Acknowledgments

The work was supported by the Ministry of Education and Science of the Russian Federation on fundamental scientific research (Agreement No. №13.6994.2017/БЧ) «Development of a methodology for determining the reliability of a heat supply system to improve energy efficiency».

References

[1] Gorshkov A and Murgul V 2018 Calculation of Heat Energy Consumption by a Typical Historical Building with a Courtyard *Adv. Intell. Syst. Comput.* **692** 577–91
[2] Kostenko V A, Gafiyatullina N M, Semchuk A A and Kukolev M I 2016 Geothermal heat pump in the passive house concept *Mag. Civ. Eng.* **68** 18–25
[3] Nefedova A, Bykova J, Kosov S and Petrichenko M 2015 Selection of individual automatic heat supply unit of ecomatic LLC vol 117(Elsevier Ltd)pp 1102–11
[4] Borodinecs A, Zemitis J, Sorokins J, Baranova D V and Sovetnikov D O 2016 Renovation need for apartment buildings in Latvia *Mag. Civ. Eng.* **68** 58–64
[5] Akhmetova I, Chichirova N and Derevianko O 2017 Revisiting heat losses calculation at district heating network *Int. J. Civ. Eng. Technol.* **8** 694–702
[6] Murgul V 2016 Assessment of the Energy - Efficient Modernization of Residential Historical Buildings in Kiev vol 73(EDP Sciences)
[7] Strogonov K, Fedyukhin A, Stepanova T and Derevianko O 2018 Estimation of Practical Significance for Application of Composite Pipes in Comparison with Metal and Polymer Materials *Adv. Intell. Syst. Comput.* **692** 1024–35
[8] Soldatenko V S, Smagin V A, Gusenitsa Y N, Gera V I and Soldatenko T N 2017 The method of calculation for the period of checking utility systems *Mag. Civ. Eng.* **70** 72–83
[9] Akhmetova I G and Mukhametova L R 2017 Innovative development of the regional economy *Innovative clusters in the digital economy: theory and practice Proc. of the Scientific and practical conference with international participation* (Saint Petersburg, May 17-22)
[10] Akhmetova I G and Chichirova N D 2016 Evalution of Thernal Insulation Type Impact on the Value of Regulatory Heat Losses in Heat and Power Systems *J. Eng. Appl. Sci.* **11** 2946–9
[11] Kiselev V G 1999 *Information and engineering support of pipeline electrochemical protection* (Saint Petersburg: SPbGUVK)
[12] Turichin G, Kuznetsov M, Sokolov M and Salminen A 2015 Hybrid Laser Arc Welding of X80 Steel: Influence of Welding Speed and Preheating on the Microstructure and Mechanical Properties vol 78(Elsevier B.V.)pp 35–44
[13] Muravyeva L and Vatin N 2016 Elaboration of the Method for Safety Assessment of Subsea Pipeline with Longitudinal Buckling *Adv. Civ. Eng.* **2016**
[14] Silva C C, De Assis J T, Philippov S and Farias J P 2016 Residual stress, microstructure and hardness of thin-walled low-carbon steel pipes welded manually *Mater. Res.* **19** 1215–25
[15] Kondrat'ev S Y and Shvetsov O V 2018 Technological and Operational Features of Drill Pipes from Aluminum Alloys 2024 and 1953 *Met. Sci. Heat Treat.* 1–7
[16] Davydov V V, Myazin N S, Logunov S E and Fadeenko V B 2018 A Contactless Method for Testing Inner Walls of Pipelines *Russ. J. Nondestruct. Test.* **54** 213–21
[17] Gritskevich M S and Garbaruk A V 2017 Influence of upstream pipe bends on the turbulent heat and mass transfer in T-junctions vol 891(Institute of Physics Publishing)
[18] Muravyeva L V 2017 Simulation modeling in pipeline safety assessment (CRC Press/Balkema) pp 213–8
[19] Loktionova E A and Miftakhova D R 2017 Fluid filtration in the clogged pressure pipelines *Mag. Civ. Eng.* **76** 214–24
[20] Kitanin É L, Smirnov Y A and Lebedev M E 2016 Development of Flow and Heat Transfer During Filling a Pipeline with Water at the Pipe Wall Temperature Below the Freezing Point *J. Eng. Phys. Thermophys.* **89** 808–14
[21] Vasilyev A, Rudskoy A, Kolbasnikov N, Sokolov S and Sokolov D 2012 Physical and mathematical modeling of austenite microstructure evolution processes developing in line-pipe steels under hot rolling **706–709** 2836–41

Experimental and numerical studies in elaboration the multi-cyclone with filter cells to processing of flue-gases of coal-fired and incineration power plants

G I Belyaeva, M G Ziganshin, R D Sukhov

Kazan State University of Architecture and Engineering, Russia, 420043, Kazan, Russia

gulnazka16@mail.ru

Abstract. Currently, in various industries for both sanitary and technological treatment of heterogeneous flows, as a rule, several devices are used that consistently provide the coarse and fine stages of gas clean up. For example, to prevent the penetration of mechanical (solid and liquid) impurities to the measurement instruments, regulation, control and automation at compressor stations of gas transmission systems the gas cleaning units are setup. They consist of two types of apparatus - cyclone dust collectors and filters. Similar gas cleaning devices are at booster compressor stations of TPPs. In this paper, we consider the effectiveness of using a multi-cyclone apparatus for such purposes, the elements of which combine both stages of cleanup. The problem of finding the optimal arrangement in the apparatus of cyclone elements with tangential inlet branch pipe is solved. A numerical model of the multi-cyclone is created and several variants of the arrangement of elements are considered. The most effective arrangement was determined taking into account the total hydraulic resistance of the apparatus and the trajectories of the motion of suspended particles.

Introduction

Coal-fire and incineration thermal power plants are the main source of air pollution with fly ash. Reduction of local concentrations of suspended solids to permissible levels can be achieved by installing efficient ash collectors. For ash trapping at the primary stage currently used the multi-cyclones, the louver-type separators, fly-ash scrubbers and on the final stage as a rule, electrostatic precipitators [1]. However, with the transition to coal-fired generation not less than 99.5% cleanup will be required. An even deeper cleanup is needed even now to the flue gases of the incineration power plants [2, 3] which are difficult to achieve in electrostatic precipitators. Therefore, it is necessary created energy-efficient and low-resource devices with filtration in a porous medium.

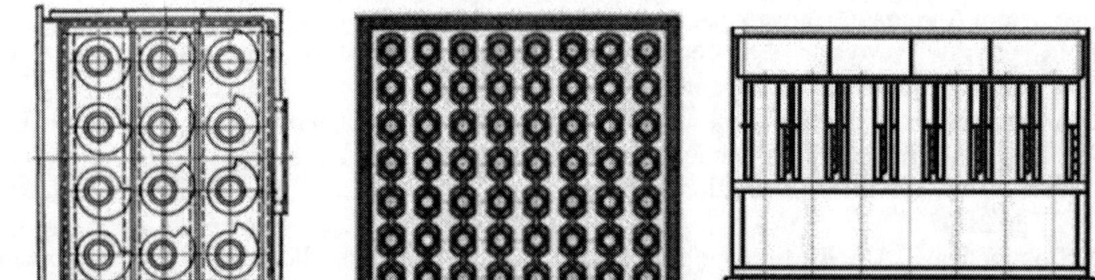

Figure 1. Examples of cells disposition in the battery cyclones with a tangential inlet: a - BCU-M [4], b - BC 2-7x (5 + 3) [5], c - CB 56 [6]

IOP Conf. Series: Earth and Environmental Science **288** (2019) 012077 doi:10.1088/1755-1315/288/1/012077

Now in the technique of dust cleaning, battery cyclones are widespread due to the simplicity of the design, reliability in operation, and low investments. However, multi-cyclones efficiently precipitate suspended particles only of medium size (from 10 μm) and larger. For smaller sizes of particles it is necessary the filtering apparatus for fine cleaning. In our paper, the problems of creating a multi-cyclone combining both stages of cleanup are considered.

The cyclone battery can be composed of cyclonic cells of various types. Taking into account the fact that the task of this work is to clean fine dust, and its adhesiveness is much higher than that of medium-sized dust, a type of return-flow cyclone elements with a tangential air supply was taken for further studies.

Figure 1 shows the arrangement of elements with a tangential inlet in the serially produced in Russia battery cyclones. It can be seen that the methods of orienting the inlet openings of the elements in different designs can be very different, while the direction of movement of the cleaned flow inside the body of apparatus is the same. This indicates that in the design of these devices, any circumstances are taken into account, except the particle trajectory in the flow of cleaned flue gases. For example, Figure 2a shows that the elements of the first line of the BCU-M multi-cyclone will be less loaded with dust, and their resistance during operation will be lower than in the following rows, which will worsen the overall cleaning degree. In the disposition of the elements as shown in Figures 2b, 2c, there remain gaps between the rows. In addition, in version b, the front rows overlap the movement of the streams to the inlets of the rear row elements. Such factors may not be of great importance for the operation of conventional multi-cyclones that process flows with a median dust diameter of 10-15 μm or more. Developing of multi-cyclones with filter elements which should target to provide a very high cleaning degree (99.5% or higher), the unsuccessful positioning of elements inlets can significantly increase the penetration of particles across the filter and nullify the efficiency of created devices.

Methods

Creating a filter cell for the multi-cyclone. A filter cell for the multi-cyclone was created. An experimental sample of the element of a multi- cyclone filter apparatus was made. A test bench was developed and assembled for its testing (Figure 2a). The diameter of the cyclone body d = 0.1 m, the diameter of the exhaust pipe d = 0.065 m. The scheme of test bench is shown in Figure 2b, the results presented in Table 1.

a
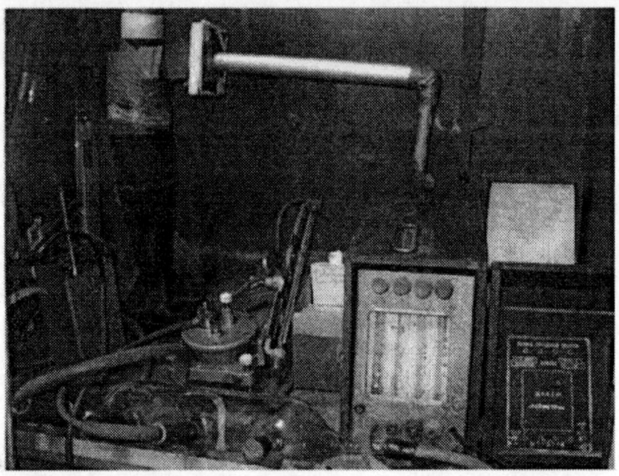
b
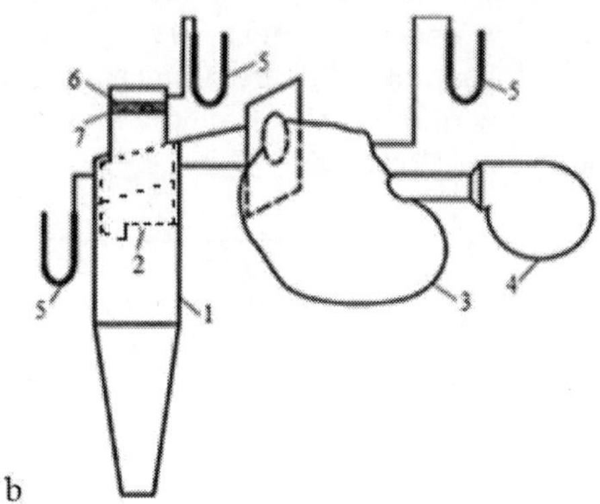

Figure 2. Photo (a) and scheme (b) of the test bench for the filter cyclonic element of the multi-cyclone: 1 - the cyclone filter element; 2 - fabric filter; 3 - container with dust (not visible in the photo); 4 - a supercharger (not shown in the photo); 5- U-shaped manometers; 6-cartridge with a grid; 7- Petryanov's fabric.

As a supercharger 4, a vacuum cleaner was used with a measuring manifold at the inlet. Air through an outlet nozzle of the vac at a pressure of about 200 Pa was supplied to container 3 with dust (talc with a median particle diameter of 10 μm and spread of 3.5). The dust particle sizes were determined in advance by sedimentation in distilled water. The mass of the dust container was determined on the analytical balance at the beginning and the end of a series of experiments with the same flow rate. Dusty flow from the container entered tangentially into the cyclone filter element 1. On the filter 2 (lavsan fabric) the dust deposited due of inertia and contact effects. The flow entered in the exhaust tube of the cyclone, where it passed through the cartridge with the Petryanov's fabric 7 and the metal grid 6, and then it removed from the system. Filters 2, 7 and grid 6 were weighed at the beginning and at the end of the test. Then, from the beginning, the middle, and the end of the filter fabric 2, samples were cut out to compare the sedimentation efficiency by the number concentration (per unit volume) of trapped particles, which determined by studying the samples under a microscope.

First, the flow rate characteristic of the filtering cyclone element was obtained; it is the dependence of its resistance on the flow rate of the emissions. When performing calibration, the flow through the element varied from a minimum of 0.001 m³/s to a maximum of 0.011 m³/s. The flow rates were calculated from the flow velocity w and the area S of the nozzle inlet d = 0.025 m in diameter. The flow velocity w was determined from the underpressure value measured by the micro-manometer at the collector, ΔP Pa, as $w = (2\ \Delta P / \rho)^{1/2}$, where the air density ρ was taken at its room temperature, which was 22-24 °C. The experiment was conducted with four combinations of flow rate (velocities) and the amount of dust in the container: with a minimum mass of dust at a minimum velocity; with a maximum dust mass at minimum velocity; with a minimum dust mass at maximum velocity; with a maximum dust mass at maximum velocity. Dust samples of the filter material were examined with a MIN-8 microscope with a 400-fold magnification. The growth of dusty sample's mass was determined on an analytical balance capable of weighing to the nearest 0.1 mg.

Detect the best location of the filtering cyclonic elements in the multi-cyclone on the basis of numerical studies of the dispersed flow motion. Since the investigation is aimed at increasing the efficiency of the dispersion flow purification with a class of suspended particles PM10, PM2.5, i.e. with size's less than 10 microns, when choosing the location of the cyclone elements, it is necessary to take into account the particle traces in the flow. The problem of finding the best location of the elements inlets of the apparatus is solved further by numerical simulation based on Computational Fluid Dynamics (CFD) methods. The RANS (Reynolds-Averaged Navier-Stokes equation) model of turbulence is used. The problem of closure of governing equations is solved with the k-Ɛ model. A 2d-model of a battery cyclone was constructed. The developed numerical model of the multi-cyclone made it possible to evaluate the influence of various factors on its hydrodynamic characteristics. In order to optimize the arrangement of the elements in the multi-cyclone body and to evaluate the efficiency of the dust collector, a numerical experiment was performed with the injection of particles into the gas flow.

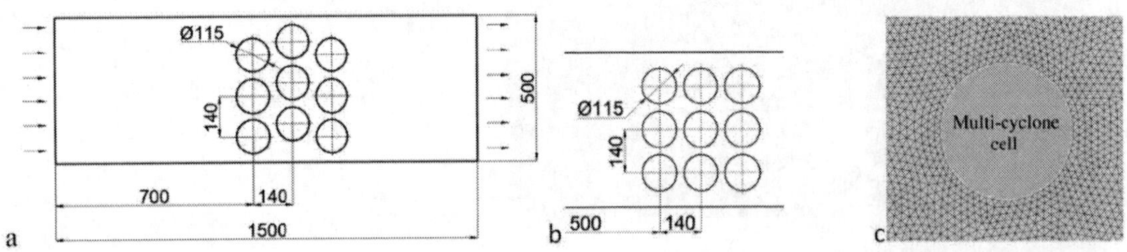

Figure 3. Geometric models of a part of the multi-cyclone bodies: a - with the staggered arrangement of cyclone elements, b - with the in-line arrangement; c is an example of a computational triangular unstructured finite element grid around one of the cyclone cells.

It can be seen from Figure 1 that for the incoming dispersed gas flow, the cells of the battery cyclone are the array of an obstacles. This constructional feature of the multi-cyclone is used in study to efficiently arranging of its filtering cyclonic elements, with taking into account the trajectories of the motion of solid particles of various sizes. Figure 3 shows an example of a geometric model of a part of the multi-cyclone body with dimensions of 1500x500 mm, with the staggered and with the in-line arrangement of obstacles - a group of 9 cylindrical elements with a diameter of 115 mm and with a spacing of 140 mm. The air flow with dust particles with dimensions from 5 to 45 μm enters the channel with velocity 0.5 m/s, and then flows around the obstacles. One part of the total mass of particles, together with air, leaves the channel, and the rest is held on obstacles. The study is carried out under standard conditions close to previous tests: T = 20 °C, P = 101325 Pa; density and dynamic viscosity of the flow are assumed to be constant: $\rho = 1{,}205$ kg / m^3, $\eta = 18{,}1 \cdot 10^{-6}$ Pa·s.

An unstructured finite element mesh based on triangular elements was created with the help of internal Gambit preprocessor tools. Here it was used because of the advantage that with its help the geometry of the region is better displayed, although a structured grid would be preferable for three-dimensional numerical studies of obstacle channels, since it requires a smaller computational resource [7]. Structured grids for studying a two-phase solid-gas system were proposed, for example, in [8], for a model with the simplest configuration for generating grids, a single spherical domain whose center coincides with the center of a spherical particle.

Before the beginning of the solution of the problem, the following boundary conditions were established: "Velocity Inlet" - uniform velocity distribution at the entrance to the channel (0.5 m / s); "Pressure Outlet" - atmospheric pressure at the exit from the channel; "Wall" is the boundary condition on the walls. The formulation of the boundary conditions had the following singularity. Impenetrable surfaces of cyclone elements performed the function of trapping particles, and the walls of the multi-cyclone bodies - the function of their reflection. Therefore, the condition "Wall" has been created as the boundary conditions of the whole discrete phase model, both for obstacles and for channel walls.

After exporting the finite element mesh in the ANSYS®Fluent software, the quality and scaling of the finite element mesh was verified (to ensure the correctness of the solution, it is necessary that the geometric dimensions of the model correspond to a scale of 1: 1). When setting up the solution of the task, the parameters of continuous and discrete phases are set. For calculations, a solver type "Based Pressure" with an "Implicit" solution scheme of the "Unsteady" task was chosen. The initial conditions for the particles are as follows: for the particle injection surface, the entrance surface of the Velocity Inlet is accepted, the type of distribution of the dispersed phase in terms of particle sizes, is taken in accordance with the Rosin-Rammler equation, with the number of fractions (Number of Diameters) 10, and with the dispersion parameter (Spread Parameter) 2.5. The material is industrial dust with a density of 2000 kg / m^3.

Results and Discussion

Experimental study of the filter cyclonic cell of the multi-cyclone.Experimental data obtained in the experiments are given in Table 1.

Tests of the filtering cyclone element sample's showed that about 75% of the dust is trapped on the filtering insert from lavsan, 20% settles on the walls. And about 5% is retained by the exit filter from Petryanov's fabric, wherein the dust, leaves a spiral trace on the surface. This indicates a fairly intense swirl of the flow in the exhaust pipe, which is the case conventional cyclones without filter inserts, as well as shown the presence of low-frequency precession of the core of a swirling flow, typical for the return-flow cyclones. Consequently, in the investigated cyclone with a constructive addition of fabric lavsan filter the nature pattern of the flow is preserved, which makes it possible to recommend its use as cyclonic cells for the multi-cyclones.

Detect the best location of the filtering cyclonic elements in the multi-cyclone on the basis of numerical studies of the motion of a dispersed flow. Figure 4 shows the results of a numerical calculation of a continuous phase flow inside a channel with filter cyclonic elements located in a staggered (Figure 4a) and in-line (Figure 4a) configuration. They show that the flow velocity reaches a maximum value of 1.8 m / s in the zones between the cyclone elements. Minimum velocities of 0.2 m / s are located in areas with increased resistance - near the walls of the channel and obstacle surfaces, as well as in the frontal and shadow areas of the obstacles. The flow characteristics in both multi-cyclone bodies by the velocity minima and maxima practically coincide, which also indicates the coincidence of the maxima and minima of the static pressures.

Table 1. Experimental data

Measured object	m_0, 10^{-3} kg	m_b, 10^{-3} kg	m_e, 10^{-3} kg	Q_{air}, 10^{-3} m^3/s	Q_{dust}, 10^{-3} kg/s	C, 10^{-3} kg/m^3
m_{dust} = *0.21, 10^{-3}* kg, t = *600 s*						
Container with dust	26.405	26.607	26.42	1.113	0.155	0.14
Filter fabric		4.137	4.245	1.113	0.09	0.08
Petryanov's fabric		0.662	0.681	1.113	0.015	0.014
m_{dust} = *1.49, 10^{-3}* kg, t = *1800 s*						
Container with dust	26.42	27.911	27.602	1.113	0.128	0.115
Filter fabric		2.998	3.208	1.113	0.087	0.078
Petryanov's fabric		0.607	0.649	1.113	0.017	0.015
m_{dust} = *0.19, 10^{-3}* kg, t = *600 s*						
Container with dust	25.443	25.635	25.488	11.3	0.245	0.022
Filter fabric		4.328	4.448	11.3	0.2	0.0179
Petryanov's fabric		0.649	0.676	11.3	0.045	0.004
m_{dust} = *1.75, 10^{-3}* kg, t = *900 s*						
Container with dust	39.523	41.276	40.978	11.3	0.414	0.0372
Filter fabric		4.341	4.426	11.3	0.118	0.106
Petryanov's fabric		0.666	0.723	11.3	0.079	0.0071

Designations: m_0 – original mass of the container; m_b, m_e – the initial and final masses of the measured objects; m_{dust}, Q_{air} –mass of dust and air flow, fed into the filtering cyclone element; Q_{dust}, C – mass loading and concentration of particles.

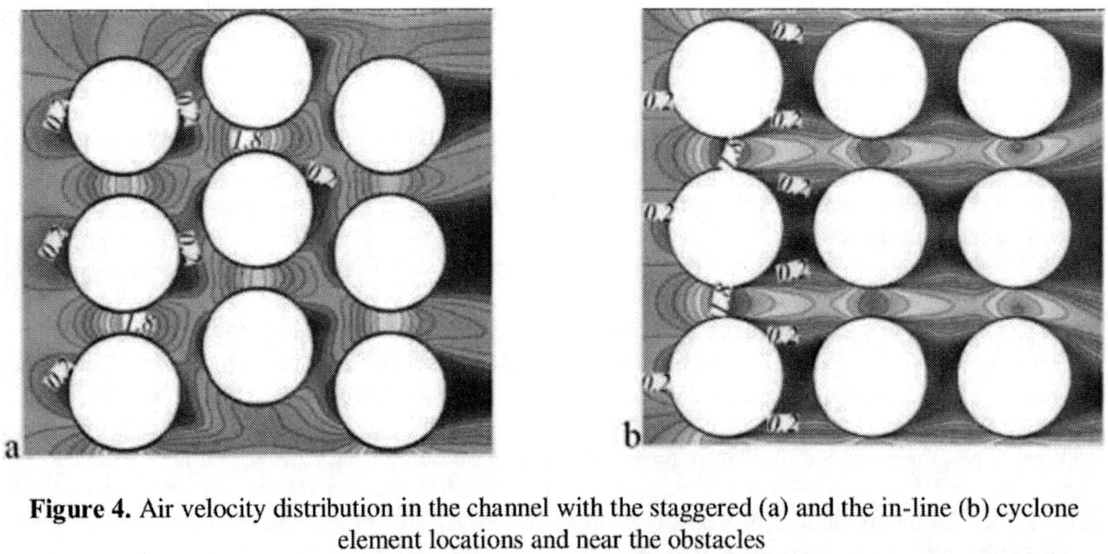

Figure 4. Air velocity distribution in the channel with the staggered (a) and the in-line (b) cyclone element locations and near the obstacles

Figure 5 presents the numerical results of calculating the trajectories of the discrete phase (industrial dust) within a channel with the staggered and the in-line arrangement of cyclone elements.

a

b

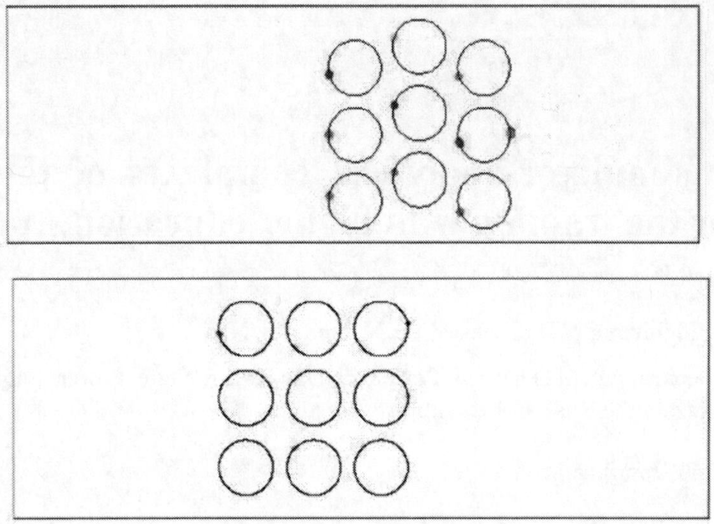

Figure 5. Results of calculation of particle trajectories in a channel: a - with the staggered arrangement of cyclonic elements; b - with the in- line arrangement of cyclonic elements; the particles color changes deep blue to red corresponded by particle sizes from 5 to 45 microns.

The results of numerical studies allowed to find out that the places of the settling of dust particles on cyclone elements do not coincide with the location of cyclonic element inlets in the serially produced multi-cyclones. You can also make a choice in favor of the effectiveness of the staggered arrangement cells, while the mass-produced multi-cyclones have a generally the in-line arrangement.

Conclusion

Thus, carried out experimental study showed the efficiency of the multi-cyclone cell with tangential inlet and the filter insert. With the numerical study obtained the hydrodynamic characteristics of disperse flows in the multi-cyclone body, and was determined the best scheme for placing its cyclonic elements with the localization of their entrances.

References

[1] Ziganshin MG, Kolesnik AA, and Ziganshin AM 2014 *Designing the devices of the dust-and-gas cleaning* (St. Petersburg LAN) p 535

[2] Tugov AN, Dick EP, and Sobolev AN 2011 Estimation of the influence of the mineral part of waste on the content of gaseous sulfur, chlorine and fluorine compounds in flue gases *Electrical Stations* **10** 15-20

[3] Tugov AN, Moskvichev VF, Fedorov LH, and Sokolov NA 2008 Methods for reducing emissions of dioxins and furans with flue gases, in the plants incinerating solid municipal waste *Electrical Stations* **12** 18-23

[4] Official site of LLC "PKF StroyServis" [available at: http://kompensator.su/catalog/ciklony/81/]

[5] Official site of the Plant of boiler equipment Официальный сайт Завода котельного оборудования [available at: http://www.zyklon.su/index.html]

[6] Official site of OJSC "Biysk Boiler Plant"Официальный сайт ОАО «Бийский котельный завод» [available at: http://www.bikz.ru/production/oborudovanie/other4/other5/cb-56/]

[7] By Prosperetti A, and Tryggvason G 2009 *Computational methods for multiphase flow* (Cambridge, New York, Melbourne, Madrid, Cape Town, Singapore, São Paulo, Cambridge University Press) p 484 [available at: www.cambridge.org/9780521847643]

[8] Bagchi P, and Balachandar S 2002 Steady planar straining flow past a rigid sphere at moderate Reynolds number *J. Fluid Mech.* **466** 365–407

IOP Conf. Series: Earth and Environmental Science **288** (2019) 012078 doi:10.1088/1755-1315/288/1/012078

Computer simulator-analytical complexes of thermal power stations for the training with higher education students

Chichirova N.D., Abasev Yu.V., Shagiev N.G.

Department of Thermal Power Plants, Kazan State Power Engineering University, Kazan, the Russian Federation

kgeu.tes@mail.ru

Abstract. The possibilities of using computer simulator-analytical complexes for training and retraining of large thermal power plants personnel are analyzed. Long-term experience of simulators using at the Thermal Power Stations Department (TPP) of Kazan State Power Engineering University (KSPU) is generalized. The requirements for computer simulators for the higher educational process are grounded.

Introduction

Automation is of particular importance for heat and power facilities. This is due to the high degree of mechanization for electrical and thermal energy production processes, as well as high pressures and temperatures in the steam-water circuits of thermal power plants, which imposes higher requirements on their regulation accuracy.

Expenditures for the fuel and energy complex development constitute a significant share in the state budget. In this regard, one of the urgent problems of the domestic economy is to increase the efficiency of large thermal power plants - thermal and nuclear power stations, powerful boiler houses.

This task solution should be facilitated by the training of highly qualified personnel who possess modern knowledge, skills and skills to work in the energy sector.

A promising area for improving the education quality in technical and technological universities can be the introduction of computer simulators in the educational process. In recent years, significant results have been achieved in the development and improvement of full-scale simulators imitating the technological processes of energy production at thermal power plants [1].

Requirements for the operational personnel of thermal power stations

The training with the computer simulators using has not yet received wide distribution in the implementation of educational programs in universities. At the same time, the existing requirements for operational personnel of thermal power plants, as required by the professional standard "Employee for the organization of operation of thermal power plant mechanical equipments" [2], provide for the ability to perform, in particular, such labor actions:

– control of operating modes and technical condition of thermal mechanical equipment, detection and recording of faults and defects of components, equipment design, according to instrument readings on the remote control board;

– carrying out emergency drills.

The same standard requires the employee skill:

– work with a personal computer and the software used at TPPs;

– determine the sequence of necessary actions in the operation performance of thermal mechanical equipment.

To train these and other labor actions and skills, it is advisable to use computer simulators, the software of which corresponds to the algorithms of control systems of technological processes on real heat power objects.

Experience of computer simulators and computer training-analytic complexes introduction in the educational process at the department of thermal power plants

At Kazan State Power Engineering University over the past decade, a significant experience has been accumulated in the successful implementation of computer simulators and computer simulators and analytical complexes in the educational process in the profile of "Thermal Power Stations" at the same department.

The first was mastered the all-mode computer simulator of the condensate power unit of supercritical pressure with a capacity of 300 MW, developed at the Ivanovo State Power Engineering University under the direction of Rabenko V.S.

On this simulator, multi-purpose classes are conducted with students studying in the areas of Bachelor's and Master's programs "Heat and power engineering and heat engineering" [3].

First, students receive theoretical knowledge, studying the content of mnemonic diagrams and the simulator management.

The second part of the training is connected with the implementation of practical tasks based on the experience of operating power units of thermal power plants. For example, one of the works is designed to acquire knowledge, skills that prevent automatic power unit shutdown in case of emergency stopping one of the two circulation technical water supply system pumps. Operational personnel must implement the least fuel-intensive transfer of the power unit to about 180 MW capacity, which will avoid an unacceptable vacuum deterioration in the turbine condenser. The power reduction should be smooth, otherwise the heat and power equipment service life is reduced.

The another work subject, also related to ensuring the continuous turbo accessory functioning, is the transfer of the increased pressure deaerator to the atmospheric regime. The main condensate feeder deaerator is characterized by high parameters (pressure 0.6-0.7 MPa and about 160 °C temperature). The transfer to atmospheric mode allows to inspect the deaerator and perform the necessary work without power unit stopping.

Finally, in the magistracy, the methods of main and auxiliary equipment thermal engineering tests can be studied using the simulator.

At present, educational methodological support is developed using the computer simulator-analytical complex of the power unit with a 450 MW combined-cycle plant (CCP-450).

The computer simulator - analytical complex of the 450 MW unit simulates the operation of the main and auxiliary equipment, control and protection algorithms, simulates control from the operator stations, is a training, pre-examination preparation and examination test for the CHP operational personnel [4].

It is designed for personnel training, and testing in the following specialties: Deputy Chief Operations Engineer, Senior Power Change Supervisor, Power Station Manager, Shift Supervisor, Senior Machinist, Power Unit Machinist, auxiliary Machinist, Electrician on duty.

Let's consider its technical characteristics [5, 6]. The CCP -450 power unit consists of two GTE-160 gas turbine units, two heat recovery boilers of type Pr-228 / 47-7.86 / 0.62-515 / 230, one steam turbine T-150-7.4, three turbogenerators, a set of auxiliary equipment (pumps, heaters, coolers, etc.) operating in a single thermodynamically connected technological cycle.

The GTE-160 gas turbine unit is a single-shaft four-stage turbo unit operating in a simple thermodynamic cycle, with an initial gas temperature of about 1060 °C and a gas temperature at the output from the turbine of 537 °C.

Electric power GTU at the calculated external conditions in accordance with GOST 20440 (outdoor air temperature 15 °C, pressure 0.1013 MPa, humidity 60%, resistance to output inlet to the compressor 1 kPa, resistance at the gases output from the gas turbine 3.3 kPa) is 155.3 MW with a efficiency coefficient of 34.12%.

Combustion products after each gas turbine are delivered to their recovery boiler. The recovery boiler has a tower layout, two high-pressure (HP) and low (LP) pressure steam pipelines with steam drums and forced and natural circulation in evaporation circuits. There is no fuel burning in the boiler, passing the exhaust gas of the gas turbine through the non-working boiler is not allowed. In the course of the gases in the boiler, the HP superheater, HP evaporator, HP economizer, LP superheater, LP evaporator and gas condensate heater are located in the boiler. Steam after superheaters of high and low pressures enters the high-pressure cylinder (into the chamber between 16 and 17 stages) of the steam turbine. Steam pressure and temperature regulation in the recovery boiler in the operating range of loads is not provided, it is designed to work with steam sliding parameters of determined by the temperature and gases flow entering the recovery boiler from the gas turbine, as well as the steam turbine operating mode.

The steam turbine T-150-7.4, with two adjustable steam samplings for heating the network water, with a nominal capacity of 161 MW at the nominal condensing mode and 129 MW at the nominal heating mode, with a rotation speed of 3000 rpm, is intended for direct drive generator , mounted on a common foundation with a turbine, and tempering to 358 MW at an estimated outside air temperature of -2 ° C. The turbine is a single-shaft two-cylinder unit.

Another simulator installed at the TPP department is simulator of the combined-cycle power unit 410 MW of Nyagan power station, which was commissioned in 2013. It has one of the highest efficiency achievable in the steam-gas cycle with condensation power generation - 58% [7].

The prototype of the CCP-410 simulator includes [4]: one gas turbine Siemens SGT5-4000F, one steam turbine SST5-3000, a heat recovery boiler Ep-270/316/46-12.5/3.06/0.46-560/560/237, SGen5-2000H turbo-generator, power unit accessories; remote objects controlled from the block control panel, including: circulating pump station, shore pump station, cooling tower with natural circulation, block gas preparation station; automated control and monitoring system SPPA-T3000 (Siemens).

Also at the department of TPP KGEU there was installed a "Plant simulator with cross-link connections for the CHPP", simulating the operation of the AMAKS software package installed on the TP-80 boiler to control the fuel supply [4].

The structure of the cross-link simulator include:

- two boiler units of TP-80 type;
- turbine unit type PT-60-130/13 with generator TV-60-2;
- turbo-unit of T-100/120-130 type with generator TVF-120-2;
- the boiler installation of the turbine PT-60;
- the heating system of the turbine T-100/120-130;
- system of main steam pipelines;
- nutrient-deaeration plant with four deaerators 0.6 MPa;
- plant circulation system;
- atmospheric deaerators 0.12 MPa for make-up.

To train students and conduct classes to improve skills and retrain workers in the energy industry a series of practical classes were developed. They provide for the fulfillment of tasks for the preparation for the start-up and start-up of individual units and aggregates, as well as for the CCP unit as a whole from the hot state and on sliding parameters from the cold state. Several lessons are associated with stop equipment.

Conclusion

The network of computer simulator classes created in recent years [3, 6, 8, 9] and the accumulated experience of conducting training sessions allows us to draw some conclusions regarding the requirements for the learning process simulators.

First of all, its maximum possible approximation to the real operating conditions of equipment in various modes is necessary. There the opportunity should be of time scaling (especially for many hours of starting and stopping operations), suspending a job, recording the current state, going back to any stage of work, having immediate messages about student mistakes made, and so on. Presence of all these simulator functions promotes increase of a degree of student independence at practical employment carrying out.

The combination of theoretical training, laboratory classes on the thermal power plant operating equipment and the practical tasks using a computer simulator implementation is necessary for the comprehensive training of future specialists to work at large thermal power plants.

Developed and successfully tested educational technologies with thermal power station computer simulators using can be used for retraining and further training of large thermal power plant workers.

Reference

[1] Actual issues of modern electric power industry simulator building development / S.I. Magid et al. // Reliability and safety of power engineering. - 2015. - №2. - P. 28-41.

[2] Order of the Labor and Social Protection Ministry of the Russian Federation No. 607н of 08.09.2015 "On approval of the professional standard" Employee for the organization of operation of thermal power plant thermal mechanical equipment "(Registered 07.10.2015 No. 39215)

[3] Thermal power plant computer simulators: Textbook. allowance / N.D. Chichirova et al., Ed. N.D. Chichirova. - Kazan: Kazan. state. power. University, 2009. - 204 p.

[4] Power station simulators for developed and implemented by JSC "TEST". URL: https://testenergo.ru/tren-zao-test/ (reference date: 04/03/2018)

[5] User's manual for the power unit simulator CCP-450 MW. - M .: JSC "Simulators of electrical stations and networks (TEST)", 2017. - 40 p.

[6] Computer CCP-450 MW unit simulator-analytical complex: workshop / under general. Ed. N.D. Chichirova. - Kazan: Kazan. state. power. Univ., 2018. - 280 p.

[7] Kolbina L. Whose is the station, whose house is this // Expert Ural. No. 39 (572). 2013. URL: http://expert.ru/ural/2013/39/chya-eto-stantsiya-chej-eto-dom/ (date of circulation: 04/03/2018)

[8] The applied tasks of the CCP-410 power unit simulator: workshop / under the general. Ed. N.D. Chichirova. - Kazan: Kazan. state. power. Univ., 2018. - 275 p.

[9] Simulator and analytical complex for a cross-link power plant: Textbook. allowance / under total. Ed. N.D. Chichirova. - Kazan: Kazan. state. power. Univ., 2018. - 221 p.

IOP Conf. Series: Earth and Environmental Science **288** (2019) 012079 doi:10.1088/1755-1315/288/1/012079

The Dynamics of transient mode dispersed jet moving drops in gas crossflow

O.V. Dolgushina, N.I. Platonov, D.M. Dolgushin

Nosov Magnitogorsk State Technical University, Lenina, 38, Magnitogorsk, Russia, 455000

helgavp@mail.ru, niplat@mail.ru, denisdolgushin@mail.ru

Abstract. It is considered the problem of the dynamics of sole drop dispersed jet in gas crossflow in active zone of contact heat exchanger with ring-gap nozzle. The equations are offered to estimate the motion trajectory of drop and residence time in active apparatus zone for transient mode 2<Re$_d$<500.

1. Introduction

Dispersed jet is formed in heat-mass-exchange apparatus with ring-gap nozzle in the process of disruption film jet at some distance from the nozzle. The contact surface of heat transfer agent dispersed jet is more than that of filmed one, so the effectiveness of heat-mass-exchange processes significantly depends on dispersed jet dynamics. There was carried out the analysis of dynamics of dispersed jet and gas [1]. It showed that the drops may be considered as separately moving from the common flow.

The model of dispersed jet moving drops in gas crossflow was put forward according to Newton resistance law [1]:

$$\frac{d\vec{w}_d}{d\tau} = -a\left|\vec{w}_d - \vec{w}_g\right|^{1-n}\left(\vec{w}_d - \vec{w}_g\right) + \vec{g}, \tag{1}$$

where

$$a = \tilde{\xi}_k \frac{1}{\tilde{\tau}}; \tag{2}$$

$\tilde{\xi}_d = \frac{A}{24}\mathrm{Re}_d^{1-n}$ – relative resistance coefficient; $\tilde{\tau} = \frac{\rho_f d_d^2}{18\mu_g}$ – time of particle dynamic relaxation, s; ρ_f – fluid density, kg/m^3; $d_к$ – drop diameter, m; μ_g – dynamic gas viscosity, Pa·s. Here the coefficients A and n are defined by drop fluid moving mode in gas flow, and depend on Reynolds number (Re$_d$). The apparatus possible modes of moving and corresponding coefficient values A and n are presented in table [2].

Equation (1) is added by initial terms for drop:

$$\tau = 0,\ r = r_0,\ z = z_0,\ w_{dr} = w_{dr0},\ w_{dz} = w_{dz0},$$

and velocity field for gas flow:

$$0 \le r \le r_h,\ z_1 \le z \le z_2,\ w_{gr} = f_1(r, z, \tau),\ w_{gz} = f_2(r, z, \tau),$$

IOP Conf. Series: Earth and Environmental Science **288** (2019) 012079 doi:10.1088/1755-1315/288/1/012079

where w_{dr} and w_{dz} – drop velocity in projections on the axis r and z correspondingly; w_{gr} и w_{gz} – gas velocity on the axis r and z correspondingly; r_h – heatexchanger radius; z_1 and z_2 – active zone apparatus vertical coordinates.

Table

The empirical coefficients of the law of resistance

№	A	n	Range on Re_d
1	24	1	$\mathrm{Re}_d \le 1$
2	25,6	1	$\mathrm{Re}_d \le 1$
3	26,3	0,8	$1 \le \mathrm{Re}_d \le 13$
4	12,3	0,5	$13 \le \mathrm{Re}_d \le 800$
5	18,5	0,6	$2 \le \mathrm{Re}_d \le 500$

2. Methods and Materials

The work presents the analytical task solution in gas flow dynamic drop for moving heat transfer agent modes from №1 till №4 (table) [1]. Moving mode №5 is transient one between modes №3 and №4, so it is necessary to examine the law of drops dynamics in this very case.

The following terms are necessary to find the solution of the given task:

– the drop at the initial moment has horizontal alignment of velocity;
– gas velocity in all points of apparatus ring gap equal to middle velocity and is directed verticaly upwards;
– the equation of drop moving along the coordinate (z or r) depends on alignment velocity of corresponding direction;
– calorific parameters of gas and fluid are the constant values.

The system of differential equations is obtained on these terms. The equations are describing velocity and trajectory of drop moving dispersed jet in gas flow

$$\frac{dw_{dr}}{d\tau} = -a\, w_{dr}^{2-n};$$
$$\frac{dw_{dz}}{d\tau} = -a\left(w_{dz} - w_{gz}\right)^{2-n} - g;$$
$$\frac{dr}{d\tau} = w_{dr};$$
$$\frac{dz}{d\tau} = w_{dz}, \tag{3}$$

the initial terms for drop:

$$\tau=0,\ r = r_o,\ z=0,\ w_{dr} = w_o,\ w_{dz} = 0; \tag{4}$$

in velocity field for gas flow:

$$r_o \le r \le r_h,\ z_1 \le z \le z_2,\ w_{gr} = 0,\ w_{gz} = w_g. \tag{5}$$

3. Results

For the transient mode moving under consideration ($2 \le \mathrm{Re}_d \le 500$) the coefficients of law of resistance takes the values A=18,5 and n=0,6 and taking into account the coefficient the equations of moving will be resembled as

$$\begin{aligned}
&\frac{dw_{dr}}{d\tau} = -a\, w_{dr}^{7/5};\\
&\frac{dw_{dz}}{d\tau} = -a\left(w_{dz} - w_{gz}\right)^{7/5} - g;\\
&\frac{dr}{d\tau} = w_{dr};\\
&\frac{dz}{d\tau} = w_{dz}.
\end{aligned} \tag{6}$$

For coordinate r it is not difficult to obtain the analytical solutions of system (6) on the terms of (4) and (5):

$$w_{dr} = w_{do} \frac{1}{\left(\frac{2}{5} a\, w_{do}^{2/5} \tau + 1\right)^{5/2}}; \tag{7}$$

$$r = r_o + \frac{5}{3} \frac{w_{do}}{a\, w_{do}^{2/5}} \cdot \left(1 - \frac{1}{\left(\frac{2}{5} a\, w_{do}^{2/5} \tau + 1\right)^{3/2}}\right). \tag{8}$$

For vertical coordinate z there are also some analytical solutions. For the law of resistance under consideration the estimated formula is obtained for falling drop time as follows:

$$\tau = \frac{5}{7g}\left(\frac{g}{b}\right)^{5/7} \left\{ \ln\left|\frac{1 - t\sqrt[5]{w_g + w_{do}}}{1 - t\sqrt[5]{w_g + w_d}}\right| + \cos\frac{5}{7}\pi \ln \frac{1 + 2t\cos\frac{1}{7}\pi\sqrt[5]{w_g + w_d} + t^2\sqrt[5]{\left(w_g + w_d\right)^2}}{1 + 2t\cos\frac{1}{7}\pi\sqrt[5]{w_g + w_{do}} + t^2\sqrt[5]{\left(w_g + w_{do}\right)^2}} + \right.$$

$$+ \cos\frac{15}{7}\pi \ln \frac{1 + 2t\cos\frac{3}{7}\pi\sqrt[5]{w_g + w_d} + t^2\sqrt[5]{\left(w_g + w_d\right)^2}}{1 + 2t\cos\frac{3}{7}\pi\sqrt[5]{w_g + w_{do}} + t^2\sqrt[5]{\left(w_g + w_{do}\right)^2}} +$$

$$+ \cos\frac{25}{7}\pi \ln \frac{1 + 2t\cos\frac{5}{7}\pi\sqrt[5]{w_g + w_d} + t^2\sqrt[5]{\left(w_g + w_d\right)^2}}{1 + 2t\cos\frac{5}{7}\pi\sqrt[5]{w_g + w_{do}} + t^2\sqrt[5]{\left(w_g + w_{do}\right)^2}} +$$

$$+ 2\sin\frac{5}{7}\operatorname{arctg}\frac{t\left(\sqrt[5]{w_g + w_d} - \sqrt[5]{w_g + w_{do}}\right)}{\sin\frac{1}{7}\pi + \frac{\left(t\sqrt[5]{w_g + w_d} + \cos\frac{1}{7}\pi\right)\left(t\sqrt[5]{w_g + w_{do}} + \cos\frac{1}{7}\pi\right)}{\sin\frac{1}{7}\pi}} +$$

$$+ 2\sin\frac{15}{7}\operatorname{arctg}\frac{t\left(\sqrt[5]{w_g + w_d} - \sqrt[5]{w_g + w_{do}}\right)}{\sin\frac{3}{7}\pi + \frac{\left(t\sqrt[5]{w_g + w_d} + \cos\frac{3}{7}\pi\right)\left(t\sqrt[5]{w_g + w_{do}} + \cos\frac{3}{7}\pi\right)}{\sin\frac{3}{7}\pi}} +$$

$$\left.+2\sin\frac{25}{7}\operatorname{arctg}\frac{t\left(\sqrt[5]{w_g+w_d}-\sqrt[5]{w_g+w_{do}}\right)}{\sin\frac{5}{7}\pi+\frac{\left(t\sqrt[5]{w_g+w_d}+\cos\frac{5}{7}\pi\right)\left(t\sqrt[5]{w_g+w_{do}}+\cos\frac{5}{7}\pi\right)}{\sin\frac{5}{7}\pi}}\right\}, \quad (9)$$

where $t=\left(\frac{g}{b}\right)^{-1/7}$, $b=13{,}95\frac{\nu_g^{0,6}\rho_g}{\rho_f d_d^{1,6}}$, ν_g – is kinematic gas viscosity, m^2/s.

The final solution of the passed distance of vertical coordinate Oz is expressed by the formula:

$$z=\frac{5}{3g}\left(\frac{g}{b}\right)^{10/7}t^3\left(\sqrt[5]{(w_g+w_{do})^3}-\sqrt[5]{(w_g+w_d)^3}\right)-w_g\tau+\frac{5}{7g}\left(\frac{g}{b}\right)^{10/7}\left\{\ln\left|\frac{1-t\sqrt[5]{w_g+w_{do}}}{1-t\sqrt[5]{w_g+w_d}}\right|+\right.$$

$$+\cos\frac{3}{7}\pi\ln\frac{1+2t\cos\frac{1}{7}\pi\sqrt[5]{w_g+w_d}+t^2\sqrt[5]{(w_g+w_d)^2}}{1+2t\cos\frac{1}{7}\pi\sqrt[5]{w_g+w_{do}}+t^2\sqrt[5]{(w_g+w_{do})^2}}+$$

$$+\cos\frac{9}{7}\pi\ln\frac{1+2t\cos\frac{3}{7}\pi\sqrt[5]{w_g+w_d}+t^2\sqrt[5]{(w_g+w_d)^2}}{1+2t\cos\frac{3}{7}\pi\sqrt[5]{w_g+w_{do}}+t^2\sqrt[5]{(w_g+w_{do})^2}}+$$

$$+\cos\frac{15}{7}\pi\ln\frac{1+2t\cos\frac{5}{7}\pi\sqrt[5]{w_g+w_d}+t^2\sqrt[5]{(w_g+w_d)^2}}{1+2t\cos\frac{5}{7}\pi\sqrt[5]{w_g+w_{do}}+t^2\sqrt[5]{(w_g+w_{do})^2}}+$$

$$+2\sin\frac{5}{7}\operatorname{arctg}\frac{t\left(\sqrt[5]{w_g+w_d}-\sqrt[5]{w_g+w_{do}}\right)}{\sin\frac{1}{7}\pi+\frac{\left(t\sqrt[5]{w_g+w_d}+\cos\frac{1}{7}\pi\right)\left(t\sqrt[5]{w_g+w_{do}}+\cos\frac{1}{7}\pi\right)}{\sin\frac{1}{7}\pi}}+$$

$$+2\sin\frac{15}{7}\operatorname{arctg}\frac{t\left(\sqrt[5]{w_g+w_d}-\sqrt[5]{w_g+w_{do}}\right)}{\sin\frac{3}{7}\pi+\frac{\left(t\sqrt[5]{w_g+w_d}+\cos\frac{3}{7}\pi\right)\left(t\sqrt[5]{w_g+w_{do}}+\cos\frac{3}{7}\pi\right)}{\sin\frac{3}{7}\pi}}+$$

$$\left. + 2\sin\frac{25}{7}\operatorname{arctg}\frac{t\left(\sqrt[5]{w_g + w_d} - \sqrt[5]{w_g + w_{do}}\right)}{\sin\frac{5}{7}\pi + \frac{\left(t\sqrt[5]{w_g + w_d} + \cos\frac{5}{7}\pi\right)\left(t\sqrt[5]{w_g + w_{do}} + \cos\frac{5}{7}\pi\right)}{\sin\frac{5}{7}\pi}} \right\}. \quad (10)$$

4. Conclusion

The kinematic drop parameters are defined according to obtained equations also for the case of changeable moving velocity of rising gas flow w_g taking place in real apparatus. To implement such an estimate all the height of apparatus was divided into small pieces where the kinematic parameters were defined on middle gas velocity. Making the calculations in some definite sequence with enough approach for practice, were defined kinematic drop specifications and lasting contact of interacting phases.

The obtained formulae are correct for the empirical law of resistance, corresponding to Reynolds number from 2 to 500 and do not occupy all the range of specifications of polydispersed flow drops of dispersed jet.

Physical and mathematical model of separate drop moving and the suggested analytical solutions allow to estimate each particle contribution into common result of interaction of dispersed jet and steam gas flow.

Formulae obtained in the current study and in the study [1] occupy all the range of drop specifications of polydispersed flow of dispersed jet formed in the result of disruption of filmed jet. The overall estimate of drops dynamics is carried out consequently by means of active zone dividing into separate parts for which one of the suggested laws of resistance is performed.

References

[1] Dolgushina O.V., Platonov N.I., Belousov V.S., Dolgushin D.M. *Dinamika kapel' dispergirovannoy strui v poperechnom potoke gaza* [Dispersed jet hydrodynamics in gas crossflow] // *Vestnik Chelyabinskogo gosudarstvennogo universiteta* [Bulletin of Chelyabinsk State University], 2012, no. 30 (284), I. 14, pp. 26-31. (in Russ.)

[2] Pazhi D.G., Galustov V.S. *Osnovy tekhniki raspylivaniya zhidkostey* [Basics of atomization techniques]. Moscow: Chemistry Publication House, 1984. (in Russ.)

IOP Conf. Series: Earth and Environmental Science **288** (2019) 012080 doi:10.1088/1755-1315/288/1/012080

Heat transfer in coiled type superheater of moisture separator-reheater of turbines at the nuclear power plant

Mikle Egorov[1], Kalyutik A.A.[1], Akhmetova I. G.[2], Makoev S.O.[1]

[1]Peter the Great St.Petersburg Polytechnic University, Institute of Energy and Transport Systems, 195251, Polytechnicheskaya, 29

[2]Federal State Budgetary Educational Institution of Higer Education Kazan State Power Engineering University, Kazan, Krasnoselskaya street, 51

mikhail.yu.egorov@gmail.com

Abstract. The reheating part of the moisture separator-reheater (MSR) of nuclear power plant (NPP) steam turbine occupies a significant part of MSR volume and requires large consumption of metal. The superheating of low pressure steam provides by the high pressure steam condensation. This steam supplies partly from turbine inlet and partly from turbine stage bleeding because of two superheater stages. The significantly less heat transfer intensity at the single-phase side of reheater comparatively with condensation side requires intensification of heat transfer from the heated-steam. For such goal the different methods can be used: for instance the cross flow of the coiled tubes and finning of straight tubes bundles. Also is perspective the usage of artificial roughness in form of small “moons” and other types of artificial roughnesses.

Introduction

In saturated steam turbines of NPPs at the high pressure casing (HPC) humidity in the HPC steam humidity reaches 10-15 %. Under such conditions effectiveness and reliability is decreased; the last blades fall under erosion and liable to failure before the expiration of the planned lifetime. At NPP the interturbine moisture separation is combined with external separation and superheating in special apparatus MSR. The technical-economic indicators of NPP as a whole depends on the effectiveness and reliability of MSR. Their masses and occupied volumes are significant, and with the growth of NPP units output the masses and sizes of MSR’s grow too. So, the increase reliability and effectiveness of MSR leads to an improvement in the overall characteristics of NPP. In the given paper the problem of steam superheating effectiveness MSR is considered for MSR operating at Russian Leningrad NPP (LAES).

Initially at the first unit of LAES the apparatus named asMSR-500 was used, which was developed in Polzunov institute and manufactured by Podolsk machinery works (ZIO). Later MSR-500 was removed and changed by MSR-500-1. Such replacement was performed after some failures SPP-500 elements, which were not related to the main functions – separation and overheating. MSR-500 was more effective in means of working process than MSR-500-1. We think that its main ideas can be used when perspective MSR develop for NPP of next generation. For new apparatus development new computation methods and operation experience appreciation will be required. These methods should be based on the modern investigations of separation and heat transfer. In the paper we presented new coiled type superheater of the same type as was used in MSR-500.

Method of reheating in moisture separator-reheater type MSR-500

In these apparatus the two stage superheating is realised. The steam heats in the spiral of helical tubes, assembled in bundles cross flowed condensing steam in two-step superheater (fig. 1).The tube bundle of first stage, which placed in the peripheral part of tube system, consists of 432 four paths spiral tubes. The heating steam is taken of stage bleeding (y<10 %; t=209°C, p=1,9 MPa [1]). The condensate outflow (t=205°C, p=1,8 MPa [1]) is performed from the tube bundle outer side. With this purpose the inner ends of the higher row of spirals, which are twisted clockwise, are welded with inner ends of low row spirals, which are twisted contraclockwise. The heating steam enters in the upper spirals, moves clockwise from the periphery to the center, passes to the low spiral row, moves through it from the center to the periphery. The tube bundle of second stage, which is placed closer to the apparatus center, is of the same construction, but consists from 216 tubes. The heat transfer is resulted by extracted inlet turbine steam condensation, which inlet temperature t=278°C, p=6,3 MPa and outlet condensate parameters equal to t=275°C, pressure p=6,1 MPa.

All coils are manufactured from stainless steel tubes 18x1,4 mm.

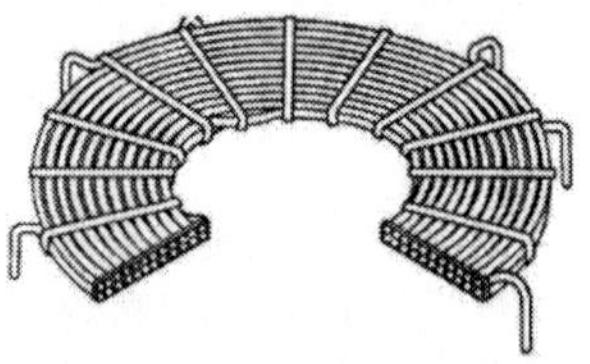

Figure 1. Two rows element of the coiled spiral tubes bundle of MSR-500.

Method of calculation of the heat transfer in a coiled type superheater

The character of cross flow in a bundle with tightly fixed tubes is shown in fig. 2. Flow and heat transfer analysis in such the bundle becomes more complicated for tubes, twisted by spirals (fig. 3).

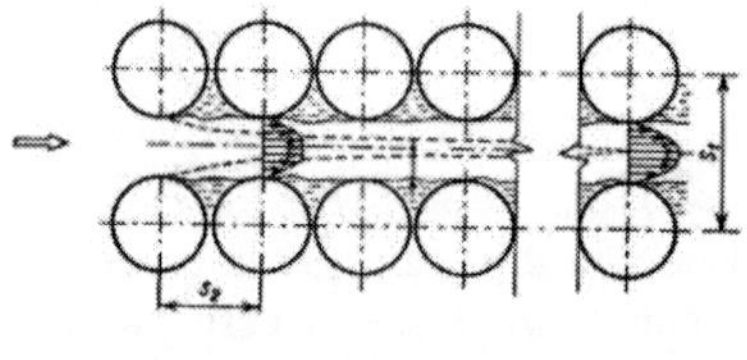

Figure 2. Sheme of flow in a “pressed” (tight) corridor-type tube bundle.

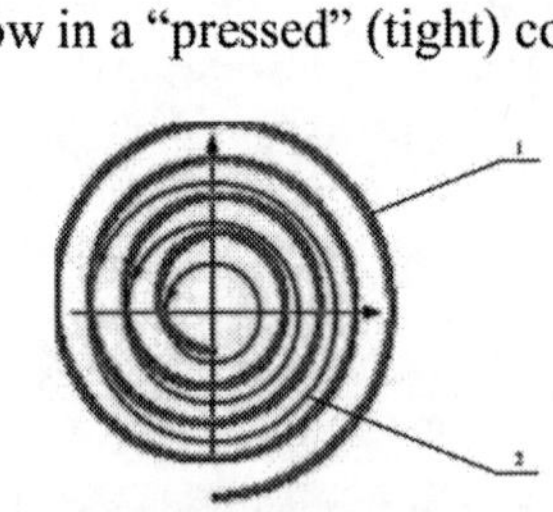

Figure 3. Sheme of actual coiled bundle geometry (1) and proposed calculation model (2).

In some works (for example, [1]) it is proposed to perform the calculations for the multirow bundle as a whole and uniform one, without taking into account the changing of heat transfer along rows. But only in multirow bundle the averaged heat transfer of whole bundle is close to heat transfer for depth row of bundle. In the construction considered here the number of rows along the depth of bundle is not a large – in MSR-500 such number is ~20 for the first stage of superheating and ~19 for the second stage. So, one can't neglect of first rows deposit to the heat transfer. Moreover, in a spiral coil the outer rows of tubes consist the larger part of heat transfer surface in comparison with deeply located rows (outer rows are located at maximal coil diameter).

We suggest to consider a spiral coil model in form of set of the concentric tube rings (fig. 3). Such a model takes into account a flow character in tight multirow bundle with simultaneously taking into account of coil geometry and deposit of outer rows. In this model in each radial section of coil is kept constant the distance between the axises of separate tubes of bundle along its depth and the distances between axises of separate tubes along bundle height is also unchangeable. So, the relative cross pitch along depth ($\sigma_1 = s_1/d$) and longitudinal (along height) pitches ($\sigma_2 = s_2/d$) are kept constant. Heat transfer coefficient α calculation are performed by the averaging of meanings of α method. Partial meaning α_1, α_2, … are obtained for separate rows (rings) with taking into account the income of each ring in overall heat transfer area,

$$\alpha_{bundle} = (\alpha_1 \cdot H_1 + \alpha_2 \cdot H_2 + \dots \alpha_j \cdot H_j)/(H_1 + H_2 + \dots H_j) \qquad (1)$$

that is the heat transfer coefficients of separate rows α_j install in formula (1) as multiplied ones by corresponding correction coefficient, which takes into account the heat transfer intensity by rows deep into the bundle.

During heat exchange process the temperature of flow changes and, consequently, meanings of the heated steam parameters change also. The account of the influence of physical properties of flow on heat transfer is connected with the choice of the conditioning temperature, by which the physical parameters are defined. Are known two methods for the taking into account the changings of physical properties with temperature. According to first one the meanings of physical properties of media are defined by the temperature of flow, and in the similarity equations are included as additional parameters. According to second method the physical properties are defined by the mean temperature between flow and wall temperature and a form of heat transfer coefficient correlation remains the same, as in case of constant meanings of physical properties. We have used the first method.

Assumptions

1. Heat losses to environment from MSR are absent. For MSR and similar apparatuses the relative heat losses are not larger, that parts of percent.
2. For non-isotermical conditions the designed thermophysical properties of heating and heated steams is each stage are taken by mean temperatures of these media.
3. Velocities of flows are equal to their averaged meanings in each stage of superheater.
4. The heat transfer surface does not have depositions; that is thermal resistance of system is constant.
5. Heat transfer from heating media by condensation inside tubes is calculated in full condensation assumption. It corresponds to the recommendation [4].

Heat transfer coefficients calculation for superheating steam by flow through coiled tubes.

We used several variants of calculations to choose the correlations corresponds conditions typical for MSR. For example, in works [10-12] are recommended to conduct heat transfer for multiflow bundle (n rows more than 10) as whole but not for rows separately; Prandtl number Pr_f is recommended to define by flow temperature, that is by meaning of bundle depth temperature. The recommended exponent at Pr_f in [10-12] is equal to 0.33 and somehow lower, than in [6] (0.35); the temperature correction factor is $(Pr_f/Pr_w)^{0.25}$.

It is known, that heat transfer intensity of the widely used in-line bundles is lower, than of staggered ones. But negligible difference between the values of heat transfer coefficients obtained when considering the tight bundle in-line and staggered approximations, confirms the validity of the assumption of the author that the true value of the heat of the investigated a dense beam should be considered to be in the range between the two values (in-line and staggered).

V.P.Isachenko in textbook [17] recommends to define heat transfer coefficient from row to row, using as determining one a mean temperature of flow. The proposed by him coefficient in the expression for Nusselt numbers is equal to C=0.26.

That remarkably increases the value of heat transfer coefficient is comparison with values, which have been used by us where C=0.2. Author [17] involves correction factor $e_s=\sigma_2^{-0.15}$, accounting an influence of relative placement of tubes in deeply located rows. It is indicated, that such a correction is received with investigation of the bundles with relatively large cross pitches (σ_1 = 1.3…2.6). From that we concluded, that for our pitch meaning (σ_1 = 1.14) correction factor e_s is not needed. For the determination of heat transfer coefficient for all bundle we are averaging the meanings of heat transfer coefficiens for separate row, using formula (1).

In the work [17] is indicated also, that by relative low extent of turbulence, particularly, by the absence of the artificial flow turbulization of incoming flow, the heat transfer intensity of the first (outer) row of in-line bundle consists of ~0.6, and of the second row ~0.9 from heat transfer intensity of the third and following rows. If a extent of turbulence of incoming flow is high the heat transfer of outer rows coincides with heat transfer of deep rows. In the work [17] are indicated also the correction coefficients for other rows, however there are not information about Reynolds numbers Re_f recommended diapason. In our calculations we have accounted, that numbers $Re_{fI} = 4.9\cdot10^4$ and $Re_{fII} = 5.7\cdot10^4$ correspond to low extent of turbulence, because the meanings of recommended in [17] correction coefficient are close to recommended by another authors [18-20] ones.

In the handbook [13] the correlation is proposed

$$Nu = 0.26\cdot Re^{0.65}\cdot Pr^{0.33} \tag{2}$$

for the determination of heat transfer for the individual row of a bundle. This correlation coincides with the given in [17] correlation. However as a difference from [17], the dependence (2) in the work [13] is recommended for the turbulent flow with indication of Re and Pr numbers diapasons Re_f=(10^3…10^5) и Pr_f=(0,7…500). Also is recommended somewhat less, than in [17] correction coefficient, taking into account the changing of heat transfer for second row of a bundle equal to 0.8. Correction $e_s = \sigma_2^{-0.15}$ on the influence of relative placement of tubes in deep rows has been taken by us equal to unity by the same reasons which was mesented above.

In the textbook [2] was proposed dependence (2) for the determination of the mean heat transfer coefficient for the bundle as a whole. That dependence of accounting of changing of heat transfer intensity in outer rows. The correction factor $e_s = \sigma_2^{-0.15}$ at coincides with given in [17] for individual row [Re_f=(10^3…10^5)]. As a difference from [17] is an absence the influence of relative placement of tubes in deep rows is introduced without indication of the diapasone of relative cross steps σ_1 which characterises the bundles for which it was obtained (as was done in [17]). For analysed by us case this factor is equal e_s = 0.98. Prandtl number is defined bundle depth distance temperature of flow; the temperature factor equals $(Pr_f/Pr_w)^{0.25}$.

Textbook [21] authors propose to perform heat transfer calculation by the dependence (2), which was obtained by V.P.Isachenko et. al [17] and was recommended for the heat transfer coefficient for each separate row. Mean liquid temperature is used as characteristic. As in [2] correction $e_s = \sigma_2^{-0.15}$ is entered but witout indication the range of σ_1. According to data [21] the correlation for first row heat transfer for Re_f>$4\cdot10^4$ and with higher exponent at Re: C=0.0266, m=0.80, n=0.33 [21]. According to early data [6] at Re~($5\cdot10^3$…$5\cdot10^4$), C=0.22, m=0.60, n=0.35.

Correlation [10] is proposed for usage in a rangle Re_f=(10^3…10^5) also in the work [18]. As in [13, 17], the calculation is proposed for individual rows of a bundle. As in [2, 21] correction $e_s = \sigma_2^{-0.15}$ has been introduced without indication of a range of σ_1. It were indicated correction coefficients for outer rows; their values concides with same ones in [17], but they are different from values of [13]. In this work is confirmed, that the flow around first row tubes do not differ practically from flow around single tube. For whole bundle is recommended the averaging the separate heat transfer coefficient, using formula (1). As a conditioning one is used mean flow temperature.

The textbook [22] has its own specific features – the difference from [2, 17, 18, 21] in coefficient e_s – it is recommended as $e_s = \sigma_2^{-0.16}$. However for our case the value of such correction practically the same as recommended ones in [2, 17, 18, 21], that is $e_s \sim 0.98$.

In [24] formula (3) is proposed to use for calculation of α for depth row,

$$Nu = 0.27 \cdot \mathrm{Re}^{0.65} \cdot \mathrm{Pr}^{0.36} \tag{3}$$

with multiplication by the coefficient c_z, accounting dependence of average heat transfer on number of rows. Multiplier in (3) is equal to 0.27, which higher, that in above considered works. In our case the increase will be equal to 35 %. The exponent at Pr in [24] equals to n=0.36, but at Re the exponetu is less (m=0.63). Recommended Re range is $10^3 \ldots 2 \cdot 10^5$.

In [24] it was marked, that for the majority of bundles the heat transfer stabilization begins from third-fourth rows. In [25] opinion offered that for the stabilized flow the heat transfer of depth rows increase with longitudinal pitch decrease (the increase 30…70 % is depended on longitudinal pitch).

For the mean heat transfer intensity in depth rows in [26] was recommended correlation (3). The recommended range of Reynolds number was expanded up to $2 \cdot 10^3 \ldots 2 \cdot 10^5$. It was marked, that in the first row the tubes are in conditions, similar single tube only if pitch is sufficiently large. In our case ($\sigma_1 = s_1/d = 1.14$ and $\sigma_2 = s_2/d = 1.11$) and correlations for a single tube in principle are not valid.

Finally we came to a conclusion, that mostly close to realised in MSR geometrical and regime parameteres are recommendations of Zhukauskas group [3, 24] – formula (4)

$$Nu = 0.308 \cdot \mathrm{Re}^{0.63} \cdot \mathrm{Pr}^{0.36} \tag{4}$$

which was obtained by data genelalization for the bundles with $\sigma_1 = (1.2 \ldots 1.4)$, $\sigma_2 = (1.09 \ldots 1.12)$ – that is mostly close to our case $\sigma_1 = 1.14$, $\sigma_2 = 1.11$. It should be marked, that the recommended in [3] value of C=0.308 leads to increase at ~14 % the heat transfer coefficient in comparison with works, where C=0.27 [20, 24, 26].

In fig. 4 are shown the heat transfer coefficients for flow of heated steam in stage of MSR SPP-500 in correspondence with [1-3, 5, 6, 9-14, 16-26].

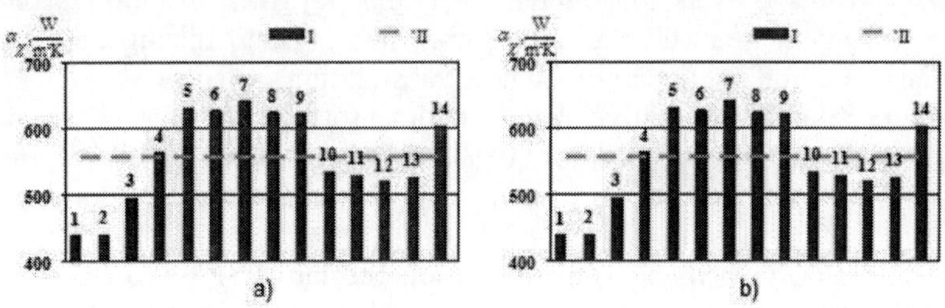

Figure 4. Heat transfer coefficient by flow of heated steam in the I (a) and II (b) stage of superheater: I–calculated values α_χ, index $\chi = 1 \ldots 14$ marks specific meanings of recommendations with correspondence to indexes and recommendations 1–Margulova T.H. [5], 2–Kutateladze S.S., Boryshanskii V.M. [6], 3–Andreev P.A.; Kirillov P.L.; Migai V.K. [10–14], 4–Budov V.M. [16], 5–Isachenko V.P. [17], 6–Kirillov P.L.; Isachenko V.P. [13, 17], 7–Isachenko V.P.; Baskakov A.P. [2, 17], 8–Isachenko V.P.; Baskakov A.P.; Kirillov P.L. [2, 17, 21], 9–Nashekin V.; Ilchenko O. [18, 22], 10–Mikheev M.A., Mikheeva I.M. [19], 11–Zhukauskas A., Ulinskas R. [3, 24], 12–Zhukauskas A.A.; Kutateladze S.S. [24, 26], 13–Zhukauskas A.A.; Tsvetkov F.F., Grigoriev B.A. [24, 20], 14–Zhukauskas A., Ulinskas R. [3]; II–averaged calculated value.

After in-line approximation let us consider staggered and mixed approximation. As already was marked, data for staggered bundles of interesting for us geometry are few. In the work [27] have been investigated regular bundle, that is with the equal relative pitches ($\sigma_1 = \sigma_2 = \sigma$) rows $\sigma = 1{,}05$ and $\sigma = 1{,}1$. These data have been compared with data for regular staggered bundles ($\sigma = 1.027...1.19$) presented in [28-31]. It was confirmed in [27], that staggered bundles with $\sigma = 1.1...1.2$ are optimal in sense of maximum of heat transfer intensity.

Authors [27] generalized their experimental data and results of other works with a correlation for regular staggered bundles:

$$Nu_f = 0.36 \cdot \{1+0.25\exp[-100(\sigma-1.18)^2]\mathrm{Re}_f^{0.6} \cdot \mathrm{Pr}_f^{0.36}\} \quad (5)$$

We have used the formula (5) and have received a Nusselt number for the first stage of MSR-500 $\mathrm{Nu}_{SP\,CI} = 281$ and for the second stage $\mathrm{Nu}_{SP\,C\,II} = 304$ that is practically the same as ones which have been obtained for in-line bundle formula (4) usage (for the first stage $\mathrm{Nu}_{SP\,C\,I} = 283$ and for the second stage $\mathrm{Nu}_{SP\,C\,II} = 303$). So we observe negligible small (less than 1 %) difference between the values, obtained by consideration of a tight bundle in-line and staggered approximations.

For actual mixed in-line staggered geometry we have taken meanings $\mathrm{Nu}_{SP\,MIX\,I} = 281$ and $\mathrm{Nu}_{SP\,MIX\,II} = 303$, as a mean values between in-line and staggered values. With such approach the heat transfer coefficients are $\alpha_{SP\,MIX\,I} = 458$ W/(m^{2}K) and $\alpha_{SP\,MIX\,II} = 605$ W/(m^{2}K), and calculated values of heat exchange surfaces are $H_{CALE\,MIX\,I} = 1423$ m^2, $H_{CALE\,MIX\,II} = 929$ m^2. The design constructive values of these surface areas were $H_{DES\,I} = 1920$ m^2, $H_{DES\,II} = 1218$ m^2 [1]. These meanings are 30 % higher. Usual values are 10-15 %.

1. Thermal calculations of superheater type of MSR-500 have next elements of the novelty in comparison with used during creation MSR-500 methodics:

1.1. The taking into account the variety of α and difference in areas of tubes each row by the definition of a mean heat transfer of a bundle.

1.2. Taking into account the temperature factor.

1.3. Introduction of corrections accounting the dependent of averaged het transfer coefficient on the quantity of row.

1.4. Comparison of the correlations, recommended by many authors for "corridor" and "chess" bundles and choice the proper correlation for actual geometry of the bundle in apparatus of MSR-500 type (mixed "corridor-chess" type).

2. As in the considered by us construction the number of rows in spiral tube coils is not a large one it is unproper to neglect the income of outer (first) rows in heat exchange, moreover that in spiral coil outer rows consists of the larger part of the heat transmission surface.

3. The difference in the quantative characteristics of a heat transfer, obtained by proposed by different authors, quite significant (up to 40 %). It means, that the conduction of additional experimental studies with real geometries of coils and preferable with steam of real parameters very desirable.

4. For conditions of the heat exchange in the superheater of SPP-500 type for the heat transfer calculations for side of heated steam (with actually defines overall heat transfer coefficient) can be recommended the correlation (fig. 5)

$$Nu_f = 0.308 \cdot \mathrm{Re}_f^{0.63} \cdot \mathrm{Pr}_f^{0.36}(\mathrm{Pr}_f/\mathrm{Pr}_w)^{0.25} \quad (6)$$

5. Calculation results testify about possibilities to reduce a reserve of heat exchange surface in MSR-500 comparatively with designed one.

6. It looks as not justified the conducted in works [2-9, 11-18, 21, 22, 24] expansion of application boundaries of V.P.Isachenko [17] correlation beyond the boundary of a range of bundles studied.

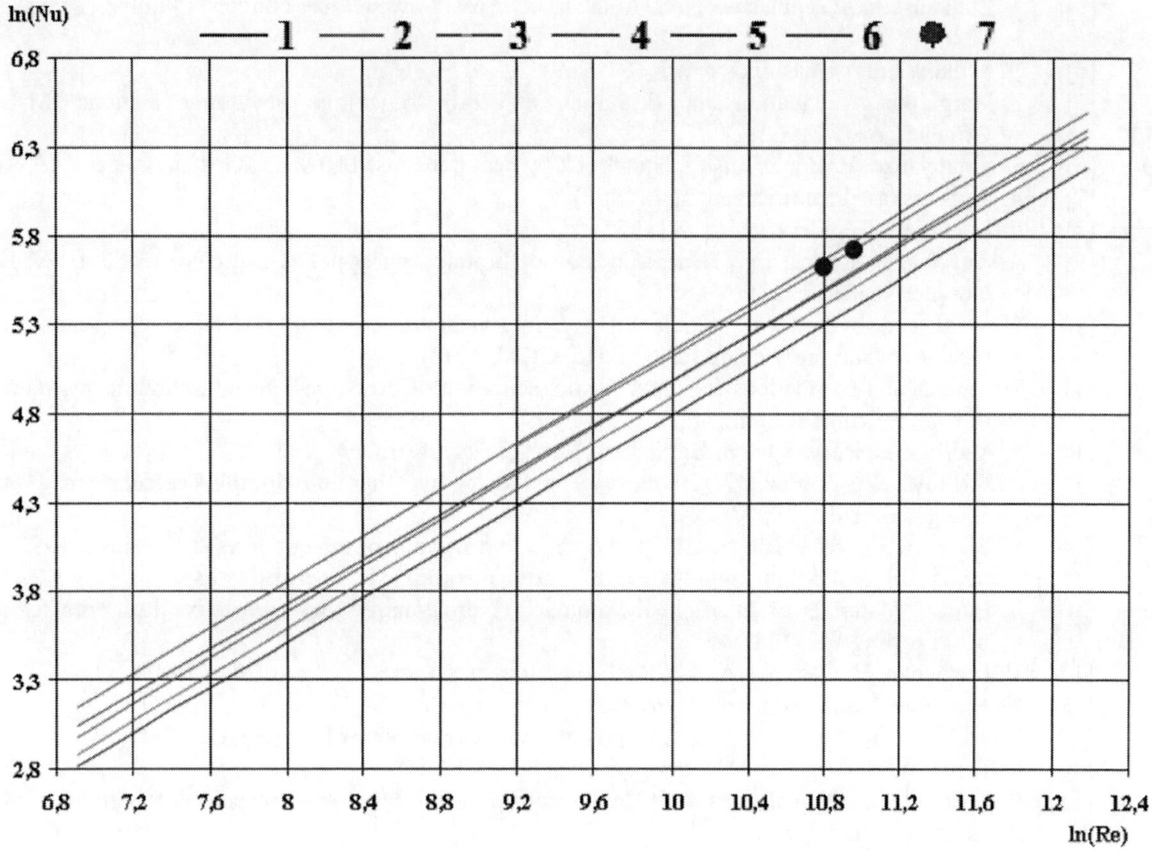

Figure 5. Non-dimensional heat transfer used for calculations: 1–$0.2c_z\,Re^{0.64}\,Pr^{0.35}$ without temperature factor, Re>$6{\cdot}10^3$; 2–$0.2c_z\,Re^{0.65}\,Pr^{0.33}(Pr_f/Pr_w)^{0.25}$ for bundle, Re=(10^3..$2{\cdot}10^5$); 3–$0.26c_z\,Re^{0.65}\,Pr^{0.33}(Pr_f/Pr_w)^{0.25}$ by rows, Re=(10^3…10^5); 4–$0.22c_z\,Re^{0.65}\,Pr^{0.36}(Pr_f/Pr_w)^{0.25}$ by rows, Re>10^3; 5–$0.27c_z\,Re^{0.63}\,Pr^{0.36}(Pr_f/Pr_w)^{0.25}$ for bundle, Re=(10^3…$2{\cdot}10^5$); 6–(6) for bundle, Re=(10^4…$2{\cdot}10^5$), σ_1=(1.2…1.4), σ_2=(1.09…1.12); 7–calculated values of I, II stages of SPP-500 type superheater.

Acknowledgments

The work was supported by the Ministry of Education and Science of the Russian Federation on fundamental scientific research (Agreement No. №13.6994.2017/БЧ) «Development of a methodology for determining the reliability of a heat supply system to improve energy efficiency».

Conclusions

There are different ways for the improving of the heat transfer effectiveness of a MSR's superheater: the usage of coiled type tube bundles, finnings and artificial roughnsses.

In the given paper are presented the methods of a calculation of the heat transfer of coiled tube bundles of a specific geometry (intermediate between corridor and chess type), based of the analysis of numerous experimental correlations by different authors.

References

[1] Design of the intermediate separastor steam reheater for turbine K-500-65/3000 (L., CKTI, 1969)
[2] A. Baskakov, B. Berg, O. Vitt. Thermotechnics (M., Energoatomizdat, 1991)

[3] A. Zhukauskas, R. Ulinskas. Heat transfer of cross flowed tube bundles (Vilnius, Mokslas, 1986)
[4] D. Volkov. Energomashinostroenie, 6 (1969)
[5] T. Margulova. Calculation and designing of steam generators of atomic stations (M-L., Gosenergoizdat, 1962)
[6] S. Kutateladze, V. Boryshanskii. Handbook by heat transmission (M-L., Gosenergoizdat, 1958)
[7] A. Zhukauskas. Teploenergetika, 4 (1955)
[8] V. Isachenko. Teploenergetika, 8 (1955)
[9] A. Gurvich, N. Kuznetsov. Thermal design of boiling aggregates (Instructive method) (M-L., Gosenergoizdat, 1957)
[10] Thermal and hydraulical calculations of intermediate separators – steam-superheaters of turbines of saturated steam for NPP (L., CKTI, 1973)
[11] Methodics and correlations for theoretical calculation of heat transfer and hydraulic resistance of heat exchange equipment (L., CKTI, 1972)
[12] P. Andreev, B. Boryshanskii, E. Firsova. Energomashinostroenie, 4 (1972)
[13] P. Kirillov, Yu. Yuriev, V. Bobkov. Handbook by thermohydroulic calculations (M., Energoatomizdat, 1990)
[14] V. Migai, E. Firsova. Heat transfer and hydraulic resistance of tube bundles (M., Nauka, 1986)
[15] Thermal design of boiling aggregates (normative method) (M., Energia, 1973)
[16] V. Budov. Materials of Int. School-Seminar "Hydrodynamic and convective heat transfer in heat exchangers" (1981)
[17] V. Isachenko, V. Osipova, A. Sykomel. Heat transmission (M., Energoizdat, 1981)
[18] V. Nashekin. Technical thermodynamics and heat transfer (M., Vicshaya shkola, 1975)
[19] M. Mikheev, I. Mikheeva. Fundamentals of heat transmission (M., Energia, 1977)
[20] F. Tsvetkov, B. Grigoriev. Heatmasstransfer (M., MEI, 2011).
[21] P. Kirillov, G. Bogoslovskaya. Heat transfer in nuclear power installations (M., Energoatomizdat, 2008)
[22] O. Ilchenko. Heat and mass exchangers thermal and nuclear power plants (Kiev, Vicsh. Shk, 1992)
[23] M. Mikheev. Izvestia AN SSSR. Energetika i transport, 5 (1966)
[24] A. Zhukauskas. Convective transfer in heat exchangers (M., Nauka, 1982)
[25] A. Zhukauskas, E. Kalinin. Intensification of heat transfer. Advances of heat transfer (Vilnius, Mokslas, 1988)
[26] S. Kutateladze. Heat transfer and hydraulic resistance (M., Energoatomizdat, 1990)
[27] M. Gotovsky, M. Belenky, A. Marinich. Trydy CKTI, 282 (2002)
[28] V. Velichko, N. Kovalenko, V. Pronin. Proc of 1th Russ. Nat. Conf. by Heat Transfer (M., MEI, 1994)
[29] A. Zhukauskas. Heat transfer of the tube bundles in cross flow of liquid (Vilnus, Mintis, 1968)
[30] Yu. Stasjuliavichus, P. Samoshka. Proc. AN LitSSR, B, 4(33) (1963)
[31] B. Boryshanskii, V. Zhinkina, E. Firsova. Trydy CKTI, 86 (1968)

IOP Conf. Series: Earth and Environmental Science **288** (2019) 012081 doi:10.1088/1755-1315/288/1/012081

Influence of the different oxidizers on the composition of sulfur products at bitumen gasification process

D.V. Ermolaev

Institute of Power Engineering and Advanced Technologies – Subdivision of the Federal Research Center «Kazan Scientific Center of Russian Academy of Sciences», PO. Box 261, 2/31 Lobachevsky st., Kazan, 420111, Russia

energoed@mail.ru

Abstract. In the article an influence of various oxidizers on the composition of sulfur products at bitumen gasification process are described. As an object of research the natural bitumen of Ashalchinskoe oil field of Tatarstan Republic (Russia) was chosen.

Introduction

At the present time the gasification process of hydrocarbon fuels belongs to clean technologies such as conversion of coal, heavy oil, peat and other solid and liquid fuels. According to data of [1] several hundred gasification plants which process the hydrocarbon fuels for production of chemicals, syngas and synthetic liquid fuels work worldwide.

One of the important sources of raw materials is high-viscosity hydrocarbons, including bitumen and oil sands. They are under study in Canada, Venezuela, Iran and other countries. In Russia a significant amount of bitumen is concentrated in the territory of the Republic of Tatarstan, where its reserves are 7 billion tons [2].

Gasification of bitumen would be allowed to produce syngas as alternative fuel source and attendant chemical products for various industries, as the methanol, ammonia or hydrogen production.

At gasification of hydrocarbon raw materials an air, oxygen, water steam or the mixtures of them use as the oxidizers. Their influence on the composition of sulfur products is not fully investigated. And this is the purpose of this work.

Materials and methods

Bitumen is a complex mixture of high-molecular hydrocarbons of oil and heteroatomic compounds, containing oxygen, sulfur, nitrogen and metals [3]. Chemical composition of bitumen is less studied than their physical, rheological and colloidal properties. Bitumen is considered like a complex of three components: oils, tars and asphaltenes [4]. Content of other substances is less 1% that is why we don't calculate them. Bitumen gasification process was considered using Texaco technology at a temperature of 1400 °C and pressure equal 1 MPa for heavy oil residues [5].

Bitumen gasification process can be considered as a complex of several stages. At first under temperature influence are heating and formation of microfractures due to water vaporization. As a result of chemical interaction between steam and surface of bitumen components under study appear light hydrocarbons. At the temperature over 900°C the oxidation-reduction reactions, steam-methane reforming and water-gas shift reactions are occurred, oxygen destroys benzene rings partially. During at this time the solid residue (a soot) from unreacted hydrocarbons and chemical substances which are components of syngas are forming [6, 7].

The final composition of produced syngas depends from the equilibrium of reaction (1):

$$CO + H_2O \leftrightarrow CO_2 + H_2 \quad (1)$$

Results and discussion

This paper presents the results of numerical studies showing the effect of various oxidizers on the composition of the sulfur products formed from depending on the oxidizer/fuel ratio. Oxygen and steam were used as the oxidizing agent. The bitumen of the Ashalchinskoye oilfield of the Republic of Tatarstan was considered as fuel, the elemental composition of which is presented in the table 1. For comparing of results the oxidizer/fuel ratio was taken equal 0.4, 0.8, 1.2 and 1.8.

Table 1. Ultimate analysis of Ashalchinskoye oilfield bitumen

Carbon	83.6 wt.%
Hydrogen	12.1 wt.%
Oxygen	0.5 wt.%
Nitrogen	0.4 wt.%
Sulfur	3.4 wt.%

Using Comsol Multiphysics software, the dependences of sulfur gasification products on oxidizer type and oxidizer/fuel ratio were obtained.

As it was written above the choice of technological parameters was taken on the base of Texaco gasification technology, which is used for utilization of heavy oil residues. Figure 1 shows the results of calculations for bitumen gasification using oxygen blow. Figure 2 shows the results of calculations for bitumen gasification using steam blow.

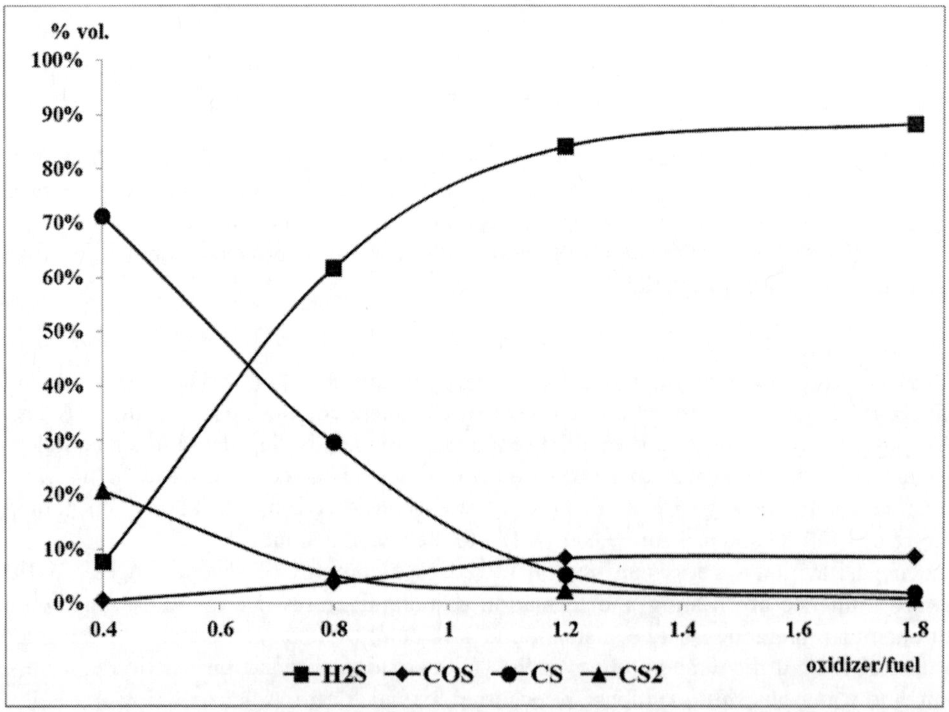

Fig. 1. Dependence of sulfur gasification products on oxygen blow

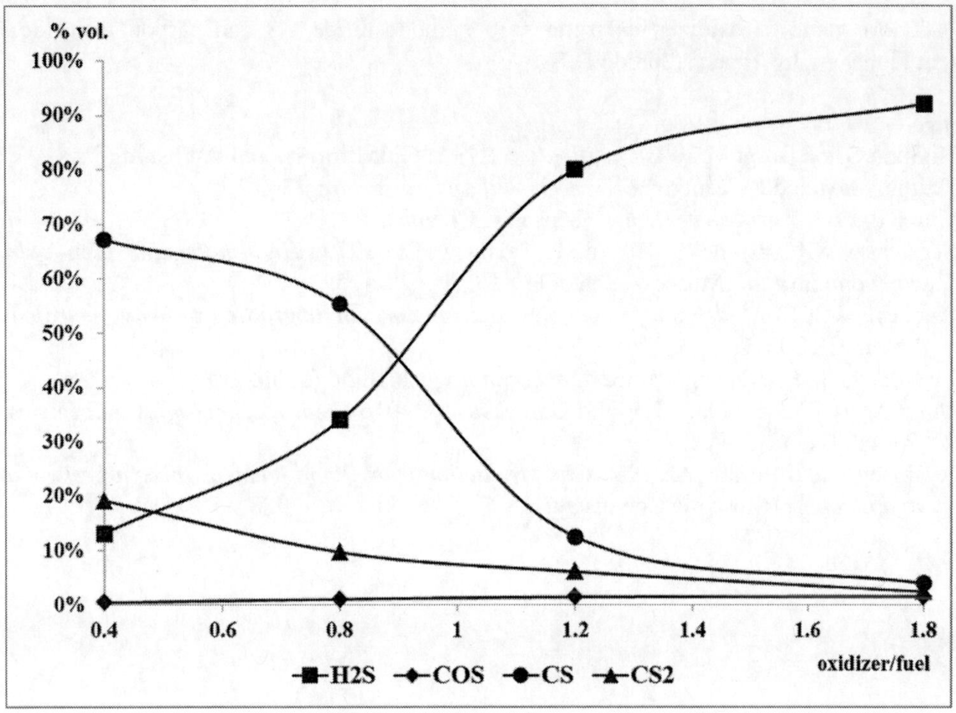

Fig. 2. Dependence of sulfur gasification products on steam blow

As you can see from figure 1, at an oxygen/bitumen ratio equal 0.4, a significant amount of sulfur from the bitumen in the gasification process proceed into the produced gas as carbon monosulfide CS and carbon disulfide CS_2. The rest of the sulfur forms a carbonyl sulfide COS and a small amount of hydrogen sulphide H_2S.

With an increase of oxygen/bitumen ratio to 0.8, a redistribution of sulfur compounds in the gas occurs: more than 60% is converted to hydrogen sulphide H_2S, about 30% of carbon monosulfide CS and rest of carbonyl sulfide COS and carbon disulfide CS_2. Further increase of oxygen/bitumen ratio leads to a significant increase in hydrogen sulphide and forming of sulfur dioxide SO_2, which amount are decreased with increasing pressure.

It can be seen from fig. 2 that with steam blow and small amount of steam/ bitumen ratio, the main amount of bitumen sulfur is converted to carbon monosulfide CS, the content of which in the produced gas decreases with increasing steam/bitumen ratio. In the same time, the amount of hydrogen sulphide H_2S is increased and at steam/bitumen ratio equal 1.8 this value is more than 90%.

The results of the calculations for the concentrations of sulfur compounds in the produced gas at steam-oxygen blow have shown that all the above-mentioned regularities remain for these conditions.

Comparing the obtained data with the results of experimental studies [8] it was obtained that in the real gasification process of sulfur liquid fuels, the sulfur contained in them is converted into the produced gas as a hydrogen sulphide H_2S (about 90%) and carbonyl sulfide COS (about 10%).

Conclusions

Based on the performed studies, the following conclusions were obtained:

- During the gasification process, the sulfur contained in the bitumen is converted in the syngas mainly in the form of hydrogen sulphide H_2S, carbonyl sulfide COS, carbon monosulfide CS and carbon disulfide CS_2;
- The distribution of sulfur amount between these compounds depends on the oxidizer/fuel ratio, temperature and pressure of gasification process;

- At low value of oxidizer/fuel ratio carbon monosulfide CS and carbon disulfide CS_2 are formed, at higher – a hydrogen sulphide H_2S.

Reference

[1] Higman C and Burgt M 2003 *Gasification* (USA: Gulf Professional Publishing)
[2] Halimov EM and Kolesnikova NV 1997 *Oil and gas geology* **3** 4-9
[3] Gun RB 1973 *Petroleum bitumen* (Moscow: Chemistry)
[4] Sergienko SR, Taimova BA and Talalaev EI 1979 *High-molecular non-hydrocarbon compounds of oil* (Moscow: Science)
[5] Ancheyta J 2013 *Modeling of processes and reactors for upgrading of heavy petroleum* (Boca Raton: CRC Press)
[6] Ermolaev DV *Petroleum Science and Technology* 2018 **36** (5) 392-397.
[7] Mingaleeva GR, Ermolaev DV and Galkeeva AA 2016 *Clean Technologies and Environmental Policy* **18** (1) 297-304.
[8] Ryabtsev II and Volkov AE 1968 *Gas production from liquid fuels for the synthesis of ammonia and alcohols* (Moscow: Chemistry)

IOP Conf. Series: Earth and Environmental Science **288** (2019) 012082 doi:10.1088/1755-1315/288/1/012082

Formation and properties of a dynamic ultrafiltration membrane

D.D. Fazullin, G.V. Mavrin

Kazan Federal University, 423812, Russia, Republic of Tatarstan, Naberezhnye Chelny, Prospekt Mira 68/19

denr3@yandex.ru

Annotation. By applying to the surface of the nylon membrane of polystyrene particles, dynamic nylon-polystyrene membranes with a final content of 3.4 % by weight were obtained. The content of polystyrene on the surface of the nylon membrane is indirectly confirmed by IR Fourier spectroscopy methods. An increase in the contact angle of the droplets of distilled water of the membrane as a result of depositing polystyrene particles on the surface of the initial membrane was revealed. According to the results of scanning electron microscopy, it was found that polystyrene particles with sizes from 0.05 to 0.2 microns are located on the surface and in pores of a nylon substrate. After applying a layer of polystyrene on the surface of the membrane, the specific productivity of the membranes decreases by an order of magnitude, due to the intensive accumulation of polystyrene particles in the pores and membrane surface. The maximum productivity of the initial and dynamic membranes is observed when the distilled water is passed through. Dynamic membrane ultrafiltration with a surface layer of polystyrene is designed to separate water-oil emulsions. The degree of separation of water-oil emulsion with a concentration of oil products of 10 g/dm^3 is more than 98 %.

Introduction

Cleaning industrial wastewater from oil products is an important task before discharging into the sewerage system or into natural water bodies. As is known, oily wastewater has a negative impact on the state of natural water: change the physical, chemical properties of water, which as a result leads to the death of hydrobionts, the disturbance of the ecosystem of the water body.

Petroleum products in waste water can be in a free state, dissolved and emulsified. If the removal of free petroleum products does not create special difficulties in the purification by sedimentation, coalescence, a whole complex of technological processes of purification is required to purify sewage from emulsified and dissolved petroleum products. At present, for the separation of emulsions, there are two fundamentally different approaches: reagent and reagent (hardware). To separate emulsions, reagent, acidic decomposition methods, coagulation methods and reagentless methods of filtration, sedimentation, centrifugation, coalescence [1], membrane separation processes are used [2-11]. And for the purification of sewage from dissolved petroleum products, adsorption [12] and membrane separation methods are used. Emulsions occur during extraction extraction of ore components and organic substances from plant raw materials, car washes, during the extraction and processing of petroleum products, and also in mechanical engineering enterprises, stable oil-containing emulsions are formed-spent lubricating and cooling liquids.

Thus, in [13], the authors obtained a superhydrophilic and superolefic membrane of polyvinylidene fluoride on the surface of which polyacrylic acid was grafted. The hierarchical micro / nanoscale structure of the surface of the membrane gives it a super-hydrophilic / superleophobic property. The membrane separates oil-in-water emulsions with a low working pressure (<0.3 bar), and the permeability of the membrane is one or two orders of magnitude higher than that of industrial filter membranes having similar parameters. Also, the resulting membrane has an excellent anti-fouling property.

Of all types of membranes, dynamic membranes have the advantage of separating water-oil emulsions. So, if the permeability of the membranes decreases, you can replace the dynamic layer with a new one, which will restore the initial performance. By using different particle sizes of the applied material or by obtaining several dynamic layers, the required pore size of the membranes can be achieved. Depending on the physico-chemical properties of the applied material, it is possible to obtain different wettability and roughness of the membrane surface.

To separate the oil-in-water emulsion, the authors of [14] obtained membranes with a dynamic layer of titanium dioxide which was formed on a porous substrate of a carbon tube. Dynamic membrane, obtained under optimal conditions, was characterized as homogeneous, dense and hydrophilic. According to the results of the experimental work, the authors found that the oil removal efficiency is 98%, and its concentration in the permeate was less than 8.3 mg / dm^3.

As a dynamic layer, a suspension of magnesium hydroxide was applied in [15], which was applied to a ceramic substrate. The membrane was used to separate the water-oil emulsion. The degree of oil removal was more than 98%, and the filtration rate was more than 100 dm^3/m^2h at an applied pressure of 100 kPa.

The authors [16] obtained a two-layer composite dynamic membrane for the separation of water-oil emulsions. The dynamic layer of the membrane was formed from kaolin and MnO_2. The microstructures of the dynamic membrane were examined with a scanning electron microscope. The results showed that the MnO_2 particles are located on the surface of the dynamic layer of Kaolin, forming a kaolin / MnO_2 two-layer composite dynamic membrane. With increasing oil concentration, the permeate flux has decreased and the oil retention coefficient has increased. In neutral or alkaline environments, the oil retention efficiency was more than 99 %. With an increase in temperature from 283 K to 313 K, the retention efficiency decreased from 99.9 % to 98.2 %, and the filtration rate increased from 120 dm^3/m^2h to 153 dm^3/m^2h.

In [17], a thin-film dynamic membrane of PAN-TFC was obtained. The efficiency of the oil-water emulsion separation was more than 99 %. Also, the authors investigated the process of surface contamination of the membrane with oil and methods of membrane regeneration. The authors decided on the need for more effective anti-fouling TFC membranes.

The main problem of cleaning oily wastewater is the separation of stable emulsified oil particles from the aqueous phase. Therefore, the purpose of this study is to increase the efficiency of the processes of ultrafiltration separation of stable emulsions of the oil-in-water type using a dynamic membrane.

Methods

A microfiltration polymer membrane of nylon with an average pore size of 0.45 μm and a diameter of 47 mm was used as the initial substrate on which the dynamic layer was applied. The dynamic layer was obtained by forming a semipermeable layer on the surface of the porous base of the suspended microparticles of polystyrene present in the filtered aqueous solution of acetone in a dynamic equilibrium with the solution. The content of polystyrene in the membrane was determined by the gravimetric method, by the weight of the membrane before and after the modification.

The particle size of the dispersed phase of emulsions and suspensions was determined by dynamic light scattering (DLS) and the ζ potential by light scattering with phase analysis (PALS) using a NanoBrook Omni analyzer.

IOP Conf. Series: Earth and Environmental Science **288** (2019) 012082 doi:10.1088/1755-1315/288/1/012082

Figure 1 shows the distribution of the particle size of the dispersed phase of a polystyrene suspension with a 25% acetone solution.

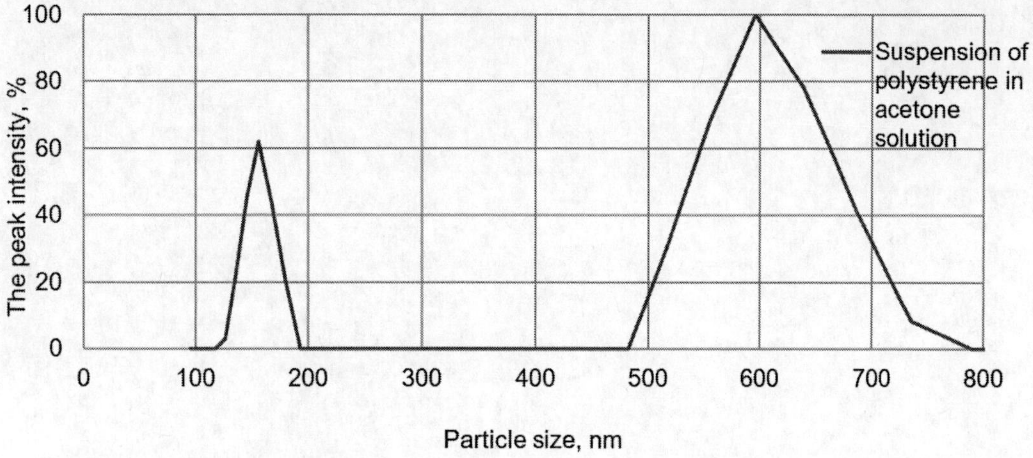

Figure 1. The graph of the distribution of the particle sizes of a dispersed phase of a polystyrene suspension in an aqueous solution of acetone

Table 1. The particle size and the ζ-potential of the dispersed phase of the suspension.

Suspension	Particle size, nm	ζ-potential, mV
Polystyrene 25 % of the p-re acetone	118 – 193, 492-790	- 7,1±0,7

The particle size of the dispersed phase of a polystyrene suspension in a 25 % acetone solution, as shown in Fig. 1 and tab. 1 is from 118 to 790 nm. To obtain a dynamic layer, use this suspension of polystyrene in an acetone-water solution.

The IR spectra of the samples were studied using an infrared FTIR spectrometer of the «Infralum FT-02» brand. IR Fourier spectrometer, allows to obtain high resolution absorption bands.

A change in the surface structure of the membranes was recorded using a scanning electron microscope of the brand «LEO-1430 VP» manufactured by «Carl Zeiss». The samples were glued to aluminum plates, the surface of the membranes was deposited with gold, cathodic deposition in argon, and viewed in high vacuum.

As the main parameters of the membrane separation of the emulsion, the specific productivity was considered, which was defined as the ratio of the amount of the filtrate formed to the product of the membrane area and the process time in terms of dm^3/m^2h, and the degree of separation of the emulsion, which was calculated as the ratio of the content of petroleum products (NP) with the help of a concentrate «KH-3», an emulsion before and after separation.

For membrane separation, a 1 % emulsion of freshly prepared coolant was used as a VME. During the separation of distilled water and emulsions, the working pressure was 0.1 MPa, the liquid temperature was – 25 °C.

Results and discussion

After applying a dynamic layer of polystyrene to the surface of the original nylon membrane, the surface of the membrane becomes hydrophobic, as determined by the increase in the contact angle of the drop of distilled water from 59.6° to 106.2° [4].

The results of investigation of the membrane surface by scanning electron microscopy with an increase of 2000 times are shown in Fig. 2.

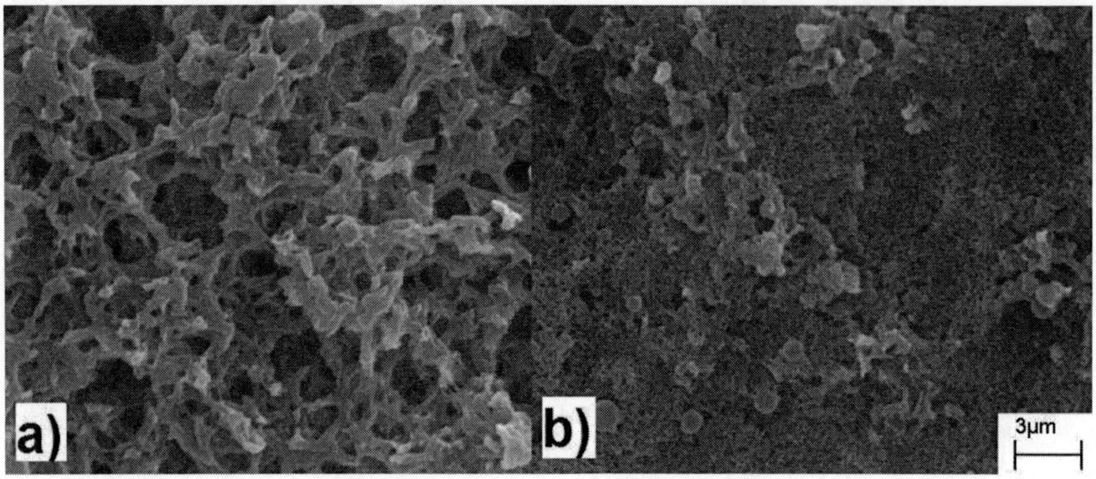

Figure 2. Microphotographs of the membrane surface: a) Initial membrane of nylon; b) the membrane after applying a dynamic layer of polystyrene (an increase of 2000 times).

According to Fig. 2, it can be seen that the polystyrene particles are located on the surface and in the pores of the membrane from nylon. In general, the dimensions of the polystyrene particles are from 0.05 to 0.2 μm.

Spectrums of the investigated membrane samples in the frequency interval 600-4000 cm 1 were obtained using the infrared FTI-08 IR spectrometer. The IR spectra of the initial and modified membranes are shown in Fig. 3.

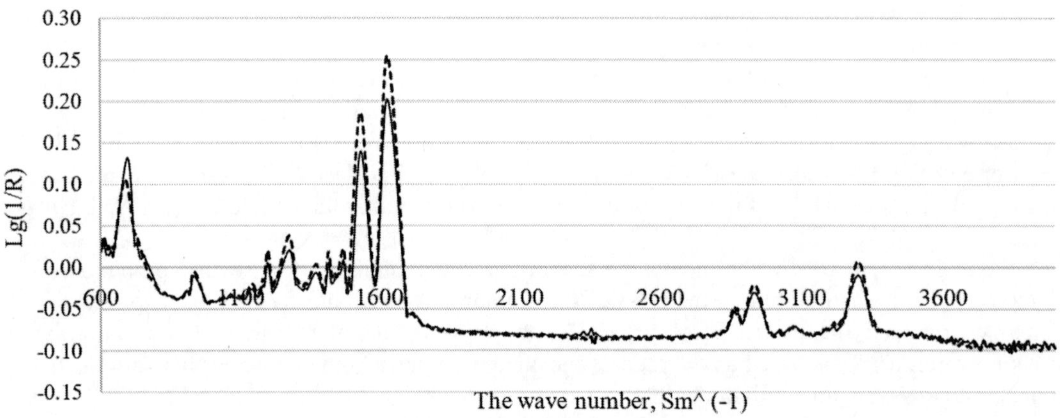

Figure 3. IR absorption spectra of the membrane samples studied: nylon (dashed line) and nylon-PS (solid line).

The characteristic absorption bands for nylon correspond to wavelengths of 1635 cm^{-1} and 1539 cm^{-1} (Fig. 3) [18]. For polystyrene, the absorption bands of 2850 cm^{-1}, 2930 cm^{-1}, 3022 cm^{-1} are characteristic [19]. After the polystyrene nylon membrane is applied to the surface, a decrease in the intensity of absorption of these bands characteristic of the nylon and an increase in the intensity of the absorption bands of 690 cm^{-1} characteristic for non-planar deformation vibrations of C-H are observed. The appearance of a weak absorption band at 753 cm^{-1} corresponds to out-of-plane deformation vibrations of 1,3 substituted C-H, groups of absorption bands of weak intensity at 2330-2370 cm^{-1} correspond to salts of amines $R_2C = NH^+$. Thus, based on the results of IR spectroscopy, in particular, the appearance of changes in the chemical structure resulting from the deposition of polystyrene on the surface of the nylon membrane, the above increase in the hydrophobicity of the nylon-PS membrane occurs.

The results of studies on the specific productivity of membranes are presented in Table. 2.

Table 2. Specific productivity of membranes

Membrane	The content of polystyrene,% (by weight)	Specific capacity of membranes, $dm^3/m^2 \cdot h$	
		for distilled water	by 1% water-oil emulsion
Nylon	-	3845	186
Nylon-PS	3,4	266	23

After applying a layer of polystyrene on the surface of the membrane, the specific productivity of the membranes decreases by an order of magnitude, due to the intensive accumulation of polystyrene particles in the pores and membrane surface. The maximum productivity of the initial and dynamic membranes is observed when the distilled water is passed through.

The results of the separation of a water-oil emulsion from an NP are shown in Table. 3.

Table 3. Degree of separation of water-oil emulsion

Membrane	The content of polystyrene,% (by weight)	Concentration of oil products, mg/dm^3		Степень очистки, %
		initial	after cleaning	
Nylon	-	10120±1012	7020±702	30,6
Nylon-PS	3,4		208±20,8	98,0

According to Table 3, it is obvious that the degree of removal of petroleum products from the water-oil emulsion using the initial nylon membrane is significantly lower than when it is separated using dynamic nylon-PS membranes. It was determined that after the deposition of a dynamic layer of polystyrene on the surface of the nylon membrane, the degree of separation of the emulsion increased to 67 %.

Conclusion

Dynamic membrane ultrafiltration with a surface layer of polystyrene is hydrophobic, designed to separate water-oil emulsions. A layer of polystyrene is formed by filtration on the surface of a nylon substrate from a suspension of polystyrene in acetone. It was determined that after depositing a dynamic layer of polystyrene on the surface of the nylon substrate, the degree of separation of the emulsion with a concentration of oil products of 10 g /dm^3 increased by 67 % and amounted to more than 98 %

Acknowledgement

The work is performed according to the Russian Government Program of Competitive Growth of Kazan Federal University.

References

[1] Fazullin D D and Mavrin G V 2017 *The Turkish Online Journal of Design, Art and Communication* 7 **Special Edition**, pp. 1686-1692.
[2] Fazullin D D, Mavrin G V, Shaikhiev I G 2016 *Petroleum Chemistry* **56** pp 454-458

[3] Fazullin D D, Mavrin G V, Shaikhiev I G 2017 *Petroleum Chemistry* **57** pp 165–171
[4] Fazullin D D, Mavrin G V, Shaikhiev I G, Nizameev I R 2018 *Petroleum Chemistry* **58** pp 145-151. DOI: 10.1134/S0965544117130047
[5] Fazullin D D and Mavrin G V 2017 *The Turkish Online Journal of Design, Art and Communication* 2017 **7 Special Edition** pp 1643-1651. DOI NO: 10.7456/1070DSE/142
[6] Yuan Gao et al 2018 *Separation and Purification Technology* **192** pp 111-126
[7] Jing Wang et al 2018 *Journal of Membrane Science* **547** pp 156-162
[8] Soobok Lee et al 1984 *Journal of Membrane Science* **19** pp 23-38
[9] Mohammadi T et al 2003 *Desalination* **157** pp 369-375
[10] Sablani S S et al 2001 *Desalination* **141** pp 269-289
[11] Alberto Lobo et al 2006 *Journal of Membrane Science* **278** pp 328-334
[12] Fazullin D D et al 2014 *Chemistry and Technology of fuels and Oils* **1** pp 88-94. DOI: 10.1007/s10553-014-0493-5
[13] Zhang W et al 2014 *Angewandte Chemie* **53 Issue 3** https://doi.org/10.1002/anie.201308183
[14] Yanqiu Pan et al 2012 *Separation and Purification Technology* **89** pp 78-83
[15] Yijiang Zhao et al 2005 *Separation and Purification Technology* **44 Issue 3** pp 212-220
[16] Tao Yang et al 2011 *Desalination* **270 Issues 1–3** pp 50-56
[17] Phuoc H H et al 2014 *Journal of Membrane Science* **452** pp 117-126
[18] Fazullin D D et al 2015 *Modern Applied Science* **9** pp 242-249
[19] Krasovskiy AN et al 2012 *High-mole. Cpd. Series A* **54** pp 851-858

Math modeling of heat and mass transfer processes during coal-water fuel gasification

A. A. Galkeeva, M. V. Pyatigina, G. R. Mingaleeva

420066, Russian Federation, Kazan, Krasnoselskaya st., 51

prujinka1990@mail.ru, pmv_83@mail.ru, mingaleeva-gr@mail.ru

Abstract. In article the current state of gasification technologies in Russia and in the world is analysed. The perspective directions of development of process – entrained flow gasification of coal-water fuel for the purpose of producing syngas for further catalytic conversion and producing chemicals are defined. The mathematical model of thermophysical characteristics calculation of process is presented. Determination results of design parametrs of a gasifier are given.

Introduction

At present, deep conversion of oil and natural gas is being carried out in Russia to produce energy and various products of chemical synthesis: methanol, ammonia, carbon monoxide, hydrogen, liquid synthetic fuels, etc. At the same time, the use of coal is limited to combustion in boilers. The technologies and plants for solid fuels combustion are morally and physically obsolete, and the process indicators do not meet the requirements of the UN for reducing emissions and technological re-equipment [1].

Based on the "Energy Strategy of Russia until 2035" project, such issues as low share of use of local fuels, development of domestic scientific and technical base of "clean" coal technologies and production of gaseous and liquid fuels, as well as competition in the domestic market through the use of cheaper raw materials require the solution [2]. Experts identify the complex use of coal as the most perspective direction [3]. This technology allows to produce heat and electric energy and various chemical products which are demanded by national economy.

Gasification is one of the most rational, energy efficient and ecologically safe directions of use of coal and coal-water fuel (CWF). Because of a possibility of a variation of syngas composition depending on a type of fuel and operating conditions it can be used in power engineering for combustion in boilers and gas-turbine plants and also as an initial component for producing a wide range of chemical products, on condition of a molar ratio of monoxide of carbon to hydrogen within 0,4-0,6 [5].

The choice of the direction of use syngas depends on its composition and operating conditions of gasification: temperature, pressure, existence and a type of oxidizer and also on reliable method of calculation of gas composition and properties, processes of a heatmass exchange in the reactor, its design. Modeling of gasificationprocess will allow to develop the gasifierunified by the type of fuels.

According to statistical data [6-7] in the world more than 150 plants making syngas which are mainly located in China, the USA and Germany are functioning. On most of it entrained flow gasifiers function that gives the chance to the plants to work on an IGCC cycle, developing electric energy, pure hydrogen and chemical products [8]. Entrained flow process is carried out at high temperatures 980-1930 °C that provides high (to 99%) conversion of carbon in gas.

Work of entrained flow gasifiersis based on experimental or operational data. At the fundamental level of a research this process is worked still out a little therefore there are no techniques of thermal, thermodynamic and aerodynamic calculation, it complicates design of gas generators when using different types of solid fuels. Thus, line heterogeneous gasification isn't captured by modern methods of modeling and a research on the basis of creation of mathematical models yet.

Math modeling

It is known that process of thermochemical conversion of coals proceeds in 3 stages. At the first stage physically bound water evaporates; at the second stage gas components of coal and the high-carbonaceous coke are formed; at the third stage gas components and the coke react with oxidizer with formation of components of syngas. The third stage is of the greatest scientific and practical interest. At thermochemical interaction of components of steam-gas mix on a surface of the solid coke syngas of various composition at regulation of operating conditions can be formed. As a rule, interaction of a gas and solid phase is described by the generalized equations of reactions between C, O_2, H_2O, CO_2, H_2, CO and CH_4 [6].

Technological parameters of processes: the fuel particle speed, the size, diffusion and heat exchange proceeding in volume of a gasifier influence finally the speed of thermochemical transformations, characteristics syngas and on the possible ways of it use.

Theoretical studies of syngas composition performed earlier by the authors [9] have shown that the optimum composition for further catalytic processing has the syngas obtained in the oxygen-free gasification of water-coal fuel.

The oxygen-free gasification of water-coal fuel is an endothermic process which requires external heat. In the method of gasification proposed by the authors, the CWF drop fall into the hot synthesis gas and, after heating, passes through all the process stages: evaporation of the CWF water, volatilization of gas components, thermal conversion of the mixture components. At each stage, the thermophysical characteristics of the process are calculated, thanks to which it is possible to determine the design parameters of the gasifier.

For determination of the amount of heat which necessary for the thermal conversion of coal-water fuel into syngas and the residence time of the drop of CWF in the gasifier the mathematical model of heat and mass transfer between drop of CWF and the flow of heating gas in the gasifier was developed [10].

$$Q_r = \sum_{i=1}^{n} (cm(T_0 - T_1))_i \tag{1}$$

$$\sum_{i=1}^{b} G_{i\_inlet} = \sum_{i=1}^{k} G_{i\_exit} \tag{2}$$

$$\frac{dQ_r}{d\tau} = \alpha S_{sur}(T_g - T_d) \tag{3}$$

$$\frac{dm}{d\tau} = \beta S_{sur}(C_{sur} - C_g) \tag{4}$$

$$m\frac{du}{d\tau} = \sum_{i=1}^{l} F_i + u\frac{dm}{d\tau} \tag{5}$$

$$L = u\tau + \frac{a\tau^2}{2} \tag{6}$$

where Q_r – the required amount of heat to evaporate water of CWF in the first stage of gasification and volatilization of gas components in the second stage, J; n – amount of rated zones of the gasifier, T_0, T_1, T_d, T_g – temperature of the heating gas at the inlet to the rated zone of gasifier, at the outlet of the rated zone, the drop of CWF and gas, respectively, K; c – specific heat of substances in reaction zones, J/kg·K; m – mass of components, kg; G – components consumption of the mix, kg/s; b – amount of initial substances; k – amount of products; τ – time of the course of certain processes (warming up the drop of CWF, mass transfer, stay of a particle of fuel in a zone), s; α – heat transfer coefficient, $W/(m^2 \cdot K)$; S_{sur} – surface area of the drop of CWF, m^2; l – almount of forces acting on particle of fuel; β – mass transfer coefficient, m/s; C_{sur}, C_g – concentration of evaporating substances at the surface and in heating gas, kg/m^3; ;∑F – sum of forces acting on a drop of CWF (on particle of fuel), H; u – drop of CWF (particle of fuel) rate, m/s; a – drop of CWF (particle of fuel) acceleration, m/s^2; L – zone length, m.

The equations 1 and 2 represent the heat and mass balance of evaporation of water and volatilization of gas components and present the laws of conservation of energy and mass. The equations 3 and 4 describe the value of heat flow and rate of mass loss and represent the equations of heat and mass transfer. The equation 5 is Meshchersky's equation and describes change of drop of CWF (or particle of fuel) rate at a variable mass of particle . The distance which is overcome by variable mass drop of CWF (or particle of fuel) at equally slow driving is determined by the equation 6.

The offered mathematical model of the processes occurring with a drop of CWF at entrained flow gasification considers process that when driving a drop its density and weight changes. At first stage water of CWF evaporates, while the droplet diameter does not change, at the outlet of the first zone drop of CWS represents a coal framework. At the second stage gas components of coal volatilize, and the coal framework turns into coke framework. At the third stage coke is destroyed and, reacting with steam-gas mix, syngas is formed.

At each stage the heat and mass transfer between a particle of fuel and gas phase takes place, at the same time process has to be provided with a necessary amount of heat for evaporation of water and gas component of coal, and and for endothermic gasification reactions. Mass transfer between a particle of fuel and gas is characterized by a diffusion coefficient which is defined through change of composition of gas in each gasification zone. To determine the geometric size of the gasifier, the residence time of the fuel particle in the gasifier and the particle rate in each reaction zone are defined.

Assumptions:

- monodisperse composition of CWF drops;
- the temperature of a gas phase in the 1st and 2nd zone of a gasifier is supported at the level of 1150K;
- the produced syngas fills entire volume of a gasifier;
- the movementof the CWF drop along the length of the gasifier is rectilinear;
- the drop of CWF moves equally slowly in each rated zone of the gasifier, drop acceleration is characterized by the rate of diffusion of gas components into the gas phase;
- at the outlet from 1st zone the steam is heated to the temperature in the gasifier, and the carbon particle is reached CWF water evaporation temperature $T_{d1} = T_{evap}$;
- in the 2nd zone the coal particle is heated to the gas components volatilization temperature, and at the outlet from 2nd zone all reaction mix is reached the temperature of the heating syngas $T_{CP2} = T_{volat} = T_{SG}$;
- in the 3rd zone the rate of the coke particle is compared to the gas flow rate $u_{CP} = u_{SG}$;
- due to the heat transfer from the heat pipes in gasifier, the syntgas temperature at the inlet to each zone is supported by constant T_{SG} =*const*.

The system of equations 1-6 for each zone corresponds to the heat and mass transfer that take place in the water evaporation and gas components volatilization. The calculation of thermochemical processes in 3rd zone is based on the determination of chemical gasification reactions heat effect and the temperature of the syngas at the outlet.

Results of studies

As a result of calculation of heat transfer in the gasification the following was determined according to the developed mathematical model.

The size of all reaction zones, and the size of the entire gasifier can be influenced by regime and technological factors.

CWF drop sizes, sprayed by the injector, exert the significant impact of the zones of water evaporation and gas components volatilization (fig. 1). It is explained by need of increase in time for course of the processes occurring with particle of fuel in the rated zones.

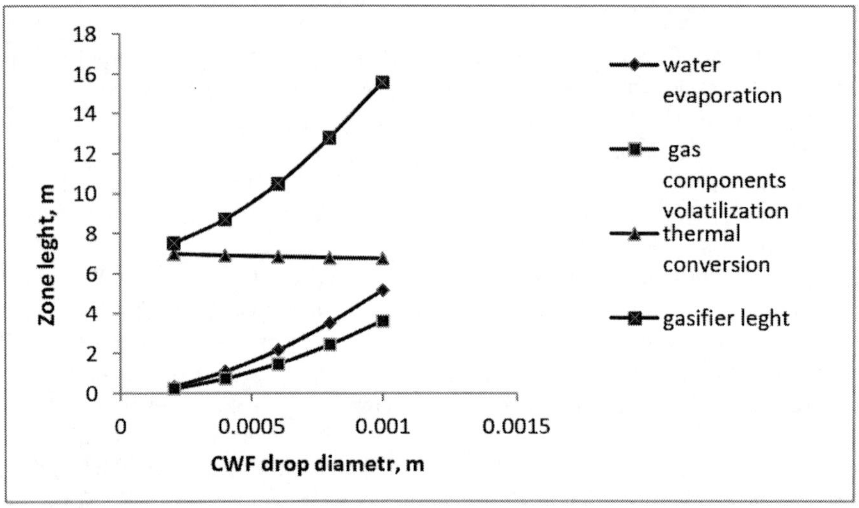

Figure 1. Dependence of length of the conditional zones of a gasifier from CWF drop sizes

At increase in rate of a drop the distance, which is overcome by CWF drop during syngas formation, respectively, and gasifier length increases. At a given gasification temperature, it is necessary to ensure a certain time of the CWF drop in the gasifier to perform its thermal conversion to syngas, regardless of the drop rate. Therefore during mathematical modeling it is necessary to determine the drop rate at which the residence time of the CWF drop in the gasifier at acceptable overall design of the gasification reactor will be provided. At a low drop rate it is required shallow spray coal-water fuel by a injector for providing necessary time for the thermochemical processes in the reactor.

At the same time varying CWF drop rate at the outlet from the injector, it is possible to reach a defined value of length of a gasifier regardless of coal brand, the problem of unification of a gas generator by type of fuel thereby is solved.

Comparison of the received results showed that with a diameter of CWF drop ~ 0,6 mm [11] to ensure the calculated time for of processes of water evaporation, gas component volatilization and thermochemical reaction components of the mix it is necessary to provide drop rate at the injector outlet of 14-18 m/s.

The reported study was funded by RFBR according to the research project No. 17-08-00295 «A».

References

[1] New Parisian climate.URL: http://www.kommersant.ru/doc/2876937
[2] «Energy Strategy of Russia until 2035». URL: http://minenergo.gov.ru/node/1026
[3] Bin Shi, Wen Xu, Erdorng Wu, Wei Wu, Po-Chih Kuo. Novel design of integrated gasification

combined cycle (IGCC) power plants with CO_2 capture. Journal of Cleaner Production. 2018. Volume 195. Pp 176-186

[4] Olkhovskii G.G. Solid fuel gasification in the global energy sector (A review). Thermal Engineering. 2015. Vol.. 62. No. 7. Pp. 465-472.

[5] Qianqian Chen, Min Lv, Yu Gu, Xiyi Yang, Zhiyong Tang, Yuhan Sun, Mianheng Jiang. Hybrid Energy System for a Coal-Based Chemical Industry. Joule. Volume 2, Issue 4, Pages 607-620

[6] Higman C., Burgt M.Gasification. Elsevierscience. 2003. Vol. 391.

[7] Slavinskaya L. Coal gasification: global trend. Neftegazovaya vertikal'. 2011. No. 8. Pp. 13-16

[8] G. Cau, D. Cocco, F. Serra. Energy and cost analysis of small-size integrated coal gasification and syngas storage power plant. Energy Conversion and Management. 56 (2012). p. 121-129

[9] Mingaleeva G.R. Ermolaev D.V., Galkeeva A.A.. Physico-chemical foundations of produced syngas during gasification process of various hydrocarbon fuels. Clean Technologies and Environmental Policy. 2016. Vol. 18. Pp. 297-304. DOI: 10.1007/s10098-015-0988-8.

[10] Galkeeva A.A., Mingaleeva G.R. Mathematical model of oxygen-free gasification of coal-water fuel. Problemy sovershenstvovaniya toplivno-ehnergeticheskogo kompleksa. Sovershenstvovanie ehnergeticheskih sistem i teploehnergeticheskih kompleksov: materialy XIII Mezhdunarodnoj nauchno-tekhnicheskoj konfencii. Saratov. November 1-3, 2016. Vol. 8. Pp. 297-300

[11] Delyagin G.N., Onishchenko A.G. The theoretical analysis of burning out of coal-water suspension taking into account its initial ash-content and humidity. Gorenie dispersnyh toplivnyh sistem. 1969. Pp. 7–18.

IOP Conf. Series: Earth and Environmental Science **288** (2019) 012084 doi:10.1088/1755-1315/288/1/012084

Analysis of the Stirling engine to identify the impact of various design parameters on the efficiency of its operation as part of an autonomous power complex based on renewable energy sources

A M Gaponenko[1], **A A Kagramanova**[1]

[1]Kuban State Technological University, Krasnodar, Russia

kagramanova@atiss.info

Abstract. The opportunity of application of Stirling engine with non-conventional and renewable sources of energy. The advantage of such use. It has been established that the four input parameters, the only parameter which can be easily changed during operation, and which effectively affects the operation of the engine is the phase difference. Dependence of work per cycle of the phase difference, called the phase characteristic, visually illustrates mode of operation of Stirling engine. The mathematical model of the cycle of Schmidt and the analysis of operation of Stirling engine in the approach of Schmidt with the aid of numerical analysis. To conduct numerical experiments designed program feature in the language MathLab. The results of numerical experiments are illustrated by graphical charts.

Numerical analysis of cycle DS in the approach of Schmidt

The calculations showed that in the perfect work of the regenerator efficiency of the cycle coincides with the Schmidt DS or efficiency of the Carnot cycle.

Dimensionless work per cycle $A/v \cdot R \cdot T_2$ depends on four dimensionless parameters: δ, τ, k, X.

Especially noticeable influence of the phase difference δ, so that its value can be controlled by the output of a heat engine: if $0 < \delta < \pi$, $A > 0$, that is, the system operates as an engine, with $\pi < \delta < 2\pi$, $A < 0$ as a refrigerating machine. Let us call the dependence of the dimensionless work from the phase difference between the phase characteristics. The maximum value of work the value of δ, which begins high, as does the kind of phase characteristics depend on the values of the three remaining parameters. Therefore, to characterize the influence of these parameters, it is convenient to represent in the graph the family of phase curves depending on any one of the three parameters at constant values of the other two. The following graph shows the family of phase characteristics at various values of the parameter k [1-6].

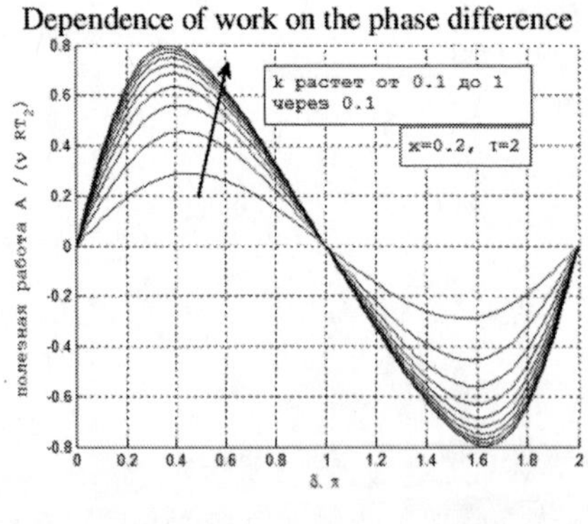

Figure 1 – Dependence of the phase characteristic from the relationship of the volume $k=Vc/V_E$ when $k \leq 1$. The arrow shows the sequence change of the curve in the collection at the specified change to the parameter k

IOP Conf. Series: Earth and Environmental Science **288** (2019) 012084 doi:10.1088/1755-1315/288/1/012084

The graph shows the influence of the phase difference δ to the value of the work per cycle. Given that the value of δ is technically easy to change in the process engine, the parameter can be used for power control of Stirling engines [7].

Many designs of Stirling engines vary the ratio of volume expansion and contraction $k = V_{c0}/V_0$.

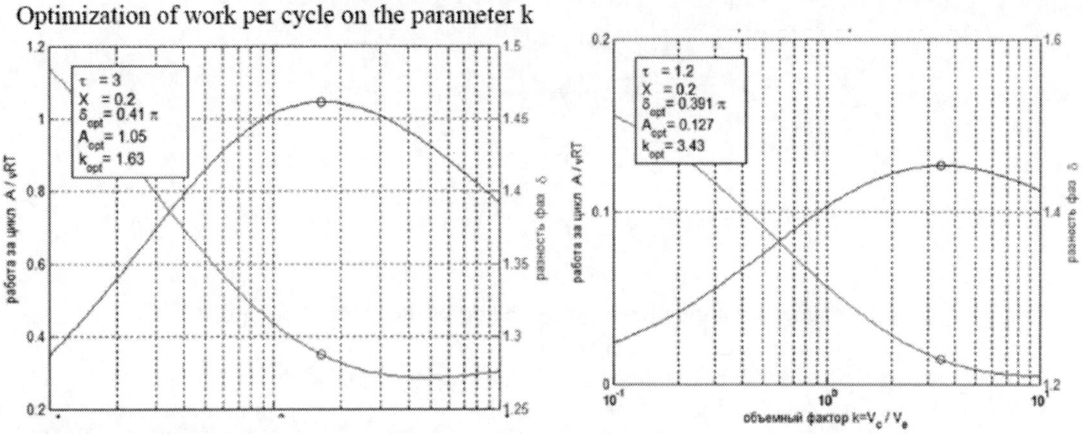

Figure 2 – Optimization of Stirling engine at the volume factor k. for different values of τ with constant value of X. Curves are shown Am (k) (in dimensionless form) and δm (k). Round markers indicate $A_{opt}(k_{opt})$ и $\delta_{opt}(k_{opt})$

From the previous picture you can see that the value of that $\delta = \delta_m$ maximizes value A is different for different values of k. This value, as well as the value Aopt, depend on two other parameters: $\tau = \frac{T_1}{T_2}$ and $X = \frac{V_D}{V_0}$.

Given that the value of δ can be adjusted in the process of operation, while, as k and X are rigidly connected with the design of the engine, and the value of τ associated with the temperature regime of work, can also be considered constant, the optimization of k can be realized as follows: by setting a specific value of τ and X, optimize And for each value of k from a fairly wide range in the neighborhood of the optimum, identifying Am(k) and δm(k), and then, from the obtained values to determine the optimum k.

The following is the result of the program MATHLAB, which uses the previously described function to perform such optimization and output in a graphic window graphics Am(k), δm(k) values Aopt, δopt and the corresponding annotations [8-10].

It follows from the figure that the optimum value of k in advance should be selected based on the expected temperature range of the engine. So for τ=3, which corresponds to T_1=900 K (627°C), T_2=300 K (27°C), kopt=1,63, while for τ =1,2, which corresponds to the temperature of the heater T_1=360 K (87°C.) at the same temperature of the refrigerator. The last case corresponds to using low-grade heat of hydrothermal reservoirs.

Each Stirling engine there is a dead volume V_D, which applies to the volume of the heater, regenerator, cooler ballast and other areas where the working fluid is not involved in the processes of compression- expansion. When designing a Stirling engine, it is important to have an idea about the impact of the proportion of dead volume to the engine. The following two drawings, made on the basis of numerical calculations explain this effect.

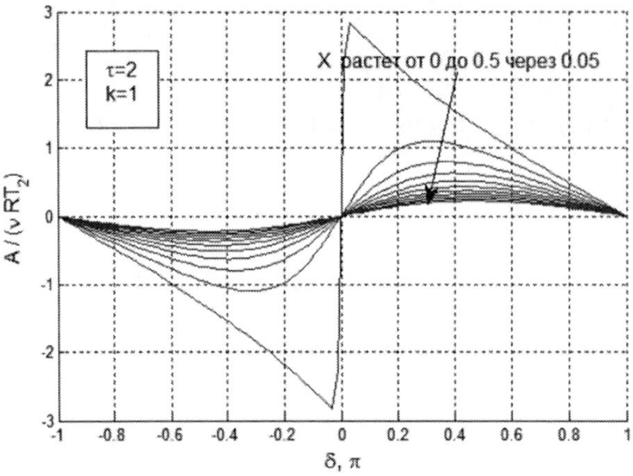

Figure 3 – Dependence of the phase characteristic of the proportion of dead volume

The figure shows that with increasing dead volume, the optimal phase shift varies almost from 0 to $\pi/2$ and at zero volume, the maximum point is unstable. With a small spontaneous decrease in δ the 2 process can go from positive to negative (the engine mode switches to the mode of refrigerating machine). From this circumstance we conclude that, although the growing proportion of the dead volume decreases maximum work, yet some of it is necessary for the stable operation of the DS.

Figure 5 shows in more detail the dependence of the optimized δ in the work and optimizes the phase difference of the proportion of dead volume X.

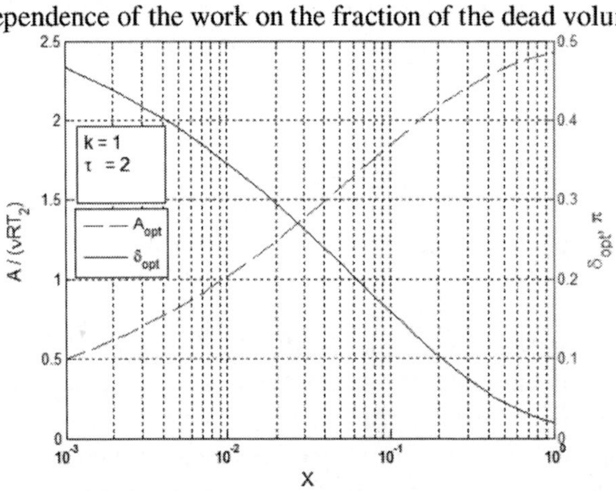

Figure 4 – dependence of the optimized dimensionless δ $A/v \cdot R \cdot T_2$ of the proportion of dead volume X and the corresponding optimizing values δopt

From figure 4 it is seen that the increase in the share of the dead volume significantly reduces engine performance.

When using DS with various sources of thermal energy, it is important to know the effect of the temperature difference between the heater and the refrigerator on engine performance. In dimensionless form, the temperature difference relative to ambient temperature (refrigerator) is expressed by the parameter τ: . To obtain the dependence from the temperature difference, an approach was used similar to that used to obtain the graphs of Fig. 4: for each τ of a certain range this value was optimized on δ and the points obtained (A_{opt}, τ) and (δ_{opt}, τ) displayed as curves on the chart. The dependence of the efficiency of the engine from the temperature difference in the dimensionless form shown in the following image.

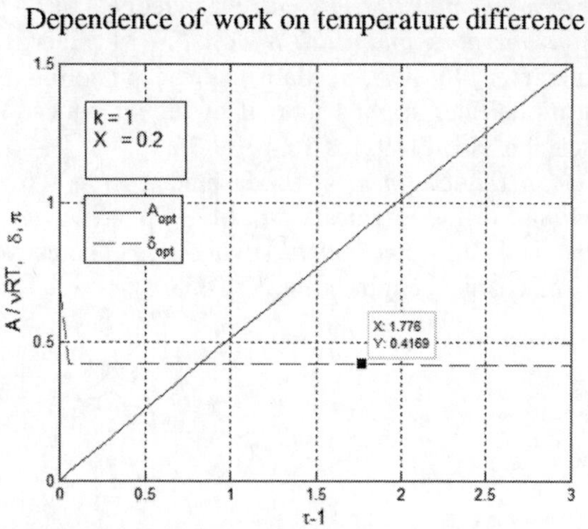

Figure 5– the dependence of the work per cycle from the temperature difference

From figure 5 it is seen that for given k and X, the work grows almost proportional to the difference of temperatures at an almost constant optimal δ.

Conclusions

Found that of the four input parameters, the only parameter that can be easily changed during operation and which effectively affects the operation of the engine, is the phase difference δ. The dependence of the cycle work from δ, called the work phase characteristics, clearly illustrates the mode of operation of the DS. Given families of phase characteristics for different modes of operation.

The paper presents the results of optimization for different temperature parameters, corresponding to high and low potential sources of heat for various values of the ratio of the volume of the compression chamber to the expansion chamber (k). The optimal value of this parameter depends on the conditions under which the calculated engine.

The results of the corresponding calculations, from which it is concluded that the decrease in the proportion of dead volume (X) leads to a significant increase in efficiency, however, too small a dead volume can lead to loss of stability of engine operation.

It is shown that the maximum in δ, the work per cycle at constant values of k, X is almost proportional to the temperature difference between the heater and the fridge.

References

[1] Gaponenko A.M., Kagramanova A.A. / Optimization of Stirling engine operation based on mathematical modeling // Advances in Engineering Research: International Conference on Actual Issues of Mechanical Engineering (AIME 2017), V.: 133. P. 285-290.

[2] Gaponenko A.M., Kagramanova A.A. 2016 Optimization of the composition of the complex of renewable energy sources with the use of cluster approach and the theory of random processes. *Polythematic online scientific journal of kuban state agrarian university-Krasnodar, KubSAU*, – №02(116). – IDA [article ID]: 1161602005. – http://ej.kubagro.ru/2016/02/pdf/05.pdf p. 94 - 109.

[3] Gaponenko A.M., Kagramanova A.A. 2016 Methodology and algorithm of mathematical analysis multifactor models optimal configuration of renewable energy sources. *International research journal.* №4(46). Vol. II. Ekaterinburg. p. 76-79.

[4] Gaponenko A.M., Kagramanova A.A. 2016 Method of determination of optimal parameters of energy efficient of power supply systems. *International research journal.* №4(46). Vol. II. Ekaterinburg. p. 80-84.
[5] Smirnov G.V. 1967 *External combustion engine*. "Knowledge", M., – p. 486.
[6] Walker G. 1978 *Machine operating on the Stirling cycle*: – M.: Energy, – p. 152.
[7] Gaponenko A.M., Kagramanova A.A. / Mathematical modeling of the Stirling engine // Scientific and educational and applied journal of higher education news. North Caucasus region. Technical science. No. 4 (192) 2016. p. 29-36.
[8] Kruglov M. G 1977 *The Stirling Engines*. – M.: Mashinostroenie, – p. 150.
[9] Reader G., Hooper P. 1986 Stirling Engines: – M.: Mir, – p. 464.
[10] Frolov K. V. 2010 (chapters. ed.) - *mechanical engineering*. The encyclopedia in forty volumes (24 volumes). – M.: mechanical engineering, – p. 458.

IOP Conf. Series: Earth and Environmental Science **288** (2019) 012085 doi:10.1088/1755-1315/288/1/012085

Mathematical modeling of Stirling engine operating modes as part of an autonomous power complex based on renewable energy sources

Aleksandr Makarovich Gaponenko[1], Aleksandra Aleksandrovna Kagramanova[1]

[1]Kuban State Technological University, Krasnodar, Russia

kaa_200110@mail.ru

Abstract. The opportunity of application of Stirling engine with non-conventional and renewable sources of energy. The advantage of such use. The resulting expression for the thermal efficiency of the Stirling engine. It is shown that the work per cycle is proportional to the quantity of matter, and hence the pressure of the working fluid, the temperature difference and, to a lesser extent, depends on the expansion coefficient; efficiency of ideal Stirling cycle coincides with the efficiency of an ideal engine working on the Carnot cycle, which distinguishes a Stirling cycle from the cycles of Otto and Diesel underlying engine. It has been established that the four input parameters, the only parameter which can be easily changed during operation, and which effectively affects the operation of the engine is the phase difference. Dependence of work per cycle of the phase difference, called the phase characteristic, visually illustrates mode of operation of Stirling engine. The mathematical model of the cycle of Schmidt and the analysis of operation of Stirling engine in the approach of Schmidt with the aid of numerical analysis. To conduct numerical experiments designed program feature in the language MathLab. The results of numerical experiments are illustrated by graphical charts.

1. Overview of possible applications and advantages of the Stirling engine

The objects of exploration, development of oil and gas wells, oil and gas extraction are often located in remote, inaccessible places far from electrical networks and highways. Autonomous power supply of these objects is currently carried out by diesel power plants, which is associated with a necessary supply of fuel and the associated large overhead. In this regard, of particular relevance is the use of non-renewable sources of energy (NRSE).

As such is known: wind energy, geothermal heat, solar radiation absorbed by either using solar panels in the form of electricity, or converted into thermal energy through solar collectors. This group should also include local energy sources such as energy of associated gas, peat, firewood, coal, coal shale, etc. Because of its specificity to ensure reliable energy, these sources must be integrated in the local energy sector, which should provide the necessary heat for heating dwellings and places of residence of staff and electricity needed for appliances and installations. Previous research of the authors [1-4] have shown the importance of optimal composition of the energy complex on the basis of RES. As part of the local energy mix should include a Converter of thermal energy into electrical energy, has a number of properties: versatility of consumed energy, great lifespan, no need regular maintenance, reliability, low noise, high power density. The only, currently known, such transducer is the generator of a Stirling is a heat engine in which the generator is combined with a Stirling engine.

Stirling engine (DS) – refers to internal combustion engines, more specifically an external supply of heat. For work DS, you only need to provide its heat exchangers the temperature difference. The heater can use the heat of the burner, steam or water heated in the geothermal or solar collectors. It is possible to sum directly the solar energy from the solar concentrator. For these cases, provided the special design of the DS with a quartz head, through which concentrated solar radiation enters directly into the expansion chamber. For these engines you can even use geothermal water pumped directly from the well, or associated oil gas, which can have high corrosion aggressiveness. As refrigerator you can use the environment, chilled water, ice or snow.

2. The principle of operation of the Stirling engine. The classic Stirling cycle. Cycle Schmidt

Stirling engine [5-9] works in a closed cycle in which the working fluid expands at higher temperatures and contracts at much lower, which ensures the positive work over the cycle.

A prototype of the real cycle is the ideal Stirling cycle consists of two isotherms and two isochor. A Stirling machine can operate in the forward loop (in this case it is the engine) and selling (in this case, it is a refrigeration machine). The expansion of the working fluid occurs at a higher temperature of the heater T1 and contraction at refrigerator temperature T2 isothermal processes. The transition of the working fluid between these temperatures is carried out during the two isochoric.

We obtain an expression for cycle efficiency. The working fluid receives heat during the isochoric heating and the isothermal expansion:

$$Q_н = q_1 + Q_1 = \nu \cdot C_V \cdot (T_1 - T_2) + \nu \cdot R \cdot T_1 \cdot \ln \frac{V_2}{V_1} \qquad (1)$$

and gives during the isochoric cooling and isothermal compression:

$$Q_x = q_2 + Q_2 = \nu \cdot C_V \cdot (T_1 - T_2) + \nu \cdot R \cdot T_2 \cdot \ln \frac{V_2}{V_1}. \qquad (2)$$

Here:

Q_1 – the amount of heat received working fluid with isothermal expansion;
q_1 – the amount of heat received working fluid during isochoric heating;
Q_2 – the amount of heat received by the worker body in an isothermal compression;
q_2 – the amount of heat received working fluid during isochoric cooling;
ν is the number of moles of a substance, C_V is the molar heat capacity at constant volume.
Work done per cycle is equal to

$$A = Q_1 - Q_2 = \nu \cdot R \cdot \ln \frac{V_2}{V_1} \cdot (T_1 - T_2) \qquad (3)$$

Thermal efficiency is

$$\eta = \frac{A}{Q_н} = \frac{\nu \cdot R \cdot \ln \frac{V_2}{V_1} \cdot (T_1 - T_2)}{\nu \cdot R \cdot T_2 \cdot \ln \frac{V_2}{V_1} \cdot T_1 + q_1} \qquad (4)$$

Stirling has offered warmth q_1получать not from the heater, and regenerator in which the heat enters into the heat exchange during the isochoric cooling of the gas. In this case, q_1 in the denominator disappears and the formula (4) has the form:

$$\eta = \frac{T_1 - T_2}{T_1}, \tag{5}$$

i.e., coincides with the efficiency of a perfect engine. From this we can draw the following important conclusions:

The work per cycle is proportional to the quantity of matter, and hence the pressure of the working fluid at ambient temperature, the temperature difference and, to a lesser extent, depends on the expansion coefficient V_2/V_1.

Efficiency of the ideal Stirling cycle coincides with the efficiency of an ideal engine working on the Carnot cycle, which distinguishes a Stirling cycle from the cycles of Otto and Diesel the underlying engine.

Schmidt, using the representation of harmonic motion of the pistons and rest of the nodes, has left unchanged the other assumptions [10]: the equality of the instantaneous pressure at any point in the volume, ideal regeneration, isterichnost expansion and contraction, the instantaneous change of the gas temperature entering the corresponding cell (perfect heat transfer), the ideal gas as the working fluid. Thus, the scheme of Schmidt, although it is close, certainly more realistic than the ideal Stirling cycle.

Let us turn to the derivation of the mathematical model of the cycle of Schmidt.

We assume that the working fluid is a perfect gas, obeying the equation Mendeleev-Clapeyron $P \cdot V = \nu \cdot R \cdot T$ how do we obtain an expression for the number of moles of the substance $\nu = \frac{P \cdot V}{R \cdot T}$ Given that all the gas is distributed in three volumes: volume V_E being at a temperature T_1, volume compression V_c, at a temperature of T_2 and dead volume V_D at a temperature of $T_D = \frac{T_1 + T_2}{2}$, express the total amount of substance:

$$\nu = \nu_1 + \nu_2 + \nu_3 = \frac{P}{R} \cdot \left(\frac{V_E}{T_1} + \frac{V_c}{T_2} + \frac{V_D}{T_D} \right), \tag{6}$$

from which we obtain the expression for the pressure:

$$P = \frac{\nu \cdot R}{\frac{V_E}{T_1} + \frac{V_c}{T_2} + \frac{V_D}{T_D}}. \tag{7}$$

In line with Schmidt's assumption of harmonic motion, consider the dependence of the volume expansion and contraction of the angle of rotation of the shaft, which is counted relative to the top dead center of the expansion piston:

$$V_E = \frac{V_0}{2} \cdot (1 + \cos\varphi); \; V_c = \frac{k \cdot V_0}{2} \cdot (1 + \cos(\varphi - \delta)),$$

where V_0 – the amplitude of volume V_E,

φ – the rotation angle of the crankshaft (phase process),

δ – the gap phase of the piston compression from the piston extension.

Also: $\tau = \frac{T_1}{T_2}$; $V_D = X \cdot V_0$; $T_D = \frac{1}{2} \cdot (T_1 + T_2)$. The expression for the pressure takes the following form:

$$P = \frac{\nu \cdot R \cdot T_2}{V_0} \cdot \frac{2}{\frac{1}{\tau} \cdot \cos\varphi + k \cdot \cos(\varphi - \delta) + \frac{2x}{(1 + \tau)}}. \tag{8}$$

The work in gas expansion is equal to

$$A_E = \oint P_E \cdot d \cdot V_E = -\nu \cdot R \cdot T_2 \oint \frac{\sin\varphi \cdot d \cdot \varphi}{\frac{1}{\tau} \cdot \cos\varphi + k \cdot \cos(\varphi - \delta) + \frac{2x}{(1+\tau)}} \qquad (9)$$

Work under the pressure

$$A_C = \oint P_c \cdot d \cdot V_c = -\nu \cdot R \cdot T_2 \oint \frac{k \cdot \sin\varphi \cdot d \cdot \varphi}{\frac{1}{\tau} \cdot \cos\varphi + k \cdot \cos(\varphi - \delta) + \frac{2x}{(1+\tau)}}. \qquad (10)$$

Here the symbol $\oint$ indicates integration cycle: $\oint \quad = \int_{\varphi=0}^{2\pi}$

Total work per cycle is equal to

$$A = A_E + A_c \qquad (11)$$

Performing the numerical integration in the above formulas, you can make a feature that allows numerically, according to the set of input parameters, the value of work per cycle. Such function was created on the MATHLAB language and used to analyze the operation of the Stirling engine in the approach of Schmidt.

3. Numerical analysis of cycle DS in the approach of Schmidt

The purpose of the analysis is to identify the dependence of the dimensionless parameters that characterize the performance of a loop $\left(\frac{P}{P_0}, \frac{V_1}{V_0}, \frac{V_2}{V_0}, \frac{A}{\nu RT_0}, \frac{Q_1}{\nu RT_0}, \frac{Q_2}{\nu RT_0}\right)$ from the dimensionless parameters characterizing the design features of the process, its mode, or as τ, k, X, δ. For the numerical analysis, a program is a function of the language MATHLAB. Using the program, you can analyze the work of the DS to determine the effect of various design parameters on its performance and determine the appropriate combination of parameters that optimizes the performance of the engine over the cycle, as will be shown below.

The Schmidt cycle

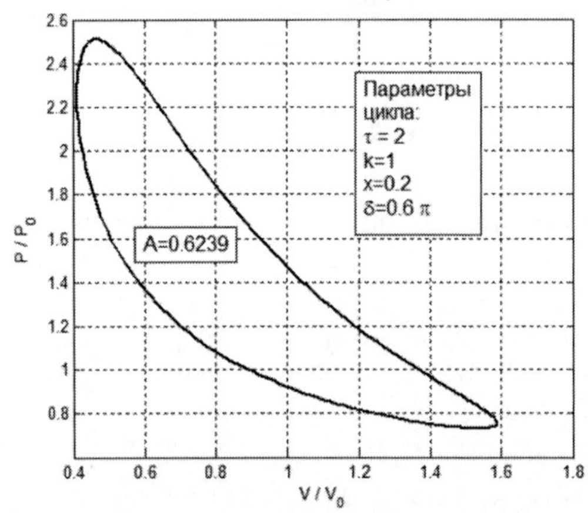

Figure 1 – Form cycle and the value of the dimensionless work for certain values of the parameters

The calculations showed that in the perfect work of the regenerator efficiency of the cycle coincides with the Schmidt DS or efficiency of the Carnot cycle.

Dimensionless work per cycle $\frac{A}{\nu \cdot R \cdot T_2}$ depends on four dimensionless parameters: δ,τ,k,X. Especially noticeable influence of the phase difference δ, so that its value can be controlled by the output of a heat engine: if $0 < \delta < \pi$, $A > 0$, that is, the system operates as an engine, with $\pi < \delta < 2\pi$, $A < 0$ as a refrigerating machine. Let us call the dependence of the dimensionless work from the phase difference between the phase characteristics. The maximum value of work the value of δ, which begins high, as does the kind of phase characteristics depend on the values of the three remaining parameters. Therefore, to characterize the influence of these parameters, it is convenient to represent in the graph the family of phase curves depending on any one of the three parameters at constant values of the other two. The following graph shows the family of phase characteristics at various values of the parameter k .

Conclusions:

As part of the local energy mix should include a Converter of thermal energy into electrical energy - generator Stirling – compact generator, combined with a Stirling engine (DS) have a number of properties: versatility of consumed energy, great lifespan, no need regular maintenance, reliability, low noise, high power density.

It is shown that with perfect regeneration and no heat losses, the efficiency of DS coincides with the efficiency of a Carnot cycle, which is the maximum possible.

It is shown that the work per cycle is proportional to the amount of substance of the working fluid, and thereby the pressure at ambient temperature.

A mathematical model of the cycle of Schmidt on the basis of which is composed of a program written in MATHLAB that enables four input parameters to analyze the work of DS to determine a suitable combination of design parameters and their influence on the efficiency of the engine. The results of such research.

Found that of the four input parameters, the only parameter that can be easily changed during operation and which effectively affects the operation of the engine, is the phase difference δ. The dependence of the cycle work from δ, called the work phase characteristics, clearly illustrates the mode of operation of the DS. Given families of phase characteristics for different modes of operation.

References:

[1] Gaponenko A.M., Kagramanova A.A. Mathematical modeling of the Stirling engine in terms of applying the composition of the power complex containing non-conventional and renewable energy / International conference "Modern problems of thermal physics and energy" 9-11 October 2017 in Moscow on the basis of the National research University " Moscow energy Institute // Journal of Physics: Conference Seriees, Volume 891, conference 1 – Режим доступа: http://doi.org/10.1088/1742-6596/891/1/012244

[2] Gaponenko A.M., Kagramanova A.A. 2016 Optimization of the composition of the complex of renewable energy sources with the use of cluster approach and the theory of random processes. *Polythematic online scientific journal of kuban state agrarian university-Krasnodar, KubSAU,* – №02(116). – IDA [article ID]: 1161602005. – Режим доступа: http://ej.kubagro.ru/2016/02/pdf/05.pdf p. 94 - 109.

[3] Gaponenko A.M., Kagramanova A.A. 2016 Methodology and algorithm of mathematical analysis multifactor models optimal configuration of renewable energy sources. *International research journal.* №4(46). Vol. II. Ekaterinburg. p. 76-79.

[4] Gaponenko A.M., Kagramanova A.A. 2016 Method of determination of optimal parameters of energy efficient of power supply systems. *International research journal.* №4(46). Vol. II. Ekaterinburg. p. 80-84.

[5] Smirnov G.V. 1967 *External combustion engine.* "Knowledge", M., – p. 486.

[6] Walker G. 1978 *Machine operating on the Stirling cycle*: – M.: Energy, – p. 152.

[7] Walker G. 1985 *Stirling Engines*: – M.: Mashinostroenie, – 408 S.

[8] Kruglov M. G 1977 *The Stirling Engines*. – M.: Mashinostroenie, – p. 150.

[9] Reader G., Hooper P. 1986 Stirling Engines: – M.: Mir, – p. 464.

[10] Frolov K. V. 2010 (chapters. ed.) - *mechanical engineering*. The encyclopedia in forty volumes (24 volumes). – M.: mechanical engineering, – p. 458.

IOP Conf. Series: Earth and Environmental Science **288** (2019) 012086 doi:10.1088/1755-1315/288/1/012086

Mixing intensification

L.R. Gaynullina, V.P. Tutubalina

Kazan State Power Engineering University, 51 Krasnoselskaya St, Kazan 420066, Russia

gainullina7819@mail.ru

Abstract. Intensification of different system mixing process, particularly in non-isothermal conditions, is a critical task for many industries, where local overheating prevention, chemical reaction acceleration or heat and mass transfer coefficient increase are required. In this regard, an experimental study of the mixing process was carried out using the transformer oil as an example. Temperature equalization time in the fluid and the mixer power consumption were used as indicators of the process efficiency. The influence of the mixer type, presence of bafflers in a vessel, and the mixer speed on the fluid mixing intensity revealed the optimal design and process parameters of the mixing units.

Introduction

In the industries such as power engineering and chemical industry, many technologies are associated with the mixing process. For different systems, especially in non-isothermal conditions, there is a matter of the mixing intensity increase that is one of the most important ways to increase the process efficiency.

The mixing process intensification enables the most qualitative chemical reaction with its acceleration, reduces the probability of local overheating of the reaction mass, increases the heat and mass transfer coefficient, etc. [1, 2]

Relation between the mixing process and hydrodynamic factors necessitates the study of hydrodynamic features of mixing, which in turn depend on the unit design, mixer profile and process parameters. One of these parameters is the fluid flow condition [3]. In laminar flow, the velocities at each point of the flow have a constant value and direction. In turbulent flow, the deformation structure of the flow is characterized by a velocity pulsation at each point of the flow, i.e. velocity variation in time at each point. Turbulent pulsations are defined not only by the velocity, but also by the turbulence scale, i.e. the mixing path length. The larger the pulsation rate, the larger the motion scale [4-6].

In general, the turbulence is formed at the fluid flow boundaries due to the interaction of the mixer with the mixed fluid and the interaction of moving flows with fixed parts of the unit, including its walls. Thus, local differences in the mixed medium structure and turbulence decay depend on the location of the turbulence sources, which in turn should be taken into account when choosing the mixer type.

The mixing efficiency is characterized by the achieved degree of homogeneity of the mixed fluid. The amount of power consumed per unit volume can be taken as a measure of the mixing intensity [7].

Experimental procedure

Experimental study of the mixing process was carried out using the transformer oil.

IOP Conf. Series: Earth and Environmental Science **288** (2019) 012086 doi:10.1088/1755-1315/288/1/012086

Different types of mixers were tested during the study: 3-blade propeller, open turbine mixer, 2-blade mixer with blades at an angle of 45° to the horizon, 2-, 3-, 4-, 6- and 8-radial-blade mixers. Mixer diameter is 75, 100, 125 and 150 mm. During the test, a thermal pulse with a power of 2.5 kJ/s was fed to a stationary fluid in the units using a heater. After 30 seconds, the heater was switched off and the mechanical mixer was rotated. The temperatures at the lower and upper points of the mixed transformer oil was measured at a given mixer speed. Temperature equalization in the upper and lower points of the mixed fluid was regarded as the end of the mixing process. The temperature equalization time, the mixer speed, the power consumed by the mixer, the thermal pulse power, the thermal pulse duration were measured during the tests.

The first series of tests was carried out to study the mixing efficiency under thermal pulse conditions, which was determined by the temperature equalization time and the consumed process power. The fluids under study were mixed using different types of mixers in a smooth-wall vessel and in a unit with four baffles. The mixer speed was n = 100 rpm. The results are presented in Tables 1 and 2.

Table 1. Comparative efficiency of mixers when mixing the transformer oil in a smooth-wall vessel (n = 100 rpm, D = 200 mm)

No.	Mixer type (dimensions, mm)	N, W	τ, sec	$N\tau$, J	Comparative efficiency with respect to	
					τ	$N\tau$
1	3-blade propeller d = 75	$1.09\cdot10^{-3}$	11.2	$7.31\cdot10^{-1}$	1.87	0.18
2	3-blade propeller d = 100	$1.14\cdot10^{-2}$	6.0	4.10	1.0	1.0
3	3-blade propeller d = 150	$0.76\cdot10^{-1}$	2.4	11.08	0.40	2.70
4	Open turbine d = 75	$5.45\cdot10^{-3}$	8.8	2.20	1.47	0.54
5	Open turbine d = 100	$5.63\cdot10^{-2}$	4.4	14.81	0.74	3.62
6	Open turbine d = 150	$3.81\cdot10^{-1}$	1.6	36.59	0.27	8.92
7	2-blade mixer d = 100, h = 40	$4.14\cdot10^{-2}$	9.0	22.37	1.50	5.46
8	2-blade mixer with blades at 45° d = 100, h = 20	$3.10\cdot10^{-2}$	10.0	18.54	1.67	4.53
9	4-blade mixer d = 100, h = 20	$5.76\cdot10^{-2}$	7.5	25.51	1.25	6.23
10	6-blade mixer d = 100, h = 20	$6.0\cdot10^{-2}$	7.0	25.11	1.17	6.17

Table 2. Comparative efficiency of mixers when mixing the transformer oil with 4 baffles (n = 100 rpm, D = 200 mm)

No.	Mixer type (dimensions, mm)	N, W	τ, sec	$N\tau$, J	Comparative efficiency with respect to	
					τ	$N\tau$
1	2	3	4	5	6	7
1	3-blade propeller, d = 75	$3.78\cdot10^{-3}$	10.2	2.35	2.12	0.35
2	3-blade propeller, d = 100	$2.29\cdot10^{-2}$	4.8	6.67	1.0	1.0
3	3-blade propeller, d = 150	$1.29\cdot10^{-1}$	2.0	15.50	0.42	2.34
4	Open turbine, d = 75	$18.98\cdot10^{-3}$	8.2	9.32	1.71	1.41

Continuation of table 2

1	2	3	4	5	6	7
5	Open turbine, d = 100	$11.31 \cdot 10^{-2}$	3.4	23.15	0.71	3.51
6	Open turbine, d = 150	$6.43 \cdot 10^{-1}$	1.4	56.11	0.29	8.45
7	2-blade mixer, d = 100, h = 40	$13.73 \cdot 10^{-2}$	8.0	65.82	1.67	9.89
8	2-blade mixer with blades at 45°, d = 100, h = 20	$4.12 \cdot 10^{-2}$	9.3	22.95	1.94	3.45
9	4-blade mixer, d = 100, h = 20	$7.65 \cdot 10^{-2}$	6.0	27.57	1.25	5.85
10	6-blade mixer, d = 100, h = 20	$5.10 \cdot 10^{-2}$	4.1	13.24	0.85	1.98

Obtained test data (Table 1 and 2) revealed the effect of two basic factors on the mixer performance: temperature equalization time (τ) and energy ($N\tau$) consumed on fluid mixing. Each of these factors taken separately can not be used to evaluate the mixer work. Together, the latter ones allow a quite complete assessment of the mixing efficiency of each mixer.

Tables 1 and 2 show that the open turbine with a diameter d = 150 mm is the most advantageous to obtain the minimum temperature equalization time in the turbulent region. However, this mixer consumes 9 times more power than a 3-blade propeller with d = 100 mm. Therefore, when comparing the mixer efficiency per unit the parameters of a 3-blade propeller with d = 100 mm were taken.

The results presented in Tables 1 and 2 show that the most effective mixers are a 3-blade propeller and an open turbine mixer. 4- and 6-radial-blade mixers with relatively good mixing quality consume more power. 2-radial-blade mixers provide insufficient mixing quality. 2-blade mixer with blades at an angle of 45° gives poor mixing quality. Baffles also reduce the temperature equalization time by 9-40 %, depending on the mixer type.

The following series of tests was carried out to study the mixer speed effect on the temperature equalization time, when mixing the transformer oil in the unit with 4 baffles (Fig.1) and in a smooth-wall unit (Fig.2).

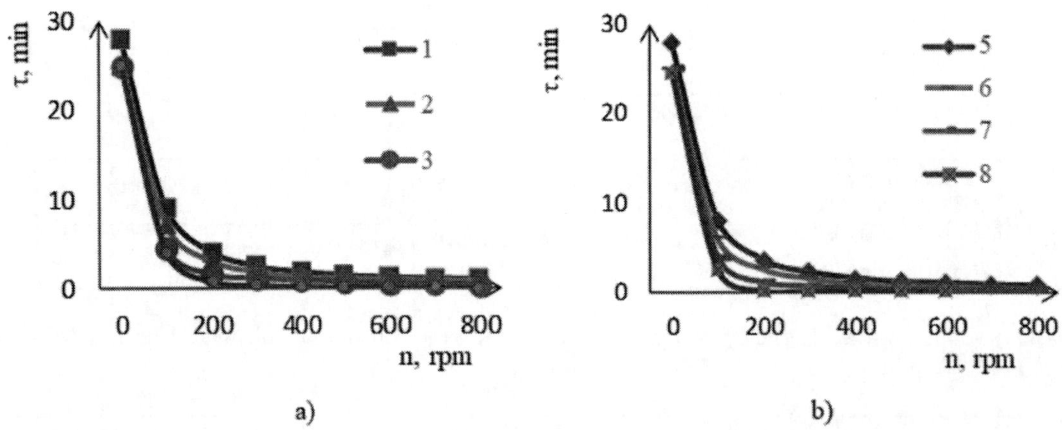

Fig. 1.* Dependence of the temperature equalization time on the mixer speed during transformer oil mixing in the unit with 4 baffles 1 – 3-blade propeller, d = 75 mm; 2 – 3-blade propeller, d = 100 mm; 3 – 3-blade propeller, d = 150 mm; 4 – 4-blade mixer, d = 100 mm; 5 – open turbine, d = 75 mm; 6 – open turbine, d = 100 mm; 7 – open turbine, d = 150 mm; 8 – 6-blade mixer, d = 100 mm, h = 20 mm.

*in all subsequent figures, the mixer types have same indication, unless otherwise stated

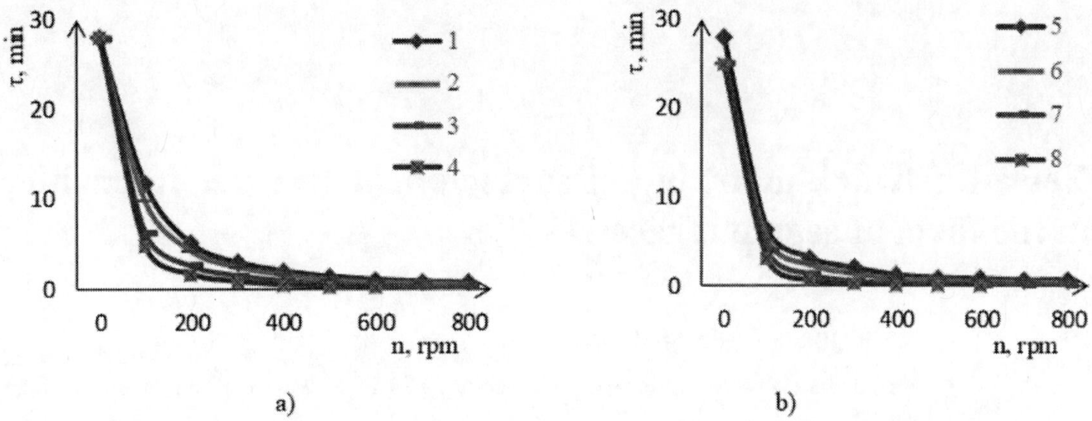

Fig. 2. Dependence of the temperature equalization time on the mixer speed during transformer oil mixing in a smooth-wall unit

In accordance with the test data shown in Figures 1 and 2, the temperature equalization time decreases with the increase in the mixer speed, by one pattern for all types of mixers both in the smooth-wall unit and in the baffled unit. The speed increase above 300 rpm does not affect the mixing process and, therefore, is impractical, since it causes an increase in power consumption. The results of experimental studies presented in Figures 1 and 2 confirm the efficiency of a 3-blade propeller and an open turbine mixer.

Conclusions

Tests determining the effect of the mixer types in smooth-wall and baffled vessels and the mixer speed on the temperature equalization time were carried out in order to increase the mixing efficiency and, as a result, accelerate the processes in transformer oil.

Studies have shown that the most effective mixers used in the tests were a 3-blade propeller with d = 150 mm and an open turbine mixer with d = 150 mm. With the best mixing quality, the time required for the temperature equalizing at the upper and lower points of vessels and the power were minimal. The optimal speed, at which the mixing efficiency increases, is 300 rpm. Further speed increase does not affect the process.

References

[1] Simulation and optimization of the chemical technology processes and systems, RASHHI-2016: *Russian-American Scientific School-Conference (May 23-25, 2016): Collection of abstracts* (Kazan: Publisher KNITU) 276 p

[2] Kovalnogov N N Magazinnik L M Fedorov R V Korolev A V 2007 *High-speed turbulent dispersed flow heat emission* (Bulletin of Ulyanovsk State Technical University) No. 1, 51-55

[3] Jaworski Z 2005 *Computational Fluid Dynamics in Chemical and Process Engineering (in Polish)* (Warszawa: Akademicka Oficyna Wydawnicza EXIT)

[4] Ashley G Smart *Exploring the extremes of turbulence* 2001 (Physics Today) 64 18 pp 18–20

[5] Adamiak I Jaworski Z 2004 *A study of turbulent flows in the Kenics static mixer* (Inz. Chem. Proc) 25 № 3/1 pp 535–541

[6] Dennis D J C F M Sogaro 2014 *Distinct Organizational States of Fully Developed Turbulent Pipe Flow* (Phys. Rev) Lett 113 234501

[7] Tutubalina V P *Fluid mixing in the vessel with a mixer* 1988 (CNIIT Neftekhim) 29 p

IOP Conf. Series: Earth and Environmental Science **288** (2019) 012087 doi:10.1088/1755-1315/288/1/012087

Neural network modeling of surface heat transfer intensifiers in the form of segment recesses

K Kh Gilfanov, R A Shakirov

Kazan State Power Engineering University, Krasnoselskaya str., 51, 420066, Kazan, Russia

shakirov.aiv@yandex.ru

Abstract. The results of neural network modeling of surface heat transfer intensifiers in the form of spherical recesses are presented, based on experimental data. The possibility and prospects of building artificial neural networks for modeling the characteristics of heat exchange surfaces are shown.

Introduction

Currently, all the generated thermal energy before its use is repeatedly converted into various heat exchangers. Therefore, the efficiency of the entire production depends on the efficiency of heat exchangers. Intensification of heat exchange and increase of energy efficiency of heat exchangers are of great interest and are of great importance for many industries.

Surface heat transfer intensifiers in the form of hemispherical (segmental) recesses and projections are widely used in heat exchange equipment. Interest in this type of intensifiers has increased after the appearance of messages on increasing the heat transfer coefficient while reducing the hydraulic resistance. [1-3] and others. They can significantly improve the efficiency of heat exchangers. It is particularly advantageous to use such heat exchangers in high-energy installations.

The design of intensified heat exchangers with optimal characteristics is complicated by the problem of generalization of research results. The analysis of literature sources shows that it is impossible to generalize the characteristics of surface heat transfer intensifiers by conventional conservation equations due to the complexity of thermal and hydro-mechanical processes. The latter is also due to the large number of design parameters of intensifiers in the form of hemispherical (segment) recesses. If it is possible to generalize the results of the research in the form of empirical formulas of the dependence of the Nusselt numbers on the Reynolds and Prandtl numbers with the involvement of the determining dimensions, then, as a rule, for a narrow class of single recesses or systems of recesses in the range of parameters of the experiments [4].

A certain way out of the situation is offered by artificial intelligence systems capable of learning or self-learning. These are artificial neural networks. Neural network modeling allows to generalize the results of the experiments of complex multiparameter processes, "looking" beyond the range of parameters in which the experiments were conducted. In neural networks, knowledge is contained in the States of many neurons and connections between them. A neural network with a certain number of inputs and outputs is constructed from neurons (nodes) connected in a certain way.

The neural network operation consists of two stages: training the network to "correct" or adequate response to the input information (input vector) and using the trained network to recognize input vectors. The last stage is often called testing. In other words, the network learns to recognize input vectors, i.e. to form output vectors corresponding to the recognized class of input vectors.

When testing (using) a trained neural network, the process of finding the nearest minimum of the objective function occurs. This restores the distorted bits of the input vector or" remembering " the unknown bits associated with the given (known) bits.

Neural network modeling

Artificial neural network is implemented using neuropace "NeuroSolutions". It contains a master of neural network architectures ("NeuralWizard"), which is used to set the architecture, select the training sample, the training criteria. When solving the task we used a version of the program with full functionality

Experimental material was used as a" teacher " [5]. Table 1 presents the relative geometric parameters of the channels and spherical recesses.

Table 1. The relative geometric parameters of the investigated channels and spherical recesses.

№	h	D	H	h/D	H/D	h/H
1	0,00071	0,00514	0,012	0,138	2,334	0,059
2	0,0015	0,00714	0,012	0,21	1,68	0,125
3	0,003	0,00916	0,012	0,327	1,31	0,25
4	0,005	0,01	0,012	0,5	1,2	0,416
5	0,00071	0,00514	0,01	0,138	1,945	0,071
6	0,0015	0,00714	0,01	0,21	1,4	0,15
7	0,003	0,00916	0,01	0,327	1,091	0,3
8	0,005	0,01	0,01	0,5	1	0,5
9	0,00071	0,00514	0,008	0,138	1,556	0,088
10	0,0015	0,00714	0,008	0,21	1,12	0,187
11	0,003	0,00916	0,008	0,327	0,873	0,375
12	0,005	0,01	0,008	0,5	0,8	0,625
13	0,00071	0,00514	0,005	0,138	0,972	0,142
14	0,0015	0,00714	0,005	0,21	0,7	0,3
15	0,003	0,00916	0,005	0,327	0,545	0,6
16	0,005	0,01	0,005	0,5	0,5	1
17	0,00071	0,00514	0,002	0,138	0,389	0,355
18	0,0015	0,00714	0,002	0,21	0,28	0,75
19	0,003	0,00916	0,002	0,327	0,218	1,5
20	0,005	0,01	0,002	0,5	0,2	2,5

Here h is the depth or height of the notch or protrusion, m; D is the diameter of the notch, ledge, m; H is the height of the channel, m.

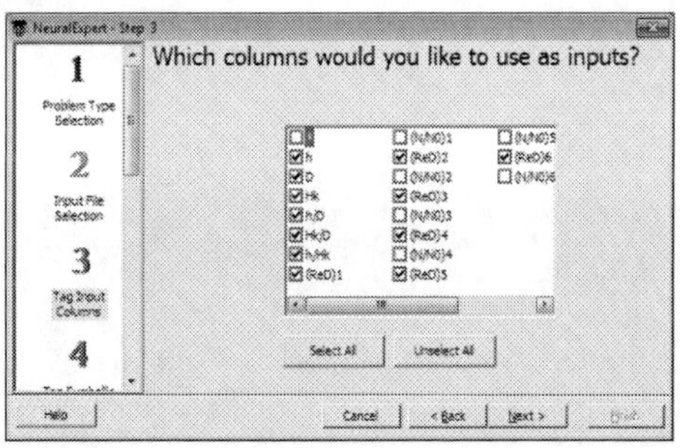

Figure 1. Setting parameters of artificial neural network

IOP Conf. Series: Earth and Environmental Science **288** (2019) 012087 doi:10.1088/1755-1315/288/1/012087

The synthesis and analysis of data on hydraulic resistance and heat transfer carried out in [5] allowed to make a sample of data for neural network modeling presented in table 2.

Table 2. Sampling data for training.

№	ReD	Nu/Nu0	ReD	Nu/Nu0	ReD	Nu/Nu0	ReD	Nu/Nu0	ReD	Nu/Nu0
1	5400	1	7800	1,03	9200	1,06	11000	1,05	15000	1,24
2	7000	1,07	7000	1,1	12000	1,25	16500	1,27	20500	1,23
3	6500	1,8	9000	1,47	12000	1,75	13400	1,67	16500	1,64
4	6200	1,44	9000	1,76	11800	2,1	13000	2,05	16000	2,4
5	3800	0,93	7800	1	9700	1,06	12100	1,07	22600	1,06
6	3500	0,93	5400	1	7000	1,07	9000	1,18	12100	1,1
7	9000	1,7	12000	1,75	13000	1,8	16000	1,76	21000	1,7
8	6200	1,92	9500	2,2	11400	2,25	17000	1,71	20000	2,5
9	5500	0,95	8000	1,04	9700	1,26	14000	1,3	16000	1,28
10	6600	1,6	5400	1,05	7000	1,27	12000	1,8	18000	1,36
11	1900	1	5000	1,5	9000	1,83	12000	1,75	13500	1,8
12	5200	2,1	6600	2,24	9200	2,05	17000	2,7	20200	2,46
13	2500	0,7	4300	0,7	6000	0,83	8500	1	11500	1,02
14	4700	1,3	7000	1,28	10000	1,3	14000	1,3	21000	1,3
15	5900	1,67	7100	1,84	10000	1,95	19000	2,23	21000	2,15
16	4800	2,9	6600	3	9900	3,14	13600	3,3	20500	3,12
17	4300	0,7	5900	0,8	8800	0,93	10200	1	20600	1,02
18	4700	1,6	7400	1,21	10800	1,41	12200	1,36	20000	1,5
19	5500	2,2	7200	2,5	10000	2,94	12400	3,18	20600	3,54
20	4400	5	7200	4,6	10800	5,3	13000	5	20700	4,84

In accordance with tables 1 and 2 is a matrix that perceived neuropace "NeuroSolutions". Next comes the process of setting up the parameters of the artificial neural network, which is shown in figure 1.

After setting up the parameters, specifying the inputs and outputs of the network and the parameters of building the network, the program completes the work by creating a network (not yet trained, just structure). The network structure is shown in figure 2.

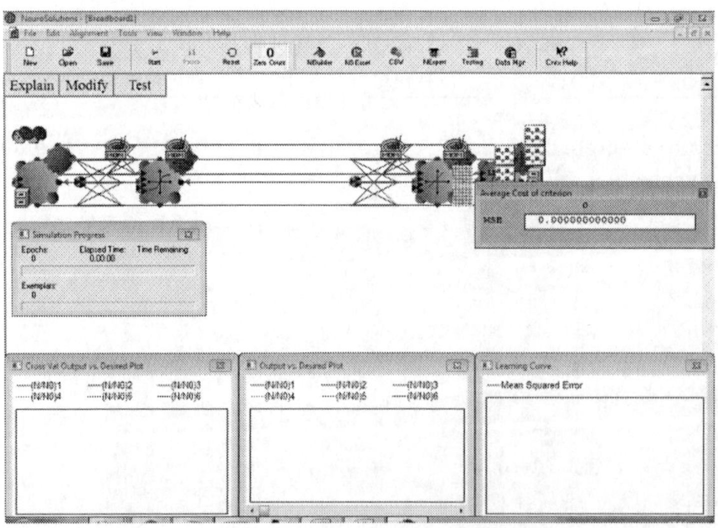

Figure 2. Structure of artificial neural network

Next, go directly to the learning process, figure 3. In the learning process, you can observe a reduction in error, which depends on the number of lessons. This can be clearly seen in figure 4 a. Also in figure 4 b are graphs showing what values gives our history network in the learning process, and what there were values actually. Thus, according to the graphs it is seen that the learning error is small and the results of the network based on the training data are practically the same as the actual values of the sample. This allows us to conclude that the network training was successful, and we can start testing i.e. calculation of the probability with which the neural network will predict the results of the experiment.

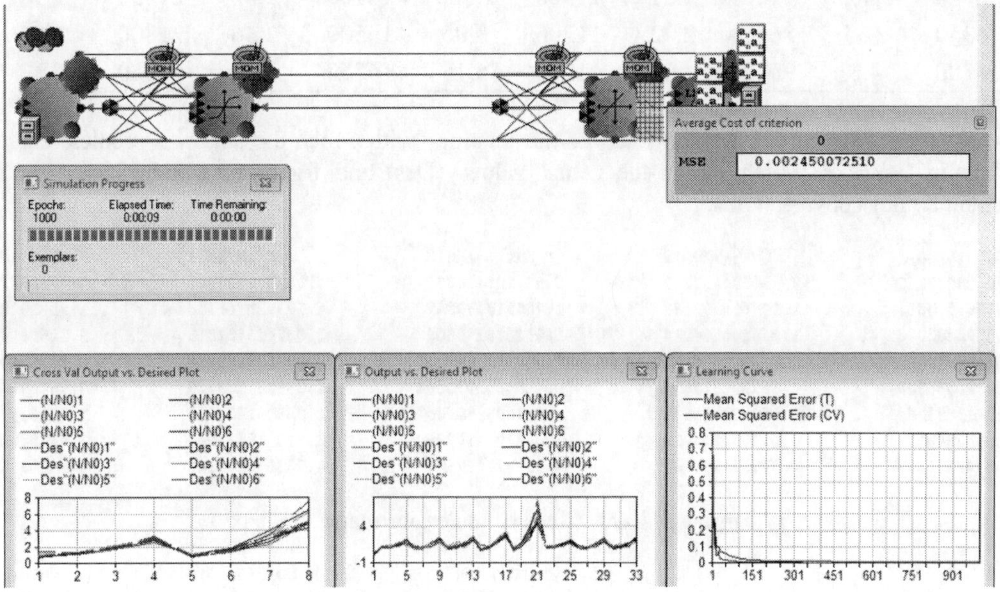

Figure 3. Neural network training

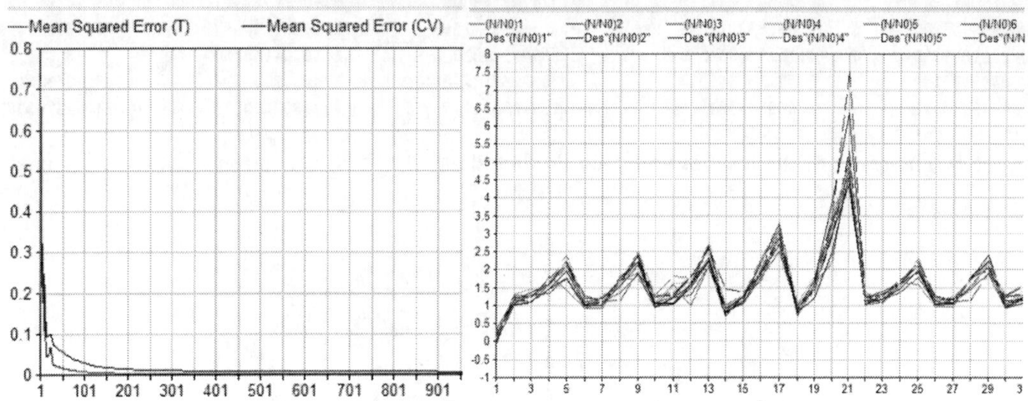

a b

Figure 4. Reduction of error depending on the number of lessons (a) and the schedule of predicted and actual values of the network (b)

As a test matrix, we use a matrix with unknown data that needs to be predicted, these are the data of Nu/Nu0 given in table 3

IOP Conf. Series: Earth and Environmental Science **288** (2019) 012087 doi:10.1088/1755-1315/288/1/012087

Table 3. Sampling data for testing

ReD	Nu/Nu0	ReD	Nu/Nu0	ReD	Nu/Nu0	ReD	Nu/Nu0	ReD	Nu/Nu0
3400	0,7	5150	0,765	7250	0,915	10000	1,01	17300	1,245
5850	1,29	8500	1,29	12000	1,3	17500	1,3	22050	1,335
6500	1,755	8550	1,895	14500	2,09	20000	2,19	22000	2,01
5700	2,95	8250	3,07	11750	3,22	17050	3,21	21650	3,06
5100	0,75	7350	0,865	9500	0,965	15400	1,01	21800	1,01
6050	1,405	9100	1,31	11500	1,385	16100	1,43	21550	1,56
6350	2,35	8600	2,72	11200	3,06	16500	3,36	21800	3,52
5800	4,8	9000	4,95	11900	5,15	16850	4,92	21750	6,17

After testing, the trained network showed the accuracy with which the data was predicted. This can be seen in figure 5, which shows the actual values (Des) and figure 6, which shows the values predicted by our network (Out).

Out (N/N0)1	Out (N/N0)2	Out (N/N0)3	Out (N/N0)4	Out (N/N0)5
0.820804297262	0.790800455221	0.897206329036	0.932779425242	0.878667382542
1.081567570644	1.149665962438	1.285918489947	1.193923961306	1.257582059310
1.665794803076	1.846195866214	1.941538378309	2.037423451818	2.106561486058
2.523158912882	2.775390838768	2.944733166008	2.988167473216	3.141591027704
0.951020036206	0.852894973636	0.968565377256	1.093561509439	0.964190719959
1.542711859785	1.552975801569	1.667093380305	1.859041654399	1.782034306406
3.058347648152	3.072930665914	3.383723411228	3.410318637966	3.436536102140
4.798449575329	4.389944988261	5.136820582679	4.783576601381	4.742795581283

Figure 5. Real values of the experimental matrix

Out (N/N0)1	Out (N/N0)2	Out (N/N0)3	Out (N/N0)4	Out (N/N0)5
0.820804297262	0.790800455221	0.897206329036	0.932779425242	0.878667382542
1.081567570644	1.149665962438	1.285918489947	1.193923961306	1.257582059310
1.665794803076	1.846195866214	1.941538378309	2.037423451818	2.106561486058
2.523158912882	2.775390838768	2.944733166008	2.988167473216	3.141591027704
0.951020036206	0.852894973636	0.968565377256	1.093561509439	0.964190719959
1.542711859785	1.552975801569	1.667093380305	1.859041654399	1.782034306406
3.058347648152	3.072930665914	3.383723411228	3.410318637966	3.436536102140
4.798449575329	4.389944988261	5.136820582679	4.783576601381	4.742795581283

Figure 6. The result of the test network (the Predicted value) Thus, we can see that the error is about 0.2 %.

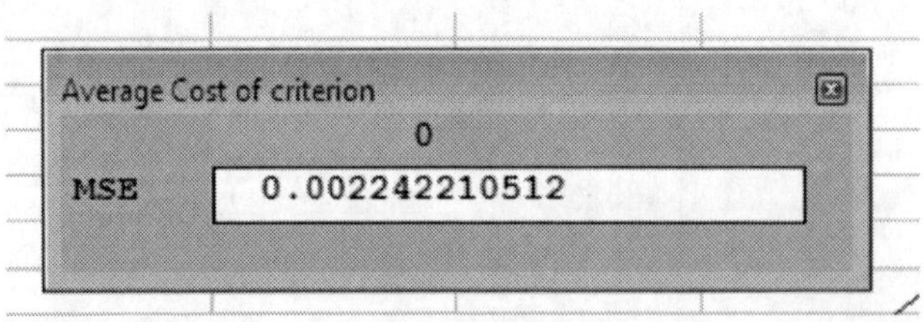

Figure 7. Network prediction error

Conclusion

Thus, the possibility of constructing artificial neural networks for modeling surface heat transfer intensifiers in the form of spherical recesses is presented. Testing of the neural network has shown an error of modeling of 0.2 %, which can be considered satisfactory, given the spread in the sample of data related to the error of the experiments. It should also be noted that for the practical use of the simulation results it is necessary to expand the range of data on the surface heat transfer intensifiers.

References

[1] Shanin Yu I, Shanin O I 2004 Intensification of heat transfer by application of spherical holes on the channel walls *Convective heat and mass transfer* (Minsk: ANB)
[2] Borisov I, Khalatov A, Kobzar S, Glezer B 2004 Comparison of thermal hydraulic characteristics for two types of dimpled surfaces ASME Paper № GT2004-54204
[3] Isaev S A 2016 Numerical simulation of the intensification of the heat exchange in a planeparallel channel with a cylindrical shallow dimple on the heated ngineering *Physics and Thermophysics* vol 89 № 5 pp 1186-1201
[4] Shchelchkov A V 2016 Intensification of vortex streams in the flow around thick lifting surfaces and due to creation of oval dimpled reliefs *7th International Conference on Vortex Flows and Vortex Models* (Rostock: *ICVFM*) p 11
[5] Gortyshov Yu F, Popov I A, Clicks A B, Ryzhkov D V 2008 Heat-Hydraulic characteristics of heat exchangers with surface intensification of heat exchange in the form of spherical recesses and protrusions (Moscow: MEI)

IOP Conf. Series: Earth and Environmental Science **288** (2019) 012088 doi:10.1088/1755-1315/288/1/012088

Thermal computer simulation of the power transformer TM-160/10

Gilfanov K.H., Nguyen Tien

Flat 71, 9 Daurskaya St. Kazan, RF 420087

kamil.gilfanov@yandex.ru

Annotation. Power transformers represent the largest part of capital investments in transmission and distribution of substations. In addition, disconnections of the power transformer cause significant economic damage to the electrical network. One of the most important parameters that determine the resource of a transformer is the temperature of the hottest point, which is determined by the density of the heat fluxes. An important task is to improve the quality of transformers, develop the most advanced technologies for their production, create and use the most modern and highly efficient materials, and reduce energy losses during their operation. The solution of these problems is impossible without deep understanding and study of the thermal regimes of the transformer.

Introduction

Power transformers represent the largest part of capital investments in transmission and distribution of substations. In addition, disconnections of the power transformer cause significant economic damage to the electrical network. One of the most important parameters that determine the resource of a transformer is the temperature of the hottest point, which is determined by the density of the heat fluxes. It was shown in [1-3] that the maximum temperature is a function of loads, overloads, cooling regimes and time. An important task is to improve the quality of transformers, develop the most advanced technologies for their production, create and use the most modern and highly efficient materials, and reduce energy losses in their work [4]. The solution of these problems is impossible without deep understanding and study of the thermal regimes of the transformer.

The increased loads of the transformer lead to operating modes with temperatures exceeded permissible values. This leads to the deterioration of insulation, decreases of the service life and reliability. Protecting the transformer from overheating due to unacceptable operating temperatures is associated either with the need to reduce the load, that is a violation of its basic function, or with the use of cooling boosting means.

The purpose of this work is to simulate the thermal conditions of the transformer, to study the distribution of heat fluxes and temperatures in the transformer under various cooling regimes and power losses.

In an according with the current standards the temperature difference of any part of the transformer is the temperature difference between this part and the cooling medium. Permissible excess temperatures of the active part of the transformer affect the internal device, dimensions, load capacity, cost and operating modes of the transformer. Control over the conformity of the temperature of the most heated point of the transformer to the permissible temperature is difficult for transformers, so its determination by experimental methods or by means of modeling is an actual task.

Calculation of power oil-filled transformer parameters

The core oil-filled three-phase transformer TM-160/10 was chosen as an object of investigation. The values of its parameters are presented in Table 1. Calculation of the main electrical quantities and geometric dimensions of the transformer, which are necessary for modeling, was performed according to the methods [5-6]. As high voltage (HV) and low voltage (LV) windings, cylindrical multilayer windings of round wire are adopted.

For the test voltage HV U_{tv1} = 35 kV, the insulation distances [7] are determined: a_{12} = 9 mm, δ_{12} = 30 mm, a_{22} = 10 mm, δ_{12} = 3 mm (Fig. 1).

Table 1. Electrical characteristics of transformer TM 160/10

Transformer type	TM-160/10
Power	160 kVA
Frequency	50 Hz
Number of phases	3
Primary voltage	(10000 ± 3×2%) V
Secondary voltage U_2	3150 V
Cooling system	oil free
Scheme and group of connection of windings	Y/Y
Short-circuit voltage	4.5 %
No-load current	2.4 %
Loss of idling	510 W
Loss of short circuit	2650 W

The core design of the transformer with the number 6, k_{cr} = 0.913 without the pressing plate was chosen. At power of 160 kVA cooling channels in the magnetic circuit are not provided. As the material of the magnetic system of the transformer, cold-rolled electrical steel grade 3404 with a density of 7650 kg/m^3 is adopted. Diameter of core rod d = 0.150 mm. The average diameter of the channel between the windings d_{12} = 0.227 m.

The winding height is l = 0.263 m. The active cross-section of the rod is S_c = 0.0153 m^2. The number of turns of LV windings per one phase is $w2$ = 327. The winding wire is with a diameter of 3.55 mm with a cross-section of 9.895 mm^2. The number of turns in the layer w_l2 = 65.48. The number of layers in the winding n_l2 = 5. The dimensions of the winding LV: internal diameter $D2_{min}$ = 180 mm, outer diameter $D2_{max}$ = 221 mm. The surface area of the LV winding is $S2$ = 0.324 mm^2. Loss of short circuit $P_{sc}2$ = 382.78 W, weight of wire of winding LV $G2$ = 18.16 kg. The heat flux density on the surface of the LV winding $q2$ = 1.183 kW/m^2. The number of turns of the HV winding at the rated voltage $wn2$ = 1038. The number of turns at one stage of the voltage winding regulation when connecting the HV winding to the star wp = 36. A multilayer cylindrical winding made of round copper wire of diameter $d1$ = 1.9 mm, section 3.14 mm^2. The number of turns in one layer is w_l1 = 109. The number of layers is n_l1 = 9.58. The internal diameter of the winding is «before- channel» $D1_{in}$ = 239 mm, «after-channel» $D1_{min}511$ = 269 mm. The outer diameter of the winding is «before-channel» $D1_{max}14$ = 261 mm, «after-channel» $D1_{max}$ = 307 mm. The total cooled surface of the winding HV $S1$ = 0.888 mm^2. Short-circuit losses $P_{sc}1$ = 514.96 W. The wire weight of the «before-channel» part of the winding $G1.14$ = 9.522 kg, the weight of the wire «after-channel» part of the winding $G1.10$ = 15.27 kg.

Computational experiment

Modeling of heat transfer and hydrodynamics was carried out in the transformer environment in the software package of the free version of Ansys 17.1 (ANSYS Free Student Product Downloads) [8-10].

In view of the fact that the exact calculation of the oil-filled transformer is limited by the computing resources of the computer, a simplified model of the TM-160/10 transformer is used.

Cylindrical surfaces that are created in the Ansys 17.1 Design Modeler geometry module are used as LV and HV windings, as well as insulation windings. Previously a sketch is created, which is then converted to a 3D shape using the Extrude operation. The models of LV and HV winding of transformer and insulation for one phase, constructed in this way, in accordance with the calculated geometry are shown in Fig. 1.

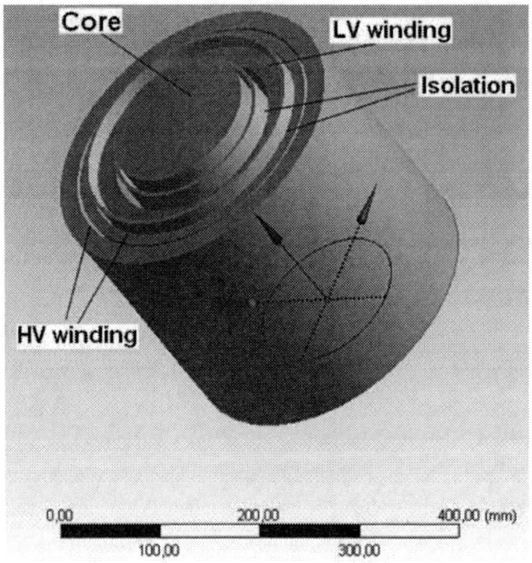

Fig. 1. Model of windings of one phase of transformer TM 160/10, created in Design Modeler of the program Ansys 17.1

Further, with the help of the Translate tool, two more phases of the transformer are created, and a core is included in the model. After this, the model is placed in the oil-filled tank.

Next, the program specifies the characteristics of the physical model. For this purpose, the characteristics of the materials of windings, rods, yokes, insulation and their environment are specified in the Steady-State Thermal module. As the material of the windings we take copper, core material is steel (Structural Steel), cooling medium is transformer oil (Engineer oil). Then in Ansys Steady-State Thermal the calculated grid area Mesh is defined, which is calculated automatically on the basis of the geometric model (Fig. 2).

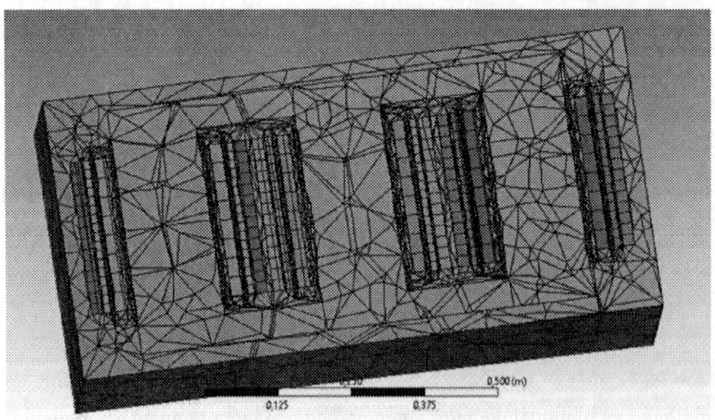

Fig. 2. The calculated grid area of the Mesh transformer (in section) in Ansys Steady-State Thermal

Short-circuit mode

The short-circuit mode is specified in Ansys Steady-State Thermal by specifying the specific heat flux q (Heat Flux). The heat flux density on the surface of the LV winding in the short-circuit mode is q_{2sc} = 1183 W/m^2, from the winding HV - q_{2sc} = 580 W/m^2.

Also for correct modeling, the condition for transferring heat by convection is introduced into the model and we denote all boundaries between the liquid and the solid with the Fluid Solid Interface tool.

From the screenshots shown, it is clear that when the air is cooled, the temperatures at which the transformer operates are significantly higher than the permissible 144…235 °C. The temperature of the hottest points corresponds to approximately 225 °C and is characteristic of the LV winding, and the core temperature t = 210 °C is also high. Due to the presence of insulation, the temperature of the HV windings is significantly lower, but it also exceeds the permissible value. The greatest density of heat fluxes, as expected, is near the windings, as the heat fluxes away from them, the intensity of the heat fluxes decreases rapidly.

Similar simulation of operation in the short-circuit mode with oil filling of the transformer was carried out. The distribution of temperatures and density of heat fluxes is shown in Fig. 3-4.

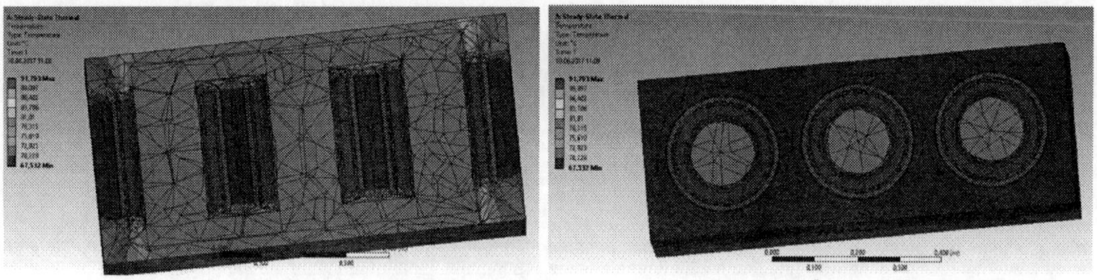

a *b*

Fig. 3. Distribution of temperatures in the transformer in the longitudinal (*a*) and transverse (*b*) sections during short-circuit operation. Cooling medium is transformer oil

The results indicate that the use of transformer oil to fill the transformer significantly reduces the temperature in the active part. The temperature distribution occupies the range of 67…91 °C. This means that the temperature of the most heated part is 91 °C and also corresponds to the winding LV.

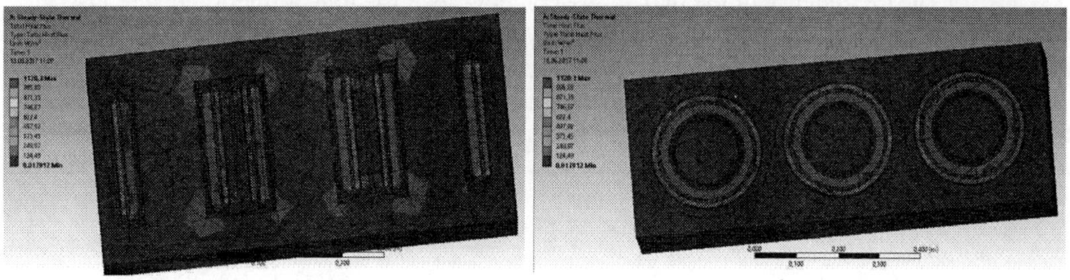

a *b*

Fig. 4. The distribution of the heat flux density in the transformer in the longitudinal (*a*) and transverse (*b*) sections when operating in the short-circuit mode. Cooling medium is transformer oil.

Dependence of the maximum temperature of the transformer on the operating mode

The service life of the transformer is determined by the maximum temperature of its windings, depending on the operating mode of the transformer. To study the dependence of the maximum temperature of the transformer TM-160/10 on the mode of its operation, simulation was performed in the range of losses from idling (510 W) to short circuit (2650 W). The corresponding temperature distribution is shown in Fig. 5.

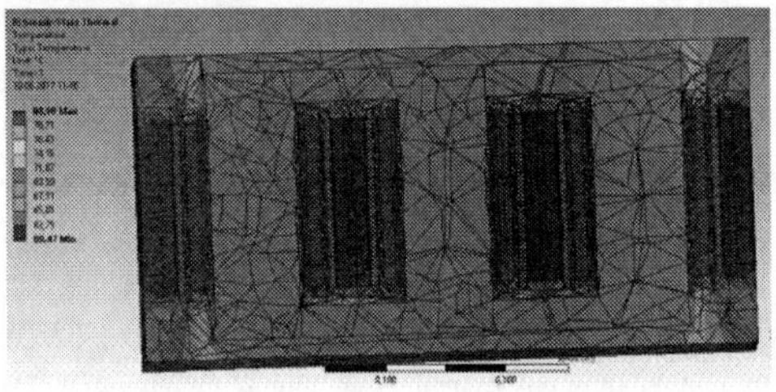

a

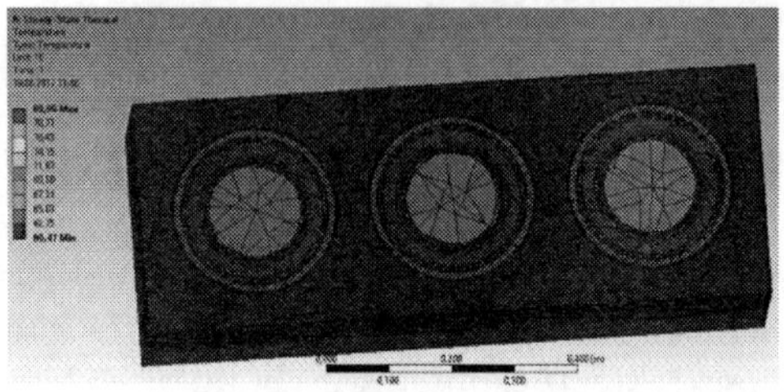

b

Fig. 5. The distribution of temperatures in the oil-filled transformer in the longitudinal (*a*) and transverse (*b*) sections with a loss of $P = 2240$ W

Based on the obtained data, the temperatures of the most heated points of the transformer are determined and the dependence of the maximum temperature of the transformer against losses is plotted (Fig. 6). The obtained dependence is well approximated by a straight line described by the equation: $t = 0{,}0261 \times P + 22{,}06$, °C.

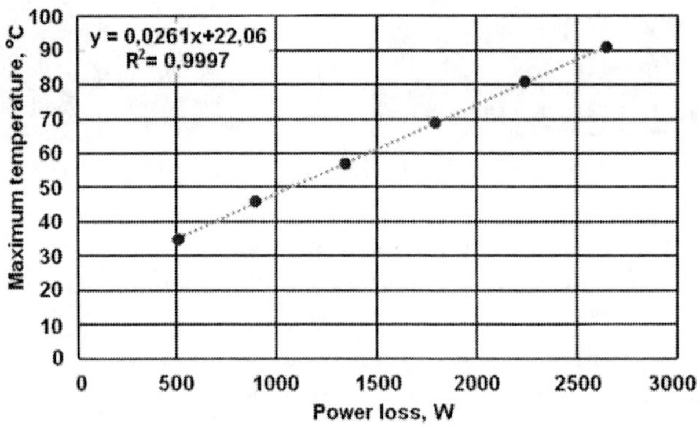

Fig. 6. The dependence of the maximum temperature of the transformer on power losses

Results

Modeling in Ansys 17.1 environment of thermal operating modes of the transformer was carried out both in the dry and in the oil-filled version. The distribution of temperatures in the active part of the transformer and near it, as well as heat fluxes, is determined. A formula is proposed for the dependence of the maximum temperature (the most heated point) on power losses in the range of idling and short circuit modes that determines its life.

References

[1] Thamir M. Abdul Wahhab, Methaq Talib Jabbar, «Design and Simulation of 11/0.4 kV Distribution Transformer Using ANSYS» Eng. & Tech. Journal, Vol. 32, Part(A), No.2, 2014. P.414-438.

[2] D. Constantin, P. M. Nicolae, and C. M. Nitu, «3D Finite Element Analysis of a three phase power transformer», Full Paper sent for EUROCON 2013 Conference, Zagreb, Croatia, 2013.

[3] Marina A. Tsili , Eleftherios I. Amoiralis , Antonios G. Kladas , Athanassios T. Souflaris (2012)

Power transformer thermal analysis by using an advanced coupled 3D heat transfer and fluid flow FEM model, International Journal of Thermal Sciences, Elsevier, 53.

[4] D. Susa, «Dynamic thermal modelling of power transformers», Ph.D.dissertation, Dept. of Electrical and Communications Eng., Helsinki Univ. of Tehnology, 2005.

[5] Watching LK Designing of power and special transformers / L.K. Watching, E.N. Pirovsky,A.V. The Pirovsky. - Kirov: Publishing house of Vyatka State University, 2006. - 246 p.

[6] Kish, L. Heating and cooling of transformers / L. Kish. Translation by M.A. Biki. - Moscow: Energy, 1980. - 180 p.

[7] Tikhomirov, P.M. Calculation of transformers / PM. Tikhomirov. - Moscow: Energoatomizdat, 1986. - 528 p.

[8] Modeling of heat exchange processes by means of a package of finite element analysis ANSYS [Electronic source] http://lib.knigi-x.ru/23raznoe/126936-1.php

[9] Yugov, V.P. Ansys 5.7 Thermal Analysis Guide/V.P. Yugov-Moscow: CADFEM, 2001.- 108 c.

[10] Ivanov, D.V. Introduction to AnsysWorkbench: A teaching method. manual for students of natural science disciplines / D.V. Ivanov, A.V. Dole. - Saratov: Amirit, 2016. - 56 p.

IOP Conf. Series: Earth and Environmental Science **288** (2019) 012089 doi:10.1088/1755-1315/288/1/012089

Dynamic packing for heat and mass exchange processes

[1]Golovanchikov A.B., [1]Merentsov N.A., [2]Topilin M.V., [3]Persidskiy A.V.

[1]Volgograd State Technical University, 400131, Russia, Volgograd, Prospect Lenin, 28

[2]«LUKOIL-Engineering VolgogradNIPImorneft» branch in Volgograd 400078, Russia, Volgograd, Prospect Lenin, 96

[3]JSC Federal Scientific and Production Centre «Titan - Barricady» 400071, Russia, Volgograd, Prospect Lenin, b/n

steeple@mail.ru

Abstract. Data of experimental study of hydrodynamics, heat and mass transfer of dynamic packing and some comparative characteristics of the packed device with packing, which have wide industrial application, are given in the article.

Introduction

When updating and engineering column apparatuses for mass exchange processes one of the key tasks is to choose optimum contact packings. Contact packings compared to plate packing have lower flow resistance and a simpler design. Besides, packed columns can work with higher gas and liquid loads, they are less sensitive to contamination if compared to plate packings.

From the point of view of design contact packings can be divided into non-regular (random) packings and regular (structured, preformed) packings. A regular packing with an individual structure consists of separate elements (rings, triangular prisms with constant or variable section by height) which are located in beds in the column tube [1]. Vertically neighbouring beds are displaced with respect to each other in order to avoid the formation of end-to-end channels. To simplify the formation of such packing its elements can be formed in advance in containers which are further on installed in the column tube. Individually formed packings are not widely spread in industry, because they considerably increase labour intensity and cost of the assembly. Compared to regular packings, random (non-regular) ones have a higher flow resistance. They are less reliable when working with contaminated media and inconvenient for maintenance.

The disadvantages of such packings are a small contact surface between the phases where heat and mass exchanges processes occur, and immobility of this surface during these processes that results in the formation of dead zones and decrease in productivity.

In order to avoid afore described disadvantages we have designed the structure of a dynamic packing with an increased surface of heat and mass exchange [2].

The packing consists of bodies that are organised in rows and one inside the other; these bodies have a form of cylindric surface. The inner rotary body is at a distance from the outer rotary body; at that the outer and the inner rings are connected by a spring. The outer diameter of the inner ring relates to the inner diameter of the outer ring as 0.7, and their heights relate to each other as:

IOP Conf. Series: Earth and Environmental Science **288** (2019) 012089 doi:10.1088/1755-1315/288/1/012089

$$\frac{h}{H} = 0.8 \div 0.9$$

where h and H are the heights of the inner and outer rings, respectively.

The packing has an additional rotary body on the outside of the outer ring at a distance from it and connected with the inner ring by a spring passing through the side wall of the outer ring.

The connection between the inner ring, the outer ring and the additional ring at the points located at equal distance from their faces facilitate the oscillation movement of the ring and of the additional rotary body under the influence of the ascending gas flow. Due to this reason smooth liquid films on the side walls of the inner ring and of the additional outer ring start oscillating with these walls, forming waves on the liquid film; their flow velocity and surface are considerably higher than the flow velocity and the surface of the smooth liquid film itself. This increases the intensity of heat and mass exchange at the phase boundary of the gas and liquid flows (Figure 1) [2].

Besides, the additional dynamic elements outside the outer ring eliminates the possibility of liquid and gas phase slip between the outer surfaces of the outer Raschig rings and involves them into an additional intensified heat and mass exchange [2]. Specific surface of the designed packing is 390 m^2/m^3, whereas standard Raschig rings of similar size have a specific surface of 190 m^2/m^3.

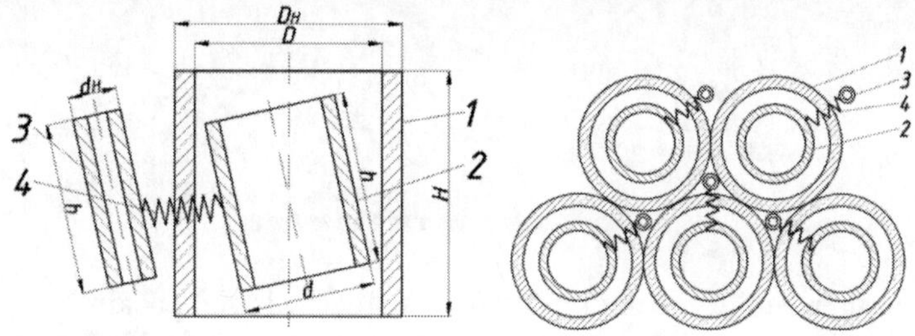

Figure 1. Designed dynamic packing for heat and mass exchange processes
1 – outer Raschig ring; 2 – inner Raschig ring; 3 – additional rotary body; 4 – spring

Experimental part

In order to perform a full range of experimental studies and compare packings having different configurations, we have designed an experimental stand (Figure 2) which is based on the modular study principle and has a cartridge system of replacement packings that provides for a quick and precise study of heat and mass exchange and flow characteristics of the packings [9].

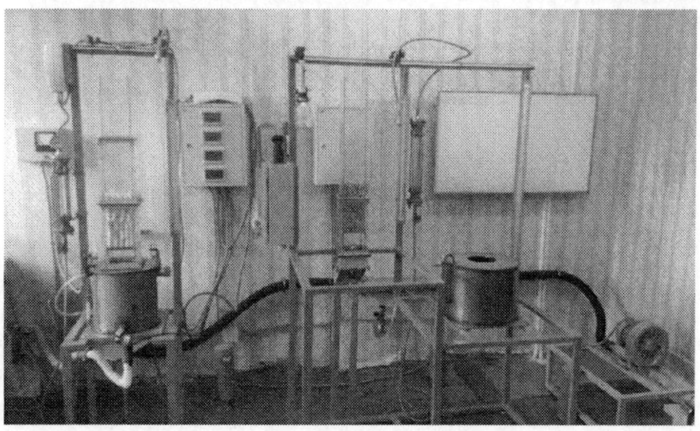

Figure 2. Experimental stand for studying characteristics of packings

To confirm the efficiency of the designed packing we have performed a series of experiments. The aim was to receive the data on its basic flow characteristics. The packings chosen for comparison were Raschig and Pall rings and a range of industrial packings [1, 2, 4, 5, 6].

Results and Discussion

Experimental data on flow resistance of dry and spray heat and mass exchange packings are given in Figure 3.

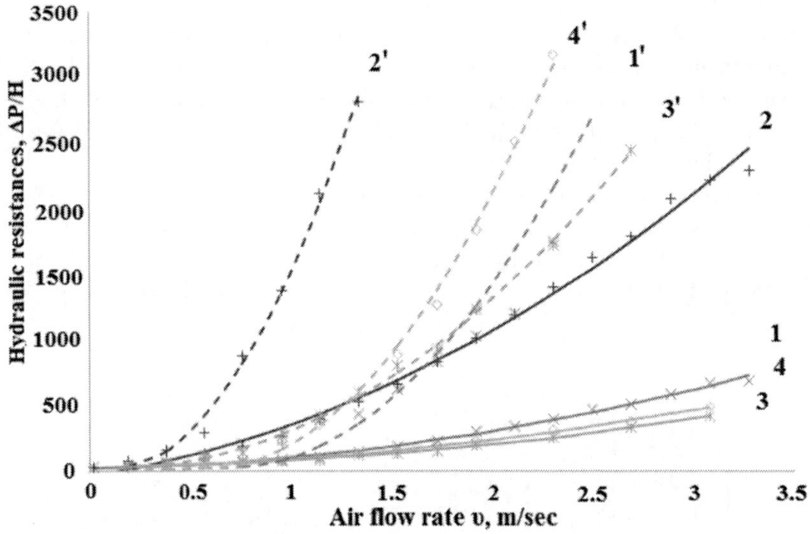

Figure 3. Flow resistance of dry (solid line) and spray (dashed line) packings vs. gas velocity in the column: 1,1' – Pall rings (25x25x1) dry, spray; 2, 2' – Raschig rings (25x25x3) not structured dry, spray; 3,3' – Raschig rings (25x25x3) structured dry, spray; 4,4' – Dynamic packing structured dry, spray (20x40x3)

In order to summarize the experimental data and prepare the comparison of the tested heat and mass exchange packing and packings that are widely spread in the industry, it is offered to use generalized equation $\lambda=f(Re_m)$ as per the method described in papers [5, 6]. Using this generalized equation we can compare their energy efficiency and describe the use of any packings having very complex configurations.

Table 1. The results of mathematical data processing of heat and mass exchange dynamic packing

α, m^{-2}	β, m^{-1}	Re	λ	l_{x1}, m	l_{x2}, m^2
		0.091	28.46		
		0.227	7.62		
$4.565\cdot10^6$	15.031	0.364	6.48	0.066	$3.2\cdot10^{-6}$
		0.456	6.23		
		0.547	5.8		
		0.638	5.25		
		0.71	5.05		

Based on the filtration curves obtained during experiments (Fig. 3) for the described heat and mass packing [5, 6] we determined linear dimensions l_1 and l_2, α and β values which are viscous and inertial factors in the Dupuit-Forchheimer equation, respectively, modified Reynolds numbers Re_m, and corresponding flow resistance factors λ, given in Table 1. This analysis was performed according to the method described in papers [5, 6]. This method helps to identify the industrial application and analyse the energy requirements for the industrial processes for any packings of any configuration. They can be summarized and are within the mode range of filtration curve $\lambda=f(Re_m)$, shown in Figure 4.

The result of the comparison of the packings is given in Figure 4.

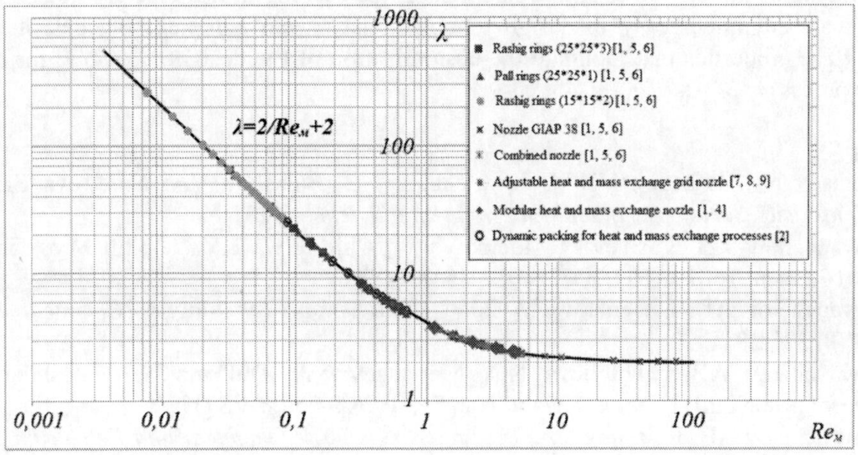

Figure 4. $\lambda=f(Re_m)$ diagram for packings with different structures [5, 6]

One of flow characteristics of packings which from our point of view is critical for identifying the industrial purposes of the packing and which we take as the basic one for our calculations [7, 8, 9] together with heat and mass transfer factors is retaining capacity, i. e. how the packing accumulates certain amount of liquid depending on the operation mode. This characteristic shows the total time of water retaining and mass transfer surface for a range of packings [7, 8, 9]. Figure 5 shows diagrams depicting the retaining capacity determined experimentally for different packings depending on gas velocity in the column.

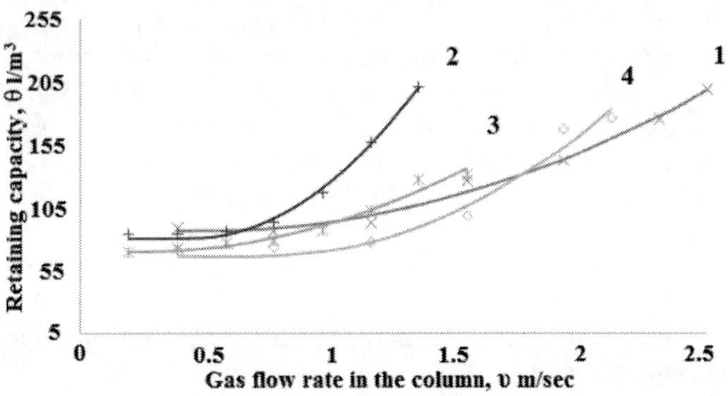

Figure 5. Retaining capacity of spray packings vs. gas velocity in the column
1 – Pall rings (25x25x1); 2 – Raschig rings (25x25x3) not structured; 3 – Raschig rings (25x25x3) structured; 4 – Dynamic heat and mass exchange packing structured (20x40x3)

Conclusions

Comparing with a range of industrial packings analysed during the experiment, the designed heat and mass exchange packing has a rather high retaining capacity that is shown in Figure 5. Based on the data on energy efficiency, flow resistance values of spray packings and a rather high retaining capacity for liquid we can conclude that this packing is promising for industry, especially for such processes as absorption and fractionation. But it should be noted that this packing can exhibit its quality only in certain resonant modes that positively influence the intensification of heat and mass exchange processes occurring in the packing, but on the other hand require fine and sensitive adjustment.

The prototype of the offered heat and mass exchange packing is manufactured of PVC tubes that makes it cheap and rather easy in design though it may seem complex. Besides, it can provide different surface properties that facilitate the customization of the designed packing for the needs of the specific process.

References

[1] Merentsov NA, Lebedev VN, Golovanchikov AB, Balashov VA, Nefed'eva EE 2018 *IOP Conference Series: Earth and Environmental Science* **(115)** 9

[2] Golovanchikov AB, Vasil'ev PS, Ljapkov AV, Chjorikova KV, Topilin MV, Tishhenko PO 2016 *Patent* 160198 RU, IPC B 01 J 19/00 VSTU

[3] Golovanchikov AB, Merentsov NA 2016 *Himicheskaja tehnologija (Chemical engineering)* **17(8)** 377-384

[4] Golovanchikov AB, Sivolobova NO, Merentsov NA, Dul'kina NA, Shishljannikov VV, Dorofeeva NI 2012 *Patent* 129450 RU, IPC F 28 S 25/08 VSTU

[5] Golovanchikov AB, Balashov VA, Merentsov NA 2017 *Chemical and Petroleum Engineering* **53(1-2)**10-13

[6] Merentsov NA, Balashov VA, Orljankina JaA 2013 *Izvestija Volgogradskogo gosudarstvennogo tehnicheskogo universiteta, Ser. Reologija, processy i apparaty himicheskoj tehnologii (The news of Volgograd state technical University, Ser. Rheology, processes and devices of chemical technology)* **1 (104)** 112-114

[7] Golovanchikov AB, Merentsov NA, Balashov VA 2013 *Chemical and Petroleum Engineering* **48(9-10)** 595-601

[8] Golovanchikov AB, Merentsov NA, Balashov VA, Orljankina JaA 2012 *Izvestija Volgogradskogo gosudarstvennogo tehnicheskogo universiteta, Ser. Aktual'nye problemy upravlenija, vychislitel'noj tehniki i informatiki v tehnicheskih sistemah (The news of Volgograd state technical University, Ser. Actual problems of management, computer science and Informatics in technical systems)* **10 (97)** 22-28

[9] Merentsov NA, Balashov VA, Golovanchikov AB, Orljankina JaA 2012 *Izvestija Volgogradskogo gosudarstvennogo tehnicheskogo universiteta, Ser. Reologija, processy i apparaty himicheskoj tehnologii (The news of Volgograd state technical University, Ser. Rheology, processes and devices of chemical technology)* **1 (88)** 78-80

[10] Kagan AM, Yudina LA, Pushnov AS 2012 *Theoretical Foundations of Chemical Engineering* **46(2)** 165-171

[11] Gorodilov AA, Berengarten MG, Pushnov AS 2016 *Theoretical Foundations of Chemical Engineering* **50(3)** 325-334

[12] Gorodilov AA, Berengarten MG, Pushnov AS 2016 *Theoretical Foundations of Chemical Engineering* **50(4)** 422-429.

[13] Mitin AK, Nikolaikina NE, Pushnov AS, Zagustina NA 2016 *Chemical and Petroleum Engineering* **52(1)** 47-52

[14] Klyushenkova MI, Kuznetsova NA, Pushnov AS, Berengarten MG, Mokrousova EA 2014 *Chemical and Petroleum Engineering* **50(7-8)** 508-512

[15] Gorodilov AA, Pushnov AS, Berengarten MG 2014 *Chemical and Petroleum Engineering* **50(1-**

2) 84-90

[16] Pushnov AS, Sokolov AS, Sidel'nikov II, Kurbatova EA, Mitrofanova EG 2014 *Chemical and Petroleum Engineering* **50(5-6)** 330-334

[17] Chizh, KV, Pushnov, AS, Berengarten MG 2014 *Chemical and Petroleum Engineering* **50(3-4)** 244-250

[18] Berengarten MG, Nevelson AO, Pushnov AS 2013 *Chemical and Petroleum Engineering* **48(11-12)** 723-729

[19] Shilin MV, Berengarten MG, Pushnov AS, Klyushenkova MI 2013 *Chemical and Petroleum Engineering* **48(9-10)** 608-614

[20] Mitin AK, Valdberg AY, Pushnov AS 2012 *Chemical and Petroleum Engineering* **48(1-2)** 50-53

[21] Tsurikova NP, Pushnov AS, Lagutkin MG, Shishov VI 2012 *Chemical and Petroleum Engineering* **48(1-2)** 3-8

[22] Kagan AM, Pushnov AS 2012 *Russian Journal of Applied Chemistry* **85(3)** 523-526

[23] Kagan AM, Yudina LA, Pushnov AS 2012 *Russian Journal of Applied Chemistry* **85(3)** 515-522

[24] Pushnov AS, Vaganov AA 2011 *Journal of Applied Chemistry* **84(9)** 1638-1641

[25] Pushnov AS, Berengarten MG, Kagan AM, Ryabushenko A 2009 *Russian Journal of Applied Chemistry* **82(4)** 723-729

[26] Tsurikova NP, Pushnov AS, Lagutkin MG 2011 *Chemical and Petroleum Engineering* **47(1)**. 97-103

[27] Petrashova EN, Lagutkin MG, Pushnov AS, Shishov VI 2011 *Chemical and Petroleum Engineering* **47(3)** 250-255

[28] Pushnov AS, Lozovaya NP, Lagutkin MG 2010 *Chemical and Petroleum Engineering* **46(1)** 3-8

Increasing the energy efficiency of the cycle of heat treatment of aluminum ingots

Gorshenin A S, Rakhimova J I, Krasnova N P

Samara State Technical University, 443100, Samara, Molodogvardeyskaya str., 244

Andersonag1@yandex.ru

Abstract. Heat treatment of aluminum ingots is a fairly energy-intensive production. Therefore, increasing energy efficiency in such processes plays an important role in reducing economic costs. The mathematical model of heat exchange occurring in the cooling chamber of aluminum ingots after homogenization annealing is considered. The model obtained made it possible to study the heat transfer in the cooling chamber. To improve energy efficiency and reduce economic costs, options for recycling hot air from the cooling chamber are offered. The estimation of economic efficiency from use of each of variants carried out.

Introduction

The existing technology of semicontinuous casting of aluminum alloys in the water-cooled crystallizer has the main technological problem - the formation of an inhomogeneous structure, micropores, cracks in the ingot due to uneven cooling of the ingot along the cross section. To solve this problem, heat treatment is carried out, one of the types of which is homogenization annealing, consisting in heating and holding ingots to a temperature higher than the dissolution temperature of the alloying components.

An important stage after homogenization of aluminum ingots is their cooling with a speed not exceeding the critical one, in order to avoid the hardening process of the alloy. For most alloy grades, this cooling temperature is 70 ° C / h. The fulfillment of this requirement is possible in several ways, but the most expedient, as shown in [1], is the cooling of aluminum ingots after homogenization annealing in a special cooling chamber, allowing to maintain a given cooling rate and increase the production of quality products.

Mathematical model of heat exchange in the cooling chamber

Application of existing methods of cooling aluminum ingots does not allow to maintain an equal cooling rate in all parts of each ingot. The existing cooling modes and methods do not take into account the geometric characteristics of the cooling chamber and the methods of placing ingots in the chamber, which allow to ensure uniform cooling and lower energy costs for the cooling process.

The solution of this problem was the development of a mathematical model of controlled convective heat transfer between cooling air and aluminum ingots. The resulting mathematical model describes the heat transfer in the cooling chamber of aluminum ingots, taking into account the geometric distribution of ingots in the cooling chamber. The mathematical model, the assumptions made to it, the description of the heat exchange process are described in more detail in [2,3].

The obtained mathematical model allowed performing an analytical study of heat exchange in the cooling chamber between air and aluminum ingots. The influence of structural factors (the diameter of

IOP Conf. Series: Earth and Environmental Science **288** (2019) 012090 doi:10.1088/1755-1315/288/1/012090

the ingot, the height of the channel between the rows of ingots) and the regime factors (the speed of the cooling air) were investigated. The study showed that maintaining a constant rate of cooling air does not allow obtaining high-quality products, since during the initial cooling period, aluminum ingots are hardened due to exceeding the critical cooling rate. Therefore, maintain a variable rate of cooling of ingots was proposed, which allowed a given cooling rate and reduce energy costs. The results of the investigations are presented in [4, 5].

Economic justification for increasing the energy efficiency of the heat treatment cycle

One of the results of the study, described above, was that the temperature of the cooling air at the outlet from the chamber is 120-130 ° C. Removing air with such a temperature into the environment is economically inexpedient, since its use will increase energy efficiency and reduce fuel costs in any technological process. Consider the options for using the heat of the cooling air for a specific heat treatment section and for the production of aluminum products (Figure 1) and the resulting fuel economy.

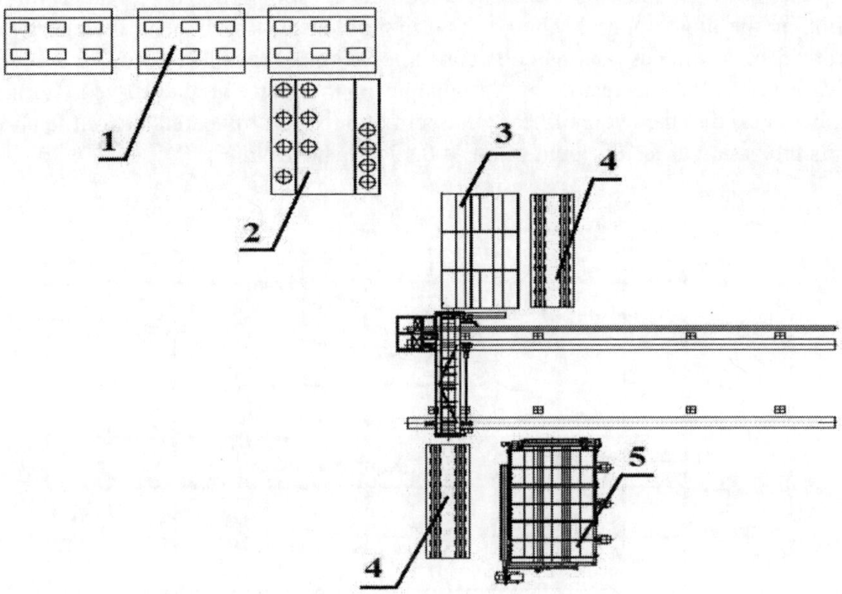

Figure 1. The plan of the heat treatment section 1-washing chamber, 2-drying chamber, 3-cooling chamber, 4-place for forming a cage, 5- homogenization furnace

The section shown in Fig. 1 consists of a homogenization furnace, a cooling chamber, and a color section of aluminum profiles, including washing, painting and drying chambers. Washing of the profiles is carried out by solutions heated by hot water from the boiler room. Drying is carried out by hot air heated from natural gas combustion products.

Thus, for this section, it is possible to utilize heated air from a homogenization furnace in the following schemes:

- supply of heated air as an oxidizer of natural gas to a homogenization furnace,
- heating the air with detergent solutions in the washing chamber,
- drying profiles in the drying chamber,
- use of heated air for air heating of the workshop.

Let's consider how in each scheme fuel consumption at recycling of heated air from a homogenization furnace is reduced and, accordingly, the energy efficiency of the heat treatment process is increased. The economic effect of the supply of hot combustion air together with fuel in the homogenization furnace will be determined as the difference in the combustion with natural gas of cold and hot air. Fuel economy during the heat treatment of 20,000 kg of aluminum ingots per hour with heating them from the initial temperature of 20 ° C to the final 550 ° C will be 9.5 m3 / h. The economic effect of the introduction of hot air heating solutions in the cleaning chamber will be determined by the reduction in natural gas consumption for heating the water in the boiler room. As a result of the application of this scheme, fuel savings can be 8.1 m3 / h. When using hot air for drying profiles, the economic effect is caused by a reduction in the gas consumption for heating the drying air, since hot air can be directly used as a drying agent. With this scheme of utilization of air, fuel economy can reach 6 m3 / h. As a result of the use of hot air for heating, it is possible to refuse traditional water heating, which is not effective in production plants. The fuel economy can be up to 5 m3 / hr when using this scheme. In all schemes, the gas flow was determined from the heat balance equation.

As can be seen from the above, any of the described schemes leads to an increase in the energy efficiency of the heat treatment process by reducing the consumption of organic fuel [6, 7]. It is only necessary to determine the most cost-effective scheme by comparing net present value (BHD). In addition to finding the most optimal scheme for utilizing hot air - improving the energy efficiency of the entire heat treatment process, let us consider the economic feasibility of using a cooling chamber in a heat treatment cycle. The results of the change in the BHD at the 10-year horizon when the cooling chamber is used to the cycle of heat treatment and the use of heated air for the above described schemes for its utilization at the discount rate E = 0.12 are shown in Fig. 2.

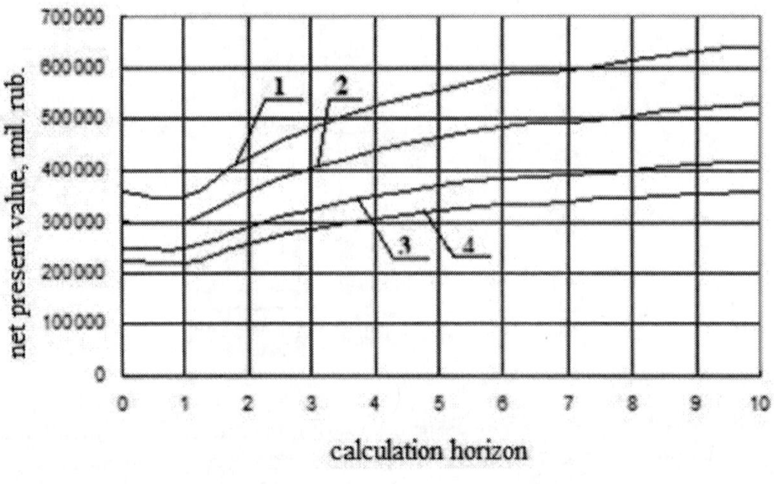

Figure 2. Change in net discounted income by years of the calculation horizon: 1 - a diagram of the use of heated air in a homogenization furnace, 2 - a diagram of the use of heated air in a washing chamber, 3 - a diagram of the use of heated air in a drying chamber, 4 - a diagram of the use of heated air for air heating.

Discounted payback period for various schemes is:

- scheme for using heated air in a homogenization furnace - 0.1 year,
- scheme of using heated air in the washing chamber - 0.16 years,
- scheme of using heated air in the drying chamber - 0.15 years,
- scheme for using heated air for air heating - 0.17 years.

The choice of the most profitable option for using heated air is determined by the highest value of net discounted income and the shortest payback period. From the considered schemes in this case, it is most expedient to use heated air for feeding to combustion in a homogenization furnace.

The conducted research shows that by utilization of hot air during homogenization of aluminum ingots it is possible to reduce fuel consumption for the heat treatment process and increase the energy efficiency of the entire process.

References

[1] Gorshenin A. S. Perfection of the cooling process of aluminum ingots by air based on modeling of controlled convective heat transfer. Thesis ... Cand.Tech.Sci .: 05.14.04 / Saratov State Technical University named after Gagarin Yu.A. Samara, 2013

[2] Gorshenin A.S., Mathematical modeling of heat transfer during cooling of a series of aluminum ingots by air. Bulletin of the Samara State Aerospace University. Academician SP Korolev (National Research University). 2012. № 2 (33). P.179-182

[3] Gorshenin A.S., Krivosheev V.E. Mathematical model of heat exchange in the "cooling air-aluminum ingot" system in a convection cooling chamber after heat treatment. Industrial energy. 2013. № 11. C.20-22

[4] Gorshenin A.S., Shchelokov A. I. Investigation of heat transfer during cooling of aluminum ingots with constant regime and design parameters for improving energy efficiency. Industrial energy. 2012. №10. C.23-25

[5] Gorshenin A.S., Krivosheev V.E. Mathematical modeling and investigation of the process of thermal processing of aluminum ingots based on controlled convective heat transfer. Bulletin of the Samara State Aerospace University. Academician SP Korolev (National Research University). 2014, No. 5-4 (47). P.137-141

[6] Pashchenko D. First law energy analysis of thermochemical waste-heat recuperation by steam methane reforming //Energy. – 2018. – T. 143. – C. 478-487

[7] Pashchenko D. I. Thermochemical recovery of heat contained in flue gases by means of bioethanol conversion //Thermal Engineering. – 2013. – T. 60. – №. 6. – C. 438-443

IOP Conf. Series: Earth and Environmental Science **288** (2019) 012091 doi:10.1088/1755-1315/288/1/012091

Control of the condition of heat exchange surfaces by free vibration method

E.V. Izmaylova, T.O. Politova, E.V. Garnyishova, M.Sh. Garifullin

Kazan State Power Engineering University, Russia, Kazan, Krasnoselskaya st., 51

evgeniya-izmailova@yandex.ru

Abstract. In power system the main heat carrier is the fresh water received from natural sources, and containing a large amount of various impurity – from the dissolved mineral salts to organic compounds. In housing and communal services and the industry cases of operation of heat-exchange devices and package boilers without special preparation of water aren't rare. During the operation of the heat-exchange equipment of impurity are allocated in a firm phase both in the form of deposits and in the form of slime. Sediments cause deterioration in a heat transfer that leads to decrease in overall performance of the equipment (to an excessive consumption of fuel, metal overheating, etc.). The radiation surfaces of modern steam generators heating are intensively warmed with a furnace torch. Density of a thermal stream in them reaches 600-700 kW/sq.m, and local thermal streams can be even higher. Therefore even short-term deterioration in coefficient of a thermolysis from a wall to the boiling water leads to significant increase in temperature of a wall of a pipe (500-600 °C and above), durability of metal can it is insufficient to sustain tension which has arisen in him. The metal damages which are characterized by emergence of blisters, fistulas are a consequence of it, and it is frequent also a rupture of pipes. The scum reduces the useful section of pipes, increasing the hydraulic resistance of pipes of the heat exchanger. Everything together it involves an excessive consumption of fuel, and according to expert data the layer of scale up to 0.2 mm thick increases fuel consumption by 3%, and a layer in 1 mm for 7%.

The serious problem of fight against sediments arises in heat exchangers of hot water supply when the section of pipes through passage and the surfaces of heat exchange almost completely grows with sediments. Overgrowing by sediments of thermal systems pipelines including reverse water supply, leads to significant increase in their hydraulic resistance, maladjustment of heating systems and big power losses by pumping of system.

There is a similar situation with power losses from sediments in air conditioning systems (table 1).

Tab. 1. Losses from sediments in air conditioning systems depending on thickness of a sediments layer

Thickness of a sediment layer (mm)	0.5	1	2	4	6	8	10	12.7
Deceleration of power of the conditioner (%)	5	9	17	23	29	34	50	56
Increase of temperature at the exit from the evaporator (°C)	0.4	0.8	1.6	3.2	4.8	6.4	8	10
Electric power overexpenditure (%)	5.8	10.6	20.2	29.4	35.6	46.8	66	76

It is possible to reduce energetic losses by timely cleaning of internal surfaces from scale and sediments. For this purpose it is necessary to have a reliable, available technique and the equipment for monitoring of sediments thickness.

For monitoring of sediments thickness on heat-exchanging surfaces it is offered to use the method of the free oscillations [1] consisting with following. In case of shock on a surface of a controlled product in it there are mechanical oscillations which will be transformed to an electrical signal by the sensor, register, spectrum analysis is carried out them [2]. Existence of sediments changes thickness, mass of heat exchange surfaces, therefore, and natural frequencies of oscillations. Having "the acoustic passport" of a new product, knowing the current oscillation frequencies of controlled surfaces it is possible to determine thickness of sediments [3].

With use of the software package of "ANSYS" [4] simulation of oscillation frequencies of the metal plate of 2 mm thickness, jammed on perimeter without precipitate and with the different thickness of sediments (0.1, 0.3, 0.7, 1, 1.5, 2.0, 3.0 mm) is carried out.

Results of frequencies calculations of natural oscillations of plates are provided in the table. 2.

Tab. 2. Results of plate measurements with different thickness of sediments

Type of a plate	clean plate	plate with a layer of 0.1 mm	plate with a layer of 0.3 mm	plate with a layer of 0.7 mm	plate with a layer of 1 mm	plate with a layer of 1.5 mm	plate with a layer of 2 mm	plate with a layer of 3 mm
Number of fluctuations form	Frequency of fluctuations [Hz]							
1	1076	1081	1096	1152	1216	1355	1526	1927
2	2002	2012	2045	2160	2284	2551	2870	3586
3	2958	2971	3013	3165	3336	3709	4164	5202
4	4431	4453	4522	4767	5033	5603	6281	7781
5	5790	5813	5894	6186	6511	7216	8062	9941
6	7542	7576	7689	8088	8525	9453	10545	11063
7	7641	7672	7778	8164	8594	9526	10647	12910
8	9542	9579	9707	10177	10697	11426	11293	13154
9	10006	10048	10190	10699	11257	11810	13121	15944
10	11440	11489	11652	11672	11574	12445	13841	16862

The analysis of the received results shows that even existence of a precipitate 0.1 mm thick leads to change of the first oscillation frequency of model on 5 Hz, and the 9th and 10th forms of oscillations on 42 and 49 Hz. Such shift of frequencies in the sound range can be determined by the spectrum analyzer of oscillations using an algorithm of the fast Fourier transform [5].

For conducting the pilot studies of influence of sediments thickness on oscillation frequencies the laboratory bench which schematic block diagram and photo is provided in fig. 1, 2 is created.

The stand consists of the registration system and processing system of signals. The device of the researched plate fixing is a part of the registration system. Excitation of plate oscillations is carried out by the mechanical drummer of pendulum type. The persistence of shock force is provided with assignment of the pendulum on the same angle. After excitation of product oscillations the analog signal from sensors of oscillations comes to the personal computer (PC) on an input of the analog-to-digital converter (ATDC). The received digital signal registers in the form of the file in a hard drive of the computer, and then is exposed to later processing.

As the sensor of oscillations in this complex the microphone is used.

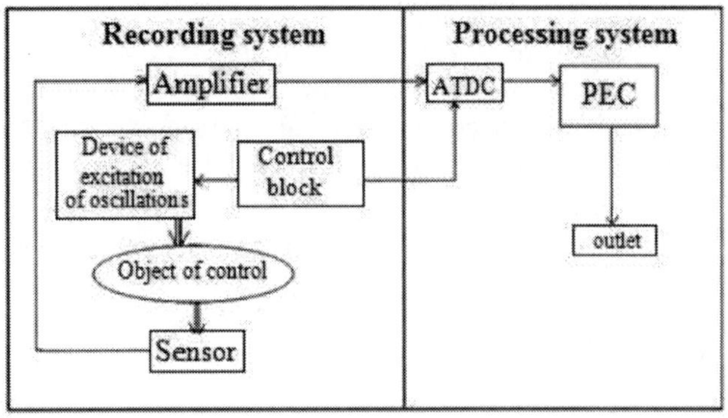

Fig. 1. Block scheme of the stand

Fig. 2. Photo of an experimental complex for a research of fluctuations at different layers of sediments, a side view

As model of a surface of the heat-exchanging equipment the steel plate 400 mm long, 160 mm wide and 2 mm thick was used.

In case of a research of acoustic characteristics with each type of a plate tenfold determination of frequencies of natural oscillations of a plate was carried out. Blow was struck to a plate, and measurement of amplitude-frequency characteristics of a response was taken by the microphone which is in break point.

After a series of experiments the layer of sediments was applied on a plate and researches repeated. Ranges of oscillations of a pure plate and plate with sediments 0.5 mm thick are provided in fig. 3.

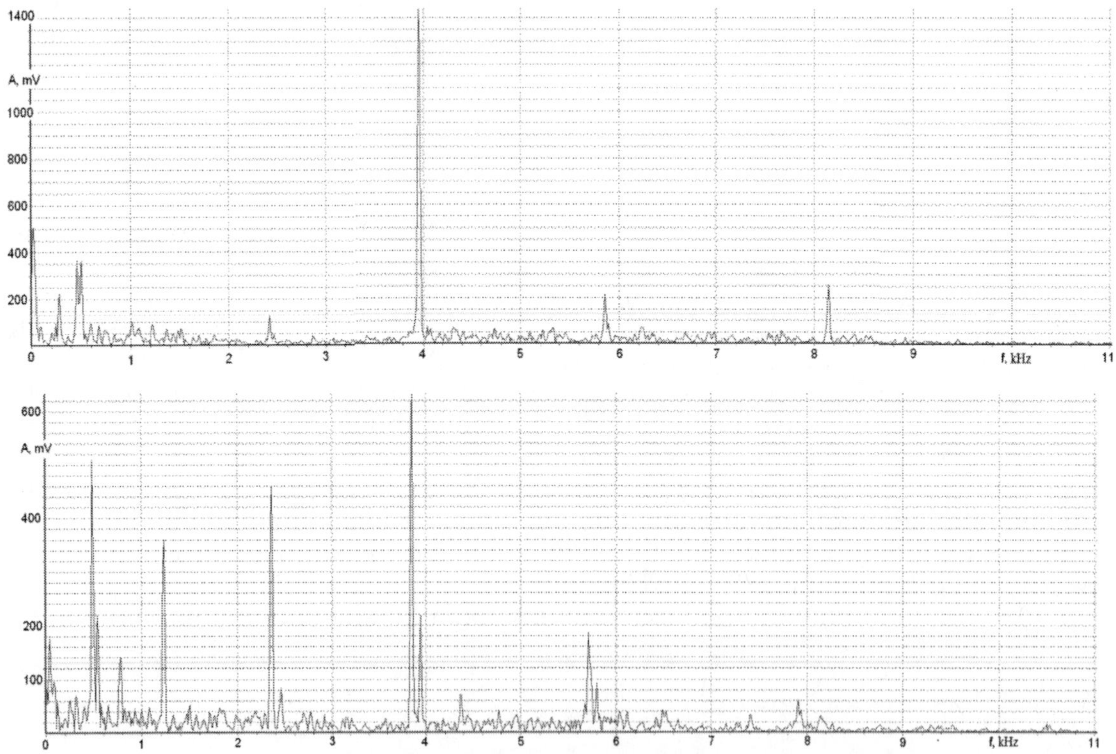

Fig. 3. Ranges of fluctuations of a clean plate (top) and with sediments 0.5 mm thick

From the analysis of ranges of the plates fluctuations presented in the figure 3 follows that sediments existence significantly changes her range of own fluctuations to plate surfaces in the range from 20 Hz to 10 kHz.

For increase in sensitivity of free fluctuations method for control of sediments it is offered to compare coefficients of r correlation of fluctuations ranges, determined by a formula:

$$r = \frac{\sum a_i a_{si} - \left(\sum a_i \sum a_{si}\right)/n}{\sqrt{\left(\sum a_i^2 - \left(\sum a_i\right)^2/n\right)\cdot\left(\sum a_{si}^2 - \left(\sum a_{si}\right)^2/n\right)}},$$

where n – the number of the compared frequencies in a range; a_i — amplitude of the checked range frequency; a_{si} – amplitude of a reference range frequency.

In fig. 4 correlation coefficients of "a pure plate" ranges and a plate with thickness of sediments of 5 mm are provided.

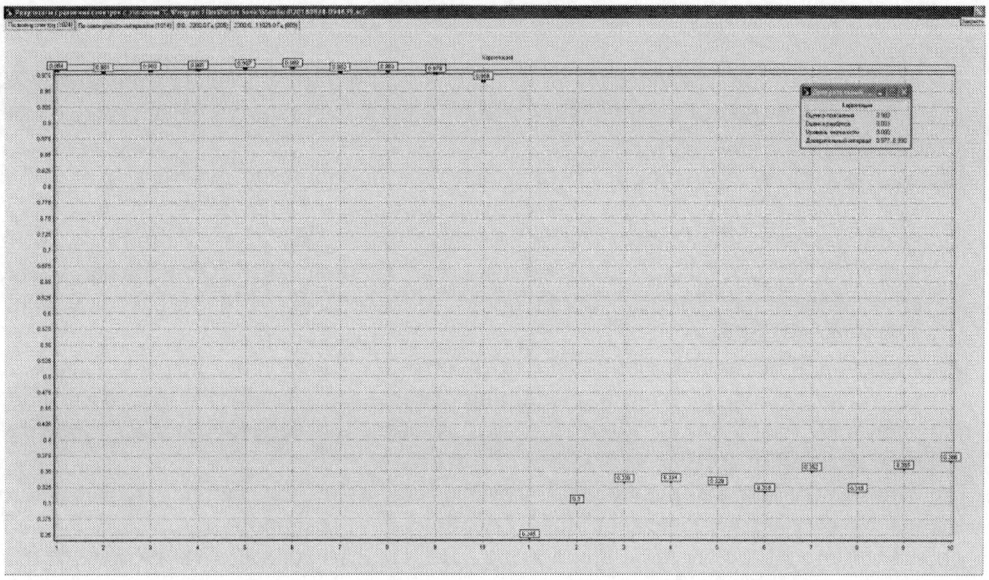

Fig. 4. Coefficients of ranges correlation of a "clean" plate and plate with sediments 0.5 mm thick

Process of record and the analysis of signals is realized in LabVIEW. The program of the analysis works as follows: signals are read out, on each signal the amplitude-frequency range is formed, ranges are compared on correlation coefficients, the received coefficients relating to one series of experiments are averaged, average coefficients are compared to border of a confidential interval.

For classification of riveted connection "suitable" or "defective" is used the approach characteristic of anomalies rejection procedures.

The analysis algorithm interprets set of the calculated values of correlation coefficient as a set of the measured values of some abstract parameter and applies the following procedure to this set of values: provision $\overline{p}$ assessment is calculated; assessment of dispersion of S as a median of absolute deviations concerning situation assessment is calculated; for the set significance value of α the confidential interval is under construction:

$$p \pm St\left(1 - \frac{\alpha}{2}, m - 2\right),$$

where – α Student's distribution quantile with freedom degrees.

In case of an exit of correlation coefficient of the compared range out of a confidential interval limits the conclusion about sediments existence on the surface becomes.

Conclusion

Existence of sediments on the surface of the heat-exchange equipment leads to the following negative consequences:

1. To decrease in the thermal productivity connected with falling of the actual coefficients of a heat transfer owing to growth of thermal resistance of heating surfaces.
2. To increase in hydraulic resistance result of reduction of their section through passage and roughness growth.
3. To loss of fuel and power of the heat-exchange equipment.

The negative effect of sediments formation can be minimized if to control of heat-exchange surfaces condition with a nondestructive control method.

Results of the conducted researches have shown that the method of free fluctuations allows to determine sediments thickness in the tenth shares of millimeter.

Reference

[1] Diagnostic system for flaw detection in the railroad car axles by acoustic parameters / Vankov Yu.V, Kazakov R.B., Pervukhin D.N. // Zavodskaya Laboratoriya. Diagnostika Materialov, 2004, 70(3), p.34-38.

[2] Acoustic spectral defectoscope for defects detection in composite materials (part 2) Vankov. Yu.V. // 2004 Pribory i Sistemy Upravleniya, EID: 2-s2.0-1542578239.

[3] Method of assessing the thermal state of the heat transfer surfaces / Saifullin E. R., Yu. V. Vankov, E. V. Izmaylova, E. R. Bazukova // LTP Coatings 2018 IOP Publishing. IOP Conf. Series: Journal of Physics: Conf. Series 1058 (2018) 012058. DOI: 10.1088/1742-6596/1058/1/012058.

[4] Determination of the oscillation frequencies of corrosion defects finite element methods in order to develop methods of acoustic monitoring of pipelines / Vankov Yu.V., Ziganshin Sh.G., Izmailova E.V., Serov V.V. 2015 // IOP Conference Series: Materials Science and Engineering DOI: 10.1088/1757-899X/86/1/012037.

[5] The effect of the size and position of transverse cracks in the natural frequencies of the rods/ Vankov Yu.V., Ziganshin Sh.G., Izmailova E.V., Zagretdinov A.R. // IOP Conference Series: Materials Science and Engineering 2015. DOI: 10.1088/1757-899X/86/1/012036.

IOP Conf. Series: Earth and Environmental Science **288** (2019) 012092 doi:10.1088/1755-1315/288/1/012092

Improving a water treatment and a heating performance of the water-to-water heat pump: misallocation and available solutions[1]

L. R. Junussova [1], S. V. Chicherin [2]

[1]Non-profit JSC Almaty University of Power Engineering and Telecommunications, Almaty, Kazakhstan.

[2]Omsk State Transport University (OSTU), Omsk, Russia

man_csv@hotmail.com

The object of our experimental investigation is a heat pump system operating at low temperatures of a low-grade heat source. The purpose of experimental research of the heat pump installation is to increase the efficiency and reliability of heat pump unit by adjusting compressor performance. To carry out the experiment we connected the reciprocating compressor with Variable Frequency Drive of AC 0..650 Hz, designed for rated capacity of 2.2 kW, as well as electricity meter. The dependence of heat load of pump condenser on the speed (rpm) is approximated by the formula and will be used to design a heat pump unit. The proposed experimental unit allowed us to determine the correlation between basic characteristics and the compressor shaft speed and determine the possible range of its regulation (50..180%). The obtained results allow us to design the water treatment plant with using heat pump unit with quality regulation systems as well.

Introduction

The most important objectives of current development of power engineering sector in the country is the increase of thermal and economic efficiency, reliability and sustainability of power engineering complex. A special place in the solution of these objectives is assigned to further development of district heating plants and distribution systems [1, 2].

One of the most effective ways of the development of heat supply sources and systems is the use of combined heat supply systems based on large and small heat power and boiler plants, matching the operation of traditional sources of heat and renewable energy sources for heat pumps, as well as autonomous power supply systems based on heat pump systems [3].

The recent research revealed the possibilities for further improving compressor characteristics by providing smooth control of their efficiency [4], improving reliability [5], reducing noise and energy consumption [6-8], as well as decreasing cost, weight and size [9]. Promising methods of regulating thermal power of air conditioning in the heat pump were proposed. The possibilities of using scroll compressors were analyzed [10]. The objectives of coordination of the evaporator and compressor

[1] The Authors thank the Government of the Russian Federation for its financial support of the Decree No. 860 August, 29 2017

operation by applying auxiliary gas cooler for simultaneous production of heat and cold by means of a heat pump were considered in our research [11, 12].

Heat pumps are used as converters of heat energy from an energy source with low temperature [13] to an energy source with higher temperature. As it follows from the description of the processes in heat pumps, its main elements are a compressor, condenser, evaporator and throttle (regulation valve) connected by a piping system for circulation of a working fluid. The combination of a heat pump and auxiliary equipment such as hydraulic machines, pipelines for supply and removal of transfer fluids (cooled and heated), power supply, control and regulation systems is a heat pump unit.

Consequently, the object of our experimental investigation is a heat pump system, operating at low temperatures of a low-grade heat source. The purpose of experimental research of the heat pump installation is to increase the efficiency and reliability of heat pump unit by adjusting compressor performance.

Materials and methods

A heat pump system with water as the lower source is proposed as the most effective one from the point of view of running costs for heating. The following conditions should be completed for this:

- water should have an appropriate quality, in terms of chemical composition - it should not cause corrosion of the evaporator of heat pump; it is necessary to obtain reliable results of its chemical and biological analysis;
- water should not contain mechanical impurities, which can increase the flow hydraulic resistance through the evaporator;
- water temperature should not be below 7°C [13];
- encroachment line should be closed to avoid the contact with air.

Ground water containing a lot of iron is used most of all as a low-grade heat source for heat pumps. When contacting with oxygen, iron comes down in the water resulting in mineral deposits in the pipes and heat pump evaporator, which increases the flow hydraulic resistance. The increase of hydrodynamic resistance of the system reduces the efficiency of a transporting pump and, consequently, leads to malfunction of the heat pump. Ground water undergoes preliminary cleaning in an ultrafiltration module before going into the evaporator of the heat pump, according to the scheme presented in fig. 1 [11, 12].

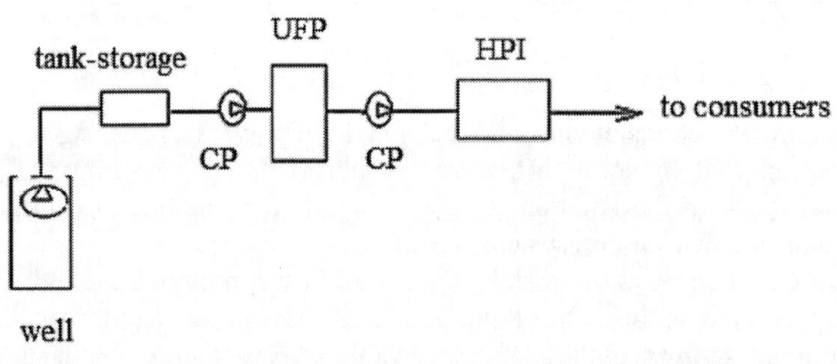

Fig. 1 Technological bloc-diagram of the water treatment plants (WTP)

We perform all experiments on a verification and calibration test bench in the flow laboratory of Almaty University of Power Engineering and Telecommunications, Kazakhstan using plastic pipes of inner diameter D = 5.0 mm (DN5) and D = 10.0 mm (DN10). Preliminary water softening is done in the ultrafiltration unit (fig. 1, UFP). Setup of the experiment is in fig. 2.

The plate pack is assembled between a fix frame plate and a movable pressure plate and compressed by tightening bolts. The plates are installed with a gasket which seals the inter plate channel and directs the fluids into alternate channels. Table 1 shows the testing conditions.

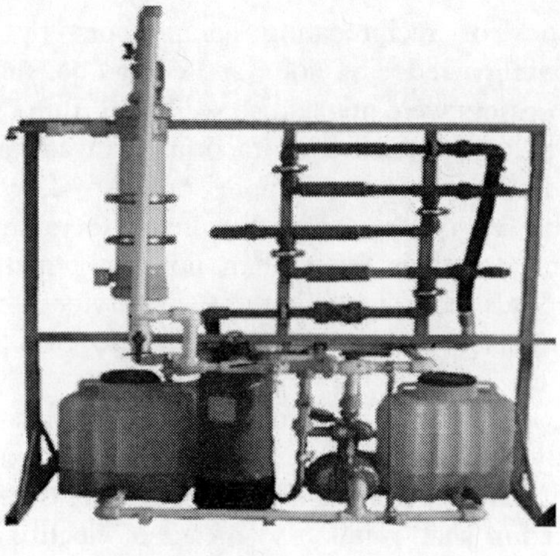

Fig. 2 Experimental setup

Table 1. Data about the setup and the elements

Outline of the test site	UFP
Water flow rate (litre per hour)	100
Number of cells	6
Machine mass, including a valve assembly filled with water (kilo)	35
Dimension XxYxZ (millimeter)	330x330x450
Pump power (kW)	0.5-2.5

To adjust the Total Dissolved Substances (TDS) setup was used and pH was controlled (Table 2).

Table 2. Chemical composition of the test solution

Test solution	Pure water characteristics (grams per litre)			Softened water characteristics (grams per litre)		
	TDS (grams per litre)	t (°C)	pH	TDS (grams per litre)	t (°C)	pH
Underground water	11.64	20	8.5..9.5	0.77	22	8.5..9.5

Prior to the testing, the solution was deaerated by purging ions (92..99%) for 2 h, because the underground water system is block with the outside, i.e., the closed system, so that the dissolved solids are very low. The unit for water pre-treatment at the ultrafiltration module allows the heat pump to operate with higher performance. When using clarified water heat pump equipment capital costs are reduced. One can get simultaneously higher quality of purified water. The ultrafiltration module can effectively collect suspended substances and the smallest particles of colloidal iron, due to the size of membrane pores 30..1000 Å. The particles up to 0.005E-06 m are removed at operating pressure of 0.2..1.0 MPa. A compressor is the most important and most expensive part of a heat pump. A heat pump ensures the thermodynamic cycle implementation and consumes a basic amount of electricity. It is impossible to achieve significant gains in efficiency and reliability of heat pumps without updating them. As for refrigerants hydrocarbons are today considered as promising alternatives to hydrofluorocarbons [10]. The reduction of pump noise impact is of great importance for consumers.

The most common type of compressor is currently the reciprocating one and its performance is studied thoroughly [9]. In recent years, the studies have identified the opportunities for further improving the characteristics of reciprocating compressors by providing smooth control of performance, enhancing reliability, reducing noise and energy consumption, weight and size, as well as cost. Other types of compressors were not studied well, and from this point of view, the question of improvement of their energy characteristics is important nowadays. An exergy analysis of vapor compression heat pumps with reciprocating compressors also shows that the largest exergy losses occur in the compressor (about 20% of the exergy supplied to the pump) [6-8].

Capacity control of compressors is most often done systematically by periodic starting and stopping [5]. However, this leads to the reduction of their service lifetime and does not permit to use fully the energy saving potential of a heat pump because of operation in off-design transient conditions.

Compressor operation depends on the mode of operation of a heat pump system. Capacity control of heat pumps can be done by reducing pressure at the compressor exit or by reducing throttle passage section or bypass of a working fluid. At present, the most progressive method of compressor capacity control is the change of speed of shaft rotation by means of electronic microprocessor device [4], so called Variable Frequency Drive (VFD). Such regulation ensures heat pump operation in design mode and reduces the number of starts of the compressor to a minimum. We have conducted an experimental study of energy characteristics of a heat pump water-water using a VFD for regulating the compressor efficiency.

To carry out the experiment we connected the reciprocating compressor with VFD "Siemensmicromaster 420" of AC in range of 0 to 650 Hz, designed for rated capacity of 2.2 kW, as well as electricity meter BFM 136, according to the schematic circuit in fig. 3.

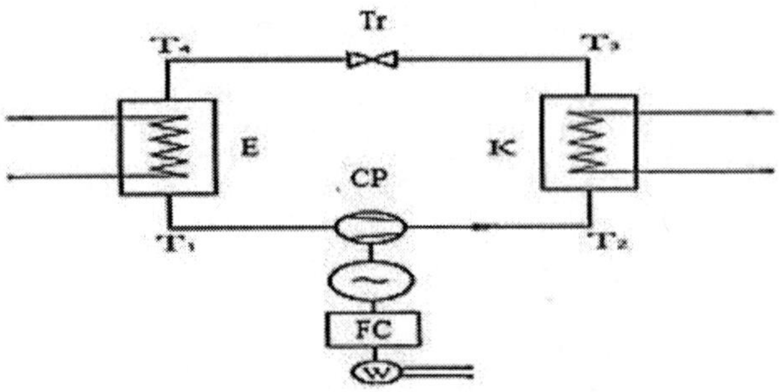

Fig. 3 Schematic circuit of the water source heat pump: CP – compressor; FC – frequency converter (VFD); K – condenser; Tr – throttle valve; E - evaporator; W - electricity meter

The energy efficiency of the heat pump is characterized by Coefficient of Performance (COP):

$$COP = \frac{Q_n}{N_{cp}}, \quad (1)$$

with Q_n – produced heat; N_{cp} – used electricity.

The compression ratio is determined by eq. 2:

$$\pi_{к} = \frac{p_{к}}{p_{и}}, \quad (2)$$

with $p_{к}$ - refrigerant vapor pressure at the compressor exit; $p_{и}$ - vapor pressure at the evaporator exit.

The dependence of COP on compression degree of the compressor is obtained by processing the experimental data in Microsoft Office Excel by means of Scatter Chart, and adding a polynomial trendline of order 2 using the least square method [14].

Case study, results and discussions

The results of experimental research of quality indicators of softened water are shown in fig. 4, 5.

The ground water after preliminary treatment in the ultrafiltration unit by means of centrifugal pump is supplied to the evaporator of the heat pump.

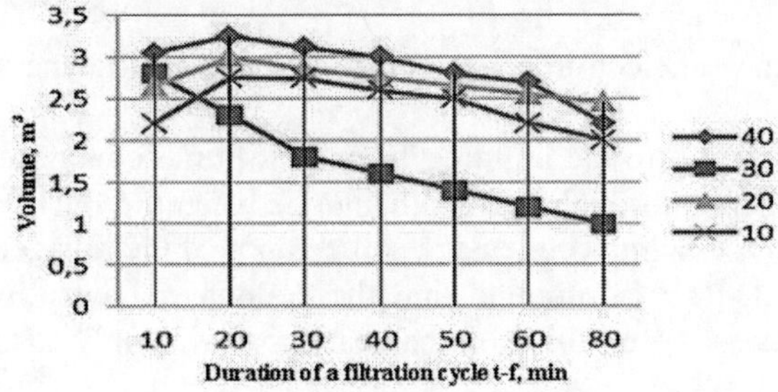

Fig. 4 Volume (m^3) profiles within a filtration cycle

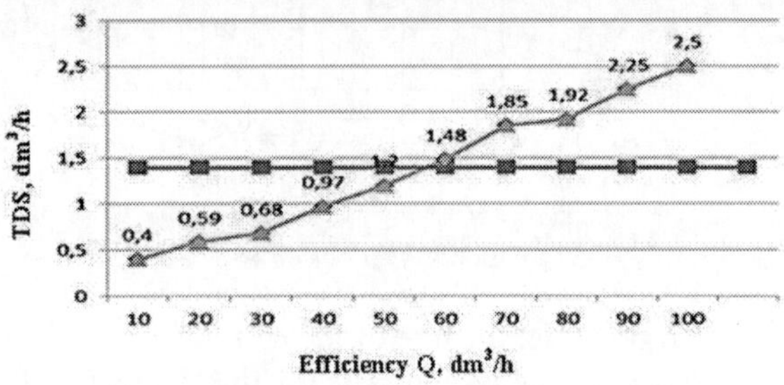

Fig. 5 Efficiency of opposite of TDS

Measured energy characteristics of a circular process in the heat pump unit allowed determining the energy performance parameters presented in the graphs (fig. 6..9). The dependence of COP on the speed (rpm) showing effective values of COP 5.55..7.23 with speed in the range of 1400 to 2150 rpm is drawn in fig. 6 based on the results of the experiments.

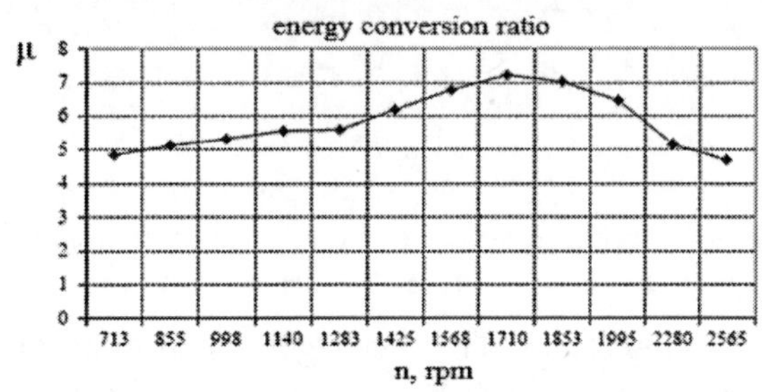

Fig. 6 COP in function of a speed (rpm)

Optimal values of compressor electromechanical η_{em} and internal efficiency $\eta_i = f(n)$ (fig. 7) correspond to the interval of variation of speed in the range of 1400 to 2150 rpm, which is determined by the intersection of curves $\eta_{em}=f(n)$ and $\eta_i = f(n)$, which corresponds to $\eta_{em} = \eta_i$ =0,95.

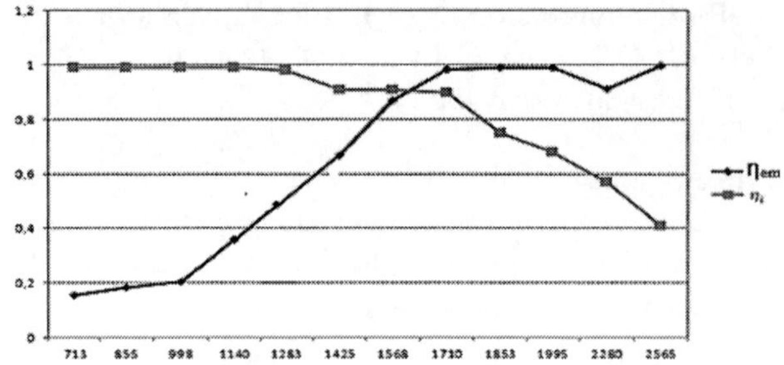

Fig. 7 Dependence of electromechanical and adiabatic efficiency on speed (rpm)

Thermal energy is transferred to the building from heat pump condenser because of condensation of a working fluid or refrigerant through heat exchange between the refrigerant and heat conductor of the heat consumer. In this case, while condensing some vapor of the refrigerant (R22) gives off heat to cold air of the room. Heated air is supplied into the condenser cavity by a fan. Fig. 8 shows the influence of changing the speed (rpm) drive on the heat capacity supply of the heat pump.

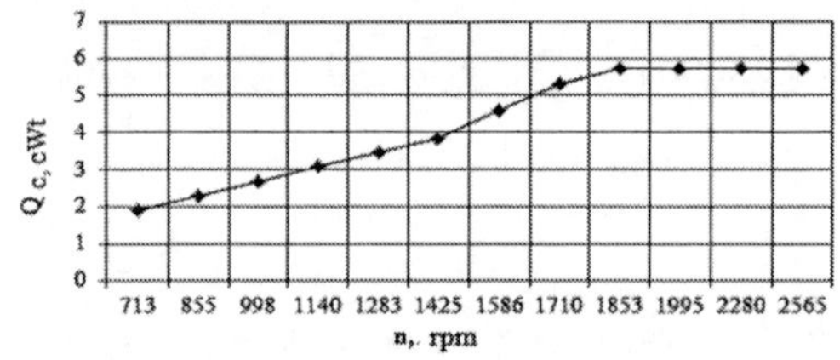

Fig. 8 Condenser heat load with a speed (rpm) on x-axis

The decrease of refrigerant compression degree of the compressor πк by an average 1.3 times leads to the increase of COP by 1.6, and on the contrary, to the decrease of electrical capacity of the compressor needed to drive it. Fig. 9 shows the dependence of COP and $N_{км}$ on compression ratio $\pi_к$. The dependence of COP on compression degree of the compressor shown in fig. 9.

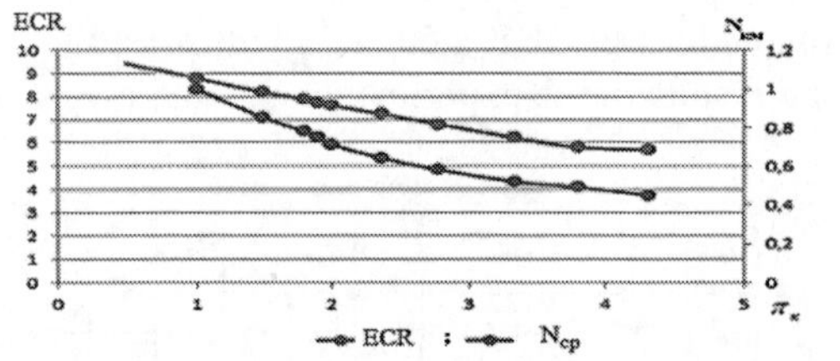

Fig. 9 Correlation between compression ratio and COP

Processing the experimental results on studying the dependence COP of heat pump on compression ratio $\pi_к$ by the least square method allowed to calculate the coefficients of regression equation:

$$COP = 6{,}83\pi_к{}^2 - 34{,}05\pi_к + 46{,}94\,. \quad (3)$$

Conclusions

The value of pump COP obtained during the experiment and the value determined by calculation conform to each other within experimental errors with the regulation of compressor capacity. The dependence of heat load of pump condenser on the speed (rpm) is approximated by the formula and will be used to design a heat pump unit.

The proposed experimental unit allowed us to determine the correlation between basic characteristics and the compressor shaft speed (rpm) and determine the possible range of its regulation (50..180%). The obtained results allow us to design the water treatment plant with using heat pump unit with quality regulation systems as well. The results will also provide the deeper control of power and temperature regimes, better matching the changing parameters of heat supply systems and the environment, as well as high efficiency in operation.

References

[1] Chicherin, S. (2016). New approach to determination of corrosion damage degree of pipeline system elements. Bulletin of the Tomsk Polytechnic University. Geo Assets Engineering. V. 327. 12. 110–115

[2] Chicherin, S. (2017). Unlocking a Potential of District Heating Network Efficient Operation and Maintenance by Minimizing the Depth of a Trench System. Bulletin of the Tomsk Polytechnic University. Geo Assets Engineering. V. 328. 9. 49–56

[3] Chicherin, S. (2018). Low-temperature district heating distributed from transmission-distribution junctions to users: energy and environmental modelling. Energy Procedia, 147, 382–389. doi:10.1016/j.egypro.2018.07.107

[4] Xia, L., Ma, Z., McLauchlan, C., & Wang, S. (2017). Experimental investigation and control optimization of a ground source heat pump system. Applied Thermal Engineering, 127, 70–80. doi:10.1016/j.applthermaleng.2017.07.205

[5] Ndiaye, D. (2016). Reliability and performance of direct-expansion ground-coupled heat pump systems: Issues and possible solutions. Renewable and Sustainable Energy Reviews, 66, 802–814. doi:10.1016/j.rser.2016.08.049

[6] Menberg, K., Heo, Y., Choi, W., Ooka, R., Choudhary, R., & Shukuya, M. (2017). Exergy analysis of a hybrid ground-source heat pump system. Applied Energy, 204, 31–46. doi:10.1016/j.apenergy.2017.06.076

[7] Dong, X., Tian, Q., & Li, Z. (2017). Energy and exergy analysis of solar integrated air source heat pump for radiant floor heating without water. Energy and Buildings, 142, 128–138. doi:10.1016/j.enbuild.2017.03.015

[8] Hu, P., Hu, Q., Lin, Y., Yang, W., & Xing, L. (2017). Energy and exergy analysis of a ground source heat pump system for a public building in Wuhan, China under different control strategies. Energy and Buildings, 152, 301–312. doi:10.1016/j.enbuild.2017.07.058

[9] Willich, C., & White, A. J. (2017). Heat transfer losses in reciprocating compressors with valve actuation for energy storage applications. Journal of Energy Storage, 14, 322–328. doi:10.1016/j.est.2017.07.024

[10] Byrne, P., Ghoubali, R., & Miriel, J. (2014). Scroll compressor modelling for heat pumps using hydrocarbons as refrigerants. International Journal of Refrigeration, 41, 1–13. doi:10.1016/j.ijrefrig.2013.06.003

[11] Junussova, L. R. (2016). Improving the Quality of Desalinated Water with Combined Water Treatment Plant Boiler. Alternative Energy and Ecology (ISJAEE), (23), 167–176. doi:10.15518/isjaee.2015.23.021

[12] Junussova, L. R. (2015). Membrane plants in water conditioning schemes at thermal power stations. European Applied Sciences. Stuttgart, Germany. 4, 60-63.

[13] Ikeda, S., Choi, W., & Ooka, R. (2017). Optimization method for multiple heat source operation including ground source heat pump considering dynamic variation in ground temperature. Applied Energy, 193, 466–478. doi:10.1016/j.apenergy.2017.02.047

[14] Xu, W., Chen, W., & Liang, Y. (2017). Feasibility study on the least square method for fitting non-Gaussian noise data. Physica A: Statistical Mechanics and Its Applications. doi:10.1016/j.physa.2017.11.108

IOP Conf. Series: Earth and Environmental Science **288** (2019) 012093 doi:10.1088/1755-1315/288/1/012093

Estimation of swirl intensity behind short twisting devices

A.I. Kadyirov, E.K. Vachagina

IPEAT FRC Kazan Scientific Center of RAS
Lobachevsky str.2/31, PO inbox 261, 420111, Kazan, Russia

e-mail: aidarik@rambler.ru

Abstract. The numerical results of the effect of short twisting devices on the damping coefficient of swirl are presented. As for twisting devices four-, six- and eight-blade inserts with a twist ratio S/D =3...5. To describe the rheological behavior of a pseudo-plastic fluid, the 0.65% aqueous carboximethyl cellulose (0.65% CMC) solutions is taken. The problem is solved by using Comsol Multiphysics.

1. Introduction

It is well known that the intensity of heat transfer under the viscous flow generally depends on the velocity distribution (convective components) [1-2]. Analysis of extensive experimental data on the flow structure obtained with various swirling methods revealed a unique relationship between the flow structure and the damping coefficient of swirl [3].

2. Problem statement

It is considered a stationary laminar three-dimensional flow of 0.65% CMC in a round tube (Fig. 1) with short twisting inserts. As for twisting devices four-, six- and eight-blades inserts with a twist ratio S/D =3...5 and length L_2=0.25D are considered. The geometric characteristics of channel are following: D = 2R= 0.025 (m); L_1= 1.5D; L_3= 16D; thickness of swirl device 0.001 (m).

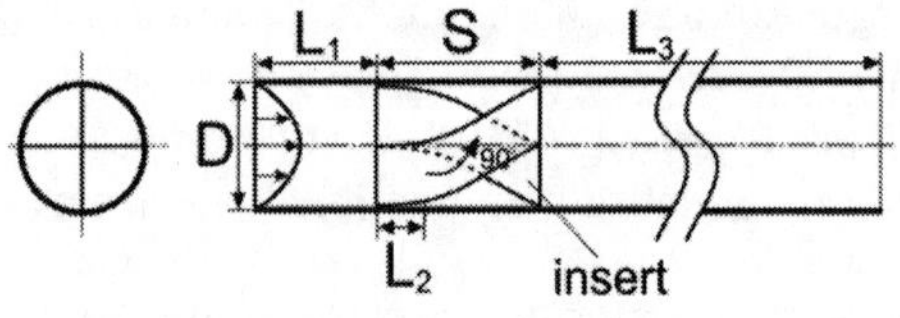

The governing equations:

$$\rho\left(\nabla\bar{V}\cdot\bar{V}\right)=\nabla\cdot\left(-p\mathbf{I}+\mu\left(\mathbf{I}_2\right)\left(\nabla\bar{V}+\left(\nabla\bar{V}\right)^T\right)\right) \quad (1)$$

$$\nabla\cdot\bar{V}=0 \quad (2)$$

where p is the pressure, Pa; $\mathbf{I}$ is the unit tensor; $\mu(\mathbf{I}_2)$ - apparent dynamic viscosity, Pa*s; $\mathbf{I}_2$ - the second invariant of rate-of-strain tensor; ρ - density of the liquid, kg/m^3; $\bar{V}$ - velocity vector.

The boundary conditions are following:

1) a developed velocity profile is taken at the entrance to the pipe [4]

$w = V_a\left(\frac{3n+1}{n+1}\right)\left[1-\left(\frac{r}{R}\right)^{(n+1)/n}\right]$, where V_a - average velocity, R - pipe radius, n - parameter of Power law model.

2) no-slip conditions are taken on the walls of the channel;

3) at the outlet of the channel there is the normal stress equal zero.

Additionally it is assumed that the mass forces are negligible. The Power law model is taken to predict the rheological behavior of 0.65% CMC with following parameters: k=0.2041($\mathrm{Pa}\cdot \mathrm{s}^n$), n=0.716. [4] To avoid uncertainty in singular point it is assumed that for values of shear rates less than 0.0001 (1/s), the viscosity sets as constant and equals to the viscosity value at a shear rate of 0.0001 (1/s).

Fig.2 shows the comparison between obtained numerical results and experimental results for laminar flow in pipe with full-length twisted tape [5]. Relative error is less than 2.5%, where f_D - Darcy friction factor.

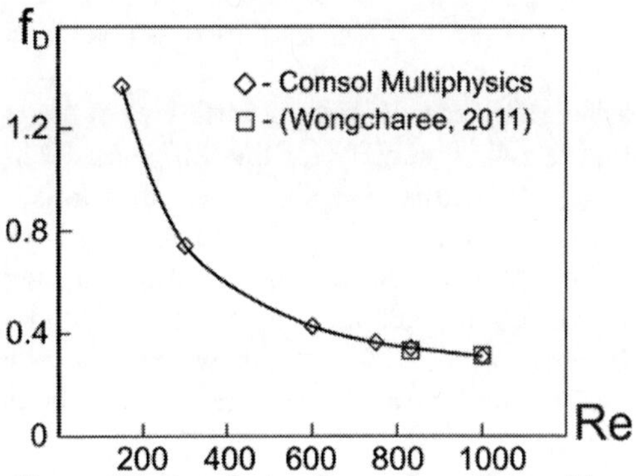

Fig. 2. Darcy friction factor for various Reynolds number

The following expression is used to calculate the damping coefficient of swirl in dimensionless form [2]

$$m(\tilde{z}) = \int_0^1 \tilde{v}_\varphi \tilde{v}_z \tilde{r}^2 d\tilde{r} \bigg/ \int_0^1 (\tilde{v}_z)^2 \tilde{r} d\tilde{r}, \tag{3}$$

$\tilde{v}_\varphi = v_\varphi / V_a$, $\tilde{v}_z = v_z / V_a$, $\tilde{r} = r/R$, $\tilde{z} = z/(R\cdot \mathrm{Re}_n)$.

The Reynolds number is calculated by using following expression [4]

$$\mathrm{Re}_n = \frac{D^n V_a^{2-n}\rho}{k' 8^{n-1}}, \quad k' = k\left(\frac{3n+1}{4n}\right)^n, \tag{4}$$

where k and n – parameters of the Power law model.

3. Results

It is observed that an increase in the number of blades leads to an increase in the swirl intensity (Fig. 3). The set of curves for a wide range of Reynolds numbers for each insert can be described an exponential law (5).

$$m(\tilde{z}) = A\exp\left(-B\tilde{z}\right) \quad (5)$$

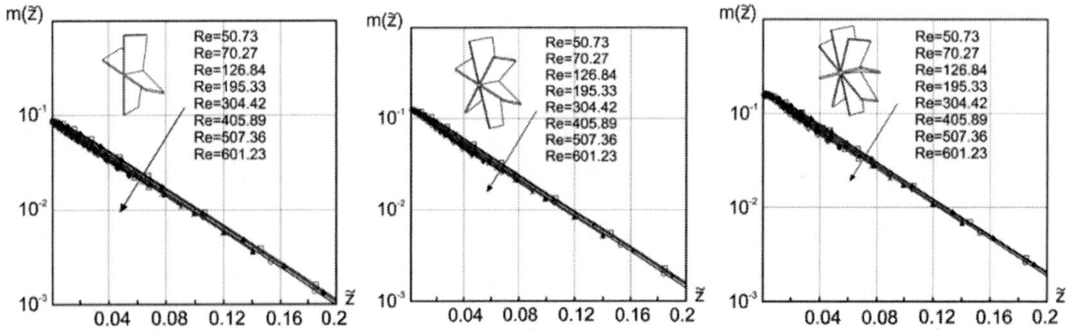

Fig. 3. Damping coefficient of swirl for various inserts (S/D =4, 0.67% CMC)

Figure 4 shows a comparison between damping coefficient of swirl for 0.67% CMC and water. In the vicinity behind the insert, the swirl intensity for the 0.67% CMC has higher values than the water. However, the trend changes after some distance. This phenomenon is due to the rheological properties of CMC, the decrease in viscosity with increasing shear rate. It can be seen that the number of blades affects only the pre-exponential factor. For water, the equations are written in parentheses and printed in italics. A directly proportional relation between the number of blades and the pre-exponential factor is found and valid for both liquids: 0.67% CMC and water. For example, a double increase in the number of blades corresponds to a double increase in the value of the pre-exponential factor. The error does not exceed 4.2%.

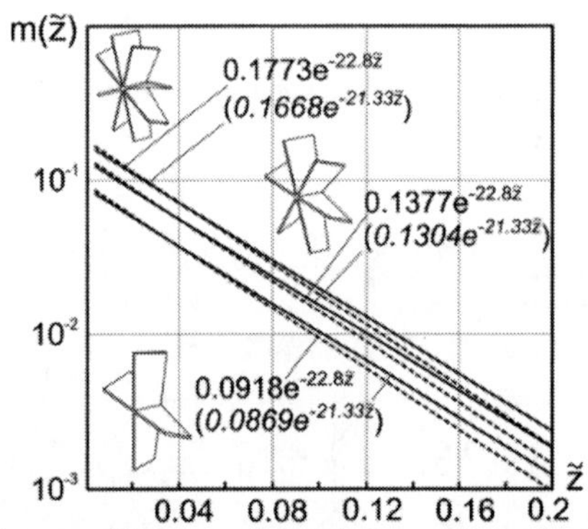

Fig. 4. Damping coefficient of swirl for various inserts (S/D = 4):
dashed line - 067% CMC; solid line – water

The inverse proportional relationship between twist ratio and the pre-exponential factor in corresponding equation (5) is established. For example $\left(5_{(S/D)}/3_{(S/D)}\right) \approx \left(0.116/0.0696\right)$ (Fig. 5). The largest value of the parameter A in equation (5) corresponds to the largest value of the damping coefficient of swirl.

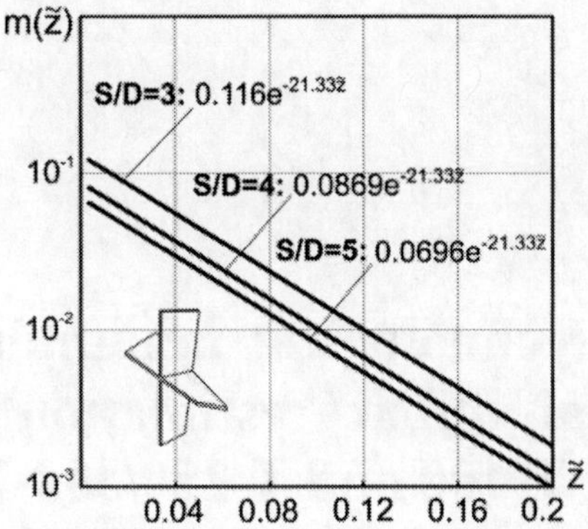

Fig. 5. Damping coefficient of swirl for various twist ratios (water)

4. Conclusion

The represented numerical results have shown that the damping coefficient of swirl can be described by an exponential equation for a wide range of Reynolds numbers. The general form of this dependence has the form $m(\tilde{z}) = A\exp(-B\tilde{z})$, where $B = 22.8$ (0.67% CMC) and $B = 21.33$ (water). The magnitude of the pre-exponential factor is inversely proportional to the twist ratio and directly proportional to the number of blades of insert.

References

[1] Manglik R M, Bergles A E 1993 *Journal of Heat Transfers*. vol 115 881-889
[2] Masaru K, Shoichiro F, Mikio A 1971 *Bulletin of the JSME* vol 14 **73** 659–670
[3] Shchukin V K, Khalatov A A 1982 Heat and mass transfer, hydrodynamics of swirled flows in axisymmetric channels (Moscow: Mechanical engineering) (*in Russian*)
[4] Wilkinson W I 1960 Non-Newtonian fluids (London – Oxford – New York – Paris)
[5] Wongcharee K, Eiamsa-ard S 2011 *International Communications in Heat and Mass Transfer* **38** 348–352

IOP Conf. Series: Earth and Environmental Science **288** (2019) 012094 doi:10.1088/1755-1315/288/1/012094

Reduction of the technological minimum by bypass method with environmental impact estimation on the example of energy blocks with T-100 and T-250/300-240 turbines

Kalyutik A.A.[1], **Grigorieva D.V.**[1] **Fedyukhin A.V.**[2], **Derevianko O.V.**[1], **Akhmetova I. G.**[3]

[1]Peter the Great St. Petersburg Polytechnic University, 194021, Russia, St.Petersburg, Polytechnicheskaya 29

[2]National Research University «Moscow Power Engineering Institute», 111250 Russia, Moscow, Krasnokazarmennaya 14.

[3]Federal State Budgetary Educational Institution of Higer Education "Kazan State Power Engineering University", Kazan, Krasnoselskaya street, 51

Kalyutik@yandex.ru

Abstract: The article describes existing methods of reducing the technological minimum with a quantitative assessment of reducing emissions of pollutants at the CHP using the method of bypass. With the help of specialized program complex the efficiency of the method has been analysed on the example of the power units at Yuzhnaya CHP-22 and CHP-21 JSC "TGC-1" with turbines T-250/300-240 and T-100-130. The article provides guidelines for the range of application of this method as well.

Statement of the problem

Thermal power plants (TPP) have a prominent place in the electric power industry of Russia. 537 TPPs with total electrical capacity of 148.6 thousand MW and 640.5 million MWh per year electric power generation operate in Russia in 2018. The annual fuel consumption of thermal power plants of general use in 2018 is estimated 265.8 million tons of reference fuel. At present, most TPPs of RF operate in the wholesale electricity and capacity market (OREM). A significant share of thermal power plants in Russia is of combined heat and power plants (CHP).

Methods of dealing with night of-peaks at CHP plants used by power companies today are based on simplified methods of selecting the main equipment, which does not provide an opportunity to find beneficial solutions and considers expected dynamics of changes in thermal loads and the uncertainty of prices for WECM insufficiently. As a result, generating companies are forced to transfer CHP power units to a reduced generation of electricity (technological minimum) mode, while maintaining the generation of thermal energy at a certain level in accordance with the schedule.

Optimization of the technological minimum at the CHP allows to reduce the fuel costs without significant investments, as well as decrease pollutant emissions [1-6].

Methods description

Method of bypassing or shutting down the high-pressure heater (HPP) can be applied to deal with night off-peak loads. Reduction of the electric power occurs due to the reduction of the steam flow to the head part of the turbine by the value of the displaced withdrawals, while the load and the parameters of the heat-extraction withdrawals remain unchanged. When the bypass channel is completely opened through the HPP, the feedwater consumption is 25-30% of its total flow. In this case, heaters work reliably with the normalized parameters of the heating steam and water. This method has the following limitations:

- regulation of a large number of valves;- difficulty in reducing steam pressure, the speed should not exceed 0.6 kgf / cm2 per minute (setting the condensate level);
- due to HPP bypass, the temperature of the feed water also decreases, while all the boiler units are designed for a certain temperature of the feed water, which means that a certain control range for the feed water must be provided.

Table 1 shows the test results of the technological minimum passage by applying the method of bypass LDPE on the T-100 block

Table 1. Test results

Characteristic	Initial	HPP bypass
Electric power, MW	77	70
Steam consumption per turbine, kg / s	91,5	79,7
Fuel consumption, kg / s	8,9	8,6
Heat supply, MW	154	154

Changing the pressure in the heating systems is done by adjusting the diaphragm or by applying completely closed position by changing the steam flow to the turbine. Maintaining the specified heat load is achieved by adjusting the proportion of the bypass of the network heaters. The scheme of the steam-turbine unit (STU) is shown in Fig. 3

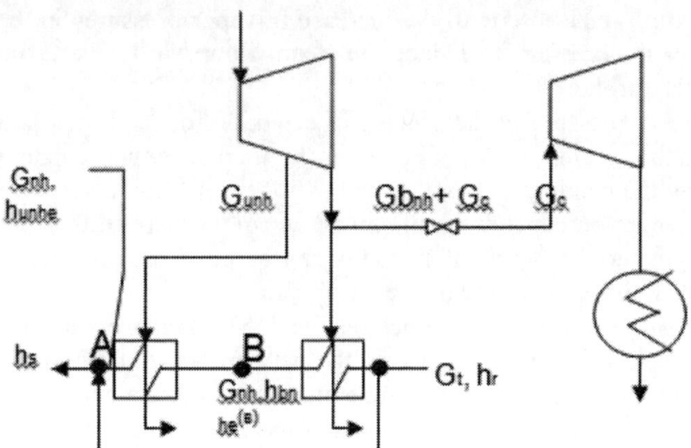

Fig. 1 The scheme of the bypass of the STU network heaters.
Gunh, Gbnh, Gc - steam consumption, respectively, in the upper (UNH) and bottom (BNH) network heaters, as well as in the condenser; Gt, Gnh, Gbl- network water flow, respectively, common, through the network heaters and through the bypass line; h_r, h_s, h_{unhe}, $h_{bnhe}^{(B)}$ – enthalpy of the network water, respectively, reverse, direct, at the exit from the UNH and at the exit from the BNH (at point B).

The turbine unit operates in the TP mode with a closed rotary diaphragm. The flow rate of steam in the condenser is a ventilation pass of steam through the diaphragm. The water flow through the network heaters is changed by bypassing a part of the flow through the bypass line. Directing part of the network water to the bypass of the network heater can be used to reduce the electric power of the turbine.

For this type of turbine units it is effective to work with counter pressure in the upper heating chamber by bypassing the network heaters. Greater flexibility in the distribution of thermal loads among network heaters can be achieved either by simultaneous flow of both heaters along the bypass line, by an individual circuit of each of the heaters, or by a combination of a common bypass with a bypass of one of the heaters. In some cases it can be limited to one heater bypass.

The thermal capacity of the heating selections and the electric power of the turbine are determined by the following formula:

$$Q_{OT} = G_{unh}(h_u - h'_{Bu}) + G_{bnh}(h_b - h'_b)$$
$$N = (G_{unh} + G_{bnh})(h_0 - h_u) + G_{lnh}(h_u - h_b) + \sum G_i H_i ,$$

where h_0, h_u, h_b is steam enthalpy , respectively, before the turbine, in the upper and lower selections.

The first term in the power expression determines the power produced by the steam in the turbine compartment before the lower selection, the second term is the capacity of the intermediate compartment, and the third term is the power of the steam chosen for regeneration.

To reduce the electrical power at a constant heat output power, it is necessary to increase the steam pressure in the upper selection (sliding counter pressure). The steam pressure in the lower selection will also increase.

In a wide pressure range pu and pb, the differences $(h_u - h'_u)$ and $(h_b - h'_b)$ remain practically unchanged. In this case, it follows from the first equation that with increasing counter pressure p_u occurs with a constant heat load Q_{OT}, the redistribution of loads between the upper and bottom network heaters, but at the same time the amount of steam consumption remains practically unchanged [7-9].

Residual $(h_0 - h_u)$ decreases due to growth h_u. The change in the power of the intermediate compartment (the second term) is a subject to additional research. The steam consumption for regeneration is somewhat reduced due to the increase in vapor pressures in the selections, so to save the sum (G_{unh}+G_{bnh}) it is necessary to reduce the steam flow rate to the turbine and thereby further reduce the power of the turbine.

The increase in steam pressure in the process of selection for the heat production allows reducing the electric power of the turbine while maintaining the thermal power capacity. The range of power reduction is limited by the maximum pressures in the upper selection determined by the manufacturer. The range of power reduction depends on the initial steam pressure in the upper selection: the lower the pressure is the greater the depth of turbine's discharge. Consequently, the depth of turbine's discharge increases with the temperature of the outside air.

The temperature of the network water behind the UNH increases due to the increase in steam pressure in the upper selection, but after mixing the network water flow after UNH with the flow of water supplied through the bypass line, the temperature of the direct supply water will be equal to its initial value. Since the total flow of the network water has not changed, the heat absorbed by the water will remain unchanged.

Thus, it can be concluded that if a part of the regeneration selections (selections for HPH) is cut off, it is possible to reduce the steam flow to the turbine by the same amount.

In heat-recovery turbines approximately 15-20% of electricity is produced by a steam of regenerative sampling, that is, the regulation by the variable regeneration coefficient changes the power within these limits [10-12]. The calculations performed for STU T-100-130 are shown in Fig. 2

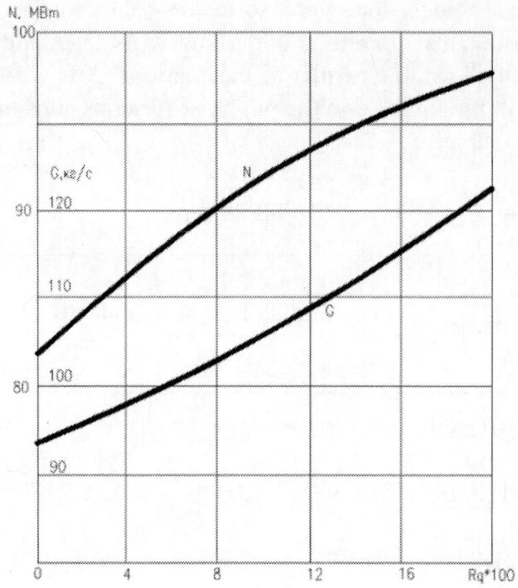

Fig. 2. Dependence of turbine T-100-130 characteristics on the coefficient of heat regeneration.

If the heat regeneration factor is changed from 0.2 to 0, the electric power is reduced from 98 to 81 MW with a decrease in steam consumption from 122 kg / s to 93 kg / s, hence this method of reducing the technological minimum is very effective for thermal stations [6,7].

STU bypass research showed that:

1. Bypassing of some part of feedwater around the STU can be an effective method of unloading.
2. Decrease in feedwater temperature in the regime of free distribution of feed water is not much higher than in the initial state, which does not affect the reliability of the pipelines and the heating surfaces of the boiler.
3. The use of this method leads to a decrease in steam consumption for the turbine and a decrease in electric power by 15-20%. The heat release from the turbine selections remaines unchanged.
4. In heat-recovery turbines approximately 15% of the electricity is produced by the steam of the regenerative sampling, that is, the regulation by the variable regeneration factor changes the power within these limits [13-15].

Based on test data, the average duration of the heating period and the duration of the night off-peak, we calculated possible fuel economy, which is 6400 tons of reference fuel. (0.68% of the average annual consumption).

To assess the possible environmental aspects of reducing the technological minimum by bypass method pollutants emission were calculated (applying LDPE bypass method and not using it), the results are given in Table 2.

Table 2.Pollutants emission

	NO		NO2		SO2	
	Emission (tons per year)	MAE (tons per year)	Emission (tons per year)	MAE (tons per year)	Emission (tons per year)	MAE (tons per year)
1working mode	511,008	510,6	2501,369	3142,1	21,097	5855,5
2 pworking mode	507,786	510,6	2484,321	3142,1	20,937	5855,5

As can be seen from Table 2, the decrease in the technological minimum did not significantly affect the amount of pollutants emission, but allowed to meet the maximum allowable emission (MAE) in NO. Table 3, 4 shows the results of calculations of fees for pollutants emission. Since the excess of MAE was small, the savings on the payment for emissions also turned out to be insignificant and amounted to 4017.37 rubles.

Table 3. Fee for pollutants emission (1 working mode)

Name of pollutants	MAE	Emission	Also within the limits of		Standard fee	Amount of payment for		
			MAE	>ПМАЕ		MAE	>MAE	Total
Nitrogen dioxide	3142,1	2501,369	2501,369	-	52	419089,37	0,00	419089,37
Nitric oxide	510,6	511,008	510,600	0,408	35	57580,36	1150,25	58730,62
sulphur dioxide	5855,5	21,097	21,097	-	21	1427,47	0,00	1427,47
					Total:	478097,20	1150,25	479247,45

Table 4. Fee for pollutants emission (2 working mode)

Name of pollutants	MAE	Emission	Also within the limits of		Standard fee	Amount of payment for		
			MAE	>MAE		MAE	>MAE	Total
Nitrogen dioxide	3142,1	2484,321	2484,321	-	52	416233,08	0,00	416233,08
Nitric oxide	510,6	507,786	510,600	-	35	57580,36	0,00	57580,36
sulphur dioxide	5855,5	20,937	20,937	-	21	1416,64	0,00	1416,64
					Total:	475230,08	0,00	475230,08

Emissions of the main pollutants (nitrogen dioxide and sulfur dioxide) from two power units of CHP were calculated the following way:

$$\text{Nitrogen dioxide emission (NO2)} = m \cdot Ql \cdot KNO2 \cdot (1-\beta) \cdot 10\text{-}3,$$

where: Ql- net calorific value of fuel, m – amount of fuel spent, KNO2 - a parameter characterizing the amount of nitrogen oxides generated per GJ of heat, kg/GJ, β - coefficient, depending on the degree of reduction of nitrogen oxides emission. Calculation of Nitric Oxide (NO) emission is similar to the calculation of Nitrogen Dioxide (NO2) emission. Determining the quantity of sulfur dioxide emission:

$$\text{Sulfur dioxide emission (SO2)} = 0{,}02 \cdot m \cdot S \cdot (1-\eta') \cdot (1-\eta''),$$

where: S - sulfur content in fuel; η′ - fraction of sulfur oxides bound by fly ash of fuel, η″ - fraction of sulfur oxides trapped in the ash collector.

As can be seen from Tables 2 and 3, the decrease in the technological minimum had no significant effect on the amount of pollutant emissions, but it allowed to meet the regulatory requirements for maximum permissible emissions, which is not always possible when the power unit is operating in nominal mode. [16-18].

Conclusion

Performed calculations make it possible to evaluate the efficiency of the LDPE bypass method for reducing the technological minimum. Given the high cost and the amount of energy consumption, this method will save a significant amount of money due to fuel economy. The reduction of pollutant emissions at CHP was not so noticeable; however, at the country level application of this method can have a significant positive effect on the ecological situation.

The problem of using this and other methods to reduce the technological minimum lies in the field of operational dispatch management. A competent distribution of loads of generating equipment will allow the most efficient use of energy resources, which in its turn will have a positive environmental impact [19, 20].

Acknowledgments

The work was supported by the Ministry of Education and Science of the Russian Federation on fundamental scientific research (Agreement No. №13.6994.2017/БЧ) «Development of a methodology for determining the reliability of a heat supply system to improve energy efficiency».

References

[1] Nizamov L.A., Chichirova N.D. Software package for optimization of modes of thermal power plant. Proceedings of the higher educational institutions. ENERGY SECTOR PROBLEMS. 2016;(1-2):64-74. https://doi.org/10.30724/1998-9903-2016-0-1-2-64-74

[2] Suter P. Research of ventilating losses in steam turbines// Report of institute of turbomachines. - Zurich, 2011. – № 4. – S. 15-28.

[3] Aminov R.Z., Garievsky M.V. The efficiency of combined-cycle CHP plant with variable electric loads, taking into account the wear and tear of equipment. *Proceedings of the higher educational institutions. ENERGY SECTOR PROBLEMS.* 2018;20(7-8):10-22. https://doi.org/10.30724/1998-9903-2018-20-7-8-10-22

[4] Koepp M., Wunsch M., Klotz E.The prospects of application of the thermal and cooling cycles in Austria// E-Wirtschaft. – Berlin, 2014. –S. 115.

[5] Loskutnikov A.A., Goryunov I.M., Bakirov F.G. Mathematical modeling of KEU on the basis of GTU and TOTE in the computer environment// Bulletin of the Ufa state aviation technical university. – 2012. - №. 2(47). – S. 127-137.

[6] Grigorieva D.V., Kalyutik A.A. Evaluation of the effectiveness of methods to mitigate technical minimum of the power unit of thermal power plant. Proceedings of the higher educational institutions. ENERGY SECTOR PROBLEMS. 2017;19(7-8):3-12. https://doi.org/10.30724/1998-9903-2017-19-7-8-3-12

[7] Gwosdz A., Leisse A., Quenders H.-J. Pulverised Coal Firing System for the Operation of Steam Generators with Low Exess Air// VGB Power Tech., - Berlin, 2015 №11. – S. 25-36

[8] Usov S.V., Kazarov S.A. // Modes of thermal power plants. L.: Energoizdat. 1985.- 240 s.

[9] Ivanov V. A. Modes powerful steam-turbine units. L.: Energoatomizdat. 1986. -248 s.

[10] Bezlepkin V.P., Mikhaylov S.Ya. The adjustment range of thermal power plants. L.: Energoatomizdat. 1990 – 168 s.

[11] ZaytsevM.V., SlabchenkoO.N., AlekhinA.A., ChupyraA.G. The choice of the variant of modernization of power units K-300-240 Zmievskaya and Zuevskaya TPP // Energy and electrification – 1996. – № 6. – S. 11-14.

[12] Slabchenko O.N., M.V. Zaytsev, A.Yu. Kozlokov, A.D. Zolotukhin. Evaluation of the Effectiveness of the LPC of turbine T-250/300-23.5 in Kharkiv CHPP-5 // Energy and thermal engineering processes and equipment. –2009. –№3. –S.41-48.

IOP Conf. Series: Earth and Environmental Science **288** (2019) 012094 doi:10.1088/1755-1315/288/1/012094

[13] SimoyuL.L., GutorovV.F., LagunV.P., BarimbergG.D. The efficiency of turbine T-250/300-240 by upgrading LPC // Teploenergetika. – 2005. – № 11. – S. 68-74.
[14] BezlepkinV.P. Regulation of the electric power of cogeneration turbines operating on heat schedule // Teploenergetika № 12. 1985. s. 49-52
[15] IvanovV.A., BorovkovV.M., VanchikovV.V., KutakhovA.G. To the question of improving the maneuverability of power plants for heat schedule // Energy, №7. 1982. S. 39-43
[16] IvanovV.A., KutakhovA.G., SerebryanikovN.I., BogomolnyyD.S. The use of CHP units for passing low electric load diagram // Teploenergetika №9. 1984. -S.10-13.
[17] Pod red. Brodov Yu.M., Kortenko V.V. Steam turbine and turbine of the Ural turbine plant. Yekaterinburg: Aparino. 2010 – 488 s.
[18] Moshkarin A.V., Kopsov A.Ya., Malkov E.S., Doverman G.I., Zaharankov E.V. New technologies in design of steam-turbine, steam-gas and vaporizing installations of thermal power plant// Tehnologii energosberejeniya na TES. -2013. - №2. – S.3-11.
[19] Kozlokov A.Yu., Goloschapov V.N., Kasilov V.I., Kasilov O.V., Ivanova L.A. Determination of a consumption of steam through TsND of the turbine T-250/300-240 during the work on the heating mode// East European magazine of advanced technologies. – 2011. – №. 6/8(54). – S. 31-36.
[20] Bistrickii L.N., Goloschapov V.N., Kozlokov A.Yu., Kasilov V.I., Shubenko A.L. Energy consumption by a turbine step at low-account operating modes// Bulletin of national technological university "Kharkiv Polytechnical Institute". Power and heat technical processes and equipment. – 2011. – №. 5. – S.60-63.

Selection of optimum inclination angle of a nozzle at circulatory heating of fuel oil in the RVS-3000 tank

R.F. Kamalov[1], J.V. Karaeva[1], R.R. Vildanov[2]

[1]Institute of Power Engineering and Advanced Technologies – Subdivision of the Federal Research Center «Kazan Scientific Center of Russian Academy of Sciences», P.B. 261, 2/31 Lobachevsky str., Kazan, 420111, Russia

[2]Federal State Budgetary Educational Institution of Higher Education «Kazan State Power Engineering University», 51 Krasnoselskaya str., Kazan, 420066, Russia

rustemran@mail.ru

Abstract. The paper presents the mathematical model and the numerical studies results of heat transfer and hydrodynamics at circulatory heating of the M100 fuel oil in the RVS-3000 tank. Calculations for various inclination angles of the nozzle in the range of 40° to -40° in increments of 10° were made. The winter storage period of fuel oil in the fuel tank is considered. Comparison of the tank heating time up to the hot storage temperature (60°C) and up to the fuel temperature before feeding to the nozzles for combustion (90°C) is performed.

Introduction

Most of the research on heat transfer and hydrodynamics at circulatory heating of the fuel oil was carried out by the staff members of the Research center for energy problems Kazan Scientific Center of Russian Academy of Sciences under the guidance of the corresponding member of the RAS Yu.G. Nazmeev [1 – 5]. The most significant results for storage of liquid organic fuel and oil products in tanks and tank fields equipped with circulatory heating systems are presented in the book [1]. This work also describes mathematical models and the numerical studies results of thermal and hydrodynamic processes for submerged plane straitened, and free jets of viscous liquid in the filled storage tanks.

In [6] heat exchange processes in the fuel oil storage tank in the presence of a heat source at the lower boundary were simulated, taking into account interaction with the environment. A comparative analysis of the dependences of mean volume temperatures in the tank versus time calculated using simplified (balanced) method and obtained as a numerical simulation result was carried out.

Of special note is the paper [7] which presents the study of circulatory heating of oil in a tank with a floating lid. When carrying out numerical studies, the k-ε turbulence model was used. For practical use, engineers are advised to adjust the spraying temperature and outflow velocity from the nozzle taking into account the Froude number.

The difficulty of carrying out the experiment for studying the circulatory heating of fuel explains the small base of experimental data on this topic. The main experimental data on the natural cooling and heating of fuel oil on the industrial scale for tanks of various volumes were obtained by Z.I. Geller [8 – 10].

Despite the widespread use of circulatory heating of fuel in the storage tanks as well as the need to improve this method of adjusting the temperature regime, currently, there is a small number of works on this topic [7].

The aim of this work is to study the influence of the inclination angle of the nozzle on the circulatory heating process of the fuel oil in the RVS-3000 tank.

The problem formulation of heat transfer at the laminar flow of fuel oil jet in the tank

When formulating the mathematical model of heat transfer at the laminar flow of the fuel oil jet the following assumptions are made:

1. nonstationarity of heat transfer processes is determined by the time dependence of the fuel oil temperature in the storage tank;
2. thermophysical properties of fuel oil such as density ρ, heat capacity c_p , thermal conductivity λ, and kinematic viscosity of fuel oil ν depend on the temperature T;
3. volume force influencing on the flow process of the fuel oil jet from the nozzle has the form:

$$\vec{F} = \alpha g \rho \left(T - T_{fuel\,oil}\right),$$

where $\alpha = \frac{\rho_{20}-\rho_0}{\rho_0(T_0-293)}$ – coefficient of volume expansion, 1/K; g – acceleration of gravity, m/s^2; $T_{fleloil}$ – fuel oil temperature at the nozzle outlet, K;

4. rheological behavior of fuel oil is Newtonian.

The primary system of equations is based on the fundamental differential equation system of energy conservation and continuum mechanics – equations of motion and continuity.

Since the supply manifolds in the fuel oil storage tanks in most cases have closely spaced nozzles the single jets flowing out of them can be replaced by a single plane jet flowing out of the plane-slot nozzle. Mathematically, this approximation allows reducing the three-dimensional formulation of the problem of heat transfer to the two-dimensional axisymmetric problem.

The geometric domain for determining the unknown variables of the problem for the fuel oil jet is shown in Fig. 1.

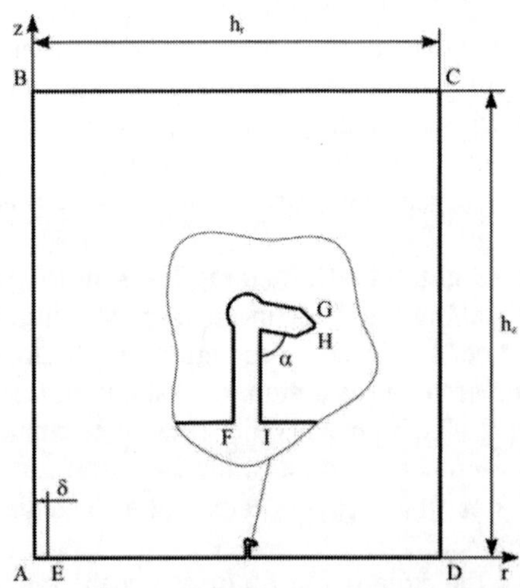

Fig. 1. The geometric domain of fuel oil jet flow in the tank

Because of the symmetry of the problem the flow is considered in the plane *ABCDE* where *r* – the coordinates of points *C* and *D*, *z* – the coordinates of points *B* and *C*, δ – outlet of the *AE* tank, h_r – radius *AD* and *BC* of the studied area, h_z – height *AB* and *DC* of the studied area, α – inclination angle of the nozzle.

IOP Conf. Series: Earth and Environmental Science **288** (2019) 012095 doi:10.1088/1755-1315/288/1/012095

Modeling of heat transfer at a laminar flowing of a fuel oil jet

The initial system of equations of motion and energy transfer describing the heat transfer process at flowing of a fuel oil jet in the general tensor formulation for the accepted assumptions has the form:

$$\rho c_p \frac{dT}{dt} = \text{div}(\lambda \text{gradT}),$$

$$\frac{d\vec{V}}{dt} = \frac{1}{\rho}\text{div}\mathbf{T} + \vec{F},$$

$$\frac{\partial \rho}{\partial t} + \text{div}\vec{V} = 0,$$

where **T** – stress tensor.

We write down the initial and boundary hydrodynamic and thermal conditions at flowing of fuel oil jet to close this system of equations.

The initial hydrodynamic condition has the form:

$$\vec{V}(t=0) = \vec{V}_0.$$

The initial temperature condition has the form:

$$T(t=0) = T_0.$$

The hydrodynamic boundary conditions:

– the velocity profile corresponding to the formed velocity profile of the Newtonian fluid in the circular pipe is specified at the inlet opening;

– at the outlet – condition of the fluid flow stabilization;

– the no-slip conditions of the fluid are specified at the solid and open boundaries of the tank.

Thermal boundary conditions:

– second-type boundary conditions – environmental heat losses – are specified at the solid and open boundaries;

– at the tank outlet we accept the predominance condition of the heat flow due to the convection compared to the heat flow due to the thermal conductivity in a direction perpendicular to the boundary;

– the temperature of the hot fuel oil fed to the tank is specified at the inlet.

The reliability of the mathematical model and the solution method was verified by comparison with the experiment results of the Z.I. Geller for circulatory heating of the cracking residues [11].

Results of numerical studies

The heating of the M100 fuel oil in the RVS-3000 tank was considered in this paper. Geometric characteristics of the tank: the diameter at the base D = 18.98 m, the height H = 11.82 m, the thickness of the tank steel walls $\delta_1 = 0.008$ m, the thickness of the tank walls insulation (mineral wool mats) $\delta_2 = 0.07$ m, the thickness of the tank aluminum coating $\delta_3 = 0.0005$ m, the volume at the maximum operating level $V = 3340$ m^3. The thickness of the concrete pad $\delta = 1.5$ m, the height of the fuel oil layer in the tank $h = 10.7$ m. The diameter of the nozzle exit section $d = 12$ mm, the height of location above the tank bottom $h_k = 0.3$ m. The nozzle inclination relative to the horizontal and vertical planes in the classical version is $\alpha = -10$ °. The supply manifold diameter $D_k = 5$ m.

The winter storage period of fuel oil in the RVS-3000 tank for Kazan was considered. The average ambient temperature for winter is 13.5 °C (January). Fuel oil in the tank is stored in the cold standby mode at $t_{cold} = 30$ °C.

Numerical studies of various inclination angles of the nozzle relative to the horizontal plane α = 40... –40 ° in increments of 10 ° were carried out. The fuel oil flow rate was specified as G_h = 60 t/h (16.67 kg/s) which is equal to productivity of the fuel oil heater PM-10-60. The fuel oil temperature at the outlet of the fuel oil heater is t_{jet} = 115 °C. The results of numerical studies of the tank heating time in winter period are shown in Fig. 2.

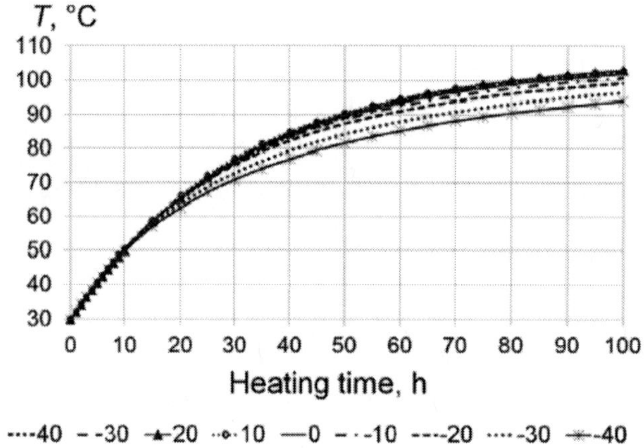

Fig. 2. The heating time of fuel oil in the RVS-3000 tank in winter period

During the first 20 hours of heating, the mean volume temperature in the tank for all inclination angles of the nozzle is practically the same. After 100 hours of heating, the mean volume temperature for inclination angles of the nozzle α = –20... – 40° is lower by 1.6 – 6.8 °C compared to the classical arrangement of the nozzle. The best heating is observed for angles higher than α = –10 °.

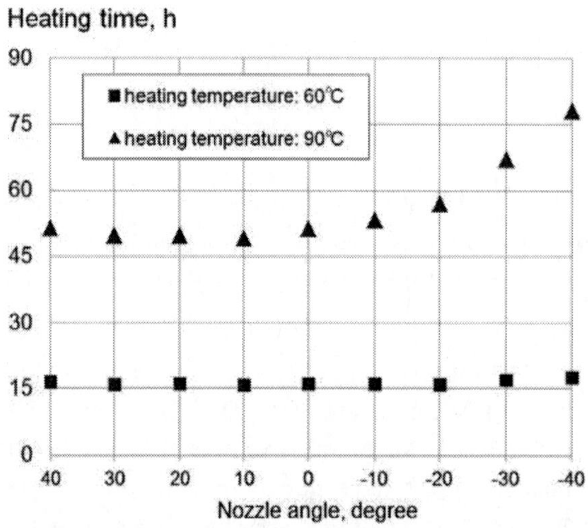

Fig. 3. The heating time of fuel oil in the RVS-3000 tank in the winter period

Fig. 3 shows the comparison of the fuel oil heating time in the RVS-3000 tank in the winter period up to the specified temperatures: 60 °C (the hot storage temperature) and 90 °C (fuel oil temperature before feeding to the nozzles for combustion).

The inclination angle of the nozzle practically does not influence the heating up to the hot storage temperature in the tank of 60 °C. The optimum nozzle angle for fuel oil heating before feeding to the nozzles for combustion is α = –10...40 °.

Conclusions

Change of the inclination angle of the nozzle in the range of – 20 ° to – 40 ° is inadvisable since a longer period of time is required for heating the tank due to an increase of heat losses through the tank bottom. Change of the inclination angle of the nozzle is advisable in the range of – 10 ° to 40 ° since a smaller period of time is required for heating the tank.

Reference

[1] Nazmeev Yu G 2005 *Heat transfer and hydrodynamics in storage systems for liquid organic fuel and petroleum products* (Moscow: Publishing house MPEI)

[2] Nazmeev Yu G, Shamsutdinov E V and Kamalov R F 2006 *Proceedings of the Russian Academy of Sciences. Power Engineering* **2** 52 – 60

[3] Shamsutdinov E V and Kamalov R F 2005 *Transactions of Academenergo* **1** 42 – 46

[4] Kamalov R F *Modelirovanie teploperenosa i razrabotka energoeffektivnih teplotehnologicheskih shem cirkulyacionnogo podogreva mazuta dlya rezervnih mazutnih hozyaystv TES. Diss. kand. tekhn. nauk* [Modeling of heat transfer and development of energy-efficient heat-engineering schemes for circulation heating of fuel oil for reserve fuel oil facilities of TPPs. Kand. eng. sci. diss.] Kazan: Kazan ScC of RAS Publ 2006 p 155

[5] Taktashev R N *Razrabotka rekomendaciy po energosberegeniyu v sisteme cirkulyacionnogo podogreva mazuta na osnove chislennogo modelirovaniya teploperenosa v rezervuarah. Diss. kand. tekhn. nauk* [Development of recommendations on energy saving in the system of circulation heating of fuel oil based on numerical modeling of heat transfer in tanks. Kand. eng. sci. diss.] Moscow: MPEI Publ 2008 p 184

[6] Kuznetsova S A and Maksimov V I 2015 *MATEC Web of Conferences* **23** 1 – 4

[7] Zhao J, Wei L, Dong H and Liu F 2016 *Case Studies in Thermal Engineering* **7** 109 – 119

[8] Geller Z I 1969 *Fuel oil as fuel* (Moscow: Energy)

[9] Geller Z I, Ashikhmin V I,. Visota K P and Shevchenko N V 1969 *Thermal Engineering* **1** 73 – 75

[10] Geller Z I and Ashikhmin V I 1966 *Power Technology and Engineering* **4** 15 – 24

[11] Zdor V O, Kamalov R F, Karaeva J V and Kadiyrov A I 2018 *Transactions of Academenergo* **1** 42 – 49

IOP Conf. Series: Earth and Environmental Science **288** (2019) 012096 doi:10.1088/1755-1315/288/1/012096

Production of biogas from poultry waste using the biomass of plants from Amaranthaceae family

J.V. Karaeva, R.F. Kamalov, A.I. Kadiyrov

Institute of Power Engineering and Advanced Technologies – Subdivision of the Federal Research Center «Kazan Scientific Center of Russian Academy of Sciences», P. B. 261, 2/31 Lobachevsky str., Kazan, 420111, Russia

julieenergy@list.ru

Abstract. The experimental results of kinetics of biogas formation for the organic waste utilization are presented for the case where combinations of technological and microbiological methods are used to enhance the process (ultrasonic pre-treatment and coenzyme with a new plant supplement). The plants from Amaranthaceae family are used as plant supplements. The influence of *Amaranthus hypochondriacus* and *Amaranthus retroflexus L.* on the process of methanogenesis for poultry wastes is analyzed. Based on the experimental results it is obtained that the maximum daily yield of biogas from substrate including the *Amaranthus retroflexus L.* and *Amaranthus hypochondriacus* are 310 ml and 252 ml respectively. For example the maximum daily yield of biogas from substrate without any plant supplements was 211 ml ("pure substrate"). Thus the biogas output rate from poultry waste using the biomass of plants from Amaranthaceae family is 47% higher than standard case. The total biogas yield using the plants supplement from Amaranthaceae family is higher than standard case.

Introduction

One of the modern ways to increase the biogas yield during fermentation is co-fermentation, i.e. joint processing of organic wastes of different origin [1]. At present time, many plant additives are known, however, the use of plants from Amaranthaceae family remains poorly studied.

Mindubaev A.Z. et al. [2] were the first who investigated the effect of *Amaranthus cruentus* on the process of methanogenesis.

It is established that the phytomass of *Amaranthus cruentus* is a stimulant for methanogenesis of hard-fermented substrates, such as beer beet and beet pulp.

The influence of amaranth on the efficiency of the process of obtaining biogas from organic waste was represented in PhD thesis by Belostotsky D.E. [3]

Yagafarova G.G. et al. [4] have developed a method for increasing the biogas yield in the fermentation of waste by applying ultrasound treatment for 4 to 8 minutes at a frequency of 22 kHz. They used shredded phytomass of crimson amaranth and aerobically stabilized activated sludge in the ratio 1:1 as complex stimulator. Unfortunately, the influence of each factor (phytomass and sludge) on the biogas production has not been analyzed separately.

The intensification of anaerobic fermentation of the wine burg and alcohol distillation is described in [5] by using natural biologically active substances (betulinol and amaranth extract) to increase the biogas yield and the content of methane in there.

Sitkey V. et al. [6] developed a technology of "dry" fermentation using silage of amaranth.

The possibility of biogas production by joint fermentation of manure and silage of amaranth *A. Caudatus* was experimentally tested in horizontal fermenter operating in the mesophilic regime.

In 2015 Adamovics A. and Dubrovskis V. represented some results on co-fermentation [7]. The silage of *Cannabis sativa L.*, *Helianthus cultus L.* and *Amaranthus L.* were used as plant supplements. Additionally, the *Metaferm* was used as the catalyst. Despite the greatest biogas yield when using *Amaranthus L.*, Adamovics A. and Dubrovskis V. proposed to use *Cannabis sativa L.*, because the latter can be used entirely, the plant contains a significant leaf mass. A significant part of the phytomass of amaranth consists of stems that are not useable for the process.

The aim of present work is the utilization of organic waste using a combination of technological and microbiological methods (ultrasonic pre-treatment and co-fermentation using a new plant supplement - *Amaranthus retroflexus L.* The feature of this study is the investigation of the methane fermentation by using *Amaranthus retroflexus L.* as plant supplement, which is recognized as a weedy plant unlike *Amaranthus Caudatus* и *Amaranthus cruentus*.

Methods

Substrate composition

As a substrate, chicken dung from a private farm was used with 29.9% of dry matter content. The dried *Amaranthus hypochondriacus* and *Amaranthus retroflexus L.* (humidity less than 5.7%) was used as the stimulator of biogas production (fig. 1).

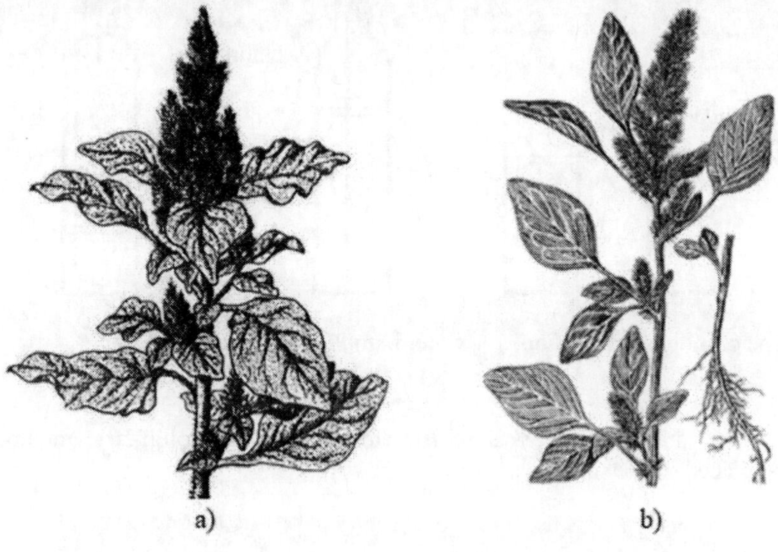

Fig. 1. Species of the Amaranthaceae family:
a) Amaranthus hypochondriacus; b) Amaranthus retroflexus L.

In three series of experiments, the chicken dung was used from one dung sample with identical characteristics.

The shredded Amaranth phytomass was added to substrates as an additional food for microflora. *Amaranthus hypochondriacus* and *Amaranthus retroflexus L.* were grown on an experimental field located in the Verkhneuslon district of the Republic of Tatarstan (harvest 2017). They were collected in the dissemination phase.

The investigated substrates are follows:

1. Control substrate composes 131 g of chicken dung which is dissolving in distilled water to reach 400 ml volume.
2. Substrate №2 composes 65.5 g of chicken dung with 9.9 g of *Amaranthus hypochondriacus*

which is dissolving in distilled water to reach 400 ml volume.

3. Substrate №3 composes 65.5 g of chicken dung with 9.9 g of *Amaranthus retroflexus L.* which is dissolving in distilled water to reach 400 ml volume.

Biomass ultrasonic pre-treatment

Before the experimental study of biogas production, all considered substrates were treated by ultrasonic wave. As an ultrasonic source, the UZTA-0.2/22-OM was used with following characteristics: power – 200 W, frequency – 20 kHz. Ultrasonic pre-treatment was conducted unde the intensity of 10 W/cm^2 during 6 minutes.

The ultrasound treatment led to homogeneous of the substrate, the stratification was not observed during the day and during the experimental study of producing biogas.

Batch experiments

The biogas production was experimentally studied in laboratory anaerobic digesters (500 ml) during 85 days. The process of methane fermentation was carried out in the mesophilic regime under a fixed temperature of 37° C (Fig. 2, 3).

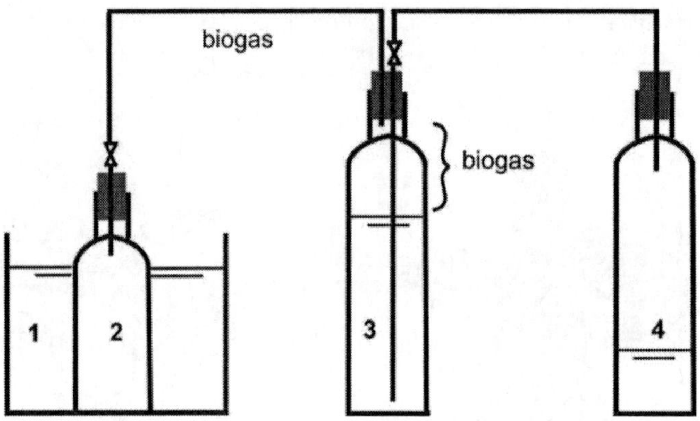

Fig. 2. Scheme of biogas production: 1 – water bath, 2 – anaerobic digester, 3 – plastic container, 4 – tank

The volume of produced biogas was registered daily by the volumetric method (measuring a volume of the displaced saline solution).

Fig. 3. The laboratory anaerobic digesters

The chemical composition of produced biogas was analyzed every 10 days by gas-liquid chromatography (GLC).

Results and discussion

Fig. 4 shows the kinetics of biogas production from control substrate and substrates including plant supplements: *Amaranthus hypochondriacus* and *Amaranthus retroflexus L.*

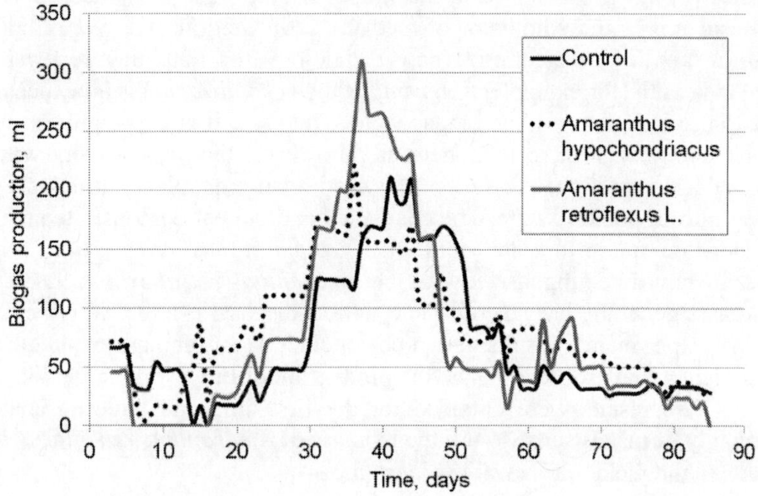

Fig. 4. Kinetics of biogas production

It can be seen that the maximum daily yield of biogas corresponds to substrate including *Amaranthus retroflexus L.* (310 ml on 37 days). Meanwhile for the substrate including *Amaranthus hypochondriacus* the maximum daily yield of biogas is registered on 41st day and reached 211 ml. After simple calculation, it is obtained that the biogas output rate from poultry waste using the biomass of plants from Amaranthaceae family is 47%, which is higher compared with results for control substrate.

Fig 5 illustrates the cumulative biogas production for investigated substrates.

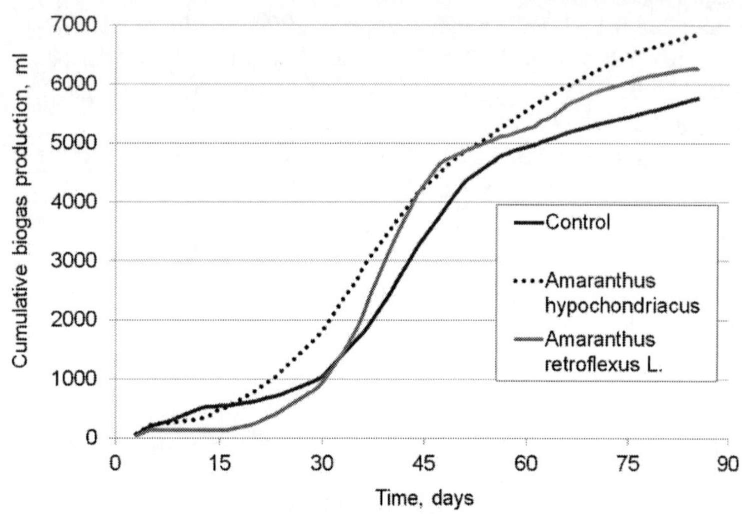

Fig. 5. Cumulative biogas production

After 85 days of holding the total biogas production from substrate №2 and substrate №3 are higher

IOP Conf. Series: Earth and Environmental Science **288** (2019) 012096 doi:10.1088/1755-1315/288/1/012096

on 19% and 9% respectively compared with the same results for control substrate. Thus it is shown that the biomass of the *Amaranthus retroflexus L.* stimulates methanogenesis like the biomass of *Amaranthus hypochondriacus*.

Conclusions

In earlier studies [2, 4, 7], it has already been shown that plant phytomass from Amaranthaceae family can be considered as the stimulator of methane fermentation. It is very important and useful result which can be applied in creating a new technology for obtaining renewable fuel. Since a cultivated type of amaranth (for example, *Amaranthus hypochondriacus*) is an expensive raw material, it is unprofitable to use it to increase the biogas yield. Therefore it is important to find affordable and cost- effective plant stimulants for methanogenesis. The aim of the present work was the study of the efficiency of using *Amaranthus retroflexus L.* (weed plant) as plant stimulants to enhance the production of biogas from poultry waste. The conducted experiment demonstrated the same enhancing effect on methanogenic microflora for both plants: *Amaranthus retroflexus L.* and *Amaranthus hypochondriacus*. It should be emphasized that the *Amaranthus retroflexus L.* is recognized as weed plant, so its phytomass generally has no cost value, and its disposal is a topical issue.

According to the experimental results, it can be concluded that biomass of *Amaranthus retroflexus L.* can be used as plant stimulator to enhance the production of biogas like a well-known *Amaranthus hypochondriacus*. These results were obtained for the first time and have no analogs in scientific literature. Accordingly, it makes sense to use the biomass of *Amaranthus retroflexus L.* (a weed plant), in practice to increase the yield of renewable biogas fuels.

Reference

[1] Hagos K, Zong J, Li D, Liu C and Lu X 2017 *Renew. Sust. Energ. Rev.* **76** 1485-96.
[2] Mindubaev AZ, Minzanova ST, Skvortsov EV, Mironov VF, Zobov VV, Akhmadullina FY, Mironova LG, Belostotsky DE and Konovalov AI 2009 *Bulletin of the Technological University* **4** 220 – 226.
[3] Belostotsky DE 2012 *Influence of amaranth on the efficiency biogas production process from organic waste* (Kazan, A.E. Arbuzov Institute of Organic and Physical Chemistry).
[4] Patent RU 2458868 (2012).
[5] Kovalev VV, Nenno VE, Bobeyka VA, Senkovskaya IA, Kovaleva OV and Unguryanu DV 2012 *Studia Universitatis* **1**(51) 160-168.
[6] Sitkey V, Gadus J, Klisky L and Dudak A 2013 *Acta Regionalia et Environmentalica* **2** 61 – 64.
[7] Adamovics A and Dubrovskis V 2015 25th Congress of the Nordic Association of Agricultural Scientists «Nordic View to Sustainable Rural Development» (Riga: University of Latvia, Main Building) pp. 385 – 389.

IOP Conf. Series: Earth and Environmental Science **288** (2019) 012097 doi:10.1088/1755-1315/288/1/012097

Determination of thermal diffusivity for rigid body at non-stationary thermal conditions

Denis Karpov[1], Vladimir Agafonov[2], Viktor Pisarenko[3], Pavel Berezin[4], Oleg Derevianko[5], Sultan Makoev[5], Mikle Egororov[5]

[1]Vologda State University, Lenina Street 15, Vologda, 160000, Russia

[2] JSC "Gazprom gazoraspredelenie Krasnodar", Stroiteley Street 23, Krasnodar, 350051 Russia

[3] LLC "Gas-transport communication and facilities", Bolshaya Posadskaya Street 9/5, Saint Petersburg, 197046, Russia

[4] LLC "Northern Alliance", Vetoshkina Street 40, Vologda, 160011, Russia

[5] Peter the Great St. Petersburg Polytechnic University, Polytechnicheskaya Street 29, Saint Petersburg, 195251 Russia

Makoev@nil-teplo.ru

Abstract. In this paper we present a new method for calculation of the thermal diffusivity of a rigid body. This method is based on contactless thermal treatment of the examined construction material by an infrared radiation source. Further we analytically find one-dimensional non-stationary temperature field of a rigid body during its heating using data from thermal elements. The calculation of the thermal diffusivity of a rigid body is made using differential heat equation.

Introduction

Thermal diffusivity a_t is a heat diffusion parameter, which is used for analytical description of non-stationary processes of heat transfer in various nature objects: rigid bodies, liquid and gaseous mediums [1].

Heat exchange theory [2,3] says that differential heat equation (in the absence of internal heat sources) is the following:

$$\frac{\partial t}{\partial \tau} = a_t \nabla^2 t \tag{1}$$

where a_t is proportionality factor in linear, homogeneous, second order partial differential equation, $\mathrm{m^2/s}$.

Thus, it is essential to know thermal diffusivity a_t to solve equation (1) and to investigate evolution of non-stationary temperature field of a rigid body in space and time. For example, the genuine solution of the heat equation (1) using heat source technique includes thermal-physical coefficient a_t [4].

In construction thermal physics finding the solution of equation (1) and determination of a_t coefficient are of current interest, as heat transfer through enclosures is always non-stationary. Ambient temperature, solar radiation intensity, wind force and direction are continuously changing, as well as indoor temperature, which is regulated by heating system and additional internal heat inputs. Fluctuations of temperature, energy and mass flows, which occur at ambient air separated by enclosures, result in variation of thermal behavior of the building itself [5].

Besides this, in construction heat engineering consideration of heat transfer at non-stationary conditions is essential when solving the following problems: determination of inner temperature fluctuation amplitudes related to irregularity of heat emission by heating system; calculation of temperature oscillation decay in enclosures caused by ambient air temperature alteration or solar radiation impact on wall surface; warming up and cooling down of massive enclosures; improving the safety of cast-in-place construction at adverse climatic conditions, reduce of energy consumption, etc [6–16]. There are a lot of literature sources concerning various methods and devices for thermal diffusivity determination of construction materials [17–25].

Classical method of a-calorimeter is based on regularities of warming up (cooling down) of material sample under study at intensive heat exchange medium with $\mathrm{Bi} \to \infty$. For regular thermal conditions the unknown value of thermal diffusivity a_t is linearly dependent on material cool down rate [26]. Cool down rate is denoted according to differential thermocouple. One of its joints is located at central area of the sample, another one is located at intensive heat exchange medium. Proportionality factor, which mathematically connects thermal diffusivity coefficient of a rigid body with cool down rate depends on geometrical parameters of the sample under study [27].

The disadvantage of this method is technical complexity of organization and carrying out theses thermal measurements. For implementation of this method one must have an a-calorimeter preheated to high temperatures in a dry cupboard and a liquid thermostat with intensive medium mixing, which provides the following condition: $\mathrm{Bi} \to \infty$.

Contact methods based on thermal non-destructive testing are the modern techniques for determination of material thermal diffusivity. The main feature of these methods, in comparison with classical ones, is the contact of temperature gauges with particular surface areas of the sample under study. Material thermal diffusivity determination is based on sample surface temperature measurements at radiative-convective warm up. These measurements are performed for two key points: edge and mid-facet. The temperature complex is calculated from a plot of sample heating. Material thermal diffusivity is calculated from the known distance between temperature measurement points taking into account temperature complex rate of change [18].

This method is nearly free of disadvantages. However, there is a need for heating chambers which fit the sample size, and this adds complexity for its practical implementation.

This work presents a special combined experimental and computational method for determination of a rigid body thermal diffusivity. This method excludes the main disadvantage of the mentioned-above techniques, namely: formation of non-stationary thermal conditions using complex and overall heat sources.

Materials and Methods

Fig.1 presents a schematic diagram for combined experimental and computational method for determination of a rigid body thermal diffusivity at non-stationary thermal conditions.

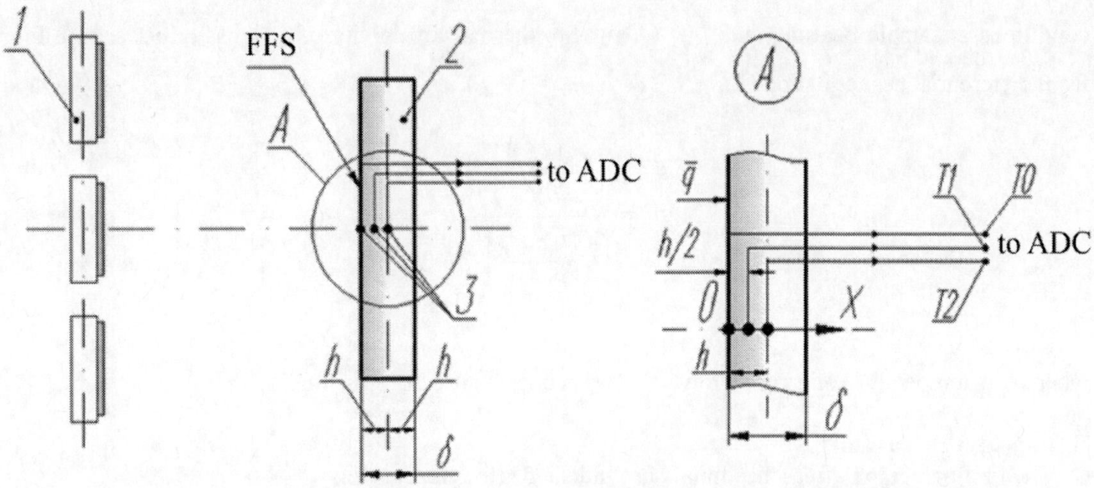

Fig.1 A schematic diagram for determination of a rigid body thermal diffusivity: 1- infra red radiation source; 2 - rigid body; 3 - a system of thermal elements

Infra red radiation source 1 is powered by electricity. The investigated rectangularly shaped rigid body 2 with $\delta = 2h$ thickness is placed at a distance from Infra red radiation source 1. The central axis of infra red radiation source 1 and rigid body 2 are coincident. The thermal elements 3 are fixed at interval $x \in [0, h]$ of a rigid body 2: $T0$ $T1$, $T2$ are placed at points with coordinates $x = 0$, $h/2$ and h, respectively. These points are connected to a PC (isn't shown here) through analog to digital converter (ADC) and converter (aren't shown here).

At initial time $\tau = 0$ the temperature field of a rigid body 2 is homogeneous and is equal to ambient air temperature. At the start point electrical energy is transferred to infra red radiation source 1, where it is converted to electromagnetic radiation of q intensity. Then it is non-contact and indestructibly transferred to a front-face surface (FFS) of the rigid body 2. Radiation energy flux q, which uniformly falls on FFS of the examined sample 2, is converted to internal energy, which is spent for heating of the whole rigid body 2. Infra red radiation source 1 uniformly irradiates FFS of the examined sample 2, consequently, temperature alteration takes place only along the axis 0X. Temperature remains the same for 0Y and 0Z directions, i.e. $\nabla_y t = \nabla_z t = 0$. Temperature alteration along the axis 0X is registered by thermal elements 3: $T0$, $T1$ and $T2$, which transfer information through ADC and converter to nonvolatile computer memory.

Non-stationary thermal conditions duration of the rigid body 2 is determined according to the expression:

$$\tau'_{max} = \frac{\delta^2}{a'_t} \tag{2}$$

where $\mathcal{E}$ is sample thickness, m; a'_t is preset thermal diffusivity, m^2/s.

Let us assume that temperature field of a rigid body 2 is known from thermal element data. So its equation is $t = t(x, \tau)$ for heating period $\tau \in [0, \tau_{nst}]$, where τ_{nst} is a segment of non-stationary thermal conditions at sample heating, and $\tau_{nst} \leq \tau'_{max}$. So, thermal diffusivity of a rigid body 2 can be found from differential heat equation (1), m^2/s:

$$a_t = \frac{\left(\frac{\partial t(x, \tau)}{\partial \tau}\right)_{x=0}}{\left(\frac{\partial^2 t(x, \tau)}{\partial x^2}\right)_{x=0}} \quad (3)$$

where t is temperature of a rigid body, $^\circ C$; x is coordinate, m; τ is time, s.

Results and Discussion

The first paragraph after a heading is not indented (Bodytext style).

Other paragraphs are indented (BodytextIndented style).

Another section of your paper

Let us use this method to determine thermal diffusivity of a silicate brick M150 (Russian State Standard GOST 379-95) with thickness $\delta = 0.12$ m ($h = 0.06$ m), Fig.2.

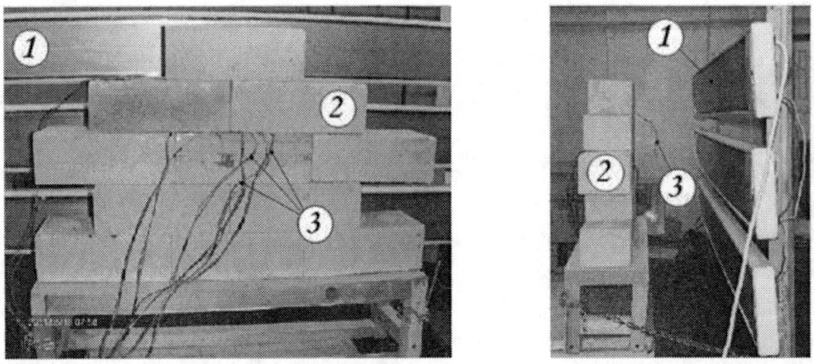

Fig. 2 Laboratory-experimental setup "energy source and receiver" for determination of construction material thermal diffusivity. 1- electrical infra red source; 2 - silicate brick, 3 - a system of chromel-alumel thermocouples

Fig. 2 shows that chromel-alumel thermocouple junctions 3: T0, T1, T2 are fixed in M150 silicate brick 2. The junctions are located along the central axis at points with coordinates $x = 0$, 0.03 and 0.06 m, respectively. Electrical infrared source 1 is Ecoline 10 R lamp of a total power 3 kW, which is located at a distance of 0.6 m from the silicate brick FFS.

The preset value of silicate brick 2 thermal diffusivity is $a'_t = 5.49 \cdot 10^{-7}$ m2/s [28]. According to the expression (2), heating period duration is $\tau'_{max} = 26218$ s. Experiment data is $\tau_{max} = 31560$ s (Fig.3); average temperature alteration in interval $x \in [0; 0.06]$ m and time range $\tau \in [26218;31560]$ s according to chromel-alumel thermocouples 3 was $\pm 2.5\ ^\circ C$. Inaccuracy of measurements is $0.60\ ^\circ C$, thus the obtained result is acceptable.

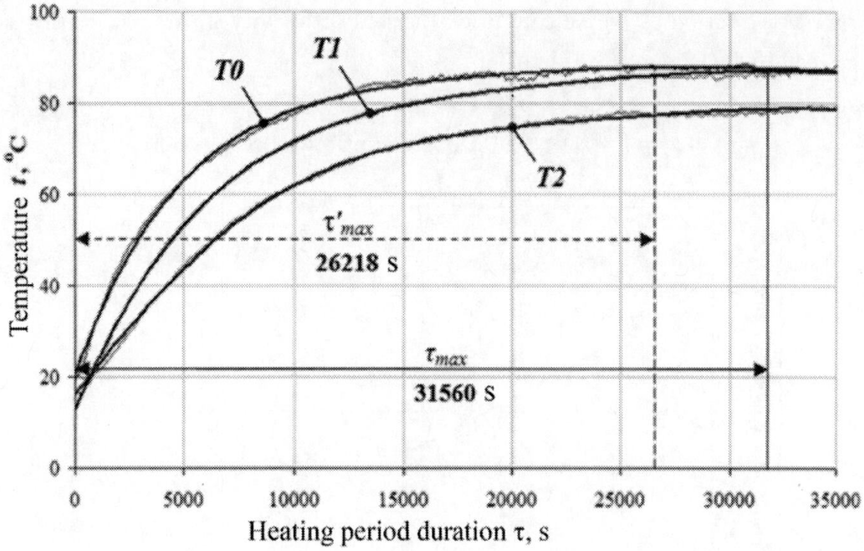

Fig 3. To the determination of non-stationary thermal conditions duration for silicate brick

Let us consider thermal conditions for silicate brick for interval $x \in [0; 0.06]$ m and time range $\tau_{nst} \in [0;15000]$ s. Fig. 4 presents temperature field of the silicate brick 2 of the form $t = t(x, \tau)$ at $x \in [0; 0.06]$ m and $\tau_{nst} \in [0;15000]$ s, drawn based on experimental data.

Functional chart, which describes heating regime for silicate brick 2 at $x \in [0; 0.06]$ m and $\tau_{nst} \in [0;15000]$ s is the following, $^{\circ}C$:

$$t = a + b\tau + cx + d\tau^2 + ex^2 + fx\tau + g\tau^3 + hx^3 + ix^2\tau + jx\tau^2, \; R^2 = 0.9927, \tag{4}$$

where $a = 22.830014$, $b = 0.010623358$, $c = -481.12022$, $d = -6.1933549 \cdot 10^{-7}$, $e = 10518.343$, $f = -0.014857501$; $g = 1.2080758 \cdot 10^{-11}$; $h = -79288.783$; $i = -0.44211261$; $j = 2.4679544 \cdot 10^{-6}$ are equation parameters.

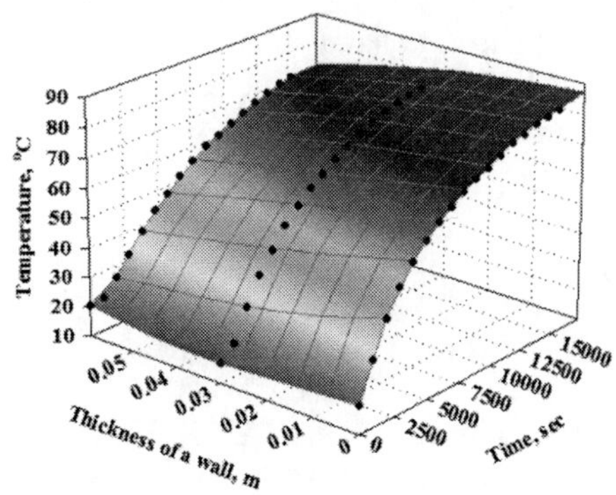

Fig. 4 Temperature field of a silicate brick for non-stationary thermal conditions

Fig. 5 presents evolution of the silicate brick thermal diffusivity in time at heating conditions at $x=0$ and $\tau_{nst} \in [0;15000]$ s of $a_t = a_t(\tau)\cdot 10^{-7}$, m²/s:

$$a_t = 5\cdot 10^{-9}\tau^2 - 0.0004\tau + 5.0896, \; R^2 = 0.9997, \quad (5)$$

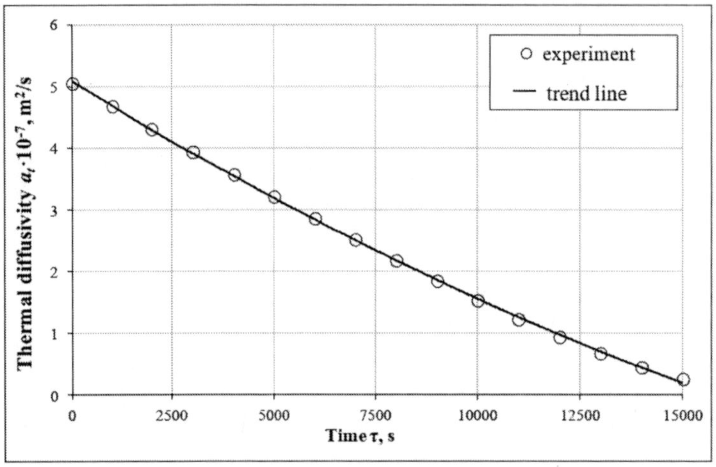

Fig. 5 Thermal diffusivity evolution in time for silicate brick

As it seen from figure 5, when thermal condition of a silicate brick approaches stationary condition, the thermal diffusivity tends to zero. According to the expression (5), the thermal diffusivity a_t of silicate brick at initial time $\tau = 0$ is equal to $5.09\cdot 10^{-7}$ m²/s (at brick temperature equal to that of ambient air $t = 19.7\ ^\circ\mathrm{C}$).

Conclusions

This work presents a new method for determination of rigid body thermal diffusivity at non-stationary thermal condition. This method takes into account advantages and disadvantages of classical techniques for determination of thermal-physical properties of construction materials. The practical implementation of this method was performed for M150 silicate brick (Russian State Standard GOST 379-95). Combined experiments and calculation result in material thermal diffusivity $5.09\cdot 10^{-7}\ \mathrm{m^2/s}$, determined for ambient air temperature. This result agrees with specified value $5.49\cdot 10^{-7}\ \mathrm{m^2/s}$ presented in [29] and data from other sources: $5.5\cdot 10^{-7}\ \mathrm{m^2/s}$ [6], $5.8\cdot 10^{-7}\ \mathrm{m^2/s}$ [28]. The developed method can be applicable in construction and heat power engineering when testing homogeneous construction units, thermal conductive and thermal insulating materials.

References

[1] Lykov A V 1972 Heat-mass exchange: reference book (Moscow: Energy)
[2] Isaev S I, Kozhinov I A and Kofanov I A et al. 1979 Theory of heat-mass exchange: University textbook ed A I Leont'ev (Moscow: Higher school)
[3] Jiji L M 2009 Heat conduction: Third edition
[4] Karpov D F, Pavlov M V, Sinityn A A and Al. E 2012 Analytical-empirical verification of experimental results for silicate brick thermal conductivity Bull. TGASU 4 210–5
[5] Bogoslovskiy V N 2006 Construction thermophysics: thermophysical basis of heating, ventilation and air conditioning: University textbook (Saint Petersburg: HVAC North-West)

[6] Fokin K F 2006 Construction heat engineering of enclosure parts of building (Moscow: HVAC Press)
[7] Zharkoy R A, Kozlobrodov A N and Nedavniy O I 7. Non-stationary heat transfer in spatial elements of outside enclosure by the example of "Velox" technology Bull. TGASU
[8] Kharkov N S 2017 Nonstationary heat and mass transfer in the multilayer building construction with ventilation channels 891
[9] Bozhokin S V and Suslova I B 2017 Comparative analysis of mathematical simulating of the non-stationary heart rate variability in diagnostic medical systems (Institute of Electrical and Electronics Engineers Inc.)
[10] Tarasova D S and Petritchenko M R 2017 Buildings quasi-stationary thermal behavior Mag. Civ. Eng. 72 28–35
[11] Zaborova D, Petrochenko M and Chernenkaya L 2016 Thermal Stability Influence of the Enclosure Structure on the Building's Energy Efficiency MATEC Web Conf. 73
[12] Tarabukina S, Simankina T, Pykhtin K and Grabovyy K 2017 Calculation of optimal outdoor enclosure in the arctic conditions vol 90
[13] Gorshkov A, Vatin N, Nemova D, Shabaldin A, Melnikova L and Kirill P 2015 Using life-cycle analysis to assess energy savings delivered by building insulation vol 117(Elsevier Ltd)pp 1085–94
[14] Penić M, Vatin N and Murgul V 2014 Double skin facades in energy efficient design Appl. Mech. Mater. 680 534–8
[15] Vatin N I, Gorshkov A S, Nemova D V, Staritcyna A A and Tarasova D S 2014 The energy-efficient heat insulation thickness for systems of hinged ventilated facades Adv. Mater. Res. 941–944 905–20
[16] Shatornaya A, Chislova M, Chislova A, Drozdetskaya M and Trubina D 2016 Ventilated Facades: Insulation Materials of Different Manufacturers vol 53(EDP Sciences)
[17] Hamdhan I N and Clarke B G 2010 Determination of thermal conductivity of coarse and fine sand soils Proc. World Geotherm. Congr. 2010 25–9
[18] Kalis H and Guseinov S 2006 Calculations of the Thermal-conductivity Coefficients for 1-D Heat Transfer Inverse Problems International Scientific Colloquium Modelling for Material Processing pp 221–6
[19] Zaborova D, Vieira G, Musorina T and Butyrin A 2018 Experimental Study of Thermal Stability of Building Materials 692 482–9
[20] Statsenko E, Ostrovaia A, Musorina T and Sergievskaya N 2018 Thermal Properties of the Building with Low Energy Consumption (LEB) Adv. Intell. Syst. Comput. 692 417–31
[21] Korniyenko S V, Vatin N I and Gorshkov A S 2016 Thermophysical field testing of residential buildings made of autoclaved aerated concrete blocks Mag. Civ. Eng. 64 10–25
[22] Kapitz M, Becker A and Wiesche S A D 2014 Influence of Thermophysical Wall Properties During Pool Boiling on Large Diamond and SiC Heaters Heat Transf. Eng. 35 472–81
[23] Statsenko E A, Ostrovaia A F, Musorina T A, Kukolev M I and Petritchenko M R 2016 The elementary mathematical model of sustainable enclosing structure Mag. Civ. Eng. 68 86–91
[24] Gorshkov A S and Rymkevich P P 2015 A diagram method of describing the process of non-stationary heat transfer Mag. Civ. Eng. 60 68–82
[25] Pukhkal V and Murgul V 2018 Effect of aerated concrete blockwork joints on the heat transfer performance uniformity vol 33(EDP Sciences)
[26] Kondrat'ev G M 1954 Regular thermal regime (Moscow: Gostehizdat)
[27] Urenev V N and Lebedev P D 1976 Heat engineering reference book: reference book: In 2 volumes (Moscow: Energy)
[28] Fokin V M, Boykov G P and Vidin Y V 2004 Basic concepts of technical thermophysics: monography (Moscow: Machine engineering)
[29] Anon 2004 Guidelines for design and construction: Design of thermal protection of buildings: SP 23-101-04: intr. 01.06.04 (Moscow: FGUP CPP)

Main trends in increasing technical and economic efficiency of underground power pipelines cathodic protection

Kiselev V.G. [1], **Kalyutik A.A.** [1], **Fedyukhin A.V.** [2], **Makoev S.O.** [1]

[1]Peter the Great St. Petersburg Polytechnic University, 194021, Russia, St.Petersburg, Polytechnicheskaya 29

[2]National Research University «Moscow Power Engineering Institute», 111250 Russia, Moscow, Krasnokazarmennaya 14.

Kalyutik@yandex.ru

Abstract. The analysis and systematization of the main factors determining the parameters of the electrochemical protection of underground metal communications (UMC) is carried out in the article. Reasons influencing current consumption and power losses in the DC circuit of the cathodic corrosion protection are discussed in detail, and practical recommendations for their reduction are given. Preparatory measures have a significant impact on anticorrosive protection of power pipelines, namely, measures to improve the quality of insulation (increasing the electrical resistance of coatings and ensuring their integrity), measures to ensure high longitudinal conductivity of pipelines, and removal of contacts of underground metal structures with other earthed structures. On the basis of the research, practical recommendations for reducing power losses and reducing current consumption in cathodic protection systems are given.

Statement of the problem

Corrosion resistance plays a significant role in energy objects design, construction and operation [1, 2, 3]. Since protection from corrosion is not only technical but also an economic problem, it is essential to take into consideration costs of specific anticorrosive measures. Chosen method of corrosion protection largely determines the reliability of power equipment operating under aggressive conditions. Due to this, measures of increasing reliability and reducing costs of combined anti-corrosion protection for underground metal communications are of special importance and will be studied in this article from the perspective of their technical and economic efficiency.

Evaluation of technical and economic efficiency of combined anticorrosive protection of UMC

In accordance with GOST 9.602 - 2005 and GOST R 51164 - 98, the criterion for the technical effectiveness of combined corrosion protection, which includes both protective anticorrosive coating and electrochemical protection (ECP), is the fulfillment of two groups of requirements:

- Requirements for the protective coating (quality of anticorrosive materials, coating technologies, properties of the finished coating);
- Requirements for the value of the protective potential without its ohmic component, which, as a rule, should be in the range from (-0.85) volts to (-1.15) volts over the copper-sulfate reference electrode during the whole service life of the object.

Within the framework of this article, we will assume that the specified technical conditions are certainly fulfilled. And, therefore, the main problem of evaluating the economic effectiveness of combined corrosion protection is the problem of optimizing the parameters of electrochemical protection, as the assessment of the economic efficiency of coatings is usually not difficult, and these data is easily accessible in the specialized literature. This approach allows us to consider the work done by the cathodic protection system to achieve the polarization potential as "useful work", that is, the work done to shift the potential of the structure from its corrosion potential φ_k to its protective potential φ_з without an ohmic component. In this case, all other costs associated with the passage of an electric current during cathodic protection of UMCs can be attributed to losses that are expediently divided into the following five groups:

- Losses associated with the conversion of an alternating current into a direct current.
- Losses associated with an excessively high protective potential of UMS, caused by insufficient protective potential adjustment system efficiency.
- Losses in the circuit of an alternating (supply) current line.
- Losses in the protective current circuit.
- Losses due to the low level of measures to reduce the protective current at UMC.

Let us look at the mentioned above losses in more detail.

The first two types of losses are closely related to the current state of the equipment in the field of direct current sources for cathodic protection and protective potential adjustment systems that are mainly determined by the market of the corresponding products. Therefore, their influence on the economic efficiency of ECP under specific design conditions is relatively simple and does not require special consideration. The third type of losses is largely determined by the interposition of UMCs and alternating current sources and, therefore, it is impossible to take them into consideration without corresponding data. On the other hand, when the required data is available, it is quite easy to calculate losses in the power line for the economic efficiency of cathodic protection, and therefore, due to the limited volume of the article, its analysis is not given here. Since losses of fourth and fifth type have a significantly greater impact on the level of costs of cathodic protection systems' construction and operation we will analyze them in more detail.

Losses in the protective current circuit and main ways to reduce them

Analysis of the processes taking place in the protective current circuit has recently received a lot of attention. In this case, various methods [4] are applied, including the method using the following formula [5]:

$$E = (E_a - E_k) + |\eta_a| + |\eta_k| + E_{э-т} + E_{cont} + E_{con.wire} \quad (1)$$

where, E – is output voltage of the cathodic protection converter. E_a – stationary anodic ground potential. E_k– stationary potential of the protected structure. $|\eta_a|$ и $|\eta_k|$ – are, respectively, absolute values of anodic and cathodic overvoltage. $E_{э-т}$ – voltage drop on the electrolyte resistance. E_{cont} - voltage drop across the contacts in the DC circuit. $E_{con.wire.}$ - voltage drop on the connecting wires (cables) of the DC circuit.

We will briefly analyze the effect of the individual components of formula (1) on the magnitude of losses in the DC circuit:

1. The effect of the $(E_a - E_k)$ value

E_a *and* E_k, are determined by the specific corrosion conditions of the anodic earthing and underground metal structure (UMS). The oscillations of these quantities usually amount to several tenths of one volt [6], and the currents flowing in the DC circuit are extremely small, which leads to an extremely insignificant effect of these parameters on the value of the power loss in the DC circuit.

2. The effect of anode overvoltage η_a.

Practical measurements [7] carried out at UMC cathodic protection in the ground show a potential drop in the range from one to two volts due to electrochemical process deceleration at an anode grounding of . This value is quite large and measures should be applied to reduce it.

3. The effect of cathodic overvoltage η_k.

If, in order to minimize power losses associated with anode overvoltage, it is required to reduce, then to solve the same problem in the protective current circuit, in contrast to andic polarization, for the quantity η_k, on the contrary, maximum growth is required with a possible small protective current [7]. In this case, the increase in power losses due to the actual cathodic polarization of the protected structure is more than compensated by the energy gain at other resistances in the DC circuit.

4. The effect of voltage drop at the contacts E_{cont} in the DC circuit.

It is known from practice that the voltage drop at all contacts in the DC circuit is usually limited to 0.05 volts. However, this drop in the potential at high protection currents can cause power losses of up to 5 watts or more, which, if the quality of the connections deteriorates, can lead to catastrophic failure of ECP systems.

5. The effect of the voltage drop at the connecting wires (cables) $E_{con.wire}$ in the DC circuit.

The length of the connecting wires in the DC circuit is determined by the geometry of the site on which the ECP system should be located and the possible harmful effect of the cathodic protection system on adjacent underground metal structures. To reduce power losses, the length of the wires (cable) under these conditions should be minimized. The choice of the optimal cross-section of the DC cable is usually performed using a special graph (see, for example, reference [4]), based on the value of the protective current, the type of cable and the life of the electrochemical protection unit or the maximum permissible voltage drop on it, should not exceed two volts.

6. The effect of voltage drop in electrolyte$E_{э-т}$.

Traditional electrochemical protection systems are characterized by such an arrangement of protected UMCs and anode groundings, where from the point of view of ohmic resistance between them it can be considered, that they are at an infinite distance from each other. Then, for the total resistance of the electrolyte$R_{э-т}$, we can write the following formula:

$$R_{э-т} = R_{\mathrm{T}} + R_{\mathrm{A}} \tag{2}$$

It immediately follows that the change in the value of $R_{э-т}$ is due, to changes in the parameters of the anodic grounding and the pipeline. Therefore, this analysis can be carried out separately, for each of these objects. Then the total power loss [8] at the ohmic resistance of the anode ground, P_{a}, with the optimal number of anodes n_{a}, will be :

$$P_k = I_s^2 \frac{R_0}{n_{\mathrm{a}}} F \frac{\rho}{\rho_0} \tag{3}$$

where, F – an interference factor of the anodes, which determines the increase in the resistance of the anode device as a result of mutual screening of individual anodes; ρ – soil resistivity in the area of anodic earthing; R_0 – resistance to spreading of a single anode, with specific resistivity of the ground ρ_o , equal to 10 Om ×m; I_s - protective current.

Along with already mentioned methods of estimating power losses on various elements of the ECP system, measures to limit the amount of protective current on the protected object [9] are of great importance as well. Their implementation dramatically reduces the consumption of electric current per unit of the protected surface.

Conclusions

- This article presents the analysis and systematization of the main factors affecting ECP system current consumption and power loss while applying UMC cathodic protection against corrosion.
- The main factors affecting current consumption and power loss in the DC circuit of the ECP system are systematized and described in detail.

References

[1] Dremicheva E.S., Zvereva E.R. Study corrosion processes of oil equipment. Proceedings of the higher educational institutions. ENERGY SECTOR PROBLEMS. 2018; 20(1-2):138-143. https://doi.org/10.30724/1998-9903-2018-20-1-2-138-143

[2] Kiselev V.G., Kalyutik A.A. Special features of the influence of corrosive elements on the localization of the processes of destroying the metal with the pipe laying in the soil. Proceedings of the higher educational institutions. ENERGY SECTOR PROBLEMS. 2015;(1-2):3-10. https://doi.org/10.30724/1998-9903-2015-0-1-2-3-10

[3] Kiselev V.G., Ruzich E.N. DIELECTRIC COATINGS AND THEIR EFFECT ON CORROSION PROTECTION OF THE OUTER SURFACE OF UNDERGROUND PIPELINES. *Proceedings of the higher educational institutions. ENERGY SECTOR PROBLEMS.* 2018;20(1-2):80-89. https://doi.org/10.30724/1998-9903-2018-20-1-2-80-89

[4] Ulrich Bette und Wolfgang Vesper. «Taschenbuch für den kathodischen Korrosionsschutz», Essen: Vulkan-Verl., - 2005. – 367 S.

[5] Agladze R.I. Applied Electrochemistry [Text] / monograph / R.I. Agladze, T.A. Vagramyan and others. / Chemistry Publishing House, Moscow, 1984.

[6] Kiselev V. G. Influence of the electric double-layer capacitance at the rate of corrosion at the phase interface /V. G. Kiselev, V. V. Sergeev and E. N. Rouzich // Corr. Rev. 2017 № Vol. 35, Issue. 1, p. 47–51.

[7] Kiselev V.G. Influence of the covering layer of underground metal structures on the efficiency of cathodic protection [Text] / V.G .Kiselev, A.A. Kalyutik // PROBLEMS OF ENERGY. NEWS OF HIGHER EDUCATIONAL INSTITUTIONS - 2016. № 5 - 6. - p. 27 - 36.

[8] Kiselev, V.G. Basic principles of underground metal structures cathodic protection design [Text] / V.G. Kiselev // Scientific and technical statements of St. Petersburg State University. - 2011. № 4 (135). - P. 111 - 116.

[9] Kiselev, V.G. Influence of preparatory measures in accordance with DIN 30676 on cathodic protection of pipelines [Text] / V.G. Kiselev, E. N. Ruzich // Scientific and technical statements of SPb GPU. - 2017. № 2. - P. 101 - 108.

IOP Conf. Series: Earth and Environmental Science **288** (2019) 012099 doi:10.1088/1755-1315/288/1/012099

Gas-filled thermal insulation use in heat supply

Krasnova NP, Gorshenin AS, Rakhimova JI

Samara State Technical University, 443100, Samara, Molodogvardeyskaya str., 244

krasnova-pt@yandex.ru

Abstract. The possibility of applying a new type of thermal insulation, which will significantly reduce heat losses during transportation of heat, increase the service life of the pipeline, and reduce the cost of heat insulation work considered. Using multi-layer thermal insulation, which is a system of closed, sealed microvolumes (pores), which are filled with carbon dioxide proposed. As a shell, high-density polyethylene is used. Such a film can be used as a ready-made material for insulation of pipelines. Depending on the temperature of the pipeline, the required number of layers of thermal insulation is applied to the pipeline.

Introduction

Reduction of thermal losses of buildings and structures, transport of coolant in pipelines of heating networks and process plants is an actual technical and economic task, especially in the Russian Federation, as most of the territory of the country belongs to climatic conditions to the northern territories. This condition requires increased expenditures of heat energy for heat supply to consumers in the residential sector and is accompanied by significant heat losses during transportation of the coolant to ensure the life of production and settlements. In both cases, a significant reduction in heat losses can be achieved by applying modern thermal insulation: for residential, industrial and administrative buildings - in the form of elements of building structures (thermal panels); for heat networks and process pipelines - in the form of a roll thermal insulation coating.

The materials used for thermal insulation are characterized primarily by the properties of thermal conductivity[1]. The lower the thermal conductivity of the material, the better it retains heat. Therefore, thermal conductivity is their passport characteristic, and the need for its measurement is relevant.

The choice of the most effective gas for the development of gas-filled insulation

As is known, gases possess the lowest coefficients of thermal conductivity. For use in gas-filled insulation, several types of gases can be used. The most common and effective are: air, carbon dioxide, oxygen and flue gases of combustion products of organic fuel[2].

The summary data of the thermal conductivity coefficient as a function of the temperature of the above gases is shown in Fig. 2.

The materials used for thermal insulation are characterized primarily by the properties of thermal conductivity. The lower the thermal conductivity of the material, the better it retains heat. Therefore, thermal conductivity is their passport characteristic, and the need for its measurement is relevant.

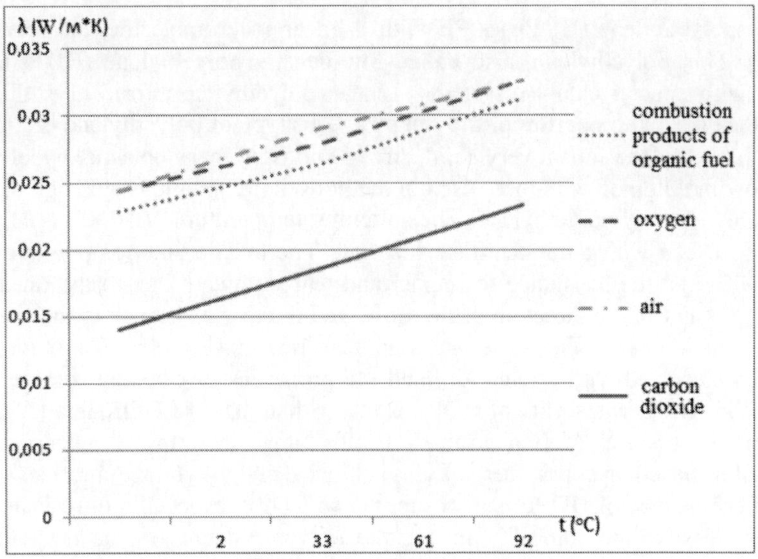

Figure 1. Dependence of the coefficient of thermal conductivity on the temperature of gaseous media.

As can be seen from Fig. 1, carbon dioxide is the lowest value of the thermal conductivity coefficient. Carbon dioxide is released in liquid form at low temperature, in liquid form at high pressure and in gaseous form. The gaseous form of carbon dioxide is obtained from waste gases in the production of alcohols, ammonia, and fuel combustion. Gaseous carbon dioxide by its properties is a non-toxic and non-explosive gas, odorless and colorless. It is proposed to use it as a gaseous medium for thermal insulation[3].

Justification of the choice of polyethylene as a shell for gas-filled insulation

As a coating for carbon dioxide, a film based on polyethylene is considered. Polyethylene is a thermoplastic polymer with a relatively low hardness, odorless and tasteless. Various research methods (microscopic, X-ray and electron diffraction, etc.) show that polyethylene has a crystal structure similar to that of normal paraffins (for example, C60H122, etc.). The degree of crystallinity of the polymer obtained by polymerization of ethylene does not reach 100%: along with the crystalline phase, amorphous is always contained. The ratio of these phases depends on the method of polymer preparation and temperature. Like high-melting waxes and paraffins, it slowly lights up and burns with a weak flame without soot. In the absence of oxygen, polyethylene is resistant to 290 ° C. Within 290-350 ° C, it decomposes into low-molecular-weight polymers of the wax type, and above 350 ° C, decomposition products are low molecular weight liquid substances and gaseous compounds - butylene, hydrogen, carbon monoxide, carbon dioxide, ethylene, ethane, etc. As can be seen from Figure 1, carbon dioxide is the lowest value of the thermal conductivity coefficient. Carbon dioxide is released in liquid form at low temperature, in liquid form at high pressure and in gaseous form. The gaseous form of carbon dioxide is obtained from waste gases in the production of alcohols, ammonia, and fuel combustion. Gaseous carbon dioxide by its properties is a non-toxic and non-explosive gas, odorless and colorless. It is proposed to use it as a gaseous medium for thermal insulation.

The mechanical properties of polyethylene depend on its molecular weight and the degree of crystallinity. With an increase in molecular weight, they improve. Crystallinity also enhances mechanical strength. In thin films, polyethylene (especially low-density polyethylene) has great flexibility and elasticity, and in thick sheets it acquires rigidity [2]. At present, polyethylene can be of high, medium and low density.

High-density polyethylene (HDPE) - PE with a linear macromolecule and relatively high density (0.960 g / cm³). This polyethylene, also called low-density polyethylene (HDPE), it is obtained by polymerization with special catalyst systems. Linear polyethylenes form crystalline regions, which strongly affect the physical properties of the samples. This type of polyethylene is commonly referred to as high density polyethylene; it is a very solid, strong and rigid thermoplastic, widely used for injection molding and blow molding of containers used in the household and industry. High density polyethylene is stronger than low density polyethylene. The softening temperature of HDPE (121 ° C) is higher than that of LDPE, so it can withstand steam sterilization. The tensile and compressive strength is higher than that of LDPE, and the resistance to impact and tear is lower. Due to the linear structure, HDPE molecules tend to orient in the direction of the flow, and the tear resistance in the longitudinal direction of the films is much lower. The differences in tear resistance in the longitudinal and transverse directions can be increased with orientation, and the properties of ribbons working on tearing will be inherent in the film. The permeability of HDPE is lower than that of LDPE, about 5-6 times, and it is an excellent barrier to moisture. Among ordinary HDPE films, in terms of moisture permeability, it is second only to films based on copolymers of vinyl chloride and vinylidene chloride.

The chemical resistance of HDPE is also superior to LDPE, especially in resistance to oils and fats. With increasing density, the solubility in organic solvents decreases, as does the permeability to solvents. HDPE is prone to environmental cracking, like LDPE, but this effect can be reduced by using high molecular weight PE grades that do not have this deficiency. Areas of application of high- density polyethylene, as a rule, coincide with the areas consuming low-density material, but the changed properties of the first, undoubtedly, improve the quality of the products produced. Thus, a film of high-density polyethylene will be stronger and more transparent, molded parts may have a smaller cross-section, and pipes and fibers will have greater strength. Increasing the melting point of new polyethylenes allows sterilization with steam. These factors, combined with the ability to regulate the properties of the products, will promote the growth of the use of polyethylenes produced on surface catalysts. Thus, the most suitable material for gas-filled insulation is a film based on high density polyethylene (low pressure). Advantages over other types of polyethylene are higher softening temperature, high tensile and compressive strength, low moisture permeability, very good weldability.

Gas-filled insulation description

The modern thermal insulation material is a complex multilayer structure and includes high density polyethylene (HDPE), the interlayer space of which is made in the form of cells filled with carbon dioxide. The process of preparing the material is the filling of the polymer hose (in this case it is high density polyethylene of high density HDPE) with carbon dioxide, after which the sleeve is welded to form hemispheres[4]. After receiving the so-called "air-bubble" film, it can be used as a ready-made material for insulation of walls of buildings and structures, as well as technological pipelines. Depending on the normalized heat flow, the required number of layers of thermal insulation is applied to the pipeline (Figure 2).

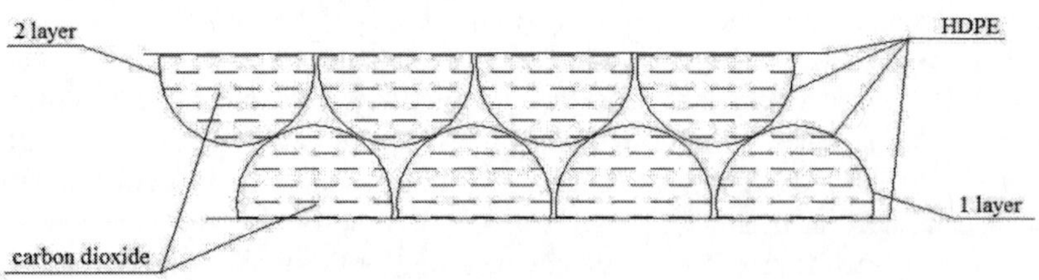

Figure 2. The scheme of gas-filled thermal insulation

Since the volume of carbon dioxide in each pore is small, the free motion that occurs in particles located in this volume can affect in the current under consideration. The motion and heat transfer depend in this case on the nature of the gas, its temperature and the dimensions of space. In this ball layer, the entire volume of gas is covered with motion. In practical calculations, it is required to determine the heat flux that passes through the gas-filled cell. The average density of the heat flux is determined by the formula:

$$q = \frac{\lambda_{eq}}{\delta} * (t_1 - t_2),$$

where λeq is the equivalent coefficient of thermal conductivity, taking into account the heat transfer as heat conductivity, and convection.

The ratio ε_eq = λ_eq / λ, where λ is the thermal conductivity of the gas, characterizes the effect of convection on heat transfer. The quantity ε_eq is a function of the complex Gr * Pr. The dependence ε_eq = f (Gr * Pr) is presented in for vertical and horizontal plane slits, annular and spherical layers filled with a gas or a dropping liquid.

In determining the similarity numbers, regardless of the shape of the interlayer, its thickness is taken as the determining dimension, and the average temperature of the gas is determined as the determining temperature.

Since the represented gas-filled insulation is multilayered, it is required to determine the overall thermal conductivity for the two layers of the HDPE film and the carbon dioxide layer. The total coefficient is determined by the formula:

$$\lambda_{ins} = \frac{\delta_{\text{HDPE}} + \delta_{\text{co2}} + \delta_{\text{HDPE}}}{\frac{\delta_{\text{HDPE}}}{\lambda_{\text{HDPE}}} + \frac{\delta_{\text{co2}}}{\lambda_{\text{co2}}} + \frac{\delta_{\text{HDPE}}}{\lambda_{\text{HDPE}}}} = \frac{0{,}0005 + 0{,}009 + 0{,}0005}{\frac{0{,}0005}{0{,}28} + \frac{0{,}009}{0{,}014} + \frac{0{,}0005}{0{,}28}} = 0{,}0155 \frac{W}{m * \text{K}}$$

This type of thermal insulation has a number of advantages: low thermal conductivity, and, as a consequence, high thermal resistance, increased moisture resistance, corrosion protection, ease of installation, durability, low cost and safety of use.

In this case, this thermal insulation material has one significant minus - the temperature range on the surface should not exceed 126°C (softening temperature of HDPE). The maximum temperature of the coolant for residential, public and administrative buildings should not be more than 95 degrees for two-pipe systems, and not more than 105 degrees for single-pipe systems.

For thermal insulation of pipelines whose temperature exceeds the softening point of HDPE, it is proposed to use gas-filled insulation over a thin layer of a known heat-insulating coating, which may be either a heat-resistant paint or a fiberglass. This layer is laid on the pipeline first and reduces the surface temperature of the pipeline to the range of application of gas-filled insulation. The next layers are laid gas-filled insulation. The number of layers of insulation depends on the surface temperature. Therefore, this thermal insulation material is perfect for thermal insulation of heating pipelines, hot and cold water supply, sewage pipelines, etc.

References

[1] Thermal insulation. Materials, structures, technologies. NTS "Stroyinform" in 2008, 444p.

[2] Physical quantities. Directory. A.P. Babichev, N.A. Babushkina, A.M. Bratkovsky and others; Ed. I.S. Grigorieva, E.Z. Meilikhova. - M.: Energoatomizdat, 1991. - 1232 p.

[3] GAS MEDIA AS MODERN THERMAL INSULATORS Uzlov PS, Krasnova NP ENERGETICS AND ENERGY SAVING: THEORY AND PRACTICE materials of the III All-Russian Scientific and Practical Conference. 2017. P. 425.

[4] NEW TYPE OF HEAT INSULATION TO REDUCE HEAT LOSS Krasnova NP, Mzhelskaya O.Yu., Shchelokov AI ,In the collection: Russia is young, The collection of materials of the VII All-Russian scientific-practical conference of young scientists with international participation. 2015. P. 136.

On low-frequency unsteady phenomena caused by cavitation in hydroturbines

P A Kuibin[1] and A V Zakharov[2]

[1] Kutateladze Institute of Thermophysics SB RAS, 1 Lavrentyev ave., Novosibirsk, 630090, Russia
[2] PJSC "Power Machines" LMZ, 18, Sverdlovskaya nab., St.-Petersburg, 195009, Russia

E-mail: kuibin@itp.nsc.ru

Abstract. When hydroturbines operate at part load or over load, intense pressure and vibration pulsations may occur. One of the instability mechanisms is due to the presence of a cavitation bubble (or cloud) behind the runner. Chen et al. [1] the phenomenon is modeled using the concept of cavitation compliance. They demonstrated the destabilizing effect of the diffuser and swirl. In the present work, the influence of two factors on the coefficients of the characteristic equation, whose solution yields natural frequencies and oscillation increments, is revealed. They are the vorticity distribution and the size of the cavitation cavity behind the runner. It is shown that the effect of swirl on stability is much smaller than that described in [1] and rapidly decreases with the growth of cavitation bubble size.

1. Introduction

Unsteady phenomena in the hydraulic part of a hydroturbine are one of main and actual problems of hydropower aggregates design, construction and operation. The description of these phenomena is complicated by the possibility of cavitation zones and the involvement of air in the water flow.

Considerable progress in modeling the instability due to the presence of the cavitation / air area behind the turbine runner was achieved by Chen et al. [1]. The authors used the concept of cavitation compliance and revealed the destabilizing effect of the diffuser and swirl. The same authors [2], considering the finiteness of the sound speed in the penstock, found the appearance of high-frequency instabilities. It follows from [1, 2] that changing the volume of the cavity, the parameters of the diffuser and the swirl intensity can control the stability of the flow and its eigen frequencies. At the same time, experimental studies [3] clearly demonstrate the effect of the gas phase on the pulsation characteristics in a vortex chamber with a precessing vortex. During transition from pure liquid to a two-phase flow, the precession frequency decreases sharply. With a further increase in the gas supply, the frequency increases monotonically. Based on the helical vortex model [4, 5], where analytical formulas for the precession frequency and the amplitude of pressure pulsations are derived, as well as analysis of the experimental data, a semi-empirical model is constructed to describe the dependence of the frequency and amplitude of pressure pulsations on the gas content [6].

It should be noted that when modeling instability, the authors of [1, 2] incorrectly interpret the Bernoulli equation and do not take into account the effects of swirl cavitation on the pressure distribution behind the turbine runner.

The aim of this paper is to develop models of the vortex flow of a fluid taking into account cavitation and to assess the effect of the vapor phase and swirl on the development of instability in the draft tube of a hydroturbine.

2. An analytical model for the instability description

Consider a system consisting of a penstock of length L_i and cross-sectional area A_i, a turbine runner (TR) and a draft tube (DT) with the area of the inlet and exit sections A_c and A_e, respectively (see figure 1). We assume that behind the turbine, at the entrance to the DT a cavity of volume V_c is formed. Chen et al. [1] derived two equations for flow rate and pressure in the system. The first equation, the continuity equation, presents the relationship between the flow rate in the penstock Q_1 and the flow rate in DT Q_2

$$Q_2 - Q_1 = \frac{dV_c}{dt} = -\rho C \frac{L_e}{A_e}\frac{d^2Q_2}{dt^2} + \rho C \frac{D-\zeta_2}{A_e^2} Q_2 \frac{dQ_2}{dt} + 2\rho C\alpha \frac{\cot\beta_2}{S}\left(\frac{Q_1}{S}\cot\beta_2 - U_2\right)\frac{dQ_1}{dt} \quad (1)$$

The second equation links the pressure at the system inlet, p_i, with the exit pressure, p_e

$$p_i = p_e + \rho\frac{L_e}{A_e}\frac{dQ_2}{dt} + \rho\frac{\zeta_2 - D}{2A_e^2}Q_2^2 + \rho\frac{L_i}{A_i}\frac{dQ_1}{dt} + \rho\frac{\zeta_T}{2A_i^2}Q_1^2 \quad (2)$$

In the system (1) – (2) the following denotations are introduced: ρ is the liquid density; $C = -\partial V_c/\partial p_c$ is the cavitation compliance; $L_e = \int(A_e/A(s))ds$ is the DT effective length; $D = (A_e/A_c)^2 - 1$ is the diffusor factor, ζ_2 is the DT loss factor. ζ_2 is assumed to be constant, although, strictly speaking, it can depend on the degree of swirl of the flow in the DT [7]. At a fixed rotation speed of TR and opening of the guide apparatus, the turbine is considered as a resistance with a constant loss factor ζ_T, which depends on the opening of the guide device. The last term of equation (1) contains the coefficient α - the pressure coefficient responsible for the swirl effect. Its definition will be discussed in the next section. β_2 denotes the angle of inclination of the blade at the exit from the TR, S is the area of the outlet cross-section of the TR, and U_2 is the peripheral velocity at the exit from the TR. In this case, the characteristic circumferential flow velocity at the outlet from the TR is

$$c_{\theta 2} = (Q_1/S)\cot\beta_2 - U_2$$

The second term on the right-hand side of equation (1), proportional to dQ_2/dt, describes the effect of the diffuser on the flow rate coefficient. With the increase in flow Q_2 the ambient pressure p_a is lowered if the diffusor factor D is greater than the loss ζ_2. The volume of the cavity increases. The second term on the left, proportional to dQ_1/dt, presents the swirl effect. This term can also be called the "flow rate coefficient", but this term is related to the flow rate Q_1 in front of the cavity. At flow rates Q_1 higher than $U_2S\tan\beta_2$, the tangential velocity $c_{\theta 2}$ and the cavity volume increase with increasing Q_1. The opposite result occurs at lower flow rates. Note that the left and right sides of equation (2) coincide with the pressure behind the runner p_a.

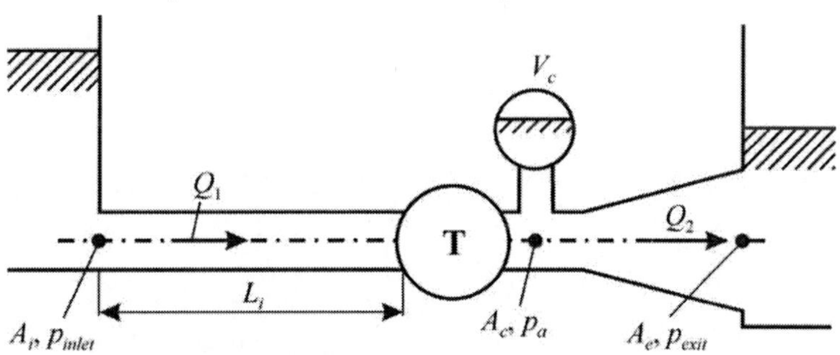

Figure 1. Scheme of the hydraulic part of the hydroturbine.

Equations (1), (2) are the basic equations for determination $Q_1(t)$ and $Q_2(t)$. For stability analysis, let us represent $Q_1 = \bar{Q}_1 + \tilde{Q}_1(t)$, $Q_2 = \bar{Q}_2 + \tilde{Q}_2(t)$ and assume that $\bar{Q}_1 \gg |\tilde{Q}_1(t)|$, $\bar{Q}_2 \gg |\tilde{Q}_2(t)|$. It is obvious that $\bar{Q}_1 = \bar{Q}_2 = \bar{Q}$, so the unsteady parts of the equations (1), (2) can be written down as

$$0 = \rho \frac{L_e}{A_e}\frac{d\tilde{Q}_2}{dt} + \rho \frac{\zeta_2 - D}{A_e^2}\bar{Q}\tilde{Q}_2 + \rho \frac{L_i}{A_i}\frac{d\tilde{Q}_1}{dt} + \rho \frac{\zeta_T}{A_i^2}\bar{Q}\tilde{Q}_1 \tag{3}$$

$$\tilde{Q}_2 - \tilde{Q}_1 = -\rho C \frac{L_e}{A_e}\frac{d^2\tilde{Q}_2}{dt^2} + \rho C \frac{D - \zeta_2}{A_e^2}\bar{Q}\frac{d\tilde{Q}_2}{dt} + 2\rho C\alpha \frac{\cot\beta_2}{S}\left(\frac{\bar{Q}}{S}\cot\beta_2 - U_2\right)\frac{d\tilde{Q}_1}{dt} \tag{4}$$

Next, we represent the nonstationary functions $\tilde{Q}_1, \tilde{Q}_2$ in the form $\tilde{Q}_1 = \tilde{Q}_{1,0}e^{i\omega t}$, $\tilde{Q}_2 = \tilde{Q}_{2,0}e^{i\omega t}$, where i is the imaginary unit. Substituting these expressions into equations (3), (4), we obtain a system of homogeneous linear equations with respect to $\tilde{Q}_{1,0}$, $\tilde{Q}_{2,0}$. Equating the determinant of the matrix of this system of linear equations to zero, we obtain the characteristic equation (5):

$$\begin{aligned}
&-\left(\frac{\rho L_i}{A_i}\frac{\rho L_e}{A_e}C\right)(i\omega)^3 \\
&-\left[\frac{\rho\zeta_T}{A_i^2}\bar{Q}\frac{\rho L_e C}{A_e} - \frac{\rho L_i}{A_i}\frac{\rho C}{A_e^2}(D-\zeta_2)\bar{Q} + 2\rho C\alpha\frac{\cot\beta_2}{S}\left(\frac{\bar{Q}}{S}\cot\beta_2 - U_2\right)\frac{\rho L_e}{A_e}\right](i\omega)^2 \\
&+\left[-\frac{\rho L_e}{A_e} - \frac{\rho L_i}{A_i} + \frac{\rho\zeta_T}{A_i^2}\frac{\rho C}{A_e^2}(D-\zeta_2)\bar{Q}^2 + 2\rho C\alpha\frac{\cot\beta_2}{S}\left(\frac{\bar{Q}}{S}\cot\beta_2 - U_2\right)\frac{\rho(D-\zeta_2)}{A_e^2}\bar{Q}\right](i\omega) \\
&+\left[\frac{\rho(D-\zeta_2)}{A_e^2}\bar{Q} - \frac{\rho\zeta_T}{A_i^2}\bar{Q}\right] = 0
\end{aligned} \tag{5}$$

Equation (5) is an equation of the third order with respect to ($i\omega$) with real coefficients. From it we can find the complex frequency $\omega = \omega_R + i\omega_I$. The real part, ω_R, is the frequency, and the imaginary part, ω_I, is the damping rate (decrement) of the perturbation. If $\omega_1 = \omega_{1R} + i\omega_{1I}$ is a solution of equation (5), then $\omega_2 = -\omega_{1R} + i\omega_{1I}$ is the second solution. Solutions ω_1 and ω_2 are essentially the same solution with the same frequency and decrement. The real part of the third solution is equal to zero, $\omega_{3R} = 0$. Thus, we have three solutions: $\omega_1 = \omega_{1R} + i\omega_{1I}$, $\omega_2 = -\omega_{1R} + i\omega_{1I}$, $\omega_3 = i\omega_{3I}$.

3. Modelling the pressure distribution behind the runner

At certain flow parameters, a steam, air-steam or air cavity may appear behind the turbine runner. We will be interested in partial or forced loads, when the flow at the inlet to the draft tube is swirled.

To study cavitational flow in the draft tube we will follow to paper [8], where models of axi-symmetrical vortices introduced in [4, 9, 10] were considered. At a given distribution of vorticity the problem on determination of the circumferential, u_θ, and axial, u_z, velocity components have been reduced to the integral of the vorticity $\omega_z(r)$.

$$u_\theta = \frac{\Gamma}{2\pi r}\Phi(r), \quad u_z = u_0 - \frac{\Gamma}{2\pi l}\Phi(r), \quad \Phi(r) = \frac{2\pi}{\Gamma}\int_0^r \omega_z(r')r'dr' \tag{6}$$

Here Γ is the total intensity of the vortex, $2\pi l$ is the pitch of helical lines, u_0 is the value of axial velocity at the vortex axis. The pressure has been found by integration

$$p = p_0 + \rho\frac{\Gamma^2}{4\pi^2}\int_0^r \Phi(r')^2\frac{dr'}{r'^3}, \tag{7}$$

where p_0 is the pressure at the vortex axis. For convenience we introduce a non-dimensional vorticity and pressure drop

IOP Conf. Series: Earth and Environmental Science **288** (2019) 012100 doi:10.1088/1755-1315/288/1/012100

$$\tilde{\omega}_z = \frac{\pi\varepsilon^2}{\Gamma}\omega_z, \quad \Delta\tilde{p} = \frac{8\pi^2\varepsilon^2}{\Gamma^2}\frac{p - p_0}{\rho} \tag{8}$$

where ε is the typical size of the vortex.

Three examples of the vorticity distributions were considered in [8]: uniform (model I), fractional-power (model II) and Gaussian (model III) distributions (see table 1).

In a liquid, when the pressure becomes less than the liquid vapor pressure, a cavitation bubble arises in the near-axis area of the vortex. The model supposed that cavity does not change along the vortex and its boundary coincides with the iso-surface of pressure equal to the liquid vapor pressure.

To apply the formula for pressure (7) it is necessary to know the constant p_0. To find it let's write down the pressure averaged over the DT entrance cross-section through the Bernoulli's integral

$$p_m = p_{tw} - \rho g H_s - \rho U_m^2/2.$$

Here p_{tw} is the tail water pressure that is equal atmospheric one, H_s is the difference of heights between the inlet cross-section and the tail water level, U_m is the flow rate mean velocity, g is the acceleration due to gravity. From other hand we calculate the pressure averaged over the cross-section

$$p_{mv} = p_0 + \frac{\rho\Gamma^2}{4\pi^2\varepsilon^2 R^2}\int_0^R r\Delta\tilde{p}dr. \tag{9}$$

The integration is performed over circular cross-section of radius R. The derived correlations for $\Delta\tilde{p}_{m0} = 8\pi^2R^2\rho^{-1}\Gamma^{-2}(p_m - p_0)$ and $\Delta\tilde{p}_{Rm} = 8\pi^2R^2\rho^{-1}\Gamma^{-2}(p_R - p_m)$ for three considered vortex models are presented in table 1. Taking $p_{mv} = p_m$ we find the pressure p_0 from equation (7) and when p_0 is less than the water vapour pressure p^{wvp}, a cavitational bubble arises. The pressure profile changes, its minimum becomes flat and instead formula (8) one has new equation:

$$p = \begin{cases} p_c, & r < r_c \\ p_c + \rho_l \dfrac{\Gamma^2}{4\pi^2}\displaystyle\int_{r_c}^{r}\Phi(r')^2\frac{dr'}{r'^3} & r \geq r_c \end{cases}, \tag{10}$$

where r_c is the radius of cavitation bubble and p_c is the pressure inside cavity.

In papers by Chen et al. [1, 2] the core pressure p_c was expressed through the ambient pressure p_a and the peripheral swirl velocity behind the runner, i.e. $u_\theta(R)$:

$$p_c = p_a - \alpha\rho u_{\theta R}^2. \tag{11}$$

For the Rankine type vortex with the core radius ε it was stated that $\alpha = (R/\varepsilon)^2 - 1/2$. Indeed, from table 1 for model I we find pressure at the tube wall

$$p_R^{\mathrm{I}} = p_0 + \frac{\rho\Gamma^2}{8\pi^2\varepsilon^2}\left(2 - \frac{\varepsilon^2}{R^2}\right).$$

If we consider the ambient pressure equal to the wall pressure, $p_a = p_R$, the cavity pressure equal to the pressure at the flow axis, $p_c = p_0$, then in view of value of the peripheral swirl velocity, $u_\theta(R) = \Gamma/2\pi R$ we obtain the same value of the coefficient α. Note that Bernoulli's equation yields some averaged pressure in the cross-section rather than pressure on a wall. To find the averaged pressure we integrate formula (11) over the tube cross-section

$$p_{mc} = p_c + \frac{\rho_l\Gamma^2}{2\pi^2R^2}\int_{r_c}^{R}\int_{r_c}^{r}\Phi(r')^2\frac{dr'}{r'^3}rdr. \tag{12}$$

The corresponding formulae for $\Delta\tilde{p}_{mc} = 8\pi^2R^2\rho^{-1}\Gamma^{-2}(p_m - p_c)$ are shown in the last line of table 1. There are three functions introduced:

Table 1. Characteristics of three models of axi-symmetrical vortices.

Model	I	II	III
$\tilde{\omega}_z$	$\begin{cases} 1, & r<\varepsilon \\ 0, & r\geq\varepsilon \end{cases}$	$\left(1+\dfrac{r^2}{\varepsilon^2}\right)^{-2}$	$\exp\left(-\dfrac{r^2}{\varepsilon^2}\right)$
$\Phi(r)$	$\begin{cases} r^2/\varepsilon^2, & r<\varepsilon \\ 1, & r\geq\varepsilon \end{cases}$	$\dfrac{r^2}{r^2+\varepsilon^2}$	$1-\exp\left(-\dfrac{r^2}{\varepsilon^2}\right)$
$\Delta\tilde{p}$	$\begin{cases} r^2/\varepsilon^2, & r<\varepsilon \\ 2-\varepsilon^2/r^2, & r\geq\varepsilon \end{cases}$	$\dfrac{r^2}{r^2+\varepsilon^2}$	$2\log 2-\dfrac{\varepsilon^2}{r^2}\left[1-\exp\left(-\dfrac{r^2}{\varepsilon^2}\right)\right]^2 +2\,\mathrm{Ei}\left(-\dfrac{r^2}{\varepsilon^2}\right)-2\,\mathrm{Ei}\left(-2\dfrac{r^2}{\varepsilon^2}\right)$
$\Delta\bar{p}_{m0}$	$2\dfrac{R^2}{\varepsilon^2}-\dfrac{3}{2}-\ln\left(\dfrac{R^2}{\varepsilon^2}\right)$	$\dfrac{R^2}{\varepsilon^2}-\ln\left(1+\dfrac{R^2}{\varepsilon^2}\right)$	$2\sum\limits_{n=2}^{\infty}\dfrac{(-1)^n\left(2^{n-1}-1\right)}{n(n-1)n!}\left(\dfrac{R^2}{\varepsilon^2}\right)^n$
$\Delta\bar{p}_{Rm}$	$\dfrac{1}{2}+\ln\left(\dfrac{R^2}{\varepsilon^2}\right)$	$\ln\left(1+\dfrac{R^2}{\varepsilon^2}\right)-\dfrac{R^2}{R^2+\varepsilon^2}$	$2\sum\limits_{n=2}^{\infty}\dfrac{(-1)^n\left(2^{n-1}-1\right)}{n\,n!}\left(\dfrac{R^2}{\varepsilon^2}\right)^n$
$\Delta\bar{p}_{mc}$	$\dfrac{r_c^4}{2\varepsilon^4}-\dfrac{r_c^2R^2}{\varepsilon^4}+\dfrac{2R^2}{\varepsilon^2}-\dfrac{3}{2}-\ln\dfrac{R^2}{\varepsilon^2}$	$\dfrac{R^2+\varepsilon^2}{r_c^2+\varepsilon^2}-\ln\left(\dfrac{R^2+\varepsilon^2}{r_c^2+\varepsilon^2}\right)-1$	$f_1\left(\dfrac{r_c^2}{\varepsilon^2}\right)+\dfrac{R^2}{\varepsilon^2}f_2\left(\dfrac{r_c^2}{\varepsilon^2}\right)+f_3\left(\dfrac{R^2}{\varepsilon^2}\right)$

$$f_1(x)=2\sum_{n=1}^{\infty}\frac{2^n-1}{(n+1)!}\frac{(-x)^n}{n},\quad f_1(x)=2\sum_{n=1}^{\infty}\frac{2^n-1}{(n+1)!}\frac{(-x)^{n+1}}{(n+1)},\quad f_3(x)=2\sum_{n=1}^{\infty}\frac{2^n-1}{(n+1)!}\frac{(-x)^{n+1}}{n(n+1)}.$$

Formula (12) can be considered as the equation for determination of the cavitational bubble radius at given mean pressure and intensity of the vortex. For the model I integration yields an equation which was previously derived by Wang et al. [11]. This be-quadratic equation can be simply resolved for r_c.

Finally, in view of equations (6, 11) we derive the following formula for the pressure coefficient of swirl

$$\alpha=\frac{2}{\Phi(R)^2}\int_{r_c}^{R}\int_{r_c}^{r}\Phi(r')^2\frac{dr'}{r'^3}r dr\,. \tag{13}$$

Obviously, the α values can be found, accordingly to equation (13), from the last line of table 1 and values of $\Phi(r)$ taken at $r=R$. In particular, for model I one has

$$\alpha^{\mathrm{I}}=\frac{r_c^4}{4\varepsilon^4}-\frac{r_c^2R^2}{2\varepsilon^4}+\frac{R^2}{\varepsilon^2}-\frac{3}{4}-\frac{1}{2}\ln\frac{R^2}{\varepsilon^2}\,. \tag{14}$$

In the numerical results described in [1] authors took $\alpha=(R/\varepsilon)^2-1/2=10$ i.e. $\varepsilon\approx 0.3086\,R$. Thus, yet at very small cavity size at given ε to R ratio we will find from equation (15) value of $\alpha\approx 8.57$ instead of 10. This means that swirl effect is essentially less than described in papers [1, 2]. Moreover, equation (15) shows that the coefficient α decreases with cavity size growth: from 8.57 at $r_c=0$ to 3.57 at $r_c=\varepsilon$ (see figure 2). For models with smooth vorticity distributions the influence of swirl is found to be yet less. In case of model II for cavity radius we have transcendent equation of type $x-\ln x=a$. This equation at $a=1$ has a single root $x=1$. At $a>1$ there are two branches with $x<1$ and $x>1$. We are interested in the upper branch and solution can be obtained numerically. Note that at $x>>1$ the approximate solution is $x\approx a+\ln a$.

The coefficient α for model II

$$\alpha^{\text{II}} = \frac{1}{2}\frac{R^2+\varepsilon^2}{R^2}\left(\frac{R^2+\varepsilon^2}{r_c^2+\varepsilon^2} - \ln\left(\frac{R^2+\varepsilon^2}{r_c^2+\varepsilon^2}\right) - 1\right). \tag{15}$$

changes from 4.83 to 1.80 when the cavity radius grows from 0 to ε. For the third model the coefficient α changes from 5.66 to 2.41 for r_c varying in the interval $[0, \varepsilon]$. The corresponding dependencies of the cavitational bubble radius and the cavitation compliance on $\Delta\bar{p}_{mc}$ were presented in [8].

An example of calculation of the disturbances frequencies and increments in dependence on flow rate for model II and different cavity size are shown in figures 3, 4 ($\varepsilon/R = 0.5$). At zero cavity size periodical oscillations are possible at $\bar{Q} < 0.566$ or $\bar{Q} > 0.690$. The wider the cavity the larger the range of flow rates where the oscillations absent.

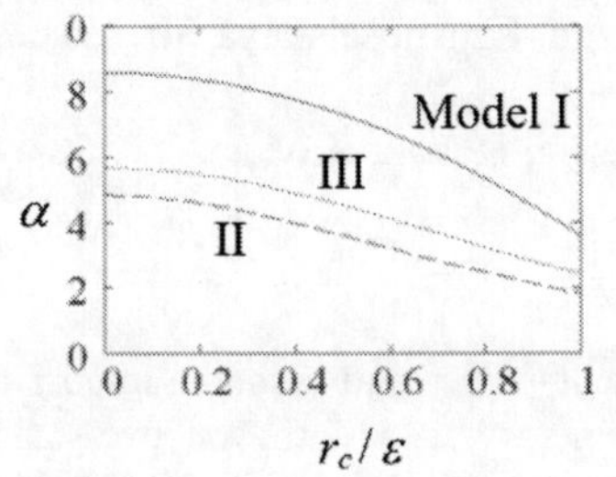

Figure 2. The pressure coefficient of swirl vs the cavity radius.

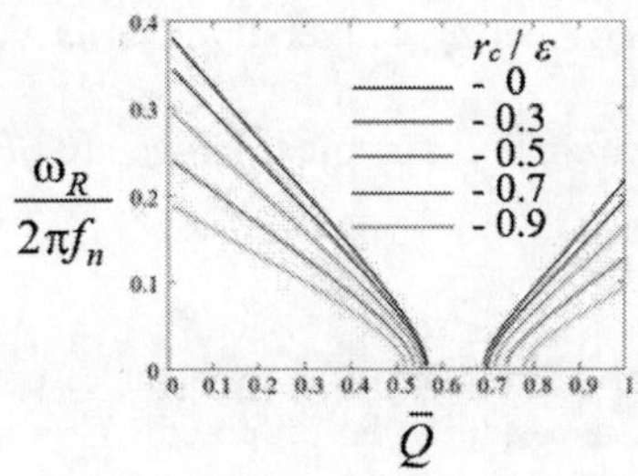

Figure 3. The frequency of unstable disturbances vs flow rate.

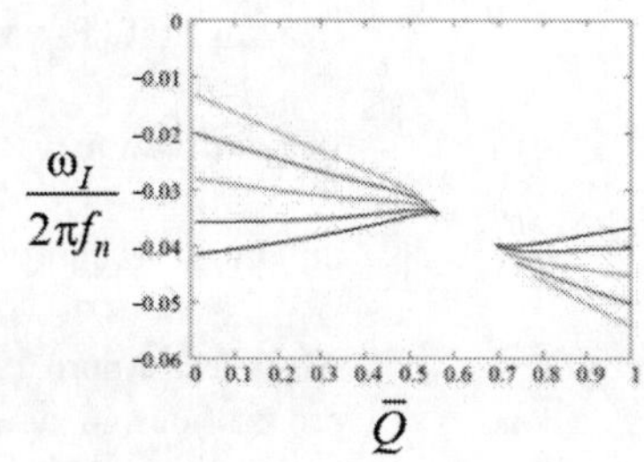

Figure 4. The increments of unstable disturbances vs flow rate.

4. Conclusion

In the paper influence of the cavitational bubble size and type of vorticity distribution on low-frequency oscillations is analyzed. A weaker influence of swirl on the stability conditions is shown in comparison with known literature data. This influence decreases with gas/vapor cavity growth. New data on instability characteristics of cavitating flow behind the hydroturbine runner are obtained.

Acknowledgments
One of the authors – P.A. Kuibin was partially supported by RFBR (project 18-58-53052).

References

[1] Chen C, Nicolet C, Yonezawa K, Farhat M, Avellan F and Tsujimoto Y 2008 *J. Fluids Eng.* **130** 041106
[2] Chen C, Nicolet C, Yonezawa K, Farhat M, Avellan F and Tsujimoto Y 2009 *J. Fluid Machinery and Systems* **2** 260
[3] Alekseenko S, Kuibin P, Okulov V and Shtork S 2010 *Heat transfer research* **41** 465
[4] Alekseenko S, Kuibin P and Okulov V 2007 *Theory of concentrated vortices: An introduction* (Berlin-Heidelberg-New York, Springer-Verlag)
[5] Kuibin P, Okulov V, Pylev I 2006 *Heat Transfer Research* **37** 675
[6] Kuibin P 2010 *Abs. 8th Euromech Fluid Mechanics Conf.*, Bad Reichenhall, Germany, S5-26
[7] Susan-Resiga R, Ciocan G, Anton I and Avellan F 2006 *J. Fluids Eng.* **128** 177
[8] Kuibin P, Pylev I and Zakharov A 2012 *IOP Conf. Series: Earth and Environ. Sci.* **15** 022001
[9] Alekseenko S, Kuibin P, Okulov V and Shtork S 1999 *J. Fluid Mech.* **382** 195
[10] Kuibin P and Okulov V 1996 *Thermophysics and Aeromechanics* **3** 335
[11] Wang X, Nishi M and Tsukamoto H 1994 *Proc. 17th IAHR Symp.*, Beijing, China

Optimal power and resource-intensive process design under uncertainty

T.V. Lapteva[1,2], K.T. Nguen[2], S.A. Laptev[1,2]

[1]Kazan State Power Engineering University, Russia, Kazan, Krasnoselskaya Str., 51

[2]Kazan National Research Technological University, Russia, Kazan, K.Marx str., 68

tanlapteva@yandex.ru

Abstract: Approach to the solution of a problem of optimal power and resource-intensive process design under uncertainty is proposed. The problem has a form of two-stage optimization problem with separate chance constraints. The approach based on the problem's functions approximations allows us to avoid calculation of multiple integrals. This allows us to reduce the problem of stochastic nonlinear programming to the usual problem of nonlinear programming.

Intrduction

It is known that the design of optimal power and resource-intensive processes occurs with partial incompleteness and inaccuracy of information, on the basis of which the designer receives a decision. For modern energy- and resource-intensive industries, uncertainties are caused by inaccuracies in the operation of the control system at associated processes, catalyst deactivation, changes in the heat exchange and mass-exchange characteristics of the apparatus. At the same time, we do not have the opportunity to regulate such impacts that affect the quality of the designed process work. We can only record such changes. Obviously, the refusal to take into account in the formulation of the optimal process design problem of the existing uncertainty can lead to a violation of the requirements for the process functioning or significantly increase the cost of the designed process.

A few approaches to account for uncertainty in optimization have been proposed [1], such as stochastic programming, robust optimization, approximate dynamic programming, and chance-constrained optimization, which is the focus of this article. In this case, the requirements for the work of production can be formulated in the form of hard or chance constraints, depending on the completeness of information on the parameters characterizing the uncertainty, and the preferences of the designer. Hard constraints imply the satisfaction of project requirements at any time during the specified period of production operation. Chance constraints can be represented as separate or joint probability constraints. This form allows us to meet project requirements with a given level of probability. Usually such a probability level is set very high [2].

Accounting for the uncertainty in the criterion of the stochastic programming problem is often presented in the form of a mathematical expectation of some function. Usually this is a function of assessing the process efficiency over the period of process operation. This period is set by the designer. A criterion of pessimism is also used, where the objective function is the worst value of the process efficiency function over the period. Often the forecast of such a criterion is mitigated by the objective function in the form of the probability boundary of the process efficiency function. The change in the operating conditions of the process is represented by the values of uncertain parameters.

It is obvious that the use of probability constraints and a criterion in the form of expectation or in the form of a probability boundary presuppose the existence of information about the law of distribution of uncertain parameters. In addition to these features, the formalization of optimal process design problems should take into account information from different stages of process life. The most common is the accounting of design and operation stages. Problem in the two-stage optimization problem (TSOP) form is used in the case of obtaining exact values of uncertain parameters from the operation stage and there is the possibility of changing the process operating mode during its operation. This impose an automated control system for the projected process. In other case the one-stage optimization problem (OSOP) form is used. Further we will assume that the uncertain parameters have a normal distribution, are uncorrelated, and their distribution functions are known. As the considered problem of designing an optimal process, we will consider TSOP with separate probability constraints (TSOPSPC).

Formulation of optimization problems with separate chance constraints

Using the objective function in the expected value form we can formulate TSOPSPC in the form:

$$\min_{d,z(\theta)} E[f(d,z(\theta),\theta)] \tag{1}$$

$$\Pr\{g_j(d,z(\theta),\theta)\le 0\}\ge \alpha_j, j=1,...,m, \tag{2}$$

where functions g_j are the left part of the inequality constraints, corresponding to design specifications, θ – n_θ-vector of uncertain parameters, d – n_d-vector of design variables, $z(\theta)$ – n_z-vector of control variables, $f(d,z(\theta),\theta)$ – function of assessing the process efficiency, α_j – a probability level. Functions $z(\theta)$ characterize the change in the process operating mode, depending on the operating conditions changing. We assume the functions f and g_j, $j=1,...,m$, are continuously differentiable. Value ranges of the parameters θ form a uncertainty region $T=\{\theta:\theta_i^L\le\theta_i\le\theta_i^U, i=1,...,n_\theta\}$.

To indicate in the problem formulation the possibility of adjusting the operating mode of the project under the changing operating conditions we use functional dependence of the control variables z on the uncertain parameters θ. The indication of this dependence is a characteristic feature of the TSOP. In case of using of the probability boundary of the function $f(d,z(\theta),\theta)$ over the selected process operation period we can formulate the problem in the following form:

$$\min_{d,z(\theta)} q \tag{3}$$

$$\Pr\{g_j(d,z(\theta),\theta)\le 0\}\ge \alpha_j, j=1,...,m, \tag{4}$$

$$\Pr\{f(d,z(\theta),\theta)\le q\}\ge \alpha_0. \tag{5}$$

In the problem (1)-(2) the objective function has the form $E_\theta[f(d,z(\theta),\theta)]=\int_T f(d,z(\theta),\theta)\rho(\theta)d\theta$, and functions of the constraints left parts have forms:

$$\Pr\{g_j(d,z(\theta),\theta)\le 0\}=\int_{\Omega_j}\rho(\theta)d\theta,\ \Omega_j=\{\theta: g_j(d,z(\theta),\theta)\le 0, \theta\in T\}.$$

Here $\rho(\theta)$ – is the probability density function. Because the parameters θ are uncorrelated, the function has the form $\rho(\theta) = \rho(\theta_1)\rho(\theta_2)...\rho(\theta_{n_\theta})$, where $\rho(\theta_i)$ – is the probability density function of the separate uncertain parameter θ_i, $i = 1,...,n_\theta$.

In the problem (3)-(5) the left part of the constraint (5) should be calculated as:

$$\Pr\{f(d,z(\theta),\theta) \le q\} = \int_{\Omega_q} \rho(\theta)d\theta, \; \Omega_q = \{\theta : f(d,z(\theta),\theta) \le q, \theta \in T\}.$$

Main issues

The optimization problem with chance constraints was introduced in a work by Charnes et al. [3]. There are many challenging aspects of solving chance constraints optimization problems. The main issues in solving chance constraints optimization problems are the computation of multiple integrals for the calculation of the probabilities of constraint satisfaction (which is very computationally intensive) and the often nonconvex nature of the feasible region of chance-constrained optimization problems. To avoid these difficulties, existing approaches for solving chance constraints optimization problems rely on solving an approximation problem. Generally, three groups of approximation methods are used to approximate chance constraints optimization problems: improvement of the Gauss quadrature, sampling-based approaches, and analytical approximation approaches. The second group of methods contains two categories of techniques: scenario approximation and sample-average approximation. Research contributions in the development of scenario approximation methods have been made by Shapiro and co-workers [4] among others. However, the scenario approximation itself is random, and its solution might not satisfy the chance constraints [5]. Sample-average approximation methods use an empirical distribution associated with random samples to replace the actual distribution and to evaluate the chance constraints. The third group of methods is divided into three categories of techniques: robust optimization, well-known probability inequalities, and the transformation of chance constraints into deterministic constraints. The approaches for obtaining robust solutions to optimization problems have been studied by Floudas and co-workers in a article [6] in which both continuous (general, bounded, uniform, normal) and discrete (general, binomial, Poisson) uncertainty distributions were considered and the framework was applied to operational planning problems. Approaches based on the transformation of chance constraints into deterministic constraints allows the calculation of multiple integrals to be avoided. However, except for a few specific probability distributions (e.g., normal distribution), it is difficult to formulate an equivalent deterministic constraint for a chance constraint. Li et al. [7] proposed a back-mapping method for solving a steady-state OSOP based on the use of the monotonic relationship between the constrained output and one of input uncertain parameters. Geletu et al. [8] proposed a new approach for solving a nonconvex OSOP with non-Gaussian distributed uncertainties. However, their approach still requires the calculation of multiple integrals. Significantly less attention has been paid to the formulation and solution of the TSOP for chemical processes with chance constraints. Ostrovsky et al. [9, 10] formulated the two-stage optimization problem with chance constraints and developed a method for solving this type of problem based on the transformation of chance constraints into deterministic constraints and the partition of the uncertainty region for the cases when the uncertain parameters have a normal distribution. However, proposed approach formulate the problem whose dimension grows exponentially from iteration to iteration. In this article, we develop a new approximate method for solving optimization problems with separate chance constraints for the case when the uncertain parameters are independent random variables. Our method is based on the transformation of chance constraints into deterministic form.

Main approximations

Consider the approximation for the functions $z_r(\theta)$, $r=1,...,n_z$. We will use linear in the parameters θ functions $\bar{z}_r^q = b_0^{q,r} + \sum_{i=1}^{n_\theta} b_i^{q,r}\theta_i^q$, where θ_i^q – i-th coordinate of some point $\theta^q \in T$. Use function $\bar{f}(d,\bar{z}^q,\theta,\theta^q)$ to approximate functions $f(d,z(\theta),\theta)$, where $\bar{f}$ is the linear part of the expansion of the function f in a Taylor series at the point θ^q:

$$\bar{f}(d,\bar{z}^q,\theta,\theta^q) = f(d,\bar{z}^q,\theta^q) + \sum_{i=1}^{n_\theta}\{(\partial f(d,\bar{z}^q,\theta^q)/\partial\theta_i)(\theta_i - \theta_i^q)\}. \quad (6)$$

Let partition uncertainty region T into subregions $T_q = \{\theta : \theta_i^{L,q} \le \theta_i \le \theta_i^{U,q}, i=1,...,n_\theta\}$, $q=1,...,Q^{(k)}$, at the k-th iteration to improve proposed approximations

$$T = \bigcup_{q=1}^{Q^{(k)}} T_q,\ T_{q_1}^{(k)} \cap T_{q_2}^{(k)} = \varnothing,\ q_1,q_2 = 1,...,Q^{(k)},\ q_1 \ne q_2. \quad (7)$$

The function $z(\theta)$ approximation has the form $\bar{z} = \{\bar{z}^q = \{\bar{z}_1^q,...,\bar{z}_{n_z}^q\}, q=1,...,Q^{(k)}\}$, and the (1)-(2) problem criteria approximation has the form [2]:

$$E_{ap}[f(d,z(\theta),\theta)] = \sum_{q=1}^{Q^{(k)}}\left(a_q f(d,\bar{z}^q,\theta^q) + \sum_{i=1}^{n_\theta}\frac{\partial f(d,\bar{z}^q,\theta^q)}{\partial\theta_i}(E[\theta_i;T_q^{(k)}] - a_q\theta_i^q)\right),$$

$$E[\theta_i;T_q^{(k)}] = \prod_{r=1}^{i-1}[\Phi(\tilde{\theta}_r^{(k)U,q}) - \Phi(\tilde{\theta}_r^{(k)L,q})]\cdot I_{iq}\cdot\prod_{r=i+1}^{n_\theta}[\Phi(\tilde{\theta}_r^{(k)U,q}) - \Phi(\tilde{\theta}_r^{(k)L,q})],$$

$$a_q = \prod_{i=1}^{n_\theta}[\Phi(\tilde{\theta}_i^{(k)U,q}) - \Phi(\tilde{\theta}_i^{(k)L,q})],\ I_{iq} = \int_{\theta_i^{(k)L,q}}^{\theta_i^{(k)U,q}} \theta_i\rho(\theta)d\theta,$$

where $\Phi(\xi)$ – is the standard distribution function, $E[\theta_i]$ is a mean value and $(\sigma_i)^2$ is the variance of parameter θ_i, $\tilde{\theta}_i^{(k)L,q} = (\theta_i^{(k)L,q} - E[\theta_i])/\sigma_i$, $\tilde{\theta}_i^{(k)U,q} = (\theta_i^{(k)U,q} - E[\theta_i])/\sigma_i$, $i=1,...,n_\theta$.

Using the $\bar{z}_r^q$ form, one can get $E_{ap}[f(d,z(\theta),\theta)] = \sum_{q=1}^{Q^{(k)}} E_{ap}[f(d,b^q,\theta)]$, where

$$E_{ap}[f(d,b^q,\theta)] = a_q f(d,b^q,\theta^q) + \sum_{i=1}^{n_\theta}\frac{\partial f(d,b^q,\theta^q)}{\partial\theta_i}(E[\theta_i;T_q^{(k)}] - a_q\theta_i^q),\ q=1,...,Q^{(k)}, \quad (8)$$

and $b^q = \{b_0^q, b_1^q,...,b_{n_\theta}^q\}$, $b_i^q = \{b_i^{q,1}, b_i^{q,2},...,b_i^{q,n_z}\}$, $q=1,...,Q^{(k)}$.

Chance constraints transformation

Let transform chance constraints into deterministic ones. Firstly, we consider the method of transforming constraints (2), which we extend below to constraints (5). We use the method for the linear in parameters θ function $g(x,\theta) = \theta_1 h_1(x) + ... + \theta_{n_\theta} h_{n_\theta}(x) - c$, $h_i(x)$, $i=1,...,n_\theta$, – arbitrary functions, c – random parameter with an expected value $E[c]$ and a variance $(\sigma_c)^2$. In [11] it is shown that probability constraint $\Pr\{g(x,\theta) \le 0\} \ge \alpha$ may be transformed into form

$$\sum_{i=1}^{n_\theta} E[\theta_i]h(x) + \Phi^{-1}(\alpha)[\sum_{i=1}^{n_\theta}(\sigma_i)^2 h_i(x)^2 + (\sigma_c)^2]^{1/2} \le E[c]. \quad (9)$$

For case of OSOP we proposed approximation of the functions g_j by the $\bar{g}_j(d,\bar{z}^q,\theta,\theta^q)$ – linear part of the their Taylor series expansion at the points $\theta^{I,j} \in T$, $j=1,...,m$ [12]. One have the form similar to the (6). We propose in [12] new form of the constraint (9) using following form of the functions $c_j = \sum_{i=1}^{n_\theta}(\theta_i^{l,j}\partial g_j(d,z,\theta^{l,j})/\partial\theta_i - g_j(d,z,\theta^{l,j}))$ and $h_{ij}(d,z) = \partial g_j(d,z,\theta^{l,j})/\partial\theta_i$,

$$[c_j - \sum_{i=1}^{n_\theta} E[\theta_i]\partial g_j(d,z,\theta^{l,j})/\partial\theta_i]/[\sum_{i=1}^{n_\theta}(\sigma_i \cdot \partial g_j(d,z,\theta^{l,j})/\partial\theta_i)^2]^{1/2} \geq \Phi^{-1}(\alpha_j)]. \quad (10)$$

Let us extend this approach to the TSOPSPC. Since (7) is hold, we can write

$$\Pr\{g_j(d,z(\theta),\theta) \leq 0\} = \int_{\Omega_{1,j}} \rho(\theta)d\theta + ... + \int_{\Omega_{Q^{(k)},j}} \rho(\theta)d\theta, \ j=1,...,m, \quad (11)$$

$$\Omega_{q,j} = \{\theta : g_j(d,z(\theta),\theta) \leq 0, \theta \in T_q\}, \ q=1,...,Q^{(k)}, \ j=1,...,m. \quad (12)$$

Let us use the same points θ^q for the expansion of functions $g_j(d,z(\theta),\theta)$ and $f(d,z(\theta),\theta)$ taking account of the $z(\theta)$ approximation in form of the function $\bar{z}^q$.

Taking into account the linear approximation of the functions $g_j(d,z(\theta),\theta)$ one can write the form of the q-th summand in the (11)

$$\int_{\Omega_q} \rho(\theta)d\theta = a_q\Phi\left(\left(c_{q,j} - \sum_{i=1}^{n_\theta}[E_q[\theta_i]\partial g_j(d,b^q,\theta^q)/\partial\theta_i]\right)\Big/\left[\sum_{i=1}^{n_\theta}[\sigma_i^q \partial g_j(d,b^q,\theta^q)/\partial\theta_i]^2\right]^{0.5}\right),$$

$$E_q[\theta_i] = \int_{T_q} \theta_i \rho(\theta)d\theta, \ \sigma_i^q = E_q[\theta_i^2] - (E_q[\theta_i])^2. \quad (13)$$

Due to the assumed independence of the uncertain parameters, we get

$$E_q[\theta_i] = \int_{T_q} \theta_i \rho(\theta)d\theta = \int_{\theta_1^L}^{\theta_1^U} \rho(\theta_1)d\theta_1 \cdot ... \cdot \int_{\theta_i^L}^{\theta_i^U} \theta_i \rho(\theta_i)d\theta_i \cdot ... \cdot \int_{\theta_{n\theta}^L}^{\theta_{n\theta}^U} \rho(\theta_{n\theta})d\theta_{n\theta}. \quad (14)$$

It is clear the expressions (13) do not depend on the search variables of the problem (1)-(2) and can be calculated before problem (1)-(2) solving. Taking into account (10) and (13), (14) we can get deterministic approximation of the constraint (2) in the form

$$\sum_{q=1}^{Q^{(k)}}\left(a_q\left(c_{q,j} - \sum_{i=1}^{n_\theta}[E_q[\theta_i]\partial g_j(d,b^q,\theta^q)/\partial\theta_i]\right)\Big/\left[\sum_{i=1}^{n_\theta}[\sigma_i^q \partial g_j(d,b^q,\theta^{j,q})/\partial\theta_i]^2\right]^{0.5}\right) \geq \Phi^{-1}(\alpha_j), \quad (15)$$

where $c_{q,j} = \sum_{i=1}^{n_\theta}(\theta_i^q \cdot \partial g_j(d,b^q,\theta^q)/\partial\theta_i) - g_j(d,b^q,\theta^q)$, $j=1,...,m$.

Using (8), (15) and expressions for the $\bar{z}_r^q$ we can write TSOPSPC in the form of the deterministic nonlinear programming

$$F^{(k)} = \min_{d,b^q} \sum_{q=1}^{Q^{(k)}} E_{ap}[f(d,b^q,\theta)] \tag{16}$$

$$\sum_{q=1}^{Q^{(k)}} a_q \frac{c_{q,j} - \sum_{i=1}^{n_\theta}[E_q[\theta_i]\partial g_j(d,b^q,\theta^q)/\partial\theta_i]}{\left[\sum_{i=1}^{n_\theta}[\sigma_i^q \partial g_j(d,b^q,\theta^{j,q})/\partial\theta_i]^2\right]^{0.5}} \geq \Phi^{-1}(\alpha_j),\ j=1,\dots,m. \tag{17}$$

We can solve problem (16) using a nonlinear programming method (for example, the sequential quadratic programming (SQP) method [13].

It is clear the way of transforming constraints (2) is easily extended to the transfomation (5) using linear approximation (6) for the function $f(d,z(\theta),\theta)$. The constraint (5) get the form

$$\sum_{q=1}^{Q^{(k)}}\left(a_q \left(s_q - \sum_{i=1}^{n_\theta}[E_q[\theta_i]\cdot\partial f(d,b^q,\theta^q)/\partial\theta_i]\right) \Big/ \left[\sum_{i=1}^{n_\theta}[\sigma_{i,q}\cdot\partial f(d,b^q,\theta^q)/\partial\theta_i]^2\right]^{0.5}\right) \geq \Phi^{-1}(\alpha_0),$$

where $s_q = \sum_{i=1}^{n_\theta}[\theta_i^q \cdot \partial f(d,b^q,\theta^q)/\partial\theta_i] - f(d,b^q,\theta^q)$.

Partition of the uncertain region

Let consider the partition rule to improve the approximation. Let us partition uncertain region T into subregions T_q, $q=1,\dots,Q^{(k)}$. We should partition the subregion T_{q^*} with the worst quality of the approximation of the function $f(d,z(\theta),\theta)$ by the function $\bar{f}(d,\bar{z}^q,\theta,\theta^q)$ and the total error of the approximation of the functions g_j by the functions $\bar{g}_j$ will be the worst. Let $d^{(k)}, b^{(k),q}$ is the problem (1)-(2) solution. To find the q^* we will use the argument of the following problem solution:

$$\max_q \max_{\theta\in T_q}[(f(d^{(k)},b^{(k),q},\theta) - f(d^{(k)},b^{(k),q},\theta,\theta^q))^2 + \sum_{j=1}^{m}(g(d^{(k)},b^{(k),q},\theta) - \bar{g}(d^{(k)},b^{(k),q},\theta,\theta^q))^2].$$

Note that the proposed approach improve the approximation of function $f(d,b,\theta)$ and functions $g_j(d,b,\theta)$, $j=1,\dots,m$, gradually, from iteration to iteration, which reduces computational costs at the first iterations.

The proposed approach was tested on a model example of the optimal design problem of a system consisting of a reactor and a heat exchanger [14]. Comparison of the obtained results with the results of known approaches [2, 9, 14] showed the effectiveness of the proposed approach, while achieving a significant reduction in time to solve the problem.

Conclusions

We proposed an approach for solving problem the design of optimal power and resource-intensive processes under uncertainty. The problem is formulated in the form of the optimization problems with separate chance constraints for cases when the uncertain parameters are independent random variables with arbitrary probability distributions. It is based on the three following steps: approximate transformation of the chance constraints into deterministic ones; approximate calculation of the expected value of the objective function with the help of the piecewise-linear approximation of the objective function; approximate representation of the control variables $z(\theta)$ as piecewise-constant functions.

To improve these approximations, partition of the uncertainty region into subregions is performed at each iteration. The proposed method excludes multiple integration. In addition, we proposed the approach for solving the TSOPSPC. Li at al. [7] considered only the method for OSOP. The computational experiment showed that the proposed approach decreases significantly the computational time of solving the TSOPSPC for the design problems of the chain of continuous stirred tank reactor and heat exchanger.

References

[1] Calfa B.A., Grossmann I.E., Agarwal A., Bury S.J. and Wassick J.M. 2015 *Computers & Chemical Engineering.* Data-Driven Individual and Joint Chance-Constrained Optimization via Kernel Smoothing **78** 51-69

[2] Ostrovsky G.M., Ziyatdinov N.N. and Lapteva T.V. 2013 *AIChE Journal.* Optimization problem with normally distributed uncertain parameters **59 Is. 7** 2471-84

[3] Charnes A., Cooper W.W. and Symonds G.H. 1958 *Manag. Sci.* Cost horizons and certainty equivalents: an approach to stochastic programming of heating oil. **4** 235-63

[4] Pagnoncelli B., Ahmed S. and Shapiro A. 2009 *J. Optim. Theory Appl.* Sample average approximation method for chance constrained programming: theory and applications. **142** 399-416

[5] Li Z., Li Z. 2015 *Comput. Chem. Eng.* Optimal robust optimization approximation for chance constrained optimization problem. **74** 89-99

[6] Li Z. and Floudas C. A. 2014 *A Comparative Theoretical and Computational Study on Robust Counterpart Optimization* (III. Improving the Quality of Robust Solutions. Ind. Eng. Chem. Res.) **53** 13112-24

[7] Li P., Arellano-Garcia H. and Wozny, G. 2008 *Comput. Chem. Eng.* Chance constrained programming approach to process optimization under uncertainty. **32** 25-45

[8] Geletu A., Hoffmann A., Kloppel M. and Li P. 2011 Eng. Optim. Monotony analysis and sparse-grid integration for nonlinear chance constrained process optimization. **43** 1019-41

[9] Ostrovsky G.M., Ziyatdinov N.N., Lapteva T.V. and Zaitsev I.V. 2011 *Chem. Eng. Sci.* Two-stage optimization problem with chance constraints. **66** 3815-28

[10] Ostrovsky G.M., Ziyatdinov N.N. and Lapteva T.V. 2013 *Comput. Chem. Eng.* Optimal design of chemical processes with chance constraints. **59** 74-88

[11] Liu B. 2002 *Physica-Verlag, Heidelberg.* Theory and practice of uncertain programming

[12] Ostrovskii G.M., Ziyatdinov N.N., Lapteva T.V. and Pervukhin D.D. 2009 *Doklady Chemistry.* One-step problems of optimization of chemical engineering systems under soft constraints. **425 Is. 1** 64-67

[13] Biegler L.T. 2010 *Nonlinear Programming Concepts, Algorithms, and Applications to Chemical Processes; MOS-SIAM Series on Optimization* (SIAM: Philadelphia, PA)

[14] Halemane K.P. and Grossmann I.E. 1983 *AIChE J* Optimal process design under uncertainty. **29 (3)** 425-33

An Approach to solve the power and resource-intensive process design problem in a one-stage optimization problem form

T.V. Lapteva[1,2], A.S. Silvestrova[1], N.N. Ziatdinov[1]

1 Kazan National Research Technological University, Russia, Kazan, K.Marx str., 68
2 Kazan State Power Engineering University, Russia, Kazan, Krasnoselskaya Str., 51

tanlapteva@yandex.ru

Abstract: Approach to the solution of a problem of optimal power and resource-intensive process design under uncertainty is proposed. The problem has a form of one-stage optimization problem with separate chance constraints. The approach based on the problem's functions approximations and approximations of the regions of the constraints satisfaction. This allows us to avoid calculation of multiple integrals. Thus we reduce the problem of stochastic nonlinear programming to the sequence of the usual problem of nonlinear programming.

Intrduction

Generally, power and resource-intensive processes are designed with the use of uncertain source data and inaccurate mathematical models. The inexactness of mathematical models arises because of the original uncertainty of chemical, physical, and economic data which are used during a process design. Therefore, it is important to design a process that guarantees the satisfaction of all design specifications either exactly or with some probability. This paper considers the issue of power and resource-intensive process optimization when at the operation stage the design specifications should be met with some probability and the control variables can be changed.

The optimization problem of a optimal process design in the case of the using an exact mathematical model and uncertain source data can be expressed as follows

$$\min_{d,z\in H} f(d,z,\theta)$$

$$g_j(d,z,\theta)\le 0,\ j=1,...,m,\ \forall\theta\in T,$$

where d – n_d-vector of design variables, z – n_z of control variables, and θ – n_θ-vector of uncertain parameters, $\theta\in T$, the region $T=\{\theta_i:\theta_i^L\le\theta_i\le\theta_i^U, i=1,...,n_\theta\}$ characterizes the changing of the operating conditions of the designed process (DP) for the operation period, $f(d,z,\theta)$ – the function of evaluation of the operational efficiency of the designed process at a time that appropriate to the operating conditions of the designed process, specified values of uncertain parameters θ. We cannot solve this problem due to an exact values of uncertain parameters θ.

The problem of the optimal design of a process under uncertainty can be formulated as follows: it is necessary to create an optimal process that would guarantee the satisfaction (exact or with some probability) of all design specifications in the case when inexact mathematical models are used and internal and external factors change during the process operation stage. Usually, the following two formulations of this problem are used:

– the formulation of the two-stage optimization problem (TSOP) takes into account possibility of the control variables change at the operation stage. Here, we suppose that at each time instant during the operation stage (a) values of all or some of the uncertain parameters can be either measured or calculated using the experimental data (thus at each time instant the process model is corrected) and (b) during the operation stage, the control variables are adjusted depending on a process state. This formulation can be used if it is possible to accurately estimate all or some of the uncertain parameters at the operation stage of process.

– the formulation of the one-stage optimization problem (OSOP) supposes that the control variables are constant at the operation stage.

The formulation of the optimization problem under uncertainty depends on the type of constraints. The constraints can be "hard" or "soft". Hard constraints must never be violated during the operation stage. Conversely, if occasional violations are allowed then the constraints are said to be soft. We consider methods of solving the one-stage optimization problems with chance constraint (OSOPCC)

1. General formulations

Let write OSOPCC with the expected value of the function $f(d,z,\theta)$ criterion during the process operation stage in the form [2]

$$\min_{d,z\in H} E_\theta[f(d,z,\theta)], \quad (1)$$

$$\Pr\{g_j(d,z,\theta)\le 0\}\ge \alpha_j,\ j=1,...,m, \quad (2)$$

where $\Pr\{g_j(d,z,\theta)\le 0\} = \int_{\Omega_j}\rho(\theta)d\theta \ge \alpha_j$, $0\le\alpha_j\le 1$, $\Omega_j=\{\theta: g_j(d,z,\theta)\le 0,\ \theta\in T\}$,

$E_\theta[f(d,z,\theta)]=\int_T f(d,z,\theta)\rho(\theta)d\theta$, $\rho(\theta)$ – is the probability density function.

The solution of problem (1)-(2) requires multidimensional integration to calculate the values of constraints (2) and the criterion in (1) at each iteration of the procedure for solving problem (1)-(2). It is computationally time-consuming operation even in the case of a small number of uncertain parameters. Currently, there are proposed a number of approaches for the economic calculation of multidimensional integrals that can be grouped into three groups:

1. Approaches based on the development of Gauss quadratures. Unfortunately, these approaches require extremely large computations with increasing dimension of the integration region.

2. A group of approaches developing ideas for methods of statistical testing. The most advanced approach in this group is currently the method of approximation of the sample-average approximation (SAA) [3]. However, these methods in turn use random variables, and such solution can lead to chance constraints violation.

3. Methods for the transformation chance constraints to deterministic form, which allows to exclude the multidimensional integration procedure. Among the methods of this group, methods for transformation linear constraints are known [4], approaches based on certain properties of constraint functions [5]. There are methods that use known inequalities in the transformation of constraints. However, such methods either impose certain requirements to the form of constraint functions, or still require multidimensional integration.

An overview of the approaches to the solution of the OSOPCC is given in [6].

2. Proposed approach basics

In [2, 7] we proposed an approach to the solution of OSOPCC (1)-(2) for independent indefinite parameters, based on our proposed method of reducing chance constraints to a deterministic form that does not depend on the form of the constraint functions and does not require multidimensional integration. However, the proposed approach required an analytic transformation of the functions of chance constraints, which is not always possible and convenient. In this paper we consider the development of the approach proposed in [2, 7], which allows to avoid preliminary analytical transformations.

2.1. Expected value quantity approximation

To reduce the computations for multidimensional integration for getting the value of the criterion for the problem (1)-(2), in [2] we used at the k -th iteration of the solving procedure the approximation of the value $E_\theta[f(d,z,\theta)]$ by a function $E_{ap}^{(k)}[f(d,z,\theta)]$, relying on the expansion of the function $f(d,z,\theta)$

$$E_{ap}^{(k)}[f(d,z,\theta)] = \sum_{q=1}^{Q^{(k)}}\left(a_q f(d,z,\theta^q) + \sum_{i=1}^{n_\theta}\frac{\partial f(d,\bar{z}^q,\theta^q)}{\partial\theta_i}(E[\theta_i;T_q^{(k)}] - a_q\theta_i^q)\right),$$

$$a_q = \prod_{i=1}^{n_\theta}[\Phi(\tilde{\theta}_i^{(k)U,q}) - \Phi(\tilde{\theta}_i^{(k)L,q})],\ E[\theta_i;T_q^{(k)}] = A_q^{(k)}\cdot I_i^{(k),q}\cdot B_q^{(k)},$$

$$A_q^{(k)} = \prod_{r=1}^{i-1}[\Phi(\tilde{\theta}_r^{(k)U,q}) - \Phi(\tilde{\theta}_r^{(k)L,q})],\ B_q^{(k)} = \prod_{r=i+1}^{n_\theta}[\Phi(\tilde{\theta}_r^{(k)U,q}) - \Phi(\tilde{\theta}_r^{(k)L,q})],$$

$$I_{iq} = \int_{\theta_i^{(k)L,q}}^{\theta_i^{(k)U,q}}\theta_i\rho(\theta)d\theta,\ \tilde{\theta}_i^{(k)L,q} = \frac{\theta_i^{(k)L,q} - E[\theta_i]}{\sigma_i},\ \tilde{\theta}_i^{(k)U,q} = \frac{\theta_i^{(k)U,q} - E[\theta_i]}{\sigma_i},$$

$$T = T_1^{(k)}\cup T_1^{(k)}\cup\ldots\cup T_{Q^{(k)}}^{(k)},\ T_q^{(k)} = \{\theta : \theta_i^{(k)L,q} \le \theta_i \le \theta_i^{(k)U,q}, i = 1,\ldots,n_\theta\},\ q = 1,\ldots,Q^{(k)},\ i = 1,\ldots,n_\theta.$$

The proposed approximation does not require the calculation of multidimensional integrals. To calculate it, one-dimensional integrals values are required. However, the values of these quantities can be obtained before the optimization procedure because the integration elements are independent of the search variables of the problem.

2.2. Approximation of the chance constraints satisfaction region

In [7] we proposed a method for reducing the chance constraints (2) to a deterministic form by approximation the T_{α_j} regions of satisfaction of constraints $g_j(d,z,\theta)\le 0$ by multidimensional rectangles R_{α_j}. For this, the hypersurfaces of the boundaries of the regions T_{α_j} obtained by the expression $\theta_{n\theta} = \varphi_j(d,z,\theta_1,\ldots,\theta_{n\theta-1})$ from the equations $g_j(d,z,\theta_1,\ldots,\theta_{n_\theta}) = 0$ were approximated by hyperplanes $\bar{\varphi}_j(d,z,\bar{\theta}_1,\ldots,\bar{\theta}_{n\theta-1})$, $\bar{\theta}_i = 0.5\cdot(\theta_i^U - \theta_i^L)$, $i = 1,\ldots,n_\theta - 1$,. As a result, we get the formulation of the problem of evaluation the criterion of the OSOPCC (1)-(2) in form (see [7])

$$\min_{d,z\in H} E_{ap}^{(k)}[f(d,z,\theta)] \tag{3}$$

$$\sum_{l=1}^{N_j^{(k)}}\left(\prod_{i=1}^{n_\theta-1}[\Phi(\tilde{\theta}_r^{(k)U,j,l}) - \Phi(\tilde{\theta}_r^{(k)L,j,l})]\cdot I_{n_\theta,j,l}\right)\ge\alpha_j,\ j = 1,\ldots,m, \tag{4}$$

where $\tilde{\theta}_i^{L,j,l} = (\theta_i^{L,j,l} - E[\theta_i])/\sigma_i$, $\tilde{\theta}_i^{U,j,l} = (\theta_i^{U,j,l} - E[\theta_i])/\sigma_i$, $E[\theta_i]$, $(\sigma_i)^2$ – the expected value and dispersion of parameter θ_i, $I_{n_\theta,j,l} = \int_{S_{n\theta,j,l}}\rho(\theta_{n\theta})d\theta_{n\theta}$, $\theta_i^{(k)L,j,l},\theta_i^{(k)U,j,l}$ – boundaries of R_{jl} regions, $l = 1,\ldots,N_j^{(k)}$, obtained by partitioning the regions R_{α_j} to improve the approximation of the functions $g_j(d,z,\theta_1,\ldots,\theta_{n_\theta})$, $i = 1,\ldots,n_\theta$, $j = 1,\ldots,m$. In (4) $S_{n_\theta,j,l} = [\theta_{n\theta}^{(k),L,j,l};\varphi_j(d,z,\bar{\theta}_1^{j,l},\ldots,\bar{\theta}_{n_\theta-1}^{j,l})]$, $\bar{\theta}_i^{j,l} = 0.5(\theta_i^{(k)L,j,l} + \theta_i^{(k)U,j,l})$, $i = 1,\ldots,n_\theta - 1$. The calculation of one-dimensional integrals $I_{n_\theta,j,l}$ is conducted according to the expressions:

$$I^{(k)}_{n_\theta,j,l} = \begin{cases} \Phi((\overline{\varphi}^{(k)}_{j,l} - E[\theta_{n_\theta}])/\sigma_{n_\theta}) - \Phi(\widetilde{\theta}^{(k)L,j,l}_{n_\theta}), & \text{if } \partial g_j(d,z,\theta)/\partial\theta_{n_\theta} \ge 0, \\ \Phi(\widetilde{\theta}^{(k)U,j,l}_{n_\theta}) - \Phi((\overline{\varphi}^{(k)}_{j,l} - E[\theta_{n_\theta}])/\sigma_{n_\theta}), & \text{if } \partial g_j(d,z,\theta)/\partial\theta_{n_\theta} \le 0, \end{cases} \quad (5)$$

где $\overline{\varphi}^{(k)}_{j,l} = \varphi_{j,l}(d,z,\overline{\theta}^{j,l}_1,\overline{\theta}^{j,l}_2,...,\overline{\theta}^{j,l}_{n_\theta-1})$ вычисляется в подобласти $R_{j,l}$, $l = 1,...,N^{(k)}_j$.

However, to calculate the values $\overline{\varphi}^{(k)}_{j,l}$, we need to express them from the equations $g_j(d,z,\overline{\theta}^{j,l}_1,...,\overline{\theta}^{j,l}_{n_\theta-1},\varphi_{j,l}) = 0$ in the subregion $R_{j,l}$, that is not always possible. Therefore, we will receive values $\overline{\varphi}^{(k)}_{j,l}$ during the execution of the optimization procedure. Let introduce the equations into the number of constraints of problem (3)-(4), add variables to the list of search variables of the problem $\overline{\varphi}^{(k)}_{j,l}$ and obtain a new form of problem (3)-(4)

$$f^{(k)} = \min_{d,z\in H,\overline{\varphi}_{j,l}} E_{ap}[f(d,z,\theta)], \quad (6)$$

$$g_j(d,z,\overline{\theta}^{j,l}_1,...,\overline{\theta}^{j,l}_{n_\theta-1},\overline{\varphi}_{j,l}) = 0,\ j = 1,...,m,\ l = 1,...,N^{(k)}_j, \quad (7)$$

$$\sum_{l=1}^{N^{(k)}_j}\left(\prod_{i=1}^{n_\theta-1}[\Phi(\widetilde{\theta}^{(k)U,j,l}_r) - \Phi(\widetilde{\theta}^{(k)L,j,l}_r)]\cdot I_{n_\theta,j,l}\right) \ge \alpha_j,\ j = 1,...,m, \quad (8)$$

where the calculation of $I_{n_\theta,j,l}$ is conducted according to (5). In the problem $mN^{(k)}_j$ of equality constraints of the form (7).

Note that in problem (6)-(8) it is assumed that for getting value $\overline{\varphi}^{(k)}_{j,l}$ is used the dependence $\theta_{n_\theta} = \varphi_j(d,z,\theta_1,...,\theta_{n_\theta-1})$ in the subregion $R_{j,l}$. However, it is not always convenient and rational to use the same undefined parameter for getting the value for various constraints $g_j(d,z,\theta) \le 0$, $j = 1,...,m$ when we solve the process design problems. Obviously, it can be used different undefined parameters for different constraints.

Let suppose that for getting a value $\overline{\varphi}^{(k)}_{j,l}$ for a constraint with a number j, we will use an undefined parameter with the number i_j, and can be executed $i_{j_1} = i_{j_2}$. In this case, the problem (6)-(8) takes the form

$$F^{(k)} = \min_{d,z\in H,\overline{\varphi}_{j,l}} E_{ap}[f(d,z,\theta)], \quad (9)$$

$$g_j(d,z,\Theta_j,\overline{\varphi}_{j,l}) = 0,\ j = 1,...,m,\ l = 1,...,N^{(k)}_j, \quad (10)$$

$$\sum_{l=1}^{N^{(k)}_j}\left(\prod_{i=1}^{n_\theta-1}[\Phi(\widetilde{\theta}^{(k)U,j,l}_r) - \Phi(\widetilde{\theta}^{(k)L,j,l}_r)]\cdot I_{n_\theta,j,l}\right) \ge \alpha_j,\ j = 1,...,m, \quad (11)$$

where $\Theta_j = \{\overline{\theta}^{j,l}_i, i \in I_j\}$, $I_j = \{i : i = 1,...,n_\theta\}/\{i = i_j\}$, $\dim I_j = n_\theta$.

The problem (9)-(11) is the problem of non-linear deterministic programming. To solve this problem, it is not necessary to conduct multidimensional integration in the process of implementing the optimization procedure. It is also not necessary to separately solve the equations $g_j(d,z,\overline{\theta}^{j,l}_1,...,\overline{\theta}^{j,l}_{n_\theta-1},\overline{\varphi}_{j,l}) = 0$ for each region to get a value. It should be noted that the derivation of the equations for the optimization level will allow us to get a faster solution of problems (9)-(11).

3. Improving chance constraints regions approximation

To improve the getting evaluation, we can use the partitioning rules for the regions T_q, $q = 1,...,Q^{(k)}$, $R_{j,l}$, $j = 1,...,m$, $l = 1,...,N^{(k)}_j$ that proposed in [2, 7]. Let $d^{(k)}, z^{(k)}$ is the problem (9)-(11) solution. We should partition the subregion T_{q^*} with the worst quality of the approximation of the function $f(d,z,\theta)$ by the function $\bar{f}(d,z,\theta,\theta^q)$. To find the q^* we will solve the problem

IOP Conf. Series: Earth and Environmental Science **288** (2019) 012102 doi:10.1088/1755-1315/288/1/012102

$$g^* = \arg\max_{q} \max_{\theta \in T_q} (f(d^{(k)}, z^{(k)}, \theta) - \bar{f}(d^{(k)}, z^{(k)}, \theta, \theta^q))^2 .$$

We should partition subregion R_{r*} the numbered $r^* = \{j, l_j^*\}$ with the worst quality of the approximation of the function $\varphi_j(d, z, \theta_1, ..., \theta_{n_\theta - 1})$ by the hyperplanes $\bar{\varphi}_{r*}^{(k)} = \varphi_{r*}(d, z, \bar{\theta}_1^{r*}, \bar{\theta}_2^{r*}, ..., \bar{\theta}_{n_\theta - 1}^{r*})$, $\bar{\theta}_i^{r*} = 0.5(\theta_i^{(k)L,r*} + \theta_i^{(k)U,r*})$, $i = 1, ..., n_\theta - 1$. To find the l_j^* we will solve the problems

$$l_j^* = \arg\max_{l} \max_{\theta \in T_l} (\varphi(d^{(k)}, z^{(k)}, \theta_1, ..., \theta_{n_\theta - 1}) - \bar{\varphi}_{j,l}^{(k)})^2 .$$

The effectiveness of the proposed approach was demonstrated on the solution of the problem of synthesis of the optimal workable one-stage of heat exchangers network (HEN). To reduce the dimension of the solving problem was used an approach to the synthesis of optimal HENs, proposed in [8].

The using of the approach became possible due to the decomposition of the multidimensional region of uncertainty into a set of two-dimensional regions. As a result, a set of problems of optimal workable HENs design on two-dimensional regions of uncertainty was obtained. By solving each of the problems of optimal workable HENs design, the level of the HEN heat and mass balance is realized in the modeling program UNISIM (Honeywell International Inc.).

The obtained optimal values of the criteria of effectiveness work of each of the operational networks were summarized in the matrix of effectiveness evaluations of networks [8]. The optimal topology of the workable HEN was obtained as a result of the solution of the assignment problem, and the problem did not contain uncertainty, it was exhausted at the level of solving separate problems of optimal workable HENs design.

Conclusions

In this paper we propose the development of the approach that we proposed earlier to solve the optimal workable CP design problems based on reducing the problem of a one-stage optimization problem with separate chance constraints to a sequence of deterministic nonlinear programming problems.

The new approach makes it possible to get rid of the procedure of the analytical solution of nonlinear equations describing the boundary of the region of satisfaction of constraints, which made it possible to approximate the constraint functions when reducing chance constraints to deterministic form. The elimination of this procedure was made possible by including these equations in the list of constraints of the solving optimization problem. This made it possible to extend the scope of the approach to cases where the analytical solution of the equations is very difficult or impossible, for example, when modeling programs are used to reduce the heat and mass balances of CP.

The efficiency of the approach was demonstrated on the solution of the problem of synthesis of an optimal workable one-stage of heat exchangers network, when the heat and mass balance of the network was calculated in the simulating program, which did not allow to obtain analytical descriptions of the boundaries of the regions for satisfaction of the chance constraints of the solving problem.

References

[1] Gupta V. and Grossmann I.E. 2011 *Computers and Chemical Engineering.* Solution Strategies for Multistage Stochastic Programming with Endogenous Uncertainties **35** 2235-47

[2] Ostrovsky G.M., Ziyatdinov N.N. and Lapteva T.V. 2013 *AIChE Journal.* Optimization problem with normally distributed uncertain parameters **59 Is. 7** 2471-84

[3] Homem-de-Mello T. and Bayraksan G. 2014 *Surveys in Operations Research and Management Science.* Monte Carlo sampling-based methods for stochastic optimization **19 Is. 1** 56-85

[4] Nemirovski A. 2012 *European Journal of Operational Research.* On safe tractable approximation of chance constraints **219** 707-18

[5] Li P., Arellano-Garcia H. and Wozny G. 2008 *Comput. Chem. Eng.* Chance constrained programming approach to process optimization under uncertainty **32** 25-45

[6] Geletu A., Kloppel M., Zhang H. and Li P. 2013 *Int. J. Syst. Sci.* Advances and applications of chance-constrained approaches to systems optimization under uncertainty **44** 1209-32
[7] Ostrovsky G.M., Lapteva T.V., Ziyatdinov N.N. and Silvestrova A.S. 2017 *Theoretical foundations of chemical engineering.* Design of chemical engineering systems with chance constraints **51 Is. 6** 961-71
[8] Ostrovskii G.M., Ziyatdinov N.N. and Emel'yanov I.I. 2016 *Theoretical foundations of chemical engineering.* Designing a heat-exchange system upon the reconstruction and synthesis of optimal systems of distillation columns **50 Is. 2** 178-87

IOP Conf. Series: Earth and Environmental Science **288** (2019) 012103 doi:10.1088/1755-1315/288/1/012103

The emergence of bifurcation phenomena during the flow of rheologically complex media in a tubular reactor

[1]S.A. Livshits, [1]N.A. Yudina, [1]T.U. Dunaeva, [2]O.V. Novikova, [1]M.S. Prudchenko,

[1]Kazan State Energy University, Kazan, Russian Federation

[2]Peter the Great St.Petersburg Polytechnic University, St.Petersburg, Russian Federation

semen19772004@mail.ru

Annotation. The mathematical model describing heat-mass transfer in the course of rheologically complex media in a tubular reactor is developed in the article. Using the example of the continuous production scheme for grade B2 polymethacrylate, solutions of the characteristic equations confirming the possibility of emergence of critical flow regimes for certain sets of control parameters were obtained.

Introduction

In 1928 Academician N.N. Semenov published a paper describing a phenomenon later called "thermal explosion in gas media". In the 30s - 40s of the XX century a great development in the study of this subject was made by such prominent scientists as: O.M. Todes, D.A. Frank-Kamenetsky and Ya.B. Zeldovich.

In 1965, Bostanjinyan S.A., Merzhanov A.G. and Khudyaev S.I. [1] equations analogous to the equations proposed by Semenov were obtained, but already for the flow of a viscous Newtonian fluid. The later investigations of the flow of viscous fluids do not give an exhaustive answer to the question of the appearance of conditions under which a progressive increase in temperature arises.

In the mathematical models available to date, the non-Newtonian properties of liquids are not taken into account and there are no methods for calculating the qualitative characteristics by means of the molecular mass distribution.

In connection with the need to solve various applied problems arising during the design and operation of thermal power plants, the task of theoretical research of heat and mass transfer processes in the laminar flow of Newtonian, decontaminating and pseudoplastic chemically reacting liquids in pipes and channels has come to the fore.

Statement of the problem

In this paper, the flow of rheologically complex media in a tubular reactor is considered. A closed mathematical model describing heat and mass transfer in the course of fluid motion inside a tubular reactor is developed, which reflects the mutual influence of hydrodynamic, heat-mass exchange, rheological and macrokinetic parameters on the polymerization process.

As a working agent (monomer) in the reactor, a polymethacrylate of grade B2 is considered, which releases heat as a result of its absolute mixing with the initiator.

IOP Conf. Series: Earth and Environmental Science **288** (2019) 012103 doi:10.1088/1755-1315/288/1/012103

In constructing the mathematical model, the following assumptions were made:

1. the flow of the laminar steady-state reaction system with the generated profile of the velocity vector at the inlet to the reactor;
2. mass forces enter the equations of motion implicitly, but through excess pressure;
3. the transfer of heat and mass along the direction of the main motion due to thermal conductivity is small in comparison with the forced transfer in the same direction;
4. the thermophysical characteristics of the system and the coefficients of mutual diffusion change insignificantly during the reaction;
5. The hydrodynamic relaxation time is much shorter than the thermal time, i.e. the profile of the velocity vector almost instantly adapts to the change in the temperature field;
6. the system under investigation does not possess elastic properties.

For the study, the system of equations of motion, energy conservation, the continuity equation and the mass transfer equations recorded for the monomer and initiator were used [2]. As a rheological model, was chosen a power-law model [4].

The system of equations under study is as follows:

$$\rho\left(\frac{\partial V_r}{\partial r}V_r+\frac{\partial V_r}{\partial z}V_z\right)=2\frac{\partial}{\partial r}\left(\mu\frac{\partial V_r}{\partial r}\right)-\frac{\partial P}{\partial r}+\frac{\partial}{\partial z}\left(\mu\frac{\partial V_r}{\partial z}\right)+\frac{\partial}{\partial z}\left(\mu\frac{\partial V_z}{\partial r}\right)-\frac{2\mu V_r}{r^2}+\frac{2\mu}{r}\frac{\partial V_r}{\partial z}$$

$$\rho\left(\frac{\partial V_z}{\partial r}V_r+\frac{\partial V_z}{\partial z}V_z\right)=\frac{\partial}{\partial r}\left(\mu\frac{\partial V_r}{\partial z}\right)+\frac{\partial}{\partial r}\left(\mu\frac{\partial V_z}{\partial r}\right)+2\frac{\partial}{\partial z}\left(\mu\frac{\partial V_z}{\partial z}\right)-\frac{\partial P}{\partial z}+\frac{\mu}{r}\frac{\partial V_r}{\partial z}+\frac{\mu}{r}\frac{\partial V_z}{\partial r}$$

$$\frac{\partial V_r}{\partial r}+\frac{V_r}{r}+\frac{\partial V_z}{\partial z}=0$$

$$V_r\frac{\partial T}{\partial r}+V_z\frac{\partial T}{\partial z}=\alpha\left(\frac{\partial^2 T}{\partial r^2}+\frac{1}{r}\frac{\partial T}{\partial r}\right)+\frac{Q_0 k_0}{c_m}e^{-E/RT}M^m J^n+\frac{2\mu I_2}{\rho c_p}$$

$$V_r\frac{\partial M}{\partial r}+V_z\frac{\partial M}{\partial z}=D_m\left(\frac{\partial^2 M}{\partial r^2}+\frac{1}{r}\frac{\partial M}{\partial r}\right)-k_0 e^{-E/RT}M^m J^n$$

$$V_r\frac{\partial J}{\partial r}+V_z\frac{\partial J}{\partial z}=D_j\left(\frac{\partial^2 J}{\partial r^2}+\frac{1}{r}\frac{\partial J}{\partial r}\right)-k_{j0} e^{-E/RT}J^s$$

$$I_2=\left(\frac{\partial V_r}{\partial r}\right)^2+\frac{1}{2}\left(\frac{\partial V_r}{\partial z}+\frac{\partial V_z}{\partial r}\right)^2+\left(\frac{V_r}{r}\right)^2+\left(\frac{\partial V_z}{\partial z}\right)^2$$

Not considering $\frac{\partial P}{\partial r}$ and taking into account $Q = \iint_\Omega V_z d\Omega = \int_0^{2\pi}\int_0^{r_1} V_z r \, d\varphi \, dr = 2\pi\int_0^{r_1} V_z r \, dr$, we have:

$$\frac{\partial P}{\partial z} = \frac{2\pi}{Q}\int_0^{r_1} r\left\{2V_z\frac{\partial}{\partial z}\left(\mu\frac{\partial V_z}{\partial z}\right) - \mu\left[\frac{\partial V_z}{\partial r}\frac{\partial V_r}{\partial z} + \left(\frac{\partial V_z}{\partial r}\right)^2\right]\right\}dr$$

Thus, in order to study the flow of a rheologically complex liquid in a tubular reactor, taking into account the Stokes assumption, we have the following system of equations:

$$\begin{cases} \frac{2\pi}{Q}\int_0^{r_1} r\left\{2V_z\frac{\partial}{\partial z}\left(\mu\frac{\partial V_z}{\partial z}\right) - \mu\left[\frac{\partial V_z}{\partial r}\frac{\partial V_r}{\partial z} + \left(\frac{\partial V_z}{\partial r}\right)^2\right]\right\}dr = \frac{1}{r}\frac{\partial}{\partial r}\left(r\mu\left(\frac{\partial V_r}{\partial z} + \frac{\partial V_z}{\partial r}\right)\right) + 2\frac{\partial}{\partial z}\left(\mu\frac{\partial V_z}{\partial z}\right) \\ \frac{\partial V_r}{\partial r} + \frac{V_r}{r} + \frac{\partial V_z}{\partial z} = 0 \\ V_r\frac{\partial T}{\partial r} + V_z\frac{\partial T}{\partial z} = \alpha\left(\frac{\partial^2 T}{\partial r^2} + \frac{1}{r}\frac{\partial T}{\partial r}\right) + \frac{Q_0 k_0}{c_m}e^{-E/RT}M^m J^n + \frac{2\mu I_2}{\rho c_p} \\ V_r\frac{\partial M}{\partial r} + V_z\frac{\partial M}{\partial z} = D_m\left(\frac{\partial^2 M}{\partial r^2} + \frac{1}{r}\frac{\partial M}{\partial r}\right) - k_0 e^{-E/RT}M^m J^n \\ V_r\frac{\partial J}{\partial r} + V_z\frac{\partial J}{\partial z} = D_j\left(\frac{\partial^2 J}{\partial r^2} + \frac{1}{r}\frac{\partial J}{\partial r}\right) - k_{j0}e^{-E/RT}J^s \end{cases}$$

$$I_2 = \left(\frac{\partial V_r}{\partial r}\right)^2 + \frac{1}{2}\left(\frac{\partial V_r}{\partial z} + \frac{\partial V_z}{\partial r}\right)^2 + \left(\frac{V_r}{r}\right)^2 + \left(\frac{\partial V_z}{\partial z}\right)^2$$

Turning to new, dimensionless functions, variables and parameters [3]

$x = \frac{r}{r_1}$, $\dot{z} = \frac{z}{L}$ – dimensionless functions of coordinates;

$\theta = \frac{E}{RT_0^2}(T - T_0)$ - dimensionless function of temperature;

$\dot{M} = \frac{M}{\tilde{M}}$; $\dot{J} = \frac{J}{\tilde{J}}$ - dimensionless and normalized concentrations of monomer and initiator;

$\dot{V}_z = \frac{\pi r_1^2}{Q}V_z$; $\dot{V}_r = \frac{L\pi r_1}{Q}V_r$ - dimensionless components of the velocity function

$\alpha = \frac{B}{E}$; $\beta = \frac{RT_0}{E}$,

and considering the power model as a rheological model:

$$\mu = \mu_0 I_2^1 = A \cdot e^{B/RT} \cdot I_2^1$$

we have:

$$\begin{cases} \dfrac{2\pi^2 r_1^4}{Q^2}\displaystyle\int_0^1 x\left\{2\dot{V}_z \frac{\partial}{\partial \dot{z}}\left(e^{-\frac{\alpha\theta}{\beta\theta+1}} \cdot \dot{I}_2^1 \cdot \frac{\partial \dot{V}_z}{\partial \dot{z}}\right) - e^{-\frac{\alpha\theta}{\beta\theta+1}} \cdot \dot{I}_2^1 \cdot \left[\frac{\partial \dot{V}_z}{\partial x}\frac{\partial \dot{V}_r}{\partial \dot{z}} + \left(\frac{\partial \dot{V}_z}{\partial x}\right)^2\right]\right\} dx = \\ \qquad = \dfrac{1}{x}\dfrac{\partial}{\partial x}\left(e^{-\frac{\alpha\theta}{\beta\theta+1}} \cdot \dot{I}_2^1 \cdot x\left(\frac{\partial \dot{V}_r}{\partial \dot{z}} + \frac{\partial \dot{V}_z}{\partial x}\right)\right) + 2\dfrac{\partial}{\partial \dot{z}}\left(e^{-\frac{\alpha\theta}{\beta\theta+1}} \cdot \dot{I}_2^1 \cdot \frac{\partial \dot{V}_z}{\partial \dot{z}}\right) \\ \dfrac{\partial \dot{V}_r}{\partial x} + \dfrac{\dot{V}_r}{x} + \dfrac{\partial \dot{V}_z}{\partial \dot{z}} = 0 \\ \dfrac{\pi r_1 R T_0^2}{QE}\left(\dot{V}_r \frac{\partial \theta}{\partial x} + \dot{V}_z \frac{\partial \theta}{\partial \dot{z}}\right) = \alpha \dfrac{R T_0^2}{E r_1^2}\left(\frac{\partial^2 \theta}{\partial x^2} + \frac{1}{x}\frac{\partial \theta}{\partial x}\right) + \dfrac{\tilde{M}^m \tilde{J}^n Q_0 k_0}{c_m} e^{-E/RT_0} e^{\frac{\theta}{\beta\theta++1}} \dot{M}^m \dot{j}^n + \dfrac{2A e^{\frac{B}{RT_0}} \pi^2 r_1^4 \mu I_2}{Q^4 \rho c_p} e^{-\frac{\alpha\theta}{\beta\theta+1}} \cdot \dot{I}_2^{l+1} \\ \dfrac{\tilde{M}\pi r_1}{Q}\left(\dot{V}_r \frac{\partial \dot{M}}{\partial x} + \dot{V}_z \frac{\partial \dot{M}}{\partial \dot{z}}\right) = \dfrac{D_m \tilde{M}}{r_1^2}\left(\frac{\partial^2 \dot{M}}{\partial x^2} + \frac{1}{x}\frac{\partial \dot{M}}{\partial x}\right) - k_0 e^{-E/RT_0} e^{\frac{\theta}{\beta\theta+1}} \dot{M}^m \dot{j}^n \\ \dfrac{\tilde{J}\pi r_1}{Q}\left(\dot{V}_r \frac{\partial \dot{J}}{\partial x} + \dot{V}_z \frac{\partial \dot{J}}{\partial \dot{z}}\right) = \dfrac{D_j \tilde{J}}{r_1^2}\left(\frac{\partial^2 \dot{J}}{\partial x^2} + \frac{1}{x}\frac{\partial \dot{J}}{\partial x}\right) - k_{j0} e^{-E/RT_0} e^{\frac{\theta}{\beta\theta+1}} \dot{j}^s \end{cases}$$

$$\dot{I}_2 = \left(\frac{\partial \dot{V}_r}{\partial x}\right)^2 + \frac{1}{2}\left(\frac{\partial \dot{V}_r}{\partial \dot{z}} + \frac{\partial \dot{V}_z}{\partial x}\right)^2 + \left(\frac{\dot{V}_r}{x}\right)^2 + \left(\frac{\partial \dot{V}_z}{\partial \dot{z}}\right)^2$$

Analytical study of the task

We expand all the functions under investigation into Taylor series in a neighborhood of the point $(0, z_0)$ on the channel axis and taking into account that the functions $\theta, \dot{M}, \dot{J}, \dot{V}_z$ are even relative x, and the function $\dot{V}_r$ is odd.

The value of the coefficients in the expansion of the function $\dot{V}_r$ is found from the continuity equation. Functions $\dot{I}_2^1$ ($1 = -\frac{1}{k}$ for polymers) and $e^{\frac{\theta}{\beta\theta+1}}$ and $e^{-\frac{\alpha\theta}{\beta\theta+1}}$ decompose into Taylor series, and the function $\dot{I}_2$ does not decompose into a series (its components decompose into series) [5].

Substituting the expansion functions for the investigated functions, we obtain a system of four algebraic equations (the continuity equation is not used, because with its help the coefficients of the Taylor series for the function $\dot{V}_r$ are obtained) [6].

From the consideration of the coefficients for the free term and at $(\dot{z} - z_0)$, adding initial conditions for the temperature, velocity, concentration of the monomer and initiator, we obtain a system of 16 algebraic equations ($k = 3, s = 1, m = 1, n = 1$).

If we substitute the boundary conditions of the first kind and renumber it θ_0 for h, we obtain the following characteristic equation for determining the temperature of the working agent:

$$\theta_1 = h - \frac{\delta \mathrm{Exp}\left(h/(1+\beta h)\right)}{4} + \frac{1}{16}\left[-\chi \mathrm{Exp}\left(\alpha h/(1+\beta h)\right) + \frac{\delta^2 \mathrm{Exp}\left(2h/(1+\beta h)\right)}{4(\beta h+1)^2}\right] -$$

$$-\frac{1}{36}\left\{\chi \mathrm{Exp}\left(\alpha h/(1+\beta h)\right)\left[(c_0-1)(1-\mathrm{Exp}(-c_1)) - \frac{\alpha\delta \mathrm{Exp}\left(h/(1+\beta h)\right)}{4(\beta h+1)^2}\right] +\right.$$

$$\left.+\frac{3\delta^3 \mathrm{Exp}\left(3h/(1+\beta h)\right)}{64(1+\beta h)^4} - \frac{\beta\delta^3 \mathrm{Exp}\left(3h/(1+\beta h)\right)}{16(1+\beta h)^3} - \frac{\chi\delta \mathrm{Exp}\left((\alpha+1)h/(1+\beta h)\right)}{16(1+\beta h)^2}\right\}.$$

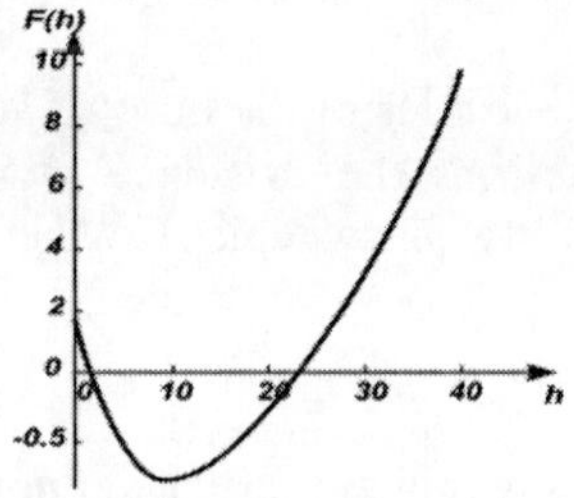

Fig.1. Graphical representation of the equation with respect to polymethacrylate grade B2 10%, taking into account the dissipative and chemical source of heat release under thermal boundary conditions of the first kind for a round tube T_0= 413 K, T_1= 418K.

In the case when thermal boundary conditions of the third kind are given on the boundary, then, after re-designation θ_0 for h, we obtain the following characteristic equation for determining the temperature of the working agent:

$$144\cdot(h-\theta_1) = 4\cdot\left\{\chi\cdot \mathrm{Exp}\left(\alpha h/(1+\beta h)\right)\cdot\left[(c_0-1)\cdot(1-\mathrm{Exp}(-c_1)) - \frac{\alpha\cdot\delta\cdot \mathrm{Exp}\left(h/(1+\beta h)\right)}{4(1+\beta h)^2}\right] +\right.$$

$$\left.+\frac{3\delta^3 \mathrm{Exp}\left(3h/(1+\beta h)\right)}{64(1+\beta h)^4} - \frac{\beta\delta^3 \mathrm{Exp}\left(3h/(1+\beta h)\right)}{16(1+\beta h)^3} - \frac{\chi\delta \mathrm{Exp}\left((\alpha+1)h/(1+\beta h)\right)}{16(1+\beta h)^2}\right\}\left(\frac{6}{Bi}+1\right)+$$

$$+9\cdot\left[\chi \mathrm{Exp}\left(\alpha h/(1+\beta h)\right)-\frac{\delta^2 \mathrm{Exp}\left(2h/(1+\beta h)\right)}{4(\beta h+1)^2}\right]\left(\frac{4}{Bi}+1\right)+36\cdot\delta \mathrm{Exp}\left(h/(1+\beta h)\right)\left(\frac{2}{Bi}+1\right)$$

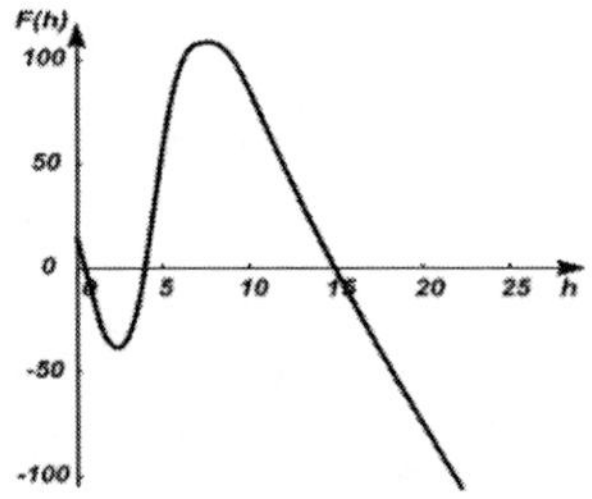

Fig. 2. Graphical representation of the equation with reference to polymetacrylate of grade B2 10%, taking into account the dissipative and chemical source of heat release under thermal boundary conditions of the third kind for a circular pipe. T_0= 413 K, T_1= 418K.

The conducted analysis of the behavior of the lines of solutions of the characteristic equations for polymetacrylate of grade B2 confirms the existence of critical flow regimes for certain sets of parameters and shows the possibility of switching to the non-calculating operating modes of heat engineering equipment.

Conclusions

1. A method for analytical investigation has been developed and a theoretical substantiation of the possibility of the appearance of bifurcation phenomena in the flow of a nonlinear viscous fluid in an infinite circular tube under thermal boundary conditions of the first and third kind is presented.
2. A technique for analytical analysis of heat and mass exchange of nonlinear viscous fluids in tubular homophase rectilinear reactors has been developed.
3. When considering the flow of polymetacrylate grade B2 in a tubular reactor, solutions of characteristic equations were obtained, analysis of which shows the possibility of switching to non-calculating operating modes of heat engineering equipment for certain values of control parameters.

References

[1] Bostanjinyan S.A., Merzhanov A.G., Khudyaev S.I. "On the Hydrodynamic Thermal Explosion" // Reports of the Academy of Sciences of the USSR, 1965, v. 163, No. 1 p. 133-136

[2] Lebedev R.V., Livshits S.A. Investigation of critical flow regimes of generalized viscous media in a tubular reactor. Proceedings of the MAI. 2011. № 46. With. 1.

[3] Kantyukov R.R., Takhaviev M.S., Lebedev R.V., Livshits S.A., Shenkarenko S.V. Analytical study on the presence of bifurcation phenomena in the flow of nonlinear viscous fluids in channels of complex geometry. // Bulletin of the Technological University. 2015. T. 18. № 4. P. 223-225.

[4] Kutateladze S.S., Popov V.I., Khabakhpasheva E.M. "Hydrodynamics of liquids with variable viscosity" // PMTF, 1966. No. 1 p. 45-49

[5] Sabitov R.R., Enikeeva S.R., Yudina N.A. Using the Pade approximant to determine the ignition conditions of a continuous infinite cylinder filled with a reactive medium under convective heat transfer // Transport and storage of petroleum products and hydrocarbons. 2017. № 1. P. 17-21.

[6] Livshits S.A., Nazmeev Yu.G., Malov K.M. Critical regimes of the nonisothermal flow of a viscous fluid in a circular tube // Fundamental research. Moscow 2005, No. 6, p.56-57.

IOP Conf. Series: Earth and Environmental Science **288** (2019) 012104 doi:10.1088/1755-1315/288/1/012104

Granulated biofuels combustion efficiency

V K Lyubov, E I Popova, A N Popov

Department of Power and Heat Engineering, Northern (Arctic) Federal University named after M.V. Lomonosov, Northern Dvina Embankment, 17, Arkhangelsk, 163002, Russian Federation

vk.lubov@mail.ru

Abstract. The article presents the results of experimental studies of the thermal and ecological performance of the Firematic 60 boiler manufactured by Herz Energietechnik GmbH. Wood pellets with diameter of 6 and 8 mm and biochar from hydrolytic lignin were used for combustion. The components of the boiler's heat balance and the emission of harmful substances were determined during the burning of the biofuels. In the course of experimental work high energy and environmental performance of the Firematic 60 boiler was reached while working on granulated fuel which gives a reason to recommend this boiler for heat supply of low-rise buildings in the conditions of the Arctic region. Organized in Onega City production of biocahar from hydrolytic lignin allowed to obtain high-quality fuel from biomass that has remained in dumps for 40-60 years. Experimental data on the change in the apparent and bulk density, mechanical durability of wood pellets thermally pretreated for 90 minutes in an electric furnace without oxygen at temperatures of 200, 250, 275 and 300 °C was obtained. It has shown that with an increase in the temperature of wood pellet torrefaction all these parameters decrease. The influence of the torrefaction process parameters of the hydrolytic lignin on the sorption properties of the biochar has been studied; sorption properties have been compared for the granules produced from the hydrolytic lignin and coniferous wood.

Introduction

An effective way of the complex solution of energy and environmental problems while providing heating load of consumers is the use of modern devices operating on solid biofuel. The research papers on highly efficient domestic and foreign heating devices positioned in the market have great importance [1,2]. Also, the research is prospective as it aims studying of torrefaction biofuel [3].

Torrefaction of solid biomass is based on its thermal treatment without air access at temperatures of 200-330 °C. As the result of this process biochar which can be used in powder form or for the production of pellets or briquettes is obtained. Torrefaction process allows to bring the characteristics of biomass closer to characteristics of coal. Different degrees of thermal treatment lead to different yields and different calorific value of the finished product [4]. Theoretical calorific value of torrefied product is between 18 and 31 MJ/kg [5] (calorific value of biochar). Black pellets have good flowability and durability, they are hydrophobic and obeyed by the same laws of grinding as coals. High energy density determines the economic feasibility of their transportation over long distances. Black pellets excel the wood pellets in all characteristics.

IOP Conf. Series: Earth and Environmental Science **288** (2019) 012104 doi:10.1088/1755-1315/288/1/012104

Purpose and objectives

Purposes of this work are:

- experimental determination of energy and environmental performance of "Firematic 60" boiler while burning wood pellets with diameter of 6 to 8 mm as well as hydrolyzed lignin black pellets;
- study of torrefaction temperature influence on transport and processing characteristics of wood pellets;
- study of torrefaction process influence on sorption characteristic of biofuel.

Results and discussion

Researches were held in Scientific and Educational Center of Energy Innovations of Higher School of Power Engineering, Oil and Gas NArFU named after M.V. Lomonosov which is connected to a district heat supply system. The reserve source of the building heat supply is the "Firematic 60" boiler made by Herz Energietechnik GmbH. The boiler is designed to operate on pellets and chips [1]. Nominal capacity of boiler is 60 kW. Detailed description of the boiler is presented in the article [6].

Multi-method research of boiler operational efficiency was carried out in three stages. In the course of the first stage the balance experiments were carried out when wood pellets were fed into the boiler furnace. Wood pellets have the diameter of 8 mm, sufficient uniform granulometric composition [7] and the thermotechnical characteristics (Table 1, tests №1,2) that correspond with requirements of current standards. At the second stage wood pellets with the diameter of 6 mm (Table 1, tests №3,4) were burned in the boiler furnace. At the third stage of research the boiler was operated on hydrolyzed lignin black pellets produced by "OAO "Bionet" (Table 1, tests №5,6). There are results of only two tests for each fuel which describe the obtained energy and environmental performance of the boiler.

Table 1. The main performance of the boiler burning wood pellets and hydrolyzed lignin black pellets

Value, dimension	No.					
	№1	№	№3	№4	№5	№6
Heat capacity (kW)	64.0	103.5	71.3	83.5	87.2	91.4
Outlet operating pressure of water (MPa)	0.27	0.27	0.22	0.24	0.30	0.30
Inlet water temperature (°C)	60.0	79.0	78.0	78.0	80.0	79.0
Moisture of pellets as received basis (%)	8.38		8.60		6.27	
Ash content of pellets as received basis (%)	0.50		0.43		2.45	
Dry-and-ash-free volatile yield (%)	85.10		84.75		63.85	
Lower calorific value as received basis (MJ/kg)	17.30		16.97		21.34	
Flue gas temperature (°C)	106.0	151.0	154.0	155.0	160.0	157.0
Excess air in flue gas	1.39	1.42	1.46	1.43	1.45	1.42
Heat loss: flue gas (%)	4.17	6.71	7.68	7.61	6.88	6.60
incomplete combustion (%)	0.02	0.01	0.01	0.01	0.00	0.00
carbon (%)	0.12	0.11	0.22	0.22	0.47	0.47
external (%)	0.47	0.29	0.42	0.36	0.34	0.33
Gross efficiency (%)	95.20	92.87	91.67	91.80	92.25	92.55
Total biofuel consumption (kg/h)	14.0	23.0	16.0	19.0	16.0	17.0
Emission of NO_x (mg/MJ)	52.0	52.0	56.0	54.0	46.0	44.0
Emission of CO (mg/MJ)	21.0	12.0	6.0	6.0	3.0	3.0
Particulate matter emission (mg/MJ)	-	-	4.84	4.39	9.59	9.34

The analysis of fuels was carried out with the use of the equipment of laboratory of thermal analysis and IKA C 2000 basic Version 2 calorimeter with LOIP FT-216-25 liquid cryothermostat. Study of the grain size distribution of granulated fuels and combustion residues was carried out with vibratory sieve shakers AS 200 Control and Microtrac S3500. The determination of velocity fields and flue gas flows was performed with Pitot tube and micromanometer of "Testo-435" precision instrument. The results of study of velocity fields were used to determine the particulate matter concentrations at the flue gas after the boiler. In order to get that done it was used external filtration method which is applied via the "OP-442 TC" impactor, dust sampling probe, filter holder, etc. "Testo-350 XL" gas analyzer was used to determine content of combustion products. Fuel consumption was determined by equation of the indirect heating balance. Experimental data analysis was performed using multi-block program-methodological complex [8]. The main results of energy survey of the "Firematic 60" boiler is presented in Table 1.

After automatic start-up the period it takes to reach the rated load does not exceed 20 minutes. After 33-38 minutes the automated control system provides inlet boiler water temperature close to the optimum value (60 °C).

The analysis of thermal conditions has shown that heat loss with flue gas is 4.17-4.68 %, but it rises when load and inlet boiler water temperature increase (Table 1). A stage fuel combustion scheme and efficient mixing of secondary air with combustible components of fuel while keeping the excess air coefficient in furnace within the range from 1.39 to 1.46 allow to reach low values of heat loss due to incomplete fuel combustion (Table 1). Values of carbon oxide concentrations corrected to excess air coefficient of 1.4 are 8.0-58 mg/Nm^3.

Carbon loss when boiler operates on all types of pellet has low range which is explained by homogeneity of grain size distribution of burning fuel, its low ash content and sufficiently high calorific value.

Prospective direction of improved biofuel production is manufacturing of black pellets [9]. Thus, pellets production from torrefied hydrolytic lignin allows to obtain (even on soft thermal treatment modes with the temperature of 180-200 °C) high quality fuel from biomass that has being remained in dumps for 40-60 years. The boiler switched from burning wood pellets with the diameter of 6 mm to hydrolyzed lignin black pellets has increased its efficiency (Table 1) despite the higher ash content of the latter (5.56 times more). In addition, minimal NO_x and CO emissions have been achieved while burning hydrolytic lignin black pellets (Table 1).

External heat loss from the boiler of rated heat capacity (60 kW) does not excess 0.5 % (that is much lower than in Russian standards [10]), it is explained by moderate overall dimensions, high quality of lining and thermal insulation of materials. Losses due to temperature of bottom ash of the boiler did not excess 0.01 % while operating on wood pellets; did not excess 0.05 % while operating on black pellets. Energy survey of the boiler burning granulated fuels has shown that it could provide the efficient operation with a load much higher than rated.

Low emission of NOx (Table 1) is explained by middle level of maximum temperatures and excess air in combustion chamber as well as two stage combustion scheme. Sulfur dioxide in combustion products while burning biofuels on each boiler mode is absent.

A stage fuel combustion scheme and intensive mixing of secondary air and combustible components of fuel make it possible to ensure boiler's efficient operation at low oxygen concentration of 4.0-6.0 %. The rise of oxygen concentration of higher than 6% leads to the increase dispersion of harmful substances to the environment. On this basis, the threshold value of the oxygen concentration in automated control system should be reduced to 4 %.

The results of soot particles emission research with the use of external filtration method under isokinetic conditions [11-13] of gas sampling have shown that soot emission factor is 1.74–5.95 g/GJ on 1 GJ of wood pellets heat (with a heat output of 72-104 kW). Soot emission factor PM 2.5 (with conversion coefficient of 0.14 [14]) is 0.243–0.834 g/GJ. When the boiler operates on hydrolyzed lignin black pellets the average soot emission factor is 2.467 g/GJ and soot emission factor PM 2.5 is 0.345 g/GJ. Research has shown that solid particles with dimension less than 15 μm are predominantly carried away to the atmosphere.

The research of efficient burning of improved fuels was carried out jointly with the research of change in apparent and bulk density as well as wood pellets durability in the process of torrefaction.

Wood pellets were supplied by "ZAO "Lesozavod 25" (Arkhangelsk city). Pellets consist of pine on 70 % and spruce on 30 %. Moisture of pellets is 6.9 %.

Wood pellets torrefaction is performed in electrical furnace without oxygen with isothermal soaking for 90 minutes at temperatures of 200, 250, 275 and 300 °C.

The bulk density is determined in accordance with GOST 32987-2014 [15]. The apparent density of samples is determined by measuring of mass and volume by external measurement. Pellets without visible defects are selected, their edges are aligned on a grinding machine. Determination of durability are performed on lignotester NHP 100 in accordance with EN 15210-1 [16]. The results are presented in Table 2.

Table 2. Characteristics of hydrolyzed lignin black pellets.

Sample	Torrefied product output (%)	Bulk density (kg/m^3)	Apparent density (kg/m^3)	Durability (%)
Starter pellets	-	597.06	1261.20	97.97
Pellets are torrefied at:				
200°C	94.13	589.97	1246.35	97.81
250°C	87.09	533.52	1127.28	93.11
275°C	78.77	514.12	1110.40	89.84
300°C	64.01	456.13	985.98	74.68

Based on the obtained data it is clear that loss of pellet mass increases as torrefaction temperature rises that is explained by thermal decomposition of main polymeric structures of the wood. Thus, hemicellulose begins to decompose at the temperature of about 190 °C. However, main breakdown of β-glucosidic bond proceeds at the temperatures ranging from 220 to 280 °C [17, 18]. Cellulose and lignin are more thermostable. Decomposition of cellulose occurs in the temperature range of 280-370 °C, and for lignin this value is in the range of 200-540 °C. During the torrefaction removal of moisture from wood occurs as well as volatiles which are formed in the process of hemicellulose breakdown and partial decomposition of cellulose and lignin. Considering the fact that quantity of lignin and cellulose into spruce timber is much higher than of hemicellulose (in 8.0 and 5.2 times accordingly) [19] the torrefaction process will be determined by the thermal decomposition of two main components (lignin and cellulose). Thus, mass of finished wood pellets while maintaining their shape and volume reduces in the torrefaction process, for which reason both the apparent and bulk densities decrease (Table 2).

Reduction of durability occurs with increasing temperature of pellets' thermal treatment (Table 2). The main reasons of it are the uneven heating of pellets into depth during torrefaction because of low thermal conductivity that leads to thermal stresses and the side-by-side process of biomass thermal decomposition which is accompanied by vapor and volatiles emissions. In this case there is the parallel motion into pellet depth of two fronts: the evaporation and thermal decomposition of biomass organic components.

In addition, studies of sorption properties evolution of hydrolyzed lignin and its torrefied product obtained at various temperatures as well as wood pellets and hydrolyzed lignin black pellets have been carried out.

Such raw materials as spruce (50 wt%) and pine (50 wt%) consisting of chips (90 wt%) and sawdust (10 wt%) were used for production wood pellets with diameter of 8 mm.

The hydrolyzed lignin was sieved on a vibratory sieve shaker Retsch AS 200 Control. Fraction of 125-500 μm was sampled for the experiment. Torrefaction of hydrolyzed lignin was performed in electrical furnace without oxygen with isothermal soaking for 90 minutes at the temperatures of 200, 250, 275 and 300 °C.

The determination of water absorption of testing samples was carried out in accordance with GOST 16483.19-72 [20]. The samples were dried preliminary until absolute dry state in a weighing bottle in an oven at the temperature of 105±2 °C. Then the sample bottles were inserted into a desiccator at the bottom of which there was a saturated sodium solution ($Na_2CO_3 \cdot 10H_2O$). Samples had been in a desiccator for 60 days. Samples were weighted after 1, 2, 3, 6, 9, 12, 20 days and then every 10 days. Absorb moisture content was determined in accuracy of 0.1 %. Results of researches are illustrated in Figure 1.

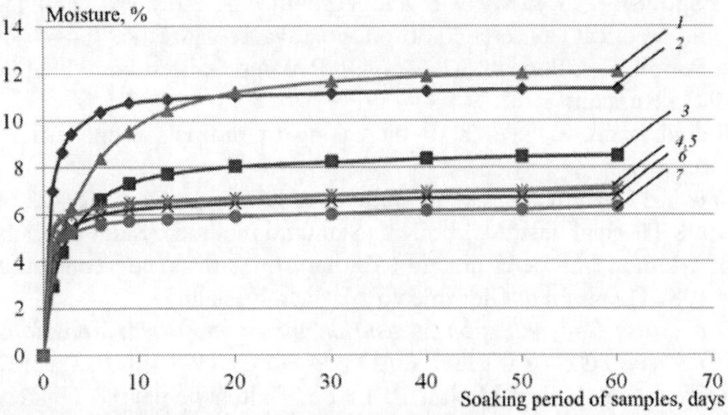

Figure 1. Absorb moisture content change: 1– wood pellets, 2 – hydrolized lignin, 3 – hydrilyzed lignin black pellets; hydrolized lignin torrefied at, °C: 4 – 250, 5 – 300, 6 – 200, 7 – 275.

Based on the obtained data wood pellets (11.9 %) and initial hydrolyzed lignin (11.2 %) have the highest absorb moisture content at the end of 60 days.

The dynamics of water absorption process in hydrolyzed lignin and its torrefied product are similar in all the cases: absorption of 80-84 % is in the first 3 days, then the velocity of this process decreases. The process of water absorption in wood pellets and black pellets proceeds more evenly, 76-86 % of all moisture are absorbed for the first 9 days. Various dynamics of water absorption process of milled samples and pellets are explained by larger surface area of the first one.

The torrefaction of hydrolyzed lignin samples leads to reduction in their sorption characteristics. Minimum value of water absorption for 60 days for hydrolyzed lignin torrefied at the temperature of 275 °C is 6.3 % which is on 4.9 % less in comparison with initial material.

Conclusions

A comprehensive energy survey has shown that the "Firematic 60" boiler provides high technical and economic performance and minimum emissions of harmful substances to the atmosphere while combusting wood pellets and hydrolyzed lignin black pellets. When burning the latter a high energy and environmental performance of the boiler has been reached wherein NOx and CO emissions were minimal. Pellets production from torrefied hydrolytic lignin allows to obtain high quality fuel from biomass that has being remained in dumps for 40-60 years.

The torrefaction process leads to the decrease in sorption characteristics of wood biomass which allows to reduce sufficiently the requirements for transportation and storage conditions of black pellets and to preserve their calorific value. However, torrefaction of finished pellets results in a reduction in their transport and processing characteristics such as apparent and bulk density and durability. Therefore, it is necessary to provide thermal treatment of initial material rather than a finished product to manufacture the pellets with high quality and processing characteristics.

References

[1] Lyubov V K, Malygin P V, Popov A N and Popova E I 2015 *Thermal Engineering* **8** 36–40
[2] Lyubov V K, Malygin P V, Popov A N and Popova E I 2014 Proceedings of International

Scintific Conference Biotehnologii v himiko-lesnom komplekse 201–205 (In Russian)

[3] Nunes L J R, Matias J C O and Catalão J P S 2014 *Renewable and Sustainable Energy Reviews* **40** 153–160

[4] Maryandyshev P A, Chernov A A, Popova E I, Eseev M K and Lyubov V K 2018 *Solid Fuel Chemistry* **52 No 3** 30–39

[5] Lyubov V K and Ivut' A E 2016 *Bulletin of CHSU* **5** 16-20 (In Russian)

[6] Popova E I, Popov A N., Lyubov V K and Varakin E A 2015 Proceedings of 1st International scientific and practical Conference Prirodopolzovanie v Arktike 464–473 (In Russian)

[7] Lyubov V K, Popov A N, Popova E I, Malygin P V and Solnyshkova L M 2017 *Lesnoy journal* **4** 149–159 (In Russian)

[8] Lyubov V K and Lyubova S V 2010 Increasing of biofuels usage efficiency (Arkhangelsk: Solti) (In Russian)

[9] Razumov E Yu and Nazipova F V 2015 Bulletin of KSTU 18 No 2 220–222 (In Russian)

[10] NPO CKTI 1998 Thermal design of boilers (Standard method) (Saint-Petersburg) (In Russian)

[11] Collection of Methods of determination of harmful substances concentrations in industrial emissions 1987 (Leningrad: Gidrometeoizdat) (In Russian)

[12] *Method of Soot Mass Concentration Measurements in Industrial Emissions and in the Air of Operational Areas* 2005 (Yaroslavl: OAO NII «TECHUGLEROD») (In Russian)

[13] Petzold A, Ogren J A, Fiebig M, Laj P, Li Sh, Baltensperger U, Holzer-Popp T, Kinne S, Pappalardo G, Sugimoto N, Wehrli C, Wiedensohler A and Zhang X 2013 *Atmospheric Chemistry and Physics* **13** pp 8365–8379.

[14] Borchsenius H and Borgnes D *Black Carbon Emissions from the District Heating Sector in the Barents Region* 2013 (Norway: NORSK ENERGI)

[15] GOST 32987-2014 2015 (EN 15103:2009) Solid biofuels – Determination of bulk density (Moscow: Standartinform) (In Russian)

[16] EN 15210-1 2009 Solid biofuels – Methods for the determination of mechanical durability of pellets and briquettes – Part 1: Pellets. European Standards (EN)

[17] Maryandyshev P A, Chernov A A, Popova E I and Lyubov V K 2016 *Solid Fuel Chemistry* **50 No 3** 167–176

[18] Maryandyshev P A, Chernov A A, Popova E I, Eseev M K and Lyubov V K 2016 *Solid Fuel Chemistry* **50 No 6** 381–389

[19] Lyubov V K, Popova E I, Shkaeva N V, Bolotova K S and Solnyshkova L M 2017 *Bulletin of CHSU* **3** 38–45 (In Russian)

[20] GOST 16483.19-72 1999 Timber – Method of determination of water absorption (Moscow: IPK Izdatelstvo standartov) (In Russian)

IOP Conf. Series: Earth and Environmental Science **288** (2019) 012105 doi:10.1088/1755-1315/288/1/012105

Research of natural biomass conversion in the Republic of Burundi

J.A. Manigomba [1], N. D.Chichirova [2], V.B.Gruzdev [2]

[1]Laboratory of Burundian State Scientific and Agrochemical Institute (ISABU), Bujumbura, Burundi.

[2]Kazan State Power Engineering University, 420066, Kazan, Krasnoselskaya, str. 51, KSPEU, Russia

amanigomba@yahoo.fr

Abstract. The paper investigates the solution of an urgent energy problem in Republic of Burundi. The pyrolysis process using a recycling of peat, wood waste, agricultural biomass, domestic and industrial waste with the objective of obtaining a secondary energy resources is proposed. Based on this method a pyrolysis gas with a ratio of CO and H2 varied from 1:1 to 1:3 is obtained. The main variable factors determining this composition are the temperature. With the increase of temperature, the pyrolysis gas increases. The number of experiments with fixed factors was chosen at least eight. The experimental setup was used to study the pyrolysis gas of dry sawdust and peat in the temperature range from 450 to 1150°C.
Key words: Thermal power stations, biomass, electric power deficiency, peat, ecology.

Introduction

The Republic of Burundi is located in the eastern part of Africa with a population of more than eight million people. The main energy source (more than 94%) for the population is wood and peat burned by open fire [1]. There is no hydrocarbon fuel in the Republic of Burundi, therefore the republic is forced to procure oil products in Tanzania, Congo and Kenya to operate its diesel power plants. But a steady rise in oil prices leads to a significant increase in the cost of produced electricity, which negatively affects both the republic economy and its energy security.

Over the past two decades, many researchers have studied the pyrolysis of biomass of various species, ranging from wood, agricultural waste, herbaceous species, subjected to various pyrolysis conditions to produce carbonized, liquid and gaseous products [2-5]. Bajus [2] investigated the pyrolysis of wood sawdust using a batch reactor under various conditions of temperature change. The results showed that wood pyrolysis products can be divided into three groups: 25.1% of the mass of the gas, 44.3% of the mass of the liquids and 30.6% of the solid residues. The main composition of liquid products: acetic acid, furfural acid formand. The main fraction of pyrolysis products of the gas are CO2, CO and CH4. The results of the experiments show that the composition of pyrolysis products strongly depends on the heating conditions [2]. Lazi et al. [3, 4] investigated the pyrolysis of agricultural residues and chips in a radiation furnace at different radiation intensities at temperatures ranging from 650 to 1000 °C. The values of the fraction of each product varied with temperature. If the temperature increased, the fraction of the solid residue initially decreased. Then, at high temperatures, there was a tendency to increase the yield of the carbonized component until a constant value was reached. The gas fraction mainly consisted of CO, CO_2, CH_4 and a small amount of H_2, C_2H_4, and C_2H_6. Gases of intermediate value were obtained from wood, while the products of the highest value came from straw. In [5], the authors studied heat transfer in a biomass layer under slow pyrolysis conditions in conjunction with the process of changing the biomass layer thickness. It was found that the endothermic dewatering process takes place at a temperature in the biomass layer from 200 to 300 °C. At temperatures ranging from 300 to 400 °C, the biomass layer is rapidly compressed due to the formation of resin from cellulose to a volume of one-third of the biomass of the initial volume.

In this paper, pyrolysis gas was obtained by using a recycling peat, wood waste, agricultural biomass, domestic and industrial waste. The studied species of biomass have practically the same elemental composition in C, H2, and O2. The study of the regularities of rice straw and wood, thermo chemical transformations, which were observed during pyrolysis and combustion, was also carried out by thermal analysis in various modes on a derivative of grade Q-1500D and experimental installations in a stationary and moving layer at different heating rates.

The investigation of obtaining pyrolysis products process

In this work, the pyrolysis of agricultural and wood waste widely distributed in Burundi is studied. Therefore, we will consider the vegetable briquettes, which have recently been widely used in small power engineering in Burundi. Briquettes and pallets can be obtained from cereals waste such as rice, sorghum, peas, beans, corn, etc. Table 1 shows the annual production volumes of some agricultural products and their wastes in Burundi [6].

Table1. Volumes of production and waste of some type's agricultural products in the Republic of Burundi

№	Sort	Production, tons / year	Waste, tons / year
1	kidney bean	220218	33065
2	Corn	123000	18450
3	Sorghum	67947	10192
4	Rice	67947	10192
5	Peas	33500	5025

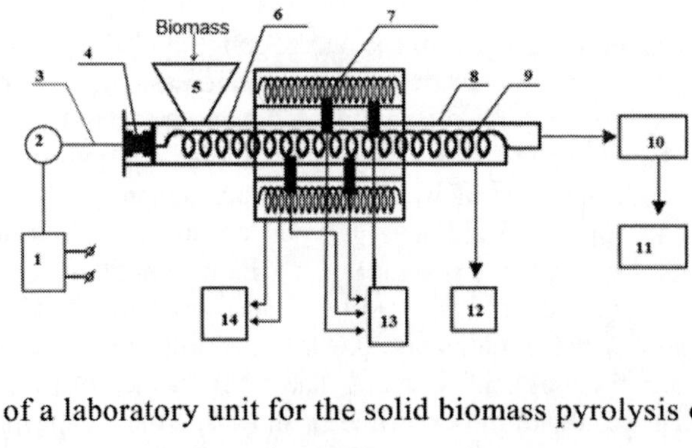

Fig.1. Block diagram of a laboratory unit for the solid biomass pyrolysis of in the moving layer

We investigated the process of pyrolysis of biomass samples in a fixed layer and in a fluidized layer. Studies in the fixed layer were carried out in a quartz glass retort with external electric heating at a heating rate of 6-9 OC / min at temperatures from 450 to 1150 ^{o}C. Pyrolysis of biomass samples in a pseudo-liquefied layer was studied in a laboratory facility consisting of a pyrolizer with a screw feeding of raw materials and external electric heating. The block diagram of the plant is shown in Fig.1.

The main elements of the flowchart are: 1– screw speed controller; 2 - motor-reducer: 3 - drive shaft of a moving screw; 4- seal block with front shaft support; 5 - receiving hopper with hermetic lid; 6- the main part of the screw body; 7- electric heater, power 1,75 kW; 8 - additional removable part of the screw body; 9-moving screw; 10 -filter-cooler of pyrolysis gas; 11-gas meter; 12 - receiving flask of solid residue; 13 - thermometers; 14–electric source of power to heater. The volatile products formed as a vapor-gas mixture in the reaction zone of this plant are discharged to the cooling and purification system 10, then to the gas meter 11, and the solid residue (coke) is collected in the receiving flask 12.

Table 2 presents data on the main components including in the studied samples organic mass. The studied species of biomass have practically the same elemental composition in C, H_2, and O_2. The study of the regularities of rice straw and wood, thermo chemical transformations, which were observed during pyrolysis and combustion, was also carried out by thermal analysis in various modes on a derivative of grade Q-1500D and experimental installations in a stationary and moving layer at different heating rates. The exponents are performed in the laboratory of the Burundian State Scientific and Agrochemical Institute (ISABU), Bujumbura. All experiments were conducted in closed rooms at an air temperature of 24-26°C. Thermocouples with a soldering diameter of 0.2 mm were used to record sample temperatures. The samples were weighted on OHAUS, PA 214C analytical weights. The samples were cooled in a laboratory desiccator. Samples were burned in a Carbolite RHF 1406 furnace at 8 different temperatures, 450, 550, 650, 750, 850,950, 1050, 1150°C. It is known that the difference between the initial weight and weight at a certain temperature, divided by the initial weight, expressed in %, gives the yield of volatile gases from studied the sample. Two methods were used to study the samples for the release of volatile gases.

Table2. The average composition of the resulting pyrolysis gas from solid household waste.

№	Pyrolysis gas composition	Component content, %	Calorific value (bottom), MJ/m3
1	Methane(CH4)	33 – 45	35,80
2	Hydrogen(H2)	12 – 28	10,80
3	Carbon monoxide(CO)	11 – 18	12,64
4	Carbon dioxide (CO2)	1,5 – 2,5	-
		Total	59,24

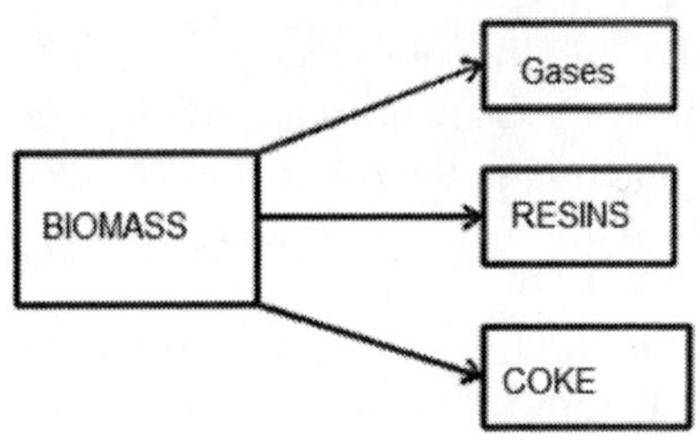

Fig. 2. Single-stage biomass pyrolysis kinetics scheme

1. A method for raw samples, when all samples without drying were placed simultaneously in a cold furnace, were heated to 450°C, held for 5 minutes, the oven furnace was turned off, all samples were taken out of it, cooled in a desiccator, weighted on an analytical balance of OHAUS, PA 214C, then again placed in the furnace, set the desired temperature of 8 values, included the furnace in the work. But since the temperature in the furnace has already decreased, in this case the combustion time could increase, since we detected the minutes 5 times from the time of the furnace reached the desired temperature. The results of the first method for determining the quantity of volatile gases yield at different temperatures, when each sample was tested in two parallel samples, and then separately from other samples, were tabulated.

2. A method for annealed samples, when the dried samples, after their initial humidity determining in the first method, each sample was placed separately from the others in a furnace preliminarily heated to the required temperature, the specimen was held for 5 minutes, the sample was taken out of the furnace, cooled in desiccator and weighted on analytical weights of OHAUS, PA 214C brand. It is known that the composition of the pyrolysis gas mainly includes CO and H_2. But depending on the method for producing pyrolysis gas, the ratio of CO: H_2 varies from 1:1 to 1:3. In direct dependence on the raw materials and the method of its processing, the paper components ratio can vary within wide limits. As a rule, the substances percentage in crude pyrolysis gas is as follows: CO - 11-18%; H_2- 12-28%; CH_4 = 33-45%; CO_2 - 1.5-2.5%. There is average composition of the pyrolysis gas obtained from biomass, and its energy value in table 2.The objective of our study is to maximize the production of pyrolysis gas as fuel from biomass and peat to ensure the operation of diesel generators of the industrial group "REGIDESO" and gas piston machines (GPM). It is known that the yield of pyrolysis gas from biomass depends on its temperature, so in our case the temperature reaction rate dependence will be considered using the Arrhenius equation.

The biomass thermal decomposition can be described on the basis of the simplified kinetic scheme shown in Fig.2 [6-8]. In accordance with this scheme, let us consider the decomposition of biomass in the high temperature region with the formation of final products, such as coke, gases and resins. The formal rate constants of these processes can be represented as the Arrhenius equation:

$$k = k_o A \exp\left(-\frac{E_a}{RT}\right). \qquad (1)$$

where k_o is a factor that does not depend on temperature, but is determined only by the type of reaction; E_a is the activation energy that characterizes the minimum energy that the particles must possess in order to enter into chemical interaction, depends on the nature of the reacting substances and independs on their temperature, J /mol; T is the temperature of the biomass, ^{O}K; R is the universal gas constant, equal to 8.31 J / mol * K.

In the calculations, carried out in this paper, the rate constants values for coke, gases and resins reaction formation were calculated on the basis of the empirical expressions presented in [9].

Results discussion

Based on the calculations results and literature analysis [10-13], it can be concluded that the main characteristic of biomass controlled thermal decomposition is a product composition its thermal conversion. Therefore, the target function in the planned experiments was the finite (at a certain point in time, corresponding to the achievement of a fixed temperature) products composition (solid, liquid and gaseous).

The main variable factors determining this composition are the temperature of the sample at the end of a single experiment, the heating rate, the biomass type, and the temperature interval. The choice of temperature changes was carried out by the method of successive approximations when the full conversion conditions were reached. The heating rate values were chosen based on the actual possibilities of the installation for industrial thermal biomass conversion. Temperature changes intervals were chosen on the basis of conditions ensuring maximum reliability of the experimental results. The results of the measurements were tabulated on the basis of which the graphs were plotted in Fig.3, when all samples were simultaneously placed in a cold furnace and heated whether to 450°C.

As the studied raw samples for pyrolysis, there were briquettes from:

Graph 1-series of 8 experiments with briquettes from rice, coffee straw and husk, initial humidity ≈ 6%;

Graph 2 - series of 8 experiments with briquettes from rice straw and wood initial humidity of 6%;

Graph 3 - series of 8 experiments with briquettes from peat A, initial humidity ≈ 6%;

Graph 4 - series of 8 experiments with briquettes from peat B, initial humidity ≈6%.

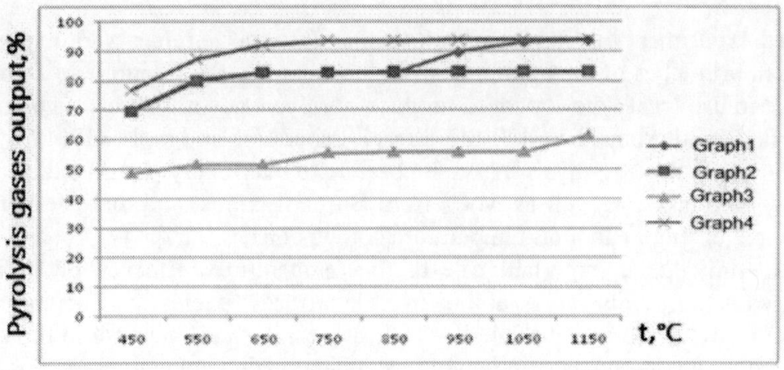

Fig.3 The dependences of the pyrolysis gas share yield from various briquettes from the temperature in the furnace.

The main types of biomass in Burundi - sawdust of Burundi palm, local peat of two deposits A and B, rice and coffee straws and husk were used for research. In preliminary experiments it was found that, despite the high requirements to the stability of the conditions for conducting a single experiment on all significant factors, some scattering of the experimental data took place. For this reason, the number of experiments with fixed factors was chosen at least eight. The experimental setup was used to study the pyrolysis of dry sawdust and peat in the temperature range from 450 to 1150 ^{O}C. It follows from Fig.3 that the total yield of the gaseous phase of pyrolysis in the region of the investigated temperatures depends significantly on the temperature in the furnace and reaches a maximum value when pyrolysis of briquettes from peat B, with a moisture content of not more than 6%. Analysis of the curves in Fig.3 shows that the briquette of rice, coffee straw and husk (series 1) with a significant increase in temperature in the furnace, has a rather sluggish yield of volatile gases, but at a temperature of 1000°C, it begins to rise sharply and reaches its maximum value of about 93%. A further temperature increase in the furnace does not lead to an increase in the volatiles yield from the sample.

A similar intensity volatile gases release is also observed in the series 2, but with a further increase in temperature over 650°C, the value of the volatile gas yield remains at 62%. Hence, it can be concluded that in series 1 with, rice, coffee straw and husks, which have an insignificantly dense structure of plant tissue and with a high air content, first the concentration of volatile gases in the temperature range (550-850°C) is stabilized, and then the stepwise conversion increase with an increase in the volatile gases yield from 82% to 93%, takes place. A series of 3 showed that the distillation of a briquette from peat A leads to a slowed down biomass decomposition reaction of, but with the temperature increase, the volatile yield continues to grow and reaches its maximum at 61% at a furnace temperature of 1150°C. Series 4 showed that the briquette from the local peat B has a more rapid rate of volatile gas evolution, which begins to increase especially from 550 to 750°C, i.e. the volatiles yield in this range was 1% for every 40°C temperature increasing in the reactor. Further temperature increase in the reactor no longer leads to an increase in the volatile gases yield from the sample. Hence it can be concluded that for peat B there is no need for high-temperature conversion and it can be satisfied with the low-temperature pyrolysis process, which is both economically and technically very beneficial for the further application of local peat type B. The obtained pyrolysis gas from type B peat has a real application as a fuel for both gas-diesel and GPM. But in order to bring it to the quality of natural gas, it is necessary to better clean both liquid (distillate) and solid fractions, which will slag the fuel supply chamber in diesel and in GPM. Nevertheless, further deep processing of pyrolysis distillate makes it possible to obtain various commodity fuel products, such as diesel fuel, gasoline, kerosene, fuel oil, solar oil, which are very relevant for the national economy of Burundi.

Conclusion

The conducted experiment of 4 series with 8 temperature measurements to determine the share of volatile gases from briquettes of the main industrial fuels used in the Republic of Burundi showed that it is quite feasible to use local type B peat to produce pyrolysis gas in sufficient volume for successful operation of the industrial group "REGIDESO" diesel -generators and in the GPM.

Fast and high-temperature pyrolysis makes it possible to maximally preserve the energy capacity of rice, coffee straw and husks, as well as wood from Burundi palms, and thereby increase the volatile gases yield for a longer time with high-temperature conversion.

Using the experiment, the important role of the temperature influence on gasification process results, especially on the pyrolysis gas output from local peat species, was determined. A threshold temperature (450°C) was also established, at which there is a sharp increase in the gas yield of from the type B peat and, as a consequence, in its volume and calorific value increase.

In general, it was determined that with an increase in the conversion temperature, an improvement in the main parameters of the resulting pyrolysis gas from local biomass species is observed, which is especially important for diesel generators and GPM, with an electricity significant deficit in the Republic of Burundi.

Acknowledgments

This work has been supported and supervised by CHICHIROVA Natalia Dmitrievna - doctor of chemical sciences, professor, director of the Institute of Heat Power Engineering, head of the department "TES". We thereby express our sincere gratitude to her for sharing his pearls of wisdom and hard work with us during the entire work.

References

[1] Z.A. Manigomba, N.D. Chichirova, V.B. Gruzdeyv, “Prospects for the use of the products of pyrolysis in diesel generators of a company «regideso» in the republic of Burundi,” *Energy sector problems*. Vol. 20, No. 2, pp. 33-40, 2018.

[2] M. M. Wright, D. E. Daugaard, J. A. Satrio, & R. C.Brown, ”Techno-economic analysis of biomass fast pyrolysis to transportation fuels,” Fuel, Vol. 89, pp.2-10, 2010.

[3] J. F. González, S. Román, J. M. Encinar & G. Martínez, “Pyrolysis of various biomass residues and char utilization for the production of activated carbons,” *Journal of Analytical and Applied Pyrolysis*, Vol. 85, No.2, pp.134-141, 2009.

[4] R. K. Sharma, J. B. Wooten, V. L. Baliga, X. Lin, W. G. Chan & M. R. Hajaligol, “Characterization of chars from pyrolysis of lignin. *Fuel*, Vol. 83, No. 12, pp. 1469-1482, 2004.

[5] M. P. McHenry, “Agricultural bio-char production, renewable energy generation and farm carbon sequestration in Western Australia: Certainty, uncertainty and risk,” *Agriculture, Ecosystems & Environment*, Vol. 129, No.3, pp. 1-7, 2009.

[6] A. V. Bridgewater, “Biomass fast pyrolysis,” *Thermal science*, Vol. 8, No. 2, pp. 21-50, 2004.

[7] D. Mohan, C. U. Pittman, & P. H. Steele, “Pyrolysis of wood/biomass for bio-oil: a critical review,” *Energy & fuels*, Vol. 20, No.3, pp. 848-889, 2006.

[8] N. Prakash, & T. Karunanithi, “Kinetic modeling in biomass pyrolysis–a review,” *Journal of Applied Sciences Research*, Vol. 4, No.12, pp.1627-1636, 2008.

[9] J. Porteiro, J. L. Míguez, E. Granada, & J. C. Moran, “Mathematical modelling of the combustion of a single wood particle,” *Fuel Processing Technology*, Vol. 87, No. 2, pp. 169-175, 2006.

[10] A. V. Bridgewater, D. Meier, & D. Radlein, “An overview of fast pyrolysis of biomass,” *Organic geochemistry*, Vol. 30, No.12, pp. 1479-1493, 1999.

[11] A. Demirbaş, “Gaseous products from biomass by pyrolysis and gasification: effects of catalyst on hydrogen yield,” *Energy conversion and management*, Vol. 43, No. 7, pp. 897-909, 2002.

[12] R. Srinivasaraghavan, T. E. Wilson, & M. J. Stark, "Pyrolysis of Industrial Wastes. In Proceedings of the 32nd Industrial Waste Conference, Purdue University may No.12, pp. 377-386, 1977.

[13] K. Tiilikkala, L. Fagernäs, & J. Tiilikkala, "History and Use of Wood Pyrolysis Liquids as Biocide and Plant Protection Product History and Use of Wood Pyrolysis Liquids as Biocide and Plant Protection Product," The open agriculture journal, 4(1), 2010

IOP Conf. Series: Earth and Environmental Science **288** (2019) 012106 doi:10.1088/1755-1315/288/1/012106

Cascade bowl-type heat and mass exchange packing with dripping irrigation mode

[1]Merentsov N.A., [2]Lebedev V.N., [3]Persidskiy A.V., [1]Balashov V.A.

[1]Volgograd State Technical University, 400131, Russia, Volgograd, Prospect Lenin, 28

[2]«LUKOIL-Engineering VolgogradNIPImorneft» branch in Volgograd 400078, Russia, Volgograd, Prospect Lenin, 96

[3]JSC Federal Scientific and Production Centre «Titan - Barricady» 400071, Russia, Volgograd, Prospect Lenin, b/n

steeple@mail.ru

Abstract. Data of experimental study of hydrodynamics, heat and mass transfer of cascade bowl-type heat and mass exchange packing with dripping spray mode and some comparative characteristics of the packed device with packing, which have wide industrial application, are given in the article.

Introduction

A large quantity of industrial packingsis known and a lot of packingsare offered to be used industrially, and the work aimed at the creation of new packings is still going on. Creation of new highly efficient packings is a difficult and time-consuming process that sometimes requires providing for necessary but colliding structural solutions [1].

Requirements set for the packings of industrial apparatuses are low flow resistance, a wide range of liquid and gas (vapour) load, ability to form a uniform velocity profile in the cross-section of the apparatus, good wettability of the packing (and in some cases hydrophobic behaviour, when it is necessary to reach dripping water flow mode), high retaining capacity for liquid, resistance to aggressive media, sufficient mechanical strength and low cost. The variety of packings used industrially is reasoned by the necessity to ensure high efficiency of the process equipment operation.

The selection of known and creation of new packings for industrial equipment is always accompanied by the necessity to assess their efficiency with regard to the specific process. Due to the specific character of the processes implemented in two-phase filtration flow it is impossible to clearly identify the most efficient type of packings for it.

The aim of the offered structure of the packing element for heat and mass exchange apparatus is to create an aerodynamically streamline configuration of the packing with the creation of good dripping liquid flow mode.

This ensures reduction of flow resistance to the ascending gas flow, decrease of energy efficiency and increase of heat and mass transfer intensity.

The result is reached by the structure where the packing element of the heat and mass exchange apparatus has a set of parallel bars on which plates with holes are placed in cascade. At that the plates are of cusp form with shaped edges and are placed with their apex down with locks placed above each plate. The cusp form of the plates of the mass exchange packing element and their installation with apex down, towards the gas flow, creates an aerodynamically streamline configuration that significantly decreases flow resistance to the ascending gas flow and energy spent on mass exchange process. The orientation of the plates in the heat and mass exchange packing elements towards the liquid flow ensures sufficient retaining capacity for liquid. This packing is universal, because depending on its orientation in space it can be applied in the process of fractionation in a splitter column and in the process of evaporation cooling. The packing type is shown in Figure 1.

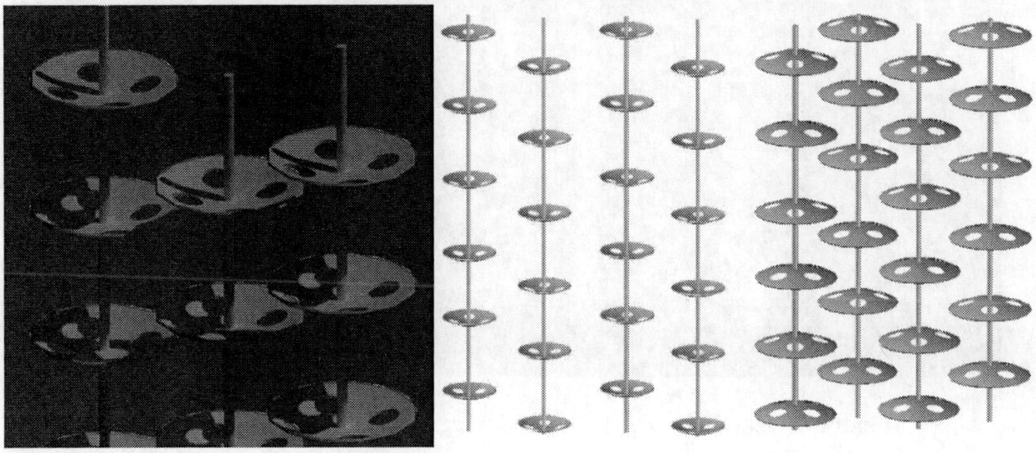

Figure 1. Structured cascade packing [2]

This packing includes parallel metal bars on which stainless steel round-shaped "cup" elements are fixed. Cups have several holes. Bars, on which the elements are fixed, pass through the middle part of the element. The elements are located concentrically with respect to each other with a certain distance from each other on the bar. The bars are attached to the frame which provides rigidity; this frame makes the assembly of the packing in the apparatuses easier, because it can be assembled by units having round or rectangular section. The plates are located in such away that drops fall down from the plate, pass the unit and hit the next plate where they break, thus, renewing their heat and mass transfer surface. These cycles repeat continuously in the whole height of the packing.

To confirm the efficiency of the designed packing we have performed a series of experiments. The aim was to receive the data on basic flow characteristics of the designed cascade packing. The packings chosen for comparison were a range of industrial packings [1-6].

Experimental part

In order to perform a full range of experimental studies and compare packings having different configurations, we have designed an experimental stand (Figure 2) which is based on the modular study principle and has a cartridge system of replacement packings that provides for a quick and precise study of heat and mass exchange and flow characteristics of the packings [1, 3].

Figure 2. Experimental stand for studying characteristics of packings

Results and Discussion

Experimental data on flow resistance of dry and spray heat and mass exchange packings are given in Figure 3.

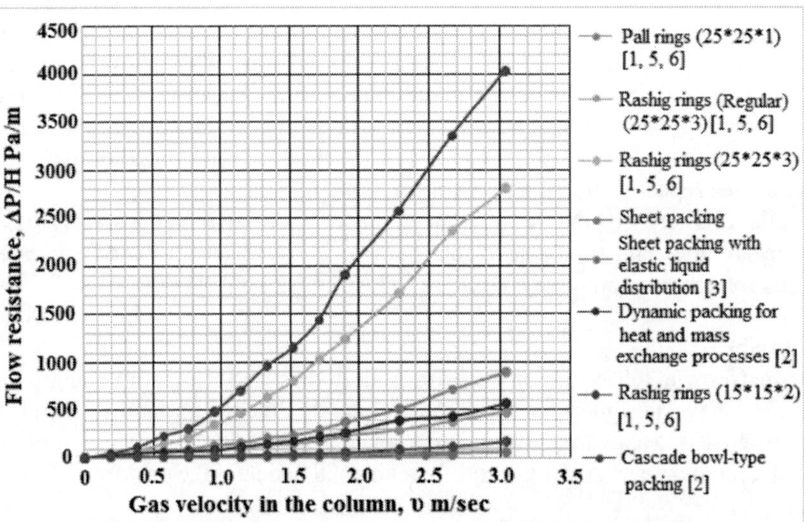

Figure 3. Flow resistance of dry packings vs. gas velocity in the column

In order to summarize the experimental data and prepare the comparison of the tested heat and mass exchange packing and packings that are widely spread in the industry, it is offered to use generalized equation $\lambda=f(Re_m)$ as per the method described in papers [5, 6]. Using this generalized equation we can compare their energy efficiency and describe the use of any packings having very complex configurations.

Based on the filtration curves obtained during experiments (Figure 3) for the described heat and mass packing [5, 6] we determined linear dimensions l_1 and l_2, α and β values which are viscous and inertial factors in the Dupuit-Forchheimer equation, respectively, modified Reynolds numbers Re_m, and corresponding flow resistance factors λ, given in Table 1. This analysis was performed according to the method described in papers [5, 6]. This method helps to identify the industrial application and analyse the energy requirements for the industrial processes for any packings of any configuration. They can be summarized and are within the mode range of filtration curve $\lambda=f(Re_m)$, shown in Figure 4.

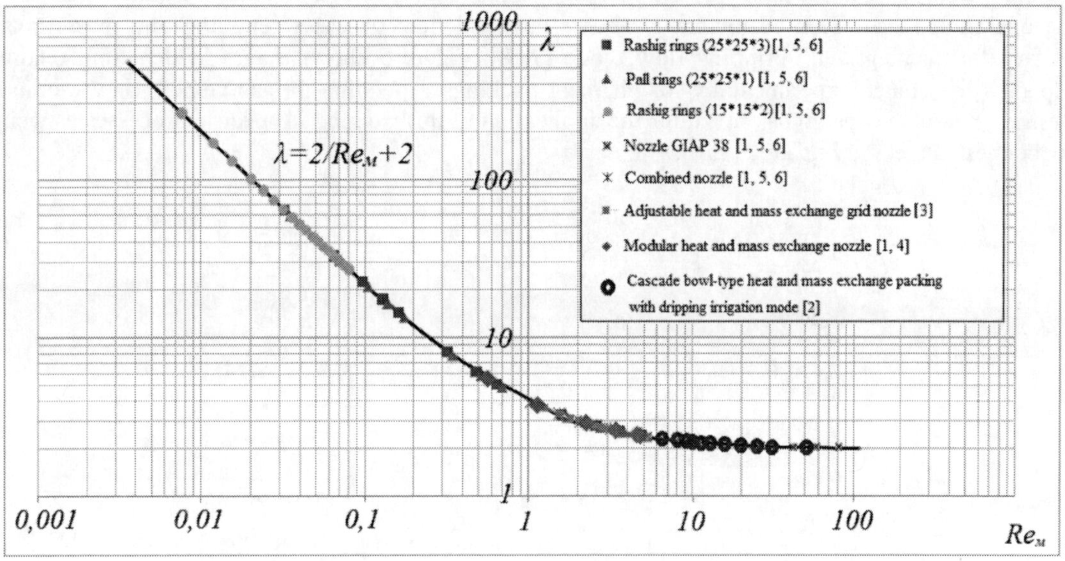

Figure 4. $\lambda=f(Re_m)$ diagram for packings with different structures [5, 6]

The analysis of the obtained experimental data on the flow dynamics of the designed cascade heat and mass exchange packing with dripping flow mode suggests that it is especially promising for evaporation cooling of water in cooling towers and in wet gas cleaning plants.

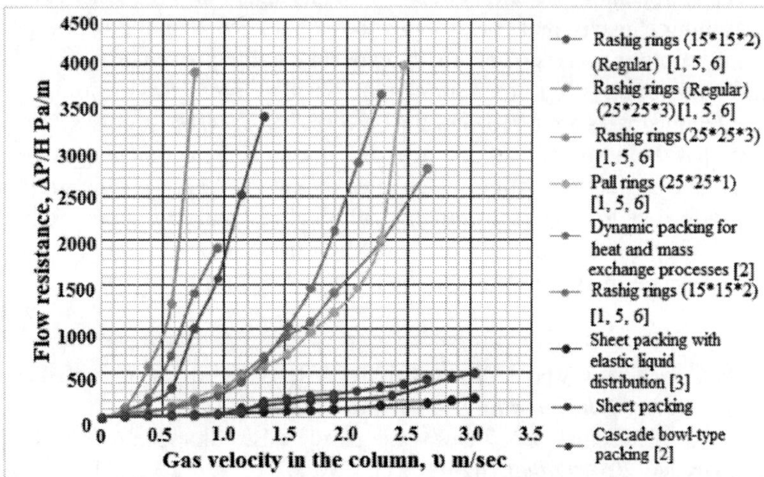

Figure 5. Relation between flow resistance of irrigation packings and gas velocity in the column

The main experimentally confirmed specific feature of the cascade bowl-type packing is the effect of oscillation of flexible cables which fix the elements. These oscillations occur even at smallest air speeds and intensify when the air flow increases. This effect occurs because the air rounds the elements in space that are fixed at a distance on flexible cables which easily lose balance position. The oscillation effect tears films on the surfaces of the packing elements, and a stable dripping flow mode is maintained in the whole height of the packing. Besides, the triggered oscillations result in heat and mass exchange processes intensification.

For packings with dripping liquid flow mode we have developed calculation methods [7, 8] where apart from heat and mass transfer coefficients there is another key parameter of the packings – retaining capacity, i. e. the ability of the packing to accumulates certain amount of liquid depending on the operation mode. This characteristic shows the total time of water retaining and mass transfer surface for packings with dripping flow mode [7, 8]. Figure 5 shows diagrams depicting retaining capacity determined experimentally for different packings depending on gas velocity in the column. Besides, retaining capacity is critical for the application of this packing in specific processes as well as the suggested method $\lambda = f(Re_m)$ [5, 6].

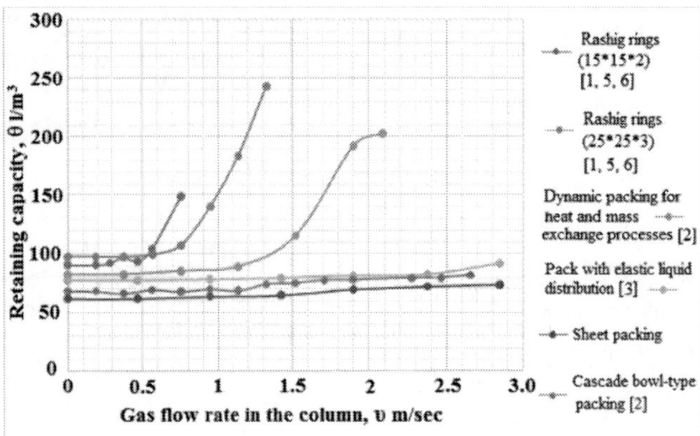

Figure 6. Retaining capacity of the irrigation packing vs. gas velocity in the column

Conclusions

Compared with a range of packings which have been studied experimentally, the designed cascade heat and mass exchange packing has a very low flow resistance that proves its energy efficiency and has satisfactory retaining capacity that corresponds to the packings suitable for evaporation cooling of circulating water in cooling towers. The main experimentally proven effect is oscillation of the elements in almost whole range of gas operation velocities that intensifies dripping flow mode and can be easily implemented in small cooling towers for local recycling water supply systems.

It is noteworthy that the designed cascade heat and mass exchange packing is very promising as a module for decontamination of circulating water in electrical field [1, 4]. This packing is very simple and can be finely adjusted for the requirements of specific heat and mass exchange process and apparatus though it may seem complex, because the structure and configuration of the packing elements as well as distances between them can be different and easily adjusted.

References

[1] Merentsov NA, Lebedev VN, Golovanchikov AB, Balashov VA, Nefed'eva EE 2018 *IOP Conference Series: Earth and Environmental Science* **(115)** 9
[2] Merentsov NA, Balashov VA, Golovanchikov AB, Shapovalov VM, Ryazanov MG, Khizhnyakov IA 2014 *Patent* 142483 RU, IPC B 01 J 19/32 VSTU
[3] Khizhnyakov IA, Ryazanov MG, H.A. Merentsov NA, Balashov VA, Golovanchikov AB 2016 *Energo- i resursosberezhenie: promyshlennost' i transport (Energy and resource saving in*

industry and transport) **1 (13)** 7-11

[4] Golovanchikov AB, Sivolobova NO, Merentsov NA, Dul'kina NA, Shishljannikov VV, Dorofeeva NI 2012 *Patent* 129450 RU, IPC F 28 S 25/08 VSTU

[5] Golovanchikov AB, Balashov VA, Merentsov NA 2017 *Chemical and Petroleum Engineering* **53(1-2)**10-13

[6] Merentsov NA, Balashov VA, Orljankina JaA 2013 *Izvestija Volgogradskogo gosudarstvennogo tehnicheskogo universiteta, Ser. Reologija, processy i apparaty himicheskoj tehnologii (The news of Volgograd state technical University, Ser. Rheology, processes and devices of chemical technology)* **1 (104)** 112-114

[7] Golovanchikov AB, Merentsov NA, Balashov VA 2013 *Chemical and Petroleum Engineering* **48(9-10)** 595-601

[8] Golovanchikov AB, Merentsov NA, Balashov VA, Orljankina JaA 2012 *Izvestija Volgogradskogo gosudarstvennogo tehnicheskogo universiteta, Ser. Aktual'nye problemy upravlenija, vychislitel'noj tehniki i informatiki v tehnicheskih sistemah (The news of Volgograd state technical University, Ser. Actual problems of management, computer science and Informatics in technical systems)* **10 (97)** 22-28

[9] Kagan AM, Yudina LA, Pushnov AS 2012 *Theoretical Foundations of Chemical Engineering* **46(2)** 165-171

[10] Gorodilov AA, Berengarten MG, Pushnov AS 2016 *Theoretical Foundations of Chemical Engineering* **50(3)** 325-334

[11] Gorodilov AA, Berengarten MG, Pushnov AS 2016 *Theoretical Foundations of Chemical Engineering* **50(4)** 422-429.

[12] Mitin AK, Nikolaikina NE, Pushnov AS, Zagustina NA 2016 *Chemical and Petroleum Engineering* **52(1)** 47-52

[13] Klyushenkova MI, Kuznetsova NA, Pushnov AS, Berengarten MG, Mokrousova EA 2014 *Chemical and Petroleum Engineering* **50(7-8)** 508-512

[14] Gorodilov AA, Pushnov AS, Berengarten MG 2014 *Chemical and Petroleum Engineering* **50(1-2)** 84-90

[15] Pushnov AS, Sokolov AS, Sidel'nikov II, Kurbatova EA, Mitrofanova EG 2014 *Chemical and Petroleum Engineering* **50(5-6)** 330-334

[16] Chizh, KV, Pushnov, AS, Berengarten MG 2014 *Chemical and Petroleum Engineering* **50(3-4)** 244-250

[17] Berengarten MG, Nevelson AO, Pushnov AS 2013 *Chemical and Petroleum Engineering* **48(11-12)** 723-729

[18] Shilin MV, Berengarten MG, Pushnov AS, Klyushenkova MI 2013 *Chemical and Petroleum Engineering* **48(9-10)** 608-614

[19] Mitin AK, Valdberg AY, Pushnov AS 2012 *Chemical and Petroleum Engineering* **48(1-2)** 50-53

[20] Tsurikova NP, Pushnov AS, Lagutkin MG, Shishov VI 2012 *Chemical and Petroleum Engineering* **48(1-2)** 3-8

[21] Kagan AM, Pushnov AS 2012 *Russian Journal of Applied Chemistry* **85(3)** 523-526

[22] Kagan AM, Yudina LA, Pushnov AS 2012 *Russian Journal of Applied Chemistry* **85(3)** 515-522

[23] Pushnov AS, Vaganov AA 2011 *Journal of Applied Chemistry* **84(9)** 1638-1641

[24] Pushnov AS, Berengarten MG, Kagan AM, Ryabushenko A 2009 *Russian Journal of Applied Chemistry* **82(4)** 723-729

[25] Tsurikova NP, Pushnov AS, Lagutkin MG 2011 *Chemical and Petroleum Engineering* **47(1)**. 97-103

[26] Petrashova EN, Lagutkin MG, Pushnov AS, Shishov VI 2011 *Chemical and Petroleum Engineering* **47(3)** 250-255

[27] Pushnov AS, Lozovaya NP, Lagutkin MG 2010 *Chemical and Petroleum Engineering* **46(1)** 3-8

IOP Conf. Series: Earth and Environmental Science **288** (2019) 012107 doi:10.1088/1755-1315/288/1/012107

Testing of the SIV method for measuring the average flow characteristics in the wake of a transverse circular cylinder

Mikheev A.N.[1], Molochnikov V.M.[1,2], Mikheev N.I.[1], Goltsman A.E.[1]

[1]Federal Research Center "Kazan Scientific Center of the Russian Academy of Sciences"

[2]Kazan National Research Technical University named after A.N. Tupolev, Kazan, Russia

vmolochnikov@mail.ru

Abstract. The results of testing the SIV method (Smoke Image Velocimetry) for two-dimensional instantaneous flow velocity vector fields measurements were presented for the case of flow in the near wake of a cross circular cylinder at Reynolds number of 3900. Measured profiles of the streamwise and transverse flow velocity components in the near wake of the cylinder at different distances were compared with the direct numerical simulation and experiments of other authors. The capabilities of SIV method were confirmed.

Introduction

A reliable prediction of the wide range of turbulent characteristics, including those which have the practical interest for different technical applications, is one of the most difficult problems in fluid mechanics. When employing the numerical modeling of complex turbulent flows, the main tool is still the method based on the solution of the Reynolds-averaged Navier-Stokes equations (RANS), which requires the use of various semi-empirical models of turbulence. Despite the continuing improvement of these models, the simulation results of separated flows and flows in the wake of bluff bodies using RANS models are still insufficient. To develop methods of numerical simulation of turbulent flows, as well as to improve turbulence models in RANS methods, reliable experimental data is needed. The most used hot-wire anemometric methods of measuring the flow velocity are contact and allow to obtain only local characteristics of the flow. The use of the PIV method and its variations makes it possible to derive the components of the flow velocity vector and their correlation characteristics in complex flows. However, they have limitations on the spatial scale.

In present paper, the results of testing the capability of the new optical SIV method for predicting the profiles of the streamwise and transverse components of the average flow velocity in the near wake of a bluff body (transverse circular cylinder) are presented. This type of flow is often used as a test to estimate the efficiency of numerical methods for modeling separated flows, testing and verification of computational codes. The largest number of studies on the cross-flow in the near wake of a circular cylinder has been performed for the Reynolds number Re = 3900. The SIV measurement results obtained in this paper are mainly compared with the available data of direct numerical simulation (DNS) and experiments. The results of the large-eddy simulation (LES) method are used in the case when there are no above-mentioned data.

Experimental setup and procedure

The wind channel and the setup arrangement used in the experiments is shown in figure 1. The fan 7 operates on the suction and drives the air in the setup. The set value of the average flow velocity in the test section is provided by the defined set of critical nozzles 6. The test section is 300-mm long 115×150 mm^2 rectangular channel. The turbulence level in the inlet of the test section does not exceed 1%.

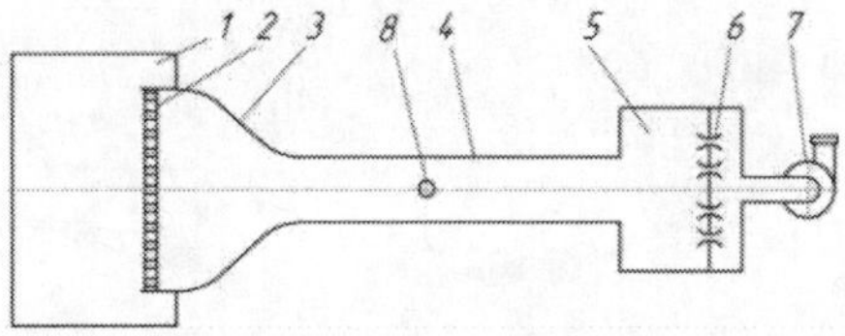

Figure 1. Schematic of the experimental setup: 1 – air-aerosol mixture preparation chamber; 2 – honeycomb; 3 – nozzle; 4 – measurement section; 5 – receiver tank; 6 – sealed baffle with set of critical nozzles; 7 – suction fan; 8 – cylinder

A circular cylinder with a diameter d = 15 mm and a length L = 150 mm was mounted horizontally on the axis of the test section at a distance of 60 mm from its inlet section. The measurements were carried out at Reynolds number Re = U d/ν = 3900 calculated by the average flow velocity U in front of the cylinder and its diameter d. The profiles of the streamwise velocity component U along the central line of the wake behind the cylinder (x-axis), as well as the profiles of the streamwise U and transverse V velocity components at distances x/d = 0.54; 1.06; 2.02; 4.0 and 10.0 from the cylinder were measured. The range of measurements for the y coordinate was y/d = -3.0 to y/d = 3.0. The velocity components in the near wake of the cylinder were measured using the SIV technique (smoke image velocimetry). A detailed description of the SIV method is given in [1, 2]. The frame resolution in the experiments was 43 × 975 pixel with the scaling factor of 0.045 mm/pixel, frame rate f = 12800 Hz and recording time of 2.0 s. The minimum interrogation window size of 11×11 pixel was used in the present measurements. Node-to-node spacing along both coordinates was equal to 4 pixels. The maximum displacement of turbulent structures between two consecutive frames was 9 pixels along the x-axis and 6 pixels along the y-axis.

Results

The profile of the streamwise velocity vector component along the central line of the near wake behind the cylinder obtained experimentally by the SIV method agrees quite well with the experimental data available in the literature and DNS results (figure 2).

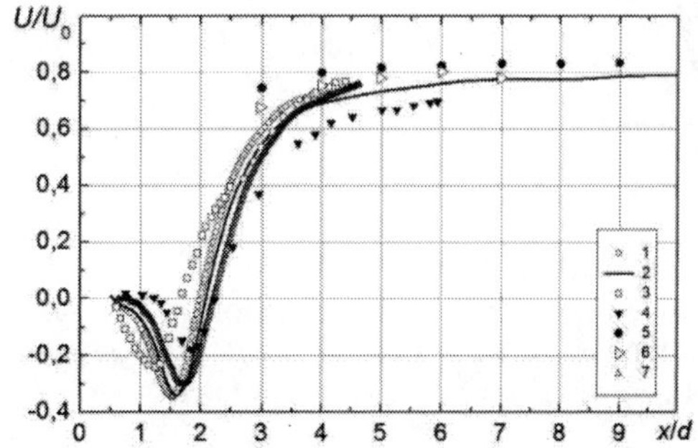

Figure 2. Streamwise velocity component along the central line of the wake behind the cylinder at Re = 3900: 1 – PIV [3]; 2 – DNS [4]; 3 – PIV [5] (private communication); 4 – PIV ((private communication, из [6]); 5 – HWA [3]; 6 – HWA [7]; 7 – present SIV measurements

IOP Conf. Series: Earth and Environmental Science **288** (2019) 012107 doi:10.1088/1755-1315/288/1/012107

The profiles of the streamwise and transverse velocity components obtained from SIV mneasurements are also in sufficient agreement with DNS profiles. Examples of respective comparisons are given in figures 3 - 8.

The profiles of the streamwise velocity component at $x/d = 0.58$ have a pronounced U-shape(figure 3), which is confirmed both by the results of the present SIV measurements and by numerical simulation using LES (Beaudan, P. & Moin, P. [9], Kravchenko AG, Moin P. [8] and the experiments of Lourenco, Shih [4]).

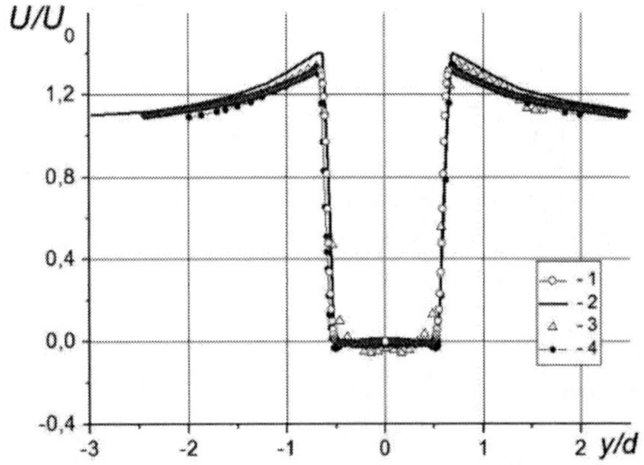

Figure 3. Profiles of the streamwise velocity component behind the cylinder at $x/d = 0.58$: 1 – SIV; 2 – LES [8]; 3 – PIV [4]; 4 – LES [9]

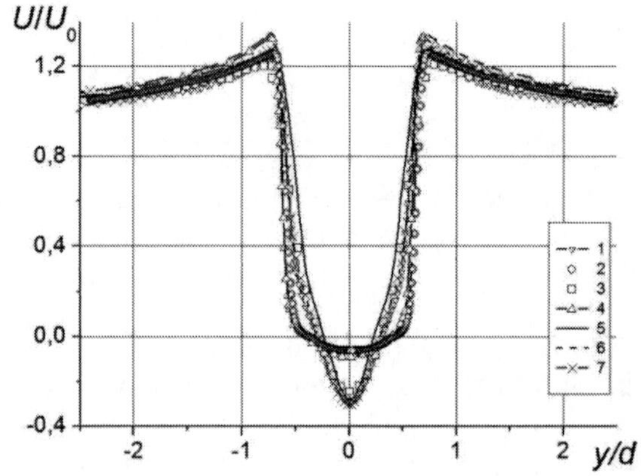

Figure 4. Profiles of the streamwise velocity component behind the cylinder at $x/d = 1.06$: 1 – DNS [6], (case I); 2 – SIV; 3 – PIV [4]); 4 – DNS [6], (case II); 5 – DNS [5] (simulation A); 6 – DNS [5] (simulation E); 7 – DNS [5] (simulation E +)

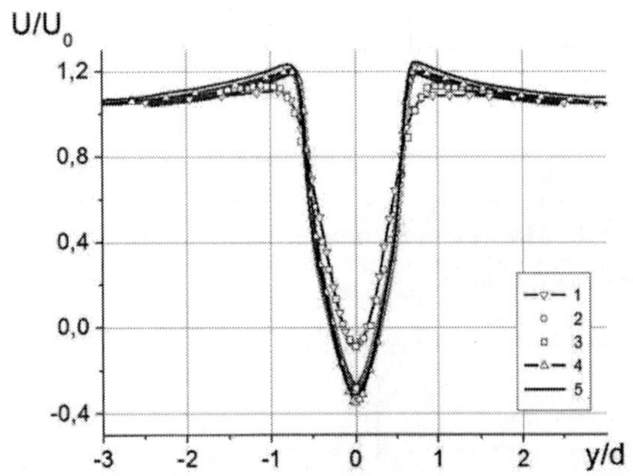

Figure 5. Profiles of the streamwise velocity component behind the cylinder at $x/d = 1.54$: 1 – DNS [6] (case I); 2 – SIV; 3 – PIV [4]; 4 – DNS [6] (case II); 5 – DNS [5] (simulation E)

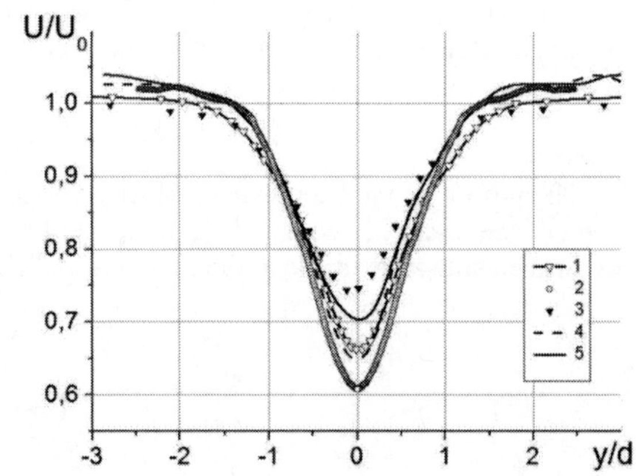

Figure 6. Profiles of the streamwise velocity component behind the cylinder at $x/d = 4.0$: 1 – DNS [6] (case I); 2 – HWA [7]; 3 – DNS [5] (simulation A); 4 – DNS [5] (simulation E); 5 – SIV

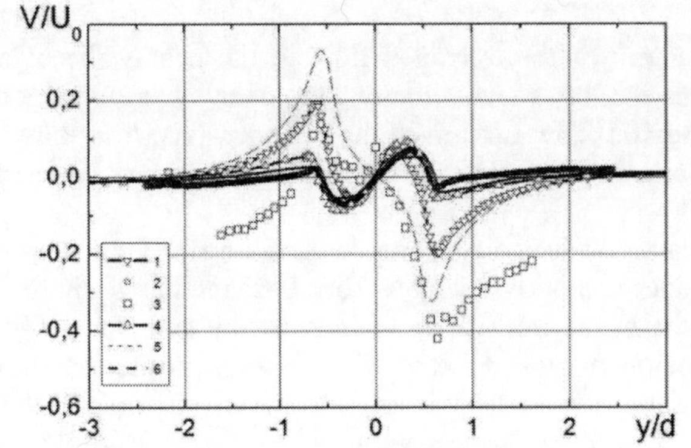

Figure 7. Profiles of the transverse velocity component behind the cylinder at $x/d = 1.06$: 1 – DNS [6] (case I); 2 – SIV; 3 – PIV [4]; 4 – DNS [6] (simulation II); 5 – DNS [5] (simulation A); 6 – DNS [5] (simulation E)

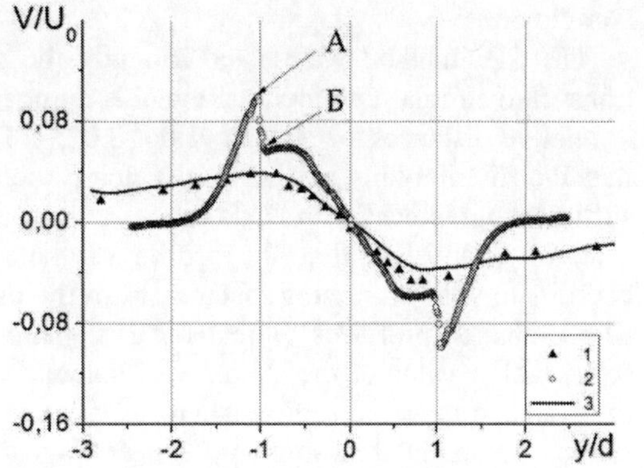

Figure 8. Profiles of the transverse velocity component behind the cylinder at $x/d = 4.0$: 1 – HWA [7]); 2 – SIV; 3 – DNS [6] (case II)

The gradual transformation from the U-shaped profile of the streamwise velocity component to the V-shape occurs (figures 3 - 5) as the distance from the cylinder increased in the range $x/d = 0.54$ – 1.54. The transition from a stagnant zone with close to zero values of the streamwise velocity component to a zone with the largest (on average) velocity of the reverse flow near the wake axis is observed right in this area of the wake, and, as the data of the flow visualization showed, the Karman vortices are generated here. The position of this area appears to be related to the size of the flow separation region behind the cylinder, which, in turn, is affected by the turbulence level of the free-stream flow, the aspect ratio (the ratio of cylinder length to its diameter), and the blockage ratio (the ratio of the cylinder diameter to the channel height). Subtle discrepancy in the values of U between DNS and experimental data using PIV and HWA and the present data (SIV measurements) results from the effect of above-mentioned factors.

The features of the distribution of the transverse velocity component at $x/d = 4.0$ obtained from the SIV data are revealed. In the vicinity of $y/d = \pm 1.0$, a sharp change in the average value of the transverse flow velocity is observed (figure 8). As some estimates show, this change corresponds to the rotation of the average velocity vector by an angle of approximately 2.5° in a section of the order of 0.1 cylinder diameter along the transverse coordinate.

An analysis of the vortex structure dynamics of the wake behind the cylinder performed using the flow visualization showed the following. In addition to moving in the direction of the free-stream flow, the large-scale vortex structures (Karman vortices) in the process of its formation shift perpendicular to this direction toward the wake axis. According to the visualization data, largest displacement is observed at a distance of about $4d$ downstream of the cylinder. The bound of the vortex forming on the lower surface of the cylinder (see figure 1) corresponds to transverse coordinate of about $y/d = +1.0$. And the vortex shedding from the upper side of the cylinder reaches the coordinate $y/d = -1.0$. The vortex periodically induces relatively large transverse velocities in this region due to the displacement across the flow. The direction of these velocities is opposite to the direction of the transverse velocity component near the bound of the wake, which causes the sharp change in the distribution of $V(y)$ (figure 8). The described feature is confirmed by oscillograms of the transverse velocity component at points A and B of the wake (marked on figure 1): the negative values of the velocity with its oscillations at the frequency of vortex formation are much greater at point B than at point A located farther from the wake axis. The average value of the transverse velocity at point B is lower than at point A due to these oscillations.

Conclusions

The SIV method was tested to study the structure of the averaged flow in the near wake of a transverse circular cylinder at Reynolds number Re = 3900. Flow velocity components were obtained at relative distances x/d = 0.54; 1.06; 2.02; 4.0 and 10.0 downstream of the cylinder axis. It is shown that the distributions of the flow velocity components derived from the SIV measurements are in sufficient agreement with DNS data.

It is established that the profile of the streamwise velocity component along the transverse coordinate with increasing distance from the cylinder smoothly changes from U-shaped to V-shaped. The U-shaped profile is typical for a stagnant zone in the immediate vicinity of the cylinder (x/d = 0.54) with a value of the streamwise velocity component close to zero. The velocity profile assumes the V-shape from the region where the velocity of the reverse flow is close to the maximum and this shape is maintained throughout the near wake.

It was found from the SIV measurements that at x/d = 4.0 in the vicinity of $y/d = \pm 1.0$, a sharp change in the average direction of the vector of streamwise velocity component by about 2.5° is observed in a section with a length of about 0.1 cylinder diameter along the transverse coordinate. This change is due to the displacement of the Karman vortices in the transverse to the wake axis direction. Also it occurs in the region of the critical (along the transverse coordinate) displacement of the vortices formed on the opposite side of the cylinder and is caused by the dynamics of this process.

This study was supported by the Russian Foundation for Basic Research (project no. 18-41-160022)

References

[1] Mikheev N I and Dushin N S 2016 *Instruments and Experimental Techniques* **59 (6)** 880.
[2] Mikheev N I, Goltsman A E, Saushin I I and Dushina O A 2017 *Experiments in Fluids* **58 (8)**, article 97.
[3] Parnaudeau P, Carlier J, Heitz D and Lamballais E 2008 *Physics of Fluids.* **20 (8)** 085101.
[4] Lourenco L M and Shih C 1993 (private communication) *Published in [9], data taken from [8].*
[5] Wissink J G and Rodi W 2008 Int. *J. Heat Fluid Flow* **29 (4)** 1060.
[6] Ma X, Karamanos G –S and Karniadakis G E 2000 *J. Fluid Mech.* **410** 29.
[7] Ong L and Wallace J 1996 *Experiments in fluids* **20 (6)** 441.
[8] Kravchenko A G and Moin P 2000 Physics of fluids **12 (2)** 403.
[9] Beaudan P and Moin P 1994 *Numerical experiments on the flow past a circular cylinder at sub-critical Reynolds number* (Tech. Rep. TF-62, Stanford University, Stanford, CA 94305).

About heat supply schemes for settlements and city districts

L R Mukhametova[1], I G Akhmetova[1], A V Fedukhin[2] , S O Makoev[3], A A Kalyutik[3]

[1]Federal State Budgetary Educational Institution of Higer Education "Kazan State Power Engineering University", Kazan, Krasnoselskaya street, 51

[2]National Research University «Moscow Power Engineering Institute», 111250 Russia, Moscow, Krasnokazarmennaya 14.

[3]Peter the Great St. Petersburg Polytechnic University, 194021, Russia, St.Petersburg, Polytechnicheskaya 29

liliyamyhametova@mail.ru

Annotation The development of heat supply schemes is particularly relevant today as according to the Federal Law of Russian Federation of July 27, 2010 No. 190-FZ "About heat supply" after December 31, 2011, the availability of a heat supply scheme is mandatory for settlements and urban districts in Russian Federation. Heat supply schemes are developed in accordance with the Order of the Government of the Russian Federation of February 22, 2012 No. 154 "On requirements for heat supply schemes, the procedure for their development and approval".

Requirements for the heat supply schemes in Russia

In the USSR, heat supply schemes for settlements were developed within the framework of the planned economy by design organizations approved by the Gosstroy of the Union republics. Those schemes were considered and approved in the manner determined by the Councils of Ministers of the Union Republics. At the same time, such targets as streamlining of further district heating and excluding the unjustified construction of small boiler stations were set for designers of the scheme in order to save fuel and reduce the number of personnel involved in heat [1–9].

Schemes were developed and used for the design and construction of thermal sources and heat networks of most large settlements, in particular, Leningrad and Moscow. However, the validity of the developed heat supply schemes ended in the 1990s, new schemes were not developed for a long time [10,11]. The reasons for this are the reduction in the total demand for thermal energy and thermal power and, as a consequence, the lack of the need for new sources of thermal energy in the 1990s and 2000s. At the same time, due to the general growth of the economy by 2007, in some regions the demand for heat energy and thermal power returned to the level of 1990s, which necessitated the development of existing heat supply systems [12–15].

In accordance with the Federal Law "About heat supply" approval of the requirements for heat supply schemes, the procedure for their development and approval are within the competence of the Government of the Russian Federation. The specified requirements are not approved by the Government of the Russian Federation yet. However, the aforementioned Law directly contains the following provisions[16]:

- The approval of heat supply schemes for settlements, urban districts with a population of five hundred thousand people or more, as well as cities of federal significance as Moscow and St. Petersburg is carried out by the federal executive body;
- Approval of heat supply schemes for settlements, urban districts with a population of less than five hundred thousand people is carried out by local government bodies of these settlements and districts;
- The scheme shall contain information on the Unified heat supplying organizations for the relevant territory;
- The scheme should identify the sources of heat supply, their loading, schedules of joint work and temperature charts;
- The scheme determines the effective heat supply zone beyond which the connection of consumers is not appropriate.

In addition, the heat supply scheme should contain measures for the development of the heat supply system, in particular, measures for the conversion of boiler stations for operation in the combined heat and power generation mode, and also, if necessary, measures for the conservation of excess heat capacity.

In terms of ensuring safety, heat supply schemes should provide the redundancy of the heat supply system, survivability and ensuring the uninterrupted operation of heat sources and heat networks.

Planning legislation provides for layouts of planned placement of capital construction projects of local importance, including heat supply facilities, within the boundaries of the settlement or the urban district.

However, the heat supply scheme is not a document of the Territorial Planning. The scheme for the locating heat supply facilities, as well as other layouts for the placement of capital construction facilities, is taken into consideration when developing a heat supply scheme for the respective territory.

The heat supply scheme is used for the following main purposes [16]:

- Definition of the heat supplying organization obliged to conclude a heat supply contract with the consumer (Unified heat supplying organization);
- To determine the possibility of connecting to the heat supply network of the capital construction site and to determine the organization obliged, if technically possible, to carry out such a connection;
- Measures for the development of the heat supply system provided by the scheme are included in the investment program of the heat supplying organization and, as a consequence, can be included in the corresponding tariff of the public utility services.

About actualization of the existing heat supply schemes

The heat supply scheme is a “living” document and must correspond to the actual rates of development of the settlement or urban district, which requires its annual updating. Currently, there is a large amount of work on the annual actualization of the existing heat supply schemes.

The main questions that arise when performing these tasks are as follows:

1. A large number of approved heat supply schemes (especially in small rural settlements) have been designed of poor quality without taking into account the requirements of the Order of the Government of February 22, 2012 N 154 “On requirements for heat supply schemes, the procedure for their development and approval”. Therefore, when updating these heat supply schemes, it is necessary to develop a new heat supply scheme that meets the specified requirements.

2. According to the requirements, the heat supply scheme must be developed for a period of not less than 15 years. Most approved heating schemes are designed only for a 15-year period. After approval of the updated heat supply scheme, the previous version of the scheme is no longer valid. Therefore, with the annual update, it is necessary to shift the billing period by one year. For example, if the initial version was developed for 2015-2030, then the estimated period of the updated version should be 2016-2031. In practice, in most cases, the actualization of the schemes occurs without a

IOP Conf. Series: Earth and Environmental Science **288** (2019) 012108 doi:10.1088/1755-1315/288/1/012108

change in the calculation period. For example, it is possible to meet the decision of the local government to update the heat supply scheme for the period of 2017-2026 years instead of 2016-2031years. Since the heat supply scheme is developed on the basis of the approved documents of the territorial planning of the settlement or the urban district, it is advisable, when updating the scheme, to adopt for the estimated period not the 15-year period, but the estimated period of the master plan.

3. The electronic model is a key element of the heat supply scheme. Modeling of hydraulic modes with the help of the model is the only way to work out perspective options for the development of the heat supply system. In addition, according to the Methodological Recommendations for the development of heat supply schemes approved by the joint Order of the Ministry of Energy and the Ministry of Regional Development of the Russian Federation of December 29, 2012 No. 565/667, with the actualization it is necessary to develop an electronic model of the "second level" with the detailing of distributing (block) heating networks to the end user.

4. The development of the heat supply system should be coordinated with other programs for the development of engineering infrastructure systems. Therefore, it is desirable that the work on the actualization of the heat supply scheme be carried out in conjunction with the Program for the Integrated Development of Utility Infrastructure Systems (PID UIS). When developing the PID UIS, among other things, the target indicators for the development of the heat supply system are determined, which allow one to assess the impact of measures provided for by the heat supply scheme on the quality and reliability of the heat supply system of a settlement or urban district.

5. Annual actualization of the heat supply scheme is necessary for the development and implementation of investment programs for heat supply or heating network organizations. In accordance with article 37 of the Order of the Government of the Russian Federation of May 5, 2014 No. 410, if the investment program does not ensure the implementation of measures to develop the heat supply system provided by the heat supply scheme, the local government within 7 days from the date of receipt of the investment program makes a decision on its return for revision.

In practice, in many cases, the investment programs of heat supply or heat network organizations are included into the heat supply scheme without any justification. As a result, the heat supply scheme turns into an expanded version of the investment program, which is unacceptable. The activities of the investment program should follow from the proposals of the heat supply scheme for the construction and reconstruction of heat energy sources, heating networks and structures, as well as from the assessment (calculation) of the reliability indicators of the heat supply system. The result of such assessment includes proposals for the replacement of worn out sections of heat networks with high values of the failure parameter and do not provide the normative reliability of heat supply.

6. As a result of the actualization of the heat supply scheme, it is recommended that a roster of changes to the current version of the scheme should be made. Also it is recommended to analyze the implementation of the envisaged measures for the development of the heat supply system for the period previous the year of actualization.

Conclusion

Most of the issues discussed concerning the quality of heat supply schemes, the estimated time frame, the coordination of heat supply schemes with other programs for the development of engineering infrastructure systems can be avoided if the heat supply scheme is updated as a part of a integrated project for the development of the territory of the municipality. In this case, PID UIS is being developed after the development and approval of the master plan and documents for the planning of the territory. And the heat supply scheme is updated in conjunction with developed PID UIS.

Acknowledgments

The work was supported by the Ministry of Education and Science of the Russian Federation on fundamental scientific research (Agreement No. №13.6994.2017/БЧ) «Development of a methodology for determining the reliability of a heat supply system to improve energy efficiency».

IOP Conf. Series: Earth and Environmental Science **288** (2019) 012108 doi:10.1088/1755-1315/288/1/012108

References

[1] VnipiEnergoprom O 2006 *RD-10-VEP. Methodological basis for the development of heat supply schemes for settlements and industrial units of the Russian Federation (Moscow)*
[2] Ovchinnikov P, Borodiņecs A and Millers R 2017 Utilization potential of low temperature hydronic space heating systems in Russia *J. Build. Eng.* **13** 1–10
[3] Ovchinnikov P, Borodiņecs A and Strelets K 2017 Utilization potential of low temperature hydronic space heating systems: A comparative review *Build. Environ.* **112** 88–98
[4] Borodinecs A, Zemitis J, Sorokins J, Baranova D V and Sovetnikov D O 2016 Renovation need for apartment buildings in Latvia *Mag. Civ. Eng.* **68** 58–64
[5] Grasmanis D, Sovetnikov D O and Baranova D V 2017 Energy performance of domestic hot water systems *Mag. Civ. Eng.* **76** 140–55
[6] Staritcyna A, Pollock E, Sokolova E and Martynenko E 2016 Energy efficiency in multi-story buildings *MATEC Web Conf.* **73**
[7] Gorshkov A and Murgul V 2018 Calculation of Heat Energy Consumption by a Typical Historical Building with a Courtyard *Adv. Intell. Syst. Comput.* **692** 577–91
[8] Anikina I D, Sergeyev V V, Amosov N T and Luchko M G 2017 Use of heat pumps in turbogenerator hydrogen cooling systems at thermal power plant *Int. J. Hydrogen Energy* **42** 636–42
[9] Trinchenko A A and Paramonov A P 2017 Introduction of low-temperature swirl technology of burning as a way of increase in ecological of low power boilers *IOP Conference Series: Earth and Environmental Science* vol 90
[10] Order of the Ministry of Energy and the Ministry of Regional Development of the Russian Federation of December 29 2012 *No. 565/667 "On approval of methodical recommendations for the development of heat supply schemes"*
[11] Akhmetova I and Mukhametova L 2017 Innovative development of the regional economy *Innovative clusters in the digital economy: theory and practiceProc. of the Scientific and practical conference with international participation* (Saint Petersburg)
[12] Gumerova E, Gamayunova O and Meshcheryakova T 2018 Energy Efficiency Upgrading of Enclosing Structures of Mass Housing of the Soviet Union **692** 432–9
[13] Akhmetova I, Chichirova N and Derevianko O 2017 Revisiting heat losses calculation at district heating network *Int. J. Civ. Eng. Technol.* **8** 694–702
[14] Gorshkov A, Murgul V and Oliynyk O 2016 Forecasted Payback Period in the Case of Energy-Efficient Activities vol 53, ed M V. (St. Petersburg State Polytechnical University, Polytechnicheskaya st. 29, St.-Petersburg, Russian Federation: EDP Sciences)
[15] Tarasova D S and Petritchenko M R 2017 Buildings quasi-stationary thermal behavior *Mag. Civ. Eng.* **72** 28–35
[16] Federal Law of Russian Federation of July 27 2010 *No. 190-FZ "About heat supply"*

IOP Conf. Series: Earth and Environmental Science **288** (2019) 012109 doi:10.1088/1755-1315/288/1/012109

Effect of ultrasonic treatment on the sorption properties of the pyrolysis product of sludge

I A Nasyrov[1], G V Mavrin[1], I G Shaikhiev[2] и V V Terentyeva[1]

[1]Kazan Federal University, 423812, Russia, Naberezhnye Chelny, Mira pr. 68/19

[2]Kazan National Research Tehnological University, 420015, Russia, Kazan, Karl Marx str., 68

chem_aleb@mail.ru

Annotation. In the article, the object of the study is a sorption material obtained by low-temperature pyrolysis from a sludge of a biological sewage treatment, and processed by ultrasound. Ultrasonic treatment is one of the methods for improving sorption characteristics by increasing the specific surface area. The adsorption capacity of pyrolysis products by methylene blue on a UNICO 2800 spectrophotometer and iodine by the titrimetric method was studied. An increase in sorption activity after ultrasonic treatment on the sorption material was found: by 2.3 times for methylene blue, 2.5 times for iodine. Sorption properties with respect to iron, copper, nickel, and zinc ions under static conditions using the atomic-emission spectrometry method with inductively coupled plasma on the Agilent 720-OES spectrometer were studied. The degree of extraction of heavy metal ions using the pyrolysis product of sludge after ultrasonic treatment reaches: for Fe – 98.7%, Ni – 84.4%, Cu – 90.5%, and Zn – 95.2%. Ultrasonic treatment with a frequency of 35 kHz for 30 minutes allows to increase the sorption capacity of the product of pyrolysis of mud sediments by Ni ions by 2.4 times, Cu - 1.2 times, Zn - 1.7 times.

Introduction

One of the effective methods of wastewater treatment from heavy metal ions is sorption purification. Such purification allows to carry out post-treatment to low concentrations of pollutant and re-use purified water in closed water circulation systems. The effectiveness of purification, depending on the sorbent used, can be 80-95%. As sorbents use activated carbon, carbon fibers, silica gels, synthetic sorbents [1]. At present, accessible and relatively cheap sorbents obtained from their wastes are increasingly being used for purification from heavy metal ions. The use of sludge after treatment with low-temperature pyrolysis as a sorption material allows to get rid of sludge accumulators and to improve the quality of wastewater treatment [2]. Sorption properties of the pyrolysis product are associated with the presence of carbon in its composition [3]. In addition, the properties of the pyrolysis product of the sludge can be changed by chemical or physical activation. One of the methods for improving the sorption characteristics of the pyrolysis product by increasing the specific surface area is ultrasonic treatment [4].

Materials and methods

The particle size distribution of the solid pyrolysis product of the sludge by size was determined by sieve analysis and using a laser particle size analyzer of the brand Microsizer 201C.

The adsorption activity by iodine of the pyrolysis product before and after ultrasonic treatment was determined in accordance with GOST 6217-74 "Active charcoal".

The adsorption activity of the pyrolysis product of sludge before and after ultrasonic treatment by the indicator methylene blue was determined in accordance with GOST 4453-74 " Coal active clarifying wood powdery".

The mineralization and specific conductivity were measured by means of the ANION-7020 conductometer.

Determination of the content of heavy metal ions and sorption properties with respect to heavy metal ions was carried out by atomic emission spectrometry using the Agilent 720-OES spectrometer.

The content of anions and cations in aqueous extract was determined by ion chromatography using a liquid chromatograph Stayer.

In order to study the sorption properties of the solid pyrolysis product, 1 g of ground sorbent and 50 ml of a model solution of the corresponding heavy metal ion with a concentration of 4-6 mg / dm^3 were placed in a series of conical flasks before and after treatment, covered with a lid and stirred on a shaker for 30 min in the conditions of temperature control (20 ± 0.1 °C). Next, the filtrate was separated from the sorbent and the initial and final concentration of the heavy metal ion was determined by atomic emission spectrometry [2].

Results and discussion

The solid pyrolysis product of sludge was used as a sorption material. It is a powder of black color with a grayish shade without foreign inclusions.

The particle distribution of the solid pyrolysis product of the sludge, determined by the sieve analysis method and using a laser particle size analyzer of the brand "Microsizer 201C", is shown in Table 1 [5].

Table 1. Particle size distribution (%)

Particle size, mm	Content,%
>5	22.3
3-5	20.9
1-3	29.3
0.5-1	9.9
0.1-0.5	12.3
0.05-0.1	1.7
0.01-0.05	1.8
0.001-0.01	1.6
0.0006-0.001	0.1
<0.0006	0.1

The granulometric analysis showed that the solid pyrolysis product of sludge contains predominantly particles with sizes from 0.1 to more than 5 mm (82.4%). The presence in the sample of large particles with dimensions of more than 1 mm is due to the content of the particles of sludge that are not decomposed during pyrolysis.

The results of the microstructure investigation by the scanning electron microscope of the brand "Jeol JSM-6390 LA" indicate that the pyrolysis product has a porous structure (Fig. 1). The porous structure allows to predict sorption properties.

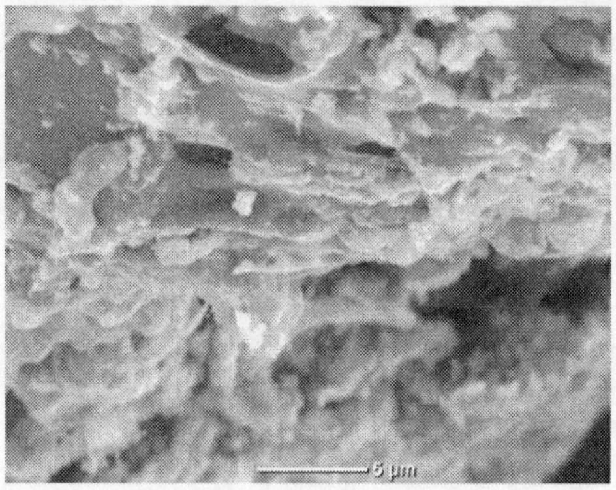

Figure 1. A microphotograph of a solid pyrolysis product of sludge

For processing as ultrasound have chosen fraction of 1-2 mm in a type of more high efficiency in dynamic conditions and rather high efficiency. The experiment was made for fractions less than 0,5 mm, 0,5-1 mm, 1-2 mm, 2-3 mm, 3 and more than mm.

The sample had the following characteristics: bulk density of the fraction (1-2 mm) - 642.8 kg/m^3, humidity 0.9%, ash content 47%, specific surface area 261.7 m^2/g, pore volume 0.156 cm^3/g (Fig. 1).

It is known that the use of ultrasonic treatment increases the capacity of solid sorbents [6, 7, 8]. When acoustically acting on the sorbent, not only the surface layer of grains changes, but also the capillary structure of the sorbent [9].

Ultrasonic treatment of the investigated material with a particle size of 1-2 mm with a predetermined volume of water was carried out in an ultrasonic bath at a frequency of 35 kHz for 30 minutes. The mass ratio of the pyrolysis-water product was 1:10, respectively, 10 g of the sample in 100 ml of water.

During the ultrasonic treatment for different times, the exposure time was determined, which ensures the maximum extraction of heavy metal ions from model solutions.

Initially, the specific electrical conductivity (SEC) and the mineralization of the aqueous solution of the processed sample after ultrasonic treatment were determined (Table 2). For comparison, Table 2 shows the mineralization and SEC data for the aqueous drawing of the pyrolysis product after a half-hour mixing of 10 g sample in 100 ml water on a shaker.

Table 2. Indices of aqueous extract and solution after treatment

sample	mineralization, mg/dm^3	SEC, µS/cm
aqueous extract of pyrolysis product	335	693
aqueous solution after ultrasonic treatment	641	1305

Ultrasonic treatment of the material leads to an increase in SEC and mineralization by a factor of 1.9. Probably, as a result of ultrasonic treatment, ions are washed from the pores by water-activated molecules of water, and the sample is demineralized.

In this case, we can observe a change in the emission of heavy metals and macro-ions into the aqueous extract from the pyrolysis product of the sludge without treatment and obtained after ultrasound treatment (Tables 3, 4 and 5).

The ultrasonic effect on the product of pyrolysis of the sludge results in isolation of heavy metals and macro-ions into the aqueous part of the working solution, including from the pores of the material. The ions that pass into the solution increase the value of the specific electric conductivity and mineralization of the aqueous phase. As a result of this process, the ion emission in the aqueous extract of the treated material decreases.

Table 3. The content of heavy metals in the aqueous extract of the pyrolysis product

sample	concentration, mg/dm^3			
	Fe	Ni	Cu	Zn
without treatment (1:50)	0.39	0.08	0.02	0.17
without treatment (1:10)	3.14	0.13	0.09	2.01
after ultrasonic treatment (1:50)	0.12	<0.01	<0.01	0.09
after ultrasonic treatment (1:10)	0.17	0.07	0.02	0.11

Table 4. Anion content in the aqueous extract of the pyrolysis product

sample	concentration, mg/dm^3				
	F^-	Cl^-	NO_3^-	PO_4^{3-}	SO_4^{2-}
without treatment (1:50)	0.535	7.13	0.03	5.41	66.5
after ultrasonic treatment (1:50)	0.301	2.32	<0.02	3.19	15.1

Table 5. Cations content in the aqueous extract of the pyrolysis product

sample	concentration, mg/dm^3				
	NH_4^+	Na^+	K^+	Mg^{2+}	Ca^{2+}
without treatment (1:50)	12.4	7.27	10.4	5.43	51.5
after ultrasonic treatment (1:50)	4.02	2.87	2.95	1.97	17.7

Assessment of adsorption was carried out on the basis of the results received by means of a technique of definition of adsorption on iodine and methylene blue.

Table 6. Sorption activity on methylene blue and to iodine

index	pyrolysis product of sludge	
	without treatment	after ultrasonic treatment
methylene blue, mg/g	16.7	38.3
iodine,%	15.4	38.1

The results obtained (Table 6) indicate an increase in sorption activity after the ultrasonic effect on the sorption material. By methylene blue sorption activity increased by 2.3 times, iodine by 2.5 times. This can be explained by an increase in the free pore volume and surface area as a result of the elution of ions from the pores.

Investigation of the sorption of heavy metal ions after ultrasonic treatment was carried out under static conditions (the ratio of the model solution and sorbent 1: 50, initial concentrations of iron, copper, nickel and zinc ions 4-6 mg/dm^3).

Comparing the concentration of ions in the initial solution with the residual (equilibrium) concentration after contact of the sorbent with the model solution, it can be concluded that adsorption capacity [9].

The results of the determination of sorption properties with respect to heavy metal ions are presented in Table 7.

The results obtained indicate an increase in the adsorption capacity with respect to nickel, copper and zinc ions after ultrasonic treatment on the pyrolysis product of sludge.

The degree of purification from heavy metal ions using the pyrolysis product of sludge after treatment is: for Fe 98.7%, Ni 84.4%, Cu 90.5%, and Zn 95.2%. This treatment allows to increase the sorption capacity of the product of pyrolysis of sludge by Ni ions 2.4 times, Cu - 1.2 times, Zn - 1.7 times.

Table 7. Efficiency of purification the model solution from heavy metal ions

heavy metal ions	initial concentration, mg/dm^3	concentration after sorption, mg/dm^3		efficiency,%	
		without treatment	after ultrasonic treatment	without treatment	after ultrasonic treatment
Fe	4.44	0.03	0.06	99.3	98.7
Ni	5.51	3.59	0.86	34.8	84.4
Cu	6.08	1.65	0.58	72.9	90.5
Zn	6.02	2.56	0.29	57.5	95.2

Thus, sonication is effective for wastewater treatment from iron, copper, zinc and nickel ions by the product of pyrolysis of sludge. The degree of sorption is almost inferior in terms of sorption to the commodity sorbent BAU [5].

Conclusions

1. As a result of ultrasonic treatment on the pyrolysis product of the sludge, heavy metals and macro-ions are released into the aqueous part of the working solution, which lead to an increase in SEC and mineralization by a factor of 1.9.

2. An increase in sorption activity after ultrasonic treatment on the sorption material was found: by 2.3 times for methylene blue, 2.5 times for iodine. This can be explained by an increase in the free volume of the pores and the surface area as a result of the washing out of the ions from the pores by the ultrasound-activated water molecules.

3. The ultrasonic effect at a frequency of 35 kHz for 30 minutes on the product of pyrolysis of sludge gives a positive result: in comparison with the untreated material, the residual concentration of iron, nickel, copper and zinc impurities decreased several times (for iron (III) - 74 times, copper II) - 10.5 times, nickel (II) - 6.4 times, zinc (II) - 20.8 times).

4. Ultrasonic treatment allows to increase the sorption capacity of the pyrolysis product of sludge by Ni ions 2.4 times, Cu - 1.2 times, Zn - 1.7 times.

5. The degree of extraction of heavy metal ions using the pyrolysis product of sludge after ultrasonic treatment reaches: for Fe 98.7%, Ni 84.4%, Cu 90.5%, and Zn 95.2%. This allows us to recommend the use of ultrasonic treatment to improve the sorption characteristics of the sorption material obtained by low-temperature pyrolysis of sludge and subsequent post-treatment of waste water from heavy metal ions.

Thus, in order to expand the name of sorption materials from carbon containing waste, an ultrasonic treatment of the solid pyrolysis product of the sludge was carried out. An increase in the sorption activity of processed products in relation to the standards of iodine and methylene blue solution and model solutions of heavy metal ions has been found.

References

[1] Lozinskaya E F, Kosolapova N I, Smorodko A V and Mitrakova T N 2015 *Chemistry of plant raw materials* **2** p 209.

[2] Nasyrov I A, Dvoryak S V and Shaikhiev I G 2016 *Acta Techn*ica **61 (4B/2016)** p 307.

[3] Nasyrov I A et al 2017 *The Turkish Online Journal of Design, Art and Communication TOJDAC* **7 SE** p 1713 DOI: 10.7456/1070DSE/150.
[4] M. Xing et al 2010 *Bioresources* **5 (3)** p 1353.
[5] Nasyrov I A et al 2017 *Journal of Fundamental and Applied Sciences* **9 (1S)** p 1615
[6] Milushkin V M and Il'in A P 2009 *Sorption and chromatographic processes* **Volume 9 Issue 2** p 308.
[7] Leykin Yu A and others 2007 *Chemical Physics* **Volume 26 (10)** p 18.
[8] Jiménez de Haro M C et al 2005 *Applied Clay Science* **Vol. 30** p 11.
[9] Sharapova A V 2013 *Bulletin of the Nizhny Novgorod University. N.I. Lobachevsky* **1** p 109.

Heat and mass exchange packing with adjustable parameters for absorption and evaporation cooling

[1]Persidskiy A.V., [2]Merentsov N.A., [3]Lebedev V.N., [2]Golovanchikov A.B.

[1]JSC Federal Scientific and Production Centre «Titan - Barricady» 400071, Russia, Volgograd, Prospect Lenin, b/n

[2]Volgograd State Technical University, 400131, Russia, Volgograd, Prospect Lenin, 28

[3]«LUKOIL-Engineering VolgogradNIPImorneft» branch in Volgograd 400078, Russia, Volgograd, Prospect Lenin, 96

steeple@mail.ru

Abstract. Data of experimental study of hydrodynamics, heat and mass transfer of packing with adjustable parameters and some comparative characteristics of the packed device with packing, which have wide industrial application, are given in the article.

Introduction

The selection of known and creation of new packings for industrial equipment is always accompanied by the necessity to assess their efficiency with regard to specific process. Specific character of different heat and mass exchange processes complicates the selection of an efficient packing for their implementation. We are trying to find a systematic approach to process the experimental data.

The aim of the work is designing and study of available heat and mass exchange packings with adjustable parameters and changeable flow modes for absorption process during selective gas cleaning and evaporation cooling of circulating water in local water supply systems at industrial enterprises.

The main task at this stage of heat and mass exchange packings study is the creation of packings providing the possibility for adjustment of their key parameters, porosity and specific surface, that consequently results in the possibility to manage flow processes. Such packings should be structured with the possibility to change their structure, or if they are non-regular, they should have elastic properties and be compressible. This idea implemented in an original engineering solution [2] to use wastes from metal treatment plants (shavings) as heat and mass exchange packings. Depending on the cutting mode shavings can have absolutely different configurations. The studied samples of shavings packing elements were obtained during lathe turning of stainless steel 12H18N10T and 08H17N14M2T (Aisi316Ti) widely used in industry.

IOP Conf. Series: Earth and Environmental Science **288** (2019) 012110 doi:10.1088/1755-1315/288/1/012110

Figure 1. Designed adjustable packing for heat and mass exchange processes with changeable packing density (Modification 1 and 2) [2]

Experimental part

In order to perform a full range of experimental studies and compare packings having different configurations, we have designed an experimental stand (Figure 2) which is based on the modular study principle and has a cartridge system of replacement packings that provides for a quick and precise study of heat and mass exchange and flow characteristics of the packings.

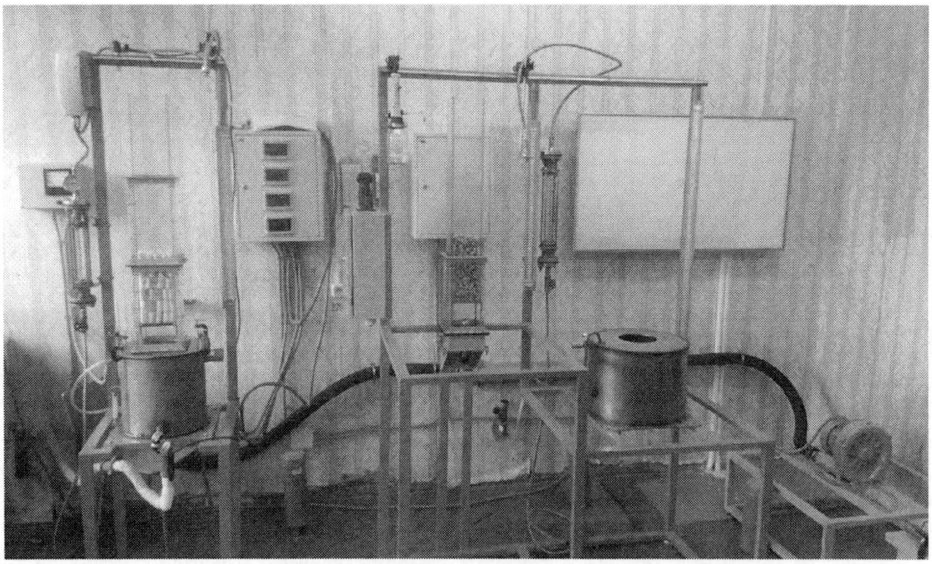

Figure 2. Experimental stand for studying characteristics of packings

Results and Discussion

Experimental data on flow resistance of dry heat and mass exchange packings are given in Figure 3.

In order to summarize the experimental data and prepare the comparison of the tested heat and mass exchange packing and packings that are widely spread in the industry, it is offered to use generalized equation $\lambda = f(Re_m)$ as per the method described in papers [5, 6]. Using this generalized equation we can compare their energy efficiency and describe the use of any packings having very complex configurations.

Based on the filtration curves obtained during experiments (Figure. 3) for the described heat and mass packing [2] we determined linear dimensions l_1 and l_2, α and β values which are viscous and inertial factors in the Dupuit-Forchheimer equation, respectively, modified Reynolds numbers Re_m, and corresponding flow resistance factors λ, given in Table 1. This analysis was performed according to the method described in papers [5, 6]. This method helps to identify the industrial application and analyse the energy requirements for the industrial processes for any packings of any configuration. They can be summarized and are within the mode range of filtration curve $\lambda=f(Re_m)$, shown in Figure 5.

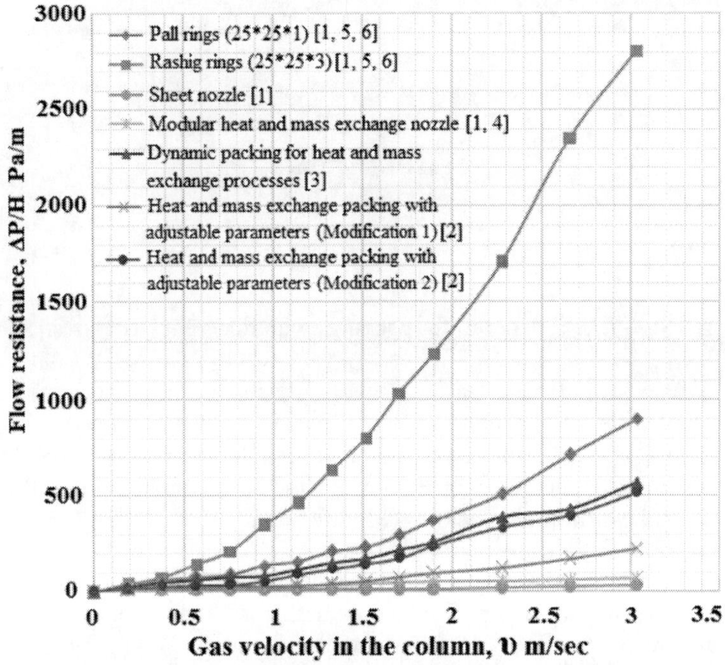

Figure 3. Relation between flow resistance of dry packings and gas velocity in the column

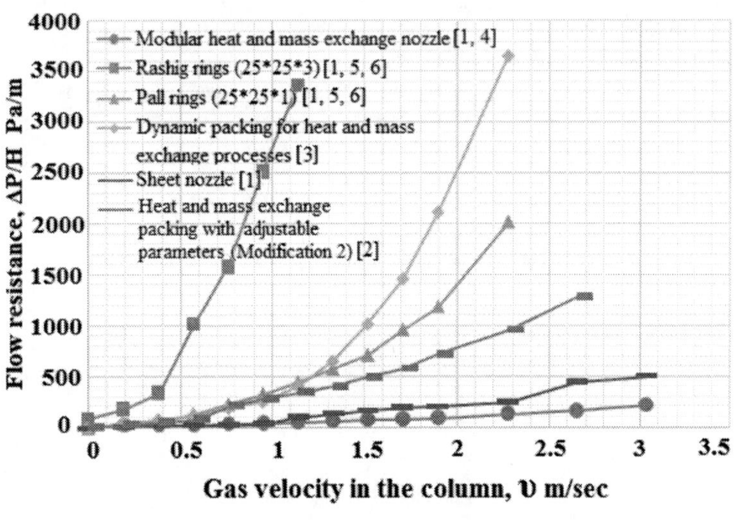

Figure 4. Relation between flow resistance of spray packings and gas velocity in the column

The result of the comparison of the packings is given in Figure 5.

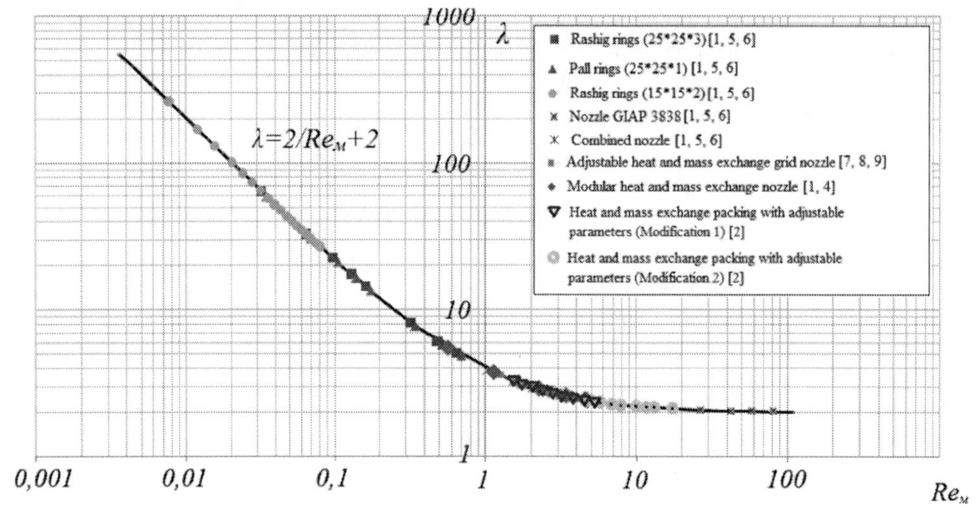

Figure 5. $\lambda = f(Re_m)$ diagram for packings with different structures [5, 6]

It is clear that wire heat and mass packings with adjustable parameters, porosity and specific surface, can have a very wide mode range because of packing density and possibility to change. The results suggest that dense adjustable packings (Modification 1) can be used for absorption, because such packings can ensure developed flow operation modes, even emulsification, though it provides a high flow resistance. While adjustable packing (Modification 2) has a very low flow resistance and implements a dripping-film liquid flow mode.

We are studying the retaining capacity of the packings with adjustable properties and we will prepare a comparative classification based on the experimental data. Retaining capacity means the capacity of the packing to accumulate a certain amount of liquid depending on the operation mode; it shows the total time of water retaining and mass transfer surface for a range of packings [7, 8]. Figure 6 shows diagrams depicting the retaining capacity determined experimentally for different packings depending on gas velocity in the column.

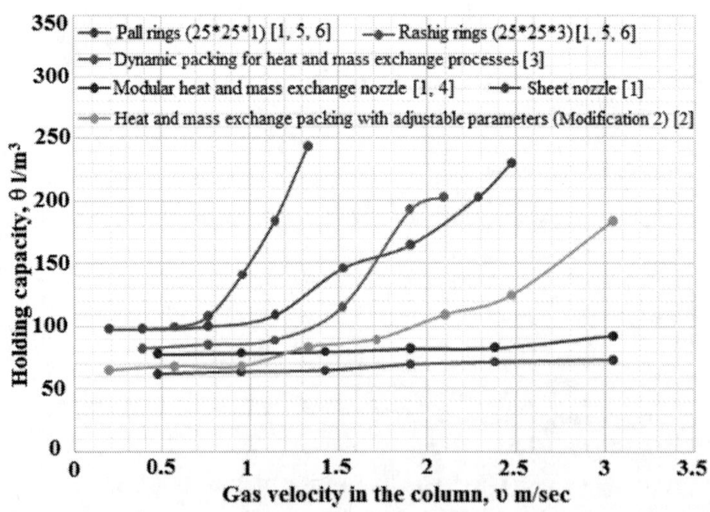

Figure 6. Retaining capacity of packing devices vs. gas velocity in the column

Correlation coefficient in all three experiments equalled 0.99 that means that the dependence $Y=A+BX+CX^2$ is direct and high. We obtained the dependence equations which are as follows:

$$Y=6{,}50072+0{,}1015X+0{,}003903659X^2 \text{ for a dry packing;}$$
$$Y=1{,}6906+1{,}0795X+0{,}018150219X^2 \text{ for irrigation packing;}$$
$$Y=410{,}736+2{,}21032X+0{,}038150271X^2 \text{ for retaining capacity (Figure 7).}$$

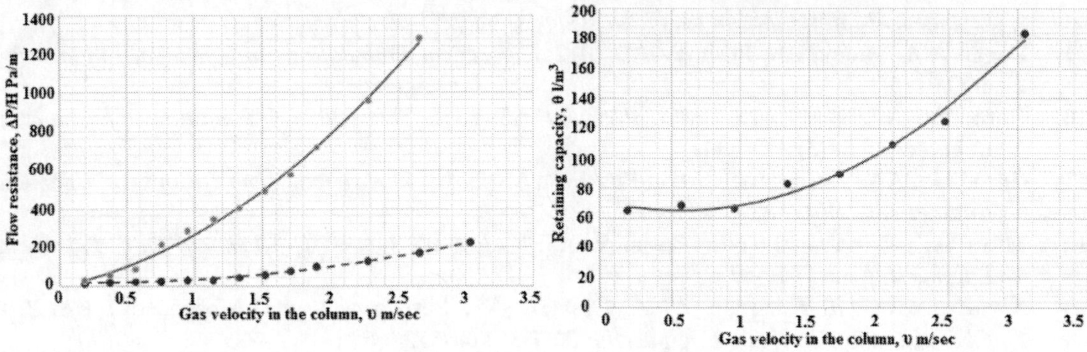

Figure 7. Deviation of experimental data from theoretical flow resistance of a dry (dashed line) and spray (solid line) studied packing and deviation of experimental data from theoretical retaining capacity of the studied packing (Modification 2) [2]

Conclusions

The obtained experimental data for the packings with adjustable porosity and specific surface and, consequently, changeable flow modes suggest that such packings are very promising for study and application, even though their analysis is complicated by conditions of creation, specific modes of cutting of metal working machines. It should be remembered that the material of such packings is waste material, though of very high quality from the point of view of material properties. Only stainless steel that is resistant to aggressive media and atmospheric wear are suitable. This is very important according to the specific character of the processes for which the packings are applicable. From our point of view ecologically and environmentally the application of such packings is an incredible process solution. The general idea and our aim is to create absolutely controllable heat and mass exchange packings that can automatically adjust to the operating flow modes; this adjustment can occur mechanically by compression and expansion, but with account of required parameters, such as heat and mass transfer coefficients, retaining capacity, response function, flow resistances.

References

[1] Merentsov NA, Lebedev VN, Golovanchikov AB, Balashov VA, Nefed'eva EE 2018 *IOP Conference Series: Earth and Environmental Science* **(115)** 9

[2] Golovanchikov AB, Vorotneva SB, Merentsov NA, Dul'kina NA, Zalipayeva OA, Sham'yanova AP 2012 *Patent* 117317 RU, IPC B 01 J 19/32 VSTU

[3] Golovanchikov AB, Vasil'ev PS, Ljapkov AV, Chjorikova KV, Topilin MV, Tishhenko PO 2016 *Patent* 160198 RU, IPC B 01 J 19/00 VSTU

[4] Golovanchikov AB, Sivolobova NO, Merentsov NA, Dul'kina NA, Shishljannikov VV, Dorofeeva NI 2012 *Patent* 129450 RU, IPC F 28 S 25/08 VSTU

[5] Golovanchikov AB, Balashov VA, Merentsov NA 2017 *Chemical and Petroleum Engineering* **53(1-2)**10-13

[6] Merentsov NA, Balashov VA, Orljankina JaA 2013 *Izvestija Volgogradskogo gosudarstvennogo tehnicheskogo universiteta, Ser. Reologija, processy i apparaty himicheskoj tehnologii (The news of Volgograd state technical University, Ser. Rheology,*

processes and devices of chemical technology) **1 (104)** 112-114

[7] Golovanchikov AB, Merentsov NA, Balashov VA 2013 *Chemical and Petroleum Engineering* **48(9-10)** 595-601

[8] Golovanchikov AB, Merentsov NA, Balashov VA, Orljankina JaA 2012 *Izvestija Volgogradskogo gosudarstvennogo tehnicheskogo universiteta, Ser. Aktual'nye problemy upravlenija, vychislitel'noj tehniki i informatiki v tehnicheskih sistemah (The news of Volgograd state technical University, Ser. Actual problems of management, computer science and Informatics in technical systems)* **10 (97)** 22-28

[9] Kagan AM, Yudina LA, Pushnov AS 2012 *Theoretical Foundations of Chemical Engineering* **46(2)** 165-171

[10] Gorodilov AA, Berengarten MG, Pushnov AS 2016 *Theoretical Foundations of Chemical Engineering* **50(3)** 325-334

[11] Gorodilov AA, Berengarten MG, Pushnov AS 2016 *Theoretical Foundations of Chemical Engineering* **50(4)** 422-429.

[12] Mitin AK, Nikolaikina NE, Pushnov AS, Zagustina NA 2016 *Chemical and Petroleum Engineering* **52(1)** 47-52

[13] Klyushenkova MI, Kuznetsova NA, Pushnov AS, Berengarten MG, Mokrousova EA 2014 *Chemical and Petroleum Engineering* **50(7-8)** 508-512

[14] Gorodilov AA, Pushnov AS, Berengarten MG 2014 *Chemical and Petroleum Engineering* **50(1-2)** 84-90

[15] Pushnov AS, Sokolov AS, Sidel'nikov II, Kurbatova EA, Mitrofanova EG 2014 *Chemical and Petroleum Engineering* **50(5-6)** 330-334

[16] Chizh, KV, Pushnov, AS, Berengarten MG 2014 *Chemical and Petroleum Engineering* **50(3-4)** 244-250

[17] Berengarten MG, Nevelson AO, Pushnov AS 2013 *Chemical and Petroleum Engineering* **48(11-12)** 723-729

[18] Shilin MV, Berengarten MG, Pushnov AS, Klyushenkova MI 2013 *Chemical and Petroleum Engineering* **48(9-10)** 608-614

[19] Mitin AK, Valdberg AY, Pushnov AS 2012 *Chemical and Petroleum Engineering* **48(1-2)** 50-53

[20] Tsurikova NP, Pushnov AS, Lagutkin MG, Shishov VI 2012 *Chemical and Petroleum Engineering* **48(1-2)** 3-8

[21] Kagan AM, Pushnov AS 2012 *Russian Journal of Applied Chemistry* **85(3)** 523-526

Control station based on synchronous motors

T.I. Petrov, A.R. Safin, I.V. Ivshin, L.V. Dolomanyuk, M.F. Nizamiev

Kazan State Power Engineering University, Kazan, Russia

tobac15@mail.ru

Abstract. The oil and gas industry is one of the leading branches of Russia, the basis of the economy. Therefore, it is obvious that increasing the technical and economic characteristics of equipment in the field of oil production is a priority for the scientific community. And there are several solutions aimed at solving this problem: use of intelligent control systems for electric drive; replacement of asynchronous motors with synchronous ones. With the integrated use of these points, it is possible to reduce energy consumption, as well as increase the percentage of extracted resource. Numerical values are individual for each particular variant, but savings of 25-35% are achieved.

1. Introduction

The introduction of intelligent energy-resource-efficient control stations for oil pumping machines is the key to solving the existing problem - optimization of oil production.

According to the Energy Strategy of Russia for the period until 2030: "The strategic goal of the state energy policy is the creation of a sustainable national innovation system in the energy sector to provide the Russian fuel and energy complex with highly effective domestic technologies and equipment, scientific, technical and innovative solutions in the amounts necessary to maintain energy security of the country ".

The development and use of a new type of energy-saving technologies in the electric power industry in Russia can be achieved only due to several conditions:

- the achievements of international science should be taken into account;
- technologies must satisfy modern trends, and correspond to forecasts in this area.

A large share in the cost of oil is the cost of electricity and maintenance of the energy complex. They reach up to 25-35% of the cost of oil.

The key tasks in the oil and gas industry are to increase the efficiency of the use of technological power equipment and to develop measures and technical means to increase the energy efficiency of the use of equipment.

At the moment, the international market offers a large selection in the field of electric drive control for rocking machines. In Russia, various solutions are also presented in the field of intelligent control systems. They are distinguished by control algorithms, data processing options, interfaces and others. It is important to note that for optimized asynchronous motors, the optimization will not reach the specified savings, due to the lack of wide possibilities for regulation, which does not allow to fully realize the advantages of systems of intelligent control systems. Also for rocking machines characterized by an uneven load regime, which reduces the useful power of induction motors.

Synchronous electric drives with permanent magnets on the rotor provide more economical consumption of energy resources, including, through the organization of feedback to the electric drives of other machines. An engine of this type is a system included in an adjustable electric drive, and is used not only for energy conversion, but also for changing the parameters of the rocking machine.

"Forecast of scientific and technological development of Russia's fuel and energy industry for the period until 2035" indicates the need, in the medium term, to optimize the cost price of heavy and viscous oil production in areas with developed infrastructure without foreign technologies. Consequently, the development and production of domestic electronics to reduce production costs with the help of "rocking machines" can be a priority.

2. Theoretical position

We should briefly dwell on the principle of synchronous machines.

The main property of synchronous machines is the field of the rotating part of the motionless relative rotor itself and performs a rotational motion together with it. The intrinsic excitation region is used in magnetic and electric machines, and the field penetrating the rotor in synchronous-reactive ones.

In the development of electric drives, the classification of synchronous machines is associated not only with the physical, design features of the machine itself, but rather with the characteristics of use, control algorithms and the construction of control stations.

A synchronous motor with permanent magnets is an electric machine with alternating current. The stator of the engine is structurally similar to the static part of induction motors containing a three-phase type of winding.

There are magnets on the moving part, depending on the position, there are only 2 types of synchronous machines: motors with magnets located on the rotor itself, in the English literature called a surface motor with permanent magnets, and motors with magnets located directly in the steel structure of the rotor. The synchronous motor is usually designed for a small power, which does not exceed hundreds of kW (7.5 - 60 kW), and is most often used in servo drives in combination with frequency converters.

2.1. Management methods

The direction of developing principles for controlling synchronous machines is determined by the improvement of the hardware component of the electric drive: the characteristics of the electric motor, the type of the converter, the presence of various sensors, and the parameters of the controllers [1].

The fundamental factors for the realization of the control station of the electric drive are the use (non-use) of the magnetic state sensors, speed and rotational angle of the rotor. For this reason, the methods are classified into a view using a rotor position sensor and a non-sensorless view, called "sensorless control".

In the standard drive on the basis of synchronous motors, a position sensor is placed on the motor shaft [2]. Types of sensors: Hall sensor, encoder, resolver, reductosin, etc. The control system is quite simple in this case, but the machine becomes less constructive, which reduces the reliability of the system.

Significant advances in the field of relevant intelligent management systems, especially the improvement of situation observers, led to the creation of electric drives that do not use rotor position sensors. In such systems, the position sensor is not available, and the necessary information for implementing the control algorithm comes from the status observer. A special role is played by such systems in those electric drive systems where the installation of a sensor is simply impossible. The advantages of systems that do not use these sensors are reliability, cheapness, and compactness.

Most control methods that do not use the rotor position sensor, described both in domestic and foreign literature, are based on mathematical models of electromagnetic processes occurring in AC motors. In such methods, the calculation of the velocity estimate is combined with the calculation of the module and the direction of the electromotive force of the moving part of the engine [3].

The most common control systems for synchronous motors are the method of direct control of torque and vector [4]. The direct control method works better for monitoring and eliminating interference, and the vector method loses response speed to changing characteristics and requires the use of more energy-consuming pulse width modulation. But it is worth noting that the vector control system also has its advantages in static modes: lower power costs and a lower ripple ratio. Consequently, this method is preferred in drives with stable load and control. Since this paper analyzes the drive for rocking machines, a choice will be made to the direction of the vector system as a method for controlling the synchronous machine, since more energy efficient modes of operation are achieved due to low levels of harmonics, in contrast to the direct control method [5].

2.2. Management system model

The model of a synchronous machine is usually considered in a rotating coordinate system (d, q), while combining the axis d along the axis of the rotor. In this case the system will look like:

$$u_d = R_s i_d + \frac{d\psi_d}{dt} - \psi_q \omega$$

$$u_q = R_s i_q + \frac{d\psi_q}{dt} - \psi_d \omega$$

$$\psi_d = L_d i_d + \psi_f$$

$$\psi_q = L_q i_q$$

$$M = \frac{3}{2} p(\psi_d i_q \times \psi_q i_d)$$

$$J\frac{d\omega_m}{dt} = M - M_H - \beta\omega_m$$

When working with an uncontrolled drive of a rocking machine, the characteristics of changing the location of the suspension point of the rod column can be determined through the geometrical dimensions of the machine elements, but it is assumed that the speed of rotation of the rocking arm will not change for a length of time equal to the period of rocking motion of the rocking machine. From this point of view, a large number of authors and in sufficient detail has investigated the kinematic system of the machine. In these studies, it was assumed that the law of the change in the location of the rocking crank (given that the rotational speed is a constant) was set, and based on this, the characteristics of the displacement of the remaining elements of the machine were determined. And based on what the dynamic forces in the links of the pumping unit were determined, and, accordingly, the electric power parameters of the drive of the machine.

When using a converter that is compatible with the swing drive, it is possible to optimize the characteristics of the suspension point. This optimization is achieved due to the fact that the frequency of rotation of the rocking engine changes, and as a result the maximum values of dynamic forces in the links of the whole machine decrease. At the same time, it is necessary to determine the law of displacement of the suspension point, as a set of laws for converting the engine speed and the movement characteristics of the rocking crank.

Thus, in the process of further development of the kinematic model of the machine, it is necessary to solve the following tasks:

1. to determine the parameters for moving the suspension point with certain rules for cranking the machine when the instantaneous value of the engine speed of the pumping unit is inconstant (let us call the direct movement);

2. determine the characteristics of the cranking motion of the rocking machine under the given laws of displacement of the suspension point (let us call the return motion).

Calculation of the characteristics of the crank motion of the machine for certain laws of displacement of the suspension point (this calculation is necessary to determine the values of the return motion)

Suppose that the law of displacement of the suspension point is given, in this case it is necessary to determine the parameters of the crank motion of the machine-the angle of rotation, the angular velocity and the angular acceleration (determine the equations of motion of the rocker crank), that is:

$$\phi_{KP} = f(S_A)$$

$$\omega_{KP} = f(S_A, v_A)$$

$$\varepsilon_{KP} = f(S_A, v_A, \alpha_A)$$

When we find the characteristics of the crank motion of the machine, we assume in the sequel that the positive direction of the displacement characteristics of the suspension point is directed downward (in the direction of acceleration of gravity). In addition, as the zero angle of the rotational motion of the rocking crank (origin), take the angle of the crank, at which the suspension point is located at the bottom dead center. In this case, the moment is considered when the crank of the rocking machine rotates clockwise.

2.3. Control station

A system simulating a pumping station control station based on high-performance synchronous motors with permanent magnets.

The model of the control station under consideration includes the following interrelated elements:

- model of a synchronous motor with permanent magnets, which describes the mechanical and electrical processes in the drive of the machine;
- block that simulates a system based on vector control, and describes the changes in the drive;
- models of kinematic problems;
- converter of values from a standard algorithm to a rotating one;
- model characterizing the "column-pump installation - liquid level";
- dynamic model of the machine, which describes the dynamic processes in the links of the installation, provided that the value of the speed of the engine shaft of the pumping unit is not constant;
- controller, necessary for implementation of sensorless control method.

3. Conclusion

The article demonstrates the relevance of this subject, based on a comparison of electric drive systems with synchronous and asynchronous motors, from an economic and practical point of view, taking into account the features of the use of electrical machines in the oil industry.

Possible areas of application of data and management methods are considered, with an overview of the solutions and advantages and disadvantages of each system. The components of the system model are included, including a synchronous motor model (rotating coordinate system) and a rocking machine (solutions of the direct and inverse kinematics problem), in the form of mathematical equations, based on which the management system is constructed.

Support

The publication of the article was carried out within the framework of the project "Creation of a series of electric drives on the basis of Russian high-performance synchronous engines for oil pumping machines using wireless data transmission systems and an adaptive control system for smart deposits", Agreement No. 074-11-2018-020 with the Ministry of Education and Science of the Russian Federation of May 30, 2018.

References

[1] Weinger A.M. Adjustable electric drives of alternating current. M .: 2009. P. 102. (in Russian)
[2] Kalachev Yu.N. Observers of the state in a vector electric drive. Moscow: 2013. P. 63.

(in Russian)
[3] Pankratov VV, Zima E.A. Energy-optimal vector control of asynchronous electric drives. Novosibirsk: Publishing House of the National Technical University, 2005. P. 120. (in Russian)
[4] Vas. P. Sensorless Vector and Direct Torque Control. Oxford: Oxford University Press, 1998. P. 376.
[5] Kalachev Yu.N. Vector regulation (practice notes). EF: 2013. P. 63. (in Russian)

IOP Conf. Series: Earth and Environmental Science **288** (2019) 012112 doi:10.1088/1755-1315/288/1/012112

Modeling of pyrolysis process of high-moisture coals

M. V. Pyatygina, A. A. Galkeeva, G. R. Mingaleeva

Kazan State Power Engineering University, Russia, Kazan, Krasnoselskaya Str., 51

pmv_83@mail.ru, prujinka1990@mail.ru, mingaleeva-gr@mail.ru

Abstract. The research is aimed at solving the fundamental scientific task of creating scientific fundamentals, deep processing of hydrocarbon raw materials (preliminary gasification, conversion, pyrolysis of solid minerals and renewable combustible resources) to produce liquid and gaseous fuels and a total increase in efficiency.

Modern lines of research in the field of thermal processing of solid fuels

In order to create efficient technological schemes and productions from the point of view of energy and resource saving, as well as increasing efficiency through the integrated use of the obtained products of processing hydrocarbon fuels, it is necessary to investigate the physico-chemical processes of decomposition of the organic mass of solid fuels, which depend to a large extent on its structure, range of temperatures, the stages of decay of compounds that make up the organic mass of raw materials and the assessment of the effect of these or other intermediate substances on the final quality and quantitative composition of the products obtained.

Investigation of the physicochemical basis for the deep processing of hydrocarbon raw materials will allow to create a methodology for calculating the technological schemes that realize these processes and to increase the energy efficiency of hydrocarbon processing.

Thermal and thermochemical processing of organic fuels is realized through such processes as gasification, pyrolysis (coking and semi-coking), hydropyrolysis, hydrogenolysis and others.

The phase and elemental compositions of brown coal and peat are studied with a wide application of such complex analysis methods as X-ray diffractometry, X-ray fluorescence and ICP-MS spectrometry supplemented with thermal gravimetry, elemental analysis of organic matter (OB) and scanning electron microscopy, Fourier transform, extraction organic solvents of different polarity, cryoscopy [1, 2]. In [3, 4], the authors noted the effect of the coal composition on structural parameters such as the presence of oxygen-containing groups, the hydrogen content in coal, and the results of thermal studies of coal and the effect of such methods of its processing as alkylation, solubility and mechanical activation on thermal degradation of coal.

Based on the obtained model of the average statistical unit (CCE) using different density and basis functionals and additional study of different CCE, calculation of IR spectra, total energy values and reactivity parameters, allowing comparison with experimental data [5, 6], the author compared on the structural characteristics of the 3D model of the so-called CCE organic mass of coal (OMC) $C_{29}H_{28}O_2$ with the molecular model of coal $C_{100}H_{79}O_7NS$. Thus, a volumetric molecular model of coal was obtained, which describes in more detail the structure of real coal, and takes into account, in contrast to the CCE OMC model, the presence in coal of such chemical elements as nitrogen and sulfur. The author also noted that, in spite of the lack of complete coincidence in the elemental composition with any of the technical coal grades given, the concentration of chemical elements in the 3D model of

$C_{100}H_{79}O_7NS$ is within the limits typical for coal, and as a result, this model in terms of structural composition best corresponds to the known experimental data for coals.

The authors of the research [7] present mathematical dependencies between the indicators characterizing the electronic and molecular structure of the organic mass of solid fuels of the series: wood - peat - brown coal - coal - anthracite. These equations are a theoretical tool for studying the processes of deep processing of solid fuels. The electronic structure of coals largely determines the variety of their properties and reactivity, as well as the features of the mechanisms of various reactions. It is shown that the parameters characterizing the electronic structure of coals have a close relationship with their molecular structure.

Confirmation of the dependence of the yield of polycyclic aromatic hydrocarbons (PAH) and phenol on the structure of coal was obtained by the authors of [8] when investigating the yield of two organic compounds (PAH and phenols) liberated during the pyrolysis of coals of various classes. So PAHs are formed not only by thermal destruction of the coal structure, but also by the combination of various low-molecular compounds with each other.

According to the author of the paper [9], in order to understand the mechanism of transformations of the organic mass of coals (OMC) and to assess the changes in their properties during the modification, an analysis of the initial stages of destruction, the most informative by the methodology of nonisothermal kinetics, which allows controlling the process from the initial stages to deep OMC conversions. The author studied the influence of methods based on the principles of chemistry of extreme effects (plasma chemistry, mechanoactivation, radiolysis, supercritical extraction, etc.) on the fuel structure change, and models of structural fragments of the investigated coals were constructed.

Thus, to date, a sufficient amount of experimental data on the composition of fuels obtained by various methods of physical and chemical analysis has been accumulated in order to obtain volumetric models of the molecular structure of fuels, as well as to find mathematical relationships between the initial fuel composition and the resulting products as a result of it pretreatment and further deep processing. Experimental data obtained by methods such as Raman spectrometry, thermogravimetry, gas-liquid chromatography, etc., are used as a basis for modeling and isolating groups of compounds, and for the mathematical description it is proposed to use numerical methods of quantum chemistry that allow obtaining a three-dimensional model of a specific fuel and structural parameters of the incoming in its composition of compounds.

The model of distributed activation energy, used for the pyrolysis of a number of materials (including coal, biomass, residual oils and kerogen), considered in [10], suggests that the thermal decomposition of numerous components is described by the distribution of activation energies.

In [11], the authors investigated the process of low-temperature pyrolysis using the Raman spectroscopy method. The change in the composition curves of different types of fuel after the pyrolysis process in a given temperature range is shown.

The author's work [12] is the closest to the concept of research. It shows a scheme for the usual pyrolysis of coal with the possible use of intermediate products. Optimal models of coal structure are considered, in the form of structural fragments and in the form of blocks of molecules of certain groups of compounds. The mechanism of pyrolysis from the point of view of molecular composition is explained. The kinetic parameters of the coal decomposition process are given, and the curves of the dependence on the process temperature are constructed. It is indicated the selection of the main products by temperature ranges.

On the basis of the analysis of the methods for studying decomposition processes of solid fuels for energy production and various chemical products, it can be concluded that attention is not sufficiently paid to the study of the chemical nature of the mechanism for the formation of groups of compounds, since the data are given only with respect to the influence of various components on the yield of the final product and general methods for calculating associated models of evaporation and combustion.

At the present stage of research in this field, not only the physicochemical aspects of thermal decomposition of hydrocarbons are revealed, but attempts are also being made to describe the mathematical and create the appropriate calculation methods.

Thus, the enthalpy of dissociation of the homolytic bonds of various bonds (CH, CC, CO and OH) for model coal compounds, which are characteristics of the functional components present in coal, were calculated in [13] using the numerical method of quantum chemistry of the two-hybrid method mPW2PLYP. The authors confirm the hypothesis of splitting the bonds of CO and CC with the formation of phenoxy radicals or benzyl radicals at the stage of initiation of coal pyrolysis.

A mathematical model based on the equations of chemical thermodynamics, which make it possible to calculate the composition of combustion products under conditions of a non-optimal fuel-oxidizer ratio, and the computational method for determining the composition of the fuel in the combustion process, were proposed in [14].

On the basis of the thermodynamic analysis of the efficiency of the use of various types of fuel and the calculation of the thermal efficiency, useful work, and the introduction of a fuel energy utilization factor, the authors of [15] obtained a result showing that low-calorie fuels, from the point of view of thermodynamics, are much more effective at steam turbine power plants , than high-calorie.

Modeling of the process of pyrolysis of water-coal suspension (WCS)

Analysis of the current state of scientific research in the field of thermal processing of hydrocarbon fuels shows that any raw material containing organic substances can be studied by the same methods as for solid fuel, since the structure also contains the basic elements - carbon and hydrogen. The difference will be only in the set of decomposition reactions and the applied regime parameters.

The physicochemical regularities of pyrolysis of water-coal suspensions of the most widespread brown coal - Irsha-Borodinsky, having the following composition: W^r=33%; A^r=9,98%; S^r=0,29%; C^r=41,2%; H^r=2,89%; N^r=0,6%; O^r=12,1%. The ratio of water to fuel: 40% - water, 60% - coal.

Pyrolysis of water-coal suspension

The principal difference between the construction of the thermal decomposition model of the water-coal suspension is that the reaction of water vapor with organic compounds of coal containing oxygen, nitrogen and sulfur is considered. According to the WCS decomposition model, moisture is first released, then products (moisture, volatile components, coke residue) are released from the WCS with a dried membrane. From the coke residue with further heating, coke is released, loose products from the heating surface, mono- and carbon dioxide.

Further, the possibility of decomposition reactions of these compounds was calculated on the basis of Gibbs energy values taking into account the bond breaking energy:

a) Primary disintegration:

Oxygen-containing: E = 230 kDj / mole - decomposition of acid carboxylic with the release of benzene, phenol, derivatives of toluene, naphthalene, and oxide carbon and water (collateral);

E = 296 kDj / mole - decomposition of i-propylphenol to form phenol and propane (collateral);

E = 377 kDj / mole - decomposition of 1,1-biphenylmethanol with the formation of biphenylmethane and water (collateral);

E = 384 kDj / mole - decomposition of crezols with the formation of phenol and methane (collateral);

E = 460 kDj / mole - decomposition of naphthols with the formation of naphthalene and water (collateral);

Sulfur-containing: E = 230 kDj / mole - decomposition of bibenzysulphide with the formation of bibenzyl and hydrogen sulphide (collateral);

Azot-containing: E = 356 kDj / mole is the degradation of 9H-carbazole, 2,3-bimethyle-1H-indole, pyrrole derivatives with the formation of biphenyl, 3-methylpropylbenzene, 1,2,4-trimethyl-3-ethylbutane, benzobutane and ammonia (collateral).

b) Second disintegration:

E=384 kDj/mole – decomposition of derivatives of toluene, naphthalene, benzene, benzobutane with the formation of benzene and methane, butane, water (collateral).

Further, for the obtained reactions according to the model (Fig. 1), the rate constant and the decomposition time are determined (Fig. 2), the yield of the components - coke, gas, tar - is estimated.

Calculation of the yield of decomposition products of organic WCS was carried out. The calculated yield of coke was 58%, coke oven gas - 38.9%, coal tar - 1.8%, crude benzene - 0.8%, ammonia - 0.1%, while for ordinary coal particles these yields are 82 %, 15%, 2.5%, 1.2% and 0.2%, respectively.

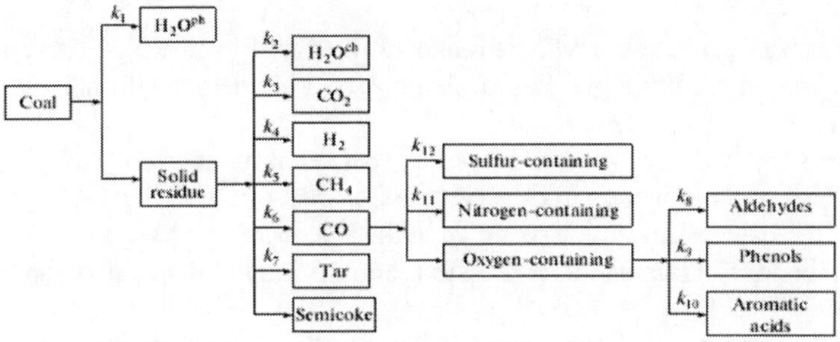

Fig. 1. Model of the kinetic parameters of the thermal decomposition (pyrolysis) of WCS

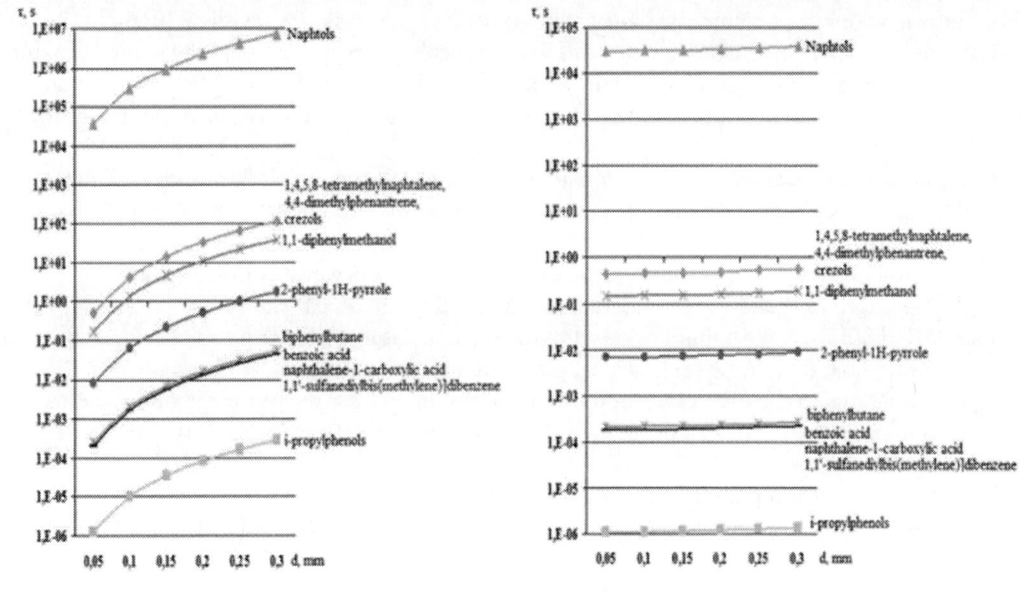

a) coal b) water-coal suspension

Fig. 2. Decomposition time of OMC compounds

Conclusions

The presence of certain substances in the mixture forming the organic mass of the fuel was revealed, modeling of the decomposition of the water-coal mixture during heat treatment was carried out.

The decomposition time of OMC for the diameter of the coal particle (d = 0.05-0.3 mm) is established, which increases with increasing diameter of the coal particle.

The greatest value of the endothermic effect is the reaction with allowance for di- and tetramethylarenes, as a result of which it is allowed to allow their withdrawal at 400 ° C and after cooling to 300 ° C of decomposition to another part of the coal organic mass to 550 ° C. And at the same time, there is a gas of methane. The first to dissociate compounds containing COOH and -S- groups.

The obtained data on the mechanism of decomposition of a drop of a coal-water suspension allow to carry out the pyrolysis process more efficiently and to regulate the yield of the target products due to the variation of the water content in the WCS and the regime parameters of the process.

References

[1] Podgaeczkij A.V., Vorob`eva I.M., Petrenko D.B., Samojlova E.K. Sostav i texnologicheskie svojstva e`nergeticheskix uglej Pavlovskogo mestorozhdeniya (Primor`e) // Otechestvennaya geologiya. №3. 2016. S. 13-20.

[2] Poloveczkaya O.S., Bojkova O.I. Ximicheskij sostav organicheskoj massy` torfa Yaroslavskoj oblasti // Uspexi sovremennogo estestvoznaniya. №2-0. 2016. S. 61-64.

[3] Kulikova M.P. Issledovanie svojstv uglya Ulug_Xemskogo bassejna // Vestnik Tuvinskogo gosudarstvennogo universiteta. №2 Estestvenny`e i sel`skoxozyajstvenny`e nauki. 2015. S. 108-118.

[4] Kulikova M.P. Issledovanie fiziko-ximicheskix svojstv Ulug-Xemskix uglej // E`nergetik. №8. 2014. S. 29-32.

[5] Poroxnov A.N. Model` srednestatisticheskoj strukturnoj edinicy OMU: analiz struktury` // Sbornik statej mezhdunarodnoj nauchno-prakticheskoj konferencii «Rol` i mesto informacionny`x texnologij v sovremennoj nauke». Saransk. 03 fevralya 2016g. S. 3-4.

[6] Poroxnov A.N. Molekulyarny`e modeli organicheskoj massy` uglya: sravnenie strukturny`x xarakteristik // Sbornik statej Mezhdunarodnoj nauchno-prakticheskoj konferencii «Nauchny`e issledovaniya i razrabotki v e`poxu globalizacii». Kirov. 05 fevralya 2016g. S. 12-14.

[7] Moskalenko T.V., Mixeev V.A. Matematicheskoe modelirovanie e`lektronnoj struktury` uglej // Vestnik Irkutskogo gosudarstvennogo texnicheskogo universiteta. №6 (89). 2014. S. 158-161.

[8] Correlation between coal structure and release of the two organic compounds during pyrolysis. Yan L., Bai Y., Zhao R., Li F., Xie K. Fuel. 2015. Vol. 145. P. 12-17.

[9] Patrakov Yu.F. Fiziko-ximicheskie prevrashheniya uglej v neizotermicheskix usloviyax i metody` ix aktivacii: avtoref. dis. … d-ra xim. nauk. Kemerovo, 2005.

[10] Approximations to the distributed activation energy model for the pyrolysis of coal. Please C.P., McGuinness M.J., McElwain D.L.S. Combustion and Flame. 2003. Vol. 133. Issues 1-2. P. 107-117.

[11] FT-Raman spectroscopic characterisation of chars from the pyrolisys of coals of varying rank. Li Xiao-jiang, Li Chun-zhu. Journal of Fuel Chemistry and Technology. 2005. Vol. 33. Issue 4. P. 385-390.

[12] Pyrolysis of Coal. Radenovic A. Journal of Chemists and Chemical Engineers. 2006. 55. (7-8). P. 311-319.

[13] A theoretical study on bond dissociation enthalpies of coal based model compounds. Li L., Fan H., Hu H. Fuel. 2015. Vol. 153. P. 70-77.

[14] Brunetkin A.I., Bondarenko A.V., Ly`syuk A.V. Opredelenie kolichestvennogo sostava neizvestnogo gazoobraznogo topliva i produktov sgoraniya na osnove izmerenny`x texnologicheskix parametrov // Trudy` Odesskogo politexnicheskogo universiteta. №1. 2014. S. 127-136.

[15] Nesenchuk A.P., Beglyak A.V., Ry`zhova T.V., Shklovchik D.I., Yarmol`chik M.A., Kachar L.I. K ocenke e`ffektivnosti ispol`zovaniya organicheskix topliv v cikle parosilovy`x ustanovok // Izvestiya vy`sshix uchebny`x zavedenij i e`nergeticheskix ob``edinenij SNG. E`nergetika. №1. 2013. S. 56-60.

IOP Conf. Series: Earth and Environmental Science **288** (2019) 012113 doi:10.1088/1755-1315/288/1/012113

Automated control system for coolant parameters with remote access

I.M. Safarov[1], D.I. Khamatkhanov[2]

[1]6 Volgogradskaya street, apt 48 Kazan, RF 420080

[2]34 Akademika Koroleva street, apt 13 Kazan, RF 344091

ildarsafarov@mail.ru

Annotation. This article proposes an automated system for monitoring and controlling the parameters of the coolant in the housing and communal services as part of the energy saving program. The proposed system is based on industrial controllers with Ethernet 100 Base-T, RS-232 and RS-485 interfaces, as well as software created in the CoDeSys software environment.

Introduction

The use of automatic control systems in various branches of the national economy makes it possible to solve increasingly complex production problems. In particular, this is also true for automatic systems for maintaining the temperature in hot water supply, return water in air intake ventilation systems, heating systems to maintain the temperature graphs.

For such systems, there are many particular solutions that allow to change the parameters of the transported medium [1]. One of them is the analog automatic control system of the shut-off and regulating valve (SORV).

Description of the problem

In this kind of systems the actuator is controlled by analog signals. The actuator's adjustment factors may depend on the actuator manufacturer, on the position sensor integrated in the valve or on the external sensor of the controlled value [2]. This leads to the inseparability of the system. It is also worth noting that not all such systems have the ability to integrate remote control. They require the use of a multiple transforming blocks that complicate the structure of the system.

The use of modern automatic control methods allows us to solve this problem, which makes it possible to substantially simplify the system. One of the solutions to this problem is the use of a proportional-integral-differentiating-regulator (PID controller) for controlling a valve with an electric drive (Figure 1) [3].

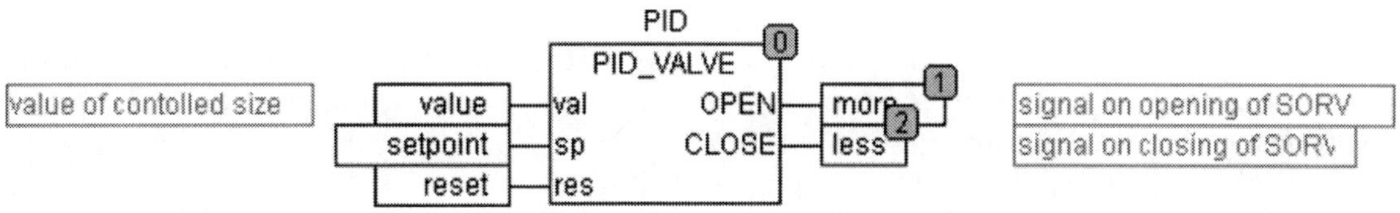

Fig. 1. PID controller for valve control

Relevance

The relevance of the proposed solution is that the system is universal and allows to remotely control the setpoint for a controlled value (temperature, drop, pressure, etc.), as well as adjusting coefficients of PID controller. This solution allows to integrate the system into the existing outdated SORV without rod position sensors by changing the program or adding a new programmable logic controller to the old system. That saves time and resources for system reinstallation. Another one advantage of this solution is fast and high-quality process control by using PID block.

Structure and description of system

The functional block (FB) PID_VALVE consists of OWEN libraries of Util.lib [5, p. 351] and PID_regulators.lib [5, p. 356] with a proprietary code (fig. 2).

By means of PID FB regulation of system's parameters is carried out. The LIN_TRAFO block linearly transforms values (range -100...100) from Y out of PID blocks to value that is acceptable for the VALVE_REG_NO_POS block by IN_VAL (range 0…100). The SORV control unit without rod position sensor VALVE_REG_NO_POS is block where instead of information about the time of rod's full speed of SORV [6] is used. DIG_FLTR is the digital filter for analog values of controlled value

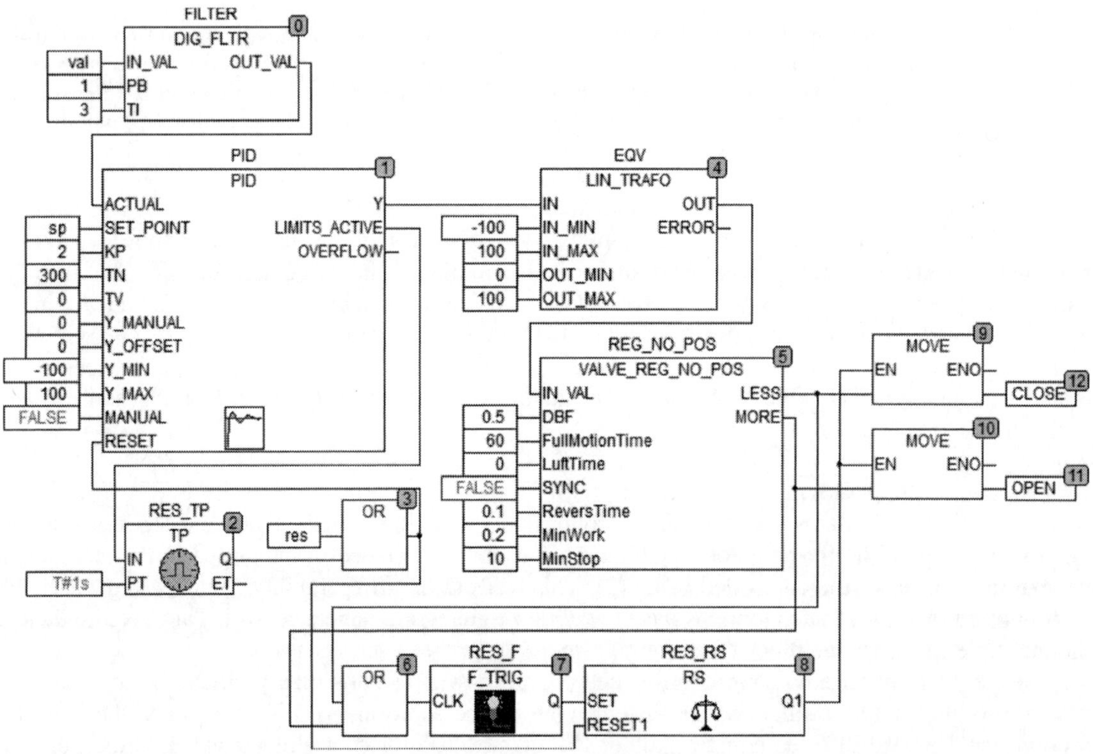

Fig. 2. Structure of the function block of PID_VALVE in the CFC editor

In fig. 3 the following variables are shown: val – the current value (pressure, temperatures, pressure difference etc); sp – setpoint (it is set in units of measure of controlled value); res – reset of an integral component of the PID-regulator.

```
    FUNCTION_BLOCK PID_VALVE
VAR_INPUT
    val, sp         :REAL; (*current value, SORV rod position, setpoint*)
    res             :BOOL;
END_VAR
VAR_OUTPUT
    OPEN, CLOSE     :BOOL; (*opening and closing signals sending to SORV*)
END_VAR
VAR
    PID             :PID;
    REG_NO_POS      :VALVE_REG_NO_POS;
    FILTER          :DIG_FLTR;
    EQV             :LIN_TRAFO;
    RES_F           :F_TRIG;
    RES_TP          :TP;
    RES_RS          :RS;
END_VAR
```

Fig. 3. Source code

Process of regulation of temperature in heating system where to the val input receives data from a temperature sensor that is sends the central heat distribution station (fig. 4) is simulated. The constant of proportionality, an integration constant and a constant of derivation of the PID block are individual and are selected empirically path [8]. The setpoint was 63 C °. Coefficients and values were taken from actual working system and used as input data.

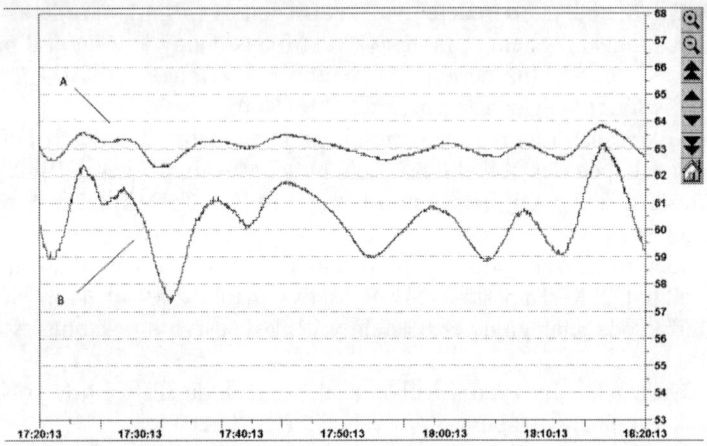

Fig. 4. The graph of temperature maintaining in the CoDeSys 2.4 environment (a counting discretization is d=250 ms) A – the output values received from a temperature sensor. B – input values

Economic justification of system

Table 1. Approximate expenses for modernization a system of one object

SORV is not changed			SORV is changed		
The required resources	Working hours	Payment (rub.)	The required resources	Working hours	Payment (rub.)
Engineer	3	600	Engineer	3	600
			Mechanic	6	1 200
			SORV		20 000
			Logistics	3	600
			Accounts department	3	600
Total		600			23 000

To calculate of compensation of the personnel participating in carrying out the following formulas was used:

$$\frac{A_S}{A_{WHM}} K_{SN} K_{AS}$$

where A_S is the average salary in the Republic of Tatarstan (31 500 rub.); A_{WHM} is average number of working hours in one month; K_{SN} is the coefficient of assignments on social needs; K_{AS} is the coefficient considering the additional salary.

Conclusion

Thus, proposed system allows to considerably reduce working personnel for serving one object (see tab. 1). Besides, the efficiency of monitoring significantly increases, and the access to data is no more than 3 minutes.

References

[1] Novikov D.A. Teoriya upravleniya organizatsionnymi sistemami. vol 2. M.: Fizmatlit, 2007. p. 604.

[2] Gurevich D.F. Truboprovodnaya armatura. Spravochnoe posobie. vol 2. L.: Mashinostroenie, 1981. p. 368.

[3] Abramov K.B. Metodika opredeleniya koeffitsientov PID-kontrollera pri modelirovanii avtomatizirovannykh system upravleniya rektifikatsionnoy kolonnoy s primeneniem paketa ChemCAD // Inženernyj vestnik Dona. 2011. №2. URL: http://www.ivdon.ru/magazine/archive/n2y2011/444.

[4] Truboprovodnaya armature s avtomaticheskim upravleniem / Gurevich D.F., Zarinskiy O.N., and Kosykh S.I.; pod red. Kosykh S.I. Leningrad: Mashinostroenie, 1982. p. 320.

[5] Rukovodstvo polzovatelya po programmirovaniyu PLK v CoDeSys 2.3. Smolensk: PK Prolog, 2006. p. 453.

[6] Kapustin N.M. Avtomatizatsiya proizvodstvennykh protsessov v mashinostroenii: Ucheb. Dlya Vuzov / Pod red. N.M. Kapustina. M.: Vysshaya shkola, 2004. p. 415.

[7] MBA8 Modul' vvoda analogoviy izmeritelniy: Rukovodstvo po ekspluatastii. M.: Atlas-Press, 2008. p. 90.

[8] Vadutov O. Nastroyka tipovykh regulyatorov po metodu Tsiglera-Nikolsa. Metodicheskiye ukazaniya k vypolneniyu laboratornoy raboty. Tomsk. 2014.

[9] Sofieva Iu.N. and Abramov K.V. Primeneniye paketa modeliruyuschikh programm ChemCAD v uchebno-trenirovochnykh kompleksakh dlya izucheniya system avtomatizatsii rektifikatsionnykh // Inženernyj vestnik Dona. 2012. №1. URL: http://www.ivdon.ru/ru/magazine/archive/n1y2012/619.

[10] Matveeva L.L. Informatsionnye tehnologii v menedjmente. M., 2010. p. 187.

[11] McGraw-Hill, Sybil P. Parker Dictionary of Scientific and Technical Terms. 6th Edition. New York: Merck, 2002. p. 2380.

[12] Grigoryuk E.N. and Bulkin V.V. Problems of Automation and Management Principles. Information Flow in Manufacturing // IOP Conference Series: Materials Science and Engineering, 2017. Volume 221, conference 1. URL: https://doi.org/10.1088/1755-1315/221/1/012006.

IOP Conf. Series: Earth and Environmental Science **288** (2019) 012114 doi:10.1088/1755-1315/288/1/012114

Quick detection of gusts and its location in the reservoir pressure maintenance system

Safin M., Shakirova K., Mirgaleeva A.

Federal State-Funded Educational Institution of Higher Professional Education "Kazan State Power Engineering University", Krasnoselskaya street, 51.

cmvorkut@mail.ru, karinashakirova@mail.ru, mirgaleeva94@mail.ru

Abstract. The article devoted to theoretical and methodical materials based on the researches conducted within trial tests reverse ultra-sonic flow meters on different areas of exploited oil fields.

Introduction to the system of maintenance of reservoir pressure

Most of the largest produce fields in Russia are exploited at a late stage of production. Residual reserves of such deposits are classified as hard-to-recover. According to various estimates, proportion of hard-to-recover reserves today exceeds half with water cut over 80%, and the average of oil recoverability factor (ORF) is between 20% and 43% in Russia. The most effective way to maintain of oil production from produce fields that are at a late stage of production is the method of waterflooding the oil deposit, which is carried out by the reservoir pressure maintenance system (RPM). The main task of RPM system is the maintenance of reservoir energy, through which fluid is withdrawn from the collector of reservoir rock through exploitation well.

The RPM system is a complex engineering infrastructure that includes facilities for preparing process fluid, pump stations, pipelines and a fund of injection wells. Such an infrastructure, together with oil reservoir into which water is pump, is an even more complex system that, when problems solving of effective injection, must be considered as a single integrated system.

Problems that arise in the MRP systems

In any complex system in which process is performed on the transformation and transfer of energy, inevitably there are losses. In the RPM system magnitude of such losses can reach significant values that affect the economy of oil and gas production enterprises, which amount to tens and hundreds of millions rubles. This article considers the issue of incident detection that are associated with gusts of pipelines, a deformation of the design of injection wells and other accident in which there are losses of water, electricity reservoir pressure of the exploited reservoir.

The solution of such a complex main task is impossible without modern measuring gauges, remote monitoring systems for the parameters of the system operation, mathematical and software.

Searching for solutions to the existing problems

Nowadays, a large number of affordable devices have appeared that allow measuring and processing the measure values of the monitored parameters (test items) with the necessary and sufficient discreteness to solve complex problems in determining process losses. This allows to differently analyze the already known physical processes in the RPM system, eliminate information gaps in determining the dynamics of physical processes and accurately determine the behavior of the parameters of the engineering infrastructure and the exploited reservoir.

Case study of the hydraulic characteristics of the bottomhole formation zone allowed to determine the complete similarity with the electrotechnical dependences of direct-current circuits [1]. It turned out that dependences of the main parameters of the "formation-well" system are identical to the electrical circuit (drawing) realized as part of integrating resistor-capacitor (RC) direct-current circuits. The given conclusion is made on the basis of application of the known laws deduced for the direct-current circuits to hydraulic processes of system "formation-well" which are below-specified [2].

The characteristic of the bottomhole pressure change during impulse function coincides with the characteristic of the electrical voltage on the capacitor plate of the integrating RC circuit. Detailed study of impulse function process on the "formation-well" system received this supposition [3]. It turned out that the behavior of the bottomhole formation zone with impulse function on the pressure of the fluid intake point at the bottom of the well is similar to the behavior of the electrolytic capacitor in the RC circuit [4].

Such an RC circuit is called an integrating network section, since this section is able to accumulate and donate a charge of electricity. The capacitor, like the bottomhole formation zone, has one interesting property – when it is discharged, it behaves almost like short-circuited, the current through it flows without limitations, approaching to infinity, and the voltage drop on it approaches to zero. When the capacitor is charged, the current becomes zero, and the voltage on it becomes equal to the voltage of the power source. The same happens with the bottomhole zone in the injection wells. When the pressure in the face takes the minimum value, the water flow in the injection well is rapidly increasing. As the formation pressure rises, the water flow decreases. It turns out interesting dependencies: for the capacitor there is a current, there is no voltage, there is a voltage there is no current, for the bottomhole formation zone the formation pressure increases, the water flow decreases, the pressure in the bottom drops, the flow increases, the parameters are different, and the regularities are the same.

An example of such a process was recorded at the remote distribution point RDP-5 at the water injection station (WIS) of the WIS-58 "Jalilneft", which is equipped with reverse flow meters capable of registering hourly integral flow readings in the memory of device. Figure 1 shows the readings of water flow devices taken at a daily time interval, within which the pumping unit on the WIS was turned off and restarted after a certain period of time.

A balance group has been established at this facility, where water pumped into the reservoir through the water lines 12 and 13 is supplied to RDP-5 from the WIS 58, and water is supplied to the injection wells via the water pipes 6,7,8 and 10. From figure 1, we see that after the unit shutdown at the WIS 58, the water flow in the injection system does not stop. After stopping the unit on the water channel 8, the direction of the fluid flow changed to the opposite direction.

The accuracy of the measured values of water flow is confirmed by the material balance equation, quantity of flooding water is equal to the quantity expended water. The average relative divergence of the balance for the day was less than one percent, which confirms the high accuracy of measuring the integral costs at hourly intervals.

Date/time	RDP-5								WIS-58				Balance group total, m3		Disbalance,%	
	6		7		8		10		12		13		Water entered	Water came out	Abs. dev	Rel. dev.
	V+	V-	V+	V-	V+	V-	V+	V-	V+	V-	V+	V-				
09.05.2013 15:39	39,931	0,000	0,006	0,000	17,753	0,000	0,012	0,004	57,893	0,000	0,002	0,000	57,899	57,702	0,20	0,34
09.05.2013 16:39	39,712	0,000	0,002	0,000	17,443	0,000	0,006	0,008	57,294	0,000	0,003	0,000	57,305	57,163	0,14	0,25
09.05.2013 17:39	38,581	0,000	0,001	0,000	16,631	0,000	4,779	0,003	59,952	0,000	0,002	0,000	59,957	60,012	-0,05	-0,09
09.05.2013 18:32	[illegible]															
09.05.2013 18:39	27,831	0,000	0,007	0,000	9,344	3,496	0,004	0,121	32,391	0,000	0,002	0,000	36,010	37,386	-1,38	-3,82
09.05.2013 19:39	10,272	0,000	0,004	0,003	0,000	5,679	0,000	0,012	4,454	0,000	0,001	0,001	10,149	10,277	-0,13	-1,26
09.05.2013 20:39	8,867	0,000	0,000	0,008	0,000	4,138	0,000	0,004	4,641	0,000	0,003	0,000	8,794	8,867	-0,07	-0,83
09.05.2013 21:39	8,363	0,000	0,000	0,007	0,000	3,261	0,000	0,002	5,011	0,000	0,000	0,006	8,281	8,389	-0,09	-1,06
09.05.2013 22:39	8,006	0,000	0,000	0,006	0,000	2,564	0,001	0,003	5,416	0,000	0,000	0,012	7,989	8,019	-0,03	-0,38
09.05.2013 23:39	8,010	0,000	0,000	0,013	0,000	2,088	0,000	0,004	5,938	0,000	0,000	0,017	8,043	8,027	0,02	0,20
10.05.2013 00:39	7,801	0,000	0,000	0,008	0,000	1,798	0,000	0,013	6,010	0,000	0,000	0,017	7,829	7,818	0,01	0,14
10.05.2013 01:39	7,603	0,000	0,000	0,012	0,000	1,574	0,000	0,010	6,034	0,000	0,000	0,015	7,630	7,618	0,01	0,16
10.05.2013 02:39	7,464	0,000	0,006	0,000	0,000	1,393	0,000	0,011	6,069	0,000	0,000	0,014	7,473	7,484	-0,01	-0,15
10.05.2013 03:39	7,243	0,000	0,006	0,000	0,000	1,347	0,000	0,011	6,100	0,000	0,000	0,013	7,358	7,262	0,10	1,30
10.05.2013 04:39	7,215	0,000	0,004	0,000	0,000	1,128	0,000	0,012	6,111	0,000	0,000	0,012	7,251	7,231	0,02	0,28
10.05.2013 05:39	7,134	0,000	0,010	0,000	0,000	1,030	0,000	0,014	6,106	0,000	0,000	0,012	7,140	7,156	-0,02	-0,22
10.05.2013 06:39	7,040	0,000	0,003	0,000	0,000	0,986	0,000	0,015	6,115	0,000	0,000	0,014	7,066	7,057	0,01	0,13
10.05.2013 07:39	6,989	0,000	0,004	0,001	0,000	0,880	0,000	0,016	6,102	0,000	0,000	0,005	6,999	6,998	0,00	0,01
10.05.2013 08:39	6,908	0,000	0,011	0,000	0,000	0,820	0,000	0,017	6,035	0,000	0,001	0,001	6,873	6,920	-0,05	-0,68
10.05.2013 09:39	6,860	0,000	0,023	0,000	0,000	0,744	0,000	0,016	6,091	0,000	0,000	0,002	6,851	6,885	-0,03	-0,50
10.05.2013 10:39	6,683	0,000	0,027	0,000	0,000	0,664	0,001	0,003	6,075	0,000	0,000	0,002	6,744	6,713	0,03	0,46
10.05.2013 11:39	6,799	0,000	0,016	0,000	0,000	0,606	0,001	0,001	6,071	0,000	0,000	0,005	6,678	6,821	-0,14	-2,14
10.05.2013 12:39	6,751	0,000	0,008	0,000	0,000	0,373	0,000	0,003	6,038	0,000	0,000	0,007	6,636	6,766	-0,13	-1,96
10.05.2013 13:04	[illegible]															
10.05.2013 13:39	33,657	0,000	0,004	0,002	15,694	0,052	0,141	0,007	32,950	0,000	0,000	0,009	33,011	31,505	1,51	2,84
10.05.2013 14:39	40,024	0,000	0,000	0,002	18,006	0,000	0,007	0,017	38,802	0,000	0,000	0,006	38,821	38,043	0,78	1,32
10.05.2013 15:39	40,104	0,000	0,001	0,001	17,941	0,000	0,000	0,013	38,728	0,000	0,000	0,005	38,744	38,142	0,60	1,02

Figure 1. Hourly values of water flow through the WIS-58.

According to the theory of electrical engineering, the reaction of a circuit of an integrating type, the equivalent of which is an RC circuit, to a single stepped action of a voltage with an amplitude U is given by the following formula:

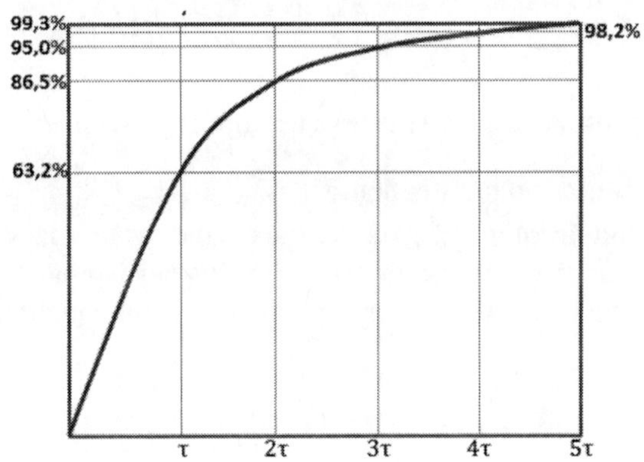

Figure 2. Graph of voltage changes on the capacitor plates.

$$U_c = U_0 + U(1 - e^{-\frac{t}{\tau}}), \qquad (1)$$

where U_0 is the initial value of voltage on the capacitor plates; U is the voltage pulse applied to the RC circuit; τ is a time constant equal to the product of the resistance to the capacitance of the capacitor,

$$\tau = RC. \qquad (2)$$

Figure 2 shows a graph of the voltage variation across the capacitor plates when the time changes, which is a multiple of the time constant τ. For the bottomhole formation zone by analogy with the RC circuit, a stepwise effect on the bottomhole pressure P_{bp} will be written similar to the mathematical dependence:

$$P_{bp} = P_{iv} + P(1 - e^{-\frac{t}{\tau}}), \tag{3}$$

where P_{iv} is the initial value of bottomhole pressure; P is the pressure pulse; t is the time; τ is the time constant, which is equal to the product of the hydraulic resistance of the bottomhole formation zone by the bottomhole zone capacity factor:

$$\tau = R_{hs} C_{bfz}, \tag{4}$$

According to the energy conservation law, the total mechanical energy that performs the work of moving a certain volume of fluid from one point of the pipeline network to another consists of the sum of the potential and kinetic energies, and this sum is constant:

$$E_{total} = E_{pot} + E_{kin} = const, \tag{5}$$

where E_{total} is the total mechanical energy, E_{pot} is the potential energy, E_{kin} is the kinetic energy.

For the energy of pressure, the energy conservation law is written in the form of the Bernoulli equation:

$$P + \rho g h + \frac{\rho V^2}{2} = const, \tag{6}$$

where $P + \rho g h$ is the hydrostatic pressure or potential energy of pressure; $\frac{\rho V^2}{2}$ is the hydrodynamic head or kinetic energy of pressure; ρ is the water density.

Considering the conditionally constant value of the water density and that the water is an incompressible liquid, for the pipeline of the RPM system having two water accounting points in which there are flow meters and pressure sensors, the Bernoulli equation will have the following form:

$$P_1 + \rho g h_1 + \frac{\rho V_1^2}{2} = P_2 + \rho g h_2 + \frac{\rho V_2^2}{2} + P_{пот} = const, \tag{7}$$

where P_1, P_2 are measured pressures at points of shipment and reception of water; V_1, V_2 are the measured values of the water velocity through the cross section of the water flow meters; $h_1 h_2$ are geometric marks of flow meters and pressure sensors installation at the points of shipment and reception of water; P_{nom} is a loss of pressure in the pipeline network between water metering points 1 and 2.

The pressure losses in the steady-state injection regime are conventionally constant, and these losses can be determined experimentally by equation (7).

The engineering infrastructure of the RPM system consists of three levels, which are conventionally depicted in Figure 3.

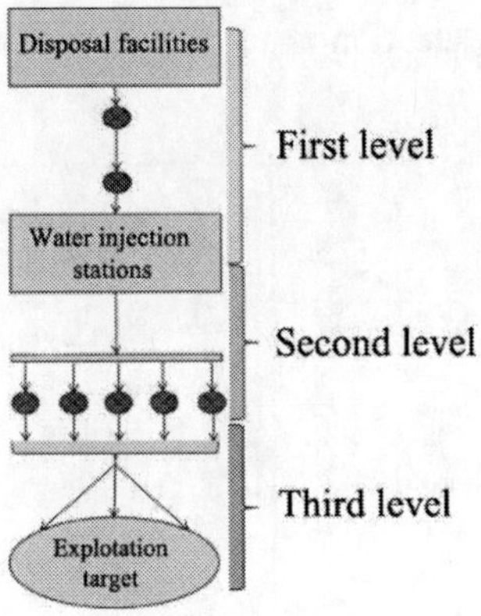

Figure 3. Three-level infrastructure of the RPM system.

The first level of the engineering network contains a balance group of water conduits with water meters that combine water treatment facilities and water injection stations.

The second level includes the balance group of water conduits with water flow meters between water injection station and manifold units at the water injection station and remote distribution points.

The third level consists of water conduits that combine manifold units and remote distribution points with injection wells. In the third group, as a rule, there are no balance groups; flow meters are installed only at the points of water shipment, there are no flow meters at the wellheads of the injection wells.

For the first level of the RPM system, having equipped water metering units with reverse flow meters and pressure sensors at the points of shipment and reception, equation (7) can determine the possible gusts of the pipeline network and also solve inverse problems to detect malfunctions of flow meters and pressure sensors.

The second level is provided by controlling the balance of the volume of the pumped liquid according to the integrated readings of the flow meters.

The third level of control is the most difficult, because there are flow meters at the points of water shipment to manifold units and remote distribution points, and there are no flow meters at the injection points at the wellheads of the injection wells. But, knowing the modes of pumping units operation at the WIS, and also controlling the injection of water in a stationary mode, using special algorithms for processing the measured values of pressures at points of the pipeline network that are common to several pipelines, it is possible, with the help of the same energy conservation law (7), with a high accuracy determine the time of occurrence of incidents, the magnitude and nature of the current changes.

Incidents are events that lead to a change in the operation of water injection in the RPM system. Incidents include: start-ups or shutdowns of pumping units; planned or unplanned changes in injection modes, including the opening or closing of individual conduits with pumping units operating at the WIS, which leads to a redistribution of costs in the water conduits and changes in pressures in common reservoirs: gusts of pipelines or disturbance of the well design.

To determine the nature of the incident, whether the incident is an accident or simply a planned measure for changing the operating mode, special algorithms are applied which, by the equation of the transient process (3), determine the magnitude of the pressure pulse and the time constant of the transient process by solving the inverse problem. For certain parameters of the amplitude of the pressure change and the time constant of the transient process, a special algorithm of the program of the remote control and monitoring system allows identifying the emergency events and changing the injection modes that do not affect the failure of the infrastructure of the RPM system, for example, the planned opening or closing of the water conduit with the WIS running.

The most difficult for automatic detection is to determine the possible disruption of the injection well design.

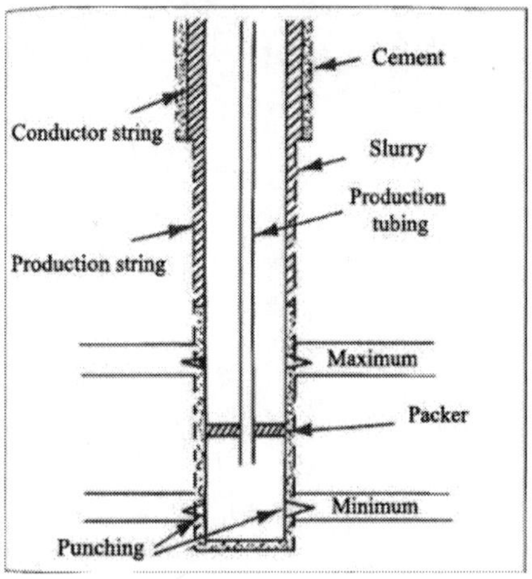

Figure 4. Design of the injection well.

Figure 4 shows the design of the injection well. The change in injection modes can be affected by the violation of the production tubing tightness, the production strings, and the packer. With such disturbances, latent pressure losses occur in the contour of the bottomhole zone of the injection well. Wells in which such disturbances occur after stopping the pumping units on the WIS begin to take water that is poured from other injection wells with greater hydrostatic pressure.

With the help of reversible counters, the direction of the water lines is directed to which water flows are directed after the pump units are stopped, and if more than one well is connected to the direction, the well that receives water after stopping the aggregates at the pump station is determined by successive shutdown of the wells. The magnitude of the pressure pulse amplitude (7) determines the nature of the problem in the injection well. To accurately determine the problem, hydrodynamic studies of the design of the injection well are carried out.

The theoretical and methodological materials outlined in this article are based on the studies carried out in the framework of pilot-scale tests of reverse ultra-sonic flow meters in different sections of the produce fields. The use of reverse flow meters in the automated system of remote control and management allowed us to identify a large number of sections of the pressure maintenance system in various OGPDs in which uncontrolled outflow of water occurs after the pumping units are stopped. In some areas, the volumes of spills reach 20% of the volume of water pumped into the reservoir. This indicates a greater efficiency compared to the methods already known [5,6,7].

The mass proportion of this problem for many OGPDs operating the field at a late stage of extracting requires a systems approach in solving the problem of determining the losses and increasing the efficiency of the MRP system's operation.

References

[1] Neprimerov N N 1962 Dissertation for the degree Doctor of Sciences *Experimental studies of some issues of underground hydromechanics and physics of the "formation-well" system* (Kazan)

[2] Neprimerov N N 1986 *Technology for optimum production of the reservoir* (Kazan)

[3] Ibragimov N G, Panarin A T, Desyatkov V K, Evtushenko S P 1996 The materials of seminar-discussion "the Concept of development of methods of increasing oil recovery" *To the question of determining the optimal period of water injection in carbonate reservoir* (Bugulma)

[4] Petrenko Y V 2016 *Theoretical foundations of electrical engineering. Transient processes in linear electric circuits: a tutorial* (Novosibirsk: publisher Novosibirsk State Technical University)

[5] Ahmed M Sattar and M Hanif Chaudhry 2008 Journal of Hydraulic Research vol 46 *Leak detection in pipelines by frequency response method*

[6] Witness Mpesha, M Hanif Chaudhry and Sarah L Gassman 2002 Journal of Hydraulic Research vol 40 *Leak detection in pipes by frequency response method using a step excitation*

[7] Jin Yang, Yumei Wen, Ping Li and Xingke Wang 2013 Urban Water Journal vol 10 *Study on an improved acoustic leak detection method for water distribution systems*

Approbation of the algorithm for combustion optimization of a multicomponent fuel with air

E R Saifullin, S A Nazarychev, V M Larionov, Yu V Vankov, K G Sadikov, Ya V Ananiev

Kazan Federal University, 18 Kremlyovskaya str., Kazan 420008, Russian Federation

mr.emilsr@gmail.com

Abstract. The article describes the laboratory setup designed to study the combustion of multicomponent fuels, in particular methane-hydrogen fuel, with air. The setup simulates a boiler with a heat exchanger. The results of experimental simulation of the initial combustion regime optimization of propane-butane fuel according to the existing algorithm are developed.

Introduction

The main fuel in the Russian Federation used in the power generation is natural gas. Due to fossil fuel reserves decrease and global energy consumption increase there is an increasing interest in using alternative fuels such as biogas, syngas and industrial hydrocarbon wastes.

In the Russian Federation there is a problem of effective utilization of petrochemical wastes. High research interest is in finding ways to utilize wastes containing methane and hydrogen. One of the ways to effective utilization these wastes is combustion in boilers. However, these wastes have a variable composition, which varies in time and depend on the processes at the petrochemical plant. The change in waste composition leads to a change of the combustion thermophysical characteristics: heat release rate, temperature and burning rate, composition of the combustion products, etc. Change in thermophysical characteristics lead to the boiler operational instability: blowout, flashback, combustion instability, fuel autoignition, etc.

There are the following methods for optimizing combustion process in boilers:

1. A method for automatically optimizing the combustion process in a boiler [1] based on controlling the air flow rate in order to eliminate carbon monoxide and maintain an optimum concentration of oxygen in the combustion products. The disadvantages of this method are the need to use gas analyzer and preset the optimum amount of oxygen in the flue gases, which is practically impossible in the case of using variable-composition fuels.
2. The method of burning process optimization [2] by measuring the temperature parameters in a flame. Disadvantages of the method are the need to use devices for contactless determination of the flame temperature and highest temperature flame point determining inaccuracy.
3. A method for automatically optimizing the combustion process by calculating the thermal efficiency of the boiler [3], which is determined from the measured values of the heat flow rate from the combustion chamber and the heat flow rate introduced by the fuel into the chamber. The disadvantage of the method is its high inertia, due to the large number of operations.

Common disadvantages of these methods: high inertia, the necessity to install additional equipment, inability to quickly respond to random changes in fuel characteristics.

The authors previously developed a method [4-7] for automatic optimization of combustion process of the variable-composition fuel in boilers. The method maintains the required heat carrier temperature in the boiler with complete combustion of fuel and minimum specific fuel consumption, based only on a change in the outlet heat carrier temperature. Preliminary, according to the developed method, the initial optimal combustion regime is achieved. Based on the change in the heat carrier temperature, the fuel and air flow rates are adjusted to eliminate the fuel underburning and air excess.

For experimental confirmation of method effectiveness, a laboratory setup simulating a boiler was developed. At the initial stage, in the absence of a source of methane and hydrogen, it is planned to conduct studies on the available propane-butane fuel in order to assess the capabilities of the laboratory setup and develop an experimental technique. Also in the experiment, the algorithm for optimizing the initial combustion regime was tested.

Description of the laboratory setup

The experiments were carried out on a laboratory setup simulating a boiler (Fig. 1). The main parts of the setup are the combustion chamber 12, gas mixer 9, flame holder 10, inspection window 11, chimney 22 and heat exchanger 13. The heat carrier (water) circulates into the heat exchanger by the pump 18. From the tank 19, water at room temperature is fed into the heat exchanger. After heating, water is fed into the tank 20.

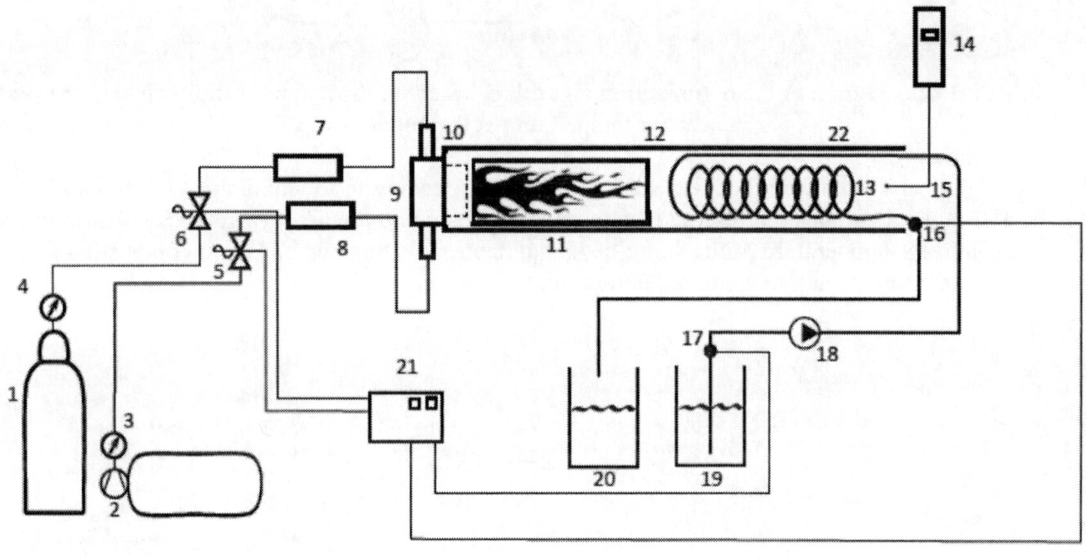

Fig. 1. Laboratory setup scheme: 1 – propane-butane tank; 2 – air compressor; 3,4 – manometer; 5,6 – gas valves; 7,8 – flow meters; 9 – gas mixer; 10 – flame holder; 11 – inspection window; 12 – combustion chamber; 13 – heat exchanger; 14 – exhaust gas analyzer; 15 – analyzer probe; 16,17 – thermocouples; 18 – water pump; 19,20 – water tanks; 21 – controller (Arduino); 22 – chimney.

Experimental procedure

The experiment was carried out according to the algorithm [8-10], based on finding the optimal fuel-air ratio by recording the change in the heat carrier temperature at the output (fig. 1) and consisted of two parts.

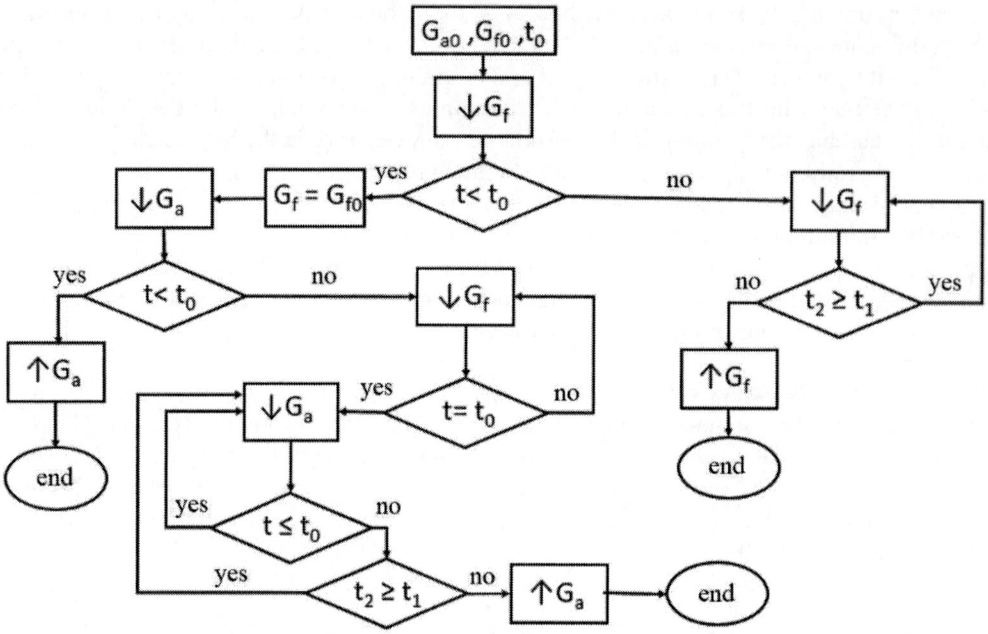

Fig. 2. Block diagram of the optimization algorithm, based on the nature of the change in the heat carrier temperature at the outlet

In the first part of the experiment, a fuel-air mixture is feeding to the laboratory setup with excess of air. The optimal combustion regime is determined by reducing the air flow rate G_a at constant fuel flow rate until the temperature of the water begins to decrease. Thus, the optimum combustion regime is achieved at the maximum temperature extremum.

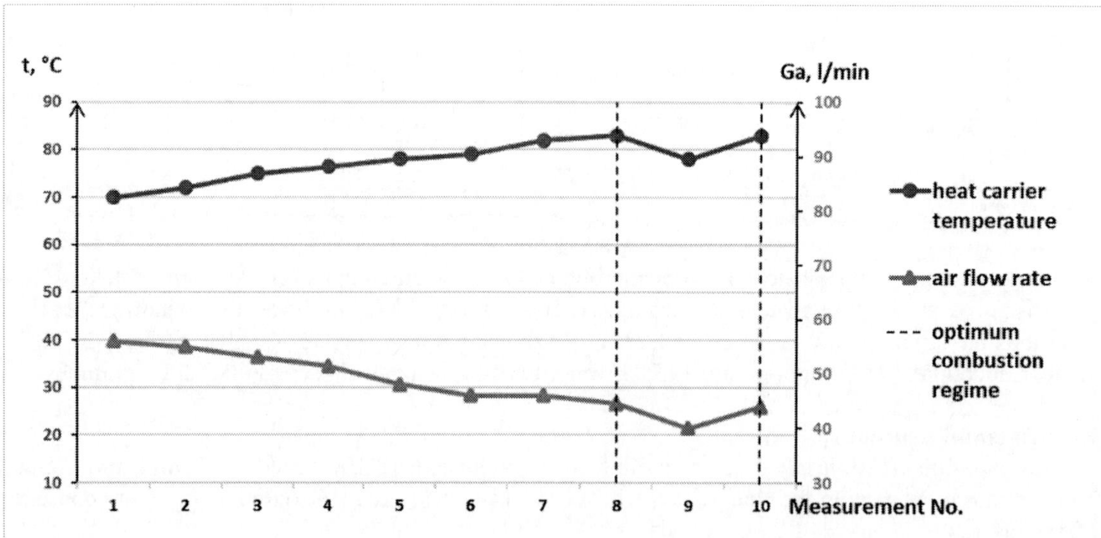

Fig. 3. Achieving the optimum combustion regime in case of excess of air

In the second part, the fuel-air mixture is feeding with an excess of fuel. To determine the optimal combustion regime, according to the algorithm, it is necessary to start reducing fuel flow rate G_f at constant air flow rate until the water temperature decreases. After decreasing the water temperature, it is necessary to return the previous fuel flow rate to achieve optimum combustion regime.

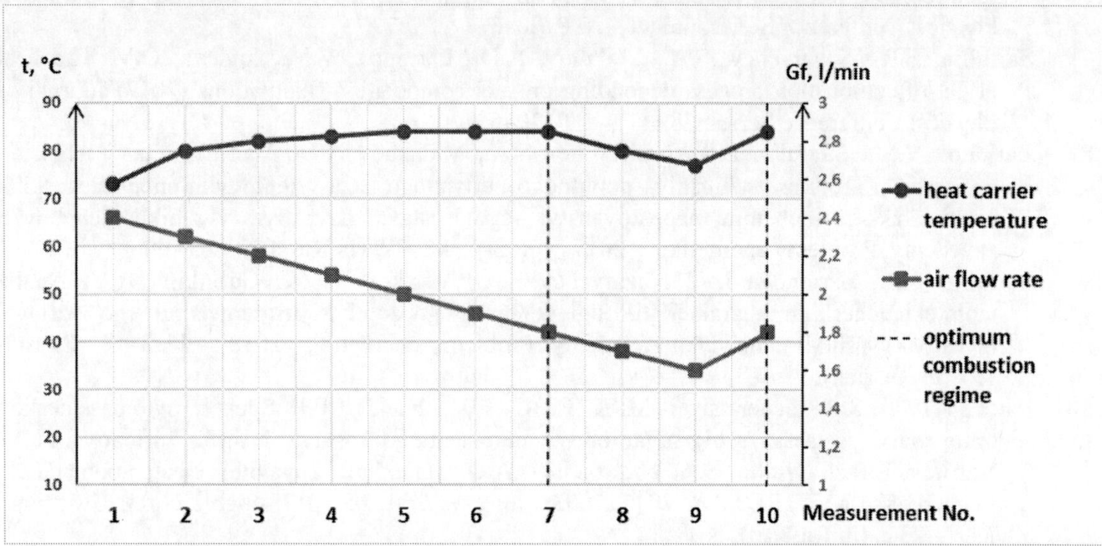

Fig. 4. Achieving the optimum combustion regime in case of excess of fuel

Conclusion

A laboratory setup simulating a boiler was created. The setup will allow studying the combustion process of variable-composition fuel in a stable mode.

The results of the experiment confirmed the previously developed algorithm for optimizing the combustion process in boilers, based on an analysis of the change in the heat carrier outlet temperature, in the case of a non-optimal initial fuel-air ratio.

Acknowledgements

The work is performed according to the Russian Government Program of Competitive Growth of Kazan Federal University. The reported research was funded by Russian Foundation for Basic Research and the government of the region of the Russian Federation, grant № 18-48-160051\18.

References

[1] A.s. 735869 SSSR, MPK5 F23N 1/02. Sposob avtomaticheskoy optimizatsii protsessa goreniya v kotle/ V.Yu Vadov, Yu.S. Denisov (SSSR). № 2431448. Zayavleno 17.12.1976. Opublikovano 25.05.1980. – 7s. (in Russian).

[2] Pat. 2493488 RF, MPK F23N 1/00. Sposob optimizatsii protsessa goreniya topliva / S. Sh. Dzantiev, B.D. Bilaonov, A. L. Rutkovskiy, V. M. Zarochentsev, A. V. Bigulov. - № 2012108906/06; Zayavleno 07.03.12 Opublikovano 10.06.2007. – 5 c. (in Russian).

[3] Pat. 2425290 RF. Sposob avtomaticheskoy optimizatsii protsessa goreniya v topke barabannogo parovogo kotla / L.S. Kazarinov, D.A. Shnayder, L.A. Koptsev, O.V. Kolesnikova, A.V. Kinash, S.V. Sedel'nikov // Byul. – 2011. – №21. – 7 s. (in Russian).

[4] Saifullin, E.R., Vankov, Yu.V. Optimization of burning process of hydrocarbon fuels with varying specific heat of combustion (2015) IOP Conference Series: Materials Science and Engineering, 86 (1), № 012006.

[5] Saifullin, E.R., Larionov, V.M., Busarov, A.V., Busarov, V.V. Optimization of hydrocarbon fuels combustion variable composition in thermal power plants (2016) Journal of Physics:

Conference Series, 669 (1), № 012037.

[6] Saifullin, E.R., Larionov, V.M., Busarov, A.V., Busarov, V.V. Thermal effect of hydrocarbon fuels combustion after a sudden change in the specific calorific value (2016) Journal of Physics: Conference Series, 669 (1), № 012043.

[7] Saifullin, E.R., Nazarychev, S.A., Malahov, A.O., Larionov, V.M., Iovleva, O.V. The heat effect of combustion process depending on fuel composition fluctuations (2017) Journal of Physics: Conference Series, 789 (1), № 012045.

[8] Larionov V.M., Sayfullin E.R., Nazarychev S.A., Malakhov A.O., Van'kov Yu.V. Algoritm optimizatsii protsessa szhiganiya poputnogo neftyanogo gaza v teplovykh energeticheskikh ustanovkakh s uchetom nepostoyanstva ego sostava. Izvestiya vysshikh uchebnykh zavedeniy. Problemy energetiki. – 2017. – № 3-4. – S. 3-9 (in Russian).

[9] Saifullin E.R., Larionova I.V., Nazarychev S.A., Van'kov Yu.V., Gazeikina A.V. Analiz teplotekhnicheskikh kharakteristik uglevodorodnogo topliva pri izmenenii ego sostava. Izvestiya vysshikh uchebnykh zavedeniy. Problemy energetiki.– 2018. – № 3-4. – S. 145-151 (in Russian).

[10] Pat. 2647940 Ros. Federatsiya: MPK F23C 1/02, F23C 1/08. Sposob avtomaticheskoy optimizatsii protsessa szhiganiya topliva peremennogo sostava / V.M. Larionov, Yu.V. Van'kov, E.R. Sayfullin, S.A. Nazarychev, A.O. Malakhov; Zayavitel' i patentoobladatel': FGAOU VO KFU, RU. – № 2017116036/06; zayavl. 04.05.2017; opubl. 21.03.2018; byul. №9. – 23 c. (in Russian).

Degassing of water with exhaust gases of the boiler

V I Sharapov[1] and R I Kamalova[1]

[1]Ulyanovsk State Technical University, st. Severny Venets, 32, Ulyanovsk, Russia

vlad-sharapov2008@yandex.ru kamalovari@oblkomhoz.ru

Abstract. A method for degassing the make-up water is described, the feature of which is the use of the exhaust gases of the boiler after the smoke exhauster as a desorbing agent. In the exhaust gases of the boiler, there is a small portion of corrosive oxygen, due to the fact that all oxygen is consumed in the combustion process in the furnace. The problem of the possible saturation of the deaerated water with carbon dioxide due to its increased content in the boiler waste gases is solved by dispensing the alkaline agent into the make-up water. The mass exchange efficiency of the make-up water degassing by the proposed method is estimated, the required specific consumption of the exhaust gases for deaeration is theoretically calculated. It is shown that the consumption of the desorbing agent, which is the exhaust gas, is sufficient for degassing practically any quantities of nutrient and make-up water.

Technology of degassing of water with waste gases of the boiler

The process of removing dissolved corrosive gases from the make-up water - deaeration - is the final stage of the technological treatment of water. Currently, the most common method of thermal deaeration, which consists in the desorption of dissolved gas when it is heated to the saturation temperature by using steam.

Water vapor is traditionally used as the desorbing agent in deaeration plants. In this case, the process of thermal deaeration is inevitably associated with the need for additional steam wastages for deaeration, as well as undesirable losses of heat with the discharge of the deaerator. Deaeration technologies significantly affect the efficiency of heat power plants [1]. In order to increase the energy efficiency of electricity generation, the process of deaeration should be carried out at the lowest possible temperature of the heat transfer medias, which provides heating of the streams of initial and deaerated water [1]. The necessity for additional steam costs for deaeration in the cycle of operation of heat power plants (HPPs), additional heat losses with deaerator dehumidifications are inevitable disadvantages of the thermal deaeration method.

In research laboratory of Heat power systems and Plants of Ulyanovsk State Technical University developed various solutions directed to increase energy efficiency and optimize the process of deaeration of the make-up water. One of the promising elaborations of development in this area is to increase the thermal efficiency of the deaeration process using non-traditional low-temperature medias as a desorbing agents [2,3].

One such solution is the use of the exhaust gases of the boiler - methane combustion products - as a desorbing agent in the deaerator [4]. The outgoing gases of the boiler practically do not contain oxygen, and the excess air values in the gas path can be assumed to be minimal provided to the organization of non-stoichiometric combustion of fuel [5]. The scheme for implementing this technology is shown in figure 1.

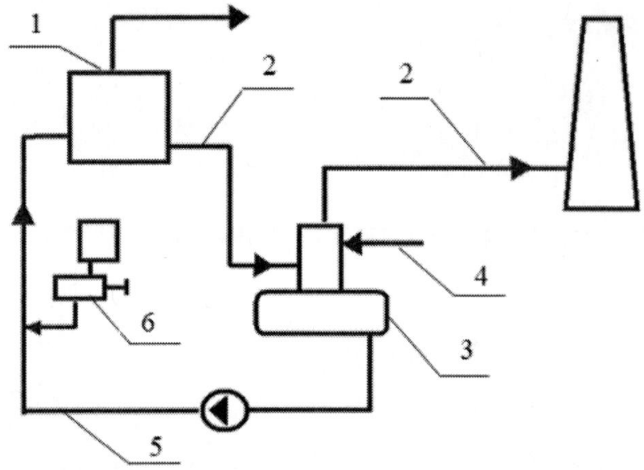

Figure 1. The scheme for degassing the make-up water with the boiler exhaust gases: 1 - steam boiler, 2 - flue gas duct, 3 - deaerator, 4 - source water pipeline, 5 - feed water pipeline, 6 - dosing system for alkaline agent.

A distinctive feature of the proposed scheme of water degassing is that the deaerator is switched on by desorbing medium into the flue gas outlet of the boiler. The outgoing gases of the boiler 1 from the flue gas duct 2 are fed to the lower part of the deaerator column 3. As they rise up inside the deaerator construction, the flue gases cross the water jets, heating the source water to the saturation temperature, and securing the release of corrosive gases and mixing them with the desorbing agent. The mixture of exhaust gases and discharged gases from the source water is removed from the top of the deaerator column through the flue gas duct to the chimney. The water thus deaerated is discharged from the deaerator by the feed pump, passes the dispenser of the alkaline agent (sodium hydroxide) 6, removal of corrosive carbon dioxide from the water. Then, the deaerated water is supplied to the economizer of the boiler 1.

The proposed technology makes it possible to increase the economics of the process of water deaeration due to elimination of steam costs for deaeration at HPPs and utilizing the heat of the boiler exhaust gases by lowering their temperature in the deaerator before being released to the atmosphere. Anticorrosion treatment of the feed and make-up water by the outgoing gases of the boiler can also be arranged in heat-generating installation in the absence of steam sources, for example in hot water boilers.

Estimation of the energy efficiency of deaeration by the exhaust gases of the boiler

To estimate the mass-exchange and energy efficiency of this solution, the values of the theoretically required consumption of the desorbing agent should be determined.

Determination of the theoretically necessary specific consumption of the desorbing agent for removal of dissolved oxygen from the water $d_{\min}^{\min}$, kg/t, is performed on the basis of the solution of the balance equations of the processes of mass transfer and heat transfer during thermal deaeration, provided that the equilibrium between the phases is reached at the exit from the deaerator [6]. It is assumed that the maximum mass-transfer and energy efficiency of the thermal deaerator is achieved with the minimum possible costs of the desorbing agent and the vapor produced by the deaerator.

The design scheme of the deaerator is shown in figure 2.

The equation for the material balance of the deaeration process can be written in the form:

$$G_{sw}X_{sw} + D_{gas}Y_{gas} = G_{dw}X_{dw} + D_{vent}Y_{vent} \tag{1}$$

where G_{sw} and G_{dw} is the consumption of the influent and deaerated waters, kg/h; D_{gas} is the consumption of exhaust gases supplied in the deaerator, kg/h; D_{vent} is the discharge of the deaerator vent steam (mixture of corrosive gases liberated from water and desorbing agent), kg/h; X_{sw} and X_{dw} are oxygen concentrations in the water at the inlet and outlet of the deaerator; and Y_{gas} and Y_{vent} are the oxygen contents in the exhaust gases at the deaerator inlet and in the vented steam at the deaerator outlet [6].

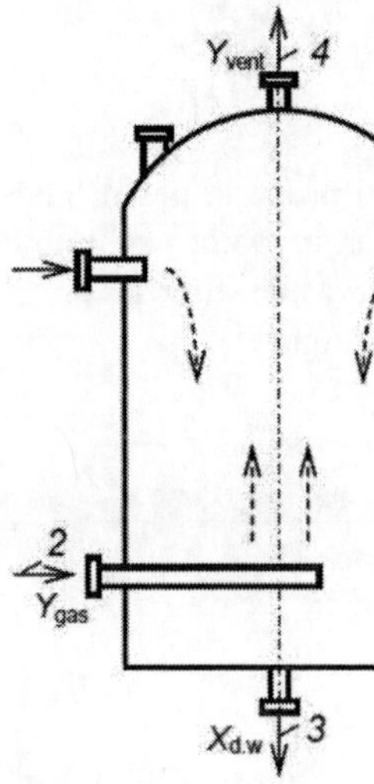

Figure 2. Scheme of deaeration column of countercurrent type: 1 - supply of source water; 2 - supply of heating steam; 3 - drainage of deaerated water; 4 - deaerator outlet tap.

Oxygen concentration Y_{gas} in gas at the deaerator input is practically zero, since the outgoing gases of the boiler do not contain oxygen because of its full consumption of the combustion process of fuel (natural gas) in the boiler furnace.

$$CH_4 + 2CO_2 = CO_2 + 2H_2O \tag{2}$$

According to the Dalton law, the total pressure of the gas and gas–steam mixtures is equal to the sum of partial pressures of gases and steams composing the mixture. From the Henry law, it follows that the concentration of the gas dissolved in water is proportional to the partial pressure of this gas over the water surface.

The oxygen concentration in the vented steam leaving the deaerator depends on the flow chart of the water and steam motion in the apparatus. At the countercurrent, mole fraction Y_{vent} of oxygen in the gas–steam mixture is

$$Y_{vent} = K_H^{O_2} X_{sw} / p \tag{3}$$

where $K_H^{O_2}$ is the Henry coefficient (phase equilibrium constant for oxygen [7]), Pa; and p is the pressure in the deaerator, Pa.

At the countercurrent flow of water and exhaust gases in the deaerator, the minimum consumption of exhaust gases is

$$D_{gas}^{\min} = G_{sw} \frac{p}{K_H^{O_2}} \frac{X_{sw} - X_{dw}}{X_{sw}}, \quad (4)$$

and its specific value is

$$d_{gas}^{\min} = \frac{D_{gas}^{\min}}{G_{sw}} \quad (5)$$

The results of calculation of the deaeration process, performed according to formulas (4) and (5) for different values of the oxygen concentration in water at the inlet to the atmospheric deaerator and the values of dissolved oxygen content in deaerated water given in GOST [8] when using as exhaust gas desorbing gases of the boiler , are shown in figure 3.

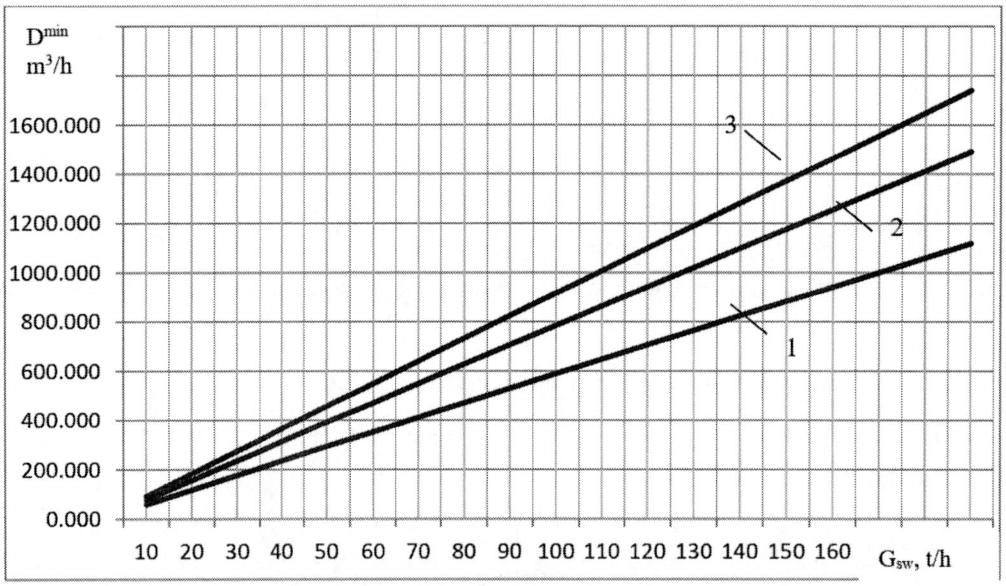

Figure 3. Theoretically required desorbent (exhaust gases) consumption at the countercurrent of water and gas as a function of the deaerator load. Quantities of $X_{s.w}$, mg/dm³: (*1*) 8, (*2*) 10, and (*3*) 12.

In this case, the theoretical flow $d_{gas}^{\min}$ corresponds to a deaerator with an infinitely large contact surface of the liquid and gas phases, which explains the relatively small amount of flue gases needed for deaeration. In practice, in deaerators mass transfer takes place at a limited contact surface of phases and for a limited time, so the equilibrium between water and steam adopted in the calculation can not be achieved.

In case of using as a stripping agent the boiler exhaust gases according to the above calculations, theoretically the required gas flow rate for deaeration will be 5-8 m³ per 1 ton of deaerated water. As follows from the obtained dependence, the required specific consumption of the desorbing agent for deaeration with the exhaust gases of the boiler exceeds the theoretically required water vapor flow rate when used as a desorbing agent.

The problem of excess carbon dioxide in the flue gases (2) and its negative effect on the quality of the de-aerated water is solved by dosing the alkaline agent (sodium hydroxide) into the pipeline of deaerated water.

Neutralization of carbon dioxide contained in feed water, alkali occurs as a reaction:

$$CO_2 + NaOH = NaHCO_3 \quad (6)$$

To maintain the pH value of the feed water is added to it an aqueous solution of ammonia, which also reacts with the free carbon dioxide by the following reaction equations [5]:

$$NH_4OH + CO_2 = NH_4HCO_3 \quad (7)$$

$$NH_4HCO_3 + NH_4OH = (NH_4)_2CO_3 + H_2O \quad (8)$$

It is advisable to introduce ammonia into chemically purified desalinated water, since this allows to protect from the carbon dioxide corrosion and the path of purified or desalted water.

To neutralize free carbon dioxide and increase the pH of feed water, organic amines are often introduced, including cyclohexylamine $C_6H_{11}NH_2$, morpholine C_4H_8NO and piperidine $C_5H_{11}N$. With alkaline properties and volatility, these corrosion inhibitors create protective oxide films on the surfaces of the feed water tract.

Conclusions

- a non-traditional technology for low-temperature deaeration of feed water with the use of boiler exhaust gases as a desorption agent is proposed;
- the efficiency of mass transfer of the water deaeration process for a new technical solution was estimated, the theoretically calculated required desorbtion agent consumption for deaeration was calculated;
- various methods for removing excess of carbon dioxide from deaerated water are considered.

References

[1] Sharapov V I 1994 *Current Problems in the Use of vacuum deaetators for Open Heat-Supply systems* (*Therm. Eng.* vol 41) pp 635-639
[2] Sharapov V I, Pazushkina O V and Kudryavtseva E V 2016 *Energy-Effective Method for Low-Temperature Deaeration of Make-up Water on the Heating Supply System of Heat Power Plants* (*Therm. Eng.* vol 63) pp 687-690
[3] Sharapov V I and Kudryavtseva E V 2016 *Energy Efficiency of Low-Temperature Deaeration of Makeup Water for a District Heating System* (*Power Tech. and Eng.* vol 50) issue 2 pp 204-207
[4] Sharapov V I, Kamalova R I, Kudryavtseva E V and Rogachev S S 2017 *Boiler installation* RF Patent 2629321 IPC F 22 B 33/18 (*Byul. Izobr.* vol 5)
[5] Roslyakov P V and Zakirov I A 2001 *Non-stoichiometric combustion of natural gas and fuel oil in heat power plants* (Moscow: MEU)
[6] Sharapov V I and Malinina (Pazushkina) O V 2004 *Determining the Theoretically Required Vapor-Venting Rate for Thermal Deaerators* (*Therm. Eng.* vol 51) pp 321-324
[7] Pavlov K F, Romankov P G and Noskov A A *Examples and tasks on the course of processes and devices of chemical technology. Textbook for high schools* (Leningrad: Chemistry)
[8] GOST (State Standart) 16860-88* 1989 *Thermal deaerators. Types, basic parametrs, acceptance, control methods* (Moscow: Izd. Stanartov)

IOP Conf. Series: Earth and Environmental Science **288** (2019) 012117 doi:10.1088/1755-1315/288/1/012117

About the intensification of convective heat transfer from combustion products to the walls of a cylindrical shaped pulsing combustion chamber

O.R. Sitnikov, G.I. Pavlov, Y.I. Khakimzyanova

Kazan National Research Technical University named after A.N. Tupolev – KAI, Russian Federation, Kazan, Karl Marx st., 10

Halmer169990@mail.ru

Abstract. Franke's critical dependence determines the theoretical relationship for the numbers Nu for pulsating and stationary flow regimes. It is based on the heat exchange intensification hypothesis due to the interaction of the pulsating velocity profile and the boundary layer near the heating surface. The article presents the results of studies to test the efficiency of the Franke's ratio for a given range of pulsed combustion chamber diameters, as well as to determine the influence of the geometric parameters of the chamber on the convective heat transfer from the combustion products to the walls of the chamber

According to the ideas of the authors [1, 2], pulsating combustion is a self-oscillatory process that arises as a result of the interaction of the combustion zone (heat release, thermal resistance and burning velocity) and flow (pressure, velocity and entropy).

According to one of the hypotheses, the intensification of heat transfer during the washing of the heating surfaces by the pulsating flow occurs as a result of the interaction of the pulsating velocity profile and the boundary layer near the heating surface. The resistance of the pulsating flow, the velocity profile and the boundary layer on the channel walls determine the mass and heat exchange between the flow and the wall. According to G. Schlichting [3], the velocity profile of the pulsating flow is determined by the pressure gradient $\delta p/\delta x$. For *cos-* vibrations $\delta p/\delta x=\rho k\ \cos(\omega\tau)$, where ρ – the density of the medium, kg/m^3; k – a constant; ω – the angular frequency, s^{-1}; τ – the time, s.

The varying velocity profile causes oscillations of the heat sink from the flow to the wall, while a significant increase in the heat transfer coefficient is observed.

Based on the temporal and spatial distribution of the velocities of the stationary and pulsating flows in the tube, and also assuming that in the turbulent flow a pulsating boundary layer arises between the wall of the tube and the core of the flow oscillating as a solid body, Frank derived the theoretical relation on the basis of the Prandtl theory [4] for Nu numbers under pulsating and stationary flow regimes:

$$\frac{Nu_{f\text{п}}}{Nu_{f\text{c}}}=\frac{1+1{,}74\left(Pr_f-1\right)Re_{f\text{c}}^{-1/8}}{1{,}02\frac{\beta^{1/8}}{b^{7/8}}\frac{Re_{f\text{п}}^{1/10}Pr}{Re_{f\text{c}}^{1/4}k}-1{,}86\frac{b^{7/8}}{\beta^{1/8}}\frac{k}{Re_{f\text{п}}^{1/10}}+1{,}02},$$

,

where $Pr_f = \nu/a$ – the Prandtl criterion; ν – the kinematic viscosity coefficient, m^2/s; a – the coefficient of thermal diffusivity, m^2/s; $Re_{fc} = W_c d/\nu$ – criterion Re for stationary mode; W_c – the average velocity of the stationary flow in the chamber, m/s; d – the diameter of the pipe, m; β and b – are constants; $Re_{fn} = A\vartheta/\nu$ – criterion Re for a pulsating flow; $A = \vartheta/\omega$ – the amplitude of pulsations, m; ϑ – the velocity amplitude, m/s; $\omega = 2\pi f$ – angular frequency of pulsation, 1/s; k – the pulsation coefficient.

As can be seen from the equation, the increase in heat transfer depends on the criteria Pr_f and Re_{fc}, the pulsation coefficient k and the Re_{fn} criterion. The effect of the Re_{fc} criterion at a constant value of W is related to the change in the pipe diameter d and the kinematic viscosity coefficient ν. Heat transfer from the pulsating flow increases with an increase in the pulsation coefficient k, and also with a decrease in the temperature of the gas flow and the value of Re_{fn}.

According to the equation for a certain tube diameter, there is an optimal relationship between the angular frequency ω, the average flow rate W and the pulsation coefficient k, in which the heat transfer is maximized.

One of the research tasks was to check the reliability of the theoretical ratio of Nusselt numbers for a given range of the chamber diameter change, as well as to determine the influence of the chamber geometric parameters on the convective heat transfer from the combustion products to the chamber walls.

The experiments were carried out with a cylindrical combustion chamber 3, opened from one side (Fig. 1). Combustion in the chamber was achieved by installing, on the closed side of the chamber, a gas injection tunnel gas type 1 with a gas burner. The thermal power of the burner at a gas pressure in front of a nozzle of 0.1 MPa was equal to 40 kW. The excess air ratio was 1.

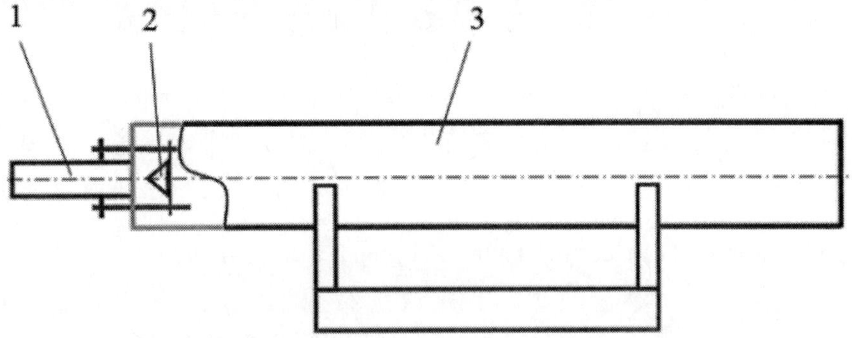

Fig.1. The scheme of the experimental setup

In chamber 3, pulsating combustion was achieved due to the installation of a combustible air mixture at the outlet of the injection gas burner 1 of a poorly flowing body - the vortex stabilizer 2 with a large cone angle.

During the experiments, the influence of the geometric parameters of the chamber on the convective heat transfer from the combustion products to the walls of the chamber was clarified. Therefore, three cylindrical pipes of different diameters were used as combustion chambers. With an unchanged length of the chamber, their internal diameters were 0.092; 0.148 and 0.206 m. In determining the convective heat transfer for the characteristic size, the diameter of the chamber was adopted.

Determination of the functions of the heat transfer coefficient α and the Nusselt number was carried out by the method described in the source [5]. For each diameter of the chamber, the rate of heating of the chamber walls, the acoustic parameters of the gas flow, and the gas flow rate in the chamber were determined. The heating rate was recorded by temperature sensors installed along the entire length of the chamber. Figure 2 shows the heating rate diagram of five sections of the chamber in the pulsed combustion regime. Measurements of the investigated signals were carried out in stationary and pulsed-burning regimes. In both cases, the local values of the heat transfer coefficient α were determined.

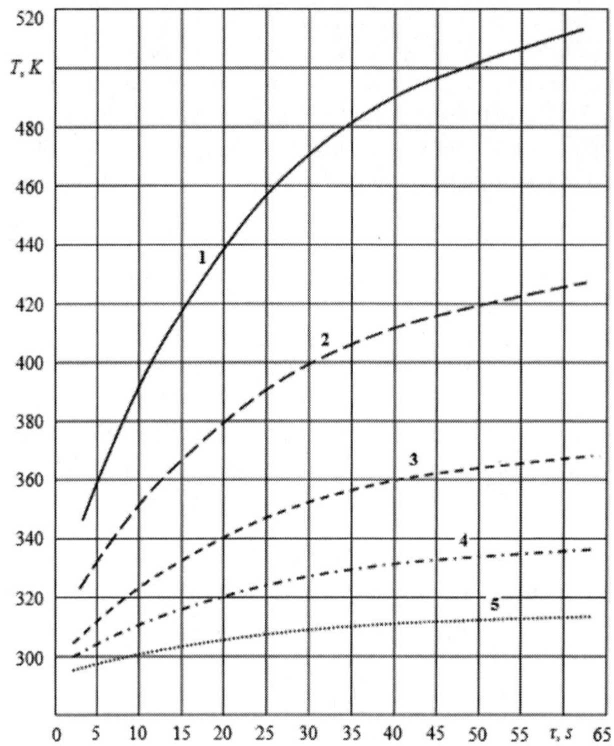

Fig. 2. Graphs of the rate of heating of the i-th section of the chamber in the regime of pulsating combustion

The results of the experiments were presented in the form of the value of the heat transfer coefficient α along the length of the chamber for the stationary and pulsating operating regimes of the three chambers. The characteristic convective heat transfer graphs are shown in Fig. 3.

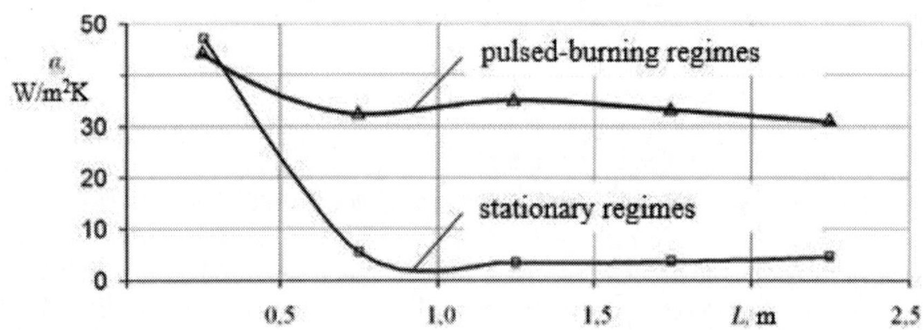

Fig. 3. Typical heat transfer graphs for a chamber with a diameter of 0.148 m

An analysis of the results of experimental studies made it possible to draw a number of conclusions.

It is established that in the regime of pulsing combustion, the value of the heat transfer coefficient α is much higher than in the stationary combustion regime. As can be seen from Fig. 3, during passing from one combustion regime to another, the heat transfer coefficient increases almost three times. It is noted that an increase in the heat transfer coefficient is associated with an increase in the velocity pulsations. In experiments with a chamber d = 0.092 m, the vibrational velocity of the gas flow reached 16 m/s. Calculation of the vibrational velocity of the gas flow during pulsating combustion was carried out on the basis of data obtained by measuring the pulsations of the gas flow pressure. The vibrational velocity in a plane traveling wave was found by the method of [6]. Figure 4 shows the characteristic graphs of the vibrational velocity of the gas flow: a solid line for harmonics with a frequency of 45 and 135 Hz in the pulsed combustion mode, an intermittent line for harmonics with the same frequencies in the steady-state combustion mode. When analyzing the diagrams of the vibrational velocity of the gas flow, it is established that in the regime of pulsating combustion considerable oscillations are generated in the chamber (up to 9 m/s). The highest values of the heat transfer coefficient α (up to 40 ... 50 W/m^{2}K) were fixed at the maximum values of the vibrational velocity. Consequently, an increase in the heat transfer coefficient α is due to a significant increase in the vibrational velocity of the gas stream.

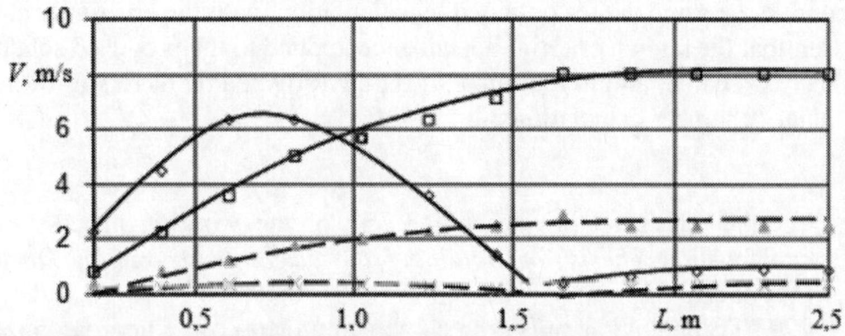

Fig. 4. Characteristic diagram of the vibrational velocity of the gas stream for a chamber with a diameter of 0.148 m

It is established that the diameter of the chamber does not significantly affect the heat transfer. Figure 5 shows the dependence of the average heat transfer coefficient $\bar{\alpha}$ average length on the chamber for the regimes of stable pulsating combustion. It is seen that as the diameter of the chamber d increases, the average heat transfer coefficient $\bar{\alpha}$ decreases insignificantly. The calculations showed an approximate dependence $\bar{\alpha} \sim d^{-0,15}$.

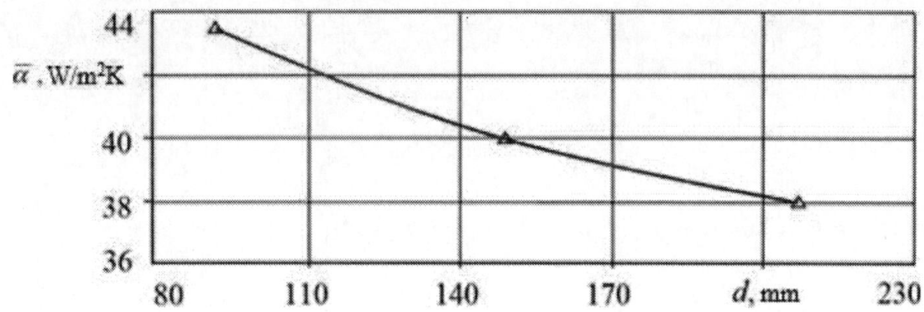

Fig. 5. Dependence of the average length heat transfer coefficient over the length of the chamber from the diameter of the chamber

Using the data obtained during the experiments and the calculated average values of the heat transfer coefficient $\bar{\alpha}$ an adequate normal equality of the Nusselt numbers was observed under pulsating and stationary gas flow regimes. The data were based on the data of the average constant speed W and the maximum oscillation amplitudes ϑ, obtained in the regime of stable pulsating combustion at the maximum heat transfer.

The obtained results of the Nusselt number ratios for different chamber diameters for pulsating and stationary operating modes are given in the table 1.

Table 1. Experimental and calculated ratios of Nusselt number values

Internal diameter of the chamber, m	Nп/Nc experimental	Nп/Nc calculated	Degree of conformity of experimental and calculated values, %
0.092	3.4	2.7	20
0.148	3.1	2.6	16
0.206	3.9	3.3	15

As can be seen from the table, the ratio of the values of Nusselt numbers for stationary and pulsating combustion regimes obtained experimentally (for chambers with internal diameter d = 0.092, 0.148 and 0.206 m) correspond to the same values obtained by calculation, with the error indicated in the table.

It allows to assert that the known criterial dependence applied to the proposed scheme of the pulsing combustion chamber is reliable, and in the range of the investigated diameters (0,092...0,206 m) it can be used for preliminary chamber calculations.

References

[1] Bychenok V.I. 2004 Thermal power engineering of the working process in apparatuses of pulsating combustion *Thesis for obtaining the scientific degree of Doctor of Technical Sciences* (Voronezh) 350p.
[2] Popov V.A. 1993 Technological pulsating combustion (Moscow: Energoatomizdat) 320p.
[3] Schlichting G. 1974 The boundary layer theory (Moscow: Nauka) 712p.
[4] Franke G. 1961 *Allgemeine Warmetechnic* V.2, P. 36-40; V.3, P. 49-59.
[5] Shenk H. 1972 Theory of Engineering Experiment (Moscow: Nauka) 382p.
[6] Kochergin A.V. 1996 Acoustic measurements (Kazan: KVAKIU) 91p.

Modeling the process of convective mass transfer in a tubular electrochemical reactor

S Y Sitnikov[1,2], L V Plotnikova[1], A F Dresvyannikov[3], N S Kiselev[1], S P Mironov[1]

[1]Kazan State Power Engineering University

[2]Kazan Federal University

[3]Kazan National Research Technological University

sjsitnikov@gmail.com

Abstract. The model, describing a processes of convective masstransfering in a tubular electrolytic cell with the much distinguishing sizes of electrodes is proposed suitable for practical calculations.
The electrochemical reactor permitting to clearing of water with the help of chemical reactions in bulk of a solution at participation of electrogenerated reagent is created.
Key words: modeling, tubular electrolytic cell, convective masstransfering, electrolysis of water, convection flow, convection in the vertical cylinder, modeling in Ansys Fluent.

Introduction

Electrochemical methods of water purification [1], the use of which is considered to be promising [2], are of particular interest in the water treatment areas of power generating objects [3]. Their advantage is the absence of secondary contamination of the treated water with ballast ions, the ability to carry out due to the electrolysis products decarbonization and disinfection of water, as well as the simplicity of hardware design and the possibility of automating the process through the use of Programmable Logic Controller.

For the electrolysis of water is usually used diaphragm (or membrane) electrolysis (DE), allowing for the separation of interelectrode spaces to carry out a variety of processes at the cathode and anode, adjust the pH in the cathode and anode chamber. DE can be operated in continuous operation for a relatively short time, then the diaphragm (membrane) is clogged with poorly soluble electrolysis products (hydroxyl compounds of multicharged ions). This leads to additional energy costs and causes the process to stop.

IOP Conf. Series: Earth and Environmental Science **288** (2019) 012118 doi:10.1088/1755-1315/288/1/012118

In addition, non-diaphragm electrolyzers – electrocoagulators (EC) are used to pre-purify water, in which dissolution of the anode takes place, followed by the formation of poorly soluble products with good coagulating ability. EC, working in a continuous mode, require special preparation of incoming water (preliminary pH change). In this regard, the task arises to develop an adequate model of the electrolyzer, which can combine the advantages of DE and EC, but does not have the above disadvantages.

The search for ways to intensify and improve the energy efficiency of the process of electrolysis of aqueous solutions involves the use of hidden possibilities. One of the approaches that can significantly change the ratio and parameters of the cathode and anode processes is the manipulation of the current density value. This approach can be implemented by using a geometric factor.

Let us consider the case of a tubular coaxial, diaphragmless electrochemical reactor with a significant (by 2-3 orders of value) difference in the area of the electrodes. The central thin cathode electrode is made of stainless steel wire grade 95X18, the outer electrode of larger area is a cylinder made of aluminum grade A95. The choice of such cell design is due to the axial symmetry of the electric field. At the same time, due to the significant difference in areas, the current density at the electrodes is also significantly different. Therefore, the process speed at the central electrode – cathode, will be much greater than the process speed at the external electrode – anode. In all further considerations, we assume that the reactor has a vertical orientation and a tubular shape, while the length (height) of the cell is much larger than its diameter.

The electrolytic cell is filled with a background electrolyte; the current of the electrolysis is close to the limit currents (i_{LIM}) of mass transfer [4]. In the presence of a background electrolyte, ions of which are present in excess compared to substances involved in electrode reactions, their migration in the electric field can be neglected. It is obvious that in this case mass transfer will be determined primarily by convective diffusion.

After switching on the electric current, at the initial stage of the process (time τ <10 s), the change in concentration determines the diffusion to the cathode. During the electrolysis, the gradient of the charge carrier concentration decreases and the diffusion current decreases. At the same time, the thermal convection caused by the heating of the solution near the central electrode is enhanced. The thermal convection flows, pushing the OH^- ions away from the cathode, are directed from the center to the periphery. At some point in time, the diffusion and convective flows have the same order, and then the convection process prevails. According to the published data [4-6], after a certain period of time, in the cell a stationary regime with a constant current intensity is established. The concentration of the reactant at the outer boundary of the diffusion layer is equal to the concentration inside the electrolyte ($C = C_0$) and is constantly maintained by convection currents that exist even in an unmixable liquid. After establishing a stationary mode of supplying the reactant to the electrode, the current does not change in time, but is determined only by the applied voltage. The entire volume of the fluid can be divided into two areas: a small area along the central electrode, inside which there is a rapid change in the concentration of the electrolysis products (the boundary diffusion layer of Levich) and the rest of the volume of the apparatus – the area of constant concentration (the Prandtl hydrodynamic layer).

Longitudinal convection

To create an adequate picture of the ongoing process, the diffusion model is taken as a basis. We assume that the flow structure is described by an expression similar to the equation of molecular diffusion [5,6]. The parameter of the model is the coefficient of longitudinal mixing D_A.

When considering convective processes, it can be excluded from consideration the upward flow, adjacent to the central electrode and due to the heating of the electrode [7,8] and the bubbling effect due to the release of hydrogen. In our case, convection and diffusion are directed in one direction and the equation for the concentration C in the longitudinal direction, along the z axis (1):

$$\frac{\partial}{\partial t}C(z,\tau) = D_A \frac{\partial^2}{\partial z^2}C(z,\tau) + V_C \frac{\partial}{\partial z}C(z,\tau), \tag{1}$$

where: D_A – coefficient of longitudinal (axial) mixing; V_C – velocity of convective flow, τ – time.

Set the initial and boundary conditions. At the initial time, the concentration of the reaction product (OH^--ions) at the central electrode 10^{-7} mol/l, so that $C(z, 0) = 10^{-7}$. The first boundary condition is the ratio $C(\infty, \tau)= 10^{-7}$ mol/l. Both of these values are very small, so with a good approximation we can assume that $C(z, 0) = C(\infty, \tau) = 0$.

The second boundary condition is the dynamics of concentration change on the surface of the «solution-air» section (z = 0; z axis is oriented along the electrodes axis) in accordance with the equation (2):

$$C(0,\tau) = C_0(1-\exp(-k\tau)), \tag{2}$$

where: k – the rate constant of the electrode reaction, depending on the time τ.

Applying the Laplace transform to the equation (2) with the above initial condition, we obtain its solution (3):

$$C(z,\tau) = C_0 \cdot e^{\left(-\frac{V_C}{D_A}z\right)}\left(1-e^{(-k\tau)}\right). \tag{3}$$

Radial convection

Equation (3), presented above, describes only the longitudinal component of the reaction product movement along the tubular reactor. In this case, it is advisable to consider its radial movement near the external electrode. Due to the complexity of the equation solution, taking into account the cylindrical geometry, we are content with a simplified case.

To describe the radial change in concentration, we make the following assumptions. Let us distinguish in a cylindrical volume of small height (the concentration change in height Δh in the first approximation is not taken into account) an area in the form of a thin parallelepiped between the axial and peripheral electrodes, to reduce the problem to the one-dimensional case. From the general equation of convective diffusion we obtain the following equation for the plane case (4):

$$v\frac{dc}{dr} = D\frac{d^2c}{dr^2}. \tag{4}$$

In order to simplify, the exact solution can be replaced by an approximate one, if one considers that, according to the conditions of the problem, $100 \cdot r_0 < R$, where r_0 is the diameter of the cathode wire, R is the diameter of the cylindrical anode (5):

$$c = c_0\left(1 - e^{\frac{v}{D}(r-R)}\right). \tag{5}$$

We take into account the following assumptions in order to simplify: the fluid is incompressible and stationary (the Peclet number proportional to the velocity of the fluid is small (Pe«1) or equal to zero).

Under the condition $v / D > 5 / r$, the exact solution differs from the approximate (5) by not more than 1%. We calculate the average concentration of the reaction product by volume (6):

$$\overline{C} = \frac{\int_0^R (2\pi r) \cdot c \cdot dr}{\pi R^2}, \tag{6}$$

where c – the concentration calculated by the formula (5). As a result of integration, we obtain an expression for the average concentration by volume (7):

$$\overline{C} = c_0\left(1 - \frac{2D}{vR} + \frac{2D^2}{v^2R^2} - \frac{2D^2}{v^2R^2}e^{-vR/D}\right). \tag{7}$$

Consider the dependence of C on v, taking into account the restrictions introduced in the derivation of formula (5) and assuming that v is numerically greater than D. It follows from (7) that the dependence is a monotonically increasing function and for large values ($v/D > 10$) the concentration is → 1. Obviously, even an insignificant increase in the convection velocity will help fill the volume of the cell with the product of the electrochemical reaction occurring on the central electrode, as confirmed by experiment [2].

FEM modeling in ANSYS

To obtain an analytical expression for convective heat and mass transfer, it is necessary to integrate the Navier-Stokes equations system, describing fluid hydrodynamics and heat transfer in it. Even with significant assumptions, this is possible only in a few simple cases. Therefore, for numerical solution of the problem of velocity distribution and convection visualization, one can use the finite element method (FEM) simulation in ANSYS.

There are two solvers for hydrodynamics problems – ANSYS CFX and ANSYS FLUENT. The main differences are in the way of integrating the equations of fluid flow and in strategies for solving equations. The ANSYS CFX solver uses a finite element grid (numerical values at grid nodes), similar to those used in strength analysis, to sample the area. Unlike ANSYS CFD, the ANSYS FLUENT solver uses a finite volume grid (numerical values at the centers of the grid cells). ANSYS FLUENT offers several approaches to the solution (density-based method, split pressure-based method, conjugate pressure-based method). In ANSYS CFX, special emphasis is placed on solving the basic equations of motion (conjugate algebraic grid).

In our case, the geometric model is created in AutoCAD and then imported into ANSYS-19.

We define the boundary conditions for the case of free convection. In the menu, select the adhesion condition "No Slip", the slip condition" Specified Shear", set the heat flux "Heat Flux " on the respective sides. In the "Reference Value" menu, the initial values of some parameters are set: density, viscosity, temperature.

Figure 1 shows the simulation result of velocity distribution of convective flow (vertical arrangement of the reactor, an aqueous solution of electrolyte, the steady-state mode).

When interpreting the results, it should be taken into account that in the case of an increase in the intensity of the process, the real flow studied in the experiment can lose axisymmetry, but in two-dimensional calculations we are based on the axisymmetric condition.

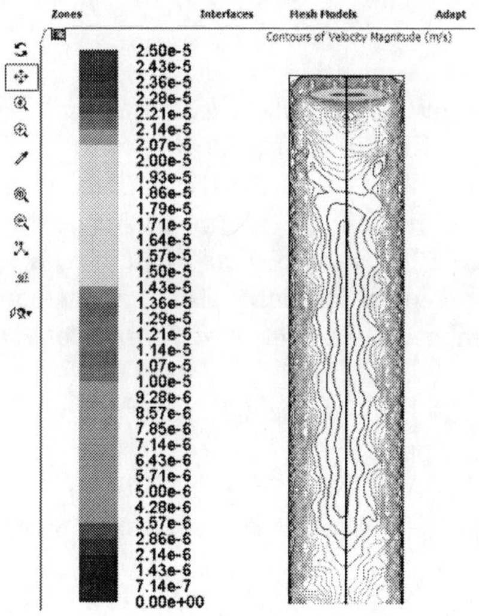

Figure 1. Simulation in ANSYS of velocity distribution of convective flows

It can be assumed that the flow rates are numerically equal to the mass transfer coefficient, with the dimension m/s = (mol/s)/(m^2·mol/m^3), where mol/s is the mass transfer rate, m^2 is the mass transfer area, m^3/mol is the molar concentration.

From figure 1 it can be seen that due to the convective mixing in a diaphragm-free cell, the electrolysis products almost completely fill the volume of the apparatus, which increases the efficiency of the apparatus. This distinguishes this scheme from traditional electrolyzers with the same area electrodes separated by a semipermeable membrane, where this effect is not observed due to carrier recombination and the presence of a diaphragm.

Summary

The above model was used by us in substantiating the efficiency of using a tubular coaxial membrane-free electrolysis cell for the electrolysis of water in order to purify it and explanation of occurring processes. When passing through the electrolyzer of the initial solution, we obtain a solution enriched with hydroxyl or hydrogen ions generated on the internal electrode. This is due to the fact that the solution enriched with the reaction products on the internal electrode and occupying almost the full volume of the cell, will be in turbulent motion and easily removed, and the solution saturated with the reaction products on the external electrode forms a laminar layer, which has an extremely low velocity (it is practically immovable and remains at the surface and in the irregularities of the external electrode).

Thus, the proposed approach makes it possible to purify water with the help of chemical reactions in the entire volume of the solution with the participation of an electrogenerated reagent, which is possible due to the hydrodynamic regime realized primarily due to the geometric configuration of the electrodes and the absence of a diaphragm.

References

[1] Sitnikov S Y and Dresvyannikov A F 2003 *Electrochemical water treatment in non-diaphragm electrolyzers* (Kazan: KSPEU Publishing Department) p 167

[2] Zhelovitskaya A V and Dresvyannikov A F 2012 *Indirect electrochemical oxidation of aromatic organic compounds* (Saarbrucken: LAP Lambert Academic Publishing) p 104

[3] Plotnikova L V, Chilikova I I, Sitnikov S Y and Efremov G I 2016 Systematic approach to the assessment of energy complex efficiency for thermal energy production with heat power saving transformer turning on *International Journal of Pharmacy and Technology* **8**(4) pp 26727-37

[4] Heifets L I and Goldberg A B 1989 Mathematical modeling of electrochemical reactors *Electrochemistry* **25** No 1 pp 3-33

[5] Sitnikov S Y, Dresvyannikov A F, Sopin V F and Sitnikova L A 2000 Mathematical model of a coaxial cell with substantially different electrode sizes *Iz. VUZ. Problems of energy* (Kazan) No 3-4 pp 112-14

[6] Dresvyannikov A F, Sitnikov S Y, Sopin V F, Dresvyannikov F N. and Sitnikova L A. 2001 Simulation of a coaxial electrochemical reactor with a central narrow-cylindrical electrode *Intra-chamber processes in power plants, acoustics, diagnostics, ecology* (Kazan) **1** p 252

[7] Sheremet M A 2010 The influence of cross effects on the characteristics of heat and mass transfer in the conditions of conjugate natural convection *Journal of Engineering Thermophysics* **19** pp 119–27

[8] Hess C F, Miller C W 1979 Natural convection in a vertical cylinder subject to constant heat flux *Int. J. Heat Mass Transfer* **22** pp 421-30

IOP Conf. Series: Earth and Environmental Science **288** (2019) 012119 doi:10.1088/1755-1315/288/1/012119

The study of the temperature field in pebble beds with volumetric heating and radial flow of coolant

Yu V Smorchkova, A N Varava, A V Dedov, A V Zakharenkov and A T Komov

National Research University "Moscow Power Engineering Institute" Russia, 111250 Moscow, Krasnokazarmennaya str., 14

Abstract. The paper presents the results of an experimental and numerical study of hydrodynamics and heat transfer in a pebble bed with internal volumetric heat release. An experimental setup was designed to investigate the hydrodynamics and heat transfer at the radial flow of a single-phase coolant. The pebble bed was set from 2.0 mm diameter sphere made of AISI 420 steel. The internal heat release model in the pebble bed is provided by high-frequency induction heating, whereby the thermal power of the high-frequency installation is 20 kW. The experiments were carried out in the range of mass flow rates of 0.1 - 0.6 kg/s. The test section simulates the fuel assembly of a nuclear reactor with microfuel. Pebbles are placed between the inner and outer perforated covers. Radial coolant flow is achieved to minimize pressure losses. The inner cover is cone-shaped and is the dispensing manifold for the coolant. The conical shape of the cover provides a uniform flow rate of the cooling liquid along the length of the test section. The temperature distribution in the pebbles were recorded.

Introduction

Recently, there have been more pressing issues about improving the energy efficiency and safety of nuclear power plants. One of the ways to solve this problem is to use a fuel assembly with microfuel (FA MF) directly cooled by a coolant [1,2]. The design of FA MF for VVER reactors was proposed in [1]. In such an assembly, the microfuel is placed between the perforated covers. Radial coolant flow is used to minimize pressure losses. Despite the fact that the geometric parameters of FA MF correspond entirely to traditional assemblies, their implementation on a large scale in nuclear power is hardly possible.

Also, there is an increasing interest in small-scale nuclear power engineering. It seems very interesting and promising to consider the use of microfuel for low-power reactors.

However, one of the difficulties that arise with the introduction of microfuel in nuclear power plants is the insufficient study of thermal-hydraulic characteristics of pebbles with internal heat release. There are no records of experimental studies on heat transfer and hydrodynamics in pebbles with radial flow coolants and internal heat release in literature.

In addition, it is necessary to resolve the issue on the choice of an optimal assembly design for low-power nuclear power plants. It is most convenient to solve such a problem using numerical simulation methods. The methods and mathematical models used in numerical calculations require verification using experimental data.

Experimental setup

Hydraulic circuit. The study was carried out on an open hydraulic circuit, which diagram is shown in Fig. 1. Distilled water was used as the working fluid, it is located in the tank (1). Circulation of the coolant is carried out by a multistage centrifugal pump Grundfos CRNE 1-4 (2). The water from the tank is injected by the pump through the filter (3) to the test section (РУ) and flows back to the tank. The hydraulic circuit is equipped with a preheater (5) and a heat exchanger (10).

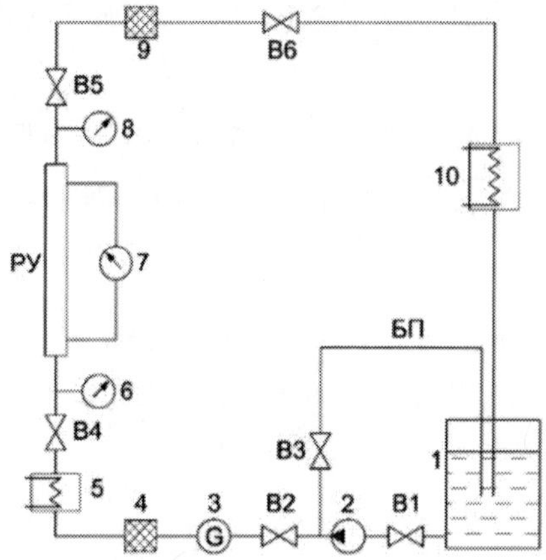

Figure 1. Hydraulic circuit

Consumption control was carried out in two ways: stepwise with the help of a frequency controller installed on the pump, and smoothly by means of a bypass line (БП) and valves (B2, B3). The flow rate through the test section was measured by an electromagnetic flowmeter (3) of the Vzlyot brand EM PROFI-212 model. To measure the pressure in the test section, the standard arrow gauges (6, 8) at the input and output of the test section and the differential pressure sensor (7) Elemer-100 were used. The experimental setup is designed for the following mode parameters: temperature in the circuit from 18 °C to 180 °C, pressure in the circuit up to 1.0 MPa, coolant flow rate (0.01-0.60) kg/s.

Test section. The plan of TS is shown in fig. 2.

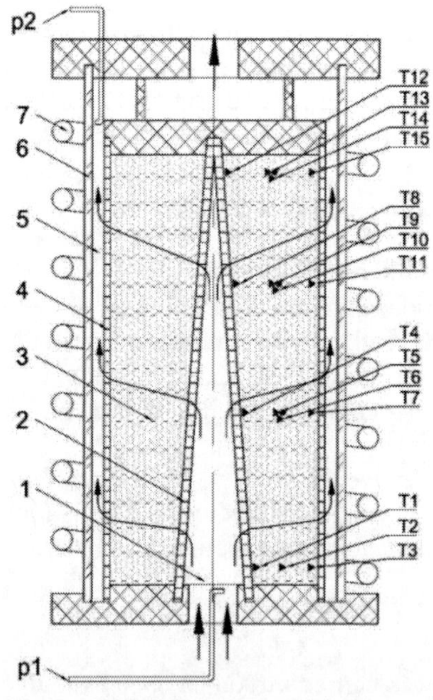

Figure 2. Diagram of the test section

The pebbles (3) are placed between the inner (2) and outer (4) perforated covers. The pebbles are filled up to a height of H = 100 mm. The coolant in such a design is fed into the dispensing manifold (1), washes over the pebbles and enters the collection manifold (5). The prefabricated collector is formed by an outer perforated cover and tube (6), which simultaneously performs the functions of external sealing of the working section. The diameter of the perforated outer cover is 47 mm, the diameter of the tube (6) is 54 mm. To ensure a uniform flow rate along the height of the pebbles, the inner perforated cover is cone-shaped. The diameter of the inlet is 15 mm. The cover is made of abs plastic using 3d printing. The perforated outer cover is a polycarbonate tube. The covers were drilled and the perforated holes were 1.5 mm in diameter. Pressure gauges were located directly in the dispensing (1) and collecting (5) manifolds at the inlet (p1) and the fluid outlet (p2) from the area occupied by pebbles.

Simulation of internal heat generation in pebbles is provided through high-frequency induction heating by means of an inductor (7), which is connected to the IHS20-60 induction heating unit. 15 cable thermocouples, installed in 4 sections along the height of the filling were placed to measure the temperature in the pebble volume. (T1-T15).

Pebbles in test sections were produced as 2.0 mm diameter spheres made of AISI 420 steel. The porosity of the pebbles $\varepsilon = 0.385$.

Experimental data

Temperature field without coolant flow. The use of induction heating to simulate the volumetric internal heating in a pebble bed is a non- standard method. Therefore, experiments on heating a pebble bed without the flow of coolant were carried out. The test section in these experiments was filled with water. The thermocouple readings were recorded during the experiments. The induction heating power was 1 kW. Temperature distribution along the height and radius of the test section is shown in Fig.3.

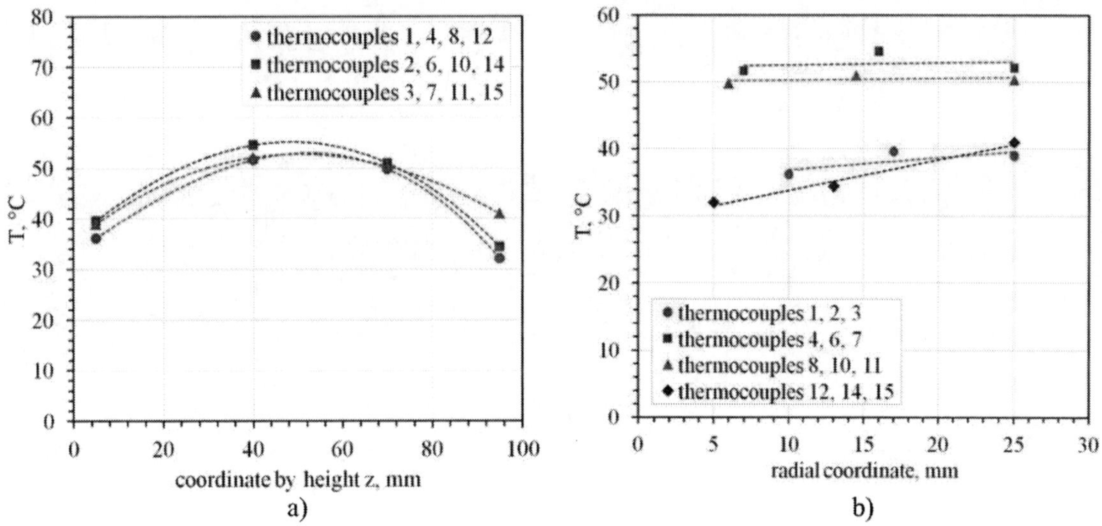

Figure 3. Temperature distribution along the height a) and radius b) of the test section

As a result of the experiments, the volume density of heat release at each point was determined by the formula:

$$q_v = \left[\varepsilon \rho_f c_{pf} + (1-\varepsilon)\rho_s c_{ps}\right]\frac{dT}{dt}$$

IOP Conf. Series: Earth and Environmental Science **288** (2019) 012119 doi:10.1088/1755-1315/288/1/012119

where – q_v is the volumetric heat release density, W/m^3, c_{pf} is the specific heat of the liquid at constant pressure, c_{ps} is the specific heat of the balls at constant pressure. The temperature at each point increases linearly with time, so the derivative dT/dt can be replaced by the temperature increment ΔT over the time τ. The distribution of the volumetric density of heat release normalized to the maximum value with respect to the dimensionless axial coordinate is shown in Fig. 4. Heat release along the height of the pebble bed has a strong heterogeneity.

When comparing the distributions of heat release along the axis in a dimensionless form with the data presented in the literature [3–5], a satisfactory agreement is observed, both qualitative and quantitative, in the central part of the backfill. At the edges of the pebble bed, the values obtained are understated. The difference between the profiles is probably due to the different geometry of the inductors.

The experimental data obtained in this paper are well described by a harmonic function:

$$q_v = q_{\max} \cos\left(2{,}5\frac{z - z_c}{H}\right),$$

where z is the coordinate of the height of the backfill, z_c is the coordinate of the center of the fill in the height

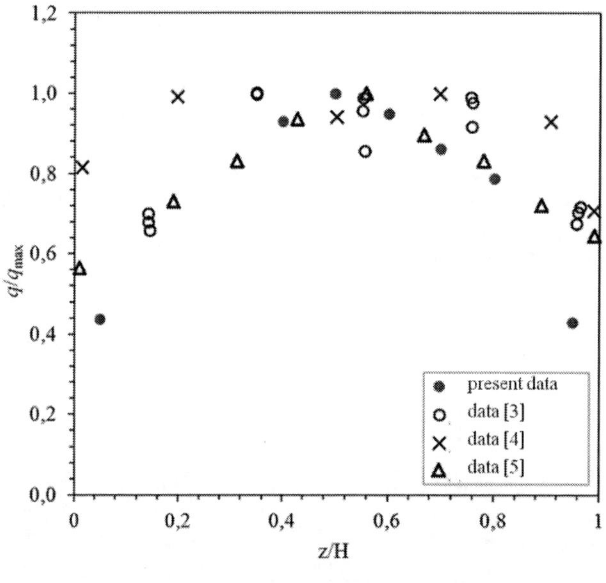

Figure 4. Dependence of the dimensionless density of volumetric heat release on a dimensionless coordinate by height

Temperature field with radial coolant flow. The heat transfer and temperature distribution in the teat section were studied for a range of coolant flow rates from 0.1 to 0.6 kg/s and volumetric heat release in the pebbles from 4.3 to 9,3·10^7 W/m^3. The distribution of temperatures recorded along the height and the radius of the pebbles are shown in Fig. 5 a), b). The results are presented for coolant flow rate 0.12 kg / s and volumetric heat release in pebbles 9,3·10^7 W/m^3. On the distribution of temperature along the height of pebbles, a maximum in the central part of the pebbles and a decrease in temperature to the ends were observed. Such a decrease in temperature is due to the peculiarities of induction heating and is due to the uneven heat release along the height. Uniform heat release is observed only in the central part of the pebbles, in the region where the thermocouples 4-7 and 8-11 are located. The graph of temperature distribution along the radius of the pebbles is as expected with a linear increase in temperature from the inner to the outer cover, which indicates uniform heating of the coolant.

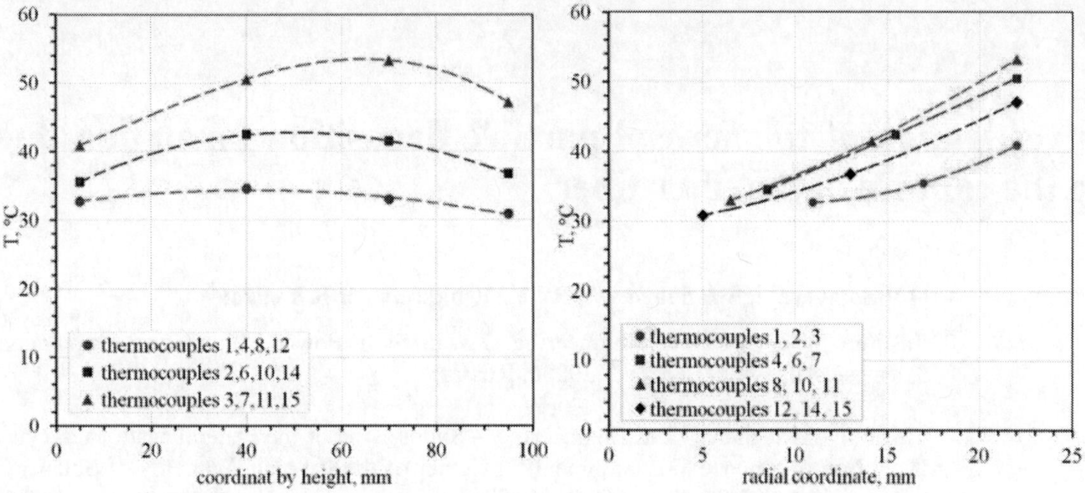

Figure. 5. Distribution of temperature over height a) and radius b) of the TS. Coolant flow rate 0.12 kg/s, volumetric heat release in the pebbles $9.3 \cdot 10^7$ W/m^3.

Conclusion

In this paper, a profile of volume heat release was obtained in a pebble bed heated by induction heating. It is shown that induction heating can be used to simulate the internal heat release in a pebble bed. Temperature distributions are obtained at the radial flow of the coolant through a pebble bed in conditions of internal heat release.

The work is supported RSF (grant № 16-19-10457)

References

[1] Ponomarev-Stepnoj N.N., Kukharkin N.E., Khrulev A.A., Degol'tsev Yu.G. and etc. Prospects of coated fuel particle application in the wwer reactors // Atomic Energy, T.86, №6, 1999. – P. 443-449.

[2] Gol'tsev A.O., Kukharkin N.E., Mosevitskij I.S., Ponomarev-Stepnoj N.N. A concept of safe tank water-moderated reactor with fuel units on the base of compact HTGR fuel rods // Atomic Energy, T.75, №6, 1993. – P. 417-423.

[3] Xianke Meng, Zhongning Sun, Guangzhan Xu Single-phase convection heat transfer characteristics of pebble-bed channels with internal heat generation // Nuclear Engineering and Design, №252, 2012. – C. 121– 127.

[4] Catton I., Jakobsson J.0. The Effect of Pressure on Dryout of a Saturated Bed of Heat-Generating Particles // Journal of Heat Transfer, №109, 1987. – C. 185-195.

[5] Naik A.S., Dhir V.K. Forced Flow Evaporative Cooling of a Volumetrically Heated Porous Layer // Int. J. Heat Mass Transfer, T.25, №4, 1982. – C. 541-552.

IOP Conf. Series: Earth and Environmental Science **288** (2019) 012120 doi:10.1088/1755-1315/288/1/012120

Investigation of the aerosol particle deposition formation due to the capture of the filter fiber

O V Soloveva[1,*], S A Solovev[2,3], R Sh Misbakhov[1], R R Yafizov[1]

[1]Institute of Heat Power Engineering, Kazan State Power Engineering University, Krasnoselskaja st. 51, Kazan, 420066, Russia

[2]Institute of Mechanics and Engineering – Subdivision of the Federal State Budgetary Institution of Science "Kazan Scientific Center of the Russian Academy of Science", Lobachevsky st. 2/31, Kazan, 420111, Russia

[3]Institute of Mathematics and Mechanics, Kazan Federal University, Kremlevskaja st. 18, Kazan, 420008, Russia

rara_avis86@mail.ru

Abstract. In this paper, an iterative model is constructed for the formation of a sediment of inertial aerosol particles on the surface of a cylindrical fiber, taking into account its effect on the gas flow and further settling of the particles. Parametric calculations are carried out for various values of the porosity of a number of fibers and the Stokes number, which determines the inertia of the particles. The shape of the sediment varies depending on the particle size. For strongly inertial particles, the thickness of the layer of settled particles is close to uniform. For weakly inertial particles with a small porosity of a number of fibers, a nonuniform distribution of the thickness of the deposited layer along the streamlined surface is observed.

Introduction

Aerosol filters are used in many devices to purify air from suspended particles. For a typical low flow rate through a filter, the deposition mechanism depends on the particle size [1, 2]. The fine particles settle on the filter elements by diffusion. Inertial precipitation and gravitation are the main processes of precipitation of large particles. The particle motion and deposition efficiency also depend on the structure of the filter. Typically, aerosol filters are a porous medium consisting of solid elements, for example, fibers [3-5] or granules [6-8], or an open cell foam material [9-12].

For small particles or in the nanosized region, the contribution of diffusion deposition increases, the intensity of which is determined by the Brownian diffusion coefficient. Investigations of diffusion deposition are discussed in [3, 8, 13]. It was shown in [14] that Brownian diffusion plays a decisive role for particles with a diameter less than 0.2 μm. For the problem of the diffusion deposition of polydisperse aerosol particles [15], analytical formulas for the efficiency of precipitation are obtained.

One of the most investigated is the inertial deposition of particles in porous media [1]. Early work on inertial deposition was devoted to the problem of flow past a gas suspension and the deposition of particles on a single cylinder. The main parameter influencing the inertial deposition is the Stokes number, therefore, in all studies, the dependence of the inertial sedimentation of suspended particles on the Stokes number is investigated [16, 17]. The notion of an effective Stokes number was introduced in [18, 19]. This made it possible to reduce the S-shaped curves of the inertial capture efficiency for the sphere and the cylinder into one universal curve in terms of the effective Stokes number. The authors of [19] obtained extended dependences not only for the efficiency of inertial capture of a cylinder in a potential flow, but also for the rates of subsidence, the concentration of particles on the cylinder surface, and the angular distribution of the quantities of interest. In [20] at Reynolds numbers of order unity, the fiber resistance forces and the trapping coefficients of inertial particles are determined. For identical values of the Stokes number, the capture coefficient is larger, the larger the Reynolds number.

In addition to inertial deposition, an important role is played by the effect of meshing particles (direct tangency), in which contact with the obstacle is determined by the geometric size of the particle and does not require that the trajectory of the particle's center cross the surface of the obstacle. Precipitation of particles due to the linkage was investigated in [21-24].

The solid elements of the filter can themselves have a porous structure. The motion of particles in a series of parallel porous cylinders was considered in [25, 26]. It was shown that the use of porous cylinders as elements of aerosol filters significantly increases the efficiency of diffusion deposition of aerosol particles. In [27], numerical calculations show that particles trapped in a porous cylinder can pass through the fibers.

During the operation of the filter and the trapping of particles on the surface of the fibers, a layer of settable particles is formed, which at the initial stage is a dendritic structure. With the further settling of the particles, the layer becomes denser and tends to a dense porous package.

The change in the form of the filter catching element affects the hydrodynamics of the flow between the fibers and the deposition of particles. In [28], the fractal form of the deposited layer, which is characteristic of the initial period of filter operation, is taken into account. Studies based on the assumption of a uniform porous layer of deposit particles on the surface of the element were carried out in [12, 29] and for cases of asymmetric precipitation in [30]. At the same time, the shape of the deposited layer can vary considerably depending on the deposition mechanism and depending on the hydrodynamics of the flow in the filter and the parameters of the particle phase [28, 31]. In the case of diffusion deposition, dendrites are formed over the entire surface of the fiber, and for the engagement mechanism, they are located on the frontal surface of the fiber. In the case of inertial collision, the particles will mainly settle near the front critical point.

In the present paper, an iterative model is constructed for the formation of a layer of inertial aerosol particles that settle on the surface of the filter fiber, taking into account its effect on the gas flow and further settling of the particles. The model includes a model of the flow of a carrier medium, the equations of motion of particles. At each iteration, the shape of the streamlined body is calculated as a result of subsidence. The thickness of the deposited layer depends on the porosity of a number of cylindrical fibers and the inertia of the particles. Parametric calculations were carried out.

Sludge formation model

We will assume that the motion of suspended particles does not affect the motion of the gas. We consider the flow past a fiber with a radius R_c of a viscous incompressible gas flow in a two-dimensional plane formulation [27, 32, 33]

$$\nabla \cdot \bar{U} = 0,$$
$$\varepsilon_c^{-2} \rho \bar{U} \cdot \nabla \bar{U} = -\nabla P + \frac{\mu}{\varepsilon_c} \nabla^2 \bar{U} - \beta \frac{\mu}{k} \bar{U}, \quad (1)$$

where $\bar{U}$ is the gas velocity, ρ is the gas density, μ is the dynamic viscosity of the gas, P is the pressure, ε_c is the porosity, k is the permeability of the porous medium, β is the coefficient determining the resistance to movement of gas in a porous medium. When calculating the external flow past a fiber ($\varepsilon_c = 1, \beta = 0$), the system of equations (1) becomes Navier-Stokes equations. When calculating the gas motion inside a porous region, the system (1) goes over into the extended Darcy-Brinkman equations. Let us choose the characteristic dimensionless parameters: for the outer flow around the fiber gas it is the Reynolds number $\mathrm{Re} = \rho U_0 2R_c / \mu$, for the gas motion inside the porous region it is the Darcy number $\mathrm{Da} = k / R_c^2$.

In Fig. 1 shows the considered region of gas flow with particles. In view of symmetry, we choose a periodic cell of height H and length $2H$. Then the packing density of cylindrical fibers is defined as $\alpha = \pi / 4h^2$, where $h = H / R_c$. To solve the system (1), we set the following boundary conditions: on the left boundary, we set the velocity of the free gas flow U_0, on the right boundary it is $\partial P / \partial n = 0$, at the upper and lower boundaries it is the symmetry conditions. The gas flow problem was solved by the finite volume method using the Fluent computational package.

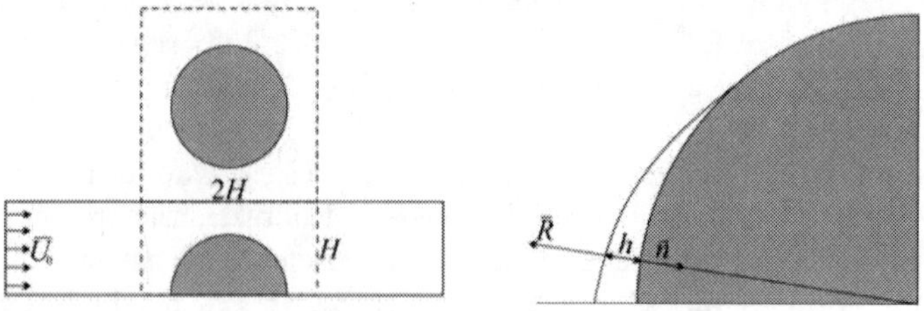

Figure 1. Scheme of calculation domain

The motion of suspended spherical particles will be considered against the background of a gas flow in the Stokes aerodynamic drag approximation

$$\frac{d\bar{V}}{dt} = \frac{\bar{U} - \bar{V}}{\tau},$$
$$\frac{d\bar{R}}{dt} = \bar{V}, \tag{2}$$

where $\bar{V}$ is the velocity of the suspended particle, $\bar{R}$ is the radius vector of the particle coordinate, $\tau = \rho_p D_p^2 / 18\mu$ is the relaxation time of the spherical particle, ρ_p is the particle density, D_p is the particle diameter. The system (2) is solved numerically by the Runge-Kutta method of the fourth order, taking into account the gas flow velocity obtained from the solution of the system (1). The characteristic dimensionless parameter in the study of particle motion is the Stokes number $\mathrm{St} = U_0 \tau / R_c$.

It is assumed that the particles settle upon contact with the surface of the fiber. For the considered problem of the motion of inertial particles, the value of the deposition efficiency E is calculated as the ratio of the number of particles reaching the surface to the total number of particles in the cross section of the cylinder far from the surface.

In this paper, we use a stationary model of gas flow. To determine the influence of build-up on the gas flow, we construct an iterative model for the formation of sediment. Let one iteration correspond to a maximum build-up thickness equal to 1% of the diameter of the cylindrical fiber. Solving the system (2), we determine the values V_n of the normal component of the velocity of the settled particles whose trajectories finish on the cylinder. Then determine the maximum thickness of the build-up as $h_{max} = 0.01 R_c$, at the point where $V_n = V_{n,max}$. In other points $h = h_{max}(V_n / V_{n,max})$. The values h are plotted along the normal to the fiber surface at the first iteration, to the current boundary of the build-up at subsequent iterations.

The normal component of the velocity of the particle V_n was chosen as the determining parameter of precipitate formation. Inertial particles colliding with a fiber can rebound [34, 35]. In the absence of electrostatic forces and the mechanism of engagement at high speeds, the particles will settle on the porous fiber at the maximum value of the normal velocity component.

Result and discussion

In the present work, calculations were made for cylindrical filaments of radius $R_c = 500$ μm, which corresponds to the fiber sizes of typical aerosol filters. The value of the velocity U_0 of the oncoming stream corresponds to the value $\mathrm{Re} = 1$. To calculate the deposition efficiency E, the start of the particles in the undisturbed flow is carried out at a distance $x = 250R_c$ from the porous cylinder. To test the calculations, gas-suspension flows were studied in a periodic row of solid cylinders. The efficiency curves for a solid-state solid cylinder at different values α of the packing density are calculated. Comparison of the obtained particle deposition efficiency curves with the efficiency curves calculated in [36] shows good agreement.

The dependence of the efficiency E on the Stokes number for various Darcy numbers for the fiber packing density parameter $\alpha = 0.01$, obtained using numerical calculations, is analyzed. The case $\mathrm{Da} = 0$ corresponds to an impermeable cylinder. In this case, the efficiency of inertial deposition decreases to very small values (close to zero) for small Stokes numbers [5]. For a porous cylinder, for sufficiently small values St, the value E remains finite. This means that we can provide a significant flow of particles to the porous cylinder, increasing its porosity. The authors of [13] used the range of Darcy numbers $\mathrm{Da} \le 0.1$ to calculate the flow of gas through a porous layer.

Calculations of sediment formation were carried out for the packing density value $\alpha = 0.01$. The buildup of buildup of inertial particles according to the proposed model for large numbers St will occur over the entire front surface of the cylinder with a uniform increase in the cross-sectional area of the fiber. Therefore, cases of small number St, where subsidence is observed only on a part of the front surface, may be of greatest interest. In Fig. 2 (a) shows the results of calculations after five iterations for the parameters $\mathrm{St} = 3$, the permeability of the cylinder $\mathrm{Da}_c = 0$, and the permeability of the deposition layer $\mathrm{Da}_h = 0$.

Five iterations determine 10% of the radius of the cylinder in the section of the greatest thickness of the sediment. The growth is formed almost on the entire front surface of the cylinder. In this case, the relative efficiency of particle capture is reduced by 4.17%. This corresponds to a decrease in the ordinate of the particle settling section.

In Fig. 2 (a) shows the calculations for the parameters of numbers $\mathrm{St} = 1.6$, $\mathrm{Da}_c = 0$, $\mathrm{Da}_h = 0$. It is seen the formation of a small growth in the front of the fiber. In this case, the deposition efficiency of the suspended particles decreases as well as in the case. On the growth, it is more clearly visible that the sediment layer at the fifth iteration is much lower in ordinate than the layer at the first iteration. This indicates the influence of build-up on the hydrodynamics of the gas flow and the motion of the particles. The relative efficiency of particle capture decreased by 12.38%. Thus, if one considers both the fiber and the resulting sediment impenetrable, that the build-up of particles results in a decrease in the efficiency of inertial deposition. Qualitatively, the pattern of formation of the build-up corresponds to [28, 31].

Let us consider the cases of the formation of a permeable build-up $\mathrm{Da}_h = 0.05$ on an impermeable fiber $\mathrm{Da}_c = 0$. We also choose the numbers $\mathrm{St} = 3$ and $\mathrm{St} = 1.6$ (Fig. 2 (b)). The formation of the build-up has a similar picture with the case $\mathrm{Da}_h = 0$. At the same time, the values E of deposition efficiency are almost unchanged in both cases. There are small increases E in the second and third iterations and the subsequent decrease. Since the permeability of the build-up is sufficiently large, the flow of gas freely passes through it, obtaining a flow similar to that of an impermeable fiber. Obviously, for the case $\mathrm{Da_h} = \mathrm{Da_c} > 0$ of formation of a built-up edge, it makes almost no contribution to the change in efficiency E.

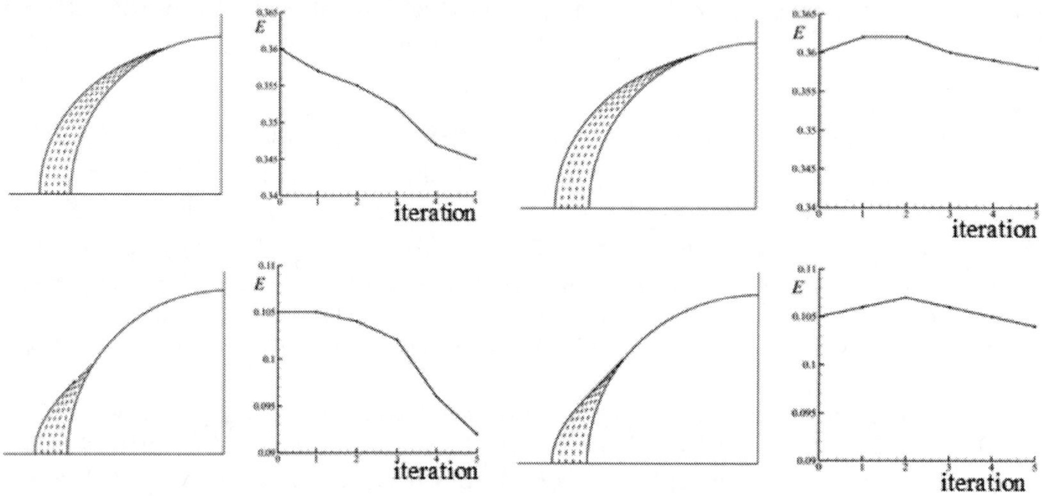

(a) (b)

Figure 2. Calculation of deposition layer formation and particle deposition efficiency at: (a) $\mathrm{Da_c} = 0$, $\mathrm{Da_h} = 0, \mathrm{St} = 3$ and $\mathrm{St} = 1.6$; (b) $\mathrm{Da_c} = 0$, $\mathrm{Da_h} = 0.05$, $\mathrm{St} = 3$ and $\mathrm{St} = 1.6$.

Let us consider the case of a flow past a strongly permeable fiber $\mathrm{Da_c} = 0.1$. In this case, the build-up built-up edge has a permeability $\mathrm{Da_h} = 0.01$ that is less than the permeability of the cylinder. The calculation result for $\mathrm{St} = 1.6$ after one iteration shows the build-up on almost the entire front surface of the fiber. In this case, after the formation of the build-up, the relative efficiency of particle capture decreased by 59.17%. That is, the built-up edge has brought the efficiency of particle capture to the efficiency of a cylinder without a build-up of the same permeability $\mathrm{Da_c} = 0.01$.

Conclusion

Analysis of the results shows that the deposition efficiency is slightly reduced in cases where the permeability of the cylinder is less than or equal to the permeability of the precipitate. Thus, even large sediment (about 10%) of inertial particles formed during long-term operation does not greatly affect the efficiency of the filter. In this case, the situation changes in the case when strongly permeable fibers are used, which are more effective, and the build-up of sedimented particles has less permeability. Numerical calculations show that a denser (but sufficiently thin) sediment changed its efficiency and brought it closer to the values, as if the entire cylinder had permeability of the deposited layer.

Acknowledgments

The reported research was funded by Russian Foundation for Basic Research and the government of the Republic of Tatarstan of the Russian Federation, grant № 18-41-160005.

The work was financially supported by the Ministry of Education and Science of the Russian Federation within the framework of the federal target program "Research and Development in Priority Areas for the Development of the Russian Science and Technology Complex for 2014-2020", an agreement on the grant in the form of a grant No. 14.574.21.0188-1 stage, unique identifier of applied scientific research (project) RFMEFI57418X0188.

References

[1] Brown R C 1993 *Air filtration. An integrated approach theory and applications of fibrous filters* (Oxford: Pergamon Press)
[2] Zagaynov V A 2010 *The inverse problem and aerosol measurements, in Aerosols -Science and Technology* / ed. by Igor Agranovski. Wiley-VCH Verlag GmbH and Co.KGaA. 241
[3] Stechkina I B and Fuchs N A 1966 *Ann. Occup. Hyg.* **9** 59
[4] Lee K W and Liu B Y H 1982 *Aerosol Sci. Technol.* **1** 147
[5] Asgharian B and Chenh Y S 2002 *Aerosol Sci. Technol.* **36** 10
[6] Wang J and Pui D Y H 2009 *J. Nanoparticle Res.* **11(1)** 185
[7] Deo S, Filippov A, Tiwari A, Vasin S and Starov V 2011 *Advan. Colloid Interface Sci.* **164(1/2)** 21
[8] Kirsh V A and Kirsh A A 2017 *Colloid J.* **79(4)** 474
[9] Brown D and Wake R C 1991 *J. Aerosol Sci.* **22** 693
[10] Hellmann A, Pitz M, Schmidt K, Haller F and Ripperger S 2015 *Aerosol Sci. Technol.* **49(1)** 16
[11] Solovev S A, Soloveva O V and Popkova O S 2018 *Rus. J. Phys Chem A.* **92(3)** 603
[12] Soloveva O V, Solovev S A, Khusainov R R, Popkova O S and Panenko D O 2018 *J. Phys.: Conf. Series.* **944** 012113
[13] Dunnett S J and Clement C F 2009 *Eng. Analys. Bound. Elements.* **33** 601
[14] Qian F, Zhang J and Huang Z 2009 *Chem. Eng. Technol.* **32(5)** 789
[15] Kwon S B, Kim H T and Lee K W 2002 *Aerosol Sci. Technol.* **36(6)** 742
[16] May K and Clifford R 1967 *Ann. Occup. Hyg.* **10** 83
[17] Israel R and Rosner D E 1983 *Aerosol Sci. Technol.* **9** 29
[18] Wang H C 1986 *J. Aerosol Sci.* **17** 827
[19] Wessel R A and Righi J 1988 *Aerosol Sci. Technol.* **9** 29
[20] Kirsh V A, Pripachkin D A and Budyka A K 2010 *Colloid J.* **72(2)** 211
[21] Nguyen X and Beeckmans J M 1975 *J. Aerosol. Sci.* **6** 205
[22] Liu G Z and Wang P K 1997 *Aerosol Sci. Technol.* **26(4)** 313
[23] Zhu C, Lin C H and Cheung C S 2000 *Powder Technol.* **112** 149
[24] Raynor P C 2008 *Aerosol Sci. Technol.* **42(5)** 347
[25] Kirsh V A 2006 *Theor. Found. Chem. Eng.* **40(5)** 465
[26] Kirsh V A 2007 *Colloid J.* **69(5)** 609
[27] Zaripov S K, Soloveva O V and Solovev S A 2015 *Aerosol Sci Technol.* **49(6)** 400
[28] Kasper G, Schollmeier S and Meyer J 2010 *J. Aerosol Sci.* **41** 1167
[29] Kirsh V A 2007 *Colloid J.* **69(5)** 615.
[30] Kirsh V A 2014 *Colloid J.* **76(4)** 435
[31] Dunnett S J and Clement C F 2012 *J. Aerosol Sci.* **53** 85
[32] Bhattacharyya S, Dhinakazan S and Khalili A 2006 *Chem. Eng. Sci.* **61** 4451
[33] Mardanov R F, Soloveva O V and Zaripov S K 2016 *IOP Conf. Ser.: Materials Sci. Eng.* **158** 012065
[34] Panfilov S V and Tsirkunov Y M 2008 *J. Appl. Mech. Tech. Phys.* **49(2)** 222
[35] Chernyakov A L, Kirsh A A and Kirsh V A 2011 *Colloid J.* **73(3)** 389
[36] Müller T, Meyer J and Kasper G 2014 *J. Aerosol Sci.* **77** 50

IOP Conf. Series: Earth and Environmental Science **288** (2019) 012121 doi:10.1088/1755-1315/288/1/012121

Regional management of district heating

N Sovetova[1], A Sinitsyn[1], E Tritenko[1], O Derevianko[2], L Mukhametova[3], A Fedukhin[4], S Makoev[2], A Kalyutik[2]

[1]Vologda State University, Lenina Street 15, Vologda, 160000, Russia

[2]Peter the Great St. Petersburg Polytechnic University, Polytechnicheskaya 29, Saint Petersburg, 195251 Russia

[3]Federal State Budgetary Educational Institution of Higer Education "Kazan State Power Engineering University", Kazan, Krasnoselskaya street, 51

[4]National Research University «Moscow Power Engineering Institute», 111250 Russia, Moscow, Krasnokazarmennaya 14.

makoev@nil-teplo.ru

Abstract. Nowadays in Russia we see rapid modernization of power engineering and transport systems. Under these conditions the issue concerning improvement of regional heat supply system energy-efficiency has gained a new importance. The aim of this study is to find new development directions of the public-private partnership (PPP), specifically from the perspective of its energy efficiency. Amid the financial crisis and economic sanctions the formation of public-private partnership allows increasing of energy efficiency of regional fuel and energy complex by means of regional energy service company creation. This paper will be useful both for central and local government and for private business.

Introduction

After the acceptance of the Federal Act No.261 "On energy saving and energy efficiency" the Russian Federation acceded the modern global trend on energy saving.

This Act concerning reduction of power consumption became a necessary measure amid the existing energy crisis and market relations. When comparing this situation with the countries of West Europe and USA, which deal with the problem of energy efficiency from the beginning of the 70s, we should study their experience of real energy economy in energetics and construction, as well as laws providing the priority for companies and citizens, which use energy-saving technologies and solutions. It should be noted, that almost simultaneously with energy crisis in USA, the first complex document - Energy Act 1992 - appeared, which denoted the basic problems and their possible solutions for energy-saving. The Government launched the program of perspective standards and other normatives, aimed at energy saving at various industry branches. This program was based on so-called "performance standards", i.e. it was oriented at the end result, which is energy economy. Enormous funding was invested in public promotion of the energy-saving program and clarification of its targets. In 2004 The USA Congress declared the successful implementation of the Energy Act 1992 and adopted a new Energy Act 2004 document. Also they developed the so-called Road Maps for various industry branches [1].

As for Germany, the Federal law initially included the public-private partnership, federal granting for modern materials and technologies usage, and various motivations. Also we should mention the "20-20-20 Program", which is the first stage of the ambitious plans of the European Union. These plans are for 90% reduction of greenhouse gases emission up to the year 2050 (in comparison to the year 1990) and energy consumption savings from 175 to 320 bln Euro annually between 2010 and 2050 (the initial investments are 270 bln Euro).

Nowadays legal regulation of the energy-saving has become even more urgent due to orientation of the countries at energy-efficiency and energy-saving. The investigation of mechanisms of government influence in the sphere of energy, estimation of efficiency of legal regulations influence and energy-efficiency increase are the main factors for lawmaking concerning private-public partnership (PPP) development. This fact is approved by gathered world experience at PPP working conditions.

Nowadays Russian heat-and-power engineering is characterized by insufficient reliability and energy-efficiency of the heat supply system, as well as danger of accidents. The majority of the enterprises have worn equipment and utilities, low production efficiency, lack of funding and the absence of concurrence. The main priorities of the Government policy for heat-supply are the following:

- Heat supply reliability according to the technical regulations;
- Heat supply and consuming energy-efficiency;
- Development of district heating networks;
- Balance the economical interests between the heat-supply organizations and consumers;
- Economically-reasonable profitability of current activity of the heat-supplying organizations and investments;
- Non-discriminatory and stable conditions for heat-supply business;
- Environmental safety;
- Safe operation of heat supply units.

The issues concerning reforming of public utilities and improvement of its performance in the Russian Federation are the key points for the following researchers: Andryushenkov A.F., Basin E.V., Vihrev YU.V., Gorbunov V.N., Demidova N.V., Kurbatov K.A., Chernyak V.Z., CHernyshov L.N., etc. A.S. Plekhanov studied experience of the European countries. V.G. Ignatov and V.I. Bytov studied the role of local self-government of housing and communal services at East Europe countries. The work of T.B. Kytakova is devoted to condominium creation experience.

The problems of energy-efficiency improvement in various fields of industry are investigated in [2-31], the estimation of Federal law on energy-efficiency applicability for heat consumption efficiency improvement is presented in [32-35]. The question of training academic and teaching staff for engineering and heat-and-power industry is of urgent interest [36-39].

Main Text

The aim of the investigation is analysis of the problems arised at investment of heat-supply system of the Vologda region and development of recommendations for improvement of regional management of this area. The object under study is Vologda region, the subject under study is Government management of the heat-supply system.

The analysis of foreign development experience was conducted [40,41]. So, the degree of government authorities participation in economic activities of the heat-and-power organizations varies from one country to another. However, the predominant activities are various types of concessions involving private investors, in particular, operation according to the two accepted models:

1. The model of the "single-buyer" The resource provider (thermal energy supplier/heating network operator) purchases thermal energy from the producers (all thermal energy sources) and sells it to the consumers under equal conditions and at equal prices. It is possible to use this model upon condition that one company is responsible for the thermal energy sale to the final consumer. The "single-buyer" model is the most common.

2. The model of "open heat networks". The thermal energy producer has a guaranteed access to the heat network upon condition that he sells the thermal energy directly to his own clients to the extent required for consumers (this model is almost not applied in practice).

When setting the tariffs for the thermal energy the following approaches are implemented:

A) "Fare construction" (this market model doesn't allow dominant suppliers to set the monopolic prices, the control for thermal energy market operation for the purpose of compliance of the existing competition law regulations is implemented by a special agency, is applicable in Finland, Denmark, Sweden, Germany, Austria, France, Great Britain, Belgium);

B) "Maximum level": the maximum tariff value is determined according to the established methodology for each company-producer, are approved by an independent national regulator. At this the resource-providing company might decrease the thermal energy price (Estonia, Latvia, Lithuania, Poland, Czech Republic, Slovakia, Hungary, Bulgaria, Macedonia);

C) "Fixed level" for the tariff setting is conducted according to the established methodology and is approved by a regulator. Here the heat-supply company cannot digress from the fixed tariff level (Russia, Belorussia, Romania, Ukraine).

Table 1 - Length of heat and steam networks along municipal areas and city districts of the Vologda region

	Total, km	Including need changing, km	Worn networks among them, km	The ratio of worn networks, %	Were changed in terms of two-pipes, km	
					total, km	Worn among them, km
At the territory of the region	**1805.0**	**794.2**	**650.6**	**36.0**	**24.0**	**20.9**
Babaevskij	35.6	17.9	17.9	50.2	0.2	0.2
Babushkinskij	5.5	0.5	0.5	9.0	0.1	0.1
Belozerskij	21.4	6.2	4.8	22.4	0.7	0.4
Vashkinskij	10.2	1.0	0.9	8.8	-	-
Velikoustyugskij	91.1	45.1	43.1	47.3	0.2	0.2
Verhovazhskij	10.1	2.6	2.5	24.8	0.2	0.2
Vozhegodskij	21.4	-	-	-	-	-
Vologodskij	145.1	25.5	21.1	14.5	2.5	2.4
Vytegorskij	20.8	10.9	10.4	50.0	0.6	0.6
Gryazoveckij	81.6	7.0	5.5	6.7	1.0	0.7
Kadujskij	83.8	36.5	26.1	31.1	0.7	0.6
Kirillovskij	28.8	12.9	12.9	44.8	0.2	0.2
Kichm.-Gorodeckij	3.6	1.2	-	-	-	-
Mezhdurechenskij	12.3	7.3	3.2	26.0	-	-
Nikol'skij	22.7	4.3	2.7	11.9	0.9	0.8
Nyuksenskij	27.4	0.8	0.8	2.9	0.1	0.1
Sokol'skij	112.0	45.2	44.8	40.0	0.9	0.7
Syamzhenskij	16.0	0.7	0.6	3.8	0.8	0.8
Tarnogskij	7.6	1.7	0.9	11.8	0.1	0.1
Totemskij	81.3	17.0	10.9	13.4	0.5	0.5
Ust'-Kubinskij	7.4	5.2	5.2	70.3	-	-
Ustyuzhenskij	43.4	26.2	20.0	46.1	0.2	0.2
Harovskij	43.3	5.2	1.2	2.8	3.0	1.1
Chagodoshchenskij	44.6	26.9	24.5	54.9	0.8	0.8
Cherepoveckij	73.2	34.2	33.5	45.8	1.7	1.7
Sheksninskij	62.5	14.4	1.0	1.6	0.1	-
Vologda	362.2	193.2	111.7	30.8	8.0	8.0
Cherepovec	330.1	244.6	243.9	73.9	0.5	0.5

At the Vologda region 803 sources of thermal energy are operating, the heat network length is 1805 km, among them 44% should be changed, the worn networks change rate is 21.8%, which is apparently insufficient for providing of pipeline wear and tear reduction, the amount of emergency situations has increased for 47%. The amount of thermal energy sources has reduced for 5% because of closure of inefficient (obsolete) ones and enlargement of heat-supply zones. These actions contribute to improvement of thermal energy production and allocation efficiency (Tables 1,2).

Table 2 - Heat losses and the amount of breakdowns in networks for municipal areas and city districts of the Vologda region

	2014			2015		
	Heat losses		Amount of breakdowns	Heat losses		Amount of breakdowns
	Total, ths. GCal	%		Total, ths. GCal	%	
At the territory of the region	**908.3**	**8.8**	**34**	**831.1**	**8.3**	**50**
Babaevskij	12.8	13.1	-	16.6	19.6	-
Babushkinskij	1.0	6.5	-	1.0	7.0	-
Belozerskij	4.4	14.1	-	7.6	13.9	-
Vashkinskij	3.8	23.8	-	3.6	24.4	-
Velikoustyugskij	53.9	21.2	4	45.5	20.6	1
Verhovazhskij	1.5	7.7	5	1.4	8.3	1
Vozhegodskij	9.4	20.2	-	8.8	19.7	-
Vologodskij	41.8	10.6	-	40.5	12.0	-
Vytegorskij	6.5	13.2	9	8.6	13.5	-
Gryazoveckij	45.0	14.0	-	43.7	14.9	-
Kadujskij	37.9	14.6	1	26.8	11.5	30
Kirillovskij	4.8	14.0	10	4.5	11.2	10
Kichm.-Gorodeckij	0.9	6.1	-	1.0	7.2	-
Mezhdurechenskij	2.9	16.3	-	2.7	16.4	-
Nikol'skij	2.5	6.0	1	2.2	5.9	-
Nyuksenskij	13.2	30.6	-	14.6	34.4	-
Sokol'skij	46.2	18.5	-	40.1	13.0	-
Syamzhenskij	3.0	14.7	-	2.5	13.0	-
Tarnogskij	3.3	16.7	-	3.1	15.8	-
Totemskij	22.8	16.9	-	20.8	15.5	-
Ust'-Kubinskij	1.5	10.5	-	3.0	21.0	5
Ustyuzhenskij	10.7	19.7	3	7.3	14.4	3
Harovskij	6.4	8.1	-	8.9	8.0	-
Chagodoshchenskij	15.3	22.3	1	13.6	22.2	-
Cherepoveckij	24.5	10.3	-	32.3	17.9	-
Sheksninskij	26.3	8.1	-	23.7	9.6	-
Vologda	71.6	8.1	-	62.0	2.9	-
Cherepovec	434.4	15.1	-	384.6	14.0	-

The specific energy costs for release of thermal energy unit from source collectors into the heat networks, as well as transportation of the thermal energy along the heat networks to the consumers characterize the energy-efficiency of the heat supply system operation (specific fuel consumption is the indicator) (kgoe/GCal).

Heat source efficiency is the value, which is inversely proportional to specific fuel consumption for 1GCal of thermal energy production taking into account energy equivalent of GCal thermal energy, expressed in kg of oil equivalent (142.86 kgoe/GCal).

For the Russian Federation territory the indicators for oil equivalent consumption per unit of the produced heat are 178.8-190.1 kgoe/GCal on the average. Consequently, the average efficiency of the boiler houses is 75.1% -79.9% (Table 3). The weighted-average efficiency of Vologda region boiler houses operation is 84%. This corresponds to average specific fuel consumption for heat release from the boiler houses collectors, which is equal to 170 kgoe/GCal.

Table 3 - Specific energy costs for release of thermal energy unit

Region	Actual fuel consumption per thermal energy unit, kgoe/GCal			boiler house efficiency, %		
	2012	2013	2014	2012	2013	2014
Russia	183.2	178.8	190.1	78.0	79.9	75.1
Central Federal District	183.4	179.4	195.1	77.9	79.6	73.2
Northwestern Federal District	168.2	166.8	165.5	84.9	85.6	86.3
South Federal District	162.5	162.5	163.7	87.9	87.9	87.3
СКФО	162.9	158.4	164.5	87.7	90.2	86.8
Privolzhsky Federal District	178.6	178.4	204.6	80.0	80.1	69.8
Ural Federal District	162.9	161.8	173.9	87.7	88.3	82.2
Siberian Federal District	227.5	211.5	213.4	62.8	67.5	66.9
Far Eastern Federal District	176.2	174.2	174.2	81.1	82.0	82.0

The highest efficiency of the thermal energy sources have Cherepovetsk region, Cherepovetsk, Velikiy Ustyug and Vologda, where this indicator exceeds 90%. For the boiler houses located at Vashkinskii, Sokol'skii, Babushkinskii, Syamzhenskii, Kirillovskii districts and Sokol town, the efficiencies are less than 50%.

After the comparison of heat supply system operation indicators (Table 4) for the Northwestern Federal District at the existing heat network length, the estimated heat losses ratio appeared to be 8.3%, the amount of breakdowns is 2.7 per 100 km.

Table 4 - Comparative characteristics of operation indicators for heat supply systems of Northwestern Federal District regions.

	the Republic of Karelia	Vologodskaya Oblast'	Novgorodskaya Oblast'	Murmanskaya Oblast'	Arkhangelskaya Oblast'	The Republic of Komi	Kaliningradskaya Oblast'
Heat networks length, km	951.3	1805.0	942.7	1068.4	1880.2	1956.8	996.5
Annual het losses (total), %	5.6	8.3	12.6	8.7	10.8	14.1	14.8
The amount of breakdowns at heat networks per 100 km	3.0	2.7	9.0	4.6	6.4	1.0	14.2

The main costs at production and allocation of the thermal energy are expenditures for fuel and power resources (42%), labour expenditures (20%), material and equipment maintenance expenditures (17-18%). The amortization expenses are 5-6%, miscellaneous expenditures are 14% (rent, expenditures for manufacturing measures, payments for negative impact on the environment, doubtful debts, etc) of the total production cost. This doesn't allow the enterprises to form their own investment sources for modernization of basic production funds.

Basing on the conducted analysis, we highlight the following problems, which hinder the development of regional heat supply system:

- Wear-and-tear of the basic production funds leads to system management losses, reliability decrease, impossibility of the required quality service provision;
- Insufficient investment for implementation of program activities, reduction of capital investments for basic production funds modernization (non-fulfillment of activities provided at the investment programs by heat supply organizations);
- due to unstable economics and absence of long-term funding, attraction of private investment to

the communal sphere possesses high risks both for potential investor and consumer;

- insufficient amount of high-efficient thermal energy sources, absence of energy-efficient projects, aimed at production cost reduction.

At modern economics conditions the majority of the countries tend to implementation of infrastructure projects with attraction of extrabudgetary funds and private investors, which results in relieve of budgetary pressures and improves quality and efficiency of industry objects exploitation. Public-private partnership (PPP) is one of the most widespread instruments for attraction of private investments for infrastructure projects [42-48].

Public-private partnership is a system of mutually-beneficial contract relationship between public and private partners. This partnership is aimed at co-funding of projects for development of socially important government-regulating industry branches of services [49,50].

According to the investigation of Association "Centr razvitiya GChP" ("PPP development Center"), nowadays there are more than 1300 projects of public-private partnership in Russia, among them are the following:

- 15 projects of the federal level;
- 191 projects of the region level;
- 1100 projects of the local level.

The analysis of the accomplished public-private partnership projects, has shown that annual average rate of investment attraction using this mechanism is rather high. It is driven by development of laws concerning concession agreements, additional guarantee instruments for investors have appeared, the institute for private concession initiative was created [51, 52].

According to the Russian Federation region rating, prepared by the "Centr razvitiya GChP" ("PPP development Center") and RF Ministry of Economic Development in 2015 [53], in 2014-2015 Vologodskaya Oblast' is a region with moderate potential for public-private partnership project development. It is on the 35 position of the rating. The subject was characterized as a region, where the initial experience is established for public-private partnership, at this the conditions for harmonic PPP development are created at the region.

During our investigation, we have denoted basic types of energy-service contracts, which are definitely attractive just for the heat supply system:

1. Energy service agreement with economies division (SharedSavings). This type of agreement implies funding of energy-efficient activities of the energy-service company with further compensation through technology-implementation economy. Savings from implementation of energy-saving technologies are divided between owner (Energy Service Customer) and energy-service company. The contract period duration is from 3 to 10 years (depending of the payback period).
2. Energy service agreement with guaranteed investment return (FirstOut). This agreement also impliesfunding of energy-efficient activities by the energy-service company with further compensation through technology-implementation economy. But in this case the contract period might be changed depending of the actual level of economy.
3. Energy service agreement with short payback time (First-Out, FirstPay-Out). This agreement implies investor cost recovery in the first place. When using this agreement the energy-service company receives all 100% of the gained economy until it will pay all expenses for project implementation.

The main stages for energy-saving project realization are shown in Figure 1.

Advantages of energy-service agreements are the following: economic efficiency increase, reduction of energy consumption at thermal energy production; release of funds at the end of the contract; absence of initial investment expenditures; the expected maximum efficiency for contractor service, as the contractor income is directly dependent on the achieved economy level, and the risk of not achieving the economy is born by the contractor.

The earlier investigations have shown [54] that this investment mechanism is practically absent at Vologodskaya Oblast'. However, this type implies implementation of projects with funds provided by the contractor with upcoming fund return at the expenses of the achieved economy for energy resources expenditures and doesn't need further funding.

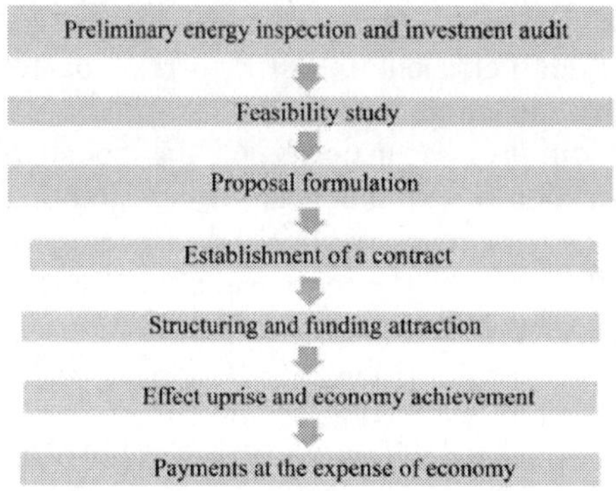

Figure 1 - Stages for energy-saving project realization

Conclusions and proposals

We believe that the following administrative mechanisms are appropriate for stimulation of the scheme implementation process and transition to the energy service contracts of the heat supply organizations. First of all, development of regional and municipal energy-service companies, as well as organizations with government and municipal contributions, which provide information and methodological suport of energy-efficiency improvement and realization of energy-saving programs at regional and municipal levels. The scheme for interaction of the potential process participants is presented in Figure 2.

Figure 2 - The scheme for interaction of participants of energy-service contract implementation process

Department of fuel and energy complex of the Vologodskay Oblast' should be denoted as the authorized agency responsible for creation of regional energy-service company, as well as the founder in Vologodskay Oblast', which is responsible for implementation of government energy-saving policy and energy-efficiency increase. This decision should be fixed by appropriate regulations and legislation.

The process for Vologodskaya Oblast' Regional energy-service company development includes three sequentially implemented steps:

Step 1: analysis of energy-service organizations, operating at subject territory; estimation of the existing conditions for activity (analysis and monitoring of the regional energy-service markets, development of object database, which are appropriate for implementation of energy-efficient activities; determination of energy-saving potential);

Step 2: Determination and development of priority company activities; estimation of funding sources; development of corporate structure;

Step 3: Formation of energy-efficient technology register, which are applicable for public service; development of methodological materials for activity organization of energy-service potential participants (government and municipal agencies, enterprises); definition of potential risks at major task implementation; development of a financial plan.

Secondly, formation of energy-efficient technology register, which are applicable for public service. In order to create a methodological basis for potential customers and investors, which are interested in attraction of funding, the regional energy-service company should carry out monitoring of agencies and enterprises with high energy-efficient potential, as well as enterprises with ineffective and expensive production and transfer of thermal energy. The analysis of production capacities of heat supply organizations, performed by the regional energy-service company, results in formation of energy-efficient technology register, which are applicable for public service at the subject territory.

We suggest to divide the activities included into the energy-efficient technology register, into the two groups: organizing, which are directly implemented by the Regional energy-saving company (energy audit, training of job responsible persons, information supply, development and actualization of heat supply schemes), and technical, which are implemented directly at the objects (replacement of worn and obsolete boiler units).

The suggested activity complex allows one to obtain not only economical effect (annual money economy of 1.8 bln rubles), but to improve the quality of the heat supply services, reduce the amount of thermal energy supply breakages for 25-30% (according to the experts opinion), reduce the amount of citizen complains up to 70%, decrease heat losses in heat networks for 1%.

References

[1] Spubin I.L., Spiridonov A.V. Zakonodatel'stvo po ehnergosberezheniyu v SSHA, Evrope i Rossii. Puti resheniya [Legislative acts on energy-saving in USA, Europe and Russia]. Vestnik MGSU. 2011. №3-1.

[2] Sinitsyn A.A. Efficiency of the new ventilation device application for premises. Life Science Journal. 2013. Vol. 10. № 4. P. 1421-1426

[3] Sinitsyn A.A. Experience of small innovative enterprises foundation in the Vologda region. Middle East Journal of Scientific Research. 2013. Vol. 16. № 10. P. 1424-1427.

[4] Zaychenko, I., Gutman, S., Kalinina, O. Adjustment of Energy Strategy of Russia to Specific Nature of Far North: Analytic Hierarchy Process. Advances in Intelligent Systems and Computing. 2018. Vol. 692. Pp. 453-462.

[5] Ilin, I., Levina, A., Iliashenko, O. Enterprise Architecture Analysis for Energy Efficiency of Saint-Petersburg Underground. Advances in Intelligent Systems and Computing. 2018. Vol. 692. Pp. 1214-1223.

[6] Statsenko, E., Ostrovaia, A., Musorina, T., Sergievskaya, N. Thermal Properties of the Building with Low Energy Consumption (LEB). Advances in Intelligent Systems and Computing. 2018. Vol. 692. Pp. 417-431.

[7] Russkov, O.V., Saradgishvili, S.E. IT-Method for uneven energy consumption planning. 2017 International Conference on Industrial Engineering, Applications and Manufacturing, ICIEAM 2017 - Proceedings. 2017. 8076414.

[8] Brunman, V.E., Vataev, A.S., Volkov, A.N., Volkov, E.A., Petkova, A.P., Kochanzhi, F.I., Plotnikov, D.G. Energy-efficient oil extraction by sucker rod borehole pumps in low-yield fields. Russian Engineering Research. 2017. Vol. 37. Pp. 378-382.

[9] Nemova, D.V., Bogomolova, A.K., Kopylova, A.I. The impact of green roofs on thermal protection and the energy efficiency of buildings. Advances and Trends in Engineering Sciences and Technologies II - Proceedings of the 2nd International Conference on Engineering Sciences and Technologies, ESaT 2016. 2017. Pp. 579-584.

[10] Shymchenko, A.V., Tereshchenko, V.V., Ryabov, Y.A., Salkutsan, S.V., Borovkov, A.I. Review of the computational approaches to advanced materials simulation in accordance with modern advanced manufacturing trends. Materials Physics and Mechanics. 2017. Vol. 32. Pp. 328-

352.

[11] Baranova, D., Sovetnikov, D., Semashkina, D., Borodinecs, A. Correlation of energy efficiency and thermal comfort depending on the ventilation strategy. Procedia Engineering. 2017. Vol. 205. Pp. 503-510.

[12] Arseniev, D.G., Shkodyrev, V.P., Yarotsky, V.A., Yagafarov, K.I. The model of intelligent Autonomous Hybrid Renewable Energy System based on Bayesian Network. 2016 IEEE 8th International Conference on Intelligent Systems, IS 2016 - Proceedings. 2016. Pp. 758-763.

[13] Zaborova, D., Petrochenko, M., Chernenkaya, L. Thermal Stability Influence of the Enclosure Structure on the Building's Energy Efficiency. MATEC Web of Conferences. 2016. 73. 2014.

[14] Staritcyna, A., Pollock, E., Sokolova, E., Martynenko, E. Energy efficiency in multi-story buildings. MATEC Web of Conferences. 2016. 73. 2009.

[15] Badenko, V., Badenko, N., Nikonorov, A., Molodtsov, D., Terleev, V., Lednova, J., Maslikov, V. Ecological Aspect of Dam Design for Flood Regulation and Sustainable Urban Development. MATEC Web of Conferences. 2016. 73. 3003.

[16] Zeb, A., De Andrade Romero, M., Baiguskarov, D., Aitbayev, S., Strelets, K. LED Lightbulbs as a Source of Electricity Saving in Buildings. MATEC Web of Conferences. 2016. 73. 2004.

[17] Starkov, V., Ovchinnikov, P., Dzampaev, T. Optimization of Energy Performance of Windows by Applying Self-Adjustable Shadings. MATEC Web of Conferences. 2016. 73. 2013.

[18] Kostenko, V., Gafiyatullina, N., Zulkarneev, G., Gorshkov, A., Petrichenko, M., Movafagh, S. Solutions to Improve the Thermal Protection of the Administrative Building. MATEC Web of Conferences. 2016. 73. 2011.

[19] Murgul, V. Methodology to Improve Energy Efficiency of Residential Historic Buildings in St. Petersburg. MATEC Web of Conferences. 2016. 53. 1046.

[20] Tarasova, D., Andreev, K., Lakić, S. The Feasibility of Energy Efficiency class Improving of a Building from Economic Point. MATEC Web of Conferences. 2016. 53. 1011.

[21] Pukhkal, V., Bieliatynskyi, A., Murgul, V. Designing energy efficiency glazed structures with comfortable microclimate in northern region. Journal of Applied Engineering Science. 2016. Vol. 14. Pp. 93-101.

[22] Perlova, E., Platonova, M., Gorshkov, A., Rakova, X. Concept project of zero energy building. Procedia Engineering. 2015. Vol. 100. Pp. 1505-1514.

[23] Vatin, N., Gamayunova, O., Nemova, D. An energy audit of kindergartens to improve their energy efficiency. Advances in Civil Engineering and Building Materials IV - Selected and Peer Reviewed Papers from the 2014 4th International Conference on Civil Engineering and Building Materials, CEBM 2014. 2015. Pp. 305-308.

[24] Tanic, M., Stankovic, D., Nikolic, V., Nikolic, M., Kostic, D., Milojkovic, A., Spasic, S., Vatin, N. Reducing energy consumption by optimizing thermal losses and measures of energy recovery in preschools. Procedia Engineering. 2015. Vol. 117. Pp. 924-937.

[25] Vatin, N., Gamayunova, O. Choosing the right type of windows to improve energy efficiency of buildings. Applied Mechanics and Materials. 2014. Vol. 633-634. Pp. 972-976.

[26] Didenko, N., Skripnuk, D. The impact of energy resources on social development in Russia. 2014. WIT Transactions on Ecology and the Environment. Vol. 1. Pp. 151-159.

[27] Penić, M., Vatin, N., Murgul, V. Double skin facades in energy efficient design. Applied Mechanics and Materials. 2014. Vol. 680. Pp. 534-538.

[28] Vatin, N., Gamayunova, O. Energy saving at home. Applied Mechanics and Materials. 2014. Vol. 672-674. Pp. 550-553.

[29] Vatin, N.I., Gorshkov, A.S., Nemova, D.V., Staritcyna, A.A., Tarasova, D.S. The energy-efficient heat insulation thickness for systems of hinged ventilated facades. Advanced Materials Research. 2014. Vol. 941-944. Pp. 905-920.

[30] Vatin, N., Petrichenko, M., Nemova, D., Staritcyna, A., Tarasova, D. Renovation of educational buildings to increase energy efficiency. Applied Mechanics and Materials. 2014. Vol. 633-634. Pp. 1023-1028.

[31] Zadvinskaya, T.O., Gorshkov, A.S. Comprehensive method of energy efficiency of residential house. Advanced Materials Research. 2014. Vol. 953-954. Pp. 1570-1577.

[32] Sinitsyn A.A., Monarkin N.N., Rogulina T.V. Federal law of power saving and opportunity of developing new the switching heat recovery devices for heat supply needs. World Applied Sciences Journal. 2013. T. 27. № 13 A. C. 361-365.

[33] Gorshkov, A., Murgul, V. Calculation of Heat Energy Consumption by a Typical Historical Building with a Courtyard. Advances in Intelligent Systems and Computing. 2018. Vol. 692. Pp. 577-591.

[34] Akhmetova, I., Chichirova, N., Derevianko, O. Revisiting heat losses calculation at district heating network. International Journal of Civil Engineering and Technology. 2017. Vol. 8. Pp. 694-702.

[35] Harmati, N., Jakšić, Z., Vatin, N. Energy consumption modelling via heat balance method for energy performance of a building. Procedia Engineering. 2015. Vol. 117. Pp. 791-799.

[36] Sinitsyn A.A. Application of the us small innovative business model in case of research and education center "Teploenergetika". Problems and trends of economy and management in the modern world Proceedings of the International Conference. 2012. P. 537-541.

[37] Tuchkevich, E., Rechinsky, A., Vysotskiy, A., Zolotova, J., Tuchkevich, V. ADN and AP programs for civil engineering students. Procedia Engineering. 2015. Vol. 117. Pp. 1142-1147.

[38] Usanova, K., Rechinsky, A., Vatin, N. Academy of construction for university applicants as a tool of university online marketing. Applied Mechanics and Materials. 2014. Vol. 635-637. Pp. 2090-2094.

[39] Arseniev, D.G., Rechinskiy, A.V., Shvetsov, K.V., Vatin, N.I., Gamayunova, O.S. Activities of Civil Engineering Institute to attract foreign students for training in civil engineering programs. Applied Mechanics and Materials. 2014. Vol. 635-637. Pp. 2076-2080.

[40] Malyshev E.A., Znamenskaya I.A. Rossijskaya i zarubezhnaya praktika gosudarstvenno-chastnogo partnerstva v sfere ehkspluatacii regional'nyh sistem teplosnabzheniya [Russian and foreign experience of public-private partnership at regional heat supply systems]. Vestnik ZabGU. 2014. № 3 (106). P. 132-139.

[41] Sedash T.N. Ispol'zovanie zarubezhnogo opyta povysheniya ehnergoehffektivnosti v rossijskoj ehkonomike [Usage of foreign experience of energy-efficiency improvement for Russian economics]. Finansovaya analitika: problemy i resheniya. 2013. Vol 9. P. 30-35.

[42] Kristensen I., McQuaid R., Scherrer W. 2015. Regional Innovation Policy and Public-Private Partnerships. -Routledge Handbook of Politics and Technology. London: Routledge. P. 249-261.

[43] Renda A., Schrefler I. Public-Private Partnerships: Models and Trends in the European Union. URL: The European Parliament, 2006. № IP/A/IMCO/SC/2005-161. P. 4-6.

[44] Pykhtin, K., Simankina, T., Karmokova, K., Zonova, A. Determination of the optimal proportions of public and private funds in project budget management. IOP Conference Series: Earth and Environmental Science. 2017. 90. 12208.

[45] Obraztsova, A., Kamenik, L. Organization of public private partnership in the renovation of urban areas. MATEC Web of Conferences. 2017. 106. 8104.

[46] Shirokova, S.V., Naidenysheva, E.G. Application of game theory in the project management of interaction between business structures and public-private partnerships. Proceedings of the 29th International Business Information Management Association Conference - Education Excellence and Innovation Management through Vision 2020: From Regional Development Sustainability to Global Economic Growth. 2017. Pp. 1150-1156.

[47] Sokolitsyn, A.S., Ivanov, M.V., Sokolitsyna, N.A., Semenov, V.P. Optimal finance distribution between public-private partnership project participants. Proceedings of the 19th International Conference on Soft Computing and Measurements, SCM 2016. 2016. Pp. 529-530.

[48] Sokolitsyn, A.S., Ivanov, M.V., Sokolitsyna, N.A. Optimal financial means and profit distribution between public and private partners in PPP project realization. Proceedings of the 28th

International Business Information Management Association Conference - Vision 2020: Innovation Management, Development Sustainability, and Competitive Economic Growth. 2016. Pp. 2122-2129.

[49] Moszoro M. 2010. Efficient Public-Private Partnerships. -Working Paper WP-884. IESE Business School -University of Navarra. 25 p

[50] Gorbunov V.N., Bajnishev S.M., Semenov A.V. Podhody k resheniyu problem investirovaniya v zhilishchno-kommunal'noe hozyajstvo na osnove gosudarstvenno-chastnogo partnerstva [Approaches for solutions of problems of investment in housing and communal services basing on public-private partnership]. Ekonomika i menedzhment innovacionnyh tekhnologij. 2014. Vol. 8 P. 56-62.

[51] O koncessionnyh soglasheniyah: federal'nyj zakon ot 21 iyulya 2005 goda № 115-FZ [Concerning the concession agreements: Federal law of July 21, 2005 No. 115-FZ].

[52] Merlin-Brogniart C. 2014. Improving Understanding of the Innovation Process in Innovation-Oriented Public-Private Partnerships. -Journal of Innovation Economics & Management. Vol. 15. Is. 3. P. 117-144

[53] Issledovanie «Razvitie gosudarstvenno-chastnogo partnerstva v Rossii v 2015-2016 godah. Rejting regionov po urovnyu razvitiya GCHP». Associaciya «Centr razvitiya GCHP», Ministerstvo ehkonomicheskogo razvitiya Rossijskoj Federacii. Moscow: Associaciya «Centr razvitiya GCHP», 2016. 36 p.

[54] Tritenko E.A., Sovetova N.P. Privlechenie investicij v kommunal'noe hozyajstvo na regional'nom urovne [Attraction of investments for housing and communal services at the regional level]. Upravlenie social'no-ehkonomicheskimi sistemami: materialy mezhdunarodnoj nauchno-prakticheskoj konferencii. M-vo obraz. i nauki RF; Vologda State University. Vologda: VoGU, 2017. Vol. 2. P.375-378.

Optimization of combined heat and power plant operating mode by means of underutilized equipment mothball

Evgeniya Sukhareva[1], Alexander Fedyukhin[1], Oleg Derevianko[2], Mikle Egorov[2], Liliya Mukhametova[3], Irina Akhmetova[3]

[1]National Research University «Moscow Power Engineering Institute», Moscow, Russia, Krasnokazarmennaya 14, Moscow, 111250 Russia

[2]Peter the Great St. Petersburg Polytechnic University, Polytechnicheskaya 29, Saint Petersburg, 195251 Russia

[3] Federal State Budgetary Educational Institution of Higer Education "Kazan State Power Engineering University", Kazan, Krasnoselskaya street, 51

derevianko@nil-teplo.ru

Abstract. This work makes it possible to link ground rules for thermal and electrical energy markets with generation plant operating. Here we consider various procedures for optimization of combined heat and power plant operating mode by means of underutilized equipment mothball. This work results in process flow diagrams for turbine hall and relative increment charts, which can be used for plant loading in the optimal way.

Introduction

The issue of managing the power plant operating modes was throughly studied and has gained great scientific and practical progress [1–9]. However, in the recent years, Russian power engineering saw far-reaching reforms. The operational-dispatch management was changed, power production became concurrent. The prices at competitive market now are not regulated by the government, but are formed based on supply and demand, and its participants compete with each other. The transformations lead to a drastic change in power system and, consequently, to a change of scientific and practical methods for its management. Now the management is performed using price signals directly targeted at generators and consumers, which are formed within the market.

A lot of power plants are now heat-underutilized, as there was a great consumer outflow in the 90's, especially those who used industrial steam bleed. As a result, for this power plant the fuel consumption increases, the product cost raises, given an impact on electrical power output.

All these aspects make our research work very challenging. The aim of this investigation was to increase the combined heat and power (CHP) plant efficiency by means of appropriate choice of its operating modes.

The following tasks were defined in it:

- Study of Moscow thermal and electrical energy markets.
- Study of optimization methods for combined heat and power plant operating modes.
- Selection of appropriate optimization method for combined heat and power plant operating mode.

IOP Conf. Series: Earth and Environmental Science **288** (2019) 012122 doi:10.1088/1755-1315/288/1/012122

- Estimation of economical efficiency of the selected method.

Modern cogeneration problems

Cogeneration based on combined electrical and thermal energy production has a number of advantages. It should be noted that special aspects of CHP plant work are connected with branch industry specialization and its specification. Cogeneration power plants produce thermal and electrical energies simultaneously. This means that its operating mode depends on demands for both product types. This aspect makes the regulation process significantly complicated.

The problem of developing objective performance indicators for combined heat and power plant at combined generation of electrical and thermal energies is still unsolved in Russia, while this question is thoroughly studied worldwide [10–18]

The degree of depreciation is characterized by combined heat and power plant equipment age, its structure is presented in Fig.1.

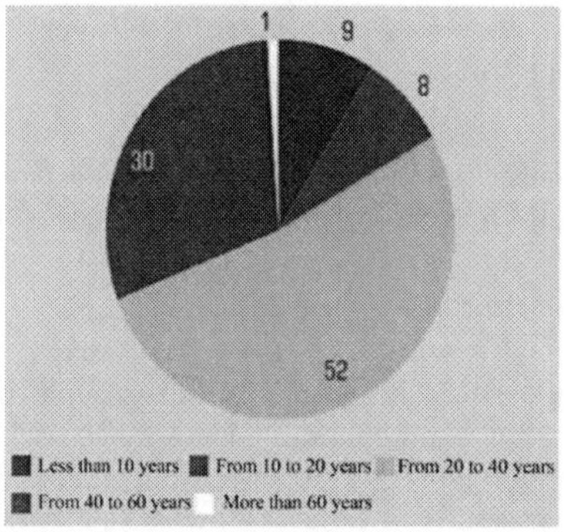

Figure 1. Age structure of the combined heat and power plant generation equipment [19]

As the primary input of power equipment was carried out in 1960-70s, nowadays it has significant functional and moral depreciation.

The usage of the worn equipment leads to:

- reducing of generation efficiency, excessive fuel consumption and additional losses;
- enhancement of workplace accidents, increasing of repair time, growth of repair costs.
- Combined heat and power plants do not have any warranties concerning capacity loads as nuclear power plants do [20] [21,22] [23] [24]. This means that market rules do not take into account technological features of combined heat and power plant operation, though their total power production is directly proportional to the heat load. Consequently, CHP plants, operating at heating mode, have some limitations for electrical load regulating and cannot perform system power-change functions.

As there are a lot of underloaded and depreciated facilities, there arises a necessity to work out a sustainable branch development scheme. It should include both modernization of the existing generation facilities (as it is more economically-viable than new construction), and new construction, where it is impossible to renovate and reconstruct the existing facilities [25].

Over the last years the spare parts support for basic combined heat and power plant equipment has improved [26] [27]. Modern multiaxes machines allow one to fabricate various blades. There are hydraulic automatics assembly units and others.

Service life extension is essentially cheaper than replacement power construction, so it is a more preferable opportunity for a large number of old CHP plants, for which there is no need for power increasing.

During modernization, the maximum technical process automation should take place, as large staff results in high semi-fixed costs.

Unlike condenser type electricity-only thermal power stations, which operate at low steam parameters, combined heat and power plants are rather competitive at thermal mode operating.

Combined heat and power plant thermal capacity decreases due to industrial production reduction, which led to closure of enterprises and steam consumption reduction. Besides this, the ratio of secondary energy resources decreases.

Russian CHP plants are characterized by an unbalance between capacity and structure of the equipment and heat and electricity market requirements. Back-pressure turbines of about 3 mln. kW total capacity stand idle [28]. This is due to absence of thermal load. Abnormal operation conditions, frequent startups and shutdowns of the equipment lead to technical depreciation and additional energy losses. And it results in additional material expenses.

Number of hours for power plant operating at installed capacity in Russia in 2014 was 4478 hours [29], in 2015 it was 4402 hours, or 50.25 % of the calendar time [30].

For industrial facilities operating hours at installed capacity (not taking into account power plants) are the following:

- for thermal power stations 4136 hours (47.21% of the calendar time);
- for nuclear power stations 7415 hours (84.65% of the calendar time);
- for hydro power stations 3354 hours (38.29% of the calendar time);
- for wind power stations 592 hours (6.75% of the calendar time);
- for sun power stations 738 hours (8.43 % of the calendar time).

Besides this, there are some regions in Russia there total power plant capacity exceeds needs of the region. This excessive electric power should be consumed somewhere. But during town and country planning every region, which imports electricity from another one, declares itself energy-deficient and develops a special program for deficit overcoming (which is actually absent) [31].

Many heating-and-power plants, despite low specific fuel consumption at cogeneration cycle, are loss-making. Typically this is due to the following reasons:

- High specific expenses for equipment exploitation and repair due to its technical depreciation and large staff of the heating-and-power plant.
- Low installed capacity utilization factor (ICUF) and mismatch between equipment composition and connected load.
- Long-term usage of combined heat and power plant equipment as a hot reserve for supplying peak electricity loads.
- Combined heat and power plant operating at condensing mode.
- Usage of understated specific fuel consumption for thermal energy production from hot-water boilers at heat tariff formation. It is accompanied by corresponding increase of specific tariffs for electricity supply (extension of cogeneration effect on combined heat and power plant segment, which is an actual fact a boiler house).
- The concurrence with hydro power plants and nuclear power plants, which compensate the major part of expenses from capacity charge at low variable expenses.
- Night marginal reduction of electricity price, so that it is less than fuel costs.
- Heat tariffs formation according to ultimate growth indices (once understated tariffs cannot be raised to the industry average level).
- Mismatch between fuel price used in heat tariffs and lack of correction during the next regulating period due to growth limitations at ultimate indices.
- Approval of understated normative values for household heat and hot water supply by local governments.
- The three latter causes have an impact only on two-production CHP plant, thus reducing it competitiveness.
- Non-payments.

- Payback of overstated or unnecessary investments.

The actual reasons for loss-making of a certain combined heat and power plant might be determined by division of its economical performing results by 3 parts:

- Cogeneration mode operation;
- Electricity supply at condensing mode.
- Heat supply at boiler house mode.

Nowadays combined heat and power plant retreat from market appears to be efficient. It is surprising, but loss-making stations appear to be efficient when retreated from the market.

Electricity produced by combined heat and power plant during condensing cycle, is expensive for most of them, and is accepted by the market only during maximum consumption periods. Approximately 30 CHP plants have marginal cost of electricity (their price bid at peak consumption is used for determination of electricity delivery costs in all huge price zone). A small combined heat and power plant at peak period might level up the whole market price for 10-15% [31].

According to the Russian Federation energy strategy up to the year 2030, the following changes of installed capacity structures will take place: The portion of nuclear power plants will increase due to commissioning of new facilities, and the portion of gaz thermal power stations will decrease due to its decommissioning. In 2015, electrical energy production by UES of Russia power plants (including that at industrial enterprises power plants) has increased against the results for the year 2014 only for 0.2%, namely:

- Thermal power plants: decrease in production for 0.9 %;
- Hydro power plants: decrease in production for 4.1%;
- Nuclear power plants: increase in production for 8.2%.

In order to provide reliable power supply, as well as failure-free and cost-effective operation of the power plant equipment, it is necessary to set rational operation modes, which take into account energy demand, technical and economical characteristics. Basic, normal operation mode is that one providing capacity according to the load chart, and supplying major power at predetermined time period.

Mothball as an approach for combined heat and power plant efficiency increase

One of the most important exploitation task is economic energy loading schedule among the power plants of the energy system and its separate units and assemblies. Simultaneously, the questions about operating units number, their startups and shutdowns should be answered.

Efficient load distribution between operating units, which provides minimum heat and fuel consumption at the power plant and energy system, is performed basing on heat consumption relative increments approach [16].

Under the conditions of financial restrictions, the most preferable approaches are repair or modernization, and mothball of the excess capacity.

The question of the unused equipment removal is essentially important at excess generating power market. Removal out of service, or excess power equipment mothball, including that used at cogeneration modes but less than 1000-2000 hours annually, together with compensation of the thermal peaks by boiler houses operating at peak modes result in amortization cost reduction, and, consequently, total expenses. The underutilized equipment mothball reduces not only operating expenses, but that ones for unit start-stop. The cogeneration remains at the same level: thermal load charts for the power plant do not change. Combined heat and power plant fully supplies heat to its consumers, whereby creating artificial power deficit.

This task can be regarded as discrete, because equipment operating mode regulation has in reality a limited accuracy, and, consequently, a finite set of variants for turbine unit loading. In order to solve this task we have developed an algorithm, which implements a process of exhaustive search for various conditions. This algorithm is presented in Fig.2.

The first step is for comparison of CHP thermal installed capacity and thermal load chart of its market. If the capacity is in excess, the operating mode of every CHP turbine unit is checked. At equipment load for more than 1000-2000 hours annually (ICUF less than 20%), the cogeneration appears to be not efficient as compared with separate production. So, this energy unit should be conserved. Further, we calculate the energy enterprise income, which are obtained by alteration of model parameters. In such a manner we examine unit mothball possibility. The obtained variants set is tested for limitations fulfillment. If one of the limitations isn't fulfilled, the corresponding variants are rejected. After this, the maximum income alternative is selected. Fig.2 presents decision-making mechanism about mothball [32].

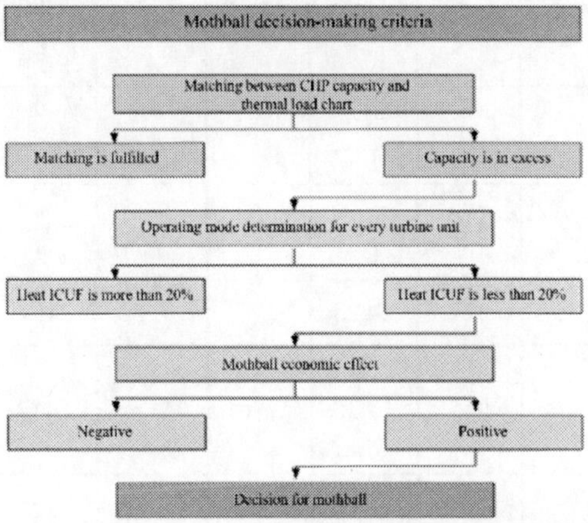

Figure 2. The mechanism for mothball problem decision.

Planning of the combined heat and power plant operation taking into account mothball

This mechanism was approved by the example of combined heat and power plant CHP-21 of Mosenergo PJSC. It is located at Dmitrovskiy District, Moscow and from 2009 is combined with CHP-28. Now the power plant has 5 power units T-110 and one ST-25. Its installed capacity is 575 MW. Annual average load is 300 MW.

Basing on generation equipment characteristics, we have calculated relative increment characteristics and process flow diagram of the turbine hall (Table 1, Fig. 3).

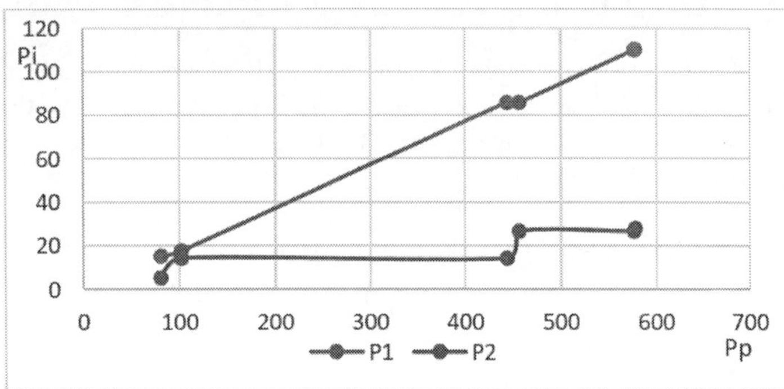

Figure 3 - Process flow diagram of the CHP-21 turbine hall

It can be seen, that plant load is well below the installed one, so the units are inefficiently loaded. 2 mothball variants are proposed: the first one is related to the absence of consumer, the second one takes into account heating load decrement. As this CHP doesn't have any industrial load, the ST turbine maintenance appears to be inefficient. So, its mothball is proposed.

Table 1 - Characteristics of turbine hall for CHP-21

Characteristic process flow points	Relative increment of turbine, GJ/MW·h	Turbine hall						Boiler house		Relative increment of plant, r_p, toe/MW·h
		T-110-130		ST-25-90		Total				
		P1, MW	Q1, GJ/h	P2, MW	Q2, GJ/h	∑P, MW	$\sum Q_{ph}$, GJ/h	$\sum Q_{bh}$, GJ/h	r_{bh}, toe/GJ	
Minimum load for boiler house	-	-	-	-	-	-	-	2221	0.0347	-
Minimum load for turbine hall	7.34	15	355	5	202	80	1981	-	-	-
Minimum CHP load	8.46	17.5	377	14.25	270	101.75	2157	2221	0.0347	0.293
1st break in turbine hall characteristic curve	8.46/ 8.6	85.88	955	14.25	270	443.65	5047	5200	0.0363	0.307/ 0.312
2nd break in turbine hall characteristic curve	8.6/ 9.63	85.88	955	26.65	376	456.05	5154	5309	0.0364	0.313/ 0.35
3d break in turbine hall characteristic curve	9.63/ 9.98	110	1187	26.65	376	576.65	6315	6505	0.0384	0.37/ 0.383
Maximum load for boiler house	-	-	-	-	-	-	-	7406	0.04	-
Maximum load for turbine hall	9.98	110	1187	28	390	578	6333	-	-	-
Maximum CHP load	9.98	110	1187	28	390	578	6333	6523.2	0.0384	0.384

The second variant involves one turbine T-110 mothball. Such mothball doesn't influence the supplied power, as these two turbines work less than 1000 hours annually. The two mothball variants are evaluated in the view of economical efficiency (Tables 2, 3). Cash outflows consist of equipment mothball expenses, and cash inflow consists of repair and amortization economy.

Table 2 - The calculation of discounted cash flow for the first mothball variant

Characteristics/Step	0	1	2	3	4	5	6	7	8	9	10	11	12
Mothball expenses	-38.4												
Repair and amortization expenses		6.1	6.1	6.1	6.1	6.1	6.1	6.1	6.1	6.1	6.1	6.1	6.1
Discount Coeff.	1	0.909	0.826	0.751	0.683	0.621	0.564	0.513	0.466	0.424	0.385	0.35	0.319
Discounted cash flow	-38.4	5.5	5.0	4.5	4.1	3.7	3.4	3.1	2.8	2.5	2.3	2.1	1.99
Discounted cash flow running total	-38.4	-32.8	-27.8	-23.2	-19.0	-15.2	-11.8	-8.7	-5.8	-3.2	-0.9	1.2	3.1

Table 3 - The calculation of discounted cash flow for the second mothball variant

Characteristics/Step	0	1	2	3	4	5	6	7	8	9	10	11	12
Mothball expenses	-88.32												
Repair and amortization expenses		18.4	18.4	18.4	18.4	18.4	18.4	18.4	18.4	18.4	18.4	18.4	18.4
Discount Coeff.	1	0.909	0.826	0.751	0.683	0.621	0.564	0.513	0.466	0.424	0.385	0.35	0.319
Discounted cash flow	-88.32	16.7	15.1	13.8	12.5	11.4	10.3	9.4	8.5	7.8	7.0	6.4	5.8
Discounted cash flow running total	-88.32	-71.5	-56.3	-42.5	-30.0	-18.5	-8.2	1.2	9.8	17.6	24.7	31.1	37.0

It is seen from Table 4, that both mothball variants are efficient. Payback period of the second variant (ST-25-90 and T-110-130 mothballs) is less than that one for the first variant (ST-25-90 mothball) by 4 months. discounted cash flow of the second variant is 11.75 times more than that one for the first variant. So it is preferable to choose ST-25-90 and T-110-130 mothballs.

Table 4 - Comparison of the proposed variants

Characteristics/Variant	1	2
Discounted cash income, mln. roubles	3.15015	37.0116
Payback period, months	7	11

It is preferable to choose ST-25-90 and T-110-130 mothballs (the second variant).

Conclusions

- Thermal and electrical energy markets of Moscow were investigated. The performed analysis shows that all combined heat and power plants operate at 50-70% of its designed operating capabilities.
- It was proved, that plants operate in a rather inefficient mode, and this deteriorates their economic performance.
- We have chosen an approach for CHP operating mode optimization taking into account power equipment mothball.
- Economic effects of the proposed methods were evaluated. Both methods are efficient. Net present value is positive for both variants.

However, despite high initial expenses for two turbines mothball, its payback period is smaller. At this, the plant fully satisfies the consumer needs. So, this variant should be recommended for further implementation.

Acknowledgments

The work was supported by the Ministry of Education and Science of the Russian Federation on fundamental scientific research (Agreement No. №13.6994.2017/БЧ) «Development of a methodology for determining the reliability of a heat supply system to improve energy efficiency».

References

[1] Saari J, Sermyagina E, Kaikko J, Vakkilainen E and Sergeev V 2016 Integration of hydrothermal carbonization and a CHP plant: Part 2 –operational and economic analysis *Energy* **113** 574–85

[2] Miroshnikova Y A 2013 Role of the system effect in the optimization of the operating modes of pumped-storage power plants *Power Technol. Eng.* **47** 233–8

[3] Ma T, Yang H, Lu L and Peng J 2015 Pumped storage-based standalone photovoltaic power generation system: Modeling and techno-economic optimization *Appl. Energy* **137** 649–59

[4] Pérez-Díaz J I, Chazarra M, García-González J, Cavazzini G and Stoppato A 2015 Trends and challenges in the operation of pumped-storage hydropower plants *Renew. Sustain. Energy Rev.* **44** 767–84

[5] Pearce J M, Denkenberger D and Zielonka H 2009 Accelerating applied sustainability by utilizing return on investment for energy conservation measures *Int. J. Energy, Environ. Econ.* **17**

[6] Anikina I D, Sergeyev V V, Amosov N T and Luchko M G 2017 Use of heat pumps in turbogenerator hydrogen cooling systems at thermal power plant *Int. J. Hydrogen Energy* **42** 636–42

[7] Fedyukhin A, Sultanguzin I, Gyul'Maliev A and Sergeev V 2017 Biomass pyrolysis and gasification comprehensive modeling for effective power generation at combined cycle power plant *Eurasian Chem. J.* **19** 245–53

[8] Nizamov L A and Chichirova N . 2016 Software package for optimization of modes of thermal power plant *Proc. High. Educ. institutions. ENERGY Sect. Probl.* **1–2** 64–74

[9] Novoselov V B, Brodov Y M, Litvinov E V, Lebedev V V and Mikhailov A G 2018 Increase efficiency of electric power stations by way of organization of additional choice of steam from steam turbines *Proc. High. Educ. institutions. ENERGY Sect. Probl.* **20** 90–9

[10] Ren H and Gao W 2010 Economic and environmental evaluation of micro CHP systems with different operating modes for residential buildings in Japan *Energy Build.* **42** 853–61

[11] Cho H, Luck R, Eksioglu S D and Chamra L M 2009 Cost-optimized real-time operation of CHP systems *Energy Build.* **41** 445–51

[12] Moradi M H, Hajinazari M, Jamasb S and Paripour M 2013 An energy management system (EMS) strategy for combined heat and power (CHP) systems based on a hybrid optimization method employing fuzzy programming *Energy* **49** 86–101

[13] Keirstead J, Samsatli N, Shah N and Weber C 2012 The impact of CHP (combined heat and power) planning restrictions on the efficiency of urban energy systems *Energy* **41** 93–103

[14] Ondeck A, Edgar T F and Baldea M 2017 A multi-scale framework for simultaneous optimization of the design and operating strategy of residential CHP systems *Appl. Energy* **205** 1495–511

[15] Syltanov M M 2010 *Optimization of CHP operating modes in regards to energy efficiency. Ph.D. Thesis in Engineering Sciences* (Moscow Power Engineering Institute)

[16] Rogalev N D, Zubkova A G and Masterova I V 2008 *Energy Economics Textbook* ed N D Rogalev (Moscow: MPEI Publishing House)

[17] Zhukov V V, Evsyukov I A and Alexandrov A S 2017 Determination of conditions for the successful start-up of Combined Cycle Power Plant (CCPP) from Black Start Diesel Generators (BSDG) *Proc. High. Educ. institutions. ENERGY Sect. Probl.* **19** 43–55

[18] Aminov R Z and Garievsky M V 2018 The efficiency of combined-cycle CHP plant with variable electric loads, taking into account the wear and tear of equipment *Proc. High. Educ. institutions. ENERGY Sect. Probl.* **20** 10–22

[19] Kolbina L 2008 Non-profitable modernization *Expert-Ural* **5**

[20] Davydov V V, Dudkin V I, Velichko E N and Karseev A Y 2015 Fiber-optic system for simulating accidents in the cooling circuits of a nuclear power plant *J. Opt. Technol. (A Transl. Opt. Zhurnal)* **82** 132–5

[21] Crossland I 2012 The economics of nuclear power *Nuclear Fuel Cycle Science and Engineering* pp 100–26

[22] Deutch J M, Kadak A C, Kazimi M S and Moniz E J 2009 Future of nuclear power *Update* 20

[23] Agafonova N D, Egorov M Y, Sergeev V V, Gotovskii M A, Kruglikov P A, Lebedev M E, Sudakov A V, Fedorovich E D and Fokin B S 2018 Heat-and-Mass Transfer Intensification in Saturated-Steam Generators in NPP with VVER as a Means for Increasing Efficiency and

Reliability *At. Energy* 1–5
[24] Obraztsov E P, Belikov A N, Menshikov N N and Popkov E N 2016 Electrical calculation code relecs in css "veb" for npp designs vol 2(American Nuclear Society)pp 1131–40
[25] Svyatkin F A, Trifonov N N, Ukhanova M G, Tren'Kin V B, Koltunov V A, Borovkov A I and Klyavin O I 2013 A modernized high-pressure heater protection system for nuclear and thermal power stations *Therm. Eng. (English Transl. Teploenerg.* **60** 635–42
[26] Eliseev V V and Piskunov V A 2016 Composite gas-turbine blades *Russ. Eng. Res.* **36** 819–22
[27] Bogov I A, Bezukhov A P, Sukhanov V A, Tolmachev V V, Novik N N, Kolenov E V, Mironova M V and Konyushin M V 2017 Full-Scale Bench for Testing Auxiliary Equipment of Steam-Turbine Units *Power Technol. Eng.* **51** 105–7
[28] Tsipylev D Y 2008 *Selection of optimum operating modes for CHP with complex equipment composition. Ph.D. Thesis in Engineering Sciences* (Moscow Power Engineering Institute)
[29] RAO UES R 2015 *Report on RAO UES of Russia functioning in 2014* (Russia)
[30] RAO UES R 2016 *Report on RAO UES of Russia functioning in 2015*
[31] Semenov V G Cogeneration in modern market conditions *Rosteplo*
[32] Sukhareva E V, Shyvalova D G and Kyrdyukova G N 2017 Planning of cogeneration power station operation by means of underutilized equipment mothball *Russ. Bus.* **4**

IOP Conf. Series: Earth and Environmental Science **288** (2019) 012123 doi:10.1088/1755-1315/288/1/012123

The Fourier-T method to solve the heat transfer problem in combustion chambers of various configurations

Toropov E.V., Lymbina L.E.

South Ural State University (National Research University), Chelyabinsk, Russia

evtor@mail.ru

Annotation. With the use of laws of conservation of different substances, the transfer equation in the combustion chambers is formulated as a partial differential equation of the second order. The solution of the obtained equation by the Fourier-T method is proposed, which helps to evaluate the influence of various factors on the space-time temperature field of the combustion chamber quickly; it is shown that the generalized Nusselt number mainly affects the solution. The developed technique allows us to estimate the temperature field of the furnace at a certain flow rate of the combustion products at the inlet of the combustion chamber as well as their composition and temperature using the analogy with heat conductivity processes. The conditions for the reliable application of the developed model are determined. Applying the theorem on multiplication of solutions, we may estimate the temperature field in the combustion chambers of various structural configurations which is an extension of a zoned method.

Introduction

Industry and science pay constant attention to the improvement of characteristics of the heat engineering equipment, since this issue is closely connected with economic and environmental performance of the units. This trend is reflected in the international program THERMY-II and in the works of leading companies producing heat power equipment. With the extensive production experience of the service of combustors of various purposes and design [1] it is not always possible to solve the problem of the efficient and safe use of various fuels quickly, as well as to change the modes of equipment operation [2-5]. The available norms for thermal calculation of combustors and modern numerical methods for their calculation require considerable time and expenditures. In addition, sometimes there is a contradiction in some methods of solving thermophysical problems including a variable separation method which has become classical in solving equations of mathematical physics. The urgency of this problem is unquestionable, and in the authors' opinion, it is necessary to solve it within fundamental laws of conservation of substances with the use of modern methods of thermophysical research of processes.

Geometric and physical conditions must correspond to the rules of the theory of similarity and modeling, the initial and boundary conditions are related to the correctness of the solution of the problem. In accordance with the statement of academician A.N. Tikhonov [6], the formulation and solution of the problem are considered to be correct, provided there is a solution which is unique and stable to small changes in the parameters of the problem. It should be noted, that the correctness corresponds to more rigid criteria than the reliability, therefore the results of solving the correct problem can be considered reliable.

When presenting the fundamentals of the method of separation of variables (the Fourier method) all publications admit "equality" of the functions, obtained as a result of separation of the main variable into a time and space function. Therefore, as a result of integrating the separated functions, there are two constants, each must include the initial temperature, and eventually the main function must include the square of the initial temperature which violates the rules of dimensional analysis. This inaccuracy is then manifested in the theorem on multiplication of solutions.

The way out of this situation is possible, if we assume immediately that the main information load about the temperature field of the bodies is carried by the second spatial derivative, and the time function is a factor reflecting the dynamics of changes in spatial temperature field. This time function can be chosen any, however, as it is shown by the analysis, spontaneous processes of heat and mass transfer (dissipative processes) are best described by an exponential with a factor of 1,0; and this factor corresponds to an exponential with power 0, that is, to the beginning of the process. In this case the index of power has a separation constant, which guarantees the reliability of mathematical description, since the initial condition is a condition for the existence and unjqueness of the solution.

Formulation of the problem and derivation of differential equation of the temperature field in the furnace

For the case under consideration when $\rho, c, \lambda = const$; $q_v = 0$, $\partial T/\partial z = 0$ for isotropic media with temperature-independent thermophysical properties the energy transfer equation for a stationary regime in the Cartesian reference system, the Fourier-Kirchhoff equation has the form

$$\rho c\left(w_x\frac{\partial T}{\partial x}+w_y\frac{\partial T}{\partial y}\right)=\lambda\left(\frac{\partial^2 T}{\partial x^2}+\frac{\partial^2 T}{\partial y^2}\right). \quad (1)$$

The law of conservation of energy in the form of a balance of heat flows for the volume of the combustion medium ∂V in the channel of the furnace of the boiler plant can be written in the form $\sum_i divq_i = 0$, where i is the heat transfer vectors for the enthalpy flow, the thermal conductivity mechanisms for the turbulent and molecular and beam mechanism respectively: $q_c = w_x \rho c T_\Sigma$; $q_\lambda = -\lambda_t \nabla T_\Sigma$; $q_r = -\lambda_r \nabla T_\Sigma$, W/m^2. This helps to transform equation (1) to the form

$$\rho c w_x \nabla T_\Sigma - div(\lambda_\Sigma \nabla T_\Sigma + \rho c D \nabla T_\Sigma) = 0 \quad (2)$$

and then, with the introduction of the relative temperature $\theta = T_\Sigma - T_w$ we go over to the canonical form of the differential equation

$$\frac{\partial^2\theta}{\partial x^2}-\frac{w_x\partial\theta}{a_\Sigma\partial x^2}+\frac{\partial^2\theta}{\partial y^2}=0. \quad (3)$$

In equation (3) the generalized temperature conductivity coefficient $a_\Sigma = \lambda_\Sigma/\rho c$, the generalized thermal conductivity coefficient λ_Σ is equal to the sum of the coefficients of molecular λ_{mol}, turbulent λ_t, mass λ_m and radiation λ_r mechanisms of the substance transfer. The coefficient of radiation mechanism is $\lambda_r = 16\sigma_0 T_\Sigma^3/3k_g$, W/m·K, where $\sigma_0 = 5{,}67\cdot10^{-8}$, k_g is the average attenuation coefficient over the combustion medium cross section, 1/m. Heat transfer is also possible together with mass transfer under conditions of temperature gradient T_Σ with mass diffusion coefficient D, which is of the order $(0{,}1\ldots1{,}0)\cdot10^{-4}$, m^2/s.

Equation (3) solution and adaptation

The separation of the variables $\theta = \theta_x \cdot \theta_y$ and the introduction of the separation constant $-k^2$ allows us to replace (3) by two ordinary differential equations and write down their solutions as

$$\theta_x'' - \theta_x' \frac{w_x}{a_\Sigma} + k^2 \theta_x = 0; \qquad \theta_x = C_1 \exp(\gamma_1 x) + C_2 \exp(\gamma_2 x); \tag{4}$$

$$\theta_y'' - k^2 \theta_y = 0; \qquad \theta_y = C_3 \sin(ky) + C_4 \cos(ky). \tag{5}$$

In the formulas (4), (5) the constants $C_{1,2,3,4}$ are determined by the single-valued conditions, and the parameters $\gamma_{1,2}$ are calculated by the formula

$$\gamma_{1,2} = \frac{w_x}{2a_\Sigma} \pm \sqrt{\frac{w_x^2}{4a_\Sigma} - k^2}, \tag{6}$$

analysis of the first level has shown that because of the small value of γ_1 for $w_x/a_\Sigma >> 1$ the constant C_1 can be ignored, and $C_2 = 1$.

With the assumption of the symmetry of the medium flow relative to the x axis in the furnace channel, the uniqueness condition is considered

$$\left.\frac{\partial \theta_y}{\partial y}\right|_{y=0} = 0, \tag{7}$$

which leads to a derivation with respect to the constant $C_3 = 0$, and the general solution of the energy equation (3) will have the form

$$\theta = \theta_x \cdot \theta_y = C_4 \exp(\gamma_2 x) \cos(ky). \tag{8}$$

A boundary condition of the third kind for the heat flow on the heat reception surface $y = \delta$

$$\left.\frac{\partial \theta_y}{\partial y}\right|_{y=\delta} = -\frac{\alpha}{\lambda_\Sigma} \theta_{y=\delta}, \qquad \text{from which} \quad \frac{k\delta}{Nu_\Sigma} = ctg(k\delta), \tag{9}$$

where the Nusselt number $Nu_\Sigma = \alpha\delta/\lambda_\Sigma$ reflects the ratio of the thermal resistances of the generalized thermal conductivity of the medium and the heat transfer on the heat reception surface: $Nu_\Sigma = (\delta/\lambda_\Sigma)/(1/\alpha)$. In formula (8) a refined solution is given in accordance with the Fourier-T method, which slightly differs from the solution in reference [7].

The only parameter affecting the solution of equation (8) is the number Nu_Σ, in the problems of thermal conductivity and diffusion such parameter are the Biot numbers Bi. Replacing the Biot number by the Nusselt number we can use the extensive reference material [8, 9] to define $C_4 = D/\theta_0$, where θ_0 is the initial temperature of the surface. In the form of the ratio of two thermal resistances to heat transfer along the y axis the Nusselt number establishes a relationship between the distributed resistance δ/λ_Σ and resistance to heat transfer concentrated on the heat-sensing wall $1/\alpha$. Therefore, to determine δ/λ_Σ it is rational to apply the dependence

$$\delta / \lambda_\Sigma = \delta \Bigg/ \left(\frac{16}{3} \frac{\sigma_0 T_\Sigma^3}{k_g} \right), \tag{10}$$

and to determine $1/\alpha$ we apply the dependence for the resulting flow on the wall

$$1/\alpha = \frac{(1/A_w - 1/2)(T_\Sigma - T_w)}{\sigma_0 (T_\Sigma^4 - T_w^4)} . \qquad (11)$$

Thus, the Nusselt number

$$Nu_\Sigma = \frac{3}{16} \frac{Bu}{(1/A_w - 1/2)} \frac{(T_\Sigma^4 - T_w^4)}{T_\Sigma^3 (T_\Sigma - T_w)} = \frac{3}{16} \frac{Bu}{(1/A_w - 1/2)} \left[1 + \frac{T_w}{T_\Sigma} + \frac{T_w^2}{T_\Sigma^2} + \frac{T_w^3}{T_\Sigma^3} \right], \qquad (12)$$

where the temperature factor is transformed to a polynomial of the third degree by algebraic transformations, with $T_w/T_\Sigma < 1$. The Bueger number $Bu = k_g ps$; k_g is the coefficient of attenuation of the radiation flow by the gases of the combustor, 1/m·MPa, p is the pressure in the combustor, MPa, s is the effective thickness of the emitting layer of gases, m.

In the solution of equation (8) the exponential factor characterizes the change in temperature along the channel length, so the definition of $\gamma_{1,2} = w_x/2a_\Sigma \pm (w_x^2/4a_\Sigma^2 + k^2)^{0,5}$ is crucial for calculating the temperature along the channel length.

According to the estimates $\gamma_2 = w_x/2a_\Sigma[1 - (1 + 4k^2a_\Sigma^2/w_x^2)^{0,5}]$, using the dependence

$$(1 + \xi)^{0,5} = 1 + 0,5\xi - 0,0625\xi^2$$

the parameter of the solution γ_2 is reduced to the form $\gamma_2 = 0,0625 w_x \xi(\xi - 2)/a_\Sigma$, where $\xi = 4k^2a_\Sigma^2/w_x^2$. The last expression makes it possible to determine the condition for the applicability of the parameter γ_2: $\xi \geq 2$.

Taking into account the periodicity of the function $ctg(k\delta)$ the solution of equation (3) in the dimensionless form is determined as the sum of particular solutions

$$\Theta = \frac{\theta}{\theta_0} = \sum_i D_i \exp(\mu_i X) \cos(\mu_i Y), \qquad (13)$$

for $\mu = k\delta$, $X = x/\delta$, $Y = y/\delta$.

The boundary condition (9) makes it possible to determine the location of the convergence point of the tangents to the temperature curves $\Theta_{X=1}$ on the heat reception surface $x = \delta$, and the ordinate of this point Y_0 is also counted from the surface $x = \delta$: $Y_0 = 1/Nu_\Sigma$ [10]. The analogy with the processes of thermal conductivity in solids helps to use the Fourier-T method in application to combustors of various shapes and flow-through heat-mass transfer units with the replacement of the Biot number by the Nusselt number, it must be taken into account that the temperature scale is directed towards the flow of the medium [11-13].

Since the ordinate Y_0 defines a figure in space, that is equidistant to the heat reception surface, any change in the latter is easily taken into account in calculating the volume temperature field of the furnace, including the change in the Nusselt number using the zone method. In this case, the uniform distribution of the flow velocity of the medium in the channel becomes unnecessary. By setting the condition for the absence of slagging in a certain section we may define the coordinate of this section by substituting in (13) $\Theta = \Theta_{sl}$.

Conclusion

On the basis of the laws of conservation of substances the energy transfer equation in the channel of the combustor is formulated in the form of a partial differential equation, which is solved by the Fourier-T method. The obtained equation of temperature variation along and across the combustor makes it possible to estimate the influence of different factors on temperature changes in combustors of various configurations and thermal stress.

References

[1] 1998 *Thermal calculation of boilers. Normative method.* – St.Petersburg: NPO CKTI-VTI.
[2] Alstom – www.alstom.com.
[3] Modeling. Software programs / ANSYS – www.ansys.com/products.
[4] Anoshin R.G., Grigoriev K.A., Lebedev B.V., Mikhailov D.N. and Starchenko A.V. 2006 Numerical study of aerothermochemical processes under various schemes of burning coal around Moscow in the furnace of the boiler БКЗ-220. *Solid fuel combustion: Proc. VI All-Russian conf., Novosibirsk.* – Ed. Institute of Thermal Physics SD RAS chapter 1 pp. 81- 87.
[5] Konyashkin V.F. 2006 Simulation of physical processes in a ring furnace with the FLUENT program. *Solid fuel combustion: Proc. VI All-Russian conf., Novosibirsk.* – Ed. Institute of Thermal Physics SD RAS chapter 1 pp. 170–177.
[6] Tikhonov A.N and Samarskiy A.A. 1977 Equations of mathematical physics (Moscow: Nauka).
[7] Toropov E.V and Osintsev K.V 2017 *Mathematical model of generalized heat transfer in the furnace of the boiler unit – the heat transfer paradigm* (Bulletin of SUSU. Series Power Engineering) vol. 17, № 1. pp. 5–12.
[8] Shvydkiy V.S., Spirin N.A., Ladygichev M.G., Yaroshenko Y.G and Gordon Y.M. 1999 *Elements of the theory of systems and numerical methods of simulation of heat and mass transfer processes* (Moscow: Intermet Engineering).
[9] Toropov E.V and Osintsev K.V 2015 *Mathematical model of heat transfer in the zone of intensive combustion of the boiler unit* (Bulletin of SUSU. Series Power Engineering) vol.15, № 4. pp. 19–25.
[10] Toropov E.V 2016 *The Systemically Structured Adaptation of Heat Transfer in Boilers* (Bulletin of SUSU. Series Power Engineering) vol.16, № 1. pp. 19 – 23.
[11] Telegin A.S., Shvydkiy V.S and Yaroshenko Y.G 2002 *Heat and mass transfer* (Moscow: ICC Akademkniga).
[12] Shorin S.N 1964 *Heat transfer* (Moscow: Vysshaya shkola).
[13] Isachenko V.P., Osipova V.A and Sukomel A.S 1981 *Heat transfer* (Moscow: Energoizdat).

Kinetics of the nitrogen oxides reaction with coke carbon in a low-temperature swirl furnace process

A A Trinchenko[1]

[1] Peter the Great St.Petersburg Polytechnic University, Polytechnicheskaya, 29, St. Petersburg, 195251, Russia

Abstract. The work is devoted to the increase of ecological parameters of organic fuel burning process in power boilers combustion chambers. The technological method of nitrogen oxides decomposition on the surface of carbon particles with the formation of environmentally safe carbon dioxide and molecular nitrogen in low-temperature swirl combustion is considered. A methodology, a mathematical model and a program for calculating the process of a fuel combustion under conditions of multiple circulation, with periodic return of fuel particles to zones with the initial concentration of oxygen, are developed. It is shown that the stepwise supply of the oxidizer to the flare leads to a decrease in the generation of toxic nitrogen oxides, the formation of reducing zones, and the decomposition of the formed nitrogen oxides on the surface of burning coke particles. It has been established that the amount of decomposed nitrogen oxides depends on the surface area of the reacting fuel particles, as well as the time of their stay in the swirl zone of the low-temperature swirl furnace. With the use of calculated results, emissions of nitrogen oxides of reconstructed boilers are reliably predicted.

1. Introduction

At present, for the production of electricity and heat, solid fuels are mainly burned in high-power pulverized-coal boilers [1–6], which have a number of drawbacks: with an increase in the unit capacity of the boiler, the dimensions, the metal consumption and the costs of fabrication and installation increase unnecessarily, harmful substances into the atmosphere, there are problems of slagging of heating surfaces. In addition, the organization of flaring requires relatively fine grinding of the fuel, which determines the presence of an expensive, difficult to operate and explosive dust preparation system.

An alternative to a pulverized-coal torch is the method of low-temperature swirl (LTS) combustion. The concept of LTS-method (LTS-technology) was proposed by Professor V.V. Pomerantsev at the Leningrad Polytechnic Institute [7] (now St. Peterburg Polytechnic University of Peter the Great) and has been further developed now [8–15].

The LTS method is based on the principle of organizing multiple circulation with respect to fuel particles with periodic return to zones with an initial concentration of oxygen. The organization of the swirl stream flows, is achieved by the interaction of burner air jets and air of lower blasting.

2. Model of fuel combustion and reaction of nitrogen oxides with coke carbon

The most complete burning of carbon particles in the LTS-combustion chamber can be described by a set of chemical reactions proposed by V.V. Pomerantsev and S.M. Shestakov [16].

The decomposition of nitrogen oxides on the coke carbon is accounted for by the reaction:

$$5.\ 2C + 2NO = N_2 + 2CO - 180.2\ \text{kJ/mol}. \qquad (1)$$

The designations of the reactive components are adopted as follows: O_2 – 1; CO_2 – 2; CO – 3; H_2 – 4; H_2O – 5, and nitrogen oxides NO – 6 and nitrogen N_2 – 7.

The change in the rate constants of chemical reactions from temperature is governed by the Arrhenius relationship:

$$k_i = k_{0i} \cdot exp[-E_i/(RT)], \qquad (2)$$

the pre-exponential factor of the i-th reaction – through the "pole" S.M. Shestakova:

$$lgk_{0i} = 0.2 \cdot 10^{-4} \cdot E_i + 2. \qquad (3)$$

Research L.A. Vulisa, V.V. Pomerantseva et al. show that between the values of the activation energy of various reactions of carbon with O_2 and CO_2 for the same coke, there is the following relationship:

$$\begin{cases} E_{C+O_2=CO} / E_{C+O_2=CO_2} = 1.1 \\ E_{C+CO_2=2CO} / E_{C+O_2=CO_2} = 2.2 \\ E_{C+H_2O=CO} / E_{C+O_2=CO_2} = 1.6 \end{cases} . \qquad (4)$$

To take into account the mass transfer near the carbon surface, the "reduced boundary film" method was used [16].

To assess the degree of influence of the reaction $CO + O_2 = CO_2$ in the burning out of a carbon particle, the criterion of N.N. Semenova is used (Se), which characterizes the ratio of the flux of matter absorbed by the homogeneous reaction of afterburning $CO + O_2 = CO_2$, to its diffusion flux:

$$Se = \left(k_4 \Delta^2 / D\right)^{1/2}, \qquad (5)$$

where k_4 – reaction rate constant; $\Delta = \delta / Nu_D$ – thickness of the reduced boundary film, m; D – diffusion coefficient, m^2/s .

As the speed of reactions $CO + O_2 = CO_2$ and $2H_2 + O_2 = 2H_2O$ increases, as well as the thickness of the boundary layer increases, oxygen is increasingly consumed to burn off CO and H_2 within the reduced film. The following cases are possible:

1. Oxygen reaches the surface of the carbon particle. This scheme is called the "burning" boundary layer scheme. It occurs at relatively low temperatures and a small thickness of the boundary layer.
2. Oxygen does not reach the surface of the carbon particle, consuming to burn off CO and H_2 within the reduced film. Such a scheme is called the "double burning" boundary layer scheme. It is possible at a high process temperature and the burning of large particles of fuel.
3. Within the given film only diffusion processes occur, and carbon monoxide and hydrogen leave the reduced film and burn in the gas stream. This case is called a circuit with a "non-burning" boundary layer. Such a scheme is typical for fine carbon particles in conditions of moderate temperatures, which is typical for pulverized-coal furnaces.

The combustion scheme of a carbon particle is determined using the N.N. Semenov criterion. At $Se \leq 0.4$, combustion of CO and H_2 can be neglected within the reduced film, and the "non-burning" boundary layer scheme can be used to analyze the combustion process. For $0.4 < Se < 2.0$ – the scheme of the "burning" boundary layer. At $Se \geq 2.0$, the "double burning" boundary layer scheme is applied. In the case where $Se > 100$, a "double burning" boundary layer is used with the CO burn-out on the burning surface inside the reduced film [17].

Analysis of the combustion of natural solid fuel particles in the LTS furnace [18], as well as the calculations performed, showed that the combustion process proceeds in the intermediate region according to the scheme of the "double burning" boundary layer (Se> 100), (the case of "wet" gasification); (figure 1).

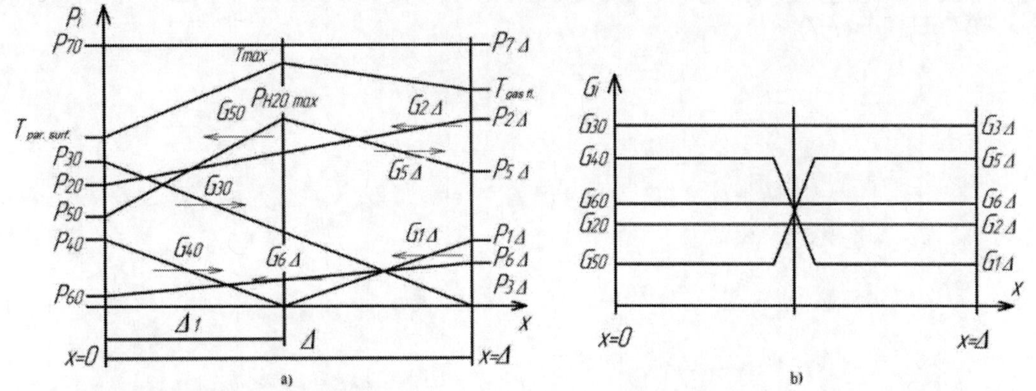

Figure 1. Distribution of partial pressures (a) and flows (b) of the components in the given film of coarse ash particle: 1 – O_2; 2 – CO_2; 3 – CO; 4 – H_2; 5 – H_2O; 6 – NO; 7 – N_2; the "0" indexes – particle surface; "Δ" – flow

Carbon coke under this scheme burns out due to the reduction reactions $C + CO_2 = 2\,CO$ and $C + H_2O = CO + H_2$. Oxygen is consumed only by reaction $2H_2 + O_2 = 2H_2O$ and does not react within the reduced film with CO (CO burns in the flue gas stream). Reactions that go without changing the volume for the case under consideration:

$$\begin{cases} 3.\ \ C + CO_2 = 2CO \\ 3'.\ C + H_2O = CO + H_2 \\ 4'.\ 2H_2 + O_2 = 2H_2O \\ 5.\ \ 2C + 2NO = N_2 + 2CO \end{cases} \quad . \tag{6}$$

The flux of carbon from the surface of the particle, taking into account the stoichiometric relationships:

$$G_c = G_{20} + 1/2G_{50} + 1/2G_{60}\ . \tag{7}$$

Expressions for partial pressures on the particle surface and component flows:

$$p_{CO_2 0} = \frac{p_{CO_2\Delta}}{1 + (k_3 \cdot \Delta / D)}; \quad G_{CO_2 0} = \frac{\alpha_D}{RT} N_3 \frac{p_{CO_2\Delta}}{1 + N_3}, \tag{8}$$

$$p_{H_2O0} = \frac{p_{H_2O\Delta} + 2p_{1\Delta}}{1 + N_{3'}}; \quad G_{H_2O0} = \frac{\alpha_D}{RT} N_{3'} \frac{p_{H_2O\Delta} + 2p_{O_2\Delta}}{1 + N_{3'}}, \tag{9}$$

$$p_{NO0} = \frac{p_{NO\Delta}}{1 + (k_5 \cdot \Delta / D)}; \quad G_{NO0} = \frac{k_5}{RT} p_{NO0} = \frac{k_5}{RT}\frac{p_{NO\Delta}}{1 + N_5} = \frac{\alpha_D}{RT} N_5 \frac{p_{NO\Delta}}{1 + N_5}. \tag{10}$$

The flow of carbon from the particle surface, (kmol/(m^2·s)):

$$G_c = \frac{\alpha_D}{RT}\left[\frac{N_3}{1 + N_3} p_{CO_2\Delta} + \frac{N_{3'}}{1 + N_{3'}}\left(p_{O_2\Delta} + 0.5 p_{H_2O\Delta}\right) + \frac{N_5}{1 + N_5} p_{NO\Delta}\right]. \tag{11}$$

3. Results of a numerical study of a low-temperature swirl furnace process

Calculations of the combustion process were carried out with reference to the combustion of milling peat in the boiler BKZ-210 (Figure 2), and the incineration of wood waste – hydrolytic lignin, in the boiler TP-35U (Figure 3).

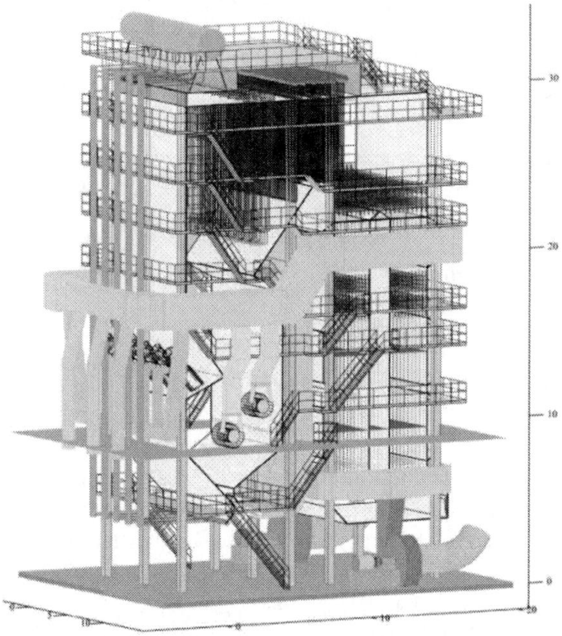

Figure 2. Boiler BKZ-210 for burning milling peat

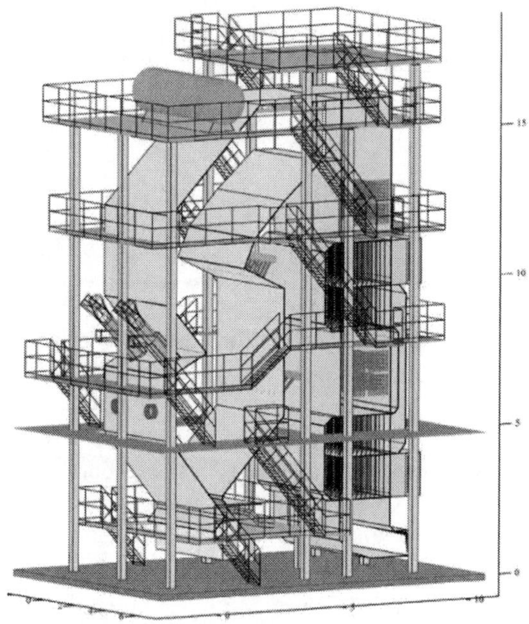

Figure 3. Boiler TP-35U for burning of hydrolytic lignin

The reacting particles motion trajectories in the LTS furnace in a known velocity field are obtained by numerically solving equation (12), which takes into account the influence of two main forces-the forces of aerodynamic drag and gravity:

$$m\frac{d\vec{V}}{d\tau}=\sum_{i=1}^{k}\vec{P_i}+(\overrightarrow{W_{fl}}-\vec{V})\frac{dm}{d\tau}. \tag{12}$$

Comparison of the calculated and experimental trajectories of particles in the LTS furnace of the BKZ-210 boiler showed their good agreement [19, 20].

The concentrations of the main reactive components for calculating the combustion process (system 6) were taken to be typical of LTS furnaces [21, 22].

As a result of fuel burnup calculations (milling peat and hydrolytic lignin) in the LTS furnace (Figure 4), the concentrations of nitrogen oxides along the combustion chamber section were determined (Figure 5), taking into account the decomposition of nitrogen oxides on the circulating coke particles surface. The concentrations obtained on the average are lower by 30 %, in comparison with the combustion of the same fuels according to the scheme of the direct-flow pulverized-coal torch [23]. Thus, low-temperature swirl combustion can be considered as a promising technological method for reducing emissions of nitrogen oxides into the atmosphere during boiler operation.

IOP Conf. Series: Earth and Environmental Science **288** (2019) 012124 doi:10.1088/1755-1315/288/1/012124

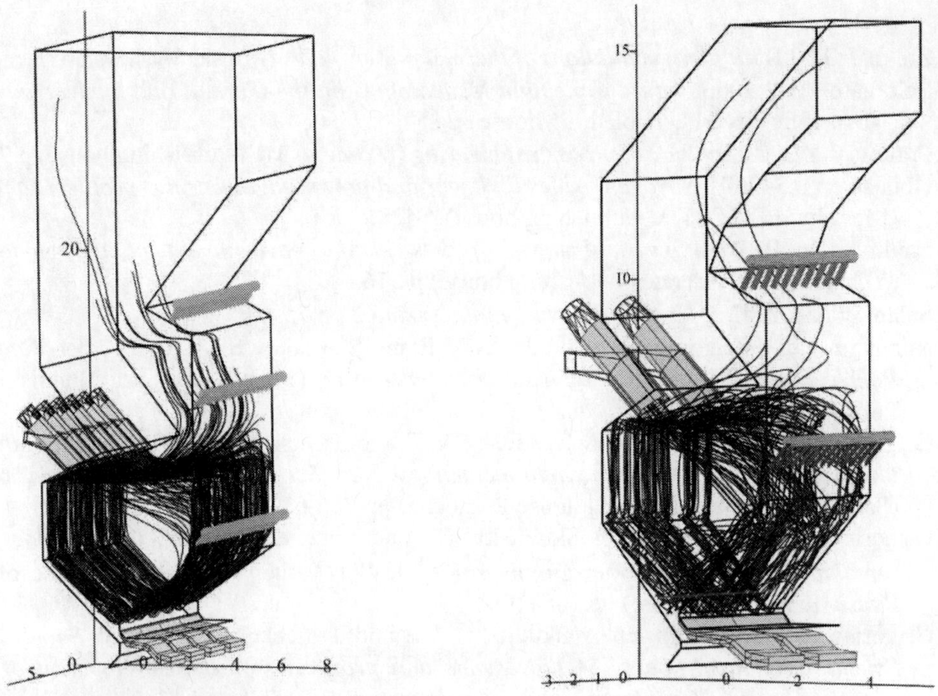

a) b)

Figure 4. Trajectories of burning particles motion in the LTS furnace:
a) – boiler BKZ-210; b) – boiler TP-35U

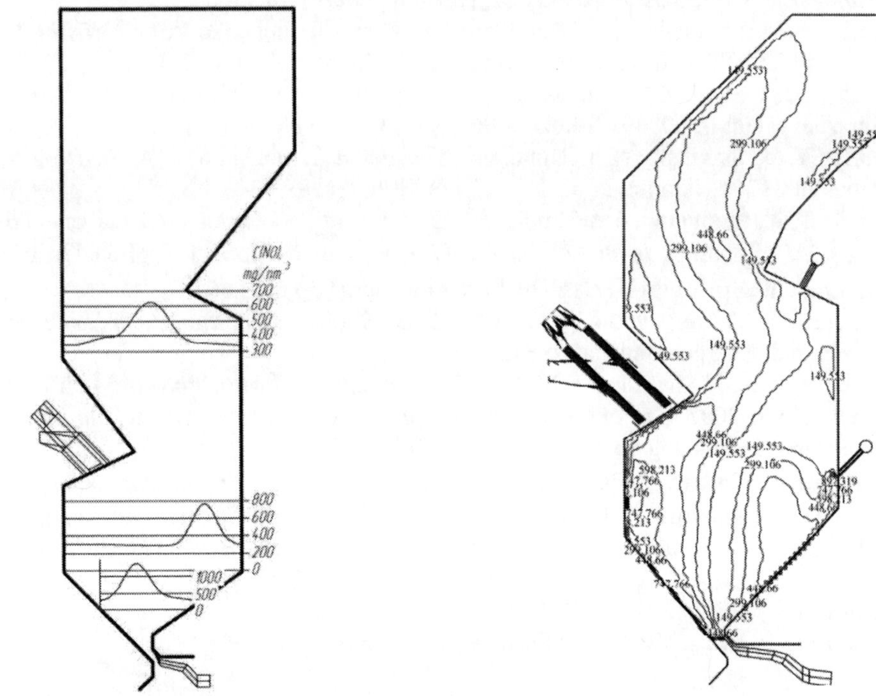

a) b)

Figure 5. Concentrations of nitrogen oxides in the LTS furnace:
a) – boiler BKZ-210; b) – boiler TP-35U

References

[1] Zhikhar G I 2015 *Boiler installations of thermal power plants* (Minsk: Vysheyshaya school) p 532

[2] Solomatov B B 2006 *Nature protection technologies on the thermal and nuclear power plants* (Novosibirsk: NGTU Publishing house) p 852

[3] Putilov V Y 2003 *Ecology of power engineering* (Moscow: MEI Publishing house) p 716

[4] Hodakov Yu S 2001 *Nitrogen oxides and heat-and-power engineeering: problems and decision* (Moscow: LLC EST-M Publishing house) p 428

[5] Shulman V L 2010 *Thermal power plants in the environment of the modern world* (Yekaterinburg: Socrates Publishing house) p 376

[6] Salamov A A 2002 *J. Heat-and-power engineeering* **2** pp 76–80

[7] Rundygin Yu A Grigoriev K A Skuditsky V E and Shestakov S M 2006 *Victor Vladimirovich Pomerantsev. To the 100 anniversary since birth* (Saint Petersburg: Publishing house Politekhn) pp 133–149

[8] Grigoriev K A Rundygin Yu A Skuditsky V E and Trinchenko A A 2003 *Formation of the technical policy of innovative high technology: Mat. Sci.-pract. conf. and workshop schools* (Saint Petersburg: Publishing house Politekhn) pp 216–222

[9] Grigoriev K A Rundygin Yu A Skuditsky V E and Trinchenko A A 2004 *Sat. Doc. Thematic seminar "Ecology in Power Engineering–2004"* (Moscow: Publishing house of the All-Russia Exhibition Centre) pp 108–109

[10] Grigoriev K A Rundygin Yu A Skuditsky V E and Trinchenko A A 2005 *Sat. Doc. IV Int. Scientific-technical. Conf. "Achievements and prospects for the development of Siberia's energy industry"* (Krasnoyarsk: Publishing house of SIBVTI) pp 144–147

[11] Grigoriev K A Rundygin Yu A Skuditsky V E and Trinchenko A A 2006 *Increasing the efficiency of production and use of energy in the Far East: Proceedings of the 4th Seminar of the Higher Schools of Siberia and the Far East on Thermal Physics and Heat Power Engineering* (Vladivostok: DVGTU Publishing house) pp 83–89

[12] Grigoriev K A Skuditsky V E Rundygin Yu A and Trinchenko A A *Swirling-type furnace* Patent of 2253801 Russia. Publ. 10.06.2005. Bulletin No. 16

[13] Grigoriev K A Skuditsky V E Rundygin Yu A and Trinchenko A A *Swirling-type furnace* Eurasian patent of 008691. Publ. 29.06.2007. Bulletin No. 3

[14] Grigoriev K A Skuditsky V E Rundygin Yu A and Trinchenko A A *Swirling-type furnace* Patent of 83761 Ukraine. Publ. 11.08.2008. Bulletin No. 15

[15] Grigoriev K A Rundygin Yu A Skuditsky V E Anoshin R G Paramonov A P and Trinchenko A A 2012 *Proceedings of the 7th Int. Symp. Coal Combustion* (Tsinghua University Press Beijing and Springer-Verlag Berlin Heidelberg) pp 662–665

[16] Pomerantsev V V Arefyev K M Akhmedov D B 1986 *Bases of the practical theory of burning* (Leningrad: Energoatomizdat) p 312

[17] Trinchenko A A and Shestakov S M 2008 *J. Sci. and tech. sheets SPbSTU* **54** pp 149–156

[18] Trinchenko A A 2007 *Mat. of the day of the young scientist at the Polytechnic University* (Saint Petersburg: Publishing house Politekhn) pp 17–18

[19] Paramonov A P Kadyrov M R Trinchenko A A 2017 *Procedia Engineering* **206** pp 546–551

[20] Paramonov A P Kadyrov M R Trinchenko A A 2017 *Procedia Engineering* **206** pp 558–563

[21] Trinchenko A A and Paramonov A P 2015 *J. Sci. and tech. sheets SPbSTU* **231** pp 72–81

[22] Trinchenko A A 2014 *J. Sci. and tech. sheets SPbSTU* **207** pp 61–70

[23] Abramov A I Yelizarov D P and Remezov A N 2001 *Increase in ecological safety of thermal power plants* (Moscow: MEI Publishing house) p 378

Increasing the efficiency of energy use in the thermal scheme of a steam boiler house

R N Valiev[1], Sh G Ziganshin[1], A V Maryashev[2]

[1]Kazan State Energy University, Kazan, Russia

[2]Povolzhsky State Technological University, Yoshkar-Ola, Russia
valievkgeu@yandex.ru

Annotation. This article discusses the issues of increasing the efficiency of energy use in the thermal scheme steamed boiler house. A positive effect is achieved by transforming the excess energy of water vapor into electricity and avoiding the use of steam at the stage of heating the raw water.

Description of the object and method ofinvestigation

The thermal scheme of the steam boiler house is shown in Fig.1. The main elements of the thermal scheme are: steam boiler 1; deaerator 2; water treatment plant 12; pumps 4, 6, 7; network preheater 5; consumers of steam 13 and network water 14.

To determine the heat-and-power efficiency of the thermal scheme, a thermodynamic analysis is carried out, which consists of four parts: information part, heat part, exergic part and analytical part. In the information part, the processing of the initial information is carried out and the following data are determined: the structure of the circuit, the characteristics of the elements, and the parameters of the flows. In the heat part, the values of the heat flows at the input and output of the circuit elements are determined. In the exergic part, the values of the exergy flows at the input and output of the circuit elements are determined. In the analytical part, an analysis is carried out with an assessment of the heat and exergic efficiency of the circuit.

Balance calculations of the elements of the thermal scheme are carried out in a package of Microsoft Excel based on the principles of decomposition calculation and methods of structural analysis [1, 2]. The thermodynamic analysis is carried out according to the procedure [3, 4]. Exergy of the substance flow is determined by the formula in accordance with[5,6]:

$$E = H - H_0 - T_0(S - S_0),\ \text{kcal/h}$$

where H, H_0, S, S_0 are flux enthalpies and entropies at the parameters of the substance flow and the environmental parameters, respectively (kcal/h); T_0 is the ambient temperature in the Kelvines, upon which it becomes impossible to heat exchange between the streams of matter in the circuit under consideration. The lowest temperature of the substance flow in the thermal scheme is T_0 = 5 °C. As the ambient pressure, we take the pressure P_0 = 1 bar.

The coefficient of heat flux is defined as the ratio E/Q [7].

IOP Conf. Series: Earth and Environmental Science **288** (2019) 012125 doi:10.1088/1755-1315/288/1/012125

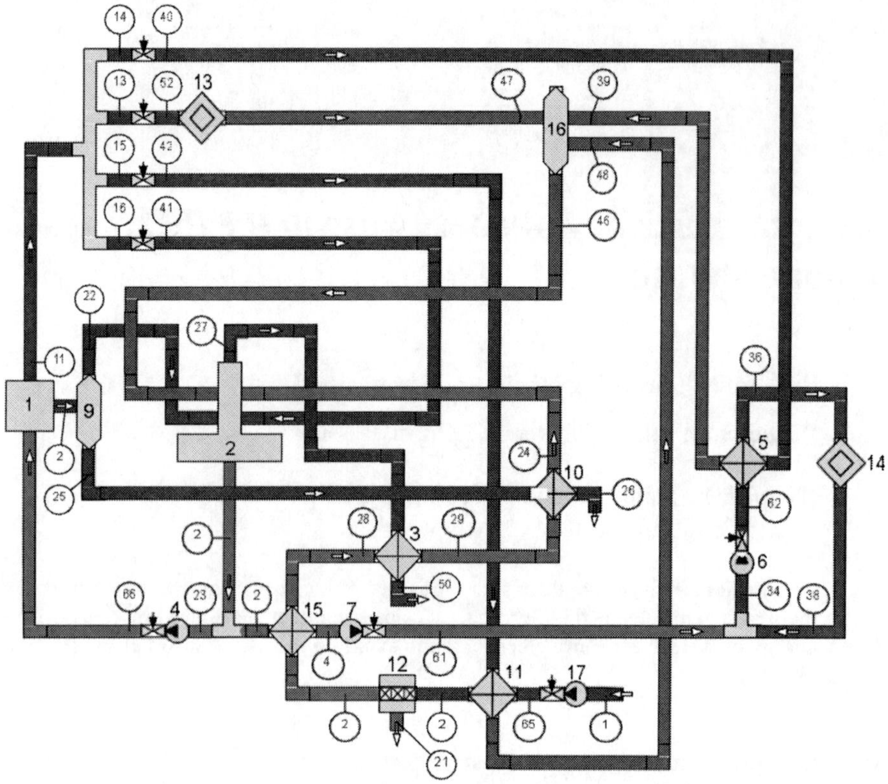

Fig.1. Thermal scheme of a steam boiler house.
Digital designations: 1 - steam boiler; 2 - atmospheric deaerator; 3 - vapor cooler; 4 – feed pump; 5 - preheater of network water; 6 - network pump; 7 - make-up pump; 8 - throttling device; 9 - continuous blowdown separator; 10 - blowdown waters cooler; 11 - preheater of raw water; 12 - water treatment plant; 13 - industrial consumers of steam; 14 - consumers of network water; 15 - deaerated water cooler; 16 - tank for collection condensate; 17 - raw water pump. The circles show the numbering of the flows (see Table 1).

Based on the results of the calculation, an information table is compiled, in which each element is represented as a node into which the flows of matter enter and exit. The results of the determination exergy of flows substances and the values of the coefficients of heat flow are given in Table 1.

Table 1. The results of the determination exergy of flows substances and the values of the coefficients of heat flow

Implemented processes	№ flow's	G, kg/h	t, °C	P, bar	H, kcal/h	S, kkal/°C/h	E, kcal/h	E/Q, в % от Q
1	2	3	4	5	6	7	8	9
Pressure increase in the pump 17	1	15679.8	5	5.9	78390	295.5	0.0	0.0
	65	15679.8	5.1	8	79965	301.3	0.0	0.0
Heating in the heat exchanger 11	65	15679.8	5.1	8	79965	301.3	0.0	0.0
	2	15679.8	35.1	7	550310	1896.0	26779.2	4.9
Cooling in the heat exchanger 11	42	800	175.7	7	532630	1293.0	173178.1	32.5
	48	800	77.8	6	62278	200.4	6732.7	10.8
Heating in the heat exchanger 15	3	15366.2	35.1	7	539310	1858.0	26272.1	4.9
	28	15366.2	49.7	6	763640	2567.0	53393.7	7.0
Cooling in the heat exchanger 15	7	6404.6	104.8	1.2	672840	2081.0	95579.0	14.2
	4	6404.6	70	1.2	448520	1461.0	43712.0	9.7
Heating in the heat exchanger 3	28	15366.2	49.7	6	763640	2567.0	53393.7	7.0
	29	15366.2	61.6	5	946310	3122.0	81690.5	8.6

Continuation of Table. 1

1	2	3	4	5	6	7	8	9
Cooling in the heat exchanger 3	27	311.7	104.8	1.2	199730	543.4	48659.7	24.4
	50	311.7	54.7	1.2	17052	56.9	1312.8	7.7
Heating in the heat exchanger 10	29	15366.2	61.6	5	946310	3122.0	81690.5	8.6
	24	15366.2	68.4	4	1051300	3432.0	100454.0	9.6
Cooling in the heat exchanger 10	25	2496	113.3	1.6	283760	867.2	43159.8	15.2
	26	2496	71.6	1.6	178720	580.7	71.6	10.0
Processes in the deaerator 2	24	15366.2	68.4	4	1051300	3432.0	68.4	9.6
	22	419.9	113.3	1.6	270320	722.3	643.8	25.7
	41	4686	157.9	1.6	3119900	8313.0	665.8	25.9
	46	83439.7	79.6	1.2	6650200	21340.0	79.7	11.1
	27	311.7	104.8	1.2	199730	543.4	640.8	24.4
Pressure increase in the pump 4	5	103600.1	104.8	1.2	10884000	33670.0	105.1	14.2
	23	97195.4	104.8	1.2	10211000	31590.0	105.1	14.2
	66	97195.4	105.4	17	10266000	31730.0	105.6	14.3
Processes in the steam boiler 1 η = 92%	66	97195.4	105.4	16	10266000	31730.0	105.6	14.3
	11	94279.6	195.2	14	62770000	145600.0	665.8	35.5
	6	2915.9	195	15	579650	1590.0	198.8	23.8
	q_1				57699620			104.0
	q_2				3461977			29.3
Condensation and cooling in the consumers of the heat 13	52	42683	175.7	7	28418000	68970.0	665.8	32.5
	47	42683	80	5	3417400	10960.0	80.1	11.1
Processes in the tank for collection condensate 16	47	38414.7	80	1.2	3075600	9866.0	80.1	11.1
	39	44225	79.4	1.2	3512300	11280.0	79.4	11.0
	48	800	77.8	1.2	62278	200.4	77.8	10.8
	46	83439.7	79.6	1.2	6650200	21340.0	79.7	11.1
Heating in the network preheater 5	62	320230	70.2	9	22494000	73230.0	70.2	9.8
	36	320230	150	8	48426000	140800.0	151.2	19.3
Condensation and cooling in the network preheater 5	40	44225	175.7	7	29444000	71460.0	665.8	32.5
	39	44225	79.4	6	3512300	11280.0	79.4	11.0
Processes in the continuous blowdown expander 9	6	2915.9	195	15	579650	1590.0	198.8	23.8
	22	419.9	113.3	1.6	270320	722.3	643.8	25.7
	25	2496	113.3	1.6	283760	867.2	113.7	15.2
Pressure increase in the pump 7	4	6404.6	70	1.2	448520	1461.0	70.0	9.7
	61	6404.6	70.2	6	449590	1464.0	70.2	9.8
Pressure increase in the pump 6	34	320230	70	4	22427000	73040.0	70.0	9.8
	62	320230	70.2	10	22494000	73230.0	70.2	9.8
Cooling in the consumers of the heat 14	36	320230	150	7	48426000	140800.0	151.2	19.3
	38	320230	70	5	22426000	73040.0	70.0	9.8

Changes in the Q and E flows of matter in the elements are determined by subtracting the input stream from the output stream. For heat exchange equipment with heat transfer through the wall, the variation for each flow separately (conditional dividing into the cooled and heated streams) is taken into account. For the heat-exchange equipment of the mixing type (the deaerator, the tank for collection condensate, the continuous blowdown separator), the required operating mode is selected (isoentropic expansion, calculating the heat load by pressure and amount of evaporation, calculating the heat load and output flow parameters by input flow parameters), and changing Q and E is taken into account as the difference between the total values of all input and output streams. For pumps, the change in Q and E is determined by the difference between the values at the inlet and outlet. The electric power consumption for the pump drive (taking into account $\eta = 0.7$) is calculated separately using the standard formula for calculating the pump power [6].

Table 2 presents the results of determining the changes in Q and E of the substance flows as it passes through the main elements of the thermal circuit of the steam boiler house. The heat efficiency ($\Sigma Q_{out} / \Sigma Q_{in} \times 100$) and the exergy efficiency of the thermal scheme (see Table 2), taking into account the elements included in the review, are 79.36% and 42.81%, respectively. The total amount of electricity spent on creating pressure in pumps with $\eta = 70\%$ is 145.047 kW.

Table 2. Results of changes in Q and E of substance fluxes

Implemented processes	№ flow's	Q_{in}, kcal/h	Q_{out}, kcal/h	E_{in}, kcal/h	E_{out}, kcal/h	ΔQ, kcal/h	ΔE, kcal/h	N_{el}, kcal/h
1	2	3	4	5	6	7	8	9
Pressure increase in the pump 17	1	78390.0		0.0		1575.0	0	-1575.55
	65		79965.0		0.0			
Heating in the heat exchanger 11	65	79965.0		0.0		470345.0	26779.20	
	2		550310.0		26779.20			
Cooling in the heat exchanger 11	42	532630.0		173178.05		-470352.0	-166445.31	
	48		62278.0		6732.74			
Heating in the heat exchanger 15	3	539310.0		26272.06		224330.0	27121.65	
	28		763640.0		53393.71			
Cooling in the heat exchanger 15	7	672840.00		95578.99		-224320.0	-51867.00	
	4		448520.0		43711.99			
Heating in the heat exchanger 3	28	763640.00		53393.71		182670.0	28296.75	
	29		946310.0		81690.46			
Cooling in the heat exchanger 3	27	199730.00		48659.66		-182678.0	-47346.90	
	50		17052.0		1312.76			
Heating in the heat exchanger 10	29	946310.00		81690.46		104990.00	18763.50	
	24		1051300.0		100453.96			
Cooling in the heat exchanger 10	25	283760.00		43159.85		-105040.00	-25350.03	
	26		178720.0		17809.82			
Processes in the deaerator 2	24	1051300.00		100453.96		-7990.00	-120946.72	
	22	270320.00		69515.13				
	41	3119900.0		808787.13				
	46	6650200.0		734921.96				
	27		199730.0		48659.66			
	5		10884000.0		1544071.82			
Pressure increase in the pump 4	23	10211000.0		1448054.65		55000.00	16059.0	-54948.80
	66		10266000.0		1464113.65			
Processes in the steam boiler 1 $\eta = 92\%$	66	10266000.0		1464113.65				
	11		62770000.0		22294458.8	53083650.0	20968451.0	
	6		579650.0		138105.90			
	q_1	57699619.6		60007604.35		-54237642.4	-58994944.1	
	q_2		3461977.17		1012660.24			
Condensation and cooling in the consumers of the heat 13	52	28418000.0		9244451.95		-25000600.0	-8865118.50	
	47		3417400.0		379333.45			
Processes in the tank for collection condensate 16	47	3075600.0		340783.81		22.00	1802.16	
	39	3512300.0		385603.25				
	48	62278.0		6732.74				
	46		6650200.0		734921.96			
Heating in the network preheater 5	62	22494000.0		2203532.75		25932000.0	7137404.50	
	36		48426000.0		9340937.25			
Condensation and cooling in the network preheater 5	40	29444000.0		9578236.25		-25931700.0	-9192633.0	
	39		3512300.0		385603.25			
Processes in the continuous blowdown expander 9	6	579650.0		138105.90		-25570.0	-25430.92	
	22		270320.0		69515.13			
	25		283760.0		43159.85			
Pressure increase in the pump 7	4	448520.0		43711.99		1070.0	235.55	-1073.91
	61		449590.0		43947.54			
Pressure increase in the pump 6	34	22427000.0		2189381.25		67000.0	14151.50	-67119.7
	62		22494000.0		2203532.75			
Cooling in the consumers of the heat 14	36	48426000.0		9340937.25		-26000000.0	-7152556.0	
	38		22426000.0		2188381.25			
Total		252252263	200189022	98626899	42223287	-52063240	-56403612	-124718
Efficiency, %		by heat: 79.36		by exergy: 42.81		by electricity: 70		

The results of the study and their discussion

To achieve the intended results, the following changes were made to the thermal scheme of the steam boiler house: a) heat exchangers in the preheating stage of raw and chemically purified water were reconnected with each other in order to make fuller use of heat; b) installed steam turbines instead of throttling devices to use the excess steam pressure after the boilers and generate electricity. The thermal scheme of the steam boiler room after the introduction of changes is shown in Fig.2.

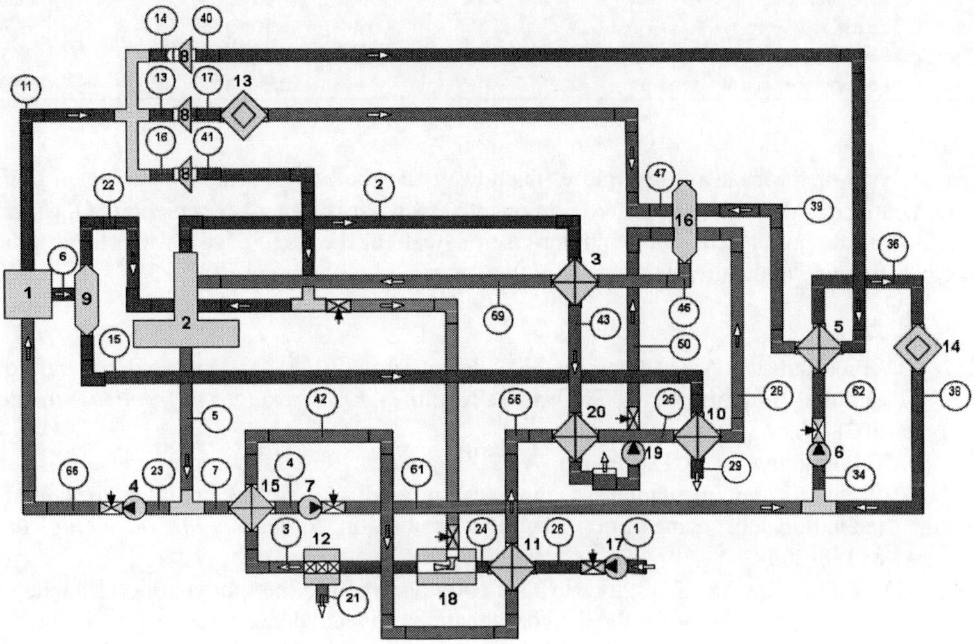

Fig.2. Thermal scheme of the steam boiler room after the changes.
Digital designations: 1÷7 - the same as in Fig.1; 8 - steam turbine; 9÷17 are the same as in Fig.1; 18 - steam utilizer; 19 - condensate pump; 20 - condensate subcooler. In the circles, the accepted numbering of the flows after the changes is given.

In connection with the organization of more complete use of low-potential heat in the boiler house scheme, the consumption of steam from boilers is reduced. The heat exchanger 11 ceases to use steam as a heat carrier and switches to the use of chemically purified water after the heat exchanger 15. A direct-acting heater 18 is used as a starting and backup device. The chemically purified water after the water purification installation 12 passes successively through the heat exchangers 15, 11, 20 and 10. All condensate streams and chemically purified water are connected in a condensate collection tank 16. Before being fed to the atmospheric deaerator 2, condensate and chemically purified water are heated in a heat exchanger 3. To generate electricity at excess steam pressure after boilers it is proposed to install steam turbines in place of the throttling devices 8. In order for steam turbines to work in the best conditions, steam should be passed through the superheater. If there is no superheater in the boiler, it is recommended to use expansion machines for working on wet steam [8].

Changes in the thermal scheme can be considered energetically effective, if as a result of their implementation, other things being equal, the consumption of fuel and energy resources will decrease or additional profit will be obtained [9]. Heat and exergic efficiency for the thermal scheme of the steam boiler house after the changes (Fig.2), were determined by the same method as for the circuit in Fig.1. Heat and exergy efficiency of the scheme after the changes: 80.72% and 50.55%, respectively. The total amount of electricity to create pressure in the pumps taking into account $\eta = 70\%$: 147.704 kW, the total amount of electricity produced in steam turbines taking into account $\eta = 75\%$: 2596.99 kW. The indicators of the heat and exergy efficiency of the thermal scheme of the steam boiler house before and after changes are summarized in Table. 3.

IOP Conf. Series: Earth and Environmental Science **288** (2019) 012125 doi:10.1088/1755-1315/288/1/012125

Table 3. Comparison of the efficiency of the thermal scheme of the boiler plant

Indicator name	Before making changes	After making changes	Increment of indicators
Heat efficiency, %	79.36	80.72	+1,36
Exergy efficiency, %	42.81	50.55	+7,74
Electric power for electric pump drive, kW	-145,047	-147,704	-2,657
Electricity produced by steam turbines, kW	0	2596,987	+2596,987

Conclusions

The analysis of the heat and exergic efficiency of the thermal scheme of the steam boiler house is carried out. Considered and proposed solutions that increase the energy efficiency of the thermal circuit. It is shown that power generation and non-use of steam at the stage of raw water heating increases the energy efficiency of the thermal scheme of the steam boiler house (see Table 3).

References

[1] L V Plotnikova, L A Kashipova and R A Ishmuratov (2016) Modeling of Heat and Power System Optimal Structure Using Software. *Journal of Engineering and Applied Sciences*, **11** (14): 3028-34

[2] L V Plotnikova, L A Kashipova, A A Zvegintsev, R A Ishmuratov and N K Nuriev (2015) Program implementation of the system analysis of the complex-structured chemical-technological scheme of petrochemical production. *Bulletin of the Technological University*, 8: 198-202

[3] R N Valiev and Yu G Nazmeev (2001) Analysis of the thermodynamic efficiency of the heat engineering scheme for the dehydrogenation of isoamylenes into isoprene in the production of synthetic isoprene rubber SKI-3. Part 1. *News of Higher Educational Establishments. Problems of energy: scientific and technical and industrial journal*, 1/2: 37-52

[4] R N Valiev and Yu G Nazmeev (2001) Analysis of the thermodynamic efficiency of the heat engineering scheme for the dehydrogenation of isoamylenes into isoprene in the production of synthetic isoprene rubber SKI-3. Part 2. *News of Higher Educational Establishments. Problems of energy: scientific and technical and industrial journal*, 3/4: 49-57

[5] G Wall and M Gong (2001) On exergy and sustainable development - Part 1: Conditions and concepts. *Exergy Int. J.*, **1**(3): 128-145

[6] I Dincer and M A Rosen (2013) Exergy: Energy, Environment and Sustainable Development (Elsevier: Oxford, UK) p 576

[7] B M Malakhov and V N Senich (1997) Thermal pollution of the environment by industrial enterprises: analyte. review (Novosibirsk: State Public Scientific and Technical Library of the Siberian Branch of the Russian Academy of Sciences-Tekhenergokhimprom-Himmashhekologiya) p 68

[8] B M Khrustalev, V N Romanyuk, V A Sednin, A A Bobich, D B Muslina and T V Bubyr (2014) To the issue of development of energy supply of industrial heat technologies and heat supply systems in Belarus. *Power engineering. Izvestiya of higher educational institutions and energy associations of the CIS*, 6: 31-47

[9] R N Valiev, Sh G Ziganshin, Yu V Vankov and R R Garipov (2017) Increasing efficiency the combined-cycle plant with a waste heat boiler due to inclusion in the scheme of the absorption transformer of the heat. *Proceedings of higher educational institutions. Problems of energy: scientific and technical and industrial journal*, 9/10: 40-50

IOP Conf. Series: Earth and Environmental Science **288** (2019) 012126 doi:10.1088/1755-1315/288/1/012126

Scheme of joint operation of an absorption heat pump and combined-cycle plant

R N Valiev[1], Sh G Ziganshin[1], A V Maryashev[2]

[1]Kazan State Energy University, Kazan, Russia

[2]Povolzhsky State Technological University, Yoshkar-Ola, Russia

valievkgeu@yandex.ru

Annotation. In this paper, we propose and consider a scheme for the combined operation of an absorption heat pump and a combined-cycle plant. A computational study was carried out and the results confirming the effectiveness of the scheme were obtained. The heat power of the gas heater of the network water due to the inclusion of an absorption heat pump is increased by 12-14% from its previous value.

1. Description of the object and method of investigation

The scheme of a combined-cycle plant without an absorption heat pump is shown in Fig.1 [1-3]. The main elements of the combined-cycle plant are: gas turbine unit 1, waste heat boiler 2 and steam turbine 3. The waste heat boiler has five heat exchange surfaces: superheater a2, evaporator b2, economizer c2, gas heater of the condensate d2 and gas heater of the network water e2. To simulate the heat consumer after gas heater of the network water, the scheme includes a water-to-water heat exchanger i2. In the scheme also identifies and numbered: deaerator 4, condenser of steam turbine 5, drum of waste heat boiler f2, cooling tower of circulating water supply 6.

The combined scheme of a combined-cycle plant and the absorption heat pump is shown in Fig.2. Absorption H2O-LiBr heat pump consists of five main elements: generator c7, condenser a7, evaporator b7, absorber d7 and heat exchanger of the solutions e7 [4-7]. The heat conversion coefficient achievable in a heat pump depends on the parameters of the heating source, on the thermodynamic cycle being realized, and on the properties of the H2O-LiBr solution [8-10]. Principal difference of the combined scheme (see Fig.2) from the scheme of a combined-cycle plant without a heat pump (see Fig.1) are the consist in the installation of an additional heating surface in the waste heat boiler and the use of heat of the circulating water for heating the heat pump evaporator. In Fig.1, before being fed into the deaerator, the condensate is heated in the gas condensate heater. In Fig.2, before being fed into the deaerator, the condensate passes successively through the absorber and the heat pump condenser and the gas water heater. Heating of the heat pump generator is carried out by means of water of the intermediate circuit from of the gas water heater.

The advantage of the combined scheme is an increase in the heat capacity of the gas heater of the network water as a result of the reduction of the heat load of the heat recovery boiler by heating the condensate. The disadvantage is the complexity of the scheme and the additional costs of acquiring and commissioning new equipment.

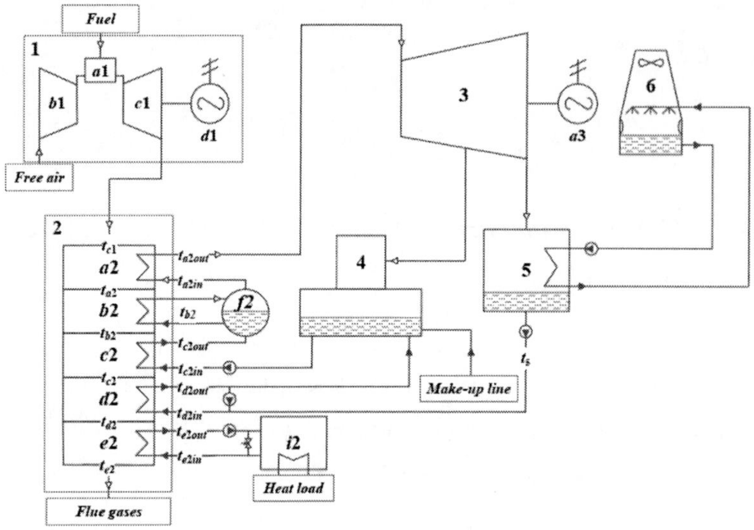

Fig.1. Scheme of combined-cycle plant without heat pump:
1 - gas turbine plant; 2 - waste heat boiler; 3 - steam turbine; 4 - deaerator; 5 - steam turbine condenser; 6 - cooling tower; a1 - combustion chamber; b1 - air compressor; c1 - gas turbine; d1 - gas turbine generator; a3 - steam turbine generator; a2 - superheater; b2 - evaporator; c2 - economizer; d2 - gas heater of the condensate; e2 - gas heater of the network water; f2 - drum; i2 - water-to-water heat exchanger. The decoding of the flow temperatures is given in Table 1.

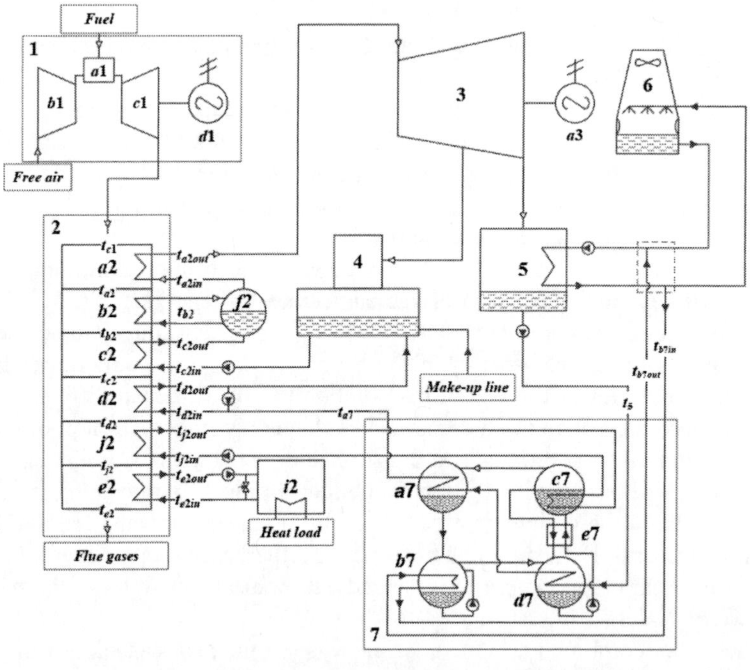

Fig. 2. Combined scheme of the combined-cycle plant and the absorption heat pump:
7 - absorption heat pump; j2 - gas heater of the water; c7 - the generator; b7 - evaporator; a7 is the condenser; d7 - absorber; e7 - heat exchanger of the solutions. The remaining symbols are the same as in Fig.1.

IOP Conf. Series: Earth and Environmental Science **288** (2019) 012126 doi:10.1088/1755-1315/288/1/012126

Table 1. Baseline and calculation results

№	Decoding of symbols and dimensions	Symbol	Fig.1	Fig.2	Fig.1	Fig.2	Fig.1	Fig.2	Fig.1	Fig.2
1	Outdoor temperature,°C	t_{ot}	-32,0	-32,0	-18,0	-18,0	-5,2	-5,2	8,0	8,0
2	Temperature combustion gas behind c1,°C	t_{c1}	518,7	518,7	526,4	526,4	532,7	532,7	538,8	538,8
3	Temperature combustion gas behind a2,°C	t_{a2}	454,0	454,0	458,4	458,4	462,0	462,0	465,4	465,4
4	Temperature combustion gas behind b2,°C	t_{b2}	255,4	255,4	255,4	255,4	255,4	255,4	255,4	255,4
5	Temperature combustion gas behind c2,°C	t_{c2}	215,7	215,7	214,8	214,8	214,0	214,0	213,3	213,3
6	Temperature combustion gas behind d2,°C	t_{d2}	166,4	183,3	164,4	181,6	162,7	180,3	161,1	179,0
7	Temperature combustion gas behind j2,°C	t_{j2}	–	173,4	–	171,5	–	169,9	–	168,5
8	Temperature combustion gas behind e2,°C	t_{e2}	110,0	110,0	110,0	110,0	110,0	110,0	110,0	110,0
9	Temperature steam outlet of a2,°C	t_{a2out}	478,7	478,7	486,4	486,4	492,7	492,7	498,8	498,8
10	Temperature steam at the input in a2,°C	t_{a2in}	245,4	245,4	245,4	245,4	245,4	245,4	245,4	245,4
11	Temperature condensate and steam in b2,°C	t_{b2}	245,4	245,4	245,4	245,4	245,4	245,4	245,4	245,4
12	Temperature condensate outlet of c2,°C	t_{c2out}	235,4	235,4	235,4	235,4	235,4	235,4	235,4	235,4
13	Temperature condensate at the input in c2,°C	t_{c2in}	159,4	159,4	159,4	159,4	159,4	159,4	159,4	159,4
14	Temperature condensate outlet of d2,°C	t_{d2out}	143,8	143,8	143,8	143,8	143,8	143,8	143,8	143,8
15	Temperature condensate at the input in d2,°C	t_{d2in}	60,0	75	60,0	75	60,0	75	60,0	75
16	Water temperature outlet of j2,°C	t_{j2out}	–	160	–	160	–	160	–	160
17	Water temperature at the input in j2,°C	t_{j2in}	–	130	–	130	–	130	–	130
18	Temperature water outlet of e2,°C	t_{e2out}	140,0	140,0	140,0	140,0	140,0	140,0	140,0	140,0
19	Temperature water at the input in e2,°C	t_{e2in}	80,0	80,0	65,0	65,0	60,0	60,0	60,0	60,0
20	Heating capacity e2, MW	Q_{e2}	8,77	9,85	8,25	9,33	7,73	8,79	7,21	8,20
21	Consumption combustion gas behind c1, kg/s	G_{c1}	146,0	146,0	142,6	142,6	138,0	138,0	132,7	132,7
22	Consumption steam through a2, kg/s	G_{a2}	18,10	18,10	18,07	18,07	17,80	17,80	17,40	17,40
23	Condensate consumption in b2, kg/s	G_{b2}	18,10	18,10	18,07	18,07	17,80	17,80	17,40	17,40
24	Consumption condensate through c2, kg/s	G_{c2}	18,47	18,47	18,44	18,44	18,16	18,16	17,75	17,75
25	Consumption condensate through d2, kg/s	G_{d2}	21,83	17,49	21,80	17,47	21,47	17,21	20,99	16,82
26	Consumption waters through j2, kg/s	G_{j2}	–	12,01	–	12	–	11,89	–	11,56
27	Consumption waters through e2, kg/s	G_{e2}	34,59	38,87	26,08	29,51	22,94	26,08	21,38	24,48
28	Heat load c7, kW	Q_{c7}	–	1550	–	1550	–	1530	–	1495
29	Heat load b7, kW	Q_{b7}	–	1088	–	1087	–	1072	–	1047
30	Heat load a7, kW	Q_{a7}	–	1153	–	1152	–	1135	–	1109
31	Heat load d7, kW	Q_{d7}	–	1490	–	1489	–	1466	–	1433
32	Heat load e7, kW	Q_{e7}	–	347,9	–	345,3	–	342,3	–	334,6
33	Temperature waters at the input in c7,°C	t_{c7in}	–	160	–	160	–	160	–	160
34	Temperature waters outlet of c7,°C	t_{c7out}	–	130	–	130	–	130	–	130
35	Temperature water at the input in b7,°C	t_{b7in}	–	35	–	35	–	35	–	35
36	Temperature water outlet of b7,°C	t_{b7out}	–	20	–	20	–	20	–	20
37	Temperature condensate at the input in d7,°C	t_5	–	39	–	39	–	39	–	39
38	Temperature condensate outlet of a7,°C	t_{a7}	–	75	–	75	–	75	–	75
39	Consumption waters through c7, kg/s	G_{c7}	–	12,01	–	12	–	11,89	–	11,56
40	Consumption condensate through d7 and a7, kg/s	$G_{d7}= G_{a7}$	–	17,49	–	17,47	–	17,21	–	16,82
41	Consumption water through b7, kg/s	G_{b7}	–	17,27	–	17,26	–	17	–	16,62

To evaluate the efficiency, the initial data were determined and the following were performed: 1) calculation of the combined-cycle plant scheme without an absorption heat pump; 2) calculation of the combined scheme of a combined-cycle plant and of an absorption heat pump; 3) a comparative assessment of the efficiency of the inclusion of an absorption heat pump in the scheme of a combined-cycle plant.

The scheme of the combined-cycle plant was calculated in a computer program designed for calculating the heat schemes of combined-cycle plants with waste heat boilers. The working algorithm of the program is based on the generally accepted methods for calculating the heat schemes of combined-cycle plants [1]. Calculation of the absorption heat pump was carried out separately from the calculation of the heat scheme of the combined-cycle plant according to the generally accepted calculation methods [5,6]. The heat load of the generator, absorber, evaporator, condenser and intermediate heat exchanger of the absorption heat pump was determined by the condensate flow rate to the deaerator. The comparability of the results was ensured by the condition that the electric power and efficiency of the gas and steam turbines, the parameters and the flow rate of the superheated steam, the temperature of the flue gases before and after the recovery boiler were adopted the same for the compared schemes and determined for four temperature values outdoor air.

2. The results of the study and their discussion

The results of interest for evaluating the effectiveness of the compared schemes are given in Table 1 and in Figures 3 and 4.

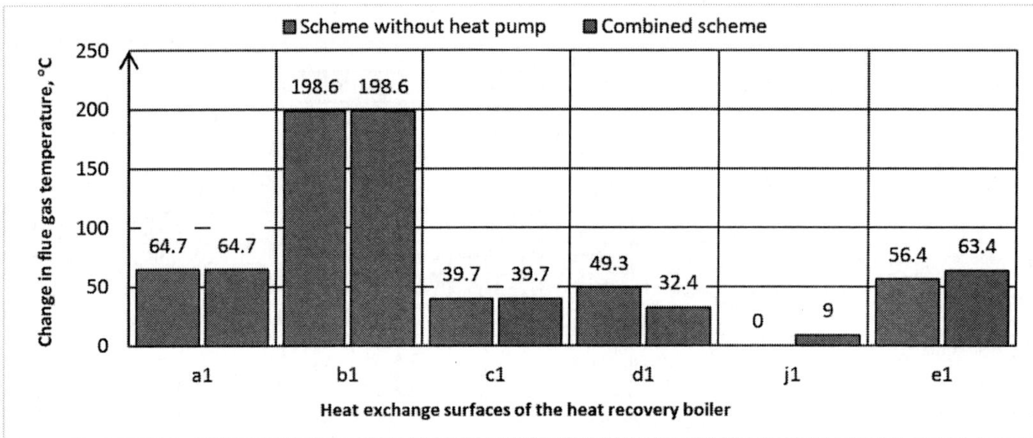

Fig.3. Change of flue gas temperature in waste heat boiler:
a2 – superheater; b2 – evaporator; c2 – economizer; d2 – gas heater of the condensate; j2 – gas heater of the water; e2 – gas heater of the network water.

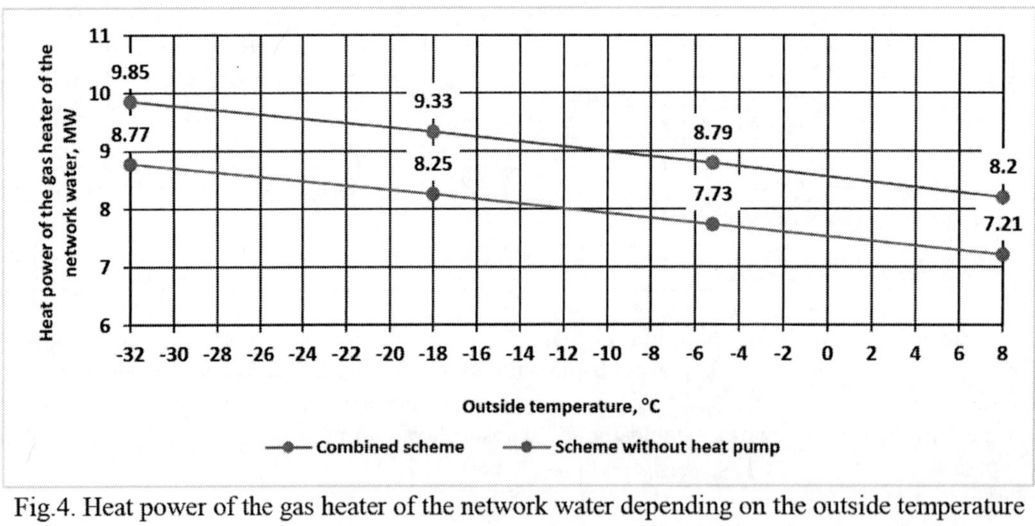

Fig.4. Heat power of the gas heater of the network water depending on the outside temperature

The histogram in Fig.3 is constructed for climatic conditions with an average outside air temperature for a heating period of -5,2 °C. It can be seen from the histogram in Fig.3 due to the preheating of the condensate in the absorber and the condenser of the heat pump, the temperature of the flue gases in the gas heater of the network water in the combined scheme changes more than in the scheme without the heat pump.

Figure 4 shows the graph of the change in the heat power in the gas heater of the network water depending on the temperature of the outside air. From the graph it follows that in the combined version of the circuit the heat power of the gas heater of the network water in the considered range of outside temperatures, with other things being equal, can be increased by 0.99 MW at the ambient temperature +8 °C, by 1,06 MW at an ambient temperature of -5.2 °C; by 1.08 MW at outdoor temperatures of -18 °C and -32 °C.

The increase in the heat power of the gas heater of the network water, other things being equal, allows to obtain additional income from the sale of heat energy, which can be estimated approximately by the formula, rubles/year:

$$\Delta D = \frac{\Delta Q}{1{,}163} \tau \cdot S \cdot \eta \qquad (1)$$

where: ΔD - income from the sale of additional heat power, rubles/year; 1,163 - conversion factor of MW to Gcal/h; ΔQ - additional realizable heat power, MW; t - the number of hours in the year during which the additional heat power is realized, h/year; S - selling price 1 Gcal of heat energy, rub/Gcal; η - efficiency coefficient of the tariff, taking into account all the additional costs and losses not included in the tariff and losses in the process of production, transportation, distribution and sale of additional heat power.

Thus, for example, with increasing heat supply to consumers from a gas heater for network water at ΔQ = 1 MW, provided that heat energy is realized during a heat energy consumption of 8400 hours/year and the cost of heat energy, S = 1626 rubles/Gcal, with a efficiency coefficient of the tariff η = 0.5, the additional annual income will be:

$$\frac{1}{1{,}163} \cdot 8400 \cdot 1626 \cdot 0{,}5 = 5872055 \text{ rubles/year.}$$

Conclusions

1. A combined scheme of a combined-cycle plant and an absorption heat pump is proposed and considered.
2. The calculation study was carried out and the results confirming the efficiency of the combined scheme in comparison with the scheme of a combined-cycle plant without a heat pump were obtained.
3. In a given temperature range of outside air, the heat power of the gas heater of the network water can be increased by 12-14% from its previous value due to the installation of the heat pump. The increase in the heat capacity of the gas heater of the network water, other things being equal, allows to obtain additional income from the sale of heat energy.

References

[1] A D Trukhny (2013) Steam-gas plants of power plants: a textbook for universities in the field of training "Power engineering" and "Heat and power engineering and heat engineering" (Izdatelstvo MEI: Moscow, RU) p 646
[2] P A Shchinnikov and et al. (2010) Comprehensive analysis of combined-cycle power plants. *Izvestiya Vysshikh Uchebnykh Zavedenii. Problems of energy: scientific and technical and*

industrial journal, 3/4: 15-20

[3] P. Ahmadi, I. Dincer, and M. A. Rosen (2011) Exergy, exergoeconomic and environmental analyses and evolutionary algorithm based multi-objective optimization of combined cycle power plants. *Energy*, **36** (10): 5886-98

[4] G V Nozdrenko and et al. (2000) Two-pipe heat supply system with absorption bromide-lithium heat pump. *News of higher educational institutions. Problems of energy: scientific and technical and industrial journal*, 5/6: 25-28

[5] A V Baranenko, L S Timofeevsky, A G Dolotov and A V Popov (2006) Absorption heat converters (SPbGUNiPT: SPb, RU) p 338

[6] L S Timofeevskii, V I Pekarev, N N Bukharin and et al. Thermal and structural calculations of refrigeration machines, heat pumps and thermal transformers. Part 1. Calculation of cycles, thermodynamic and thermophysical properties of working substances (SPbGUNiPT: SPb, RU) p 260

[7] L V Galimova LV (2004) Efficiency of energy-saving systems based on absorption thermo-transformers (Astrakhan. State Technical University, author's abstract. dis. Dr. Tech. sciences: Astrakhan, RU) p 32

[8] C Keil, S Plura, M Radspieler, C Schweigler (2008) Application of customized absorption heat pumps for utilization of low-grade heat sources. *Applied Thermal Engineering*, **28** (16): 2070-76

[9] V N Romanyuk and A A Bobich (2015) Absorption heat pumps at the thermal power station of the Belarusian Energy System by the example of Mozyrskaya CHP. *Energy and Management*, 1: 13-20

[10] V N Romanyuk and A A Bobich (2016) Justification of the parameters of absorption bromide-lithium heat pumps for utilization of secondary energy resources at the CHP with the help of a passive experiment and determination of appropriate changes in various estimates of the operation of the power system. *Energy and Management*, 1: 14-23

Assessment of transport losses of heat at change of properties of thermal isolation

Yu.V. Vankov[1], **E.R. Bazukova**[2], **A.S. Gavrilov**[3]

[1,2,3] Kazan state power engineering university, 420066, Krasnoselskaya st., 51, Kazan, Russian Federation

[1]yvankov@mail.ru, [2]ell17@mail.ru

Summary. Research is devoted to increase in system effectiveness of transportation of thermal energy. The control system of thermal insulation properties realized in the environment of LabVIEW is developed. Assessment of transport losses of heat taking into account real service conditions of pipelines and heterogeneity of thermal insulation properties is carried out. Influence of extent of aging of thermal isolation on increase in size of density of a thermal stream is shown. Calculation of specific thermal losses of an elevated steam line taking into account thermal destruction of isolation is carried out.

Introduction

There are 10% of oil production in the world are extracted in the Russian Federation. Explored reserves of "lungs" of oil decrease and increase stocks of high-viscosity oil and natural bitumens. The known technologies of production of high-viscosity oil and natural bitumens are energy-intensive, a large amount of steam is used in them. Reduction of power losses at production of high-viscosity oil and bitumens is one of the most important tasks of oil production.

In process to transportation of steam from a source to the well there are thermal losses to which reduction apply different types of thermal insulation. At operation of steam lines the indicator of the long-term properties of isolation confirming her service life at the level of the established requirements is important.

It is known that the materials applied in Russia can lead to increase in coefficient of heat conductivity by 20-30% already for the first time 5 years of operation [1-3]. For example, producers guarantee that polyurethane foam works at a temperature of 150 °C. However at a temperature of 120 °C there is his destruction (figure 1).

Timely assessment of thermal insulation properties of heat conductors with the subsequent her replacement will allow to save energy resources significantly.

Methods

In spite of the fact that in modern scientific literature much attention is paid to the questions connected with definition of thermal losses now there is no method of calculation of thermal losses considering heterogeneity of thermal insulation on length of the pipeline and the factors leading to decrease in heat-shielding properties of a covering: aging, change of thickness, sagging, partial lack and moistening of isolation.

Figure 1. End face of two-layer thermal isolation (basalt + polyurethane foam).

Check of a condition of pipelines isolation is under operating conditions offered to be carried out by means of the device of which operation the method of an additional wall is the cornerstone. It consists in the fact that the additional wall with the known thermal resistance λ_0 / δ_0 densely nestles on a surface of a wall which thermal losses need to be defined (figure 2).

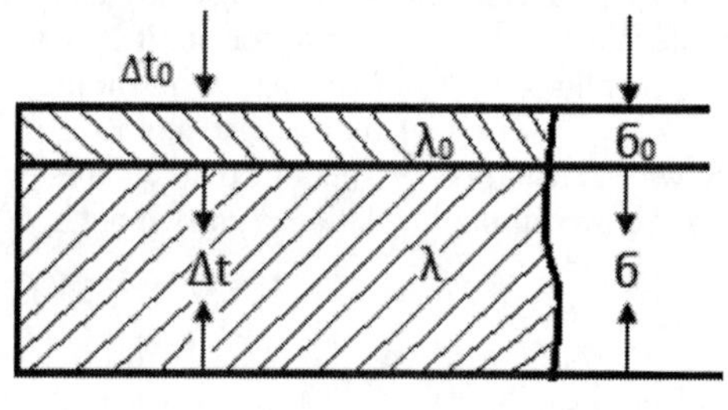

Figure 2. Illustration of an "additional" wall method.

Measuring temperature drop in an additional wall, it is possible to find the thermal stream passing through it from the equation

$$q = \frac{\lambda_0}{\delta_0} \Delta t_0 .$$

If the thermal resistance of an additional wall in comparison with the thermal resistance of the main isolation is small, then at the established thermal state the same thermal stream will pass also through the main wall. This stream also makes thermal losses of the pipeline.

The converter of the device represents a belt length equal to perimeter of the surveyed pipeline (figure 3). In places of a belt, to the corresponding 3, 6, 9 and 12 hours there are sensors of a thermal stream.

Sensitive element of the thermopair sensor is the hyper battery of copper-constant couples executed on galvanic technology. On square centimeter of the sensor about 1500 thermocouples are fitted.

Their total signal is proportional to the difference of temperatures on thickness of the sensor and, therefore, to a thermal stream through him. Density of a thermal stream is defined by the sensor by multiplication of a signal by the calibration factor having a value of 5 to 40 W/m^2·mV. Thickness of thermal stream sensors – 2 mm.

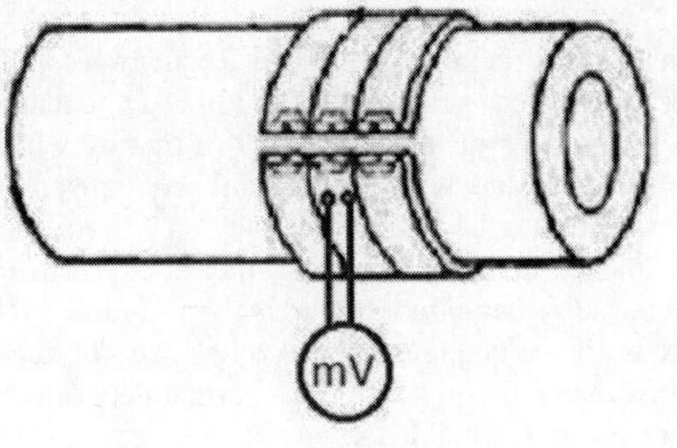

Figure 3. Converter arrangement on the surveyed pipeline.

For registration and information processing, coming from sensors of a thermal stream, the program, in a development environment of virtual LabVIEW devices is created.

Functionality of the virtual device is defined by his block diagram which is graphic realization of an algorithm.

The block diagram of the developed program is represented in figure 4.

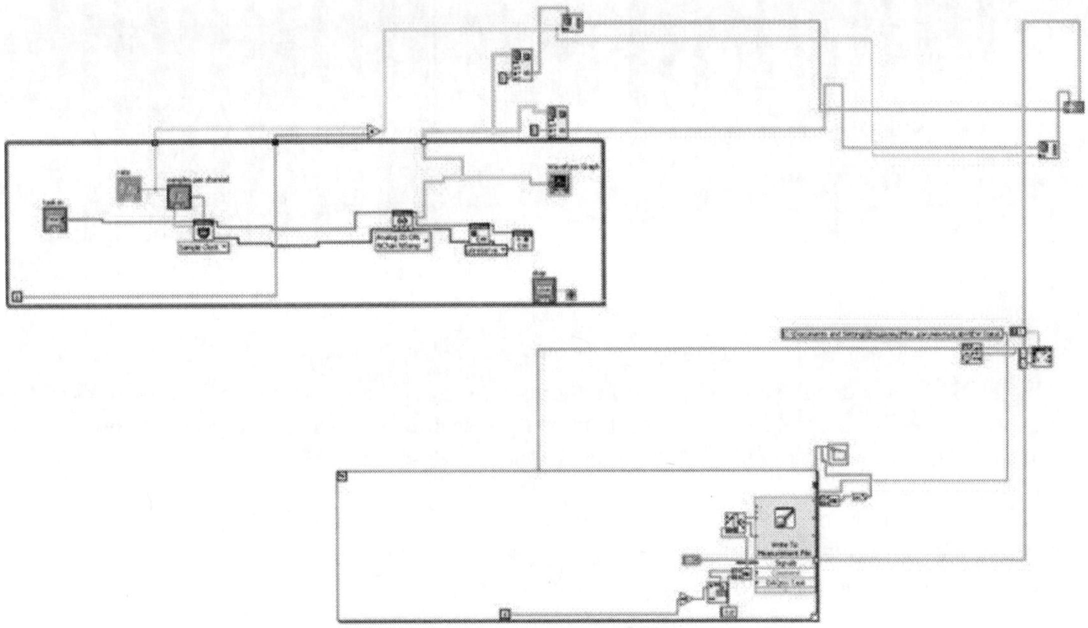

Figure 4. Block diagram of the developed program.

Before carrying out researches calibration of measuring system has been carried out. As the model device serially released measuring instrument of a thermal stream ITP-MG4.03/X(Y) "Stream" was used.

For assessment of thermal losses, during aging isolation, measurements with four models of isolation with different extent of destruction are executed:

- a sample No. 1 - not damaged isolation (100%);
- a sample No. 2 - 90% of an initial state;
- a sample No. 3 - 50% of an initial state;
- a sample No. 4 - 30% of an initial state.

Conclusion

By the technique given to Construction Norms and Regulations 41-03-2003 "Thermal isolation of the equipment and pipelines" (updated version of 61.13330.2012) calculation of specific thermal losses of steam lines of elevated laying d-76 of mm and d-159 of mm by which steam with a temperature of 194 °C of initial thermal isolation and with destruction was transported has been made. Results of calculation are presented on the chart (the figure 5).

Difference in thermal losses, during accounting thermal destruction and without her account, significantly. In case of thermal destruction of thermal insulation of loss when transporting steam increase by 39%. This circumstance acquires special relevance in connection with increase in energy costs. From the figure 5 it is visible that operation of thermal network in the conditions differing from design leads to essential increase in thermal losses.

The received results testify to need of accounting of real service conditions at assessment of thermal energy losses.

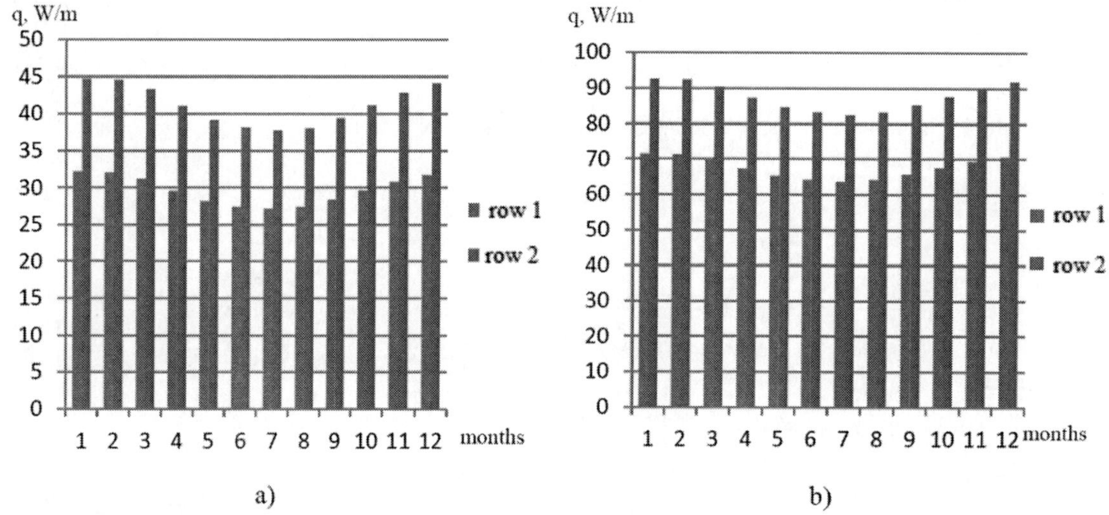

Figure 5. Results of calculation of specific thermal losses of a steam line a) d-76 mm, b) d-159 mm. A row 1 – specific thermal losses without temperature destruction of PPU-isolation. A row 2 – specific thermal losses at increase in coefficient of heat conductivity owing to temperature destruction for 20%

Timely definition of "problem" places by means of the developed system will allow to provide implementation of energy saving programs . Considerable economy of energy resources can be reached due to local repair and insulating works on concrete sites of pipelines with abnormal thermal losses.

References

[1] Smorodova O V 2016 Tool assessment of dynamics of mineral-cotton thermal isolation aging *Innovative science* **7-8** pp. 90-92

[2] Bazukova E R, Vankov Yu V 2015 Thermal losses at deterioration in properties of isolation in use *Engineering bulletin of Don* **3** ivdon/ru/ru/magazine/archive/n3y2015/3249.

[3] Dementiev A G 2000 Assessment of polyurethane foam service life for thermal insulation *Penopoliuretan* **2** pp. 38-41

[4] Gerashchenko O A 1971 *Basics of thermometry* (Kiev: Naukova thought) p. 191

[5] Vankov Yu V, Bazukova E R, Islamova S I, Busarov A V 2017 Estimation of heat transport losses during aging isolation *Innovative, information and communication technologies* **1** pp. 572-576

IOP Conf. Series: Earth and Environmental Science **288** (2019) 012128 doi:10.1088/1755-1315/288/1/012128

Mathematical simulation of the thermal processes in the heating networks insulation using thin-film covering

Zakirova I.A., Chichirova N.D.

Kazan state power engineering university, Kazan, Russian Federation

iazakirova@mail.ru, ndchichirova@mail.ru

Annotation. The necessity of application of computer technologies in the educational process with increasing requirements to the higher technical education quality is shown. The possibility of the program complex using of finite element modeling in the learning process, in the numerical study of the convection effect on heat exchange in the porous insulation of heat network pipelines using thin-film coatings, is considered.

Introduction

In the modern world, with growing demands on the quality of higher technical education, there is a rapid and widespread introduction of digital technologies into educational and research processes. The future engineer should be able to navigate in the modern software, to master the basic methods of working with programs for calculating various processes in their field in accordance with the training profile.

In the educational process, specialized mathematical complexes for scientific and engineering calculations, such as Mathcad, MatLab, Maple, MuPAD, GNU Octave, Scilab, Maxima, etc., can be used.

Along with theory teaching, calculation methods and techniques of heat transfer process experimental studies in the preparation of qualified engineering personnel in various specialties in the direction of heat power engineering, it is important to expand the possibility of heat exchange theory applying in engineering practice using the computer technology. At the same time, the student, the future engineer, must understand the physical features of the processes and be able to describe the phenomenon under investigation mathematically.

The introduction of computer technologies in all fields of science and technology led to the widespread use of numerical methods for solving problems in mathematical physics. One of the universal methods for solving boundary value problems is the finite element method (FEM). There is a large number of finite element modeling software packages, such as ANSYS, ELCUT, Femlab, FEMM, FlexPDE and others.

Description of finite element modeling program complex

From existing packages for solving differential equation systems, in the learning process, you can use the FlexPDE package [1]. Because, one of the advantages of this package is the availability of the student version, which is available for free distribution, when everyone can receive a key for its activation by e-mail, registering on the developer's site. This software product is designed to build scenario models, solving differential equations by the finite element method. The solution of the equations system is represented graphically. The system of differential equations can be stationary or non-stationary. In the framework of a single problem, stationary and nonstationary equations can be considered simultaneously. The equations can be nonlinear.

This program contains a complete set of functions necessary for solving the system of partial differential equations: - editing function for scenarios preparing; - generator of finite element grids; - function of finite elements selection for solution; - a graphical function to represent a graph of the results [2].

Advantages of this program is the ability to specify any number of geometric areas for a solution with various material properties, while it is fairly simple to use, and therefore it is of value for educational purposes.

In addition, it is designed to solve partial differential equations in an arbitrary form, which allows students to formulate a mathematical model independently.

The script for solving problems in the FlexPDE package is usually small in volume and easily compiled. It includes information about the desired functions, solved equations, the geometry of the domain boundary, the boundary conditions and the graphs of the required dependencies.

Application of the FlexPDE finite element simulation software package in the numerical study of thermal processes in the insulation of pipelines of heating networks using thin-film coatings

In work [3] results of experimental researches on application of a thin-film covering on a surface of a heat insulating design of heat network pipelines are presented. To theoretically the energy-saving effect from thin-film coating substantiate, a numerical study of thermal processes in thermal insulation was carried out. As an example of the FlexPDE package application, one can consider the results of solving the problem of complex heat transfer in the porous heat-insulating structure of the heat network pipeline taking into account natural convective processes (Figure 1a).

The convective phenomena in porous media study is of interest for solving both engineering and scientific problems. The most complete review of experimental and theoretical papers is presented in [4]. In [5], nonstationary regimes of natural convection in a closed porous cylindrical cavity were considered. Also, in a number of works, convective currents in annular layers of a saturated porous medium with an internal or external heat source under conditions of a closed shell have been studied [6, 7]. We note the paper [8], in which the effect of natural convection on heat exchange in porous annular layers under the condition of a closed outer shell was numerically investigated.

The heat exchange process in the pipeline insulation of heating networks has a stationary steady state, in view of the long-term regime throughout the entire heating season. The transition regime is insignificant during the periods of the heat supply system start-up and shutdown.

With a quasi-steady-state thermal state, the amount of heat flowing from the heat carrier to the concentric cylindrical heat insulation surface passing through any point of the field is equal to the amount of heat leaving this concentric surface to the external medium.

The structure of the insulated pipeline of the heat network consists of several layers in the thickness of the concentric arrangement (Figure 1a), in this case consisting of mineral wool (3) as an insulating layer and a covering layer consisting of glass-reinforced plastic (GRP) with a thin-film coating applied on its surface (4). Hot water (1) is used as the coolant. The temperature of the coolant throughout the entire model area under consideration remains unchanged. The layer adjacent to the heat carrier is the steel pipe wall (2).

The simulation area is a porous medium between two horizontal coaxial cylinders and consists of the following zones (Figure 1b):

I - is the basic insulating layer having a porous character and consisting of mineral wool, where R_0 is the inner radius of the main insulation layer corresponding to the outer diameter of the steel pipe outer wall, R_1 is the main insulation layer outer radius, with permeability κ_1, thermal conductivity λ_1 and layer thickness $\Delta R_1 = R_1 - R_0$;

II - a covering layer consisting of GRP and a thin-film rubber coating applied on its surface, where R_2 is the outer radius of the thermal insulation structure, with permeability K_2, thermal conductivity λ_2 and layer thickness $\Delta R_2 = R_2 - R_1$;

III - the air environment, where T_3 is the average ambient temperature, α_3 is the heat transfer coefficient between the surface of the thin-film coating and the surrounding air.

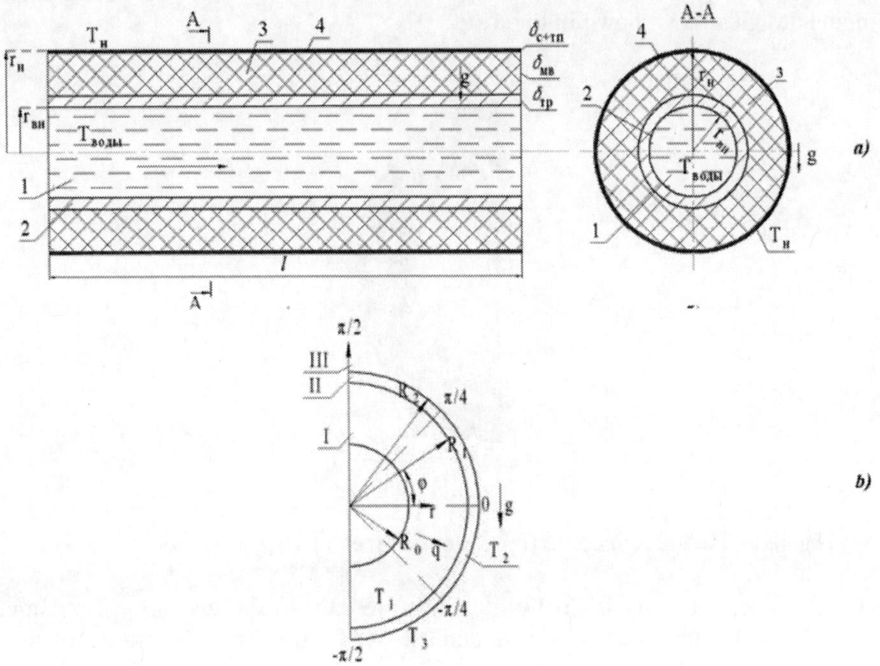

Figure 1. Pipeline model of the: a) physical model of the pipeline; b) a two-dimensional model of a pipeline thermal insulation design

To calculate the flow field and heat transfer under steady-state conditions, the convection equation was used in the Boussinesq approximation, where the surface force was replaced by an equivalent volume resistance force in accordance with Darcy's law [9].

$$\mu\bar{v}/k = -\nabla p + \rho g \beta \Delta T, \qquad \mathrm{div}\, \mathrm{v} = 0, \qquad \sigma(\bar{v}\nabla)T = \lambda^{*}\nabla^{2}T$$

The heat exchange in the pipeline insulation proceeds identically along the length in the radial direction and the main change in the heat flux occurs in the plane of the cross section, so the heat exchange simulation can be carried out in two-dimensional coordinates.

Because of the symmetry with respect to the vertical axis passing through the center of the annular layer [9], which represents the heat-insulating construction of the pipeline (Fig. 1b), the mathematical model can be applied for half the annular region.

There is (Figure 2) the script element for solving this problem in the FlexPDE package describing the equations of the considered mathematical model:

```
EQUATIONS     {PDE's, one for each variable}
p:  DX(vx)+DY(vy)=0 !p=pinit
u: div(l1*grad(u))=roCp*(vx*DX(u)+vy*DY(u)) {one possibility } !u=uinit
phi: phi=(arctan((y-dR)/(x+0.00001))+Pi/2)!*180/pi
then
 psi:  div(grad(psi)) + w = 0! solve streamline equation separately from velocities
```

Figure 2. The FlexPDE program script element (p is the pressure, u is the temperature

IOP Conf. Series: Earth and Environmental Science **288** (2019) 012128 doi:10.1088/1755-1315/288/1/012128

Finite element analysis begins with the discretization of the problem under investigation and divides it into mesh cells, called finite elements, interconnected at the node points, and a finite-element grid of the model design area is shown in Figure 3.

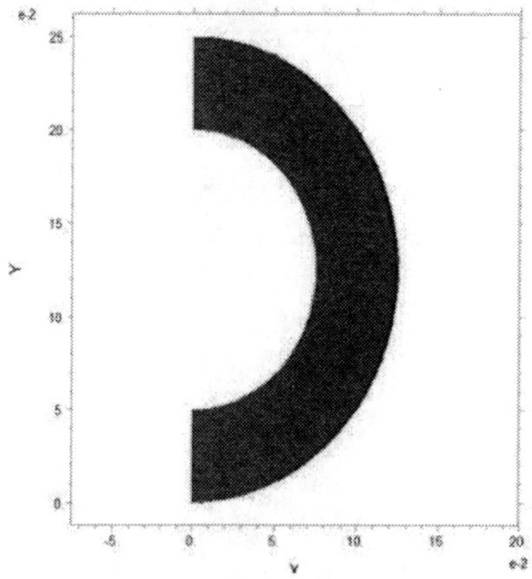

Figure 3. Two-dimensional finite element mesh of the model design area

The results of numerical analysis in the FlexPDE program in the insulation layer under natural convection conditions, for the pipeline with a diameter of 57 mm and a thickness of the insulating layer of 60 mm are shown in Figure 4.

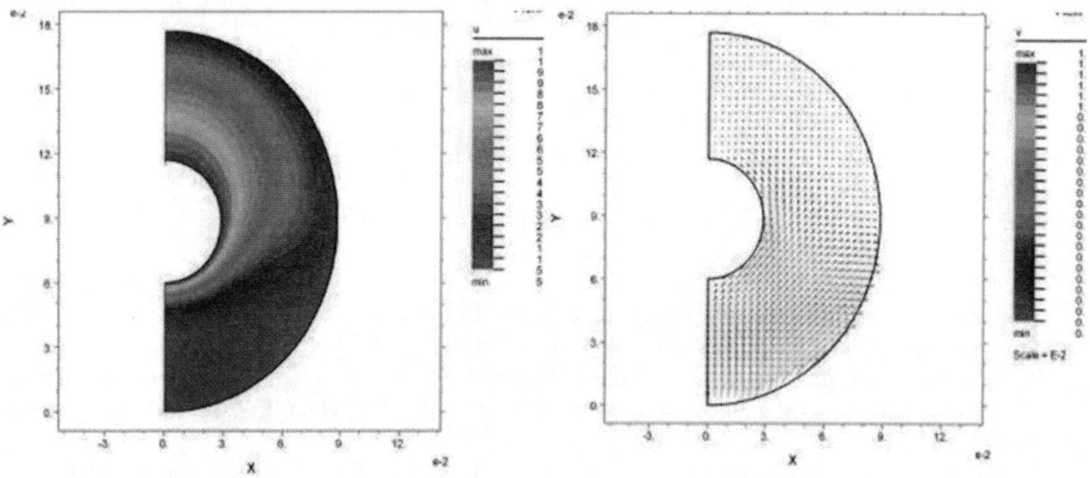

a) *b)*

Figure 4. Results of numerical analysis a) distribution of temperature fields, ° C; b) velocity distribution, m/s.

Conclusion

It should be noted that developed by the teachers mathematical model for the thermal processes calculation in the thickness of the heat insulation structure of the heating network pipelines for the FlexPDE package, which simulates physical processes, can be used for teaching students and graduate students. Thus, they will be able to compare the experimental data with the calculated ones. In addition, the package will be very useful for students and graduate students in the performance of research and development.

Reference

[1] www.pdesolutions.com (date of the application 27.04.2018)

[2] Dvoretsky S.I., Ermakov A.A., Ivanov O.O., Akulinin E.I. Computer modeling of processes and devices for food, bio- and chemical technology in the FlexPDE environment: Proc. allowance / Tambov: Tamb. state. tech.un-ty, 2006. 72 pp.

[3] Nield D.A. Convection in porous media. 3rd ed. / D.A. Nield, A. Bejan. N.Y./ Springer-Verlag, 2006.

[4] Zakirova I.A., Chichirova N.D. Experimental determination of effectiveness of thermal insulation of heating networks involving application of thin-film coatings. – Safety and Reliability of Power Industry. 2017. vol. 10, no. 2, pp. 148-154.

[5] Trifonova T.A., Sheremet M.A. Numerical studies of nonstationary regimes of conjugate natural convection in a porous cylindrical region (Darcy-Boussinesq model). - Computer studies and modeling, 2013. Vol. 5, No. 2, p. 179-191.

[6] Trofimova A.V., Tsibulin V.G. Convective motions in the porous annular sector. Applied Mechanics and Technical Physics. 2011. T. 52, No. 3, p.116-125.

[7] Bessonov O.A., Brailovskaya V.A. Spatial model of thermal convection in the gap between horizontal coaxial cylinders with anisotropic porous filling // Izv. RAS. Mechanics of fluid and gas. 2001. № 1. p. 145-155.

[8] Brailovskaya V.A., Petrazhitsky G.B. Thermal and laminar convection of liquid in the annular region for a given heat flux. – Applied Mechanics and Technical Physics, 1977, No. 3, p. 109-114.

[9] Brailovskaya V.A., Petrazhitsky G.B., Polezhaev V.I. Natural convection and heat transfer in porous interlayers between horizontal coaxial cylinders. - Applied Mechanics and Technical Physics, 1978, No. 6, p. 90-96.

IOP Conf. Series: Earth and Environmental Science **288** (2019) 012129 doi:10.1088/1755-1315/288/1/012129

Development of quality monitoring devices for industrial water in heat supply systems

R.S. Zaripova[1], E.A. Saltanaeva[1], N.G. Bikeeva[1], E.V. Priimak[2]

[1] Kazan State Power Engineering University,

[2] Kazan National Research Technological University

Russia, Kazan, Fuchik str., 72-97 zarrimma@mail.ru

Annotation. The article is devoted to the development of basic principles and the creation of prototypes of devices for monitoring the rigidity of industrial water in the heat supply systems of the power system. The proposed approaches are based on the electrochemical principle, according to which the industrial water is an electrolyte with a certain chemical composition and electrical conductivity. Based on this principle, new methods and measuring devices are being developed to determine the rigidity of industrial water in heat supply systems by continuously measuring the concentration of alkali and alkaline earth ions.

Relevance

At present, the development of effective methods and tools for the analysis of elements is one of the most important tasks the solution of which the development of many branches of the national economy depends on. The urgency of work in this direction is determined by fast-growing need of sophisticated science and technologies in clean and ultrapure materials, as well as monitoring of the environment and the World Ocean.

Among a wide variety of analytical methods, more and more attention is paid in our country and abroad to one of the most promising areas of electroanalysis - the method using membrane sensors based on ion-selective electrodes. Membrane sensors are devices that allow you to quickly and accurately determine the chemical composition of the environment in which the sensor is placed. It can be placed directly into the process solution, where it will acquire a potential, depending on the composition of the solution. Membrane technology is used in areas where traditional methods are inapplicable or ineffective. Convenience and simplicity of work with ion-selective electrodes promote their wide distribution and application in the most various areas.

Fields of application of membrane sensors are medicine, biology, soil science, oceanology, analysis of environmental pollution. They are used in the control of petroleum products, large chemical aggregates, nuclear reactors, technological solutions in many industries. At thermal power plants, the use of membrane sensors allows to automate the processes of water chemistry control, as well as automatically control the ionic composition of the industrial water at different stages of its processing. Such wide application of sensors is due to their ability to determine the concentration of the relevant components without disturbing the integrity of the object. The more selective the electrode is for a given ion, the wider its application in different media [1].

Advantages of membrane sensors are relative simplicity, compactness of hardware design, cheapness, speed, high sensitivity, selectivity, ecological purity, continuity of the measurement process and the possibility of its automation. But with frequent and long-term use of an ion-selective membrane in the sensor (concentration measurement takes a long time), there is a rapid "poisoning" and clogging of the membrane with ions of other metals. As a consequence, the threshold of its sensitivity to the determined ions decreases, and its shelf life decreases. Elimination of these shortcomings by reducing the time of the measuring process and, consequently, increasing the accuracy of the measurement will make it possible to obtain more reliable data on the composition of the solution. In connection with this, the development of a fast-acting method for nondestructive testing of parameters of the natural environment, substances, materials and products using membrane sensors is an urgent task.

It is known that the rigidity of industrial water along with the level of free oxygen is one of the two most important factors that affect the efficiency of the heat transfer process of a water coolant and the service life of equipment in heat supply structures of various types (thermal power plants, thermal power stations, local systems). An increase in the level of stiffness associated with an increase in the concentration of Ca^{2+} and Mg^{2+} ions in the aqueous medium will promote the formation of a scale in the pipeline system of a closed heat exchange cycle and, thereby, reduce the efficiency of the heat transfer process. This can lead to significant energy losses in water heating systems as well as in case of violation of the high-speed regime of the water coolant. To avoid emerging problems, it is necessary to develop effective methods and devices for the continuous monitoring of the water-chemical balance of the aquatic environment. In this context, the proposed project is relevant and aimed at addressing specific energy conservation objectives.

To determine the concentration of ions of alkaline and alkaline earth metals, at present, the use of membrane sensors based on ion-selective electrodes is promising. They have a high sensitivity and an acceptable measurement speed. In connection with this, the development of such sensors is an urgent task.

Methodology of research

The membrane sensor is a galvanic half-cell consisting of an ion-selective membrane, an internal contacting solution and an internal reference electrode (Figure 1). For practical convenience, they are all located in the same housing. Another half-cell is formed by an external reference electrode immersed in a standard electrolyte. The membrane is an elastic transparent film with a thickness of 0.1-1 mm made of a plasticized polymer (matrix) with an ionophore embedded in it. Ionophore is a structure that passes ions from the solution into an internal solution. Crown ethers, calixarenes, and podands are used as ionophores. The polymer matrix is made of polyvinyl chloride and plasticizers [2].

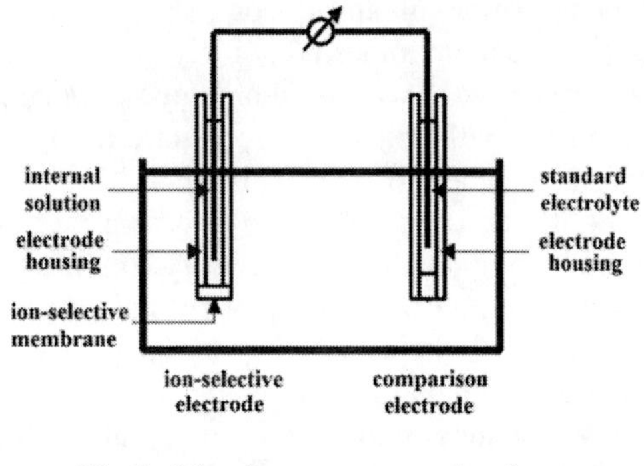

Fig.1. Membrane sensor circuit

The use of such sensors is based on the measurement of membrane potentials. The membrane sensor is an ion-selective electrode consisting of an internal semi-reference element immersed in an internal solution, a membrane and a reference electrode consisting of a comparison semi-element

IOP Conf. Series: Earth and Environmental Science **288** (2019) 012129 doi:10.1088/1755-1315/288/1/012129

immersed in a standard electrolyte. Both electrodes are placed in the test solution. The principle of the sensor is based on the fact that the membrane transmits ions of only one metal, preventing the penetration of other ions. Consequently, a potential difference appears at the boundaries of the membrane, which corresponds to the measured concentration [3].

The ion-selective membrane transmits ions of only one metal, preventing the penetration of ions of other metals. With the appropriate composition and structure of the membrane, its potential depends only on the concentration of this ion on both sides of the membrane. No other process that takes place in the membrane affects the membrane potential. Consequently, a potential difference appears at the boundaries of the membrane, which corresponds to the measured concentration. If the ion-selective membrane passes only the detectable ions, then at a constant temperature the electrode potential can be expressed by the Nernst equation.

The membrane is a film with a thickness of 0,3-0,5 mm made of a plasticized polymer (matrix) with an ionophore embedded in it. Crown ethers, calixarenes, and podands are used as ionophores. In the sensor being developed, the polymer membrane is made of polyvinyl chloride and plasticizers.

An equivalent scheme of the membrane sensor is proposed (Figure 2). It consists of a circuit R_iC and an ideal voltage source U_c. R_i is the output resistance of the sensor. Thus, the sensor is a capacitor that, when immersed in a solution, is charged through R_i to E, and when it is removed from the solution it is discharged.

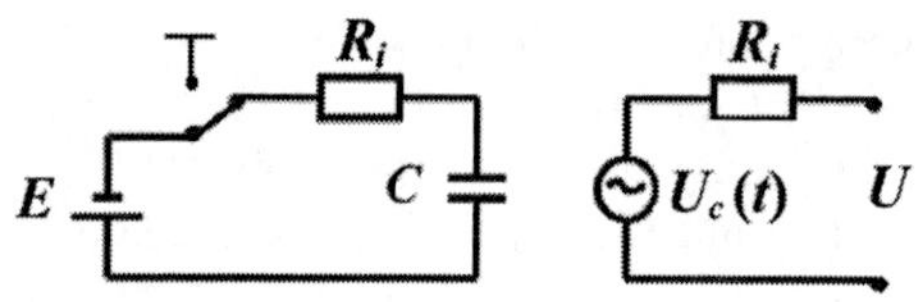

Fig.2. Equivalent Circuit of Membrane Sensor

Description

Observance of the water-chemical regime is an essential component of the normal operation of a thermal or nuclear power plant, as well as any local heat and water supply system. Under conditions of elevated temperature and pressure, strict limitations on the ionic composition of the water coolant must be carried out to reduce the formation of deposits and its corrosive properties in order to extend the life of the equipment used. This requires careful water treatment and, consequently, continuous monitoring of the chemical composition of the aqueous solution entering the cycles of power plants or water supply and heating systems.

To determine the ionic composition of an aqueous solution, membrane sensors based on ion-selective electrodes with replaceable membranes are widely used, which make it possible to determine the concentration of ions of various types with a high degree of accuracy. For example, the limit for measuring impurity concentrations with ion-selective electrodes for industrial water is $10^{-6} \div 10^{-8}$ mol/l, while for feed water of steam generators, about 10^{-5} mol/l.

It is known that the membrane sensor is a first-order measuring system with a transient operating mode. Therefore, a standard measuring method in laboratory conditions involves the selection of water samples from the pipeline with a certain periodicity, the possible preparation of the sample and the measurement itself after the sensor has reached a stationary operating mode in a minute time interval.

This method is well tested and will be very effective when the ionic composition of the water coolant flowing in the pipeline deviates markedly from the allowable on spatial scales, which are determined by the speed of the water flow and the total time including the period of the measuring process and the duration of the individual measurement. If the scale of deviations is much less than the specified value, there is always a danger that one can not "see" a noticeable violation of the water-chemical regime that will take place in the interval between adjacent measurements. To avoid such a situation, it is necessary to make the measuring process continuous. To do this, the membrane sensor should be placed in a bypass line with possible preparation of the water flow (filtration of solid particles, establishment of the required pH-factor and ionic strength of the solution, masking of interfering ions) and then transmit the digitized signal for remote processing in the appropriate information measuring environment.

With this approach, ion concentration monitoring of the water flow will be conducted at time scales determined by the signal-to-noise ratio of the analog signal of the membrane sensor and the sampling rate of the analog-to-digital converter, for example, a microcontroller or National Instruments interface board. If the change in the ionic composition of the aquatic environment occurs at times significantly exceeding the time constant of the sensor, which is a few seconds, then it manages to enter the stationary mode of operation. Otherwise, it is necessary to take into account the peculiarities of the transient mode of operation of the membrane sensor during the measuring process.

In determining the variable concentration of ions in the liquid that is washing the sensor, the lag effects associated with the time over which the concentration of ions of a given type on the surface of the ion-selective membrane will correlate with the value in the flow will influence the measurement process. This time depends on the thickness of the fixed diffusion layer near the surface of the membrane sensor (Nernst layer), which is determined by the flow regime of the water flow in the flow system. Thus, with continuous electrochemical analysis of the water flow, a number of features of the measurement process that require preliminary consideration arise in order to determine the optimal measurement regimes and design an appropriate information and measurement system.

The advantage of the developed information-measuring system based on the membrane sensor is a wide range of detectable concentrations. It has been experimentally established that the system can be used to determine the concentrations of alkali and alkaline earth ions in the range from 1 to 10^{-6} mol/l.

Results

Continuous monitoring of the hardness of industrial water based on the measurement of the concentration of ions Ca^{2+}, Mg^{2+} and other metals is carried out using an information-measuring system based on a membrane sensor. To determine the ionic composition of an aqueous solution, membrane sensors based on ion-selective electrodes with replaceable membranes are widely used, which make it possible to determine the concentration of ions of various types with a high degree of accuracy.

With the help of the proposed high-speed method for determining ion concentration in an aqueous medium on the basis of ion-selective electrodes, it is possible to continuously monitor its chemical composition. The accuracy of continuous measurements is comparable to the accuracy of traditional discrete measurements, which are carried out in the stationary mode of operation of the membrane sensor. This will make reliable information about the ionic composition of water reliable and will allow monitoring its correct changes in real time. Note that the corresponding information-measuring system is sufficiently simple in its design and does not require significant monetary costs in the implementation.

The innovative component of the project is determined by the desire to create inexpensive and competitive devices for the continuous monitoring of the water and chemical balance of the coolant in heat supply systems, so that the costs of creating, operating and maintaining these monitoring devices are significantly less than the economic losses caused by energy overexposure.

References

[1] Zaripova R S and Belavin V A 2006 *The number of higher educational institutions. Problems of energetics* vol 3-4 (Kazan: Kazan State Power Engineering University) pp 93-98

[2] Belyaeva L R, Zaripova R S, Petrushenko Y Y and Popov E A 2011 *The number of higher educational institutions. Problems of energetics* vol 1-2 (Kazan: Kazan State Power Engineering University) pp 119-126

[3] Plotnikova L V, Chilikova I I, Sitnikov S Y and Efremov G I 2016 Systematic approach to the assessment of energy complex efficiency for thermal energy production with heat power saving transformer turning on *International Journal of Pharmacy and Technology* **8**(4) pp 26727-37

IOP Conf. Series: Earth and Environmental Science **288** (2019) 012130 doi:10.1088/1755-1315/288/1/012130

Analysis of thermodynamic efficiency of the fuel preparation systems with an intermediate hopper at thermal power plants

J N Zatsarinnaya[1], A G Logacheva[2], A A Solovyova[3]

[1,2,3] Federal State Budgetary Educational Institution of Higher Education "Kazan State Power Engineering University", Kazan, Krasnoselskaya Street, 51

[1]waysubbota@gmail.com, [2]logacheva.alla@yandex.ru, [3]solovyova.alena@mail.ru

Annotation. The thermal and thermodynamic calculation of closed and open loop individual systems of coal preparation at TPP with different types of mills reveals the possibilities of increasing the efficiency of the technological scheme by utilizing the heat removed from the system and thus increasing the thermodynamic efficiency.

Coal as a fuel and its processing at thermal power plants

The main advantages of using solid fuel as one of the main energy carriers are its wide distribution, convenience and simplicity of storage, and low cost. The availability of coal reserves in the world are enough to last at average 230 years, in Russia - more than 500 years [1].

However, solid fuels require complex and expensive preparation for combustion. Despite a large number of scientific developments aimed at increasing the efficiency of particular elements of fuel preparing systems, the evaluation of their performance is mainly carried out experimentally or as a result of industrial tests [2].

Types of coal preparation systems

The systems of fuel (coal) dust preparation are divided into two types: central and individual [3].

The systems of fuel dust preparation are divided into two types: central and individual.

Central systems of dust preparation are intended for preparing and supplying fuel for several boilers of the power plant. The central system of dust preparation can operate independently.

Individual systems of dust preparation are designed for preparation and supply of dust for one boiler. At power plants in Russia individual systems with an intermediate dust hopper prevail. Such systems are equipped with ball mills or hammer mills, which is explained by the peculiarity of the coals burned in Russia (high content of ash and moisture). In these systems, the presence of the hopper contributes to a more stable operation of the boiler irrespective of the dust preparation system, since dust is supplied from the intermediate hopper.

In other countries roller mills have been widely used [4].

Individual coal preparation system

The individual coal preparation system at TPP consists of several main stages (the scheme is shown in Fig. 1) [5].

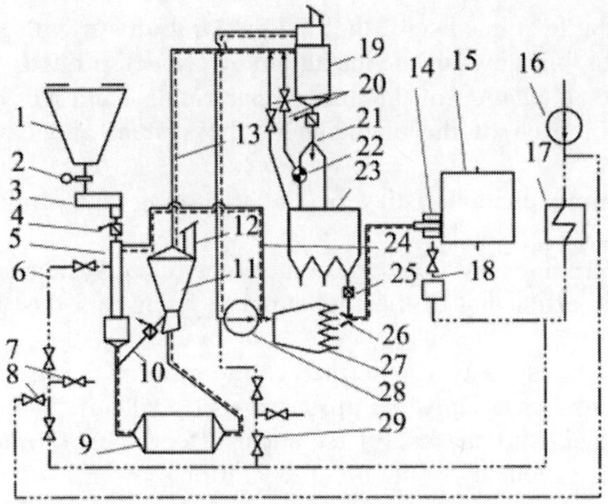

Fig. 1 – Individual fuel preparation system at thermal power plant

The Fig.1 shows: 1 - boiler hopper, 2 - a shut-off gate, 3 - coal feeder, 4 - flasher valve, 5 - pipe-dryer, 6, 7, 8 - hot air duct, air additive and cold air ducts, 9 - mill, 10 - dust return tide, 11 - separator, 12 - explosion valve, 13 - dust line, 14 - burner, 15 - boiler, 16 – primary air fan, 17 - air preheater, 18 - air heater, 19 - cyclone, 20 – drain pipes , 21 - swing gate, 22 - reversible screw, 23 - pulverized coal bin, 24 - recirculation line, 25 - dust feeder, 26 - dust mixer, 27 - primary air distributor, 28 - mill fan, 29- hot air duct.

Coal enters the territory of the power plant usually in railway cars, after unloading of which with the help of car dumpers, it can be supplied either directly to the fuel supply line or conveyed to the warehouse. In winter pre-heating of the fuel in the defrosting construction (overall housing) is provided. The grinding of fuel to the size of pieces of about 25 mm each is carried out in crushers of coarse or fine crushing.

Prepared this way, coal enters the pulverizing system of the boiler room, which is a drying and grinding unit consisting of a mill, separator, cyclone and mill fan. Hot air, flue gases or a mixture of both may be used as a drying agent. Dust preparation systems may have closed loop or open loop modification, i.e. the spent drying agent can be fed into the furnace of the boiler or, after cleaning, discharged into the atmosphere. Coal dust from the intermediate hopper is transported by means of the feeder into the mixer, where hot air blows through the hot blast fan. The mixture of hot air and coal dust then is supplied to the furnace of the boiler. The hot air from the hot blast fan is the primary air necessary for burning of the fuel.

Analysis of fuel preparation systems efficiency

In conveyed research individual dust preparation systems with an intermediate dust hopper equipped with ball mill, roller mill and hammer mill with drying of fuel by air and combustion products were considered.

Calculations were carried out to identify the possibility of improving the presented technological scheme in order to minimize energy costs for the process of solid fuel preparation for combustion. Such approach to the evaluation of system efficiency makes it possible to compare different types of systems [6].

Approach to system efficiency analysis

The analysis of the operation of the solid fuel preparation system was carried out with determination of thermodynamic parameters of the system; in particular the exergy of the flows supplied to the system and taken from it was evaluated. The evaluation of the thermodynamic efficiency of the system was carried out in several stages.

1) On the basis of the heat balance of the fuel preparation system, the thermal characteristics of the incoming and outgoing flows of the installation were determined.

2) The aerodynamic calculation of the dust preparation system was carried out, as a result of which the resistances of each of the elements of the system and of the whole system were determined.

3) The thermodynamic characteristics of the incoming and outgoing flows of the dust preparation system were determined.

4) The values of internal losses of exergy in the elements of systems were determined.

5) The total exergic efficiency of the fuel supply system and dust preparation system were determined.

Exergic efficiency of the system was defined as the ratio of the useful technical effect of the system to the spent. In this case, only the production of coal dust of the required humidity and temperature may be considered the useful technical effect. The formula below represents the calculation of the exergic efficiency of the fuel preparation system,

$$\eta_e = \frac{E_{пв}''}{E_{п.}' + E_{в.}' + L_1' + E_{х.в}' + \Sigma L_{тп}' + E_{са}' + L_2' + E_{г.в}' + L_3'},$$

$E_п'$ – exergy of the steam supplied to the air preheater; $E_в'$ – exergy of the air; L_1' – power of fans of the defrosting device; $E_{са}'$ – exergy of drying agent; $\Sigma L_{тп}'$ – power of fuel supply mechanisms; L_2' – power of the mill and mill fan; $E_{уп}'$ – exergy of coal dust at the outlet of the drying and grinding plant; $E_{qт}$ – exergy of the heat flow spent on fuel heating and evaporation of moisture; $E_{г.в}'$ – exergy of hot air supplied to the mixer; $E_{пв}''$ – exergy of the dust-air mixture at the outlet of the mixer; L_3' – power of the hot blast fan.

Exergic efficiency calculation results

The results of calculations of exergic efficiency calculations for fuel preparation system equipped with ball mill, roller mill and hammer mill with drying of fuel by air and combustion products are presented in Table 1.

Table 1. The results of exergic efficiency calculations for different types of fuel preparation system

Type of dust preparation system		Exergic efficiency of the system, %
Individual system of dust preparation, equipped with ball mill:	under depletion	19,9
- open loop modification with drying by air:	under pressure	19,5
- open loop modification with drying by combustion products:	under pressure	11,6
- closed loop modification with drying by air:	under pressure	31,4
closed loop modification with drying by combustion products:	under pressure	42,2
Individual system of dust preparation, equipped with roller mill:	under depletion	18,3
- open loop modification with drying by air:	under pressure	16,7
Individual system of dust preparation, equipped with hammer mill: - open loop modification with drying by air:	under pressure	22

Conclusion

The results of calculations show that the system for preparation of fuel for combustion is imperfect. It can be seen that systems operating under depletion have an exergy efficiency higher than systems operating under pressure. Fuel preparation systems equipped with hammer mill have the highest efficiency, which is explained by the small consumption of the drying agent and the small specific grinding capacity of the mills.

One of the ways to increase the efficiency of the system is the utilization of the heat of the flows outgoing from the system. A significant amount of heat is lost with the condensate and heated air drained from the defrosting device, with a drying agent spent in the dust preparation system, which could be used to preheat the fuel.

References

[1] Plakitkina L S 2013 *The state-of-the art and future trends of coal production in the period up to 2030 in the main basins and coalfields of the Far Eastern Federal District* (Moscow, Nauchno-proizvodstvennaya kompaniya «Gemos Limited»)

[2] Dincer I, Midilli A and Kucuk H 2014 *Progress in Exergy, Energy, and the Environment* (Springer International Publishing)

[3] Mingaleeva G R, Zatsarinnaya J N and Cyshevskij R V 2004 *5th International Symposium "Resource efficiency and energy saving" Proc.* p 55

[4] Taillon J and Blanchard R E 2015 *Energy* vol 88 pp 56-57

[5] Zatsarinnaya J N, Mingaleeva G R and Nazmeev J G 2005 *Sovremennye naukoemkie tekhnologii* vol 5 p 54

[6] Dubrovskiy V, Sedelnikov N, Zubova M and Dihnova A 2016 *EPJ Web of Conferences "Thermophysical Basis of Energy Technologies 2015"* p 1017

IOP Conf. Series: Earth and Environmental Science **288** (2019) 012131 doi:10.1088/1755-1315/288/1/012131

Heat supply system computer laboratory stand development

Zatsarinnaya, Yu. N.[1,2], Staroverova N. A.[1], Volkova M.M.[1], Galymullina S.I.[2], Khakimzyanov R. K.[1]

[1]Kazan national research technological University, Kazan

[2]Kazan state power engineeringUniversity, Kazan

waysubbota@gmail.com nata-staroverova@yandex.ru shadana@mail.ru

Abstract. The process of developing a virtual laboratory stand of automatic temperature control in the closed heat supply system is discussed in detail in the article.
The virtual laboratory stand of the temperature control system can be easily replicated. It provides a significant reduction of material and time costs for its creation and maintenance. This distinguishes it from traditional laboratory equipment

Hot water is necessary for hygienic and household needs. The water temperature varies from 50 to 70 degrees in this system. In a simple form, the local hot water-supply system is a water heating installation and a process piping that supplies heated water to the points of consumption [8].

Hot water systems are connected to the heating network through water-to-water heat exchangers. Several schemes are applied for the inclusion of heaters in two-pipe networks with simultaneous connection of heating and hot water supply systems: pre-wired, parallel, two-stage sequential, two-stage mixed, two-stage mixed with a flow limiter. Installation of storage tanks for load balancing hot is necessary in a number of cases. They are also used as a reserve in case of a break in the supply of coolant. Reserve tanks are installed in hotels with restaurants, baths, laundries, shower screens in production, etc. Therefore, the parallel circuit can be without a battery, with a lower storage tank and with a top storage tank.

The circulation pump installed on the supply line is present in the coolant supply circuit. A reserve pump of the same power is placed around the working pump in accordance with the rules.

This scheme uses the heat energy of the return line of the heating system, as well as the heat of the water flowing through the closed DHW circuit. This water is mixed with cold water from the water pipe. The mixture enters the second stage heat exchanger only after this. The PID controller regulates the outlet temperature due to the valve on the coolant supply line. The scheme allows you to significantly save, removing the excess load from the boilers and using the maximum heat available. It should be noted that filters are installed at the inlet to the heat exchanger. Reliable and durable operation of the unit depends on this. Sensors that are needed to monitor the flow, pressure and temperature in the system are also installed. The implementation of automatic alarm is due to them [1].

The modern education system contributes to the strengthening of student's theoretical knowledge and acquiring the necessary skills in the chosen direction through laboratory work. The issue of purchasing equipment for laboratory work is acute in many universities. We have developed a virtual laboratory temperature control stand in the heat supply system in the modeling environment

OmegaLand. Virtual stands provide an opportunity to model production processes and control the assimilation of knowledge. Spent on laboratory work time is reduced at the same time. This is due to the use of computer effects.

The task of creating a laboratory stand is very relevant for many universities. The task of creating a laboratory stand is very relevant for many universities. This makes it possible to equip all educational institutions, regardless of their geographical location and new training equipment. Also, it does not affect the economic effect of the purchase of equipment

There are various software tools that allow you to create virtual stands. The OmegaLand simulation system was used to create this virtual stand [2].

Each OmegaLand product contains several "modules" and VMspace. Modules provide an independent function. VMspace is the mechanism for systematically connecting these modules.

For example, the visual Modeler dynamic modeling (VM) package. It classifies functional modules as basic functional modules, optional functional modules, and user functional modules. Basic functional modules include the necessary fundamental modules to build almost any application system. Optional functional modules may be selected depending on the task. Both basic and optional function modules are products of Omega Simulation [3]. User-defined function modules are provided and added by the user. The term "application" is used to indicate the software in the format of performance In OmegaLand. It includes functions as part of a real application system, while the term "module" is used to include mechanisms that also generate each application [4]

In addition, the exact models created in the Visual Modeler dynamic modeling program can be implemented into the OmegaLand system as applications. The process plant model thus created can be executed on its own in the process model module (that is, in the Visual Modeler). This was done during the development of the stand.

The virtual laboratory stand of the closed system of heat supply consists of a plate heat exchanger and its piping system.

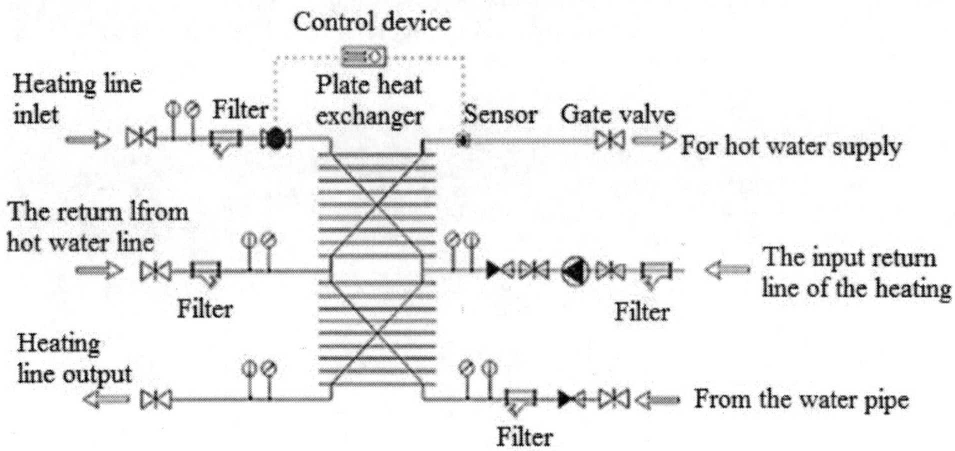

Fig. 1 Two-stage plate heat exchanger process diagram

The creation of the schema begins with adding components used in the technological process in the module Visual Modeler.

The ability to add elements to the diagram and enter initial parameters of the model is provided after adding components and creating the system.

Further, the addition and configuration of valves, sensors, heat exchangers, PID controller is carried out.

Creating and setting up a mnemonic that can be used in the simulator is made in the Graphic module. It is necessary to connect the valves with pipes, to supply sensors, pumps and PID controller in the appropriate places.

Discrete sensors and PID controller with valves should also be connected by a dotted line in accordance with the scheme of Visual Modeler [5]. The final view of the scheme, built in Graphic Builder, is shown in figure 2.

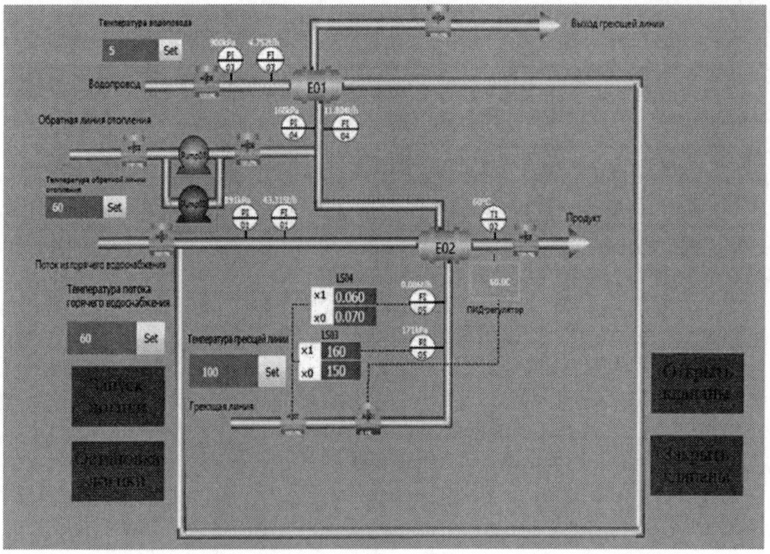

Fig 2. The final scheme of the process in Graphic Builder

Linking icons to the Visual Modeler is the next step in creating a mimic.

As a result, the virtual laboratory bench shown in Figure 3, based on a two-stage heat exchanger strapping scheme, was created on the basis of a virtual process model.

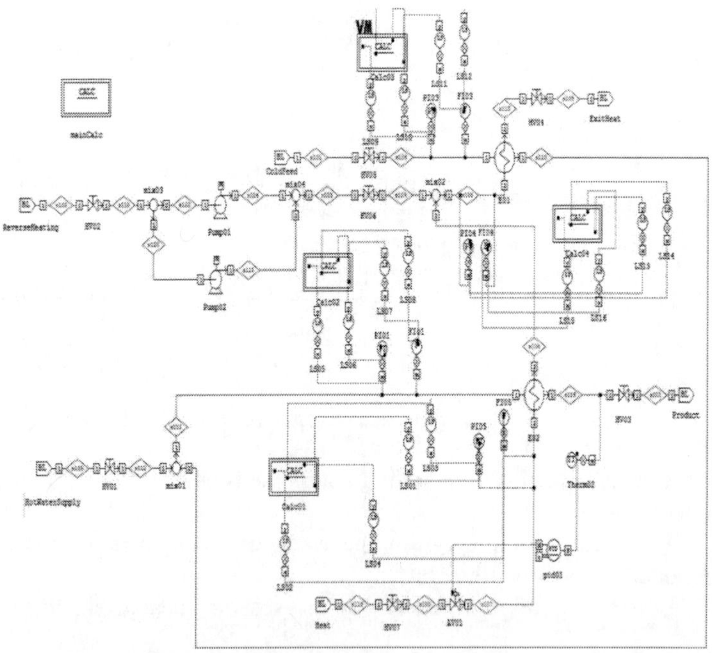

Fig. 3 - Virtual laboratory stand layout

The scheme consists of the heat exchangers E01 and E02, Pump01 pumps and spare pump Pump02, shut-off valves HV01, HV02, HV03, HV04, HV05, HV06, HV07, one control valve AV01, PID pid01, modules for the implementation of the logic of automatic protection and alarm Calc01, Calc02, Calc03, Calc04, mainCalc, flow sensors FI01, FI02, FI03, FI04, FI05, pressure gauges PI01, PI02, PI03, PI04, PI05, discrete sensor LS01, LS02, LS03, LS04, LS05, LS06, LS07, LS08, LS09, LS11, LS12, LS13, LS14, LS15, LS16.

Tap water comes from the ColdFeed inlet and is sent to the heat exchanger E01. There, it is initially heated by the heat of the mixture of the coolant that exits the heat exchanger E02 and the heat of the reverse heating flow that exits the ReverseHeating element. Further, water is mixed with the return flow of hot heating coming from HotWaterSupply. After that, water is sent for reheating to the heat exchanger E02 with the heat carrier coming out of the Heat element. After that, the heated water goes to the outlet, to the consumer. The PID regulates the flow of the heat transfer medium from the Heat element, changing the degree of openness of the AV01 valve, depending on the setpoint values. Discrete sensors receive a signal with sensors and costs.

Conclusions

The quality of education received by a student is determined not only by the level of theoretical training, but also by the ability to use the knowledge gained in practice.The student acquires practical skills through laboratory works and seminars, which are provided specifically in the curriculum. However, laboratory stands that involve physical modeling of the technological process, even in a simplified form, mean serious financial costs for their creation and maintenance in a working condition.In recent years, Information technology has been increasingly used, in particular in the organization of laboratory workshops. This is done to improve the level of practical training of scientific and technical specialists. The laboratory virtual stand presented in this article can easily be replicated, it provides a sign

References

[1] ZatsarinnayaYu.N., Rakhmatullin RR, Rizvanova GI Information transport bus of enterprises (ESB) in distributed energy companies / Yu. N. Zatsarinnaya, RR Rakhmatullin, GI Rizvanova // Bulletin of Kazan Technological University - 2013. - № 5 - P. 278-280.

[2] Development of a model of automatic pump capacity control based on modern automation equipment / Staroverova NA, ZatsarinnayaYu.N., Gerke A.R., Lira A.V. // Bulletin of Kazan Technological University. . - 2017. - № 15, стр.114-118

[3] Application of the possibilities of virtual laboratories in the educational process of a technical college // Sadanova BM, Oleinikova AV, Alberti IV, etc. // N. Young Scientist. - 2016. - №4. - P. 71- 74.

[4] Manual for working with the program OmegaLand / ITK - Tokio / Japan: 2006 - 130c.

Methodological questions of intellectual multisensor analyzers construction

Zhitkov A.N[1]., Fedyukhin A.V.[2], Filaretov G.F.[1], Chervova A.A.[1], O.V. Derevianko[3]

[1]National Research University «Moscow Power Engineering Institute», 111250 Russia, Moscow, Krasnokazarmennaya 14.

[2]V. A. Trapeznikov Institute of Control Sciences, Russian Academy of Sciences, Russia, Moscow, Profsoyuznaya, 65.

[3]Peter the Great St. Petersburg Polytechnic University, 194021, Russia, St.Petersburg, Polytechnicheskaya 29

gefefi@yandex.ru, fedyukhinav@yandex.ru

Abstract. In this paper we consider actual questions, connected with construction and usage of multisensor analyzers of "electronic nose" or "electronic tongue" types. Despite a large amount of papers concerning various aspects of this theme, there is a significant lack of methodological works among them. This work presents a discussion on general key problems of improvement of intellectual multisensor analyzers, namely the problems of optimization of sensor system, selection of algorithm means, evaluation of identification ability of the analyzer, typification of the utilized project design solutions at its construction. Typical examples of intellectual multisensor analyzers are presented, which are successfully used in various areas, including that for control and monitoring of water quality.

Introduction

Intellectual multisensor analyzers are referred to scientific devices of the new generation, which are based on combined usage both of achievements in microelectronics, chemical material science, sensors, and modern information technologies. The main purpose of this type of analyzers is express-analysis of content of complex air or water mediums, express classification (diagnostics) of the analyzed medium state or state of the object under control aiming at determination of its consumer properties [1–11]. Actually, these analyzers form a separate class of information-measuring systems. The main feature of these systems is usage of teaching procedures. This circumstance allows one to say that these systems might be considered as technical means, which simulate olfactive or taste systems of living organisms. That's why they are called "Artificial nose" in case of analysis of air (gas) medium or "Artificial (electronic) tongue" in case of liquid mediums [8].

Despite a large amount of papers concerning various aspects of this theme, there is a significant lack of methodological works among them, which consider general key problems of improvement of intellectual multisensor analyzers. We refer here to the problems of optimization of sensor system, selection of algorithm means, evaluation of identification ability of the analyzer, typification of the utilized project design solutions at its construction. This consideration will be performed using the results of long-term investigations of the National Research University "MPEI" in the area of methodological and program and algorithmic provision of multisensor systems and its' usage primarily for liquid medium analysis [12].

Synthesis of optimum sensor system

It should be noted immediately that consumer characteristics of quality identifying devices for various application areas are primarily governed by characteristics of the sensor system (sensor block). Generally, the sensor block contains a massive of sensors of rather high dimension *n* (typical *n* values are in the range from 10 to 30).

The process of sensor block creation, as a rule, is iteration and includes some stages:

a) Initial synthesis of sensor system. It is carried out on the basis of general considerations of the character of the analyzed medium, its possible chemical components, ions, etc. The obtained from the initial synthesis stage sensor system must be of substantial excess character, i.e. contain a large number of sensors, which sometimes exceed the run-time version.

b) Experimental investigation of the synthesized system using the typical samples of the production which will be further analyzed. The aim of the investigation is to determine operating capability of the system, its principal suitability for solving the considered applied problem. In the case of positive answer to this question, the obtained experimental material might be used at the further stages of sensor block development.

c) Determination of sensor informativeness and choice of its optimum (minimum) amount. It is valuable to carry out this stage using the specially designed technique, which provides two-stage procedure for selection of the most informative sensors [13]. Basing on the results of the first stage, the sensor ranking for its informativeness takes place. The low-sensitive and likely-responsive sensors are drop out. The first stage is devoted to informativeness check of the sensors selected at the first stage by direct usage of classification algorithms, which are laid down in the analyzer.

d) Tests of the optimized sensor set in conditions which are close to practical implementation. At positive result of the tests, one might say that the functional sensor block of the certain analyzer was created.

Algorithmic provision of the multisensor analyzers

It is known, that the key feature of the considered multisensor analyzers is that they operate using the principle of preliminary teaching on samples. The end result of the teaching experiment is the training sampling of the volume *N*. It contains data set came from *n* sensors of the sensor block in every experiment with indication of identity of the sample to a certain product (class). In an *n*-dimension factor space (sensor space) the elements of the training sampling are displayed as a certain cloud on *N* points.

A large number of various classification algorithms are known [14–17]. They are used for identification of separating borders between certain classes basing on the training sampling.

Nowadays in practice the most popular are the following algorithms:

a) The main components method (MCM) in two-dimensional case, when all experimental points are displayed on the plane. Usually elliptical areas are displayed simultaneously, which characterize the degree of its variation for each class (Fig.1). These ellipsis define some attraction areas of points, which are attributed to different classes. They denote, where with probability *p* may appear the next point, if it is attributed to this class (as a rule, p=0.9 or p=0.95 is chosen).

This method has a good visibility and interpretability of the obtained classification results.

b) Multi-layer perceptron with sigmoidal activation functions and the amount of hidden layers from 1 to 3 and output layer of m neurons according to the amount of identifying classes. The results are represented as column diagrams, which show the values of output network signals y_j, $j = 1, 2,\ldots, m$. At this the sample is referred to f class if the following condition is fulfilled: $0.75 \leq y_f \leq 1.0$, the others $y_j \leq 0.25$. If these conditions are not fulfilled, the classification is considered as unreliable.

IOP Conf. Series: Earth and Environmental Science **288** (2019) 012132 doi:10.1088/1755-1315/288/1/012132

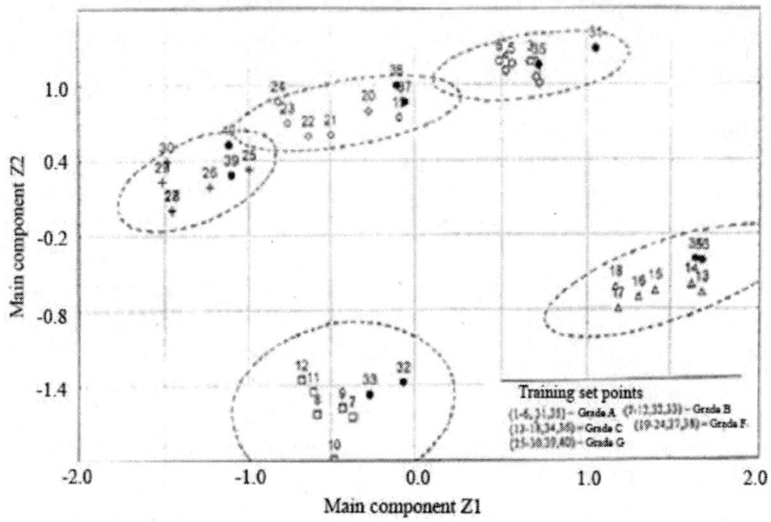

Fig.1. Classification using MCM

It should be noted that both these methods have an essential weakness: if at analyzer usage there is a possibility for appearance of new class samples, which are not included into the training sampling, then the whole recognition algorithm should be taught again. Hence the interest for ANN, based on ideas of adaptive resonance theory (ART), which are free of the above-said limitation.

Evaluation of the identification ability of the multisensor analyzers

This evaluation is carried out for the purpose of reliability assessment of the identification results, performed by the analyzer during solving a certain applied task. As an identification reliability factor one usually use probability P of right classification of samples of all classes. P value is set by the user, usually $P = 0.95$. Solution of this task might be traced to a standard problem of examination of statistical hypothesis about the value of probability P_0 of wrong classification at the given level of criterion $\alpha = 1 - P$. Without getting into details of elementary mathematical transformations, we will write the resulting formula for critical boundaries of amount of cases for wrong classification *N*fault to the total amount of experiments N, which were carried out during the tests:

$$Nfault / N = u_{1-\alpha} \sqrt{\frac{P_0(1-P_0)}{N}} + P_0,$$

where $u_{1-\alpha}$ is quantile of normal distribution. This formula is an approximate one, but gives rather precise results at N>20. If, for example, N=25, $P_0 = P = 0.95$, $u_{0.05} = 1.64$, when using the above formula we will obtain the critical boundary for the amount of faults at identification Nfault=3.07. Consequently, if during the tests we revealed more than 3 faults, we conclude that it has insufficient identification ability.

It should be noted that for some applied tasks it is important to check identification ability of the analyzer for revealing of new production types, which were not represented during its teaching.

Typification of the project design solutions at construction of multisensor analyzers

Fig.2 presents an example of general view of multisensor analyzer of "Electronic tongue" type, intended for identification of various bottled drinking water types and discovery of counterfeit products (the amount of brands in the initial variant is $m = 6$).

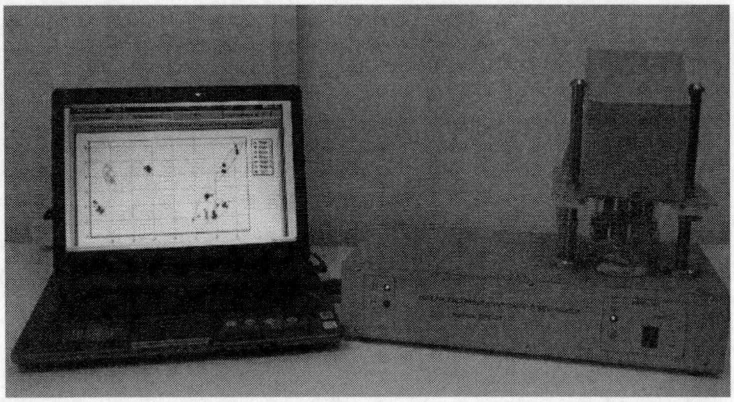

Fig.2 General view of multisensor analyzer of "Electronic tongue" type

The analyzer consists of a sensor block of 10 ion-selective electrodes, which were selected according to the method described above. Algorithmic provision includes statistical algorithm (MCM with further usage of linear discriminate analysis), neural network algorithm (four-layer perceptron with amount of neurons at the layers $n_1 = n$, $n_2 \approx n/2$, $n_3 \approx n_2 + m/2$, where n = 10) and self-teaching ART-algorithm. At combined usage of all three algorithms the decision on attribution to one class or another is made according to the principles of majority logic. The analyzer has successfully passed all necessary tests and further showed high efficiency at solving of tasks for identification of various water mediums.

Besides this, it might be useful for organizing of monitoring process for drinking water quality, which is obtained after filtration of water of poor quality. For this purpose we may recommend to use three samples as initial ones used at the teaching stage: low-quality water from its source; the same water after purification using a certain filter (at initial stage the filtering element is new); and a reference water sample (for example, distilled water). The latter one is used for creation of a reference set on the plane of the first major components for comparison with other samples. The further work of the analyzer in the operation mode allows one to estimate changes in water quality and discover how good is the filtering system operation, as well as how quickly do the filter characteristics change and when the filtering element should be changed of regenerated.. It is clearly seen from gradual shift of the representation points which correspond to the filtered water at the plane of two first major components from the initial state (new filtering element) sidewards the points corresponding to water from source, as it is shown in Fig.3.

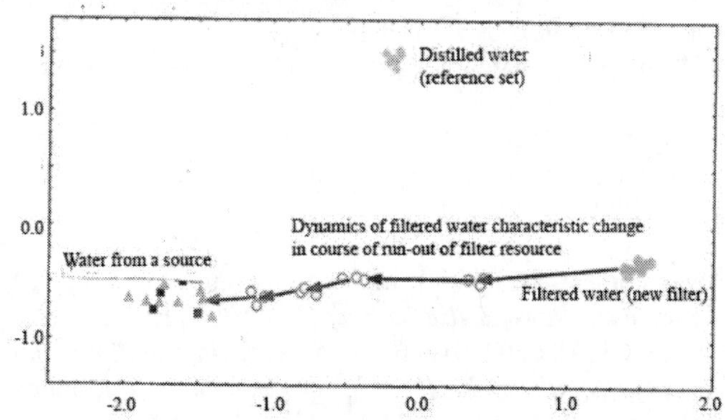

Fig. 3 Monitoring of water quality change using multisensor analyzer

The idea of typification was also used at construction of other multisensor analyzers. Among them is analyzer of "electronic nose" type for identification of perfume productions, multisensor analyzer of vodka authenticity (MAP-V), which received the Certificate of Russia Gosstandard for confirmation of measurement type with registration in State Register of Measuring Equipment.

Conclusions

For successful implementation of intellectual multisensor analyzers one should clearly understand its place and role among the other techniques and analytical equipment intended for solving of these tasks. They can't replace the organoleptic tests, chromatographic analysis, mass-spectrometry methods, etc. Its major advantage is relatively low cost of analysis, usage simplicity, objectivity, rapidness and absence of sample preparation. So up to now its main purpose is express-analysis, preliminary rejection of production according to its quality, including field tests. It is clear that in the course of gaining experience and increasing of its possibilities, the area of its application will gradually increase.

References

[1] Yaroshenko I, Kirsanov D, Kartsova L, Sidorova A, Borisova I and Legin A 2015 Determination of urine ionic composition with potentiometric multisensor system *Talanta* **131** 556–61
[2] Debus B, Kirsanov D, Yaroshenko I, Sidorova A, Piven A and Legin A 2015 Two low-cost digital camera-based platforms for quantitative creatinine analysis in urine *Anal. Chim. Acta* **895** 71–9
[3] Yaroshenko I, Kirsanov D, Kartsova L, Sidorova A, Sun Q, Wan H, He Y, Wang P and Legin A 2016 Exploring bitterness of traditional Chinese medicine samples by potentiometric electronic tongue and by capillary electrophoresis and liquid chromatography coupled to UV detection *Talanta* **152** 105–11
[4] Shkodyrev V P and Yagafarov K I 2018 The approach to emergency situation prediction in dynamical systems using neural networks vol 2018–Febru (Association for Computing Machinery) pp 27–32
[5] Averin I A, Igoshina S E, Karmanov A A, Pronin I A, Moshnikov V A and Terukov E I 2017 Simulation of the sensor response of vacuummeters with sensitive elements based on multicomponent oxide nanomaterials with the fractal structure *Tech. Phys.* **62** 799–806
[6] Baryshnikova M V, Filatov L A, Petrov A S and Alexandrov S E 2015 CVD Deposited Titania Thin Films for Gas Sensors with Improved Operating Characteristics *Chem. Vap. Depos.* **21** 327–33
[7] Mileshko L P, Kotov V N, Korolev A N, Cherepahin I I and Scherbinin I P 2009 *Multisensor systems: textbook* (Taganrog: Izd-vo TTI YUFU)
[8] Vlasov Y G, Legin A V and Rudnickaya A M 2008 Electronic tongue is a system of chemical sensors for water analysis *Ros. him. zh.* **LII** 101–12
[9] Filaretov G F and Zhitkov A N 1999 Application of artificial neural networks in sensor systems *Datchiki i Sist.* **5** 2–9
[10] Kashaev R S, Kozelkov O V. and Kubango B E 2017 On-line flow proton magnetic resonance analyzers for oil-well liquid control on GOST P 8.615-2005 GSI *Proc. High. Educ. institutions. ENERGY Sect. Probl.* **19** 137–51
[11] Kashaev R S, Kozelkov O V., Malev N A and Kubango B E 2017 Automatic control of rotary pump using measurements of oil-well liquid by on-line proton magnetic resonance analyzer *Proc. High. Educ. institutions. ENERGY Sect. Probl.* **19** 119–31
[12] Filaretov G F and Zhitkov A N 2001 Analyzer of "Artificial nose" type *Materialy XIII nauchno-tekhnicheskoj konferencii "Datchiki i preobrazovateli informacii sistem izmereniya, kontrolya i upravleniya" (Datchik – 2001)* (MGIEHM) pp 301–3
[13] Filaretov G F 2001 Method for synthesis of sensor block of analyzer of "Artificial nose" type *Materialy XIII nauchno-tekhnicheskoj konferencii "Datchiki i preobrazovateli informacii*

sistem izmereniya, kontrolya i upravleniya" (Datchik – 2001) (MGIEHM) pp 303–5

[14] Utkin L V, Chekh A I and Zhuk Y A 2016 Binary classification SVM-based algorithms with interval-valued training data using triangular and Epanechnikov kernels *Neural Networks* **80** 53–66

[15] Sonkin K, Stankevich L, Khomenko Y, Nagornova Z, Shemyakina N, Koval A and Perets D 2016 Neurological classifier committee based on artificial neural networks and support vector machine for single-trial EEG signal decoding **9719** 100–7

[16] Stankevich L and Sonkin K 2016 Human-robot interaction using brain-computer interface based on EEG signal decoding **9812 LNCS** 99–106

[17] Utkin L V and Zhuk Y A 2017 Interval SVM-Based Classification Algorithm Using the Uncertainty Trick *Int. J. Artif. Intell. Tools* **26**

IOP Conf. Series: Earth and Environmental Science **288** (2019) 012133 doi:10.1088/1755-1315/288/1/012133

Decrease of pressure losses in elbow fittings of ventilation systems of thermal power plant buildings

A M Ziganshin[1], E E Beljaeva[1], K I Logachev[2] and O A Averkova[2]

[1]Thermal power engineering, Gas supply and Ventilation Department, Kazan State University of Architecture and Engineering, Russia, 420043, Kazan, Russia

[2]Heat and Gas supply and Ventilation Department, Belgorod State Technological University named after V.G. Shukhov, 308012, Belgorod, Russia

amziganshin@kgasu.ru

Abstract. The paper presents the results of a numerical study of the flow in the elbows of the ventilation systems in buildings of thermal power plants. Elbow fittings, as well as other shaped elements of ventilation systems are the main causes of energy losses. A method is proposed for reducing such losses through the use of an insert profiling the sharp edge of the elbow along the outline of the vortex zone. For a wide range of practical dimensions, numerical models of "sharp" elbows at 90° have been studied, a good correspondence between pressure losses in them, numerical and known experimental data has been shown. The outlines of the vortex zones, as well as the lengths of the influence zones of the elbow on the flow parameters upstream and downstream, are determined. Next, numerical models of profiled elbows are constructed and studied; a significant reduction in pressure losses (up to 64%) and lengths of influence zones (up to 45%) as compared with unprofiled ones is shown. Dependence has been found to determine the local resistance coefficient of the profiled elbow, which can be used in the design of ventilation systems.

Introduction

The main energy losses of ventilation systems in buildings of thermal power plants are related with local resistance of various shaped parts of ductworks - elbows, tees, cross junctions and other duct fittings. The cause of the loss is the change in the flow direction in this disturbing elements (DE) and in the formation of vortex zones (VZ) due to separation of flow on the sharp edges.

Due to inertia, the flow begins to deform long before the shaped element and completes the reorganization far beyond it. Such places of transformation of the flow structure are called influence zones (IZ) upstream and downstream of disturbing element. Available data about the length of IZ are poor and often have an experimental character [1]. This lack of information makes difficult to calculate the local resistance coefficients and, accordingly, local pressure losses in the ventilation systems during its design. A little more is known about the outlines of vortex zones in the elbows [2, 3]. However, information on the three-dimensional outlines of the VZ for elbows of different depths for different flow regimes is lacking.

One of the most advanced methods for studying hydro- and aerodynamic phenomena today is a numerical simulation. At present, there are many works on the profiling of DE using the methods of computational fluid dynamics (CFD), for example, [4, 5].

In this paper, we propose a solution for reducing the local resistance coefficients (LRC) and pressure losses in the ventilation ductwork by placing a special profiling insert inside the "sharp" shaped element, which completely repeats the outline of the free streamline separating the main flow in the channel and the region of VZ [6]. The outlines of the VZ and the lengths of the influence zones are also given.

Methods

To implement the proposed solution at the initial stage, the task was set is to determining the LRC of "sharp" elbows to compare the obtained results with reliable data of other authors [1]. To determine the dependence of LRC of the DE on its geometric characteristics, a number of problems for flow in a three-dimensional channel with an 90° elbow in a wide range of dimensions have been solved (Figure 1). The width of the channel before the bend (b_0) (Check: under the text b0 is often called the height) took the values 0.1 m and 0.2 m, after (b_1) - 0.02; 0.06; 0.1; 0.2 and 0.3 m. The depth of the channel (a_0) was 0.025; 0.05; 0.1; 0.2 and 0.4 m and ∞ (two-dimensional task). The input boundary ***ABCD*** was modeled using the "velocity inlet" boundary condition (BC), which defines a uniform velocity profile v = 10 m/s; at the output boundary ***EFGH***, the gauge pressure was assumed to be zero, the remaining boundaries were set as a solid wall.

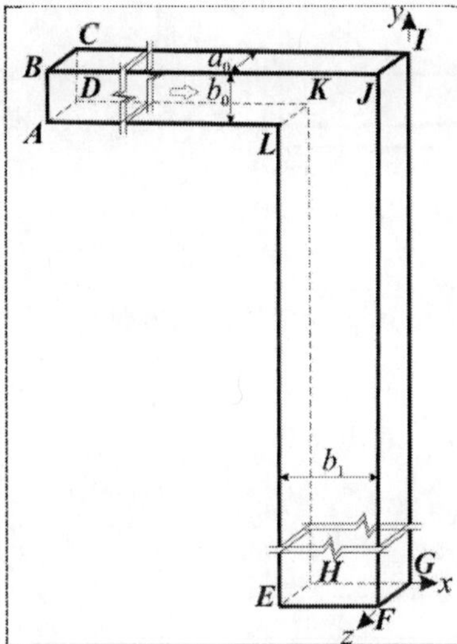

Figure 1. Geometry of the calculated area of unprofiled "sharp" elbow (for the case $b_1/b_0 = 2$ and $a_0/b_0 = 4$).

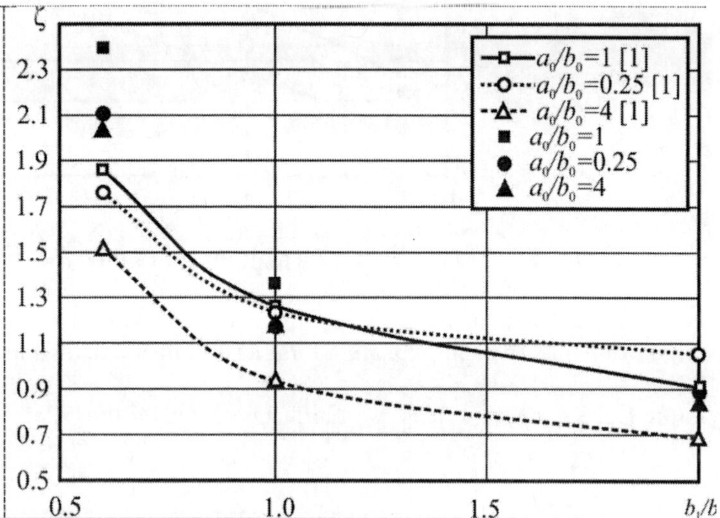

Figure 2. Dependence of the LRC of "sharp" elbow on the dimensions b_1/b_0 and a_0/b_0.

Figure 2 shows the dependence of the LRC of "sharp" elbow on the dimensions b_1/b_0 and a_0/b_0. The obtained results allow to make a conclusion about the decrease of elbows LRC with the increase of the channel area after the elbow, which agrees well with the data of [1] (the difference from them is within 1% -26% for different b_1/b_0). This is explained by the fact that in a such bend the flow velocity in the outlet section of the elbow reduces, and respectively reducing loss of pressure in the disturbing element.

Based on the results of the numerical calculation, the minimum value of the LRC is obtained for the maximum investigated depth $a_0/b_0 = 4$, and it decreases with increasing b_1/b_0. This also agrees with [1] and is explained by a decrease in pressure loss in the local resistance due to a decrease in the flow velocity in the outlet section of the elbow. In [1] are mentioned two factors that affect the LRC in such elements with increasing channel size after the elbow: increase of the vortex zone, which is accompanying with an increase in energy losses, and a simultaneous decrease in the velocity, which leads to a decrease of the LRC. It should be noted that for $a_0/b_0 = 1$, for larger values of b_1/b_0, the LRC is somewhat larger than for a smaller in depth elbow with $a_0/b_0 = 0.25$, although the vortex zone increases with decreasing the depth of channel (Figure 6).

Results and Discussion

Influence zones

In Figure 3, the icons show the lengths of the influence zones (IZ) before (l_b) and after (l_a) the elbow, depending on the size of the channel after bend and the depth of the channel.\

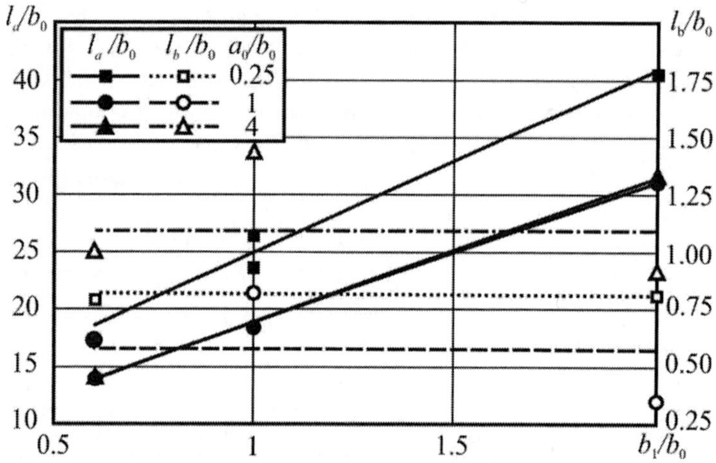

Figure 3. Dependence of the lengths of the IZ on b_1/b_0 and a_0/b_0 for the "sharp" elbow.

The lines shows the results of approximation - solid for l_a/b_0, dotted for l_b/b_0. Dependences of the lengths of the IZ and approximating formulas for their calculations are given in the table 1.

Table 1. Dependence of lengths of influence zones for the "sharp" elbow.

a_0/b_0	l_a/b_0	l_b/b_0
0.25	$12.596b_1/b_0+6.218$	0.583
1	$12.212b_1/b_0+6.513$	0.8
4	$15.962b_1/b_0+9.013$	1.1

It follows from Figure 3 that when the channel size after elbow (b_1) increases, the IZ after the elbow is lengthened due to the increase in the diffusive effect, which increases the flow separation and vortex formation [1]. In addition, it should be noted that the greater the depth of the channel a_0, the more length of IZ after elbow. Dependences of the IZ length before the elbow on the change of the channel size b_1/b_0, for the number of numerical experiments carried out, are not observed (Figure 3). However, it can be noted that as the channel depth a_0 increases, the IZ before the elbow becomes longer.

Vortex zones

To study the dependence of the vortex zone length on the flow regime, three tasks were solved in the self-similar modes with velocities: 10 m/s ($Re = 6.7\cdot10^4$), 80 m/s ($Re = 5.3\cdot10^5$) and 1000 m/s ($Re = 6.7\cdot10^6$) for the channel with $a_0/b_0 = 0.25$ and $b_1/b_0 = 1$. The outlines of the obtained vortex zones are shown in figure 4, where the two-dimensional case ($a_0/b_0 = \infty$) is shown with a solid line. It can be seen that with the Re increases, the size of VZ changes insignificantly - on the main line the outlines completely repeat each other; some difference is observed in the area where the VZ is touching the wall of the elbow.

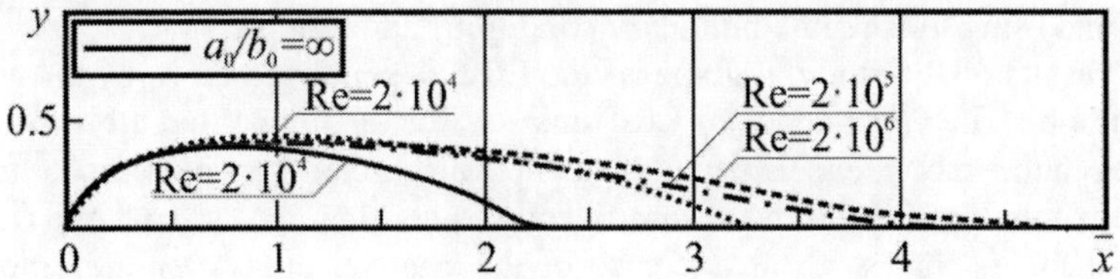

Figure 4. The lengths of the vortex zones for elbow with $a_0/b_0 = 0.25$, $b_0/b_1 = 1$ for Re from $2\cdot10^4$ to $2\cdot10^6$.

Figure 5 shows the dependence of the dimensionless length of the vortex zone x_0/b_0 on the channel depth a_0/b_0. In this figure the lengths of the vortex zone for a depth $a_0/b_0 = 0.25$ for a different Re number are shown as individual icons.

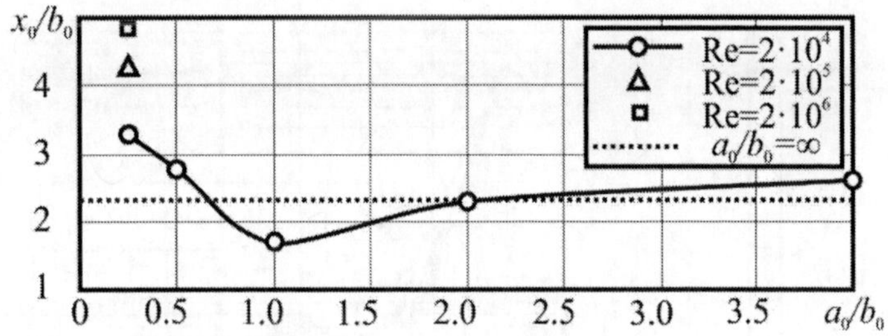

Figure 5. Dependence of the dimensionless length x_0/b_0 of the vortex zone on a_0/b_0.

Figures 4 and 5 shows that when the depth of the channel increases, the length of the VZ decreases. Perhaps this is due to the smaller turning radius of the flow: if the depth of the channel a_0 is several times greater than its width b_0, then the change in the flow direction in the elbow will be smoother.

Figure 6 shows the outline of the VZ of elbows with different depths, determined along the center of the channel. As preliminary study showed, significant differences in the shape of the three-dimensional vortex zone along the channel depth are observed only near the walls [5].

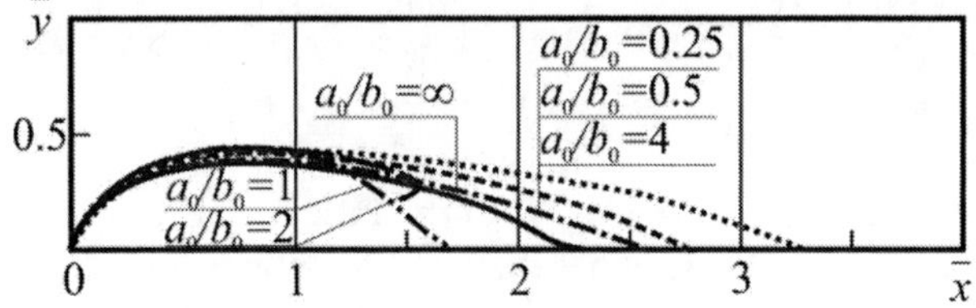

Figure 6. Outlines of vortex zones for elbows with depths $a_0/b_0 = 0.25 \div 4$.

It can be seen from figure 6 that for elbows with different sizes of a_0 practically along the entire length of the vortex zone its height and outlines agree well with the outline of the separation zone obtained for the two-dimensional model.

Modeling of efficient profiled elbow

The above-defined outlines of the vortex zones that forms due to separation of flow at the inner edge (taken from the two-dimensional model) were used to create computer models of the profiled efficient elbows (Figure 7). Although profiling of the external elbow corner also leads to a decrease in the LRC, but it is less significant [1, 5], and therefore was not considered here.

At the next stage solved a number of tasks on the flow of air in a channel with a profiled "sharp" elbow with an angle 90° (Figure 7) for the same range of channel sizes b_1/b_0 and a_0/b_0 as for the unprofiled ones, and using those same boundary conditions and models.

Based on the results of the study, a graph of the LRC dependence on b_1/b_0 for various depth a_0/b_0 was plotted (Figure 8). The dependence of LRC on b_1/b_0 for the unprofiled elbow of one depth - a_0/b_0 = 1 and according to the known experimental data [1] is also given for comparison here. It can be seen that the resistance of the profiled elbow is significantly lower than the "sharp" one (reduction to 64%). It can also be concluded that, with an increase of the channel area after the elbow, a decrease of elbows LRC is observed, which agrees well with the data of [1]. This is due to the fact that at the outlet section of the elbow, the velocity of air decreases, reducing the pressure loss of the disturbing element. As the depth of the channel increases, the elbow resistance also drops. For three-dimensional elbows, the minimum LRC is also obtained for a_0/b_0 = 4. In addition, the plot of ζ for the two-dimensional model lies below the others, i.e. the LRC for this case is minimal. This indicates that the profile of the profiling insert, taken from the outline of the vortex zone in the two-dimensional elbow, is the best for two-dimensional model (i.e. $a_0 = \infty$) and slightly worse for the elbows with other depths.

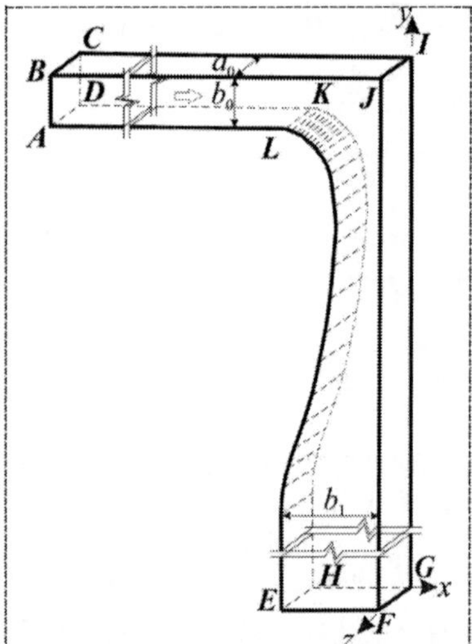

Figure 7. Geometry of the computational domain of the profiled elbow (for the case $b_1/b_0 = 2$ and $a_0/b_0 = 4$).

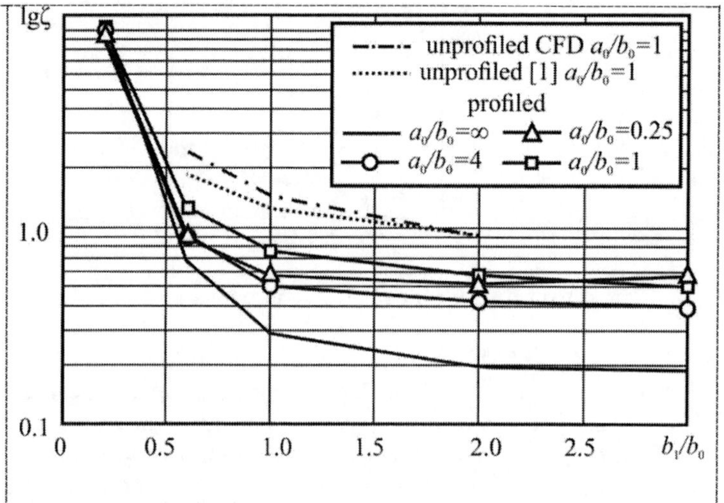

Figure 8. Dependence of LRC on the ratio of the sizes b_1/b_0 and a_0/b_0 for the profiled elbow.

Figure 9 shows the dependence of the lengths of the influence zones before l_b/b_0 (dotted line) and after the elbow l_a/b_0 (solid line) on the channel size after the elbow and channel depth for the profiled elbows. At this graph, for comparison, data for unprofiled elbow with depth a_0/b_0 = 1 are shown.

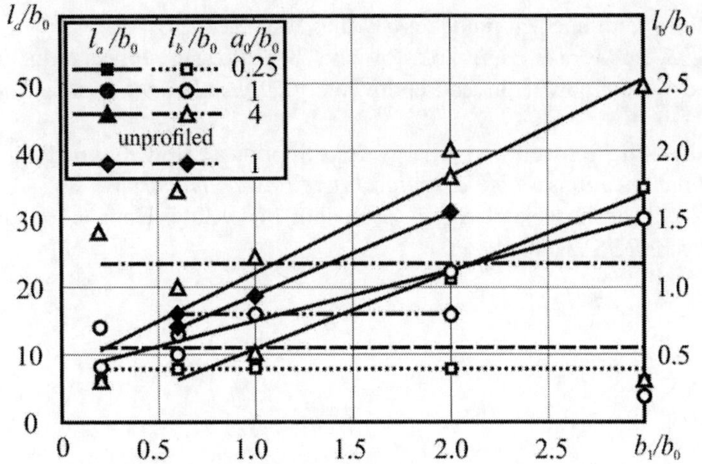

Figure 9. Dependence of lengths of the IZ on b_1/b_0 and a_0/b_0 for profiled elbows.

The results of IZ study in profiled elbows, approximating with formulas are given in the table 2.

Table 2. Dependence of IZ lengths for the profiled elbows

a_0/b_0	l_a/b_0	l_b/b_0
0.25	$11.434b_1/b_0$-0.741	1.18
1	$7.525b_1/b_0$+7.339	0.55
4	$14.251b_1/b_0$+7.719	0.4

The increase of the IZ length after the elbow occurs in the same way as for the unprofiled elbows, with the increase in the size of the channel after turn (b_1), and with increasing channel depth a_0, the IZ after elbow lengthened. Dependences of the length of the IZ before elbow on the change of the size of channel b_1/b_0 are also not observed, while increasing the depth of the channel a_0, the length of the IZ before elbow increases. Comparison of IZ of the unprofiled and profiled elbow showed that during profiling the IZ decreases to 45.5% before elbow and from 7.7 to 40.9% after it, depending of dimensions of elbow.

Conclusion

Thus, a detailed numerical investigation of the flow in the "sharp" elbow allowed to determine the outlines of the vortex zones formed when the flow is separate from the inner edge of the 90° elbow, the lengths of influence zone and the dependence of the local resistance coefficient on the wide range of its width and depth. Further, based on this data the numerical models of profiled elbows was created and the dependences of LRC and IZ on the dimensions of elbow were determined. Showing a significant reduction of pressure losses and a decrease in the length of the influence zones of such kind profiled elbows, which demonstrate their high energy efficiency.

Acknowledgments

The reported research was funded by the Russian Science Foundation, grant No. 18-79-10025.

References

[1] Idel'chik I E and Steinberg M O 1994 *Handbook of Hydraulic Resistance* (Boca Raton, FL: CRC Press) p 790
[2] Heskestad G 1971 Two-Dimensional Miter-Bend Flow *J. Basic Eng.* **93** 433–43
[3] Lichtarowicz a. and Markland E 2006 Calculation of potential flow with separation in a right-

angled elbow with unequal branches *J. Fluid Mech.* **17** 596

[4] Ziganshin A M, Beljaeva E E and Sokolov V A 2017 Pressure losses reduction with profiling of sharp elbow and elbow with dead-end *News of Higher Educational institutions. Construction* 108–16

[5] Ziganshin A M 2015 Reduction of energy consumption of flow by profiling of fittings in power plants communications *Safety & Reliability of Power Industry* 63–8

[6] Ziganshin A M, Aleshchenko I S and Ziganshin M G 2016 Connection fitting with profiling inserts *Patent RUS* 2604264

Numerical modelling of energy efficient exhaust orifices and hoods in ventilation and air conditioning systems in buildings and facilities of thermal power plants

A M Ziganshin[1], K E Batrova[1], K I Logachev[2] and D N Mingazeeva[3]

[1]Thermal power engineering, Gas supply and Ventilation Department, Kazan State University of Architecture and Engineering, Russia, 420043, Kazan, Russia

[2] Heat and Gas supply and Ventilation Department, Belgorod State Technological University named after V.G. Shukhov, 308012, Belgorod, Russia

[3]Industrial Engineering Department, LLC "Velesstroy" Building Construction Company,125047, Moscow, Russia

amziganshin@kgasu.ru

Abstract. The paper presents the results of numerical simulation of energy-efficient inlet parts of exhaust ventilation systems in the form of the last side orifice and an exhaust hood. The profiling of the inlet sharp edges is used, which is made according to the outlines of the vortex zones that generates when the flow is separated at this edges. For this, the non-profiled geometry of the exhaust orifices is first modelled, where the outlines of the vortex zones are determining. Then a numerical model with profiling is created and the energy losses of such an improved shaped elements is determined. A significant reduction in pressure loss has been obtained: for the exhaust hood more than 40% and 30% for the last side orifice.

Introduction

Ventilation and air-conditioning systems at industrial facilities of thermal power plants are a complex structure, constructed to provide normative sanitary and hygienic requirements of meteorological conditions in the working area of the premises - relative humidity, speed and air purity. The main elements of the systems include air ducts, which can have, depending on the purpose and location of the serviced premises in the buildings or thermal power plant (TPP) facilities, very complex configurations with a large number of fittings and various functional devices.

For example, [1] prescribes the provision in the engine and boiler rooms the multizoneventilation systems. At the same time, in the boiler rooms of the TPP main buildings operating on gaseous fuels, supply of fresh air must be provided in the amount of three times the air exchange per hour. Cable floors and cable tunnels that pass inside and outside buildings must be equipped with mechanical ventilation without recirculation with the removal of air from each compartment of the cable rooms outside of the building. In this case, the exhaust ducts can be combined into collectors, and the systems must be equipped with automation, which intensifies the ventilation to avoid exceeding the temperature in the premises above 35 °C. In the transformer chambers and in the rooms of current-limiting reactors, the supply and exhaust systems also operate with the function of temperature control, and in the air of storage battery rooms ventilation should ensure the content of sulfuric acid aerosols within the maximum permissible concentration (MPC) and hydrogen within the explosion-proof concentration. At the same time, the air ducts of the exhaust ventilation systems have functional devices with hoods and orifices for local exhaust of air from the process equipment, exhaust openings and other structural elements.

In [2] it was shown that the energy consumptions in the communications of TPPs significantly reduce the efficiency of innovative solutions introduced at the station. Thus, for example, a typical main building of a small 350 MW cogeneration plant has an area of about 11,000 m^2 [3]. Having determined in accordance with the instructions [1] air flow rate in the premises of the main buildings of TPPs operating on gaseous fuels with an air change rate equal 3 and having a room height of 6 m, we obtain an air flow rate of 198,000 m^3/h. To supply such flow rate at a pressure of 8 kPa, 55x8 = 440 kW is required. Having estimated that the same amount is needed for the exhaust system and the energy consumption for maintaining the ventilation system of the entire station is twice as high, we will get the energy consumption for the system of 1.76 MW, which is 0.5% of the generated electric power of the TPP. For comparison, it can be recalled that widely developed in recent years generating systems for super-overcritical pressure above 38 MPa and a temperature of more than 700 °C allow to increase the generation efficiency by only 5%. Consequently, a decrease in the pressure losses in the ventilation system can make a tangible contribution to the improvement of the station's performance in energy efficiency.

The results of numerical simulation of flow to the last (at the duct end plug) side orifice and hood (Figure 1) of the exhaust ventilation system are given in the paper. Flows to the hoods have been studied for a long time; there are works both for determining the pressure losses of such devices [4] (and containing recommendations for their design) [5, 6], and for determining the velocity field in front of the hood [7, 8]. In this case, both experimental methods of research and numerical modeling were used. There is much less work to study the currents in front of the side orifice, although such a device in the exhaust ventilation systems is also very common. One can note the paper [9], where the resistance of such anorifice is determined, for the case of a circular channel. It is known that in such devices, when the air flow is detached from the sharp entrance edge, a separation zone arises that causes local pressure losses. If the inlet section of the exhaust hole is profiled according to the outlines of the vortex zone, as suggested in [10] for the duct ventilation elements - elbows, confusers and diffusers, vortex formation can be reduced or eliminated, and therefore the resistance of such devices can be reduced.

At the first stage of creating such energy-efficient exhaust devices, it is necessary to determine the outlines of the formed vortex zones (VZ). Works on determining the outlines of vortex zones at the inlets of the hoods is very few. The outlines of VZ are determined in [11, 12] for air flow to a circular hoods with an angle of inclination of the hood flange 90° by the discrete vortices method. For the circular hoods with a large range of sizes and angles of flanges, the results of experimental and numerical studies of both separation zones and velocity fields are given in [13]. Numerical studies were carried out using the methods of discrete vortices and computational fluid dynamics. The results are shown to be in good agreement with the experimental data. For slot hoods the work [14] is known, where the VZ parameters are determined by the method of conformal mappings. The found outlines of the VZ are limited by one angle of inclination of the flange, but there is some information on the length and depth on which the VZ develops in the hood. Thus, there is no systematic information about the outlines of the vortex zones formed due to flow separation at the inletsof the last side orifice and the flat exhaust hood.

Numerical study of unprofiledexhaust orifices

In this paper, we present the results of a numerical study of the flow, in order to determine the outline of vortex zones and the subsequent modeling of energy-efficient inlets of exhaust ventilation systems with the profiling insert. The study is carried out for one case of the geometry of each type of considered exhaust orifices. For the exhaust hood - the length of the flange is equal to the half-width of the channel ($d = b = 0.1$m, $\alpha = 30°$, figure 1*a* - the symmetrical half of the computational domain is

IOP Conf. Series: Earth and Environmental Science **288** (2019) 012134 doi:10.1088/1755-1315/288/1/012134

shown), the dimensions of the computational domain are: height - 1m, width - 2m, channel length 2m, for the last side orifice - channel width H = 0.16m, orifice dimension h = 0.3 m, dimensions of the computational domain (Figure 1*b*): height - 1m, width - 1.5m, channel length - 6m. Boundary conditions (BC) for the hood: ***AF***–velocity outlet v_0 = 5 m/s (Re> 50000), ***FE***– symmetry BC; ***BCDE***-external free boundaries (gauge pressure is set to zero); ***ABGI*** - impermeable walls. BC for the side orifice: at the outlet ***AJ*** –mass flow rate G = 1kg/s (Re> 50000); external boundaries ***BCDEF***–external free permeable boundaries (gauge pressure is set to zero), on all solid walls (***AI***, ***GF***, ***FJ***) - BC "wall".

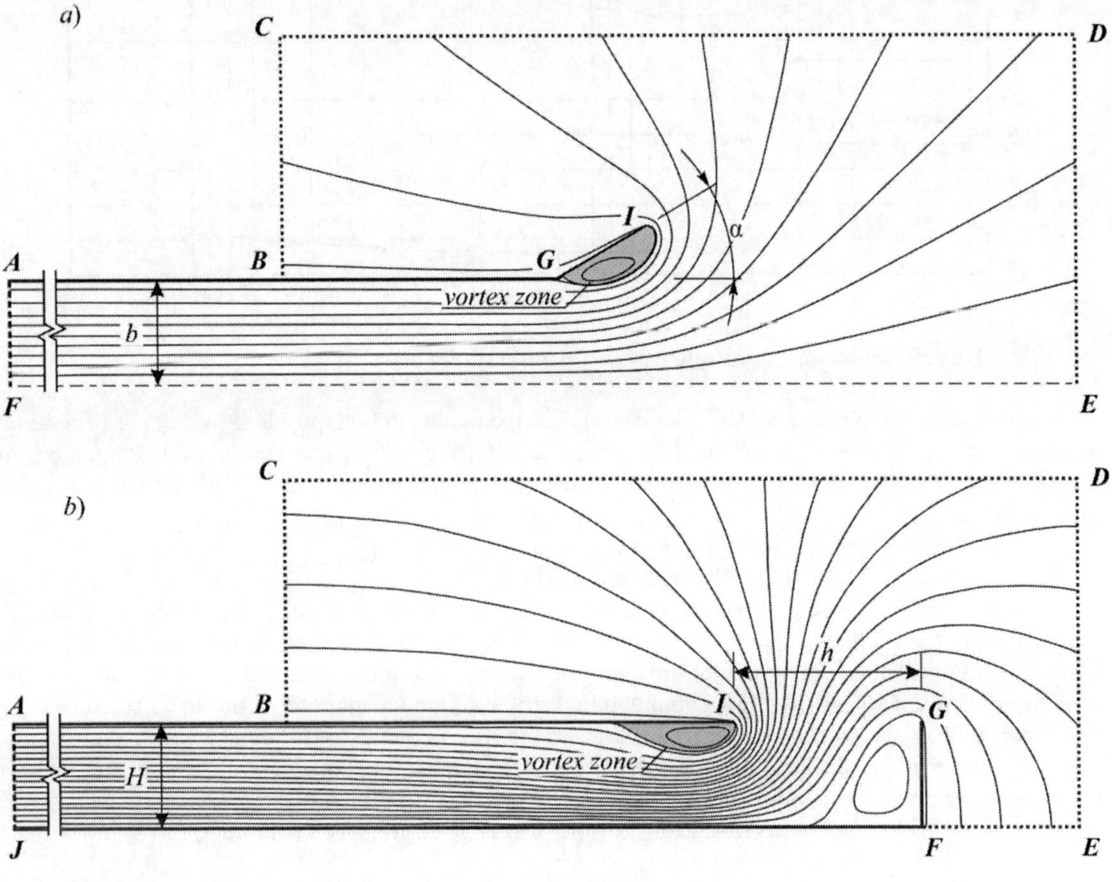

Figure 1. Geometry of the computational domain and a flow streamlines to the exhaust hood (*a*) and the last side orifice(*b*).

The tasks are solved in a two dimensional formulation, with ANSYS® Fluent Software using RANS models, which are the most suitable for modeling such flows based on the results of preliminary studies– "standard" k-ε for the side orifice and the"Reynolds Stress Model" for the hood. For modeling a boundary layer "Enhanced Wall Treatments" is used. For each task, a grid convergence study is carried out – the grid is refined in stages, with solving and determining of a local resistance coefficient (LRC) at each stage.

$$\zeta=(P_b - P_e - \Delta P_{fr})/P_d,$$

here $P_b$is the initial excess pressure assumed to be zero; P_e is the final pressure at the end of the duct (at the ***AF*** boundaryfor hood and ***AJ*** forthe side orifice) [Pa]; $\Delta P_{fr}= R_{fr}\cdot l$is a pressure losses in the duct due to the friction [Pa]; $P_d= \rho v_0^2/2$ is a dynamic pressure in the duct [Pa]; R_{fr} is a unit friction losses

IOP Conf. Series: Earth and Environmental Science **288** (2019) 012134 doi:10.1088/1755-1315/288/1/012134

of pressure in the duct determined by processing the numerically computed field of total pressures in the duct [Pa/m]; and *l* is the length of the duct [m] (Figure 1).

Further figure 2 shows the dependence of LRC on the dimensionless distance *y*+, which characterizes the degree of refinement of the computational grid.

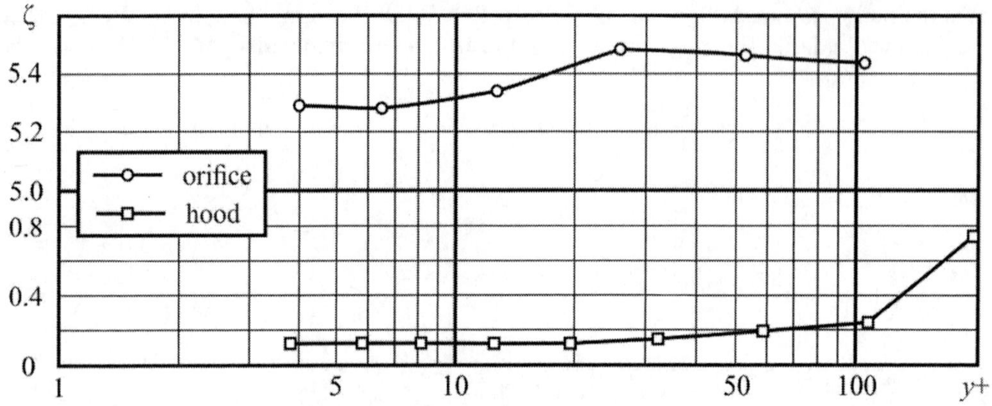

Figure 2. Verification of the tasks

It can be seen that on the last two calculated grids, the results in both problems no longer depend on the degree of grid refinement, and the LRC equal 5.29 for the side orifice and 0.121 for the hood. In this case, the cells have minimum dimensions of the order of 0.005 mm, and their total number is more than 2 million.

Modeling of energy efficient profiled exhaust orifices

Next, a numerical model of energy-efficient exhaust orifices wasbuilt, where the inlet part is profiled using the outlines of the vortex zones (see figure 1). Boundary conditions and models were used the same as in the study of non-profiled elements. Also, all solved tasks werestudied to grid convergence. Similarly, for the last computational grids, the independence of the local resistance on the dimensions of the calculation cells is characteristic. As a result, it was found that for the exhaust hood, the decrease of LRC was 42% ($\zeta = 0.07$), and for the last side orifice – 30% ($\zeta = 3.7$). In this case, the streamlines for the profiled exhaust hood shows that after the profiling insert a secondary VZ appeared, which, although much less, leads nevertheless to pressure losses (Figure 3).

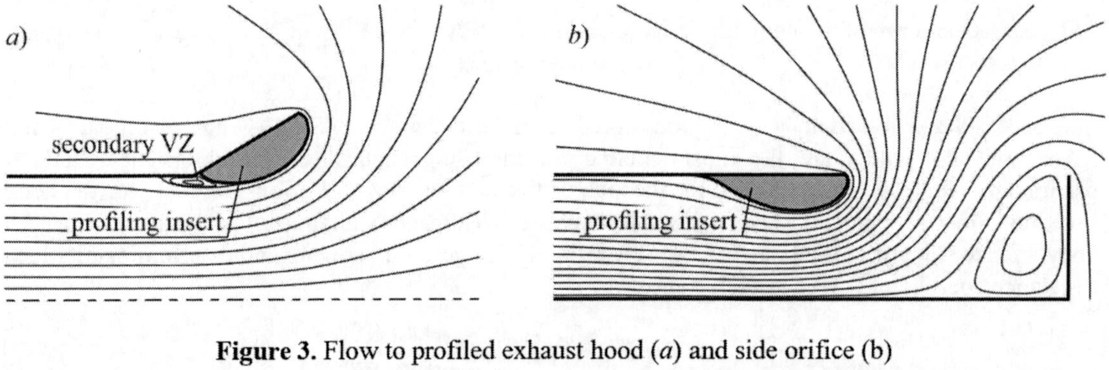

Figure 3. Flow to profiled exhaust hood (*a*) and side orifice (b)

Therefore, further it is interesting to simulate an exhaust profiledwith the outlines of the secondary VZ. For the case of last side orifice, such a secondary VZ did not observed. Thus, we can talk about the possibility of reducing the pressure loss, and hence the energy consumption of the exhaust ventilation systems by profiling its input elements in the form of the exhaust hood and the last side orifice.

References

[1] SP 90.13330.2012. 2012 *Thermal power stations* (Moscow: Ministry of Regional Development of Russia) p 113

[2] Ziganshin A M 2015 Reduction of energy consumption of flow by profiling of fittings in power plants communications *Safety & Reliability of Power Industry* 63–8

[3] Kupcov I P and Ioffe J R 1985 *Design and construction of thermal power plants* (Moscow: Jenergoatomizdat) p 408

[4] Figueroa C E 2011 Hood entry coefficients of compound exhaust hoods. *J. Occup. Environ. Hyg.***8** 740–5

[5] DallaValle J M 1952 *Exhaust hoods* (New York: The Industrial Press) p 146

[6] Logachev I N and Logachev K I 2014 *Industrial Air Quality and Ventilation* (CRC Press) p 417.

[7] Cascetta F 1996 Experimental evaluation of the velocity fields for local exhaust hoods with circular and rectangular openings *Building and Environment***31** 437–49

[8] Betta V, Cascetta F, Labruna P and Palombo A 2004 A numerical approach for air velocity predictions in front of exhaust flanged slot openings *Building and Environment***39** 9–18.

[9] Khanzhonkov V I and Davydenko N I 1959 Resistance of the lateral openings of the pipeline end section *Industrial aerodynamics***15** 38-46

[10] Ziganshin A M, Aleshchenko I S and Ziganshin M G 2016 Connection fitting with profiling inserts *PatentRUS*2604264

[11] Logachev K I, Averkova O A, Tolmacheva E I, Logachev A K and Dmitrienko V G 2016 Modeling of Air and Dust Flows in the Range of Action of a Round Suction Funnel Above an Impermeable Plane. Part 1. A Mathematical Model and Algorithm for its Computer Implementation *Refractories and Industrial Ceramics***56** 679–83

[12] Logachev K I, Averkova O A, Tolmacheva E I, Logachev A K and Dmitrienko V G 2016 Modeling of Air and Dust Flows in the Range of Action of a Round Suction Funnel Above an Impermeable Plane. Part 2. Characteristics of Separation Region and Efficiency of Capture of Dust Particles1 *Refractories and Industrial Ceramics***57** 103–7

[13] Logachev K I, Ziganshin A M, Averkova O A and Logachev A K 2018 A survey of separated airflow patterns at inlet of circular exhaust hoods *Energy Build.***173** 58–70

[14] Posokhin V N, Salimov N B and Safiullin R G 2003 On the form of the vortex zones at the entrance to the exhaust hood *News of higher educational institutions. Problems of power engineering* **3-4** 39–47

IOP Conf. Series: Earth and Environmental Science **288** (2019) 012135 doi:10.1088/1755-1315/288/1/012135

Method of calculation of power efficiency by optimization of work of heat supply systems

Yu.N. Zvonareva[1], Yu.V.Vankov

Department of Industrial power system and systems of heat supply

Kazan State Power Engineering University-KGEY, Kazan, Russian Federation

e-mail: [1]Skulinaun@mail.ru

Abstract. In this work results of mathematical modeling and results of the pilot studies conducted at the laboratory stand for the system of heat supply of Kazan, the development of power of the large cities directed to development of an integrated approach at a research of problems are integrated. By means of the program Zulu complex the mathematical model intended for optimization of work of power systems during their operation has been created.Results of the conducted researches have allowed to analyse influence of stage-by-stage introduction at consumers of automatic metering stations and weather regulation (AITP) on a hydraulic operating mode of system of heat supply for large municipal associations . In this work the main criteria are also considered, the algorithm and a method of calculation of power efficiency is developed by optimization of systems of municipal heat supply by introduction of the automated individual thermal points (AITP). The questions of influence of the made technical solutions on financial and economic and investment indicators are raised. The considered actions for energy resources saving are investigated for the purpose of reduction of volumes of consumption or more rational use of fuel and energy resources.

1. Introduction

Management of the systems of development, transportation and distribution of energy carriers is a basis of increase in power and economic efficiency of economic entities of the industry and the sphere of housing and communal services.

Realization of the having potential of energy saving requires complex interaction of elements of external and internal engineering systems on the way from a warmth source to the consumer. Economy of energy is reached due to optimization of a holiday of thermal energy, regulation of the hydraulic mode [1,2], introduction of the automated thermal points (AITP) which are a link of internal and external thermal networks [3].

The tool of the solution of the main problems which are available in concrete thermal network are the computer models giving the chance to define the hydraulic and thermal mode of system [4].

Modeling of the modes of a holiday of heat by means of the electronic scheme allows to analyze profitability of options of loading of the heating equipment from the point of view of minimization of costs of transfer of heat, specific fuel consumption on a holiday of thermal and electric energy taking into account opportunities of thermal network for transportation of the heat carrier.

Earlier authors have analysed influence of stage-by-stage introduction of automatic metering stations and regulation on the most remote from the central heat distribution station the consumers connected according to the elevator scheme [5].

Optimum consumption of thermal energy at respect for the required comfort in rooms of buildings is the main objective of energy saving in the sphere of municipal heat supply to which much attention is paid [6] today.

Comparison of result, the research (electronic modeling of operating modes and pilot studies) conducted in various ways is presented in the figure 1.

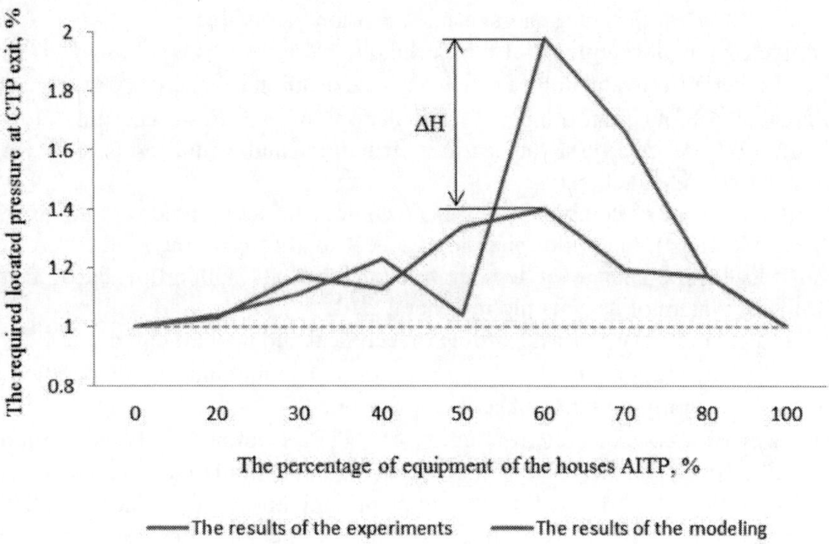

Figure 1 The required located pressure at CTP exit depending on stages of introduction of ITP

Apparently from the schedule, considerable influence on the hydraulic mode of system of heat supply, renders the level of equipment of consumers of the AITP system of 60%. At this value significant increase in the located pressure on an entrance to the considered contour, therefore, and increase in pressure in the giving pipeline is required. Increase in pressure of network water can lead to emergence of the emergency or transition hydraulic processes caused by fluctuation of pressure in the system of heat supply of the city in general [7]. It can't be allowed from the point of view of reliability of the operated systems of heat supply which is one of the major factors at implementation of the contractual relations between the heatsupplying organizations and consumers of thermal energy [8].

The difference ΔH between the received values during an experiment and modeling in our opinion is explained by an error of devices.

The most optimum from the point of view of quality of heat supply and power overall performance of system is providing AITP more than 80% of consumers of total number. At the same time the set thermal and hydraulic mode in the system of heat supply remains, there is no disbalance of system and the required temperature condition directly at the consumer is exposed [9].

2. Goals and objectives

The method of calculation of energetic efficiency is developed for increase in energy saving taking into account step-by-step implementation in the systems of municipal heat supply of automated metering stations and regulations of thermal energy and also for maintenance optimum, from the point of view of saving of energy resources, the heathydraulic mode by optimization of systems of municipal heat supply by implementation of the automated personal thermal points (APTP) (further the Technique).

Use of this Technique allows to estimate technical and economic efficiency of the actions directed to saving and optimum use of energy resources.

The technique is developed for the closed systems of heat supply.

It can be applied as to quarterly calculation of systems of heat supply of the large cities, and to calculation of diagrams of heat supply of small settlements.

The purpose of development of the Technique is complex assessment of efficiency, the energy saving actions offered to implementation in natural and monetary value.

The main objectives by development of the Technique were:

- choice of criteria for evaluation of efficiency of execution of an energy saving action;
- development of an algorithm of calculation of criteria according to saving from implementation of energy saving actions in natural and value terms, based on a technical entity of the executed actions.

The main criteria in case of compilation of this technique selected indices of the expenditure of the heat carrier (network water), indices of change of a heat load of customers, in case of oscillations of temperature of outside air, change of temperature in locations and influence of these criteria on hydraulic stability of system of heat supply in general [10].

Algorithm of calculation of criteria according to technical and economic indices:

- According to metering devices the main sentries and annual indices of operation of the studied system of heat supply for each subscriber are defined.
- The number of customers for installation of AITP is defined with the optimum percent of equipment received earlier as a result of the carried-out simulation [11].
- The composition, the cost of the equipment and costs of project installation works is determined by directories of vendors and suppliers of personal thermal points.
- For each of the selected customers the potential of saving of thermal energy is defined with implementation of AITP.
- Taking into account change of a heat load on customers, hydraulic calculation of the considered system of heat supply is carried out.
- The economic efficiency from implementation of schedulable energy saving technologies is defined and the payback period of the offered actions is defined [12].

3. Results of work

By this technique an assessment of power efficiency of the realized energy saving actions (introduction of AITP) on the example of the residential quarter No. 38 to Kazan has been given.

The efficiency of any implemented projects and actions by optimization of systems of heat supply is defined by a ratio of expenses and the results considered actions.

Assessment of commercial effectiveness was carried out I taking into account the following requirements:

- • the capital investments put in calculations are indexed according to price indexes on capital investments and elements of their technological structure;
- • the cost of design installation works is accepted according to settlement prices of the partner companies (producers and suppliers of the equipment).

As the main criterion of commercial effectiveness the discounted payback period (return of the capital), internal standard of profitability, the index of profitability (profitability), the net discounted income is accepted [13, 14].

IOP Conf. Series: Earth and Environmental Science **288** (2019) 012135 doi:10.1088/1755-1315/288/1/012135

The discounted period of payback (DPB) is the time demanded for a covering of all discounted capital expenditure at the expense of the gained income.

$$Inv = \sum_{t=1}^{DPB} \frac{CF_t}{(1+r)^{t-1}},$$

Inv - initial investments; CFt - a pure cash flow of month of t; r - monthly discount rate.

The net discounted income (NPV) considers dynamics of change of profit during the settlement period.

$$NPV = \sum_{t=1}^{Tp} \frac{CF_t}{(1+r)^{t-1}} - Inv,$$

Tp – the settlement period.

The index of profitability (PI) – the relation of the given effects to the size of the discounted capital investments:

$$PI = \frac{\sum_{t=1}^{Tp} \frac{CF_t}{(1+r)^{t-1}}}{Inv},$$

The internal standard of profitability (IRR) – characterizes intensity of return of the spent means,

By results of calculation the summary table of the key financial performance of implementation of the project on introduction of AITP for a quarter No. 38 is made and the schedule of payback is constructed.

Table 1. Key financial performance.

Indicator	Value
Simple period of payback of tentative expenses (PB), month.	31
The discounted period of payback of tentative expenses (DPB), month.	35
The net present value (NRV), rub.	51,55
Profitability index (PI)	3,45
Internal rate of return (IRR),%	66,6

Ratio of the discounted and simple period of payback of initial expenses are reflected in the figure 2.

According to the carried-out calculations the annual expected effect of realization of actions for optimization of systems of heat supply will allow to save on average up to 30% consumed and that isn't less important, the paid thermal energy. The payback period of the incurred capital expenditure on average doesn't exceed three and a half years.

Thus, proceeding from the analysis of indicators of economic efficiency the introduced actions in this case are economically expedient.

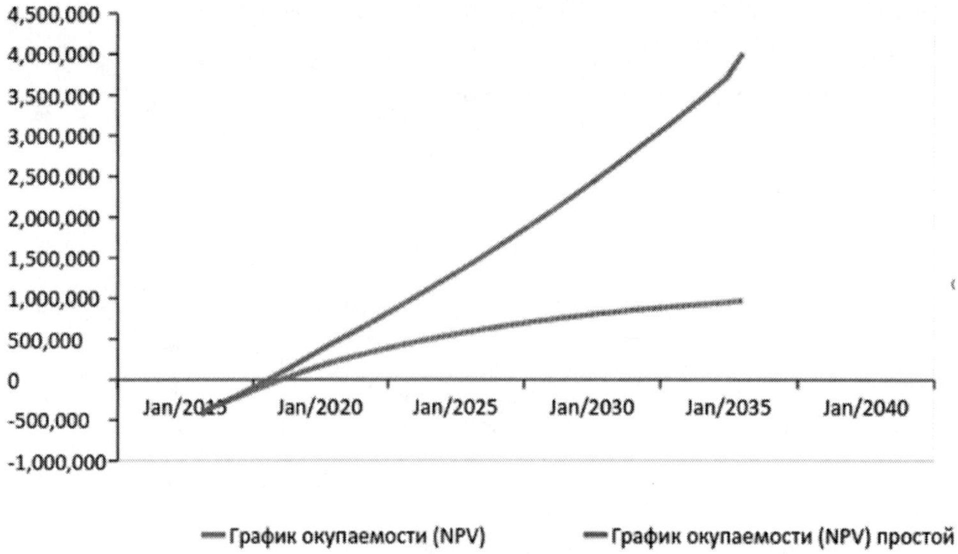

Figure 2. Schedule of payback (NPV)

4. Conclusion

At management of the modern systems of heat supply the issues of providing consumers the required heat carrier parameters connected with installation of the set pressure, distribution of streams, etc. are resolved.

Comparison of the results received at various ways (mathematical modeling, a laboratory experiment, a calculation method) allows to give more exact assessment to power efficiency of the introduced energy saving actions

For making decision by investors (the managing director heatgenerating and transporting the heat carrier to the companies) on expediency of participation in implementation of the project on equipping by automatic metering stations both regulations of consumers of thermal energy, and their influence on hydraulic stability of all system in general except technical efficiency, for detailed assessment, it is also necessary to make calculation of commercial effectiveness of the offered actions.

On the basis of the technical and economic indicators received when calculating for the offered technique the conclusion about efficiency and expediency of introduction of the Project and also about reliability and non-failure operation of work of system of heat supply is drawn when holding actions for energy resources saving for the purpose of reduction of volumes of consumption or more rational use of fuel and energy resources.

References

[1] Popov. V.E. Investments into infrastructure and economic growth: regional aspect//Regional economy. – 2009. - No. 1. – pp. 59-69.

[2] Zvonareva Yu.N.. Optimization of modes of jont operations of two heat sources for improving economic efficiency. Texnika i texnologiya: materialy XII mezhdunarodnoj nauchno-prakticheskoj konferencii: publishing house «Sputnik +», 2014.

[3] Shein I.S, Izvekov A.V., Collaboration of two sources of heat supply//Reliability and safety of power. 2012. No. 2 (17).

[4] Dikul. L.O. Priority directions of venture investment activity in industrial sector of economy//Economy, statistics and informatics. – 2011. - No. 4. – pp.30-35.

[5] Chistovich A. How to warm the cities. The Russian Expert review No. 2 2006, - http://www.protown.ru/information/articles/3325.html

[6] Zvonareva Yu.N., Vankov Yu.V., Management of the systems of heat supply// Radio electronics, electrical equipment and power: XXIII mezhdunarodnoj nauchno-prakticheskoj konferencii T. 3. M.: MEI publishing house, 2017. – 448 p.

[7] Madorskij B.M., Shmidt V.A.. Operation of central heat distribution stations, systems of heating and hot water supply. Moscow, 1971. 168 p.

[8] Shein I.S., Izvekov A.V.. Some questions of optimization of functioning of city thermal knot//Radio electronics, electrical equipment and power. M.: MEI, 2010.

[9] Sazanov B.V. Thermal power plants: Manual for technical schools of heat power specialties / M.: Energy, 1974. – 224p.

[10] Zvonareva Yu.N., Vankov Yu.V., Nazarychev S.A.. Estimation of economic effect for consumers in terms of adoption of automatic junctions of acquisition of heat energy. Inzhenernyj vestnik Dona, 2015, №4 URL: iv don.ru/ru/magazine/archive/n4y2015/3315.

[11] Employment of Danfoss automation at heating points of systems of central heat supply of buildings. Benefit. RB.00.H8.50.-M.: OOO «Danfoss», 2014, pp. 63.

[12] Zvonareva Yu.N., Vankov Yu.V., Work of system of heat supply at stage-by-stage introduction of the automated individual thermal points // News of higher educational institutions. Problems power engineering specialists / Publishing house: KGEU (Kazan). - 2017. - T. 19 - № 1-2.

[13] Drigo M.F.. Methodology of an assessment of economic efficiency investment decisions//Management in Russia and abroad, 2008. - No. 1. – pp.18-24.

[14] Pyrkov V.V. Modern thermal points. Automatic equipment and regulation. K.: II DP «Takispravi», 2007. 252 p.

AUTHOR INDEX

AUTHOR INDEX

AUTHOR INDEX

AUTHOR INDEX